BUILDING CONSTRUCTION

BUILDING CONSTRUCTION

Principles, Materials, and Systems

Second Edition

Madan Mehta
University of Texas at Arlington

Walter Scarborough
Hall Building Information Group, LLC

Diane Armpriest
University of Idaho

Original edition, entitled Building Construction: Principles, Materials, & Systems, 2nd Edition, By Mehta, Madan; Scarborough, Walter; Armpriest, Diane, published by Pearson Education, Inc, Copyright © 2013 by Pearson Education, Inc.

Indian edition published by Pearson India Education Services Pvt. Ltd. Copyright © 2016.

ISBN 978-93-325-7509-7

First Impression, 2016

Published by Pearson India Education Services Pvt. Ltd, CIN: U72200TN2005PTC057128.
Formerly known as TutorVista Global Pvt. Ltd, licensees of Pearson Education in South Asia

Head Office: 7th Floor, knowledge Boulevard, A-8(A) Sector-62, Noida (U.P) 201309, India

Registered Office: 4th floor, Software Block, Elnet Software City, TS -140, Block 2 & 9
Rajiv Gandhi Salai, Taramani, Chennai, Tamil Nadu 600113.Fax: 080-30461003,
Phone: 080-30461060, www.pearson.co.in email id: companysecetary.India@pearson.com

Printed in India by Sai Printo Pack Pvt. Ltd.

FOREWORD

The pedagogical philosophy of this book continues to be robust and unique. Therefore, it is not surprising that the first edition of the book was well received. The current (second) edition retains the book's original approach but has been substantially enhanced to bring it up-to-date with the current knowledge base and includes full-color photographs, photo-realistic renderings, and line art. I expect that it will become the leading text on construction materials and systems, and I am honored to recommend it to the academic and professional communities with renewed confidence.

As the Chairman and CEO of one of the largest, international architecture and engineering firms, I have come to appreciate the tremendous amount of knowledge today's architects, engineers, and constructors need to produce functional, economical, aesthetically pleasing, and high-performance buildings. Contemporary design and construction professionals cope, on a daily basis, with an exploding amount of information, mesh together a growing range of products, work out increasingly complex assembly details, and coordinate several diverse specialties.

In this second edition of *Building Construction: Principles, Materials, and Systems*, the authors have once again utilized their long experience and diverse expertise to improve upon the first edition and present the complexity of building construction in a more accessible volume. It clearly provides the basics of building science as applied to the art of transforming materials and systems into constructible buildings. Principles that influence building performance provide the background necessary to understand why, as well as how, buildings are assembled as they are.

The book appropriately addresses each of the primary building assemblies—foundations, walls, floors, ceilings, and roofs—and how they join, seal, and integrate with other components. The performance of building enclosures and systems is reviewed in detail, which enhances the reader's understanding of the comprehensive, integrated nature of the building design and construction processes. Almost all building materials and systems have been covered in depth.

The book is unique among the available books on the subject because it is a joint effort of three authors—two of whom are engaged full time in academia and another who has an extensive background in the profession. Together, the authors' combined expertise in architecture, engineering, and construction disciplines provides a holistic treatment of the subject. Although written primarily to educate students of architecture, engineering, and construction, the book will continue to serve as a reference for practitioners.

An exhaustive work, *Building Construction: Principles, Materials, and Systems* uses text and concepts, photographs, and detailed drawings to convey the construction assembly techniques, theory, and technology inherent in architecture. It also highlights the building professionals' involvement, as stewards of the environment, in producing sustainable, purposeful, and high-performance buildings on a life-cycle basis.

H. Ralph Hawkins, FAIA, FACHA, MPH, LEED AP
Chairman and CEO
HKS Inc.

INTRODUCING THE SECOND EDITION OF THE GROUNDBREAKING *BUILDING CONSTRUCTION: PRINCIPLES, MATERIALS, AND SYSTEMS*

Groundbreaking

The *first edition* was groundbreaking in its organization. The first book to focus initially on principles of construction, it gave students a greater understanding before moving on to materials and systems.

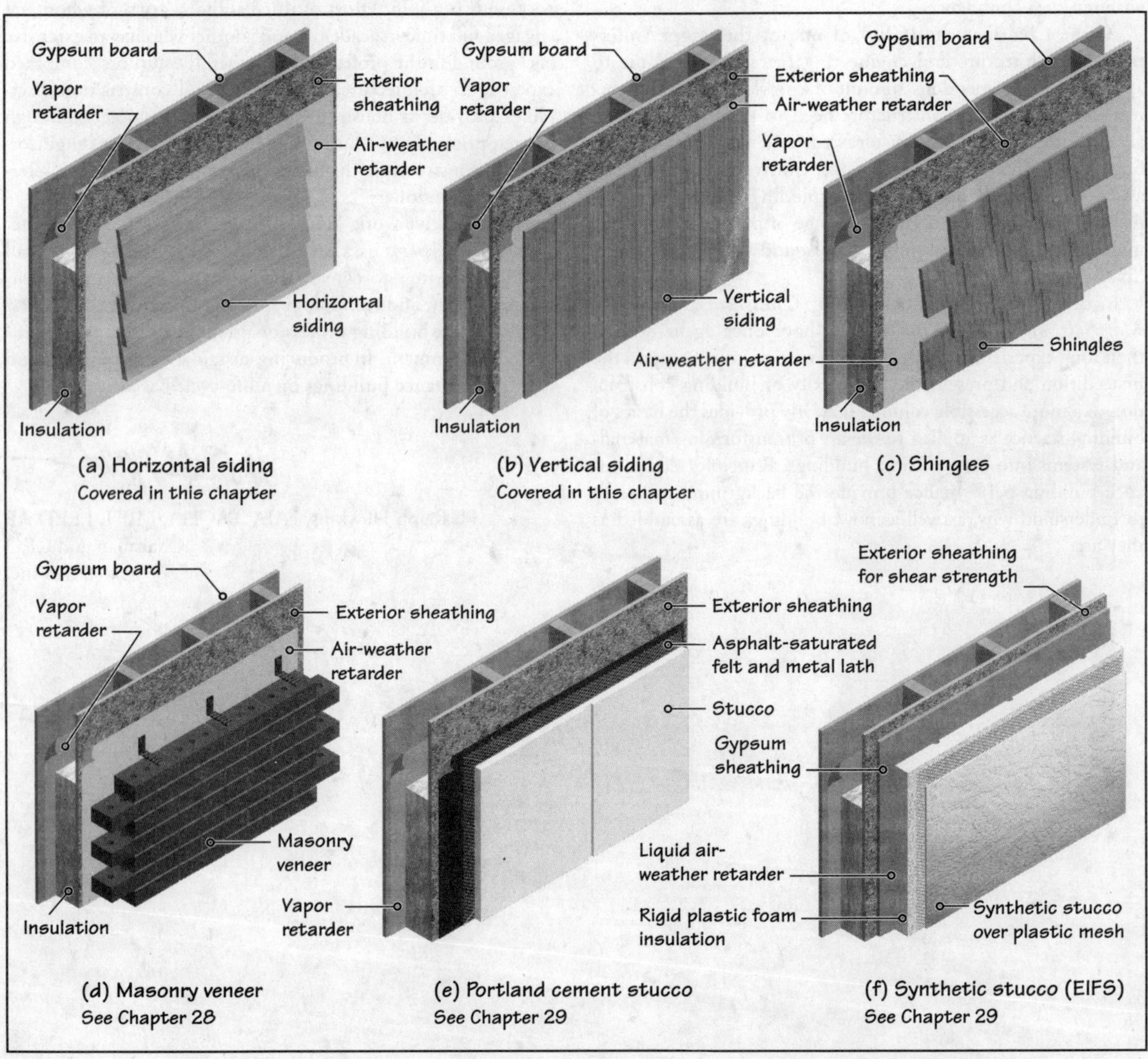

(a) Horizontal siding
Covered in this chapter

(b) Vertical siding
Covered in this chapter

(c) Shingles

(d) Masonry veneer
See Chapter 28

(e) Portland cement stucco
See Chapter 29

(f) Synthetic stucco (EIFS)
See Chapter 29

Engaging

The *second edition* makes building construction more engaging and accessible than ever before.

Hundreds of photos and illustrations make the material even more engaging.

With the chalk lines in place, framing the ground floor walls can begin, Figures 5, 6, and 7. Temporary braces stabilize the walls until the frame is complete and the exterior walls are sheathed. When double-height spaces are required, continuous or double studs are used to frame the high segments of the wall.

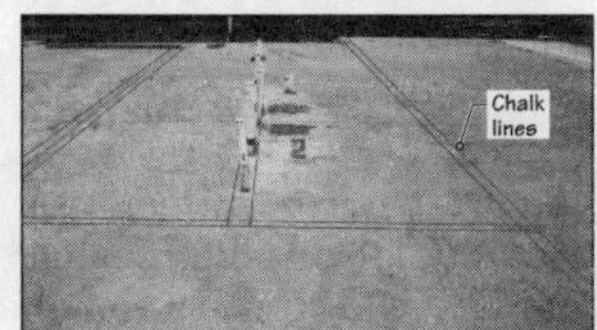

FIGURE 4 Chalk lines on the slab indicate the location of the walls. Note that plumbing pipes align and are located within the thickness of the wall.

FIGURE 5 Wall assembly under construction.

FIGURE 6 Walls are typically framed on the slab or floor platform and subsequently tilted up into position.

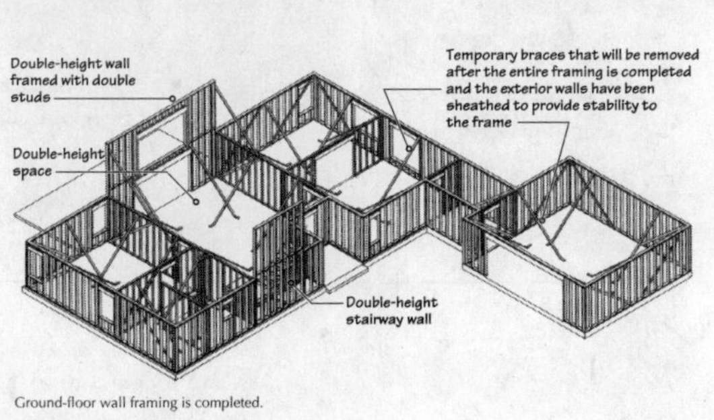

FIGURE 7 Ground-floor wall framing is completed.

(Continued)

Principles in Practice demonstrates practical applications of key concepts.

New Content

- Streamlined content designed to be more accessible
- More end-of-chapter review questions
- Expanded coverage on sustainable manufacturing of modern steel
- New coverage of building information modeling (BIM) and integrated project delivery (IPD)
- Coverage of all-precast concrete construction

Support

No book of this class would be complete without a robust supplement package for instructors and students.

For Instructors

- Instructor's Manual
- PowerPoint Presentations
- Test Bank

For Students

- Homework and Classroom Assignment Manual

PREFACE

We gratefully acknowledge the overwhelming response of the academic and professional communities to the first edition of this text in recognition of the text's unique pedagogical philosophy and distinguishing features. We hope that this full-color, updated second edition will continue to meet or exceed the expectations of the readers.

As stated in the preface to the first edition, building construction is a society's most dynamic enterprise, rooted in the inherent desire of humans to continually improve their habitat. The impact of this dynamism on building construction is that new products and construction systems are added to the existing stock in a never-ending process. In addition to new materials, new versions of traditional materials are proliferating. Consequently, today's designers and constructors face an unprecedented challenge as they strive to make well-informed decisions in the face of an expanding number of competing products and systems available for a given application.

The most accessible sources of information on building products often come from those who produce them. Although product testing is generally conducted by independent agencies, the results come to us through the filter of the manufacturers' vested interests and aggressive sales strategies, slanting the information unduly in favor of a specific product.

How do we, as designers and constructors, develop a critical faculty that enables us to sift facts from exaggeration and relevance from insignificance? How do we deal with the immense regional and international diversity of construction? How do we learn to function successfully in an increasingly litigious environment?

The core philosophy of this text is embedded in the belief that the best way for today's designers and constructors to respond successfully to the challenges just cited is for them to be sufficiently grounded in the principles of construction and the related systems. After all, the principles will be sustained, just as the conventional materials and practices will become dated all too soon. Equally important is the belief that architects, engineers, and constructors, well versed in the principles governing the performance of materials and systems, can produce a more wholesome and sustainable built environment.

PARTS 1 AND 2

The study of principles alone is not sufficient to master or fully grasp a subject. The principles must be illustrated by current and traditional practices in order to connect with real-world design and construction. *Building Construction: Principles, Materials, and Systems* aims to do just that. Therefore, the text is divided into two parts:

- Part 1, consisting of 10 chapters, deals primarily with the principles of building materials' and building assemblies' performance.
- Part 2, consisting of 27 chapters, deals primarily with specific materials and assemblies.

The experience of two of the book's authors, who have served in academia for many years, indicates that by introducing the basic principles common to the performance of most materials early in the course, it is possible to preclude or reduce repetition when progressing from one material or system to another. For example, thermal insulation is provided differently in different assemblies, but the thermal behavior of various assemblies is entrenched in the same basic principles. Similarly, all building assemblies must have some measure of fire endurance. Therefore, it makes sense to deal with the general properties and performance requirements of materials and systems in advance of their specifics.

However, there cannot be an absolute separation between the principles and their application. Some reiteration of the fundamentals as one proceeds through various materials and systems is unavoidable—in fact, pedagogically necessary. Additionally, there are certain principles that are limited only to one or two materials or assemblies. Those principles must be discussed in pertinent chapters.

DISTINGUISHING FEATURES

The book has several distinctive features. Each chapter is divided into convenient segments that pause with a set of multiple-choice (*Practice Quiz*) questions. Their purpose is to help the reader gain a broad understanding of the subject and assimilate its highlights. Answers to practice quizzes are provided in the end sheets of the book to facilitate self-testing by the readers. Testing for a detailed understanding of the chapter is accomplished through *Review Questions* provided at the end of each chapter.

A large number of *Margin Notes* expand on critical topics or topics that would stimulate the reader to pursue additional self-directed inquiry. Where a critical topic needs expanded coverage, beyond that given in the main body of the chapter or in margin notes, it is provided in the *Expand Your Knowledge* sections. Almost every chapter has one or more such sections.

Several chapters conclude with a section entitled *Principles in Practice*. These sections provide the opportunity for the interested reader (graduate students, practicing architects and engineers) to pursue in-depth study of the applications of construction principles to design issues without creating an undue burden in the main body of the chapter.

An entire chapter (Chapter 10) is devoted to the *Principles of Sustainable Construction* in Part 1 of the text, and this is followed up with a special section called *Focus on Sustainability* in various chapters of Part 2 of the text.

Appendix B: Preliminary Sizing of Structural Members provides rules of thumb for determining approximate dimensions of structural members of conventional wood, steel, masonry, and concrete structures as one consolidated reference—a handy guide in design studios for students and practitioners.

The book is copiously illustrated. Two-dimensional and three-dimensional line art supplement the text descriptions.

ABOUT THE AUTHORS

Madan Mehta, B.Arch., M.Bdg.Sc., Ph.D., P.E., is a faculty member at the School of Architecture, University of Texas at Arlington, and teaches courses in construction and structures. He was previously the Director of the Architectural Engineering Program at King Fahd University, Saudi Arabia. A licensed professional engineer (Texas), Fellow of the Institute of Architects (India), and Member of the American Society of Civil Engineers, he has worked in India, Australia, the United Kingdom, Saudi Arabia, and the United States. With academic credentials in both architecture and engineering, he ran a comprehensive architecture/engineering practice while working as a faculty member at the Delhi School of Architecture, and he worked for a large general contractor in the United States during a leave of absence. He is the author of several full-length books and monographs on building construction, architectural structures, and architectural engineering.

Walter R. Scarborough, CSI, SCIP, AIA, is Vice President and Regional Manager for Hall Building Information Group, LLC. He is a specifications consultant and registered architect (Texas) with over 35 years of comprehensive technical architectural experience in specifications, document production, and construction contract administration. He has produced documents and administered construction for a large number and variety of building types. Previously the Director of Specifications for 10 years for one of the largest architectural firms in the world, he was responsible for building sciences research, manager of a department of specifiers, and master specification development and maintenance, in addition to being the specifier for major healthcare, sports, detention, municipal, and commercial projects, some valued in the hundreds of millions of dollars. He is active in the Construction Specifications Institute (CSI) at the local level (past president, secretary, and technical director) and national level (Education Committee and Practice Guide Task team), holds several CSI certifications, is Chairman of the Institute's Education Committee, was awarded CSI's prestigious J. Norman Hunter Memorial Award for advancing building sciences and specifications, and is the revision author for CSI's *Project Delivery Practice Guide* and its associated education program.

Diane Armpriest, M.L.A., M. Arch., is Associate Professor and Chair, Faculty of Architecture and Interior Design, College of Art and Architecture, University of Idaho. Before joining the faculty in 2001, she worked as an architectural project manager, and as a project developer and construction manager for neighborhood nonprofit housing providers. Her teaching and research interests include the pedagogy of architectural building construction technology, the expression of structure and materials in Northwest regional architecture, and the relationship between building and site. Previously, she was Associate Professor of Landscape Architecture at the University of Cincinnati. Highlights of her work there include research in resource-efficient design and construction and working with students on design-build projects.

CONTENTS

PART 1: PRINCIPLES OF CONSTRUCTION

CHAPTER 1

An Overview of the Building Delivery Process (How Buildings Come into Being)

CHAPTER OUTLINE

Building construction is a complex, significant, and rewarding process. It begins with an idea and culminates in a structure that may serve its occupants for several decades, even centuries. Like the manufacturing of products, building construction requires an ordered and planned assembly of materials. It is, however, far more complicated than product manufacturing. Buildings are assembled outdoors by a large number of diverse constructors and artisans on all types of sites and are subject to all kinds of weather.

Additionally, even a modest-sized building must satisfy many performance criteria and legal constraints, requires an immense variety of materials, and involves a large network of design and production firms. Building construction is further complicated by the fact that no two buildings are identical; each one must be custom-built to serve a unique function and respond to its specific context and the preferences of its owner, user, and occupant.

Because of a building's uniqueness, we invoke first principles in each building project. Although it may seem that we are "reinventing the wheel," we are in fact refining and improving the building delivery process. In so doing, we bring to the task the collective wisdom of the architects, engineers, and contractors who have done so before us. Although there are movements that promote the development of standardized, mass-produced buildings, these seldom meet the distinct needs of each user.

Regardless of the uniqueness of each building project, the flow of activities, events, and processes necessary for a project's realization is virtually the same in all buildings. This chapter presents an overview of the activities, events, and processes that bring about a building—from the inception of an idea or a concept in the owner's mind to the completed *design* by the architects and engineers and, finally, to the actual *construction* of the building by the contractor.

Design and construction are two independent but related and generally sequential functions in the realization of a building. The former function deals with the creation of the *documents,* and the latter function involves interpreting and transforming these documents into reality—a building or a complex of buildings.

The chapter begins with a discussion of the various personnel involved in a project and the relational framework among them. Subsequently, a description of the two major elements of design documentation—construction drawings and specifications—is provided. Finally, the chapter examines some of the methods used for bringing a building into being, referred to as the *project delivery methods.* From the owner's perspective, these methods are called *project acquisition methods.*

The purpose of this chapter, as its title suggests, is to provide an overall, yet distilled, view of the construction process and its relationship with design. Although several contractual and legal issues are discussed, they should be treated as introductory. A reader requiring additional information on these topics should refer to texts specially devoted to them.

1.1 PROJECT DELIVERY PHASES

The process by which a building project is delivered to its owner may be divided into the following five phases, referred to as the *project delivery phases.* Although there is usually some overlap between adjacent phases, they generally follow this order:

- Predesign phase
- Design phase
- Preconstruction phase
- Construction phase
- Postconstruction phase

1.2 PREDESIGN PHASE

During the *predesign phase* (also called the *planning phase*), the project is defined in terms of its function, purpose, scope, size, and economics. This is the most crucial of the five phases, and is almost always managed by the owner and the owner's team. The success or failure of the project may depend on how well this phase is defined, detailed, and managed. Obviously, the clearer the project's definition, the easier it is to proceed to the subsequent phases. Some of the important predesign tasks are:

- *Building program definition*
- *Economic feasibility assessment,* including the project's overall budget and financing
- *Site assessment and selection,* including verifying the site's appropriateness and determining its designated land use (Chapter 2)
- *Governmental constraints assessment,* for example, building code and zoning constraints (Chapter 2) and other legal aspects of the project
- *Sustainability rating*—whether the owner would like the project to achieve the U.S. Green Building Council's (USGBC's) Leadership in Energy and Environmental Design (LEED) certification at some level (see Chapter 10)
- *Design team selection*

BUILDING (PROJECT) PROGRAM

This includes defining the activities, functions, and spaces required in the building, along with their approximate sizes and their relationships with each other. For a house or another small project, the program is usually simple and can be developed by the owner without external assistance. For a large project, however, where the owner may be an institution (such as a corporation, school board, hospital, religious organization, or governmental entity), developing the program may be a complex exercise. This may be due to the size and complexity of the project or the need to involve several individuals—a corporation's board of directors, for example—in decision making. These constituencies may have different views of the project, making it difficult to create a consensus.

Program development may also be complicated by situations in which the owner has a fuzzy idea of the project and is unable to define it clearly. By contrast, experienced owners tend to have a clear understanding of the project and generally provide a detailed, unambiguous program to the architect.

It is not unusual for the owner to involve the architect and a few other consultants of the design team in preparing the program. In this instance, the design team may be hired during the predesign phase. When the economic considerations of the project are paramount, the owner may also consult a construction cost analyst.

Whatever the situation, preparing the program is the first step in the project delivery process. It should be spelled out in writing and in sufficient detail to guide the design, reduce the liability risk for the architect, and avoid its misinterpretation. If a revision is made during the progress of the project, the owner's written approval is necessary.

NOTE

Building (Project) Program

The American Institute of Architects (AIA) Document B141, *Standard Form of Agreement Between Owner and Architect*, defines the building program as "the owner's objectives, schedule, constraints and criteria including space requirements and relationships, special equipment, flexibility, expandability, systems, and site requirements."

1.3 DESIGN PHASE

The *design phase* begins after the selection of the architect. Because the architect (usually a firm) may have limited capabilities for handling the broad range of building-design activities, several different, more specialized consultants are usually required, depending on the size and scope of the project.

In most projects, the design team consists of the architect, civil and structural consultants, and mechanical, electrical, plumbing, and fire-protection (MEPF) consultants. In complex projects, the design team may also include an acoustical consultant, roofing and waterproofing consultant, cost consultant, building code consultant, signage consultant, interior designer, landscape architect, and so on.

Some design firms have an entire design team (architects and specialized consultants) on staff, in which case the owner will contract with a single firm. Generally, however, the design team comprises several different design firms. In such cases, the owner typically contracts the architect, who in turn contracts the remaining design team members, Figure 1.1.

Thus, the architect functions as the prime design professional and, to a limited degree, as the owner's representative. The architect is liable to the owner for his or her own work and that of the consultants. For that reason, most architects ensure that their consultants carry adequate liability insurance.

In some projects, the owner may contract some consultants directly, particularly a civil consultant (for a survey of the site, site grading, slope stabilization, and site drainage), a geotechnical consultant (for investigation of the soil properties), and a landscape architect (for landscape and site design), Figure 1.2. These consultants may be engaged before or at the same time as the architect.

Even when a consultant is contracted directly by the owner, the architect retains some liability for the consultant's work. This liability occurs because the architect, being the prime design professional, coordinates the entire design effort, and the consultants' work is influenced a great deal by the architectural decisions. Therefore, the working relationship

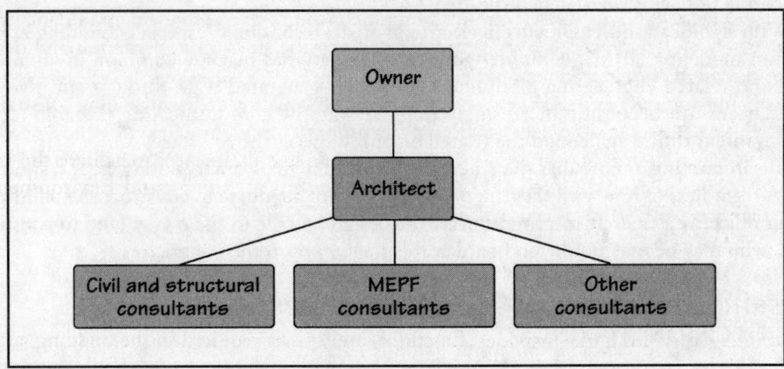

FIGURE 1.1 Members of a typical design team, and their interrelationships with each other and the owner in a traditional contractual setup. A line in this illustration indicates a contractual relationship between parties. ("MEPF consultants" is an acronym for mechanical, electrical, plumbing and fire consultants.)

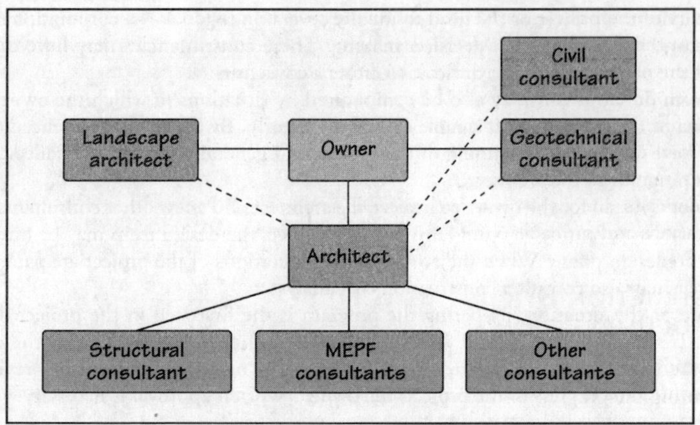

FIGURE 1.2 Members of a typical design team, and their interrelationships with each other and the owner in a project where some consultants are contracted directly by the owner. A solid line in this illustration indicates a contractual relationship between parties. A dashed line indicates a communication link, not a contract.

between the architect and an owner-contracted consultant remains essentially the same as if the consultant were chosen by the architect.

In some cases, an engineer or another professional may coordinate the design process. This generally occurs when a building is a minor component of a large-scale project. For example, in a highly technical project such as a power plant, an electrical engineer may be the prime design professional.

In most building projects, the design phase consists of three stages:

- Schematic design stage
- Design development stage
- Construction documents stage

SCHEMATIC DESIGN (SD) STAGE—EMPHASIS ON DESIGN

The *schematic design* gives graphic shape to the project program. It is an overall design concept that illustrates the key ideas of the design solution. The major player in this stage is the architect, who develops the design scheme (or several design options) with only limited help from the consultants. Because most projects have strict budgetary limitations, a rough estimate of the project's probable cost is generally produced at this stage.

The schematic design usually goes through several revisions, because the first design scheme prepared by the architect will rarely be approved by the owner. The architect communicates the design proposal(s) to the owner through various types of drawings—plans, elevations, sections, freehand sketches, and three-dimensional graphics (isometrics, axonometrics, and perspectives). For some projects, a three-dimensional scale model of the entire building or the complex of buildings, showing the context (neighboring buildings) within which the project is sited, may be needed.

With significant developments in electronic media technology, especially building information modeling (BIM), computer-generated imagery has become common in architecture and related engineering disciplines. Computer-generated walk-through and flyover simulations are becoming increasingly popular ways of communicating the architect's design intent to the owner and the related organizations at the SD stage.

It is important to note that the schematic design drawings, images, models, and simulations, regardless of how well they are produced, are not adequate to construct the building. Their objective is merely to communicate the design scheme to the owner (and to consultants, who may or may not be on board at this stage), not to the contractor.

DESIGN DEVELOPMENT (DD) STAGE—EMPHASIS ON DECISION MAKING

Once the schematic design is approved by the owner, the process of designing the building in greater detail begins. During this stage, the schematic design is developed further—hence the term *design development* (DD) stage.

While the emphasis in the SD stage is on the creative, conceptual, and innovative aspects of design, the DD stage focuses on developing practical and pragmatic solutions for the exterior envelope, structure, fenestration, interior systems, MEPF systems, and so forth. This development involves strategic consultations with all members of the design team.

Therefore, the most critical feature of the DD stage is decision making, which may range from broad design aspects to details. At this stage, the vast majority of decisions about products, materials, and equipment are made. Efficient execution of the construction documents depends directly on how well the DD is managed. A more detailed version of the specifications and probable cost of the project is also prepared at this stage.

Construction Documents (CD) Stage—Emphasis on Documentation

The purpose of the *construction documents* (CD) stage is to prepare all documents required by the contractor to construct the building. During this stage, the consultants and architect collaborate intensively to work out the "nuts and bolts" of the building and develop the required documentation, referred to as *construction documents*. All of the consultants advise the architect, but they also collaborate with each other (generally through the architect) so that the work of one consultant agrees with that of the others.

The construction documents consist of the following:

- Construction drawings
- Specifications

Construction Drawings

During the CD stage, the architect and consultants prepare their own sets of drawings, referred to as *construction drawings*. Thus, a project has architectural construction drawings, civil and structural construction drawings, MEPF construction drawings, landscape construction drawings, and so on.

Construction drawings are dimensioned drawings (usually computer generated) that fully delineate the building. They consist of floor plans, elevations, sections, schedules, and various large-scale details. The details depict a small portion of the building that cannot be adequately described on smaller-scale plans, elevations, or sections.

Construction drawings are the drawings that the construction team uses to build the building. Therefore, they must indicate the geometry, layout, dimensions, types of materials, details of assembling the components, colors and textures, and so on. Construction drawings are generally two-dimensional drawings, but three-dimensional isometrics are sometimes used for complex details. Construction drawings are also used by the contractor to prepare a detailed cost estimate of the project at the time of bidding.

Construction drawings are not a sequence of assembly instructions, such as for a bicycle. Instead, they indicate what every component is and where it will be located when the building is completed. In other words, the design team decides the "what" and "where" of the building. The "how" and "when" of the building are entirely in the contractor's domain.

NOTE

Working Drawings and Construction Drawings

The term *working drawings* was used until the end of the twentieth century for what are now commonly referred to as *construction drawings*.

Specifications

Buildings cannot be constructed from drawings alone, because there is a great deal of information that cannot be included in the drawings. For instance, the drawings will give the locations of columns, their dimensions, and the material used (such as reinforced concrete), but the quality of materials, their properties (the strength of concrete, for example), and the test methods required to confirm compliance cannot be furnished on the drawings. This information is included in the document called *specifications*.

Specifications are written technical descriptions of the design intent, whereas the drawings provide the graphic description. The two components of the construction documents—the specifications and the construction drawings—complement each other and generally deal with different aspects of the project. Because they are complementary, they are supposed to be used in conjunction with each other. There is no order of precedence between the construction drawings and the specifications. Thus, if an item is described in only one place—either the specification or the drawings—it is part of the project, as if described in the other.

For instance, if the construction drawings do not show the door hardware (hinges, locks, handles, and other components) but the hardware is described in the specifications, the owner will get the doors with the stated hardware. If the drawings had precedence over the specifications, the owner would receive doors without hinges and handles.

EXPAND YOUR KNOWLEDGE

Relationship Between Construction Drawings and Specifications

Construction Drawings	Specifications
Design intent represented graphically	Design intent represented with words
Product/material may be shown many times	Product/material described only once
Product/material shown generically	Product/material identified specifically, sometimes proprietary to a manufacturer
Quantity indicated	Quality indicated
Location of elements established	Installation requirements of elements established
Size, shape, and relationship of building elements provided	Description, properties, characteristics, and finishes of building elements provided

Generally, there is little overlap between the drawings and the specifications. More importantly, there should be no conflict between them. If a conflict between the two documents is identified, the contractor must bring it to the attention of the architect promptly. In fact, construction contracts generally require that before starting any portion of the project, the contractor must carefully study and compare the drawings and the specifications and report inconsistencies to the architect.

If the conflict between the specifications and the construction drawings goes unnoticed initially but later results in a dispute, the courts have in most cases resolved it in favor of the specifications—implying that the specifications, not the drawings, govern the project. However, if the owner or the design team wishes to reverse the order, it may be so stated in the owner-contractor agreement.

THE CONSTRUCTION DOCUMENT SET

Just as the construction drawings are prepared separately by the architect and each consultant for their respective portions of the work, so are the specifications. The specifications from various design team members are assembled by the architect in a single document called the *project manual.* Because the specifications consist of printed (typed) pages (not graphic images), a project manual is a bound document—like a book.

The major component of a project manual is the specifications. However, the project manual also contains other items, as explained later in this chapter.

The set of construction drawings (from various design team members) and the project manual together constitute what is known as the *construction document set,* Figure 1.3. The construction document set is the document that the owner and architect use to invite bids from prospective contractors.

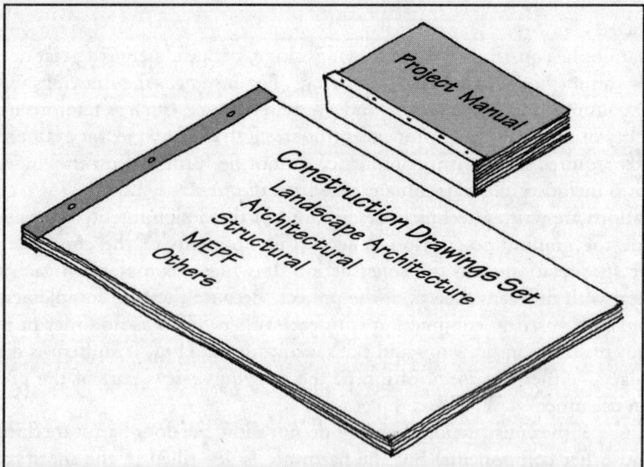

FIGURE 1.3 A construction document set consists of a set of architectural and consultants' construction drawings plus the project manual. The project manual is bound in a book format.

EXPAND YOUR KNOWLEDGE

Surety Bonds

The purpose of a surety bond is to ensure that if the contractor fails to fulfill contractual obligations, there will be a financially sound party—referred to as the *surety* (also called the *guarantor* or *bonding company*)—available to take over those unfulfilled obligations. The bond is, therefore, a form of insurance that the contractor buys from a surety, generally a corporation.

There are three types of surety bonds in most building projects (A few others may be required in some special projects):

- Bid bond
- Performance bond
- Payment bond

Bid Bond

The purpose of the *bid bond* (also called the *bid security bond*) is to exclude frivolous bidders. It ensures that, if selected by the owner, the bidder will be able to enter into a contract with the owner based on the bidding requirements, and that the bidder will be able to obtain performance and payment bonds from an acceptable surety.

A bid bond is required at the time the bidder submits the bid for the project. If the bidder refuses to enter into an agreement or is unable to provide the required performance and payment bonds, the surety is obliged to pay the penalty (bid security amount), usually 5% of the bid amount, to the owner.

Performance Bond

The *performance bond* is required by the owner before entering into an agreement with the successful bidder. The performance bond ensures that if, after the award of the contract, the contractor is unable to perform the work as required by the bidding documents, the surety will provide sufficient funds for the completion of the project.

A performance bond protects the owner against default by the contractor or by those for whose work the contractor is responsible, such as the subcontractors. For that reason, the general contractor will generally require a performance bond from major subcontractors.

Payment Bond

A *payment bond* (also referred to as a *labor and materials bond*) ensures that those providing labor, services, and materials for the project—such as the subcontractors and material suppliers—will be paid by the contractor. In the absence of the payment bond, the owner may be held liable to those subcontractors and material suppliers whose services and materials have been invested in the project. This liability exists even if the owner has paid the general contractor for the work of these subcontractors and material suppliers.

Pros and Cons of Bonds

The bonds are generally mandated for a publicly funded project. In a private project, the owner may waive the bonds, particularly the bid bond. This saves the owner some money because although the cost of a bond (the premium) is paid by the contractor, it is in reality paid by the owner since the contractor adds the cost of the bond to the bid amount.

Despite their cost, most owners consider the bonds (particularly the performance and payment bonds) a good value because they eliminate the financial risks of construction. The bid bond is unnecessary in an invitational bidding method where the owner knows the contractor's financial standing and the ability to perform. However, where uncertainty exists, a bid bond provides excellent prequalification screening of the contractor. Responsible contractors generally maintain a close and continuous relationship with their bonding company, so the bonding company's knowledge of a contractor's capabilities far exceeds that of most owners or architects (as an owner's representative).

Whether or not the release of the bidding documents is restricted, the procedure stated earlier ensures that all the bidders are similarly qualified with respect to financial ability, experience, and technical expertise. Because all bidders receive the same information and are of the same standing, the competition is fair; therefore, the contract is generally awarded to the lowest qualified bidder.

DBB Method—Competitive Sealed Proposal

This method is very similar to competitive sealed bidding and is commonly used for publicly funded projects. The difference between competitive sealed bidding and competitive sealed proposal methods is that the owner's selection of the general contractor is not based on price alone but also on such other criteria as the contractor's past experience, safety record, proposed personnel, project schedule, and so on. To ensure fairness, the advertisement and bidding documents must provide the details of the selection criteria, with relative weightings assigned to each criterion.

DBB Method—Invitational Bidding

Invitational bidding, also called *closed bidding*, is another variation of the DBB method that is generally used for quasi-public and some private projects. In this method, the owner preselects general contractors who have demonstrated, based on their experience, resources, and financial standing, their qualifications to perform the work. The selected contractors are then invited to bid for the project, and the contractor with the lowest bid is then awarded the contract. The architect (as the owner's representative) may be involved in the prescreening process.

PRACTICE QUIZ

Each question has only one correct answer. Select the choice that best answers the question.

11. The MasterFormat has been developed by the
- **a.** Construction Specifications Institute.
- **b.** American Society for Testing and Materials.
- **c.** American National Standards Institute.
- **d.** American Institute of Architects.
- **e.** Associated General Contractors of America.

12. The MasterFormat consists of
- **a.** 20 divisions.
- **b.** 30 divisions.
- **c.** 40 divisions.
- **d.** 50 divisions.
- **e.** none of the above.

13. In the MasterFormat, Division 02 refers to
- **a.** General Requirements.
- **b.** Existing Conditions.
- **c.** Masonry.
- **d.** Metals.
- **e.** none of the above.

14. In the MasterFormat, Division 04 refers to
- **a.** General Requirements.
- **b.** Existing Conditions.
- **c.** Masonry.
- **d.** Metals.
- **e.** none of the above.

15. In the MasterFormat, windows are part of
- **a.** Division 05.
- **b.** Division 06.
- **c.** Division 07.
- **d.** Division 08.
- **e.** none of the above.

16. In the MasterFormat, roofing is part of
- **a.** Division 05.
- **b.** Division 06.
- **c.** Division 07.
- **d.** Division 08.
- **e.** none of the above.

17. In the MasterFormat, flooring is part of
- **a.** Division 05.
- **b.** Division 06.
- **c.** Division 07.
- **d.** Division 08.
- **e.** none of the above.

18. In the traditional competitive bidding method, the owner has separate contracts with the general contractor and the subcontractors.
- **a.** True
- **b.** False

19. Who is responsible for ensuring the safety of workers on the construction site of a typical building project?
- **a.** The architect
- **b.** The structural engineer
- **c.** The general contractor
- **d.** The owner
- **e.** All of the above collectively

20. In the traditional competitive bidding method for a building, there is generally
- **a.** one general contractor.
- **b.** one general contractor and one subcontractor.
- **c.** one general contractor and several subcontractors.
- **d.** several general contractors and several subcontractors.

21. A surety bond is provided by the owner to the general contractor.
- **a.** True
- **b.** False

22. The three surety bonds used in a typical construction contract are
- **a.** bid bond, contract award bond, and completion bond.
- **b.** prescreening bond, award bond, and completion bond.
- **c.** bid bond, performance bond, and completion bond.
- **d.** bid bond, performance bond, and payment bond.
- **e.** none of the above.

NOTE

Construction Documents and Contract Documents

The terms *contract documents* and *construction documents* are used interchangeably. Although both sets of documents are essentially the same, the construction documents become contract documents when they are incorporated into the contract between the owner and the contractor.

1.8 CONSTRUCTION PHASE

Once the general contractor has been selected and the contract awarded, the construction work begins, as described in the *contract documents*. The contract documents are virtually the same as the bidding documents, except that the contract documents are part of a signed legal contract between the owner and the contractor. They generally do not contain Division 00 of the MasterFormat.

In preparing the contract documents, the design team's challenge is to efficiently produce the graphics and text that effectively communicate the design intent to the construction professionals and the related product suppliers and manufacturers so that they can do the following:

- Propose accurate and competitive bids
- Prepare detailed and descriptive submittals for approval
- Construct the building with a minimum number of questions, revisions, and changes

SHOP DRAWINGS

The construction drawings and the specifications should provide a fairly detailed delineation of the building. However, they do not describe it to the extent that fabricators can produce building components directly from them. Therefore, the fabricators generate their own drawings, referred to as *shop drawings*, to provide the higher level of detail necessary to fabricate and assemble the components.

Shop drawings are not generic, consisting of manufacturers' or suppliers' catalogs, but are specially prepared for the project by the manufacturer, fabricator, erector, or subcontractors. For example, an aluminum window manufacturer must produce shop drawings to show that the required windows conform with the construction drawings and the specifications. Similarly, precast concrete panels, stone cladding, structural steel frame, marble or granite flooring, air-conditioning ducts, and other components require shop drawings before they are fabricated and installed.

Before commencing fabrication, the fabricator submits the shop drawings to the general contractor. The general contractor reviews them, marks them "approved," if appropriate, and then submits them to the architect for review and approval. Subcontractors or manufacturers cannot submit shop drawings directly to the architect.

The review of all shop drawings is coordinated through the architect, even though they may actually be reviewed in detail by the appropriate consultant. Thus, the shop drawings pertaining to structural components are sent to the architect and then to the structural consultant for review and approval. The fabricator generally begins fabrication only after receiving the architect's review of the shop drawings.

The review of shop drawings by the architect is limited to checking that the work indicated therein conforms with the overall design intent shown in the contract documents. Approval of shop drawings that are later discovered to deviate from the contract documents does not absolve the general contractor of the responsibility to comply with the contract documents for quality of materials, workmanship, or the dimensions of the fabricated components, Figure 1.9.

XYZ
Architects
Any City

Review is only for the limited purpose of checking for conformance with information given and the design concept expressed in the Contract Documents. Review is not conducted for the purpose of determining the accuracy or completeness of other details such as dimensions or quantities or for substantiating instructions for installation or performance of equipment or systems designed by the Contractor, all of which remain the responsibility of the Contractor. Review or acceptance of a specific item shall not indicate approval of an assembly of which the item is a component. Review neither extends nor alters any contractual obligations of the Architect or Contractor and review shall not relieve the Contractor of responsibility for any deviation from the requirements of the Contract Documents.

- ☐ Accepted as Noted
- ☐ Revise, Resubmit
- ☐ Product Substitution, Submit in accordance with Section 01630
- ☐ Value Engineering, Submit in accordance with Section 01630
- ☐ Not Reviewed; Submittal not required by contract documents
- ☐ Reviewed for Architectural Issues Only
- ☐ Reviewed for Project Close Out Requirements Only

By: Date:

FIGURE 1.9 A typical stamp used by architects to indicate the result of review of shop drawings and other submittals.

Mock-Up Samples

In addition to shop drawings, full-size mock-up samples of one or more critical elements of the building may be required in some projects. This is done to establish the quality of materials and workmanship by which the completed work will be judged. For example, it is not unusual for the architect to ask for a mock-up of a typical area of the curtain wall of a high-rise building before the fabrication of the actual curtain wall is undertaken. Mock-up samples go through the same approval process as the shop drawings.

Other Submittals

In addition to shop drawings and any mock-up samples, some other submittals required from the contractor for the architect's review are:

- Product material samples
- Product data
- Certifications
- Calculations

1.9 CONSTRUCTION CONTRACT ADMINISTRATION

The general contractor will normally have an inspection process to ensure that the work of all subcontractors is progressing as indicated in the contract documents and that the work meets the standards of quality and workmanship. On smaller projects, this may be done by the *project superintendent*. On large projects, a team of quality-assurance inspectors generally assists the contractor's project superintendent. These inspectors are individuals who, by training and experience, are specialized in their own areas of construction—for example, concrete, steel, or masonry.

Additional quality control is required by the contract through the use of independent testing laboratories. For instance, structural concrete to be used on the site must be verified for strength and other properties by independent concrete testing laboratories.

Leaving quality control of materials and performance entirely in the hands of the contractor is considered inappropriate. It can render the owner vulnerable to omissions and errors in the work, and it places an additional legal burden on the contractor. Therefore, the owner usually retains the services of the project architect to provide a third-party level of scrutiny to administer the construction contract. If not, the owner will retain another independent architect, engineer, or inspector to provide construction contract administration services. The contracting community favors this third-party oversight of its work.

ARCHITECT'S OBSERVATION OF CONSTRUCTION

The architect's role during the construction phase has evolved over the years. There was a time when architects provided regular supervision of their projects during construction, but the liability exposure resulting from the supervisory role became so adverse for the architects that they have been forced to relinquish this responsibility. Instead, the operative term for the architect's role during construction is *field observation* of the work.

The observational role still allows the architects to verify that their drawings and specifications are transformed into reality just as they had conceived. It also provides a sufficient safeguard against the errors caused by the contractors' misinterpretation of contract documents in the absence of the architects' clarification and interpretation.

The shift in the architect's role to observer of construction also recognizes the important and entirely independent role that the contractor must play during construction. This recognition provides full authority to the general contractor to proceed with the work in the manner that the contractor deems appropriate. This reinforces the earlier statements that:

- The architect determines the *what* and *where*..
- The contractor determines the *how* (means and methods) and *when* (sequence) of construction.

In other words, daily supervision or superintendence of construction is the function of the contractor—the most competent person to fulfill this role. The architect provides periodic observation and evaluation of the contractor's work and notifies the owner if the work is not in compliance with the intent of the contract documents. This underscores the division between the responsibilities of the architect and the contractor during construction.

Note that by providing observation, the architect does not certify the contractor's work. Nor does the observation relieve the contractor of its responsibilities under the contract. The contractor remains fully liable for any error that has not been discovered through the architect's observation. However, the architect may be held liable for all or part of the work observed, should the architect fail to detect or provide timely notification of work not conforming with the contract documents. This omission is known as *failure to detect.*

Because many components can be covered up by other items over days or hours, the architect should visit the construction site at regular intervals, as appropriate to the progress of construction. For example, earthwork covers foundations and underground plumbing, and gypsum board covers ceiling and wall framing. Observing the work after the components are hidden defeats the purpose of observation.

On some projects, a resident project architect or engineer may be engaged by the architect, at an additional cost to the owner, to observe the work of the contractor. Under the conditions of the contract, the contractor is generally required to provide this person with an on-site office, water, electricity, a telephone, and other necessary facilities.

INSPECTION OF WORK

There are only two times during the construction of a project that the architect makes an exception to being an observer of construction. At these times, the architect inspects the work. These inspections are meant to verify the general contractor's claim that the work is (a) substantially complete and (b) has been completed and hence is ready for final payment. These inspections, explained in Section 1.10, are referred to as:

- Substantial completion inspection
- Final completion inspection

PAYMENT CERTIFICATIONS

In addition to construction observation and inspection, there are several other duties the architect must discharge in administering the contract between the owner and the contractor. These are outlined in the box "Summary of Architect's Functions as Construction Contract Administrator." Certifying (validating) the contractor's periodic requests for payment against the work done and the materials stored at the construction site is perhaps the most critical of these functions.

An application for payment (typically made once a month unless stated differently in the contract) is followed by the architect's evaluation of the work and necessary documentation to verify the contractor's claim. Because the architect is not involved in day-to-day supervision, the issuance of the certificate of payment by the architect does not imply acceptance of the quality or quantity of the contractor's work. However, the architect has to be judicious and impartial to both the owner and the contractor and perform within the bounds of the contract.

CHANGE ORDERS

There is hardly a construction project that does not require changes after construction has begun. The contract between the owner and the contractor recognizes this fact and includes provisions for the owner's right to order a change and the contractor's obligation to accept the *change order* in return for an equitable price adjustment. Here again, the architect performs a quasi-judicial role to arrive at a suitable agreement and price between the owner and the contractor.

1.10 POSTCONSTRUCTION (PROJECT CLOSEOUT) PHASE

Once the project is sufficiently complete, the contractor will ask the architect to conduct a *substantial completion inspection* to confirm that the work is complete in most respects. By doing so, the contractor implies that the work is complete enough for the owner to occupy the facility and start using it, even though there might be cosmetic and minor work yet to be completed.

The contractor's request for substantial completion inspection by the architect should include a list of incomplete portions of the work (to be completed), referred to as the *punch list.* The punch list, which is prepared by the contractor, is used by the architect as a checklist to review all work, not merely the incomplete portions of the work. If the architect's inspection discloses incomplete items not included in the contractor's punch list, they are added to the list by the architect.

The substantial completion inspection is also conducted by the architect's consultants, either with the architect or separately. Incomplete items discovered by them are also added to the punch list. If the additional items are excessive, the architect may ask the contractor to complete the selected items before rescheduling the substantial completion inspection.

SUBSTANTIAL COMPLETION—THE MOST IMPORTANT PROJECT DATE

Before requesting a substantial completion inspection, the contractor must submit all required guaranties and warranties from the manufacturers of equipment and materials and the specialty subcontractors and installers used in the building. For instance, the manufacturers of roofing materials, windows, curtain walls, mechanical equipment, and other materials warrant their products for specified time periods. These warranties are in addition to the standard one-year correction period between the owner and the contractor.

The warranties are to be given to the architect at the time of substantial completion for review and transmission to the owner. Because the obligatory one-year correction period between the owner and the contractor, as well as other extended-time warranties, begin from the date of substantial completion of the project, the substantial completion date is an important project closeout event. That is why the contractor is allowed a brief time interval to complete the work fully after the successful substantial completion inspection.

Before seeking a substantial completion inspection, the contractor is required to secure a *certificate of occupancy* (Chapter 2) from the authority having jurisdiction over the project—usually the city where the project is permitted and built. The certificate of occupancy confirms that all appropriate inspections and approvals, required by the authority having jurisdiction, have taken place and that the site has been cleared of the contractor's temporary facilities so that the owner can occupy the building without obligations to any authority.

After substantial completion, the contractor is no longer liable for the maintenance (cleaning and upkeep), utility costs, insurance, and security of the project. These responsibilities and liabilities are transferred to the owner.

FINAL COMPLETION INSPECTION

After the contractor carries out all the corrective work identified during the substantial completion inspection and so informs the architect, the architect (with the assistance of the

NOTE

Summary of the Architect's Functions as Construction Contract Administrator

- Observe construction
- Inform the owner of the progress of work
- Guard the owner against defects and deficiencies
- Review and approve shop drawings, mock-up samples, and other submittals
- Prepare change orders, if required
- Review correspondence between the owner and the contractor and take action if required
- Prepare certificates of payment
- Make the substantial completion inspection
- Make the final completion inspection
- Review manufacturers' and suppliers' warranties and other project closeout documentation and forward them to the owner
- Make a judicious interpretation of the contract between the owner and the contractor when needed

consultants) carries out the final inspection of the project. If the final inspection passes, certification for final payment is issued by the architect, which entitles the contractor to receive the final payment from the owner.

Before the certification for final payment is executed by the architect and, finally, by the owner, the owner receives the record documents, keys and key schedule, equipment manuals, and other specified necessities. Additionally, the owner receives all legal documentation to indicate that the contractor will be responsible for claims made by any subcontractor, manufacturer, or other party with respect to the project.

RECORD DOCUMENTS (AS-BUILT DOCUMENTS)

As previously stated, minor changes are often made during the construction of a project. These changes must be recorded for the benefit of the owner should the owner wish to alter or expand the building in the future. Therefore, after the building has been completed, the contractor is required to provide a set of *record drawings* (previously known as *as-built drawings*). These drawings reflect the changes that were made during the course of construction by the contractor.

In addition to record drawings, *record specifications,* as well as a set of approved shop drawings, are usually required to complete the record document package delivered to the owner.

1.11 DESIGN-NEGOTIATE-BUILD PROJECT DELIVERY METHOD

In the design-bid-build (DBB) method, discussed thus far, the architect designs the project and prepares the bidding documents. After the bidding documents have been completed, the architect assists the owner in selecting the general contractor, which is done through competitive bids (sealed bids, sealed proposals, or invitational bids).

Once construction begins, the architect visits the site to observe the work in progress, advises the owner whether the work conforms with the contract documents, and acts on the general contractor's requests for periodic payments to be made by the owner. In other words, the architect functions (in a limited sense) as the owner's representative and provides professional service from the inception to the completion of the project.

Apart from the fact that the DBB method is simple and well understood, it has several other advantages: (a) there is a single point of responsibility for construction, (b) the contractor is selected through aggressive and open competition, and (c) the project's scope and cost are fully defined before construction starts.

A major disadvantage of the DBB method is the absence of the contractor's preconstruction (design-phase) services. A delivery method that addresses this concern uses a negotiated contract and is called the *design-negotiate-build* (DNB) project delivery method.

The DNB method is used when the owner knows of one or more reputable, competent, and trusted general contractors. The owner simply negotiates with these contractors concerning the overall contract price, time required for completion, and other important details of the project. The negotiations are generally conducted with one contractor at a time, and after negotiations with all selected contractors are complete, the owner analyzes the bids and selects a general contractor.

A major advantage of the negotiated contract is that the general contractor can be on board during the design (or predesign) phase. This helps the owner ensure that the architect's design is realistically constructible. In many situations, the contractor may advise the architect of simpler, less expensive, or more sophisticated building systems to realize the architect's design intentions.

Additionally, because the contractor is the one who is most knowledgeable about construction costs, budget estimates can be obtained at various stages during the design phase. This means that *value engineering* can proceed throughout the design phase instead of being undertaken at the end of this phase or during construction, as in the DBB method of project delivery.

Because the vast majority of owners have to work within a limited budget, the negotiated contract is a popular delivery method for private projects. The services offered by the contractor during the design phase of a negotiated contract are referred to as the contractor's *preconstruction services.*

The negotiated contract is not devoid of competition, because the general contractor obtains competitive bids from numerous subcontractors and material suppliers. Because the general contractor is selected during the SD or DD stage, the bids from some or all subcontractors can be obtained earlier, which may shorten the project delivery time.

NOTE

Value Engineering

Value engineering (VE) is the science of obtaining balance among the cost, reliability, and performance of a product or project. VE is commonly used in building projects when the project's cost exceeds its budget. It relates to the substitution for a product or assembly that is part of the original design with an alternative that provides equivalent performance, reliability, and aesthetics at a lower cost.

1.12 CONSTRUCTION MANAGEMENT-RELATED PROJECT DELIVERY METHODS

In the delivery methods so far discussed (design-bid-build and design-negotiate-build), the role of the architect remains essentially the same: the architect designs the project, helps the owner select the contractor, and provides construction administration services during the construction phase as the owner's representative. In the 1970s, in response to cost overruns and time delays caused by lack of realism in the design of several projects, owners began to seek the assistance of the contracting community during the design phase of the project. This approach became more common as project complexities grew, giving birth to an entirely new profession called *construction management.*

Construction Manager as the Owner's Agent—CMAA Method

The project delivery method in which a construction manager (CM) is included is referred to as the *construction manager as agent (CMAA) method.* In this method, the owner retains a CM as the owner's agent to advise on such issues as cost, scheduling, site supervision, site safety, construction finance administration, and overall building construction.

Note that the CM is not a contractor, but a manager who plays no entrepreneurial role in the project (unlike the general contractor, who assumes financial risks in the project). In most CMAA projects, the owner hires the CM as the first step. The CM may advise the owner in the selection of the architect and other members of the design team as well as the contracting team.

The birth of the CMAA delivery method does not mean that there is no construction management in design-bid-build, or the negotiated contract method. It exists, but it is done informally and shared by the design team and the general contractor.

The introduction of a CM on the project transfers various functions of the general contractor (in a traditional method) to the CM. Thus, in the CMAA method, the general contractor becomes redundant. Therefore, there is no general contractor in this method.

In the CMAA method of project delivery, the owner awards multiple contracts to various trade and specialty contractors, whose work is coordinated by the CM. Thus, the structural framework of the building may be erected by one contractor, masonry work done by another, interior drywall work by yet another, and so on.

Each contractor is referred to as the *prime contractor,* who may have one or more subcontractors, Figure 1.10. The task of scheduling and coordinating the work of all the contractors and ensuring site safety—undertaken by the general contractor in the DBB and DNB methods—is done by the CM in the CMAA method. Additionally, the CM administers the contracts between the prime contractors and the owner. Note, however, that because the CM is only an agent (employed to administer the contract on behalf of the owner), all the financial risks and other liabilities in the project are assumed by the owner.

Thus, the owner, by assuming part of the role of the general contractor, eliminates the general contractor's markup on the work of the subcontractors. The owner may also receive a reduction in the fee charged by the architect for contract administration. Although these savings are partially offset by the fee that the owner pays to the CM, there can still be substantial savings in large but technically simple projects.

The CMAA project delivery method is particularly attractive to owners who are knowledgeable about the construction process and can participate fully in all of its aspects, from bidding and bid evaluation to the closeout phase.

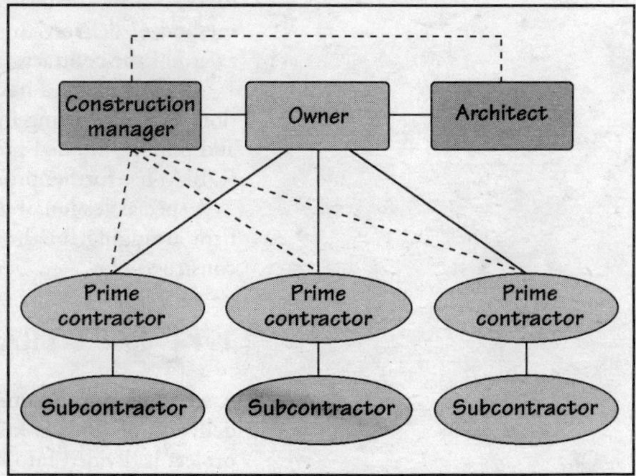

FIGURE 1.10 Contractual relationships between various parties in the CMAA method of project delivery. A solid line in this illustration indicates a contractual relationship between parties. A dashed line indicates a communication link, not a contract.

Construction Manager at Risk (CMAR) Method

A disadvantage of the CMAA method lies in the liability risk that the owner assumes, which in the design-bid-build method is held by the general contractor. This means that there is not the same incentive for the CM to optimize efficiency as when the CM carries financial risks.

Additionally, in the CMAA method, there is no single point of responsibility among the various prime contractors. Each prime

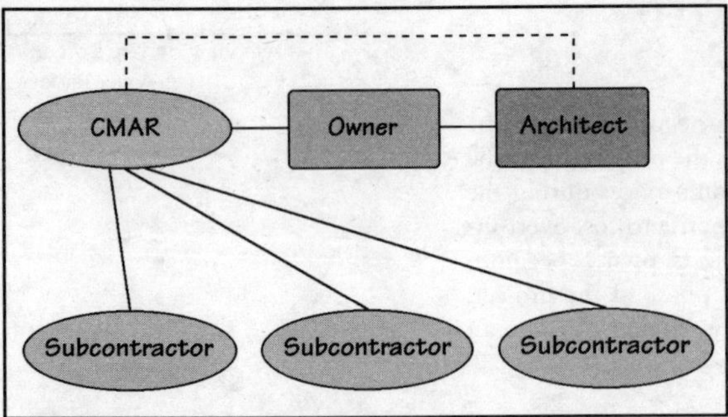

FIGURE 1.11 Contractual relationships between various parties in the CMAR method of project delivery. A solid line in this illustration indicates a contractual relationship between parties. A dashed line indicates a communication link, not a contract.

contractor has a direct contract with the owner. Consequently, the CM has little leverage to ensure timely performance. The owner must therefore exercise care in selecting the CM because the cost, timeliness, and quality of the ultimate product are heavily dependent on the expertise of the CM.

In response to the preceding concerns, the CMAA method has evolved into what is known as the *construction manager at risk* (CMAR) method. In this method, the roles of the CM and general contractor are performed by one entity, but the compensation for these roles is paid separately by the owner.

In the CMAR method, the owner contracts with a CMAR company (a) to provide construction management services during the design phase of the project for a professional fee and (b) to work as the general contractor of the project. Thus, the CMAR company works with the architect during the design phase to develop construction documents that will meet the owner's budget and schedule. In doing so, the CMAR company functions as the owner's representative. The relationships between the various parties in a CMAR project delivery method are shown in Figure 1.11.

After the drawings are completed, all the work is competitively bid by subcontractors and the bids are opened in the owner's presence. The work is normally awarded to subcontractors with the lowest bids. In working as the general contractor, the CMAR company assumes all responsibilities for subcontractors' work and site safety. The CMAR method is being increasingly used for publicly funded projects such as schools, university residence halls, and apartments buildings.

1.13 DESIGN-BUILD (DB) PROJECT DELIVERY METHOD

A project delivery method that integrates design and construction activities into a single entity is called the *design-build (DB) method.* In this method, the owner awards the contract to one firm, which designs the project and also builds it, either on a cost-plus-profit basis or a lump-sum basis. In many ways, this method resurrects the historic *master-builder method,* in which there was no separation between the architect and the contractor. The design-build firm is usually a general contractor, which, in addition to providing construction capabilities, has a design team (of architects and engineers) within the organization or a closely allied separate organization.

The DB method has the advantage of integrating design and construction, thus fostering teamwork between the design team and the contractor throughout the project. It can provide a reduction in change orders for the owner, faster project completion, and a single source of responsibility. The major disadvantage is that the owner does not receive the protection provided by the checks and balances inherent in delivery methods with separate design and construction responsibilities. Consequently, once the contract has been awarded to a DB firm, the owner loses much of the control over the project. Therefore, for the DB method of delivery to succeed, the end result must be meticulously defined prior to the award of the contract.

The DB method has been in existence for decades in single-family residential construction. It is now being increasingly accepted in commercial construction—for both private and publicly funded projects. The establishment of the Design-Build Institute of America (DBIA) has further promoted the method.

A special version of the DB method, referred to as the *turnkey method,* consists of the DB firm arranging for the land and financing for the project in addition to designing and constructing it.

1.14 FAST-TRACK PROJECT SCHEDULING

A scheduling technique that can be used to save project delivery time with most project delivery methods is known as the *fast-track scheduling technique.* In this technique, the project is divided into multiple segments, and each segment of construction is awarded to different contractors through negotiations. The division of construction into segments is such that the segments are sequential. Thus, the first segment of the project may be site construction (site development, excavations, and foundations), the second segment may be structural framing (columns, beams, and floor and roof slabs), and the third segment may

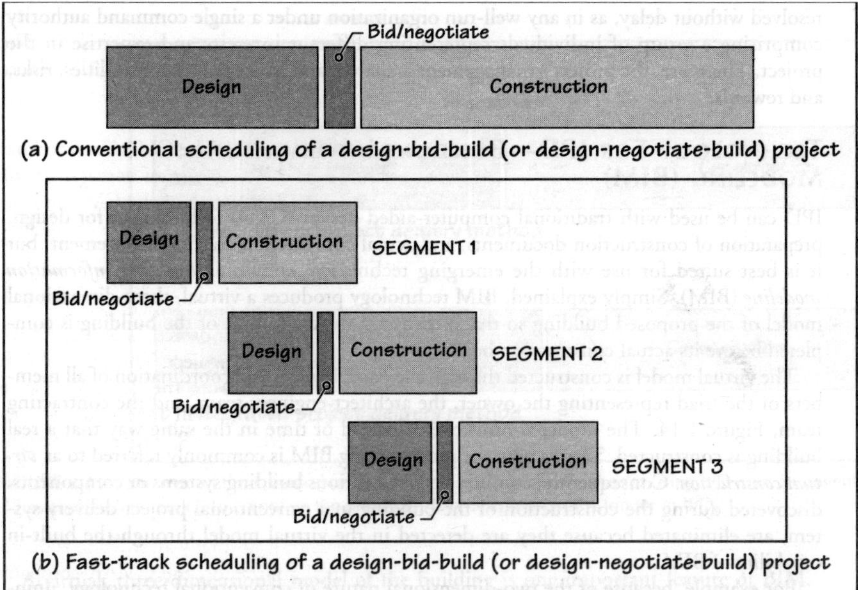

FIGURE 1.12 In fast-track scheduling, a project is segmented into parts, which overlap in time. As shown in this illustration, segmentation shortens project delivery in comparison with the design-bid-build or design-negotiate-build method. Fast-track scheduling can also be applied to other project delivery methods.

be the exterior enclosure, interior finishes, and project closeout that takes the project to completion.

Sequential segmenting of the project saves time because the earlier segments of the project can be constructed while the construction documents for the later segments are still in progress, resulting in overlapping design and construction processes, Figure 1.12.

Fast-track sequencing requires a great deal of coordination between segments. It also requires a commitment from the owner that the decisions will not be delayed and, once made, will not be changed.

1.15 THE INTEGRATED PROJECT DELIVERY (IPD) METHOD

The *integrated project delivery* (IPD) method is the ultimate in promoting harmony, collaboration, and integration among all team members who contribute to the project. While the members of the triad (owner, architect-engineer team, and contracting team) are separated into three distinct entities in the design-bid-build or CMAR method, and into two distinct entities in the design-build method, they are integrated fully into one entity in the IPD method, Figure 1.13.

In fact, the IPD method involves not simply the integration of the three major entities but of all those who contribute to the project (owner, architect, engineers, general contractor, subcontractors, fabricators, material suppliers, etc.). All participants come on board during the design phase or as soon as their expertise is needed. The entire delivery process, from inception to completion, is open across participants, with continuous sharing of knowledge.

The central underlying philosophy of IPD is across-the-board, trust-based collaboration in a zero-blame and zero-litigation environment. Differences and disputes are

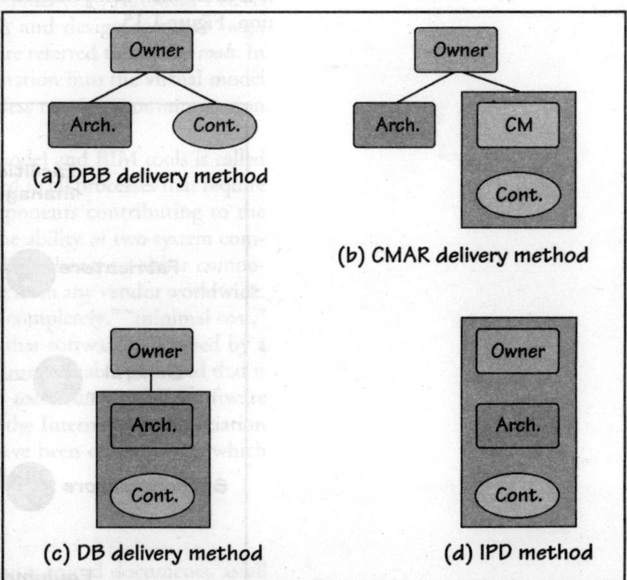

FIGURE 1.13 This illustration shows the relative integration among three major entities—owner, architect (Arch.), and contractor (Cont.)—in four major project delivery methods. Note that there is limited integration among the three entities in the DBB or CMAR method, partial integration in the DB method, and (supposedly) full integration in the IPD method. (The term *architect* implies the entire design team, which includes the architect and the architect's consultants.)

CHAPTER 2

Construction Regulations and Standards

CHAPTER OUTLINE

The primary requirement of a building is that it should be safe and healthy. Visual appeal and economic viability, though important, are secondary requirements. To deliver a safe and healthy building is primarily the responsibility of the design and construction professionals—architects, engineers, and builders. However, as with all public health and safety issues, the design and construction of buildings are regulated by numerous federal, state, and local laws. The most important of these laws are contained in a document called the *building code.*

A building code is enforced by local jurisdictions, such as cities or municipalities, under police powers granted to them by the state. No building may be constructed unless it meets the requirements of the building code of the local jurisdiction.*

Although a few building code provisions are based on traditional construction practices, most provisions are firmly grounded in scientific and quantifiable data of construction performance. There is, therefore, a strong link between code requirements and the science of construction.

Ongoing technological developments in the building industry require that building codes be constantly reviewed and revised. This task is usually beyond the resources of most local jurisdictions. It is, therefore, handled by an independent agency—the model code organization—whose primary responsibility is to develop, maintain, and publish the building code and the other related codes.

*There are often unregulated areas in a state or county where buildings may be constructed without conforming to any building code. Rural communities and areas outside urban boundaries are generally unregulated.

This chapter begins with a discussion of the objectives of a building code, followed by a description of the organizational principles and contents of the current model building code. Finally, the chapter deals with *zoning ordinances* and other important laws that affect building construction.

2.1 OBJECTIVES OF A BUILDING CODE

The objectives of a building code are to ensure that all new construction and renovated buildings provide a minimum level of safety, health, and welfare to the occupants and the public at large. Although under no legal obligation to do so, the owner or the designer may choose to exceed the requirements of the code. For the sake of economy, however, a large majority of buildings are designed to satisfy only the minimum requirements.

A building code does not regulate aspects of design that relate to a building's appearance. It deals with the issues of a building's performance. Therefore, aesthetics, color, and form-related attributes are outside the purview of building codes.

Additionally, the code protects the safety, health, and welfare not only of the owner of the project but also of the general public, because the interests of the owner, the general public, and the building occupants may be at variance with one another. What may be in the best interest of the owner may not be in the best interest of the public or the occupants of the building.

For this reason, building construction is regulated by an impartial authority, such as the state, county, or city. It is the responsibility of the regulatory authority to ensure that the interests of all concerned parties are protected. Although design and construction professionals generally dislike having their work policed by an external authority, building codes have one major benefit for them all—that of liability protection. If the building has been designed and constructed in accordance with the building code and other applicable regulations, the design and construction professionals are exposed to a substantially lower liability risk.

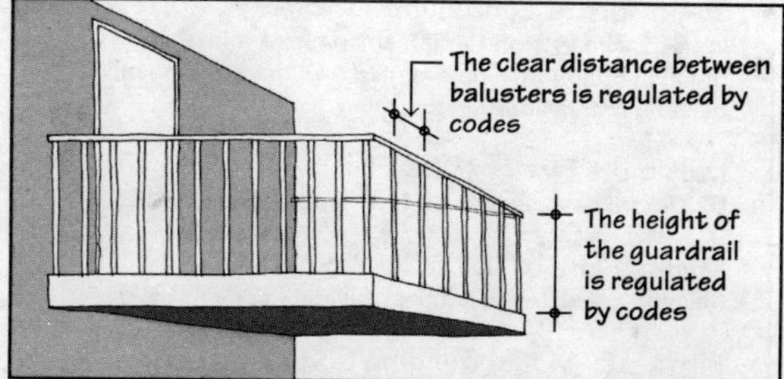

FIGURE 2.1 In addition to its structural integrity, the height of a balcony's guardrail as well as the clear space between the balusters is regulated by building codes.

In more specific terms, a building code regulates the following aspects of building design and construction:

- Life safety
- Fire safety
- Structural safety
- Health and welfare
- Property protection

LIFE SAFETY

Although both fire safety and structural safety are essentially life safety issues, the reverse is not always true. The term *life safety* has its own independent existence in building codes, because several safety regulations in codes are related neither to fire safety nor structural safety. For example, a guardrail on a balcony, apart from being structurally adequate, must provide protection from the danger of falling over the top of the rail or from between its vertical or horizontal members, Figure 2.1.

Thus, the regulations relating to the height of a guardrail and the clear space between its members are life safety regulations with no relationship to fire safety or structural safety. Similarly, the handrail on a staircase, apart from being structurally strong, must provide adequate grippage. A handrail whose cross section is either too large or too small will not provide the required safety. Therefore, building codes prescribe maximum and minimum dimensions of a handrail in addition to its clearance from the adjacent wall, Figure 2.2. These regulations are life safety regulations, as distinct from structural safety and fire safety regulations.

The relationship between the treads and risers of a staircase, the dimensional uniformity of the treads and risers, and the slope of a ramp are some other examples of life safety issues in codes, Figure 2.3. Additionally, accessibility regulations requiring that a building be easily accessible to individuals with disabilities are, in many ways, life safety regulations.

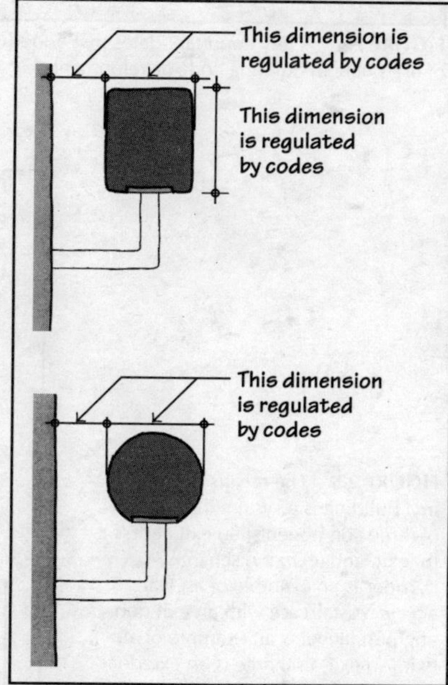

FIGURE 2.2 To ensure adequate grippage, the building code regulates various dimensions of a staircase handrail.

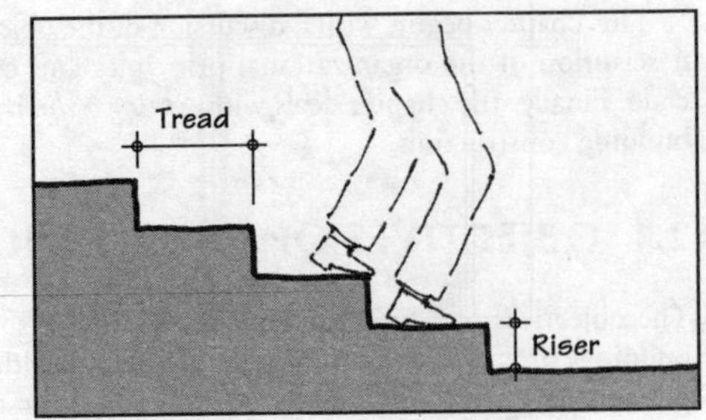

FIGURE 2.3 The dimensions of the treads and risers of a stair are regulated by building codes. Additionally, each tread must be of the same dimension and profile. The same applies to the risers; see Section 35.1.

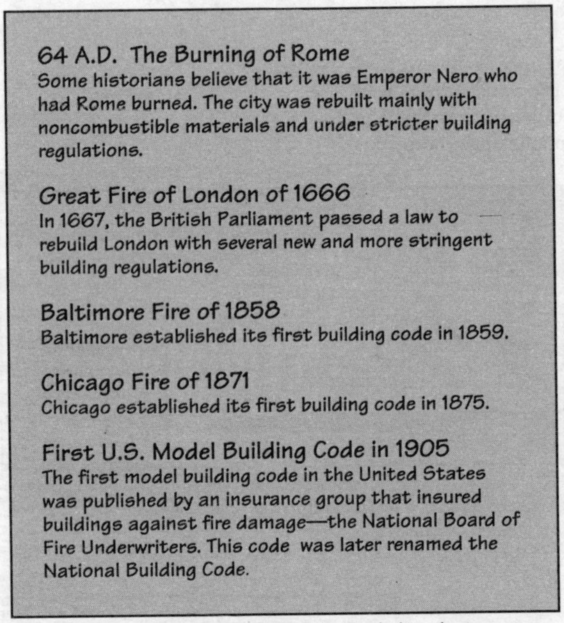

64 A.D. The Burning of Rome
Some historians believe that it was Emperor Nero who had Rome burned. The city was rebuilt mainly with noncombustible materials and under stricter building regulations.

Great Fire of London of 1666
In 1667, the British Parliament passed a law to rebuild London with several new and more stringent building regulations.

Baltimore Fire of 1858
Baltimore established its first building code in 1859.

Chicago Fire of 1871
Chicago established its first building code in 1875.

First U.S. Model Building Code in 1905
The first model building code in the United States was published by an insurance group that insured buildings against fire damage—the National Board of Fire Underwriters. This code was later renamed the National Building Code.

FIGURE 2.4 A few important dates that underline the importance of fire safety in building code development.

FIRE SAFETY

Fire safety regulations in building codes are among the most important regulations. If one were to separate structural regulations from building codes, the bulk of the remaining regulations would relate, in one way or another, to fire safety issues. In fact, the history of building codes is replete with instances in which the world's prominent cities either promulgated building codes for the first time or drastically revised their existing ones after the outbreak of major fires, Figure 2.4.

Obviously, fire safety regulations relate to the use of fire-resistant materials and construction. As we will see in Chapter 7, the types of building construction, as defined by the codes, are directly related to the fire resistance of the major components of the building—walls, floor slabs, roofs, beams, columns, and so on. We will also observe in that chapter that smoke plays a predominant role in the safety of occupants during a fire. Therefore, fire safety regulations are, in fact, fire and smoke safety regulations.

One of the most important sets of fire safety regulations deals with the means of egress from the building (escape routes) should a fire occur. Building codes regulate various aspects of the means of egress system in a building, including the width and height of exit enclosures, fire resistance of materials used therein, illumination levels, exit signs, and so on, Figure 2.5.

Similarly, there are code regulations for exit doors, including height and width, fire resistance, panic hardware, and direction of

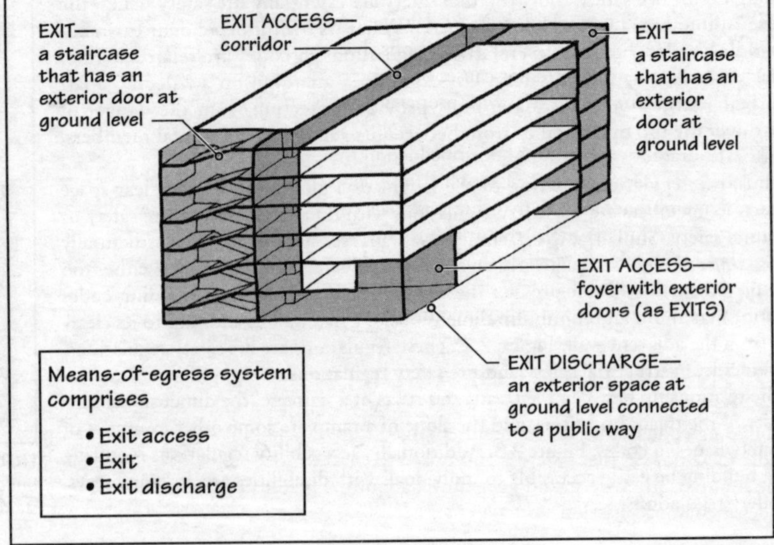

FIGURE 2.5 The means of egress in a building is a system that consists of three components: (a) exit access, (b) exit, and (c) exit discharge. A corridor is an example of an exit access. A staircase with an exit door at ground level is an example of an exit. An exit discharge is an exterior space at ground level that is accessible to a public way. Building codes regulate various aspects of the means-of-egress system for fire safety reasons.

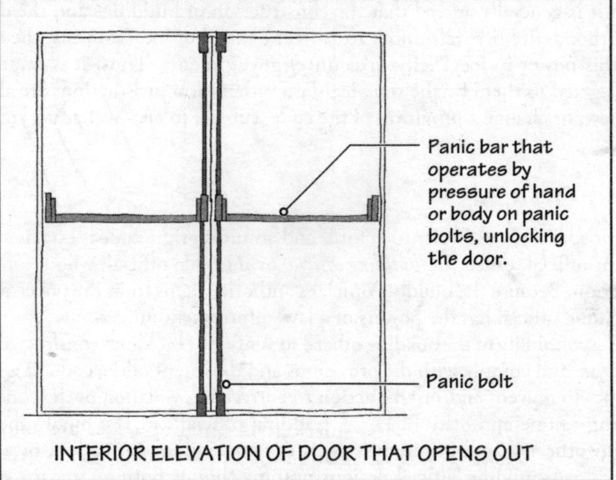

FIGURE 2.6 Panic hardware on exit doors is regulated by codes to ensure fire safety in buildings.

swing, Figure 2.6. The regulations governing the provision of fire and smoke detection, fire and smoke alarms, fire suppression, fire-extinguishing systems (automatic sprinklers and standpipes), and so on, are parts of the fire safety regulations.

STRUCTURAL SAFETY

Structural safety is obviously one of the primary objectives of a building code. A building code contains several chapters that provide detailed regulations dealing with the structural design of buildings. As stated previously, structural regulations are the most numerous and account for the largest volume of the building code document.

HEALTH AND WELFARE

Although the main preoccupation of a building code is with traditional concerns of life safety, fire safety, and structural safety, it also deals with human health and welfare issues. Therefore, the building code contains provisions related to lighting, ventilation, sanitation, temperature, and noise control.

For the same reasons, the minimum dimensions of habitable rooms are regulated by the codes. Regulations pertaining to energy conservation and accessibility for individuals with disabilities are also included. In fact, energy-conservation regulations are increasingly becoming integral to building design.

PROPERTY PROTECTION

Safeguarding a building against loss or damage (property protection) is indirectly ensured through life safety and fire safety regulations, because the burning of the building or its structural collapse are two major causes of property damage. However, the codes also contain several regulations that deal with property protection through requirements for materials' durability.

These durability regulations are embedded in the codes through materials and construction standards (see Section 2.7) to which the codes refer. For instance, the codes require that roof membranes and many other materials conform to relevant ASTM specifications. These specifications contain requirements for the durability of materials. Several other durability requirements, such as those for the decay of structural and nonstructural wood from termite and fungal attack, degeneration of materials due to freeze-thaw action, and corrosion of metals, are expressly stated in the codes.

2.2 ENFORCEMENT OF A BUILDING CODE

A building code is a legal document enforced through the police powers of the state. It is under these powers that a state is authorized to enact legislation for the safety of its citizens. The health, traffic safety, and general welfare of the public are also promoted by the state under the same legislation.

NOTE

Means of Egress

A means of egress in a building is a system that consists of the travel route taken by a building occupant to escape from the interior of the building to a public way. The route, which must remain unobstructed at all times, passes through the following three elements: (a) exit access, (b) exit, and (c) exit discharge.

Exit Access: An exit access is an interior space that leads to an exit. Thus, a corridor leading from a room to an exterior door at the ground level or to a staircase that has an exterior door opening at the ground level is an exit access. A room or other interior space is also part of the exit access.

Exit: An exit consists of an exterior door at the ground level or a staircase that opens at the ground level.

Exit Discharge: An exit discharge is an exterior or interior space immediately beyond the exit at the ground level that leads to a public way. It may be a courtyard, plaza, or any other open-air space adjoining an exit door. An exterior door is an exit only if it opens onto an exit discharge. An exterior door opening into an open-air courtyard that is closed on all sides is not an exit.

A portico or an open verandah at the ground floor that connects an enclosed staircase to the street is also an exit discharge.

Because it is generally agreed that the construction of buildings and the development of neighborhoods are best left under the direct control of local citizens, the state usually delegates this power to local (city or county) governments. Thus, it is under the police powers delegated to them by the state legislature that local jurisdictions are able to enact, adopt, review, or change a provision of the code, subject to any overriding state or federal legislation.

NOTE

Alternative Titles for the Building Official

Some jurisdictions call the building code enforcement official the *director of the building department* or *manager of building safety*. Some very small jurisdictions may have a part-time employee to enforce the codes and may designate that person as the *building inspector*.

BUILDING OFFICIAL

The local jurisdiction's authority to enforce and administer the code is exercised through an enforcement official, called the *building official* or the code official, who is an employee of the jurisdiction. Because the building official's authority stems from the police powers of the state, a building official has the powers of a law enforcement officer.

It is the responsibility of the building official to verify that buildings constructed in the jurisdiction are safe and comply with the provisions and the intent of the code. It is the building official's duty to prevent and/or take action to correct any violation of the code.

According to one authority [2.1], "A building official . . . is a highly specialized law enforcement officer whose prime mission is the prevention, correction, or abatement of violations. . . . A building official designs nothing, builds nothing, repairs nothing. His responsibility is merely to see that those persons who are engaged in these activities do so within the requirements of the law."

In order to effectively discharge duties, a building official must have a thorough knowledge of the building code, related city regulations, and the science of building construction. Law enforcement responsibilities require that a building official's dealings with the public be fair and impartial.

As the head of a city building department, the building official is usually assisted by several other functionaries (plans examiners and building inspectors) to carry out such tasks as plan reviews and field inspections. In a small city or jurisdiction, the building official may be the only person performing these tasks. Some cities (particularly smaller cities) may outsource some or all of a building official's functions to an outside consultant—a noncity employee.

NOTE

Plans for a Building Permit

The word *plans* is generally used for what really is a set of drawings that fully describe the building to be constructed. The set generally includes plans, elevations, sections, and details.

BUILDING PERMIT APPLICATION

Job Address

Legal Descr.	Lot No.	Blk.	Tract

Contractor	Mail address	Zip	Phone
Permit Applicant (Engineer/Architect/Owner)	Mail address	Zip	Phone

Use of Building	Number of floors

Work use (Check one)

New construction	Rehab
New construction-shell	Repair
Remodel	Fire repair
Interior finish	Demolition
Addition conversion	Move
	Swimming pool
	Spa/hot tub
	Misc. uses

Square feet in area	Square feet in garage	Type of slab Post-tensioned	Rebar

NOTICE

Separate permits are required for electrical, plumbing, air conditioning and signs. This permit becomes null and void if work or construction authorized is not commenced within 120 days or if construction or work is suspended or abandoned for a period of 120 days at any time after work is commenced. Applicant is the owner and/or has the owner's consent to do the requested work. I hereby certify that I have read and examined this application and know the same to be true and correct. All provisions of law and ordinances governing this type of work will be complied with whether specified herein or not. The granting of permit does not presume to give authority to violate or cancel the provisions of any other state or local ordinances regulating construction or the performance of construction.

Plan check fee	Permit fee
Type of const.	Occupancy group & Div.
Size of bldg (Total sq ft)	No. of stories
Use zone	Max. Occ. load
Fire sprinklers reqd. Yes	No
Special approvals	
Zoning	
Health Dept.	
Fire Dept.	
Soil report	
Other (specify)	

WHEN PROPERLY VALIDATED (IN THIS SPACE) THIS IS YOUR PERMIT

FIGURE 2.7 A typical building permit application.

BUILDING PERMIT

The general procedure followed in administering the code is to require that the design of a proposed building comply with the provisions of the code before official permission to commence construction is granted by the city. Before granting approval for construction in the form of a *building permit*, the city requires the submission of a building permit application, along with copies of plans for the proposed building. A typical building permit application form is shown in Figure 2.7.

If the plans are in accordance with the building code and other applicable laws, the approved building permit and plans are returned to the owner. After receiving them, the owner may commence construction. If the plans do not comply, the owner is notified by the building official, and the plans must be revised until conformance with the code is achieved.

Once construction begins, periodic and progressive inspections by the building official or the building official's representative—the building inspector—ensure that the construction meets code requirements. The building is inspected at several stages during construction. Typically, these stages are (1) prefoundation (to inspect sewage disposal and water-supply lines), (2) post-foundation, (3) after the erection of the structural frame, (4) after insulation and vapor retarder installation, (5) mechanical, electrical, and plumbing (MEP) rough-in, (6) MEP final, and (7) final building completion.

Every inspection concludes with a report generated by the inspector to indicate whether or not construction is progressing in conformance with the code. Some cities color-code is the most extensive part of the code and includes a separate chapter for each structural material—concrete, masonry, steel, and wood.

Nonstructural Materials Provisions deal with the use of nonstructural materials, such as aluminum, glass, gypsum board, and plastics.

Building Services Provisions deal with electrical, mechanical, and plumbing aspects relevant to architectural design (excluding technical aspects of these systems, which are covered in specialty codes).

Miscellaneous Provisions These deal with miscellaneous concerns, such as construction in the public right of way, site work, demolition, and existing structures.

2.6 APPLICATION OF A BUILDING CODE

Experienced design and construction professionals realize that a building code is a large, complex document. Therefore, they consider the essential features of the code early in the design process and gradually introduce its details as the design process progresses.

Because of the code's complexity, only a brief overview of how a building code is applied to a proposed building is presented here. The most important features of this process, subdivided into eight steps, are illustrated in Figure 2.14.

Step 1 in the application of the code is to determine the occupancy classification of the building. The next step (*Step 2*) is to determine the type of construction to be used. *Step 3* determines the building's frontage and whether or not automatic sprinklers will be provided.

Step 4 involves determining the maximum allowable area and the maximum allowable height of the building. This is related to the degree of hazard present in the building. The fundamental premise of a building code is to have almost the same degree of life-safety risk in all buildings, referred to as *equivalent risk theory*. Thus, if the hazard present in the building is small, the allowable area and height of the building can be large. Conversely, if the hazard present is large, the allowable area and height of the building must be small.

The hazard posed by a building is a function of its occupancy classification, construction type, frontage, automatic sprinklers, area, and height. The more fire-resistive the construction type and (or) the greater the frontage, the lower the hazard. The provision of fire detection and suppression systems (e.g., automatic sprinklers) also reduces the hazard. The larger and (or) taller the building, the greater the hazard it poses.

Other steps in the application of the building code include reviewing the conformity of the building to the detailed provisions of its particular occupancy (*Step* 5) and type of construction (*Step* 6) and reviewing requirements for egress and accessibility (*Step* 7). Finally, the building must be reviewed for conformity with structural engineering requirements (*Step* 8).

OCCUPANCY CLASSIFICATION

Of the various factors that pose hazards in a building (as mentioned previously), its occupancy is most critical. Occupancy-related hazard in a building depends primarily on the following:

NOTE

Factors That Determine the Code-Allowed Area and Height of a Building

The following factors determine the allowable area and height of a building (see Principles in Practice at the end of this chapter):

- Occupancy classification
- Type of construction
- Frontage (open spaces)
- Automatic sprinklers

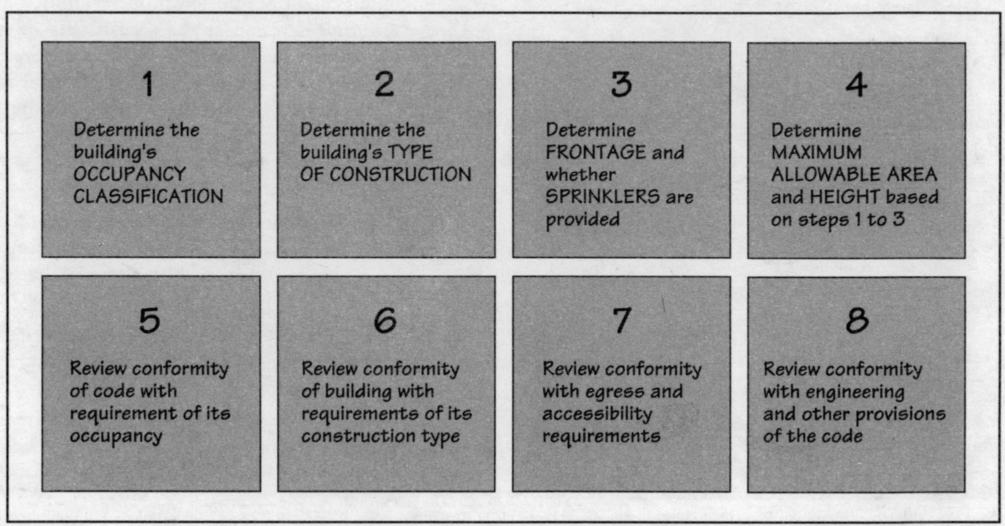

FIGURE 2.14 Steps involved in the application of a building code to a proposed building.

Concentration of Occupants Buildings that accommodate a large number of people, such as stadiums, assembly halls, auditoriums, and houses of worship, present a greater hazard than buildings with a smaller concentration of occupants, such as individual dwellings and apartment buildings.

Fuel Content Buildings that contain flammable materials, such as automobile repair garages, paint and chemical stores, and woodworking mills, are more hazardous because they contain more combustibles than other buildings.

Mobility of Occupants Buildings such as nursing homes, hospitals, day care centers, and prisons have occupants whose mobility is severely limited. Such buildings present a great deal of hazard because their occupants cannot exit in the event of a fire or some other emergency.

Familiarity of Occupants with the Building The greater the occupants' familiarity with the building, the lower the hazard. This conclusion is based on the premise that in the event of a fire or some other emergency, the occupants will be able to exit a building easily if they are familiar with it. A hotel or office building is, therefore, more hazardous than an individual dwelling.

The International Building Code classifies buildings into 10 occupancy groups. Each occupancy group is given a letter designation, such as A, B, . . . , U and is further classified into subgroups, called *divisions.* For example, the Assembly occupancy is divided into five divisions, which are designated as A-1, A-2, A-3, A-4, and A-5, Table 2.1.

TABLE 2.1 OCCUPANCY CLASSIFICATIONS

Generally decreasing hazard →

A Assembly	**A-1** Movie theatre, TV and radio stations, concert hall, etc.	**A-2** Banquet hall, restaurant, nightclub, etc.	**A-3** House of worship, gymnasium, lecture hall, museum, etc.	**A-4** Skating rink, swimming pool, tennis court, etc.	**A-5** Stadium, grandstand, amusement park, etc.
B Business	Bank, professional offices (architects, attorneys, doctors, etc.), beauty shop, fire and police stations, post office, educational institution beyond grade 12, etc.		**A room or space used for assembly purposes by fewer than 50 persons and as an accessory to another occupancy is considered to belong to that occupancy. If the number of persons using that space is 50 or more, it is considered an Assembly occupancy. For example, a lecture or classroom for fewer than 50 students in an educational building (Occupancy E if up to grade 12 or Occupancy B if beyond grade 12) is classified as Occupancy E or Occupancy B. A classroom for more than 50 students in the same building is classified as Occupancy A-3.**		
E Educational	Educational buildings up to grade 12 with 6 or more persons at any one time and day care centers for at least 5 children $2\frac{1}{2}$ years or older.				
F Factory	**F-1** Automobile, bicycles, bakery, clothing, furniture, etc.	**F-2** Beverages, brick and masonry, glass, gypsum, etc.			
H High-Hazard	**H-1** Structures with explosives, unstable chemicals, etc.	**H-2** Structures with *combustible* dust, flammable gas, etc.	**H-3** Structures with aerosols, flammable solids, etc.	**H-4** Structures with toxic materials, etc.	**H-5** Structures with semiconductor fabrication facilities, etc.
I Institutional	**I-1** Convalescent home, assisted living center, etc.	**I-2** Hospital, nursing home, 24-hour day care center, etc.	**I-3** Prison, detention center, correction center, etc.	**I-4** Adult care facility with at least 5 people, etc.	
M Mercantile	Department store, drugstore, market, salesroom, motor vehicle service station, etc.				
R Residential	**R-1** Hotel, motel, boarding house, etc.	**R-2** Apartment, dormitory, monastery, fraternity, etc.	**R-3** Single-family or two-family dwelling unit, etc.	**R-4** Assisted living for more than 5 but fewer than 16 persons.	
S Storage	**S-1** Aircraft hangar, furniture, clothing, grain, leather, etc.	S-2 Cement, glass, parking garage (open or enclosed), etc.			
U Utility	Agricultural building, carport, greenhouse, livestock shelter, etc.				

Source: Adapted from the International Building Code (2009), published by the ICC. The information given in this table is incomplete. For complete and precise information, refer to the source cited.

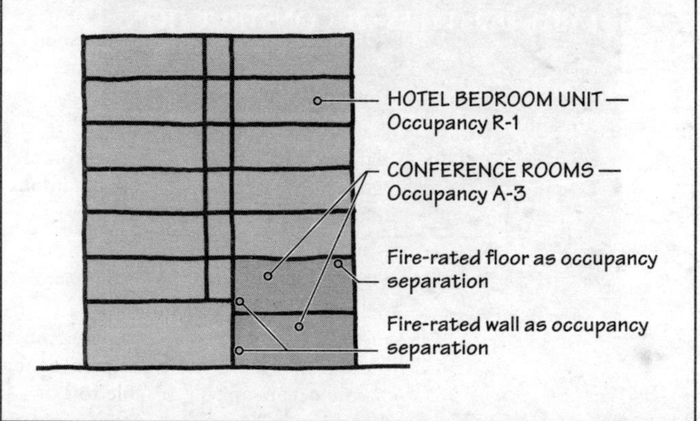

FIGURE 2.15 The separation between two different occupancies in a building is called an *occupancy separation*. In this illustration, the two different occupancies shown are hotel bedrooms (Occupancy R-1) and conference rooms (Occupancy A-3). An occupancy separation may be a wall, a floor, or both, each of which must have the minimum fire resistance rating specified by the code. If occupancy separations are not provided (i.e., if nonseparated occupancies are provided), the entire building will be treated as belonging to the more hazardous occupancy.

Generally, the hazard present in occupancies decreases with increasing division number. Thus, Occupancy A-2 is generally less hazardous than Occupancy A-1, A-3 is less hazardous than A-2, and so on.

MIXED OCCUPANCY AND OCCUPANCY SEPARATION

A building may consist of one occupancy or more than one occupancy. The latter is referred to as a *mixed-occupancy building*. Mixed-occupancy buildings are fairly common. For instance, a hotel may consist of bedrooms and a few conference rooms. Hotel bedrooms are Occupancy R-1, whereas conference rooms are classified as A-3, Table 2.1. Similarly, an office block may consist of individual offices, a day care center, a gymnasium, and a few retail shops. The individual offices are Occupancy B, a day care center is Occupancy E, a gymnasium is Occupancy A-3, and retail shops are Occupancy M.

Because of the variable degree of hazard present in a mixed occupancy, the code requires that each occupancy be separated from the other occupancy by an occupancy separation of a specified minimum fire resistance. For example, the code stipulates that in a hotel building consisting of R-1 and A-3 Occupancies, the separation between these two occupancies must have a minimum fire resistance of 2 h, Figure 2.15. The occupancy separation (which is a fire-rated element) may be a wall, floor, or both. The objective is to separate the two occupancies vertically as well as horizontally.

If the required occupancy separation is not provided, the entire building will be treated as belonging to the more hazardous group, adversely affecting the allowable area and height of the building. In the example of Figure 2.15, the entire building will be treated as Occupancy A-3, because this is more hazardous than Occupancy R-1. The allowable area and height of the building will be as permitted for Occupancy A-3.

If the required occupancy separation is provided, the building will be considered one building with two distinct occupancies, and each occupancy can have its own allowable area and height. Additional details of the allowable area and height of a building (including a mixed-occupancy building) are given in Principles in Practice at the end of this chapter.

TYPES OF CONSTRUCTION

The building's type of construction also creates its own hazard. The more fire-resistive the construction, the smaller the hazard. The International Building Code classifies construction into five basic types—I, II, III, IV, and V. This classification is based on the degree of fire resistance of various structural components of the building, such as floors, roof, columns, beams, exterior walls, and interior partitions.

Type I is the most fire-resistive construction, and Type V is the least fire-resistive. Each of these types is further subdivided into two subtypes, except Type IV (which has no subtype), Table 2.2. A detailed description of the construction types is given in Chapter 7.

NOTE

Accessory-Use Space

Some buildings will have one dominant occupancy and one or more minor occupancies. If the total floor area of the minor occupancies is less than 10% of the total area of the building on any floor, the minor occupancies may be treated as an *accessory-use space*. An occupancy with an accessory-use space may be considered a single (dominant) occupancy, requiring no separation between the occupancy and the accessory-use space.

For example, consider a high-rise apartment building with a beauty shop and a restaurant on the ground floor. The dominant occupancy of the building is R-2, Table 2.1. The beauty shop is Occupancy B, and the restaurant is Occupancy A-2. If the total area of Occupancy B and Occupancy A-2 is less than 10% of the total ground-floor area of the building, these two occupancies may be considered accessory-use spaces.

Exceptions to this policy are classrooms or assembly spaces used for fewer than 50 persons. Such spaces need not be limited to 10% of the building's area, provided that they are accessory to that occupancy.

Incidental-Use Space

Some minor occupancies may be more hazardous than the dominant occupancy. Such occupancies are treated not as accessory-use spaces but as *incidental-use spaces*. For example, a covered parking garage in a high-rise office building is an incidental-use space. Incidental-use spaces require a fire-rated separation from the main occupancy. The International Building Code lists the spaces that should be treated as incidental-use spaces and gives the corresponding fire-rating requirement of the separation.

TABLE 2.2 TYPES OF CONSTRUCTION

Decreasing fire resistance ↓		Decreasing fire resistance →	
	Type I	Type I(A)	Type I(B)
	Type II	Type II(A)	Type II(B)
	Type III	Type III(A)	Type III(B)
	Type IV	No subclassifications	
	Type V	Type V(A)	Type V(B)

PRACTICE QUIZ

Each question has only one correct answer. Select the choice that best answers the question.

11. A prescriptive building code provision is easier to enforce than a performance provision.
a. True **b.** False

12. Performance-type building code provisions inhibit the use of innovative building materials and construction systems compared with prescriptive type provisions.
a. True **b.** False

13. Writing and periodically updating the building code is generally done by
a. the individual city.
b. the individual state.
c. the federal government.
d. a specialized independent agency.

14. The International Building Code is published by the
a. U.S. government.
b. International Conference of Building Officials.
c. Building Officials and Code Administrators International.
d. International Code Council.
e. United Nations Agency on Building Codes.

15. The International Building Code is generally updated every
a. 2 years. **b.** 3 years.
c. 4 years. **d.** 5 years.
e. 10 years.

16. The adoption of a model building code in the United States is done at the level of
a. each city. **b.** each state.
c. either (a) or (b).

17. The only code published by the International Code Council is the International Building Code.
a. True **b.** False

18. The first step in applying the building code to a project is to
a. determine if the building is provided with an automatic sprinkler system.
b. determine the width of open spaces around the building.
c. determine the occupancy classification of the building.
d. determine the number of exits from the building at the ground floor.

19. The total number of occupancies (excluding the divisions) as specified in the International Building Code is
a. 6. **b.** 8.
c. 10. **d.** 12.
e. none of the above.

20. Which of the following is not a recognized building code occupancy?
a. Assembly **b.** Educational
c. Business **d.** Office
e. Institutional

21. A hotel building is a
a. business occupancy. **b.** residential occupancy.
c. hotel occupancy. **d.** hospitality occupancy.
e. none of the above.

22. The International Building Code divides the types of construction into
a. Types U, V, X, and Y. **b.** Types 1, 2, 3, 4, and 5.
c. Types I, II, III, IV, and V. **d.** Types I, II, III, IV, V, and VI.

23. The most fire-resistive type of construction is
a. Type I(B). **b.** Type 5(A).
c. Type I(A). **d.** Type V(A).

2.7 CONSTRUCTION STANDARDS

Standards are the foundations of modern building codes. They contain technical information that addresses (a) the properties of a building product or a component, (b) test methods to determine the properties of materials, and (c) the method of installation or construction.

The word *standard* is used here as a qualifier for words such as the *properties, test methods,* or the *method of installation.* Thus, a standard, in fact, refers to one of the following:

Standard specification This deals with the quality of materials, products, and components.

Standard test method This determines the particular performance of a product or system through a test including methods of sampling and quality control.

Standard method of practice—construction practice, installation practice, or maintenance practice This includes specific fabrication, installation, and erection methods of a component, including its maintenance.

A standard should not be confused with a building code provision. Although a code specifies the design criteria or the required properties of a component, the standard specifies the procedures and equipment required to verify the criteria or measure the properties. For instance, a building code will give the required fire resistance (e.g., 2 h) for a wall, roof, floor, and so on. The construction standard prescribes the test procedure and equipment required to measure the fire resistance of that element.

A fully prescriptive code does not require any standards because it does not refer to any performance criterion or property. By contrast, a performance-type code must rely heavily on the use of standards. As the trend toward performance-type codes has grown, the use of standards in building codes has increased accordingly.

ORGANIZATIONS THAT WRITE STANDARDS

The development of standards is beyond the scope of most local jurisdictions or a model code organization because many standards are referenced by a model code. For example, the International Building Code references nearly 500 standards produced by numerous organizations. The tasks of formulating and updating the standards are, therefore, performed by organizations that have the necessary facilities and expertise for obtaining and evaluating performance data. These organizations may be divided into four types: trade associations, professional societies, governmental organizations, and organizations whose primary purpose is to create, maintain, and publish standards.

Trade Associations A trade association is an association of manufacturing companies making the same product. For example, the American Wood Preservers' Association (AWPA) is an organization whose membership comprises industries that treat lumber, plywood, and other wood products with preservatives and fire-retardant chemicals. Some other important trade organizations involved in writing standards are the National Roofing Contractors' Association (NRCA), Brick Industry Association (BIA), National Concrete Masonry Association (NCMA), and Tile Council of America (TCI).

The function of a trade association is to coordinate the tasks of member companies and to protect their interests. It deals with all matters that affect the specific industry, such as collecting statistics, dealing with tariffs and trade regulations, and disseminating information about the products to design and construction professionals, particularly emphasizing their benefits over those of any competing or alternative product.

In order to ensure quality and competitiveness, most trade associations publish standards for the products they serve, to which all member companies must conform. Several trade associations also have certification programs to grade and stamp the product. AWPA is such an organization, and an AWPA grade stamp on wood and wood products is regarded as a mark of quality and standardization.

Professional Societies The professional societies, such as the American Society of Civil Engineers (ASCE) and the American Society of Heating, Refrigeration and Air Conditioning Engineers (ASHRAE), International Code Council (ICC) are another source of standards, although not extensive. These organizations develop standard specifications and testing procedures as professional obligations. For example, the ASCE's publication *Minimum Design Loads for Buildings and Other Structures,* ASCE 7 Standard, is heavily referenced by the building codes.

Governmental Organizations Examples of such organizations include the U.S. Department of Housing and Urban Development (HUD), Federal Emergency Management Association (FEMA), U.S. Department of Commerce (DOC), and Consumer Product Safety Commission (CPSC).

ORGANIZATIONS WITH STANDARDS PRODUCTION AS THEIR PRIMARY PURPOSE

By far the largest and most frequently referenced standards-producing organizations are those whose main function is to produce standards. In countries other than the United States, this task is normally handled by a single umbrella organization at the national level. In Britain there is the British Standards Institution (BSI), in Australia the Standards Association of Australia (SAA), and in Germany the Deutsches Institute fur Normung (DIN).

ZONING MAP

A *zoning map* is based on a comprehensive master plan of the city that takes into account the present and future use of land under the city's jurisdiction. The map shows the district to which a piece of land belongs, and the zoning text specifies the types of buildings that can be built in that district. For instance, in a single-family dwelling district, the zoning ordinance will permit the construction of single-family residences and several other facilities that may be needed in that district, such as fire stations, local houses of worship, community centers, parochial schools, playgrounds, and parks.

ZONING TEXT

The *zoning text* (as distinct from the graphical component—the zoning map) forms the bulk of the zoning ordinance. In addition to specifying the land use, a zoning text gives the development standards for the city's land. These standards include the maximum ground coverage, maximum floor area ratio, minimum setbacks, maximum height, and number of stories that may be built on a piece of land. In some cases, the types of exterior building envelope materials may be regulated.

Ground Coverage Ground coverage refers to the percentage of land area that may be covered by the building. For instance, for an office building in a downtown area, the ground coverage permitted may be as high as 100%, whereas for a single-family dwelling in a suburban location, the maximum ground coverage may be as low as 25%.

Floor Area Ratio *Floor area ratio* (FAR) refers to the total built-up area (sum of the built-up areas on all floors) divided by the total land area. An FAR of 2.0 means that the total allowable built-up area of the building on all floors is equal to twice the land area. Thus, an FAR of 2.0 is achieved by a two-story building covering the entire piece of land, or a four-story building covering 50% of the land area, or a six-story building covering one-third the land area, and so on.

Ground coverage and FAR are powerful zoning restrictions that control the overall volume of buildings in a district without greatly jeopardizing design freedom. Volume control is an indirect means of aesthetic control of the development. It also controls population density, traffic volume, environmental pollution, and open spaces.

Restrictions placed on the height or the number of stories under a zoning ordinance also affect population density and traffic volume in the same way as ground coverage and FAR, but the main purpose of height restriction is to be sure that the buildings are within the capabilities of fire-fighting equipment of the city. In some cities, height restrictions may be imposed to maintain views of existing buildings or to preserve the aesthetic character of a district.

Setbacks Setbacks refer to the distances a building must be set back from lot lines, thereby specifying the minimum sizes of open areas around the building. Historically, the purpose of a front setback was to reserve land for any future widening of the street. In modern times, setbacks provide greater privacy, help to insulate buildings from traffic noise and fumes, and otherwise enhance the aesthetic appeal of neighborhoods.

ZONING ORDINANCE REVIEW AND ADMINISTRATION

The formulation and enforcement of a zoning ordinance are usually delegated by the state to local authorities. Like the building code, the zoning ordinance is enforced under the police powers of the state.

Each local authority has several decision-making bodies that contribute to the formulation and review of the zoning ordinance. Theoretically, the city council is the highest zoning ordinance authority. For practical purposes, however, the zoning ordinance is under the charge of the *zoning and planning commission* of the city, which is typically a body consisting of several local citizens. The responsibilities of this commission include the formulation, development, and review of the zoning plan.

The zoning commission is (ideally) composed of individuals who have distinguished themselves by demonstrating outstanding and unselfish interest in civic affairs. They are appointed by the city council to serve for a fixed term and receive no remuneration for their work.

The initial formulation and subsequent reviews of the city's zoning plan by the zoning commission must go through a public hearing process. This allows all citizens to participate in city planning and the zoning decision-making process.

Although the zoning commission and the city council are the bodies responsible for zonal plan review, the day-to-day work related to zoning administration is carried out by the city's planning department. Thus, the owner of a piece of land who desires to rezone (i.e., change the land use of) the property would submit the request to the city's planning department, which would review it, make a recommendation, and forward it to the zoning commission for a decision. If the zoning commission agrees to rezoning, a public hearing is arranged. After the public hearing, the zoning commission makes its decision, which is usually in the form of a recommendation to the city council. The city council may support the zoning commission's recommendation or hold its own public hearing on the matter before voting.

BOARD OF ZONING APPEALS

The implementation of land use control involves more than a mere formulation or review of the zoning plan. It also involves the enforcement of zoning ordinances that give rise to problems of application and interpretation. These problems are referred to the board of zoning appeals, also called the *board of adjustment.* This board consists of several members, usually appointed by the mayor or the city council.

The primary responsibility of the board of adjustment is to hear appeals on actions taken by city officials who administer zoning regulations and to make special exceptions to zoning regulations. Special exceptions generally refer to granting variances to setbacks, ground coverage, FAR, and so on, whose literal enforcement may cause unnecessary and undue hardship to the owner. Variances granted by the board must not be contrary to the public interest and must not act to relieve the owner of self-created hardship. The meetings of the board are open to the public.

2.10 BUILDING ACCESSIBILITY—AMERICANS WITH DISABILITIES ACT (ADA)

The Americans with Disabilities Act (ADA) is a 1990 federal law that came into effect in January 1992. The objective of the act is to ensure equal opportunity to individuals with disabilities. Part of the act is devoted to regulating access to buildings so that they are usable by individuals with disabilities.

The act requires that all new construction and alterations to existing construction provide barrier-free access routes on the site, into, and within all building spaces. Single-family dwellings, multifamily dwellings, and buildings funded by religious organizations, such as churches and temples, are exempt from ADA's accessibility provisions.

The concept of barrier-free access existed for nearly two decades before the promulgation of ADA. The ANSI (A117.1) standard entitled "Standard for Accessible and Usable Buildings and Facilities" had been adopted by several states and local jurisdictions.

In 1969, the U.S. federal government passed the Architectural Barriers Act (ABA), which states that if federal funds are spent in the construction of a facility, that facility must be designed to be accessible to the physically handicapped according to the Uniform Federal Accessibility Standard (UFAS). In 1980, UFAS provisions were absorbed into ANSI 117.1. The difference between ABA and ADA is that ADA is much wider in scope. Additionally, ADA is mandated for application to all buildings except those mentioned earlier, regardless of the source of funds used in constructing them.

ADA is a comprehensive legislative act that addresses four major concerns: Title I pertains to employment discrimination against individuals with disabilities. Title II refers to the accessibility of public transportation lines, such as railroads and buses. Title III deals with accessibility in buildings, and Title IV covers accessible telecommunication facilities, such as telephones for the hearing impaired or individuals with other handicaps, and automatic teller machines.

Only Title III and Title IV of ADA are of concern in the design of buildings. The provisions of Title III are detailed in ADA Accessibility Guidelines (ADAAG). The International Building Code also has provisions relating to accessibility (in a separate chapter). However, it must be recognized that a building code and ADA are two different legislative provisions. A building code is a local jurisdiction's law, whereas ADA is a federal law. The owner of the building is required to meet the provisions of both laws.

ADA's accessibility requirements relate primarily to the design of entrances, doors, stairways, ramps, changes of level, sidewalks, elevators, drinking fountains, toilet facilities, door width, doorknobs, and so on.

PRINCIPLES IN PRACTICE

Code-Allowable Area and Height of a Building (*Continued*)

AREA INCREASE DUE TO FRONTAGE

The basic allowable area per floor (Figure 1) may be increased for frontage. The frontage increase is available only if more than 25% of the building's perimeter fronts a public way or an accessible open space at least 20 ft wide. The open space must be accessible from a street or fire lane at all times. The area increase per floor due to frontage is given by the following equation:

$$\text{Area increase due to frontage} = A_b I_f \qquad \textbf{Eq. (1)}$$

where A_b = basic allowable area from Figure 1
I_f = a coefficient, whose value is given by the expression

$$I_f = \left[\frac{F}{P} - 0.25\right]\frac{W}{30} \qquad \textbf{Eq. (2)}$$

where F = the part of the building's perimeter that fronts on a public way or an accessible open space having a minimum width of 20 ft. Thus, the part of the perimeter that is included in F should satisfy the following two conditions: (a) it should front a minimum 20 ft wide open space within the lot or a minimum 20 ft wide public way, and (b) the open space within the lot should be accessible directly to a public way or through a minimum 20 ft wide open space provided within the lot.

P = perimeter of the entire building. If $(F/P) \leq 0.25$, that is, the accessible frontage is less than or equal to 25% of the total perimeter of the building, set $I_f = 0$, which means that no area increase due to frontage is available.

W = width of the public way or accessible open space, not less than 20 ft; in other words, for $W < 20$ ft, set $W = 0$. Also, if $W > 30$ ft, set $W = 30$ ft, that is, $W/30 = 1.0$. In other words, an accessible open space greater than 30 ft does not yield any additional area increase.

AREA INCREASE DUE TO SPRINKLERS

The basic allowable area per floor (Figure 1), may be increased if the entire building is provided with automatic sprinklers. The area increase per floor due to sprinklers is given by

$$\text{Area increase due to sprinklers} = A_b I_s \qquad \textbf{Eq. (3)}$$

where $I_s = 2.0$ (which implies a 200% increase) for a multistory building
$I_s = 3.0$ (which implies a 300% increase) for a single-story building

HEIGHT INCREASE DUE TO SPRINKLERS

If automatic sprinklers are provided throughout the building, the maximum height of the building may be increased by 20 ft, and the maximum number of stories may be increased by one. Thus, from Figure 1, the maximum allowable height of Occupancy B with Type III(A) construction is 65 ft, not exceeding five stories. If this building is provided with automatic sprinklers, the maximum allowable height is 85 ft, and the maximum number of stories allowed is six.

TOTAL ALLOWABLE AREA PER FLOOR

Thus, the total area per floor including the increases due to frontage and sprinklers, $A_{\text{per floor}}$, is given by

$$A_{\text{per floor}} = \begin{array}{l}\text{basic allowable area, } A_b \text{(from Figure 1)} + \\ \text{area increase due to frontage (from Eq. 1)} + \\ \text{area increase due to sprinklers (from Eq. 3)}\end{array} \qquad \textbf{Eq. (4)}$$

TOTAL ALLOWABLE AREA ON ALL FLOORS

The total allowable area on all floors, $A_{\text{all floors}}$ is given by

$$\begin{array}{l}(A_{\text{all floors}}) = 3.0(A_{\text{per floor}}) \text{ for a building with three or more floors} \\ (A_{\text{all floors}}) = 2.0(A_{\text{per floor}}) \text{ for a building with two floors}\end{array} \qquad \textbf{Eq. (5)}$$

Thus, if $A_{per\ floor}$ for an eight-story building = 25,000 ft^2, the maximum total allowable area on all floors for this building is $A_{all\ floors}$ = 75,000 ft^2. If the building is a two-story building, $A_{all\ floors}$ = 50,000 ft^2.

Examples 1 and 2 illustrate the use of this procedure.

EXAMPLE 1 (ALLOWABLE AREA OF AN UNSPRINKLERED ONE-STORY BUILDING)

Determine if the plan of the furniture warehouse shown in Figure 2 meets the building code restrictions on allowable area and height. The building is one story high (total height = 30 ft), and the type of construction used is Type II(B). No automatic sprinklers are provided.

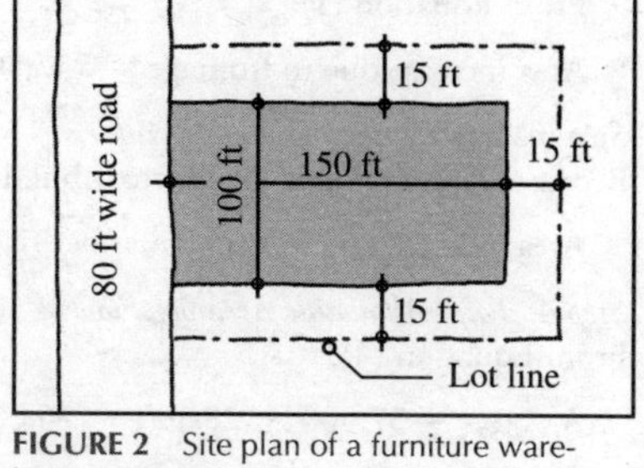

FIGURE 2 Site plan of a furniture warehouse, referring to Example 1.

SOLUTION

From Table 2.1, a furniture warehouse is Occupancy S-1.

Step 1: From Figure 1, the basic allowable area is A_b = 17,500 ft^2, and the basic allowable height = 55 ft.

Step 2: Area Increase Due to Frontage
Only one (100 ft) side of the building fronts a public way. Because the other three open spaces are less than 20 ft wide, they do not count as accessible open spaces. Hence

$$F = 100 \text{ ft} \quad \text{and} \quad P = 100 + 100 + 150 + 150 = 500 \text{ ft}$$

Hence, (F/P) = (100/500) = 0.2. Because (F/P) is less than 0.25, meaning that less than 25% of the building's perimeter fronts an accessible open space, I_f = 0. From Equation (1),

$$\text{Area increase due to frontage} = 0$$

Step 3: Area Increase Due to Sprinklers
Because no sprinklers have been provided,

$$\text{Area increase due to sprinklers} = 0$$

Step 4: Allowable Area per Floor
From Equation (4),

$$(A_{per\ floor}) = 17{,}500 + 0 + 0 = 17{,}500 \text{ ft}^2$$

Step 5: Allowable Area on all Floors
Because the building is one story high, the allowable area on all floors is 17,500 ft^2.

Area and Height Provided
The area provided is 150 × 100 = 15,000 ft^2, which is less than the allowable area of 17,500 ft^2. Therefore, the area provided is OK.

The height of the building is 30 ft, which is less than the maximum allowable height of 55 ft (Figure 1), which is also OK.

EXAMPLE 2 (ALLOWABLE AREA OF A SPRINKLERED MULTISTORY BUILDING)

An architect's site plan of a six-story office building (80 ft tall) is shown in Figure 3. Determine if the area and height of the building conform to the building code. The type of construction used is Type II(A). The entire building is provided with an approved automatic sprinkler system.

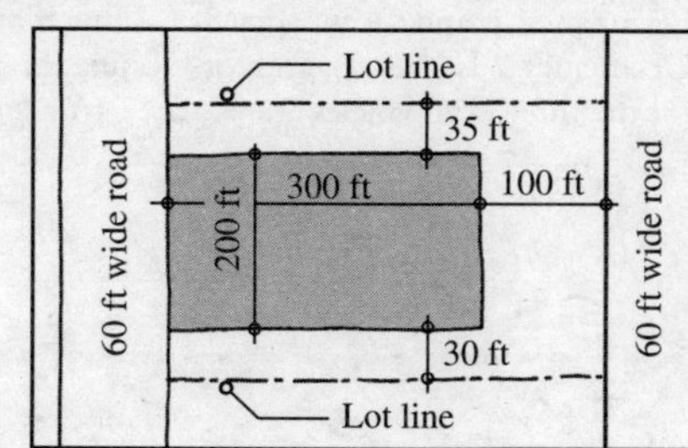

FIGURE 3 Site plan of an office building, referring to Example 2.

SOLUTION

From Table 2.1, an office is Occupancy B.

Step 1: From Figure 1, the basic allowable area per floor is A_b = 37,500 ft^2, the basic allowable height is 65 ft, and the number of stories is five.

(Continued)

PRINCIPLES IN PRACTICE

Code-Allowable Area and Height of a Building (*Continued*)

Step 2: Area Increase Due to Frontage
The building has open spaces on all four sides, and their width (W) exceeds 20 ft. Hence,

$$F = 200 + 300 + 200 + 300 = 1{,}000 \text{ ft} \quad \text{and} \quad P = 1{,}000 \text{ ft}$$

Hence, $(F/P) = (1{,}000/1{,}000) = 1.0$. Because $W \geq 30$ ft on all sides of the building, set $W = 30$. Hence, $W/30 = 1.0$. From Equation (2):

$$I_f = (1.0 - 0.25)(1.0) = 0.75$$

From Equation (1):

$$\text{Area increase due to frontage} = (37{,}500)(0.75) = 28{,}125 \text{ ft}^2$$

Step 3: Area Increase Due to Sprinklers
Because the building is a multistory building, $I_s = 2.0$. From Equation (3),

$$\text{Area increase due to sprinklers} = (37{,}500)(2.0) = 75{,}000 \text{ ft}^2$$

Step 4: Total Allowable Area per Floor
From Equation (4):

$$A_{\text{per floor}} = 37{,}500 + 28{,}125 + 75{,}000 = 140{,}625 \text{ ft}^2$$

Step 5: Total Allowable Area for the Entire Building
From Equation (5), the total allowable area for the entire building is

$$A_{\text{all floors}} = 3.0(140{,}625) = 421{,}875 \text{ ft}^2$$

Area Provided
The area provided per floor is $300 \times 200 = 60{,}000$ ft^2, which is less than the allowable area ($A_{\text{per floor}}$) of 140,625 ft^2. Therefore, the area provided per floor is OK.

The total area provided on all floors in the building is $6(60{,}000) = 360{,}000$ ft^2, which is less than the allowable area ($A_{\text{all floors}}$) of 421,875 ft^2. Therefore, it is OK.

Height Provided
The basic allowable height for occupancy is B = five stories (Figure 1), with a maximum height of 65 ft. Because the building is provided with automatic sprinklers, the maximum allowable height is $65 + 20 = 85$ ft, and the number of stories allowed is $5 + 1 = 6$. Thus, the provided height of 80 ft and six stories is OK.

ALLOWABLE AREA OF A MIXED-OCCUPANCY BUILDING

The allowable area and height determinations given so far refer to a single-occupancy building. For a mixed-occupancy building, two cases must be distinguished.

First, if the building is not provided with required occupancy separations (an *unseparated mixed-occupancy building*), the allowable area and height are calculated by assuming that the entire building belongs to only one occupancy. These calculations are made for each occupancy provided. The smallest area and the smallest height obtained from the calculations are the allowable area and height of the building.

Second, if required occupancy separations have been provided, the building is treated as a duly *separated mixed-occupancy building.* For such a building, the sum of the respective floor areas provided for each occupancy divided by their allowable floor areas at each floor must not exceed 1.0.

For example, consider a building with three different occupancies. Let us call these occupancies Occupancy 1, Occupancy 2, and Occupancy 3. Assume that the floor area provided at any one floor for Occupancy 1 is $(A_{\text{prov}})_1$, for Occupancy 2 is $(A_{\text{prov}})_2$, and for Occupancy 3 is $(A_{\text{prov}})_3$. Assume further that the respective allowable areas for each floor for the three occupancies are $(A_{\text{allow}})_1$, $(A_{\text{allow}})_2$, and $(A_{\text{allow}})_3$; then the following equation must be satisfied:

$$\frac{(A_{\text{prov}})_1}{(A_{\text{allow}})_1} + \frac{(A_{\text{prov}})_2}{(A_{\text{allow}})_2} + \frac{(A_{\text{prov}})_3}{(A_{\text{allow}})_3} \leq 1.0 \qquad \textbf{Eq. (6)}$$

Each question has only one correct answer. Select the choice that best answers the question.

35. If a single-story building is provided throughout with automatic sprinklers, the allowable area per floor of the building may be increased by
a. 100%.
b. 150%.
c. 200%.
d. 250%.
e. 300%.

36. If a multistory building is provided throughout with automatic sprinklers, the allowable area per floor of the building may be increased by
a. 100%.
b. 150%.
c. 200%.
d. 250%.
e. 300%.

37. If automatic sprinklers are provided throughout the building, the building's height may be increased by
a. 20 ft above the basic allowable height.
b. 40 ft above the basic allowable height.
c. 60 ft above the basic allowable height.
d. 100 ft above the basic allowable height.
e. none of the above.

38. If the maximum allowable area per floor for a building is 20,000 ft^2 and the maximum allowable height is 11 floors, what is the maximum allowable area for the entire building?
a. 40,000 ft^2
b. 60,000 ft^2
c. 100,000 ft^2
d. 220,000 ft^2
e. none of the above

REVIEW QUESTIONS

1. What is the primary purpose of building codes? Explain.
2. What is the difference between a prescriptive and a performanc type code provision? Explain with the help of an example.
3. List the model code organizations that existed in the United States prior to the year 2000 and the building codes published by each.
4. Describe the relationship between a building code and construction standards. List three important standards organizations in the United States.
5. What information can you derive from a given designation of an ASTM standard such as E 119-95a?
6. List the first five steps you will follow in ascertaining that a building conforms to the provisions of the building code.
7. List at least three codes published by the International Code Council.
8. What do the terms *ground coverage* and *floor area ratio* (FAR) mean? Are they part of building code provisions? Explain.

CHAPTER 3

Loads on Buildings

CHAPTER OUTLINE

Buildings are subjected to several types of loads, Figure 3.1. Two main classifications of loads are

- Gravity loads
- Lateral loads

Gravity loads are caused by the gravitational pull of the earth and act in the vertical direction. Therefore, they are also referred to as *vertical loads*. Gravity loads include the materials and components that comprise the buildings, as well as people, rainwater, snow, furniture, equipment, and all that is contained within the building. Gravity loads are further classified as *dead loads* and *live loads*, as explained in subsequent sections.

The two primary sources of *lateral loads* on buildings are wind and earthquakes. The effect of each is to create loads in the lateral (other than vertically downward) direction. For example, wind creates horizontal forces on a wall as well as vertically upward forces (suction) on a flat roof. The main effect of earthquake ground motion is to create horizontal forces in buildings, although a small amount of vertical force may also exist. Additional examples of lateral loads are earth pressure on basement walls, water pressure on tank walls, and loads caused by blasts and moving vehicles or equipment.

In this chapter, we examine the nature of different types of loads on buildings. The treatment given here is brief and qualitative and sets the stage for a detailed investigation of the topic by an interested reader.

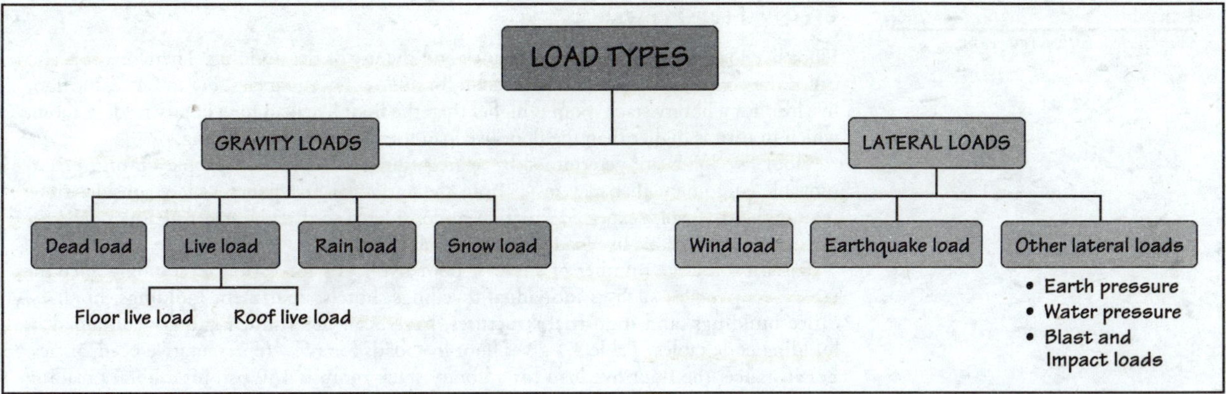

FIGURE 3.1 Some of the important loads in buildings.

Units of Measurement for Loads

In the U.S. system of units, a load is expressed in pounds (lb). The pound is a small unit for measuring building loads. Therefore, the unit kilopound (kip) is commonly used; 1 kip is equal to 1,000 lb. Thus, 2.5 kips = 2,500 lb.

When the load is distributed over a surface such as a floor or roof, it is generally expressed in pounds per square foot (psf), Figure 3.2. A beam is treated as a linear element, because its width is generally much smaller than its length. Therefore, the load on a beam is generally expressed in terms of pounds per foot length of the beam (lb/ft). If the load on the beam is large, it is generally stated as kilopounds per foot (kips/ft). The load on a column is expressed in terms of pounds (lb) or kilopounds (kips).

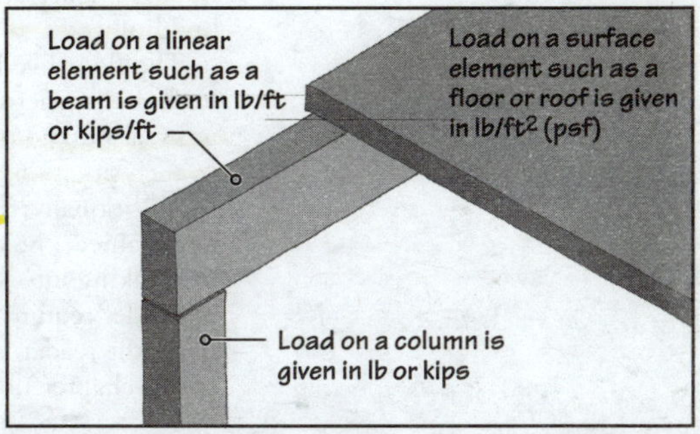

Load on a linear element such as a beam is given in lb/ft or kips/ft

Load on a surface element such as a floor or roof is given in lb/ft^2 (psf)

Load on a column is given in lb or kips

FIGURE 3.2 The loads on various elements of a building are expressed in different units.

3.1 DEAD LOADS

Dead loads are always present in a building; that is, they do not vary with time. They include the weights of the materials and components that comprise the structure. The dead load of a component is computed by multiplying its volume by the density of the material. Because both the densities and the dimensions of the components are known with reasonable accuracy, dead loads in a building can be estimated with greater certainty than other load types. Densities of some of the commonly used materials are given in Table 2 in Principles in Practice at the end of this chapter.

The dead load for which a building component is designed includes the self-load of the component plus the dead loads of all other components that it supports. For example, the dead load on a column includes the weight of the column itself plus all the dead load imposed on it. In Figure 3.3, the dead load on a column is the weight of the column plus the dead load from the beams and slab resting on it. Similarly, the dead load on a beam is the weight of the beam itself plus the dead load from the slab that it supports. The dead load on the slab is only the self-weight of the slab. However, if the slab supports a floor finish, ceiling, light fixtures, or plumbing and electrical pipes, their weights must be included in the dead load acting on the slab.

NOTE

Loads in the SI System of Units

The unit of measurement for load in the SI system of units is the newton, which is abbreviated N. A newton is even smaller than a pound (1 lb = 4.45 N). Therefore, the unit kilonewton (kN) is commonly used.

For the load on a floor, the unit newton per square meter (N/m^2) is used instead of pound per square foot (psf). A newton per square meter is a pascal (Pa), and a kN/m^2 is a kilopascal (kPa). Thus, 1.0 N/m^2 is synonymous with 1.0 Pa, and 1.0 kN/m^2 with 1.0 kPa; 1 psf = 47.88 Pa.

For the load on a linear element, the unit is newton per meter (N/m), or kilonewton per meter (kN/m) is used. The load on a column is expressed in newtons (N) or kilonewtons (kN).

(For conversion from the U.S. system of units to the SI system of units, and vice versa, see Appendix A.)

3.2 LIVE LOADS

While the dead load is a permanent load on the structure, *live load* is defined as the load whose magnitude and placement change with time. Such loads are due to the weights of people (animals, if the building houses animals), furniture, movable equipment, and stored materials. As shown in Figure 3.1, live loads are further divided into

- Floor live load
- Roof live load

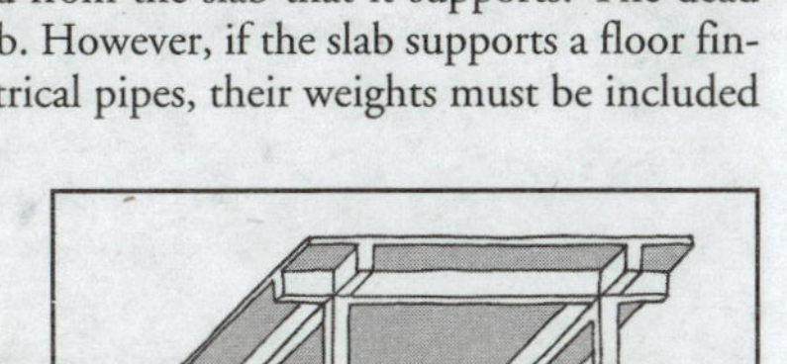

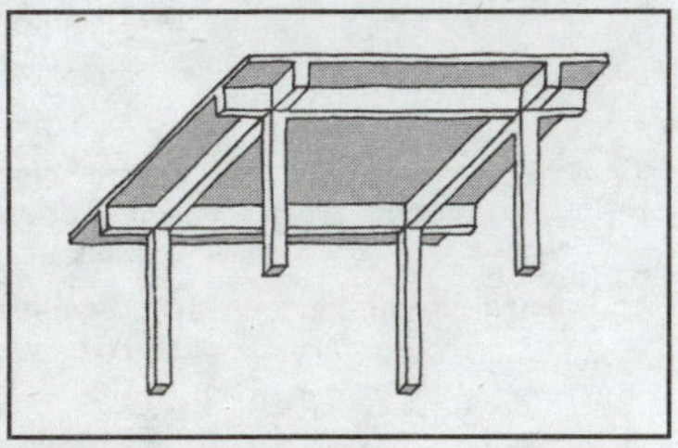

FIGURE 3.3 The dead load on the column of a building includes the weight of all components that it supports, such as the beams and floor slab plus the self load of the column.

FLOOR LIVE LOAD

Floor live load depends on the occupancy and the use of the building. Therefore, it is also called the *occupancy load*, and it is different for different occupancies. For instance, the floor live load for a library stack room is higher than the floor live load for a library reading room, which in turn is higher than the floor live load for an apartment building.

Floor live loads are determined by aggregating the loads of all people, furniture, and movable equipment that may result from the particular occupancy. Safety considerations require that the worst expected situation be considered so that the structure is designed for the maximum possible live load that may be placed on it.

Based on a large number of surveys, floor live loads for various commonly encountered occupancies, such as individual dwellings, hotels, apartment buildings, libraries, office buildings, and industrial structures, have been determined and are contained in building code tables. Table 3.1 gives floor live loads for a few representative occupancies. For instance, the floor live load for a library stack room is 150 psf, for a library reading room is 60 psf, and for an apartment building is 40 psf. Note that these are the minimum floor live loads for which the building must be designed. Even if the actual live load on a floor is smaller, the building must be designed for the minimum live load specified by the building code.

The floor live load values in Table 3.1 are conservative. In most situations, the actual loads are smaller than those given. However, the architect or the engineer must recognize unusual situations that may lead to a greater actual load than the one specified in the code. In such situations, the higher anticipated load should be used.

Additionally, if the live load for an occupancy not included in building code tables is to be obtained, the architect or the structural engineer must determine it from first principles, taking into account all the loads that may be expected on the structure. Most building codes require that such live load values be approved by the building official. (At this stage, the reader is advised to study the examples given in Principles in Practice at the end of this chapter.)

ROOF LIVE LOAD

The live load on a roof takes into account the weight of repair personnel and temporary storage of construction or repair materials and equipment on the roof. Roof live load is generally given as 20 psf acting on the horizontal projected area of the roof, Figure 3.4.

TABLE 3.1 MINIMUM FLOOR LIVE LOADS FOR SELECTED OCCUPANCIES

Occupancy	Description	Load (psf)
Access floor system	Office use	50
	Computer use	100
Auditoriums	Fixed seating areas	60
	Movable seating areas	100
	Stage areas and enclosed platforms	125
Garages	Passenger cars	40
Hospitals	Wards and rooms	40
	Operating theaters	60
Libraries	Reading rooms	60
	Stack rooms	150
Manufacturing	Light	125
	Heavy	250
Offices	General office	50
	Corridors	80
	Lobbies	100
Residential dwellings	Dwellings and hotel guest rooms	40
	Habitable attics	30
	Corridors, public spaces, and balconies	100
Schools	Classrooms	40
	Corridors and lobbies	80
Stores	Light	125
	Heavy	250

Note: For authoritative information, refer to the building code of the jurisdiction.

This applies to a roof that will not be used in the future as a floor. If the building is expected to be extended vertically in the future, the roof live load is the floor live load for the anticipated future occupancy. Additionally, if the roof is to be landscaped, the load due to the growth medium and landscaping elements must be considered as dead load on the roof.

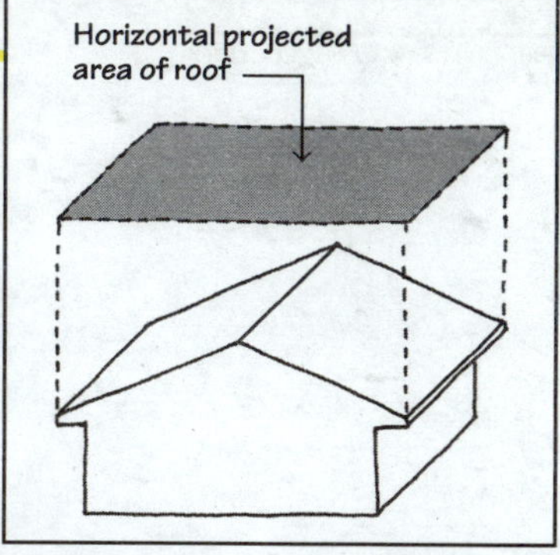

FIGURE 3.4 Horizontal projected area of a pitched roof.

3.3 RAIN LOAD

Although roofs are designed to have adequate drainage so that no accumulation of water occurs, loads resulting from accidental accumulation of melted snow or rainwater must be considered as a possibility. Drains may be blocked by windblown debris or hail aggregating on the roof or the formation of ice dams near the drains.

Long-span, relatively flat roofs are particularly vulnerable to rainwater accumulation because, being flexible, they deflect under the weight of water. This deflection leads to yet more accumulated water, causing additional deflection, which increases accumulation. If adequate stiffness is not provided in the roof, the progressive increase of deflection can cause excessive load on the roof. Water accumulation has been the cause of complete collapse of several long-span roofs.

Generally, roofs with slope greater than $\frac{1}{4}$ in. to 1 ft (1:48 or, say, 1:50 slope) are not subjected to accumulated rainwater unless roof drains are blocked. Building codes mandate $\frac{1}{4}$ in. to 1 ft as the minimum slope required for roofs, which, apart from providing positive drainage, also helps to increase the life and improve the performance of roof (waterproofing) membranes.

Building codes also require that in addition to primary drains, roofs must be provided with secondary (overflow) drains. The secondary drains must be at least 2 in. above the primary drains so that if the primary drainage system gets blocked, the secondary system will be able to drain the water off the roof, Figure 3.5.

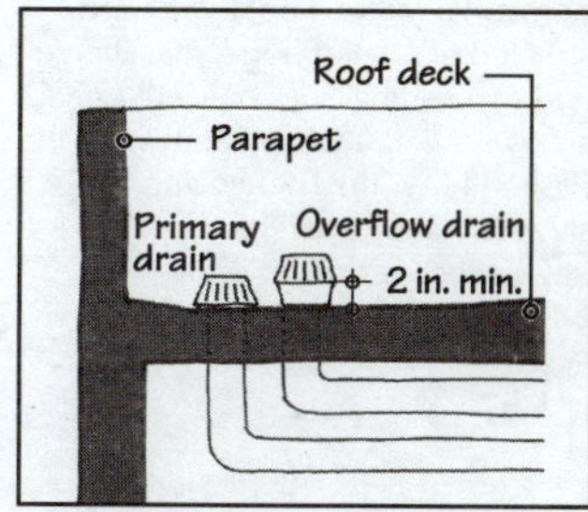

FIGURE 3.5 A roof with a parapet must be provided with two independent drainage systems.

Water accumulation generally occurs on roofs that are provided with a parapet. In the absence of a parapet, water accumulation will generally not occur. Therefore, secondary drains are not required on unparapeted roofs or on steep roofs.

PRACTICE **QUIZ**

Each question has only one correct answer. Select the choice that best answers the question.

1. The two main categories under which all building loads may be classified are
   a. dead loads and live loads.
   b. gravity loads and lateral loads.
   c. terrestrial loads and celestial loads.
   d. first-class and second-class loads.

2. The loads on a floor are generally expressed in
   a. lb/ft^2.
   b. lb/ft.
   c. lb.
   d. lb/ft^3.
   e. none of the above.

3. The loads on a beam are generally expressed in
   a. lb/ft^2.
   b. lb/ft.
   c. lb.
   d. lb/ft^3.
   e. none of the above.

4. The loads on a column are generally expressed in
   a. lb/ft^2.
   b. lb/ft.
   c. lb.
   d. lb/ft^3.
   e. none of the above.

5. Which of the following loads can be estimated with greatest certainty?
   a. Dead loads
   b. Live loads
   c. Wind loads
   d. Earthquake loads
   e. Snow loads

6. The floor live load in a residential occupancy is generally assumed to be
   a. 40 psf.
   b. 10 psf.
   c. 20 psf.
   d. 5 psf.
   e. none of the above.

7. Which of the following occupancies has the highest floor live load?
   a. Classrooms
   b. Hotel guest rooms
   c. Library stack rooms
   d. Library reading rooms
   e. Offices

8. The roof live load in a building is generally
   a. less than the floor live load.
   b. greater than the floor live load.
   c. the same as the floor live load.

9. Roof live load is a function of the building's occupancy.
   a. True
   b. False

10. In which of the following roofs will you consider the effect of rain load?
    a. Roof with a parapet
    b. Roof without a parapet

3.4 WIND LOAD BASICS

Although wind loads are primarily horizontal, they also exert an upward force on horizontal elements such as flat and low-slope roofs. Resistance against upward wind force is provided by anchoring the building to its foundations. Resistance against horizontal loads requires

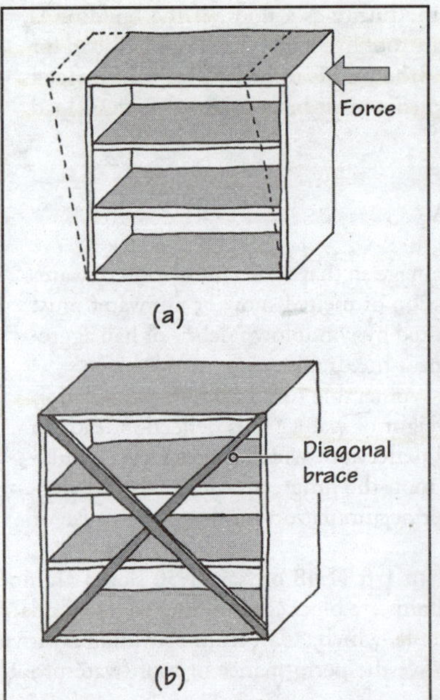

FIGURE 3.6 **(a)** When pushed by a horizontal force, a book rack constructed of several horizontal shelves and two vertical supports will move sideways (i.e., rack), as shown by the dashed lines. **(b)** The racking of shelves can be prevented by diagonal braces (as shown) or by sheet bracing.

anchorage to foundations and the use of stiffening elements. These stiffening elements are commonly referred to as *wind-bracing elements.*

The wind-bracing requirement for a building is precisely the same as the requirement for either diagonal bracing or sheet bracing of a bookshelf. If a book rack is constructed of several shelves mounted on two side supports, it will be adequate to carry the gravity load from the books, but the assembly will be unstable. When acted upon by a horizontal force from either direction, it will move sideways (or *rack*), as shown by the dotted lines in Figure 3.6(a).

Racking of the bookshelf can be prevented by diagonally bracing one of its faces, Figure 3.6(b). An alternative method of stiffening the book rack is to apply a sheet material over the entire face instead of diagonal braces. Sheet bracing generally provides greater lateral stiffening than diagonal bracing.

For buildings, several methods of wind bracing are used, depending on the functional and aesthetic constraints of the building and the magnitude of wind loads. Because wind loads increase with the height of a building, stiffening the building against wind loads becomes increasingly important as the height of the building increases. In fact, the design of tall buildings is heavily dominated by wind-bracing requirements. Architects have often used this structural requirement as a forceful aesthetic expression of building facades in several buildings. Two such examples are shown in Figures 3.7 and 3.8.

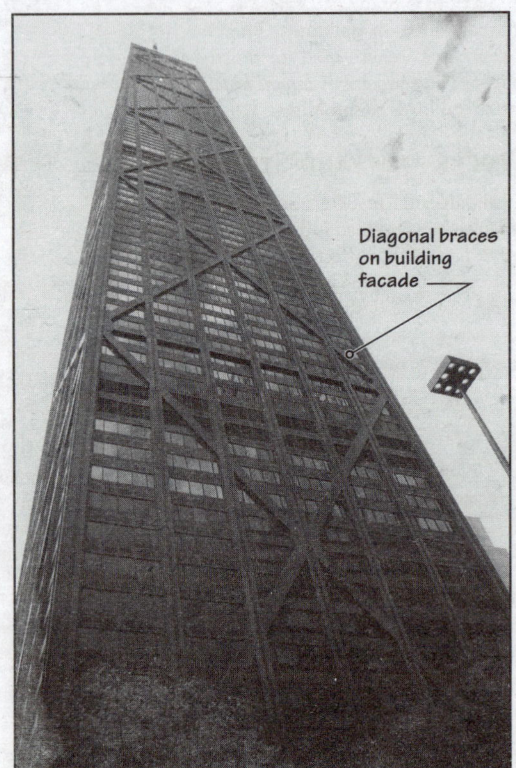

FIGURE 3.7 The lateral load resistance in the 100-story John Hancock Center, Chicago, has been provided by diagonal braces on its exterior facade. The building was designed by the architecture and engineering firm Skidmore, Owings and Merril (SOM) and was completed in 1970.

FIGURE 3.8 Diagonal braces used on the exterior facade of the 52-story New York Times Building, New York City, provide lateral load resistance. The building was designed by Renzo Piano Building Workshop in collaboration with Fox and Fowle Architects and was completed in 2007.

In both the John Hancock Center and the New York Times Building, the lateral load resistance has been provided by diagonal braces on the building's exterior—a structural feature that makes a bold architectural statement. However, diagonal braces may also be used in the building's interior. Additionally, several other means of wind and earthquake load resistance are available and are described in the Principles and Practice section of Chapter 19.

As we will see in Sections 3.8 and 3.9, there is much similarity between the effects of wind load and earthquake load on buildings. Therefore, many of the structural strategies used for wind load resistance are also used for resisting earthquake loads. Thus, the more commonly used term for wind bracing is *lateral bracing* because both wind load and earthquake load represent lateral loads on buildings.

TORNADOES, HURRICANES, AND STRAIGHT-LINE WINDSTORMS

The wind speed is fundamental to determining wind loads on buildings. It is the most important factor, because the wind load is directly proportional to the square of wind speed. In other words, if other factors that affect wind loads are ignored, doubling the wind speed quadruples the wind load. Tripling the wind speed makes wind load nine times larger.

Obviously, buildings must be designed for the maximum probable wind speed at that location. The highest wind speeds are usually contained in a tornado—a rotating funnel-shaped column of air that produces strong suction, Figure 3.9.

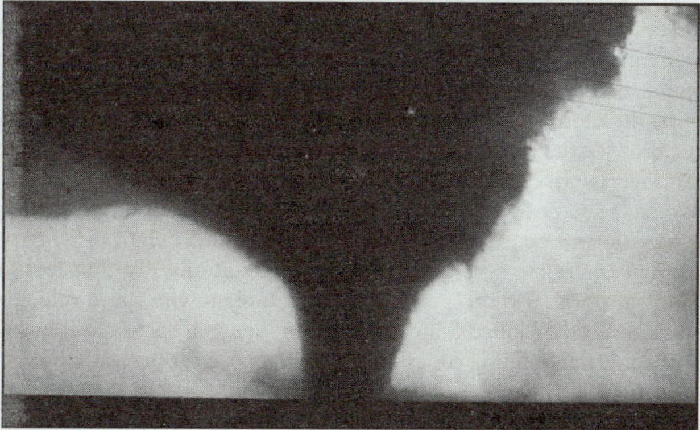

FIGURE 3.9 Tornado in Seymour, Texas, on April 10, 1979.
(Photo obtained from the World Wide Web using the following URL; http:/www.photolib.noaa.gov/bigs/nssl0068.jpg)

NOTE

A Note on Tornadoes

The suction produced in a tornado funnel lifts objects into the air like a giant suction pump. If a tornado hits a water body such as a lake or a sea, a waterspout is formed. Nearly 60% of all tornadoes are weak in intensity. However, 38% are strong enough to topple walls and blow roofs off buildings. Nearly 2% of tornadoes are so violent that they destroy virtually everything in their path.

Actual measurements of wind speeds in a tornado are not available because the area of ground struck by a tornado is relatively small. Posttornado damage observations have revealed that an average tornado has a life span of nearly 30 min and covers a nearly 10-mi-long by 0.25-mi-wide area of ground, that is, 2.5 mi^2.

Thus, the probability of a tornado hitting a particular building or intercepting a permanently located wind speed–measuring instrument, such as that installed at an airport, is very small. Additionally, the destructive power of a tornado is so large that whenever a tornado has hit such measuring instruments, it has completely destroyed them.

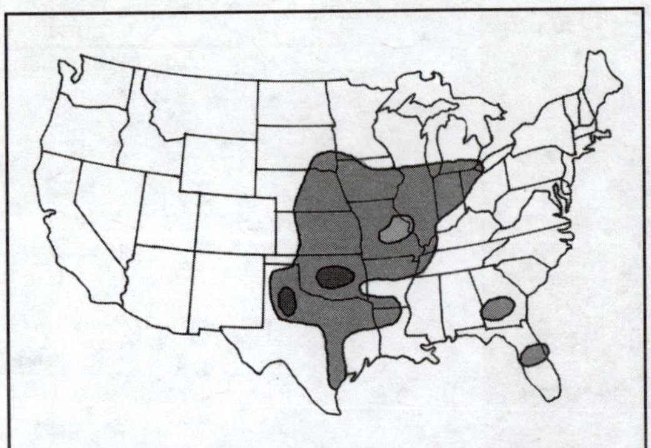

FIGURE 3.10 An approximate representation of tornado-prone regions of the United States. The darker the color of the region, the greater the tornado activity. However, no region of the United States is absolutely tornado-free. For hurricane-prone regions of the United States, see Figure 3.11.

Because of the extremely small probability that a violent tornado will hit a particular building, design wind speeds adopted by building codes for various locations do not reflect the wind speeds obtained in tornadoes. In spite of the fact that design wind speeds do not account for tornadoes, buildings have to be designed to resist tornadoes in tornado-prone locations. For this purpose, sufficient empirical information is available to design and construct tornado-resistant buildings. Figure 3.10 shows tornado-prone areas in the United States.

After tornadoes, hurricanes represent the next most severe wind phenomenon. Known as *typhoons* or *tropical cyclones* in Asia, hurricanes pose greater life safety and economic hazards to buildings and urban infrastructure than tornadoes. Like a tornado, a hurricane is a rotating funnel-shaped wind storm, except that the funnel in a hurricane is much larger than that of a tornado.

An average hurricane within which strong winds exist has a diameter of about 500 mi and has an average lifespan of 10 days. Depending on the location, winds of up to 180 mph may occur within a hurricane. Most damage from hurricanes occurs in coastal areas and is caused primarily by the uplifting of roofs, windborne debris, and uprooted trees or branches of trees. Tidal surges caused by hurricanes can be more devastating than wind-created damage.

As a hurricane travels inland, it loses energy and degenerates into a tropical rain storm. In the United States, hurricanes develop over warm areas in the Atlantic Ocean, Caribbean Sea, Gulf of Mexico, and northeastern Pacific Ocean. Although hurricane-induced property damage has increased over the years in the United States due to the migration of population toward coastal areas, loss of life has decreased due to improvements in forecasting and warning systems.

After tornadoes and hurricanes, local windstorms (e.g., thunderstorms) are the next category of violent wind activity. Because they are nonrotating or nonspinning, they are also referred to as *straight-line winds*. Being a local phenomenon, a windstorm strikes a much smaller area of ground than a hurricane. However, the frequency of occurrence of windstorms is much higher. Consequently, the damage caused by windstorms is substantial. In determining the design wind speed for a location, we consider both the hurricane activity and the straight-line wind activity.

Design (Basic) Wind Speed

Wind tends to be gusty, implying that the wind speed varies continuously. Because wind speed–measuring instrumentation has a finite response period over which it averages the wind speed, instantaneous wind speed measurements are not possible. The shortest averaging interval used in wind speed–measuring instruments in the United States is 3 seconds. Because building design must be based on maximum wind speeds, we refer to the maximum wind speed as *peak 3-s gust speed*.

Peak 3-s gust speeds for several selected locations are continuously recorded in the United States. The peak 3-s gust speed that is not likely to be exceeded for 700 years at a location is defined as the *basic wind speed* for that location. Basic wind speed, also called *design wind speed*, is used in determining the design wind loads on buildings.

The period of 700 years is called the *return period*. This return period is used for determining the design wind speed for designing buildings with an average risk occupancy, such as individual homes, hotels, and apartment buildings. A longer return period (1,700 years) is used for determining the design wind speed for occupancies of above-average risk, such as emergency shelters, broadcasting stations, and police stations. A shorter return period (300 years) is used for below-average risk occupancies, such as farm buildings for animal shelters. These return periods may seem excessive, but they include a statistically determined safety margin.

By international agreement, wind speeds are recorded at a height of 33 ft (10 m) above ground. Although there are several other recording stations, wind speeds are typically recorded at local airports. A wind speed map giving design wind speed values (with a 700-year return period) for the United States is shown in Figure 3.11. Map values are to be used as a general guide. Local building codes must be referenced for precise values.

Induced Pressure and Suction

Being a fluid, wind exerts pressure on building surfaces. The direction of wind pressure is always perpendicular to the building surface. Thus, on a flat roof, wind pressure is vertical.

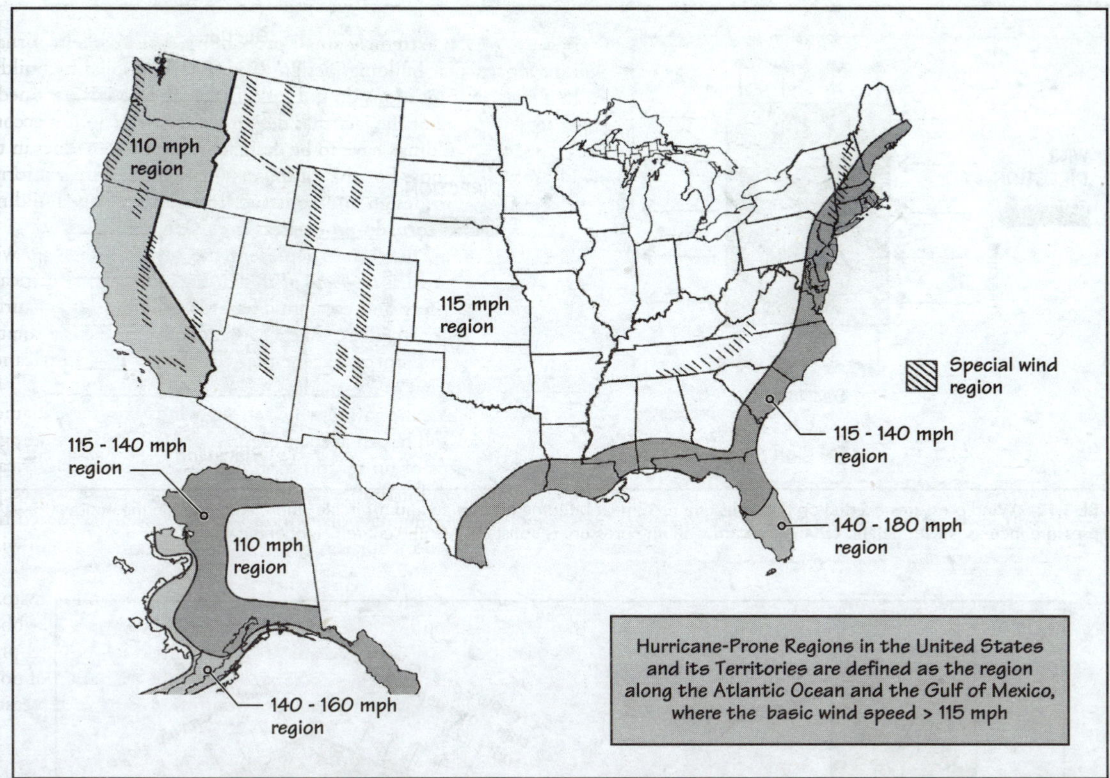

FIGURE 3.11 A peak 3-second gust wind speed (basic wind speed) map of the United States in miles per hour (mph) for the design of buildings with average risk (return period of 700 years), such as homes, offices, restaurants, and so on.
Source: Adapted from the American Society of Civil Engineers, *Minimum Design Loads for Buildings and Other Structures*, SEI/ASCE 7-10 (2010). This map presents information that has been highly simplified and must be treated as conceptual. For accurate information, refer to the source stated.

On a vertical wall, it is horizontal. Whatever the direction of pressure, it acts either toward the surface or away from the surface (suction).

On a rectangular building, wind causes inward force on the windward wall and suction on the other walls (leeward wall and side walls), Figure 3.12(a). The largest pressures occur on the windward and leeward walls. Because a given wall may be a windward or a leeward wall, depending on wind direction, it has to be designed for both the maximum (inward) pressure and maximum suction.

Wind tunnel tests have shown that on the windward wall, wind pressure varies with height, increasing as the height increases. On the leeward wall, there is no appreciable change in pressure with respect to height. Hence, uniform pressure is assumed to act on the leeward wall, Figure 3.12(b).

As for a roof, a flat or near-flat roof is subjected to suction. A pitched roof has to be designed for suction as well as downward pressure. This is because if the wind blows parallel to the ridge, the roof is under suction, Figure 3.13(a). If the wind blows perpendicular to the ridge, the leeward slope is subjected to suction. The windward slope of the roof, however, comes under suction if the roof slope is small, which turns to downward (toward the roof) pressure as the slope increases, Figure 3.13(b).

Spatial Variation of Wind Pressure

Wind pressure on a building's facade not only varies with height above ground, but it also varies from point to point at the same height above ground. For example, significantly larger pressures are obtained at the edges and corners of a building facade than in the central part (field). In other words, a corner window is subjected to greater pressure than a window in the middle of the wall. Similarly, wind pressures on the corners of a roof are greater than those along the perimeter of the roof, which in turn are greater than those in the middle of the roof.

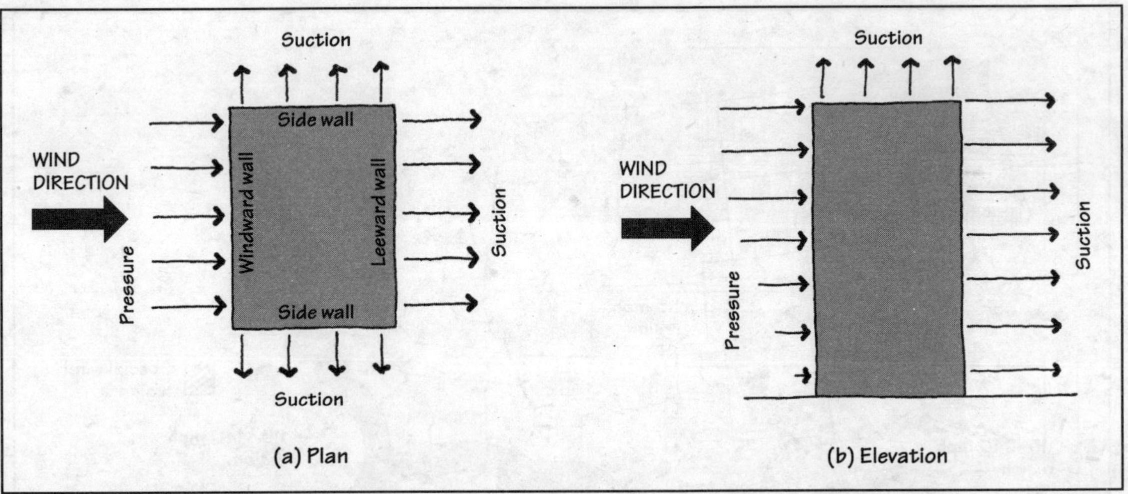

FIGURE 3.12 Wind pressures exerted on the walls of a rectilinear building **(a)** in plan and **(b)** in elevation. Note that on the windward wall, wind pressure increases with height. On the leeward wall, the pressure is constant over the entire height of the wall.

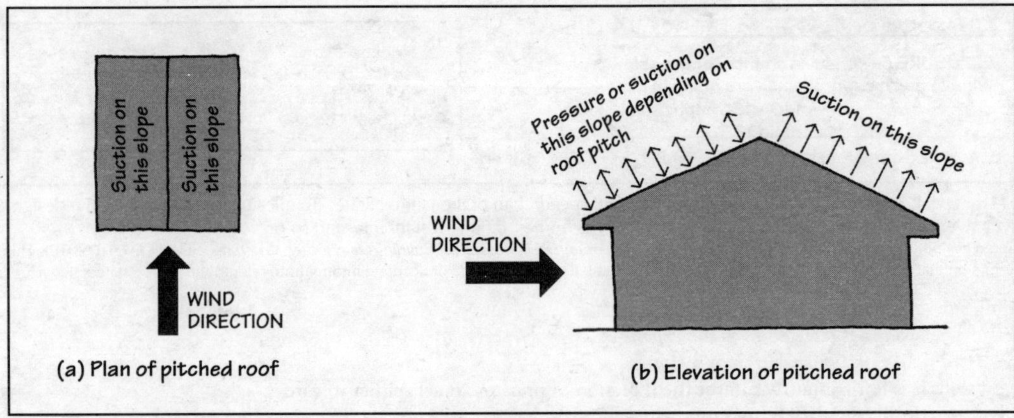

FIGURE 3.13 Wind pressures on a pitched roof.

3.5 FACTORS THAT AFFECT WIND LOADS

The predominant effect of wind speed on wind pressure has already been mentioned. Other important factors that affect wind pressure are:

- Height above ground
- Exposure classification of the site
- Enclosure classification of the building

Height Above Ground

Through our daily experience, we know that wind speed increases with height above the ground. An open window on a higher floor of a multistory building feels breezier than one on a lower floor. Consequently, wind pressures are higher on a tall building than on a short one.

Exposure Classification of the Site

Wind pressures on a building also depend on the neighborhood in which the building is located. A building located in a vast open stretch of land, with no other buildings or obstructions in its neighborhood, is subjected to greater wind pressures than the same building located in a densely built-up area. This is because the wind speed in an open stretch of land is higher than in a built-up or highly wooded area.

EXPOSURE B—A downtown location

EXPOSURE B—An urban or suburban location

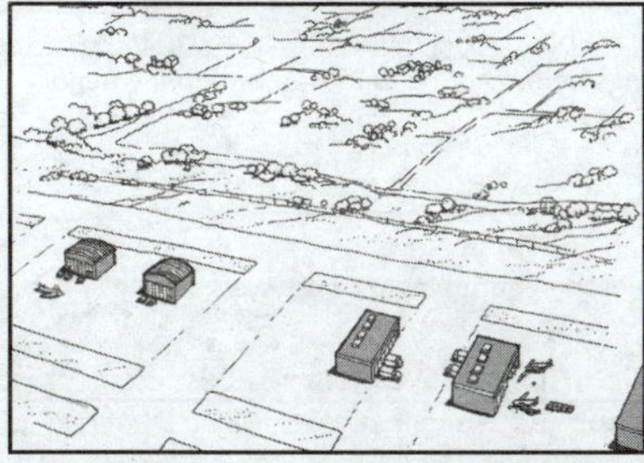

EXPOSURE C—An open country or grassland. Wind speed measurements are made in Exposure C.

EXPOSURE D—A flat, unobstructed site facing a large water body

FIGURE 3.14 Site exposure categories—B, C and D. A densely built downtown location was earlier categorized as Exposure Category A, but it was discovered that, due to the tunneling effect, wind pressures in some parts of a downtown area are equal to those of an urban/suburban location (Exposure Category B). Therefore, Category A was deleted.

Building sites are classified into three types, based on the roughness of the ground. The roughness of the ground is referred to as the *exposure category* of the site. The three exposure categories to which a particular site may belong are B, C, or D. *Exposure Category B*, or simply *Exposure B*, refers to an urban, suburban, or wooded area with several closely spaced obstructions, Figure 3.14. A large city center with several high-rise and closely spaced buildings also belongs to Exposure B.

Exposure C refers to open terrain with scattered obstructions involving heights generally less than 30 ft. An airport is a typical example of Exposure C. *Exposure D* refers to flat, unobstructed ground facing a large body of water (a large lake- or seafront). Everything else being the same, wind pressures are highest in Exposure D, followed by Exposure C, and least in Exposure B.

Enclosure Classification of the Building

Building envelopes are not perfectly airtight. They are subject to air infiltration and exfiltration. Additionally, they have windows and doors, which may not be fully sealed. Because of air infiltration through the openings, the interior of a building is always subjected to some internal pressure in addition to the pressure on the exterior envelope of the building.

Internal pressure is particularly critical in buildings that have large openings on one side or two adjacent sides, because these tend to "cup" the wind, creating a "ballooning" effect, Figure 3.15. Buildings in which the ballooning effect is large are called *partially enclosed buildings*. They include buildings in which the area of openings in one wall is much greater than that in the remaining walls, such as aircraft hangars, warehouses with open dock doors, or garages that are open only on one side.

A building that is not partially enclosed is called an *enclosed building*. Most buildings belong to this (enclosed building) category. Everything else being the same, a partially enclosed building is subjected to greater wind pressure than an enclosed building. If the exterior doors and windows in a building shatter in a windstorm, an enclosed building may become a partially enclosed building, increasing the wind pressure on the building for which it has not been designed. Therefore, the design of the building must ensure that its

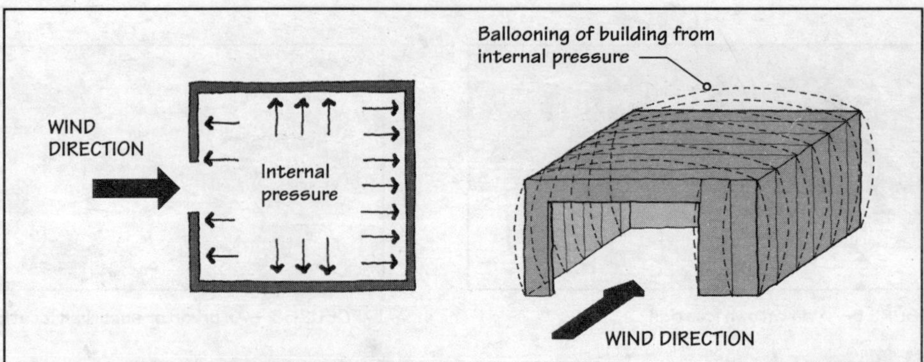

FIGURE 3.15 Ballooning of a building with a relatively large opening in one wall. Such a building is referred to as a *partially enclosed building*. A building that is not partially enclosed is called an *enclosed building*. Most buildings belong to the enclosed building category.

EXPAND YOUR KNOWLEDGE

Important Facts About Wind Loads

1. Wind Load Is Expressed in Terms of Wind Pressure
Wind loads on buildings are generally expressed as wind pressure, that is, in pounds per square foot (psf).

2. Wind Pressure Is Proportional to the Square of Wind Speed
Everything else being the same, the wind pressure on a building component is proportional to the square of wind speed. More precisely:

$$p = \left[\frac{V}{20}\right]^2$$

where: p = wind pressure in pounds per square foot (psf)
V = wind speed in miles per hour (mph)

Thus, if V = 100 mph, p = 25 psf. If V = 90 mph, p = 20 psf. For V = 10 mph, p = 0.25 psf.

3. Magnitude of Wind Pressures on Low-Rise Buildings
Because wind speed refers to that measured at a height of 33 ft (10 m) above the ground, the approximate value of wind pressure on the components of most low-rise buildings (one to three stories tall) can be obtained by the simplified equation just given. This is an extremely important result, as it helps us understand the performance of air and vapor barriers, covered in Chapter 6.

4. Wind Pressure on a Building Component Equals the Difference Between Inside and Outside Air Pressures
The wind pressure on a building component, such as a wall, window, or roof is, in fact, the difference between inside and outside air pressures. If the component is subjected to suction under wind loads, it implies that the inside pressure is greater than the outside pressure. Conversely, if the outside pressure is greater than the inside pressure, the component will be subjected to push-in pressure. The inside pressure in a building is generally the atmospheric pressure. The outside pressure changes with wind speed.

envelope remains intact under the worst expected windstorm. Many buildings have suddenly collapsed (or suffered major damage) under a storm immediately after the collapse of one or more of their exterior doors or windows (see Figure 3.25).

3.6 ROOF SNOW LOAD

A roof is designed for either the roof live load or the snow load, whichever is greater. The reason is that in the event of full snow load on a roof, the roof is not likely to be accessed by a repair or construction crew, who would impose additional live load on it. Like the roof live load, snow load is also expressed in terms of the horizontal projected area of the roof (Figure 3.4).

Fundamental to determining roof snow load is the ground snow load of a location. Ground snow load values with a 50-year return period for selected U.S. locations are given in Table 3.2. These values should be treated as approximate. For authoritative data, the building code or the city's building department should be consulted.

Factors That Affect Roof Snow Load

The factors that affect snow load on a roof are

- *Ground snow load at the location*
- *Roof slope* The greater the slope of the roof, the smaller the snow load. For a flat or near-flat roof, the snow load should theoretically be equal to the ground snow load.

TABLE 3.2 APPROXIMATE GROUND SNOW LOADS FOR SELECTED LOCATIONS

Location	Snow load psf	Location	Snow load psf
Phoenix, Arizona	5	Jackson, Mississippi	5
Little Rock, Arkansas	10	Albany, New York	35
Atlanta, Georgia	10	Raleigh, North Carolina	15
Sprigfield, Illinois	20	Harrisburg, Pennsylvania	25
Des Moines, Iowa	25	Dallas, Texas	10
Topeka, Kansas	25	Anchorage, Alaska	50

Note: The snow load information given is approximate and has been adapted from the American Society of Civil Engineers' *Minimum Design Loads for Buildings and Other Structures*, ASCE 7-10 (2010). For accurate values, the local building code should be referenced.

However, records of measurements of snow deposits on roofs have shown that less snow is typically present on flat roofs than on the ground because wind blows the snow off the roofs and some snow melts due to heat escaping from the heated interiors. Snow load on flat roofs in open areas (i.e., areas that are not obstructed by wind-shielding elements such as higher structures, terrain, and trees) can be as low as 60% of that on the nearby ground.

- *Wind exposure classification of the site (Exposure B, C, or D)* Everything else being the same, Exposure D has the lowest snow load, due to the high wind speeds, and Exposure B has the highest snow load.
- *Warm roof or cold roof* An example of a warm roof is that of a continuously heated greenhouse, and an example of a cold roof is that of a typical residential structure in which the attic space is ventilated, which keeps the roof cold. An unheated building is another example of a structure with a cold roof. Snow load is smaller on a warm roof than on a cold roof.
- *Risk category of building's occupancy* This factor is essentially the same as for the wind loads. Above-average risk occupancies, such as emergency shelters, police stations, fire stations, and radio and television studios, are designed for greater snow load than average-risk occupancies.

PRACTICE **QUIZ**

Each question has only one correct answer. Select the choice that best answers the question.

11. Highest wind speeds are usually obtained in
a. hurricanes. **b.** wind storms.
c. thunderstorms. **d.** tornadoes.

12. An average tornado strikes a larger area on the ground than a hurricane.
a. True **b.** False

13. The design wind speed is also referred to as the
a. standard wind speed. **b.** official wind speed.
c. uniform wind speed. **d.** straight-line wind speed.
e. basic wind speed.

14. The design wind speed for an average-risk building at a location has a probability of occurrence of
a. once in 300 years. **b.** once in 200 years.
c. once in 100 years. **d.** once in 50 years.
e. none of the above.

15. The design wind speed for a location is the peak
a. 60-s gust speed. **b.** 10-s gust speed.
c. 5-s gust speed. **d.** 3-s gust speed.
e. 1-s gust speed.

16. For most of the United States, the design wind speed for an average-risk building is
a. 95 mph. **b.** 105 mph.
c. 115 mph. **d.** 125 mph.
e. 150 mph.

17. For wind blowing along one of the two major axes of a rectangular building, the windward wall is subjected to positive (toward-the-wall) pressure.
a. True **b.** False

18. For wind blowing along one of the two major axes of a rectangular building, the leeward wall is subjected to positive (toward-the-wall) pressure.
a. True **b.** False

19. For wind blowing along one of the two major axes of a rectangular building, the side walls are subjected to positive (toward-the-wall) pressure.
a. True **b.** False

20. On a leeward wall, the wind pressure
a. increases with height above the ground.
b. decreases with height above the ground.
c. is almost constant with height above the ground.

21. With respect to wind pressure, a building site is classified in terms of
a. Exposures A, B, C, and D. **b.** Exposures A, B, and C.
c. Exposures X, Y, and Z. **d.** Exposures P, Q, R, and S.
e. Exposures B, C, and D.

22. The design snow load on a roof is a function of
a. ground snow load of the location.
b. roof slope.
c. wind exposure classification of the site.
d. all of the above.
e. both (a) and (b).

3.7 EARTHQUAKE LOAD

A severe earthquake is one of the most terrifying natural events a person can experience. Regardless of where it occurs, it makes instant headlines all over the world. In the past, earthquakes killed a large number of people. Today, earthquakes pose a smaller threat to life than in the past (at least in developed nations, and discounting the December 2004 tsunami in the Indian Ocean and the January 2010 Haiti earthquake) because of our improved knowledge of constructing earthquake-resistant structures. Statistics indicate that the current annual loss of life resulting from earthquakes is much smaller than that due to hurricanes, building fires, floods, or automobile accidents.

Ground shaking, the phenomenon most commonly associated with earthquakes, causes damage to buildings and other infrastructural facilities, Figure 3.16. Records show that the earthquakes of 1811 and 1812 in Missouri shifted the course of the Mississippi River considerably [3.1]. In addition to ground shaking, a major earthquake can produce several related problems, such as landslides, surface fractures (Figure 3.17), soil liquefaction (Section 11.3), tsunamis, and fires.

TSUNAMIS

A tsunami is caused by the physical displacement of the seabed during an earthquake, which creates waves of up to 50 ft in height. So far, most tsunamis have occurred in the Pacific Ocean, with Japan witnessing the largest number of them—hence their Japanese name. However, the 2004 tsunami that occurred in the Indian Ocean was the most devastating in terms of the loss of human lives.

FIGURE 3.16 Damage to railroad tracks on the Puerto Barrios wharf caused by the 1976 Guatemala earthquake.
(Photo obtained from the World Wide Web using the following URL: http://libraryphoto.cr.usgs.gov/htmllib/batch31/batch31j/batch31z/batch31/gueq0024.jpg)

GEOLOGICAL EXPLANATION OF EARTHQUAKES

The cause of earthquakes is fairly well established. However, we are still far from perfecting a methodology to accurately predict the location and time of earthquake occurrence.

Geologists believe that the earth's crust is divided into several individual segments called *crustal* or *tectonic plates,* Figure 3.18. These solid plates, floating on a molten mantle below, are constantly in motion relative to each other. The motion is so slow that it is significant only in geological time. The relative motion in the earth's crust causes compressive, tensile, or shear forces to develop, depending on whether the plates press against each other, pull apart, or slip laterally under one another. When these stresses exceed the maximum capacity of the crust to absorb them, a fracture at the crust occurs. It is this fracturing, or slippage, which is always sudden, that produces a shock wave known as *seismic motion.*

FIGURE 3.17 Fissures in the ground caused by the 1989 Loma Prieta earthquake, San Francisco Bay area (Richter magnitude 7.0).
(Photo obtained from the World Wide Web using the following URL: http://pubs.usgs.gov/dds/dds-29/tif/img0087.tif)

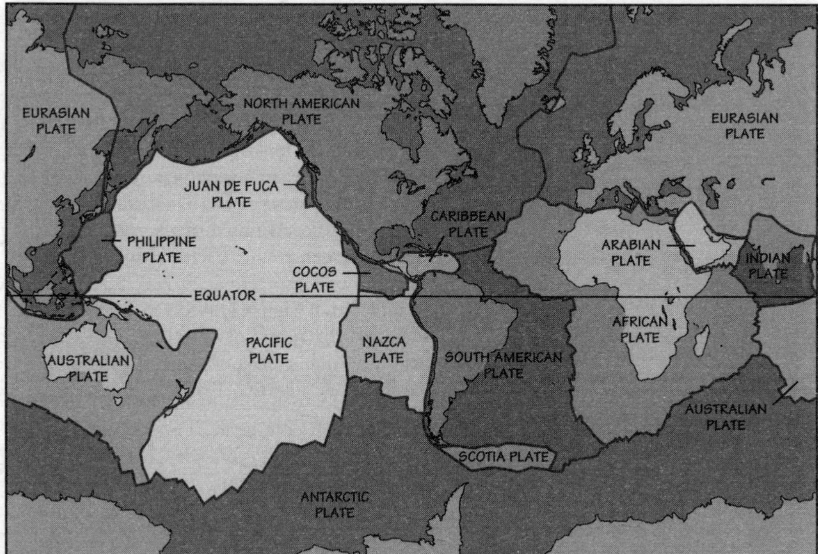

FIGURE 3.18 Map of the earth with an approximate depiction of major tectonic plates. (Illustration by the U.S. Geological Survey.)

The plane where the fracture occurs is called a *fault.* The location where the fault originates is called the *focus,* which is generally inside the earth's crust. The point directly above the focus on the earth's surface is called the *epicenter,* Figure 3.19.

The fracture on the surface of the earth may or may not be noticeable or may appear as a minor fissure, as shown in Figure 3.17. The 1906 San Francisco (California) earthquake left a deep ground rupture extending hundreds of miles. Known as the *San Andreas fault,* it stands as a distinct landmark created by an earthquake.

FREQUENCY OF OCCURRENCE AND LOCATION OF EARTHQUAKES

An earthquake may occur at any location. In fact, mild earthquakes occur quite frequently: Several thousand a day occur in various regions of the earth, Table 3.3. They go unnoticed because they may occur in remote places or with a magnitude that is below the threshold of human perception.

The regions of the earth in proximity to plate boundaries are susceptible to more frequent and more severe earthquakes. Nearly 90% of all earthquakes occur in the vicinity of plate boundaries, because the plate boundaries represent lines of greater weakness in the earth's crust. Observe in Figure 3.18 that the Pacific plate and the North American plate meet along the West Coast of the United States—a region known for its high seismic activity.

The remaining 10% of earthquakes occur because of the faults within the plates. Called *intraplate earthquakes,* they occur far away from plate boundaries. Earthquakes that have

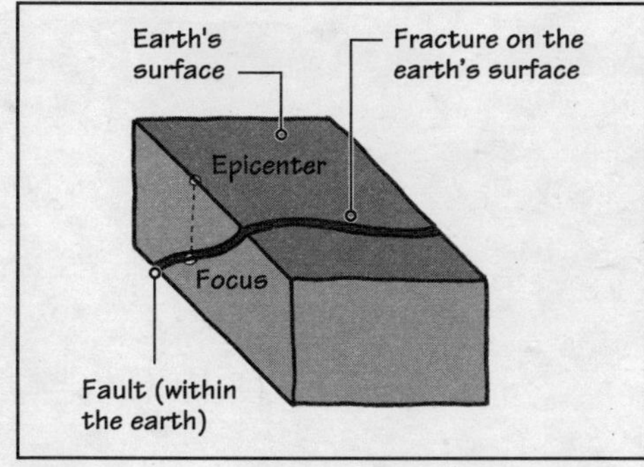

FIGURE 3.19 Fault, focus, and epicenter of an earthquake.

TABLE 3.3 WORLDWIDE FREQUENCY OF EARTHQUAKE OCCURRENCE

Descriptor	Richter magnitude	Average frequency per year
Great	≥8.0	0–1
Major	7–7.9	17
Strong	6–6.9	134
Moderate	5–5.9	1,319
Light	4–4.9	13,000
Minor	3–3.9	130,000
Very minor	2–2.9	1,300,000

Source: EQ Facts and Lists, National Earthquake Information Center, U.S. Geological Society Web site (http://www.usgs.gov).

occurred in the Eastern and Midwestern United States are intraplate earthquakes. Intraplate earthquakes are infrequent but may be as severe as plate boundary earthquakes.

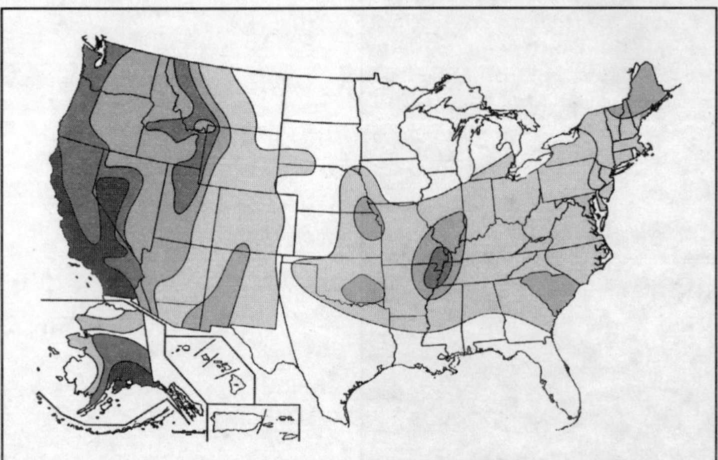

FIGURE 3.20 Map of the United States showing approximate variation of seismic activity. The darker the area, the greater the seismic activity. White areas have little or no seismic activity.

Because seismicity is a geographical activity, it is possible to delineate areas that are seismically more active than others. For the United States, this activity is shown in Figure 3.20, in which the degree of seismic activity is denoted by the darkness of the gray color. Thus, unshaded areas of the country have little or no seismic activity, and the darkest areas are seismically most active.

EARTHQUAKE INTENSITY—RICHTER MAGNITUDE

For the general public, the scale most commonly used to describe the intensity of an earthquake is the Richter scale. The Richter scale was devised in 1935 by Professor Charles Richter of the California Institute of Technology. The scale begins at zero but has no upper limit. However, the severest earthquake recorded to date on this scale is 9.0.

According to the Richter scale, the total energy released by an earthquake (E) is proportional to the Richter magnitude (R), as given by the following expression:

$$E \propto 10^{1.5R}$$

The symbol $\propto$ is the sign of proportionality, implying that earthquake energy is proportional to $10^{1.5R}$. Because $10^{1.5}$ is approximately equal to 32, this expression implies that an increase of 1.0 on the Richter scale gives a 32 times increase in the energy output of the earthquake. Thus, an earthquake measuring 8.0 on the Richter scale is nearly 32-fold more powerful than one measuring 7.0 and nearly 1,000 times more powerful than one measuring 6.0. (Note: $32 \times 32 =$ approximately 1,000.)

An earthquake of up to 2.0 on the Richter scale goes unnoticed by humans; 4.0 to 5.0 may cause slight damage to poorly designed buildings. At 7.0, the damage is usually extensive in poorly designed buildings. An earthquake with a magnitude of 6.5 and above is considered significant. Table 3.4 lists some of the major earthquakes that have occurred in various parts of the world in the past 100 years. (Major earthquakes may be measured by their Richter magnitude, number of fatalities, or loss of property.)

TABLE 3.4 SOME OF THE MAJOR EARTHQUAKES SINCE 1900

Year	Location	Richter magnitude	Loss of life	Year	Location	Richter magnitude	Loss of life
1905	Kangra, India	8.6	19,000	1952	Hokkaido, Japan	8.6	600
1906	Columbia and Ecuador	8.9	1,000	1952	Alerce, Chile	8.5	2,000
1906	San Francisco, USA	8.3	700	1960	Alaska, USA	8.5	115
1906	South Central Chile	8.6	1,500	1964	Moro, Philippines	8.0	6,500
1907	Karatag, Tajikistan	8.0	12,000	1976	Tangshan, China	7.5	255,000
1908	Messina, Italy	7.2	100,000	1976	Sunda, Indonesia	8.0	200
1920	Gansu, China	7.8	200,000	1977	Bucharest, Romania	7.2	1,600
1923	Kanto, Japan	7.9	143,000	1985	Mexico City	8.1	5,000
1927	Tsinghai, China	7.9	200,000	1990	Kobe, Japan	6.8	6,400
1932	Gansu, China	7.6	70,000	1995	Turkey	7.7	17,000
1933	Honshu, Japan	8.9	3,000	1999	India	7.7	20,000
1935	Quetta, Pakistan	7.5	60,000	2004	Sumatra (tsunami)	9.0	283,000
1939	Erzincan, Turkey	8.0	33,000	2005	Azad Kashmir, Pakistan	7.6	80,000
1945	Iran and Pakistan	8.2	4,000	2010	Haiti	7.0	200,000
1948	USSR	7.9	110,000	2010	Chile	8.8	600
1950	India and China	8.6	1,500				

Source: U.S. Geological Survey Earthquake Hazards Program (http://earthquake.usgs.gov) and other sources.

3.8 FACTORS THAT AFFECT EARTHQUAKE LOADS

The ground motion caused by an earthquake consists of random vibrations. Like any other vibration, this motion is also characterized by its amplitude and frequency. Although the Richter scale gives the total energy released, it provides no information about the characteristics of ground motion (frequency and duration of vibrations) and the interaction of ground vibration with the vibrational characteristics of the building and the underlying soil. Therefore, the Richter scale cannot be used in determining earthquake loads on buildings. Instead, the following factors must be considered:

- Ground motion
- Building's mass and ductility of the structural frame
- Type of soil
- Building occupancy's risk category

GROUND MOTION

To understand how an earthquake impacts a building, imagine a building just before the earthquake. Every part of the building is in static equilibrium. As soon as the earthquake occurs, the ground is displaced horizontally. Because the building has inertia, it takes a finite period of time for the horizontal displacement to travel to the upper parts of the building. In the meantime, the building is subjected to deformations, as indicated by its deformed shape, Figure 3.21. Thus, although there are several differences between the effects of an earthquake and a windstorm on a building (see Section 3.9), the overall building deformation produced by both is similar.

An earthquake results in ground acceleration due to an abrupt ground motion. This acceleration produces an inertial force in the building. This force, which tends to oppose the ground motion, is referred to as the *total earthquake load* on the building. The magnitude of the total earthquake load can be obtained from Newton's second law of motion. According to this law, when a body is accelerated, the inertial force (F) acting on it is equal to the product of the acceleration and the mass of the body:

$$F = ma$$

where m is the mass of the body and a is the acceleration imparted to it.

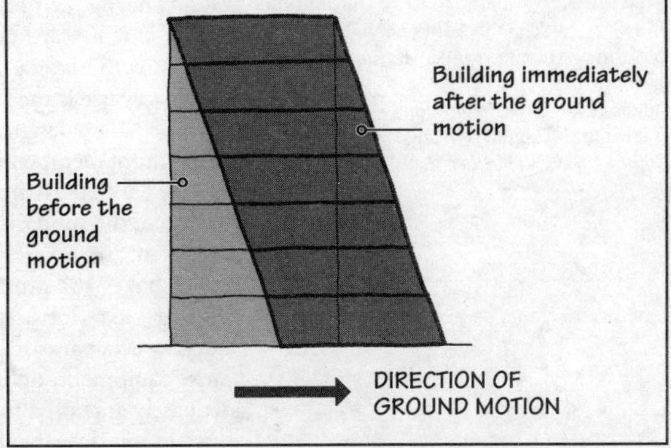

FIGURE 3.21 Deformation of a building resulting from ground movement. Observe that the deformation caused by an earthquake is essentially similar to that caused by wind. Therefore, the structural provisions for earthquake and wind load resistance are essentially similar. In the building shown, similar deformation would have occurred from wind blowing from the right-hand side.

When Newton's law is applied to buildings, m refers to the weight of the building and a refers to the acceleration of the ground. If ground acceleration is small, the earthquake load is small. This is obvious because if the ground moves very slowly (with negligible acceleration), the building will simply go along with this motion as one unit, and the opposing inertial force (the earthquake load) will be small. In the case of a sudden and swift motion (large acceleration), the lower part of the building moves horizontally, whereas the upper part remains in its original position (see Figure 3.21). In an effort to maintain the status quo, a large opposing (inertial) force is produced in the building—that is, the total earthquake load on the building is large.

BUILDING'S MASS AND DUCTILITY OF THE STRUCTURAL FRAME

From Newton's second law of motion, the other factor that affects the magnitude of the earthquake load is the weight of the building. Lighter buildings attract a smaller earthquake load than buildings constructed of heavy materials. Thus, concrete frame and masonry structures attract a greater earthquake load than wood frame or steel frame structures. Residential structures with wood siding or stucco finishes attract a smaller earthquake load than those with brick or stone veneer.

Indeed, because of its light weight, the earthquake load on a tent is negligible even in a severe earthquake, Figure 3.22. A tent structure has the additional advantage of its built-in resistance to

FIGURE 3.22 A tent resists earthquake forces admirably due to its light weight and its ability to absorb deformations.

FIGURE 3.23 Damage to an unreinforced masonry building in Santa Monica from the Northridge (California) earthquake on January 17, 1994.
(Photo obtained from the World Wide Web using the following URL: http://www.ngdc.noaa.gov/hazard/img/200res/19/19398.tif)

absorb structural deformations (ductility)—an important structural property for earthquake regions.

Buildings constructed of heavy materials with little or no capacity to absorb deformations are among the most hazardous buildings in a seismically active zone. Such buildings include those whose structural system consists of unreinforced masonry structures (i.e., bricks, stone or concrete block walls with no vertical reinforcement), Figure 3.23.

Soil Type and Risk Category of the Building's Occupancy

The effect of soil on earthquake load is described in Section 11.3. The risk category of the building's occupancy has the same importance as for wind and snow loads; that is, occupancies with above-average risk must be designed for a higher earthquake load than occupancies with average or below-average risk.

3.9 WIND VERSUS EARTHQUAKE RESISTANCE OF BUILDINGS

As discussed in Section 3.4, many of the structural strategies used for wind load resistance and earthquake resistance are essentially similar. Buildings are, therefore, generally designed to resist either earthquake loads or wind loads, whichever causes the worst effect (greater stresses). This is based on the assumption that there is a negligible probability of maximum wind speeds occurring at the same time as an intense earthquake. In regions of low seismic activity, wind loads govern the design of the lateral resistance of buildings; in regions of intense seismic activity, the reverse is the case.

The design of lightweight envelope components, such as glass curtain walls and roof membranes, is governed by wind loads even in highly seismic regions. (Remember that earthquake loads are influenced by the weight of the component.)

Despite the similarities between the provisions for wind load and earthquake load resistance, there are several differences in the details. An important difference is that as an earthquake shakes the ground on which the building rests, it affects all components of the building—exterior as well as interior components. Damage to the building's contents and injury to occupants may occur by falling interior contents, Figure 3.24. Thus, all components, equipment, and fixtures in a building located in a seismically active zone must be adequately anchored to remain intact during an earthquake.

Wind, on the other hand, acts on the building envelope. It is through the envelope that wind loads are transmitted to the building structure. If the envelope remains intact, the building will generally retain its overall structural integrity in a storm. Thus, the wind damage in a building usually begins with the damage to envelope components, such as the roof, exterior walls, exterior doors, windows, and other cladding elements. Once the building changes from fully enclosed to partially enclosed (see Figure 3.15), the wind loads on the

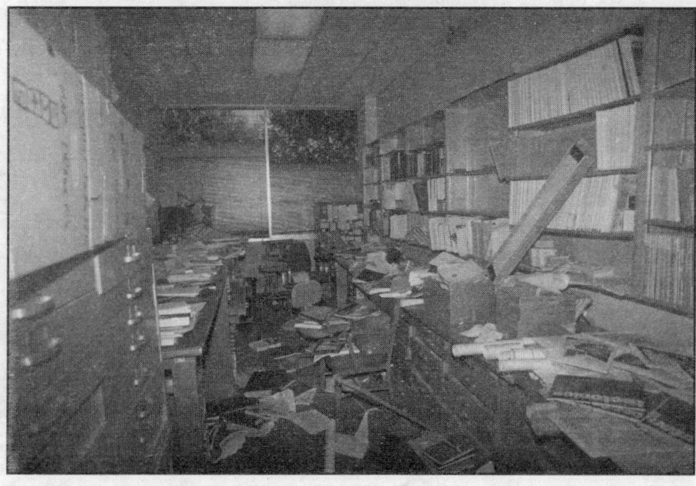

FIGURE 3.24 Effect of earthquake shaking (1989 Loma Prieta earthquake) on the contents of a building's interior. Note that the building envelope is intact. Greater damage could have occurred if the interior fixtures were inadequately anchored.
(Photo obtained from the World Wide Web using the following URL: http://pubs.usgs.gov/dds/dds-29/tif/img0040.tif)

building increase due to the ballooning effect, which may result in additional damage or collapse of the structure, damage to interior contents, and injuries to occupants, Figure 3.25.

FIGURE 3.25 Damage to the glass curtain wall of the 37-story Bank One Building, Fort Worth, Texas, by a forceful tornado on March 28, 2000. As soon as the glass in the curtain wall shattered, substantial damage to the building's interior (ceiling, drywall, electrical fixtures, etc.) occurred.

Distinct Design Approaches

The more important difference between wind and earthquake resistance lies in their basic design approaches. For wind load resistance, the structure is designed so that under the action of maximum wind loads, it remains elastic—that is, the structure is designed to suffer no permanent deformation under the worst expected storm.

The design approach is different for earthquake load resistance. The loads on a building produced by the worst expected earthquake for the location are so large that if the building were designed to remain elastic after the earthquake, it would be prohibitive in cost. Therefore, the earthquake loads for which a building is designed may be smaller than the maximum earthquake loads expected on it.

The underlying design philosophy is that the building should remain elastic when resisting minor earthquakes. In the event of an intense earthquake, part of the earthquake's energy may be dissipated in permanently deforming the building, but the building must otherwise remain intact to provide complete safety to its occupants.

Permanent deformations in a structure are possible only if the structure has the ability to sustain such deformations. As we will see in Chapter 4, materials that can sustain permanent deformation prior to failure are called *ductile* materials. Ductility is an essential property for buildings located in seismic zones but is not a requirement for resisting wind loads.

A building must not deform permanently even in the most severe windstorm. Additionally, it must be rigid enough to reduce the deflections caused by strong winds. The swaying of the upper floor of tall buildings under wind loads must be controlled to remain within acceptable limits of human tolerance.

The reverse is true for earthquake loads. Buildings must be able to deform, even permanently, to absorb the energy delivered to them by the earthquake. If brittle materials are used, they will fail under earthquake forces. Several reinforced concrete frame buildings with (brittle) unreinforced masonry infill walls between frames have collapsed or suffered serious damage under earthquake forces, Figure 3.26. Such buildings would generally be unharmed by violent storms.

FIGURE 3.26 A typical pattern of damage to unreinforced masonry infill walls in buildings with a reinforced concrete frame structure. If the infill walls shown in this illustration had been reinforced horizontally and vertically, they would have suffered minor (or no) damage depending on the earthquake's intensity.

PRINCIPLES IN PRACTICE

Dead Load and Live Load Estimation

TRIBUTARY AREA OF A BUILDING COMPONENT

In computing the dead loads and other loads on a component, the concept of *tributary area* is important to understand. The tributary area for a component is the area (or areas) of the building that *contributes* load on that component.

For example, in a floor system that consists of wood planks supported by parallel beams, the tributary area for a beam is the area of the floor that lies halfway to the adjacent beam on the left and halfway to the adjacent beam on the right. In Figure 1, the tributary area for beam A is shown by the colored area. All the load that is placed on this colored area must be carried by beam A.

The tributary area for a column in a building is similarly obtained. In the framing plan of the building shown in Figure 2, the tributary areas of an interior column X, an exterior column Y, and a corner column Z have been colored. Thus, column X receives all the load placed on the colored area of the floor surrounding it. This is a rectangular area with dimensions $0.5(L_1 + L_2)$ by $0.5(W_1 + W_2)$. Thus, if L_1 and L_2 are 12 ft and 18 ft, respectively, and W_1 and W_2 are each 13 ft, the tributary area for column X is 15 ft × 13 ft.

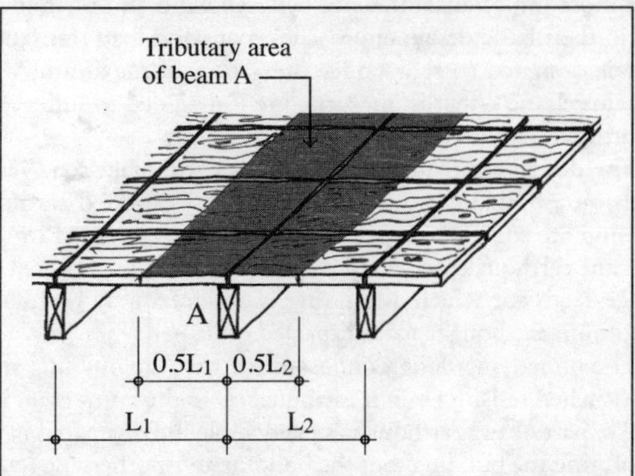

FIGURE 1 The tributary area of beam A is the colored area. The width of this area = $0.5(L_1 + L_2)$, where L_1 and L_2 are the center-to-center distances between beams.

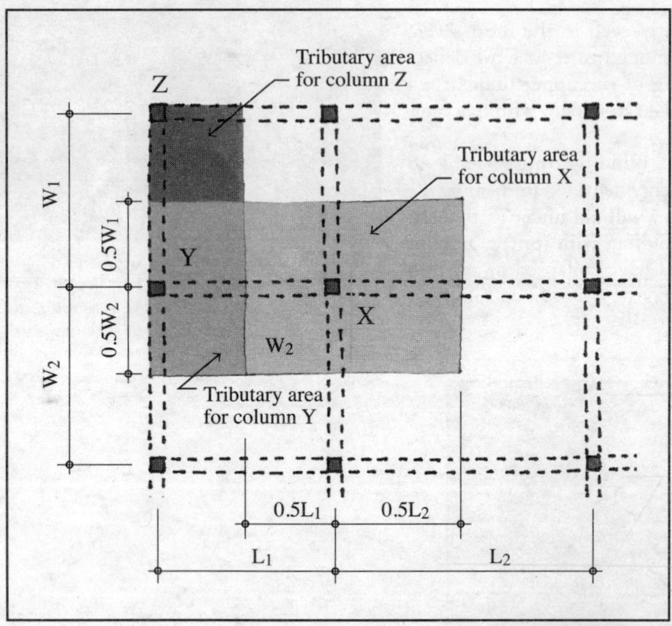

FIGURE 2 Tributary areas for columns X, Y, and Z.

EXAMPLE 1 (ESTIMATING THE DEAD LOAD ON A BEAM)

Determine the dead load on beam A of Figure 3, where each (wood) beam is 4 in. × 16 in. in cross section and is spaced 6 ft on center. The (wood) floor planks are $2\frac{1}{2}$ in. thick. Assume that the density of wood is 40 lb/ft^3 (pcf).

SOLUTION

Because the load on a beam is generally expressed in pounds/foot, we will isolate a 1-ft length of beam A and determine all the dead load on this length. What holds for this 1-ft length also holds for the rest of the beam.

PRINCIPLES IN PRACTICE

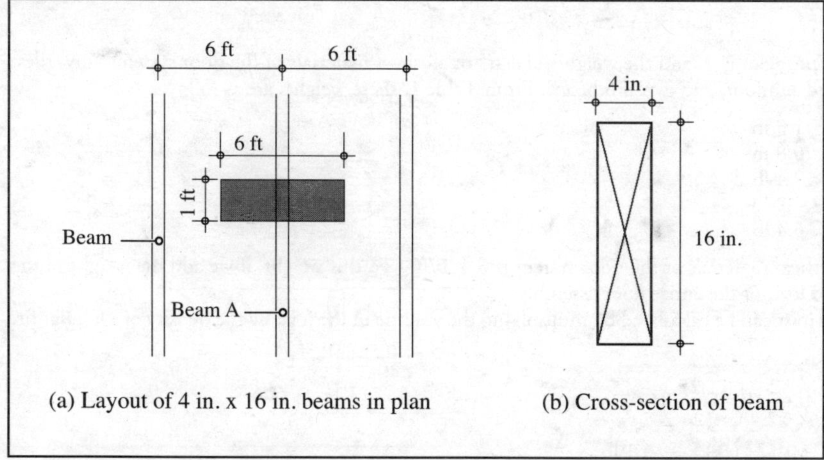

FIGURE 3 Refers to Example 1.

$$\text{Volume of a 1-ft-long beam} = \left(\frac{4}{12}\right)\left(\frac{16}{12}\right)(1) = 0.44 \text{ ft}^3$$

$$\text{Weight of a 1-ft-long beam} = 0.44(40) = 17.6 \text{ lb}$$

From Figure 3, the tributary area of a 1-ft-long beam is 1 ft $\times$ 6 ft = 6.0 ft^2. Hence, the weight of 6 ft^2 of floor planks is borne by a 1-ft-long beam.

$$\text{Volume of 6.0 ft}^2 \text{ planks} = (6.0)\left(\frac{2.5}{12}\right) = 1.25 \text{ ft}^3$$

$$\text{Weight of planks} = 1.25(40) = 50.0 \text{ lb}$$

The total dead load on beam A = 17.6 + 50.0 = 67.6 lb/ft.

EXAMPLE 2 (ESTIMATING THE DEAD LOAD OF A TYPICAL WOOD FRAME RESIDENTIAL FLOOR)

Determine the dead load of a typical residential floor constructed of 2 $\times$ 12 (nominal) Douglas fir floor joists spaced 16 in. on center. The subfloor consists of $\frac{3}{4}$-in.-thick plywood and a $\frac{1}{4}$-in.-thick hardboard underlayment, Figure 4. Vinyl tiles are used as floor finish, and the ceiling consists of $\frac{1}{2}$-in.-thick gypsum wallboard.

Note: A (nominal) 2 $\times$ 12 floor joist has actual dimensions of $1\frac{1}{2}$ in. $\times$ $11\frac{1}{4}$ in.

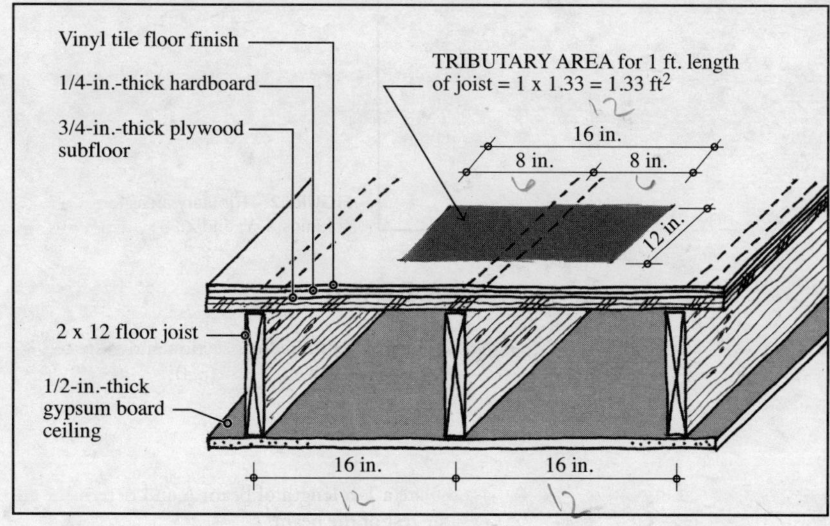

FIGURE 4 Floor and ceiling construction in a typical wood frame residential building—refers to Example 2.

(Continued)

PRINCIPLES IN PRACTICE

Dead Load and Live Load Estimation (*Continued*)

SOLUTION

A simple method for solving this problem is to add the weights of 1 ft^2 of all sheet materials of the floor system: vinyl tiles, hardboard underlayment, plywood subfloor, and gypsum board. From Table 1, these weights are as follows:

Vinyl tiles	1.0 lb
Hardboard (2 × 0.4)	0.8 lb
Plywood (6 × 0.4)	2.4 lb
Gypsum board (4 × 0.55)	2.2 lb
Total weight	6.4 lb

Thus, the total load due to all sheet materials of the floor system is 6.4 lb/ft^2. To this weight, if we add the weight of the floor joist, we will obtain the dead load of the entire floor assembly.

The weight of a 1-ft-long floor joist can be calculated by multiplying the volume of the joist by the density of Douglas fir, which is 34 lb/ft^3, Table 2.

$$\text{Volume of a 1-ft-long joist} = \left(\frac{1.5}{12}\right)\left(\frac{11.25}{12}\right)(1) = 0.12\ ft^3$$

$$\text{Weight of a 1-ft-long joist} = (0.12)(34) = 4.1\ lb$$

TABLE 1 APPROXIMATE (SURFACE) DENSITIES OF SELECTED MATERIALS

Component	Weight lb/ft^2	Component	Weight lb/ft^2
Roof Coverings		Plywood or hardboard (per $\frac{1}{8}$-in. thickness)	0.4
Asphalt shingles	2.0	Linoleum, vinyl, or asphalt tile ($\frac{1}{4}$ in. thick)	1.0
Clay tiles	12–18	Marble or slate or terrazzo (per 1-in. thickness)	12
Gravel-covered built-up roof	4.0	Granite	15
Insulation (per 1 in. thickness)		Ceilings	
Fiberglass	0.1–0.2	Gypsum board (per $\frac{1}{8}$-in. thickness)	0.55
Extruded or expanded polystyrene, or polyurethane or polyisocynurate	0.2	Plaster on wood lath	8.0
Floors and Floor Finishes		Acoustic fiber tile	1.0
Hardwood flooring, $\frac{7}{8}$-in. thick	4.0	Mechanical duct allowance	4.0

TABLE 2 APPROXIMATE (VOLUME) DENSITIES OF SELECTED MATERIALS

Material	Weight, lb/ft^3	Material	Weight, lb/ft^3
Metals		Wood	
Cast iron	450	Oak	47
Copper	556	Douglas fir	34
Lead	710	Hem fir	28
Steel	490	Southern pine	37
Zinc	450	Wood Products	
Cement and Concrete		Plywood	36
Concrete (normal weight)	145	Particleboard	45
Reinforced concrete (normal weight)	150	Gypsum Products	
Structural lightweight concrete	85–115	Gypsum wallboard	50
Insulating concrete	25–50	Lath and plaster	55
Portland cement	94	Insulating Materials	
Lime	45	Fiberglass	1.0–2.5
Masonry		Rockwool	1.0–2.5
Clay brick	110–130	Extruded or expanded polystyrene	2.0
Concrete block (lightweight)	105–125	Polyisocyanurate	2.0
Concrete block (normal weight)	135	Glass (flat glass)	160
Masonry mortar or grout	130	Earth	
Stone		Clay, silt, or sand (dry packed)	100
Granite	165	Crushed stone	100–120
Limestone	165	Water	63
Marble	170	Ice	57
		Air	0.075

PRINCIPLES IN PRACTICE

From Figure 4, a 1-ft-long floor joist carries 1.33 ft^2 of floor.

Therefore, the weight of floor joist that contributes to 1 ft^2 of the floor is $\left(\frac{4.1}{1.33}\right) = 3.1$ lb.

Hence, the total dead load of 1 ft^2 of floor is $6.4 + 3.1 = 9.5$ lb.

This is an important result because the dead load of a conventional wood frame floor is assumed to be 10 psf.

EXAMPLE 3 (ESTIMATING THE TOTAL GRAVITY LOAD ON A TYPICAL WOOD FRAME RESIDENTIAL FLOOR)

Determine the total gravity load (dead load + live load) on the floor of Example 2. The floor is part of an apartment building.

SOLUTION

The dead load of the floor is 10 psf, Example 2.

The live load on a residential floor is 40 psf, Table 3.1.

Hence, the total gravity load (dead load + live load) is $10 + 40 = 50$ psf.

EXAMPLE 4 (ESTIMATING THE DEAD LOAD OF A WOOD FRAME RESIDENTIAL FLOOR WITH LIGHTWEIGHT CONCRETE TOPPING)

In many multifamily dwellings, a lightweight concrete topping is used over a wood floor to increase the sound insulation between floors. Determine the dead load of such a floor, assuming $1\frac{1}{2}$-in.-thick lightweight concrete topping. The density of lightweight concrete is 100 pcf.

SOLUTION

The dead load of a typical residential wood floor is 10 psf, Example 2.

Volume of 1 sq ft of concrete topping $= 1.0 \times 0.125 = 0.125$ ft^3 (Note that $1\frac{1}{2}$ in. $= 0.125$ ft)

Weight of 1 sq ft of concrete topping $= 0.125 \times 100 = 12.5$ lb

Hence, the total dead load of the floor is $10 + 12.5 = 22.5$ psf.

PRACTICE QUIZ

Each question has only one correct answer. Select the choice that best answers the question.

23. According to the geologists, the earth's surface is composed of several segments that are called
- **a.** earth segments.
- **b.** surface plates.
- **c.** tectonic plates.
- **d.** geological segments.
- **e.** segmental plates.

24. The region of the United States that is relatively more earthquake prone is the
- **a.** northern United States.
- **b.** southern United States.
- **c.** eastern United States.
- **d.** western United States.
- **e.** none of the above.

25. The number of mild (minor to very minor) earthquakes occurring on the earth is almost
- **a.** 1 or 2 per year.
- **b.** 10 to 20 per year.
- **c.** 100 per year.
- **d.** 1,000 per year.
- **e.** several thousand per year.

26. An earthquake measuring 7.0 on the Richter scale is
- **a.** twice as strong as a 6.0 earthquake.
- **b.** 10 times stronger than a 6.0 earthquake.
- **c.** 100 times stronger than a 6.0 earthquake.
- **d.** 500 times stronger than a 6.0 earthquake.
- **e.** none of the above.

27. Everything else being the same, a building constructed of heavyweight materials is subjected to a greater earthquake load than a building constructed of lightweight materials.
- **a.** True
- **b.** False

28. A reinforced concrete beam measures 12 in. × 24 in. in cross section. If the density of reinforced concrete is 150 lb/ft^3, the weight of the beam is
- **a.** 450 lb/ft.
- **b.** 350 lb/ft.
- **c.** 300 lb/ft.
- **d.** 300 lb.
- **e.** 300 lb/ft^3.

29. In a building whose structural system consists of columns, beams, and floor and roof slabs, the columns are spaced at 40 ft on center in one direction and 25 ft on center in the other direction. What is the tributary area for an interior column?
- **a.** 1,000 ft^2
- **b.** 1,000 ft
- **c.** 500 ft^2
- **d.** 250 ft^2
- **e.** None of the above

PRACTICE QUIZ (Continued)

30. In a building whose structural system consists of columns, beams, and floor and roof slabs, the columns are spaced at 40 ft on center in one direction and 25 ft on center in the other direction. What is the tributary area for a corner column?
    a. 1,000 ft^2
    b. 1,000 ft
    c. 500 ft^2
    d. 250 ft^2
    e. None of the above

31. The total approximate dead load of a floor in a typical wood frame residential building is
    a. 50 psf.
    b. 30 psf.
    c. 10 psf.
    d. 5 psf.
    e. none of the above.

32. The total approximate gravity load on a floor in a typical wood frame residential building is
    a. 50 psf.
    b. 30 psf.
    c. 10 psf.
    d. 5 psf.
    e. none of the above.

REVIEW QUESTIONS

1. What is the difference between dead load and live load? Explain with the help of examples.
2. What is the approximate wind speed in a tornado? Explain why tornado wind speeds are not considered in determining the design wind speed for a location.
3. Using sketches and notes, explain the types of wind pressure that occur on a rectangular building.
4. Using sketches and notes, explain how roof slope affects wind pressure on a roof.
5. Explain the differences between wind Exposure Categories B, C, and D.
6. Using sketches and notes, explain the ballooning effect in buildings.
7. What is the Richter scale? What is the difference between the amounts of energy released by two earthquakes that are 8.0 and 6.0 on the Richter scale?
8. What types of construction are (a) suitable and (b) unsuitable in an earthquake-prone region?
9. Explain the major differences between designing buildings for earthquake resistance and for wind resistance.

CHAPTER 4

Load Resistance (The Structural Properties of Materials)

CHAPTER OUTLINE

The history of the development of architectural form coincides with the evolution of human understanding of the structural properties of materials and their application to the design and construction of buildings. Every civilization has found it necessary to understand the strength and limitations of indigenous materials in order to design safe and durable structures. The small spans of beams used in Greek temples, as shown in Figure 4.1, were due mainly to the designer's instinctive understanding that stone is a brittle material that is weak in tension and would break if the span is too long or the beam size is too small.

Stone's high strength in compression was exploited in columns and walls, but the column spacings and the sizes of openings in walls had to be small. Although the main reason for providing capitals at the tops of columns in Greek temples was nonstructural, the capitals provided greater structural safety by reducing beam spans even further.

The discovery of the arch—a form that carries loads primarily in compression—enabled early builders to span greater distances in stone (or brick) than was previously possible. The arch led to the development of the vault, which is essentially a three-dimensional version of the arch, and subsequently to the dome—a form obtained by rotating the arch about a center point.

The arch, the vault, and the dome dominated architecture for centuries as the primary form-giving elements in buildings because the major construction materials available at the time, such as stone and brick, were brittle materials. The use of wood, which is less brittle than stone, was limited to either minor buildings or minor components of buildings because of its lack of durability. Sooner or later, wood structures were either consumed by fire or destroyed by termites.

Masonry buildings, on the other hand, are extremely durable, and some have survived for thousands of years. Therefore, the formal, intellectual and technological use of brittle

FIGURE 4.1 Parthenon, Acropolis, Greece. Observe the relatively small span of beams in response to the weakness of stone in bending.
(Photo by Brent Wong/Shutterstock)

materials became highly perfected over time. Gothic cathedrals are excellent examples of how the limitations in the tensile strength of brittle materials were overcome to produce magnificent and technically daring structures, Figure 4.2.

The Industrial Revolution brought many new developments in architectural form. The availability of wrought iron and, later, cast iron had already revived the beam as a horizontal spanning element because of iron's high tensile strength. Large openings became possible without the use of arches, and the progression from load-bearing masonry buildings to the skeleton frame did not require much imagination. After steel became commercially available (in the mid-nineteenth century), the frame structure replaced load-bearing masonry as the predominant structural system.

The invention of new materials—prestressed concrete, high-strength steels, reinforced concrete, and concrete masonry—led to the development of new structural forms. In fact, the modern movement and subsequent developments in architecture owe a great deal to the exploitation of the structural properties of materials.

In this chapter, we examine some of the important properties that influence the strength of materials. Emphasis is placed on properties that relate directly to basic structural materials, such as steel, concrete, brick, stone, and wood. However, the principles examined have general application.

Structural properties of materials and the behavior of structures under loads cannot be divorced from each other. They are so interrelated that often the distinction between the two is vague at best. Ductility and brittleness, elasticity, and plasticity, which are properties of materials, are also used to describe the behavior of an entire structure or components of a structure. Therefore, structural behavior is discussed in this chapter where necessary to describe the structural properties of materials.

FIGURE 4.2 Bourges Cathedral, Bourges, France. Buttresses and flying arches were some of the structural innovations of the time in combating the inherent weakness of stone in tension.
(Photo courtesy of Dr. Jay Henry)

4.1 COMPRESSIVE AND TENSILE STRENGTHS OF MATERIALS

When a force acts on a member (such as a building component), the member develops an internal resistance to the applied force. The intensity of internal resistance to the applied external force is called the *stress.* If the applied force is large, the internal resistance is large, and so is the stress. If the applied force is small, the stress developed in the member is also small.

If a member is unable to develop any resistance to an applied force, the stress in the member is zero. For instance, if we hold a cable (or rope) between two hands and push it inward, that is, put a compressive force on the cable, we observe that it does not resist the applied force and simply gives in, Figure 4.3(a). The (compressive) stress in the cable in this case is zero.

If, on the other hand, the cable is pulled, that is, a tensile force is applied on the cable, the cable tends to resist the force and becomes taut, implying the existence of tensile stress in the cable, Figure 4.3(b). The stress in the cable increases as the applied tensile force is increased.

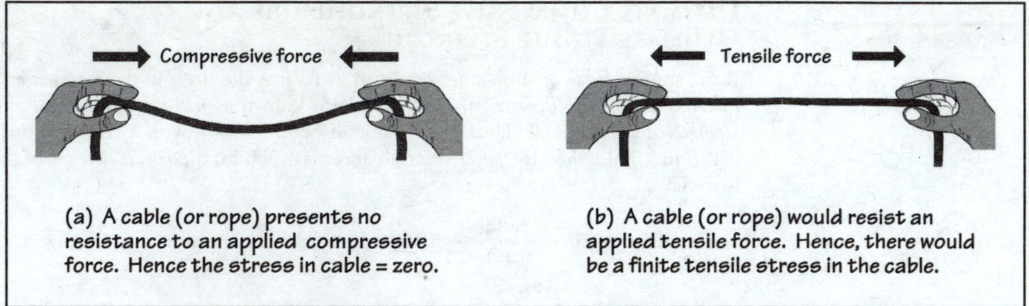

Compressive force
Tensile force
(a) A cable (or rope) presents no resistance to an applied compressive force. Hence the stress in cable = zero.
(b) A cable (or rope) would resist an applied tensile force. Hence, there would be a finite tensile stress in the cable.

FIGURE 4.3 Stresses in a cable under **(a)** an applied compressive force and **(b)** an applied tensile force.

From this example, we learn that the stress can either be compressive or tensile, depending on the type of external force. If the external force is compressive, the stress created in the member is *compressive stress* (or simply *compression*), and if the external force is tensile, the stress created in the member is *tensile stress* (or simply *tension*). A column or wall in a typical building is in compression, Figure 4.4(a). In a simple truss made of two rafters and a ceiling joist, the rafters are in compression and the ceiling joist is in tension, Figure 4.4(b). In a suspension bridge, the vertical pylons are in compression and the suspension cables are in tension, Figure 4.5.

Quantitatively, the stress is defined as the force acting on a unit cross-sectional area (1 in.^{2} or 1 ft^2) of the member. That is why we referred to the stress as the *intensity* of internal resistance. The symbol commonly used for stress is f. Thus,

$$f = \frac{\text{force}}{\text{area}} = \frac{P}{A} \qquad \textbf{Eq. (1)}$$

where P = force and A = cross-sectional area of the member.

If the force is expressed in pounds and the cross-sectional area is in square inches, the unit of stress is pounds per square inch (psi). If the force is expressed in kips (1 kip = 1,000 lb), the unit of stress is kips per square inch (ksi).

For instance, if the force applied on the cable of Figure 4.3(b) is 500 lb and if the diameter of the cable is 0.5 in. (cross-sectional area = 0.196 in.^{2}), the tensile stress in the cable is

$$f = \frac{500}{0.196} = 2{,}550 \text{ psi} = 2.55 \text{ ksi}$$

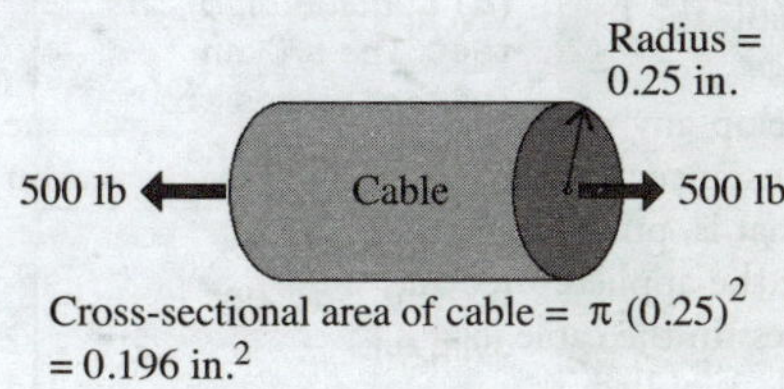

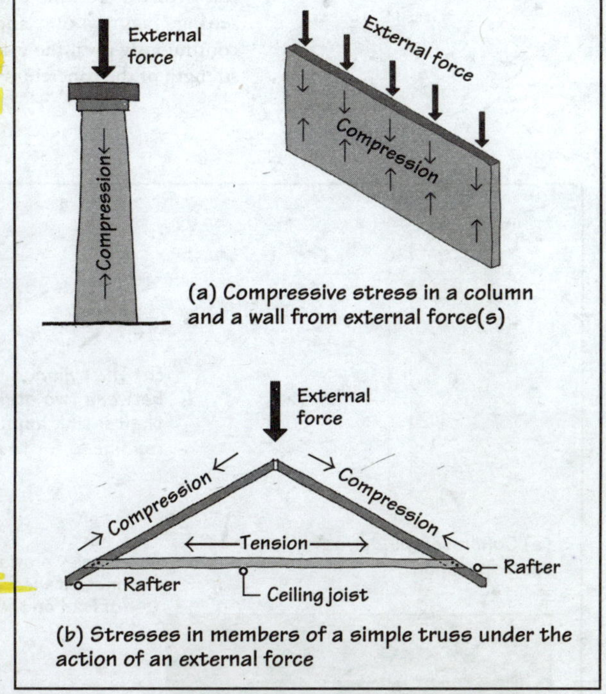

FIGURE 4.4 Compressive and tensile stresses in building components.

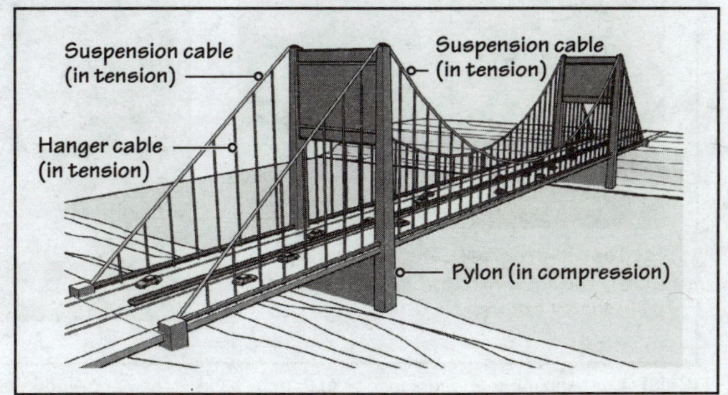

FIGURE 4.5 Tensile and compressive stresses in the members of a suspension bridge.

ULTIMATE COMPRESSIVE STRENGTH AND ULTIMATE TENSILE STRENGTH

If the applied force on a member results in its failure, the stress in the member at failure is referred to as the *ultimate stress,* which is also known as *ultimate strength* or simply the *strength* of the material. Thus, if the cable of Figure 4.3(b), with a cross-sectional area of 0.196 in.^{2}, fails when the applied tensile force is 1,000 lb, the *tensile strength* of the cable's material is

$$\text{Tensile strength} = \frac{1{,}000}{0.196} = 5{,}100 \text{ psi} = 5.1 \text{ ksi}$$

TESTING OF MATERIALS TO OBTAIN THEIR ULTIMATE STRENGTH

Standards such as those mentioned in Chapter 2 prescribe the approved test methods for determining the tensile and compressive strengths of various materials. For example, if we wish to determine the compressive strength of concrete, we perform a standard test. This test involves crushing a 6-in.-diameter cylinder of concrete 12 in. high in a controlled setting, Figure 4.6(a) and 4.6(b). As the load is applied, it is measured and recorded continuously until the cylinder is crushed (fails), Figure 4.6(c). The ultimate compressive strength of the concrete is obtained from the load that caused the failure.

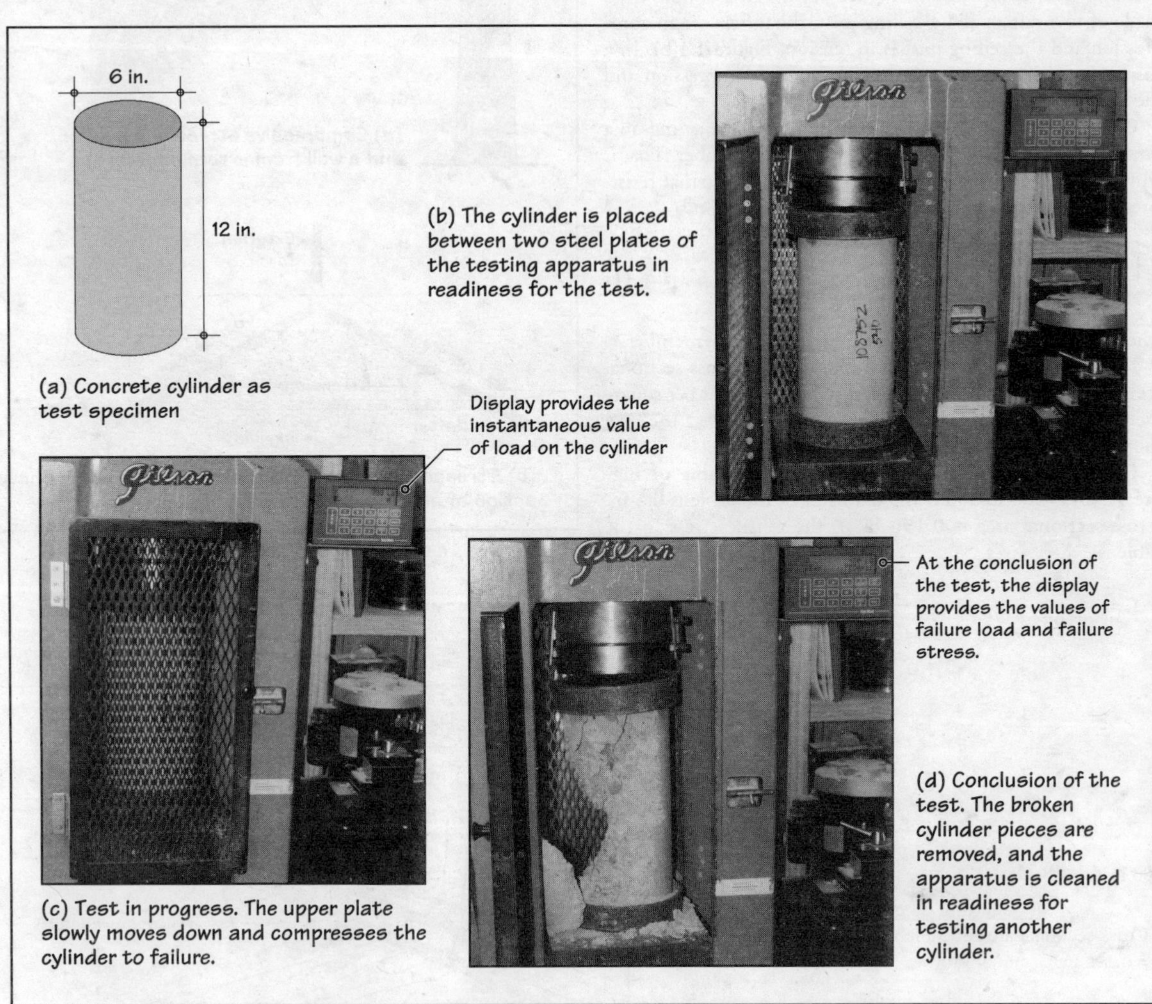

6 in.
12 in.
(a) Concrete cylinder as test specimen
(b) The cylinder is placed between two steel plates of the testing apparatus in readiness for the test.
Display provides the instantaneous value of load on the cylinder
(c) Test in progress. The upper plate slowly moves down and compresses the cylinder to failure.
At the conclusion of the test, the display provides the values of failure load and failure stress.
(d) Conclusion of the test. The broken cylinder pieces are removed, and the apparatus is cleaned in readiness for testing another cylinder.

FIGURE 4.6 Apparatus and test specimen (6 in. × 12 in. concrete cylinder) used for determining the compressive strength of concrete.

For instance, in Figure 4.6, let us assume that the cylinder failed when the load reached 115 kips. Then the compressive strength of concrete would be

$$\text{Compressive strength} = \frac{\text{load at failure}}{\text{area of cylinder}} = \frac{115}{28.27} = 4.1 \text{ ksi}$$

$$= 4{,}100 \text{ psi}$$

The method and test setup (machine) for determining the compressive strength of brick or concrete block masonry is similar. The specimen used is a masonry prism of a specified size made from several bricks or blocks with mortar joints, Figure 4.7.

The tensile strength of concrete or masonry is very low compared with its compressive strength—approximately 10% of its compressive strength. Thus, if the compressive strength of concrete is 4,000 psi, its tensile strength is only about 400 psi. That is why the tensile strength of concrete is generally ignored in the design of concrete structures; that is, we assume that the tensile strength of concrete is zero.

We typically determine the tensile strength of steel using a round bar specially machined at the ends to provide a grip for the machine to pull the bar to failure. A commonly used tensile-testing machine is shown in Figure 4.8. The compressive strength of steel is approximately equal to its tensile strength, provided that the steel member is prevented from buckling (see Section 4.9).

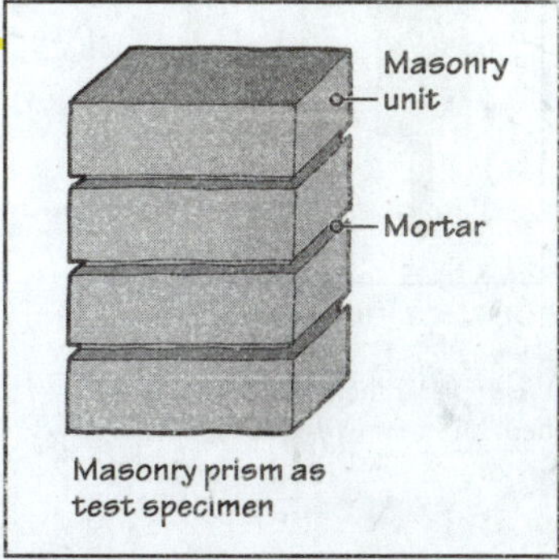

FIGURE 4.7 Test specimen used for determining the compressive strength of masonry.

4.2 DUCTILITY AND BRITTLENESS

Stress is nearly always accompanied by deformation of the member. The exception to this rule is a *rigid body*. A rigid body is defined as one that does not deform at all under the action of loads. In other words, a rigid body will not stretch, shorten, bend, or change shape regardless of the magnitude of the load placed on it. In the real world, a rigid body does not exist, because all materials deform when loaded. Actual bodies, such as building components, will deform under loads, although the deformations in them are usually too small to be visible to the human eye.

The deformation caused by a compressive or a tensile stress is simply the change in the length of the member. Tensile stress causes the member to elongate, and compressive

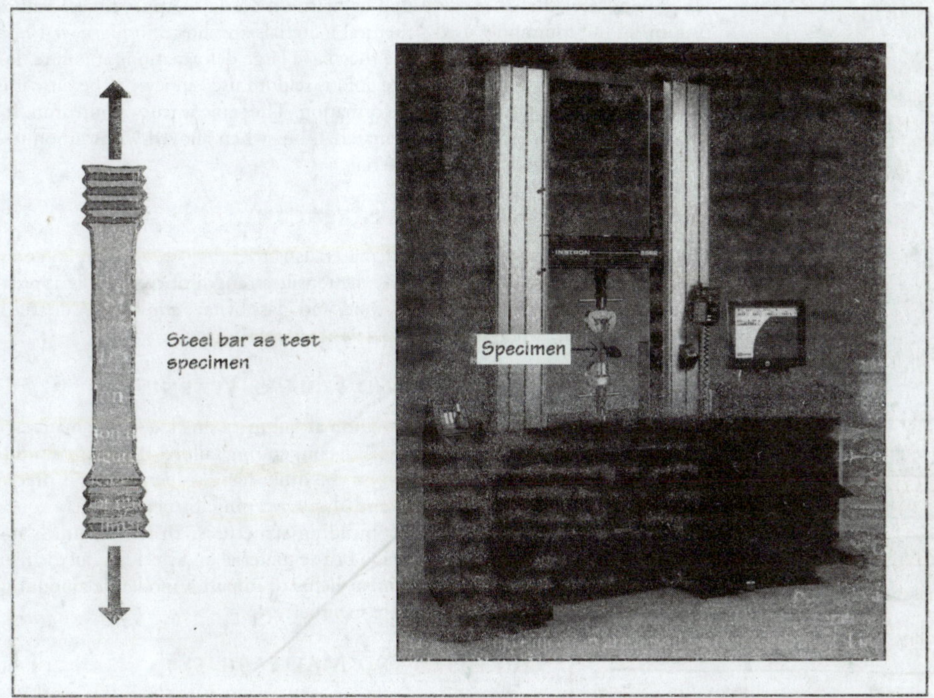

FIGURE 4.8 A typical Universal Testing Machine (UTM) for measuring the tensile strength of a steel specimen. (Photo courtesy of Instron® Corporation)

stress causes it to shorten. We are interested in both the absolute value of the change in length and its relative value. The relative change in length, defined as the change in length divided by the original length, is called the *strain.*

Thus, in Figure 4.9, if the original length of the bar is L, which—under the action of a tensile force—becomes L + δ, then the strain in the bar, denoted by the Greek letter ε (epsilon), is given by

$$\varepsilon = \frac{\text{change in length}}{\text{original length}} = \frac{\delta}{L} \qquad \textbf{Eq. (2)}$$

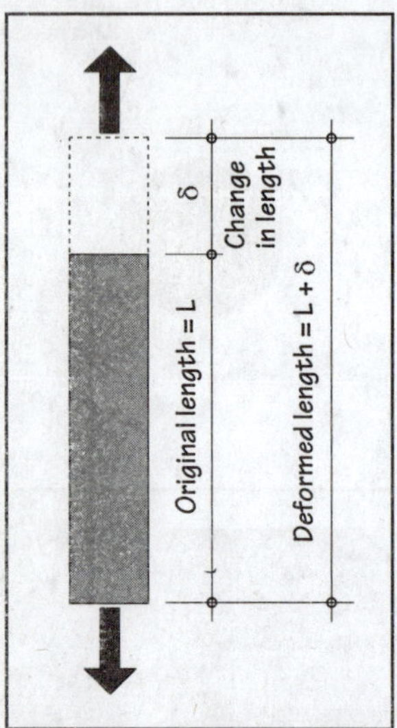

FIGURE 4.9 If a bar, whose original length is L, stretches to length (L + δ) under the action of a force, then the strain in the bar, $\varepsilon = (\delta/L)$.

From Equation (2), if a 10-ft (120-in.)-high column shortens by 0.25 in. under a load, the strain in the column is

$$\text{Strain } (\varepsilon) = \frac{0.25}{120} = 0.002$$

Because strain is the ratio of two lengths, it is simply a number with no units of measure. The strain caused by a tensile force is called *tensile strain*; that caused by a compressive force is called *compressive strain.* The strain in the element at failure is referred to as the *ultimate strain.*

The ultimate strain for most building materials is usually small. For example, the ultimate compressive strain of concrete or masonry is approximately 0.003. It is, therefore, more convenient at times to express strain as a percentage. For example, a strain of 0.003 is equal to 0.3% strain. The ultimate strain for mild steel (the term *mild steel* is defined later) is nearly 0.35 (or 35%), which is fairly large. In other words, for mild steel, a 100-in.-long steel bar will elongate by 35 in. at failure; that is, its length will become 135 in.

Materials that produce large deformations before failure are called *ductile materials.* Conversely, materials that do not deform much before failure are called *brittle* materials. The corresponding properties are called *ductility* and *brittleness.* There is no general agreement as to the limiting value of the ultimate strain that distinguishes a ductile material from a brittle material, but we generally regard a material as brittle if its ultimate strain is less than or equal to 0.5%. A ductile material is one whose ultimate strain is greater than 0.5%.

NOTE

Brittle and Ductile Materials

A brittle material is one for which ultimate strain ≤0.5%.

A ductile material is one for which ultimate strain >0.5%.

Materials such as brick, stone, concrete, and glass have an ultimate strain of less than 0.5% and are, therefore, classified as brittle materials. Approximate values of ultimate strains of the commonly used structural materials are shown in Figure 4.10.

Most metals are ductile, because they have large deformations at failure. Pure metals are more ductile than their alloys. Pure gold is seldom used in jewelry because it is very ductile and, hence, prone to damage by deformation. The same is true of pure iron. It is too ductile and soft for use as a construction material, but when alloyed with carbon to form steel, it becomes a useful construction material.

BRITTLENESS AND TENSILE STRENGTH

One of the characteristics of brittle materials is that they are stronger in compression than in tension. As stated in Section 4.1, the tensile strength of concrete is approximately 10% of its compressive strength. Brick, stone, and glass behave similarly. A ductile material such as steel has equal strength in compression and tension.

DUCTILITY AND FAILURE WARNING

Because its deformation at failure is small, a brittle material does not give any visual warning of its impending failure. Its failure is sudden. Failure of a ductile material, on the other hand, is gentle and is preceded by excessive deformation and ample warning. Structural safety concerns mandate ductile failure in building structures. Brittle failure is not permitted. Because concrete is a brittle material and steel is a ductile material, the use of steel reinforcement helps to impart a moderate amount of ductility to reinforced concrete members.

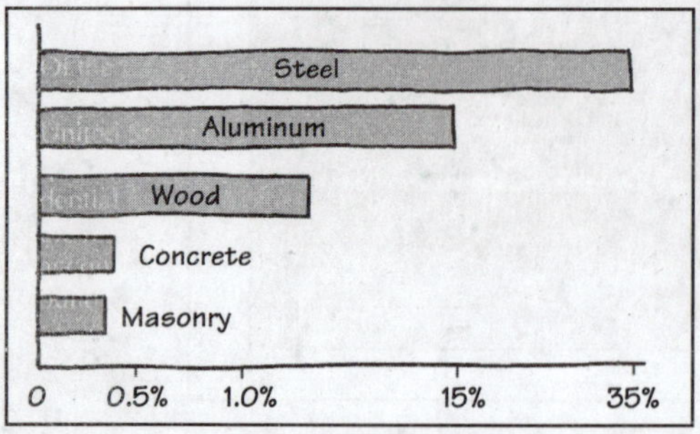

FIGURE 4.10 Approximate values of ultimate strain for selected materials (not to scale).

DUCTILITY AND MALLEABILITY

The terms *ductility* and *malleability* are used interchangeably in lay literature, although, strictly speaking, they represent different properties of materials. As stated previously, ductility is the property of a material to deform extensively prior to failure. Malleability, on the other hand, is the property that allows a material to be shaped by hammering, forging, pressing, and rolling. Ductile materials are generally malleable, but they need not be. For instance, cast iron is a ductile material (according to the definition of ductile material given earlier), but it cannot be converted into shapes by hammering or pressing. Cast-iron shapes are obtained by casting molten iron into molds.

PRACTICE QUIZ

Each question has only one correct answer. Select the choice that best answers the question.

1. The material in an arch is primarily in
   a. compression. b. tension.
   c. both tension and compression. d. shear.
   e. none of the above.
2. When a cable or rope is compressed by a force, what is the compressive stress in it?
   a. Equal to the force applied
   b. Force divided by the cross-sectional area of rope
   c. Zero
   d. Infinite
   e. None of the above
3. In a simple three-member truss consisting of two rafters and a tie, each member is in
   a. compression. b. tension.
   c. either tension or compression. d. shear.
   e. both tension and compression.
4. In a suspension bridge, which member(s) is (are) in compression?
   a. Suspension cable
   b. Hanger cables
   c. Support pylons
   d. Both suspension cable and hanger cables
   e. None of the above
5. A rectangular column measuring 12 in. × 12 in. in cross section carries a load of 18 kips. What is the stress in the column?
   a. 18 kips b. 18 ksi
   c. 2.25 ksi d. 125 psi
   e. None of the above
6. The test specimen used for determining the compressive strength of concrete in the United States is a
   a. cube measuring 12 in. × 12 in. × 12 in.
   b. cube measuring 6 in. × 6 in. × 6 in.
   c. prism measuring 6 in. × 6 in. × 12 in.
   d. cylinder measuring 12 in. in diameter × 6 in. high.
   e. cylinder measuring 6 in. in diameter × 12 in. high.
7. When we determine the strength of concrete using the test specimens, we determine the concrete's
   a. ultimate compressive stress. b. ultimate tensile stress.
   c. ultimate shear stress. d. modulus of elasticity.
   e. all of the above.
8. In relation to its compressive strength, the tensile strength of concrete is
   a. much higher. b. slightly higher.
   c. nearly the same. d. much lower.
   e. slightly lower.
9. A 20-ft-high column shortens by 0.6 in. under a load. What is the resulting strain in the column?
   a. 0.0020 b. 0.0020 in.
   c. 0.0025 in. d. 0.0025
   e. None of the above
10. Which gives greater warning before failure?
    a. A brittle material
    b. A ductile material

4.3 YIELD STRENGTH OF MATERIALS

Because a load causes deformation, stress and strain occur simultaneously in a member. Generally, the strain increases as the stress increases. However, there are situations where stress may exist without strain or strain without stress. For example, a beam resting on two end supports will be subjected to a change in its length as the temperature of the beam increases or decreases. In this case, the strain occurs without any stress.

But if the same beam is clamped between the two end supports so that it cannot change its length, the beam will be subjected to stresses as its temperature changes—a case in which the stress occurs without the strain. The clamped beam will be subjected to compressive stress when its temperature rises and to tensile stress when its temperature falls.

The relationship between stress and strain conveys a great deal of information about the structural properties of the material and is unique to each material. This relationship is usually given in a graphical form, called the *stress-strain diagram.* In the stress-strain diagram, the stress is generally represented on the vertical axis and the strain on the horizontal axis.

The stress-strain diagram of one of the most important structural materials—low-carbon steel—is shown in Figure 4.11. The diagram is obtained by subjecting a round steel bar of standard length and diameter to an increasing tensile force. This is done in a tensile-testing machine (Figure 4.8), which is fully automated and gives the stress-strain diagram directly, thus avoiding the need to calculate stresses and strains and to plot them manually.

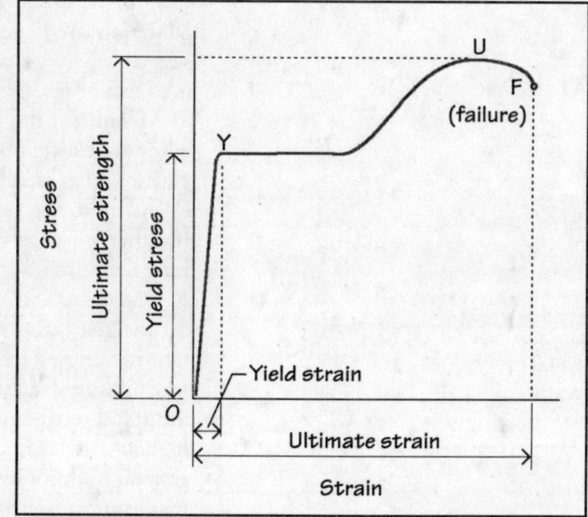

FIGURE 4.11 Stress-strain diagram for low-carbon (mild) steel in tension (not to scale).

NOTE

Low-Carbon (Mild) Steel

Low-carbon steel is known as *mild steel* because of its higher ductility and the consequent ease with which it can be bent to shape and rolled into various cross-sectional shapes. High-strength steels are comparatively more brittle. They contain more carbon than mild steel. Therefore, they cannot be bent to shape or rolled into cross-sectional shapes with the same ease.

The diagram of Figure 4.11 begins at the origin (point O) because the strain must obviously be zero when the stress is zero—a no-load condition. It consists of three important points; Y, U, and F.

Point Y is called the *yield point,* and the stress corresponding to this point is called the *yield stress.* Yield stress is an important property of steel. A few types of steel are distinguished from each other by their yield stress values. For instance, A-36 steel is so called because its yield stress is 36 ksi (the letter *A* in A-36 symbolizes a ferrous metal, as per ASTM classification (Table 2.3). Steel used as reinforcing bars in concrete is called grade 40 or grade 60 steel, depending on whether its yield stress is 40 ksi or 60 ksi.

YIELD STRESS AND ULTIMATE STRENGTH

In region OY of Figure 4.11, the relationship between stress and strain is a straight line, and stress is proportional to strain. Beyond point Y, the deformation in the bar increases substantially with little or no increase in the load, and the stress-strain diagram becomes (almost) horizontal. In other words, the steel yields after reaching point Y like a plastic material (such as modeling clay).

The yielding of steel produces a major rearrangement of the atoms in its crystalline structure, making steel stronger and harder, so the bar is able to sustain a greater load. This effect, referred to as *strain hardening,* is similar to that obtained by mechanical working on steel, such as cold rolling or hammering. We experience strain hardening when we take a piece of steel wire and bend it between our fingers. The wire becomes harder and stronger after bending, and it is more difficult to bend again at the original bend point.

Further increases in the load take the diagram of Figure 4.11 to point U, the point of maximum stress. The maximum stress is called the *ultimate stress,* or the *ultimate strength* of steel. Observe that the stress-strain diagram drops down between point U and the failure point F, indicating a loss of strength near failure. This is due to the manner in which the materials are tested and indicates the inability of testing equipment to keep the load on a rapidly disintegrating material.

Although point U represents the ultimate strength of steel, for all practical purposes it is the yield stress that is considered as the limiting stress in steel. Because steel deforms excessively after yield and because excessive deformation is to be avoided in a structure, we limit the actual stress in steel members to remain well below the yield stress. Therefore, although not absolutely correct, yield stress is also referred to as the *yield strength* of steel.

The yielding of the material is not peculiar to steel. It occurs in several metals and is an extremely useful property, because it provides ductility. Additionally, the strain hardening that follows yield makes the metal stronger.

4.4 ELASTICITY AND PLASTICITY

A material is called an *elastic material* if it is fully recoverable from deformation after the load is removed. The property responsible for the elastic behavior of the material is called *elasticity.* The converse of elasticity is called *plasticity.* Thus, a material that does not recover from its deformation after the removal of the load is called an *inelastic,* or *plastic, material.* Materials are generally elastic only up to a small strain value—approximately 0.5%. As stated earlier, the ultimate strain in brittle materials such as concrete, brick, stone, and glass is of this order (0.5%). Therefore, brittle materials are generally elastic up to failure.

Metals, on the other hand, can sustain large deformations, usually greater than 10%. They are elastic up to a certain stress level, and if loaded beyond that stress value, they are plastic—that is, their deformation beyond that point is permanent. For example, steel is elastic up to the yield point (between points O and Y in Figure 4.11) but plastic thereafter. In other words, if the stress in a steel member remains below the yield stress, the member is fully recovered from any deformation after the load is removed. Materials that are elastic up to a certain stress value and plastic thereafter are called *elastic-plastic materials.* Thus, low-carbon steel is an elastic-plastic material. The same applies to aluminum—another important metal used in buildings, generally in windows and glass curtain walls.

Building structures are designed so that the stresses in materials remain within the elastic limit and so that no inelastic deformations occur under design loads. Because they are permanent, inelastic deformations affect the appearance of structures. The exception to this general philosophy is made in the design of structures located in highly seismic areas. The maximum intensity of earthquake loads in such areas is so high that if structures were designed to remain elastic, member sizes (columns, beams and floor slabs) would have to be so large as to make the structure highly uneconomical.

Thus, as stated in Section 3.9, the structural design philosophy for buildings located in seismic zones is that the building components should deform elastically under low to moderately intense earthquakes so that after the earthquake event, the building regains its original shape and size. However, under severe earthquake loads, the structure is designed to deform inelastically (plastically) without compromising its safety or integrity. The structure may become unusable and unserviceable because of plastic deformations but will remain integral to provide life safety.

4.5 MODULUS OF ELASTICITY

The ratio of stress to strain in a material is called the *modulus of elasticity.* Thus,

$$\frac{\text{Stress}}{\text{Strain}} = E \qquad \textbf{Eq. (3)}$$

where E denotes the modulus of elasticity. A measure of the material's stiffness, the modulus of elasticity is a property of a material. The larger the value of E, the stiffer the material and the less it will deform under a given load.

On the stress-strain diagram, the value of E is represented by the slope of the diagram. The greater the slope, the larger the value of E and, hence, the stiffer the material. Thus, in Figure 4.12, the value of E is larger for Material 1 than for Material 2. For a *rigid* material, the value of E would be infinite. This is obvious from Equation (3), because for a rigid material, the strain is zero regardless of the value of stress. Therefore, the stress-strain diagram for a rigid material is a vertical line because on a vertical line, the strain is zero.

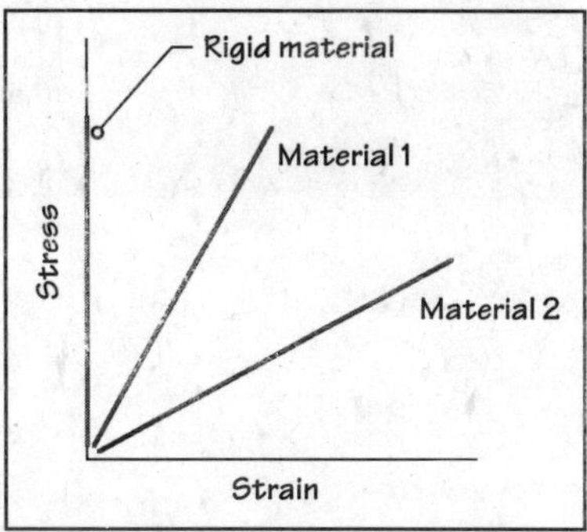

FIGURE 4.12 The slope of a material's stress-strain diagram gives the magnitude of the modulus of elasticity of the material, E. The greater the slope, the greater the value of E. Thus, in this illustration, Material 2 has the lowest value of E.

The concept of the modulus of elasticity as a material property (constant for a material) assumes a linear stress-strain diagram. If the stress-strain diagram is a curve, the modulus of elasticity is not constant but varies with the stress level. Most construction materials fall into this category, that is, they do not have a linear stress-strain diagram. The exception is steel, which has a linear stress-strain diagram only up to the yield point. Because yield stress is, practically, the limiting stress in steel, steel may be considered as having a linear stress-strain diagram for all practical purposes.

The modulus of elasticity of a material that does not have a linear stress-strain diagram is obtained from the straight line whose slope is approximately equal to the slope of the initial part of its stress-strain diagram. This approximation is justified on the basis that in actual buildings, the materials are usually loaded to low- or moderate-stress levels so that the slope of the initial part of the diagram is a good approximation to the modulus of elasticity of the material.

Figure 4.13 shows the stress-strain diagram for concrete. Because the diagram is nonlinear, the modulus of elasticity is given by the slope of straight line OA, where A represents 40% of the compressive strength of concrete. Thus, if the strength of concrete is 4,000 psi, point A represents a stress level of 1,600 psi.

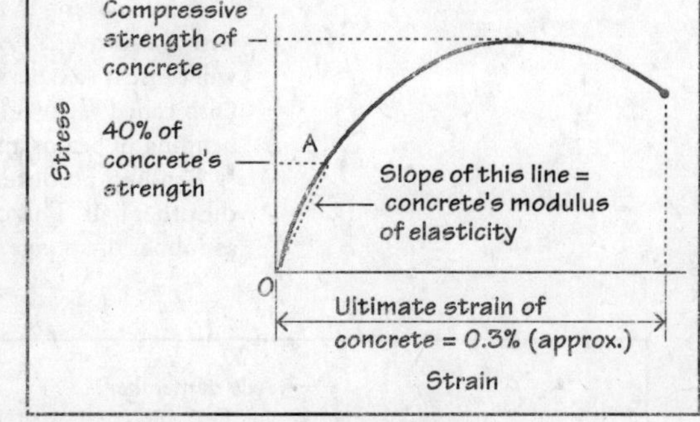

FIGURE 4.13 Stress-strain diagram of concrete.

UNITS OF MODULUS OF ELASTICITY

From Equation (3), the units of the modulus of elasticity are the same as those of stress, because strain has no units. In other words, the units of E are psi or ksi (pascals in the SI system). The E value of materials is usually a large number—millions of pounds per square inch (gigapascals in the SI system, where 1 GPa = 10^9 Pa).

For a soft rubber, the value of E is a relatively small number. Of the commonly used materials, diamond has the highest value of E (greatest stiffness). That is why diamond-tipped tools are used to cut window glass, masonry, and concrete. Table 4.1 gives the modulus of elasticity values of a few selected materials.

MODULUS OF ELASTICITY AND STRENGTH OF MATERIALS

Being a measure of the stiffness of the material, modulus of elasticity determines how much a material will deform under loads. Because deformation control is a structural design criterion, modulus of elasticity is an important material property. It determines how much a

NOTE

Artificial Diamond

The diamond commonly used on the tips of cutting tools, such as saw blades and glass-cutting wheels, is artificial. Artificial diamond, first patented by Edward Acheson in 1926, was carborundum—a trade name for silicon carbide—made from clay and carbon. Since then, several other metal carbides, such as tungsten carbide and titanium carbide, have been produced. Pure diamond is 100% crystalline carbon.

TABLE 4.1 MODULUS OF ELASTICITY OF SELECTED MATERIALS

Material	Modulus of elasticity ($\times 10^6$ psi)
Wood	1.0–2.0
Brick masonry	1.0–3.0
Concrete	3.0–8.0
Steel	29.0
Aluminum alloys	10.5
Window glass	10.4
Diamond	170
Rubber	0.001

beam or slab will deflect under a given load or how much a tall building will sway sideways under the action of wind.

For most materials, the modulus of elasticity increases with the strength of the material. Higher-strength species of wood, for example, have higher moduli of elasticity than lower-strength species. The same is true of concrete and masonry. The greater the concrete strength, the greater its modulus of elasticity. However, steel has the same modulus of elasticity regardless of its strength. In other words, high-strength steels have the same modulus of elasticity as low-carbon (mild) steel.

Steel is the stiffest of all construction materials; that is, it has the highest value of E (29×10^6 psi). Its modulus of elasticity is based on the straight-line portion of the stress-strain diagram, points O to Y in Figure 4.11. Because, unlike other materials, the modulus of elasticity of steel does not increase with strength, high-strength steel is not as commonly used in structural steel sections as low-carbon (mild) steels.

4.6 BENDING STRENGTH OF MATERIALS

So far, we have discussed situations in which either tensile or compressive stress is present in a member. For example, in the suspension bridge of Figure 4.5, the pylons are in compression and the cables and hanger bars are in tension. In such cases, the external force acts along the axis of the member, as shown in Figure 4.14(a). The presence of the axial force may be obvious, as in Figure 4.5, or not directly obvious, as in the rafter-joist assembly of Figure 4.4(b).

Because the external force is axial, the stress created in the member is referred to as an *axial stress.* The axial stress is uniform over the entire cross section, as shown in Figure 4.14(a), and may be either compressive or tensile. However, if the external force acts perpendicular to the axis of a member, as seen in Figure 4.14(b), the member bends, subjecting itself to both compressive and tensile stresses simultaneously.

A force acting in a downward direction on a beam that is supported at the ends bends it with a concave *water-holding curvature,* like the inside of a saucer. An upward-acting force will cause it to have a convex *water-shedding curvature,* like the outside of a sphere. Bending (also called *flexure*) is a predominant action in building structures. Gravity loads produce bending in beams and slabs, and lateral loads cause bending in walls and columns.

Bending produces compression in one-half of a member's cross section and tension in the other half. This can be demonstrated by bending a member made of a soft material such as rubber or sponge. If we mark two vertical lines, PQ and RS, on one of the longitudinal

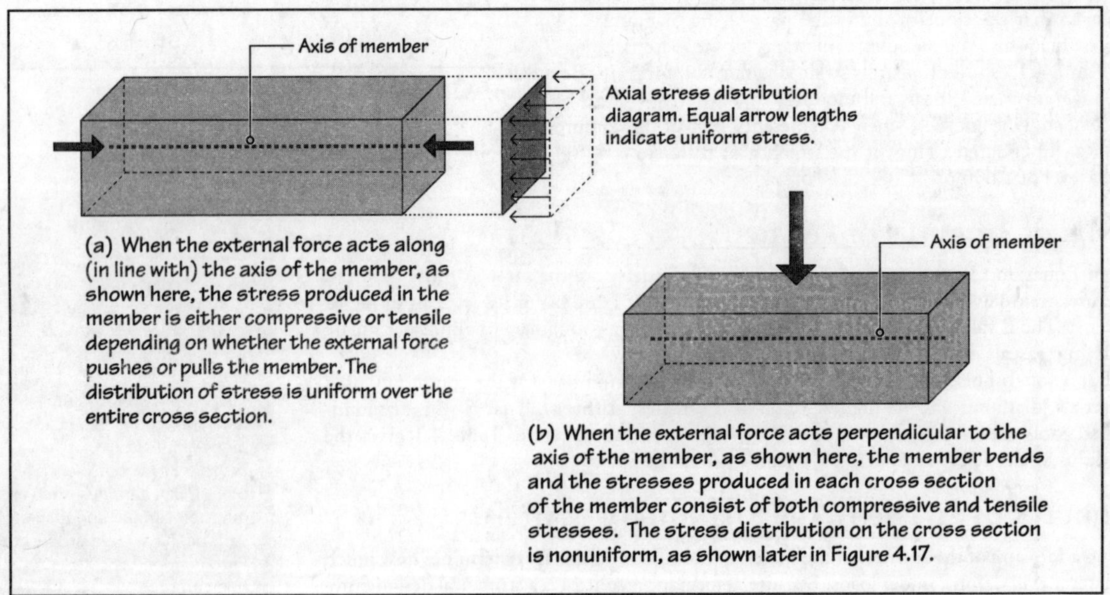

Axis of member
Axial stress distribution diagram. Equal arrow lengths indicate uniform stress.
(a) When the external force acts along (in line with) the axis of the member, as shown here, the stress produced in the member is either compressive or tensile depending on whether the external force pushes or pulls the member. The distribution of stress is uniform over the entire cross section.
Axis of member
(b) When the external force acts perpendicular to the axis of the member, as shown here, the member bends and the stresses produced in each cross section of the member consist of both compressive and tensile stresses. The stress distribution on the cross section is nonuniform, as shown later in Figure 4.17.

FIGURE 4.14 The orientation of the external force with respect to the axis of the member and the types of stress produced in the member.

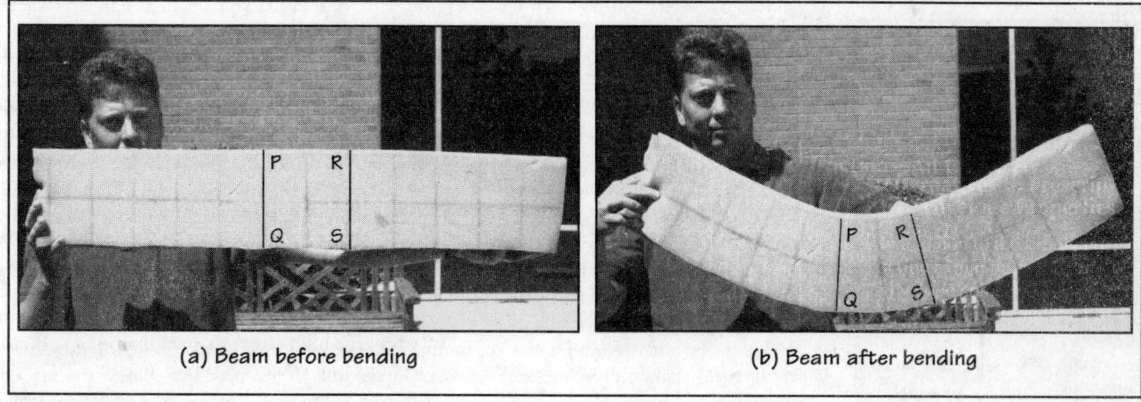

FIGURE 4.15 Demonstration of bending in a beam. In a beam bent with a water-holding curvature, as shown in **(b)**, the upper half of the beam is in compression and the lower half of the beam is in tension.

faces of a sponge beam and bend it to produce a concave curvature, as shown in Figure 4.15, we see that the lines have become closer at the top and farther apart at the bottom. Fibers of the beam represented by line PR have become shorter in length, and fibers on line QS have become longer.

Further examination of the deformed shape of rectangle PQSR shows that maximum contraction takes place along line PR, the extreme top surface of the beam. The maximum elongation is produced along line QS, the extreme bottom surface of the beam. In other words, compressive stresses are maximum at the top surface, and tensile stresses are maximum at the bottom surface of the beam.

The change from compressive to tensile stresses occurs on the surface that neither contracts nor stretches; that is, this surface maintains its original dimension and is, therefore, unstressed. This surface is the neutral plane, commonly referred to as the *neutral axis* of the beam, Figure 4.16. Thus, the neutral axis is a line on the beam's cross section where the stresses are zero.

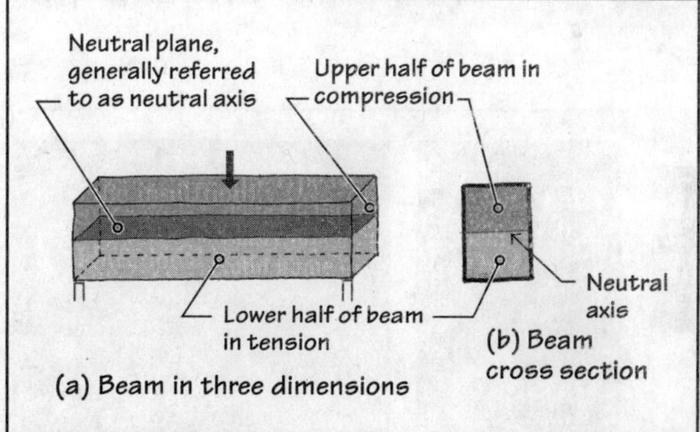

FIGURE 4.16 The location of the neutral axis in a beam under bending stresses. Note that the neutral axis is, in fact, a neutral plane. It is called the neutral axis because we generally draw a beam in two dimensions—in cross section—in which the neutral plane is shown as a line.

Improving the Bending Strength of Members

If we isolate a small length of the beam, such as length PQSR in Figure 4.15, and examine the distribution of stresses on cross sections represented by lines PQ and RS, we see that the distribution is as shown in Figure 4.17. The term *bending stress* (or *flexural stress*) is used to describe this distribution of stresses—compressive stresses on one side of the neutral axis and tensile stresses on the other side.

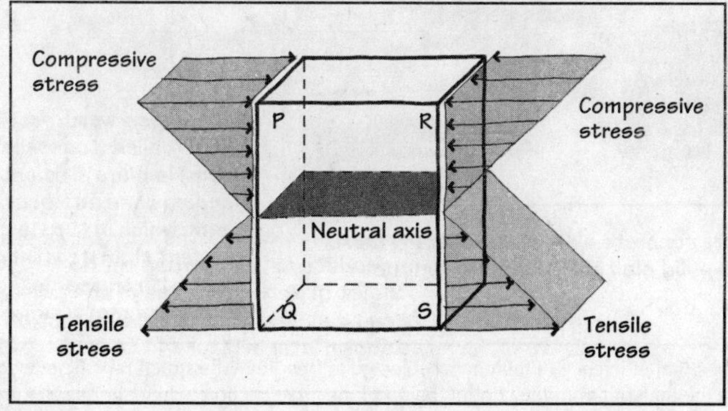

FIGURE 4.17 Stress distribution on a small length, PQSR, of the beam in Figure 4.15.

NOTE

Structural Efficiency

A structurally efficient shape is one that can withstand a greater load for the same amount of material. For example, a beam with an I-section (Figure 4.18) is structurally more efficient because it can withstand a greater load than a rectangular beam of the same cross-sectional area and span. A hollow-core concrete slab (Figure 4.19(b)) is structurally more efficient than a solid slab with the same amount of material.

The stress distribution shown in Figure 4.17 represents stresses in the beam when it bends with a water-holding curvature. If the beam bends with a water-shedding curvature, the stresses will simply be reversed; that is, tensile stresses will be created in the upper half of the beam and compressive stresses in the lower half of the beam.

Stresses are unequally distributed throughout the beam's cross section in bending. They are concentrated at the top and bottom of the cross section, but the central part of the cross section is relatively stress free—an inefficient use of the material. By contrast, the axial stress distribution is uniform over the entire cross section, as shown in Figure 4.14(a). Thus, structural configurations that generate axial stresses in members use the material efficiently.

In suspension structures, where members are either in compression or tension (Figure 4.5), loads are carried very efficiently. That is why suspension structures are popular for long-span structures, such as, bridges.

Architects and engineers have overcome the structural inefficiency in bending by developing the I-shaped cross section. This configuration places most material at the top and bottom of the beam where stresses are greatest, thus making it more efficient than a (solid) rectangular section. I-shaped beams in steel, wood, and concrete are commonly used in buildings and civil engineering structures, Figure 4.18.

Some of the other strategies used in response to the structural inefficiency in bending are tubular cross sections, hollow-core concrete slabs, and open-web-steel I-shape beams, Figure 4.19.

FIGURE 4.18 I-section beams are commonly used in steel, wood, and concrete.

(a) A tubular (hollow structural) section is more efficient in bending than a solid rectangular section of the same cross-sectional area.

(b) A hollow-core concrete slab is structurally more efficient than a solid slab with the same amount of material.

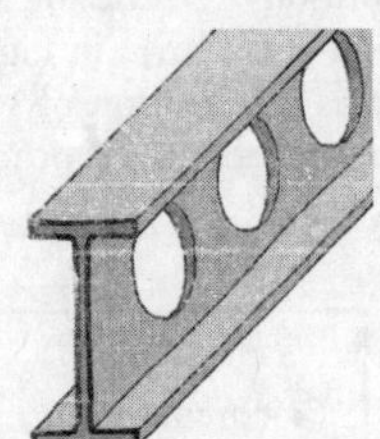

(c) An open-web I-section beam (called a castellated beam) is more efficient in bending than an I-section beam, which in turn is more efficient than a rectangular beam with the same amount of material.

FIGURE 4.19 Some strategies used to improve the structural efficiency of members subjected to bending. The structural efficiency of a member subjected to bending improves by removing the material from the central region of the cross section (where the stresses are low) and placing it closer to the extremes edges of the cross section (where the stresses are high).

BENDING STRENGTH OF BRITTLE MATERIALS IS LOW

Because bending creates tensile as well as compressive stresses, brittle materials, being weak in tension, are also weak in bending. Ductile materials, such as steel, which have equal strengths in tension and compression, have a higher bending strength. Reinforced concrete is a composite material that exploits the moderately high compressive strength of concrete and the high tensile strength of steel.

Steel is placed in those locations of a reinforced concrete member where tensile stresses are present. For instance, in a beam supported at two ends, tensile stresses occur in the lower half of the beam's cross section. Therefore, steel reinforcing bars are placed along the length of the beam at the bottom of the beam's cross section, Figure 4.20. (Also see Section 22.5 for additional details.)

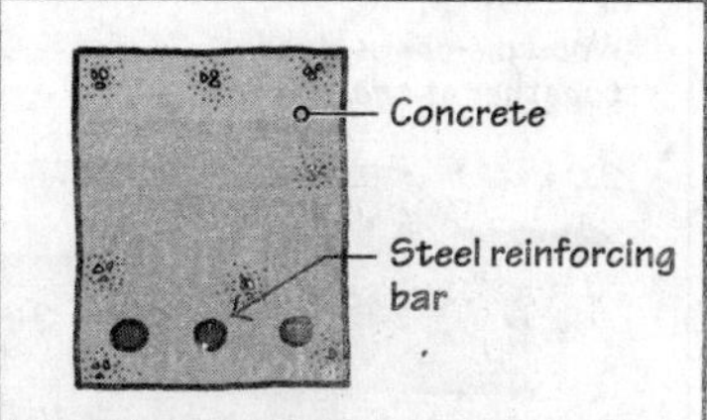

FIGURE 4.20 Section through a typical reinforced concrete beam. Note that steel reinforcing bars are used in the region of the cross section that is subjected to tension.

PRACTICE **QUIZ**

Each question has only one correct answer. Select the choice that best answers the question.

11. In a typical stress-strain diagram of a material, the stress is generally plotted along the vertical axis.
- **a.** True
- **b.** False

12. What is the yield stress of grade 60 reinforcing steel?
- **a.** 60 lb
- **b.** 6,000 lb
- **c.** 6,000 psi
- **d.** 60 psi
- **e.** None of the above

13. The stress-strain diagram of mild steel is nearly a straight line up to the yield point.
- **a.** True
- **b.** False

14. Low-carbon (mild) steel is
- **a.** an elastic material.
- **b.** a plastic material.
- **c.** an elastic-plastic material.
- **d.** none of the above.

15. Concrete is
- **a.** an elastic material.
- **b.** a plastic material.
- **c.** an elastic-plastic material.
- **d.** none of the above.

16. Modulus of elasticity refers to how
- **a.** strong a material is.
- **b.** stiff a material is.
- **c.** elastic a material is.
- **d.** serviceable a material is.
- **e.** none of the above.

17. The units of modulus of elasticity are
- **a.** psi.
- **b.** ksi.
- **c.** Pa.
- **d.** GPa.
- **e.** all of the above.

18. The bending stresses along the neutral axis of a beam are
- **a.** minimum.
- **b.** maximum.
- **c.** zero.
- **d.** none of the above.

19. Which of the following beam cross sections is structurally more efficient in bending?
- **a.** Rectangular (solid) section
- **b.** I-section

20. Which of the following beam cross sections is structurally more efficient in bending?
- **a.** Rectangular (solid) section
- **b.** Tubular section

4.7 SHEAR STRENGTH OF MATERIALS

So far, we have discussed the strength of materials under axial stress and bending stress. Both of these stresses—axial stress (which can be compressive or tensile) and bending stress (a combination of compressive and tensile stresses)—are *normal stresses,* implying that the stress acts perpendicular (normal) to the cross section of the member (see Figure 4.14(a) and 4.17). A second type of stress, called *shear stress,* acts tangential to the cross section of the member. In fact, normal stress and shear stress are the only two basic types of stress. Various types of stress that exist in a member belong to either normal stress or shear stress, Figure 4.21.

Shear stress is caused by a force that is tangential to the cross section of the member. Because it is tangential, a shear force tends to produce a sliding effect on the body. Shear stress is, therefore, associated with the resistance of the layers within the body to sliding over each other, Figure 4.22. Like normal stress, shear stress is also defined as the shear force divided by area and has the same units as normal stress—psi or ksi.

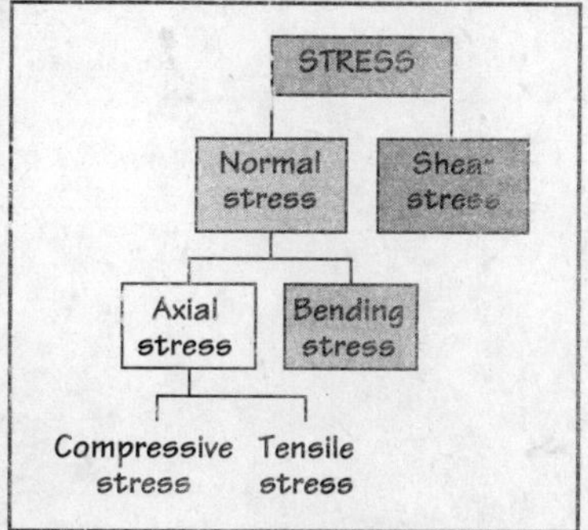

FIGURE 4.21 Types of stresses that may exist in a structural member.

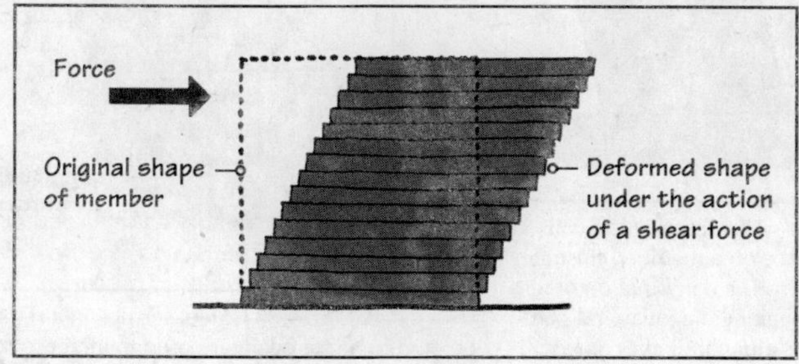

FIGURE 4.22 The deformation caused by a shear (tangential) force on a member.

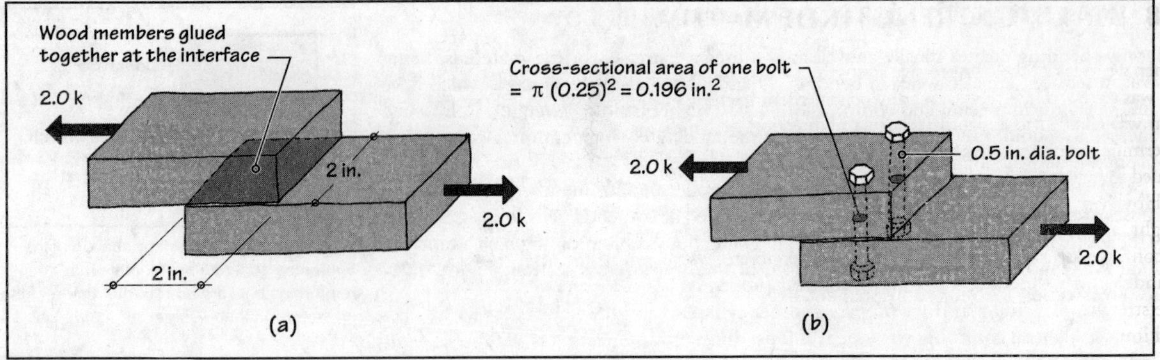

FIGURE 4.23 The magnitude of shear stress equals the shear force divided by the area that resists the shear force (sliding action).

Consider an assembly of two wood members glued together at the interface, as shown in Figure 4.23(a). If a force pulls on the assembly, axial tension is developed in each member. In addition to the axial stress, the assembly develops shear stress to resist sliding at the interface. The interface area of the assembly that resists shear has been colored. For example, if this area is 2.0 in. by 2.0 in. (4.0 in.^{2}) and if the force acting on the assembly is 2.0 kips, then the shear stress developed at the interface is

$$\text{Shear stress} = \frac{\text{force}}{\text{area}} = \frac{2.0}{4.0} = 0.5 \text{ ksi} = 500 \text{ psi}$$

If the members are joined together with bolts, as shown in Figure 4.23(b), instead of being glued together at the interface, the shear force will now be resisted by the cross-sectional area of the bolts. For example, if there are two 0.5-in.-diameter bolts, the shear stress in bolts is

$$\text{Shear stress} = \frac{2.0}{2(0.196)} = 5.1 \text{ ksi}$$

Low Shear Strength of Brittle Materials

As shown in Figure 4.22, a shear force causes an angular deformation in an element, changing the element's shape, not its dimensions. Thus, a rectangular element becomes an oblique element. This introduces tension along one diagonal of the member and compression along the other, Figure 4.24. Because of the tension created by shear, brittle materials—which are weak in tension—are also weak in shear. Steel, on the other hand, has a fairly high shear strength, because it has the same (high) strength in tension as well as compression.

In the concrete beam illustrated in Figure 4.20, the longitudinal steel reinforcement placed at the bottom of the beam is adequate to resist tensile stresses resulting from bending, but additional steel is required to resist shear. Steel stirrups are therefore used in concrete beams to resist shear stresses, Figure 4.25. (See Section 22.5 for additional details.)

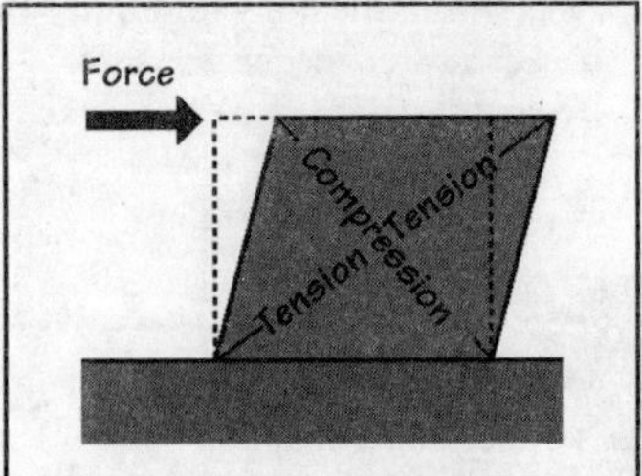

FIGURE 4.24 The deformation caused by a shear force on a member illustrates that shear stress is a combination of tensile stress along one diagonal and compressive stress along the other diagonal of the member.

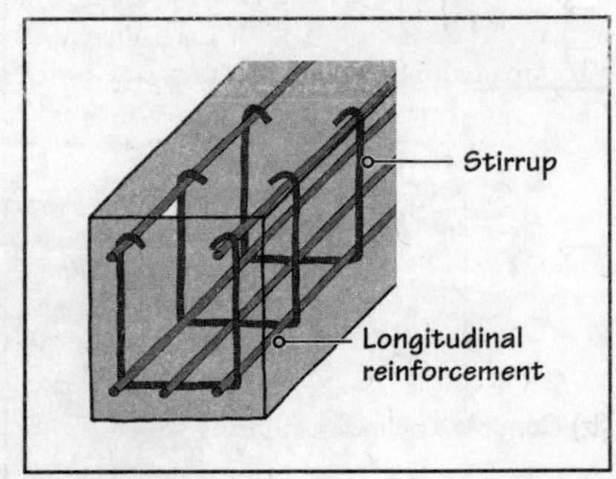

FIGURE 4.25 Steel stirrups in a reinforced concrete beam are used to increase the beam's shear strength.

4.8 BEARING STRENGTH OF MATERIALS

When we speak of the compressive strength of a material, we imply that a specimen of the material was tested to failure without any confinement in the lateral direction; that is, the test specimen was allowed to expand freely in the direction perpendicular to the load. This is how we determine the compressive strength of concrete in a standard compression test. As the cylinder is loaded to failure from the vertical direction, it is free to bulge out horizontally (see Figure 4.6).

However, if a cylinder is tested with horizontal confining stress throughout its entire height, as shown in Figure 4.26, we observe that the cylinder fails at a higher load than an unconfined cylinder made from the same concrete. For example, assume that a concrete cylinder under the standard compression test (no lateral confinement) tests to a compressive strength of 3,500 psi. If we prepare another cylinder from the same batch of concrete and impose a lateral confining stress of 1,000 psi during the compression test, we find that the cylinder tests to a compressive strength of nearly 9,000 psi. When the lateral confinement is increased to 2,000 psi, a cylinder from the same batch of concrete tests to a compressive strength of nearly 13,000 psi, Figure 4.27.

Confining compressive stress exists in deep and thick columns and walls. In such elements, the material in the interior portions is not free to expand laterally because of the confining effect of the exterior mass. Consequently, the interior material posts a higher compressive strength than the material on the exterior of such elements.

This fact is also used in the design of a column footing. The concrete footing generally has a much larger area than the cross-sectional area of the column. Therefore, as the column delivers the load to the footing, the concrete immediately under the column experiences the confining effect of the mass of the concrete surrounding the column base. Thus, the concrete in the footing exhibits greater overall compressive strength than that obtained from the standard compression test. This state of compressive stress, created at the surface of contact between the column and the footing, is referred to as the *bearing stress,* and the corresponding strength of the material is called the *bearing strength* of the material.

In general, when one element delivers a compressive load on the supporting element, the surface area of contact between the two elements is called the *bearing area.* If the bearing area is larger than the area of the supporting surface, a confining effect occurs, and the stress created in the bearing area is referred to as the *bearing stress.* Therefore, bearing stress is the (compressive) load divided by the bearing area (contact area) between the two elements.

Bearing failure occurs when the load causes the bearing stress in the material to exceed its bearing strength. Usually, bearing failure does not result in complete crushing of the material but rather partial crushing, or what is referred to as *local crushing.* Local crushing in a structural member is similar to the crushing (or settlement) of ground directly under the pointed heel of a shoe.

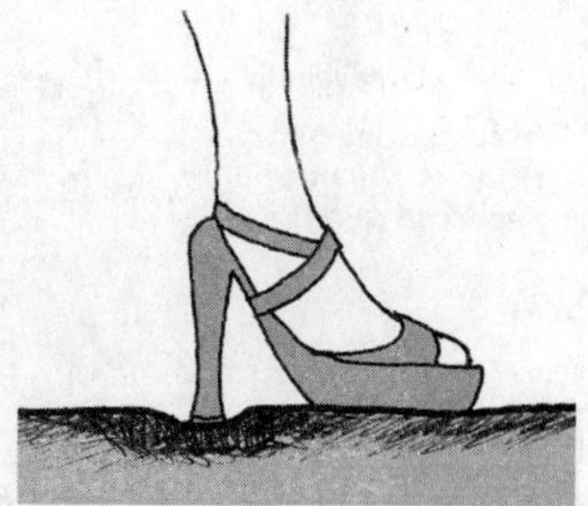

Local crushing or settlement (bearing failure)

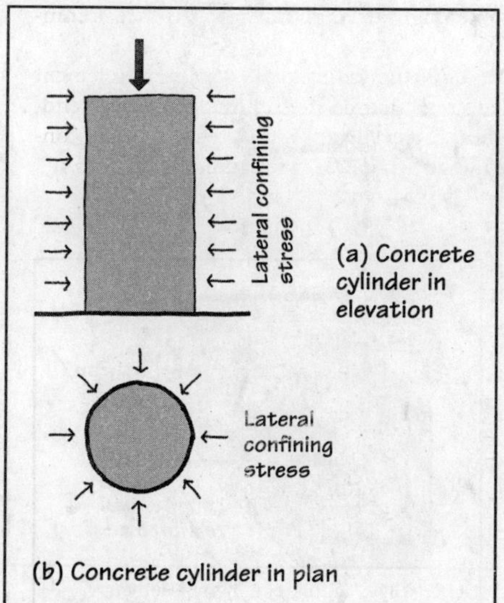

FIGURE 4.26 A concrete cylinder (shown in plan and elevation) tested to failure under lateral confinement.

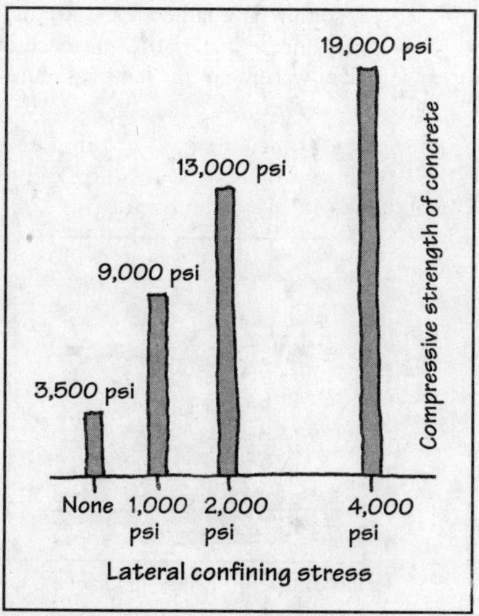

FIGURE 4.27 The effect of lateral confinement on the compressive (crushing) strength of concrete. The compressive strength of (a given batch of) concrete increases as the confining stress on it is increased.

BEARING PLATES

In building structures, a steel plate is generally used to prevent bearing failure (local crushing) that could occur when one element delivers a concentrated compressive load to another element. For example, if a wide-flange steel column is placed directly on a concrete footing, a bearing failure (local crushing) in the concrete could result, because concrete's bearing strength is much lower than the compressive stress in the steel column.

The provision of a steel plate, with a larger surface area, at the base of the column distributes the column load over a larger area, reducing the stress on the concrete. Because the column bears on the steel plate, the steel plate is referred to as the *bearing plate,* Figure 4.28. For the same reasons, a bearing plate is also used when a steel beam rests on a concrete column or wall, Figure 4.29.

Referring to Figure 4.28 again, the column is said to "bear" on the bearing plate; the bearing plate "bears" on the concrete footing, and the footing "bears" on the soil below. The bearing stress of the materials must be checked at each interface to ensure that it does not exceed the bearing strength of the material.

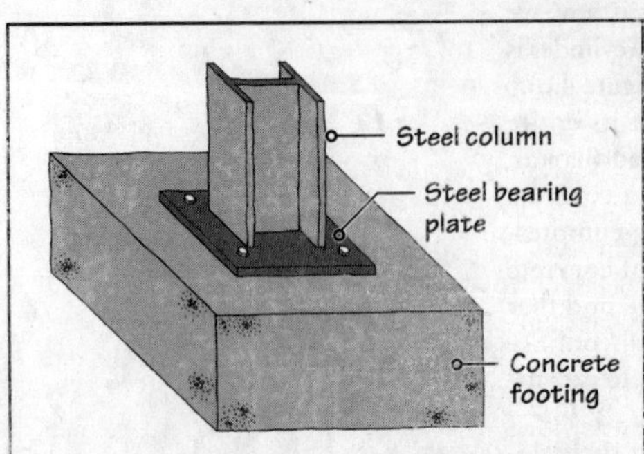

FIGURE 4.28 Bearing plate under a steel column. In this illustration, the column is said to "bear" on the bearing plate. The bearing plate is said to "bear" on the footing, and the footing is said to "bear" on the soil.

BEARING CAPACITY OF SOIL

Because the soil directly under a footing is confined by a large mass of the surrounding ground, the compressive strength of the soil is referred to as the *bearing strength* or, more commonly, as the bearing capacity of the soil (see Section 11.4). Observe that the *bearing strength* of a material is essentially its compressive strength plus the effect of confinement, when present. If there is no confinement, the bearing strength of the material is the same as the compressive strength of the material.

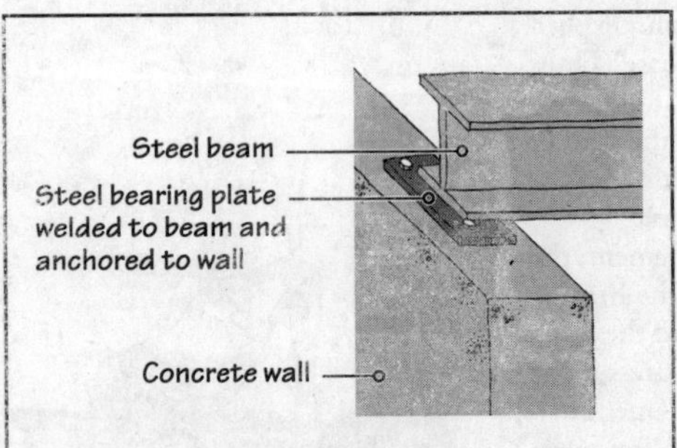

FIGURE 4.29 Bearing plate under a steel beam.

4.9 STRUCTURAL FAILURES

The structural failure of a member can occur in several ways, depending on the type of stress in the member. Because stresses in a member can either be compressive, tensile, or a combination of tensile and compressive stresses, compressive and tensile failures are the two basic types of structural failure.

COMPRESSIVE AND TENSILE FAILURES

A compressive failure is sudden and catastrophic, Figure 4.30. A tensile failure (particularly of a ductile material) is preceded by necking and elongation of the member, giving sufficient warning of the failure, Figure 4.31.

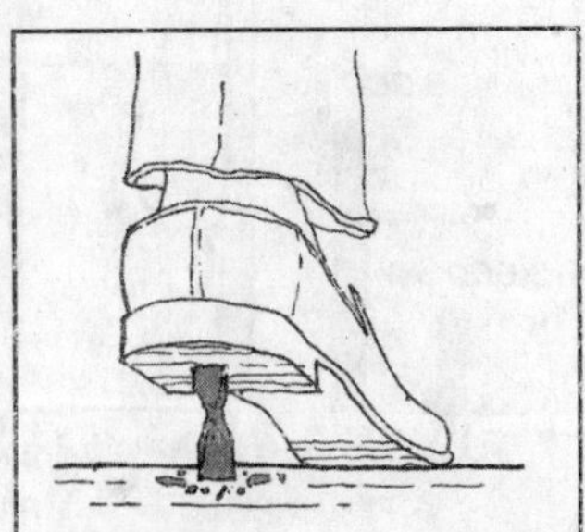

FIGURE 4.30 Failure mode of a member under axial compression (compressive failure). This mode of failure occurs in a short, squat member made of brittle material such as concrete, brick, or stone. Compressive failure is abrupt.

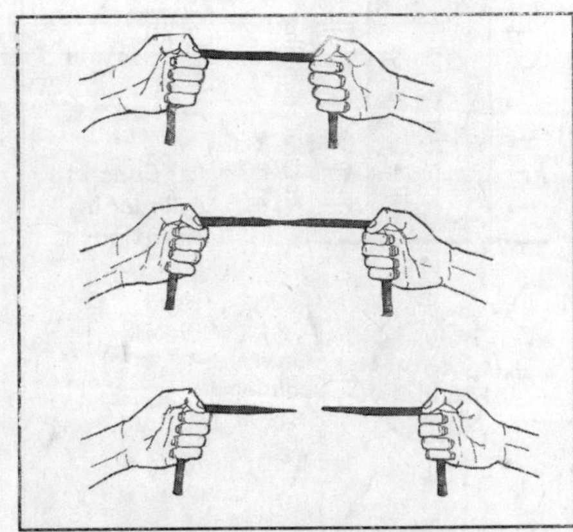

FIGURE 4.31 Failure of a member in tension—tensile failure. If the member is made of a ductile material, such as steel or aluminum, tensile failure is gradual and is generally preceded by "necking" of the member, and hence a large deformation.

BENDING FAILURE

Bending (stress) failure may either be sudden or gradual, depending on whether the member fails in compression or tension. The same applies to shear failure, because, like bending stress, shear stress is also a combination of compressive and tensile stresses.

BUCKLING FAILURE

A type of failure that is entirely different from the failures just described is buckling failure. Buckling failure occurs in a long, slender member subjected to axial compressive stress. (A slender member is one whose length is much greater than its cross-sectional dimensions.) Although a short, squat member will fail by the crushing of the material (as shown in Figure 4.30), a long, slender member will bend under an axial compressive load. The bending of a member under an axial compressive load is called *buckling.*

Buckling is similar to bending. The difference between bending and buckling is that bending is produced by loads that are perpendicular to the axis of the member and buckling is caused by axial loads. A more significant difference between the two, however, is that the bending of a member takes place under all magnitudes of loads, however small. Buckling occurs when the load increases to a certain critical value, called the *critical load.*

Therefore, unlike bending, buckling takes place abruptly. The buckling of a member is recognized as one of the failure mechanisms, although it usually does not lead to a complete collapse of the structure. By buckling under the load, the member becomes unstable but will generally not fracture. In some situations, however, the instability caused by buckling in one member may lead to excessive stresses in other members, thereby causing a progressive collapse of the structure.

Buckling can be easily demonstrated by taking a plastic ruler and subjecting it to increasing axial compression. The more slender the ruler, the smaller the compressive force needed to produce a buckling failure in it, Figure 4.32.

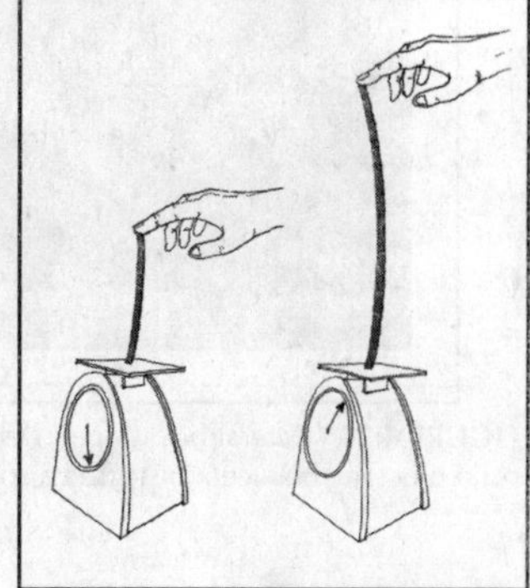

FIGURE 4.32 Buckling of a slender member. The more slender the member, the smaller its buckling load.

Buckling is an important consideration in the design of columns and load-bearing walls. Steel columns are more prone to buckling than concrete columns because they are more slender. Even slender beams and floor joists are prone to buckling because of the presence of compressive stresses resulting from their bending. Remember, bending is a combination of tensile and compressive stresses.

To strengthen a member against buckling failure, we can either reduce its slenderness—that is, increase its cross-sectional dimension—or brace it at intermediate point(s) along its length or height. Usually, the latter approach (bracing the member) is used, because increasing the cross-sectional dimension is uneconomical. For instance, studs in a wood light frame wall must be solidly blocked if it is determined that they may buckle under loads, Figure 4.33. Slender beams and joists require similar bracing, Figure 4.34, because compression is produced in their top fibers as they bend. This compression may cause the slender beams and joists to buckle sideways, a phenomenon known as *lateral buckling.*

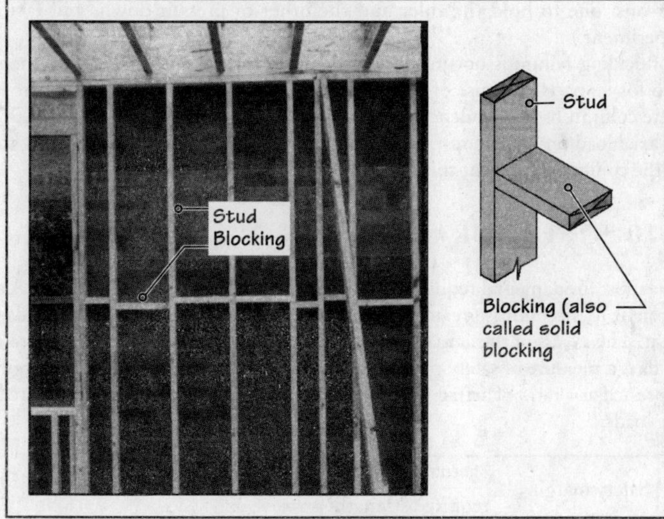

FIGURE 4.33 Solid blocking in a wood frame wall to prevent the buckling of studs.

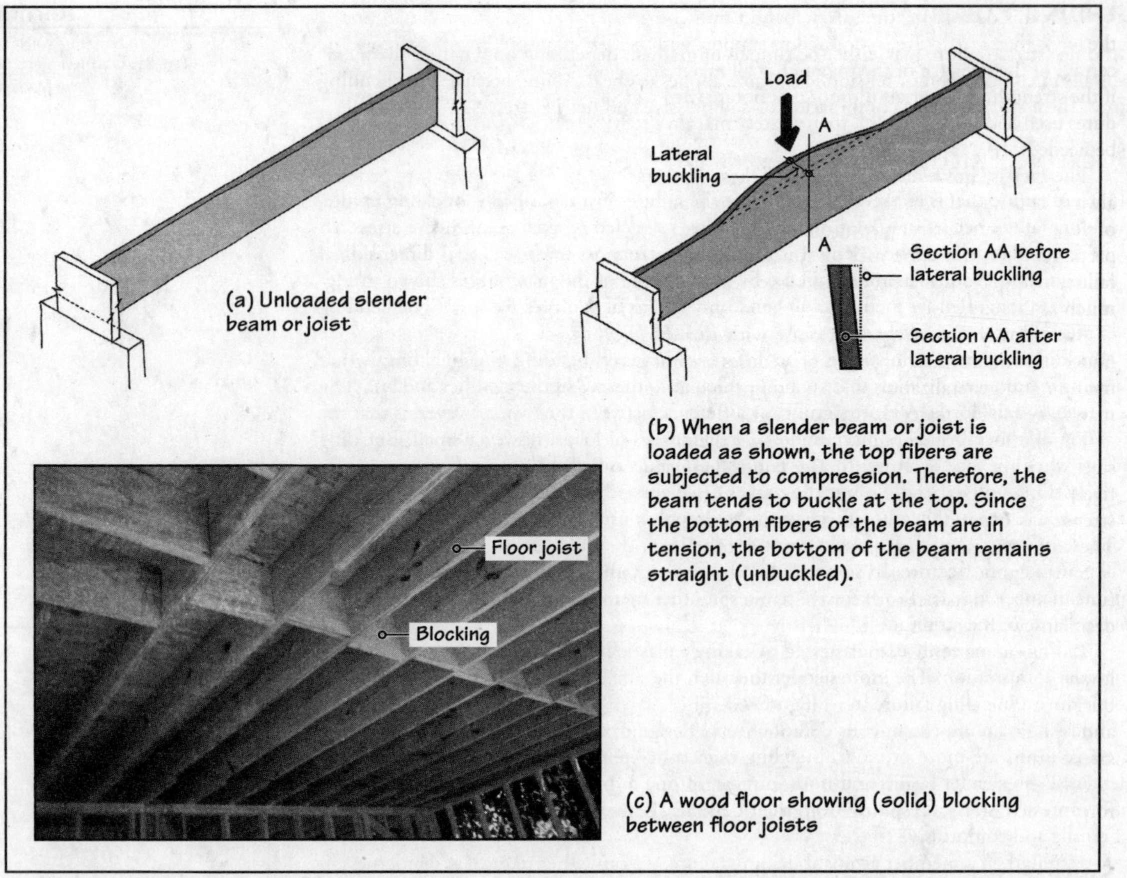

Load

A

Lateral buckling

A

Section AA before lateral buckling

Section AA after lateral buckling

(a) Unloaded slender beam or joist

(b) When a slender beam or joist is loaded as shown, the top fibers are subjected to compression. Therefore, the beam tends to buckle at the top. Since the bottom fibers of the beam are in tension, the bottom of the beam remains straight (unbuckled).

Floor joist

Blocking

(c) A wood floor showing (solid) blocking between floor joists

FIGURE 4.34 Lateral buckling of slender beams or joists. In wood floor joists (which are quite slender), lateral buckling is prevented by using solid blocking between joists (see also Section 15.6).

Lateral buckling can be verified by taking a 12-in.-long plastic ruler, using it as a horizontal beam, and subjecting it to a vertical load, as shown in Figure 4.34(b). We will see that the ruler will not bend downward but horizontally at the top. (Two persons, one to hold the ruler and the other to press it down, will facilitate the experiment.)

Buckling failure is not just a consideration for the entire structural member but also for its parts. Because of their slenderness, the vertical reinforcing bars in a concrete column have a tendency to buckle and break through the concrete cover under an axial load on the column. Horizontal ties are, therefore, provided along the height of the column to prevent this failure mode, Figure 4.35.

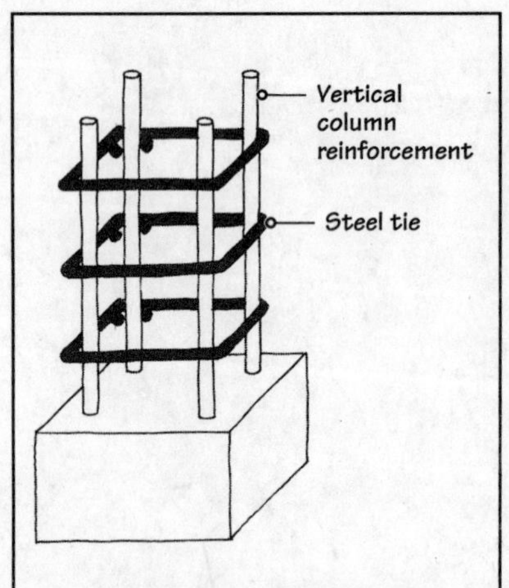

FIGURE 4.35 Horizontal ties in a reinforced concrete column prevent the buckling of vertical reinforcement due to the compressive load on the column.

4.10 STRUCTURAL SAFETY

The most fundamental requirement of a structure is that it should be safe. A structure is safe if its strength is greater than the strength required to carry the loads imposed on it. The excess of the actual strength of the structure over that required to carry the loads is a measure of safety provided in the structure. Thus, the safety margin may be expressed as a ratio of actual strength to required strength—strength required to carry the loads:

$$\text{Safety margin} = \frac{\text{actual strength}}{\text{required strength}} \quad \textbf{Eq. (4)}$$

With this definition, the safety margin must be greater than 1.0. If we could ascertain the loads that a structure is required to support with perfect certainty, if we were absolutely certain of the strength of the material, if all materials gave sufficient warning before failure, if the strength of materials did not deteriorate with time, and if the structural design procedures used had no ambiguity, then a safety margin slightly greater than 1.0 is all that would be needed.

The fact is, however, that none of these "ifs" can be ignored in practice. There are always uncertainties in the values of design loads used, and there is usually a great deal of variation in the strength of materials. For example, if several concrete cylinders were prepared from the same mix of concrete under identical conditions and were tested to failure using the same procedures and testing equipment, the results could vary by as much as 20%.

Additionally, materials deteriorate with time, and their deterioration is not the same under all conditions. Our design procedures are not exact and involve considerable approximation and generalization. Considering these uncertainties, it is necessary that the safety margin be much greater than 1.0.

The greater the uncertainty, the greater the safety margin needed. In a material such as steel, which is manufactured under controlled conditions, the safety margin used is relatively low. Concrete, on the other hand, is virtually always a site-manufactured material. Its strength is dependent on placing, compacting, and curing, all of which occur on site. Therefore, the safety margin used for concrete is higher.

Similarly, because wood is a naturally grown material, there is considerable variation (uncertainty) in its strength for the same specie and grade. The safety margin required in designing with wood is even higher than that required for concrete.

The safety margin is also dependent on the type of stress present. Usually, the margin is lower for bending stress or axial tension than for axial compression, shear, bearing, or buckling. The reason is that failure of a member in axial compression, shearing, bearing, and buckling is relatively more abrupt than in bending or axial tension, requiring a higher safety margin.

A large safety margin leads to uneconomical design. A judicious compromise is, therefore, needed between adequate safety and economic considerations.

Structural Safety—Allowable Stress

There are two ways in which safety is incorporated in a structure. One way is to limit the stress in members caused by loads to a value that is less than its failure stress. This limiting stress is called the *allowable stress.* The stress in the material caused by loads must be less than or equal to the allowable stress. The ratio of the failure stress to the allowable stress is the safety margin, which is commonly called the *factor of safety:*

$$\boxed{\text{Factor of safety} = \frac{\text{failure stress}}{\text{allowable stress}}} \qquad \textbf{Eq. (5)}$$

For example, if the failure stress of a material is 2,000 psi and if a factor of safety of 2.0 is used, the allowable stress of that material is $2{,}000/2.0 = 1{,}000$ psi.

Structural Safety—Load Factor and Resistance Factor

An alternative method of incorporating safety in structures is to inflate the value of loads on the structure, that is, to design the structure for loads that are greater than those that will actually be present on the structure. This method is more sophisticated because the loads can be inflated based on the degree of uncertainty present in their determination. Load inflation is achieved by multiplying the loads by suitable multiplication factors, called the *load factors.* Because a load factor represents inflation in the value of a load, a load factor is greater than 1.0.

The load factor is relatively small for dead loads, because the uncertainty in their determination is small. The load factor is larger for live loads because of greater uncertainty in their determination.

Because there is some uncertainty in the determination of the strength of materials, the strength of the material is reduced by a factor called the *resistance factor.* The resistance factor is less than 1.0. This method of structural design is, therefore, referred to as the *load resistance factor method,* as opposed to the *allowable stress method* described previously.

PRACTICE QUIZ

Each question has only one correct answer. Select the choice that best answers the question.

21. The primary purpose of longitudinal steel bars in a reinforced concrete beam is to overcome the weakness of concrete in tension.
a. True   b. False

22. A material that is weak in tension is
a. strong in shear.
b. weak in shear.
c. no relationship between the shear and tensile strengths of a material.

23. A material that is weak in tension is
a. strong in bending.
b. weak in bending.
c. no relationship between the tensile and bending strengths of a material.

24. The stirrups in a reinforced concrete beam are used to
a. increase the strength of the beam in bending.
b. reduce the deflection of the beam.
c. increase the durability of the beam.
d. increase the strength of the beam in shear.
e. none of the above.

25. The bearing strength of a material is closely related to its
a. bending strength.   b. tensile strength.
c. compressive strength.   d. modulus of elasticity.
e. none of the above.

26. In a steel column that rests on a concrete footing, the area of the bearing plate used is generally
a. greater than the cross-sectional area of the column.
b. equal to the cross-sectional area of the column.
c. smaller than the cross-sectional area of the column.

27. The primary purpose of ties in a reinforced concrete column is to
a. prevent the buckling of vertical reinforcement in the column.
b. prevent the crushing of concrete.
c. increase the shear resistance of the column.
d. none of the above.

28. The horizontal blocking in wood studs helps prevent
a. compressive failure of studs.
b. shear failure of studs.
c. tensile failure of studs.
d. buckling of studs.

29. The allowable stress in a material is generally
a. equal to its ultimate stress.
b. less than its ultimate stress.
c. greater than its ultimate stress.
d. none of the above.

30. In most materials, the safety margin for compressive stress and bending stress is the same.
a. True   b. False

REVIEW QUESTIONS

1. What are the units of stress in the U.S. system of units and the SI system of units?
2. A 1-in.-diameter steel bar is subjected to a tensile force of 30 kips. Determine the stress in the bar.
3. If the bar in Problem 2 is 10 ft long, determine the elongation in the bar. The modulus of elasticity of steel is 29×10^6 psi.
4. Sketch the stress-strain diagram of low-carbon (mild) steel, showing all of its important parts.
5. Explain the modulus of elasticity. Which property of the material does it represent? Give the approximate values of the modulus of elasticity of steel, concrete, and wood.
6. Explain the difference between the elasticity and modulus of elasticity of a material.
7. Using sketches and notes, explain how we increase the bending efficiency of building components.
8. Explain why a brittle material is weak in shear.
9. Explain the bearing strength of a material. Which other property of the material is bearing strength related to?
10. Explain the concept of buckling. What measures do we adopt to prevent buckling failure of building components?
11. If the ultimate strength of a material is 6,000 psi and the factor of safety is 2.5, determine the allowable stress of the material.

CHAPTER

5 Properties of the Envelope–I (Thermal Properties)

CHAPTER OUTLINE

In Chapters 3 and 4, we discussed loads on buildings and the structural properties of materials. The focus was on the structural system—the components that work together to support and give form to a building. This chapter and the following chapter focus on the performance properties of the building's enclosure. (In practice, several different names are used for the building's enclosure, including *envelope* and *skin.* In this text, we generally use the term *building envelope.*)

The building envelope is one of the most important parts of a building, from both the aesthetic as well as the performance perspective. Aesthetically, it is where the design team spends most of its time and effort and where the construction community provides the greatest amount of quality control. From the performance perspective, the role of the envelope is to separate the interior of the building from the exterior environment: heat and cold (i.e., thermal), air, water vapor, (liquid) water, noise, dust, insects, and so on. In addition to being structurally adequate, the envelope must mediate between the exterior and the interior of the building so that required interior environmental conditions can be achieved. In other words, the envelope must control the flow of heat/cold, air, vapor, water, noise, dust, and so on.

Of the environmental parameters mentioned, four of them—thermal, air, vapor, and water—impact the design of the envelope most significantly. They are also interrelated because the flow of air or vapor through the envelope impact the building thermally, and vapor and water are essentially the same material but in different states (gas and liquid, respectively). In this chapter, properties related to heat flow through the envelope are discussed. The next chapter deals with properties related to air and water vapor flow through the envelope. (Chapter 27 deals with rainwater infiltration through the envelope.)

5.1 BUILDING DESIGN AND THERMAL COMFORT

Providing thermal comfort in buildings has been an important design objective from the earliest times. The considerations of climate and manipulation of resources to produce a comfortable building have always had a profound influence on building/shelter design. The Eskimos' igloo and the adobe houses of the Middle East are some of the early examples of comfort-conscious designs in two extremes of climate.

Figure 5.1(a) and 5.1(b) shows daily (24-h) variations of temperatures inside a typical igloo and an adobe dwelling, along with the respective outdoor temperatures [5.1]. The relative stability of the indoor air temperature and the large difference between the indoor and outdoor temperatures in both the igloo and the adobe dwelling attest to the success of these designs in providing thermal comfort.

Water bodies in open courtyards of the hot, dry regions of northern India and wind scoops on roofs in the hot, humid climates of the Middle East are design solutions that provided cooling by manipulating the form and orientation of buildings to respond to the challenges of the local site and climate, Figure 5.2.

Twentieth-century advances in heating, ventilating, and air-conditioning (HVAC) equipment made it possible to achieve interior comfort independent of building form and climate. Advances in materials and their mass production allowed designers to build larger spaces and taller buildings. These factors, in conjunction with modern ideas about design, resulted in buildings that conformed to specific design aesthetics rather than considerations of site and climate. Similar design solutions were used everywhere in the world, because the HVAC systems allowed any design solution to be adapted to any climate.

The resulting design freedom was fully exploited by architects to the extent that an all-glass exterior became a recognized facade formula, particularly for commercial buildings. Although the glass envelope was capable of resisting moisture, it tended to collect heat, increasing the requirements for cooling, even in moderate climates. The situation was acceptable because energy prices were low and the cost of operating buildings was small, but it changed dramatically due to the 1973 Arab-Israeli war when the oil-producing and exporting (OPEC) countries placed an embargo on the sale of oil and quadrupled oil prices.

With the increase in energy prices, the cost of heating and cooling buildings became the single most important recurring liability. At the same time came the realization that the earth's fossil fuel reserves were rapidly being depleted. Consequently, energy conservation became an important design objective, which led to a substantial reorientation of architectural thinking.

Consideration of thermal properties and thermal behavior of buildings assumed greater importance. The building industry reacted by producing materials and equipment that are thermally more efficient, and the design community responded by insulating the building envelope against heat gain and heat loss. In fact, today's building envelope is a multilayered, often complex assembly of materials and components.

As shown in Figure 5.3, an energy-conserving building design is a function of several variables. It includes the consideration of the outside climate, site conditions, building orientation and shape, building services (air conditioning, water supply, lighting, and waste management), and thermal properties of the building envelope.

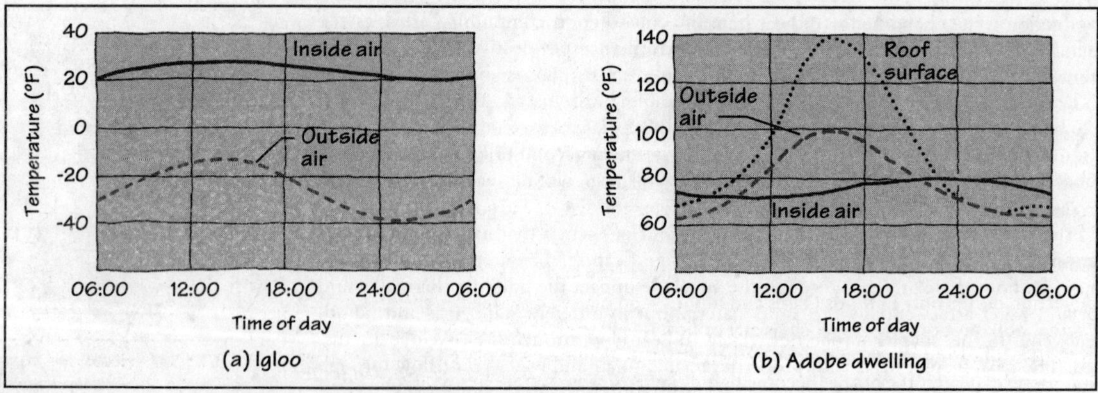

FIGURE 5.1 Daily variations of temperature in **(a)** an igloo and **(b)** an adobe dwelling.

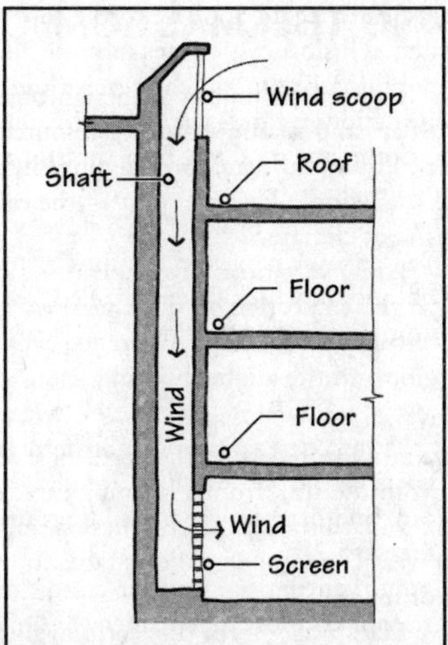

FIGURE 5.2 Section through a wind scoop showing the penetration of breeze from a vertical shaft into a house interior. Wind scoops were used extensively in coastal areas of the Middle East.

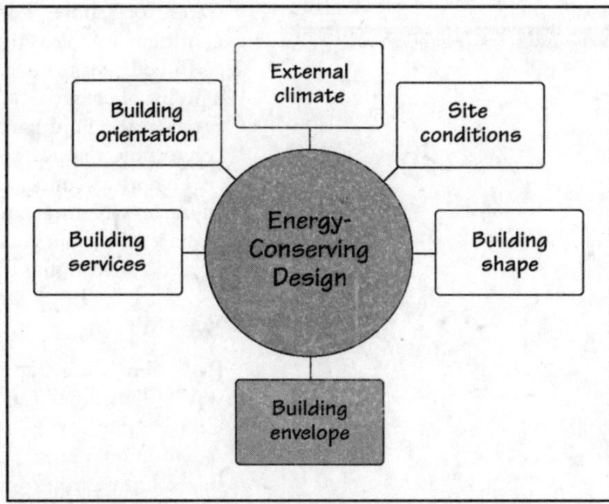

FIGURE 5.3 Factors that affect energy-conserving building design. The focus of this chapter is on the elements of the energy-conserving building envelope.

In this chapter, only the thermal properties of materials and assemblies that influence the design and performance of the building envelope are covered. A discussion of insulating materials is provided at the end of the chapter.

UNITS OF ENERGY

Because thermal properties of materials are related to energy flow through them, an understanding of the units of measurement of energy is necessary. In the SI system of units, energy is measured in Joules (J), and the rate of energy flow through a material is measured in Joules per second (J/s), which is called Watts (W). Both the Joule and the Watt are small units. Therefore, the kilo-Joule (kJ) and kilowatt (kW) are commonly used. In the U.S. system of units, the unit for energy is the British thermal unit (Btu), and the corresponding unit for rate of energy flow is Btu per hour (Btu/h).

5.2 CONDUCTION, CONVECTION, AND RADIATION

From our daily experience, we know that if two objects, which are at different temperatures, are placed in contact with each other, the temperature of the warmer object decreases and that of the cooler object increases. This phenomenon occurs because heat flows from an object at a higher temperature to that at a lower temperature, Figure 5.4. Heat transfer will continue until the temperature of the two objects becomes equal. There are three modes by which heat transfer can take place: conduction, convection, and radiation.

CONDUCTION

Conduction is the phenomenon of heat transfer between the molecules of a substance that are in contact with each other. It is the only mode of heat transfer within solids and is characterized by physical contact between the molecules in the substance. When one end of a solid substance is heated, the molecules in the neighborhood of the heated end begin to vibrate about their mean positions with a higher average velocity. Because of the physical contact between the molecules of the substance, the increased vibrational energy in the heated molecules is transferred to adjacent cooler molecules, which in turn transfer their increased energy to still cooler molecules in their neighborhood, and so on.

CONVECTION

Convection occurs only in fluids (gases and liquids) and may be described as energy transfer by actual bulk motion of a gas (such as air) or liquid (such as water). A familiar example of convection is heating water on a stove. When water in a container is placed on a stove, the water particles nearest the burner become lighter in density as they are heated. Being lighter, these particles rise to the top and are replaced by colder (and hence heavier) particles of

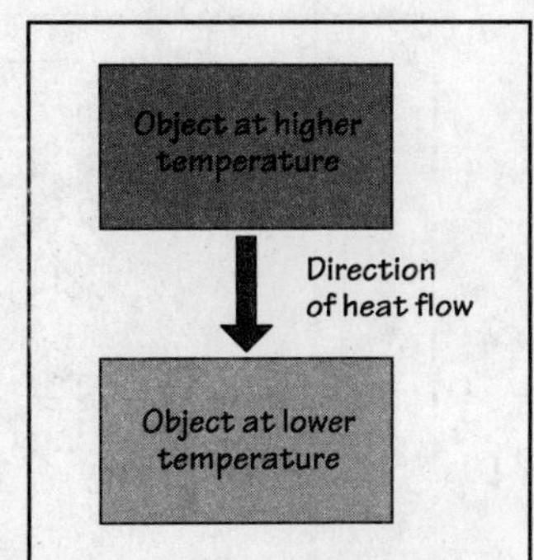

FIGURE 5.4 Heat flows from an object at a higher temperature to an object at a lower temperature.

water from above. The colder particles are heated again and rise to the top, and the cycle continues as long as the water is being heated. This process creates motion within the fluid.

In addition to the transfer of energy by bulk motion, which creates fluid flow, a small amount of energy is also transferred by conduction in fluids because of the physical contact between the fluid particles. However, the dominant mode of heat transfer in fluids is by convection, that is, the movement of fluid particles.

The process of heat transfer described by the heating of water on a stove is called *natural convection.* Natural convection results from density changes in the fluid. Convection may also occur without changes in the density of fluid, in which case it is called *forced convection.* Forced convection is produced by fans, pumps, and wind motion.

Radiation

Radiation is the transfer of energy between two objects in the form of electromagnetic waves. Thus, the heat and light received from the sun, from a burning candle, or from an electric lamp reach us by electromagnetic radiation. Unlike conduction and convection, radiation does not require a medium to travel. This property allows radiation from the sun to reach the earth through the vast realm of space.

It is important to appreciate that heat radiation is a form of electromagnetic radiation, and all electromagnetic radiation is transformed into heat when absorbed by an object. Electromagnetic radiation can exist in a variety of wavelengths, each with different characteristics. Some of the segments of electromagnetic radiation are infrared, visible (light), ultraviolet, X-rays, and gamma rays.

Thus, although the radiation from the sun consists of ultraviolet, visible, and infrared radiations, all three components of solar radiation are converted to heat when they are absorbed by a building surface. Similarly, all radiation from an electric lamp in a room is converted to heat as it gets absorbed by the occupants, furniture, room surfaces, and room air.

Simultaneity of Conduction, Convection, and Radiation

During the process of heat transfer through a building envelope, all three modes usually come into play simultaneously. Thus, to reduce heat transfer through the envelope, all three modes must be considered, and the heat transfer potential of each mode must be reduced.

Consider the section through the brick cavity wall shown in Figure 5.5. Energy from the sun reaches the wall by radiation and is partly absorbed and partly reflected by the wall's exterior surface. The absorbed energy then travels by conduction through the wall (from surface A to surface B) until it reaches the air space.

Inside the air space (from surface B to surface C), the heat-transfer mode changes to radiation and convection. The subsequent mode is conduction through the inner layer of the wall (from surface C to surface D). Finally, the energy is transferred from the inner surface of the wall (surface D) to indoor air, room surfaces, furniture, and occupants by convection and radiation.

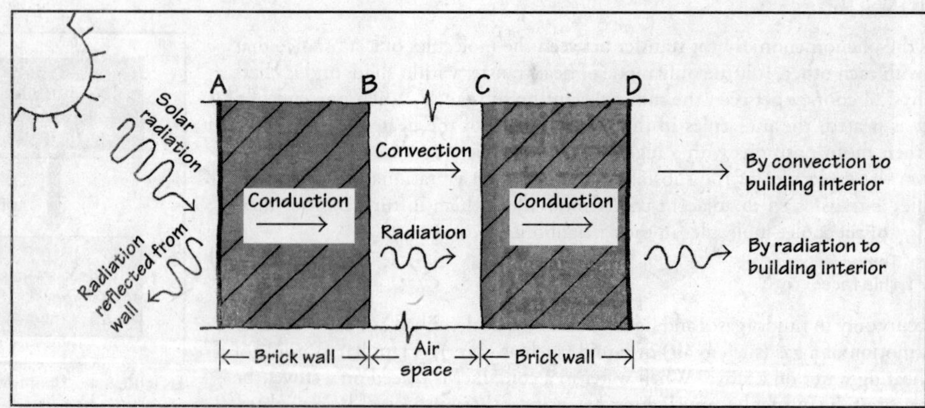

FIGURE 5.5 Modes of transfer of solar heat through a brick cavity wall.

5.3 R-VALUE OF A BUILDING COMPONENT

The term R is called *thermal resistance*, or simply the *R-value*. It is the measure of the ability of a component to resist the flow of heat through it. The R-value is related to heat flow primarily by conduction.

To understand the concept, consider a rectangular plate (such as a wall or roof) of thickness L, as shown in Figure 5.6. Let the surface temperatures of the two faces of the plate be t_1 and t_2. If t_2 is greater than t_1, heat will travel from face 2 to face 1. Experiments have shown that the rate of heat transfer by conduction from face 2 to face 1 is given by

$$q_c = \frac{(t_2 - t_1)}{\rho L} \qquad \text{Eq. (1)}$$

where q_c is the rate of heat conduction in Btu/h through 1 ft^2 of the plate (Btu/(h-ft^2)).

Equation (1) shows that the rate of heat conduction through the plate is directly proportional to the difference between the surface temperatures of the two faces of the plate. If there is no difference between the surface temperatures—that is, if $(t_2 - t_1)$ is zero—no heat conduction will occur. Additionally, the rate of heat conduction is inversely proportional to the thickness, L, of the plate. This means that the greater the thickness of the plate, the smaller the amount of heat conducted through it.

The term ρ in Equation (1) is called *thermal resistivity*. Thermal resistivity is a property of the material and is, therefore, constant for the material. If the value of ρ is large, the rate of heat conduction through the plate is small. Such a material is called a *thermal insulator* or an *insulating material*. Conversely, if the value of ρ is small, the rate of heat conduction through the plate is large. Such a material is called a *thermal conductor*.

If we now replace ρL in Equation (1) by R, it becomes

$$q_c = \frac{(t_2 - t_1)}{R} \qquad \text{Eq. (2)}$$

Again, the term R refers to thermal resistance, and a component with a high R-value is more insulating (resistant to the flow of heat) than a material with a low R-value. It is the R-value that is generally quoted in referring to the insulating value of a component or assembly. Thus, we speak of the R-value of a component, such as R-10 or R-11. Most insulating products are labeled with their R-values, Figure 5.7.

From Equation (2), the units of R-value are (ft^2-°F-h)/Btu. The units of ρ are [(ft^2-°F-h)/(Btu-in.)], Table 5.1. Because both of these units are rather long and complex, we usually omit the units in practice—a practice that we will follow in this text.

When the R-value of a component is given in the SI system of units, it is usually referred to as the *RSI-value*. Thus, if the R-value of a component is R-10, its RSI-value is 1.76,

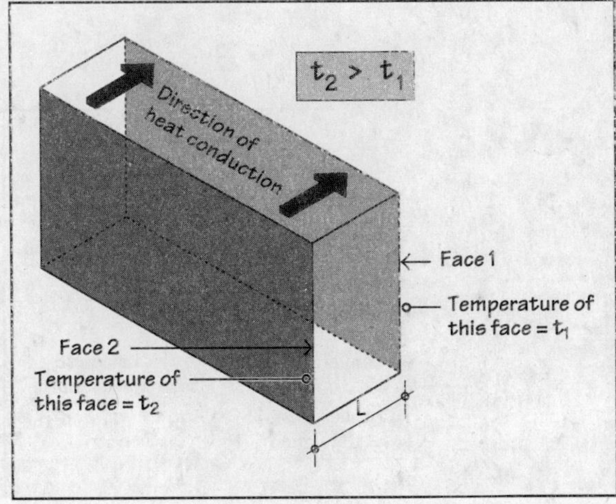

FIGURE 5.6 Heat transfer by conduction through a plate of thickness L.

(a)

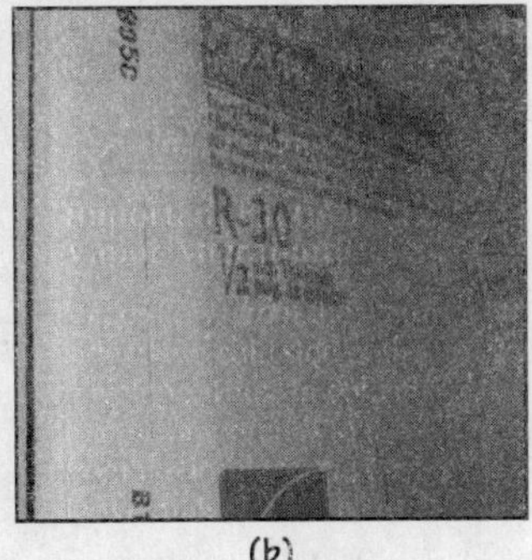

(b)

FIGURE 5.7 (a) R-11 fiberglass insulation in batt form. (b) R-3 extruded polystyrene insulation in board form.

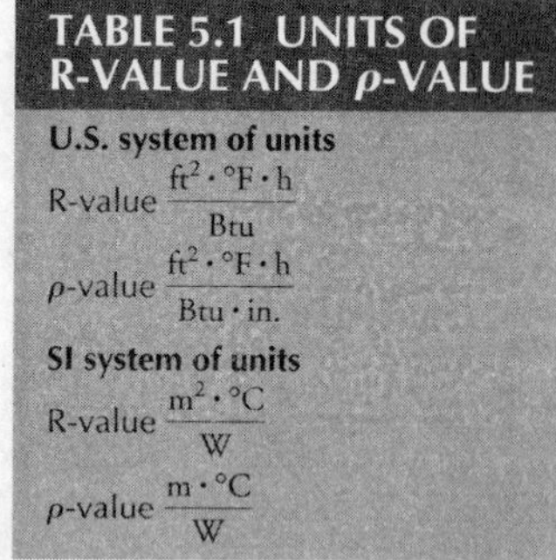

TABLE 5.1 UNITS OF R-VALUE AND ρ-VALUE

U.S. system of units	
R-value	$\frac{ft^2 \cdot °F \cdot h}{Btu}$
ρ-value	$\frac{ft^2 \cdot °F \cdot h}{Btu \cdot in.}$
SI system of units	
R-value	$\frac{m^2 \cdot °C}{W}$
ρ-value	$\frac{m \cdot °C}{W}$

To convert an R-value to an RSI-value, multiply the R-value by 0.176.

which is referred to as RSI-1.76. Table 5.2 also gives the conversion factors for R-values and ρ-values.

DIFFERENCE BETWEEN R-VALUES AND ρ-VALUES

It is important to note again that the ρ-value is a property of the material, whereas the R-value is a property of the component of a certain thickness. Because they are related to each other by the relationship $R = \rho L$, where L is the thickness of the component, the ρ-value is, in fact, the R-value of a component whose thickness is 1 in. Thus, if we know the ρ-value of a material, we can determine the R-value of a component of any thickness. For example, if the ρ-value of a material is 2.5, then the R-value of a 6-in.-thick component made of that material is $2.5 \times 6 = 15.0$; that is, the R-value of this component is 15.

Because the ρ-value is a material property, it is the ρ-value that is used for comparing the insulating effectiveness of materials. Again, a material with a higher ρ-value is a better thermal insulator. The ρ-values of some commonly used materials are given in Table 5.2.

TABLE 5.2 ρ-VALUES OF COMMONLY USED BUILDING MATERIALS

Material	Resistivity (ρ-value) $(ft^2 \cdot °F \cdot h)/(Btu \cdot in.)$
Metals	
Steel	$0.0032 = 3.2 \times 10^{-3}$
Copper	3.2×10^{-4}
Aluminum	3.2×10^{-3}
Ceramic materials	
Clay bricks	0.20
Concrete (normal weight)	0.15
Concrete (structural lightweight)	0.25–0.35 (depending on density)
Insulating concrete (perlite or vermiculite)	1.70
Concrete masonry	Depends on the type of concrete and cell insulation
Limestone	0.15
Sandstone	0.18
Glass sheet	0.14
Plaster	0.35
Gypsum wallboard	0.60
Portland cement plaster	0.30
Wood and coal	
Softwoods (solid lumber or plywood)	0.9
Fiberboard	2.4
Wood charcoal	2.2
Coal	0.85
Insulating materials	
Granulated cork	3.0
Vermiculite (loose fill)	2.1
Perlite (loose fill, density 5.0 pcf)	3.0
Perlite (loose fill, density 10.0 pcf)	2.4
Expanded perlite board	2.8
Fiberglass (loose fill, assuming all voids are filled)	3.5
Fiberglass (batt or blanket)	3.5
Mineral wool (rock wool)	3.5
Expanded polystyrene (EPS) board (bead board)	4.0
Extruded polystyrene (XPS) board	5.0
Polyurethane board (laminations on both sides)	6.5
Polyisocyanurate (ISO) board (laminations on both sides)	6.5
Foamed-in-place polyicynene	3.6
Gases	
Air	5.6
Argon	8.9
Carbon dioxide	9.9
Chloro-fluoro-carbon (CFC) gas	16.5
Hydro-chloro-fluoro-carbon (HCFC) gas	15.0
Water	0.24

ρ-value = R-value of 1-in.-thick material

To convert a ρ-value to a ρSI-value:	**Multiply the ρ-value by 6.93.**
To convert an R-value to an RSI-value:	**Multiply the R-value by 0.176.**

Note: The values given in this table are representative values. Consult manufacturers' data for precise values.

Example 1
(R-Value from ρ-Value)

Determine the R-value of (a) a 6-in.-thick concrete wall, (b) a 6-in.-thick fiberglass blanket, and (c) a 0.25-in.-thick glass sheet.

Solution

From Table 5.2, the ρ-values of concrete, fiberglass, and glass sheet are 0.15, 3.5, and 0.14, respectively.

a. For a 6-in.-thick (normal-weight) concrete wall, $R = \rho L = (0.15)(6) = 0.9$.
b. For a 6-in.-thick fiberglass blanket, $R = \rho L = (3.5)(6) = 21.0$.
c. For a 0.25-in.-thick glass sheet, $R = \rho L = (0.14)(0.25) = 0.035$.

Note that although the concrete wall and the fiberglass blanket are both of the same thickness, the fiberglass blanket is nearly $21.0/0.9 = 23$ times more effective as a thermal insulator than the concrete wall. In other words, it will take a nearly $23 \times 6 = 138$-in.-thick concrete wall to provide the same thermal insulation as a 6-in.-thick fiberglass blanket. Also note that the R-value of a sheet of glass is virtually zero.

EFFECT OF DENSITY AND MOISTURE CONTENT ON THE ρ-VALUE

As a general rule, a low-density material has a high ρ-value. Being fibrous, granular, or cellular in structure, a low-density material has air trapped within its voids. Because air has a high ρ-value, a low-density material is a better insulator than a high-density material. In fact, still air is one of the best thermal insulators available. Circulating air cannot be a good insulator because the circulation of air increases convective heat flow, thereby reducing the ρ-value.

In other words, the basic requirement of an insulating material is that the solid constituents of the material must divide its volume into such tiny air spaces that air cannot circulate within them. This characteristic ensures that convective heat transfer is virtually negligible. It also implies that all insulating materials are low-density materials.

Another important parameter that affects the ρ-value is the moisture content of the material. Water has a much lower ρ-value than air (ρ-value of air = 5.6 and ρ-value of water = 0.24, Table 5.2). Because a moist material has water trapped inside its pores in place of the air, its ρ-value is lower than that of a dry sample.

It is, therefore, important to ensure, in the detailing of buildings, that insulating materials remain dry throughout their useful life spans. This fact is important because most insulating materials due to their location, are particularly vulnerable to water absorption from rain. In cold climates where condensation of water vapor within insulating materials is an additional hazard, vapor retarders are installed in roof and exterior wall assemblies (see Sections 6.5 to 6.7).

ENTRAPMENT OF HCFC GAS INSTEAD OF AIR

Although most insulating materials contain small volumes of air to provide high insulating values, some plastic foam insulations (or simply *foamed plastics,* discussed at the end of the chapter) contain a gas that is more insulating than air. The gas commonly used is hydro-chloro-fluoro-carbon (HCFC) gas, which is nearly 2.5 times more effective than air as an insulation (see Table 5.2).*

Foamed plastics that trap HCFC are manufactured by using HCFC gas as a foaming agent in a liquid polymer. From Table 5.2, the ρ-value of bead board (a foamed plastic with air inside the beads, also called *EPS board*) is 4.0, whereas the ρ-value of polyisocyanurate board, which contains HCFC, is 6.5.

Note that plastic foam insulations that trap HCFC have a closed cellular structure to retain the gas within the material, Figure 5.8. Despite their cellular structure, some of the trapped HCFC gas finds

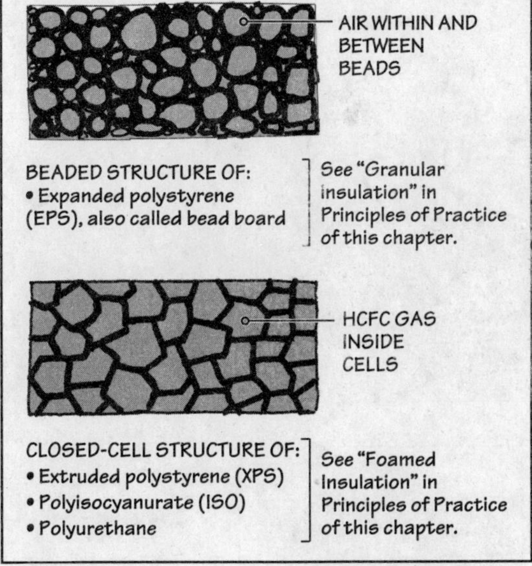

FIGURE 5.8 Cellular structure of various plastic foam insulations.

*The gas used earlier as a foaming agent was chloro-fluoro-carbon (CFC) gas. However, it was found to deplete the ozone layer in the upper atmosphere of the earth (see Chapter 10). HCFC gas is less destructive of the ozone layer, and hence it has replaced CFC gas. The insulating potential of HCFC is only slightly lower than that of CFC.

its way into the atmosphere. Therefore, air is able to permeate into the cell cavities of these foams, driving some HCFC gas out from the edges and ends of a board.

The steady outward migration of HCFC gas means that the R-value of such plastic foams decreases with time. Therefore, in referring to the R-value of a plastic foam with HCFC gas, we generally quote its *stabilized,* or *aged,* R-value.* The stabilized R-value is lower than the R-value immediately after the plastic foam's manufacture. In most cases, the R-value of such a material stabilizes after six months, after which the decrease in R-value is almost negligible.

PRACTICE QUIZ

Each question has only one correct answer. Select the choice that best answers the question.

1. Apart from the climate of the location, the other factors that affect energy-conscious building design are
   a. building orientation and building shape.
   b. building envelope.
   c. site conditions.
   d. all of the above.
   e. (a) and (b).

2. Which of the following events provided a major push toward energy-conserving building design?
   a. Depletion of energy resources during the Second World War
   b. Freedom gained by many Asian and African countries around 1950
   c. Arab-Israeli war of 1967
   d. Arab-Israeli war of 1973

3. The unit for the measure of energy in the U.S. system of units is the
   a. calorie.
   b. joule.
   c. Btu.
   d. watt-hour.
   e. all of the above.

4. The unit for the measure of energy in the SI system is the
   a. calorie.
   b. joule.
   c. Btu.
   d. watt

5. Within a solid, the mode of heat transfer is by
   a. conduction.
   b. convection.
   c. radiation.
   d. both (a) and (b).
   e. both (b) and (c).

6. Within an air space (cavity), the mode of heat transfer is primarily by
   a. conduction.
   b. convection.
   c. radiation.
   d. both (a) and (b).
   e. both (b) and (c).

7. Air is a good thermal insulator, provided that it exists in large volumes.
   a. True
   b. False

8. The R-value of a 2-in.-thick material is 4.0. What is the R-value of the same material if its thickness is 1 in.?
   a. 1.0
   b. 2.0
   c. 6.0
   d. 8.0
   e. None of the above

9. The R-value of a building assembly in the SI system is referred to as the
   a. R-value (SI).
   b. SI-value.
   c. RS-value.
   d. RSI-value.
   e. none of the above.

10. A wet or moist material provides a higher R-value than if the same material is dry.
    a. True
    b. False

11. One of the main reasons for using HCFC gas in some insulating materials is that it
    a. makes the material noncombustible.
    b. makes the material more economical.
    c. makes the material more resistant to decay.
    d. increases the strength of the material.
    e. none of the above.

12. Insulating materials that contain HCFC gas within their cells have a higher R-value than those that contain air within their cells.
    a. True
    b. False

13. Which of the following insulating materials contains HCFC gas within their voids?
    a. Extruded polystyrene
    b. Polyisocyanurate
    c. Polyurethane
    d. All of the above
    e. Only (b) and (c)

14. For the same thickness, which of the following materials has the highest R-value?
    a. Extruded polystyrene
    b. Expanded polystyrene
    c. Polyisocyanurate
    d. Fiberglass
    e. Gypsum

15. For the same thickness, which of the following materials has the smallest R-value?
    a. Extruded polystyrene
    b. Expanded polystyrene
    c. Polyisocyanurate
    d. Fiberglass
    e. Gypsum

16. The R-value of a 1-in.-thick extruded polystyrene board is 5.0. If the R-value of a 1-in.-thick brick wall is 0.2, how thick should the brick wall be to equal the R-value of 1-in.-thick extruded polystyrene?
    a. 10 in.
    b. 5 in.
    c. 50 in.
    d. 2.5 in.
    e. None of the above

17. The concept of *aged R-value* is applicable to
    a. all insulating materials.
    b. insulating materials that contain HCFC gas.
    c. fiberglass insulation.
    d. perlite board insulation.
    e. all building materials.

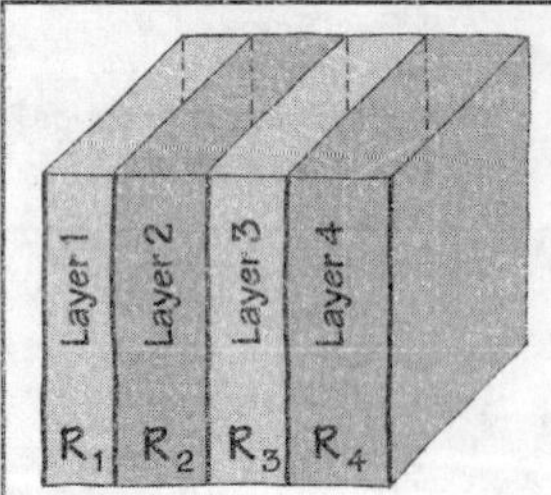

FIGURE 5.9 A multilayer assembly.

5.4 R-VALUE OF A MULTILAYER COMPONENT

Usually a wall or roof assembly consists of several layers of different materials, as shown in Figure 5.9. For example, the wall shown in Figure 5.10 consists of three layers of materials: a layer of brick, a layer of insulation, and a second layer of brick. Using the laws of physics,

*The Polyisocyanyrate Manufacturers Association (PIMA) refers to the stabilized R-value as *long-term thermal resistance* (LTTR).

it can be shown that the total resistance of a multilayer component, shown in Figure 5.9, is obtained by adding the R-values of the individual layers:

$$R_t = R_1 + R_2 + R_3 + R_4 + \cdots \qquad \textbf{Eq. (3)}$$

where R_t is the total resistance of the multilayer component and R_1, R_2, R_3, . . . , are the resistances of layer 1, layer 2, layer 3, and so on, respectively. Note that Equation (3) is valid only if there is no thermal bridging in the assembly (see Example 5 and the box entitled "Thermal Bridging in Building Assemblies" later in the chapter).

Example 2
(Estimating the R-Value of an Assembly)

Calculate the R-value of a wall assembly that consists of 2-in.-thick extruded polystyrene (XPS) insulation sandwiched between an 8-in.-thick (nominal) brick wall and a 4-in.-thick (nominal) brick wall, as shown in Figure 5.10.

Note: The actual thickness of an 8-in. nominal brick wall is $7\frac{5}{8}$ in., and that of a 4-in.-thick brick wall is $3\frac{5}{8}$ in. (see Section 24.5).

Solution

From Table 5.2, the ρ-values of brick wall and extruded polystyrene insulations are 0.2 and 5.0, respectively. The total R-value of the assembly is determined as follows:

Element	R-value
4-in.-thick brick wall	3.625(0.2) = 0.725
2-in.-thick polystyrene board	2.0(5.0) = 10.0
8-in.-thick brick wall	7.625(0.2) = 1.525

$$R_t = 0.725 + 10.0 + 1.525 = 12.25$$

Because an R-value is given by the nearest whole number, the wall is an R-12 wall.

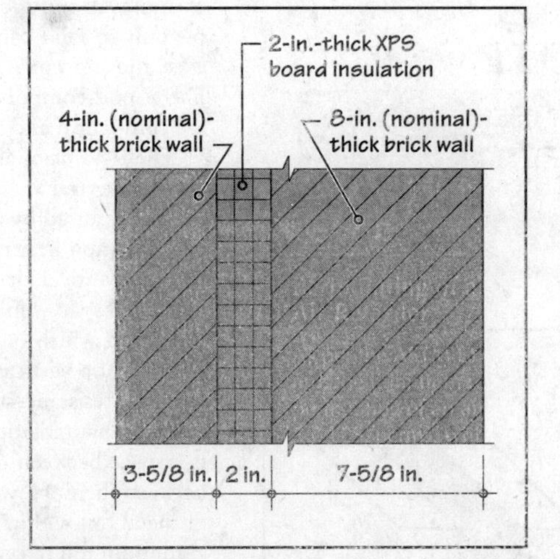

FIGURE 5.10 A brick wall assembly with sandwiched insulation.

SURFACE RESISTANCES

In addition to the layers of solid materials comprising an assembly, there are two invisible layers that contribute to its total R-value. These layers are thin films of air of nearly zero velocity that cling to the inside and outside surfaces of an assembly, Figure 5.11. Because each air film has a certain thickness, it increases the total R-value of the assembly.

The R-value of each film is called the *film resistance* or, more commonly, the *surface resistance.* Thus, there is an inside surface resistance, R_{si}, and an outside surface resistance, R_{so}, corresponding to the inside and outside air films, respectively. Because the surface resistances are always present, Equation (3) is modified as follows:

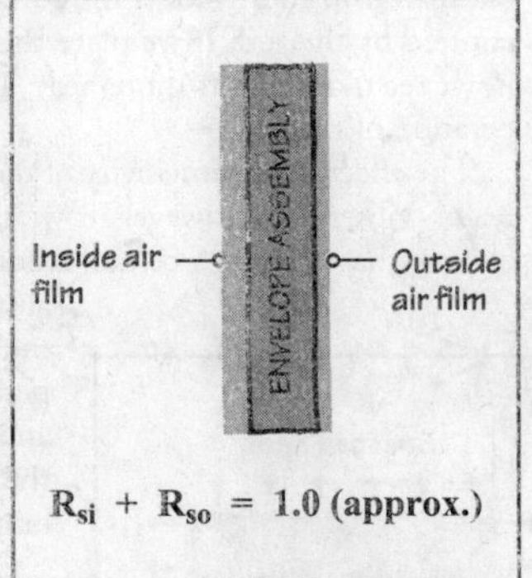

FIGURE 5.11 Inside and outside air films on the surfaces of an assembly add to its R-value. The R-value of the inside air film (called *inside surface resistance,* R_{si}) is approximately 0.7. The R-value of the outside air film (called *outside surface resistance,* R_{so}) is approximately 0.2. Both air films together add an R-value of 0.9 (approximately 1.0) to the total R-value of the assembly.

$$R_t = R_{si} + R_1 + R_2 + R_3 + R_4 + \cdots + R_{so} \qquad \textbf{Eq. (4)}$$

It is the total resistance obtained by including the two surface resistances that is responsible for *air-to-air heat transfer* through an envelope assembly. Because the inside air velocity is usually smaller than the outside air velocity, the inside film is thicker than the outside film. A good approximation of R_{si} is 0.7, and that of R_{so} is 0.2 (in the U.S. system of units). Thus, $R_{si} + R_{so}$ may be taken as approximately equal to 0.9, or, say, 1.0.

Note that the total R-value of an assembly, obtained from Equation (4), is typically rounded to the nearest whole number (integer).

NOTE

R-Value of a Single Sheet of Glass

The R-value of a single sheet of glass is R-1, which is due mainly to the internal and external film resistances.

NOTE

R-Value of an Air Space in a Wall (Wall Cavity)

The R-value of a vertical (wall) cavity is approximately equal to R-1.

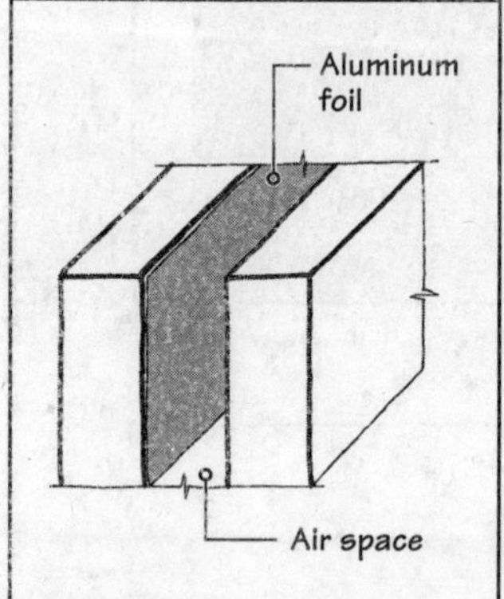

FIGURE 5.12 Radiation heat transfer across an air space is reduced if one surface of the air space is lined with a bright metal such as aluminum foil. Thus, the R-value of an (aluminum) foil-lined air space is greater than that of an unlined air space.

Because both surface resistances provide a total R-value of only 1.0, they do not add significantly to the total R-value of a typical roof or wall assembly (see Example 3 later in this chapter). However, they contribute greatly to the R-value of a glazed window or a roof skylight, because a single sheet of glass has an extremely small R-value—virtually zero. See Example 1. In fact, the total R-value of a single sheet of glass is 1.0, which is provided almost entirely by the inside and outside surface resistances.

5.5 SURFACE EMISSIVITY

So far, we have discussed conductive heat transfer—the dominant mode in most building components. The other two modes, convection and radiation, occur only in air spaces that exist in wall and roof assemblies and attics. Theoretically, the concept of R-value is applicable only to solid components in which conduction is the only mode of heat transfer. However, the concept is extended to air spaces also, and an air space is assigned an R-value just like a solid component, although the primary modes of heat transfer through an air space are convection and radiation.

The R-value of an air space varies with its width, its orientation (whether it is a horizontal or a vertical air space), and the direction of heat flow. The change in the R-value of an air space caused by a change in its width is a complex function of conduction, convection, and radiation heat transfers within the space.

For a vertical air space (wall cavity), the R-value increases with an increase in the width of the air space, up to a value of 1.0. This value applies to an air space that is $\frac{3}{4}$ in. wide. Beyond a width of $\frac{3}{4}$ in., the R-value of the air space does not increase. Therefore, the R-value of a vertical air space is generally assumed to be 1.0. This takes into account the two film resistances within the cavity and both convection and radiation heat transfers.

Note that radiation heat transfer is unaffected by the width of the cavity. As long as there is a space between the two opposite surfaces of a cavity, radiation heat transfer will occur between them. However, radiation heat transfer can be altered (reduced) by including a layer of metal foil within the air space, Figure 5.12. This is generally achieved by laminating an aluminum foil to one of the two surfaces of the air space, referred to as *lining the air space.*

R-VALUE OF A FOIL-LINED VERTICAL AIR SPACE

Aluminum foil reduces radiation heat transfer due to its low *emissivity.* Emissivity is a property of the surface of an object that refers to its potential to emit radiation. In fact, the magnitude of radiation emitted by an object is directly related to its emissivity and its temperature.

The effect of temperature on the radiation emitted by an object is well known. If we place an iron rod inside a fire for a short while and then withdraw it, we can feel the heat emitted by the rod. If we place the same rod in the fire for a long time and then withdraw it, we see that it emits more heat. The increase in radiated heat is due to the rise in the temperature of the rod.

The effect of the emissivity of the object on the amount of radiation emitted by it is not as readily observed. However, if we take radiation measurements on a heated object, we will observe that it emits a certain amount of heat. If the same object is covered all around with aluminum foil, we will observe that the heat emitted by it is much less, Figure 5.13. Restaurants commonly keep a baked potato wrapped in aluminum foil so that it will emit less heat and thereby cool more slowly.* In this case, the foil reduces the emission of heat. (Note that the term *emission* is used only with radiation, not with conduction or convection.)

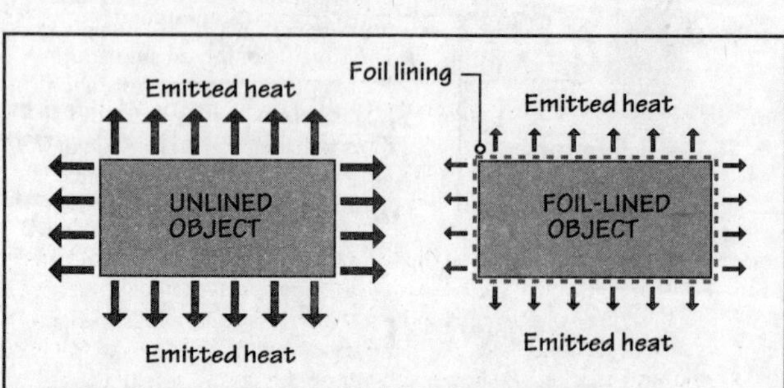

FIGURE 5.13 Radiation emitted by two identical objects that are at the same temperature. The unlined object emits more radiation than the same object that is lined (wrapped) with aluminum foil.

The emissivity of a material lies between 0 and 1. Most building materials (brick, concrete, wood, plaster, gypsum board, steel, etc.) have a high emissivity—approximately 0.9. Polished metals, on the other hand, have a low emissivity. A highly polished metal, such as stainless steel or aluminum, has an emissivity of only 0.05. This means that if an object made of brick, concrete, wood, and the like is covered with aluminum foil, the heat emitted by it will be (0.05/0.90) = 0.06, that is, 6% of the heat emitted by the same object without the foil covering.

*The aluminum foil also helps retain the moisture within the potato.

Note that emissivity is a property of the surface, not of the bulk of the material. Thus, changing the surface characteristic of an object changes its emissivity. For instance, plywood's emissivity is nearly 0.9, but if that plywood is laminated with aluminum foil, its emissivity becomes only 0.05. Aluminum foil is the metal that is commonly used in building assemblies because of its relatively low cost.

Lining a surface of the air space with foil implies that the heat emitted by that surface is small; hence, the amount of heat transferred to the opposite surface is small. This increases the R-value of the air space. Thus, the R-value of a typical foil-lined air space in a wall is approximately 2.5 (an increase of 1.5 over that of an unlined air space).

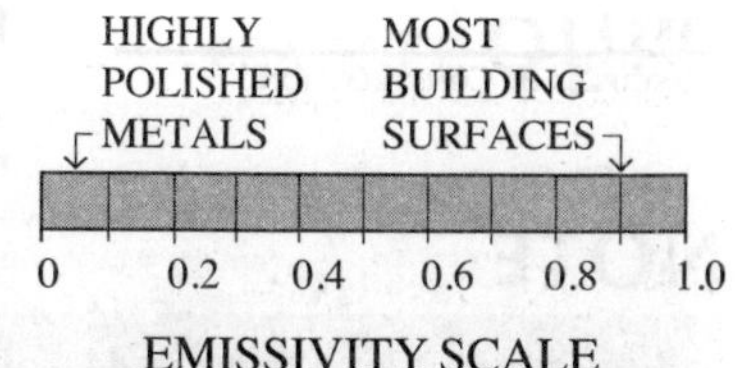

Location of Low-Emissivity Material (Aluminum Foil) in an Air Space

A fundamental fact related to emissivity is that *the emissivity of a surface is always equal to its absorptivity.** Because most building materials have an emissivity of 0.9, their absorptivity is also 0.9, which means that they absorb 90% of the radiation falling on them. A foil-lined surface, on the other hand, absorbs only 5% of the radiation falling on it, because its absorptivity (or emissivity) is 0.05. Its reflectivity is, therefore, 0.95; that is, it reflects 95% of the radiation falling on it. That is why aluminum foil is also referred to as a *reflective insulation* or a *radiant barrier*.

This fact implies that the location of the foil in an air space—on the warmer or cooler side of the space—has a negligible effect on the R-value of the air space. Thus, in Figure 5.14, surface 1 emits a large amount of heat because its emissivity is high (equal to 0.9). When this heat falls on surface 2, most of it is reflected back because surface 2 is lined with aluminum foil with a reflectivity of 95%. Consequently, net heat transfer from surface 1 to surface 2 is reduced, increasing the R-value of the air space (in comparison with that of an unlined air space).

If surface 1 is lined with a low-emissivity material (aluminum foil), the heat emitted by it toward surface 2 is small, Figure 5.15. Therefore, the heat flow across the air space is small, thereby increasing the R-value of the air space. Thus, in either case (Figure 5.14 or 5.15), the end result is the same.

NOTE

R-Value of a Foil-Lined Wall Cavity

The R-value of a vertical (wall) cavity lined with aluminum foil is approximately equal to R-2.5.

NOTE

R-Value of a Horizontal Air Space

The R-value of an unlined horizontal air space, such as an attic, is approximately 2.5. The R-value of a horizontal air space lined with aluminum foil is approximately 6.5.

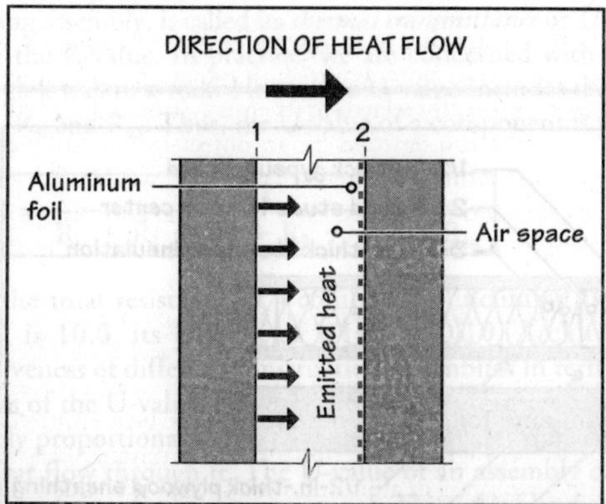

FIGURE 5.14 In this illustration, the wall on the left side of the air space is warmer than the wall on the right side. Therefore, surface 1 of the air space emits heat to surface 2. Because surface 2 is lined with aluminum foil (with an emissivity of 0.05, i.e., a reflectivity of 95%), 95% of the heat falling on surface 2 is reflected back to surface 1. This reduces net heat transfer from surface 1 to surface 2, thus increasing the R-value of the air space.

If there is no foil lining on surface 2, its emissivity is 0.9 (reflectivity of 0.1, or 10%). Thus, only 10% of the heat falling on surface 2 is reflected back, resulting in relatively greater heat transfer from surface 1 to surface 2, and hence a lower R-value of the air space.

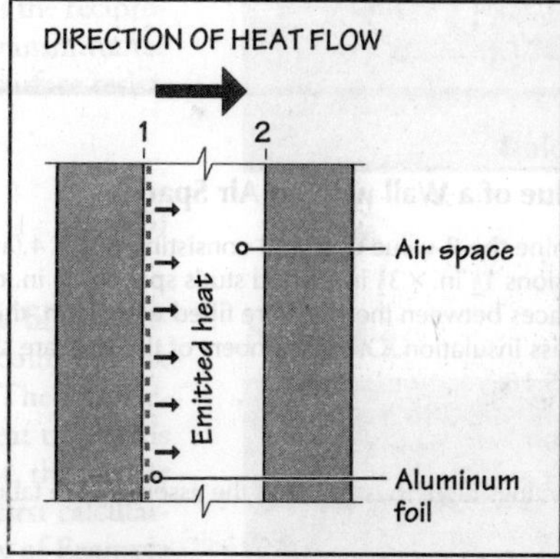

FIGURE 5.15 In this illustration also, the wall on the left side of the air space is warmer than the wall on the right side. Therefore, surface 1 of the air space emits heat to surface 2. However, because surface 1 is lined with aluminum foil (with an emissivity of 0.05, compared with 0.9 of an unlined surface), it emits much less heat to surface 2 (compared with the assembly of Figure 5.14). Therefore, heat transfer from the left-side wall to the right-side wall is inherently low.

In other words, radiation heat transfer through a foil-lined air space is approximately the same regardless of whether the foil is on the warmer side (as shown here) or on the cooler side of the air space, as shown in Figure 5.14.

*Emissivity and absorptivity of a surface are equal if the radiation falling on the object is approximately of the same wavelength as the radiation emitted by it. The wavelength of radiation emitted by an object is a function of its temperature. Because the temperature range of building surfaces is small, the equality of emissivity and absorptivity of a surface is true for all practical purposes.

Question 1—where to insulate—is easily answered. Obviously, the insulation is required only in those components of the building through which heat flow would occur. These include all components of the building envelope: exterior walls, roof, floor above an unconditioned basement or crawl space, a concrete slab supported on the ground, the walls of a conditioned basement, and so on, as shown in Figure 5.20. Interior walls and floors need no insulation unless they are exposed to an unconditioned space.

Question 2—the amount of insulation required—is a function of two factors:

1. Economics and sustainability concerns
2. Climate of the location.

Economics and sustainability: From a purely economic standpoint, the amount of insulation required in the envelope is a function of the price of energy and the cost of providing and installing the insulation. If energy prices are low, the cost of heating and/or cooling a building is low. In that case, a lower level of insulation is economical. On the other hand, if energy prices are high, a greater amount of insulation is justified. This is precisely why the levels of insulation in buildings increased after the abrupt rise in world energy prices in 1973. Another equally important role of insulation is to improve the sustainability of buildings because increased levels of insulation generally conserve energy in heating and cooling of buildings, thereby reducing air pollution, global warming, and the buildings' carbon footprints (see Chapter 10).

Climate of the location: Climate affects the insulation requirement in the following way. In a severe climate, a greater amount of insulation is needed than in a moderate climate. In a temperate climate, little or no insulation may be required. From an economic standpoint again, there generally is an optimum level of envelope insulation for a particular climate. This is defined as that level of insulation at which the amount saved in energy consumption over the entire life of the building exceeds the cost of providing and installing the insulation.

For determining the insulation requirements of a building envelope, two climate parameters of the location are used: (a) exterior air temperature and (b) exterior humidity level. Based on the exterior air temperature, the United States has been divided into eight *temperature zones*, ranging from zone 1 to zone 8. Zone 1 represents a location that is warm

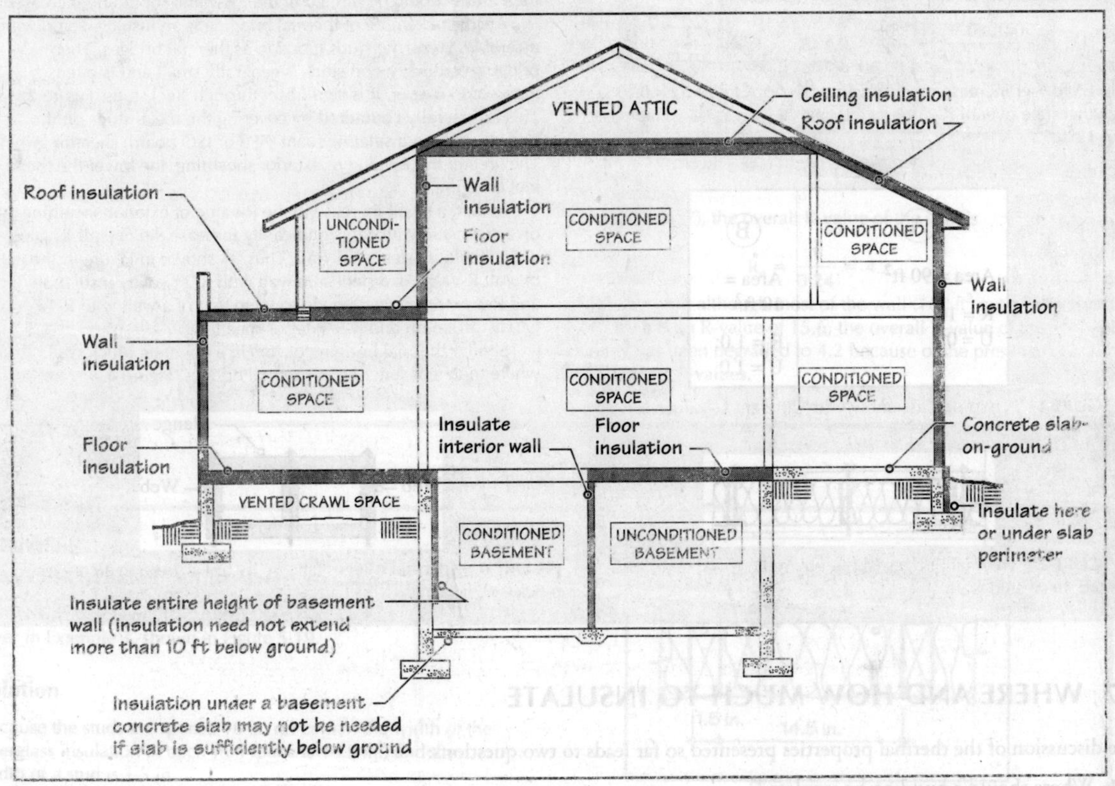

FIGURE 5.20 Location of insulation in a typical building with (thermally) conditioned and unconditioned spaces. Insulated components of the building have been colored.

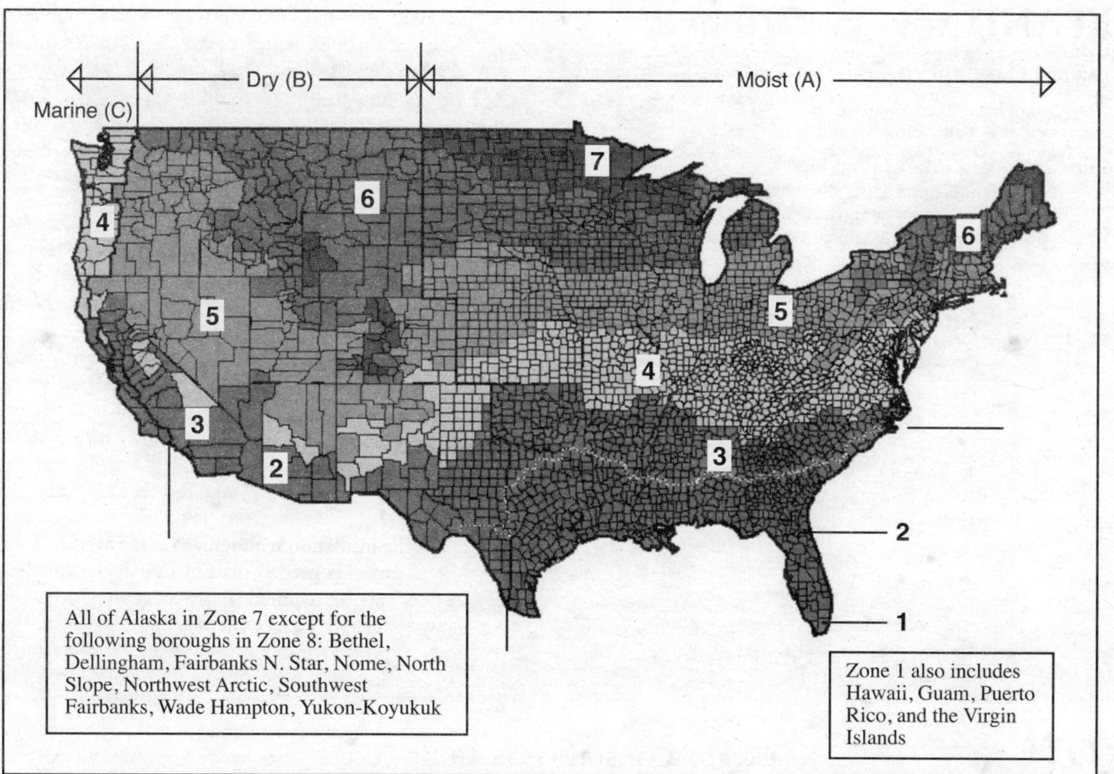

FIGURE 5.21 Climate zone map of the United States developed by the United States Department of Energy (DOE) and adopted by the International Energy Conservation Code (IECC).
Source: U.S. Department of Energy web site: http://resourcecenter.pnl.gov/cocoon/morf/ResourceCenter/dbimages/full/973.jpg

throughout the year (and hence cooling dominated), and zone 8 refers to a location that is subarctic (hence, heating dominated). Zones 2 through 7 fall within the two extremes represented by zones 1 and 8.

The energy use in a building is dependent not only on the exterior air temperature of a location but also on the exterior humidity level. To account for exterior humidity, the United States has been divided into three *humidity divisions*: Division A (moist), Division B (dry), and Division C (marine). Temperature zones and humidity divisions have been used by the United States Department of Energy (DOE) to develop a comprehensive *climate zone map* of the United States, Figure 5.21. The DOE's (hygrothermal) climate zone map is referenced by the codes to prescribe the thermal properties of a building's envelope and other energy-efficiency requirements.

EXPAND YOUR KNOWLEDGE

Climate Classification—The Climate Zones

The climate classification of the United States represented by the climate zone map of Figure 5.21 was developed specifically for prescribing energy efficiency measures in buildings. This classification is not valid for other purposes, such as studying the relationship between climate and agriculture, climate and animal diversity/behavior, and so on. Other climate classifications exist for these purposes.

Only two climate parameters—*outside humidity* and *outside air temperature*—were considered important for the classification, disregarding other climate parameters such as solar radiation, precipitation, wind speed, and so on. As previously stated, according to this classification, a location lies within one of the eight temperature zones (zone 1 to zone 8), and within one of the three humidity regions (A, B, and C).

Humidity Regions: Of the three humidity regions, two regions—moist (region A) and dry (region B)—each cover approximately 50% of the continental United States. The division between the two regions is indicated by a virtual straight line that distinguishes the moist eastern half of the United States from the dry western half. A thin strip along the Pacific Coast has been designated as region C (marine) to recognize its mild summer climate, requiring no or little summer cooling in a typical building.

Temperature Zones: The eight temperature zones of climate classification are based on two fundamental indices related to the outside air temperature: *heating degree days* (HDD) and *cooling degree days* (CDD). Both HDD and CDD have numerical values for a location, which have been found to correlate fairly well with the annual energy consumption of a typical building. (A reader

(Continued)

EXPAND YOUR KNOWLEDGE

Climate Classification—The Climate Zones (*Continued*)

interested in learning more about HDD and CCD is referred to books that deal with these two indices.)

Climate Zones of the United States

With 8 temperature zones and 3 humidity regions, a total of (8 × 3 =) 24 climate zones are possible. However, in the United States only 15 zones exist, as shown in the following table (also in Figure 5.21). The reason for developing 24 zones is to make climate classification applicable globally (outside the boundaries of the United States) and to allow future modifications.

For example, in the International Energy Conservation Code, the thermal properties required of building envelope components are based primarily on temperature zones (disregarding the humidity regions, except the Marine region). Both humidity regions and temperature zones are, however, needed for the specification of some aspects of the HVAC equipment.

Fifteen Climate Zones of the United States

Climate zone	Zone description	Climate zone	Zone description
1A	Very hot-moist	4C	Mixed-marine
2A	Hot-moist	5A	Cool-moist
2B	Hot-dry	5B	Cool-dry
3A	Warm-moist	6A	Cold-moist
3B	Warm-dry	6B	Cold-dry
3C	Warm-marine	7	Very cold*
4A	Mixed-humid	8	Subarctic*
4B	Mixed-dry		

*Subclassification based on humidity regions was considered unnecessary for very cold and subarctic temperature zones (zones 7 and 8, respectively) because exterior humidity impacts energy consumption only if the building is cooled.
Source: R. S. Briggs, R. G. Lucas, and T. Z. Taylor: "Climate Classification for Building Energy Codes and Standards", Technical Paper Final Review Draft, Pacific NW National Laboratory, March 26, 2002.

NOTE

Energy Use in the United States

A total of 98.33 quads of energy were used in the United States in the year 2003 (1 quad = 1 quadrillion Btu = 10^{15} Btu). Of this, 38.8 quads were used in buildings, representing nearly 39% of the total energy use. Industry and transportation accounted for 33% and 28%, respectively.

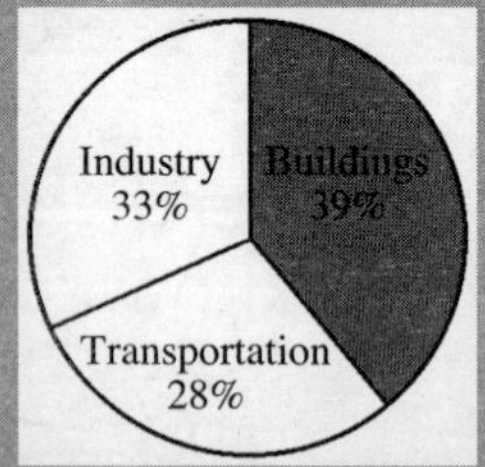

Of the total 38.8 quads of energy used in buildings in the United States in 2003, 21.3 quads were used in the residential sector. In other words, the residential sector accounted for nearly 55% of building energy use.

Source: Summary Tables, *Annual Energy Outlook 2005 with Projections to 2025*. U.S. Department of Energy, DOE/EIA Report No. 0383 (2005).

Energy Conservation Code

An important document regulating energy efficiency in buildings in the United States is the *International Energy Conservation Code* (IECC). This comprehensive document deals with all aspects of energy use in buildings, including the amount of insulation in the opaque (unglazed) portion of the building envelope, thermal properties of fenestration (windows and doors), air infiltration, lighting levels, efficiency of HVAC and other energy-consuming equipment, and so on.

The discussion presented here focuses on residential occupancies (up to three stories in height). (Interestingly, more energy is used in residential buildings than in commercial and industrial buildings combined.) Additionally, our discussion in this chapter is limited to the amount of insulation in opaque (nonglazed) portions of the building envelope, a critical factor affecting energy use in residential buildings. The IECC's requirements for other parts of the building envelope (doors, windows, glazing, and so on) are covered in Chapters 30 and 31.

The corresponding role of the envelope in commercial (or industrial) buildings is relatively less critical because the energy use of such buildings tends to be intermittent and is impacted more by lighting, occupant load, manufacturing processes, and so on. Therefore, the IECC provisions for envelope insulation in nonresidential buildings are less stringent than those for residential buildings. However, because of the architectural complexity of commercial and industrial buildings, energy-performance-modeling software is generally required in the design of most such buildings.

Insulation Requirements for Residential Buildings

The IECC requirements for minimum R-values of opaque (nonglazed) portions of wall and roof assemblies of residential buildings (constructed of wood frame), as a function of climate zones, are given in Figure 5.22. Note that the R-values given refer only to cavity insulation. They do not include the R-value contribution from other components of the assembly such as gypsum board, interior and exterior surface resistances, and so on.

Note also that the code does not require a higher R-value of the opaque portions of walls if the fenestration area (window and door areas) in the walls is large. That is, the prescribed R-value of the opaque portions of walls is independent of the wall fenestration, as long as the fenestration meets the maximum U-value (minimum R-value) and other required thermal properties. Similarly, the required R-value of a residential roof-ceiling assembly is also independent of the area of skylights.

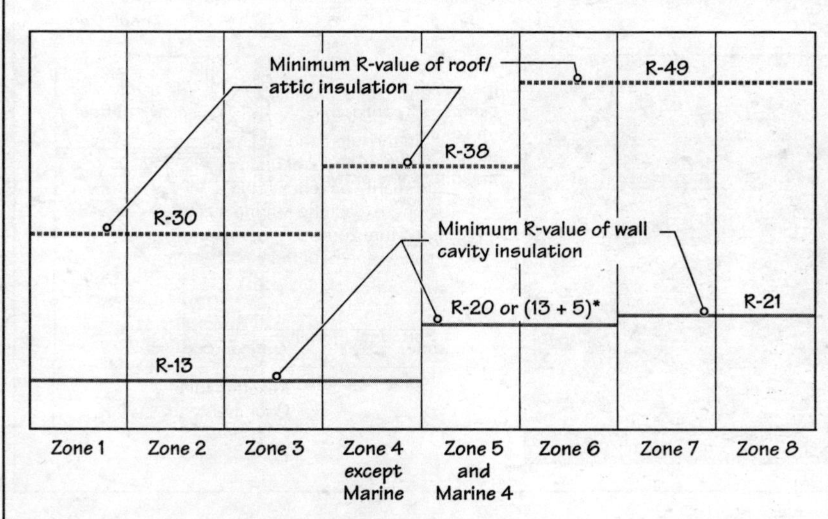

FIGURE 5.22 Minimum insulation requirements for the walls and roof of residential buildings (single-family and multifamily dwellings of up to three stories) constructed of wood light frame. The values given should be regarded as approximate. The exact values should be obtained from the International Energy Conservation Code (IECC), 2009.
*(13 + 5) means R-13 cavity insulation plus R-5 continuous exterior insulating sheathing.

In other words, the code provisions for residential buildings do not consider the effect of thermal bridging caused by fenestration, which generally has a much lower R-value than the opaque parts of walls or roofs. This apparent anomaly was introduced to simplify the code's enforcement. (Simpler though less stringent code provisions lead to better enforcement, yielding greater energy efficiency.) Examples 6 and 7 illustrate the use of Figure 5.22.

Example 6
(Minimum Required R-Value of Insulation in Walls)

Determine the minimum required R-value of insulation in cavities between the studs of a wood-frame, single-family dwelling in Chicago, Illinois. The area of the opaque portions of the walls is 1,500 ft^2. The total glazed area is 300 ft^2.

Solution

Chicago is in climate zone 5. The minimum required R-value of insulation in wall cavities for climate zone 5 is R-20, Figure 5.22. The relative areas of glazing and opaque wall are unimportant, as previously explained. However, the IECC's provisions for the maximum U-value of glazing (and other thermal properties of fenestration) must be met. (We will see in Chapter 31 that the maximum permissible U-value of glazing for climate zone 5 is 0.35.)

Example 7
(Minimum Required R-Value of Insulation in a Roof-Ceiling Assembly)

Determine the required minimum attic (roof-ceiling assembly) insulation for a wood-framed single-family dwelling in Chicago, Illinois.

Solution

From Figure 5.22, the minimum R-value of attic insulation for climate zone 5 is R-38.

INSULATION IN FLOORS OVER AN UNCONDITIONED BASEMENT OR CRAWL SPACE

A floor over an unconditioned basement or crawl space should be insulated. For a wood joist floor, the insulation is placed between joist cavities in full contact with the underside of the subfloor, Figure 5.23. Figure 5.24 gives the minimum R-value of such floor insulation.

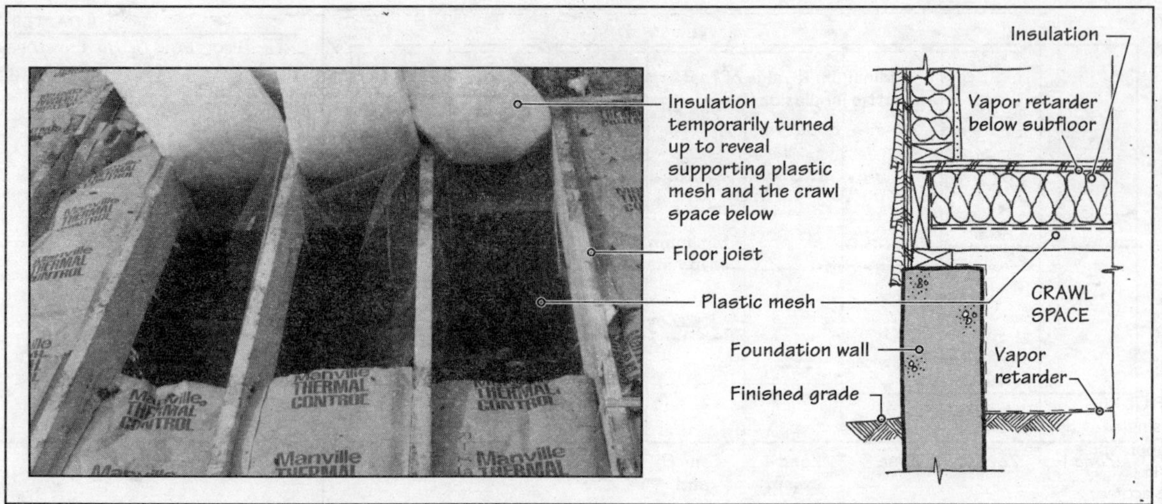

FIGURE 5.23 Insulation in a floor over a vented crawl space.

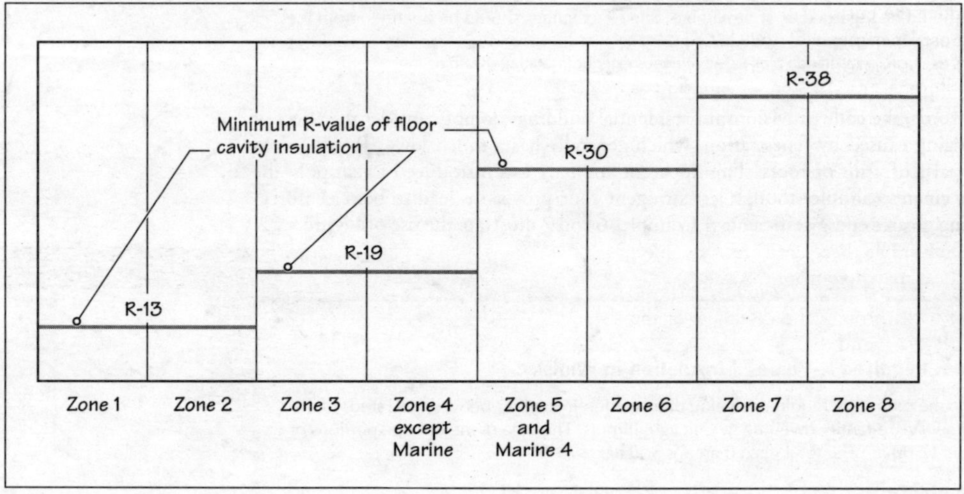

FIGURE 5.24 Minimum insulation requirements for the floors of residential buildings (single-family and multifamily dwellings of up to three stories) constructed of wood light frame. The values given should be regarded as approximate. The exact values should be obtained from the International Energy Conservation Code (IECC), 2009.

Insulation in Walls of a Conditioned Basement or Crawl Space

The walls of a conditioned basement must be insulated throughout the basement wall's height, but the insulation need not extend more than 10 ft below ground. Depending on the construction type, the insulation may be placed toward either the interior or the exterior of (a concrete or concrete masonry) basement wall.

The walls of a conditioned (unventilated) crawl space should be insulated similarly, and the insulation should extend from the floor overhead to the crawl space ground and then at least an additional 2 ft vertically into the ground or horizontally by the same amount. The minimum R-values of basement and crawl space wall insulation are given in Figure 5.25. With basement or crawl space wall insulation provided, there is obviously no need to insulate the floor above.

> **NOTE**
>
> **Frost-Protected Shallow Foundations**
>
> The principles used in reducing the loss of interior heat through a slab-on-ground are also used in designing frost-protected shallow foundations (FSPF) in cold climates. See Chapter 12.

Insulation Under a Concrete Slab-on-Ground

If the floor is a ground-supported concrete slab (a *slab-on-ground,* also called a *slab-on-grade*), it needs to be insulated only along its exposed perimeter. There is no need to place

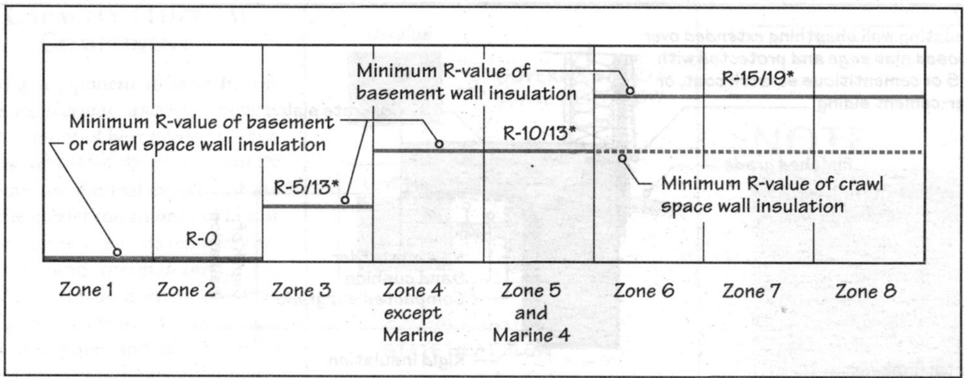

FIGURE 5.25 Minimum insulation requirements for conditioned basement and conditioned crawl space walls of residential buildings (single-family and multifamily dwellings of up to three stories) constructed of wood light frame. The values given should be regarded as approximate. The exact values should be obtained from the International Energy Conservation Code (IECC), 2009.
*In ($\frac{5}{13}$, $\frac{10}{13}$ or $\frac{15}{19}$), the first value refers to R-value of continuous wall insulation and the second value refers to the R-value of cavity insulation. Either of the two is acceptable.

insulation under the entire slab because the heat loss from a slab-on-ground occurs only from the exposed perimeter of the slab. The portion of the slab away from the exposed perimeter does not contribute much to heat loss, as explained shortly.

From Section 5.3, the thermal resistance of an element is directly proportional to its thickness, L. In fact, the quantity L in Figure 5.6 refers to the length of the heat-flow path through the element. The longer the path, the greater its resistance to heat flow.

A simplified visualization of heat loss from a slab-on-ground shows that there are three possible paths along which heat may flow from the slab:

1. Vertical path through the thickness of the slab and into the ground
2. Horizontal path through the slab
3. Through-the-ground paths (shown by the curved lines in Figure 5.26)

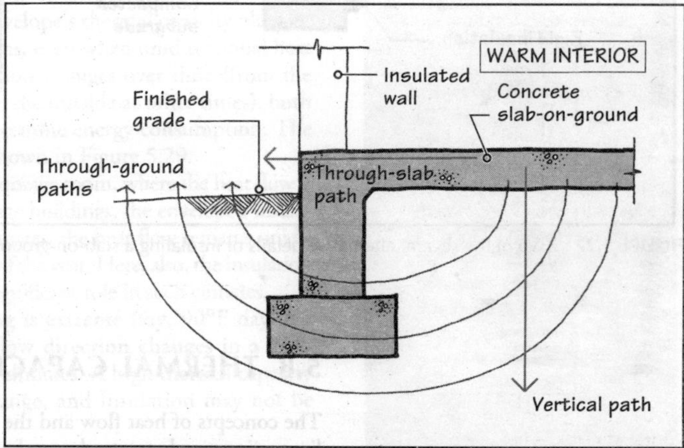

FIGURE 5.26 Various heat flow paths through a slab-on-ground (also called *slab-on-grade*).

The vertical path represents an extremely long path (as long as the diameter of the earth) and, therefore, an extremely large resistance to heat flow, implying that almost no heat will flow in that direction. Heat losses through the other two paths are functions of their respective lengths. Because the region of the slab away from an exposed edge has long path lengths, it does not contribute much to heat loss from the slab. Only the region that is close to the exposed edges affects heat loss. Therefore, a slab-on-ground can be adequately insulated around its exposed perimeter. Very little is gained by insulating under the entire slab.

The standard practice is to insulate under the slab with either horizontal or vertical insulation or a combination of both (vertical insulation is generally preferred). Two commonly used details for insulating a slab-on-ground are shown in Figure 5.27. In both details, the location of the insulation is meant to insulate the horizontal and curved heat flow paths. The minimum R-value of the insulation is a function of the climate zone of the location, as given in Figure 5.27. Note that insulation under a slab-on-ground is not required in climate zones 1 to 3.

NOTE

Properties of Below-Ground Insulation

Insulation exposed to the ground, such as in basement and crawl-space walls and under slabs-on-ground, must be rigid and have a closed-cell structure. As shown in Figure 5.8, XPS and ISO meet both requirements. Their closed-cell structure helps to resist water absorption, which is necessary so that they retain their insulating efficacy. (Remember, water is a poor thermal insulator.)

Another required property of such insulation is termite resistance. XPS and ISO, being soft, are subjected to termite degradation because termites bore through them (but do not consume the foam as food). Boron-treated XPS and ISO are available and are claimed by their manufacturers to be termite resistant. Another termite-resistant insulation is cellular glass (see Principles in Practice at the end of this chapter). Cellular glass derives its termite-resistant property from its high compressive strength so that termites are unable to bore through it.

5.9 THE MOST EFFECTIVE FACE OF THE ENVELOPE FOR INSULATION

The placement of insulation on the outside of the assembly (compared with its location in the middle of the envelope section or toward the interior of the building) is more effective for the following reasons:

- It reduces energy consumption because it eliminates (or reduces) thermal bridging in the envelope.
- It moderates the temperature of structural components, reducing their expansion and contraction, thereby increasing their life span.

Only when a building is intermittently heated or cooled, such as a church hall, the interior placement of insulation may be considered. The interior placement of insulation allows faster heating or cooling of interior air because it reduces energy loss to heat or cool the structural components of the building.

PRACTICE QUIZ

Each question has only one correct answer. Select the choice that best answers the question.

27. The U.S. Department of Energy's climate zones are based on
a. the exterior air temperature.
b. the exterior humidity.
c. precipitation.
d. (a) and (b).
e. (b) and (c).

28. The coastal region of the western United States lies in
a. climate zone 1(A).
b. climate zone 2(A).
c. climate zone 3(A).
d. climate zone 4(A).
e. none of the above.

29. The U-value and R-value of an assembly are related to each other by which relationship?
a. $U = R$
b. $U + R = 1$
c. $(U)(R) = 1.0$
d. $U^2 + R^2 = 1.0$
e. None of the above

30. A building assembly with a higher U-value is thermally more insulating than one with a lower U-value.
a. True
b. False

31. Which of the following metal stud wall assemblies gives a higher overall R-value? In both assemblies, the total amount of insulation is R-15.
a. Assembly A has an exterior insulating sheathing of R-5 and within-stud insulation of R-10.
b. Assembly B has an exterior insulating sheathing of R-3 and within-stud insulation of R-12.

32. In a cold climate, a concrete slab-on-ground must be insulated under its entire area.
a. True
b. False

33. The walls of a heated basement must be insulated over the entire depth of the wall.
a. True
b. False

34. In which of the following buildings will heat flow occur unidirectionally?
a. Continuously heated or cooled buildings
b. Intermittently heated buildings
c. Intermittently cooled buildings
d. All of the above
e. None of the above

35. For energy conservation, materials with a high thermal capacity are effective in climates in which the heat flow through the building envelope is
a. bidirectional within a 24-h period during a large part of the year.
b. unidirectional within a 24-h period during a large part of the year.
c. absent during a large part of the year.

36. Insulation is most effective if it is placed
a. on the outside face of the assembly.
b. on the inside face of the assembly.
c. in the center of the assembly.

PRINCIPLES IN PRACTICE

Insulating Materials

A large variety of insulating materials are currently used in buildings. They can be classified in several ways based on their configuration, physical structure, or combustibility. If classified on the basis of configuration, insulating materials are classified as either *rigid insulation* or *flexible insulation.* Rigid insulation is commonly used in flat roofs and may be in the form of rigid boards or poured-in-place lightweight concrete. Flexible insulation is commonly used in stud walls and attic spaces and may be in the form of blankets, batts, or loose-fill insulation. Graphically, we distinguish between the rigid and flexible insulations, as shown in Figure 1.

A classification that is more suitable for describing various insulations is based on the insulation's physical structure. According to this classification, insulating materials may be divided into the following three categories:

- Fibrous insulation
- Granular insulation
- Foamed insulation

Commonly used materials in each category are shown in Table 1.

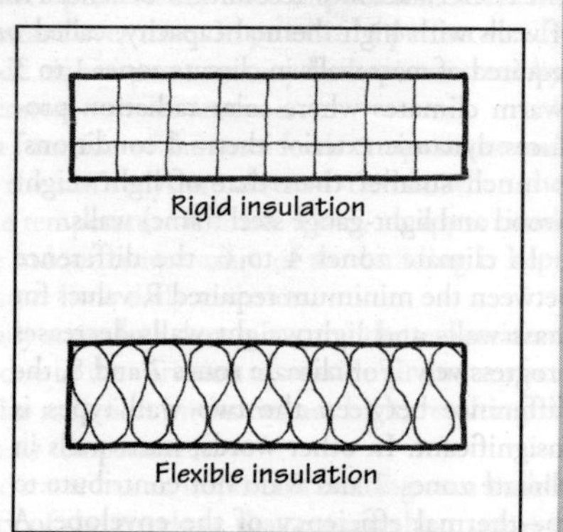

FIGURE 1 Graphical notations for rigid and flexible insulations for use in cross sections of building assemblies.

PRINCIPLES IN PRACTICE

FIBROUS INSULATION

Fibrous insulating materials derive their high thermal resistivity from the air contained between the fibers. The fibers can be of mineral base or cellulosic base. Three types of mineral fibers are generally used: (a) glass fibers, called *fiberglass,* (b) fibers obtained from natural rock, called *rock wool,* and (c) fibers obtained from slag, called *slag wool.*

Fiberglass is made by fiberizing molten glass, and rock wool and slag wool are made by fiberizing molten rock and slag, respectively. The fibers are then sprayed with a binder and fabricated into the finished product. Fiberglass, rock wool, and slag wool insulations are fire resistant, moisture resistant, and vermin resistant.

Cellulosic fibers are derived from recycled newspaper, wood, and sugarcane. They are usually treated for fire resistance, minimization of smoke contribution, fungal growth, and decomposition by moisture.

Fiberglass is the most commonly used fibrous insulation and is available in the form of

- Batts
- Blankets
- Semirigid boards

NOTE

Rock Wool and Slag Wool

Rock wool and slag wool are together referred to as *mineral wool.* Although both mineral wool and fiberglass are noncombustible, mineral wool is more fire resistant than fiberglass because it has a higher melting-point temperature. Slag is a waste product obtained from the manufacture of iron, see Chapter 18.

Batts and blankets are normally used as wall insulation. Semirigid boards are also used in walls, but in locations where greater rigidity is needed, such as in spandrel areas of glass curtain walls. Batts and blankets are similar in appearance, composition, and density. The only difference between them is that blankets are available in rolls, whereas batts are precut from rolls into standard dimensions.

Apart from other sizes, both batts and blankets are manufactured in standard widths to fit 12-in., 16-in., and 24-in. center-to-center spacing between wood or metal studs. They are available either unfaced, Figure 2(a), or faced, Figure 2(b). Facing usually consists of asphalt-impregnated paper, called *building paper,* which functions as a vapor retarder (see Section 6.6). Building paper–faced insulations are also available with projecting flanges for stapling to wood studs, Figure 3. Flanges may be stapled either to the faces of studs (face stapled) or to the sides of studs (inset stapled, as shown in Figure 2(b)).

Unfaced insulation is particularly useful as additional insulation in attic spaces where faced insulation already exists or in ceiling spaces where rigid ceiling tiles provide support to the insulation. Unfaced batt insulation is also available to pressure fit within stud spaces, which may be used when a cover material has already been placed on the outside face of the assembly. If a vapor retarder is required in such an assembly, it must be provided separately, Figure 4.

Glass wool, mineral wool, or cellulosic fiber insulation may also be used as loose-fill insulation. These are usually blown into air spaces by pneumatic blowers. Blown-in insulation, Figure 5, is popular in retrofit applications (older buildings) where either no insulation or inadequate insulation existed previously. However, the potential for condensation within the insulation must be examined in a retrofit application, because the addition of insulation may create condensation where none existed earlier. Refer to Principles in Practice at the end of Chapter 6.

TABLE 1 COMMONLY USED INSULATING MATERIALS

Physical structure	Configuration	Insulating material
Fibrous insulation	Batts, blankets, and semirigid boards	Fiberglass Rock wool Slag wool
	Loose-fill (blown in)	Cellulosic fibers Fiberglass Rock wool
Granular insulation	Rigid boards	Perlite board Expanded polystyrene (EPS)
	Insulating concrete	Perlite concrete Vermiculite concrete
Foamed insulation	Rigid boards	Plastic foams Extruded polystyrene (XPS) Polyisocyanurate (ISO) Cellular glass
	Insulating concrete	Foamed concrete
	Foamed-in-place insulation	Foamed-in-place polyurethane

(Continued)

PRINCIPLES IN PRACTICE

Insulating Materials (*Continued*)

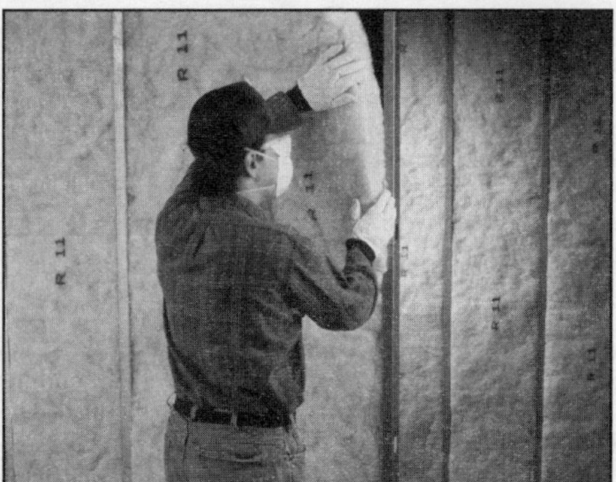

FIGURE 2(a) Unfaced fiberglass insulation installed between studs. The insulation will be covered with a vapor retarder (see Figure 4) before gypsum board is installed on the studs. (Photo courtesy of CertainTeed Corporation)

FIGURE 2(b) Kraft paper faced fiberglass insulation installed between studs. (Photo courtesy of CertainTeed Corporation)

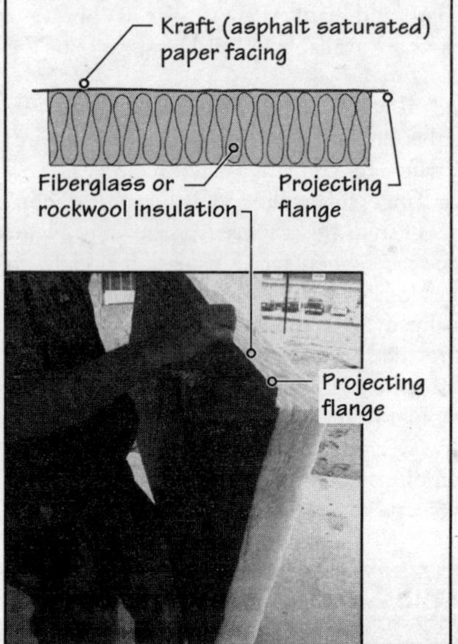

FIGURE 3 Projecting paper flanges in kraft paper–faced fibrous insulation.

FIGURE 4 A clear polyethylene sheet vapor retarder being installed over unfaced fiberglass insulation between wood studs.

GRANULAR INSULATION

In materials with a granular structure, the air voids are contained inside tiny hollow beads or granules. Three types of granules are in common use

- Expanded perlite granules
- Expanded vermiculite granules
- Expanded polystyrene (EPS) granules

Perlite is a glassy volcanic rock that is expanded into granules by heat treatment. The expansion of perlite takes place due to the 2% to 6% moisture present in the crude perlite rock. When heated suddenly to a temperature of nearly 1,600°F, the water in crushed perlite particles is converted to steam. The pressure of the steam expands the perlite particles, creating small air-filled granules, Figure 6. The process of expanding perlite is, in fact, similar to that of making popcorn.

Expanded vermiculite granules are made from mica. The process of making vermiculite granules is similar to that of making perlite granules.

FIGURE 5 Blown-in insulation in an attic.
(Photos courtesy of CertainTeed Corporation)

Expanded polystyrene granules (also called *polystyrene beads*), popular as packing material, are used to make EPS insulation boards, as described in the following paragraph. Of the three granular materials, polystyrene granules are combustible, whereas perlite and vermiculite granules are noncombustible.

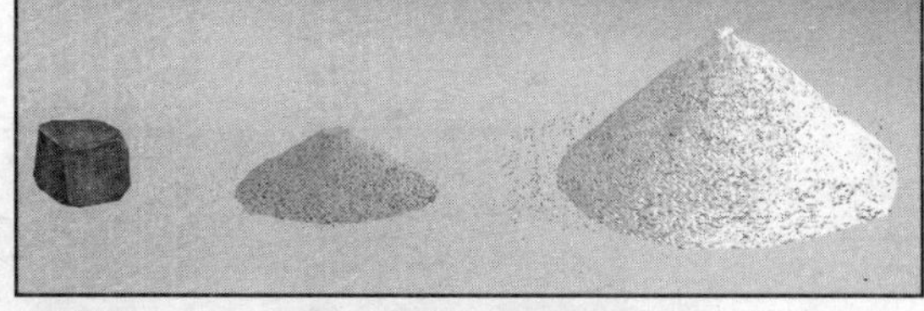

FIGURE 6 Three stages of perlite production show the great increase in volume that takes place on expansion by heating of the perlite rock. Although this illustration refers to perlite rock, it applies to all rocks that contain some water and are commonly used in expanded state (see Figure 21.14).

Perlite Board and Expanded Polystyrene (EPS) Board

Expanded perlite is also available as a rigid board. Called *perlite board,* it is made with a combination of expanded perlite granules and mineral and cellulosic fibers. The fibers help increase the strength of the board. Asphalt is also added to the combination to improve water repellency and rot resistance. Perlite board is widely used as an insulation over low-slope (flat) roof decks. It is noncombustible and can withstand the high temperature of hot asphalt.

Expanded polystyrene (EPS) boards—also referred to as *bead boards*—are made by packing polystyrene granules (beads), nearly $\frac{1}{8}$ in. diameter, in a mold and fusing them together under heat and pressure. The molded material is simply a large rectangular block, referred to as a *bun,* which is sliced into boards of required thickness, Figure 7. An EPS board is essentially of the same material as that used in picnic coolers.

Although the compressive strength of an EPS board is almost the same as that of a perlite board, it is not as widely used in low-slope (flat) roof insulation because of its combustibility and its inability to withstand high temperatures. It is, however, commonly used in exterior insulation and finish systems (EIFS) wall systems (see Chapter 29).

EPS and Insulating Concrete Forms (ICF)

An interesting use of EPS is in hollow blocks, which are used as formwork for concrete. Referred to as *insulating concrete forms* (ICF), ICF blocks are stacked and reinforced with steel bars, and their cells are filled with concrete to form an insulated concrete wall, Figure 8.

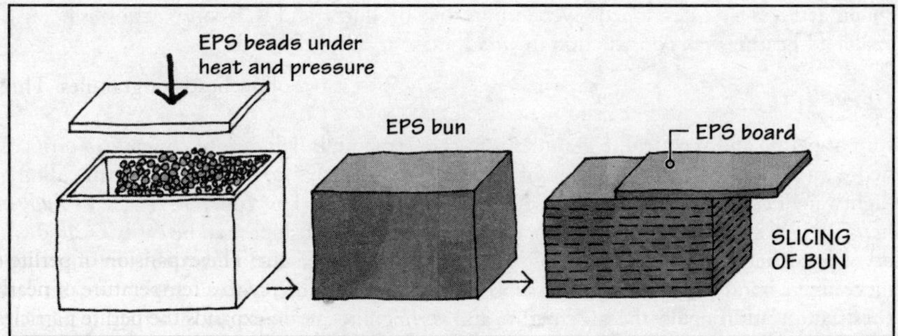

FIGURE 7 Manufacturing process of EPS boards.

(Continued)

PRINCIPLES IN PRACTICE

Insulating Materials (*Continued*)

Fire retardant treated EPS panels with interior plastic ties constitute an ICF unit. The units are used as formwork for an insulated concrete wall, as shown in (b). Plastic ties, which connect the two opposite EPS panels of a unit, are permanently embedded in the wall's concrete.

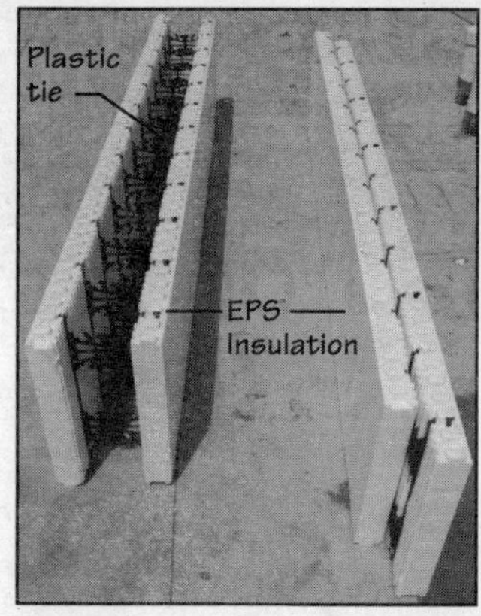

(a) ICF units in open and collapsed profiles. The interior ties are hinged so that the units can be collapsed for shipment. They are opened when laid in the wall.

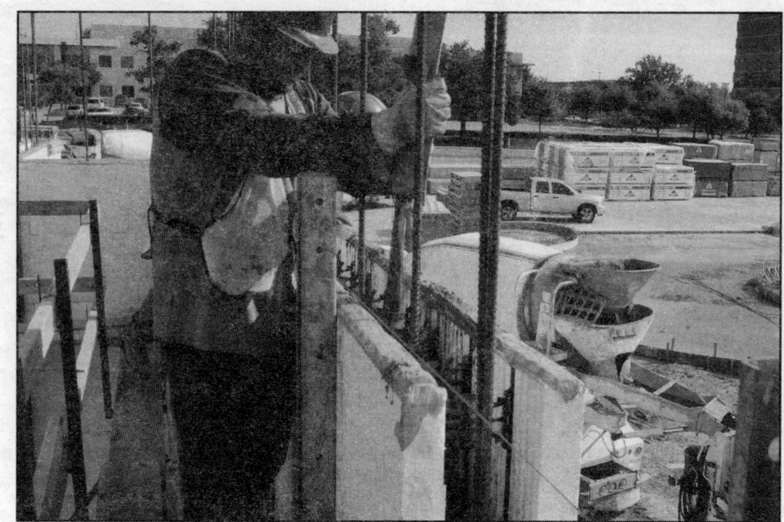

(b) Concrete being placed in an ICF wall

FIGURE 8 Insulating concrete form (ICF) walls are used as exterior walls for residential and commercial buildings. The wall's high R-value makes ICF construction an attractive alternative where energy conservation is critical.

When EPS is used as permanent concrete forms, ICF construction eliminates the use of conventional wood or steel formwork. ICF manufacturers have developed several innovative products, and ICF construction is being used as an alternative to traditional load-bearing wall construction in wood, masonry, and concrete.

Insulating Concrete

Another application of perlite and vermiculite granules is in (nonstructural) lightweight concrete. Vermiculite or perlite granules, combined with portland cement in a ratio of 1 (portland cement): 4 to 8 (perlite or vermiculite) granules by weight, makes a lightweight concrete, which has good insulating properties. This concrete is called *insulating concrete.*

Insulating concrete is well suited for insulation on a flat roof because roof slopes can be easily created with wet concrete. The other method of providing slope on a flat roof is to use tapered board insulation, which is far more labor intensive. Because insulating concrete bonds well to most roof substrates, such as reinforced concrete slab or steel deck, it provides high-wind-uplift resistance. Additionally, because perlite and vermiculite are inorganic materials, insulating concrete is noncombustible and provides a high degree of fire resistance to the roof.

The R-value of insulating concrete is lower than those of other insulations (e.g., plastic foams). However, the advantages of insulating concrete, such as its fire resistance, easy sloping to drains, and good bonding to substrate, can be combined

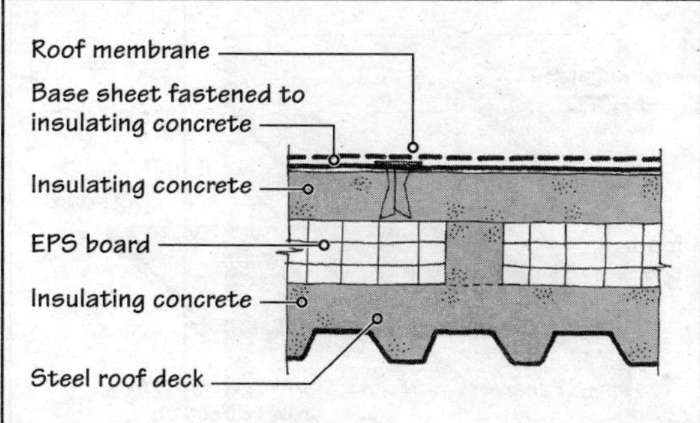

FIGURE 9 Cross section through an insulating concrete roof deck with sandwiched EPS boards.

FIGURE 10 Insulating concrete being poured over a roof deck.

with the higher R-value of EPS boards to create a hybrid assembly. In this assembly, insulating concrete is poured below and above the EPS boards so that the EPS boards are sandwiched between the two layers of concrete, Figures 9 and 10.

To structurally integrate the two layers of concrete and the EPS boards into a monolithic whole, the EPS boards are provided with holes, Figure 11. Insulating concrete is particularly well suited as insulation over steel roof decks, see Chapter 33.

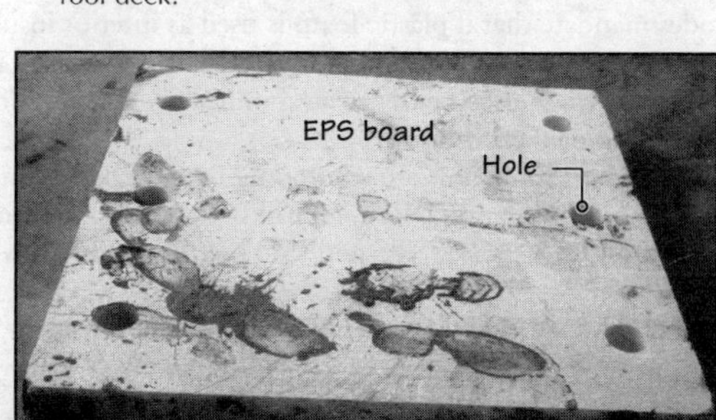

FIGURE 11 Because EPS boards are sandwiched between two layers of concrete, the holes in the EPS boards structurally integrate the layers.

FOAMED INSULATION

The most commonly used insulating foams are synthetic (plastic) foams, such as *extruded polystyrene* and *polyisocyanurate.* Both are used as rigid boards, and their physical structure is cellular with closed cell cavities. The cavities contain HCFC gas.

The boards are made by the extrusion process. This process consists of forcing a semiliquid material through an aperture (die). This is precisely the same concept as forcing toothpaste out of a tube, the mouth of the tube being the die. Polystyrene board obtained from the extrusion process is called *extruded polystyrene board* (XPS) to distinguish it from the molded EPS (beaded) board.

The extrusion process is complex and requires a large capital outlay in manufacturing. By contrast, the molding process is simpler and less expensive. That is why XPS boards are costlier than EPS boards but provide a higher ρ-value and higher compressive strength.

XPS boards are made by mixing liquid polystyrene with liquid HCFC (as the blowing agent) and forcing the mixture through a die, Figure 12. XPS was first manufactured by the Dow Chemical Company in the United States during the 1940s under the trade name Styrofoam. However, today, several manufacturers make XPS boards and distinguish their products from each other by using different colors.

The form of HCFC gas used as a blowing agent in the production of polyisocyanurate boards (commonly referred to as *ISO boards*) vaporizes at room temperature, so the liquid mixture of polymer and blowing agent is spread between the top and bottom facers to give ISO boards their required shapes. ISO boards are, therefore, manufactured with facers on both sides. XPS boards, on the other hand, do not require facers. The process of manufacturing ISO boards is shown in Figure 13.

The facers used on ISO boards retard the migration of HCFC gas out of the foam. There-

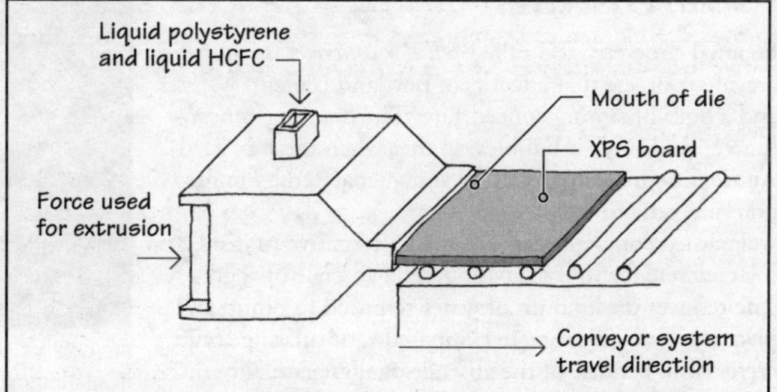

FIGURE 12 Manufacturing process of XPS boards.

(Continued)

PRINCIPLES IN PRACTICE

Insulating Materials (*Continued*)

fore, ISO boards have higher thermal resistivity than XPS boards. Facers consist of aluminum foil, glass fiber–reinforced polyester, plaster boards, wood fiber boards, and so on.

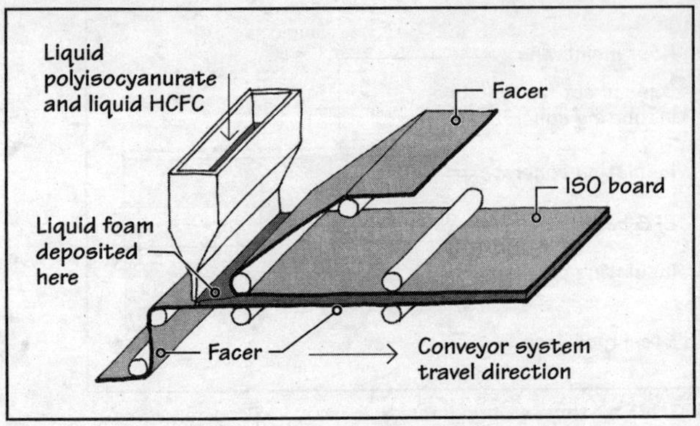

FIGURE 13 Manufacturing process of ISO boards.

USES AND LIMITATIONS OF PLASTIC FOAM BOARDS

The closed cellular structure of XPS and ISO boards implies that they have high resistance to water and water vapor penetration. An EPS board, being a beaded product, is relatively more permeable to water and water vapor than an XPS or ISO board (see Figure 5.8).

Plastic foams are combustible and require protection from flames and high temperatures. They should not be used toward the interior of a wall. Building codes mandate that if plastic foam is used as interior insulation, it must be covered with a minimum of $\frac{1}{2}$-in.-thick gypsum board or an equivalent thermal barrier. Plastic foams are resistant to fungal growth and chemical decomposition but can be destroyed when used below ground in heavily termite-infested soils. Polystyrene is sensitive to daylight and will deteriorate after prolonged exposure.

XPS and ISO boards are commonly used as insulation on flat or low-slope roof decks. ISO board is relatively more common as roof insulation because it can better withstand the high temperature of hot asphalt than XPS and is also more fire resistant. Other uses of XPS and ISO boards are in wall sheathing, basement walls, and insulation under slabs-on-grade.

CELLULAR (FOAMED) GLASS BOARDS

Another rigid-board foam product is cellular glass, which is made by expanding molten glass into a closed-cell structure and cooling it to form boards. Cellular glass boards have a much higher compressive strength than XPS or ISO boards. They are also impermeable to water vapor and water and are suitable as insulation in roofs that carry heavy loads, such as roof plazas and rooftop parking areas, and as perimeter insulation under concrete slabs-on-grade under heavy loads.

FOAMED-IN-PLACE INSULATION

Foamed-in-place insulation is formed by spraying a liquid chemical (or a mixture of two liquid chemicals) in position. The liquid foams, when sprayed, expand to nearly 30 times the original volume of the liquid. After expansion, the foam solidifies into a closed-cell structure in a few minutes. The advantage of foamed-in-place insulation is that it can be provided in preexisting cavities of odd shapes where other types of insulation would be less desirable or impossible to use.

A commonly used foamed-in-place insulation, called *polyicynene* by its manufacturer, is used for insulating stud cavities, Figure 14. The open-cell structure of the foam traps air and expands as it is sprayed into voids. The manufacturer claims that it fills the voids, providing a good air barrier.

FIGURE 14 Spraying of polyicynene insulation in stud cavities. The insulation expands and cures (hardens) quickly when it is shaved flush with the studs.
(Photo courtesy of Icynene, Inc.)

FOAMED CONCRETE

Foamed concrete (also called *cellular concrete*) is a nonaggregate concrete that consists of portland cement, water, and a liquid foaming concentrate. The foaming concentrate creates tiny air bubbles, so that when the mix hardens, it contains a matrix of air voids separated by pure portland cement walls.

Foamed concrete is an excellent alternative to insulating concrete for roof insulation. The absence of aggregate reduces the amount of water required to pump and place the foamed concrete compared to insulating concrete. Also, because of the absence of aggregate, the ρ-value of foamed concrete is slightly higher than that of insulating concrete.

Each question has only one correct answer. Select the choice that best answers the question.

37. Which of the following plastic foam insulations contains air within its voids?
- **a.** Extruded polystyrene (XPS) board
- **b.** Expanded polystyrene (EPS) board (bead board)
- **c.** Polyisocyanurate board (ISO board)

38. Which of the following insulating materials is available in the form of rigid boards?
- **a.** Fiberglass
- **b.** Rock wool
- **c.** Insulating concrete
- **d.** Extruded polystyrene
- **e.** Vermiculite

39. Which of the following insulating materials is available in the form of batts?
- **a.** Fiberglass
- **b.** Expanded perlite
- **c.** Insulating concrete
- **d.** Extruded polystyrene
- **e.** Vermiculite

40. Which of the following insulations is combustible?
- **a.** Fiberglass
- **b.** Insulating concrete
- **c.** Perlite
- **d.** Extruded polystyrene
- **e.** Vermiculite

41. In which of the following locations is blown-in insulation most commonly used?
- **a.** Stud wall cavities
- **b.** Masonry wall cavities
- **c.** Top of a flat roof
- **d.** Attic spaces

42. Lightweight insulating concrete is generally used as insulation in
- **a.** walls.
- **b.** steep roofs.
- **c.** foundations.
- **d.** all of the above.
- **e.** none of the above.

REVIEW QUESTIONS

1. Using the appropriate table, determine the individual R-values of the following materials:
   a. 4-in. nominal (actual $3\frac{5}{8}$-in.) thickness brick wall
   b. $\frac{5}{8}$-in.-thick gypsum board
   c. 2-in.-thick extruded polystyrene board
2. Draw a plan of the following wall assembly and then determine its R-value. Starting from the outside face of the wall, the assembly consists of the following layers:
   4-in. nominal ($3\frac{5}{8}$-in. actual) brick veneer
   2-in. air space (cavity)
   0.75-in.-thick extruded polystyrene sheathing
   2×4 studs with 3.5-in.-thick fiberglass insulation
   0.5-in.-thick gypsum wallboard

   (Note: In calculating the total R-value, ignore the presence of wood studs and assume that the entire layer consists of fiberglass insulation.)
3. What is the U-value of the assembly in Question 2?
4. What is the aged R-value? To which materials does this concept apply and why?
5. Determine the overall U-value of a roof whose opaque portion, with an area of 2,500 ft^2, has an R-value of 35. There is a skylight in the roof, which has an area of 100 ft^2, and its R-value is 2.0.
6. Using the tables given in the text, determine the minimum required R-value of cavity insulation for the opaque portions of the walls of a single-family home in
   a. climate zone 4(C).
   b. climate zone 7.
7. If the walls of the building in Question 6 consist of 15% glazed area, determine the overall R-value of the walls. Assume that (a) the components of the opaque portions of the wall other than cavity insulation provide an R-value of 2.0 and (b) the R-value of the glazing is 2.5.
8. Determine the minimum required R-values of the roof-ceiling assemblies in Question 6.
9. Using sketches and notes, explain why it is unnecessary to insulate under an entire concrete slab-on-ground.
10. What are high-thermal-capacity materials? In which climates are these materials effective?
11. Explain why exterior placement of insulation in the building envelope is preferable. In which situations would you prefer inside placement of insulation?

CHAPTER 6

Properties of the Envelope–II (Air and Water Vapor Control)

CHAPTER OUTLINE

Thermal insulation, discussed in Chapter 5, controls only the flow of energy through a building envelope, and energy flow is only one of the several environmental factors that must be controlled in the design of a building envelope. The other important environmental factors, intimately related to energy, are (a) air, (b) water vapor, and (c) liquid water. In this chapter, we will focus on air and water vapor control. Issues related to water infiltration control are covered in Chapter 27, immediately preceding the chapters (Chapters 28 and 29) that deal with construction of exterior wall cladding.

AIR

Outside air often contains dust, pollen, ozone, and other pollutants. Inside air is typically subjected to mechanical conditioning (heated/cooled, humidified/dehumidified, and filtered). The unregulated flow of outside air to the interior not only introduces pollutants but also increases energy consumption because the outside air entering the envelope uncontrolled is not subjected to such conditioning.

The entry and exit of such air (called air *infiltration* and *exfiltration*, respectively) take place through cracks, gaps, voids, unsealed joints, and at the interface between envelope assemblies. It is important to note that the amount of air infiltration must equal exfiltration. This fact ensures that the interior spaces are neither pressurized nor depressurized. Therefore, collectively, we refer to air infiltration and exfiltration as *air leakage.* Air leakage is essentially unwanted ventilation, which must be reduced to the extent possible.

WATER VAPOR

Water and water vapor (collectively called *moisture*) are essentially the same material but in different states—water is a liquid, and water vapor is a gas. Water vapor generally occurs in combination with air. The leakage of air through the envelope is, therefore, always accompanied by the leakage of water vapor.

The movement of water vapor through the envelope assembly can cause significant problems if it converts to (liquid) water through condensation. Water wets the insulation and reduces its R-value, corrodes metals, promotes insect infestation, creates dampness giving rise to mold and fungi, peels off paints, and reduces the strength of cellulose- and gypsum-based materials. Controlling water vapor flow (in order to prevent its condensation within the envelope) is, therefore, essential.

LEAKAGE AND DIFFUSION

Before proceeding, it is important to understand that leakage is one of the two means by which a gas can migrate into and out of its container. The term *leakage* has already been defined as the migration of a gas (such as air and water vapor) through gaps, cracks, and voids in a container (building envelope). The migration of air and water vapor can also occur even when the envelope is fully sealed (without any gaps and voids, so that there is no leakage). Such migration is called *diffusion.* Although diffusion and leakage lead to the same end result, their mechanisms are different. Diffusion refers to the movement of a gas through the body of the container (envelope), whereas leakage refers to its movement through (unintended) holes in the envelope—for example, at joints between components.

Difference between leakage and diffusion of gas (or air) through a balloon.

Consider a rubber or latex balloon inflated with air or some other gas. Sooner or later, the air (or gas) will migrate out of the balloon. If the balloon wall is thicker, it may take a few days longer for the balloon to deflate than if it is thinner, but deflation will eventually occur. It will happen even if the mouth of the balloon (the only leakage site) is fully sealed. Obviously, the migration of air or gas in this situation occurs through the body of the balloon, not through any hole. This migration is called *diffusion.*

The rate at which diffusion occurs through a building component is called its *permeability.* Permeability is a property of (a) the component's material and (b) the pressure difference (air pressure or vapor pressure difference) across the component. The greater the pressure difference, the greater the permeability of the component. To compare the permeability of two materials or components, we must test them under the same pressure difference. As we will see later in this chapter, building materials and components have different permeability values for air and vapor, and some materials, such as glass and metals, are impermeable to both air and vapor.

To reduce the passage of air or vapor through the envelope, both leakage and diffusion must be controlled. Sealing the cracks, holes, and gaps reduces leakage but not diffusion. Diffusion reduction requires additional strategies. In most contemporary buildings, however, leakage accounts for most air and vapor migration through the envelope. Sealing the envelope is, therefore, fundamental to controlling the passage of air and vapor.

6.1 AIR LEAKAGE CONTROL

Two factors affect air leakage in a building:

- Leakage area
- Pressure difference between inside and outside air caused by (a) wind and (b) temperature.

Leakage Area: The leakage area is a measure of the relative tightness of the envelope. It is the area of cracks, tears, holes, and openings in the envelope and is a function of the type of construction, design, and workmanship. The leakage sites in the envelope usually occur at the joints between various components, such as exterior doors, windows, skylights, fireplaces, electrical outlets, plumbing, and duct penetrations.

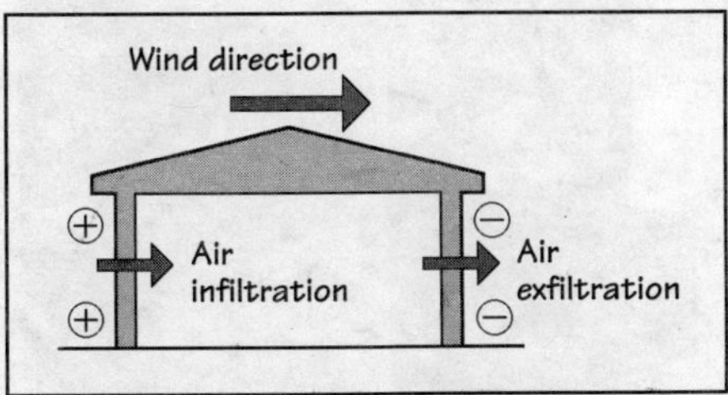

FIGURE 6.1 Air infiltration and exfiltration through a building caused by the pressure difference created by wind. Because wind creates positive (+) pressure on the windward face of a building and suction, that is, negative (–) pressure on the other faces, air infiltration and exfiltration occur through the building as shown.

Air Pressure Difference Due to Wind: We observed in Section 3.4 that the windward face of a building is under positive pressure and the other faces are under suction (negative pressure). Therefore, air infiltrates through the windward face and exfiltrates through the nonwindward faces of the building, Figure 6.1. This process is accelerated with increasing wind speed.

Air Pressure Difference Due to Temperature: Another factor that affects the inside-outside air pressure difference is the difference between the inside and outside air temperatures. Because the density of air increases with decreasing temperature, cool (more-dense) air exerts higher pressure than warm (less-dense) air. Thus, during the cooling season, cool inside air has higher pressure than warm outside air. This leads to continuous exfiltration of cool air and infiltration of warm outside air through the envelope, Figure 6.2(a). During the heating season, the process is reversed, Figure 6.2(b).

Because the inside-outside pressure difference is primarily a function of the outside climate (wind speed and air temperature), there is little that can be done to control it. Thus, the only means of reducing air leakage in a building is to reduce the leakage area in the building envelope. This requires sealing all joints between building components (joints between exterior sheathing panels, between door/window frame and wall, between door frame and door, etc.). Sealants between fixed components are typically nonhardening synthetic compounds (Chapter 9). *Weather stripping,* made from a resilient, compressible material, is used to seal gaps around operable components.

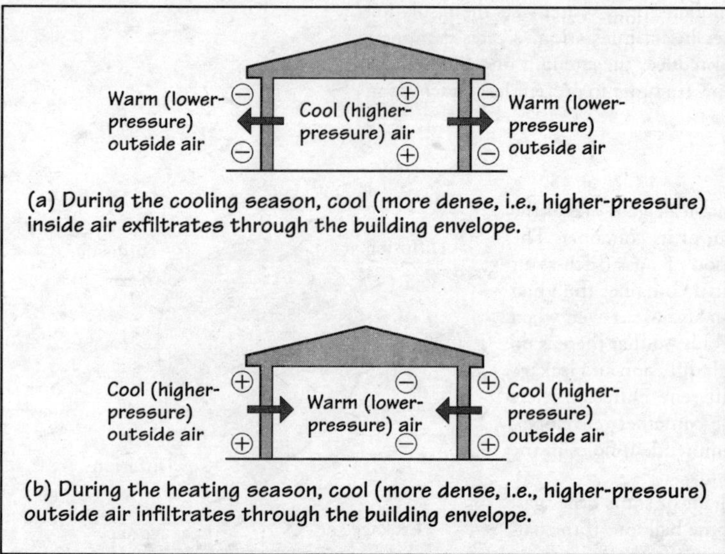
Warm (lower-pressure) outside air
Cool (higher-pressure) air
Warm (lower-pressure) outside air
(a) During the cooling season, cool (more dense, i.e., higher-pressure) inside air exfiltrates through the building envelope.
Cool (higher-pressure) outside air
Warm (lower-pressure) air
Cool (higher-pressure) outside air
(b) During the heating season, cool (more dense, i.e., higher-pressure) outside air infiltrates through the building envelope.

FIGURE 6.2 Air leakage (infiltration and exfiltration) through a building caused by the temperature difference between the inside and outside air. The plus (+) sign indicates relatively greater pressure, and the negative (–) sign indicates lower pressure.

6.2 AIR DIFFUSION CONTROL

In addition to controlling air leakage through the envelope (by using sealants, gaskets, and weatherstripping), we need to reduce air diffusion through the envelope's material(s). This is achieved with the help of an air diffusion barrier in the envelope. Because an air diffusion barrier does not completely stop diffusion, it is really not a barrier but a retarder, commonly called an *air retarder.* It has fairly low air permeability.

NOTE

1 mil = $\frac{1}{1,000}$ in. = 0.001 in.

An air retarder is generally used only on the exterior walls. This is because most leakage sites occur in the walls of the envelope, Figure 6.3. There are two reasons for not using an air retarder in roofs: (a) roofs have a much smaller surface area than the walls—a difference that increases with increasing building height—and (b) waterproofing materials used on roofs are good air retarders.

An air retarder typically consists of a 5- to 10-mil-thick plastic sheet that is wrapped over the exterior walls before installing the exterior wall cladding, Figure 6.4 (see also Figure 6.9). The air retarder is microperforated. The perforations allow very little air to pass through but provide a high degree of permeability to water vapor. Depending on the manufacturer, an air retarder is made from a perforated plastic sheet or plastic fibers that are either woven or spunbonded into a sheet. (See Section 6.7 for the reasons why an air retarder must be perforated.)

The use of an air retarder, as a *continuous (sheet) wrap* over the exterior walls of wood or light-gauge steel frame buildings, has become standard practice, Figure 6.5. Air retarder wraps are secured to exterior wall sheathing with staples or fasteners, and the joints between them are lapped and sealed with a self-adhesive tape. To reduce the number of joints, air retarders are made in long, wide rolls; the roll width generally covers a one-story height.

Because of its low air permeability and continuity, a properly installed air retarder wrap precludes the need to seal the joints between individual wall sheathing panels. An air retarder is, therefore, effective against both air leakage and air diffusion.

FIGURE 6.3 Typical leakage sites in a wood frame or light-gauge steel frame building. Notice that most leakage sites in a building are in the walls.

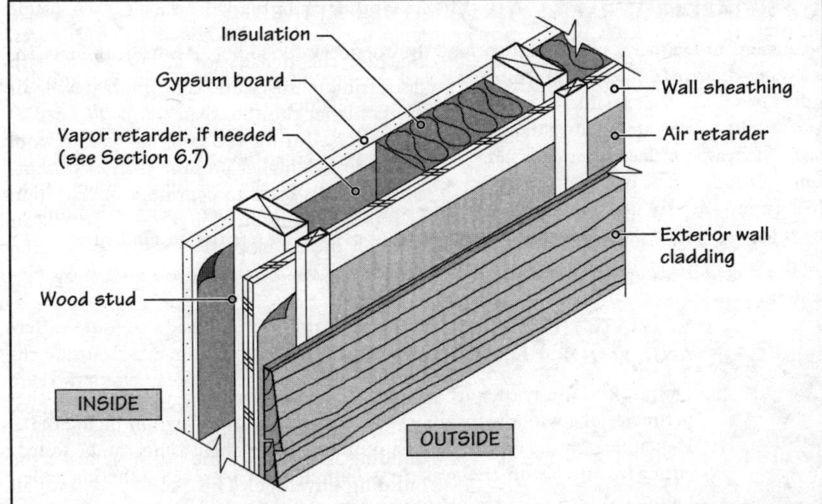

FIGURE 6.4 Location of a wrap-type air retarder in a wall assembly framed with wood or light-gauge (cold-formed) steel.

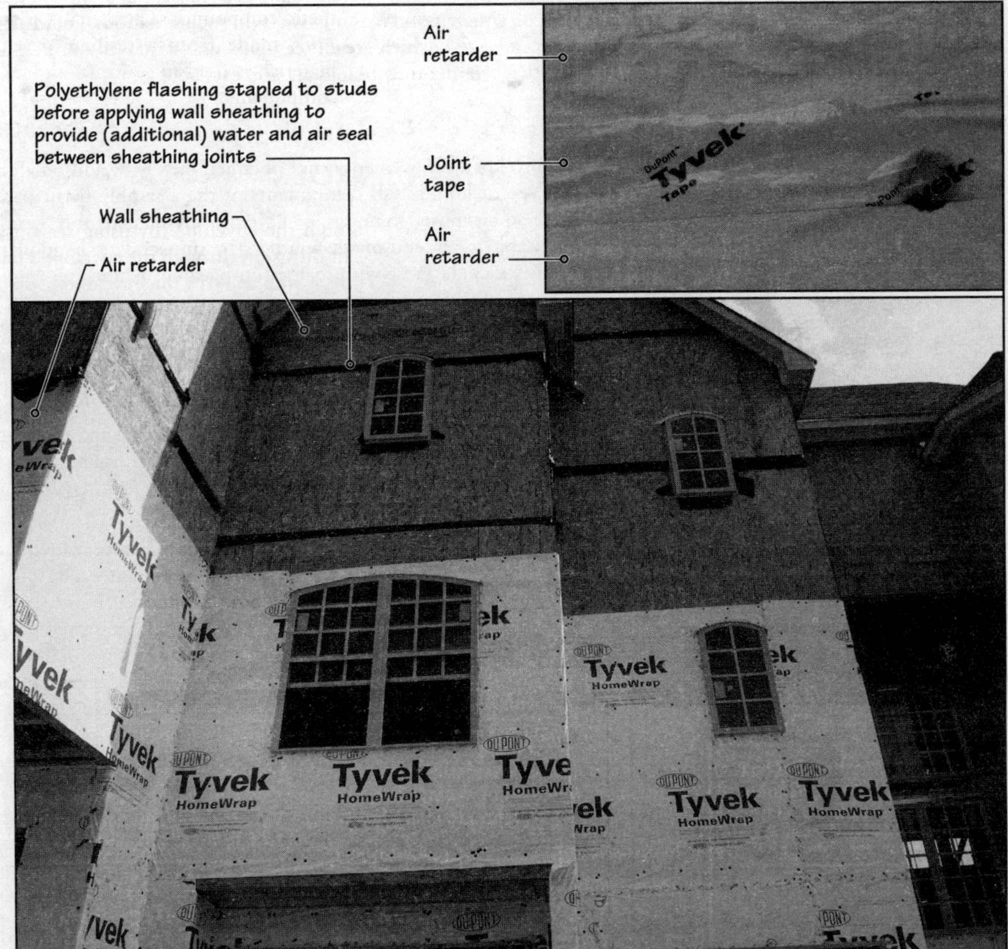

FIGURE 6.5 An air retarder wrapped over the exterior wall sheathing of a three-story wood frame condominium building. The photograph was taken while the installation of the air retarder was in progress. After installation of the air retarder over all of the exterior sheathing is completed, exterior cladding will be installed. The magnified part of the air retarder shows the self-adhering tape between the air retarder joints.

AIR RETARDER WRAP AS AIR-WEATHER RETARDER

Because an air retarder is immediately behind the exterior wall cladding, it is made to function as a water-resistive (*hydrophobic*) membrane as well—to prevent the passage of rainwater into the wall. Therefore, an air retarder is, in fact, an air-water retarder (generally called an *air-weather retarder*). Its microperforations, although permeable to water vapor, do not allow (bulk) water to pass. The reason is that (because of surface tension) several water molecules cohere together to form relatively large volumes (e.g., drops). On the other hand, the molecules of water vapor (being a gas) are separated from each other. Consequently, while the water vapor passes through, water (drops) cannot pass through the microperforations of an air retarder. In summary:

An air-weather retarder resists the passage of air and water but permits the water vapor to go through.

NOTE

Important Properties of a Wrap-Type Air Retarder

1. *Air Permeabilty:* The most important property of an air retarder is its air diffusion rating (air permeability). The air permeability of a wrap-type air retarder is generally required to be less than 0.004 cubic foot per minute per square foot (cfm/ft^2) at a pressure difference of 1.57 pounds per square foot (psf). This value is the permeability of a $\frac{1}{2}$-in.-thick gypsum board. In other words, a $\frac{1}{2}$-in.-thick gypsum board has been chosen as the standard against which the air permeability of other materials is determined.
2. *Breathability:* An air retarder wrap is required to be *vapor permeable*, referred to as *breathable*. It must have a minimum vapor permeability of 5 perms, with no limit placed on its maximum value. (*Perm* is a measure of vapor permeability of a material, discussed later in this chapter.)
3. *Water Resistance:* An air retarder must be water resistant.
4. *Tensile Strength:* Because an air retarder completely separates the interior air from the exterior air, it must have sufficient tensile strength to resist the pressure differences between the inside and outside air. A typical air retarder wrap obtains much of its strength from the rigid backing to which it is fastened, such as wall sheathing.

SEALING THE EXTERIOR WINDOW AND DOOR PERIMETER

Despite the use of the air retarder, the joints between the wall and the perimeter of a window (or door) can leak air (see the arrows in Figure 6.3). Depending on the type of detail around the jamb, these joints can be sealed either by an elastomeric sealant (Chapter 9) or by a self-adhering tape, Figure 6.6.

LIQUID AND PEEL-AND-STICK AIR-WEATHER RETARDERS

Wrap-type air retarders are generally used on walls with nailable exterior sheathing, such as wood frame or light-gauge steel frame walls. They cannot be used on concrete and masonry walls. In such situations, liquid air-weather retarders are used, which are either rolled on or sprayed on. Peel-and-stick air-weather retarder membranes are also available.

AIR LEAKAGE AND DIFFUSION—ASSEMBLY PROPERTIES

The control of air movement is not obtained merely by add-on air retarders (wrap-type or liquid). All components of the assembly participate in this process, some more, some less. Therefore, if the assembly is inherently impermeable to air, add-on air retarders are unnecessary. Examples of such assemblies include those with foamed-in-place insulation (see Figure 14 in Principles in Practice, Chapter 5) and SIP assemblies (Chapter 17). In these assemblies, the insulation, with its closed-cell structure, has very low air permeability and clings snugly to framing cavities. However, such assemblies generally require a water-resistive membrane.

Other assemblies belonging to this category are insulated metal panels and glass-aluminum curtain walls. Because both metal and glass are impermeable to air, these assemblies only need leakage control by joint seals and gaskets; diffusion control (through the use of an air retarder) is not needed.

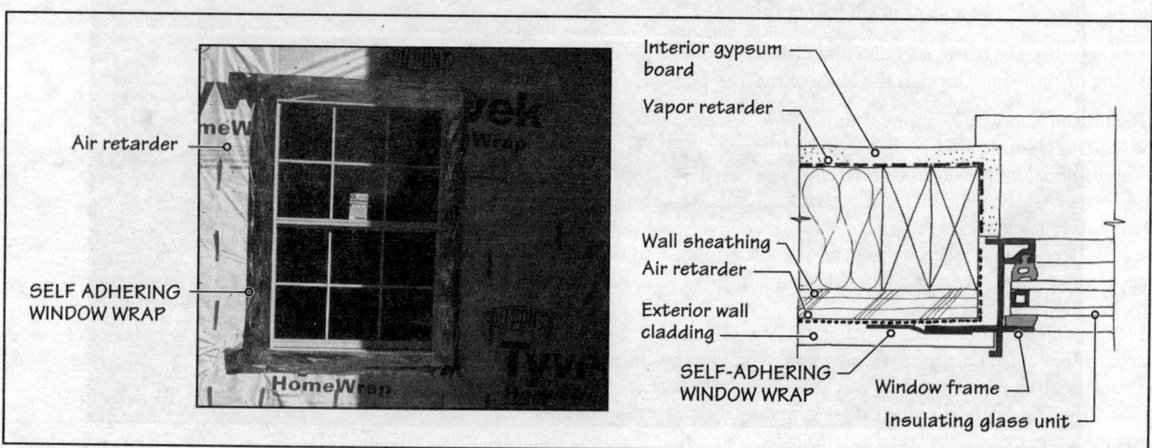

FIGURE 6.6 A self-adhering tape on the outside of a window (referred to as *window wrap*) to reduce air leakage through the joint between the window and the wall, photographed before the application of exterior cladding.

Each question has only one correct answer. Select the choice that best answers the question.

1. Air leakage through a building envelope is a function of
   - **a.** the area of holes, cracks, and openings in the envelope.
   - **b.** wind speed.
   - **c.** inside-outside air temperature difference.
   - **d.** all of the above.
   - **e.** (a) and (b).
2. Wrap-type air retarders and liquid air retarders can be used interchangeably in an assembly.
   - **a.** True
   - **b.** False
3. A wrap-type air retarder is available in many varieties, but in all of them, it is made from
   - **a.** an unperforated plastic sheet.
   - **b.** a perforated plastic sheet.
   - **c.** an asphalt-coated paper.
   - **d.** any one of the above.
   - **e.** none of the above.
4. An air retarder is generally placed
   - **a.** toward the warm side of the envelope assembly.
   - **b.** toward the cold side of the envelope assembly.
   - **c.** on the exterior face of interior gypsum board.
   - **d.** on the exterior face of an exterior wall assembly.
   - **e.** any one of the above.
5. An air retarder is generally used on
   - **a.** the walls.
   - **b.** the ceiling.
   - **c.** the roof.
   - **d.** all of the above.
6. The primary purpose of sealants and weatherstripping is to
   - **a.** reduce vapor diffusion through the envelope.
   - **b.** reduce air leakage through the envelope.
   - **c.** reduce air diffusion through the envelope.
   - **d.** all of the above.
   - **e.** (a) and (b).
7. The use of wrap-type or liquid air retarders is recommended in all heated or cooled buildings.
   - **a.** True
   - **b.** False
8. Breathability of an air retarder refers to its
   - **a.** permeability to air, water vapor, and water.
   - **b.** permeability to air and water vapor.
   - **c.** permeability to air.
   - **d.** permeability to water.
   - **e.** none of the above.

6.3 WATER VAPOR IN AIR

Almost all air contains some water vapor. Absolutely dry air, that is, air without water vapor, is rare. It is used in insulating glass units (see Chapter 30) and in laboratories for special purposes. Water vapor content in outside air is usually higher in coastal areas than inland locations.

Like air, water vapor is a gas. Consequently, it exerts pressure on the surfaces of the enclosure containing it. However, although air and water vapor are thoroughly mixed together, the pressure exerted by water vapor is independent of that exerted by air. In other words, air pressure and vapor pressure act separately on enclosure surfaces, Figure 6.7. This fact has an important consequence:

> *The diffusion of air and the diffusion of vapor through a building envelope are independent of each other and are controlled by their own individual pressure differences across the envelope. In other words, the diffusion of air is related to the air pressure difference across the envelope, and the diffusion of vapor is similarly related to the vapor pressure difference.*

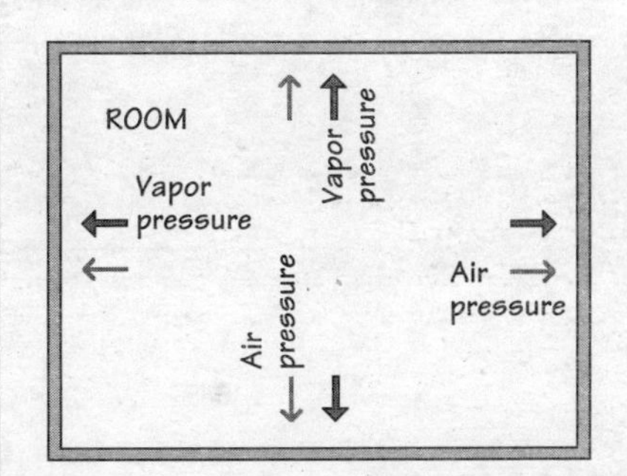

FIGURE 6.7 Although thoroughly mixed together, air and vapor exert pressure on the boundaries of a room independent of each other. This fact is as per Dalton's Law of Partial Pressures.

MOVEMENT OF WATER VAPOR THROUGH ASSEMBLIES

Like air, vapor has two means of migrating through a building assembly:

- Vapor leakage (through holes and cracks in the assembly)
- Vapor diffusion(through the body of the assembly)

Because air and water vapor are thoroughly mixed, vapor and air leak through the envelope together. Controlling air leakage also controls vapor leakage.

Vapor diffusion, however, is different. As stated in the introduction to this chapter, diffusion is measured by the material's (or assembly's) permeability, which is a function of the pressure difference across the material (or assembly). Because air pressure and vapor pressure act independently of each other, air permeability and vapor permeability are not related to each other. That is why it is possible to make an air retarder, which has low air permeability but high vapor permeability.

Not just air retarders but, in general, all *building assemblies are more vapor permeable than air permeable.* This means that water vapor can pass (leak or diffuse) through walls and roofs with greater ease than can air. The reason is that the vapor pressure difference between the inside and the outside is generally much higher than the corresponding air pressure difference. This fact explains why it is easy to make air retarders with high vapor permeability (breathable). (In Section 6.7, we will see why air retarders need to be vapor permeable.)

NOTE

Important Fact About Vapor Migration

Vapor flows from the warm side to the cold side of the envelope. Thus, in cold climates, vapor tends to flow from the inside to the outside. In warm, humid climates, vapor tends to flow from the outside to the inside. In other words, heat flow and vapor flow are generally in the same direction.

EXPAND YOUR KNOWLEDGE

Why Water Vapor Can Flow Through an Assembly with Greater Ease than Air

Typical Inside-Outside Air Pressure Difference

As stated in Section 3.5 (Expand Your Knowledge section), the pressure difference between the inside and outside air is approximately equal to $(V/20)^2$, where the pressure difference is in pounds per square foot (psf) and V is the outside wind speed in miles per hour (mph). Thus, if V = 20 mph (a fairly windy condition), the air pressure difference between the inside and outside = 1.0 psf. On a moderately windy day with 10 mph wind speed, the corresponding air pressure difference is only 0.25 psf.

Typical Inside-Outside Vapor Pressure Difference

Vapor pressure is related to the amount of water vapor present in air. Air cannot contain an unlimited amount of water vapor. When the air contains the maximum amount of water vapor it can possibly hold, it is referred to as *saturated air,* and the corresponding vapor pressure is referred to as the *saturation vapor pressure*.

The saturation vapor pressure increases with air temperature, as shown in the accompanying table. Because the relative humidity (RH) of air is directly related to the saturated air, the vapor pressure of air can be obtained by

Vapor pressure of air = (RH/100)(vapor pressure of saturated air)

Saturation Vapor Pressure as a Function of Air Temperature

Temperature (°F)	Saturation vapor pressure (psf)
10	4.5
30	11.5
50	25.5
70	52.5
90	100.5

From the table, if the air temperature = 70°F and RH = 45%, the vapor pressure of air = (45/100)52.5 = 23.6 psf. Similarly, for air at 10°F and 80% RH, the vapor pressure = (0.80)(4.5) = 3.6 psf. These two estimates of vapor pressure are for typical inside and outside conditions, respectively, in cold climates during the heating season, giving a typical inside-outside vapor pressure difference of 20.0 psf.

Compare this vapor pressure difference of 20 psf with the typical air pressure difference of approximately 0.25 psf. For the air pressure difference to equal 20 psf, the wind speed must be approximately 90 mph—an extremely rare occurrence. *Summarizing, therefore, it is because of the large vapor pressure difference across the envelope that water vapor flows through the envelope much more easily than air.*

Note that the RH of about 45% is the one commonly maintained in mechanically controlled interior environments. Air with a high RH (above 70%) feels moist, promotes fungal growth, and is uncomfortable and unhealthy. Air with a low RH (30% or less) dries human skin, creates static electricity in carpets, and aggravates respiratory problems.

NOTE

Dew Point and Vapor Impermeability of the Surface Necessary for Condensation

Note that condensation occurs only on surfaces through which water vapor cannot pass (permeate). In other words, water vapor will continue to move freely until it encounters a surface that interrupts its movement. At that location, condensation of vapor is possible, provided that the temperature of the surface is less than the air's dew point. Thus, because window glass is impermeable to vapor, condensation may occur on it if its temperature is below the dew point of surrounding air. Condensation does not occur in an air space even if its temperature is below the dew point.

6.4 CONDENSATION OF WATER VAPOR

Consider air at a certain temperature and relative humidity (RH). If no moisture is added to or subtracted from this air and its temperature is decreased, its relative humidity will increase. If the decrease in temperature continues, a temperature will be reached at which the RH of the air is 100%. The temperature at which the air's RH becomes 100% (i.e., when the air becomes saturated) is called its *dew point temperature,* or simply the *dew point* of air. If the temperature of the air is decreased below the dew point, the water vapor in air converts to (liquid) water—a phenomenon known as *condensation.*

Condensation occurs commonly in nature. In heated buildings, condensation is often observed during the winter on the inside surface of window glass. Such condensation is more pronounced in more humid interiors, such as indoor swimming pools, aerobic centers, bathrooms, and kitchens.

CONCEALED CONDENSATION AND SURFACE CONDENSATION

Apart from condensing on the surfaces of a window glass or any other cold surface, warm interior air can also condense inside a wall or roof assembly. If the vapor can move into a wall or roof assembly, it will condense where the temperature of the assembly is at or below the dew point of the migrating vapor. The condensation of vapor inside an envelope assembly is referred to as *concealed condensation,* as opposed to *surface condensation,* which occurs on the envelope's surface, such as a window glass.

Although both surface and concealed condensation are undesirable, the latter causes more problems. As stated earlier, concealed condensation wets the interior of an assembly, accelerating metal corrosion, wood decay, mold and fungi growth, and other problems. Therefore, the control of concealed condensation is particularly important.

Note, however, that water vapor by itself is not damaging to a building assembly. It becomes damaging when it converts to water.

6.5 CONTROL OF CONDENSATION

Surface condensation can be prevented by simply increasing the R-value of the assembly. The increase in the R-value of the assembly increases the temperature of the interior surface during winter, reducing the potential for condensation at the interior surface. Condensation occurs on the inside surface of a single glass sheet during winter because its R-value is quite low (R = 1). If the single glass sheet is replaced by an insulating glass unit (with an R-value of, say, R-3 or R-4), condensation on the inside glass will not occur unless the interior air is very humid and/or the outside air temperature is very low.

In general, if the R-value of the envelope is low, the dew point of air occurs at the surface of the envelope, leading to surface condensation. If the R-value of the envelope is raised, the dew point shifts to within the body of the envelope, Figure 6.8. This leads to concealed condensation, which should be avoided. (See Principles in Practice at the end of this chapter to determine the location of the dew point in an envelope assembly.)

CONTROL OF CONCEALED CONDENSATION—THE USE OF A VAPOR RETARDER

Concealed condensation occurs only if the water vapor is able to enter the envelope assembly and is then unable to exit. Therefore, a two-part strategy is used to control concealed condensation.

1. The first part of the strategy is to prevent the entry of water vapor into the assembly. This is achieved by using a vapor barrier in the envelope. A vapor barrier is a material that is impermeable to vapor. It is also impermeable to air because if air leaks through it, so will vapor. Thus, a vapor barrier is both an air barrier and a vapor barrier. Note that the term *vapor retarder* is more appropriate, because most commercial vapor barriers are not absolutely vapor impermeable.
2. The second part of the strategy, discussed in Section 6.7, is to facilitate any vapor that has entered the assembly to escape freely to the outside.

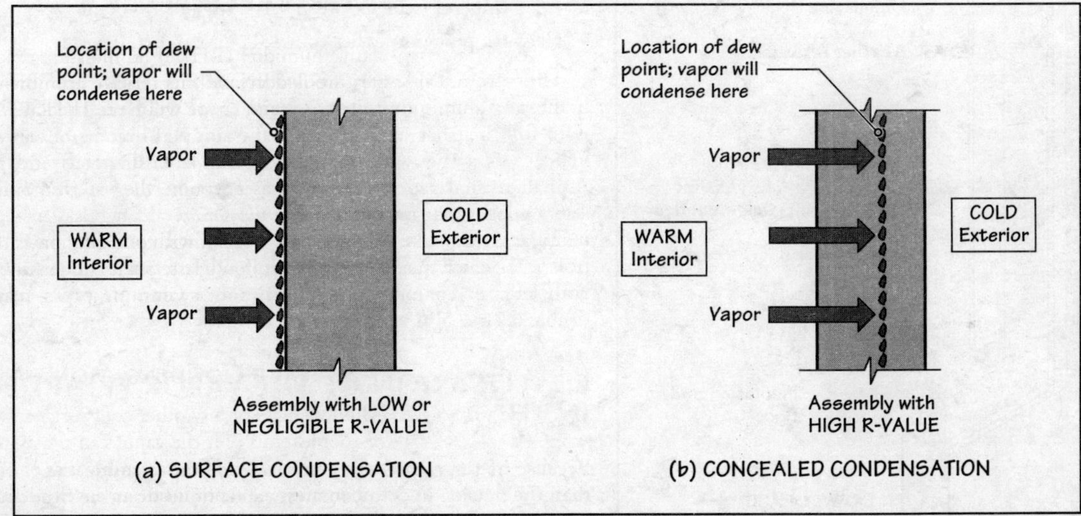

FIGURE 6.8 The effect of the R-value of an assembly on the location of the dew point during a heating season. **(a)** If the assembly has a low or negligible R-value (such as a single sheet of glass), the interior surface will be cold. Consequently, the dew point is likely to occur on the interior surface of the assembly, where the warm interior air will condense. **(b)** If the assembly has a high R-value, the interior surface will be warm. Therefore, the dew point will occur somewhere inside the assembly. Note that the condensation will occur only if the interior vapor can permeate (leak or diffuse) into the assembly. If vapor cannot permeate into the assembly, there will be no condensation within the assembly (see Example 2 in Principles in Practice).

TABLE 6.1 APPROXIMATE PERM RATINGS OF SELECTED MATERIALS (IN THE U.S. SYSTEM OF UNITS)

Component	Perm rating (perm)	Component	Perm rating (perm)
Aluminum foil (unpunctured)	0.0	Brick masonry, 4 in. thick with tooled mortar joints	4.0
Aluminum foil on gypsum board	0.1	Concrete block masonry, 8 in. thick with tooled mortar joints	6.0
Built-up roofing, 3- to 5-ply	0.0	Unpainted gypsum board	6.0
15-lb asphalt felt	4.0	Latex or enamel paint on gypsum board	3.0
Polyethylene sheet, 4 mil thick	0.08	Building paper, grade B	1.0
Polyethylene sheet, 6 mil thick	0.06	Exterior oil paint, 3 coats on wood	0.3–1.0
Polyethylene sheet, 8 mil thick	0.03	Commercial air retarders	15–60

NOTE

Units of Permeance (Perm Rating)

U.S. System of Units

1 perm = 1 grain of water vapor passing through 1 ft^2 of a material in 1 h under a vapor pressure difference of 1 in. of mercury. (1 lb = 7,000 grains.)

SI System of Units

1 perm = 1 mg of water vapor passing through 1 m^2 of a material in 1 s under a vapor pressure difference of 1 Pascal.

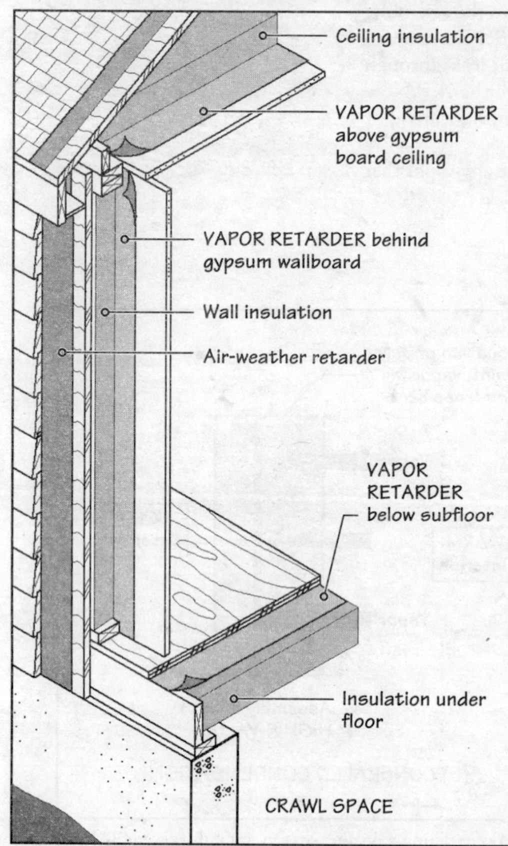

FIGURE 6.9 As per the codes, a vapor retarder (Class I or II) is required on all interior surfaces of the envelope in climate zones 5, 6 , 7, 8 and marine 4. The objective is to prevent interior vapor from migrating into the envelope assembly, where it could condense.

6.6 MATERIALS USED AS VAPOR RETARDERS

The rate at which water vapor diffuses through a material is measured by its permeability. The unit of vapor permeability is called the *permeance,* or simply the *perm.* A material with a higher perm value (or rating) is more vapor permeable. If the perm rating is zero, the material is impermeable to vapor diffusion. Such a material is a perfect vapor retarder—a vapor barrier—provided that it is free of holes, cracks, and unsealed joints.

Apart from being a property of the material, the perm rating is a function of the material's thickness. A greater thickness of the same material has a lower perm rating. Therefore, when the perm rating is given, the thickness of the component must be stated. The perm ratings of selected materials (in the U.S. system of units) are given in Table 6.1. For instance, the perm rating of a 4-mil-thick polyethylene sheet is 0.08; a 6-mil-thick polyethylene sheet's perm rating is 0.06.

CLASSIFICATION OF VAPOR RETARDERS

Vapor retarders are classified into three types:

Class I: perm rating ≤0.1 perm

Class II: perm rating between 0.1 perm and 1.0 perm

Class III: perm rating between 1.0 perm and 10.0 perm

Thus, from Table 6.1, a polyethylene sheet, built-up roof membrane, and aluminum foil are Class I vapor retarders. In fact, aluminum foil, if it has no perforations, has a perm rating of zero, and hence it is a (perfect) vapor barrier, as are all metals and glass. Asphalt-treated paper, referred to as *building paper* or *kraft paper,* is also a good vapor retarder (Table 6.1). Several fiberglass insulation manufacturers make fiberglass batts faced with building paper (Section 5.7), which qualifies as a Class II vapor retarder. Gypsum board with latex or enamel paint qualifies as a Class III vapor retarder (Table 6.1).

6.7 LOCATION OF THE VAPOR RETARDER IN THE ASSEMBLY

Because of the greater concentration of vapor inside the building than the outside in cold climates, vapor flows from the inside to the outside in these climates. The codes, therefore, require that a vapor retarder be installed toward the interior side of the building envelope in cold climates (climate zones 5, 6, 7, and 8 and marine 4), Figure 6.9. The objective is to contain the vapor within the building interior so that it will not migrate into the envelope assembly, where it could condense.

To achieve this objective, all penetrations through the vapor retarder and all vapor retarder joints should be fully sealed. Additionally, the vapor retarder should be of (almost) zero perm rating so that the vapor does not diffuse through it. Although the codes allow either Class I or Class II vapor retarders, a Class I vapor retarder is preferred. A 4-mil-thick polyethylene sheet is the most commonly used material, installed immediately behind the interior gypsum board, as shown in Figure 6.9 (see also Figure 6.4). Building paper is another common choice, which, as stated previously, is also available as a facing on fibrous insulation.

In mixed and warm climates, the use of vapor retarder is not mandated by the codes. However, a Class III vapor retarder may be used in climate zones 3 and 4 in the same way as for colder climates. In climate zones 1 and 2, the vapor retarder may be omitted. A vapor retarder is also unnecessary in buildings that are not conditioned or insulated.

BREATHABILITY OF THE ASSEMBLY BEYOND THE VAPOR RETARDER

Because the vapor retarder is not a (perfect) vapor barrier and sealing the leaks may be imperfect, some vapor is bound to migrate into the assembly, where it may condense if not allowed to escape out of the assembly. To ensure its escape, the assembly beyond the vapor retarder should be vapor permeable (breathable), Figure 6.10. That is why the air retarder, which is placed on the exterior face of the walls, is required to be vapor permeable. Vapor permeability of the air retarder is achieved through its perforations. If the air retarder is not perforated, any vapor that travels through the vapor retarder will be trapped between the vapor retarder and the air retarder and will condense there. To further facilitate the escape of vapor out of the assembly, a vented air space between the air retarder and the exterior wall cladding is desirable (see Figure 6.4).

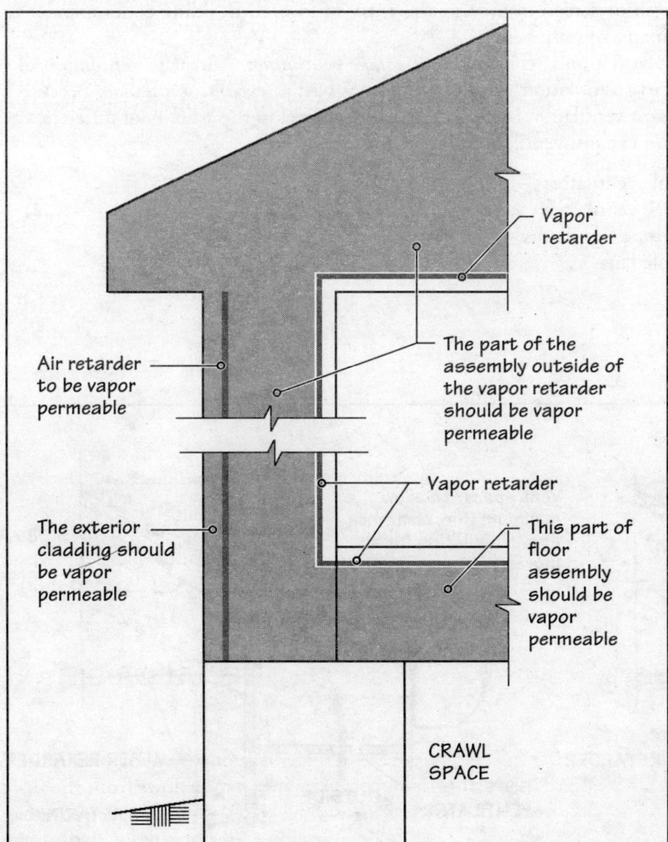

FIGURE 6.10 Because a vapor retarder encloses an interior space completely, as shown, interior vapor will not permeate into the envelope assembly. However, to ensure that any unintended small amount of interior vapor, which may migrate into the envelope assembly, does not condense there, the assembly beyond the vapor retarder should be vapor permeable (breathable).

NOTE

Vapor Retarder and Climate Zones

As per the building codes, a Class I or II vapor retarder is mandated on the interior side of an envelope assembly in cold climates—climate zones 5, 6, 7, 8 and marine 4. A vapor retarder is not mandated in other climate zones.

NOTE

Perm Rating of an Air Retarder

As per the ASTM E1677 standard, the minimum required perm rating of an air retarder is 5.0 U.S. perms, which is 50 times the maximum perm rating of a Class I vapor retarder. The perm rating of commercial air retarders is much greater than 5.0, making them quite breathable.

What about the interior vapor that migrates beyond the vapor retarder into the ceiling? Because the roof cover materials that are not vapor permeable, vapor must be removed by venting the attic space. Without adequate attic ventilation, the vapor will be trapped in the attic, where it may condense. For the same reason, the crawl space should also be ventilated.

6.8 IMPORTANCE OF ATTIC VENTILATION

Ventilating the attic is necessary for reasons other than condensation control. In cold climates, ventilation prevents the formation of ice dams at projecting roof eaves. If the attic is not ventilated, it remains relatively warm because of the entry of heat into the attic from the warm interior, but the eave overhangs are cold. This melts the snow in the middle of the roof, which freezes again at the eaves, forming ice dams, Figure 6.11(a). The problem is worse in a roof-ceiling assembly with inadequate insulation.

Ice dams can substantially increase the load at eave overhangs and gutters. They also prevent water from draining off the roof, which may cause roof leakage. Adequate ceiling insulation, coupled with ventilation of the attic, helps to eliminate ice dams as well as the buildup of water vapor, Figure 6.11(b).

Attic ventilation is also required in warm climates to reduce heat transmission from the attic into the interior of the building. The temperature of the air in an unventilated attic becomes much higher than that of the outside air because of the ability of air and the materials of the roof to store heat. This is particularly critical during the summer.

Temperature differences between the outside air and the attic air in excess of 40°F have been reported in unventilated or inadequately vented attics. The higher attic temperature increases heat transfer to the interior of the building. As shown in Figure 6.11(b), attic ventilation can be provided through openings in the soffit. The openings must be covered with a mesh or screen to prevent the entry of insects. Building codes mandate a certain minimum area of soffit vents.

Soffit ventilation is considered as *intake ventilation.* Adequate ventilation of the attic requires cross ventilation, requiring intake as well as *exhaust ventilation.* Because warm air rises, exhaust ventilation is required at a higher level in the attic. Four different alternatives are used for exhaust ventilation, Figure 6.12:

- Gable ventilators
- Ridge ventilators
- Turbine ventilators
- Gable fans

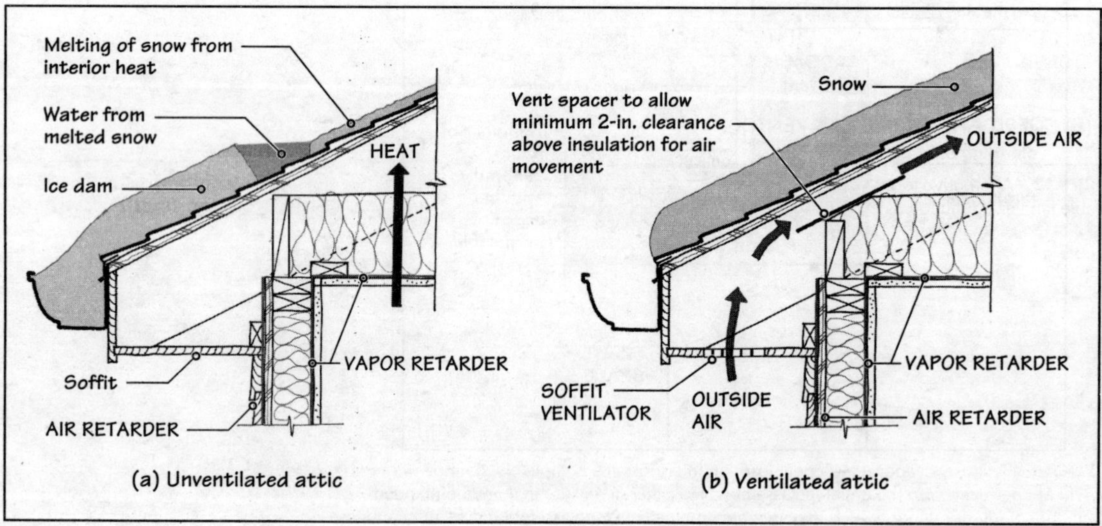

FIGURE 6.11 The effect of attic ventilation on the formation of ice dams at the overhanging eave of a sloping roof.

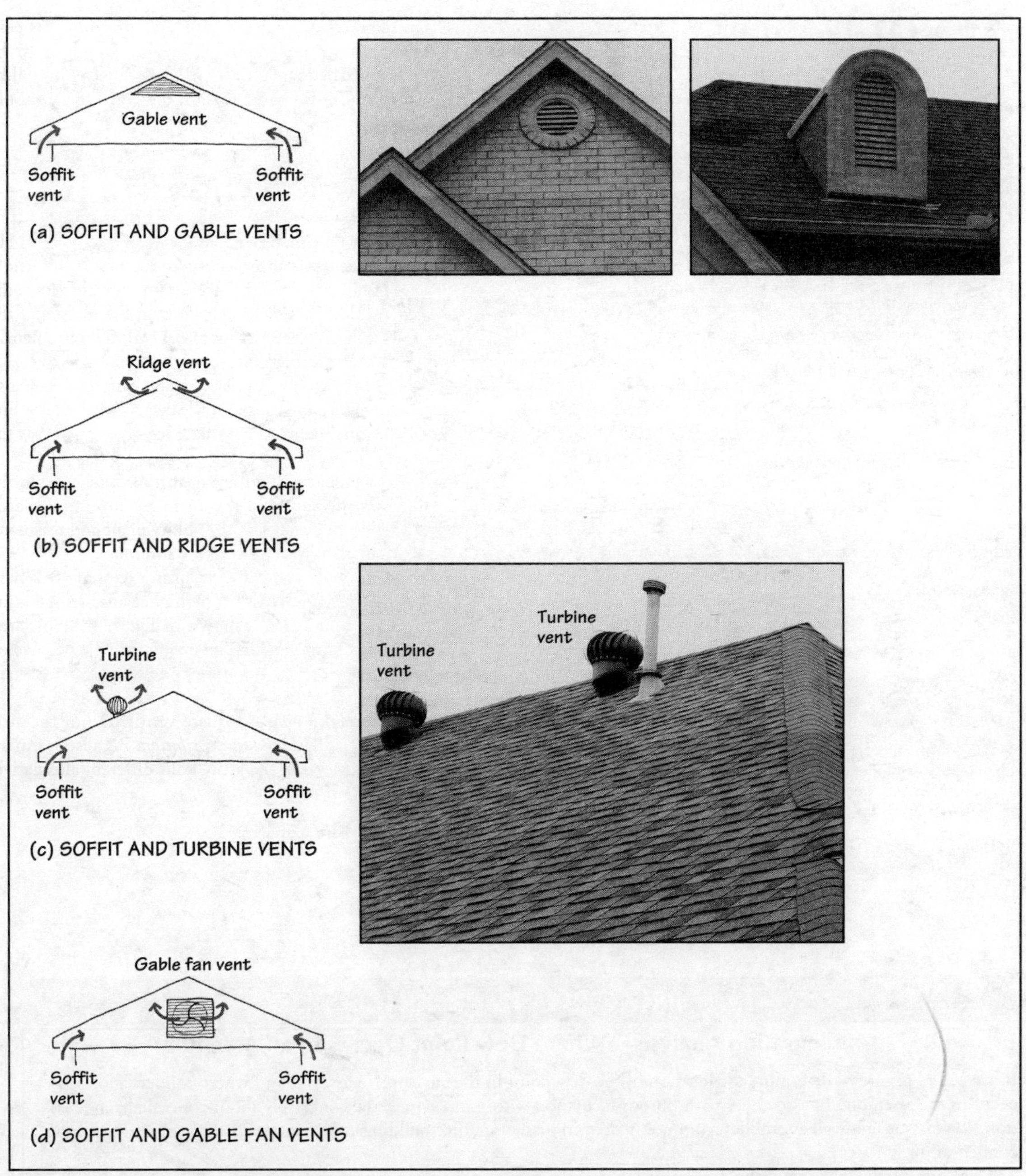

FIGURE 6.12 Alternative forms of attic ventilation. Soffit vents function as intake ventilators, while gable vents, ridge vents, turbine vents, and gable fan vents function as exhaust ventilators. Any one of the four exhaust ventilation options may be used along with soffit vents.

EXPAND YOUR KNOWLEDGE

Vapor-Tight Construction and Attic Ventilation

Although preventing condensation in an attic by providing attic ventilation is perhaps the safest approach and one that is currently mandated by building codes, it is not without disadvantages. In extremely cold climates with frequent blowing snow and rain, venting can cause serious problems by permitting snow and rain to infiltrate the vents. Experience in cold Canadian climates has demonstrated that if indoor humidity levels are controlled and the ceiling is provided with a well-installed vapor retarder with an extremely low vapor permeance (vapor-tight construction), it is possible to prevent attic condensation without providing attic ventilation.

A major advantage of eliminating attic ventilation is that it improves the effectiveness of attic insulation. However, if attic ventilation is not provided, the problem of ice dam formation should be investigated or prevented by eliminating eave overhangs.

PRACTICE QUIZ

Each question has only one correct answer. Select the choice that best answers the question.

9. In a typical dwelling in the northern United States, the vapor pressure difference between the inside and outside air is generally

- **a.** the same during the heating and cooling seasons.
- **b.** smaller during the heating season than the cooling season.
- **c.** greater during the heating season than the cooling season.

10. The dew point of air is always

- **a.** less than or equal to the air's temperature.
- **b.** greater than or equal to the air's temperature.
- **c.** dew point is unrelated to the air's temperature.

11. In heated buildings, the vapor flow is generally

- **a.** from the inside to the outside of the envelope.
- **b.** from the outside to the inside of the envelope.

12. Concealed condensation occurs inside an envelope assembly during a heating season if

- **a.** interior vapor cannot permeate the assembly.
- **b.** interior vapor can permeate the assembly but cannot exit it.
- **c.** the R-value of the assembly is high.
- **d.** (a) and (b).
- **e.** (b) and (c).

13. Which of the following materials has the lowest perm rating?

- **a.** Plastic sheet
- **b.** Metal sheet
- **c.** Brick wall
- **d.** Concrete wall
- **e.** All of the above materials have the same perm rating.

14. In which of the two assemblies, one with a negligible R-value (assembly A) and the other with a high R-value (assembly B), is concealed condensation likely to occur?

- **a.** Assembly A
- **b.** Assembly B

15. Perm rating is a property that measures

- **a.** how effectively the air diffuses through the material.
- **b.** how effectively water vapor leaks through the material.
- **c.** (a) and (b).
- **d.** neither (a) nor (b).

16. In general, the greater the thickness of a material, the greater its perm rating.

- **a.** True
- **b.** False

17. For a material to be considered a Class I vapor retarder, its perm rating in the U.S. system of units must be

- **a.** greater than 2.0.
- **b.** less than or equal to 2.0.
- **c.** greater than 1.0.
- **d.** less than or equal to 1.0.
- **e.** none of the above.

18. A vapor retarder is mandated in the building envelope

- **a.** in all climate zones.
- **b.** only in climate zones 3 to 8.
- **c.** only in climate zones 4 to 8.
- **d.** only in climate zones 5 to 8.
- **e.** none of the above.

19. A vapor retarder is perforated so that

- **a.** water may pass through the pores.
- **b.** water vapor may pass through the pores.
- **c.** air may pass through the pores.
- **d.** none of the above.

20. Ventilation of an attic space is generally considered important in

- **a.** cold climates only.
- **b.** moderately cold climates only.
- **c.** hot climates only.
- **d.** all climates.
- **e.** none of the above.

21. Soffit vents alone are not adequate to provide attic ventilation.

- **a.** True
- **b.** False

22. When gable vents are provided to ventilate an attic space,

- **a.** ridge vents are also necessary.
- **b.** turbine vents are also necessary.
- **c.** soffit vents are also necessary.
- **d.** no other vents are necessary.

23. An ice dam is likely to occur

- **a.** in the middle of a sloping roof.
- **b.** at the eave of a sloping roof.
- **c.** in the middle of a flat roof.
- **d.** at the edge of a flat roof.

PRINCIPLES IN PRACTICE

Condensation Analysis—Where Dew Point Occurs in an Assembly

It is often necessary to determine the location of the dew point in the envelope to assess if and where condensation may occur in the assembly. In assemblies with fibrous insulation within its cavities, the concern is about concealed condensation. In glass curtain wall assemblies, commonly used in high-rise office buildings, the concern is about (interior) surface condensation.

Determining the location of the dew point requires determining the temperature variation through the assembly. A graphical representation of temperature variation is called the *temperature gradient*, which is a line passing through the cross section of the assembly. Because condensation is generally a cold-climate phenomenon, the temperature gradient line slopes down toward the outside, because the temperature is highest on the inside, and reduces gradually toward the outside of the assembly.

From Equation (2) in Chapter 5, we see that the amount of heat (q) passing through an element is a function of its R-value and of the temperature drop, $(t_2 - t_1)$, across the element, that is, $q = (t_2 - t_1)/R$. Thus, $(t_2 - t_1) = qR$.

In a multilayer assembly, the amount of heat that enters the first layer of the assembly must exit the last layer so that the amount of heat (q) flowing through each layer is the same. This implies that the temperature drop across a layer is directly proportional to the R-value of the layer. In other words, the greater the R-value of the layer, the greater the temperature drop across that layer.

Consider an assembly consisting of three layers (layers 1, 2, and 3), say with R-values of R-1, R-3, and R-6, respectively, so that the total R-value of the assembly = 10. Assume that the total temperature drop across the three layers is 50°F. Then the temperature drop across layer 1 is 5°F, across layer 2 it is 15°F, and across layer 3 it is 30°F, as shown by the temperature gradient line superimposed on the assembly's cross section in Figure 1.

The temperature gradient line is more easily obtained if the cross section of the assembly is drawn based not on the physical dimensions of each layer, but on the R-value of each layer. In that case, the temperature gradient line is one continuous straight line. The temperature gradient line of the assembly of Figure 1, obtained from this simplified procedure, is shown in Figure 2.

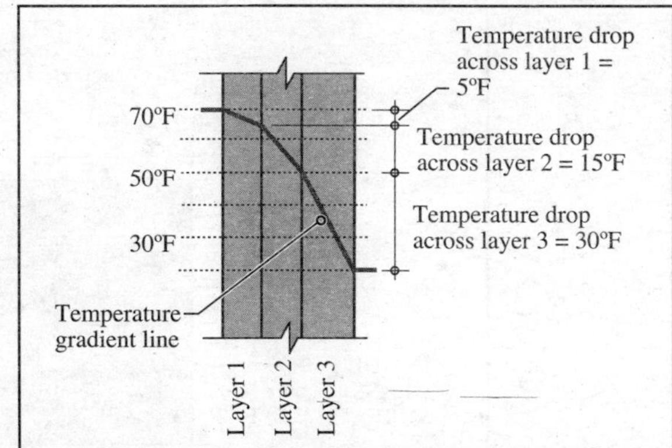

FIGURE 1 Temperature gradient line superimposed on an assembly's cross section, consisting of three layers. The thickness of each layer is based on its actual physical dimension. Note that the temperature gradient line is not a straight line.

LOCATION OF THE DEW POINT (DP) IN THE ASSEMBLY

Once the temperature gradient line has been obtained, the dew point (DP) can be located from Table 1 (at the end of this section), which provides the DP of air for various temperature and RH values. For example, the DP of air at 70°F and 30% RH is 37°F. The following examples illustrate the procedure.

EXAMPLE 1

Determine the location of the DP on a glazing consisting of a $\frac{1}{4}$-in.-thick glass sheet. Assume that the inside air film (surface) resistance, $R_{si} = 0.7$, and the outside air film resistance, $R_{so} = 0.2$. Assume that the inside and outside air temperatures are 70°F and 10°F, respectively, and the inside RH = 45%.

SOLUTION

The R-value of the glass sheet is zero. Therefore, the total R-value of the assembly considering inside and outside air films = $R_t = 0.7 + 0 + 0.2 = 0.9$. Figure 3 shows the temperature gradient through the assembly's cross section (the thickness of each layer is equal to its R-value). From Table 1, the DP of the inside air (at 70°F and 45% RH) = 47.5°F

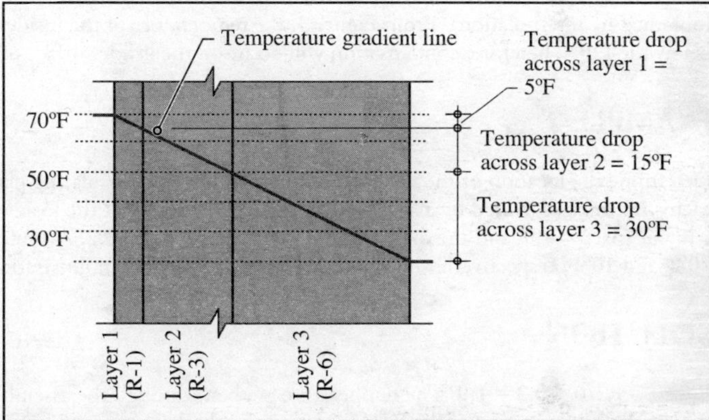

FIGURE 2 Temperature gradient line superimposed on the cross section of the assembly in Figure 1, but with the thickness of each layer proportional to its R-value. Note that the temperature gradient line is a straight line.

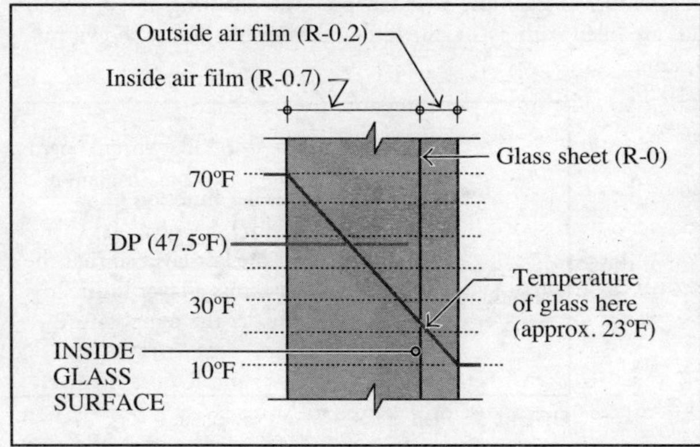

FIGURE 3 Temperature gradient in the assembly of Example 1. The thickness of each layer is proportional to its R-value.

(Continued)

Condensation Analysis—Where Dew Point Occurs in an Assembly (*Continued*)

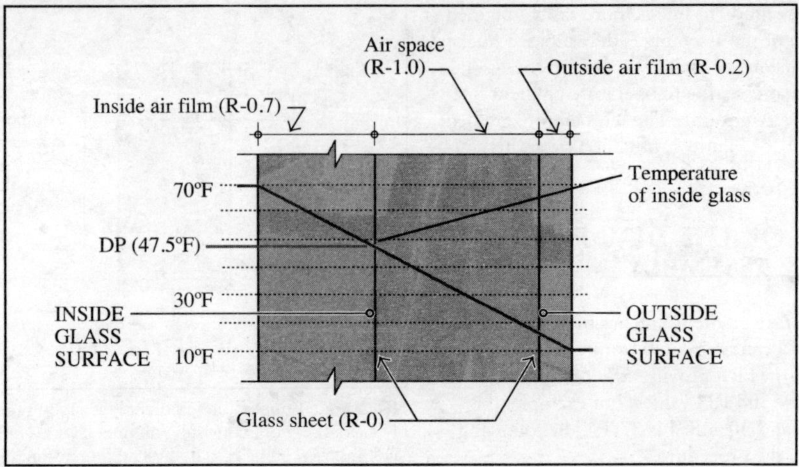

FIGURE 4 Temperature gradient in the assembly of Example 2. The thickness of each layer is proportional to its R-value.

(obtained by interpolation). From Figure 3, the temperature of the inside surface of the glass (approximately 23°F) is well below the DP. Therefore, condensation will occur on the inside surface of the glass.

EXAMPLE 2

Determine the location of the DP for a glazing with (sealed) insulating glass units (IGU), consisting of two glass sheets, each $\frac{1}{4}$ in. thick separated by a $\frac{1}{2}$-in.-wide air space. Assume that the inside air film resistance, $R_{si} = 0.7$, the R-value of the $\frac{3}{8}$-in. air space = 1.0, and the outside air film resistance, $R_{so} = 0.2$. Assume that the inside and outside air temperatures are 70°F and 10°F, respectively, and the inside RH = 45%. (For an illustration of an IGU, see Figure 30.19.)

SOLUTION

$R_t = 0.7 + 1.0 + 0.2 = 1.9$. The temperature gradient through the assembly is shown in Figure 4. As shown, the temperature of the inside surface of the IGU is approximately the same as the DP of the interior air. Therefore, some condensation is likely to occur on the glass. Also note that because an IGU is a sealed unit, the air (or water vapor) cannot permeate it. Therefore, there will be no condensation within the IGU.

EXAMPLE 3

Determine the location of the DP in the brick veneer wall assembly of Example 3 (in Chapter 5), consisting of 2 × 4 wood studs spaced 16 in. on center. The spaces between the studs are filled with $3\frac{1}{2}$-in.-thick fiberglass insulation, as shown in Figure 5 (which is a copy of Figure 5.16 in Chapter 5). Assume that the inside and outside air temperatures are 70°F and 10°F, respectively, and the inside RH = 45%.

SOLUTION

From Example 3 (Chapter 5), the R-values of various layers of the assembly are

Inside air film resistance = 0.7
$\frac{1}{2}$-in.-thick gypsum board = 0.3
$3\frac{1}{2}$-in.-thick fiberglass insulation = 12.2

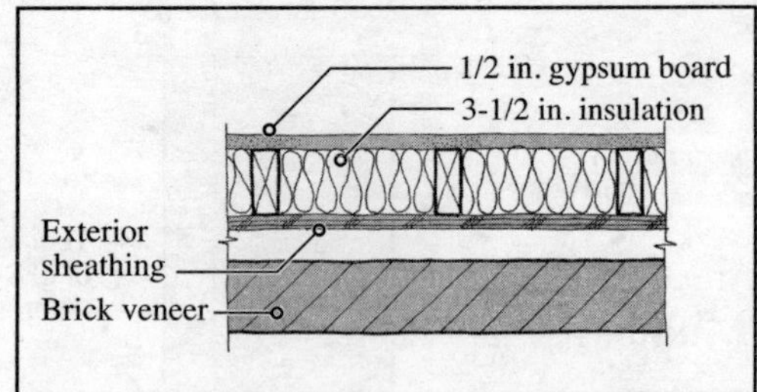

FIGURE 5 Plan of the wall of Example 3.

PRINCIPLES IN PRACTICE

$\frac{1}{2}$-in.-thick plywood = 0.5
2-in.-wide air space = 1.0
$3\frac{5}{8}$-in.-thick brick veneer = 0.7
Outside air film resistance = 0.2

Figure 6 shows the temperature gradient line superimposed on the cross section of the assembly based on the R-values of layers. Note that DP (47.5°F) is located within the insulation. Therefore, condensation will occur within the insulation if inside vapor can permeate to it.

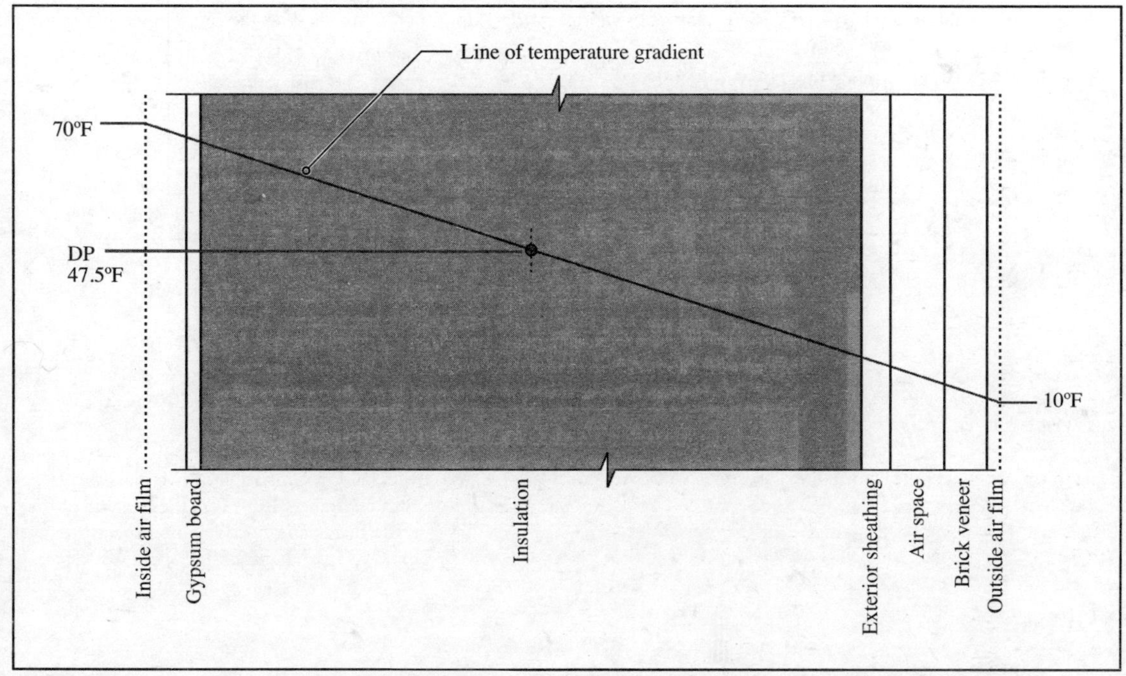

FIGURE 6 Temperature gradient in the wall of Example 3. The thickness of each layer is proportional to its R-value.

TABLE 1 DEW POINT (DP) AS A FUNCTION OF AIR TEMPERATURE AND RH

	Air temperature (°F)				
RH (%)	50	60	70	80	90
100	50	60	70	80	90
90	47	57	67	77	87
80	44	54	64	73	83
70	40	50	60	69	79
60	36	46	55	65	74
50	33	41	50	60	69
40	27	35	45	53	62
30	21	29	37	46	54
20	13	20	28	35	43
10	6	9	13	20	27
0	This air will not condense.				

REVIEW QUESTIONS

1. Using a sketch, explain the difference between leakage and diffusion. To which material (solid, liquid or gas) does the concept of diffusion apply?
2. Explain why an air retarder is called an air-weather retarder. Apart from a wrap-type air retarder, which other types of air retarder are commonly used?
3. Using a three-dimensional sketch, show the locations of the air retarder and vapor retarder in a wall assembly.
4. Explain why the micro-perforations in an air retarder allow water vapor to pass through but neither (liquid) water nor air.
5. Explain the difference between surface condensation and concealed condensation and state in which situations they might occur.
6. Explain why it is much easier for vapor to flow through building assemblies than air.
7. Explain the difference between various classes of vapor retarders. In which climates does the building code require the use of a vapor retarder?
8. Explain why an attic space must be ventilated.

CHAPTER 7

Fire-Related Properties

CHAPTER OUTLINE

Fires that could engulf an entire neighborhood or a city are rare these days.* Because of sustained improvements in fire safety and zoning regulations and advances in fire detection and suppression equipment used in buildings, fires today are generally limited to individual buildings or a small group of buildings. If fire safety considerations were less stringent than those currently in use, fire of the severity shown in Figure 7.1 would have not only caused the total destruction and collapse of this building, but would also have produced a domino effect that would have burned and destroyed a large area of the city.

Despite various improvements in fire-safe design and construction, the frequency of building fires and the resulting property losses are still substantial enough to require continued research into materials and construction systems. Even though they are gradually decreasing, the statistics of deaths and injuries resulting from fires are still appalling. Fire continues to be the single largest killer of building occupants in the United States. By comparison, natural disasters, such as earthquakes, hurricanes, and floods, account for a much smaller number of lives lost, and deaths caused by structural failures of buildings are rare. For instance, in the United States, approximately 2,700 people (excluding firefighters) died and several times more were injured in building fires in 2009 [7.1].

Because fire is the biggest hazard to life safety in modern buildings, building codes recognize this fact by making fire protection an important objective. For example, the classification of buildings in various occupancy groups, as provided in building codes, is based primarily on the degree of fire hazard present in buildings. The maximum permissible building area and height are also based on the ability of various construction

*Although building fires are fairly common in contemporary times, fires such as the Great Fire of London (1666), which lasted for several days and nights and destroyed virtually half of the city of London, are, thankfully, a thing of the past.

FIGURE 7.1 Four floors of the First Interstate Building in Los Angeles, California, gutted with fire. The building is still standing and is in service. Its survival is ascribed primarily to its fire-safe construction. (Photos courtesy of W.R. Grace & Co. Both photos are copyrighted by W.R. Grace & Co.)

assemblies to withstand fire. In fact, if the structural engineering provisions are disregarded in a building code, then of the remaining regulations, nearly two-thirds relate to fire safety considerations.

Factors That Affect Fire Safety in Buildings

Fire safety in buildings is a function of several variables, which may be grouped under the following three categories: (a) architectural design, (b) fire protection systems, and (c) public education, Figure 7.2. Fire-safe architectural design, an extensive area of study in itself, is outside the scope of this text. Briefly, it relates to the design features of a building that reduce the fire hazard and assist in *speedy evacuation* of people in the event of a fire.

Fire protection of a building is of two types—*active fire protection* and *passive fire protection.* Both have their distinct roles, and a building may have one or both types. Active fire protection relates to extinguishing or suppressing the fire after it occurs by using automatic sprinklers. In addition to sprinklers, the active system includes fire detection, an alarm system, hydrants, standpipes, and other firefighting equipment.

Passive fire protection is an integral part of the building's materials and construction, whose purpose is to increase the fire endurance of building assemblies, contain ignition, and retard the spread of fire. More specifically, passive fire protection includes compartmentalizing the building into smaller volumes, using fire-resistive barriers (walls, floors, roofs, etc.), and choosing interior finishes whose flame- and smoke-generating characteristics are safe.

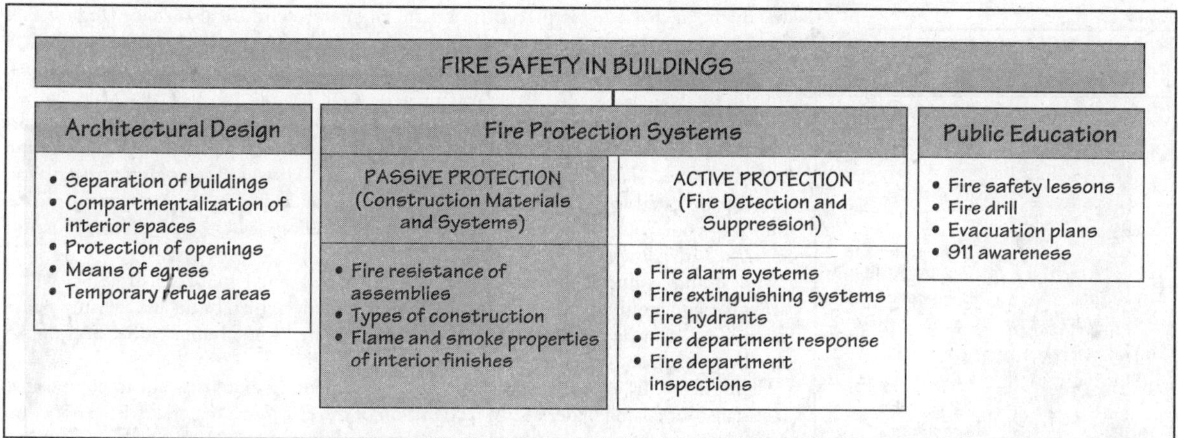

FIGURE 7.2 Factors that affect fire safety in buildings. In this chapter, we focus only on passive protection systems.

Although both active and passive protection systems are important, we focus only on passive fire protection systems and the principles and properties related to them in this text.

7.1 FIRE CODE AND BUILDING CODE

In addition to the building code, a building is regulated by the jurisdiction's fire code. As previously stated, fire safety is one of the prime objectives of a building code. However, once the building has been built and occupied, it must be properly maintained to remain safe against fire and other hazards. The regulations that cover aspects of fire safety in a building during its use and occupancy are included in the *fire prevention code,* or simply the *fire code.*

The fire code regulates such items as the location, maintenance, and installation of fire protection appliances; maintenance of the means of egress; and the storage of combustible and (or) hazardous materials. For example, the storage of flammable liquids in a dry cleaning establishment is regulated by the fire code, but the construction of the enclosure to store the flammable liquids and the requirement for an automatic fire suppression system are contained within the building code.

The building code and the fire code are two arms of a jurisdiction's building safety ordinances. They are enforced by the jurisdiction's *building official* and *fire official,* respectively. Because the distinction between fire-safe construction and fire prevention can be subtle and, at times, indistinguishable, there is always a certain amount of overlap between the fire code and building code provisions. The fire department and the building code department of a city are expected to work in close cooperation because the aims of both are the same—to ensure public safety.

The principal fire code in the United States is the Fire Prevention Code, published by the National Fire Protection Association (NFPA). However, in order to ensure compatibility and correlation between the building code and the fire code, the International Code Council publishes its own fire code, called the International Fire Code (Figure 2.12).

7.2 COMBUSTIBLE AND NONCOMBUSTIBLE MATERIALS

In the context of fire-safe construction, building materials must be distinguished either as *combustible materials* or as *noncombustible materials.* Metals used in construction (iron, steel, aluminum, copper, etc.) and concrete, brick, stone, gypsum, and so on, are noncombustible materials. Wood, paper, and plastics are combustible materials.

The distinction between a combustible and a noncombustible material is generally, although not always, obvious. For example, it is not immediately clear whether fire-retardant-treated wood is a combustible or a noncombustible material. Additionally, some materials may contain a small combustible content, which may or may not contribute appreciably to a fire, such as concrete made with polystyrene beads as aggregate. Are these materials to be considered combustible or noncombustible materials?

A precise definition of noncombustibility is, therefore, necessary. This is provided for in the ASTM E136 Test. In this test, a predried specimen of the material measuring

$1\frac{1}{2}$ in. by $1\frac{1}{2}$ in. by 2 in. is placed in a 3-in.-diameter cylindrical furnace, Figure 7.3. Before the material is placed in the furnace, the furnace temperature is raised to 1,382°F (750°C). The furnace has a transparent lid to allow the inspection of the specimen during the test.

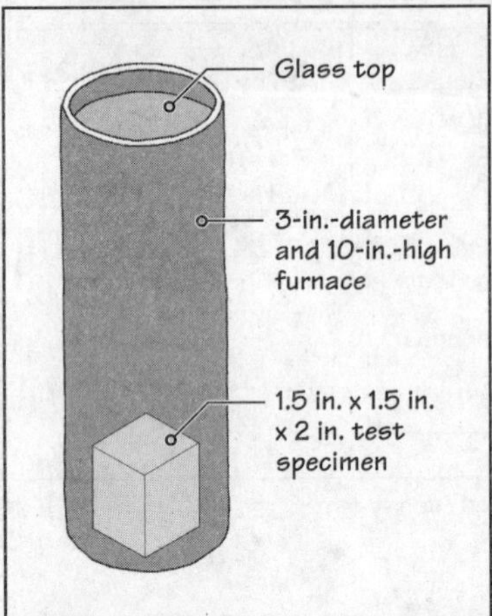

FIGURE 7.3 ASTM E136 test setup to determine whether a material is combustible or noncombustible.

The specimen is reported as noncombustible if it satisfies the conditions laid down in the test. Fire-retardant-treated lumber and plywood do not pass the ASTM test and are, therefore, classified as combustible materials. However, recognizing that fire-retardant-treated wood makes a reduced contribution as fuel in the early stages of fire, building codes permit its use in some limited situations where only noncombustible materials are allowed.

For example, non-load-bearing partitions in Type I and Type II construction, which are otherwise required to be of noncombustible materials (metal stud and gypsum board assemblies), may be constructed of fire-retardant-treated wood framing. (See Section 7.5 for the types of construction.)

The ASTM E136 test is applicable only to elementary materials, not to composites, laminated, or coated materials. In recognition of the fact that a material with a noncombustible core (as per the ASTM E136 test), but with a thin facing of a combustible material or a combustible paint, does not contribute greatly to fire, building codes classify such a material as a noncombustible material, Figure 7.4.

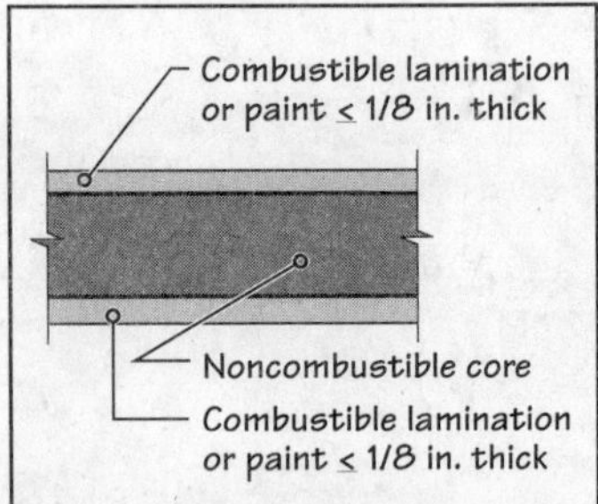

FIGURE 7.4 A material with a noncombustible core but with a combustible lamination less than or equal to $\frac{1}{8}$ in. (3 mm) thick is regarded by the building codes as a noncombustible material.

The constraints placed on the combustible paint or laminate are that its thickness shall not exceed $\frac{1}{8}$ in. and its flame spread rating shall not exceed 50 (see Section 7.8 for the explanation of flame spread rating). Thus, gypsum board, which has a thin paper lining, is classified as a noncombustible material because the thickness of the combustible paper lining is less than $\frac{1}{8}$ in.

NONCOMBUSTIBILITY OF A MATERIAL AND ITS ABILITY TO WITHSTAND FIRE

Note that noncombustibility refers only to the fact that a noncombustible material will not add fuel to the fire. Noncombustibility is not related to the ability of the material to withstand fire. For example, wood is a combustible material and steel is a noncombustible material, but a structure constructed of heavy sections of wood is better able to withstand fire than one constructed of unprotected steel members, as explained later in this chapter.

7.3 PRODUCTS GIVEN OFF IN A BUILDING FIRE

The two products given off in a building fire (referred to as *products of combustion*) are heat and smoke. Smoke consists of (a) fire gases, (b) solid particulate matter (fine particles of solid matter suspended in the atmosphere), and (c) liquid particulate matter (condensed vapors dispersed in the atmosphere as tiny liquid droplets), Figure 7.5. Although

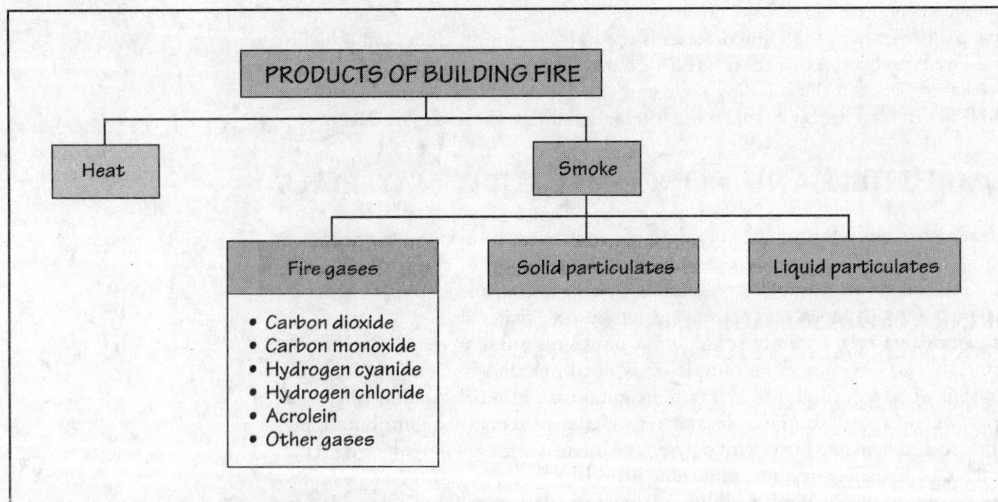

FIGURE 7.5 Products of combustion in a building fire.

fire gases in smoke are generally transparent, the solid and liquid particulate matter form an opaque cloud and diminish visibility in a fire.

Both products of combustion—smoke and heat—are responsible for human deaths and injuries. The statistics based on autopsies conducted on people who died in building fires indicate that, because of smoke's toxicity, smoke inhalation accounts for nearly 65% of all deaths. Burns and heat account for 30%, and the remaining 5% of deaths are caused by emotional shock and heart failure, Figure 7.6.

The toxicity of smoke is due to gases produced in a fire. The types of gases depend on the type of material that burns. Carbon dioxide and carbon monoxide are produced in almost all fires. If a sufficient supply of oxygen (air) is available during the fire, the carbon that is present in almost every combustible material, such as wood, cotton, silk, wool, and plastics, is converted to carbon dioxide. This type of combustion is called *complete combustion.* However, most fires take place under conditions of incomplete combustion, where the amount of oxygen available is less than that required for complete combustion. In such a case, carbon monoxide (CO) is produced.

Both carbon dioxide (CO_2) and carbon monoxide are toxic to humans. However, carbon monoxide is considered the major threat to human life in building fires. Although less toxic than some of the other fire gases, it is generally present in such large quantities that its overall toxic effect is most devastating. It is estimated that nearly 50% of fatalities caused by smoke inhalation occur due to the effect of carbon monoxide.

Carbon dioxide is also present in large quantities in a building fire, but its toxicity is considerably lower than that of carbon monoxide. Other gases produced in a fire may include hydrogen cyanide, hydrogen chloride, or acrolein. Hydrogen cyanide is produced from the burning of wool, silk, leather, rayon, and plastics containing nitrogen. Hydrogen chloride is produced from polyvinyl chloride (PVC). Acrolein is produced from wood and paper.

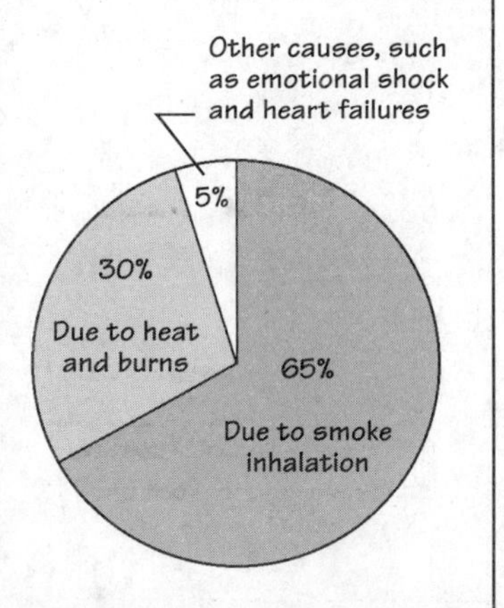

FIGURE 7.6 Causes of deaths in a building fire.

EFFECT OF SMOKE ON VISIBILITY

The solid and liquid particulate matter in smoke reduces visibility. Although reduced visibility has no direct effect on fire fatalities, it impedes the escape of a building's occupants and prolongs their exposure to toxic gases and heat. This often results in panic conditions, which are known to have caused several deaths in building fires. Smoke also has irritating effects on eyes and lungs, and this may result in serious medical problems.

PRACTICE QUIZ

Each question has only one correct answer. Select the choice that best answers the question.

1. Based on recent statistics, approximately how many people are killed in building fires in the United States every year?
   a. 10,000 b. 1,000
   c. 100 d. None of the above

2. The design of the means of egress system in a building is a part of
   a. active fire protection. b. passive fire protection.
   c. basic fire protection. d. upgraded fire protection.
   e. none of the above.

3. The automatic sprinkler system in a building is a part of
   a. active fire protection. b. passive fire protection.
   c. basic fire protection. d. upgraded fire protection.
   e. none of the above.

4. The fire code is one of the most important chapters in a building code.
   a. True b. False

5. A material whose core is noncombustible but is covered with a combustible lamination or paint is considered a noncombustible material, provided that the thickness of the paint or lamination does not exceed
   a. 1 in. b. $\frac{1}{2}$ in.
   c. $\frac{1}{4}$ in. d. $\frac{1}{8}$ in.
   e. $\frac{1}{16}$ in.

6. Statistics based on autopsies conducted on people who died in building fires indicate that most deaths occur due to
   a. burns. b. smoke.
   c. emotional shock. d. heart failure.

7.4 FIRE-RATED ASSEMBLIES AND COMPARTMENTALIZATION OF A BUILDING

One of the important design strategies employed to reduce the fire hazard in a building is to subdivide it into small compartments so that the fire is limited to the compartment of its origin. This strategy is based on the fact that the fire will either extinguish itself after a while, due to the lack of oxygen in the compartment, or take some time to spread to other compartments. During this period, the occupants can evacuate to a place of safety.

Requirements of a Fire-Rated Assembly

The concept of compartmentalization, as shown in Figure 7.7, assumes that the boundary elements of the compartment (walls, floors, and roofs) will function as barriers against the spread of fire to adjacent compartments of the building. We refer to such barriers as *fire-rated assemblies.* To meet this criterion, a fire-rated assembly must satisfy the following three requirements:

1. A fire-rated assembly should be able to perform its structural function without collapse; that is, it should be able to sustain the loads for which it has been designed throughout the duration of the fire.
2. A fire-rated assembly should remain fire-tight; that is, it should develop no cracks during the duration of the fire. The purpose of this requirement is to ensure that smoke and flames will not spread to adjacent compartments.
3. The temperature of the unexposed face of a fire-rated assembly during a fire should be so low that the heat received by radiation and (or) conduction through the assembly will not ignite combustibles in adjacent compartments.

Fire-Resistance Rating

When exposed to a fire for a sufficient length of time, all building assemblies, regardless of their material, will eventually crack, disintegrate, and collapse. Thus, a fire-rated assembly can satisfy the stated requirements for a limited duration only. The ability to endure fire for the duration for which it satisfies all the stated requirements is called the *fire-resistance rating* or simply the *fire rating* of the assembly.

In other words, the fire rating of an assembly is its ability to confine the fire to the enclosure of its origin for a particular duration. Accordingly, it is measured in units of time, that is, in hours or fractions thereof. Thus, a 1-h fire rating means that the fire-rated assembly can endure a typical building fire for at least 1 h. A 2-h fire rating means that it can endure fire for at least 2 h.

Note that by virtue of its definition, the fire rating is a property of the assemblies, such as walls, floors, and roofs, and not simply of the materials of which the assemblies are constructed. Thus, the fire rating of a single material, such as concrete or wood, is meaningless.

Openings in a Fire-Rated Assembly

Obviously, a fire-rated assembly can function as a fire barrier only if the openings in the barrier (such as doors and windows) have an equivalent fire rating. Thus, in addition to the three requirements of a fire-rated assembly, the fourth requirement is that the openings therein should also be fire-rated. Equivalent fire rating of an opening does not mean that its fire rating should be equal to that of the assembly in which it is provided. Generally, the fire rating required of a door or window is less than the fire rating of the wall in which it is located.

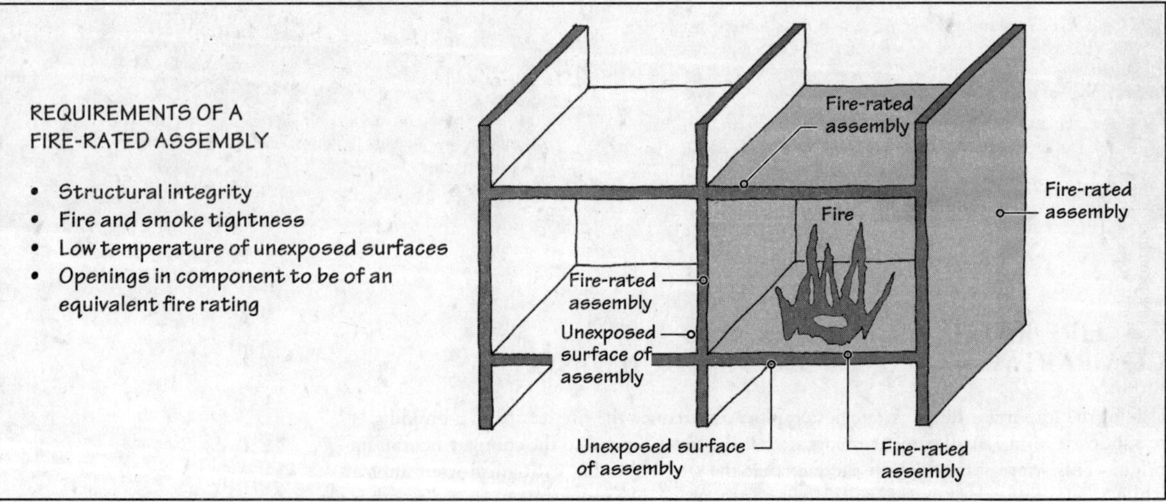

FIGURE 7.7 Compartmentalization of a building by fire-rated assemblies. A fire-rated assembly may be either vertical or horizontal.

7.5 TYPES OF CONSTRUCTION

As stated in Chapter 2, building construction is classified by building codes under the following five types:

- Type I
- Type II
- Type III
- Type IV
- Type V

The classification is based on the fire ratings of the various critical assemblies of the building. The higher the fire rating of these assemblies, the more fire resistive the type of construction. Type I construction is the most fire resistive, and Type V construction is the least fire resistive. For a given occupancy (Chapter 2, Principles in Practice), building codes allow greater area and greater height for a more fire-resistive type of construction.

As shown in Table 7.1, the assemblies whose fire rating is considered in determining the type of construction classification of a building are the structural elements—floor-ceiling assemblies, structural frame (columns, beams, girders, and trusses), bearing walls, and roofs. This highlights the importance of the building's structural integrity in fighting a fire.

Nonbearing exterior walls or interior partitions are not included in determining the building's type of construction, although their fire rating is a consideration in building design and is generally related to the building's type of construction. The fire rating of doors and windows is also not included, but it may be required to satisfy other code provisions.

As Table 7.1 shows, the fire-rating requirement of an assembly is generally given by the number of hours. Thus, the building codes require these assemblies to be rated as 0, 1, 1.5, 2, or 3 h. Number 0 in Table 7.1 implies that there is no requirement for a fire rating; that is, the assembly may be a *nonrated assembly.* Because a rated assembly in Table 7.1 must have a minimum rating of 1 h, an assembly with a rating of 59 min or less is considered a nonrated assembly, that is, one with a fire rating of zero. The maximum fire rating required of any assembly by building codes is 3 h (Table 7.1).

Types I, II, III, and V are further divided into subtypes A and B. Thus, there are nine types of construction classifications. More precisely, therefore, Type I(A) is the most fire-resistive and Type V(B) is the least fire-resistive construction type.

NOTE

Types of Construction—An Overview

Noncombustible Construction–Type Group

Type I and Type II All envelope elements (exterior bearing or nonbearing walls and roof), interior partitions, and the structural frame must be of noncombustible materials such as steel, concrete, or masonry.

Noncombustible/Combustible Construction–Type Group

Exterior (bearing or nonbearing) walls, must be of noncombustible materials in both Types III and IV.

Type III Structural frame and interior partitions may be of combustible materials (wood light-frame), noncombustible materials, or a combination thereof.

Type IV (HT) The structural frame must be of heavy sawn lumber or heavy glue-laminated lumber sections without concealed spaces. That is why Type IV construction is called *heavy timber* (HT) construction. To qualify as heavy timber, the minimum sizes of structural frame members are specified by the code. Interior partitions may be of heavy sawn or glue-laminated lumber, or may be of 1-h-rated construction in any material—combustible, noncombustible, or a combination thereof.

Combustible Construction–Type Group

Type V All structural elements, exterior walls, and internal partitions may be of combustible materials (wood frame), noncombustible materials, or a combination thereof.

TABLE 7.1 MINIMUM FIRE-RATING REQUIREMENTS (IN HOURS) FOR VARIOUS TYPES OF CONSTRUCTION

	Type of Construction		Building's Structural Elements				
			Structural frame (columns, girders, and trusses)	Floor construction (slabs, beams, and joists)	Roof construction (slabs, beams, rafters, and joists)	Bearing walls: Exterior	Bearing walls: Interior
Noncombustible group	Type I	A	3	2	1.5	3	3
		B	2	2	1	2	2
	Type II	A	1	1	1	1	1
		B	0	0	0	0	0
Noncombustible/Combustible group	Type III	A	1	1	1	2	1
		B	0	0	0	2	0
	Type IV	HT	HT	HT	HT	2	1 or HT
Combustible group	Type V	A	1	1	1	1	1
		B	0	0	0	0	0

Notes:
1. **Nonbearing exterior walls and interior partitions** in the noncombustible group (Types I and II) must be of noncombustible materials.
2. **Minimum fire rating of exterior nonbearing walls** is a function of separation distance (distance measured from the wall to the center of the street or to the line between two adjacent lots). If the separation distance is greater than 30 ft, the walls need not have any fire rating.
3. Minimum fire rating of interior partitions depends on the purpose they serve (e.g., corridor walls, tenant separation walls, guest room separation walls in hotels or motels.)

Source: International Building Code (2009), published by the International Code Council.

The five types of construction are categorized under three groups:

- *Noncombustible group*—consisting of Types I (A and B) and Types II (A and B). In these types of construction, all envelope and structural assemblies must be of noncombustible materials (concrete, steel, or masonry).
- *Noncombustible/combustible group*—consisting of Types III (A and B) and Type IV. In these types of construction, the exterior walls must be of noncombustible materials, whereas other assemblies of the building may be of combustible or noncombustible materials. A combustible type of construction generally refers to one whose structural frame consists of wood or wood-based products.
- *Combustible group*—consisting of Types V (A and B). In these types of construction, all envelope and structural elements may be of combustible or noncombustible materials.

Although not expressly mentioned, Types III (A and B) and Types V (A and B) are of wood light-frame construction. The difference between them is that, in Types III (A and B), the exterior walls are required to be of noncombustible materials. In Types V (A and B), the exterior walls may be of combustible or noncombustible materials.

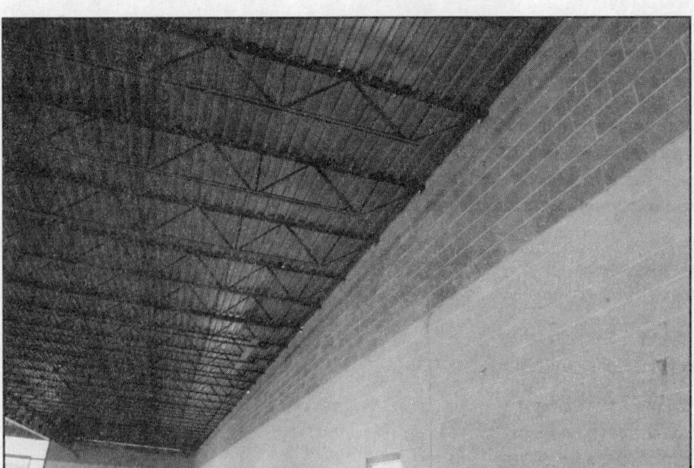

FIGURE 7.8 An example of Type II(B) construction. In this building, the load-bearing walls are of masonry, and other structural elements (columns, roof joists, and roof deck) are of steel. Therefore, it is a noncombustible type of construction—Type I or Type II. However, because the roof assembly is not protected against fire, its fire rating is (much) less than 1 h, qualifying as a nonrated assembly. Therefore, it is a Type II(B) construction.

Note that the fire rating required of various members in Type II(B) and Type V(B) is the same—zero. The difference between the two is that in Type II(B), all envelope and structural elements must be of noncombustible materials. Generally, such construction consists of masonry walls, unprotected steel columns, beams, and roof deck, Figure 7.8. Type V(B) consists of wood light-frame construction. Most single-family dwellings are of Type V(B), in which the envelope and structural elements need not be fire rated (zero fire rating). A typical example of Type V(B) construction is shown in Figure 7.9.

Type IV is called *heavy timber* (HT) construction. In HT construction, floors, roofs, and the structural frame are of heavy sections of wood without any added gypsum board or similar fire protection. In other words, in HT construction, the wood members are exposed without any concealed spaces such as those that occur in a conventional wood light-frame building. For example, in a conventional wood light-frame floor, concealed

FIGURE 7.9 Most single-family and multifamily dwellings in North America are built of wood frame construction, as shown here (see also Chapters 15 and 16). The wood members are typically protected against fire with $\frac{1}{2}$-in.-thick gypsum board on both sides, which gives the wall, floor, and roof assemblies a fire rating of nearly 40 min. Since the fire rating is less than 1 h, such assemblies meet the requirements of Type V(B) construction—a nonrated construction.

spaces occur between floor joists, Figure 7.10. Similar spaces occur in conventional wood light-frame walls and attics. Such spaces are absent in HT construction.

HT construction is slow burning and is assumed to provide an equivalent of a 1-h fire rating. Thus, the overall fire endurance of HT construction is generally considered equivalent to that of Type III(A). An example of Type IV (HT) construction is shown in Figure 7.11.

A typical example of Type I(A) construction is a cast-in-place reinforced concrete building, Figure 7.12. It is relatively easier to obtain a Type I(A) or Type I(B) classification in a reinforced concrete building compared with a steel frame building because a steel frame building requires add-on fire protection, while a reinforced concrete building has inherent protection provided by concrete.

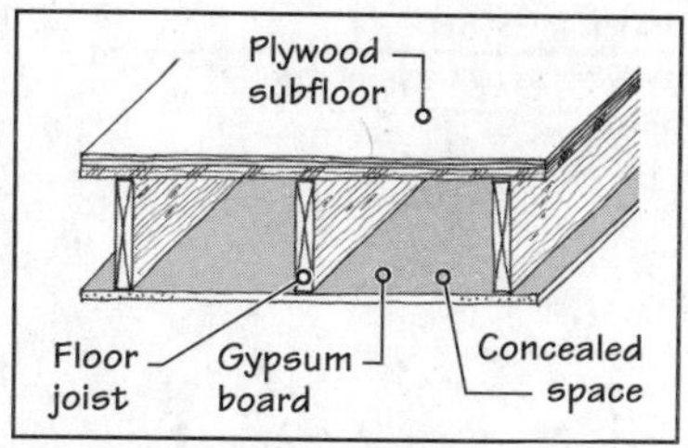

FIGURE 7.10 Floor assembly in a typical wood light-frame construction (Type V construction) illustrates concealed spaces that occur in walls, floors and roofs with such construction.

IDENTIFYING THE TYPE OF CONSTRUCTION

For a building to be classified as a given type of construction, the fire ratings of various assemblies must either be equal to or greater than those given in Table 7.1. The fire ratings given in Table 7.1 are minimum fire ratings. A building cannot be designated as

FIGURE 7.11 An example of heavy timber construction—Type IV. Notice heavy sections of wood with no protective covering and no concealed spaces.

FIGURE 7.12 A cast-in-place reinforced concrete building (under construction). It is not possible to precisely judge its type of construction [Type I(A), Type I(B) or Type II(A)] without examining the construction drawings of the building because the fire rating of a cast-in-place reinforced concrete building is a function of the dimensions of the structural members and the concrete cover over steel reinforcement. However, such a building is likely to be a Type I(A) or Type I(B) construction, probably Type I(A).

belonging to a particular type of construction unless it meets or exceeds all the fire-rating requirements for that type of construction.

Note again that only the structural assemblies of a building determine its type of construction. These include (interior and exterior) load-bearing walls, the structural frame (columns, beams, trusses, etc.), floor assemblies, and roof assemblies. In these assemblies, only the part that is structural is considered in determining whether the assembly is combustible or noncombustible. The nonstructural parts of the assembly (insulation, doors, windows, trims, etc.) are ignored because the building code allows several combustible finishes in Types I and II construction.

For example, a roof assembly consisting of a reinforced concrete slab (which is noncombustible) topped by plastic foam insulation and a polymeric roof membrane (which are both combustible) is considered a noncombustible roof assembly. Similarly, a noncombustible structural floor with a combustible floor finish, such as carpet or vinyl tiles, is regarded as a noncombustible floor.

Determining the Type of Construction

Example 1

Using Table 7.1, determine the type of construction of a building in which the materials and fire ratings of different components are as follows:

Structural frame—steel columns and steel girders	2-h rating
Exterior bearing walls	None provided
Interior bearing walls	None provided
Floor assemblies—steel deck on steel beams with concrete topping	1-h rating
Roof assembly—steel deck	1-h rating

Solution

Because all components are of noncombustible materials, the building is either Type I or Type II construction. Comparing the fire ratings of the building's components with those of Table 7.1, the building is classified as Type II(A).

Example 2

What changes must be made in the fire ratings of components to upgrade the construction of Example 1 to Type I(B)?

Solution

Examining the fire ratings required for Type I(B) construction in Table 7.1, we see that if the fire rating of the floor assemblies of Example 1 is increased to a minimum of 2 h, it can be classified as Type I(B) construction.

Example 3

Determine the type of construction for the building in which the materials and fire ratings of different components are as follows:

Structural frame—columns and girders	None provided
Exterior bearing walls—concrete masonry	2-h rating
Interior bearing walls—wood studs and gypsum board	1-h rating
Floor assemblies—wood joists and plywood floors	1-h rating
Roof assembly—wood rafters and plywood sheathing	1-h rating

Solution

Because some components are of combustible materials, the building will be classified as Type III, Type IV, or Type V. Type IV is ruled out since the roofs and floors are not constructed of heavy timber. Comparing the given fire ratings with those of Table 7.1, we observe that the building is of Type III(A) construction. Note that the exterior bearing walls are of noncombustible materials, as required for this type of construction. If nonbearing exterior walls are provided, they also must be of noncombustible materials.

Example 4

If the exterior bearing walls of Example 3 are of wood studs and gypsum board, but with the same (2-h) fire rating, and all other components remain unchanged, what will be the revised construction type?

Solution

Because the exterior bearing walls are of a combustible material, the type of construction will be Type V(A).

PRACTICE QUIZ

Each question has only one correct answer. Select the choice that best answers the question.

7. Which of the following properties are required of a fire-resistive building component?
   a. Structural integrity during the fire
   b. Freedom from cracks during the fire
   c. Resistance to the change of color during the fire
   d. All of the above
   e. (a) and (b)

8. Fire rating is a property of
   a. the material of which a component is made.
   b. the thickness of the component.
   c. the entire assembly comprising the component.
   d. weight of the component.
   e. all of the above.

9. Fire rating is measured in units of
   a. time.
   b. temperature increase per hour.
   c. heat released by the component in a fire test.
   d. pounds per square inch.
   e. all of the above.

10. Building codes divide the types of construction into
    a. Types 1 to 5.
    b. Types I to V.
    c. Types 1 to 6.
    d. Types I to VI.
    e. none of the above.

11. Noncombustible types of construction are
    a. Types 1, 2, and 3.
    b. Types 1 and 2.
    c. Types I, II, and III.
    d. Types I and II.
    e. Type I.

12. Combustible types of construction are
    a. Types 4, 5, and 6.
    b. Types 4 and 5.
    c. Types IV, V, and VI.
    d. Type VI.
    e. Type V.

13. Among the various types of construction included in a building code, which is the most fire-resistive type?
    a. Type I(a)
    b. Type V(A)
    c. Type V(b)
    d. Type HT
    e. None of the above

14. The total number of types of construction, including their subclassifications, in a building code is
    a. 6.
    b. 7.
    c. 8.
    d. 9.
    e. 10.

15. Which type of construction is commonly used in a typical single-family dwelling in the United States?
    a. Type I(a)
    b. Type V(A)
    c. Type V(b)
    d. Type HT
    e. None of the above

16. The fire rating of building components is generally obtained from fire tests, not through calculations.
    a. True
    b. False

7.6 FIRE-STOPPING OF PENETRATIONS AND FIRE-SEALING OF JOINTS

Walls, floors, and roofs in buildings are often penetrated by HVAC ducts, pipes, electrical conduits, and communication wiring. If these assemblies (walls, floors, and roofs) are to function as effective barriers to the spread of fire, the space around the penetrations must be thoroughly sealed. The sealing of the space around a penetration is called *fire-stopping.*

An unsealed space will allow heat, smoke, and flammable gases to move from one side of the barrier to the other, compromising the fire rating of the barrier. The fire rating of fire-stops is generally required to match the fire rating of the barrier in which they are installed.

Fire-stopping materials must obviously be noncombustible. Several are in common use. They consist of mineral wool, sealants, and foams, Figure 7.13. Foamed fire-stop is a material that expands at room temperature and fills the space. Fire-stop materials also help to seal spaces against air and sound transmission.

Just as penetrations through a fire-rated building element are required to be fire-stopped, joints in them (such as expansion joints and control joints, Chapter 9) are also required to be fire rated.

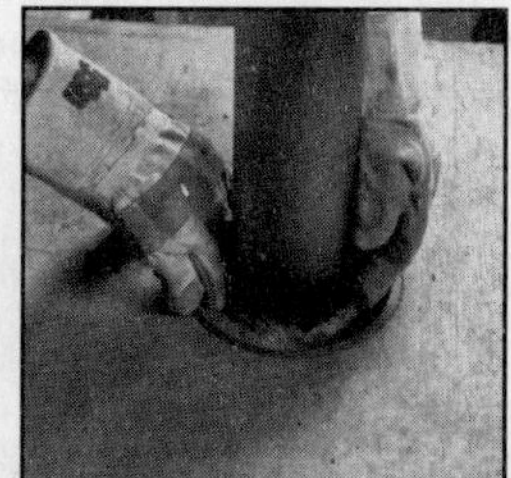
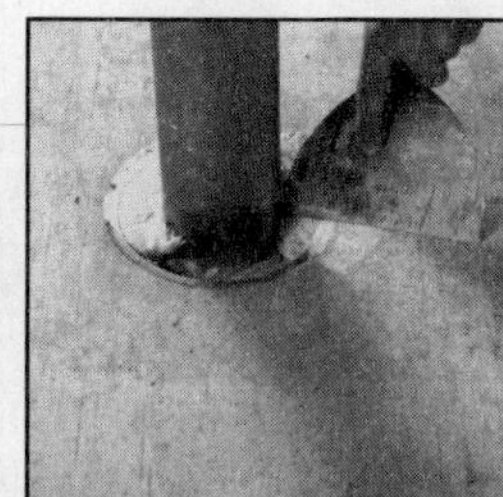
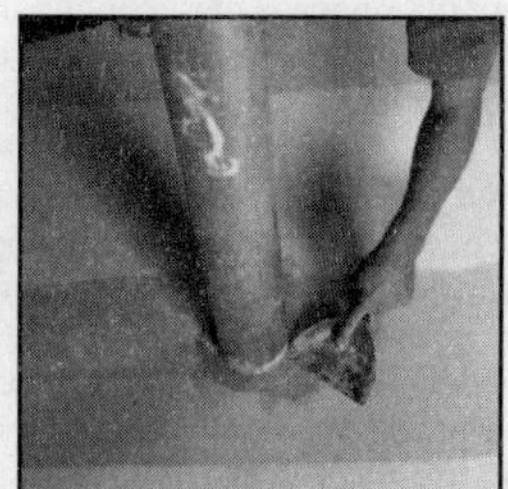

FIGURE 7.13 A typical fire-stopping around a penetration in a floor generally consists of high-density mineral wool pressure fitted into the void, followed by a semiliquid, fire-resistant sealant troweled over mineral wool packing. (Photo courtesy of U.S. Gypsum Company)

BUILDING CODE REQUIREMENTS FOR FSI AND SDI

Based on FSI and SDI values, building codes classify interior finishes into three classes: Class A, Class B, and Class C, Table 7.3. The objective of this classification is to regulate the use of interior finish materials according to the fire safety requirements of spaces. Because Class A interior finishes are the least hazardous, building codes mandate that they be used in enclosed vertical exit ways—staircases and elevators. A minimum of Class B interior finish material is required in other exit ways—corridors, lounges, foyers, and so on. Class C finishes may be used in areas such as individual rooms, Figure 7.15.

TABLE 7.3 CLASSIFICATION OF INTERIOR FINISHES

Flame spread class	FSI	SDI
Class A	0–25	<450
Class B	26–75	<450
Class C	76–200	<450

There are a few exceptions to these rules, for which the applicable building code should be consulted. For example, Class C materials are allowed to be used in all areas of a single-family dwelling, even in stair enclosures and corridors. Additionally, the provision of automatic sprinklers in a building allows Class B finishes to be used in vertical exit areas where Class A is generally needed.

No classification is given to interior finishes with an FSI greater than 200 and an SDI greater than 450, because these materials are not allowed as interior finish materials except in utility occupancies. Only finishes applied on walls and ceilings are regulated for FSI and SDI values. Floor finishes such as carpets and floor tiles are not rated for FSI or SDI.

NOTE

Pill Test

The pill test simulates the effect of a lighted cigarette, match, or fireplace ember on a carpet or rug. In the test, a methenamine tablet is placed on the carpet or rug specimen, and the tablet is ignited with a match. If the specimen burns more than 3 in. in any direction from the location of the tablet, the specimen fails the test.

FIRE AND FLOOR FINISHES

Because flames spread vertically upward, floor finishes do not get involved in a building fire until fairly late in the burning process. Therefore, most traditional nonfibrous floor finishes, even the combustible finishes (such as wood, vinyl, and linoleum), are exempt from any fire-related properties. However, the fire properties of carpets and rugs are regulated, and they must pass the *pill test.*

FIRE AND ROOF COVERINGS

Although roof coverings (required for the waterproofing of roofs) are exterior to the building, their fire properties are regulated because their burning can pose a hazard to an adjacent building or an adjacent part of the same building. The reverse may also occur in that the burning embers of roof-covering material on a building may land on an adjacent building. Roof-covering materials are classified as Class I, Class II, or Class III (as per the ASTM E108 test), Class I being the most fire resistive and Class III the least fire resistive. Building codes mandate a higher class of roof-covering material for a more fire-resistive type of construction.

FIRE RESISTANCE AND SURFACE FLAMMABILITY

Note that the fire rating of barriers and the surface-burning properties of interior finishes and roof coverings are entirely independent and unrelated properties. Whereas the fire rating is a measure of the structural integrity of a barrier during a fire, the surface burning properties (e.g., FSI and SDI) measure the fire hazard of a finish material. For example, HT construction may give the equivalent of 1 h or more of fire rating, but it presents some

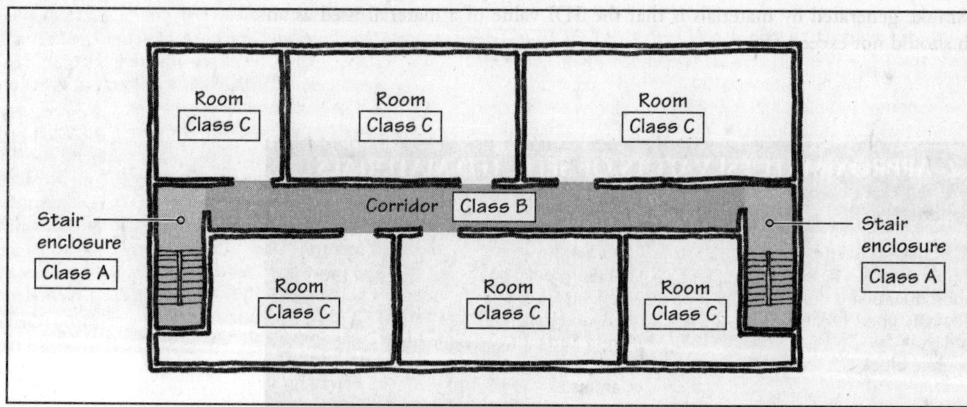

FIGURE 7.15 General building code requirements for the class of interior finishes.

degree of hazard as an interior finish. A thin steel plate, on the other hand, is a poor fire barrier with a virtually zero fire rating, but it presents no hazard as an interior finish material.

7.8 IMPORTANCE OF ACTIVE FIRE PROTECTION

As previously stated, active fire protection (through the use of automatic sprinklers) is an important part of fire-safe construction. Building codes encourage the use of automatic sprinklers by allowing greater area and height for a building that is sprinklered (see Chapter 2, Principles in Practice) in comparison with a building that is not sprinklered. For some occupancies, however, the use of automatic sprinklers is mandated by building codes, even when the passive fire protection provided in the building is of a high order (e.g., Type I(A) or Type I(B) construction).

In addition to permitting an increase in the area and height of the building, building codes allow a reduction in fire-resistance rating for some occupancies if the building is sprinklered. Lowering of the class of interior finish materials (from Class A to Class B or from Class B to Class C) is also permitted in some situations if the building is sprinklered.

The trade-off between active and passive protection (i.e., downgrading the type of construction of the building for the use of sprinklers) is a matter of debate among experts. Proponents of passive systems (which include the manufacturers of spray-on fire protection, fire-stopping materials, gypsum board, etc., and the design and construction community in general) believe that passive fire protection is foolproof. Being an integral part of a building's type of construction, passive protection is forever present and requires little or no maintenance, whereas an active system (being an electromechanical system) may fail or malfunction due to any one of several reasons—power outage, inadequate water supply, corrosion of pipes and fittings, poor system maintenance, and so on.

Proponents of active systems (e.g., the National Fire Sprinklers Association) claim that their systems are tested after initial installation and routinely maintained thereafter throughout the building's existence. The statistics of injuries and fatalities from building fires clearly illustrate the role sprinklers play in adding to the fire safety of buildings. It is for this reason that the use of sprinklers is being mandated by codes in situations where none was required earlier.

For example, the 2009 International Residential Code (IRC) requires that all new one- and two-family dwellings and townhouses in the United States constructed after January 1, 2011, shall be equipped with automatic sprinklers in the hope that this will reduce the number of fatalities in one- and two-family dwellings in the United States. This IRC provision is highly controversial among stakeholders because although it has obvious life-safety and property protection benefits, it also has the drawback of increasing the initial construction and long-term maintenance costs. Consequently, several states have adopted 2009 IRC with the sprinkler mandate, while others have adopted 2009 IRC without the sprinkler mandate.

NOTE

Fatalities from Building Fires and Natural Disasters in the United States

Fire continues to be a major killer of humans in buildings. As stated at the beginning of this chapter, the annual average number of fatalities from building fires in the United States is approximately 3,000. By comparison, the corresponding average number of fatalities from all natural disasters (hurricanes, snowstorms, tornadoes, floods, earthquakes, etc.) in the United States is approximately 200.

An extremely revealing detail of building-fire fatalities is that (of the total 3,000) nearly 70% occur in one- and two-family dwellings; 12% occur in other residential occupancies (multifamily dwellings and hotels). Commercial buildings account for fewer than 8% of the fatalities.

PRACTICE QUIZ

Each question has only one correct answer. Select the choice that best answers the question.

17. The term *fire-stopping* refers to
- **a.** a fire-rated wall with a minimum fire rating of 3 h.
- **b.** a fire-rated floor with a minimum fire rating of 3 h.
- **c.** a fire seal in a non-fire-rated wall or floor.
- **d.** a fire seal in a fire-rated wall or floor.
- **e.** none of the above.

18. The term *FSI* refers to
- **a.** fire safety index.
- **b.** fire and smoke index.
- **c.** fire-speed index.
- **d.** flame-spread index.

19. SDI measures
- **a.** visibility through smoke.
- **b.** toxicity of smoke.
- **c.** rate of smoke generation.
- **d.** amount of carbon dioxide in smoke.
- **e.** all of the above.

20. Based on their fire properties, interior finishes are divided into
- **a.** Classes I, II, III, and IV.
- **b.** Classes A, B, C, and D.
- **c.** Classes W, X, Y, and Z.
- **d.** Classes A, B, and C.
- **e.** Classes X, Y, and Z.

21. The fire properties that determine the class of interior finishes are
- **a.** fire-resistance rating and FSI.
- **b.** FSI and SDI.
- **c.** fire-resistance rating and SDI.
- **d.** fire-resistance rating, FSI, and SDI.

22. The maximum value of SDI recognized by building codes is
- **a.** 450.
- **b.** 300.
- **c.** 150.
- **d.** 50.
- **e.** none of the above.

REVIEW QUESTIONS

1. List the factors that affect fire safety in buildings.
2. Discuss the differences between a fire code and a building code.
3. Explain why we need a test to determine whether a building material is combustible or noncombustible when we generally know which material is combustible and which is noncombustible.
4. What is the difference between Type III(A) and Type IV construction? Explain.
5. Determine the type of construction of a building with the following data (Table 7.1):
   Structural frame; structural steel; 2-h rated
   Floors; steel joists, steel deck, and concrete topping; 1-h rated
   Roof; steel deck; 2-h rated
   Exterior bearing walls; none provided
   Interior bearing walls; none provided
6. What type of construction would the building in Problem 5 be if the floor's fire rating is increased to 2 h?
7. Discuss the difference between active and passive fire protection of buildings.
8. Discuss the pros and cons of using active fire protection for one- and two-family homes and townhouses.

CHAPTER 8 Acoustical Properties of Materials

CHAPTER OUTLINE

Acoustical and noise-control concerns exist in almost every modern building. An acoustical consultant is usually required for the design of an auditorium or a concert hall or for the solution of a complicated noise problem. However, commonplace acoustical issues in most buildings, such as selecting the appropriate interior finishes for a lecture room, controlling noise in a busy dining hall, or isolating a hotel bedroom from street noise, are fairly elementary and are resolved by design professionals without a consultant's help.

This chapter provides an understanding of the acoustical properties of materials and assemblies that are routinely used in buildings. It begins with a discussion of the basic characteristics of sound waves (frequency, speed, and wavelength), leading to the decibel scale—a unit for rating the loudness of sound. This is followed by a discussion of the sound-insulation and sound-absorbing properties of materials.

8.1 FREQUENCY, SPEED, AND WAVELENGTH OF SOUND

Sound is the human ear's response to pressure fluctuations in the air caused by vibrating objects. For example, a tap on a wall produces sound because the tap makes the wall vibrate back and forth. The back-and-forth motion of the wall is transferred to air particles in direct contact with the wall. These particles then transfer their vibratory motion to the neighboring particles, and so on.

The domino-type transfer of vibratory motion from particle to particle is what we refer to as a *sound wave*. The number of back-and-forth cycles that an object (or the air particles)

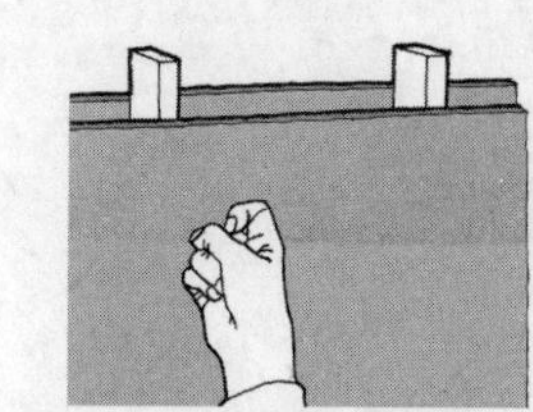

A tap on a wall makes the wall vibrate, which in turn makes the air particles vibrate, producing sound waves.

moves in 1 s in a sound wave is called *sound frequency.* Its unit is cycles per second (c/s), which is also termed *hertz* (Hz) after the Austrian physicist Heinrich Hertz. Vibrations with frequencies lying between 20 Hz and 20,000 Hz (i.e., 20 kHz) are audible. This range of frequencies is called the *audible frequency range.*

Subjectively, the frequency of a sound is perceived as its pitch. A high-pitched sound means that it has a high frequency. A female voice is slightly higher-pitched than a male voice.

Sounds in our environment do not generally consist of individual frequencies (a single note or pure tone), such as that produced by a tuning fork. Most sounds are complex; that is, they consist of a continuum of several frequencies. Thus, the human speech consists of all frequencies ranging from nearly 200 Hz to 5 kHz. The male voice peaks at about 400 Hz, and the female voice peaks at about 500 Hz. The range of frequencies in music is larger than the range for speech.

Frequency of sound is an important acoustical concept because the properties of building materials and construction assemblies are frequency dependent, meaning that building products vary in how they transmit or absorb sound in its many frequencies.

NOTE

Cycles per Second and Hertz

Hertz (Hz) and cycles per second (c/s) are used interchangeably, that is,

1 Hz = 1 c/s

1,000 Hz = 1 kilohertz = 1 kHz

NOTE

Speech Sounds

Human speech lies in the frequency range of 200 Hz to 5 kHz. However, most of the speech energy lies in the range of 250 Hz to 2 kHz.

NOTE

Infrasonic, Ultrasonic, and Supersonic

Frequencies below 20 Hz are called *infrasonic* frequencies. They are not heard, but they are perceived by humans as vibrations.

Frequencies above 20 kHz, referred to as *ultrasonic* frequencies, are also not heard by humans.

An object traveling at a speed greater than the speed of sound is said to be traveling at *supersonic* speed.

Speed of Sound

The speed of sound in the air has been measured as 1,130 ft/s (344 m/s) and is independent of sound frequency or loudness. Because sound travels from particle to particle in a medium, a medium is required for the existence of sound. In other words, sound cannot be produced or travel in a vacuum. This is in contrast with light and heat waves, which, being electromagnetic in nature, do not require a medium for travel.

Wavelength of Sound

As a sound wave travels, it creates excess pressure (called *compression*) and reduced pressure (called *rarefaction*) in space, just as water waves produce *crests* and *troughs.* The distance between adjacent compression peaks or adjacent rarefaction peaks in the air at an instant of time is called the *wavelength.* The frequency, wavelength, and speed of wave motion are related to each other by the following simple relationship:

$$\boxed{\text{Speed} = \text{frequency} \times \text{wavelength}} \qquad \textbf{Eq. (1)}$$

Thus, the wavelength of sound at 100 Hz is 1,130/100 = 11.3 ft. At 1 kHz, the wavelength is only 1.13 ft. At 10 kHz, the wavelength is 0.113 ft, that is, approximately 1 in. Wavelength is not as commonly used in describing material properties as frequency. However, as Equation (1) shows, it is easy to convert from frequency to wavelength and vice versa.

8.2 THE DECIBEL SCALE

The physical quantity associated with the loudness of sound is sound pressure, which can be expressed in pounds per square foot (psf) but is generally expressed in pascals (Pa), a unit in the SI system. The minimum sound pressure to which our ear responds, the *threshold of audibility,* is 0.00002 Pa. The sound pressure that corresponds to the sensation of pain in the ear is approximately 20 Pa (0.4 psf). This is still a very small pressure compared to the atmospheric pressure (101,300 Pa = 2,100 psf) under which we all live.

Although the sound pressures are small, their range is extremely large. Therefore, it is more convenient to use the *decibel* scale in expressing sound pressure. The decibel scale is a logarithmic scale that compares a given sound pressure with the reference sound pressure of 0.00002 Pa, the threshold of audibility.

To distinguish sound pressure (expressed in pascals) from that expressed in decibels, we call the latter the *sound pressure level* (SPL). In other words, the unit for SPL is the decibel (dB). The threshold of audibility is 0 dB, and the threshold of pain is 130 dB. Figure 8.1 gives the approximate SPLs of common environmental sounds.

Obtaining convenient numbers is not the only reason for the use of the decibel scale. The decibel scale also corresponds more directly to the ear's perception of loudness. For instance, a change of 1 dB in SPL is hardly perceived by the human ear. The minimum change in SPL that is just perceptible is 3 dB. A 5-dB change is quite noticeable, and a 10-dB change is perceived as a substantial change.

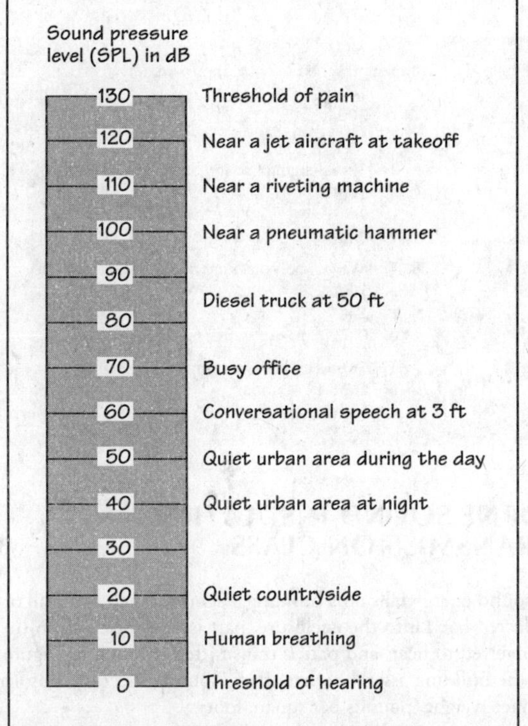

FIGURE 8.1 Commonly occurring sounds and their decibel ratings.

8.3 AIRBORNE AND STRUCTURE-BORNE SOUNDS

In building acoustics, we distinguish between two types of sound based on the sound's origin. The reason for the distinction is that building components behave differently with respect to them. The two types of sound are

- Airborne sound
- Structure-borne sound

Most sounds in buildings are airborne sounds, such as the sounds generated by human conversation and musical instruments, Figure 8.2. Fans, motors, machinery, airplanes, and automobiles are some of the other sources that produce airborne sound.

FIGURE 8.2 Examples of airborne sound sources.

Structure-borne sound is produced by an impact of some sort on building components—walls, floors, roofs, and so on. The impact causes building elements to vibrate, and as they vibrate, they radiate sound. Because it is impact related, structure-borne sound is also referred to as *impact sound.* Thus, when a nail is driven into a wall or a person walks on a floor, structure-borne sound is produced, Figure 8.3. Other examples of structure-borne sounds are vibrating machinery rigidly connected to a floor, plumbing pipes attached to a wall, and the slamming of a door.

In other words, a structure-borne sound originates in an impact- or vibration-producing source that is in contact with a building component. The building component works as an amplifier of the sound generated by the impact or vibrating source. The impact or vibrating source itself may not create much sound.

For example, a vibrating water tap does not create much sound of its own, but if it is rigidly connected to a wall, its sound is greatly amplified because of the vibrations it produces in the wall. Similarly, the sound produced by a string instrument, such as a guitar, is amplified severalfold by the wooden body (sounding board) on which the strings are mounted.

Some sources can produce both airborne and structure-borne sounds. Once the structure-borne sound is produced by a building component, it becomes airborne sound and reaches the receiver as such.

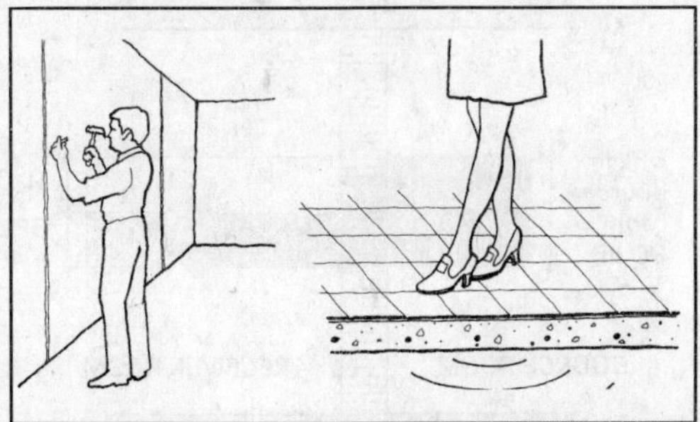

FIGURE 8.3 Examples of structure-borne sound sources.

PRACTICE QUIZ

Each question has only one correct answer. Select the choice that best answers the question.

1. Cycles per second is also called a
   - a. pascal.
   - b. newton.
   - c. hertz.
   - d. hulen.
   - e. none of the above.
2. Sound that is audible to human beings has a frequency lying between
   - a. 2 kHz and 20 kHz.
   - b. 200 Hz and 20 kHz.
   - c. 100 Hz and 800 Hz.
   - d. 20 Hz and 20 kHz.
   - e. 50 Hz and 50 kHz.
3. The pitch of a sound is related to
   - a. sound frequency.
   - b. sound energy.
   - c. sound speed.
   - d. all of the above.
   - e. none of the above.
4. Sound speed in air is approximately
   - a. 10 ft/s.
   - b. 100 ft/s.
   - c. 500 ft/s.
   - d. 1,000 ft/s.
   - e. 1,100 ft/s.
5. The wavelength of sound at a frequency of 1,000 Hz is approximately
   - a. 10 ft.
   - b. 5 ft.
   - c. 2 ft.
   - d. 1 ft.
   - e. 0 ft.
6. The wavelength of sound at a frequency of 100 Hz is approximately
   - a. 10 ft.
   - b. 5 ft.
   - c. 2 ft.
   - d. 1 ft.
   - e. 0 ft.
7. The sound created by human speech is
   - a. airborne sound.
   - b. structure-borne sound.
   - c. both (a) and (b).
   - d. neither (a) nor (b).

8.4 AIRBORNE SOUND INSULATION—SOUND-TRANSMISSION CLASS

When airborne sound energy falls on a building assembly (such as a wall or ceiling), part of the energy is reflected back into the enclosure, part is absorbed within the material of the assembly and converted to heat, and part is transmitted through it, Figure 8.4. An important property of the building assembly that affects sound reflection, absorption, and transmission is its surface weight (pounds per square foot).

Massive, heavyweight assemblies, such as thick masonry or concrete walls, are good sound insulators because they reflect back most of the sound. Stated differently, heavyweight assemblies are poor transmitters because a relatively small fraction of sound energy goes through them. Conversely, lightweight assemblies, such as stud walls and wood floors, are generally poor sound insulators because a relatively large amount of sound energy goes through them. However, as described later, the sound-insulating properties of lightweight assemblies can be improved by a few special provisions.

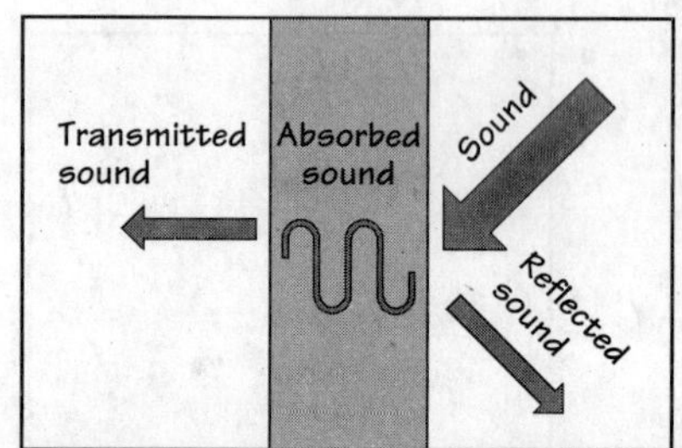

FIGURE 8.4 When a sound falls on a building assembly, part of it is reflected, part is absorbed, and the remaining part is transmitted through the assembly.

The airborne sound insulation of assemblies is measured by a quantity called *sound-transmission loss* (TL). TL is defined as the loss in sound-pressure level that occurs as the sound passes through the assembly. For example, if the sound-pressure level on the source side of a wall is 90 dB and it is 40 dB on the receiver side, the TL of the wall is 50 dB, Figure 8.5. If the sound-pressure level on the source side is 70 dB, then the sound-pressure level on the receiver side of the same wall is 20 dB.

The greater the TL of an assembly, the greater the sound insulation provided by it. An assembly with a TL of 50 dB is a better sound insulator than one with a TL of 40 dB. The TL of an assembly varies with frequency, generally increasing with increasing frequency. In other words, building assemblies generally provide greater sound insulation at high frequencies than at low frequencies.

Because of the variation of TL with frequency, TL cannot be used to compare the sound-insulating efficacy of one assembly with another. For such a comparison, a single-number index is required. This is provided by the quantity called *sound-transmission class* (STC). The STC of an assembly is its average TL over frequencies ranging from 125 Hz to 4 kHz. The greater the STC, the more sound insulating the assembly. Note that the unit decibel is omitted in quoting the STC value in order to distinguish STC from TL. Thus, we use STC 54, not STC 54 dB, in stating the sound insulation of an assembly. Whole numbers are used in quoting the STC.

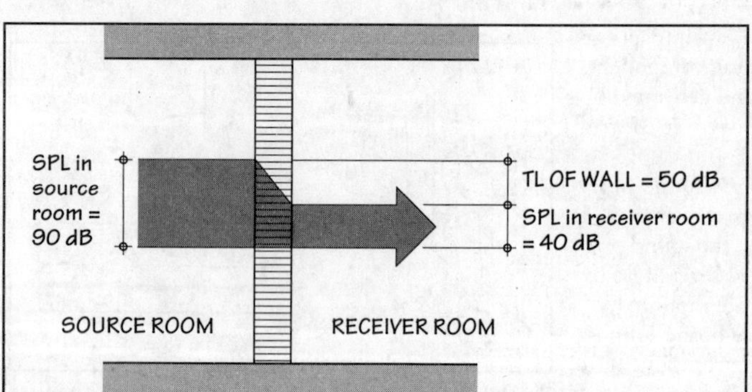

FIGURE 8.5 Definition of the transmission loss (TL) of an assembly.

SUBJECTIVE PERCEPTION OF STC VALUES

STC is an important acoustical index. Manufacturers routinely provide the STC values of their assemblies.

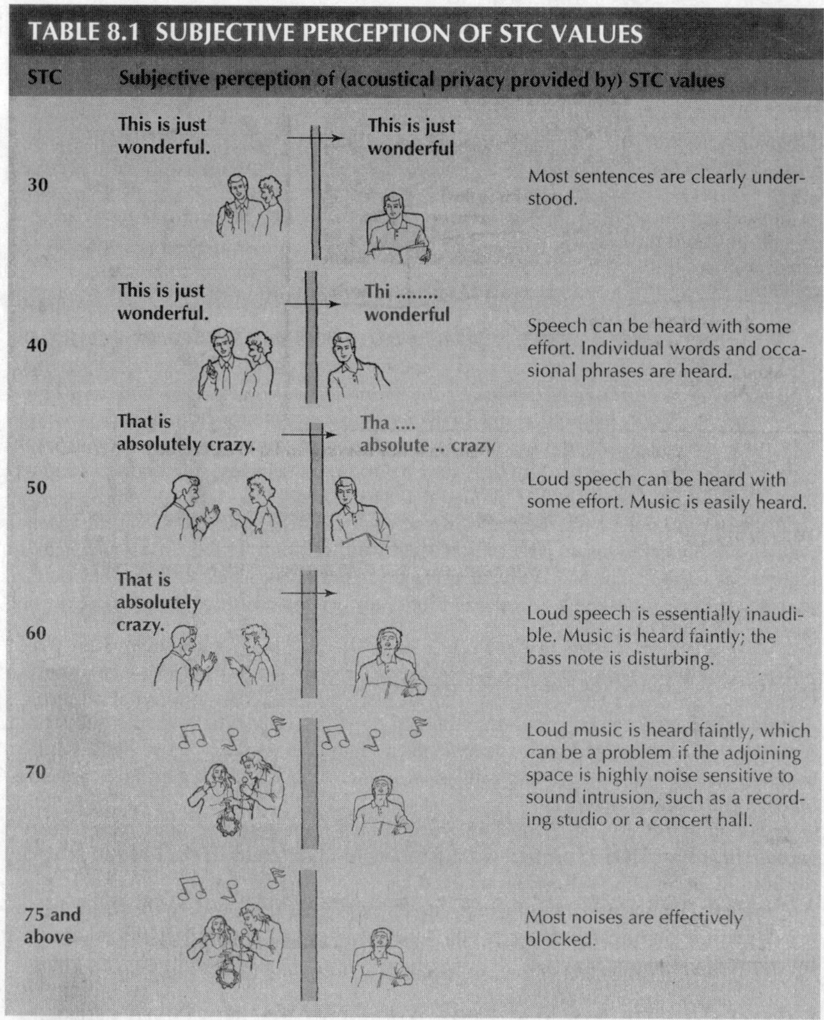

TABLE 8.1 SUBJECTIVE PERCEPTION OF STC VALUES

STC	Subjective perception of (acoustical privacy provided by) STC values		
30	This is just wonderful.	This is just wonderful	Most sentences are clearly understood.
40	This is just wonderful.	Thi .... wonderful	Speech can be heard with some effort. Individual words and occasional phrases are heard.
50	That is absolutely crazy.	Tha .... absolute .. crazy	Loud speech can be heard with some effort. Music is easily heard.
60	That is absolutely crazy.		Loud speech is essentially inaudible. Music is heard faintly; the bass note is disturbing.
70			Loud music is heard faintly, which can be a problem if the adjoining space is highly noise sensitive to sound intrusion, such as a recording studio or a concert hall.
75 and above			Most noises are effectively blocked.

This table assumes a reasonably quiet environment in the receiving room.

Building codes require that the party wall separating two dwelling units (condominiums or apartments) must have a minimum STC of 50. A higher value is usually required as a good design practice. But how much higher? In other words, how much acoustical privacy does an STC 50 assembly or an STC 55 assembly provide? Table 8.1 attempts to answer these questions. As illustrated in this table, an assembly with an STC of 50 will allow loud speech to be heard through the assembly by a curious listener.

IMPROVING THE STC VALUE OF LIGHTWEIGHT ASSEMBLIES

As stated earlier, the TL (and hence the STC) of an assembly can be increased if it is made heavier. However, this not only adds to the cost of the assembly but also increases the dead load on the structure. That is why lightweight assemblies, such as steel or wood stud walls lined with gypsum board on both sides, are commonly used.

The STC of a wall with 2 × 4 wood studs and $\frac{1}{2}$-in. gypsum board on each side is only 37. This is an extremely low STC value and, as shown in Table 8.1, provides little acoustical privacy. The following are commonly used methods of increasing the STC of stud-framed walls:

- Add fibrous insulation (e.g., fiberglass or mineral wool) within stud cavities.
- Decouple the gypsum board layer on one side of the assembly from the remaining part of the wall.
- Use multiple gypsum board layers to increase the weight (mass) of the assembly.

TABLE 8.2 APPROXIMATE STC VALUES OF SELECTED WOOD STUD ASSEMBLIES WITH $\frac{1}{2}$-IN.-THICK GYPSUM BOARD ON BOTH SIDES OF STUDS

Gypsum board support system	Without cavity absorption Layers of gypsum board 1 + 1	1 + 2	2 + 2	With cavity absorption Layers of gypsum board 1 + 1	1 + 2	2 + 2
2 × 4 wood studs	37	40	43	40	43	46
2 × 4 wood studs, resilient channel (or clip) on one side	40	45	49	50	53	57
2 × 4 wood studs, resilient channel (or clip) on both sides	41	46	51	49	53	58
2 × 4 wood stud wall (staggered studs)	41	48	52	50	54	58
Double-stud wall with 1-in. space between studs	46	53	57	57	61	63

As shown in this table, using a resilient channel (or clip) on one side of a wall is adequate. Using the channel (or clip) on both sides of the wall gives a negligible increase in STC value.

Fibrous insulation helps to retard the vibration of air within the wall cavity, thereby absorbing the sound. Decoupling reduces the transfer of vibrations from one side of the wall to the other. Using multiple gypsum board layers increases the surface weight of the wall. Table 8.2 shows the STC values of a few selected lightweight assemblies.

Decoupling is one of the most effective means of increasing the STC of a lightweight wall. It is achieved by using either a staggered-stud wall or a double-stud wall assembly, Figures 8.6 and 8.7. The STC value of a staggered-stud wall is lower than that of a double-stud wall because the staggered-stud wall is not fully decoupled due to the presence of common top and bottom plates.

Resilient Channels and Resilient Clips

A more commonly used means of decoupling is to use a resilient channel or a resilient clip. A resilient channel is a light-gauge steel member, Figure 8.8 (see also Figure 20.6). Resilient channels are fastened to studs (typically at 24 in. on center), and the gypsum board is fastened to the resilient channels.

In fastening the resilient channel, its free end must be toward the top, so that when the gypsum board is fastened, the free end pulls away from the studs. If the free end is toward the bottom, decoupling will not be achieved.

A resilient clip is a more recently introduced device as an alternative to the resilient channel. It consists of a light-gauge steel clip with a rubber pad, Figure 8.9. The rubber pad functions as a vibration absorber.

Importance of Air Seals for Airborne Sound Insulation

Because sound travels through the air, it is important that an assembly contain no voids if it is required to have a high STC. Remember that if air can go through the assembly, so will sound. It is futile to increase the surface weight of the assembly or use resilient channels, clips, or staggered studs if air can leak through the assembly. Therefore, the assembly should be fully sealed at all edges (perimeter) and contain no air-leakage sites.

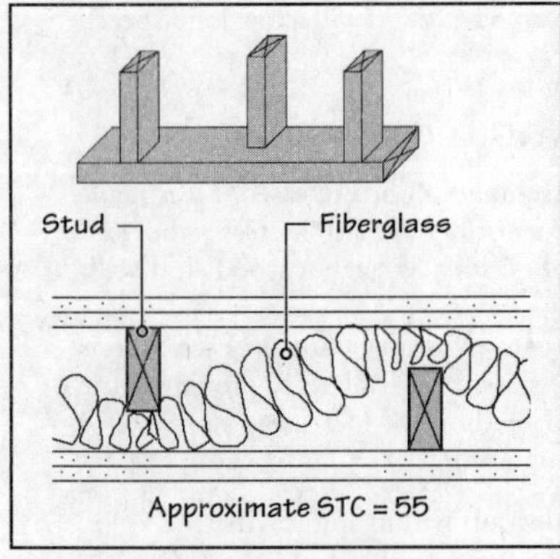

FIGURE 8.6 Staggered-stud wall assembly with cavity absorption (fiberglass or mineral wool).

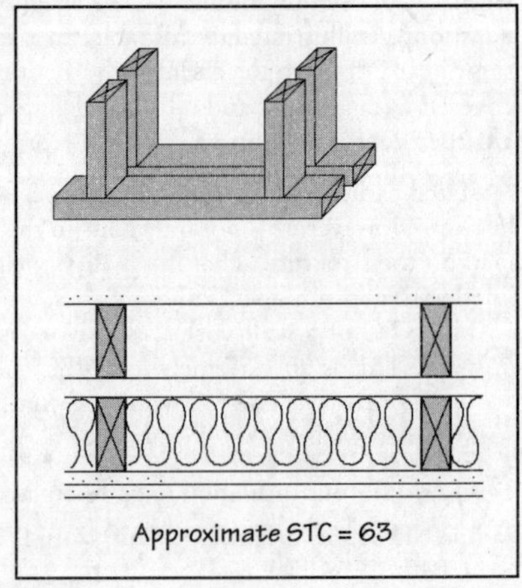

FIGURE 8.7 Double-stud wall assembly.

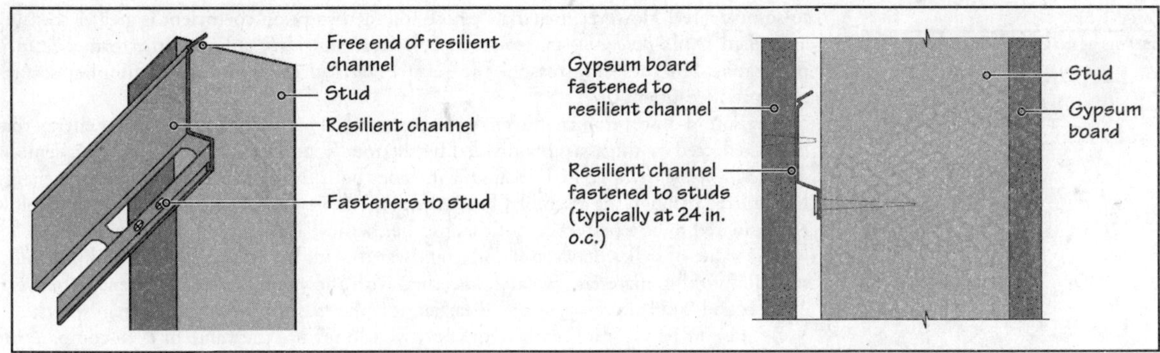

FIGURE 8.8 Resilient channel's profile and use.

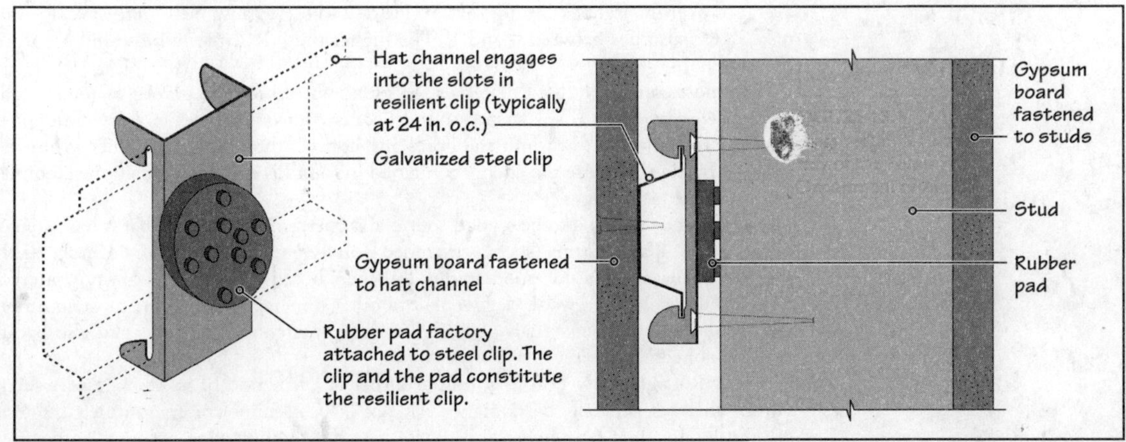

FIGURE 8.9 Resilient clip and its use.

8.5 STRUCTURE-BORNE SOUND INSULATION—IMPACT INSULATION CLASS

An effective way to reduce structure-borne sound transmission is to reduce the vibrations at the source (before they become structure-borne) by absorbing them through the use of resilient materials. For example, the most common structure-borne sound in buildings is produced by the floors as a result of footsteps, furniture movement, or the vibration of machinery supported by the floor. The best way of reducing the transmission of this sound through the floor is to dampen the impact through a soft covering over the floor. The measure used to quantify structure-borne sound insulation is *impact insulation class* (IIC). As with STC values, the unit decibel is not used with IIC values.

A soft covering, such as carpet backed by a foam underlayment, increases the IIC substantially. For example, the IIC of a bare 6-in.-thick concrete floor is 25. The same floor covered with a carpet and underlayment gives an IIC of 85—an increase of 60. Note that a carpet does not improve the STC of a floor. The STC of a 6-in.-thick concrete slab is 55, regardless of whether the floor is bare or covered with a carpet.

The improvement in the IIC of wood floors by using a carpet is less pronounced because of the inherent resilience of a wood floor. For example, a typical plywood floor on wood joists and a gypsum board ceiling directly attached to the studs has an IIC of 34 without the carpet and 55 with the carpet, Figure 8.10. Building codes require a minimum IIC of 50 for a floor-ceiling assembly between two independent dwelling units, such as in an apartment building.

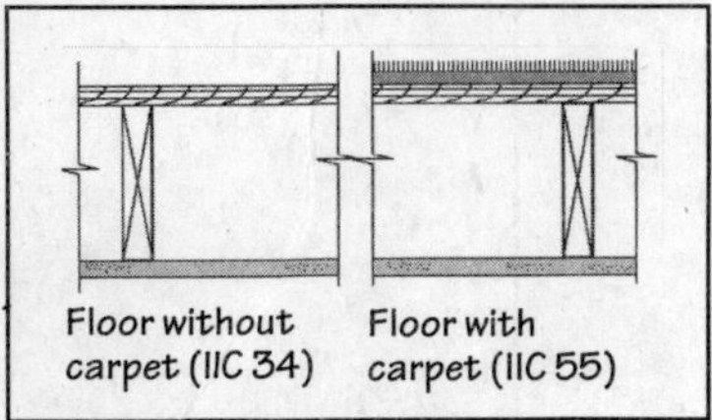

FIGURE 8.10 IIC of a wood floor without and with carpet and underlayment.

8.6 SOUND ABSORPTION—NOISE-REDUCTION COEFFICIENT

While the sound originating from outside a receiving room is attenuated (reduced) by using sound-insulating construction, that originating from within the room can be reduced only by using sound-absorbing materials. All materials and objects absorb sound to some degree. However, materials whose sound-absorption coefficient is greater than 0.2 are called *sound-absorbing materials,* or *acoustical materials,* although the former term is preferable. For the same reason, the term *acoustical treatment* usually implies sound-absorptive treatment.

The sound-absorption coefficient of an assembly, α, is defined as the sound energy that is not reflected by the assembly divided by the total sound energy falling on the assembly. The term *not reflected* is used because it includes both the absorbed energy and the energy transmitted through the assembly. From the point of view of an enclosure, an open window is considered to be a perfect acoustical absorber because it does not reflect any sound.

The value of α lies between 0 and 1 and varies with the frequency of sound. For most sound-absorbing materials, we are concerned with the value of α for frequencies between 250 Hz and 2 kHz because, as noted earlier, it is the frequency range of human speech.

Because of its variation with frequency, we cannot use the value of α to compare the sound-absorbing efficacy of one material with another. Therefore, we use a single-number metric called the *noise-reduction coefficient* (NRC), which is the value of α for a material averaged over the frequency range of 250 Hz to 2 kHz. Because NRC is the average value of α, NRC also lies between 0 and 1. The higher the NRC, the more sound absorptive the material.

The most commonly used materials for sound absorption are porous materials, called *porous sound absorbers.* They absorb sound because the air particles in the vicinity of the material go back and forth into the pores and convert their vibrational energy into heat through friction. Because the energy contained in sound is extremely small, the amount of heat so created is negligible.

For a porous material to be a good sound absorber, it is important that it has interconnected pores. Plastic foams (such as extruded polystyrene, polyisocyanurate, etc.), used as thermal insulators, are not good sound absorbers because their pores are not interconnected. Fiberglass, rock wool, slag wool, and other mineral fibers are the most commonly used sound-absorbing materials because they have interconnected pores and also because they are noncombustible.

The NRC value of a porous sound absorber is a function of its thickness. A thicker porous absorber generally has a higher value of NRC. Manufacturers routinely quote the NRC value for their sound-absorptive materials. A commonly used porous absorber is the ceiling *tile,* generally a mineral fiberboard product, in which the perforations can take various patterns, Figure 8.11. Other materials, particularly fiberglass wrapped in a perforated fabric, is commonly used to absorb sound, thereby reducing ambient noise levels. Figure 8.12 is an example of low-height partitions in an open-plan office faced with fabric-wrapped fiberglass.

FIGURE 8.11 Various types of perforations used in ceiling tiles made from mineral fiberboard (see Chapter 37).

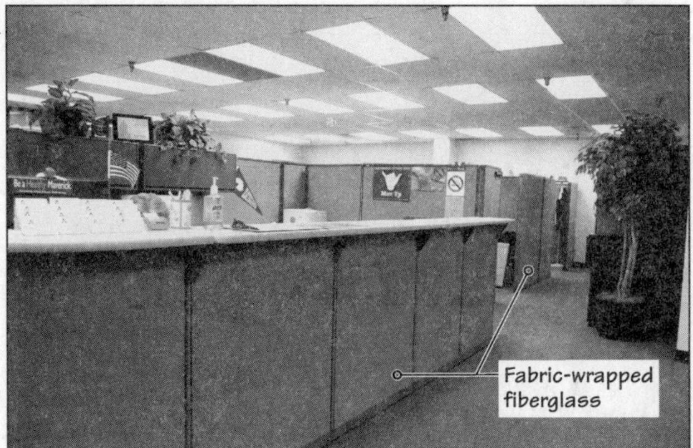

FIGURE 8.12 Low-height partitions in a large open-plan office are usually made of fabric-wrapped fiberglass to provide sound absorption, reducing the ambient noise level.

PRACTICE **QUIZ**

Each question has only one correct answer. Select the choice that best answers the question.

8. Compared to a lightweight building element, a heavyweight element is generally a poor transmitter of airborne sound.
a. True **b.** False

9. Most of the noise produced in a busy restaurant or dining hall is
a. airborne sound. **b.** structure-borne sound.

10. Sound transmission class (STC) is a measure of the
a. sound-absorbing property of a material.
b. sound-insulating property of a material with respect to airborne sound.
c. sound-insulating property of a material with respect to structure-borne sound.
d. sound-reflecting property of a material.

11. For a party wall between two dwelling units, such as two apartments, the building codes require its minimum STC value to be
a. 100. **b.** 75.
c. 65. **d.** 55.
e. 50.

12. A resilient clip consists of a small light-gauge steel member with a rubber pad fastened to it.
a. True **b.** False

13. A resilient channel is used to attenuate airborne sound, whereas a resilient clip is used to attenuate structure-borne sound.
a. True **b.** False

14. A sound barrier with an STC of 50 will generally
a. block all speech and music, however loud.
b. block loud speech but not loud music.
c. block loud music but not loud speech.
d. not block music but will block speech sufficiently to be heard only by a curious listener.

15. The measure that is used for structure-borne sound insulation is
a. transmission loss.
b. impact isolation index.
c. impact insulation class.
d. structural insulation index.

16. A commonly used porous sound absorber is
a. gypsum board. **b.** aluminum foil.
c. microperforated plastic. **d.** oriented strandboard.
e. fiberglass.

17. The noise-reduction coefficient (NRC) of a material is evaluated between the frequency range of 250 Hz and 2 kHz because
a. it is virtually impossible to evaluate this property outside the above frequency range.
b. the human ear is insensitive to noise outside the above frequency range.
c. most of the sound energy in the noise produced in buildings lies in the given frequency range.
d. most of the sound energy in human speech lies in the given frequency range.
e. none of the above.

18. The NRC of a material lies between
a. 0 and 100. **b.** 1 and 100.
c. 0 and 10. **d.** 1 and 10.
e. none of the above.

CHAPTER 9

Principles of Joints and Sealants (Expansion and Contraction Control)

CHAPTER OUTLINE

Building joints result from our need to work with materials that can be easily and efficiently fabricated, transported, and assembled on site. The joints also allow field adjustments in assemblies during construction. Additionally, visible connections resulting from the joints provide the opportunity to develop ideas of scale, pattern, color, texture, and so on, on building planes and surfaces, Figure 9.1.

The most important issue in the design of building joints is that the dimensions of building components are constantly changing, not only with respect to their original dimensions, but also relative to each other. For example, window glass expands and contracts in response to external temperature changes. The changes in glass dimensions are usually different from the corresponding changes in the frames that hold the glass. Yet we still need to be able to make a secure connection between the window glass and the frame.

STATIC JOINTS AND MOVEMENT JOINTS

Over the years, two different strategies have been developed for joining building components. We could provide a nonmoving joint (connection) between the components, called a *static joint.* In the case of the window glass and frame, a static joint would most likely result in broken glass, a broken frame, or both. The alternative is to provide a connection that allows both materials to move independently and at the same time hold the window glass securely in the frame. This movable connection is called a *dynamic joint* or, more commonly, a *movement joint.*

Static joints typically connect materials that are the same or similar. For instance, the mortar joints between units of masonry, the joint between two sheets of gypsum board, and most structural joints provide nonmoving connections between components. By their

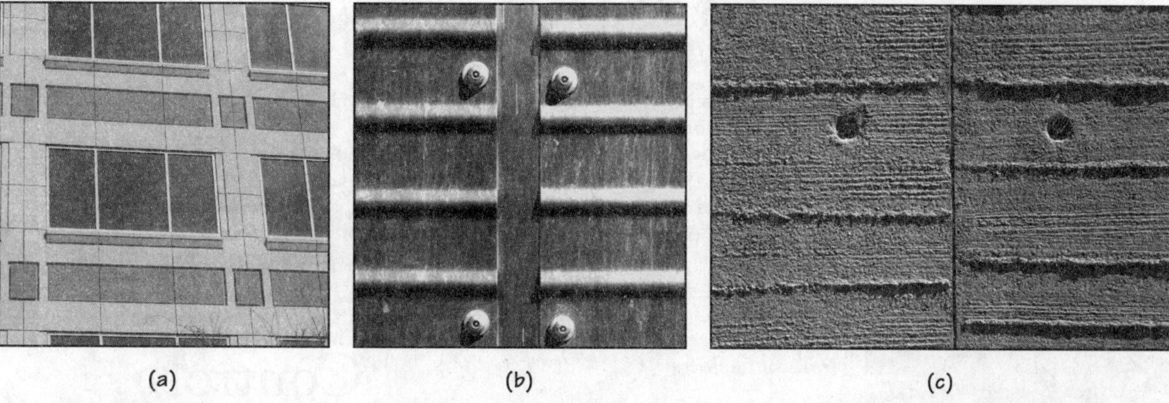

FIGURE 9.1 **(a)** Joints are a common feature in buildings, particularly on building facades. Some of these joints are aesthetic joints to provide surface relief, rhythm, patterns, and so on, while others are provided for performance reasons related to building design and (or) construction. **(b)** The fasteners used in the joint in the copper cladding of Burton Barr Central Library, Phoenix, Arizona, contribute to the rhythm and pattern in the cladding. **(c)** The textured surface and the bold horizontal lines in the exterior concrete walls of the same library building relieve the monotony of a large concrete surface.

nature, the movement of components in a static joint is restricted, so the stress created by such restraint must be counteracted by increased strength of the components.

A well-designed movement joint, on the other hand, has no movement stress. Therefore, it is used at the connection of materials that move at different rates. This implies that movement joints are placed where dissimilar materials meet. They may also be needed to subdivide a component into smaller sizes in order to reduce internal stresses. It is for this reason that a concrete slab-on-grade is subdivided into smaller areas with a gridwork of joints. It is also the reason that a masonry wall—an assembly of static (mortar) joints—must be provided with movement joints at intervals.

The focus of this chapter is on movement joints. Because a movement joint is generally required to be sealed against air, water, and noise transmission (particularly the joints that occur in the external envelope of the building), we will also examine the commonly used sealants and sealant backup materials. Issues related to preventing movement of rainwater through wall assemblies and their joints are discussed specifically in Chapter 27.

NOTE

Construction Joint

A special type of static joint that is provided between two concrete placements is called a *construction joint* or *cold joint* (Chapter 22).

9.1 TYPES OF MOVEMENT JOINTS

Based on the purpose they serve, movement joints are classified as

- *Building joints*—joints between different parts of the building as a whole
- *Component joints*—joints between individual components of an assembly

A building joint divides the entire building into two or more separate buildings. Based on the purpose for which it is provided, it is classified as either a *building separation joint* or a *seismic joint.*

A joint between two adjacent building components can be an *expansion joint,* a contraction joint (commonly called a *control joint*), or an *isolation joint.* A complete list of joint types in a building is given in Figure 9.2. Joint types are discussed in detail in the following sections.

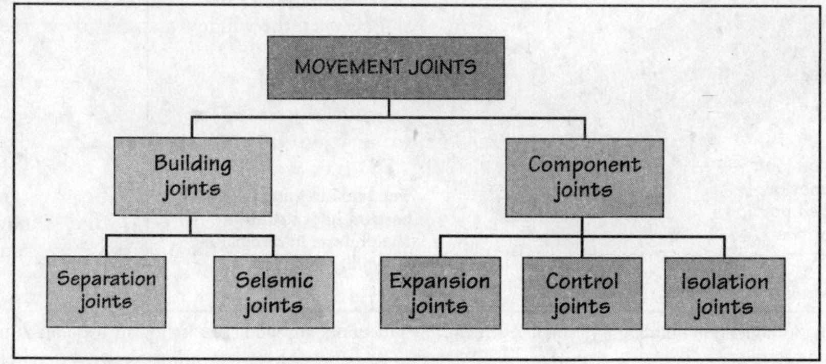

FIGURE 9.2 Types of movement joints in a building.

9.2 BUILDING SEPARATION JOINTS AND SEISMIC JOINTS

NOTE

Building Separation Joint (or Building Expansion Joint)

The term *building expansion joint* is often used for building separation joint in some literature. This is not entirely correct, because a building separation joint accommodates more than mere expansion and contraction of a building.

NOTE

Spacing Between Building Separation Joints

The suggested spacing of 250 ft between building separation joints should be regarded as a rough guide. Most buildings of a simple rectangular shape up to 300 ft in length do not need separation joints. On the other hand, a building with major deviations from a rectangular shape may need separation joints at less than 250 ft on center.

A building separation joint runs continuously throughout the entire building from floor to floor and from face to face. It divides a large and geometrically complex building into smaller, individual, and geometrically simplified buildings, which can move independent of each other. This joint, typically 1.5 to 2 in. wide, accommodates the cumulative effect of various types of movements that occur in a building as a whole, as distinct from the movements that occur in individual components of the building.

A building separation joint prevents the stresses created in one part of the building from affecting the integrity of the other part. A building separation joint is needed in large buildings. As a rough guide, it should be provided at 250-ft (75-m) intervals, Figure 9.3. In addition, good practice requires the provision of building separation joints at the following discontinuities:

- Where a low building mass meets a tall mass. This prevents problems caused by differential expansion and contraction (including creep), in addition to any differential settlement of foundations. Minor changes in height should be disregarded.
- Where the building changes direction, such as in an L-shaped or T-shaped building. Minor changes in direction should be ignored.
- Where the building's structural material changes, such as where a steel frame building meets a concrete frame. This prevents problems due to differential movement between disparate structural systems.

As far as possible, a building separation joint should divide a large building into separate smaller buildings of simplified geometries, such as rectangles. The division into separate buildings requires the use of two columns and two beams at a building separation joint, Figures 9.4 to 9.6.

The duplication of columns and beams at a building separation joint is an ideal solution. An alternative solution that is commonly used is shown in Figures 9.7 and 9.8, in which a single column with a bracket is used.

Seismic Joint

A seismic joint is similar to but also different from a building separation joint. Building separation joints are introduced to accommodate moderate building movements caused by temperature and moisture changes, creep, and foundation settlements. Although they can accommodate small vertical movement, they are designed to accommodate mainly horizontal movement, that is, in the direction perpendicular to the joint.

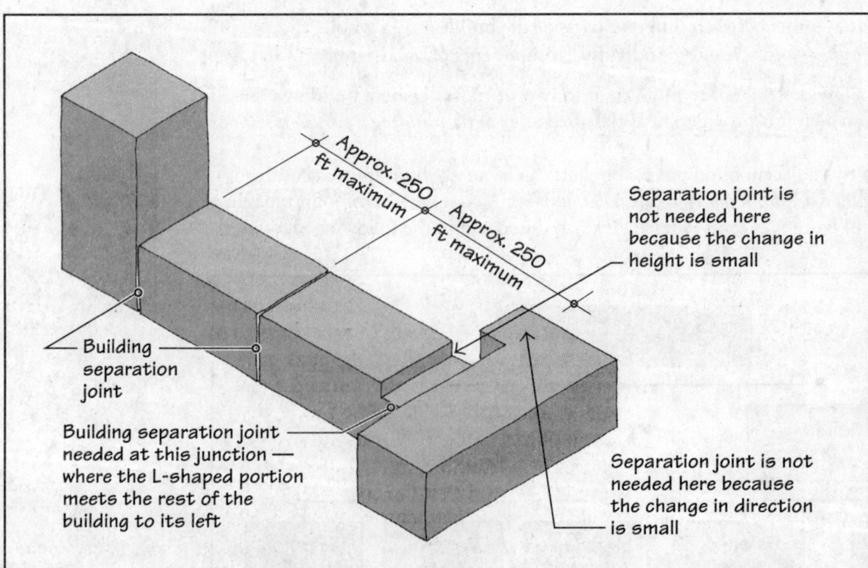

FIGURE 9.3 A building separation joint should divide the building into smaller, structurally independent sections (as if they are separate buildings) so that each section is geometrically simple.

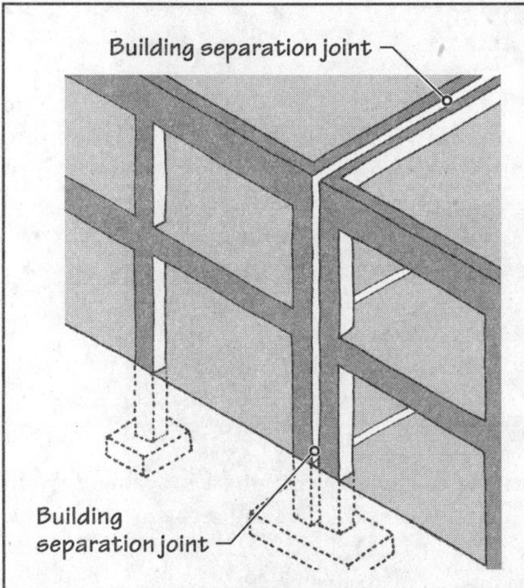

FIGURE 9.4 A building separation joint runs through the entire building from the ground floor (or the basement, if provided) up to the roof. Two columns are often used at the separation joint. However, a combined footing may be used for the two adjacent columns.

FIGURE 9.5 Building separation joint with two columns in a reinforced concrete building. Because the building is under construction, daylight is showing through the joint between the columns. In a completed building, the joint is sealed and covered with a joint cover, as is the case here for the separation joint in the floor.

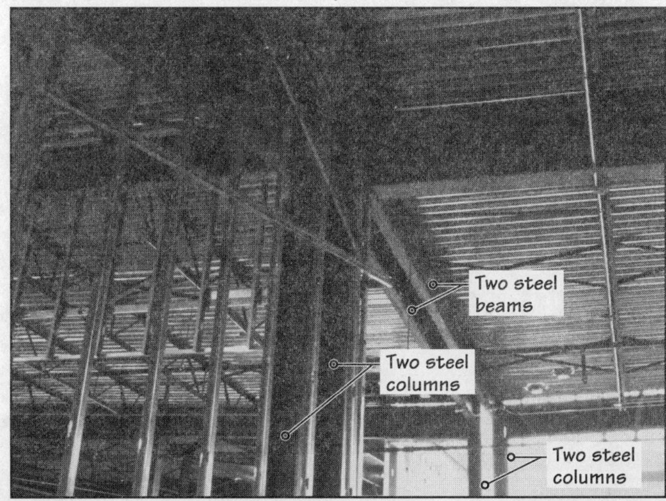

FIGURE 9.6 Building separation joint with two columns and two beams in a steel frame building. Because the building is under construction, daylight is showing through the joint between the columns. In a completed building, the joint is sealed and covered with a joint cover.

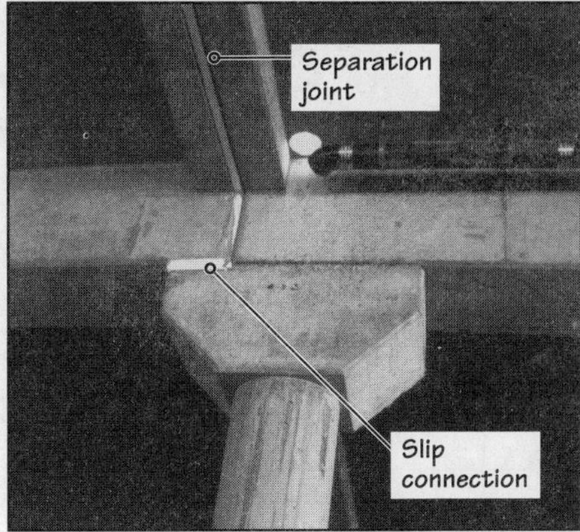

FIGURE 9.7 Building separation joint using a single column in a reinforced concrete building (see Expand Your Knowledge in Chapter 23 for details). Observe the provision of a beam on each side of the separation joint.

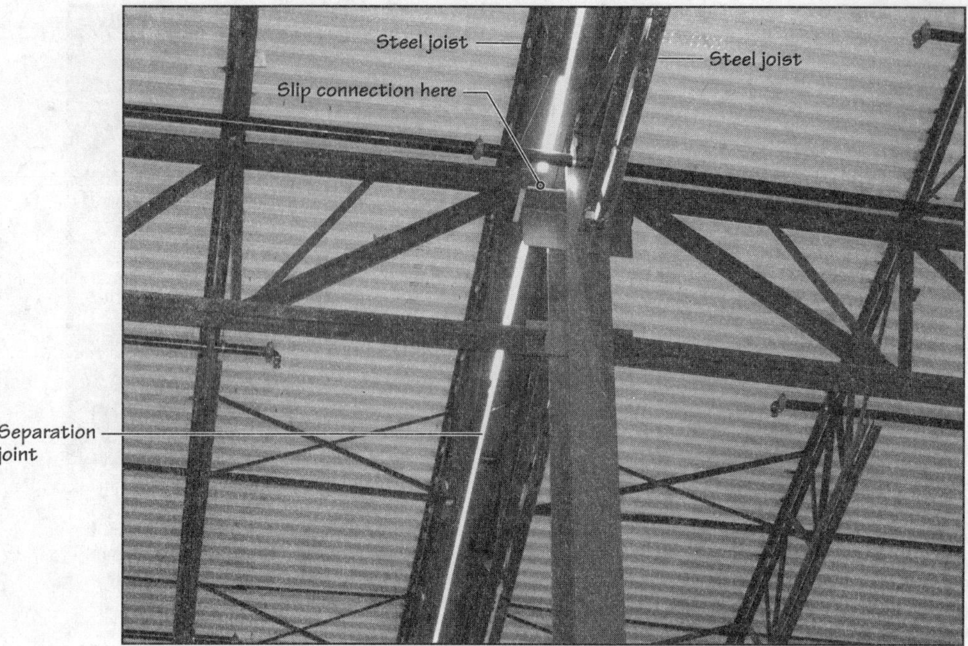

FIGURE 9.8 Building separation joint in a steel frame building using a single column. Because the building is under construction, daylight is showing through the separation joint. In a completed building, the joint is sealed and covered with a joint cover (see Chapter 19 for details of the separation joint). Observe the provision of a steel joist on each side of the separation joint.

Like building separation joints, seismic joints are also provided where there are major dissimilarities in building form. However, the purpose of seismic joints is to ensure that one section of the building does not collide with the adjacent section during an earthquake. A seismic joint must accommodate simultaneous movements in horizontal as well as vertical directions, that is, in all three principal directions.

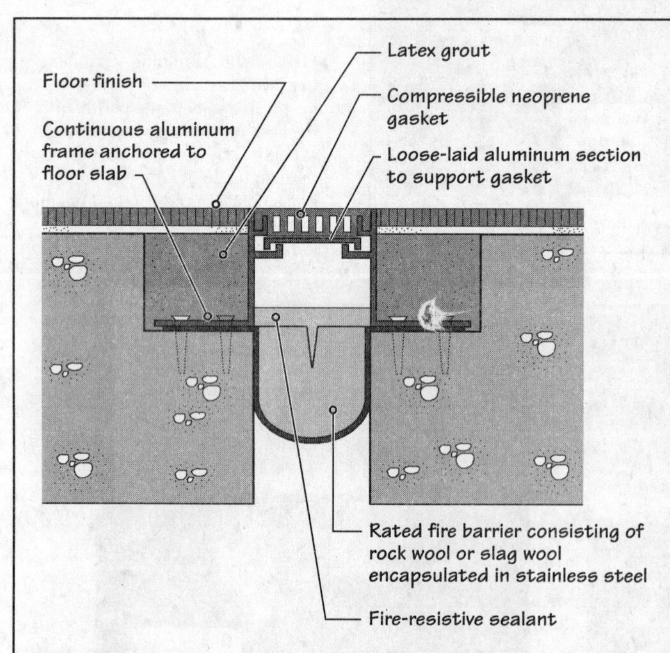

FIGURE 9.9 A typical detail of a building separation joint cover at the floor level. Similar covers are provided in interior and exterior walls and roofs. Joint cover manufacturers' details and specifications should be strictly followed for structural, fire rating, and other performance reasons.

A seismic joint is generally much wider than a building separation joint. In highly seismic locations, a seismic joint may be a few feet wide for a tall building. The width of a building separation joint is the same at each floor. The width of a seismic joint, on the other hand, should increase with height. For aesthetic and architectural reasons, however, a constant width is generally used.

Using double columns at a seismic joint is the preferred solution. However, a single column with seated connections (Figures 9.7 and 9.8) that are adequately restrained to prevent the seat from sliding off the support is acceptable.

Covers Over Building Separation or Seismic Joints

Building separation and seismic joints must be covered. Various proprietary products are available to cover the joints at the roof, floors, and walls. A simple floor joint cover consists of a compressible gasket between two metal sections. Each metal section is anchored to the floor slab, Figure 9.9.

A more elaborate floor joint cover that allows greater movement is used for seismic joints. It is important to ensure that the joint has the same fire resistance as the floor, wall, or roof in which it occurs. Once again, proprietary fire-resistive joint materials (typically consisting of ceramic insulation encapsulated in a stainless steel sheet) are available to use within the joint. Building separation joints at the roof are discussed in Chapter 33.

9.3 MOVEMENT JOINTS IN BUILDING COMPONENTS

Movement in a building component can be caused by several phenomena. Therefore, many factors must be considered in sizing and detailing movement joints in building components:

- Thermal movement
- Moisture movement
- Elastic deformation and creep
- Construction tolerances and other considerations

Although thermal movement and elastic deformation occur in all materials, the same is not true of creep and moisture movement. Steel does not creep, and its dimensions are not affected by moisture variations. Thermal movement is reversible if the component is unrestrained. Moisture movement, on the other hand, may or may not be reversible, depending on the material, Table 9.1. Creep deformation is irreversible; that is, creep is a permanent deformation.

In addition to the factors just stated, construction and material tolerances and several other factors must be considered in sizing building component joints.

Because of the multiplicity of factors and their intricate interrelationships, the theory of sizing movement joints can be fairly complex. Therefore, in practice, joint dimensions are based on well-acknowledged rules of thumb. In most situations, the rules of thumb are established by material manufacturing associations. For instance, the Brick Industry Association (BIA) provides rules of thumb for the size and spacing of joints in brick walls. Similarly, the Portland Cement Association (PCA) and the American Concrete Institute (ACI) provide rules of thumb for joints in concrete slabs and walls.

TABLE 9.1 TYPES OF MOVEMENT IN SELECTED MATERIALS

Building material	Thermal	Elastic deformation	Creep	Reversible moisture	Irreversible moisture
Steel	X	X	—	—	—
Concrete	X	X	X	X	—
Concrete masonry	X	X	X	X	—
Brick masonry	X	X	X	—	X
Wood	X	X	X	X	—

X indicates the presence of particular movement in the material and—indicates the absence of such movement.

EXPAND YOUR KNOWLEDGE

Importance of Movement Joints in Building Components

The importance of providing an adequate space for the movement of components may be appreciated by considering the expansion of a linear member such as a beam. Assume that the original length of a beam is L. Assume further that, under certain temperature and moisture changes, its length increases by an amount δ. If the beam is fully restrained at its ends so that it cannot increase in length, it will be subjected to axial compression. Consequently, the beam will tend to buckle (bend upward or downward), Figure 9.10. If the deflection of the beam is denoted by h, we find that h is several times larger than the beam's unrestrained extension, δ.

To better visualize this fact, consider the following illustrative scale model. An approximately 4-ft-long sheet made of a flexible material (e.g., acrylic) has been cut to fit snugly

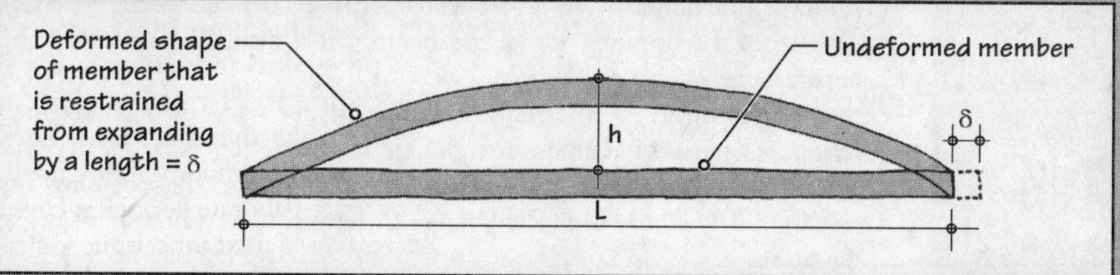

FIGURE 9.10 Deformation of a member that is prevented from expanding. Note that restraint against even a small expansion (δ) can cause a large lateral deformation (h) in the member.

(Continued)

EXPAND YOUR KNOWLEDGE

Importance of Movement Joints in Building Components (*Continued*)

between two raised ends, Figure 9.11(a). The raised ends restrain the expansion of the sheet. If we slightly shorten the length of the sheet (say, by approximately $\frac{1}{16}$ in.) on each end by placing a 25-cent coin between the sheet and the raised end, we find that the sheet curves up substantially (approximately 2 in.), Figure 9.11(b).

Apart from the fact that such a large deformation is visually unacceptable in buildings, it also creates structural problems. A component made of a ductile material (e.g., metal or plastic) will buckle in such a situation. On the other hand, materials such as concrete, masonry, gypsum drywall, and glass will simply break under such a large deformation because, being weak in tension, these (brittle) materials cannot resist the tensile stress created by excessive bending.

This simple model highlights the importance of providing movement joints. A similar situation would occur if the beam were to contract instead of expanding. Contraction creates axial tensions, and most brittle materials will fail under relatively small contraction because their tensile strength is low.

If the component is free to move due to the provision of movement joints, no stress will be created in the component. However, although movement joints provide a convenient solution to expansion and contraction, they can be avoided (where the situation so requires) by increasing the strength and stiffness of the component. For example, the buckling of the acrylic sheet in Figure 9.11(b) can be prevented by increasing the strength (i.e., thickness) of the sheet or replacing it by a stiffer material.

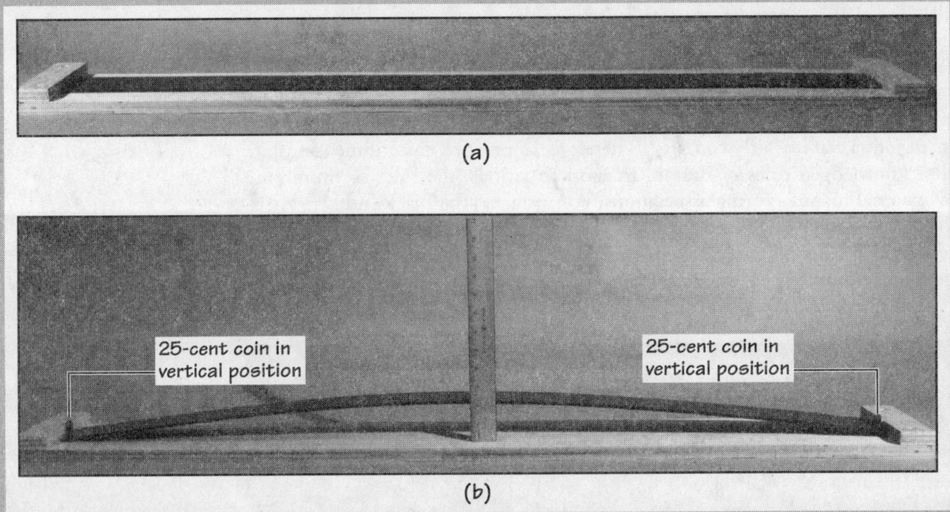

FIGURE 9.11 (a) An acrylic sheet in an undeformed state. (b) Deformation of the same sheet (nearly 4 ft long) shortened by approximately $\frac{1}{16}$ in. at each end by placing 25-cent coins. Note that the (lateral) deformation of the sheet (approximately 2 in.) is much larger than the longitudinal shortening (approximately $\frac{1}{8}$ in.).

Despite the general use of rules of thumb, a basic understanding of the theory of movement of building components is important and is presented in the following sections. This information is critical when the designer must depart from the rules of thumb. It is also important in situations where rules of thumb do not exist.

NOTE

Units of α

The units of L and δ_t are inches, and the unit of (ΔT) is °F. Substituting these units in Equation (1), we get

$$\text{in.} = \alpha(\text{in.})(°\text{F})$$

So that the units on both sides of the equals sign are consistent in this equation, the units of α must be

$$\frac{1}{°\text{F}}, \text{ that is, } °\text{F}^{-1}$$

9.4 THERMAL MOVEMENT

Thermal movement is generally the most critical movement because it occurs in all components, particularly those exposed to the exterior climate, such as exterior walls, cladding, roof membranes, slabs, and paving. It is also the largest type of movement in most components.

To determine the thermal expansion or contraction of a component, assume that its length is L and the temperature change it experiences is Δt(°F or °C). If the component is not restrained, its length will increase (or decrease) so that the new length will be $(L + \delta_t)$ or $(L - \delta_t)$, where δ_t is the change in length. The value of δ_t is given by the following equation:

$$\boxed{\delta_t = \alpha L(\Delta t)} \qquad \textbf{Eq. (1)}$$

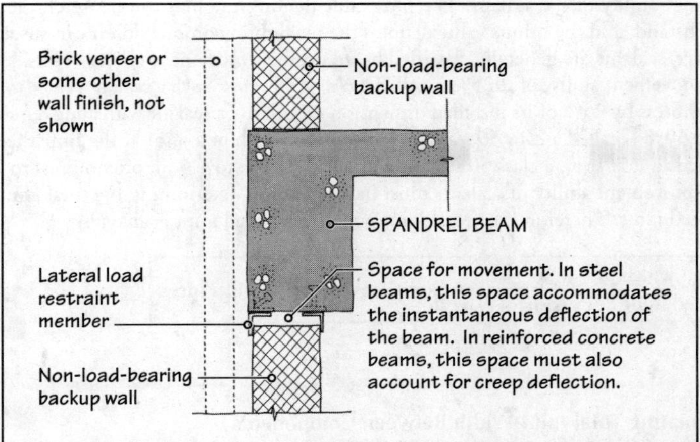

FIGURE 9.16 The connection between the spandrel beam and the non-load-bearing wall below must account for the beam's deflection (see also Figure 28.6).

components and is, therefore, irreversible. In other words, even if the load is removed, creep deformation will remain.

Elastic deformations are routinely determined for all types of structures as part of structural calculations. For concrete structures, creep deformation is also determined. From the architectural viewpoint, the most critical deformation occurs in spandrel beams. Because spandrel beams have exterior walls anchored to them, it is necessary to provide flexibility in the anchorage of exterior walls so that neither the elastic nor the creep deflection of spandrel beams will cause additional stresses in exterior walls.

If a non-load-bearing wall is placed directly under a spandrel beam, an adequate space between the beam and the top of the wall should be provided to allow the beam to deflect under the load, Figure 9.16. If adequate space is not provided, the load from the beam will be transferred to the wall, damaging it.

NOTE

Spandrel Beam

A spandrel beam is a beam that runs on the outside edge of a floor or roof of the building; see Figure 9.16.

9.7 TOTAL JOINT DIMENSION

In addition to thermal, moisture, elastic, and creep deformations, there are several other factors that should be considered in determining the joint dimensions. One such important consideration is material tolerances to recognize that building components vary from their specified dimensions. For example, it is not uncommon to see brick or concrete masonry units that are undersized or oversized. Noticeable differences in dimensions from those stated on the drawings can occur in prefabricated concrete curtain wall panels.

In addition to material or component tolerances, construction tolerances should be taken into account. For instance, an expansion joint in a masonry wall may be shown as a vertical joint on the drawings, but the constructed joint is unlikely to be absolutely vertical. Construction tolerances are a function of quality control during construction. Smaller tolerances require good construction practices and supervision. Poor supervision demands larger tolerances. Therefore

$$\text{Width of joint} = \begin{array}{l}\text{width based on}\\ \text{temperature, moisture,}\\ \text{and other movements}\end{array} + \text{tolerances} \qquad \textbf{Eq. (3)}$$

Additionally, foundation settlement and story drift (see Principles in Practice in Chapter 19) due to wind or earthquake loads are important. Some chemical processes may also cause movement. Steel expands as it corrodes. Excessive corrosion of steel can cause cracking and sometime spalling of concrete. Because water expands on freezing, the freezing of water causes similar effects in materials that absorb water, such as brick, concrete, concrete masonry, and wood.

EFFECT OF A SEALANT ON JOINT DIMENSION

Most joints in building envelope components are filled with elastomeric sealants. The sealants have a finite ability to cyclically stretch and compress, which is given as the plus-minus movement ability of the sealant. The plus value denotes the maximum stretch the sealant can withstand, and the minus value denotes the maximum compression it can sustain.

In fact, sealants are generally classified based on their movement ability. Thus, a sealant with a movement ability of ±25% is called a *class 25 sealant.* Such a sealant is able to stretch and compress by 25% of its installed dimension. Similarly, a sealant with a movement ability of ±50% is called a *class 50 sealant.* Thus, if the width of a joint at the time the sealant is applied to it is 1 in., a class 50 sealant will safely stretch to 1.5 in. or compress to 0.5 in.

The movement ability of sealants must be included in determining the total joint width of a sealed joint. Therefore, the total joint width of a sealed joint is given by

$$\text{Joint width of sealed joint} = \frac{100}{\text{sealant class}}\,[\text{joint movement}] + \text{tolerances} \qquad \textbf{Eq. (4)}$$

Estimating Total Joint Width Between Components

Example 5

Continuous vertical expansion joints are to be provided in a brick veneer wall at 25 ft on center. Determine the total joint width, assuming that the joints are sealed with a class 50 sealant.

Solution

From Examples 2 and 3, the thermal and moisture movement of a 25-ft-long wall is 0.108 + 0.090 = 0.198 in. From Equation (4), total joint width is

$$\frac{100}{50}(0.198) = 0.396 \text{ in.}$$

This joint width does not include tolerances. In practice, the expansion joints in a brick veneer wall should be $\frac{1}{2}$ in. (0.5 in.) wide.

9.8 PRINCIPLES OF JOINT DETAILING

In detailing a movement joint in a component, it is important to distinguish between an *expansion joint* and a *control* (*shrinkage*) *joint.* Because the joint width in an expansion joint will become smaller over time, the filler in an expansion joint (if used) must be elastomeric to allow unrestrained movement of the components.

The filler in a control joint, on the other hand, may be elastomeric or nonelastomeric. If a nonelastomeric filler is provided in a control joint, the detailing of the joint must allow unrestrained shrinkage of the components.

The difference in the detailing of an expansion joint and a control joint can be further appreciated by examining typical joint details used in brick masonry and concrete masonry walls. As stated earlier, brick masonry walls expand over time after construction as the bricks absorb water from the mortar and the atmosphere. Therefore, brick masonry walls require expansion joints. By contrast, concrete masonry walls shrink after construction as the water used in manufacturing concrete masonry units and in the preparation of mortar evaporates. Therefore, concrete masonry walls require control joints.

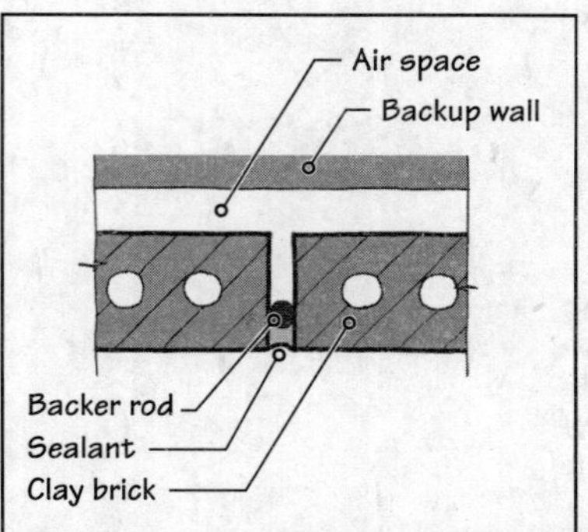

FIGURE 9.17 Detail of an expansion joint in a brick veneer wall. The joint is filled with a compressible backer rod and an elastomeric sealant (see also Figure 28.16).

Figure 9.17 shows the detailing of an expansion joint in a brick veneer wall. Note that the joint is filled with a compressible backer rod and an elastomeric sealant.

Figure 9.18 shows a typical control joint detail in concrete masonry walls. The joint is filled with masonry mortar to provide shear key between the two parts of the wall. To ensure that the mortar does not adhere on both sides of the joint and restrain the shrinkage of the wall, asphalt-saturated paper is placed on one side of the joint to break the bond.

In explaining the difference between a control joint and an expansion joint, we considered moisture-induced movement in brick and concrete masonry walls, which is largely permanent. On the other hand, to accommodate reversible movement, such as that created by temperature changes, requires either an open joint or one filled with an elastomeric sealant (Figure 9.17). In other words, an open joint or an elastomerically filled joint is a versatile joint.

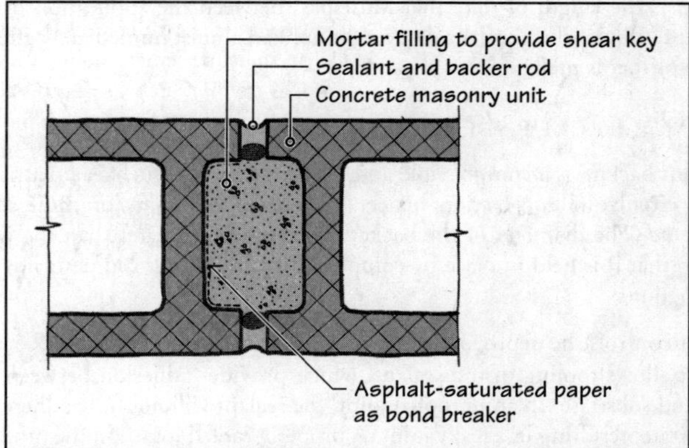

FIGURE 9.18 Detail of a control joint in a concrete masonry wall (see also Figure 25.25).

Note that because temperature-induced movement is almost universal, expansion joints are provided in most components. Control joints, on the other hand, are provided in concrete and concrete masonry components. Detailed discussions of joint detailing are provided in later chapters. For example, construction joints, control joints, expansion joints, and isolation joints used in concrete construction are covered in Chapter 22.

9.9 COMPONENTS OF A SEALED JOINT

The primary purpose of sealing a joint is to prevent water penetration. However, control of air leakage, dust penetration, and noise transmission are additional benefits of sealing a joint. An unsealed joint may also get filled with incompressible materials over time, resulting in joint failure.

Although joint sealant is the most important component of a sealed joint, these other components must also be carefully selected:

- Substrate
- Primer
- Sealant backup
- Bond breaker

SUBSTRATE

For a sealed joint to function effectively, it is important that the sealant be fully adhered to the surfaces of components meeting at the joint (called the *substrate*). A nonadhering sealant will obviously result in a leaky joint.

The adhesion of the sealant to the substrate depends on the chemical compatibility of the sealant with the substrate material. Because all sealants are not compatible (or equally compatible) with all substrate materials, it is important to obtain compatibility information from the sealant manufacturer. This is particularly vital because substrate materials such as concrete and masonry are sometimes treated with water-repellent chemicals, and metals are sometimes treated with paints and other protective coatings. These treatments may inhibit sealant adhesion.

Some substrate materials may require special joint preparation, such as the application of primers to make the substrate compatible with the sealant. In any event, some preparation will be required on all substrates, including cleaning the surfaces of loose particles, contaminants, frost, ice, and so on.

PRIMER

The purpose of a primer is to improve the adhesion of a sealant to the substrate. Some sealants require primers on all types of surfaces, whereas others require primers on a few types of surfaces. Most sealant manufacturers require primers for concrete and masonry substrates to stabilize the substrate by filling the pores and strengthening weak areas.

Some sealants require different primers for different substrates, which is an important consideration in the selection of a sealant if disparate materials are to be joined. Another consideration is the length of time that must pass between the application of the primer and the sealant. Obviously, a sealant that can be applied almost immediately after the application of the primer is preferable.

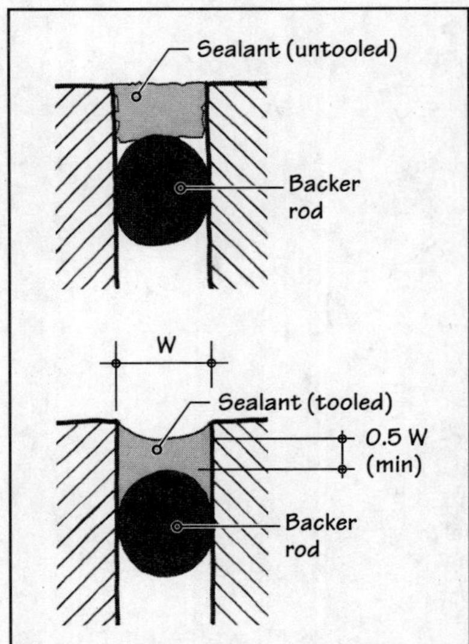

FIGURE 9.19 Untooled and tooled sealants in joints. Note that the depth of the sealant must be at least half the width of the joint.

SEALANT BACKUP—THE BACKER ROD

A sealant backup is a compressible and resilient material, such as a plastic foam. It is usually circular in cross section; hence, it is called a *backup rod* or, more commonly, a *backer rod.* The diameter of the backer rod should be greater than the width of the joint so that it is held in place by compression. The backer rod performs the following functions:

- It controls the depth and shape of the sealant.
- It allows tooling of the sealant, which provides adhesion between the sealant and substrate. In an untooled joint, the sealant will not fully adhere to the substrate, resulting in a leaky joint. A properly tooled joint, on the other hand, will have full contact with the substrate, Figure 9.19. For the joint to be tooled, the backer rod must be under sufficient compression not to slide under the pressure of tooling.
- It acts as a temporary joint seal until the sealant is applied.

For the backer rod to function as a secondary seal, it must remain elastic throughout its life and not develop significant compression set. It should also be compatible with the sealant. Therefore, a backer rod approved by the sealant manufacturer should be used.

BOND BREAKER

For the sealant to function effectively, it should be bonded to only two opposite surfaces of the substrate so that it comes under axial tension or axial compression when the joint moves. If a joint is similar to the one shown in Figure 9.20(a), a bond breaker is necessary on the rear side of the joint to prevent sealant failure, shown in Figure 9.20(b).

A bond breaker is required only if the third surface is hard and unyielding, such as concrete, metal, masonry, or any other inflexible backup. A bond breaker is not required with a flexible backup that will not significantly restrict the freedom of sealant movement. Thus, a bond breaker is not required in a conventional joint with a foam backer rod, such as that shown in Figure 9.19.

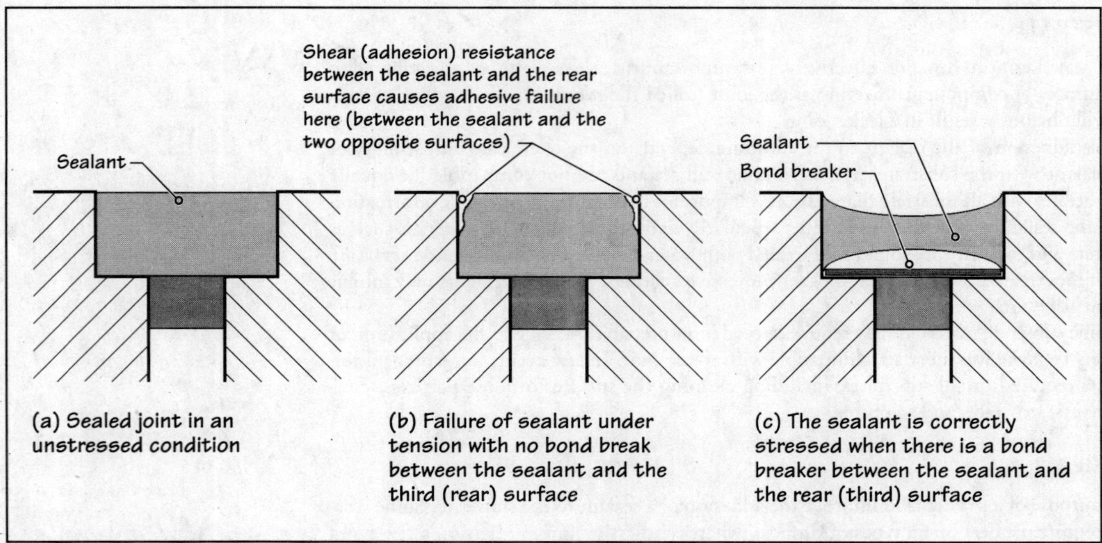

FIGURE 9.20 An ideal sealed joint should only be bonded to two opposite surfaces so that the sealant is subjected to either tension or compression. A bond between the sealant and the third (in this case, the rear) surface may cause sealant failure, as shown in **(b)**, because of the adhesion of sealant to the third surface. Therefore, where a third surface exists in a sealed joint, a bond breaker should be used, as shown in **(c)**.

9.10 TYPES AND PROPERTIES OF JOINT SEALANTS

Joint sealants may be divided into three categories:

- Preformed tapes
- Caulks
- Elastomeric sealants

Preformed tapes are available in rolls. They function as seals only if they are under pressure; they are commonly used in window glazing, door jambs, and gypsum drywall.

Caulks are doughlike materials and are the first generation of sealing compounds. Glazing putty is an example of the earliest caulks used in buildings. It consists of nearly 12% linseed oil and 88% chalk (calcium carbonate) that hardens as it cures (oil evaporation), leading to cracking, loss of elasticity, and, hence, loss of its sealing ability. It is rarely used today.

TYPES OF ELASTOMERIC SEALANTS

Elastomeric sealants are synthetic materials (polymers) and are the ones most commonly used in contemporary construction. The five commonly used synthetic sealants are

- Polyisobutylene
- Acrylics
- Polyurethane
- Polysulfide
- Silicone

Polyisobutylene and acrylic sealants have very low movement ability and are used as caulking materials in joints that do not move or move very little, such as the static joints between aluminum window frames. Polyisobutylene (referred to as *butyl caulk*) has excellent resistance to water-vapor transmission. Therefore, it is commonly used as the primary seal in an insulating glass unit. The secondary seal in an insulated glass unit is provided by the silicone sealant because of its durability and good adhesion (Section 30.6).

Polyurethane sealant has high abrasion resistance. Therefore, it is the sealant of choice in horizontal joints subjected to foot traffic. It has good compatibility with a wide variety of substrates—another factor in its favor.

Polysulfide sealant has excellent chemical and weathering resistance. It is the sealant of choice for swimming pools, wastewater treatment plants, water-treatment plants, and so on, where a durable watertight seal is required. Silicone sealant has high strength and exceptional resistance to ultraviolet radiation. It is the only sealant that can be used in structural glazing.

Table 9.3 gives a comparative overview of three of the most commonly used sealants—polyurethane, polysulfide, and silicone—and the following sections provide a description of the criteria that must be considered in their selection.

MOVEMENT ABILITY OF SEALANTS

The most important property of a sealant is its ability to withstand cyclic joint movements (Section 9.7). Currently available sealants are classified into three categories: low-range sealants, medium-range sealants, and high-range sealants.

A low-range sealant has limited movement ability, of the order of ±5% or less. These sealants include (a) oil-based caulks and (b) butyl or acrylic caulks. Low-range sealants should be used only in a nonmoving joint. Their principal advantage lies in their low cost.

Medium-range sealants have a movement range of up to ±12.5%, and high-range sealants have a movement range larger than ±12.5%. Silicone sealants normally have the highest movement capability, up to ±50%.

STRENGTH AND MODULUS OF SEALANTS

A sealant must be able to withstand movement stresses. Most joint failures occur due to excessive tensile stress. Tensile failure may occur either at the substrate-sealant interface (failure due to inadequate adhesion) or within the sealant (failure due to inadequate cohesion).

Modulus refers to the modulus of elasticity of sealant, a measure of the stiffness of a material (Chapter 4). A high-modulus sealant is stiffer than a low-modulus sealant and is desirable where joint movement is small. Thus, a high-modulus sealant is required in structural glazing, where the stiffness of the joint between the glass and the aluminum frame is important. A low-modulus sealant, on the other hand, is desirable where joint movement is

NOTE

Modulus of a Sealant

From Equation (3) in Chapter 4, the stress in a material is the product of the modulus of elasticity of the material and the strain:

$$\text{Stress} = \text{modulus of elasticity} \times \text{strain}$$

Thus, stress in a material is directly proportional to its modulus of elasticity. In other words, a low-modulus material will be subjected to lower stress than a high-modulus material if the strain (deformation) of the two materials is the same.

TABLE 9.3 COMPARATIVE OVERVIEW OF COMMONLY USED SEALANTS

	Type of sealant		
Sealant property	Polyurethane	Polysulfide	Silicone
Movement ability	Up to ±25%	Up to ±25%	Up to ±50%
Water vapor permeability	Fair	Good	Poor
Abrasion resistance	Excellent	Poor	Fair
Chemical resistance	Fair	Excellent	Fair
Durability	Good	Good	Excellent
Substrate compatibility	Excellent	Good	Incompatible with some substrates
Paintability	Good	Good	Poor

large and the sealant is required purely for sealing purposes. A low-modulus sealant will be under lower stress due to movement.

NOTE

Joints in Building Components—Lapped Joints

There are several movement joints in buildings that function because of the overlap between building components. Such joints generally do not need to be sealed. Some examples of lapped joints are joints between exterior wall siding (Chapter 16), joints between roof shingles or tiles, and joints between underlayment felts below the shingles (Chapter 34).

Tooling Time, Cure Time, and Temperature Range

Tooling time refers to the time (usually measured in minutes) that must be allowed to elapse before the sealant changes from a liquid state to a semisolid state so that it can be tooled. *Cure time* refers to the time it takes for the sealant to be hardened to its final elastomeric state.

Application temperature range refers to the ambient air temperature range within which the sealant can be applied. *Performance temperature range* is the range over which the sealant will maintain its properties after it has cured.

Life Expectancy of Sealants

Life expectancy is the time (in years) after which the sealant may have to be reapplied. Most high-grade sealants, such as the structural silicones, are quoted by their manufacturers to have a life expectancy of 20-plus years.

One-Part or Two-Part Sealants

Sealants can either be one-part or two-part sealants. One-part sealants are easier to use but generally take longer to cure and have long tack-free time. Longer curing time means a greater delay in tooling and possible damage to the sealant if the joint moves excessively before the sealant has fully cured. Long tack-free time means that the sealant will tend to attract more dirt.

PRACTICE QUIZ

Each question has only one correct answer. Select the choice that best answers the question.

11. Creep deformation occurs in
- **a.** concrete members.
- **b.** steel members.
- **c.** aluminum members.
- **d.** all of the above.

12. Creep deformation is caused by
- **a.** live loads.
- **b.** dead loads.
- **c.** wind loads.
- **d.** all of the above.
- **e.** none of the above.

13. A spandrel beam in a building is a beam that
- **a.** has a floor slab or roof slab on both sides.
- **b.** runs on the outside edge of a floor or roof.
- **c.** is partially embedded in the ground.
- **d.** is supported by a continuous wall below it.
- **e.** supports a continuous wall above it.

14. A non-load-bearing wall that runs under a spandrel beam should be adequately separated from the beam to allow the beam to move freely in the
- **a.** vertical direction.
- **b.** horizontal direction.
- **c.** both (a) and (b).

15. In determining the total width of a movement joint between building components, we account for the dimensional and construction tolerances. Poor or inadequate quality control on a construction site
- **a.** increases construction tolerances.
- **b.** decreases construction tolerances.
- **c.** increases dimensional tolerances.
- **d.** decreases dimensional tolerances.
- **e.** none of the above.

16. A joint between two walls has been determined to be 0.5 in. if the joint is not to be filled with a sealant. If the same joint is to be filled with a class 50 sealant, its width should be
- **a.** 1 in. minimum.
- **b.** $2\frac{1}{2}$ in. minimum.
- **c.** $\frac{1}{4}$ in. minimum.
- **d.** $\frac{1}{4}$ in. maximum.
- **e.** none of the above.

17. Which of the following materials does not represent a joint sealant?
- **a.** Silicone
- **b.** Polysulfide
- **c.** Polyisobutylene
- **d.** Polyvinyl chloride

18. Which of the following joint sealants has the highest abrasion resistance?
- **a.** Polyisobutylene
- **b.** Acrylic
- **c.** Polyurethane
- **d.** Silicone

19. Which of the following joint sealants has the highest movement ability?
- **a.** Polyurethane
- **b.** Acrylic
- **c.** Polyisobutylene
- **d.** Silicone

20. The sealant in a joint works best when it is subjected to
- **a.** tension, compression, and shear.
- **b.** tension, compression, shear, and bending.
- **c.** tension or shear only.
- **d.** tension or compression only.

REVIEW QUESTIONS

1. Using sketches and notes, explain the various types of movement joints used in buildings.
2. Explain the relative suitability of a static joint versus a dynamic (movement) joint in buildings.
3. With the help of a sketch, explain why and where building separation joints should be provided.
4. Explain the differences between a seismic joint and a building separation joint.
5. What is creep, and how does it affect the movement of building components? Explain.
6. Explain why a low-modulus sealant is generally preferred over a high-modulus sealant.
7. Silicone, polyurethane, and polysulfide are commonly used as joint sealants. Explain their relative suitability for various applications.
8. Discuss the benefits of tooling a joint sealant.

CHAPTER 10

Principles of Sustainable Construction

CHAPTER OUTLINE

Before the beginning of the twentieth century, buildings contributed little to environmental degradation. Building construction was largely craft based, and most buildings were built by human labor or low-tech tools, occasionally using simple machines. The use of locally produced materials was the norm, and the amount of energy used in mining raw materials and converting them into finished products (referred to as *embodied energy*) was low.

In addition to low embodied energy, the energy used in a building's operation and maintenance (lighting, heating, cooling, water supply, and waste disposal) was low. Air conditioning had not yet been invented. The dominant source of interior illumination was daylight, and the levels of artificial lighting in buildings were modest.

To ensure that buildings were comfortable and healthy, architects spent a great deal of design effort in creating climate-responsive buildings through appropriate site planning, building orientation, fenestration design, and the use of appropriate landscape materials (see the introduction to Chapter 5).

Discovery of Mechanical Air Conditioning

Around the beginning of the twentieth century, several technical discoveries changed the way buildings were constructed and used. The most significant change was the invention of air conditioning (by Willis Carrier in 1902). This produced a ducted system of comfort conditioning using warm or chilled air. By the middle of the twentieth century, most new buildings in North America were fully air conditioned—a technology that spread not only to buildings throughout the world, but also to forms of transportation that carried people and perishable goods (trains, buses, automobiles, etc.). The result was a hefty increase in the world's energy use.

PASSENGER ELEVATORS AND SKYSCRAPERS

Another major development was the elevator developed by Elisha Graves Otis, who publicly demonstrated its safety in 1854 in New York City. The elevator's safety and the subsequent improvements in its efficiency and speed provided the necessary stimuli for the construction of tall buildings. Initially, Chicago and New York City—and later other urban centers throughout the world—began constructing tall buildings, giving birth to an entirely new building form, the skyscraper.

The amount of energy required for the construction of a square foot of a skyscraper is far greater than that required for the construction of a low-rise building. This is due not only to the use of heavy construction equipment (excavators, bulldozers, graders, loaders, cranes, etc.) required for the construction of a skyscraper, but also to the disproportionately large amount of materials needed in its structural frame.

In addition to a large amount of embodied energy, tall buildings consume large amounts of energy for operation and maintenance. On the other hand, because tall buildings yield denser urban habitations, they can reduce the amount of energy required for urban infrastructure—roads, water supply, sewage, and waste-disposal facilities.

The twentieth century also witnessed several major sociopolitical changes that led to further escalation in the world's energy use. The independence of several Asian and African countries and their industrialization led to the expansion of international trade and travel and increased building and infrastructural construction. The population explosion resulting from medical breakthroughs that drastically reduced infant mortality and increased adult life spans was another major reason for the increased use of energy.

USE OF EARTH'S RESOURCES BY BUILDINGS

The escalation of energy use continued unabated until the oil embargo of 1973. This event made the world suddenly realize its excessive dependence on energy and the pace at which the earth's finite energy resources were being depleted. An even more significant realization was how the world's excessive use of energy was irreparably damaging the environment and endangering people's health.

For example, the thinning of the ozone layer in the earth's atmosphere due to the use of ozone-depleting chemicals was well established during the latter part of the twentieth century. Global warming, caused by the emission of carbon dioxide and other gases, was no longer a theoretical prediction but a reality that could be proved by irrefutable measurements.

By this time (around the 1970s), the world was beginning to be aware that we were abusing the planet's resources and that unless something was done soon, the damage to our ecosystem could be irreversible. We also realized that energy is only one of the several resources needed for our survival, and, for humankind to have a reasonable future, all resources, such as land (including landfills), water, air, and other materials, must be conserved and protected against depletion, pollution, and degradation.

This realization led to the concept of pursuing an *ecologically benign,* or *environmentally friendly,* existence. The term *sustainable development* eventually stuck because it included the dimension of time. Another commonly used term is *green development,* which emphasizes its association with the environment.

NOTE

Embodied Energy

Embodied Energy in a Building

The embodied energy in a building is the amount of energy consumed by all the processes involved in the production of a building—from the acquisition of raw materials to the delivery and placement of the final product in the building. It includes the energy used in mining, manufacturing, transportation, construction, and administrative functions.

Embodied Energy in a Product

The embodied energy in a material or product is the amount of energy used in its production, including the administrative functions related to its production. It does not include the energy used in the product's transportation to the construction site and its installation in the building.

EXPAND YOUR KNOWLEDGE

Ozone Depletion

The ozone layer lies in the earth's atmosphere—between 10 and 20 mi (18 and 35 km) from the earth's surface. It is in this atmospheric region that the ultraviolet (UV) component of solar radiation converts atmospheric oxygen to ozone. (An oxygen molecule consists of two atoms of oxygen, i.e., O_2, and an ozone molecule consists of three atoms of oxygen, i.e., O_3.)

The other components of solar radiation convert the ozone molecule back to an oxygen molecule, so there is a perpetual cycle of conversion from oxygen to ozone and back to oxygen. If the cycle is not disrupted, a constant thickness of the ozone layer is maintained in the atmosphere at all times. The ozone layer prevents the harmful UV-B rays from reaching the earth by absorbing them. In humans, UV-B rays cause skin cancer and eye cataracts and weaken the immune system.

Substances such as chlorine react aggressively with ozone and disrupt the ozone-oxygen cycle, decreasing the thickness of the ozone layer. Several modern-day chemicals contain chlorine, such as methyl chloroform (used in cleaning solvents, adhesives, and aerosols), carbon tetrachloride (used in dry cleaning), and chlorofluorocarbons (CFCs), used as freon in refrigeration equipment.

Since the 1970s, scientists had warned that the ozone layer was being depleted, but it was not until 1987 that measurements revealed an unexpectedly large ozone-depleted hole over the Antarctic. In response to the revelation, new production of ozone-depleting substances was banned under the 1987 international agreement referred to as the *Montreal Protocol*. Despite this agreement (and provided that all countries seriously adhere to it), the chemicals that are already in use will eventually find their way into the atmosphere, causing further damage.

The first serious worldwide discourse on sustainable development is generally traced to the (1987) United Nations publication entitled the *Brundtland Report* [10.1], named after Gro Brundtland, prime minister of Norway, who chaired the UN World Commission on Environment and Development. The gist of the report is best described by the following excerpt:

> *. . . the challenge faced by all is to achieve sustainable world economy where the needs of all the world's people are met without compromising the ability of the future generations to meet their needs.*

Since the publication of the Brundtland Report, sustainability has become an all-encompassing discipline because truly sustainable development must include all facets of human activity—agriculture, manufacturing, transportation, buildings, and infrastructure.

This chapter begins with a discussion of the fundamentals of sustainability as applied to building design and construction. Subsequently, it deals with the issues of sustainability as applied to materials and construction.

It must be noted that although sustainable materials and construction practices have been growing, they are still in the early stages and are likely to go through extensive evolution as we learn from experience. It is expected that sustainability issues will become mainstream and will be given the same importance in buildings as structural safety, fire protection, and energy conservation. A step in this direction has already been taken in the United States through the draft formulation of the first sustainable building design code by the International Code Council in November 2010, known as the *International Green Construction Code* (IGCC).

NOTE

Resource Consumption by Buildings

Among the five major sectors of the U.S. economy (agriculture, transportation, manufacturing, infrastructure, and buildings), the building sector consumes a large share of the earth's resources,* as buildings:

- Use 36% of total energy
- Use 30% of raw materials
- Use 12% of potable water
- Produce 30% of total waste
- Emit 30% of greenhouse gases

See also the box entitled "Some Interesting Energy Use Data" in Section 5.11.

*Source: Gregory Kats, et al. "The Costs and Benefits of Green Buildings," A Report to California's Sustainable Building Task Force, October 2003.

10.1 FUNDAMENTALS OF SUSTAINABLE BUILDINGS

As noted earlier, sustainable building design and construction is not a new concept. It is an idea that reemerged in the 1970s and gathered momentum after the 1973 oil embargo. At that time, the focus was limited to energy conservation and the use of alternative energy sources in buildings. As awareness grew that the problem was larger than one of just energy use, a wide range of related environmental issues were also incorporated in the thinking. The work of the architects and designers who propagated this thought came to be known by the more inclusive term *sustainable design*.

More specifically, sustainable design recognizes that buildings are major consumers of resources and continually generate waste and pollution. The processes used for manufacturing materials and the construction of buildings also use resources and produce waste and pollution.

The land used under the buildings and related infrastructure (roads, bridges, water supply, sewage-treatment plants, etc.) disrupts the ecosystems. Several interior materials,

EXPAND YOUR KNOWLEDGE

Greenhouse Gases (GHG) and Global Warming

The earth's atmosphere contains a number of gases—nitrogen (approximately 79%), oxygen (approximately 20%), carbon dioxide (approximately 0.04%), and several others in small amounts. Some of the atmospheric gases, particularly carbon dioxide, are responsible for producing the greenhouse effect on the earth. This allows the sun's radiation to pass through the atmosphere but traps the earth's long-wave radiation. The effect is similar to that produced by glass in a greenhouse (see the box entitled "Some Important Facts About Radiation" in Chapter 30).

In addition to carbon dioxide, there are several other gases, such as methane, nitrous oxide, and water vapor, that are responsible for the greenhouse effect. They are collectively referred to as the *greenhouse gases* (GHG). The greenhouse effect is useful because, in its absence, the earth would have been much colder, making life more difficult in several regions.

However, due to the rapidly increasing use of fossil fuels, the amount of carbon dioxide released in the atmosphere has increased substantially. Additionally, the use of nitrogen-based fertilizers for farming has increased the release of nitrous oxide. Consequently, the greenhouse effect has become more pronounced in the past century and has raised the average temperature of the earth.

Scientists believe that if the present trend continues, the earth's temperature could rise further, melting most of the polar ice and causing sea levels to rise so that low-lying countries, such as the Netherlands and Bangladesh, could be submerged. Cities such as New York, London, and Bangkok may also be at risk. Perhaps the most frightening part of global warming is the chain reaction that could ensue, upsetting ecosystems and causing increased flooding, severe droughts, heat waves, and so on.

However, despite two major United Nations Framework Conventions (the 1997 Kyoto Convention and the 2009 Copenhagen Convention), there is little agreement among major industrialized nations on curbing GHG emissions because of serious disagreement between the developing and developed worlds. While the developed world believes that emission curbs must be shared by all countries, the developing world believes that, being the major emitters and the original creators of the problem, the developed world must currently carry most of the burden.

finishes, and furnishings emit unhealthy gaseous products, causing harm to humans. The goals of sustainable architecture, therefore, are as follows:

Integrated Site Design: Promote development of the built environment that minimizes its impact on the natural systems and processes of the site or restores and improves sites that have been poisoned or altered greatly over time. Strategies include minimizing site disruption, increasing development density, minimizing the buildings' footprints, using pedestrian-friendly neighborhoods, developing links to public transportation, using landscaping that conserves water and reduces the heat island effect; and so on.

Water Conservation: Use water-conservation strategies that reduce storm water runoff and introduce water-harvesting techniques to increase local aquifer recharge; reduce or limit the use of potable water for landscaping; use low-flow plumbing fixtures, water-efficient appliances and heating, ventilation, and air-conditioning (HVAC) equipment; and so on.

Energy Conservation and Atmosphere Protection: Minimize energy use through energy-efficient HVAC, lighting, and other equipment; increase the use of renewable energy sources; reduce atmospheric ozone depletion, and so on.

Resource Efficiency: Reuse existing buildings; design long-lasting buildings that can be adapted for changing uses over time; reduce construction waste and implement construction waste management; increase the use of durable and reusable materials; use materials with greater recycled content; use locally or regionally produced products; and so on.

Indoor Environment: Maintain good indoor air quality; increase ventilation effectiveness; reduce the emissions of volatile organic compounds (VOCs) and other contaminants by interior materials; increase the use of daylighting of interiors; and so on.

Experts [10.2] claim that although a sustainable building generally has a higher initial cost, the financial payback through energy savings and lower waste-disposal and water-consumption costs is rapid. Additionally, there are several financial payback features that cannot be easily quantified: lower employee health costs and greater productivity due to the building's healthier indoor environment (cleaner air, increased amount of daylight, and greater tenant control of temperature and lighting).

10.2 ASSESSING THE SUSTAINABILITY OF BUILDINGS

Without standards to guide the design and construction for sustainability, it is difficult for a designer to make decisions that successfully address the numerous diverse and interrelated sustainability issues encountered in a project. For that reason, several organizations have worked for years to develop objective, practical, and fair systems to evaluate the sustainability of buildings of various types.

One organization that has become a key player in the United States is the U.S. Green Building Council (USGBC). It was formed in 1993 by a diverse group of individuals representing various interest groups, such as building owners, architects, engineers, constructors, environmentalists, building material manufacturers, utility companies, financial and insurance experts, and local, state, and federal governments.

The USGBC has devised a system to rate the sustainability of a building's design and construction. The rating system is voluntary, is based on consensus, and provides a third-party, independent measure of a building's sustainability. It is called the *Leadership in Energy and Environmental Design (LEED®) Green Building* rating system. Starting in 1998 as a pilot version (LEED 1.0), the rating system has gone through a number of updates and refinements (LEED 2.0, LEED 2.1, and LEED 2.2). The current version, introduced in 2009, is LEED 3.0, also called LEED v3.

NOTE

Major Differences Between the LEED v3 Rating System and Its Prior Version

- The total number of points has been raised to 100 in LEED v3 from 69 in the previous version (64 in five topical categories and 5 for innovation and design). The topical categories, across which 100 points are distributed, remain unchanged at five.
- According to the USGBC, the point redistribution across categories in LEED v3 is based on a more rigorous scientific foundation than in its previous version.
- The relative weights of three categories have been increased: Energy and Atmosphere—from 27% to 35%; Sustainable Sites—from 22% to 26%; and Water Efficiency—from 8% to 10%. The relative weights of Indoor Environmental Quality and of Materials and Resources have been reduced.
- A new nontopical Bonus Points category has been introduced in LEED v3 with a total of 10 points. The Innovation and Design category that existed as a separate category in the previous version has been moved to the Bonus Points category, which also includes Regionalization—a new concept that did not exist in the previous version.
- Although the green building certification levels—Certified, Silver, Gold, and Platinum—remain the same, the minimum number of points required to achieve these levels has been rationalized to 40%, 50%, 60%, and 80%, respectively. The larger point gap between the Gold and Platinum levels emphasizes the higher bar set for Platinum-rated buildings.
- A more predictable development cycle is proposed to be introduced by USGBC for future changes in the LEED rating system.

Because different building types pose different challenges, there are several variations in the rating system, which include

- LEED for new construction and major renovations (LEED-NC)—a comprehensive rating for the entire building, most commonly sought.
- LEED for existing buildings: operations and maintenance (LEED-EB)—applies to existing buildings to improve their impact on the environment by selecting sustainable equipment.
- LEED for commercial interiors (LEED-CI)—applies to tenant improvement of facilities.
- LEED for core and shell (LEED-CS)—applies to new buildings in which none or little interior work is done by the owner.
- LEED for schools (LEED-S)—a comprehensive rating system like LEED (NC) but recognizes the special sustainability requirements of schools (kindergarten to 12th grade).
- LEED for healthcare (LEED-HC)—a comprehensive rating system that recognizes the special sustainability requirements for health-care facilities.
- LEED for homes (LLED-H)—a comprehensive rating system for homes.
- LEED for neighborhood development (LEED-ND)—sustainable urban and city planning.

The LEED rating systems are based on the performance of a building (or a neighborhood) under the following five topical categories, as shown here and in Figure 10.1. The maximum possible point score in each category is

- Sustainable Sites (SS)—26 points
- Water Efficiency (WE)—10 points
- Energy and Atmosphere (EA)—35 points
- Materials and Resources (MR)—14 points
- Indoor Environmental Quality (EQ)—15 points

A building can score points in all five categories, with the total number of points in all categories adding up to a maximum of 100. The measure of a building's sustainability is determined by the sum of all points it scores, and the total score determines the *LEED certification level,* as shown in Figure 10.1. A building with a total score of 39 points or less is not considered a *green* building per the LEED rating system.

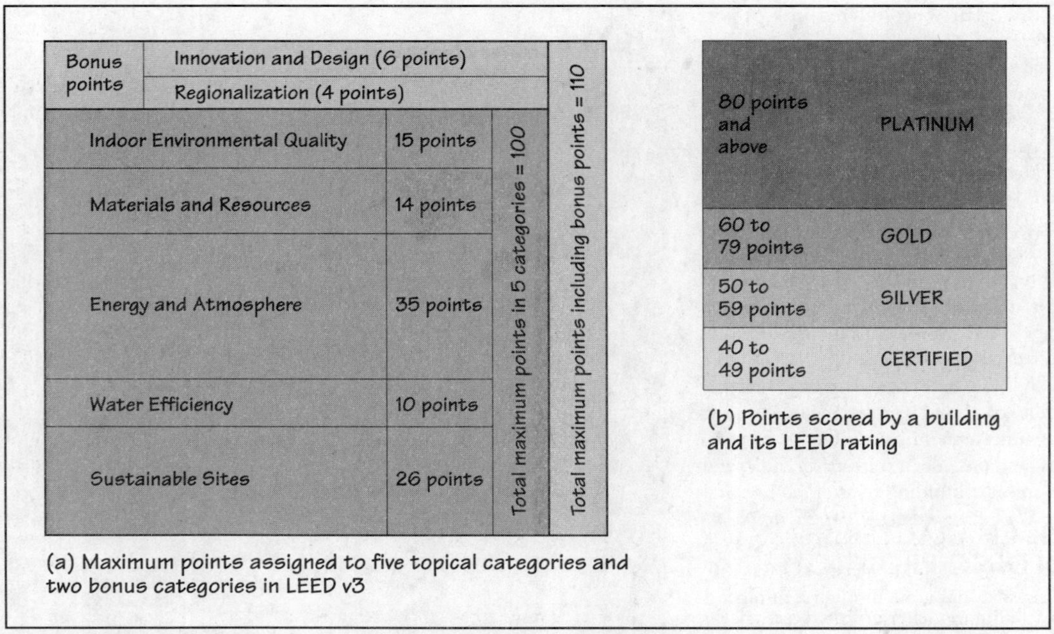

Bonus points
Innovation and Design (6 points)
Regionalization (4 points)
Indoor Environmental Quality 15 points
Materials and Resources 14 points
Energy and Atmosphere 35 points
Water Efficiency 10 points
Sustainable Sites 26 points
Total maximum points in 5 categories = 100
Total maximum points including bonus points = 110
(a) Maximum points assigned to five topical categories and two bonus categories in LEED v3

80 points and above PLATINUM
60 to 79 points GOLD
50 to 59 points SILVER
40 to 49 points CERTIFIED
(b) Points scored by a building and its LEED rating

FIGURE 10.1 **(a)** Relative importance of five topical and two bonus categories in the LEED v3 rating system. **(b)** Point scores required by a building for four LEED ratings: Certified, Silver, Gold, and Platinum.

In addition to the 100 points distributed across five topical categories, the LEED v3 rating system provides the opportunity to earn 10 bonus points (6 points for innovation and design and 4 points for regionalization—consideration of the special environmental requirements of a region). The 10 points are over and above the regular 100 points to help a project achieve certification through design innovation and consideration of the specific regional environment. Thus, the maximum possible number of points a project may score is 110. For example, a building that scores 54 points (out of 100) from the five topical categories and 6 bonus points (out of 10) will earn a total score of 60 points, changing its LEED rating from Silver to Gold.

Each topical category is further subdivided into several subcategories, which delineate the requirements of that subcategory and the number of points that can be earned by meeting the stated requirements. Additionally, each category has a few prerequisites that must be met before any points can be assigned in that category.

For example, a prerequisite in the Sustainable Sites (SS) category is that the construction of the building must include measures for *construction activity pollution prevention.* This requires controlling soil erosion, waterway sedimentation, and airborne dust generation during construction. Similarly, in the Energy and Atmosphere (EA) category, one of the prerequisites is the *commissioning of the building energy systems* by an independent commissioning authority. The commissioning authority appointed to satisfy the prerequisite in the EA category may also be used as the overall commissioning authority for the project.

Overall Commissioning Authority

To ensure that the building's performance after completion is as intended and designed (with respect to energy use, envelope design, water use, waste management, indoor air quality, etc.), the LEED rating system recommends that the building be commissioned. *Commissioning* is a systematic evaluation to verify that the basic building components and systems have been installed and calibrated to function as intended.

Engaging an independent *commissioning authority* (also called a *commissioning agent*) helps the owner obtain an objective evaluation of the design and construction team's work. Engaging a commissioning authority early in the project-delivery process (preferably during the design phase) ensures that the postcompletion verification strategies used by the authority are clear to the design and construction teams.

PRACTICE **QUIZ**

Each question has only one correct answer. Select the choice that best answers the question.

1. The embodied energy in a building is a measure of the energy consumed
- **a.** to extract raw materials from the earth and manufacture a finished building product.
- **b.** in the administrative functions related to raw material extraction and product manufacturing.
- **c.** to transport the finished product to the building site and install it in the building.
- **d.** all of the above.
- **e.** (a) and (b).

2. The increased use of a few modern-day chemicals has increased the thickness of the ozone layer.
- **a.** True
- **b.** False

3. Global warming has been caused mainly by the
- **a.** thickening of the ozone layer in the atmosphere.
- **b.** increase in the emission of greenhouse gases.
- **c.** decrease in the cloud cover over the earth.
- **d.** deforestation of the earth's surface.
- **e.** all of the above.

4. One of the major U.S. organizations that deals with sustainable architectural design is the
- **a.** USEBC.
- **b.** USFBC.
- **c.** USGBC.
- **d.** USHBC.
- **e.** none of the above.

5. The total number of topical categories in the LEED-NC rating system is
- **a.** 10.
- **b.** 8.
- **c.** 5.
- **d.** 4.
- **e.** 2.

6. Per the LEED-NC rating system, a building may receive recognition as a green building at the following levels:
- **a.** Platinum, Gold, Silver, and Bronze
- **b.** Gold, Silver, Bronze, and Certified
- **c.** Gold, Silver, and Bronze
- **d.** Gold, Silver, and Merit
- **e.** Platinum, Gold, Silver, and Certified

7. A building that receives a total score of 55 points in the LEED–NC rating system will be recognized at the
- **a.** Sustainable level.
- **b.** Certified level.
- **c.** Silver level.
- **d.** Gold level.
- **e.** Green level.

10.3 CHARACTERISTICS OF SUSTAINABLE BUILDING PRODUCTS

Because building materials constitute a large part of the environmental burden created by a building, the use of green building materials and products is one of several constituents that make a building sustainable. Extracting materials from the earth and processing them into a finished product require energy and water resources and produce waste, some of which may be hazardous.

Some products give off toxic gases after installation. Others require cleaning with chemicals that may do likewise. Postconsumer disposal of products consumes landfills, some of which may pollute groundwater.

Materials whose overall environmental burden is low are referred to as *green materials.* The relative greenness of a material is based on the same basic determinants as for the building as a whole. More specifically, the greenness of a product is a function of the following factors:

- Renewability
- Recovery and reusability
- Recyclability and recycled content
- Biodegradability
- Resource (energy and water) consumption
- Impact on occupants' health
- Durability and life-cycle assessment of product

NOTE

Current Laws on Emissions and Manufacturing

Several federal and state laws currently control toxic emissions from products during and after manufacturing. Laws also exist to control, regulate, and monitor the entire raw-material use, manufacturing, and use cycle of products. The goals of sustainability, however, aim for even higher standards.

Renewability

The *law of conservation of matter* states that all matter on the earth and within its atmosphere can neither be destroyed nor created. In other words, whatever existed on the earth at the dawn of time will always exist. Its physical or biological state may, however, change, either through natural or human-made processes. The basic elements of which a material is composed continue to exist forever on the earth or in its atmosphere. For example, when iron corrodes, it becomes iron oxide. The amounts of iron and oxygen on the earth remain unchanged in this transformation.

The transformation of matter (contained within both physical and biological realms) from one state to the other and in various combinations is cyclical. In other words, matter cycles back and forth within physical realms and also from the physical to biological realms.

For example, there is a constant amount of water on the earth, held within the oceans, in the atmosphere, and in other terrestrial entities—both living and nonliving. The processes of evaporation, condensation, consumption, and disposal of water simply move it from one state to another and from one realm to another. There may be drought in one region and excessive rainfall in another region, but the total quantity of water on the earth and in its atmosphere remains constant.

FOCUS ON SUSTAINABILITY

Reduce, Reuse, and Recycle—Three Tenets of Sustainability

Reduce, reuse, and recycle are considered to be the three most important tenets of sustainable construction. These tenets are listed in order of their importance. The first tenet (reduce), in fact, far outweighs the other two in importance.

Although a great deal of stress is currently being placed on reusability and recyclability, the same level of concern is lacking for appropriate sizing of buildings, automobiles, and other items of human consumption. Regardless of how successful we are with reusability and recyclability, sustainability will not be achieved without seriously addressing the *reduction* tenet.

A worrying statistic about buildings is the gradual increase in the average size of new homes built in the United States during the past three decades. In 1970, a typical new house was 1,500 ft^2. In 2005, the corresponding size was 2,300 ft^2—an increase of approximately 50% [10.3]. At the same time, the average family size has become smaller.

Thus, all materials are theoretically renewable. However, some materials renew over a short-duration cycle, whereas the cycles of others are long, and for some materials the cycle may extend over millions of years. Materials that have a short renewal cycle and require limited processing input to convert them to a usable form are referred to as *renewable materials.* Conversely, materials that have long renewal cycles and require large processing input are called *nonrenewable materials.* A renewable material is greener than a nonrenewable material.

The renewal cycle of forests is brief—usually 25 to 50 years. To transform trees into wood products requires a small amount of additional resources. Wood is, therefore, a renewable material. Perhaps one of the most renewable building materials is adobe. We can dig earth from the ground, shape it into adobe bricks, and construct buildings with them with a negligible amount of processing. When an adobe building is no longer needed, it can be demolished, and its material can be reprocessed for use in a different building with zero renewability duration. Alternatively, the material may be returned to the earth in almost the same state in which it was first obtained.

NOTE

Difference Between Renewal and Recycling

The difference between the renewal and recycling of a material is subtle. *Renewal* refers to the process of recycling that occurs in nature. *Recycling* is a deliberate process.

Metals, stone, glass, and plastics are examples of nonrenewable materials because they take much longer to renew and require excessive processing resources to convert them into usable states. Take the example of steel again. If left unprotected, it rusts. Rust is essentially the same as iron ore but in a highly diluted form. To obtain iron from the rust—which, by mingling with the other constituents of the earth's surface, becomes further diluted in iron concentration (i.e., becomes a low-grade iron ore)—requires enormous processing energy and produces enormous waste and pollution. Natural geological processes can provide iron ore with a high iron concentration, but these processes would take millions of years.

RECOVERY AND REUSABILITY

An effective way of greening a building is by using materials that have been recovered from the demolition of existing buildings. This reduces raw-material extraction and also reduces the burden on landfills. Obviously, salvaged materials must be obtained from nearby sources (to reduce transportation) and must be of usable quality, and their remaining life spans must meet the requirements of their reuse. The economics of recovery and reuse are more favorable with durable materials, such as bricks, natural stone, steel, and aluminum, than with materials with short life spans.

It is expected that as the use of salvaged materials becomes an accepted part of architectural practice, demolition recovery will become a more important commercial enterprise. The role of manufacturers in this area cannot be overemphasized. As the manufacturing industry moves further toward sustainability and assumes responsibility for the entire life cycle of the materials it produces, salvage and reuse could become part of the manufacturers' responsibilities.

In this scenario, the manufacturer retains ownership of the product, which is transferred to the consumer for only a certain period. At the end of that period, the buyback of the product by its manufacturer is mandated, and it is to be refurbished, remanufactured, or recycled as needed. The ultimate disposal of the product in a landfill also remains the manufacturer's responsibility. This should make product manufacturing and its use in buildings approach a closed-loop system, Figure 10.2.

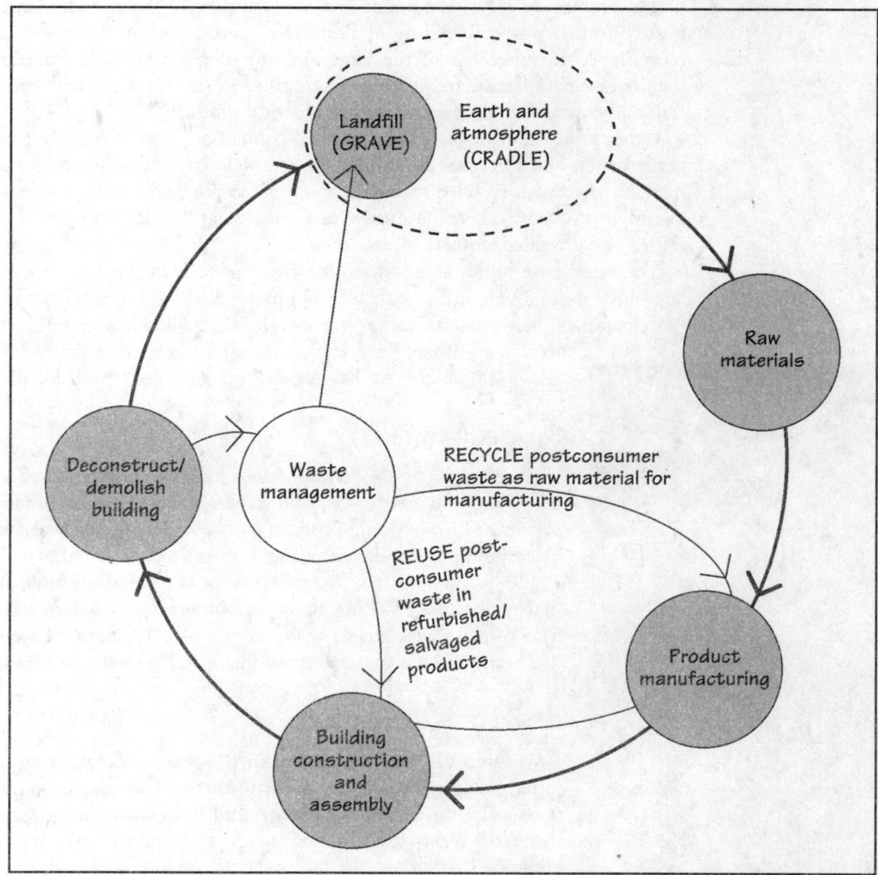

FIGURE 10.2 A closed-loop (cradle-to-cradle) building product life cycle. Note that in an ideal closed-loop life cycle, any waste sent to a landfill should be biodegradable so that it becomes food for living creatures, and all emissions into the atmosphere should be benign.

As environmental legislation makes waste management and disposal more costly, manufacturers will be increasingly forced to make their products greener. Removing hazardous contents, if any, before disposal and ensuring rapid degradability of materials (e.g., computers, furniture, or carpets) in landfills will be the manufacturers' responsibility.

An obvious extension of demolition recovery is the concept of *building deconstruction*, which implies that when a building has served its useful life, it is not demolished but deconstructed into its components. The deconstructed components can be refurbished if necessary and sold for use in another building.

Conceptually, deconstruction is the reverse of construction. Only those components of a deconstructed building that cannot be reused are disposed of as waste.

RECYCLABILITY AND RECYCLED CONTENT

The greater the amount of recycled content in a material, the greener it is. Recycling is generally classified into two types: *internal recycling* and *external recycling*.

Internal recycling is the reuse of materials that are by-products or waste products from manufacturing. The brick- and glass-manufacturing industries and several others have used internal recycling for a long time. Broken or defective bricks or glass are reused as raw materials in the manufacturing process. Ready-mix concrete manufacturing plants recycle wastewater reclaimed from washing concrete mixers after they return for recharge and delivery.

External recycling, on the other hand, is reclaiming a material for use in the manufacturing process of the same or a different product after it has become obsolete or unnecessary in its present application. This is referred to as *postconsumer waste recycling*.

For example, concrete obtained from a demolished building can be crushed and used as a drainage layer under a new concrete slab-on-grade in place of crushed stone. In some cases, crushed old concrete may be used as coarse aggregate in new concrete. Similarly, reinforcing or structural steel recovered from a demolished building can be melted and processed as new steel.

The amount of recycled content in a material depends on the economics of procuring post consumer waste. Recycling of aluminum (beverage) cans became successful early in recycling history because of the large amount of resource input (particularly energy) required to manufacture new aluminum from its ore, as compared with procuring recycled cans and converting them into a new or identical product.

Other products with significant recycled content are steel, carpets, gypsum board, rubber, plastics, and fiberglass insulation. Increasingly larger number of products are being manufactured with recycled materials, and their recycled content is also increasing. Consequently, recycling has become a major and diverse industry that employs a large number of skilled and unskilled workers.

The number of highly skilled scientists and engineers in the recycling industry is quite large because manufacturing with recycled materials is technically as complex as (if not more complex than) manufacturing from virgin materials. For example, one of the major recycling problems in the glass industry has been to remove color from used glass—a problem that has engaged the brains of several top glass scientists.

TABLE 10.1 EMBODIED ENERGY IN SELECTED MATERIALS

Material	Embodied energy MJ/kg	MJ/m^3
Adobe	0.42	819
Bricks	2.5	5,170
Concrete block	0.94	2,350
Concrete (4,000 psi)	1.3	3,180
Stone	0.79	2,030
Lumber	2.5	1,380
Plywood	10.4	5,720
Particle board	8.0	4,400
Aluminum (virgin)	227	515,700
Aluminum (recycled)	8.1	21,870
Steel (virgin)	32.0	251,200
Steel (recycled)	8.9	37,200
Copper (virgin)	70.6	631,164
Glass (float)	15.9	37,550
Gypsum board	6.1	5,890
Fiberglass insulation	30.3	970
Extruded polystyrene	117	3,770
Cellulose insulation	3.3	112

BIODEGRADABILITY

Everything else being identical, biodegradable materials, such as adobe, wood, and paper, are greener than nonbiodegradable materials such as metals and plastics. Thus, building product packaging, if required, consisting of paper-based products should, therefore, be preferred.

Ideally, all waste sent to landfills should be biodegradable and nontoxic so that it becomes food for plants and other living creatures. Some ecologists have coined the axiom *waste must equal food*. Waste that cannot be converted to a biodegradable material should be recycled over and over again.

RESOURCE CONSUMPTION

Two resources that are critical in product manufacturing are energy and water. Both are, therefore, determinants of a material's greenness. As stated previously, the amount of energy used in manufacturing a material is called its *embodied energy*. Embodied energy is generally measured in megajoules per kilogram (MJ/kg) or MJ/m^3.

Table 10.1 gives approximate values of the embodied energy in selected building materials [10.4]. Because all materials are not equal in terms of their use and properties, embodied energy cannot be used directly in comparing the energy efficiency of one material with that of another. However, it is a good measure for comparing the energy efficiency of similar products—those that belong to the same application group. For example, embodied energy may be used in comparing plywood with oriented strandboard, one carpet with another carpet, or one type of gypsum board with another.

Regardless of the materials' embodied energy, the use of locally produced materials reduces the building's embodied energy by reducing transportation energy use. It also improves the local economy.

Data for total water used in manufacturing a product, similar to embodied energy data, are not generally available.

IMPACT ON OCCUPANTS' HEALTH

The relative greenness of a material is also a function of its impact on human health. A number of modern building materials, particularly adhesives used in wood products and floor finishes (carpets, linoleum, and vinyl and rubber floorings), paints, sealers, and sealants, emit volatile organic compounds, which are harmful to humans beyond a certain level of concentration. These emissions adversely impact indoor air quality (IAQ) and, hence, the health of building occupants and are regulated by the U.S. Environmental Protection Agency (EPA).

Poorly ventilated spaces with high moisture contents and the use of fibrous materials may support the growth of mold when the materials become wet. Mold has been known to cause human health problems.

In addition to the impact of materials on building occupants, their impact on the earth and its biosphere must be accounted for, because some materials will either leach toxic materials during their use or degrade into hazardous substances in landfills.

DURABILITY AND LIFE-CYCLE ASSESSMENT (LCA) OF A PRODUCT

A product's durability is closely related to its sustainability. A material with high embodied energy but a long service life may be more sustainable than one with low embodied energy but a short service life. Additionally, a material may not emit hazardous compounds when installed in the building, but the chemicals required for its manufacture or periodic maintenance may do so. Therefore, the overall ecoburden of a product should be determined based on its expected service life. In other words, a life-cycle assessment (LCA) of a product's greenness is more important than its initial greenness.

The LCA of a product evaluates the environmental impact of the product over its entire life cycle, including the product's production, transportation, use, reuse, and its final disposal. Environmental impact measures include global warming, ozone depletion, production of pollutants and toxic waste, resource consumption, land use, and so on.

10.4 ASSESSING THE SUSTAINABILITY OF BUILDING PRODUCTS AND ASSEMBLIES

The growing public acceptance of and demand for sustainable building design and construction has brought building material and product manufacturers under increasing pressure to develop green products. Manufacturers that do not respond to this demand know that they are at risk of marginalization.

The public demand for green products has been accompanied by two important developments: (a) the issuance by the FTC of the "Guides for the Use of Environmental Claims" (16 CFR Part) and (b) the development of metrics to assess the sustainability (greenness) of products. The objective of the FTC's guidelines is only to ensure that the green claims made by manufacturers avoid consumer deception by refraining from overstatement and the use of language that may lead to unintended implications.

There are two types of metrics (objective methods of measuring performance) to assess the greenness of building products and assemblies:

1. Metrics based on a single attribute or a few selected environmental attributes
2. Metrics based on a comprehensive, life-cycle assessment (LCA) of all important environmental attributes

It is important to note that the use of green products (and sustainable construction in general) is largely voluntary at the present time. The situation is rapidly evolving, however, as more and more U.S. states and cities introduce sustainability requirements for new

NOTE

Environmental Marketing Claim Guides for Products

As the green building movement became popular with owners, designers, and constructors, almost every product manufacturer began to claim its product as green (greenwashing). In an attempt to protect consumers against deceptive marketing practices, the U.S. Federal Trade Commission (FTC) issued "Guides for the Use of Environmental Marketing Claims" in 1992, popularly known as the "Green Guides." It was revised in 1998. Another major revision has been completed, is out for public comment (as of early 2011), and is expected to be finalized for conformance by the industry.

The "Green Guides" provides comprehensive guidance to manufacturers on how to formulate their green marketing claims, including labels, logos, emblems, and promotional materials in graphical, text, digital, and electronic formats, and so on, in a manner that has an acceptable, unambiguous, transparent and independent, third-party substantiation. It does not deal with or establish any specifications or measurement standards for the greenness of products. The "Guides" applies to the product, its components, and its packaging.

construction. It is also important to note that any acceptable sustainability metric must be transparent, independent, and substantiated by third-party evaluation that is based on a recognized evaluation standard. (Even a third-party evaluation must be transparent, resulting from the use of a recognized standard.)

Subjective claims of greenness are unacceptable because almost every product has some green virtues. For example, manufacturers of products made from metals (steel, aluminum, copper, etc.) claim that their products are green because metals can be fully recycled. Manufacturers of wood-based products claim greenness because wood is rapidly renewable and is a biodegradable material. Masonry product manufacturers promote their products for reasons of reuse, recyclability and zero VOC emissions, and so on.

10.5 ASSESSING PRODUCT SUSTAINABILITY BASED ON A SINGLE ATTRIBUTE OR A LIMITED SET OF ATTRIBUTES

It may appear at first that sustainability metrics based on a single attribute or a limited number of environmental attributes have little use because they are not comprehensive. This is not true provided that the metric is significantly relevant to the product's performance. For example, the most important environmental attribute of paints and adhesives is VOC emissions. Emission ratings of these products, therefore, provide valuable information for facility executives, who may use these ratings along with several other factors (such as color, cost, etc.) in making their selection. Similarly, ratings of windows and glass curtain walls based on thermal performance (i.e., their impact on energy consumption) are useful metrics because energy conservation is arguably the single most important sustainability attribute.

An additional advantage of single- or limited-attribute metrics is that they can be provided in the form of labels or marks (the Energy Star label, the Green Seal label, etc.), which are easy to comprehend and, therefore, better disposed to adoption. Table 10.2 provides an overview of some of the commonly used metrics for assessing the greenness of building products based on single or limited environmental attributes.

ENERGY STAR LABEL

The Energy Star label was introduced by the U.S. Environmental Protection Agency (EPA) in 1992 to recognize energy-efficient computers. The program's success led to its extension to other product areas and its acceptance by several other countries. Thus, Energy Star is now an international energy-labeling system that rates and labels

- appliances (refrigerators, air conditioners, washing machines, and so on)
- products (light bulbs and fixtures, office products, building products, and so on)
- homes
- commercial and industrial buildings

TABLE 10.2 COMMONLY USED METRICS FOR ASSESSMENT OF SUSTAINABILITY OF BUILDING PRODUCTS BASED ON A SINGLE ATTRIBUTE OR A LIMITED SET OF ENVIRONMENTAL ATTRIBUTES

Rating label or certificate	Rating organization and its headquarters	Attributes considered in rating	Building products rated
Energy Star label	U.S. Environmental Protection Agency (EPA) Washington, D.C.	Energy efficiency	Doors, windows, and skylights; appliances, HVAC equipment, lighting, exit signs, etc.
Green Seal certification mark	Green Seal Washington, D.C.	Several product-specific attributes	Paints, coatings, adhesives, windows, and doors
Greenguard certificate	Greenguard Environmental Institute (GEI) Marietta, Georgia	Indoor air quality	Adhesives, paints, and wallpapers
Green Label and Green Label Plus	Carpet and Rug Institute (CRI) Dalton, Georgia	Indoor air quality	Carpets, rugs, and cushions
Certified Wood under Forest Stewardship Council (FSC) rules	SmartWood Richmond, Vermont Scientific Certification Systems (SCS) Emeryville, California	(a) Sustainability of forests from which wood is derived (b) Chain-of-custody certification of wood-based products	Wood used in any product

Among the building products, only those used in the building envelope are Energy Star rated, such as windows, skylights, and exterior doors; insulation, radiant barriers, and air sealing products; and roofing products. Because different climate zones in the country have different energy efficiency requirements, the Energy Star label on a door, window, or skylight shows the climate zone(s) for which the label is appropriate. An Energy Star–labeled door, window, or skylight certifies that its performance either equals or exceeds the requirements of the International Energy Conservation Code (see Chapter 5).

A comprehensive Energy Star label is available for homes and commercial/industrial buildings. The label for new homes requires the energy analysis of the proposed home through the Home Energy Rating System (HERS)—a software package developed by the Residential Energy Services Network (RESNET). The analysis is performed by RESNET-approved raters based on the proposed home's construction drawings. The analysis yields a HERS index that determines if the proposed home meets the requirements of the Energy Star label and, if not, what improvements are needed.

The analysis is followed by inspections during construction and tests of air leakage after completion, called *blower tests*. Two blower tests are performed; one test determines the air leakage in the building envelope, and the other test determines the leakage in the HVAC duct system.

Homes built to the minimum requirements of the International Energy Conservation Code (IECC) receive a HERS rating of 100. A HERS index greater than 100 indicates that the home does not meet the code requirements. Many older homes built prior to IECC's adoption belong in this category. A HERS index of less than 100 indicates that the proposed home is more energy efficient than the minimum code requirements. Each 1-point decrease in the HERS index implies 1% greater energy efficiency. Thus, a HERS index of 90 means that the proposed home is 10% more efficient than the minimum code-required home. To receive an Energy Star label, a proposed home must have a HERS index of at least 85 in climate zones 1 to 5 and an index of 80 in climate zones 6 to 8, Figure 10.3.

Green Seal Label

The Green Seal organization has developed fairly stringent standards for measuring the greenness of several building products (such as paints, coatings, doors, and windows) based on a number of environmental attributes that are particularly relevant to the products. However, it is best known for its seals on residential and commercial cleaning products (hand cleaners, carpet cleaners, floor cleaners, and so on).

Greenguard Certificate

Greenguard certification focuses primarily on indoor air quality to certify products such as paints, adhesives, sealants, and interior furnishings and is provided by the Greenguard Environmental Institute (GEI), which is particularly well known for certifying building construction that meets standards for mold prevention.

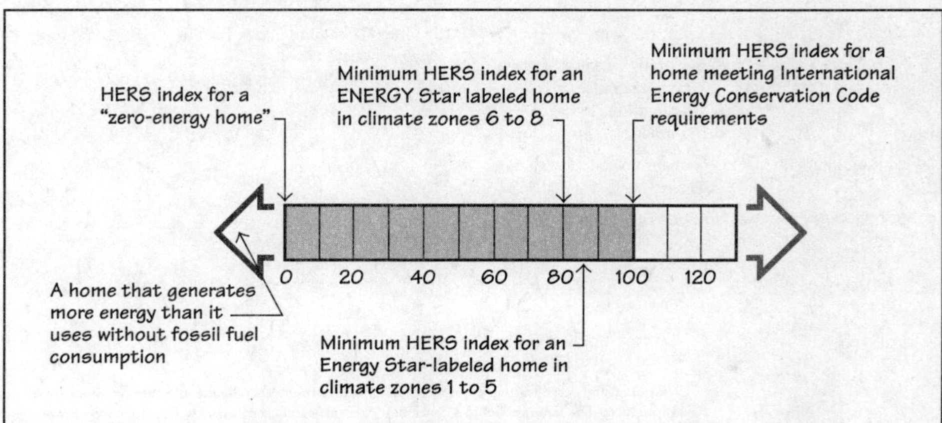

FIGURE 10.3 HERS index and Energy Star label requirements for homes.

GREEN LABEL AND GREEN LABEL PLUS

Green Label is the mark assigned by the Carpet and Rug Institute (CRI) to carpets, rugs, and cushion materials that have low VOC emissions. Green Label Plus is a more stringent mark than Green Label.

CERTIFIED WOOD LABEL

The Certified Wood label is carried by wood products that have been produced by manufacturers according to the guidelines promulgated by the Forest Stewardship Council (FSC). The FSC guidelines prescribe responsible management practices for forest growth and harvesting that meet the social, economic, and environmental needs of the present and future generations. The FSC maintains a number of independent certifiers around the world.

The two major FSC-accredited certifiers in the United States are SmartWood and Scientific Certification Systems. (See Chapters 13 and 14 for additional details.) Note that FSC's Certified Wood label refers only to the environmental management aspects of forests. It does not include other sustainability concerns, such as embodied energy or emissions of VOCs by engineered wood products.

An additional certificate for wood or products that have some wood content is the chain-of-custody certificate. It is an inventory-control system that traces a wood product from its origin to the consumer to provide quality assurance.* *Chain of custody,* a term borrowed from criminology, refers to the entire chain of processes through which wood originating from a certified forest is transformed through various processes into a finished product, including its distribution and sale. The product may be an all-wood or a composite wood product.

SUSTAINABILITY VERIFICATION OF PRODUCTS BY THE INTERNATIONAL CODE COUNCIL

The sustainability evaluation of building products by the International Code Council, called the Sustainable Attributes Verification and Evaluation (SAVE) program, is limited to evaluating a product for the following attributes: recycled content, bio-based content, regional manufacturing, certified wood content, solar reflectance and emissivity of roofing products, and VOC emissions from paints and adhesives.

PRACTICE QUIZ

Each question has only one correct answer. Select the choice that best answers the question.

8. Which of the following building materials is most renewable?
- **a.** Steel
- **b.** Aluminum
- **c.** Natural stone
- **d.** Wood

9. The embodied energy of a product is a good index for comparing the energy efficiency of
- **a.** one material with another material regardless of the type of material.
- **b.** similar materials—materials that belong to the same group.

10. Weight for weight, which of the following materials has the highest embodied energy when made from virgin raw materials?
- **a.** Steel
- **b.** Aluminum
- **c.** Natural stone
- **d.** Lumber

11. The durability and sustainability of a material are related to each other.
- **a.** True
- **b.** False

12. The Certified Wood label considers all aspects of sustainability of wood.
- **a.** True
- **b.** False

13. The life-cycle assessment (LCA) of a product refers to its cost over its entire service life.
- **a.** True
- **b.** False

14. The Energy Star labeling program applies to
- **a.** appliances.
- **b.** products, such as windows and doors.
- **c.** buildings.
- **d.** all of the above.
- **e.** only (a) and (b).

15. Green Label and Green Label Plus programs apply to
- **a.** appliances.
- **b.** products, such as windows and doors.
- **c.** homes.
- **d.** all of the above.
- **e.** none of the above.

16. The HERS index is used in rating for the
- **a.** Green Seal label.
- **b.** Certified Wood label.
- **c.** Green Seal Plus label.
- **d.** Energy Star label.
- **e.** none of the above.

*Chain-of-custody certification: A wood product manufacturer generally manufactures certified as well as regular (noncertified) products. The same applies to a distributor, who may store certified as well as noncertified wood products. Therefore, a chain-of-custody certificate implies that a Certified Wood product is what it has been certified for and that it has not been mixed with noncertified products during manufacturing distribution, or sales.

10.6 ASSESSING THE SUSTAINABILITY OF BUILDING PRODUCTS AND ASSEMBLIES BASED ON A COMPREHENSIVE SET OF ATTRIBUTES

Unlike labels or certificates based on single- or limited-attribute metrics, a comprehensive assessment of sustainability is obtained only by evaluating the greenness of a product over its entire life—life-cycle assessment (LCA). This complex assessment is feasible only through computer-based analysis. The two more commonly used software programs for this purpose are

- Building for Environmental and Economic Sustainability (BEES) software—developed by the U.S. National Institute of Standards and Technology (NIST) to determine the greenness of building products
- EcoCalculator for Assemblies—developed by the Athena Institute (Ontario, Canada) to determine the greenness of building assemblies

BUILDING FOR ENVIRONMENTAL AND ECONOMIC SUSTAINABILITY (BEES) SOFTWARE FOR BUILDING PRODUCTS

BEES is a Windows-based program that determines the LCA of building products along with their life-cycle cost (LCC). The two features, environmental performance (through LCA) and economic performance (through LCC), have been combined because while LCA should be an important consideration in product selection, it is generally not the sole criterion in practice and has to be balanced with the cost of the product.

The software is, however, flexible and allows the determination of LCA and LCC independent of each other or in any user-defined combination by assigning different weights to LCA and LCC. For example, LCA and LCC can be given equal weights (i.e., 50% each), or 60% LCA and 40% LCC, and so on. The weights can also be set at 100% for LCA and 0% for LCC, or vice versa, to obtain individual values of LCA and LCC. Thus, the software generates an LCA score, an LCC score, and an overall score of the product, Figure 10.4.

The overall environmental burden of a product over its entire life cycle (cradle-to-gate, that is, beginning with raw materials to processing to the finished product as delivered at the gate of the manufacturing facility) is accounted for by considering 12 parameters, referred to as *environmental impact categories* in the LCA part of BEES. These categories are shown in Figure 10.5, along with the relative weights of each category considered in BEES. For example, global warming has a weighting of 29%, human health, 13%, and so on.

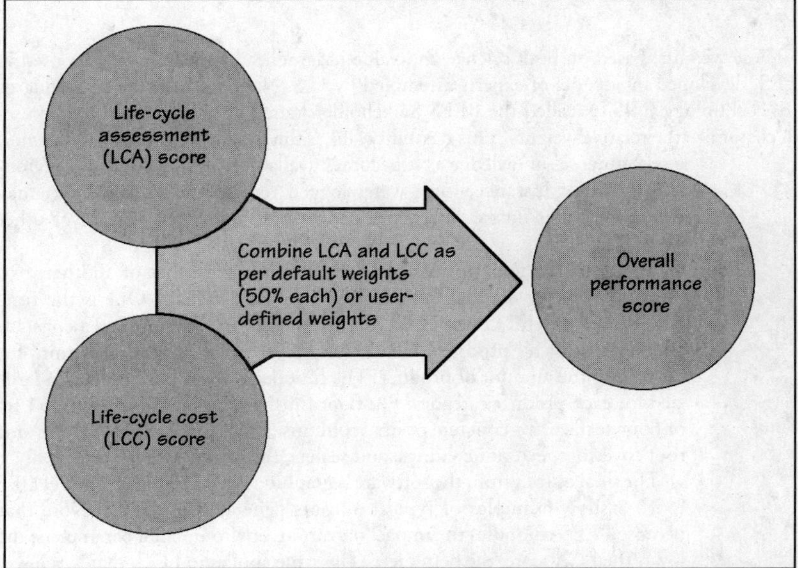

FIGURE 10.4 The output from BEES software for a product consists of an LCA value and an LCC value, and a combination of the two values provides the overall performance of the product.

NOTE

More About BEES

BEES 4.0, the current BEES software, is a fourth-generation product that includes environmental and economic performance data on 230 building materials and products. The software is downloadable free from the Web, including the accompanying manual. The LCA part of BEES is based on the International Standards Organization (ISO) Standard 14040-06, "Environmental Management—Life Cycle Assessment—Principles and Framework," and the LCC is based on ASTM Standard E917-05, "Standard Practice for Measuring Life-Cycle Costs of Buildings and Building Systems."

The LCC provided by BEES includes the initial cost of the product plus any future costs needed for its maintenance and replacement. The future costs are converted to the present value so that the total LCC of the product is given in present-value terms. Because the future costs are generally higher than the present cost due to inflation, they must be discounted to obtain their present value. The discount rate, which is related to the inflation rate, is set at 3% in BEES, but it can be changed to any desired value by the user to reflect an estimate of the future inflation rate.

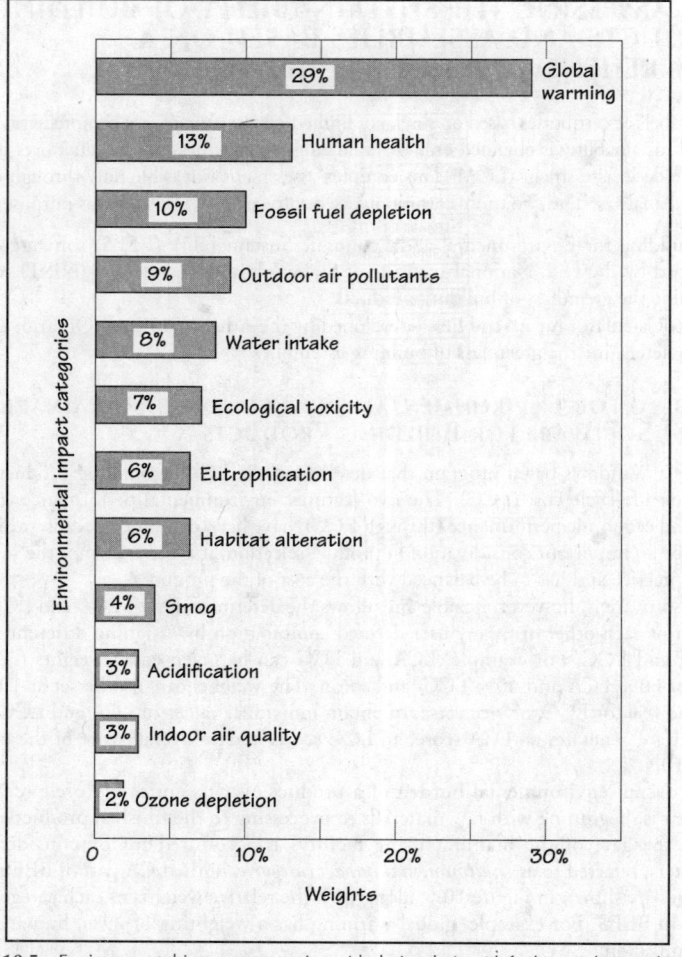

FIGURE 10.5 Environmental impact categories with their relative (default) weights used in BEES. The software allows the user to redistribute weights across categories so long as the aggregate weight equals 100.

These weights, based on both science and value judgments, are default weights used in BEES, developed by a panel of experts assembled by U.S. National Institute of Standards and Technology (NIST), called the BEES Stakeholder Panel. However, the software user can customize the relative weights. This flexibility allows the program to accommodate new environmental knowledge as it becomes available and to account for regionalization. For instance, for a water-logged region, the software user may delete the water intake category and distribute its weight (8%) over other categories.

To compare the performance of one product with that of another, two factors have been standardized in BEES across products. One is the time period over which LCA and LCC are computed. In BEES, this is a constant of 50 years for all products. The other factor is the "functional unit" for measuring the amount of product. The functional units have been standardized for each product category. For floor finishes, the functional unit is 1 ft^2 of floor surface; for concrete beams, columns, and slabs, the unit is 1 yd^3; for roof coverings, exterior coatings, and sealers, the unit is 100 ft^2; and so on.

The final output from the software is graphic as well as tabular. Figures 10.6 to 10.8 show examples of typical outputs generated by BEES. Note that because BEES computes the impact on various environmental parameters, the lower the LCA score, the better it is. The same applies to LCC, that is, a lower LCC score is a better score. This similarity between LCA and LCC provides an easy way of combining the scores.

NOTE

Eutrophication, Acidification, and Habitat Alteration

Eutrophication refers to the addition of mineral nutrients to the soil or water, such as nitrogen and phosphorus, which, in large quantities, reduce ecological diversity by increasing algae growth that leads to the lack of oxygen, killing species such as fish. *Acidification* refers to compounds released in air in a gaseous state or as solid particles, which, when dissolved in water, produce acids (referred to as *acid rain*) that adversely affect soil, buildings, trees, and other living creatures. *Habitat alteration* refers to the use of land by humans that may adversely impact threatened and endangered species.

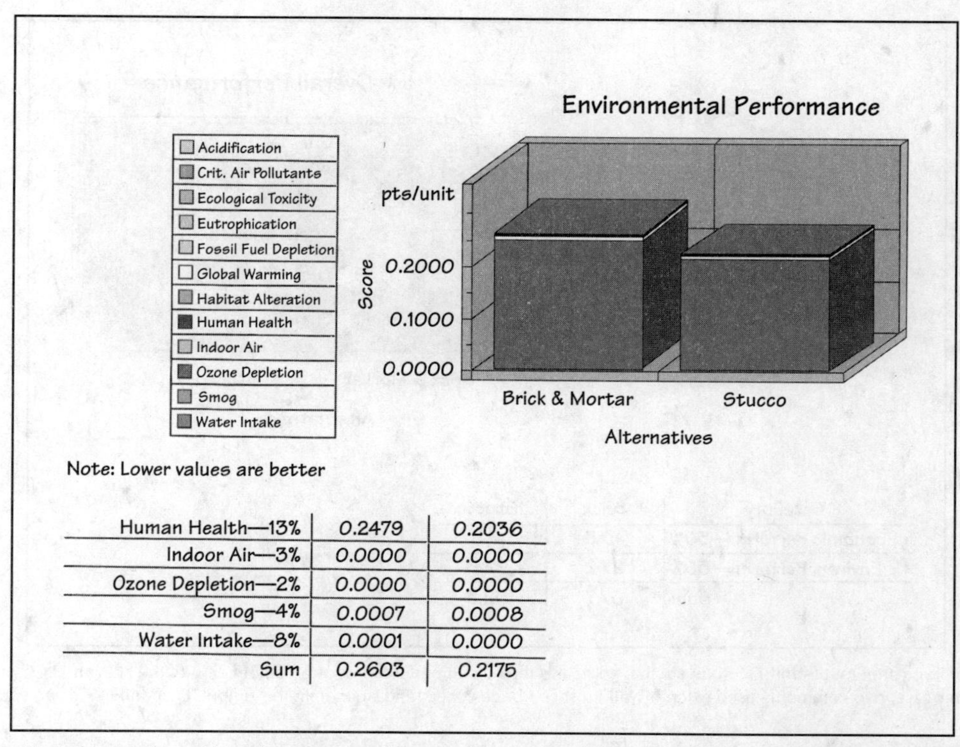

Human Health—13%	0.2479	0.2036
Indoor Air—3%	0.0000	0.0000
Ozone Depletion—2%	0.0000	0.0000
Smog—4%	0.0007	0.0008
Water Intake—8%	0.0001	0.0000
Sum	0.2603	0.2175

FIGURE 10.6 Typical output from BEES software that compares the environmental performance in terms of the LCA of two materials (in this case, two commonly used exterior wall finishes—brick veneer and stucco).

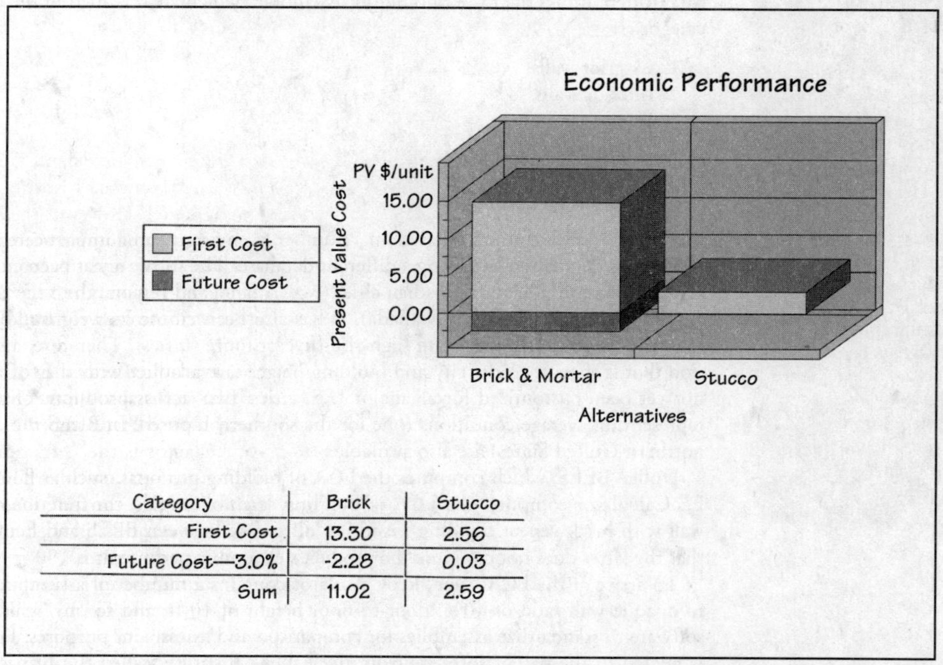

Category	Brick	Stucco
First Cost	13.30	2.56
Future Cost—3.0%	-2.28	0.03
Sum	11.02	2.59

FIGURE 10.7 Typical output from BEES software that compares the economic performance in terms of the LCC of two materials (in this case, two commonly used exterior wall finishes—brick veneer and stucco).

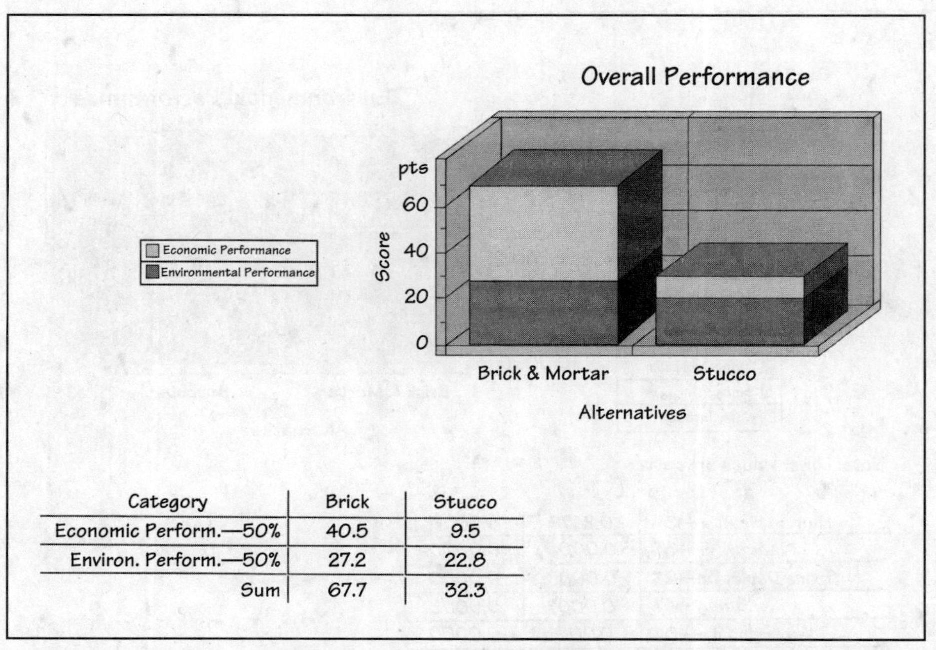

Category	Brick	Stucco
Economic Perform.—50%	40.5	9.5
Environ. Perform.—50%	27.2	22.8
Sum	67.7	32.3

FIGURE 10.8 Typical output from BEES software that compares the overall performance as a combination of the LCA and LCC of two materials (in this case, two commonly used exterior wall finishes—brick veneer and stucco). In this output, LCA and LCC have equal weights (50% each).

NOTE

Limitations of BEES

The versatility of BEES is dependent on the size of its database. At the present time, the database includes mostly generic products. The environmental impact data of these products are based on the average of similar products manufactured by various manufacturers. Thus, the software does not provide the performance differences between products from competing manufacturers. However, the database is being constantly expanded, and data on proprietary products are being added. As the software gains greater acceptance from design and construction professionals, its database is expected to grow.

ATHENA ECOCALCULATOR FOR ASSEMBLIES

EcoCalculator is Mac- and Windows-compatible software (developed on the Microsoft Excel platform) that is downloadable free of charge from the Web site of the Athena Institute. It determines the LCA of building assemblies, which are classified in the following six categories:

1. Exterior walls
2. Interior walls
3. Roofs
4. Windows
5. Intermediate floors
6. Columns and beams

The LCA of assemblies is based on a number of environmental impact categories similar to those used in BEES but using a different database. The software has been customized for several different U.S. and Canadian cities (e.g., Atlanta and Pittsburgh in the United States and Toronto and Calgary in Canada). It has also been customized for building height—low-rise (up to four stories) and high-rise (five or more stories). Therefore, a software version that is specific to the city and building height is available. Because EcoCalculator has not yet been customized for all major U.S. cities, two regional versions of the software, representing average conditions (one for the southern United States and the other for the northern United States) are also available.

Unlike BEES, which computes the LCA of building products (such as floor coverings), EcoCalculator computes the LCA of building assemblies (e.g., concrete masonry backup wall with brick veneer cladding). Another difference between BEES and EcoCalculator is that the latter does not compute the product's economic performance.

To arrive at the LCA value, EcoCalculator contains a number of assumptions (glazing-to-opaque-wall ratio of 40%, floor-to-floor height of 10 ft, and so on), which allows the software to standardize assemblies for comparison and assessment purposes. EcoCalculator is related to the parent software from the Athena Institute, called the Impact Estimator, which computes the LCA of (entire) buildings.

PRACTICE QUIZ

Each question has only one correct answer. Select the choice that best answers the question.

17. Which of the following environmental impact categories has the highest weight in the default weights set in BEES software?
- **a.** Human health
- **b.** Fossil fuel depletion
- **c.** Outdoor pollutants
- **d.** Global warming
- **e.** Ozone depletion

18. Which of the following environmental impact categories has the lowest weight in the default weights set in BEES software?
- **a.** Human health
- **b.** Fossil fuel depletion
- **c.** Outdoor pollutants
- **d.** Global warming
- **e.** Ozone depletion

19. In comparing various products based on their LCA scores, the product with the lowest score is the worst.
- **a.** True
- **b.** False

20. The time period used in BEES for determining the LCC of a product is
- **a.** 25 years.
- **b.** 50 years.
- **c.** 75 years.
- **d.** 100 years.
- **e.** none of the above.

21. The life-cycle cost (LCC) of a product obtained from BEES is its
- **a.** equivalent future cost.
- **b.** equivalent present cost.

22. EcoCalculator for assemblies has been developed by the
- **a.** National Institute of Standards and Technology (NIST).
- **b.** American Society of Testing and Materials (ASTM).
- **c.** International Code Council (ICC).
- **d.** International Standards Organization.
- **e.** none of the above.

23. EcoCalculator for assemblies uses the same database as BEES.
- **a.** True
- **b.** False

REVIEW QUESTIONS

1. In the LEED v3 rating system, list the categories that are used to measure the sustainability of buildings.
2. Explain the role of commissioning in the LEED rating system for new construction.
3. List various factors that determine the sustainability of a building product.
4. Using a sketch and notes, explain an ideal closed-loop product life cycle.
5. Explain which materials are considered renewable and which are considered nonrenewable and why.
6. Using a sketch and notes, explain the HERS index and Energy Star label requirements for homes.
7. Describe the important features of BEES and its current limitations.

CHAPTER 11 Soils and Excavations

CHAPTER OUTLINE

As stated in Chapter 1, an important part of the predesign phase of a building is site evaluation. This is a multifaceted evaluation. Some facets impact the construction materials and methods employed for the building significantly, while others play only a minor role. For example, the cost of land, the land use, and the site's accessibility to infrastructural services, though important, have a marginal impact on the construction aspects of the building. By contrast, several other site evaluation facets impact the construction aspects of the building substantially. These facets can be categorized as

- Surface investigation of the site
- Subsurface (i.e., below-surface) investigation of the site

Surface investigation involves making a preliminary judgment about the site's suitability for the proposed building. The first part of surface investigation is a visual assessment of the site's topography, vegetation, storm water drainage pattern, foundation systems used in nearby buildings and their performance, and so on. If visual assessment concludes that the site is potentially suitable for the proposed building, the second part of the surface investigation—an engineering land survey of the site—is undertaken.

The land survey provides physical measurements of the site's boundaries, topography, drainage patterns, encroachments, easements, ownership, legal history, and so on, and is a critical component of the surface investigation. If the land survey analysis is favorable for the project, the next step is subsurface investigation of the site.

Subsurface investigation deals with conditions below the ground surface to determine the requirements for the foundations and excavations. Subsurface conditions have a significant influence on the building design, construction materials, structural system, construction cost, and schedule. For example, it is more expensive and time-consuming to excavate in a rocky stratum or a stratum with a high water table. This chapter deals with aspects of subsurface investigation that impact foundation design and construction.

SOIL—A STRUCTURAL MATERIAL

The structural integrity of the building foundation is largely a function of the earth material (i.e., the soil) on which it rests. Because of its geological history, soil is a fairly complex material. Its complexity is embedded in its chemical composition, physical characteristics, variations in the depth below ground, and variations from location to location. The structural aspects of soils, as distinct from the soil's botanical characteristics (of interest to farmers), constitute a branch of study known as *soil mechanics*, or, more recently, as *geotechnical engineering* or *geostructural engineering*.

This chapter provides a brief and simplified discussion of soils' geotechnical characteristics that are essential for an architect or a constructor to know in order to make preliminary assessments and to communicate with geotechnical and structural consultants. It begins with a discussion of soil classification, leading to the methods used for field and laboratory investigations of soil properties. Because all foundation systems require some type of excavation, various methods of excavation and earthwork are presented. This is followed by a discussion of the support systems used to prevent the failure of deep excavation faces and trenches. The chapter concludes with a brief discussion of excavation dewatering.

11.1 CLASSIFICATION OF SOILS

If organic matter, which is present only in small quantities in most locations, is ignored, the earth's crust consists mainly of mineral (i.e., inorganic, noncombustible) matter. The earth's mineral matter is generally classified as *rocks* and *soils*.

In rocks, the mineral particles are firmly bonded together. Soil consists of either individual particles or a conglomerate of several easily separable particles that have resulted from the weathering of rocks (see the margin note). In most locations, the top layer of the ground consists of soil; rock generally occurs deep under the surface.

In most situations, the soil has adequate strength and other characteristics to support a building. When this is not the case, building foundations must bear on bedrock. (Rocks are classified as *igneous rocks, sedimentary rocks,* and *metamorphic rocks;* see Chapter 25.)

GRAVELS, SANDS, SILTS, AND CLAYS

There are a number of characteristics that must be considered in determining the ability of a soil to support building loads. One important characteristic is soil classification based on the size of soil particles. The size of soil particles is measured by passing a dried soil sample through a series of sieves, each with a standardized opening size (see Figure 11.4). Soil particles that are retained in a No. 200 sieve are called *coarse-grained soils* (>0.075 mm), and soils that pass through this sieve are considered to be *fine-grained soils*.

As shown in Figure 11.1, coarse-grained soils are further divided into *gravels* or *sands*. Gravels have particles that are retained on a No. 4 sieve but are smaller than 3 in. Particles larger than 3 in. are called *cobbles* or *boulders*. Sand particles pass through a No. 4 sieve but are retained on a No. 200 sieve.

Fine-grained soils are divided into *silts* and *clays*. Particles smaller than 0.075 mm but larger than 0.002 mm are classified as silt. Particles smaller than 0.002 mm are classified as clay.

UNIQUENESS OF CLAY (SWELLING AND SHRINKING)—THE EXPANSIVE SOILS

Apart from the particle size, a major factor that distinguishes clay from other soil constituents is the particle shape. Gravel and sand particles are approximately equidimensional; that is, they are approximately spherical or ellipsoidal in shape. This is because gravels and sands are the result of mechanical weathering.

Clay particles, on the other hand, are the result of chemical weathering. They are, therefore, not equidimensional but have flat, platelike shapes. Because of their flat particle shape, the surface-area-to-volume ratio of clays is several hundred or thousand times greater than the corresponding ratio for gravels and sands. The characteristics of silt particles lie between those of clay and sand.

The behavior of clayey soils is greatly influenced by the electrostatic forces that develop between platelike surfaces. In the presence of water, these forces are repulsive, which increases the space between plates. Therefore, in the presence of water, clayey soils swell, and as water decreases (i.e., when they dry), they shrink.

Soils that are predominantly clayey are unstable because they expand and contract, depending on the amount of water present in them, and are referred to as *expansive soils*. If

NOTE

Rock-Soil Cycle

In the early stages of the cooling of the earth from its molten state, the earth's crust consisted only of rocks. Abrasion caused by the movement of water and air on the earth's surface disintegrated rocks into smaller soil particles. Fatigue resulting from thermal expansion, contraction, freezing, and thawing of rocks enhanced their disintegration.

In addition to *mechanical weathering*, just described, rocks also went through *chemical weathering*, caused by their reaction with atmospheric oxygen, plant life, and various chemicals transported by water.

Both weathering processes continue to this day. As the disintegrated particles are carried by water and deposited in valleys, lakes, and deltas, they are eventually converted back to rock due to the pressure of overlying material. Thus, rock and soil formations are parts of a perpetual cycle referred to as the *geological cycle*.

Bedrock

Bedrock is continuous layer of native rock that has not weathered to soil. The depth of bedrock below the earth's surface may vary from a couple of feet to 100 ft or more.

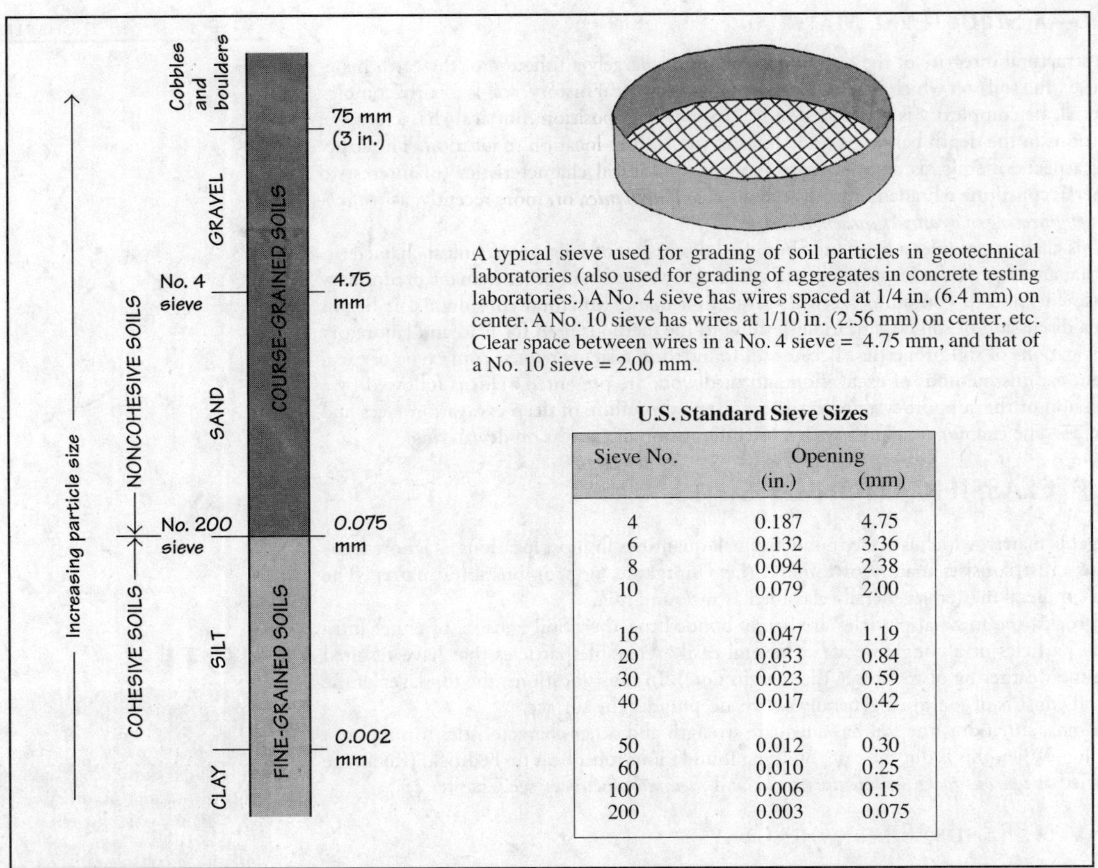
Cobbles and boulders
75 mm (3 in.)
GRAVEL
No. 4 sieve
4.75 mm
COURSE-GRAINED SOILS
SAND
NONCOHESIVE SOILS
Increasing particle size
No. 200 sieve
0.075 mm
COHESIVE SOILS
SILT
FINE-GRAINED SOILS
0.002 mm
CLAY
A typical sieve used for grading of soil particles in geotechnical laboratories (also used for grading of aggregates in concrete testing laboratories.) A No. 4 sieve has wires spaced at 1/4 in. (6.4 mm) on center. A No. 10 sieve has wires at 1/10 in. (2.56 mm) on center, etc. Clear space between wires in a No. 4 sieve = 4.75 mm, and that of a No. 10 sieve = 2.00 mm.

U.S. Standard Sieve Sizes

Sieve No.	Opening (in.)	Opening (mm)
4	0.187	4.75
6	0.132	3.36
8	0.094	2.38
10	0.079	2.00
16	0.047	1.19
20	0.033	0.84
30	0.023	0.59
40	0.017	0.42
50	0.012	0.30
60	0.010	0.25
100	0.006	0.15
200	0.003	0.075

FIGURE 11.1 The distinction between coarse-grained and fine-grained soils and between gravels, sands, silts, and clays. The term *cobbles* is used for pieces of naturally rounded rocks between 3 and 12 in. The term *boulders* describes similar material greater than 12 in. For U.S. standard sieve sizes, see American Society for Testing and Materials (ASTM), "Standard Specifications for Wire Cloth and Sieves for Testing Purposes," ASTM E11-04.

foundations supported on expansive soils are not designed to account for the soil's instability, significant structural damage may occur, particularly to lightweight buildings, such as wood light frame buildings. (See the section entitled "Expansive Soils" in Principles in Practice at the end of this chapter.)

NOTE

Organic Soils

When a soil consists largely of organic matter such as fully or partially decayed plant matter (peat), it is called *organic soil.* Organic soils are highly compressible and unsuitable for building foundations. They are typically dark in color and have a characteristic odor of decay. Additionally, they may have a fibrous texture due to bark, leaves, grass, branches, or other fibrous vegetable matter.

Organic soils form the top layer of soil and are typically removed before construction begins. If the soil characteristics are desirable, the soils are often stockpiled and redistributed after construction is complete.

COHESIVE AND NONCOHESIVE SOILS

One factor that distinguishes fine-grained from coarse-grained soils is the property of cohesion. Fine-grained soil particles adhere to each other in the presence of water and are, therefore, called *cohesive soils.* Coarse-grained soils are typically single-grained, lacking cohesiveness, and are referred to as *noncohesive soils.*

Clayey soils are more cohesive than silts. It is due to its cohesiveness that a potter can mold objects from wet clay, and excavated (cut) faces in a clayey soil can remain vertical without collapsing for a greater depth than other soil types. Even shallow excavations in noncohesive soils must be sloped for stability or shored by a support system.

11.2 GEOTECHNICAL INVESTIGATIONS—SOIL SAMPLING AND TESTING

Field exploration of below-ground soil conditions is an important first step before designing a building's foundations. The objectives of this exploration and sampling are to determine the:

- *Engineering properties of the soil* at various depths, such as the soil's strength (shear strength, unconfined compressive strength, triaxial compressive strength, and so on). These properties help determine the soil's bearing capacity.

- *Particle-size distribution of the soil* to assess the stability of cut faces, required excavation support, and the soil's drainage characteristics.
- *Plasticity index of the soil* to assess if the soil is unstable (shrinks and swells with changes in its moisture content); see Principles in Practice at the end of the chapter.
- *Nature of the excavation* that will suit the soil. (For instance, excavation in rock is more complicated and expensive than excavation in an ordinary soil.)
- *Depth of the water table* to plan for dewatering of excavations (if required) and its effect on existing construction.
- *Compressibility of the soil* to assess potential foundation settlement.

Two methods are generally used for field exploration: (a) the test pit method and (b) the test boring method.

TEST PIT METHOD

The test pit method consists of digging pits or trenches that are large enough for visual inspection of the soil and for procuring samples for laboratory testing by getting inside the pit or trench. It is considered to be the more reliable method because it allows direct inspection and assessment of several properties of the undisturbed soil through field tests with simple instruments.

There are several limitations to the test pit method. It is more expensive, and the cost increases substantially as the depth of the pit increases. Also, if the water table is high, the depth of inspection is limited.

TEST BORING METHOD

The test boring method, which is more commonly used, involves obtaining soil samples by boring into the ground using a truck-mounted, power-driven hollow stem fitted at the end with either a cutting bit or a fluted drilling auger. A typical drilling operation is shown in Figure 11.2.

(a)

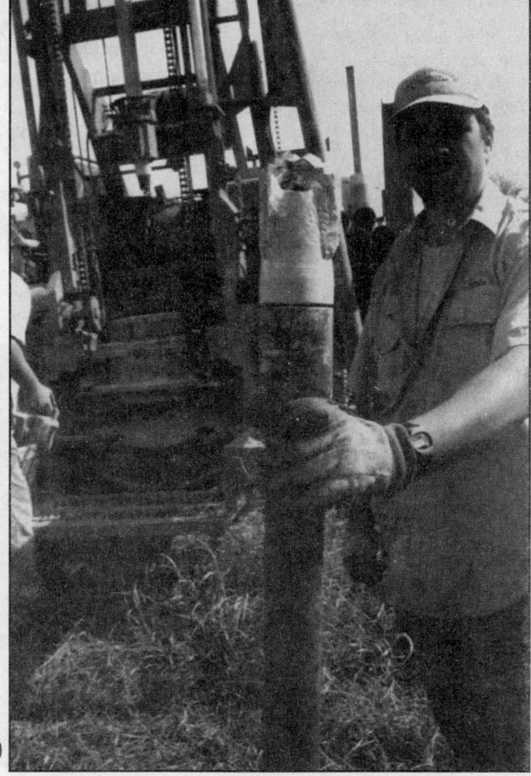

(b)

FIGURE 11.2 **(a)** A truck-mounted drilling machine bores a hole in the ground with a cutting tip at the end of a hollow pipe stem. During the drilling operation, compressed air is pushed through the hollow stem to suck the drilled soil out to keep the boring clean. After drilling to the required depth, the drilling bit is removed and a sampling tube is attached to the pipe stem to bring out a soil specimen. The process is repeated for various depths below ground. **(b)** A hollow pipe stem with a typical cutting bit.

FIGURE 11.3 Soil samples obtained from borings through the use of a sampling tube are wrapped in plastic, labeled with the date of sampling, location of boring at the site, depth, and so on, and sent to a geotechnical laboratory.

When the boring reaches the depth at which a sample is required, the cutting bit is withdrawn and a standard steel sampling tube (called a *Shelby tube*) is pushed into the soil through its full height and the core sample is removed. (A tube that opens up into two semicircular halves, called a *split-spoon,* is also used instead of the Shelby tube. From a rocky soil, core samples are cut out.)

Soil specimens obtained from sampling tubes are wrapped in plastic, labeled with the required information, and transported to the testing laboratory, Figure 11.3. Several test borings are generally required across a large site, and the data obtained from various borings are coordinated. The test boring method also allows determination of the soil's density and strength by impacting the bottom of the boring with a standard amount of impact energy. The test, called the *standard penetration test,* is performed at several elevations in a boring.

Laboratory Testing of Soil Samples

Soil samples obtained from pits or borings are subjected to a number of laboratory tests. The sieve analysis determines the particle-size distribution of the soil. This is used to classify the soil (see Principles in Practice at the end of this chapter). It also provides important information about several engineering properties of the soil. Sieve analysis is performed on an oven-dried part of the specimen by placing the sample on the uppermost sieve of a sieve stack, consisting of sieves of different sizes, Figure 11.4.

As shown in Figure 11.4, the bottom of the sieve stack is a No. 200 sieve. This sieve separates silts and clays from sand. In other words, sieve analysis is not performed to separate silts from clays because their particle size is too small to be separated by sieving. Instead, the soil sample retained on the No. 200 sieve is soaked in water, and the solution is agitated and then allowed to sediment (i.e., settle down). Because the heavier particles settle first, the rate of sedimentation provides information about the percentages of silts and clays.

Other tests include determining the soil's moisture content, dry density, liquid limit, plastic limit, compressive strength (unconfined, i.e., axial compressive strength, and confined, i.e., triaxial compressive strength), shear strength, and so on.

The data obtained during boring—from visual inspection of the soil, a standard penetration test, water table elevation, and so on—are immediately recorded on the testing laboratory's standard data sheet, referred to as the *boring field log.* After subsequent laboratory testing of samples, a final *boring log* is prepared. Each boring, therefore, results in its own boring log. One such log is shown in Figure 11.5.

NOTE

Standard Penetration Test

The standard penetration test consists of striking the sampling tube with a hammer that delivers a standard amount of energy with each blow (a 140-lb weight falling over the tube through a height of 30 in.). The number of blows required to drive the tube through a depth of 6 in. is recorded.

The total number of blows for the second and third 6 in. of penetration (i.e., a total 12 in. of penetration) is called the *N-value* (or the *blow count* or the *standard penetration resistance*). The first 6 in. of penetration is ignored because it occurs in the disturbed part of the soil. A maximum of 50 blows are used, and if the tube penetrates less than 6 in. in 50 blows, the corresponding penetration distance is recorded. The test is generally performed every 5 ft of depth into the soil; see Figure 11.5.

The N-value is used as a rough index of the soil's bearing capacity, particularly for sandy soils. For a precise determination of the bearing capacity, other soil properties must be used along with the N-value.

11.3 SOIL TYPE AND EARTHQUAKES

In seismically active areas, different types of soils react differently to earthquake vibrations. Assume two types of subsurface conditions—one consisting of highly dense soil (e.g., bedrock as an extreme case) and the other consisting of loosely consolidated soil (soft soil), Figure 11.6. Assume further that both of these subsurface conditions are subjected to the

(a)

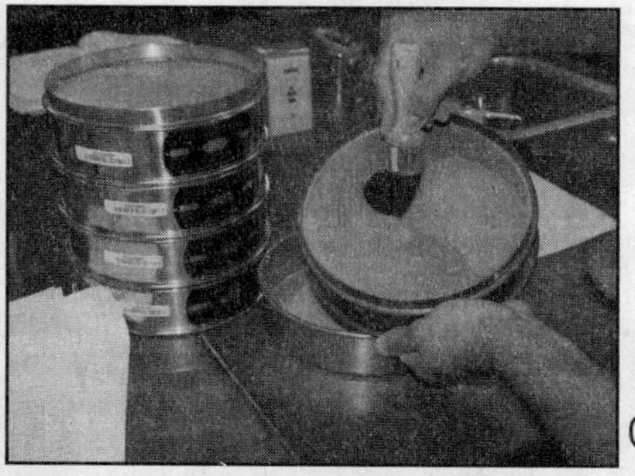

(b)

FIGURE 11.4 **(a)** A typical stack of sieves used to determine the grain-size distribution of oven-dried soil specimens. The bottom of the stack is not a sieve but a container. The sieve immediately above the container is generally a No. 200 sieve. The sieve stack is shaken per the standard procedure, either manually or by placing it on a motorized shaker. The amount of soil retained on each sieve is weighed. The amount obtained from the bottom container is the amount that passes through a No. 200 sieve. A graph showing the percentages of soil passing through various sieves is plotted. **(b)** After sieve analysis is performed for a specimen, the sieves are cleaned and restacked for analyzing the next specimen. (Photos courtesy of the Geotechnical Laboratory of the University of Texas at Arlington)

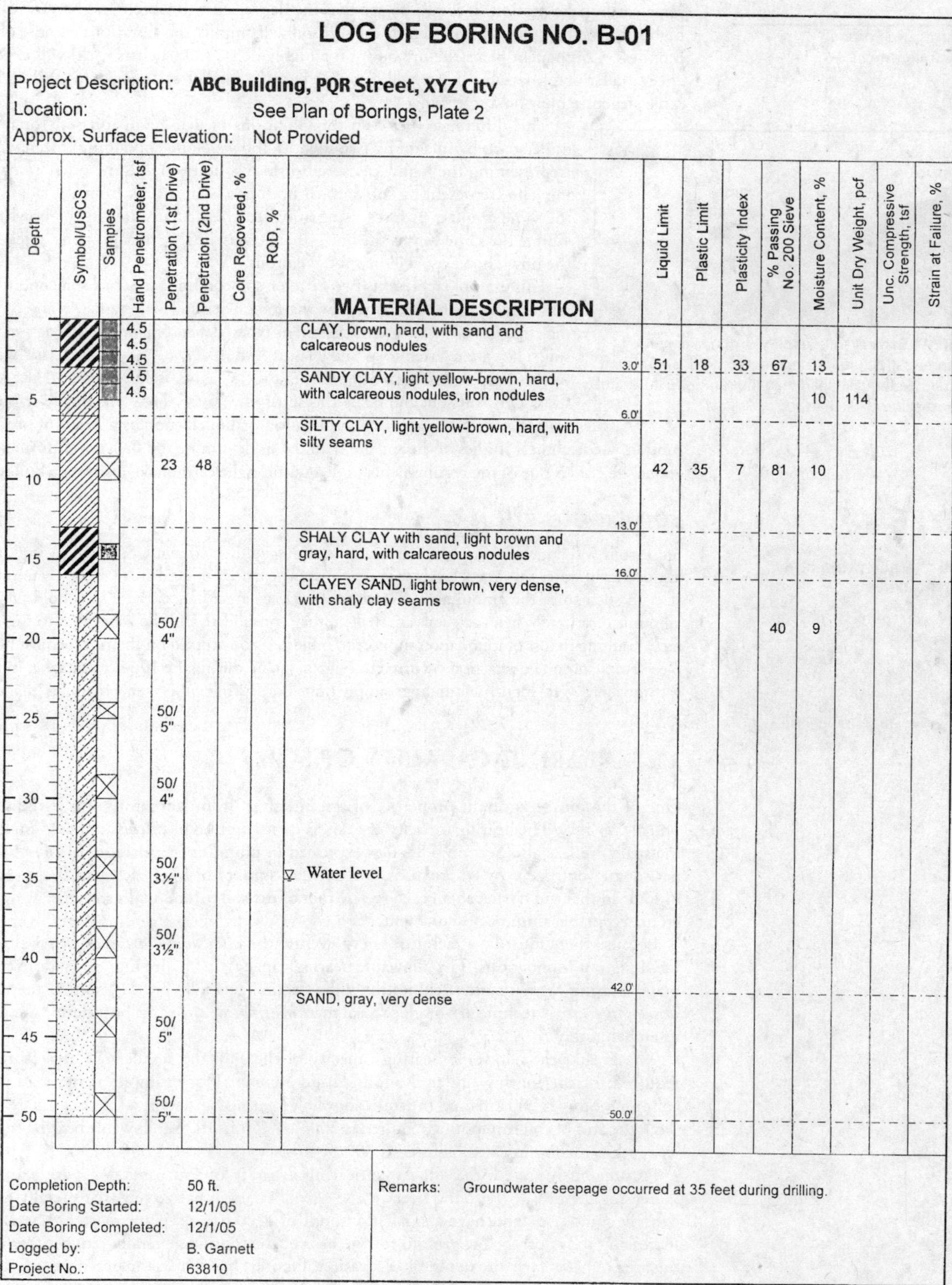

LOG OF BORING NO. B-01

Project Description: **ABC Building, PQR Street, XYZ City**
Location: See Plan of Borings, Plate 2
Approx. Surface Elevation: Not Provided

Depth	Hand Penetrometer, tsf	Penetration (1st Drive)	Penetration (2nd Drive)	Core Recovered, %	RQD, %	MATERIAL DESCRIPTION	Liquid Limit	Plastic Limit	Plasticity Index	% Passing No. 200 Sieve	Moisture Content, %	Unit Dry Weight, pcf	Unc. Compressive Strength, tsf	Strain at Failure, %
0–3.0'	4.5 4.5 4.5					CLAY, brown, hard, with sand and calcareous nodules	51	18	33	67	13			
3.0'–6.0'	4.5 4.5					SANDY CLAY, light yellow-brown, hard, with calcareous nodules, iron nodules					10	114		
6.0'–13.0'		23	48			SILTY CLAY, light yellow-brown, hard, with silty seams	42	35	7	81	10			
13.0'–16.0'						SHALY CLAY with sand, light brown and gray, hard, with calcareous nodules								
16.0'–42.0'		50/ 4"				CLAYEY SAND, light brown, very dense, with shaly clay seams				40	9			
		50/ 5"												
		50/ 4"												
		50/ 3½"				Water level								
		50/ 3½"												
42.0'–50.0'		50/ 5"				SAND, gray, very dense								
		50/ 5"												

Completion Depth: 50 ft.
Date Boring Started: 12/1/05
Date Boring Completed: 12/1/05
Logged by: B. Garnett
Project No.: 63810

Remarks: Groundwater seepage occurred at 35 feet during drilling.

FIGURE 11.5 A typical boring log. (Courtesy of Kleinfelder, Fort Worth, Texas) (See Principles in Practice at the end of this chapter for an explanation of some of the terms used in this log.)

same earthquake motion. It is easy to see that the dense soil will vibrate less because of its high stiffness than the soft soil. In fact, the soft soil will amplify the vibrations because of its limpness. Consequently, a structure built on a highly dense soil (such as rock) will experience smaller deformations (hence will be subjected to a smaller earthquake load) than the same structure built on a soft soil.

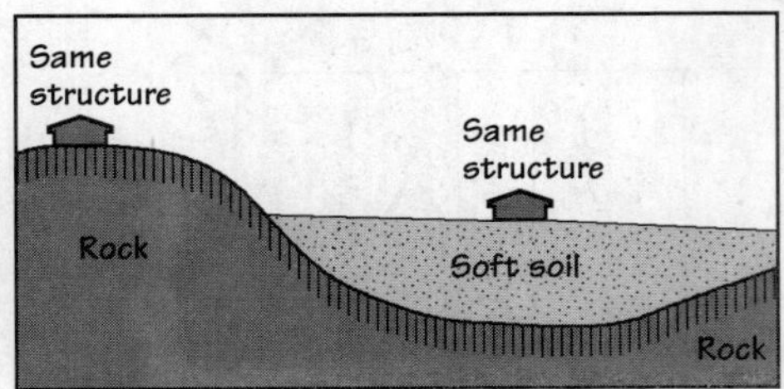

FIGURE 11.6 Soft soils amplify earthquake motion. Therefore, a structure located on a soft soil is subjected to a larger earthquake load than the same structure situated on a stiff soil, such as rock.

This difference (between the vibrations of dense and soft soils) may be understood by imagining two identical bowls—one containing frozen water (representing the highly dense subsurface condition) and the other containing jello (representing soft soil). If both bowls are struck from the side with the same amount of force using a mallet, the frozen water in the bowl will vibrate the same way as the bowl, that is, very little. The vibrations of jello in the bowl, however, will be more vigorous.

This example represents two extreme soil conditions. Actual soil conditions range anywhere between these two extremes. Seismic load standards and building codes classify soils into six different types, depending on the stiffness of the soil. They are referred to as Site Classes A, B, C, D, E, and F. Site Class A represents hard rock; Class B represents rock; Class C is soft rock or dense soil; Class D is stiff soil; and Class E is soft soil. Class F represents highly sensitive soils that may experience consolidation or liquefaction as a result of ground motion. Geotechnical studies of the soil are required to determine the site class. Information about site class is one of the requirements in determining the earthquake loads on a building.

NOTE

Soil Classification and Earthquakes

Class A Hard rock
Class B Rock
Class C Soft rock (dense soil)
Class D Stiff soil
Class E Soft soil
Class F Highly sensitive soil

SOIL LIQUEFACTION

Some soils will liquefy under seismic vibrations, losing all their strength. Soil liquefaction usually occurs in water-saturated, sandy soils in which the particles of sand are of relatively uniform size. In such soils, the amount of interparticle space is large; hence, these soils hold greater amounts of water. When such soils are shaken, they consolidate and the water rises to the surface, resulting in loss of foundation support to buildings constructed on them. The most notable example of soil liquefaction occurred in Nigata, Japan, during the 1964 earthquake. In this earthquake, several high-rise buildings simply tipped over while otherwise remaining intact.

11.4 BEARING CAPACITY OF SOIL

One of the most important properties of a soil derived from laboratory tests is the soil's *bearing capacity*. The bearing capacity of a soil is its strength to bear loads imposed on it by the structure (see also Section 4.8). It is expressed in pounds per square foot (psf) or kips per square foot (ksf). In other words, the bearing capacity of a soil determines the maximum load that can be placed on each square foot of the soil before it fails structurally or has an unacceptable amount of settlement.

Because there must be a factor of safety, we use the *allowable bearing capacity* value in the design of foundations. The allowable bearing capacity is smaller than the failure (ultimate) bearing capacity and includes the safety factor (generally 3.0). Note that the term *bearing pressure* is technically more precise than *bearing capacity*, but both terms are used interchangeably.

The greater the allowable bearing capacity of the soil, the smaller the area (spread) required for the footing under the foundation element. For example, if the load on a column footing is 200 kips and the soil's allowable bearing capacity is 4 kips per square foot (ksf), the area of column footing required is $200/4 = 50\ ft^2$. If the allowable bearing capacity is 5 ksf, the footing area required is $200/5 = 40\ ft^2$ (1 kip = 1,000 lb).

As previously stated, the allowable bearing capacity of soil generally increases with increasing depth below ground because the deeper strata of native soil are generally more densely compacted and have a smaller amount of decomposed plant matter. Therefore, increasing the depth below ground for the base of the footing generally reduces the footing area but increases the depth of excavation, Figure 11.7. Thus, the bearing capacity of the soil is generally evaluated at various levels below ground to establish the most suitable level for the base of the footing—to balance the footing area with the amount of excavation needed.

DEPTH OF THE FROST LINE

A soil property that is even more critical than the soil's bearing capacity is its stability. As previously stated, an unstable soil (such as a predominantly clayey soil) is unfit for

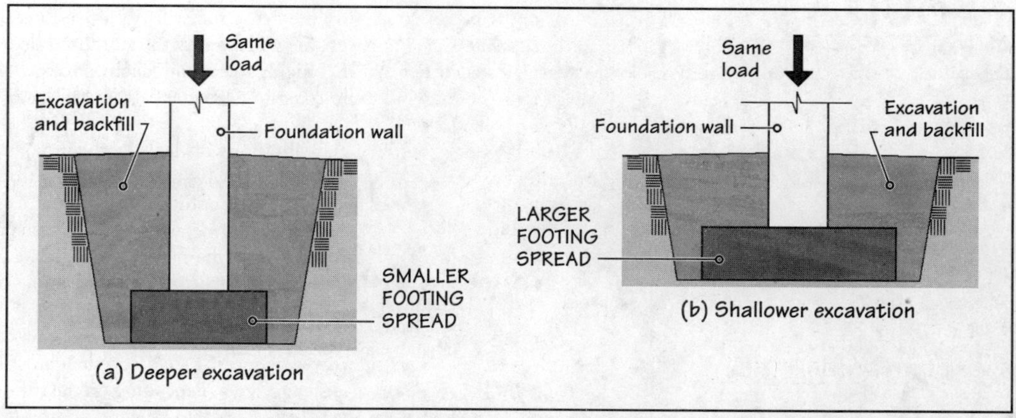

FIGURE 11.7 The bearing capacity of soil at a site generally increases with the depth below ground. Consequently, the footing area (spread) required is smaller if the base of the footing is located deeper in the ground. However, going deeper into the ground increases the excavation depth, somewhat negating the benefit of the smaller footing area. A balance between these two opposite effects is generally sought unless there are other considerations.

foundation support. In such soils, the foundation elements must penetrate deep into the ground to a depth where the soil is stable and has sufficient bearing capacity.

Another factor (apart from the soil's clay content) that produces instability is the freezing of the ground, which is defined by the *frost depth*. The frost depth (also called the *freezing depth*) at a location is the maximum depth into the ground beyond which the water in the soil will not freeze. The frost depth is larger in colder climates. In the continental United States, the frost depth varies from approximately 6 in. in southern regions to approximately 90 in. in northern regions. The building departments of municipalities provide more accurate information.

The base of a footing must extend below the frost depth. If the base of the footing is above the frost depth, the freezing of water in soil will cause the soil under the footing to heave, damaging the footing. In Canada and the northern United States, the depth below ground for the base of footings for lightweight structures is generally controlled by the frost line, not by the allowable bearing capacity of the soil.

Frost depth is not an issue with bedrock because of the absence of water. For the same reason, it is not an issue with a soil that does not hold water, such as free-draining soils with gravel or coarse sand.

NOTE

Relationship Between Foundations and Footings

As previously stated, the load-bearing capacity of the earth's surface generally increases with depth below the ground. Therefore, unless the ground surface at a building site consists of stable, hard rock, the structural elements of a building (such as load-bearing walls and columns) must be taken below the ground—to a stratum where the soil is stable and has adequate strength to carry the superimposed loads. The part of the structural system below the ground floor of the building (below the basement floor if the building has a basement) that carries the load into the ground is called the *building foundation system* (see also the introduction to Chapter 12).

The soil on which the load-bearing walls and columns rest is generally weaker than the materials used to construct the walls and columns. Therefore, the walls and columns generally need to be widened at their bases to ensure that the bearing pressure on the soil is less than its bearing capacity. These widened parts are called the *footings*. Note that footings are part of the foundation system. (A detailed discussion of foundation systems is presented in Chapter 12.)

Presumptive Bearing Capacity of Soil

The allowable bearing capacity of a soil should be obtained from geotechnical investigations of the site. However, its approximate value, based on the particle size of the soil at the location (without geotechnical investigation) is allowed to be used in situations where

- the building is small;
- adequate information about the soil from adjoining areas is available;
- the site does not contain fill of an unknown origin; and
- the soil is known to be stable (nonexpansive).

Table 11.1 gives approximate (presumptive) allowable bearing capacity values, as stated in the International Building Code. The code values are highly conservative and, hence, their use generally results in overdesign of foundation systems. In larger structures, it is, therefore, more economical to seek formal subsurface soil investigation.

TABLE 11.1 PRESUMPTIVE ALLOWABLE BEARING CAPACITY OF SOILS

Class of material	Allowable bearing capacity
Crystalline bedrock	12.0 ksf
Sedimentary and foliated rock	4.0 ksf
Soil types GW and GP	3.0 ksf
Soil types SW, SP, SM, SC, GM, and GC	2.0 ksf
Soil types CL, ML, MH, and CH	1.5 ksf

For symbols of soil types, refer to the Unified Soil Classification System in Principles and Practice at the end of this chapter.
Source: International Code Council, *International Building Code*, 2009, Chapter 18.

PRACTICE **QUIZ**

Each question has only one correct answer. Select the choice that best answers the question.

1. Soil and rock consist essentially of the same matter.
   a. True **b.** False
2. In the commonly used soil classification system, *sand and silt* are classified as
   a. coarse-grained soils.
   b. fine-grained soils.
   c. organic soils.
   d. inorganic soils.
   e. none of the above.
3. The four commonly used classifications of soils arranged in order of decreasing particle size are
   a. gravel, sand, silt, and clay.
   b. gravel, silt, sand, and clay.
   c. gravel, silt, clay, and sand.
   d. gravel, sand, clay, and silt.
   e. none of the above.
4. Which of the following statements is correct?
   a. Gravel particles have a platelike shape.
   b. Sand particles have a platelike shape.
   c. Silt particles have a platelike shape.
   d. Clay particles have a platelike shape.
5. Cohesive soils are those in which the dry soil particles tend to separate from each other when mixed with a small amount of water.
   a. True **b.** False
6. To obtain soil-sampling tubes from below ground, which soil-testing method is commonly used?
   a. Test pit method
   b. Test boring method
   c. Test pressure method
   d. Any one of the above
   e. None of the above
7. During soil testing, soil samples are obtained from
   a. one central location on the site but from different depths below ground.
   b. several locations on the site but from the same depth below ground.
   c. several locations on the site and from several depths below ground.
   d. none of the above.
8. All testing of soil samples is generally done at the site
   a. because modern soil sampling equipment is fully equipped with soil testing facility.
   b. because the delay in bringing soil samples to the laboratory falsifies test results.
   c. because the vibration of soil samples caused by transportation falsifies test results.
   d. all of the above.
   e. none of the above.
9. In terms of soil behavior in earthquakes, building codes classify a construction site based on soil conditions. The primary property of soil that is used in this classification is
   a. the stiffness/density of soil.
   b. the size of soil particles.
   c. the swelling and shrinkage characteristics of soil.
   d. whether the soil is native to the site.
   e. none of the above.
10. The bearing capacity of a soil refers to
    a. the density of soil.
    b. the modulus of elasticity of soil.
    c. the compressive strength of soil.
    d. whether the soil is native to the site.
    e. none of the above.
11. The bearing capacity of a soil is generally expressed in terms of
    a. pounds or kips per square foot.
    b. pounds or kips per square inch.
    c. pounds or kips per cubic inch.
    d. pounds or kips per inch.
    e. pounds or kips.
12. If the bearing capacity of the soil is high, the footing area required is large.
    a. True **b.** False
13. The maximum presumptive bearing capacity value of bedrock provided by the International Building Code is
    a. 18.0 ksf.
    b. 16.5 ksf.
    c. 15.0 ksf.
    d. 12.0 ksf.
    e. None of the above.
14. The building codes do not mandate geotechnical investigations for all building construction sites.
    a. True **b.** False

NOTE

Types of Fill—Backfill, Grading Fill, Select Fill, and Engineered Fill

Backfill refers to the material excavated from the site that is used to fill the excavated trenches or pits after constructing the foundations or basement. Unless the excavated material is unsuitable, backfill involves returning the excavated material to its original location.

Grading fill (also called *fill*) is soil that is used for grading work—to bring the finished ground surface to the required contours and slope for drainage, landscaping, erosion control, aesthetics (land form), and so on, as per the site design. Generally, the material not used as backfill is used as grading fill unless it is unsuitable.

Select fill is a soil with required properties that is imported to a given site, placed, and compacted to produce the desired result. For example, a select fill may consist of a coarse-grained, stable soil to replace the existing expansive (clayey) soil at a site.

Engineered fill is soil that is specially prepared, placed, compacted, and supervised per the geotechnical engineer's design and specifications. For example, it may be soil mixed with crushed angular aggregate of a required grading to give a controlled void space to increase storm water retention and reduce runoff.

11.5 EARTHWORK FOR EXCAVATIONS AND GRADING

Site preparation in advance of construction generally includes

- Fencing the site from adjacent public or private property
- Locating and marking existing underground utility lines so that they will not be damaged during construction
- Demolishing unneeded existing structures and utility lines
- Marking trees to be saved, and removing unneeded trees, shrubs, topsoil, and extraneous landfill, if present

EXCAVATION

Excavation is the first step of construction. It refers to the process of removing soil or rock from its original location, typically in preparation for constructing foundations, basements, and underground utility lines and for grading of the ground surface. Excavated material required for backfill or grading fill is stockpiled on the site for subsequent use.

Unneeded material is removed from the site for appropriate disposal. Excavations are generally classified as

- Open excavations
- Trenches
- Pits

Open excavations refer to large (and often deep) excavations, such as for a basement. *Trenches* generally refer to long, narrow excavations, such as for footings under a wall or utility pipes. *Pits* are excavations for the footing of an individual column, elevator shaft, and so on.

The depth of excavation depends on the type of soil and the type of foundation. For a slab-on-ground, the depth of excavation is small, whereas the depth of excavation for a basement is directly related to the number of basement floors. For trenches and pits, the excavation must reach soil with sufficient bearing capacity to support the load. If the soil with the required bearing capacity and stability is not present at a reasonable depth, deep foundation elements, which do not require conventional excavation (see Chapter 12), are used.

Grading

Grading involves moving earth from one location of the site to another and changing the existing land surface to the desired finished surface configuration as per the site plan and drainage plan. On most sites, grading is separated into rough grading and finish grading. Rough grading is done along with excavations for foundations, basements, and utility trenches. Finish grading is generally done toward the end of the project as per the landscape design.

Both excavation and grading require various types of power equipment, such as excavators, compactors, and heavy earth-moving equipment (front-end loaders and backhoes), some of which are shown in Figure 11.8.

(a) Excavator and hauler

(b) Trencher

(c) Compactor

(d) Mini compactor (rammer), generally used for compacting small areas

FIGURE 11.8 Some of the equipment used for earthwork (excavation, hauling, and compaction).

11.6 SUPPORTS FOR OPEN EXCAVATIONS

Excavations in the soil generally require some type of support to prevent cave-ins while the foundation system or basement walls are constructed. The simplest excavation support system consists of providing adequate slope in the excavated (cut) face so that it is able to support itself, Figure 11.9. This is feasible only if the site is large enough to accommodate sloped excavations. Because the amount of excavated material is large, sloped excavations also require a larger site area for stockpiling the material until it is used as grading fill or backfill.

The maximum natural slope at which a soil will support itself must be determined from soil investigations. Excavation in coarse-grained soils requires a shallower slope than excavation in fine-grained soils. A sloped excavation may either be uniformly sloped or stair-stepped, called a *benched excavation,* Figure 11.10. Benches increase slope stability. They are also easier to compact. Compacted benches further enhance slope stability.

Self-supporting sloped excavations cannot be provided where the site area is restricted or adjoining structures are present. In these cases, the excavation must consist of vertical cuts. In cohesive soils, shallow vertical cuts (generally 5 ft or less in depth) may be possible without any support system. Deeper vertical cuts must be provided with a support system. These support systems may be temporary (removed after the construction is complete) or permanent. Some of the commonly used methods of supporting deep vertical cuts in the soil are

- Sheet piles
- Cantilevered soldier piles
- Anchored soldier piles
- Contiguous bored concrete piles
- Secant piles
- Soil nailing
- Bentonite slurry walls

EXCAVATION SUPPORT FOR DEEP VERTICAL CUTS—A TOP-DOWN CONSTRUCTION SYSTEM

All the support systems just listed involve the construction of an earth retention system. Because the excavated cuts are constructed in vertical segments, proceeding from the top and moving downward, the system of construction is referred to as *top-down construction.* In this system, the earth pressure is resisted by the excavation support system, and a separate basement wall is constructed (where needed) abutting the excavation support, Figure 11.11(a). Thus, the basement wall is not subjected to any earth pressure.

This is in contrast to the construction of a conventional basement wall, where the excavation for the entire height of the wall is made first (e.g., using the sloped excavation of Figure 11.9) and the basement wall is constructed subsequently. In this case the earth

FIGURE 11.9 An example of a self-supporting open excavation for a basement.

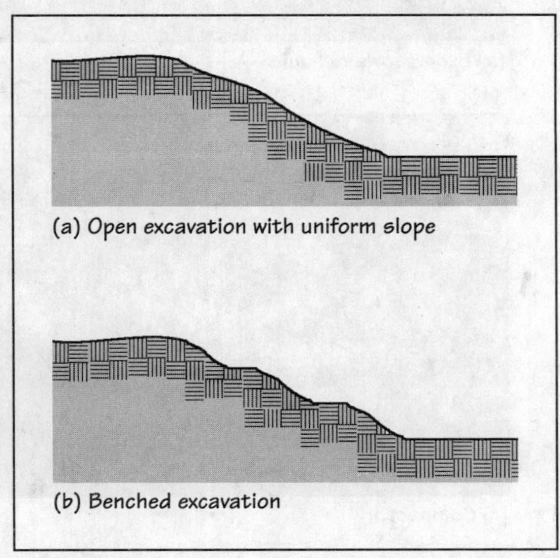

FIGURE 11.10 Two alternative methods used for self-supporting open excavations.

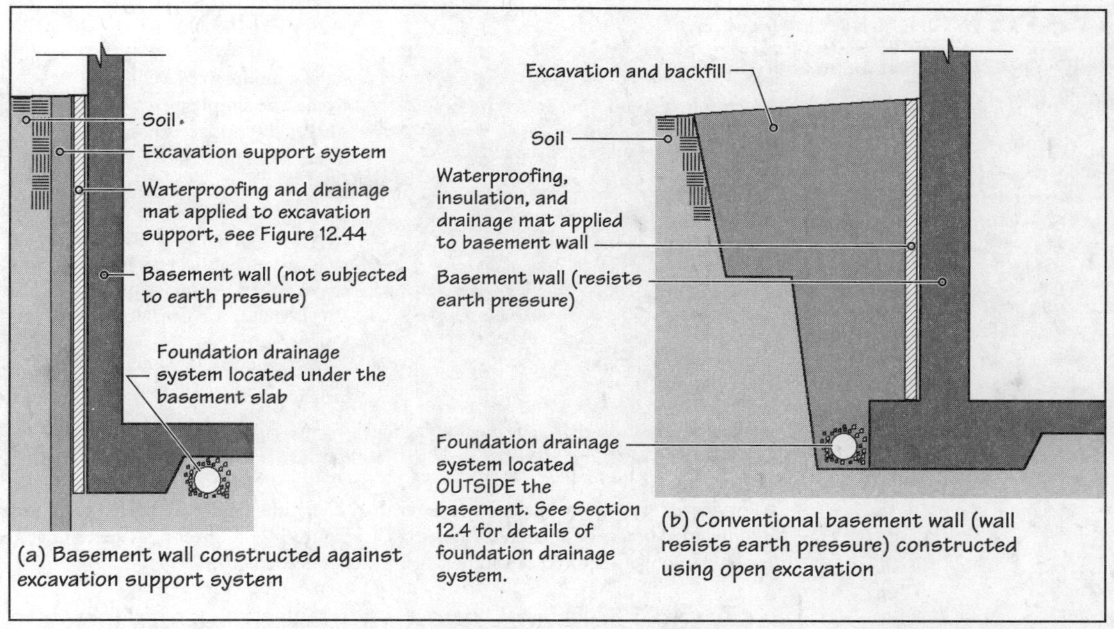

FIGURE 11.11 Two alternative ways of constructing a basement: **(a)** basement wall not subjected to earth pressure (the earth pressure is resisted by an independent excavation support system), and **(b)** earth pressure resisted by the basement wall. Provisions for the drainage of subsoil water are essential in both cases. Several details (e.g., waterproofing of the basement slab) are not shown for clarity. Refer also to Figures 12.43 to 12.45.

retention is provided by the basement wall, which is constructed from the bottom up, Figure 11.11(b). In top-down construction, the earth retention is provided by the excavation support system, and is the system of choice for deep excavations and tight sites.

In both cases, it is important to provide sufficient drainage to drain water from the soil behind the basement wall (in conventional construction) or behind the excavation support system, because wet soil exerts significantly greater pressure than dry soil.

Excavation Support Using Sheet Piles

For depths of up to about 15 ft, vertical sheets of steel, referred to as *sheet piles,* can be driven into the ground before commencing excavations. Sheet piles consist of individual steel sections that interlock with each other on both sides. The interlocks form a continuous barrier to retain the earth. Structurally, sheet piles function as vertical cantilevers and, therefore, must be buried in the soil for sufficient depth below the bottom of the excavation. (Sheet piles are also used as permanent barriers against water, e.g., harbor walls, flood-protection walls, etc.)

Sheet piles are available in many cross-sectional profiles. The most commonly used profile is a Z-section, Figure 11.12. The sections are driven into the ground one by one using either hydraulic hammers or vibrators, Figure 11.13.

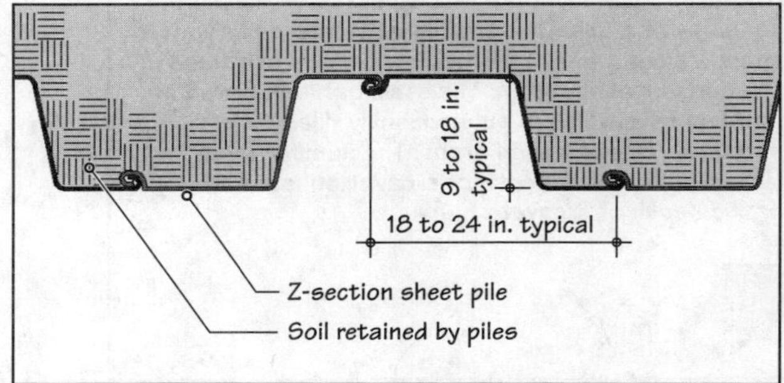

FIGURE 11.12 Steel sheet piles with a Z-shaped profile. Adjacent sections interlock with each other. Several other sectional profiles are also available.

FIGURE 11.13 Steel sheet piles being driven into the soil using a diesel pile driver. (Photo courtesy of R. I. Carr)

EXPAND YOUR KNOWLEDGE

Top-Down Construction for an Entire Building

The top-down method of construction is applicable not only to excavation supports but also to the construction of an entire building. The method allows the construction of the basement and the above-ground structures simultaneously and is, therefore, also called the *up-and-down method* of construction. It reduces construction time, particularly for buildings with several basement floors, commonly used in downtown metropolitan areas.

In this method, perimeter basement walls are constructed first, using secant piles or concrete diaphragm walls (described in Section 11.6). Next, drilled piers are constructed up to the ground floor level at locations indicated for building columns in the plans (see Section 12.4 for the construction of drilled piers). The piers function as supports for the above-ground columns and as columns for the basement structure. The ground floor is now constructed, leaving some bays of the ground floor structure open to allow access for basement excavation.

After the construction of the ground floor structure, columns and other structural elements of the building for the upper floor are constructed while the excavation for the first basement level continues simultaneously. The excavation for the basement exposes the piers and the diaphragm walls. As previously stated, the piers function as basement columns, and the diaphragm walls function as outer basement walls.

Because the vertical support elements for the first basement floor are already in place, this floor is now constructed, again leaving a few access openings for the excavation of the lower basement levels. Note that the floor of each basement acts as lateral support for the diaphragm walls. The process described continues for additional above-ground and below-ground levels.

For deeper excavations (generally greater than 15 ft), sheet piles are braced with horizontal or inclined braces or anchored with tiebacks, Figure 11.14. Sheet piles are removed after they are no longer required or can be left in place if needed.

Excavation Support Using Cantilevered Soldier Piles

One of the disadvantages of sheet piles is the noise and vibration created in driving them, particularly in stiff soils where the vibratory method is ineffective and hydraulic hammers must be employed. An alternative to sheet pile excavation support is the soldier pile system. In this support system, H-shaped steel columns (called *soldier piles* or *H-piles*) are placed in the ground. The piles are placed in predrilled circular holes approximately 6 to 8 ft on

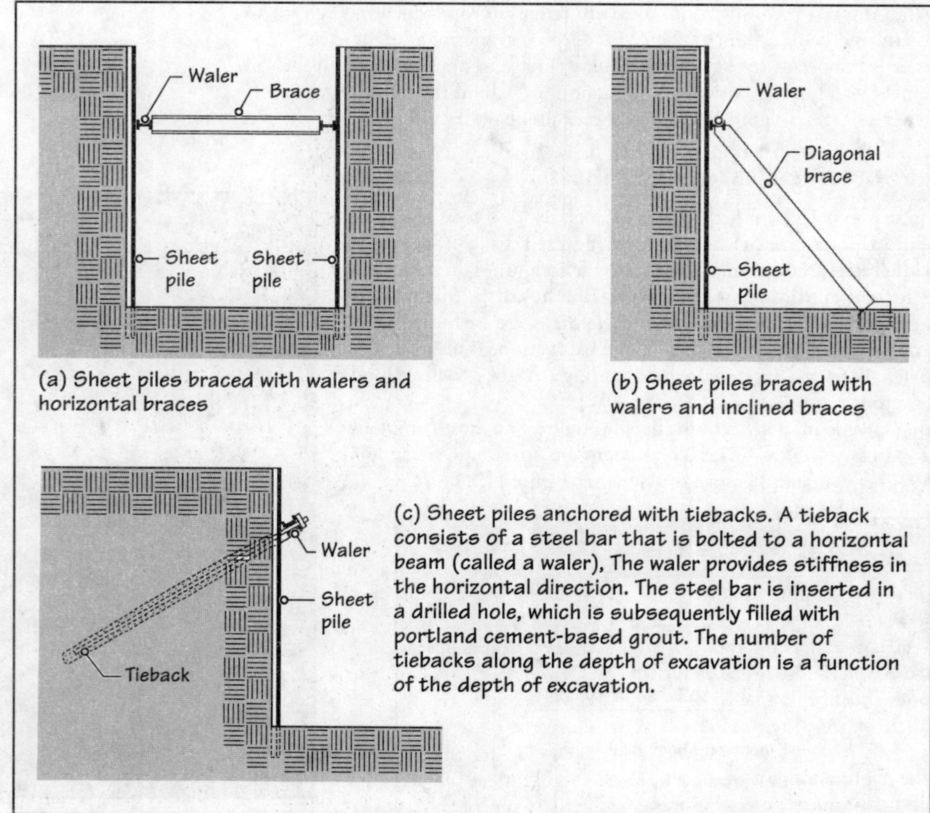
Waler
Brace
Sheet pile
Sheet pile
(a) Sheet piles braced with walers and horizontal braces
Waler
Diagonal brace
Sheet pile
(b) Sheet piles braced with walers and inclined braces
Waler
Sheet pile
Tieback
(c) Sheet piles anchored with tiebacks. A tieback consists of a steel bar that is bolted to a horizontal beam (called a waler), The waler provides stiffness in the horizontal direction. The steel bar is inserted in a drilled hole, which is subsequently filled with portland cement-based grout. The number of tiebacks along the depth of excavation is a function of the depth of excavation.

FIGURE 11.14 **(a)** and **(b)** Two alternative ways of bracing sheet piles in deep excavations—horizontal and inclined braces. **(c)** Sheet piles with tiebacks.

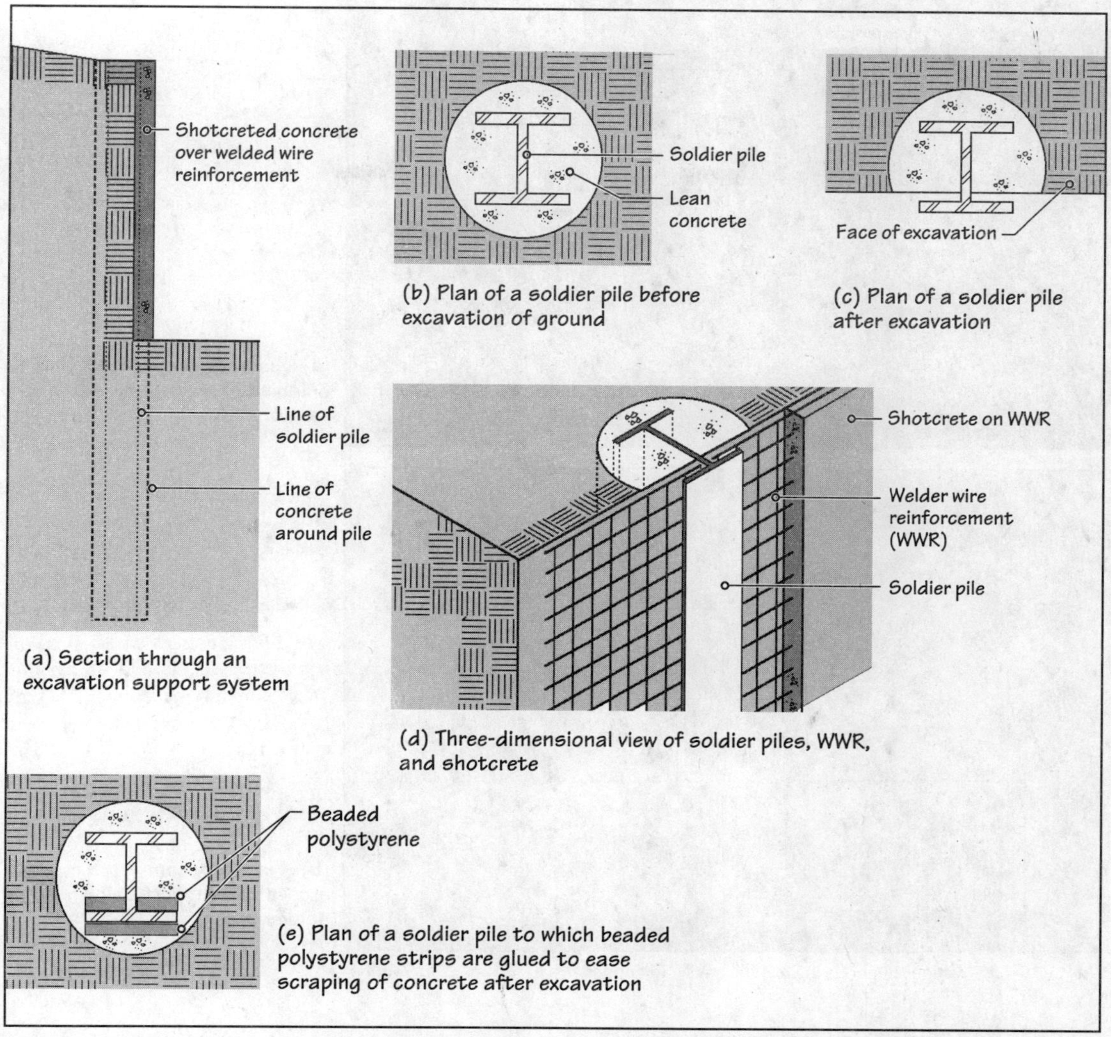

(a) Section through an excavation support system

(b) Plan of a soldier pile before excavation of ground

(c) Plan of a soldier pile after excavation

(d) Three-dimensional view of soldier piles, WWR, and shotcrete

(e) Plan of a soldier pile to which beaded polystyrene strips are glued to ease scraping of concrete after excavation

FIGURE 11.15 Steps showing the construction of cantileverd soldier pile excavation support with welded wire reinforcement (WWR) and shotcrete.

center. After the piles are placed, the holes are filled with lean concrete (about 1500 psi strength), Figure 11.15(b). Excavation of the ground abutting the piles is commenced after the concrete around the piles has gained sufficient strength.

The ground is excavated in vertical segments 5 to 6 ft in height that expose the encased piles. Part of the concrete encasing the piles is now scraped to make way for the welded wire reinforcement (WWR) placed between the flanges of adjacent H-piles, Figure 11.15(c and d). The leanness of concrete allows the scraping to take place. Some earth retention contractors glue beaded polystyrene strips to the front and back sides of the outer flange of the piles to simplify the scraping of concrete, Figure 11.15(e). (See Section 21.13 for a discussion of WWR.)

After the WWR is placed, concrete is sprayed over the reinforcement under high velocity and pressure (referred to as *shotcrete* or *gunite*). The shotcreted concrete is floated to provide a smooth finish, Figure 11.16. Pressure-treated, rough-sawn horizontal lumber members, called *lagging*, are an alternative to shotcrete between soldier piles (see Figure 11.17).

Structurally, the shotcrete on WWR (or lumber lagging) between the adjacent piles functions as a wall that transfers earth pressure horizontally to the piles. The piles behave as vertical columns cantilevered from the ground. Therefore, like the sheet piles, soldier piles must be buried sufficiently below the bottom of the excavation to provide the required cantilever action, as shown in Figure 11.15(a).

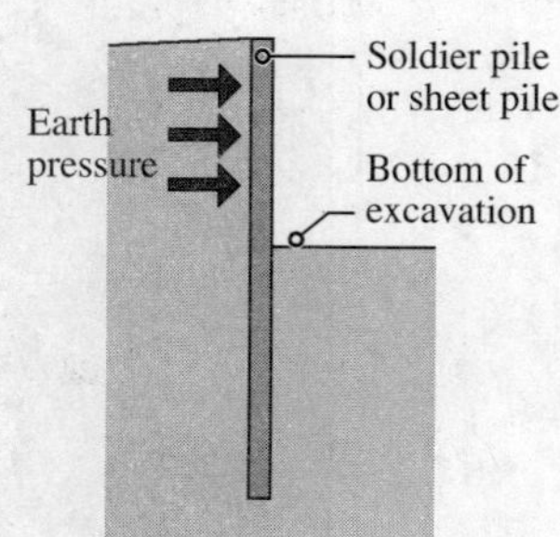

Soldier piles or sheet piles must penetrate sufficiently into the ground below the bottom of the excavation in order to resist earth pressure as vertical cantilevers.

(a) Shotcreting over WWR and finishing of an already shotcreted part of a wall

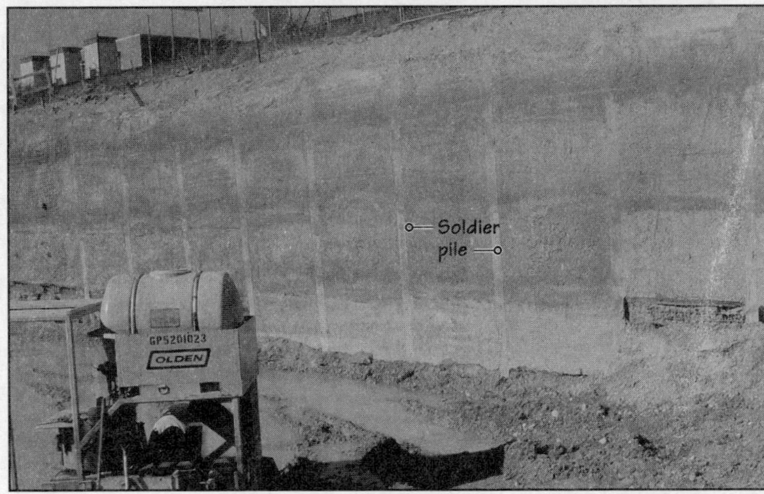

(b) A partially completed excavation support using soldier piles

FIGURE 11.16 Excavation support using cantilevered soldier piles.

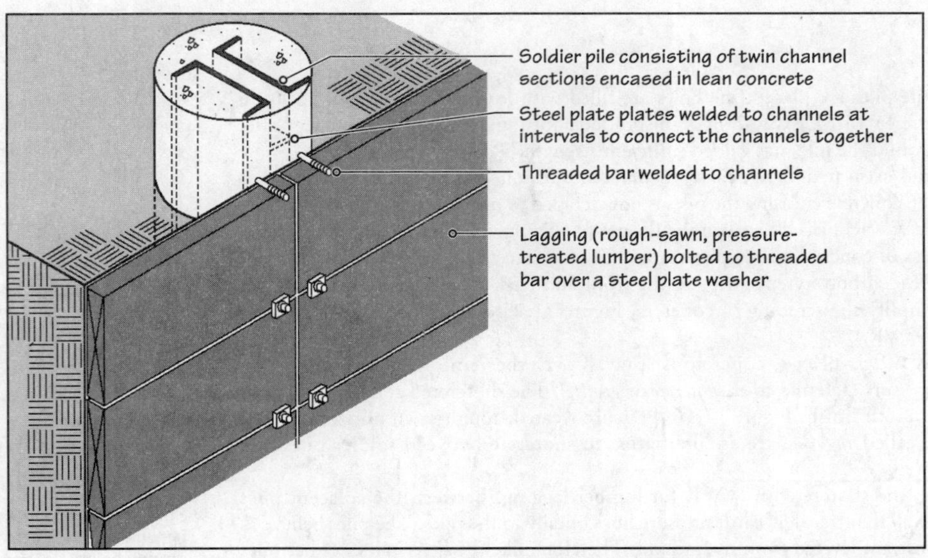

FIGURE 11.17 Soldier pile support system with twin channel sections and lumber lagging between piles. Lumber lagging may be replaced by welded wire reinforcement and shotcrete.

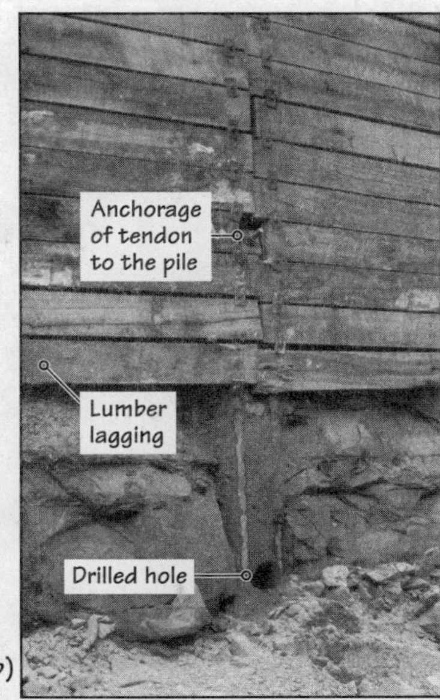

FIGURE 11.18 **(a)** The hole for a tieback is drilled through the space between the twin steel channels comprising the pile. **(b)** A high-strength steel tendon is now placed into the hole, and the hole is filled with portland cement-based grout. After the grout has gained sufficient strength, the tendon is stressed to the required tension and anchored to the pile as shown.

Excavation Support Using Anchored Soldier Piles

The use of a cantilevered soldier pile system is uneconomical beyond a depth of approximately 15 ft because of the increase in pile cross section. For deeper excavations, an anchored soldier pile system is employed, which is similar to the cantilevered pile system except that the piles are tied back (anchored) into the ground. The commonly used vertical support members for this system consist of two steel channels with a space between them. The channels are connected together with steel plates welded at intervals in this space, Figure 11.17.

Drilling for tieback anchors is done through the space between the twin C-sections of piles, Figure 11.18. After a tieback hole has been drilled, steel bars or high-strength steel tendons are placed in the hole, and the hole is grouted, Figure 11.19. The bars or tendons are subsequently anchored to the piles, Figure 11.20. If high-strength tendons are used,

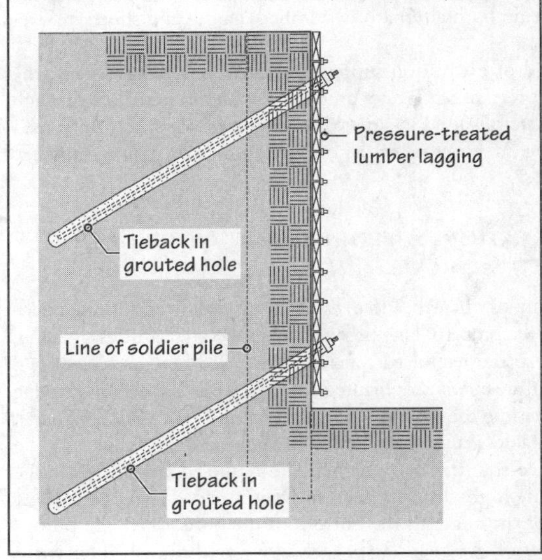

FIGURE 11.19 Section through an anchored soldier pile support system with lumber lagging. Tiebacks are installed as the excavation proceeds. This is followed by bolting the lagging to the piles. The number of tiebacks required is a function of the depth of excavation and the type of soil.

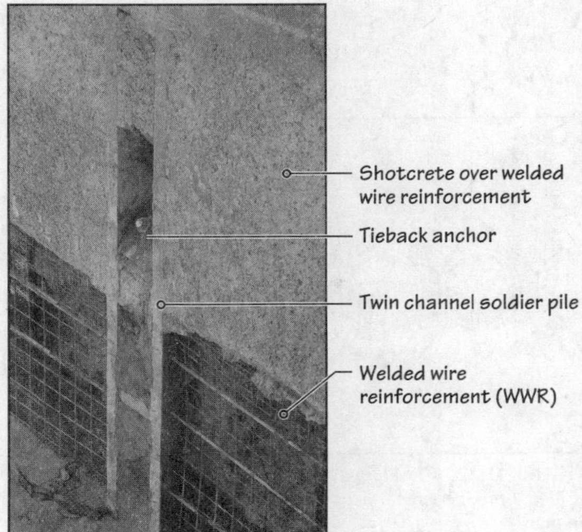

FIGURE 11.20 Tieback anchor detail in a anchored soldier pile support system with WWR and shotcrete.

(a)

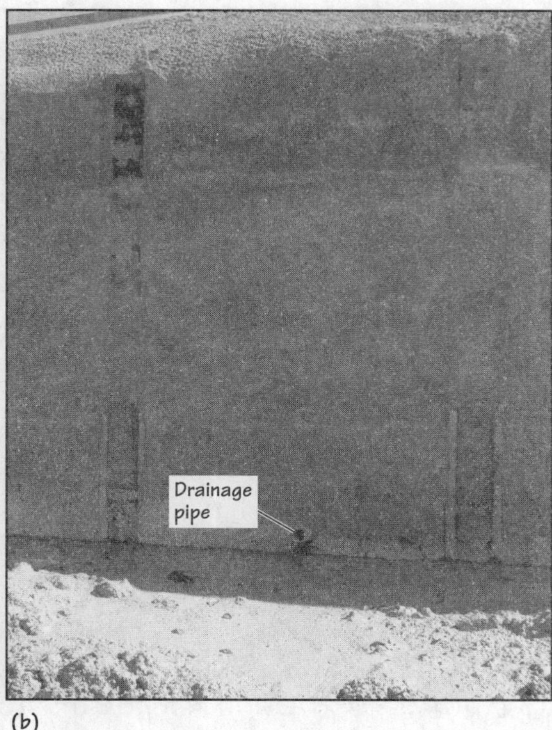

(b)

FIGURE 11.21 **(a)** Anchored pile excavation support system with lumber lagging. **(b)** Anchored pile excavation support system with shotcrete and welded wire reinforcement. Tieback anchors are not visible because they have been shotcreted over. The drainage pipe drains water from behind the support system. The water is collected in the sump located in the interior of the basement from where it is pumped out. All excavation support systems need such internal drainage.

pressure grouting of holes and posttensioning of tendons are done after the grout has gained strength. The soil in which posttensioned tiebacks are used must have adequate holding power to withstand the pull-out force on the tendons.

The earth retention between the anchored piles can be provided by shotcrete over WWR, as shown in Figure 11.20. Alternatively, pressure-treated lumber lagging, attached to the piles through threaded bolts, may be used (as shown in Figures 11.17 and 11.18). Overall views of anchored pile systems using both alternatives (lumber lagging and shotcrete over WWR) are shown in Figure 11.21.

Whereas the sheet pile method of excavation support is relatively water resistant, the soldier pile and lagging system allows a much greater amount of water to percolate through the lagging. Therefore, the pile system is used in situations where water percolation is relatively small. In both cases, however, dewatering of the excavation may be required to keep the excavation dry (see Section 11.7).

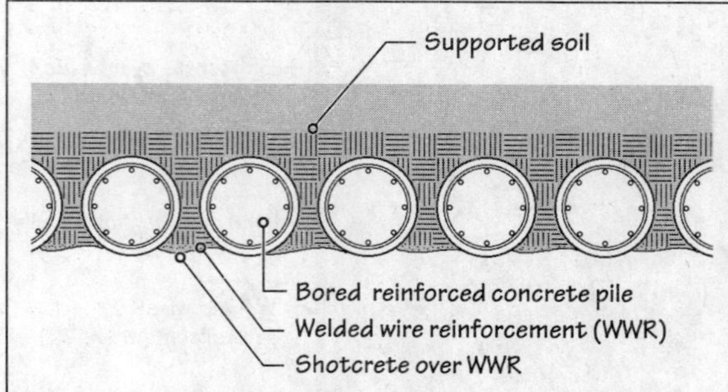

FIGURE 11.22 Closely spaced drilled (bored) concrete piles shown in plan.

Excavation Support Using Contiguous Bored Concrete Piles

In situations where the (deep) excavation is close to an adjacent building or the property line, tiebacks cannot be used. In this situation, closely spaced reinforced concrete piles, called *contiguous bored piles* (CBPs), are often used, Figure 11.22. Each pile is made by screwing an auger into the ground. The auger, called a *continuous flight auger*, has a hollow stem in the middle of a continuous spiral drill.

Once the drill has reached the required depth below the ground, high-slump concrete is pumped down the hollow stem of the auger to the bottom of the bore. Once the pumping starts, the auger is progressively withdrawn. The withdrawing auger brings the soil from the bore to the surface, where it is removed. Thus, the sides of the bore are supported at all times by the soil-filled auger or concrete. Immediately after the entire bore has been concreted, a reinforcement cage is lowered in the concrete-filled bore.

FIGURE 11.23 Contiguous bored piles used as supports for a deep basement excavation in a tight urban location. In areas farther away from the existing building or property line, the soldier pile and lagging system is used. Note that as the excavation proceeds, the lower part of drilled concrete piles will also be shotcreted as shown here for the upper part.

The depth of CBPs is limited by the length of the reinforcement cage that can be lowered through the concrete. (The reinforcement cage is specially designed to have adequate stiffness to penetrate into fresh concrete.) CBPs up to 100 ft deep have been constructed. They are generally 18 to 36 in. in diameter, depending on the depth of the excavation.

After the piles have been constructed and the excavation proceeds, the piles and the (small) soil spaces between them are shotcreted to reduce water percolation through the cut face, Figure 11.23.

Excavation Support Using Secant Piles

A major shortcoming of CBPs is the gaps between piles and the consequent lack of water resistance of the excavation support. This problem is overcome by the use of the modified version of CBPs called *secant piles.*

Secant piles essentially consist of two sets of interlocking contiguous piles. The first set, called the *primary piles,* is bored and concreted in the same way as the CBPs. The center-to-center distance between the primary piles is slightly smaller than twice their diameter.

After the primary piles are constructed, the *secondary piles* are bored at mid-distance between the primary piles, which also bores through part of the primary piles, Figure 11.24(a). The boring for secondary piles is done before the concrete in the primary piles has gained full strength to reduce wear on the blades of the auger. The secondary piles are concreted and reinforced in the same way as the CBPs. In most situations, only the secondary piles are reinforced, Figure 11.24(a). This precludes accidental cutting of any misaligned reinforcement in the primary piles. Thus, the unreinforced primary piles provide a water-resistant layer between the secondary piles.

Where greater bending strength of the support system is required, the primary piles are also reinforced, generally with a wide-flange steel section, and the secondary piles are reinforced with a circular steel reinforcement cage, Figure 11.24(b).

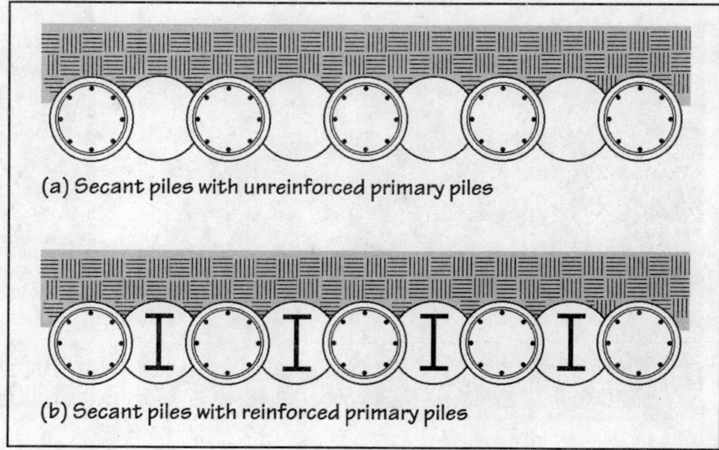

(a) Secant piles with unreinforced primary piles
(b) Secant piles with reinforced primary piles

FIGURE 11.24 Excavation support through secant piles.

EXCAVATION SUPPORT USING SOIL NAILING

Soil nailing is a means of strengthening the soil with closely spaced, inclined steel bars that increase the cohesiveness of the soil and prevent the soil from shearing along an inclined plane. The inclined bars are almost perpendicular to the possible shearing plane. In other words, the steel bars connect imaginary inclined layers of the earth into a thick block that behaves as a gravity-retaining wall when excavated, Figure 11.25.

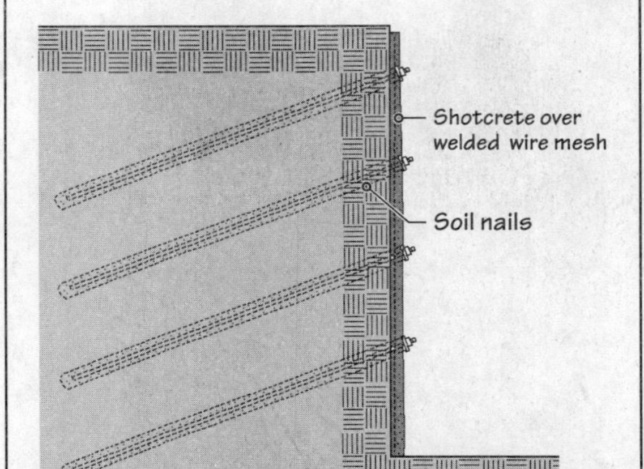

FIGURE 11.25 Section through a soil-nailed excavation support.

The process of soil nailing consists of the following steps:

- The soil is first excavated 5 to 7 ft deep, depending on the ability of the cut face to remain vertical without supports.
- Holes are drilled along the cut face at 3 to 4 ft on centers so that one hole covers approximately 10 to 15 ft^2 of the cut face, Figure 11.26(a).
- Threaded steel bars (approximately 1 in. in diameter) are inserted in the holes. The length of the bars is a function of the soil type but is approximately half the final depth of excavation. The bars protrude a few inches out of the holes.
- The holes are grouted with concrete, Figure 11.26(b).
- WWR is placed over the wall and tied to the protruding bars.
- A layer of shotcrete is applied to the mesh.
- Plates and washers are inserted in the protruding bars and locked in position with a nut.
- A second layer of shotcrete may be used if the soil-nailed wall is the finished wall, or a cast-in-place concrete wall may be constructed against it.
- These steps are repeated with the next depth of cut.

Figure 11.27 shows a soil-nailed wall that is almost complete. Note that soil nails are not posttensioned.

EXCAVATION SUPPORT USING BENTONITE SLURRY AS TRENCH SUPPORT

Another excavation support system, commonly used in situations where the underground water table is relatively high, is a reinforced concrete wall. Construction of such walls is done by excavating 10-ft- to 15-ft-long discontinuous trench sections down to bedrock, called *primary panels.* The width of the trench sections is the required thickness of the concrete wall. So that the soil does not collapse, the trench is continuously kept filled with bentonite slurry as the excavation proceeds. (Bentonite slurry is a mixture of water and bentonite clay, which pressurizes the walls of the trench sufficiently to prevent their collapse during excavation.)

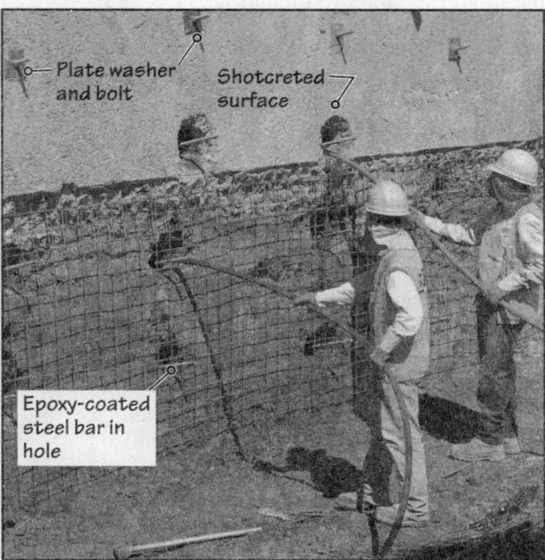

FIGURE 11.26 **(a)** Drilling of holes in excavated soil. **(b)** Grouting of holes after the epoxy-coated steel bar (soil nail) has been placed in the drilled holes.

FIGURE 11.27 View of a soil-nailed excavation support. (Photo courtesy of Schnabel Foundation Company)

Special excavation equipment is used to extract soil through the slurry-filled trench. After the excavation for the entire primary panel is complete, a reinforcement cage is lowered into the trench. Concrete is then placed in the trench panel using two or more *tremie pipes*, typically one at each end of the panel. Concrete is placed from the bottom up, and the discharge end of the tremie is always buried in concrete. A tremie pipe is generally an 8-in.- to 10-in.-diameter steel pipe with a hopper at the top, Figure 11.28.

As concreting proceeds, the slurry is pumped out from the top of the trench and stored for later use. After the primary panels have been constructed, excavation for secondary panels (between the primary panels) is undertaken in the same way as for the primary panels. To provide shear key and water resistance between primary and secondary panels, a steel pipe is embedded at the end of each primary panel prior to its concreting. These pipes are removed after the concrete in the primary panels has gained sufficient strength.

The tremie pipe method of concrete placement requires great care and expertise, particularly the initial placement of concrete, which is generally a richer mix. The concrete must also be placed slowly so that it does not get too diluted by the slurry.

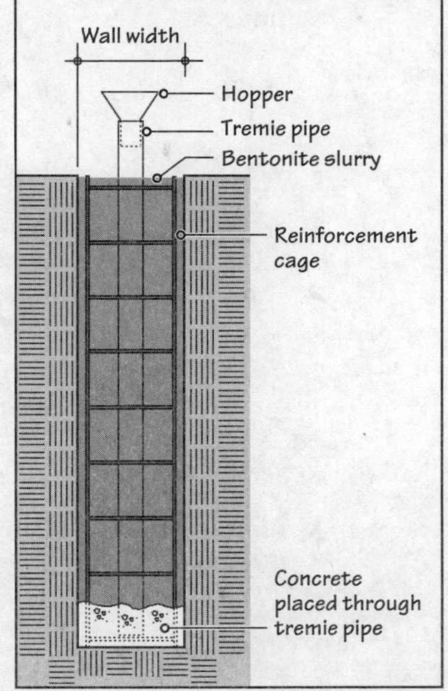

FIGURE 11.28 Trench support through bentonite slurry.

11.7 KEEPING EXCAVATIONS DRY

It is important to keep excavations free from groundwater. Even a small amount of water saturation in excavated areas impedes further excavation, is unsafe for workers, and is harmful to equipment and machines. Additionally, if the amount of groundwater is reduced, so is the pressure exerted by the soil on excavation supports—sheet piles, soldier piles, and so on.

Groundwater control in an excavation consists of two parts: (a) preventing surface water from entering the excavation through runoff and (b) draining (dewatering) the soil around the excavations so that the groundwater level falls below the elevation of proposed excavation. Two commonly used methods of dewatering the ground are *sump pumps* and *well points*.

DEWATERING THROUGH SUMPS

Sump dewatering consists of constructing pits (called *sumps*) within the enclosure of the excavation. The bottom of sumps must be located below the final elevation of the excavation. As the groundwater from surrounding soil percolates into the sump, it is lifted by automatic pumps and discharged away from the building site, Figure 11.29. The number of required sumps is a function of the excavation area. The method of discharge of pumped water must meet with the approval of local authorities.

DEWATERING THROUGH WELL POINTS

Sump dewatering works well in cohesive soils, where the percolation rate is slow and where the water table is not much higher than the final elevation of the base of the excavation.

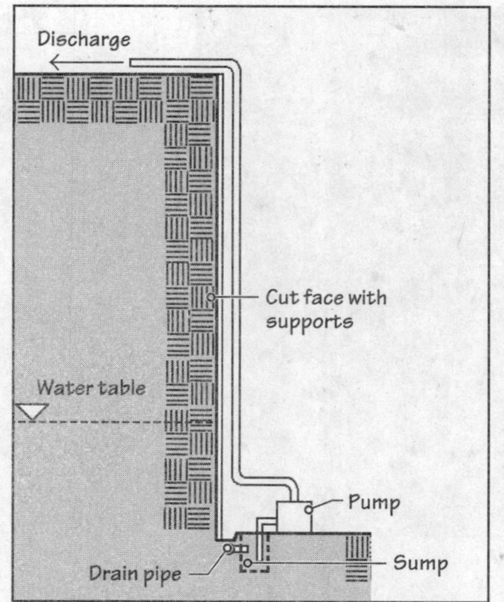

FIGURE 11.29 Sumps for groundwater control in an excavation.

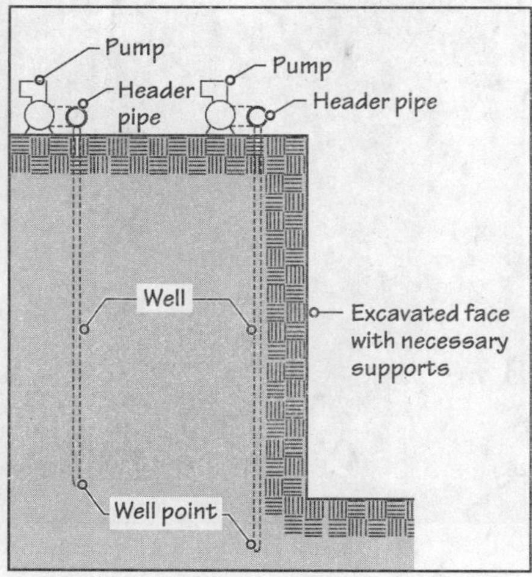

FIGURE 11.30 Section through an excavation showing two rings of well points.

A more effective dewatering method uses forced suction to extract groundwater. This is done by sinking a number of vertical pipes with a screened end at the bottom (called *well points*) around the perimeter of the excavation. The well points reach below the floor of the excavation and are connected to large-diameter horizontal header pipes at the surface.

The header pipe is connected to a vacuum-assisted centrifugal pump that sucks water from the ground for discharge to an appropriate point. For a very deep excavation, two rings of well points may be required. The well points in the ring farther away from the excavation terminate at a higher level than those close to the excavation, Figure 11.30.

Whereas the sump method of dewatering does not greatly affect the existing water table, dewatering by well points can lower the water table considerably. The effect of this on the adjoining buildings must be considered because it can cause consolidation and settling of the foundations of existing buildings on some types of soils.

Dewatering of excavations can be fairly complicated and generally requires an expert dewatering subcontractor for large and complicated operations.

PRACTICE QUIZ

Each question has only one correct answer. Select the choice that best answers the question.

15. Grading at a site refers to
- **a.** evaluating the existing subsurface conditions of a site.
- **b.** evaluating the groundwater conditions at a site.
- **c.** compacting the soil to required density.
- **d.** all of the above.
- **e.** none of the above.

16. Engineered fill refers to the soil that is
- **a.** specially formulated to provide the required properties.
- **b.** placed per the geotechnical engineer's specifications.
- **c.** compacted per the geotechnical engineer's specifications and inspection.
- **d.** all of the above.
- **e.** none of the above.

17. A benched excavation is generally used
- **a.** on open suburban sites.
- **b.** on tight downtown sites.
- **c.** where the excavation depth is less than 5 ft.
- **d.** where dewatering of the site is not required.

18. Sheet piles are used
- **a.** as shallow foundations in buildings.
- **b.** as deep foundations in buildings.
- **c.** as formwork for concrete walls.
- **d.** as excavation supports.
- **e.** none of the above.

19. Which of the following is not used for excavation supports?
- **a.** Sheet piles
- **b.** Soldier piles
- **c.** Soil nailing
- **d.** Precast concrete piles
- **e.** Contiguous bored concrete piles

20. In the excavation support system using soldier piles, the soldier piles consist of
- **a.** sheet steel.
- **b.** sheet aluminum.
- **c.** structural steel sections.
- **d.** reinforced concrete.
- **e.** precast concrete.

21. In the soldier pile and lagging system, the lagging generally consists of
- **a.** pressure-treated lumber.
- **b.** pressure-treated plywood.
- **c.** structural steel sections.
- **d.** sheet steel.
- **e.** none of the above.

22. An alternative to lagging that is commonly used with the soldier pile system of excavation support is
- **a.** pressure-treated lumber.
- **b.** pressure-treated plywood.
- **c.** structural steel sections.
- **d.** sheet steel.
- **e.** shotcrete over welded wire reinforcement (WWR).

23. Tiebacks used with the anchored soldier pile system of excavation support consist of
- **a.** prestressing tendons in grouted holes.
- **b.** structural steel angle sections in grouted holes.
- **c.** reinforcing steel bars in grouted holes.
- **d.** prestressing tendons or reinforcing steel bars in grouted holes.
- **e.** none of the above.

24. Contiguous bored piles are constructed using
- **a.** structural steel sections.
- **b.** site-cast reinforced concrete.
- **c.** precast concrete.
- **d.** precast, prestressed concrete.
- **e.** none of the above.

25. In terms of design and construction, secant piles are related to
- **a.** sheet piles.
- **b.** soldier piles.
- **c.** contiguous bored piles.
- **d.** none of the above.

26. The center-to-center distance between secant piles is generally
- **a.** 5 to 8 ft.
- **b.** 8 to 12 ft.
- **c.** 10 to 15 ft.
- **d.** 15 to 20 ft.
- **e.** none of the above.

27. Soil nailing refers to
- **a.** a cut face of excavation supported by closely spaced steel bars or angles.
- **b.** a cut face of excavation supported by closely spaced steel bars and shotcrete.
- **c.** a cut face of excavation supported by closely spaced steel bars and shotcrete over WWR.
- **d.** a cut face of excavation supported by closely spaced steel bars in grouted holes and shotcrete over WWR.
- **e.** none of the above.

28. Dewatering of an excavation, as described in this text, can be done by the
- **a.** pit method.
- **b.** sump method.
- **c.** well-point method.
- **d.** (a) and (b).
- **e.** (b) and (c).

PRINCIPLES IN PRACTICE

Unified Soil Classification

The soil's classification (into gravel, sand, silt, and clay), described in Section 11.1, is too broad to be useful in building construction, because subsurface deposits containing pure gravel, pure sand, and so on, are extremely rare. Generally, we find soil mixtures that contain two or more of the four soil materials in varying proportions. Therefore, a more detailed classification is required. The classification commonly used is called the *Unified Soil Classification System* (USCS). In the USCS, the distinction between coarse- and fine-grained soils is as follows.

- *Coarse-grained soils* are soils that contain more than 50% of material retained on a No. 200 sieve.

  Coarse-grained soils are further subdivided into *gravelly soils* and *sandy soils.* Gravelly soils are those in which 50% or more of the coarse fraction is retained on a No. 4 sieve. In sandy soils, more than 50% of the coarse fraction passes through a No. 4 sieve.

  Gravelly soils are further subdivided into *clean gravels* and *dirty gravels.* Similarly, sandy soils are further subdivided into *clean sands* and *dirty sands.*
- *Fine-grained soils* are soils that contain 50% or more of material passing through a No. 200 sieve. Fine-grained soils are further subdivided into Group L soils and Group H soils. This distinction is based on the plasticity of the soil, which is measured by the soil's *liquid limit.* A soil with *low* plasticity (liquid limit ≤50%) belongs to Group L soils. A soil with *high* plasticity (liquid limit >50%) is classified as a Group H soil.

Additional details of USCS are given in Table 1.

SYMBOLS USED IN THE USCS

USCS uses the following symbols to designate various soils:

G for gravel

S for sand

M for silt

C for clay

W for well-graded particles, that is, when particles of various sizes are present, giving a relatively denser soil

P for poorly graded soils, that is, soils with a very little (i.e., almost uniform) gradation of particle sizes. (Such soils contain more void space and are, therefore, less dense.)

L for low plasticity (liquid limit ≤50%)

H for high plasticity (liquid limit >50%)

O for organic soils

(Continued)

PRINCIPLES IN PRACTICE

Unified Soil Classification (*Continued*)

TABLE 1 UNIFIED SOIL CLASSIFICATION SYSTEM (USCS)

Major divisions			Group symbols	Group description
Coarse-grained soils >50% material retained on No. 200 sieve	**Gravelly soils** >50% of coarse fraction retained on No. 4 sieve	**Clean gravels** Fines <5%	GW	Well-graded gravels and gravel-sand mixtures, little or no fines
		In field test, no stain on wet palm	GP	Poorly graded gravels and gravel-sand mixtures, little or no fines
		Dirty gravels Fines >12%	GM	Silty gravels, gravel-sand-silt mixtures
		In field test, stain on wet palm	GC	Clayey gravels, gravel-sand-clay mixtures
	Sandy soils ≥50% of coarse fraction passes through No. 4 sieve	**Clean sands** Fines <5%	SW	Well-graded sands and gravelly sands, little or no fines
		In field test, no stain on wet palm	SP	Poorly graded sands and gravelly sands, little or no fines
		Dirty sands Fines >12%	SM	Silty sands, sand-silt mixtures
		In field test, stain on wet palm	SC	Clayey sands, sand-clay mixtures
Fine-grained soils ≥50% material passes through No. 200 sieve	**Silts and clays** Liquid limit ≤50%		ML	Inorganic silts, very fine sands, rock flour, silty or clayey fine sands
			CL	Inorganic clays of low to medium plasticity, gravelly clays, sandy clays, silty clays, lean clays
			OL	Organic silts and organic silty clays of low plasticity
	Silts and clays Liquid limit >50%		MH	Inorganic silts, micaceous or diatomaceous fine sands or silts, elastic silts
			CH	Inorganic clays of high plasticity, fat clays
			OH	Organic clays of medium to high plasticity
Highly Organic Soils			Pt	Peat, muck, and other highly organic soils

Note: Gravelly or sandy soils with fines between 5% and 12% carry a hyphenated designation such as GW-GM, GW-GC, and so on.
Source: Adapted from American Society for Testing and Materials (ASTM): "Standard Practice for Classification of Soils for Engineering Purposes (Unified Soil Classification System)," ASTM D2487, with permission.

Two symbols are used in conjunction to designate a soil. Thus, GW designates a well-graded gravel, GP designates a poorly graded gravel, ML designates a silty soil with a liquid limit less than or equal to 50%, and so on.

PLASTIC LIMIT AND LIQUID LIMIT OF FINE-GRAINED SOILS

The liquid limit of soil (a measure of the soil's plasticity) indicates the behavior of soil in the presence of water. If we take a sample of completely dry fine-grained soil and put a little water in it, it will become slightly moist, or semidry.

If the amount of water in the soil sample is increased, it will become plastic (puttylike), implying that it can be rolled into a rope between the palms of both hands. The ratio of the weight of water to the dry weight of the soil (expressed as a percentage moisture content) when a soil just changes from a semidry state to the plastic state is called the *plastic limit* (PL) of the soil.

If we keep adding water to the soil sample, a stage is reached when the soil loses its plasticity and turns into a slurry, that is, a liquid, Figure 1. The moisture content of soil when it just turns into a liquid is called its *liquid limit* (LL).

PRINCIPLES IN PRACTICE

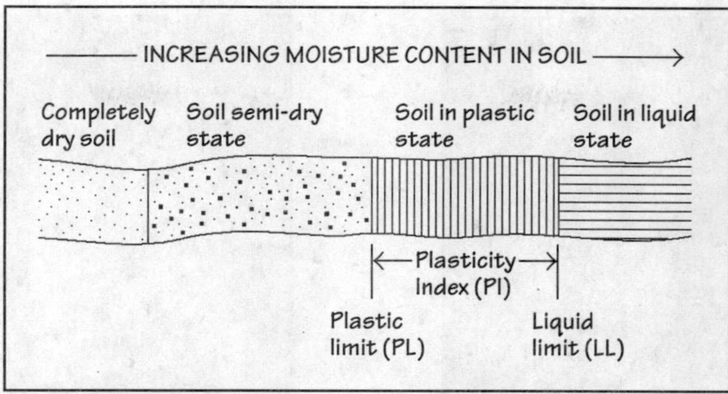

FIGURE 1 Four states of a fine-grained soil with increasing moisture content.

Gravels and sands do not become plastic. They change from a solid state to a liquid state. That is why the LL and PL concepts apply only to fine-grained soils—silts and clays.

Another important soil property is the plasticity index (PI), defined as

$$PI = LL - PL$$

For example, if a soil has a PL of 35% and an LL of 60%, its PI is $60 - 35 = 25\%$. Between two soils with the same LL, the one with the larger PI value retains more water and is, therefore, more susceptible to swelling and shrinkage due to changes in moisture.

EXPANSIVE SOILS

Some fine-grained soils expand with an increase in moisture content and shrink when the moisture content decreases. Foundations in such soils, referred to as *expansive soils*, are constructed to ensure that the structure is not adversely affected by soil movement.

A rough guide to the swell-shrink susceptibility of a fine-grained soil is provided by the following indices:

Highly expansive soil (high swell-shrink potential): LL > 50% and PI > 30%.

Medium expansive soil (medium swell-shrink potential): LL between 25% and 50% and PI between 15% and 30%.

Stable soil (low swell-shrink potential): LL < 25% and PI < 15%.

Expansive soils are found in many regions of the United States, such as California, Nevada, Texas, Arizona, and Maryland.

TESTING FOR LIQUID LIMIT OF FINE-GRAINED SOILS

A liquid-limit testing apparatus consists of a metal cup mounted on a hard rubber base, Figure 2. The motor attached to the cup raises it to a prescribed height and releases it for a free fall on the rubber base.

To determine the LL, the soil specimen is mixed with different amounts of water to give several batches of the specimen. A batch is placed in the cup and flattened to the required spread and thickness in the prescribed manner. A groove is then cut in the flattened batch with a standard grooving tool (shown in the foreground in Figure 2).

The number of falls of the cup that close the groove in a batch are recorded. A graph relating the moisture content of the soil with the number of falls for each batch is plotted. From this graph, the LL is obtained by reading the amount of moisture content that would be required to close the groove with 25 falls.

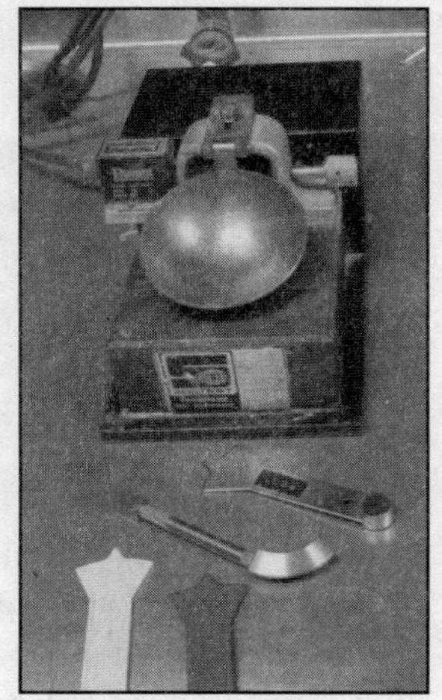

FIGURE 2 A liquid limit testing apparatus. (Photo courtesy of the Geotechnical Laboratory of the University of Texas at Arlington)

TESTING FOR PLASTIC LIMIT OF FINE-GRAINED SOILS

A plastic-limit test involves taking a prescribed weight of dry sample and mixing it with water until it is almost saturated. The sample is rolled back and forth between two plastic plates until a thread nearly $\frac{1}{4}$ in. in diameter is formed, Figure 3.

(Continued)

PRINCIPLES IN PRACTICE

Unified Soil Classification (*Continued*)

FIGURE 3 Rolling moist soil into a thread. (Photo courtesy of the Geotechnical Laboratory of the University of Texas at Arlington)

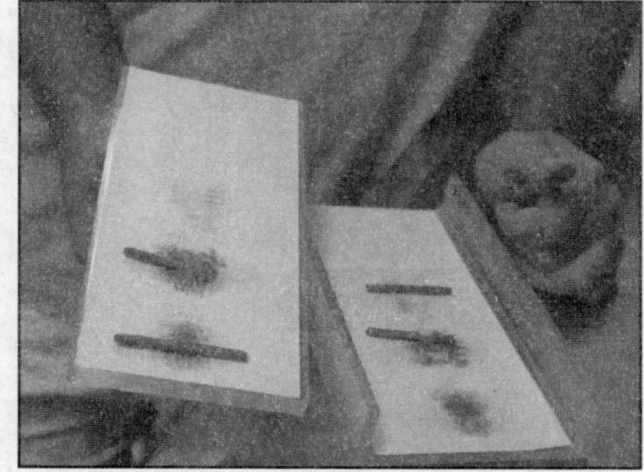

FIGURE 4 As the moist soil thread is progressively dried while rolling, it breaks. (Photo courtesy of the Geotechnical Laboratory of the University of Texas at Arlington)

The thread is then balled up and rolled again and then again. Each rolling operation is performed under a fan so that the specimen gets progressively drier. A stage is reached when the thread breaks into short lengths, Figure 4. The percentage moisture content in the specimen at this stage is the PL of the soil.

Note that both the LL and PL tests are performed on the fraction of fine-grained soils that passes through a No. 40 sieve.

PRACTICE **QUIZ**

Each question has only one correct answer. Select the choice that best answers the question.

29. In the Unified Soil Classification System, the symbol M stands for
- **a.** medium-grained soil.
- **b.** moist soil.
- **c.** soil of medium plasticity.
- **d.** none of the above.

30. In the Unified Soil Classification System, the symbol GP stands for
- **a.** poorly graded gravel.
- **b.** pea gravel.
- **c.** peat and gravel.
- **d.** soil with good performance.
- **e.** none of the above.

31. Which of the following soils has the lowest liquid limit?
- **a.** Clay
- **b.** Silt
- **c.** Sand
- **d.** Mixture of silt and clay
- **e.** Mixture of silt and sand

32. The expansiveness of a soil is determined from the values of
- **a.** plastic limit and liquid limit.
- **b.** liquid limit and plasticity index.
- **c.** plastic limit and plasticity index.
- **d.** none of the above.

REVIEW QUESTIONS

1. Explain what is included in surface investigation and subsurface investigation of a site.
2. Explain the difference between (a) fine-grained soils and coarse-grained soils, (b) sand and gravel, and (c) silt and clay.
3. Explain why clayey soils are unstable.
4. Explain the difference between how a stiff, dense soil performs during an earthquake compared with a soft, loosely compacted soil.
5. Explain why the wall and column footings must be placed below the frost line.
6. Explain where self-supporting basement excavation may be used.
7. List various excavation support systems commonly used.
8. List the steps needed in the construction of a soil-nailed excavation support system.
9. Using sketches, explain the difference between cantilevered soldier pile and anchored soldier pile excavation support systems.

CHAPTER 12

Below-Grade Construction (Foundation Systems and Basements)

CHAPTER OUTLINE

As shown in Figure 12.1, a building may be considered to comprise a (a) superstructure, (b) substructure, and (c) foundation system. In this classification, the *superstructure* generally refers to the part of the building that is above the ground level. The *substructure* refers to basements, if provided, and the *foundation system* refers to the structure that lies below the substructure. Where a basement does not exist, a building has only a superstructure and foundations.

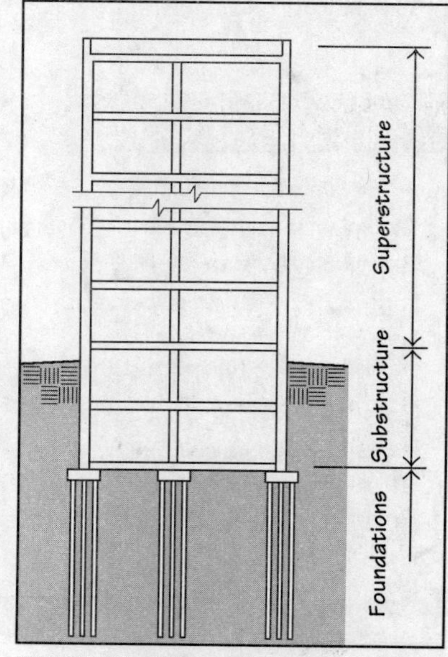

FIGURE 12.1 Three parts of a structure.

SHALLOW AND DEEP FOUNDATIONS

The choice of foundation type is governed by the characteristics of the soil near the surface and the pressure exerted on it by the loads on the building. Foundations are classified as shallow or deep. *Shallow foundations* extend a relatively short distance below the ground and bear directly on the upper soil stratum. Because of their lower cost, they are preferred over deep foundations. However, because the soil's bearing capacity is low at smaller depths from the surface, shallow foundations are generally limited to low- to mid-rise buildings.

Deep foundations are used where soil of adequate strength is not available close enough to the surface for the use of shallow foundations. Deep foundations consist of either piles or piers and are generally used for high-rise buildings. However, where the soil near the surface is unsuitable (e.g., subject to swell and shrink phenomenon), deep foundations are utilized for low-rise buildings as well so that the loads are transferred to a suitable stratum.

In other words, the choice of a foundation system for low to mid-rise buildings is governed by the type of near-surface soil. If the near-surface soil is stable and has adequate strength, shallow foundations are appropriate. On the other hand, if the near-surface soil is either weak or unstable (i.e., expansive), deep foundations must be utilized. In the case of lightly loaded low-rise buildings, shallow foundations can be designed for use on expansive soils.

This chapter begins with a broad description of shallow and deep foundations, followed by a more detailed study of the systems. The study presented in Sections 12.7 and 12.8 classifies building foundation systems into two categories: (a) foundation systems on stable soils and (b) foundation systems on expansive soils. Most foundation systems suitable for expansive soils may also be used where near-surface soils have a low bearing capacity or are otherwise unsuitable.

FOUNDATION MATERIALS

The materials commonly used for foundations are concrete (reinforced with either reinforcing steel or posttensioning tendons) and reinforced masonry (concrete masonry units). Less commonly used materials are plain (unreinforced) concrete, unreinforced masonry, and preservative-treated lumber and plywood. Our discussion is limited to reinforced concrete, posttensioned concrete, and reinforced masonry foundations.

NOTE

Foundation Failures Due to Expansive Soils

As described in Section 11.1, expansive soils are present throughout United States, but predominate in some states. Experts claim that damage to foundations not adequately designed for expansive soils totals several billion dollars in the United States every year. This damage receives little media attention because it is not life-threatening, like fires, earthquakes, hurricanes, and so on.

12.1 SHALLOW FOUNDATIONS

There are two types of shallow foundations:

- Foundation systems consisting of footings (see Figure 12.2)
- Monolithic concrete foundation under the building's entire footprint (see Figure 12.4)

FOUNDATION SYSTEMS WITH FOOTINGS

The widened base of a column or wall is called the *footing.* Widening of the base is essential because the strength of the soil supporting the column or wall is generally lower than the strength of the material used for the column or wall. Thus, the primary purpose of a footing is to distribute the superimposed load on a large area of the soil so that the pressure on the soil is less than or equal to the soil's strength (bearing capacity). Most footings are constructed of reinforced concrete, but plain concrete footings may be used for lightly loaded walls. Commonly used foundation systems with footings are:

- Footings under perimeter foundation walls and interior piers (short columns) with an elevated ground floor, Figure 12.2(a)
- Footings under columns and load-bearing walls with a concrete slab-on-ground, Figure 12.2(b)
- Isolated column footings with or without a basement, Figure 12.2(c)

TYPES OF FOOTINGS

As shown in Figure 12.3, footings are of the following types:

- *Continuous wall footings,* also called *strip footings,* are commonly used where the superimposed loads are linear, generally from a load-bearing wall, Figure 12.3(a).
- *Isolated* (independent) *footings* are used where the superimposed load is a point load—for example, from a column, Figure 12.3(b). The bearing capacity required under a column footing is generally higher than that under a wall footing because of the concentrated load on the column. Sometimes, particularly with steel columns, a stepped footing (with a pedestal) is used to save concrete, Figure 12.3(c).
- A *combined footing* is a combination of two isolated column footings. It is used where two or more adjacent columns are closely spaced and heavily loaded. Combining them into one footing reduces the excavation cost and distributes the load over a larger area, Figure 12.3(d). A combined footing is also used where an exterior column must be placed within the property line or adjacent to an existing building, Figure 12.3(e).

NOTE

Pier, Drilled Pier and Bored Pier

The term *pier* in building construction generally means a column-like element that terminates a short distance above the ground and supports the structure above it. It may be of two types:

- A stub (short) masonry or concrete column constructed on a footing.
- A reinforced concrete column (buried in the soil) and supported on rock or firm soil deep in the ground. In this case, the pier is constructed by drilling a hole in the ground into which a steel reinforcement cage is lowered, followed by the placement of concrete in the hole.

To distinguish between the two types of piers, we have called the first type a *pier,* and the second type a *drilled pier* in this text. The terms *bored pier, bored pile,* and *caisson* are also used in place of *drilled pier.*

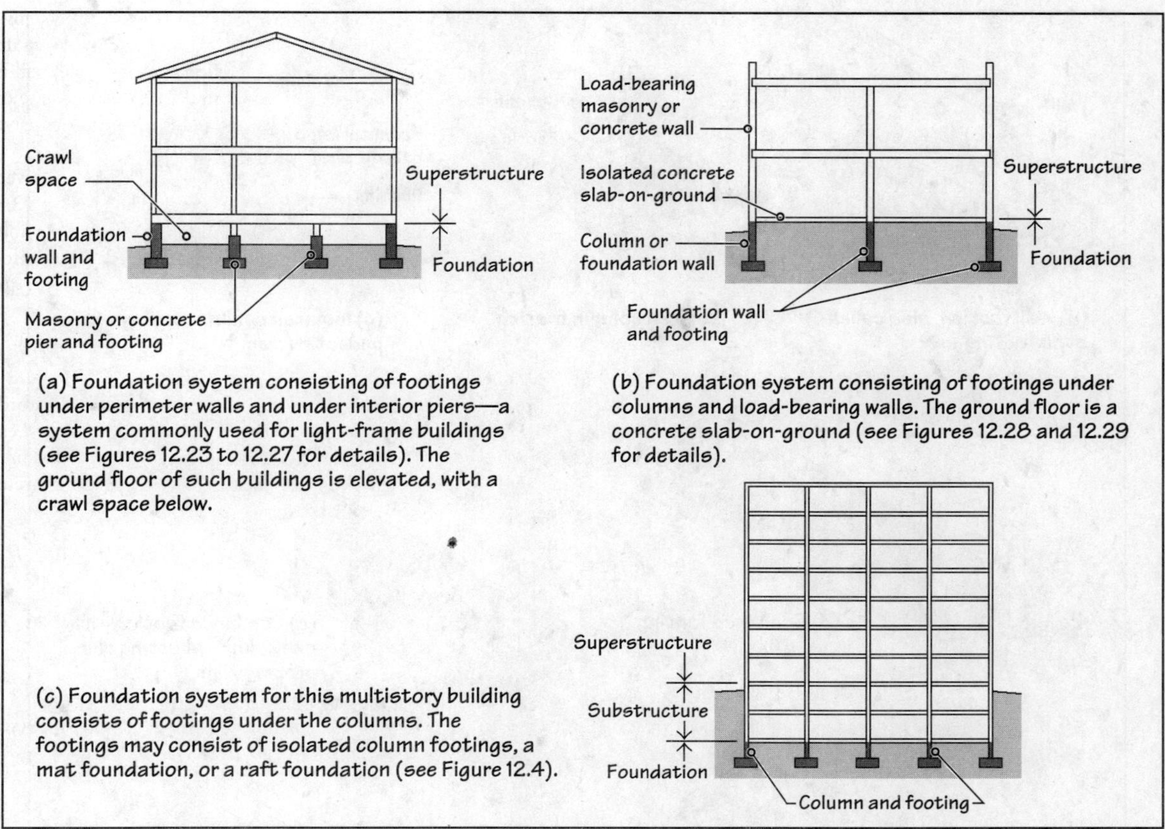

FIGURE 12.2 Commonly used shallow foundation systems with footings.

Two alternatives to a combined footing are the *strap footing,* Figure 12.3(f) and the *cantilevered footing,* Figure 12.3(g), in which a cantilevered grade beam spans between the two isolated column footings.

Monolithic Concrete Foundation Under the Building's Entire Footprint

The foundation system consisting of a large monolithic concrete base, generally covering the entire footprint of the building, can be of three types:

- A *slab-on-ground foundation* is the most widely used monolithic foundation system because of its low cost and ease of construction. It is the system of choice for low-rise, light frame residential or commercial buildings, such as single-family homes, multifamily homes, motels, and professional offices. The slab functions as a foundation system and also as the ground floor of the building, Figure 12.4(a). The slab-on-ground may be a reinforced concrete slab or a posttensioned concrete slab.
- In a *mat foundation,* also called a *mat footing,* all columns and walls of a building bear on one large, thick reinforced concrete slab, Figures 12.4(b) and (c). This foundation is used where the bearing capacity of the soil is low (even deep below the surface) so that independent column footings become so large that only small, unexcavated areas remain between the footings. If the excavation required for isolated footings in a building is more than 50% of the footprint of the building, it is generally more economical to use a mat foundation.

  A mat foundation is also used where the bedrock is so deep that the use of a deep foundation system (drilled piers and piles) is uneconomical. Because of the large

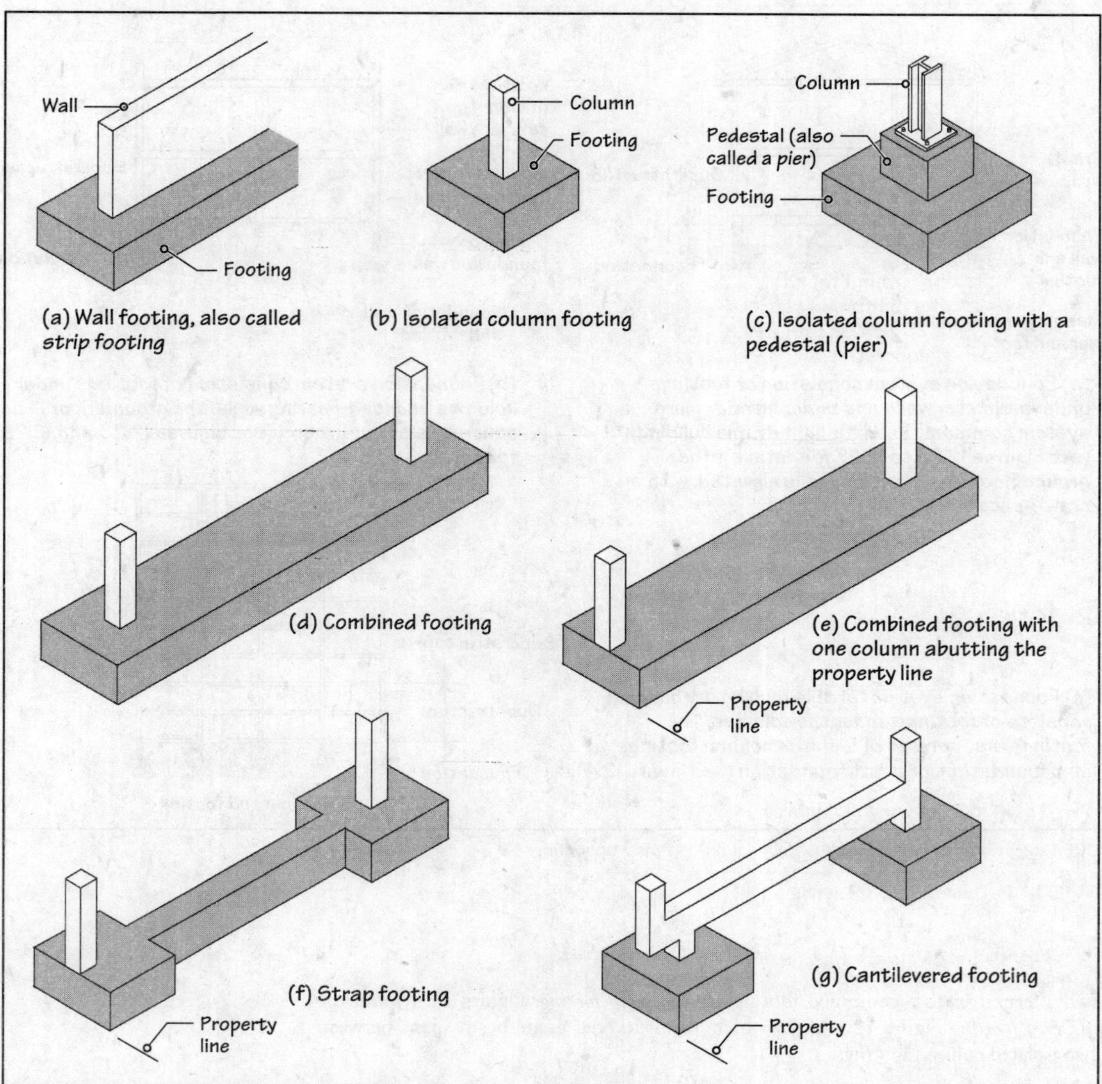

FIGURE 12.3 Commonly used footing types.

thickness of a mat foundation, it is sufficiently rigid to distribute loads from individual columns evenly on the underlying soil.

- A *raft*, also called a *floating foundation*, is a type of mat foundation. It consists of a hollow mat formed by a grid of thick reinforced concrete walls between two thick reinforced concrete slabs, Figure 12.4(d). In this case, the weight of the soil excavated from the ground is equal to the weight of the entire building, so that the pressure on the soil is unchanged from the original condition, making the building float on the soil.

12.2 DEEP FOUNDATIONS

Deep-foundation systems include piles and drilled piers that are like slender columns buried in the ground. They are prevented from buckling because they are confined by the soil. Piles and drilled piers transfer the load either to bedrock or to soil of high bearing capacity while passing through unsuitable soil conditions.

Piles are generally steel, timber, or concrete elements driven into the ground, except concrete piles, which can also be site-cast in predrilled holes. Site-cast piles are called *drilled piers* or *caissons.* Thus, contiguous bored concrete piles (shown in Figure 11.23) can be called drilled piers. Another term used for drilled piers is *bored* or *drilled piles.*

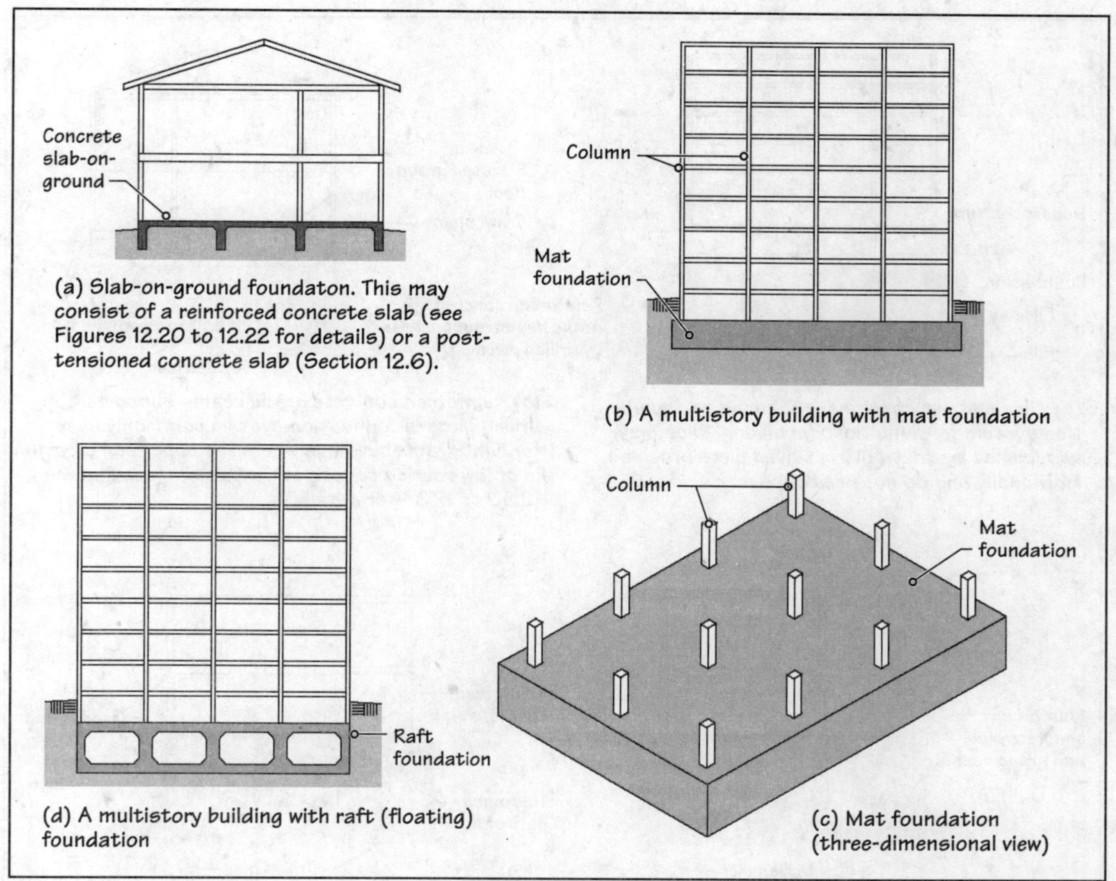

FIGURE 12.4 Foundation systems comprising a large, monolithic reinforced concrete base, generally covering the building's entire footprint.

Thus, drilled piers are always of reinforced concrete, while piles (used for foundations, distinct from sheet piles used for excavation support) may consist of reinforced concrete, steel, or timber.

As previously stated, deep foundations are used where shallow foundations cannot be used because the pressure imposed by the structural elements is too large (e.g., under the columns of a tall building) for the soil's bearing capacity at shallow depths, Figure 12.5(a).

Deep foundations are also used in low-rise, lightly loaded buildings in situations where the soil near the surface is unstable. Figure 12.5(b) shows the foundation system of a low-rise building consisting of reinforced concrete grade beams supported on drilled piers—commonly used in unstable soil conditions. The ground floor of this building is supported on grade beams with a crawl space under the floor, similar to that of Figure 12.2(a).

Figure 12.5(c) shows a foundation system that is an alternative to that of Figure 12.5(b), where the ground floor consists of a concrete slab supported on grade beams that bear on drilled piers. Figure 12.5(d) shows another alternative in which the ground floor slab is supported on enlarged concrete caps that bear on drilled piers.

Because the settlement of deep foundation systems is relatively small, deep foundations are also used in buildings that house items or instruments sensitive to even small amounts of settlement, which is likely to occur in shallow foundations.

12.3 PILES AS DEEP FOUNDATIONS

As previously mentioned, steel, precast concrete, and wood are common pile materials. The material selected depends on availability, cost, below-grade environment, type of soil, the load to be supported, and the equipment required to drive the piles. Steel and concrete piles are generally used under heavier loads. Wood piles are limited in their load-bearing

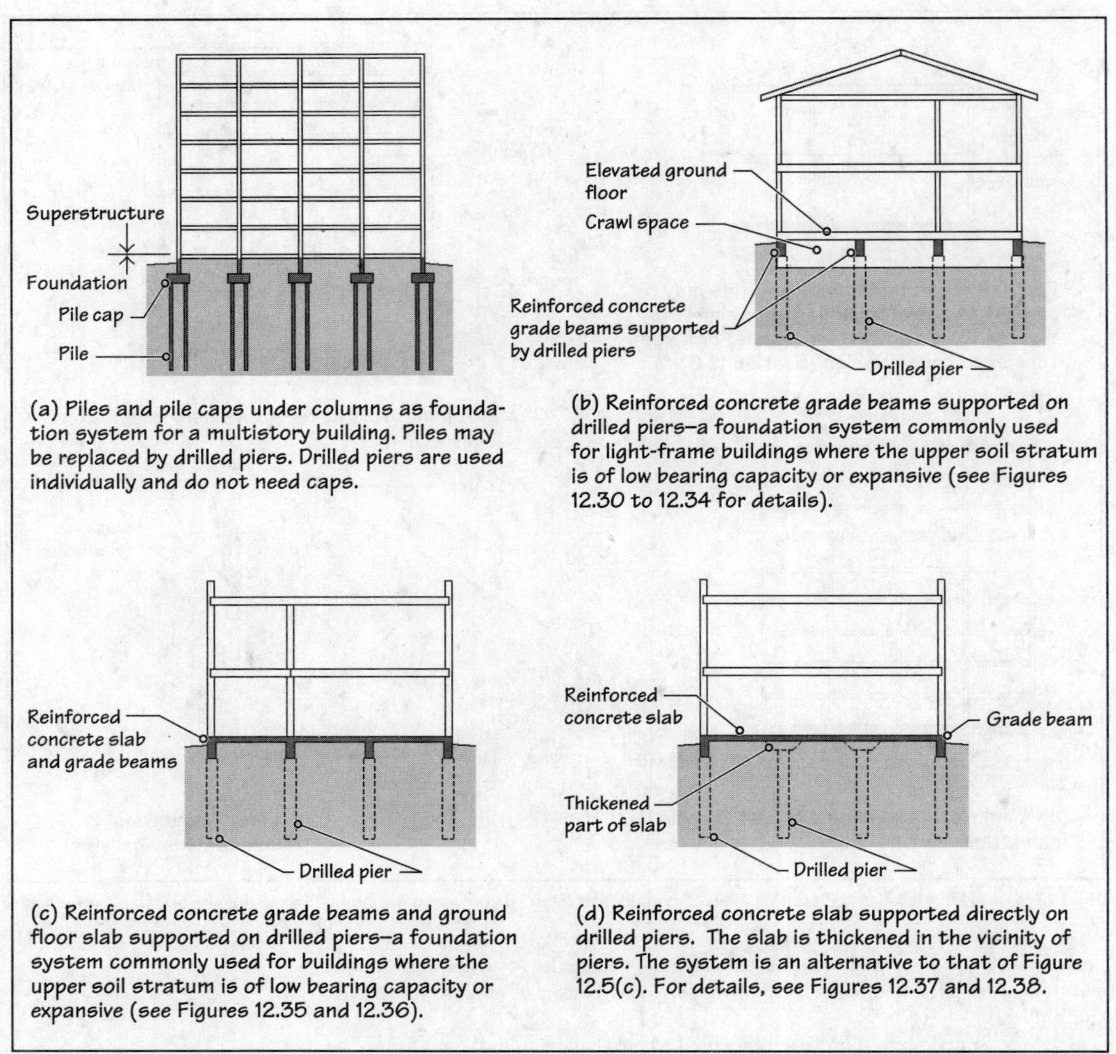

Superstructure
Foundation
Pile cap
Pile
(a) Piles and pile caps under columns as foundation system for a multistory building. Piles may be replaced by drilled piers. Drilled piers are used individually and do not need caps.
Elevated ground floor
Crawl space
Reinforced concrete grade beams supported by drilled piers
Drilled pier
(b) Reinforced concrete grade beams supported on drilled piers—a foundation system commonly used for light-frame buildings where the upper soil stratum is of low bearing capacity or expansive (see Figures 12.30 to 12.34 for details).
Reinforced concrete slab and grade beams
Drilled pier
(c) Reinforced concrete grade beams and ground floor slab supported on drilled piers—a foundation system commonly used for buildings where the upper soil stratum is of low bearing capacity or expansive (see Figures 12.35 and 12.36).
Reinforced concrete slab
Grade beam
Thickened part of slab
Drilled pier
(d) Reinforced concrete slab supported directly on drilled piers. The slab is thickened in the vicinity of piers. The system is an alternative to that of Figure 12.5(c). For details, see Figures 12.37 and 12.38.

FIGURE 12.5 Commonly used deep foundation systems.

capacity because of the nature of the material and the limitations on the cross-sectional area of tree trunks.

Steel piles may be H-shaped or hollow pipes. Hollow steel piles are filled with concrete after being driven. Steel piles are covered with protective coatings if used in corrosive environments. Concrete piles are made of precast concrete and are generally solid. They are subject to attack by the sulfur present in some soils, requiring the use of sulfate-resistant portland cement (Chapter 21). Concrete piles can be provided with a steel tip extension for easier penetration. Wood piles are treated with creosote or chromated copper arsenate (CCA) as a preservative. (For a discussion of CCA, see Section 13.15.)

Piles support loads through two mechanisms. Piles that support loads primarily through friction created between the surface of the pile and the soil are called *friction piles,* Figure 12.6(a). They are generally tapered to a narrow cross section at the bottom to facilitate driving. Piles that transfer most of the load to the bottom stratum, with very little through friction, are called *tip-bearing* or *end-bearing piles,* Figure 12.6(b). For piles driven in weak soils, the end reaction from tip bearing may be ignored.

Because they are driven, piles are smaller in cross section than drilled piers, and the load capacity of an individual pile is also small. Piles are, therefore, used in clusters of three or more at 2 to 3 ft on center and support a reinforced concrete cap called a *pile cap.* A pile cap functions like an isolated column footing or a continuous strip footing under a wall, Figure 12.7. Two or more columns can be placed on a large pile cap.

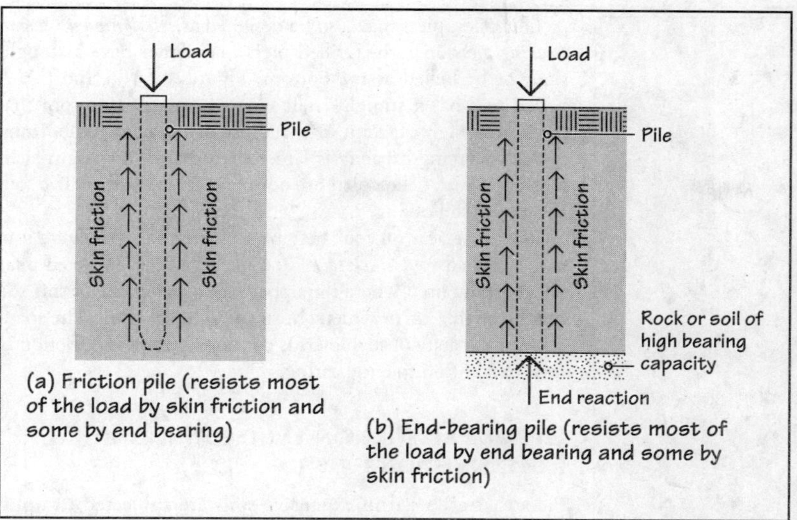

FIGURE 12.6 Friction piles and end-bearing piles.

DRIVING OF PILES

Although pile-driving machines, Figure 12.8, come in different types, depending on the type of soil and the pile material, they basically consist of a heavy weight placed between guides so that it can move up and down in a straight line. The driving force comes from the repeated hammering action as the weight is dropped on the pile head, raised, and dropped again, using diesel, hydraulics, or compressed air for the operation.

12.4 DRILLED PIERS AS DEEP FOUNDATIONS

Drilled piers are often used individually (as opposed to a cluster of piles) because they can be made almost as large as required. The drilled pier diameter generally varies between 18 and 36 in. Larger-diameter piers (6 ft in diameter or larger) are used for bridges or high-rise structures. Another reason for using individual piers is that their vertical alignment is almost assured. (Because piles are driven, alignment is uncertain. Therefore, an individual pile is seldom used.)

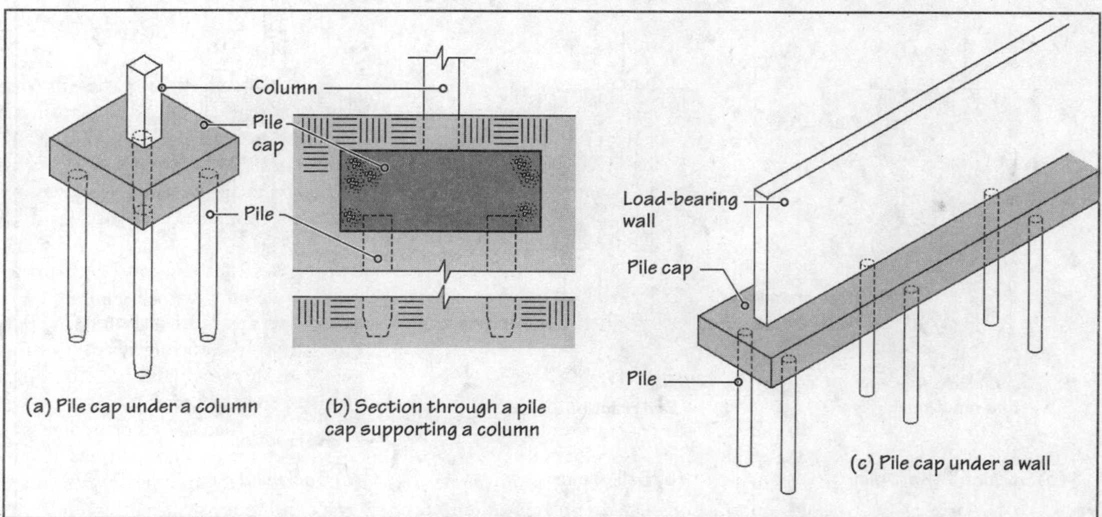

FIGURE 12.7 Reinforced concrete pile caps over a cluster of piles.

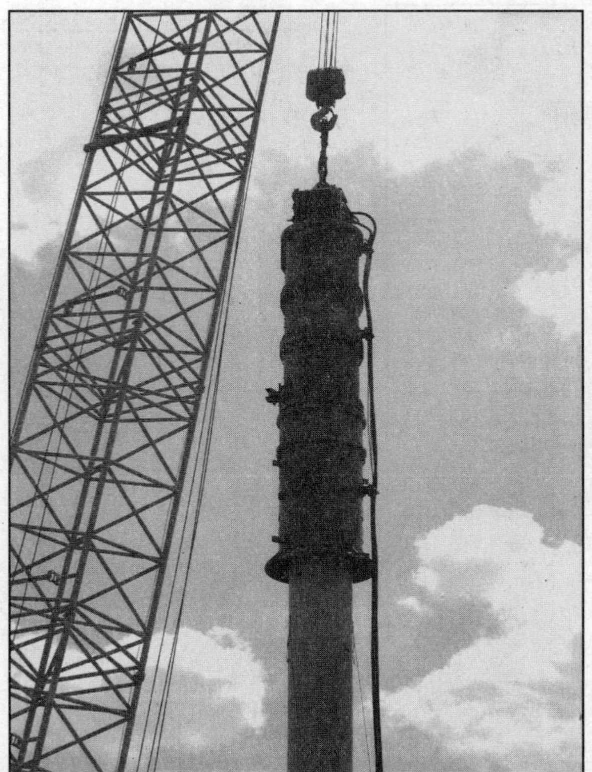

FIGURE 12.8 A round concrete pile being driven by a hydraulic impact hammer. (Photo courtesy of Pile Driving Contractors Association [PDCA], Orange Park, Florida)

Like piles, piers may also be designed as *friction piers* or *end-bearing piers* or both. Drilled piers may either have a straight shaft or be belled at the bottom, Figures 12.9(a) and (b). A belled pier has a straight shaft with an enlarged bottom that bears on a high-capacity soil. Because of the enlarged bottom, its load-carrying capacity is larger than that of a straight-shaft pier. However, belled piers are not practical in noncohesive soils because of soil caving.

Piers that bear on rock may be socketed (i.e., penetrate into rock, referred to as *rock sockets*), Figure 12.9(c). Socketed piers are generally used where the upper stratum consists of soft soil that is unable to provide significant skin friction. The load-carrying capacity of such piers is developed from skin friction in rock in addition to end bearing.

Uplift Pressure on Drilled Piers in Expansive Soils

Piers that are used in expansive soils are subjected to uplift pressure caused by soil expansion. In such soils, the uplift pressure on the pier from the upper, expansive zone of the soil (referred to as the *active zone*) must be countered by skin friction from the lower, stable (inactive) zone of the soil, Figure 12.10. Where the (lower) stable zone is not sufficiently deep, the pier may need to be socketed in the rock to provide the necessary anchorage against uplift. A belled pier may also be used for the same reason.

Construction of Drilled Piers

Piers are generally constructed by first drilling a hole of predetermined diameter into the soil (and later into the rock for a socketed pier). This is accomplished by a truck-mounted rig furnished with a drilling auger. An auger consists of a continuous blade formed into an open helix (drill). Different augers are used for cutting into soil and rock. Soil augers are simple blade-type augers, Figure 12.11. Augers used for cutting into rock or hard soil are provided with carbide-tipped teeth, Figure 12.12.

As the auger rotates, the soil or rock cut by the auger collects in the spaces between the auger helix. When the space gets filled, the auger is brought up and moved away from the

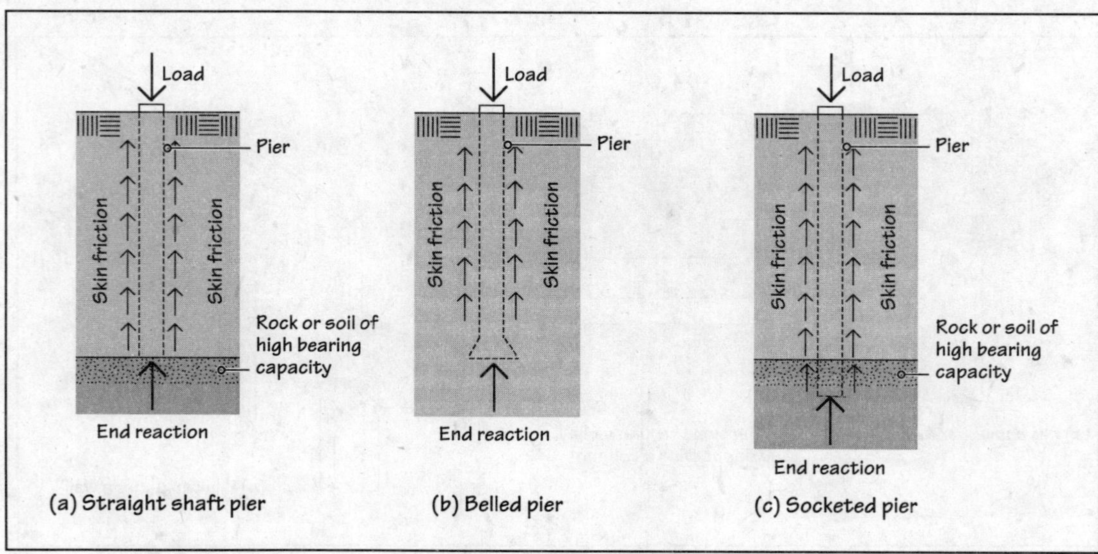

FIGURE 12.9 Types of drilled (bored) piers.

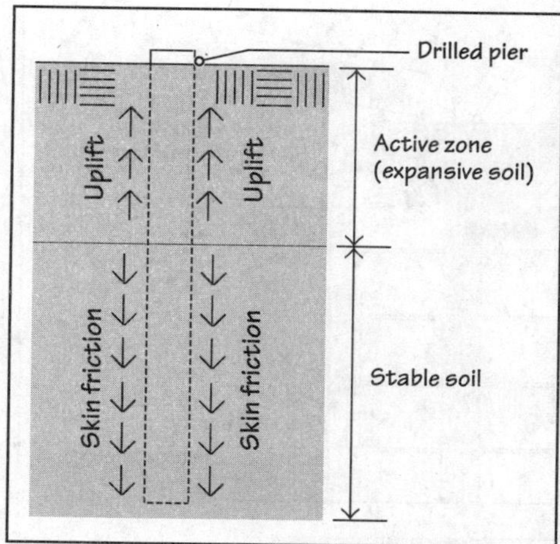

FIGURE 12.10 An expansive soil creates uplift on a drilled pier as it swells. This uplift force must be counteracted by skin friction and (or) end anchorage, particularly if the dead load on the pier is small.

FIGURE 12.11 A typical auger for drilling a hole in soil.

hole. The cut material is then spun off (the auger drill is rotated vigorously in the opposite direction), and the auger is lowered again into the hole for additional cutting.

Another common pier-drilling device is a bucket-type auger, Figure 12.13. It consists of a bucket with cutting edges. As the auger is rotated, it cuts the soil and deposits the cut material in the bucket. When the bucket is full, it is brought up and emptied, and the process is repeated for additional cutting.

After the hole has been drilled to the required depth, a steel reinforcement cage is lowered into the hole, Figure 12.14, and concrete is placed in the hole. Placement of concrete is generally done using a tremmie, which is a steel pipe topped by a funnel, Figure 12.15. The tremmie takes the concrete close to its intended location and prevents separation of concrete particles, which would occur if the concrete were dropped from a distance. The tremmie is gradually brought up as the hole is filled with concrete.

If the drilled hole is deep, and (or) if there is a possibility of the soil caving in, a temporary steel casing is inserted in the hole as the drilling advances, Figure 12.16. A steel casing is a hollow tube whose diameter is slightly smaller than that of the drilled hole. The casing is gradually extracted from the hole as the concreting of the pier advances.

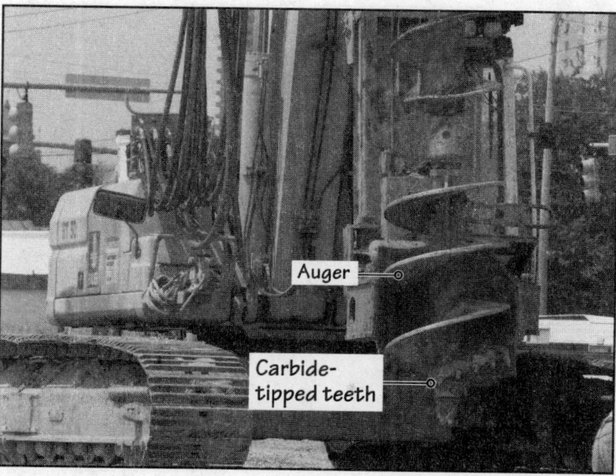

FIGURE 12.12 A typical rock auger with carbide-tipped teeth. (Photo courtesy of Jeffrey Machines, Inc., Birmingham, Alabama)

FIGURE 12.13 A typical bucket-type auger for drilling in soil.

FIGURE 12.14 A pier hole with reinforcement, ready to receive concrete. Steel reinforcing dowels or steel anchor bolts will be placed before concreting; see Figure 12.18. (Photo courtesy of Andrew Kocher)

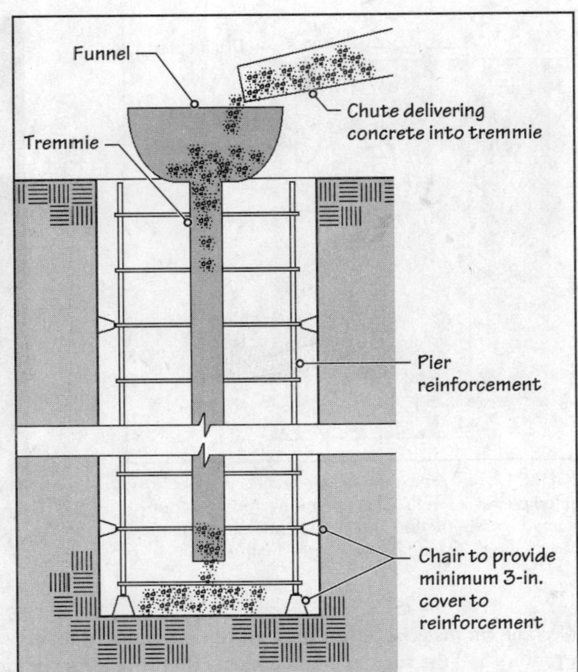

FIGURE 12.15 The use of a tremmie for placing concrete into a pier hole.

Steel casing is also used in water-bearing soils to prevent the entry of water into the hole. The bottom of the casing is sealed by spinning it into the rock (rock does not contain water) while applying a downward vertical force, Figure 12.17. After sealing of the hole, the water is removed, reinforcement is lowered, and the concreting operation is begun. The small amount of water left in the hole floats on top of the concrete because it is lighter than concrete. This water is removed after the entire hole has been concreted. The casing is now gradually removed from the hole while the concrete is still green. Some additional concreting is required during this operation to fill the void created by the removal of casing.

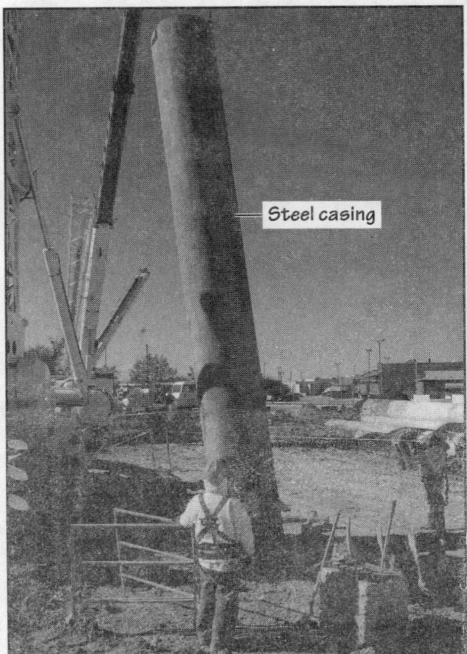

FIGURE 12.16 A steel casing being lowered into a pier hole.

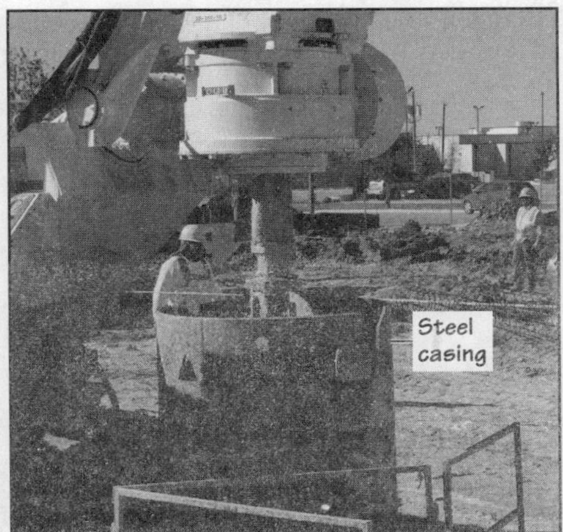

FIGURE 12.17 Steel casing being rotated to push it into hard soil (or rock) to seal its bottom against the entry of water.

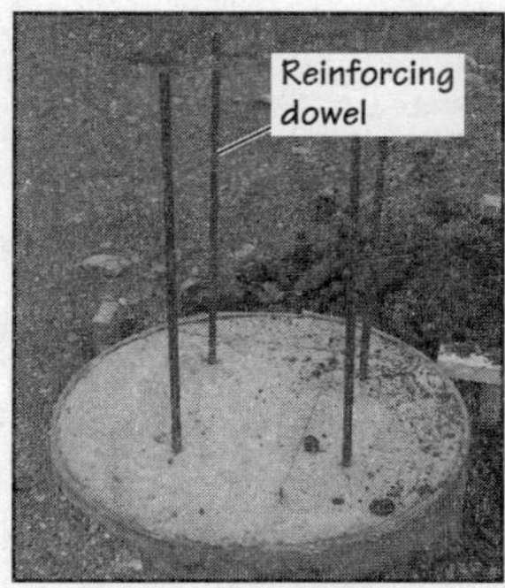

(a) Pier to support a concrete column.

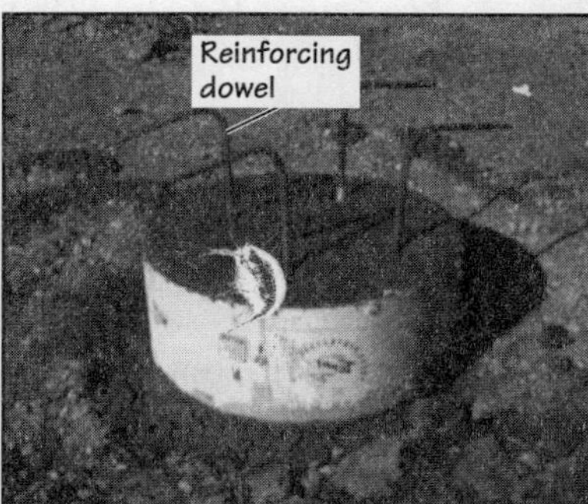

(b) Pier to support a slab-on-ground or grade beam.

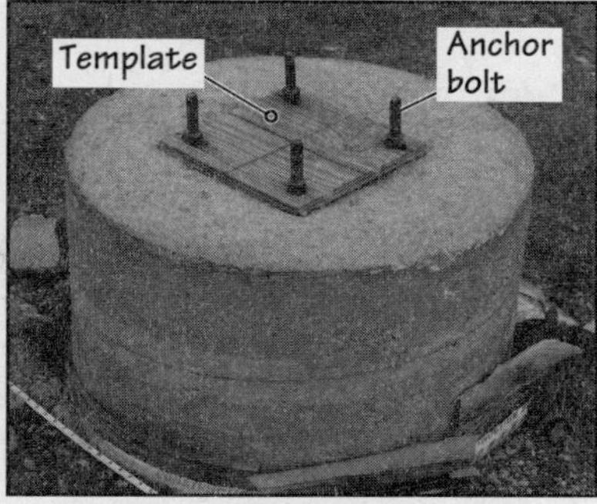

(c) Pier to support a steel column. The (removable plywood) template shown here is for accurate positioning of anchor bolts to match the holes in the column base plate.

FIGURE 12.18 Steel anchor bolts or reinforcing dowels are provided in the pier for anchorage to the structural member to be supported by the pier.

Drilled Pier Termination

Where a pier is used to support another concrete member (a grade beam or concrete column), steel reinforcing dowels that extend beyond the top of the pier are placed in the member before concreting is completed, Figures 12.18(a) and (b). For a pier that supports a steel column, the pier top is provided with steel anchor bolts, Figure 12.18(c).

Piles versus Drilled Piers

Piles and piers have their own unique applications. Piles are ideally suited for marine and coastal sites, where drilled piers are not suitable because of the presence of water, a high water table, or the presence of a sandy soil with cave-in potential. As the pile is driven, an assessment of the soil's bearing capacity is obtained from the blow count and the resulting pile penetration. Piles can be used immediately, with no wait time for the concrete to cure. However, pile driving is noisy and requires heavier equipment. It also disturbs the surrounding area with vibrations caused by the driving operations.

Piers are particularly suited for urban locations where vibrations and noise from pile driving are unacceptable. They can also support a much larger load than a pile and, unlike piles, do not require capping.

The technology required for piers is less complex; hence, drilled piers are generally more economical than driven piles. For this reason, piers are popular for foundation systems for low-rise buildings in expansive (clay) soils.

12.5 FOUNDATION SETTLEMENT

All foundation systems require a determination of the bearing pressure created by the loads on the underlying stratum (soil or rock). This bearing pressure must be less than the stratum's allowable bearing capacity (see Section 11.4). In addition to having an adequate bearing capacity, foundations must resist settlement. This is generally of greater concern for shallow foundations because they bear on a more compressible stratum than deep foundations.

Immediate and Consolidation Settlement

Settlement occurs as the soil compresses under the loads. The greater the soil's compressibility, the greater the settlement. In coarse-grained soils, settlement is immediate; that is, it occurs as the load is applied. Therefore, most of the settlement in coarse-grained soils occurs during the construction of the building. In fine-grained (particularly clayey) soils, part of the settlement is immediate, and the remainder (called *consolidation settlement*) occurs over a period of several months or years. Consolidation settlement occurs when the water held between clay particles is squeezed out by the load, and the soil consolidates and shifts.

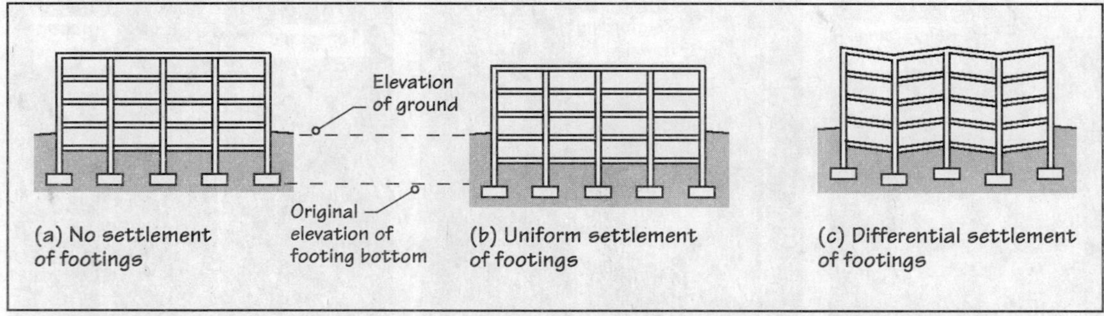

FIGURE 12.19 Uniform and differential settlement of footings.

UNIFORM AND DIFFERENTIAL SETTLEMENT

Settlement may be uniform over the entire structure or nonuniform (referred to as *differential settlement*). In uniform settlement, the entire structure settles by the same amount. In the case of excessive settlement, the ground floor of a building may sink to below the finished ground level, Figure 12.19(b).

One of the most frequently cited examples of the sinking of building foundations comes from Mexico City, which is sited on a soft soil. The increase in the city's population has led to the excessive use of underground water, resulting in the gradual lowering of the ground surface. The ground floor of the famous Palacio de Bellas Artes (Palace of Fine Arts) in Mexico City (designed by architect Federico Mariscal) has sunk by about 14 ft since it was completed in 1934. Fortunately, the settlement has been uniform, so the resulting problems were easily solved.

Differential settlement is more serious. It leads to deformation of the structural frame, imposing stresses on the structure for which it has not been designed and possible structural failure, Figure 12.19(c). Building codes specify the maximum permissible amount of differential settlement of footings.

Properly designed and constructed foundation systems with a monolithic base (mat or raft foundations) have negligible differential settlement due to their rigidity. Differential settlement is an important design consideration in foundation systems consisting of multiple footings because of the local variation in soil's compressibility, uneven compaction, and uneven bearing pressure induced on the soil by the superimposed loads. Therefore, footing dimensions are determined not merely on the bearing capacity of the soil but also on footing settlement—to ensure that all footings in the building have approximately the same amount of settlement.

PRACTICE **QUIZ**

Each question has only one correct answer. Select the choice that best answers the question.

1. A raft foundation is a shallow foundation system.
   a. True **b.** False
2. Which of the following foundation systems does not extend, generally uninterrupted, for the entire footprint of the building?
   a. Slab-on-grade foundation **b.** Mat foundation
   c. Raft foundation **d.** Pier or pile foundation
3. A strip footing is generally used under a
   a. column. **b.** wall.
   c. drilled pier. **d.** pile.
   e. all of the above.
4. A strap footing is generally used where
   a. the soil is expansive.
   b. the soil has low bearing capacity.
   c. the columns in the buildings are closely spaced.
   d. the columns in the building abut the property line.
   e. none of the above.
5. A deep foundation system consists of a
   a. slab-on-ground foundation. **b.** mat foundation.
   c. raft foundation. **d.** pier or pile foundation.
   e. all of the above.
6. Which of the following foundation items must be integrated under a concrete cap?
   a. Isolated column footings **b.** Wall footings
   c. Driven piles **d.** Drilled piers
   e. All of the above
7. Drilled piers used as foundation elements consist of
   a. cast-in-place reinforced concrete.
   b. cast-in-place unreinforced concrete.
   c. precast concrete.
   d. structural steel.
   e. wood.
8. Which of the following foundation system has large voids?
   a. Concrete slab-on-ground **b.** Mat foundation
   c. Raft foundation **d.** Strip footing
   e. None of the above

PRACTICE **QUIZ** *(Continued)*

9. Pile foundations are commonly used in coastal areas.
   a. True
   b. False
10. Drilled piers transfer building loads mainly by end bearing, while piles do so by both end bearing and skin friction.
    a. True
    b. False
11. In expansive soils, drilled piers are subjected to
    a. a downward force in the expansive region of the soil, which generally occurs well below the ground surface.
    b. an uplift force in the expansive region of the soil, which generally occurs well below the ground surface.
    c. an uplift force in the expansive region of the soil, which generally occurs in the upper soil layer.
    d. any one of the above, depending on the local geology.
    e. none of the above.
12. The augers used for boring pier holes generally use
    a. a hammering operation.
    b. a vibratory operation followed by a hammering operation.
    c. a rotary operation with a helix-type blade cutter.
    d. a rotary operation followed by a hammering operation.
    e. none of the above.
13. A tremmie is used during the drilling of a pier to
    a. prevent the caving in of the soil.
    b. bring the cut soil out of the hole.
    c. bring up the groundwater that collects at the bottom of the pier.
    d. seal the pier hole against percolation of water.
    e. none of the above.
14. Where a tremmie is used in a pier, casing is unnecessary.
    a. True
    b. False
15. In a deep foundation system, foundation settlement is generally larger than in a shallow foundation system.
    a. True
    b. False
16. Settlement of foundations causes stresses in the structural frame of the building when the settlement is
    a. caused by live loads.
    b. caused by dead loads.
    c. caused by lateral loads.
    d. large and uniform over the entire foundation area.
    e. nonuniform over the foundation area.

12.6 FOUNDATION DRAINAGE

Most building assemblies are adversely impacted by water. Water accelerates wood decay and promotes the growth of mold and fungi. Specifically with respect to foundations, water in the soil increases lateral pressure on basement and crawl space walls, corrodes reinforcement in concrete and masonry, and causes heaving of soil during freezing weather. Additionally, excess water can cause flooding of basements and crawl spaces and provide a breeding ground for mosquitoes. Clayey soils are particularly vulnerable to excess water because of their swell-shrink characteristic.

Two types of water can affect a buildings: (a) *surface water* from rain and snow melt and (b) *subsurface water* (also called *groundwater*). Surface water control is obtained by sloping the finished ground away from the building. It is important that sufficient slope is provided to divert water quickly away from the building so as to reduce its potential to seep into the ground and add to the groundwater. A 4% slope ($\frac{1}{2}$ in. per foot), provided over the first 10 ft from the building, is generally considered adequate.

CONTROL OF GROUNDWATER—FRENCH DRAIN

While the control of surface water is important in all buildings, the control of groundwater is generally required only in buildings with a basement or crawl space—to reduce water pressure on basement and crawl space walls. This is achieved by installing a drainage system all around the foundation. The drainage system consists of a perforated drain pipe placed in a narrow trench and surrounded with a drainage medium consisting of clean (preferably washed), crushed stone or gravel. A filter fabric, made of a synthetic material, encapsulates the drainage medium. The filter fabric captures fine soil particles and prevents them from clogging the voids in the drainage medium and pipe perforations. A number of terms, such as *french drain, trench drain,* and *footing drain,* are used for this drainage system.

The pipe, generally 4 to 6 in. in diameter, made of polyvinyl chloride (PVC) or corrugated polyethylene, is perforated on one edge. The perforated edge is placed face down to attract the lowest level of water into the pipe. The pipe must terminate in a storm water drain or in daylight. Alternatively, the pipe may terminate in a sump pit, from where a self-actuated pump lifts the water into a storm drain.

It is important that the drainage system be installed a little above the bottom of the foundation. If installed below the foundation, it may tend to wash away the soil below the foundation. The details of foundation systems given in Sections 12.7 and 12.8 illustrate the concept of foundation drainage.

12.7 DETAILS OF FOUNDATION SYSTEMS ON STABLE SOILS

Some of the commonly used foundation systems for low-rise buildings on stable soils are

- Reinforced concrete slab-on-ground foundation
- Exterior foundation walls, interior piers (short columns), and an elevated ground floor
- Wall and column footings and ground floor comprising a reinforced concrete slab

Reinforced Concrete Slab-on-Ground Foundation

As conceptually diagramed in Figure 12.4(a), a concrete slab-on-ground foundation functions as the ground floor of the building and also as the foundation system for the structural elements of the building. It is easily constructed because it requires little excavation. Excavated trenches function as formwork for concrete work, making it extremely cost effective for low-rise, lightly loaded buildings. Its limitation is that it can only be used in warm climates where the frost depth is small (see also the discussion of frost-protected shallow foundations in Section 12.9).

The thickness of the slab is primarily a function of the loads on the slab and the type of soil. A fairly stable soil with low swell-shrink potential (such as a sandy soil) provides a stratum of uniform strength. Over such a stratum, a 5-in.-thick slab with nominal reinforcement, with the perimeter and intermediate beams under load-bearing walls, is adequate.

If the soil has some swell-shrink material that is susceptible to changes in moisture content, the slab needs to be designed as a raft so that it has the required stiffness to float on the soil (move up and down as one thick plate with minimal bending and minor cracking). This requires a thicker slab and a greater amount of reinforcement. In some cases, the slab thickness required may be so great that it is more economical to provide a ribbed slab (also called a *waffle slab*) in place of a slab of constant thickness.

In other words, a ribbed slab can be structurally as strong and stiff as a slab of constant thickness but uses a smaller amount of concrete and steel. A ribbed slab derives its strength and stiffness from its geometry, which comprises a slab intercepted by intermediate beams in both directions, spaced 8 to 10 ft apart, in addition to perimeter beams.

Figure 12.20 shows the excavation for a ribbed concrete slab-on-ground. This figure further highlights the ease with which a ribbed slab-on-ground foundation can be constructed. The construction of a uniformly thick slab-on-ground foundation is even simpler. Figures 12.21 and 12.22 further explain the construction of a ribbed reinforced concrete (RC) slab-on-ground foundation. (The construction details relating to an RC slab-on-ground are presented in Chapter 22.)

FIGURE 12.20 Excavation for a typical reinforced concrete slab-on-ground foundation showing the network of trenches in both directions. The trenches serve as formwork for concrete beams.

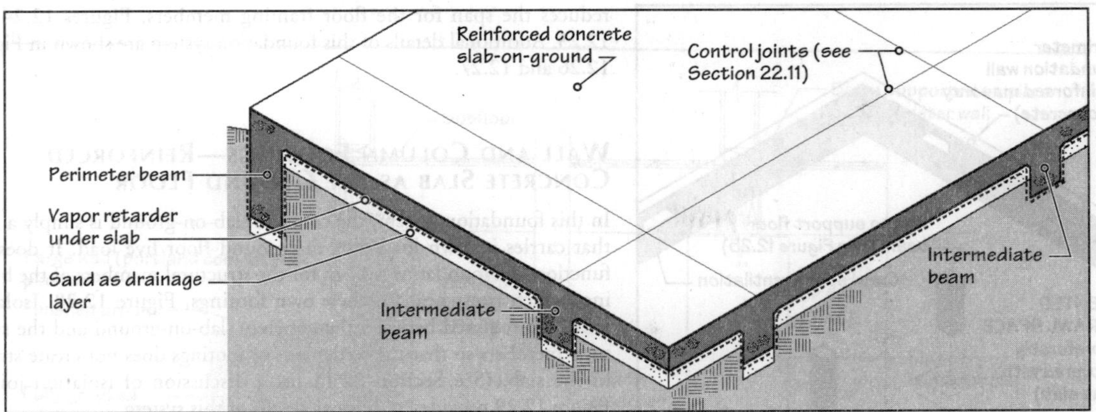

FIGURE 12.21 A typical reinforced concrete slab-on-ground foundation used as a foundation system for low-rise light-frame buildings. The slab is generally a ribbed slab, that is, a slab stiffened with perimeter and intermediate beams in both directions.

Exterior Foundation Walls, Interior Piers (Short Columns), and an Elevated Ground Floor

This foundation system, shown conceptually in Figure 12.2(a), is commonly used for light-frame buildings in cold climates, where, because of the deeper frost line, a slab-on-ground foundation cannot be used. It consists of strip footings under load-bearing exterior walls, Figure 12.23. As previously stated, the bottom of the footings must be below the frost line.

The ground floor framing members span from exterior wall to exterior wall, creating a crawl space below. If the distance between the opposite exterior walls is large, interior piers (short columns) are provided to support an intermediate beam. The beam

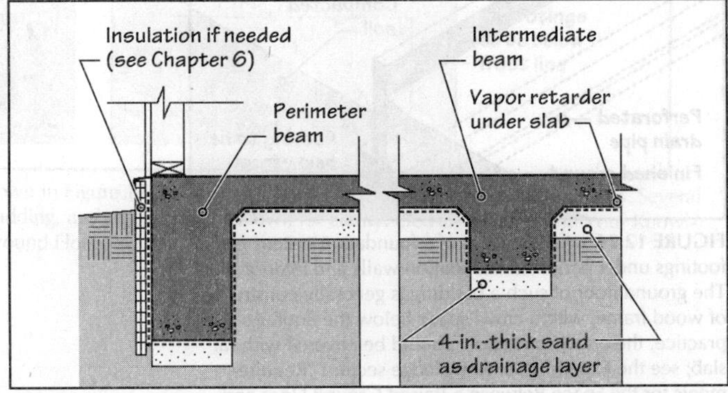

FIGURE 12.22 Part section through the slab shown in Figure 12.21. This detail focuses only on foundation items. Other items, such as exterior wall insulation, wall cladding, interior drywall, and so on, are not shown for clarity.

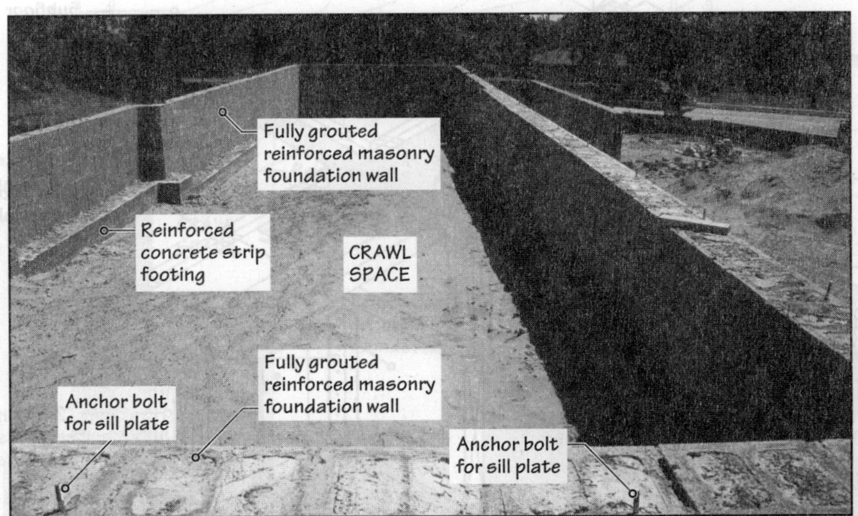

FIGURE 12.23 The foundation system of this single-family house consists of perimeter foundation walls. In this building, the distance between the opposite foundation walls is small. This allows the ground floor framing members to span from exterior wall to exterior wall, precluding the need for intermediate support(s) of floor framing members (see Figure 12.24). The wall footings are partly exposed because adequate soil-bearing capacity was available at shallow depth below the ground.

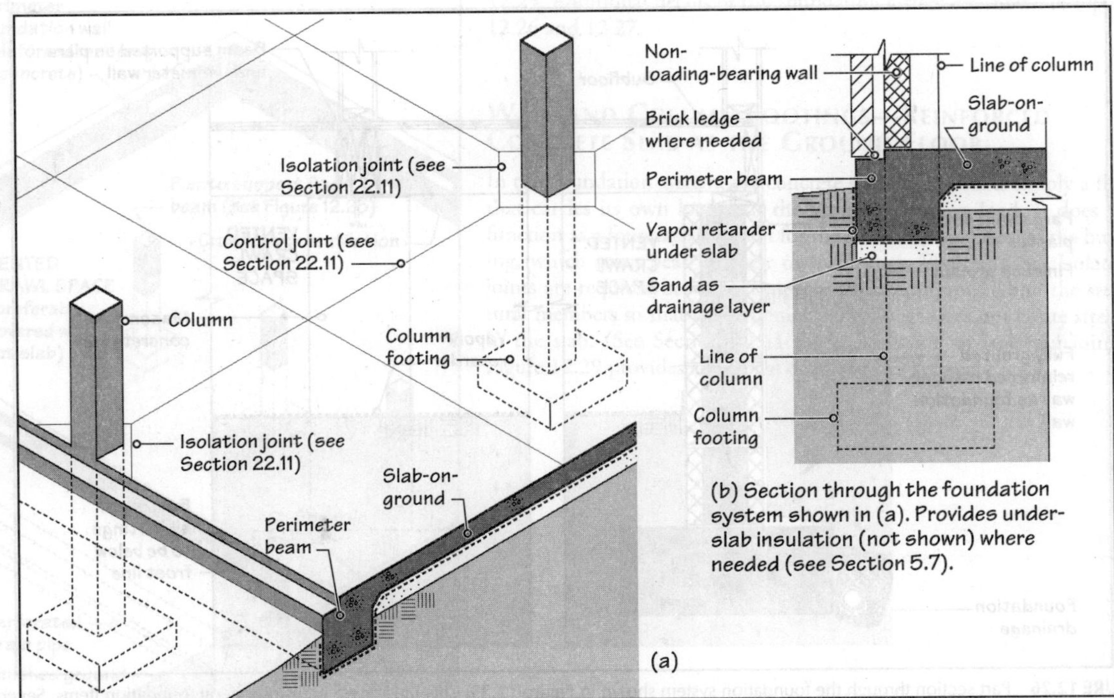

FIGURE 12.28 Foundation system for a building with isolated columns and non-load-bearing perimeter walls. Non-load-bearing walls are supported on a reinforced concrete slab-on-ground. Note that isolation joints are required between the columns and the slab-on-ground.

EXPAND YOUR KNOWLEDGE

Requirements for the Space Between a Raised Ground Floor and the Underlying Soil

As described in this chapter, the ground floor structure in many buildings is not supported on the ground but is raised above it, creating a space between the ground floor and the underlying soil. This space may be made accessible or left inaccessible, depending on the floor's structural frame or the requirements of any mechanical, electrical, and plumbing (MEP) systems installed under the floor.

Accessibility of Below-Floor Space

A raised ground floor structure made of wood or steel framing members is liable to deteriorate due to the buildup of moisture in the space below the floor. (Wood is impacted by fungal decay and steel by corrosion. Termite infestation is an additional concern for wood frame members.) With such a ground floor structure, an accessible crawl space is essential because it allows inspection and repair of the floor structure.

If the floor structure consists of cast-in-place reinforced concrete, the above concerns do not exist. Therefore, the space under a raised reinforced concrete floor need not be accessible.

An accessible below-floor space is referred to as *crawl space*. Building codes prescribe minimum dimensions of access opening for an accessible crawl space.

Control of Moisture Buildup in Below-Floor Space

Apart from the effect of below-floor moisture on the structural members, its entry (in the form of water vapor) into the space above the ground floor is a concern. Therefore, adequate control of below-floor moisture is required, regardless of the type of structural frame. Building codes require that the build up of moisture in this space be controlled. Two alternatives are provided: (a) ventilate below-floor space or (b) provide a vapor retarder.

Minimum ventilation requirements, and the protection of ventilators against the entry of insects into below-floor space, are prescribed by the building codes. If a vapor retarder is used for moisture control, it must cover the entire surface below the floor. The joints between the vapor retarder membranes must be lapped and sealed, and the membrane must be turned up and anchored to the perimeter foundation walls, columns, and other obstructions.

Generally, if the space below the ground floor is accessible (i.e., a crawl space), both alternatives (ventilation and a vapor retarder) are recommended for redundancy and improved performance.

Rat Slab and Crawl Space

To make the crawl space easier to crawl on, a good practice is to provide a 2-in.-thick, lightly unreinforced concrete slab over a vapor retarder. The slab also helps to deter pests, such as mice and roaches, and hence is commonly known as a *rat slab*.

12.8 DETAILS OF FOUNDATION SYSTEMS ON EXPANSIVE SOILS

Commonly used foundation systems for low-rise buildings on expansive soils are listed below. As stated in the introduction to this chapter, most of these systems may also be used where soils have a low bearing capacity or are otherwise unsuitable (e.g., have excessive organic content).

- Posttensioned concrete slab-on-ground foundation
- Grade beams and drilled piers with (elevated) ground floor over a crawl space
- Grade beams and drilled piers with a suspended concrete ground floor slab
- Drilled piers with a suspended concrete ground floor slab

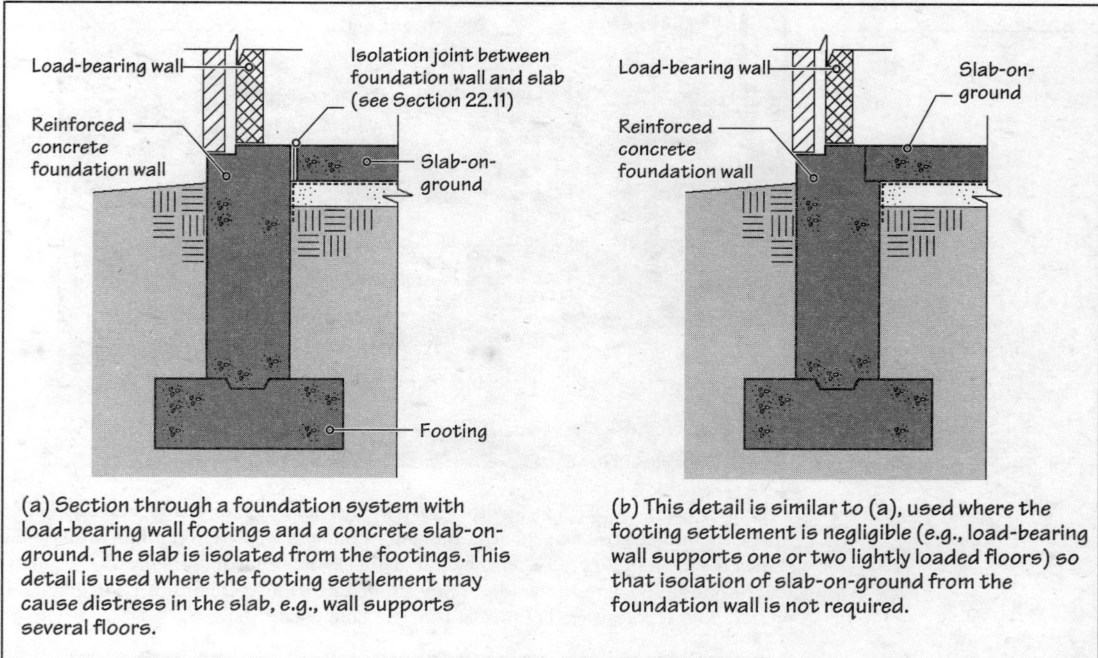
Load-bearing wall
Isolation joint between foundation wall and slab (see Section 22.11)
Reinforced concrete foundation wall
Slab-on-ground
Footing
(a) Section through a foundation system with load-bearing wall footings and a concrete slab-on-ground. The slab is isolated from the footings. This detail is used where the footing settlement may cause distress in the slab, e.g., wall supports several floors.
Load-bearing wall
Slab-on-ground
Reinforced concrete foundation wall
(b) This detail is similar to (a), used where the footing settlement is negligible (e.g., load-bearing wall supports one or two lightly loaded floors) so that isolation of slab-on-ground from the foundation wall is not required.

FIGURE 12.29 Foundation systems for buildings with load-bearing wall footings and concrete slab-on-ground floors.

Posttensioned (PT) Concrete Slab-on-Ground Foundation

If the soil is highly expansive, the slab-on-ground must be made so stiff that it floats on the soil as a raft—the entire foundation responding to the soil's swelling and shrinkage as one monolithic unit. Although a reinforced concrete slab-on-ground can be designed to accomplish this objective, it becomes quite uneconomical (even a ribbed reinforced concrete slab is uneconomical) compared with a post tensioned (PT) concrete slab-on-ground.

A PT concrete slab-on-ground foundation is similar to a ribbed reinforced concrete slab-on-ground foundation except that high-strength steel reinforcements (called *tendons*) are used in place of normal steel. A PT slab also consists of perimeter beams and intermediate beams in both directions, and a finished PT concrete slab-on-ground looks identical to that of a reinforced concrete slab-on-ground foundation (shown in Figures 12.20 to 12.22).

Posttensioning introduces compression in the slab, which not only stiffens the slab but also prevents the development of cracks that are inherent in a reinforced concrete slab. Thus, a PT concrete slab-on-ground does not require control joints. (The construction of PT concrete slabs-on-ground is discussed in Chapter 22.)

Grade Beams and Drilled Piers with (Elevated) Ground Floor Over a Crawl Space

A foundation system that is considered more reliable on expansive soils than a PT concrete slab-on-ground is one that uses drilled piers to transfer building loads to a stable stratum while passing through the expansive zone (see Figure 12.10). In this foundation system, the piers terminate a small distance above the ground level where they support a network of grade beams.

The grade beams support the ground floor structure of the building. Any superstructure columns and load-bearing walls are also supported on the piers, Figure 12.30. Because of the elevated ground floor, this foundation system results in an (accessible) crawl space.

Grade beams may be of concrete or steel, and the ground floor may be framed with wood, steel, or concrete, depending on the building's occupancy, durability, and fire-resistance requirements. Perimeter grade beams are generally of concrete and also serve as retaining walls for backfill. Figure 12.30 shows a foundation system where the interior grade beams are of structural steel supported by steel columns bearing on drilled piers. The ground floor framing is of steel as well.

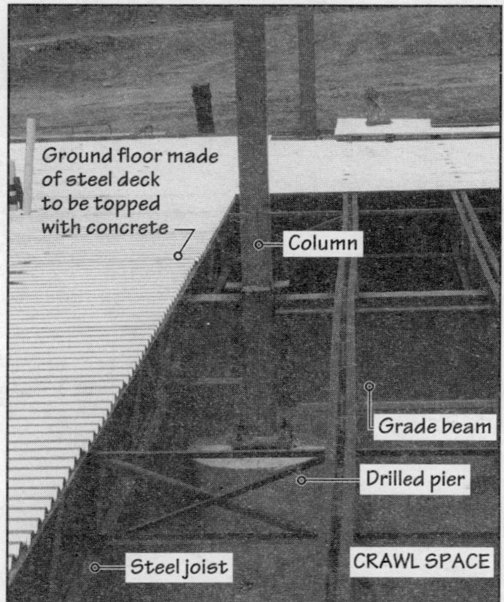
Ground floor made of steel deck to be topped with concrete
Column
Grade beam
Drilled pier
Steel joist
CRAWL SPACE

FIGURE 12.30 Foundation system of a building on expansive soil in which the columns and grade beams are supported on drilled piers. In this building, the perimeter grade beams (not shown) are made of reinforced concrete. Interior grade beams are made of structural steel. The ground floor of the building is framed with steel joists and steel deck.

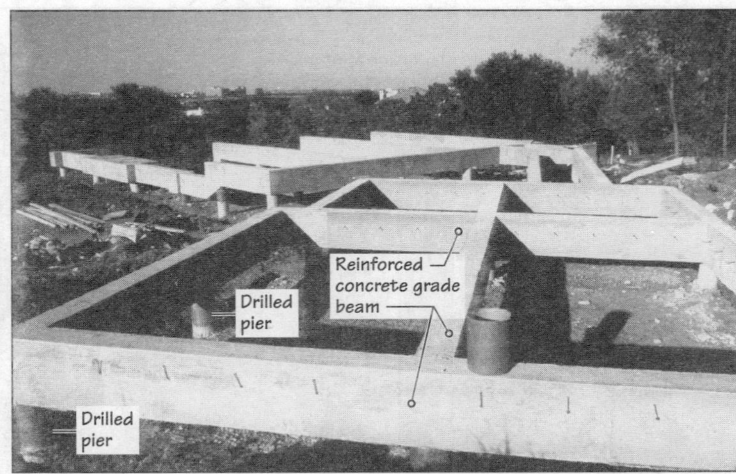

FIGURE 12.31 Foundation system comprising reinforced concrete grade beams supported by drilled piers. Grade beams support the ground floor framing, as shown in Figure 12.32. A crawl space is created between the ground floor framing and the underlying soil.

Figures 12.31 and 12.32 show the foundation system of a residential building with cast-in-place reinforced concrete grade beams and wood floor. A section through this foundation system is shown in Figure 12.33. The foundation system shown in Figure 12.34 is that of a school building where the grade beams are of cast-in-place reinforced concrete and the ground floor is constructed of precast concrete hollow-core slabs.

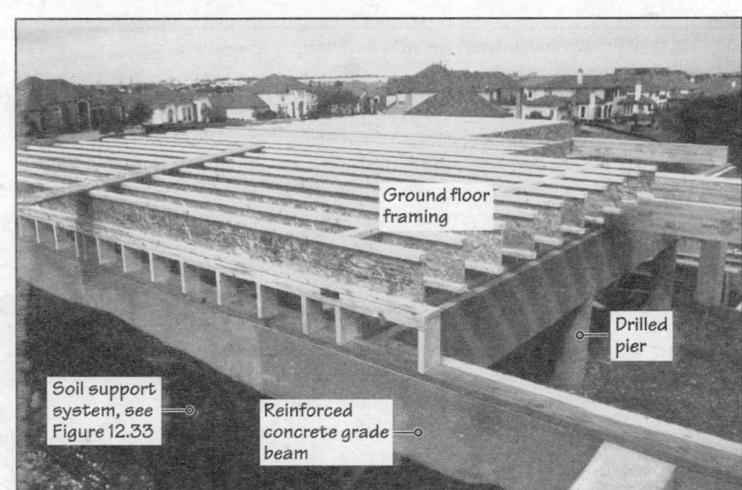

FIGURE 12.32 This figure shows the progress in the construction of ground floor framing (consisting of wood light-frame members) over the foundation system shown in Figure 12.31.

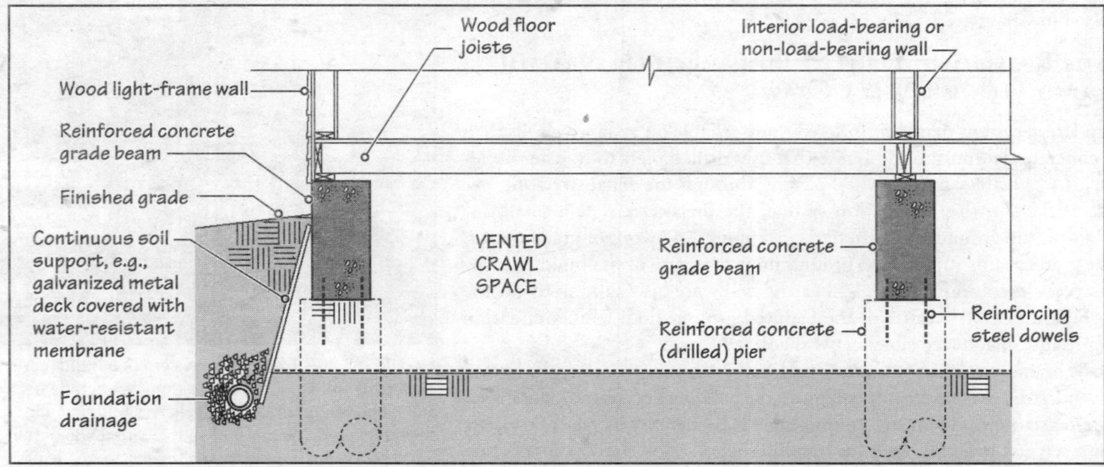

FIGURE 12.33 Part section through the foundation system shown in Figure 12.32. Note that the focus of this detail is only on foundation items. For clarity, several other items are not shown.

FIGURE 12.34 Foundation system comprising drilled piers and reinforced concrete grade beams. The ground floor consists of precast concrete hollow-core slabs. A layer of concrete will be placed over the entire floor system (grade beams and hollow-core slabs). Steel stirrups projecting from grade beams are intended to integrate all floor elements into one monolithic concrete floor.

Grade Beams and Drilled Piers with a Suspended Concrete Ground Floor Slab

The provision of a crawl space has the advantage of providing easier access for repair and maintenance of below-floor utility lines and HVAC ducts (where provided). However, it raises the ground floor elevation, requiring additional earthwork around the building. Therefore, where a crawl space is unnecessary, the foundation system shown in Figure 12.33 is modified to the one shown in Figure 12.35.

In this system, reinforced concrete grades beams, supported on drilled piers, are partially buried in the ground. The formwork for grade beams between the piers is provided by cardboard boxes (referred to as *void boxes*). The void boxes bear on the soil and are strong enough to withstand the weight of concrete and other construction loads. However, they decompose in time, leaving a clear space between the underside of the grade beams and the soil. This allows the soil to swell and shrink without imposing any pressure on the beams.

The grade beams support a cast-in-place reinforced concrete slab. The formwork for the slab is also provided by void boxes. Thus, after the void boxes have decomposed, a space is created between the slab and the soil, and the slab functions as a suspended slab. Note that the space between the soil and the slab is inaccessible.

The construction of this foundation system is shown in Figures 12.36(a) and (b). After the piers have been constructed, the grade beams are cast, but only partially. The

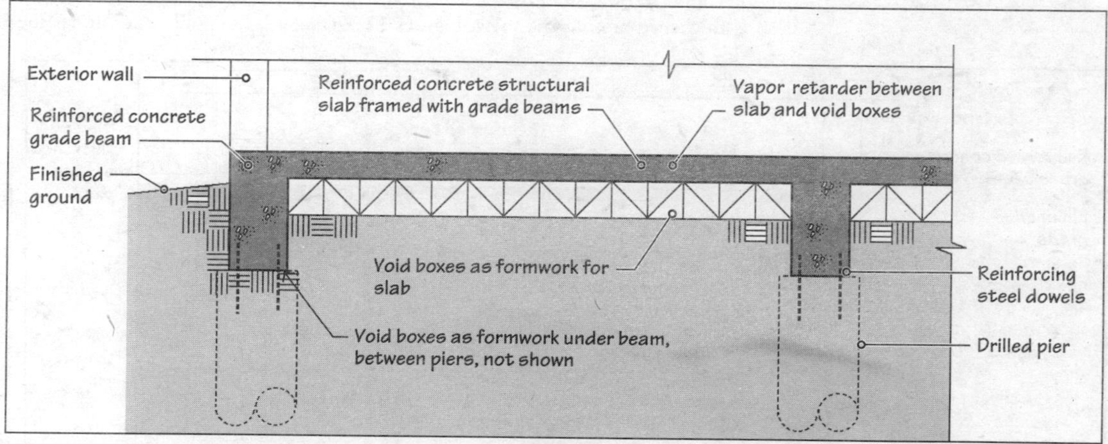

FIGURE 12.35 Foundation system comprising grade beams and drilled reinforced concrete piers without a crawl space. The ground floor slab is not ground-supported but is supported by grade beams as a suspended slab. Void boxes, made of cardboard, are used as formwork under the slab and grade beams. Void boxes decompose shortly after construction, leaving an (inaccessible) space under the slab and grade beams. This space allows the soil to swell and shrink without affecting the grade beams and slab.

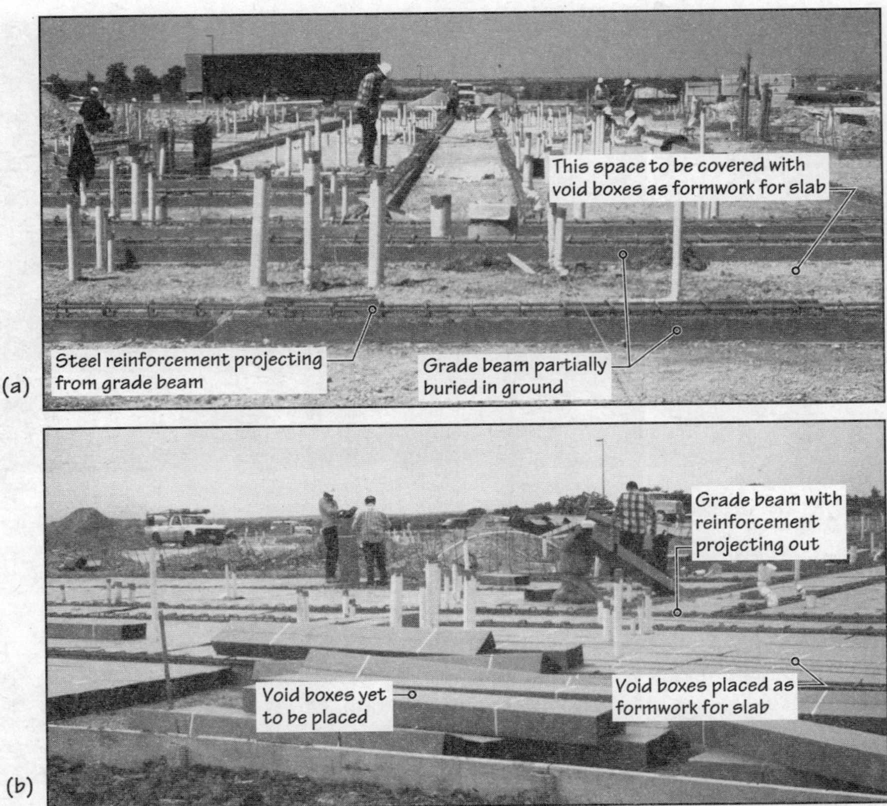

FIGURE 12.36 Construction of the foundation system shown in Figure 12.35. In **(a)**, the grade beams have been partially completed. Figure **(b)** shows preparation for the construction of the slab-on-ground between the grade beams.

upper portion of the grade beams is not cast at this time, but is cast along with the floor slab.

After the partial concreting of grade beams, void boxes are laid as formwork for the slab, followed by reinforcement and concrete placement. The floor slab and (partial) grade beams are structurally integrated through stirrups that can be seen projecting out in Figure 12.36(a). This foundation system is structurally identical to an elevated reinforced concrete floor structure used in a multifloor reinforced concrete building (see Chapter 23).

Drilled Piers with a Suspended Concrete Ground Floor Slab

Another foundation system uses a structural slab supported directly on drilled piers. The slab is thickened around the piers, Figures 12.37 and 12.38. Void boxes under the slab

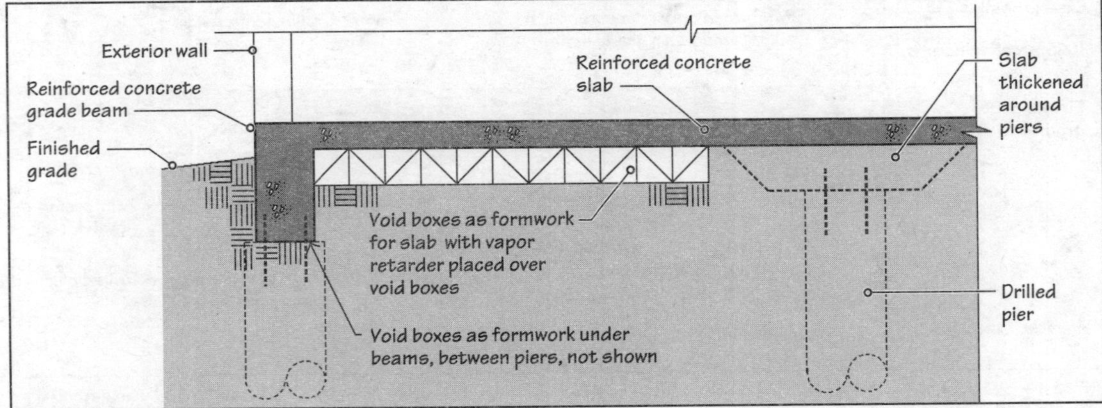

FIGURE 12.37 Foundation system in which the concrete slab-on-ground is supported directly on pier caps. Grade beams are generally provided at the perimeter of the building to support exterior walls. Structurally, this ground floor slab is similar to the elevated flat slab described in Chapter 23.

function as the formwork. Grade beams are provided at the perimeter only to support exterior wall loads. Structurally, this foundation system is identical to an elevated flat slab floor (see Chapter 23).

12.9 FROST-PROTECTED SHALLOW FOUNDATIONS

Until recently, a concrete slab-on-ground could not be used in cold climates because it was not deep enough to extend below the frost line. Thus, the foundations, even for light-frame buildings, had to be deeper than those required by the bearing capacity of the soil.

The introduction of the *frost-protected shallow foundation* (FPSF) system has made it possible to use a concrete slab-on-ground foundation in climates where the frost line is deep. An FPSF system incorporates insulation around the perimeter of the slab-on-ground foundation and uses the heat escaping from a heated building to keep the temperature of the slab above the ground's freezing temperature (32°F). In other words, an FPSF system effectively raises the depth of the frost line under a building's foundation, Figure 12.39.

A section through a typical frost-protected concrete slab-on-ground foundation is shown in Figure 12.40. Both vertical and horizontal insulation must be used, and it must extend around the entire footprint of the building, Figure 12.41.

The arrangement of insulation in an FPSF, shown in Figure 12.40, appears similar to insulating the perimeter of a concrete slab-on-ground for energy conservation (Section 5.11). However, there is a fundamental difference between the workings of the two insulation systems. The purpose of below-slab insulation for energy conservation is to reduce the loss of interior heat into the ground.

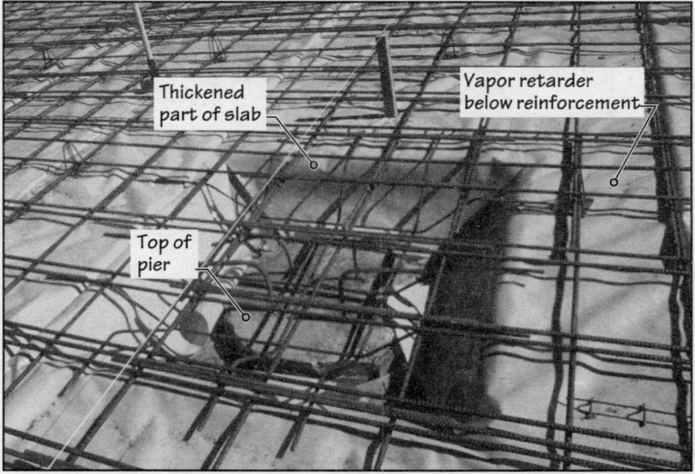

FIGURE 12.38 The foundation system of Figure 12.37. Note that the slab is thickened around the perimeter.

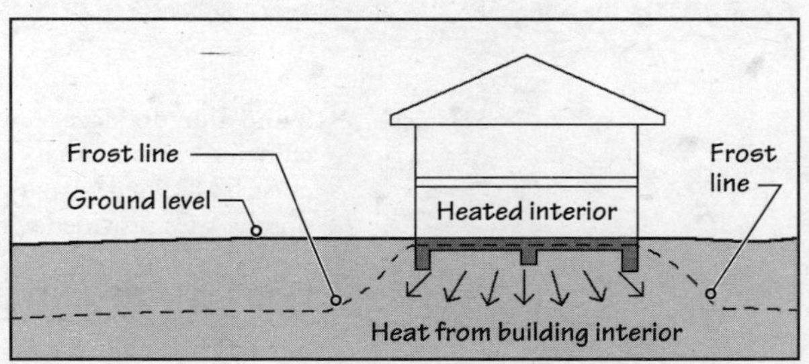

FIGURE 12.39 In a frost-protected shallow foundation (FPSF), the heat from the building's interior raises the frost line under the building foundation. The purpose of below-ground perimeter insulation is to ensure that the heat from the building's interior escapes vertically under the foundation, not sideways.

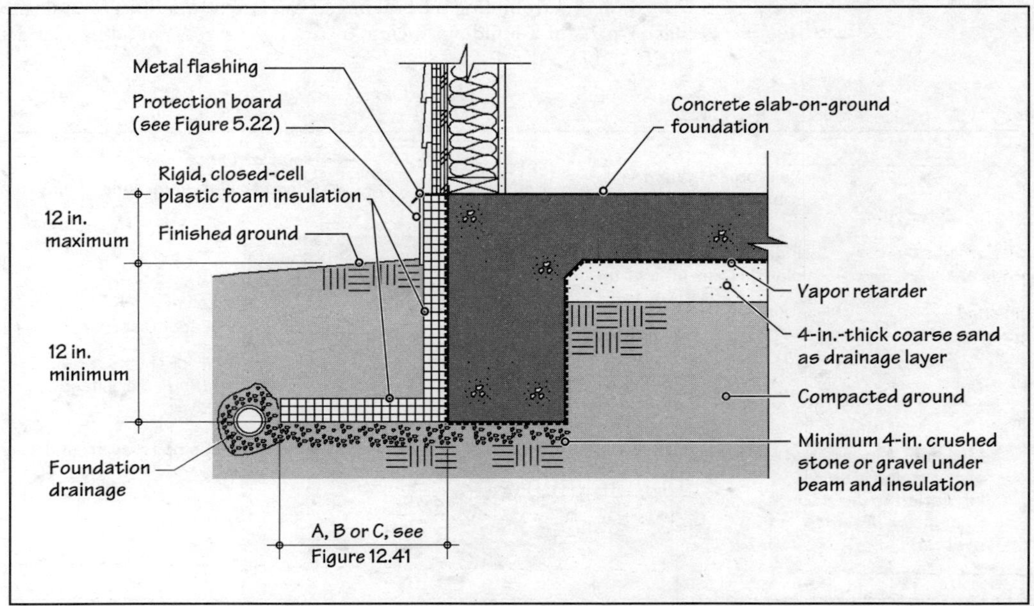

FIGURE 12.40 Section through a frost-protected concrete slab-on-ground.

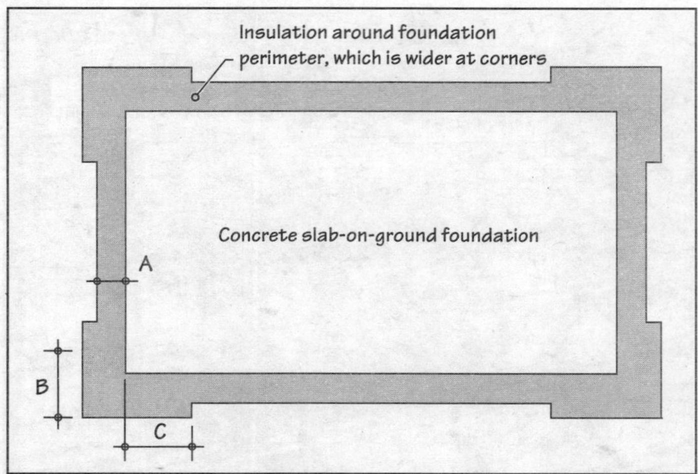

FIGURE 12.41 Plan of a frost-protected concrete slab-on-ground. For dimensions (A, B, and C) and R-values of insulation, see the International Residential Code. (For additional details, refer to the *Revised Builder's Guide to Frost Protected Shallow Foundations* by the National Association of Home Builders.)

The purpose of insulation in an FPSF is not reduce the loss of interior heat into the ground, but to direct it to escape vertically into the foundation, and to educe its sideway (horizontal) escape. This fact emphasizes that if an FPSF is used, interior heating is mandatory during freezing weather to prevent foundation failure by frost heave. However, experience has indicated that brief periods of power shutdown do not adversely impact an FPSF [12.1].

The height and width of insulation in an FPSF, and the associated R-values, are a function of the local climate. The International Residential Code provides detailed requirements for the design and construction of an FPSF. In a properly designed FPSF, the depth of the foundation below the finished ground need not exceed 16 in. even in the coldest parts of the United States.

Foundation Drainage: An Important Requirement for an FPSF

While perimeter insulation is key to the success of an FPSF, the other important requirement is foundation drainage. A well-drained foundation moves water away from the foundation, reducing the potential for frost heave. Additionally, water from roofs and other sources must be drained away from the building. That is why an FPSF is suited to moderately sloping sites provided with suitable surface water drainage.

Limitations and Additional Requirements for an FPSF

In considering the use of an FPSF, the following points should be kept in mind:

- An FPSF system does not apply to permanently frozen ground (permafrost).
- As the term *frost-protected shallow foundation* implies, it is applicable to all shallow foundation systems, including those with foundation walls and strip footings, Figure 12.42.
- The insulation used in an FPSF must be closed-cell and resistant to deterioration by water. Extruded polystyrene (see Chapter 5) best fits this requirement.
- Although an FPSF system depends on the heat from the building's interior to raise the frost line surrounding the building, it can also be modified for unheated buildings or parts of a building, such as a parking garage. (For additional details, see [12.1] and [12.2]).

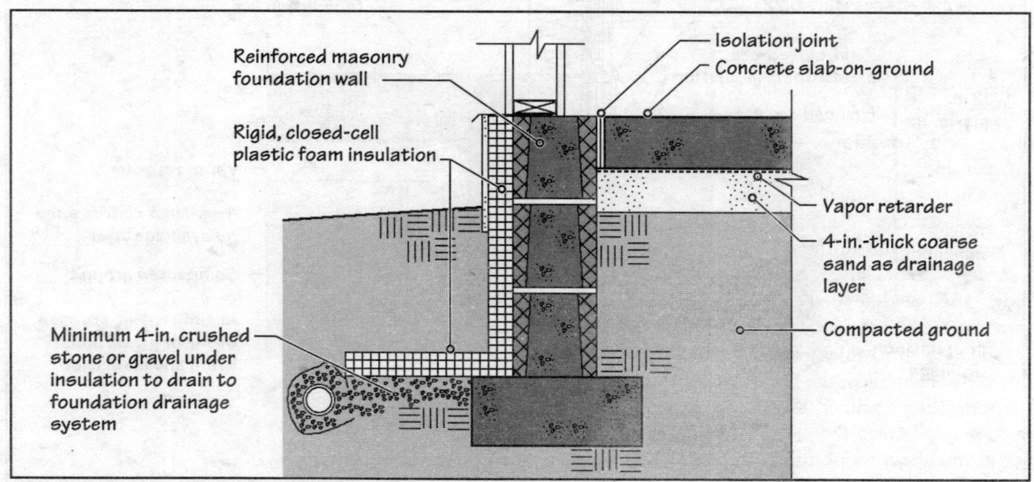

FIGURE 12.42 Section through a frost-protected foundation comprising concrete footing under a masonry foundation wall.

12.10 BASEMENT CONSTRUCTION AND WATERPROOFING

The term *basement* refers to an enclosed space that is below the ground floor of the building. A residential basement is generally one floor deep, with a floor height of 8 to 10 ft, with the top of the basement wall terminating approximately 2 to 3 ft above the finished ground level. On a sloping site, the basement floor may be above the ground level on one or two sides, referred to as a *walk-out* or *daylight basement.* Residential basements are more common in colder regions of the United States, where the frost line is deep so that a basement can be economically provided without much additional excavation. Commercial basements can be several floors deep, particularly in buildings located in downtown areas.

CONDITIONED AND UNCONDITIONED BASEMENTS

A basement can be conditioned (to maintain a constant indoor temperature) or unconditioned, where the indoor temperature fluctuates with the outdoor temperature. For a conditioned basement, only the basement walls (not the floor) need to be insulated as per the requirements of the local energy code. Generally, the insulation need not extend beyond 10 ft below the ground (see Section 5.7).

Basement walls may be insulated from the inside or the outside. In a multilevel commercial basement, the insulation is required only for the upper basement level. Therefore, in such basements, interior insulation is common. In residential basements, which are only one level deep, outside placement of insulation (over the entire depth of the basement) is common. Note that outside insulation is more effective (see Section 5.9).

The walls of an unconditioned basement should be insulated the same way as those of a conditioned basement. However, a lower R-value insulation may be justified. The ceiling of an unconditioned basement should be insulated as per the local energy code. Generally, this insulation is placed between the ground-floor framing members.

STRUCTURAL CONSIDERATIONS

The major structural components of a residential basement are the walls and the floor. Cast-in-place reinforced concrete is the most commonly used material for both walls and the floor of contemporary basements, although reinforced masonry and pressure-treated wood may be used for the walls. Basement walls must be designed to withstand lateral earth pressure and gravity loads from the upper floors. A section through a typical reinforced concrete residential basement is shown in Figure 12.43. Wall footings consist of strip footings and are cast first, followed by basement walls and the basement slab.

BASEMENT WATERPROOFING

Waterproofing is obviously the most important consideration for basements. It is applied to both the basement walls and the floor. For the walls, it is applied on the outside; for the floor, it is applied under it so that it forms a continuous, unbroken barrier to the entry of water. In earlier times, the use of layers of hot coal-tar pitch or hot asphalt alternating with fiberglass felt to yield a 3- to 5-ply membrane was the most common basement-waterproofing technique—a technique identical to that used for built-up roofing today (Chapter 33).

Although a hot-tar or asphalt system is easily applied on a horizontal surface, its application on a vertical surface is difficult. Safety concerns are critical in underground waterproofing because of the limited working space available. Therefore, a hot-applied waterproofing system has largely been replaced by cold-applied systems.

Commonly used cold-applied, below-grade waterproofing systems are similar to single-ply roof membranes and consist of rubberized asphalt or thermoplastic sheets. The sheets, usually 60 mil (1.5 mm) thick, are available in self-adhering rolls with a release paper. Before the sheets are applied, the basement wall and floor surfaces are cleaned and primed with the manufacturer's primer. Because the sheets are self-adhering, they form a continuous waterproof membrane.

In place of membrane waterproofing, liquid-applied elastomeric compounds may be used, which can be sprayed on, rolled on, or brushed on. They are particularly attractive for complex surface formations but are prone to application errors.

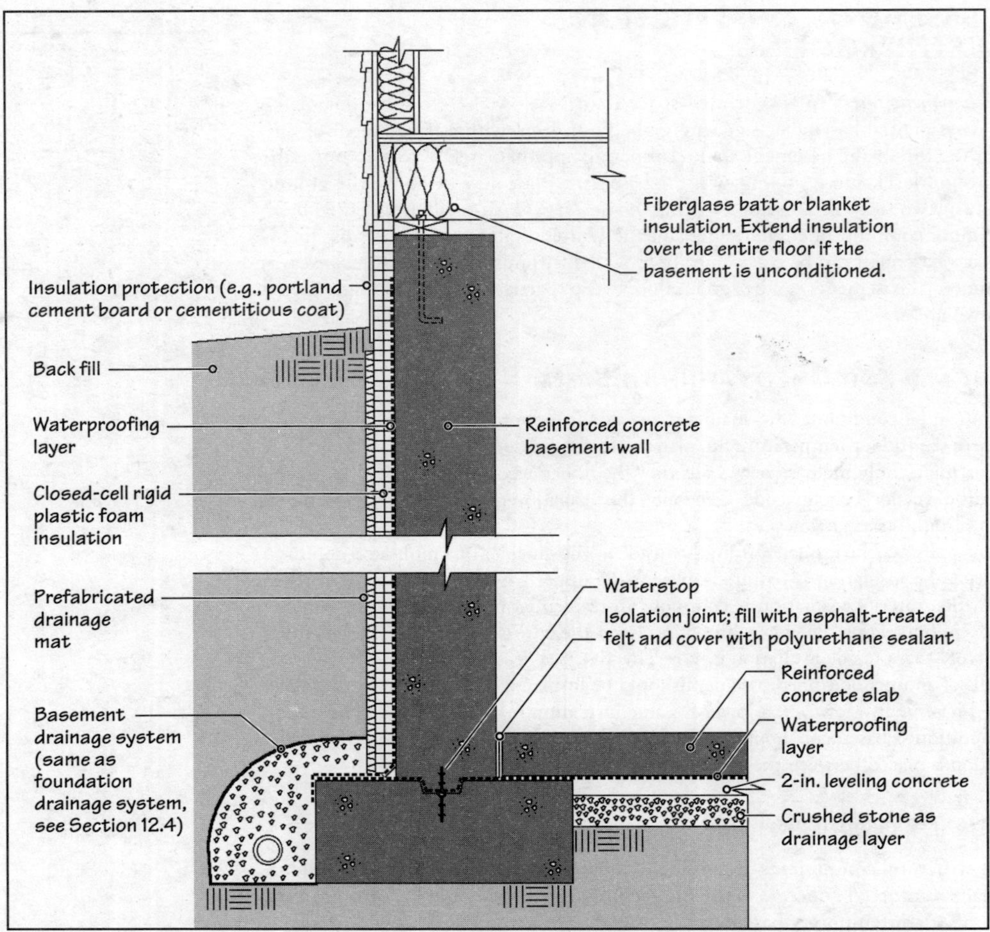

FIGURE 12.43 Section through a typical residential basement constructed with an open excavation system (see Figure 11.9). Although a reinforced concrete wall is shown here, it can be replaced by a reinforced masonry wall.

Basement Drainage

The basement drainage system is an important component of basement waterproofing. Its purpose is to collect, drain, and discharge groundwater away from the building. The drainage system consists of drainage mats and a drain pipe. The drainage mats have a thick open-weave structure that allows the subsoil water to drain downward by gravity. They eliminate or reduce water pressure acting on a below-grade wall and also protect the waterproofing layer and/or insulation from damage that might be caused by the backfill.

In a typical basement, waterproofing is applied directly to (the outside of) the basement wall, followed by insulation and, finally, the drainage mat, Figure 12.43. Some manufacturers provide the drainage mats laminated to extruded or expanded polystyrene board.

The drain pipe used under for basement drainage is identical to the one used for foundation drainage—a 4- to 6-in.-diameter perforated pipe set within a bed of crushed stone that allows the water to seep into the pipe (Section 12.6).

Positive-Side, Negative-Side, and Blind-Side Waterproofing

The waterproofing layer shown in Figure 12.43 is called *positive-side waterproofing* because it has been applied on the side of the wall with direct exposure to water. *Negative-side waterproofing* refers to waterproofing that is applied from the opposite side, that is, from the interior. Negative-side waterproofing is generally used in remedial applications.

Soil Treatment Under Foundations Against Termites

The soil under and around building foundations must be treated with a termicide to prevent damage by termites. Because termites live in underground colonies, the treated soil creates a toxic barrier where the termites perish in trying to reach the building.

The treatment consists of spraying a waterborne chemical into the soil before constructing the foundations. The chemicals are regulated by the U.S. Environmental Protection Agency (EPA) and should be applied by a specialized and bonded company. The efficacy of the treatment is of a limited duration depending on the chemical and its concentration; it may need to be reapplied a few times during the building's life. Therefore, the treatment done before the building's construction is called *pretreatment*.

Although pretreatment is critical for wood structures, it must be applied to all buildings, regardless of the materials used in their structural frame (steel, concrete, masonry, etc.). It may be omitted in climates where termite infestation is not a concern, such as extremely cold or extremely arid climates.

Pretreatment of soil is only one of several measures needed to protect the building against termites. Additional measures are needed particularly for wood frame buildings, as discussed in Section 13.14.

Positive-side waterproofing is more effective and is more commonly used. It also protects the wall against water seepage, mold growth, metal corrosion, and so on. It requires adequate space outside a below-grade wall for applicators to work, which is later backfilled. Positive-side waterproofing, therefore, remains accessible after its completion.

Blind-side waterproofing refers to waterproofing applied to the outside of the wall that becomes inaccessible after the wall's construction. It is used in situations where the basement wall is supported by an excavation-support system such as soldier piles and lagging, which are left in place. In this situation, the drainage mat and waterproofing layer are applied to the excavation-support system and the basement wall is constructed against it.

Blind-side waterproofing manufacturers provide the drainage mat and waterproofing layer integrated into one membrane. In this composite material, the waterproofing layer faces the interior; that is, fresh concrete for the basement wall is placed against the waterproofing layer. The drainage mat is in contact with the excavation-support system. Figures 12.44 and 12.45 illustrate blind-side waterproofing application.

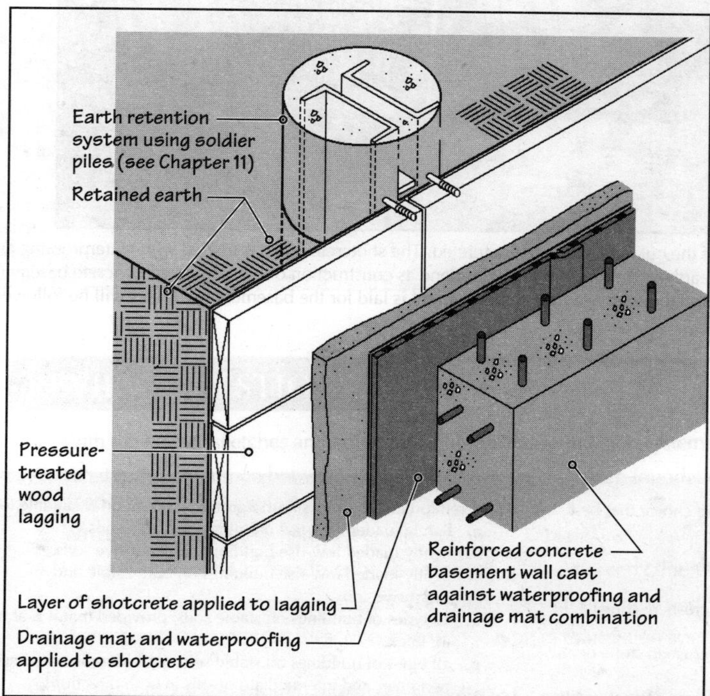

FIGURE 12.44 Waterproofing of a reinforced concrete basement wall cast against a soldier pile and lagging system of excavation support (blind-side waterproofing).

CHAPTER 13

Materials for Wood Construction–I (Lumber)

CHAPTER OUTLINE

There are many structural and nonstructural applications for wood products in building construction. Because *wood* is a general term, it is important to clarify the terminology for its various uses.

The term *lumber* applies to wood products derived directly from logs through sawing and planing operations only, with no further manufacturing except cutting to length or finger jointing two or more pieces (see Section 13.6 for a discussion of finger jointing). In other words, only solid pieces of wood are classified as lumber. Hence, the terms *lumber, solid lumber, solid sawn lumber,* and *sawn lumber* are used synonymously.

In recent years, manufactured wood products—products manufactured with altered or transformed wood fibers—have been developed. These products are identified by their individual names and are not called lumber. Plywood was one of the earliest of these products.

Glue-laminated lumber, parallel strand lumber, and laminated veneer lumber are products that have the same or similar profiles as solid lumber. Wood trusses and I-joists are other examples of manufactured wood products. These products have been formulated primarily to conserve the decreasing supplies of high-quality solid lumber members.

This chapter and the following four chapters deal with wood construction. The properties and uses of various wood products are presented first. Lumber is discussed in this chapter and manufactured wood products in Chapter 14. In practice, lumber and manufactured wood products are used together in the same building. Wood light-frame construction, a common construction system utilizing lumber and manufactured wood products, is described in Chapter 15.

Chapter 16 deals with frequently used exterior and interior finishes in wood light-frame buildings. Chapter 17 deals with panelized wood frame construction in what is referred to as *structural insulated panel* (SIP) construction.

13.1 INTRODUCTION

A cursory review of architectural history reveals that stone masonry was the construction system of antiquity, yet sufficient evidence suggests that the use of wood in buildings could have preceded it. Historic stone structures survived due to their relative durability, whereas wood structures of antiquity were either consumed by fire or destroyed by wood-consuming organisms.

However, many centuries'-old surviving wood buildings indicate that wood is a fairly durable material, provided that it is protected from fire and biological destruction. An example illustrating wood's durability is shown in Figure 13.1. Many other examples exist in several European countries.

Wood—An Easily Renewable Material

One of the major advantages of wood as a building material is its renewability. Theoretically, all materials are renewable. However, wood is a building material that is renewed within the human time span. Trees for most softwood construction lumber are harvested within 25 to 40 years. Other naturally occurring building materials (stone, iron, aluminum, copper, etc.) are also renewed, but within geological time spans—in thousands or millions of years.

In order to optimize wood's renewablility, sufficient forest land must be available to grow trees, and the land must be carefully managed (lumber farming). Unfortunately, due to high population densities, the availability of land for lumber farming is severely limited in most parts of Africa, Asia, Europe, and Latin America. Coincidentally, forest-management skills are also relatively underdeveloped in these countries.

FIGURE 13.1 John Ward House, Salem, Massachusetts, built in the seventeenth century and still standing. (Photo courtesy of Dr. Jay Henry)

EXPAND YOUR KNOWLEDGE

Photosynthesis and Respiration

The Processes That Make Wood a Rapidly Renewable Material

The growth of trees and plants occurs due to a natural process called *photosynthesis*. In this process, the water drawn from the soil and the environment reacts with the carbon dioxide in the air to produce a basic sugar (glucose), releasing oxygen. This reaction requires a tree or plant as the medium and the presence of light energy from the sun, hence the term *photosynthesis*, whose literal meaning is "to synthesize with the help of light." In other words

$$\text{Carbon dioxide} + \text{water} \xrightarrow[\text{energy}]{\text{Light}} \text{glucose} + \text{oxygen}$$

$$6CO_2 + 6H_2O \xrightarrow[\text{energy}]{\text{Light}} C_6H_{12}O_6 + 6O_2$$

Glucose is subsequently converted into other organic molecules that comprise the tree—cellulose, hemicellulose, and lignin. Cellulose* and hemicellulose, which are chemically similar to glucose, constitute nearly two-thirds to three-quarters of the weight of wood in a tree. Cellulose and hemicellulose form the walls of wood cells, and lignin is the glue that binds the cells together.

Although cellulose, hemicellulose, and lignin constitute the bulk of the wood, small quantities of other minerals and chemicals are also transported into the tree.

Photosynthesis is the reverse of *respiration*—a process by which the food that humans and animals eat is converted into sugar. This sugar is subsequently metabolized into carbon dioxide and water, and energy is released in this process:

$$\text{Glucose} + \text{oxygen} \longrightarrow \text{carbon dioxide} + \text{water} + \text{energy}$$

$$C_6H_{12}O_6 + 6O_2 \longrightarrow 6CO_2 + 6H_2O + \text{energy}$$

Photosynthesis and respiration, being the reverse of each other, form an ecological cycle on which all terrestrial life depends.

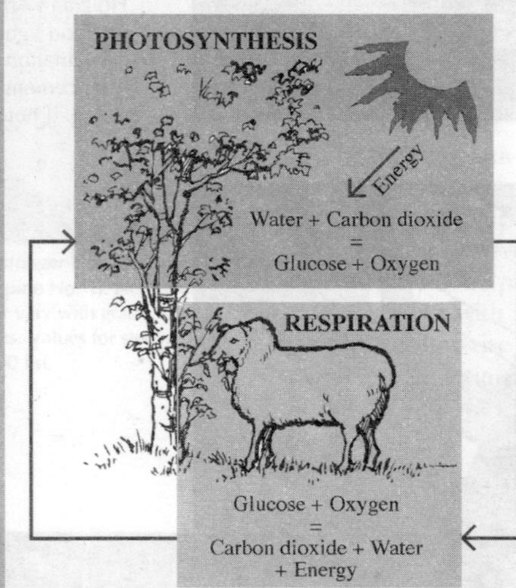

The tree obtains water from the soil and carbon dioxide from the atmosphere and converts them into sugar (contained in plant life). This reaction, in which oxygen is released, requires the presence of energy that is supplied by the sun.

Animals, humans and other creatures consume the sugar in plants. With the help of oxygen that they breathe, they convert these substances into carbon dioxide (that they exhale) and water is given off in animal and human waste.

*Chemically, a cellulose molecule is represented by $(C_6H_{10}O_5)_n$. The subscript n indicates that a cellulose molecule is a polymer consisting of several $C_6H_{10}O_5$ monomers linked together end-to-end as the loops in a chain. Typically, in a cellulose polymer, n may be as high as 10,000, which gives polymer lengths of nearly 0.01 mm. The orientation of the polymer chains along the long axis of the tree provides many of the basic properties of wood.

13.2 GROWTH RINGS AND WOOD'S MICROSTRUCTURE

A tree consists of three essential parts: (a) root structure, (b) trunk, and (c) branches and the leaf system. The root structure stabilizes the tree, and the trunk provides the strength required to support the branches and leaves. In addition to providing stability and strength, the other important function of the root structure and trunk is to conduct water mixed with food (nutrients) from the soil to the leaves and branches.

As water moves through the tree and reaches the leaves, it is converted to cellulose by *photosynthesis*. The transportation of water from the soil is accompanied not only by food, but also by several other dissolved minerals and chemicals in small quantities. Water transportation requires the microscopic structure of a tree to consist of tiny hollow tubes, referred to as *cells*. To transport food efficiently, the cells must be oriented along the long axis of the tree—vertically in the trunk and along the length of an individual branch. The cells are visible only when magnified several times under a microscope.

Figure 13.5 shows part of the unmagnified cross section through the trunk of a tree. The outermost dark-colored layer is the bark, followed by several increasingly smaller rings toward the center of the trunk. These rings are referred to as *annual rings*, because a new

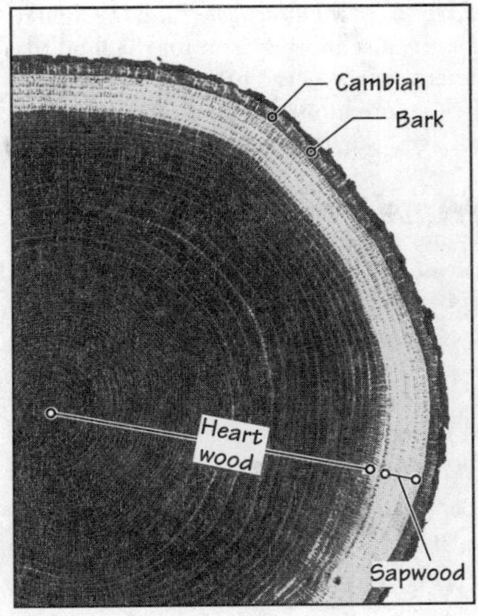

FIGURE 13.5 A cross section through the trunk of a tree. The cambium layer, in which cell division takes place, lies between the bark and the outermost growth ring. The interior dark-colored region is the heartwood, and the light-colored region is the sapwood. The radial lines (which are rather faint in this diagram) are the rays (Section 13.3). (Photo courtesy of the Forest Products Laboratory, U.S. Department of Agriculture)

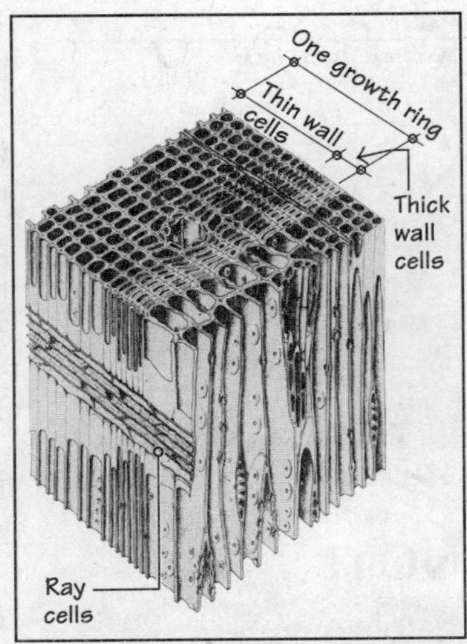

FIGURE 13.6 A magnified view of the cross section of a tree trunk. (Photo courtesy of the Forest Products Laboratory, U.S. Department of Agriculture)

ring is typically added every year. *Growth ring* is, however, a better term because in some trees, more than one ring may be added per year.

The addition of growth rings occurs in the *cambium* layer, which lies between the bark and the outermost growth ring (Figure 13.5). The cells in the cambium layer divide and subdivide, producing the cells of the growth rings and the bark. As each growth ring is added, the girth of the tree increases.

Earlywood and Latewood

The cells added during the wet season have thinner walls and larger cavities. During the dry season, when the tree has to transport less food, the cell walls are thicker and the cell cavities are smaller. The cells added during the wet and dry seasons are called *earlywood* and *latewood*, respectively.

It is the difference between the cells of earlywood and latewood that distinguishes each ring, Figure 13.6. The distinct growth rings give wood its characteristic grain structure.

In some species the difference between earlywood and latewood is less distinct than in others. For instance, in basswood, all cells have a similar wall thickness and cavity size, and the annual rings are indistinct. This yields wood with a relatively flat and even grain structure. By comparison, Southern yellow pine, Figure 13.7, has highly differentiated cell sizes, resulting in a pronounced grain structure.

Heartwood and Sapwood

As the tree ages and increases in girth, not all cells are needed to conduct food. As this happens, walls and cavities of the cells formed when the tree was younger (i.e., the rings closer to the center of the trunk) are impregnated with several chemical substances. This process increases the tree's mass and strength, which is required to support its increasing height.

The obvious manifestation of the chemical deposition is a change in the color of the wood. Thus, the growth rings added in the trunk earlier during the life of the tree are darker in color than those added later (Figure 13.5).

NOTE

Earlywood and Latewood

The terms *springwood* and *summerwood* were used earlier for earlywood and latewood, respectively, which erroneously represented a connection with a calendar season of the year. In many tropical areas, which remain wet throughout the year, the growth of a tree occurs throughout the year.

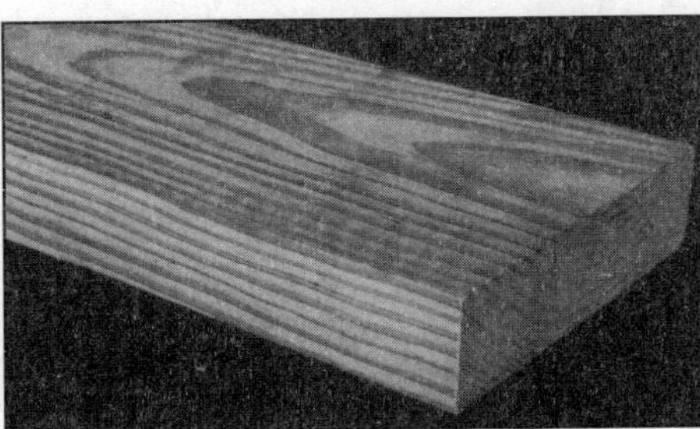

FIGURE 13.7 A piece of southern yellow pine illustrating its pronounced grain structure. By comparison, a piece of basswood, commonly used for architectural model making, has a flat, even grain structure.

The darker-colored region of the trunk is referred to as the *heartwood*, and the lighter-colored region is called the *sapwood*. The word *sap* is just another term for the food that the tree conducts; hence, the term *sapwood* merely expresses the function of the light-colored outer growth rings in relation to the dark-colored inner rings. The change from sapwood to heartwood is gradual, and as the tree ages, the heartwood portion in the tree increases.

The chemical deposition in many commercial trees is toxic to fungi. Therefore, the heartwood of most commercial trees is relatively more decay resistant than the sapwood. In some species, the decay resistance of the heartwood is significant. Before the introduction of pressure-treated wood, the heartwoods of redwood and cedar were commonly specified for lumber elements resting directly on concrete or masonry foundations, such as sill plates and sleepers.

NOTE

Naturally Decay-Resistant Wood

Heartwood of redwood and cedar is naturally decay resistant and is recognized as such by the building codes.

Contemporary building codes also recognize the decay resistance of the heartwood of redwood and cedar by allowing them to be used where pressure-treated wood is required. The use of pressure-treated wood is, however, more common due to its greater decay and termite resistance.

NOTE

The Term *Species*

In this text, the term *species* is used in its customary sense, not in the botanically correct sense. In the botanist's language, a particular species is designated by the combination of *genus* (generic name) and *species* (specific name). For instance, the botanical name of Eastern white pine is *Pinus strobus*, in which *Pinus* represents the genus and *strobus* represents the species. Similarly, the botanical name of Ponderosa pine is *Pinus ponderosa*. For the sake of simplicity, the term *species* in this text includes both botanical terms—genus and species. Note also that the term *species* is used in both singular and plural forms.

13.3 SOFTWOODS AND HARDWOODS

Lumber is divided into two broad categories—*hardwoods* and *softwoods*. The hardwoods are generally denser than the softwoods. Because density is a major determinant of the strength of wood, hardwoods are generally stronger than softwoods.

The distinction between softwoods and hardwoods is, however, not based on the density of the wood, because several hardwoods are lighter than softwoods, and vice versa. For instance, balsa, among the lightest and weakest of woods, is a hardwood. On the other hand, slash pine and longleaf pine are softwoods but are denser than many hardwoods, Figure 13.8.

The distinction between softwoods and hardwoods is based on their botanical characteristics. Softwood-producing trees, in general, do not bear flowers and have a single main stem, and most of them are evergreen, with leaves that are needlelike (i.e., conical; hence, softwood trees are also called *conifers*), Figure 13.9. Pines, firs, spruces, cedars, hemlocks, and so on, are softwoods.

Hardwood-producing trees are generally flowering trees, have broad leaves, and are typically deciduous, shedding and regrowing leaves annually. Oak, walnut, birch, elm, teak, mahogany, rosewood, and so on, are hardwoods. Figure 13.10 shows a typical hardwood tree.

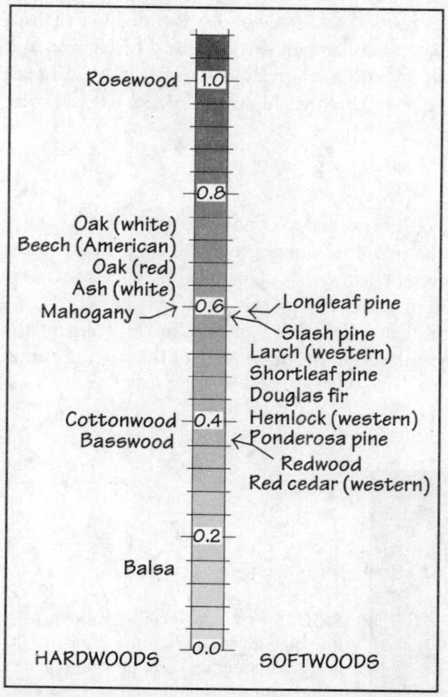

Softwoods in the United States come from two main regions, the western wood region and the southern pine region, and to a small extent from the northern and northeastern region, Figure 13.11. In addition to indigenous softwoods, a great deal of softwood lumber in the United States is imported from Canada. Most hardwoods in the United States are from the eastern region, ranging from New York to Georgia.

FIGURE 13.8 Specific gravities of selected softwoods and hardwoods.

NOTE

Specific Gravity

The specific gravity of a material is an index of its density relative to the density of water, that is, the density of the material divided by the density of water. Thus, if the density of a material is less than the density of water, its specific gravity is less than 1.0. Such a material will float on water. Conversely, if the density of a material is greater than the density of water, its specific gravity is greater than 1.0, and it will sink in water.

Because the density of water is 62.5 lb/ft^3, a wood whose specific gravity is 0.5 has a density of 31.25 lb/ft^3.

MICROSTRUCTURE OF SOFTWOODS AND HARDWOODS

Although basically similar, there is some difference between the cellular structures of softwoods and hardwoods. In softwoods, all longitudinal cells are of the same type and almost the same size, Figure 13.12. These cells perform two functions: (a) provide strength to the tree and (b) conduct food from the ground up.

FIGURE 13.9 Softwood (pine) trees in the Willamette Industries forest, Ruston, Louisiana.

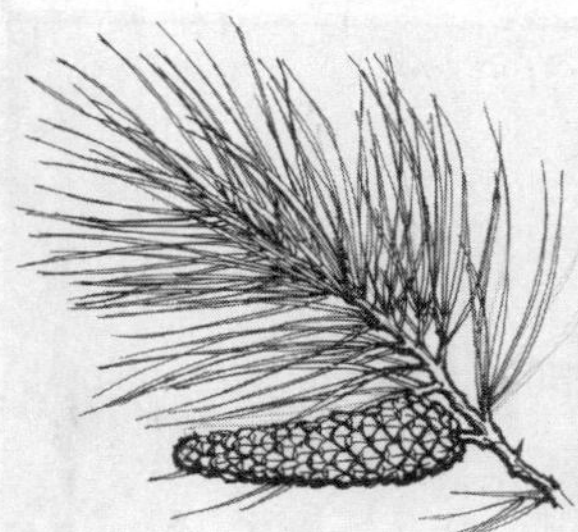

Pine cone with its scaled surface

Broad leaves of red oak

FIGURE 13.10 A hardwood tree (red oak).

The cellular structure of hardwoods is more complex. Unlike softwoods, the longitudinal cells of hardwoods are of two types—one type with small cavities and the other with large cavities—with two distinct functions. The smaller cells in hardwoods provide strength to the tree, and the cells with larger cavities, referred to as *vessels,* conduct food, Figure 13.13. In many hardwoods, the vessels can be seen under an ordinary magnifying glass.

In addition to the longitudinally oriented cells, both softwoods and hardwoods have transverse cells called *rays.* Rays are perpendicular to the longitudinal cells. They provide radial transfer of food, as well as transverse strength.

Ray cells are far more prominent in hardwoods than in softwoods. Due to the prominence of ray cells and the diversity of their longitudinal cells, most hardwoods show a more interesting grain structure than do softwoods.

Wood's Anisotropicity

Despite the differences previously described, the microstructures of both softwoods and hardwoods are essentially similar and consist of a cluster of longitudinally oriented hollow

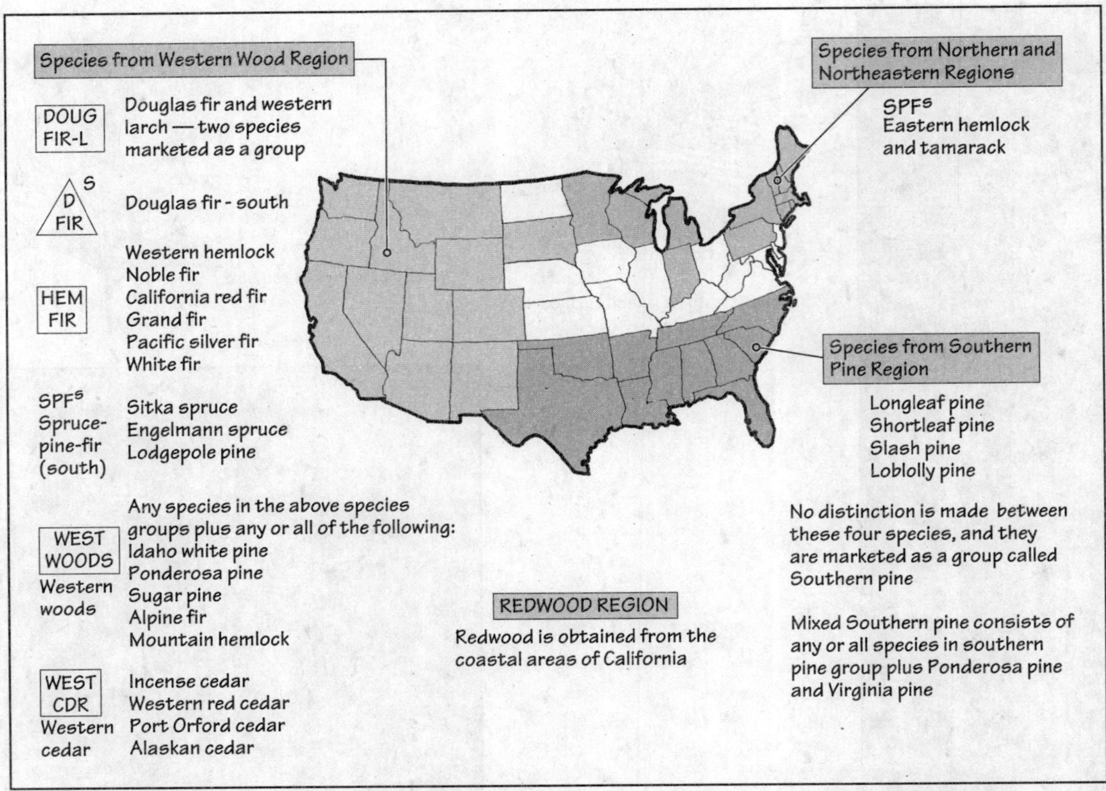

FIGURE 13.11 Map of the major softwood-producing regions in the United States showing the states included in each region.

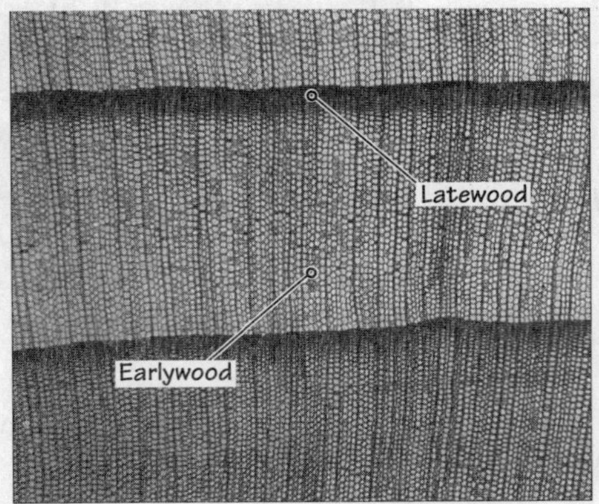

FIGURE 13.12 Magnified view of a softwood's growth rings showing the cellular structure. The lighter-colored region, with larger cell cavities and thinner walls of a growth ring, represents earlywood. The darker-colored region with smaller cell cavities and thicker walls represents latewood. (Photo courtesy of the Forest Products Laboratory, U.S. Department of Agriculture)

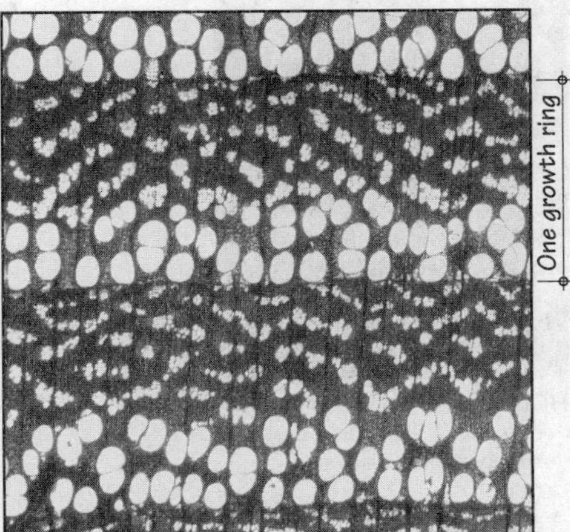

FIGURE 13.13 Magnified view of a hardwood's growth rings showing that it has a more complex microstructure than that of softwood. (Photo courtesy of the Forest Products Laboratory, U.S. Department of Agriculture)

tubes. This cluster resembles a bundle of drinking straws glued together and reinforced with a few transversely oriented tubes (ray cells), Figure 13.14. This simplification clarifies some of the important properties of wood.

For instance, a piece of lumber is much stronger along the grain (parallel to the axis of cells) than across the grain (perpendicular to the axis of cells), Figure 13.15. In fact, lumber's tensile strength across the grain is so small that it is neglected, as if the individual straws were not glued together. Thus, unlike steel and concrete, which are isotropic materials (having the same properties in all directions), wood is an anisotropic material.

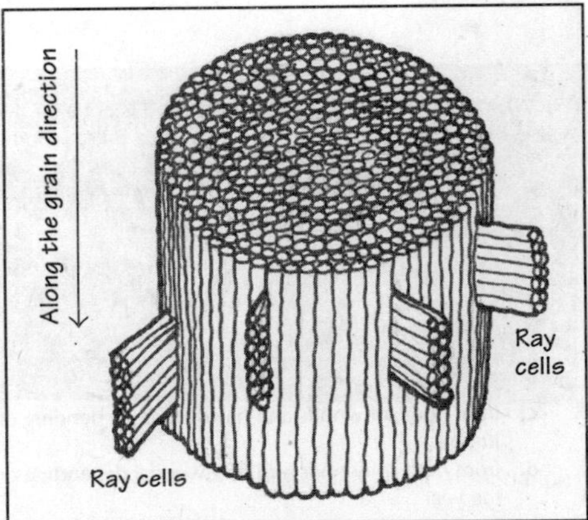

FIGURE 13.14 A highly simplified version of the cellular structure of wood—consisting mainly of hollow tubes oriented along the length of the member, resembling a bundle of straws glued together.

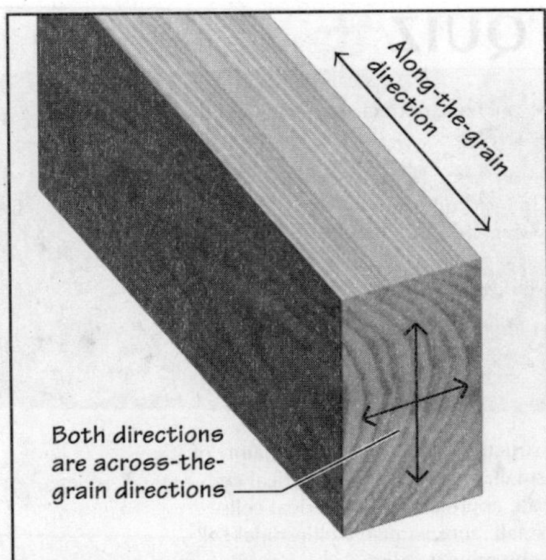

FIGURE 13.15 Along-the-grain and across-the-grain directions in lumber.

The weak glue bond between individual cells (drinking straws) explains why a lumber piece will easily split along the grain when nailed near an end, Figure 13.16. For the same reason, wood is weaker in shear along the grain than across the grain, because individual cells can easily slip (slide) against each other.

As we will see in Section 13.5, a piece of lumber shrinks substantially across the grain because the cell walls become thinner as water leaves them. However, along the grain (lengthwise), the departure of water has very little effect. Thus, lumber shrinks negligibly along the grain.

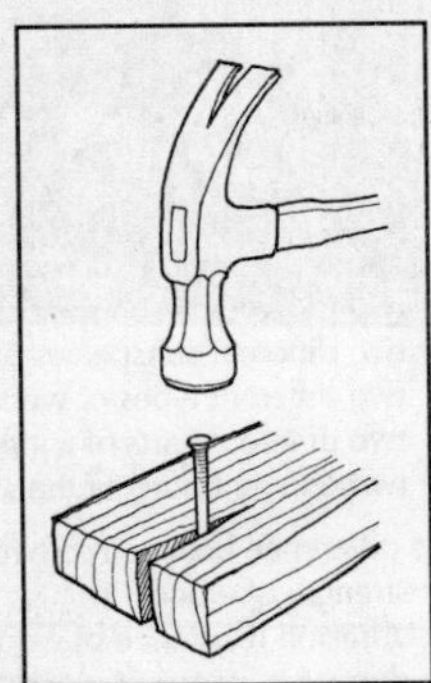

FIGURE 13.16 Lumber splits easily when nailed close to an end because of its low tensile strength across the grain resulting from the weak glue bond between fibers.

Uses of Softwoods and Hardwoods

Softwood trees mature for harvesting two to three times faster than hardwood trees. Nearly 75% of North American forests contain softwoods. Because of their faster growth and, hence, their relative abundance, softwoods are commonly used for structural framing of wood buildings in North America. Thus, in most buildings, floor joists, rafters, ceiling joists, studs, sheathing, and so on, are of softwood lumber.

Hardwoods are commonly used for finish flooring, where their higher densities (giving greater abrasion resistance) are useful. Because hardwoods have a more interesting and varied grain structure, they are also used for wall paneling, trims, cabinets, furniture, and so on. Table 13.2 gives some of the major softwood and hardwood species used in building construction in North America.

TABLE 13.2 MAJOR SOFTWOOD AND HARDWOOD LUMBER SPECIES USED IN NORTH AMERICA

Softwood lumber		Hardwood lumber
Framing lumber (studs, floor and ceiling joists, rafters, headers, etc.)	**Roof shingles, fencing, etc.**	**Finish flooring**
Southern pine, western lumber, north and northeastern lumber, Canadian lumber (See Figure 13.11)	Decay-resistant woods such as redwood, western red cedar, white cedar, and southern cypress	Red oak and white oak are most common. Sugar maple, pecan, hickory, and teak are also used.
Siding, paneling, fascia boards, etc.	**Finish flooring**	**Paneling and molding**
Almost all species used for framing can be used for these nonstructural applications.	Denser species such as southern pine and douglas fir	Ash, beech, hickory, red oak, white oak, sugar maple, pecan, black walnut, and teak
	Doors, windows, and cabinets	
	Ponderosa pine, sugar pine, and Idaho white pine	

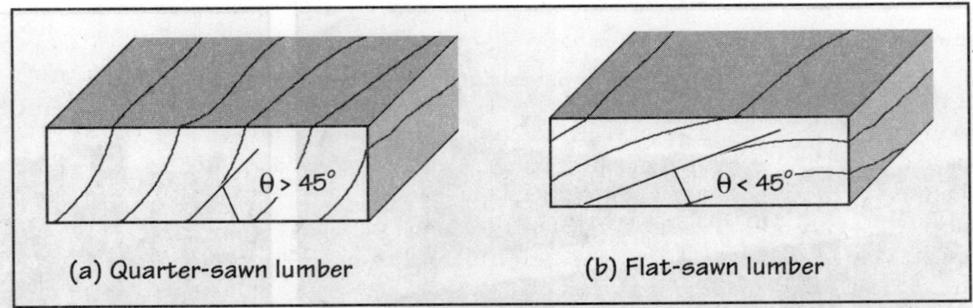

FIGURE 13.21 Theoretical definitions of quarter-sawn and flat-sawn lumber members.

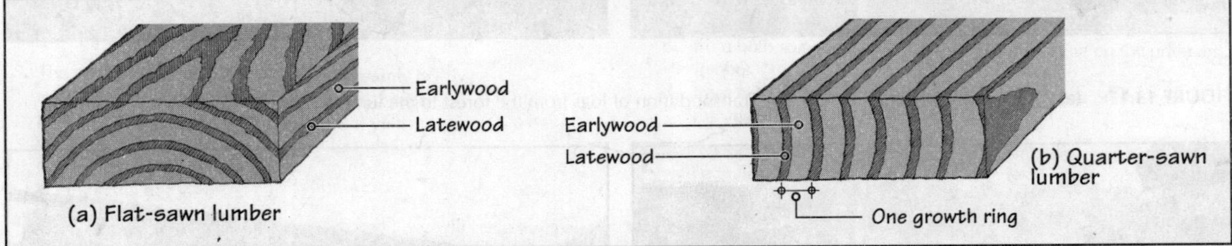

FIGURE 13.22 The distribution of denser latewood and lighter earlywood in quarter-sawn lumber is more even than in flat-sawn lumber.

In a radial-sawn piece farther from the log's center, the angle between the rings and the wider faces of the cross section is less than 90°. Some lumber experts define a quarter-sawn piece as the one whose growth rings make an angle of 45° or more with the wider face of its cross section, Figure 13.21 (even if the piece has been obtained from flat sawing). Thus, some of the pieces in Figure 13.19 are classified as quarter-sawn. If the angle is less than 45°, the piece is classified as flat-sawn.

Quarter sawing is more complex and usually more wasteful. Hence, most framing lumber is obtained by flat sawing. As we will see later in this section, quarter-sawn lumber is more dimensionally stable. It resists wear and abrasion more evenly because denser latewood is more uniformly exposed on the surfaces of the lumber piece, Figure 13.22. Quarter-sawn lumber is commonly specified for high-grade finish floors.

13.5 DRYING OF LUMBER

After sawing, the lumber is dried and subsequently surfaced smooth before being shipped to lumber yards for sale. Although the drying of wood begins from the time the tree is cut and converted to logs, the moisture content prior to sawing far exceeds that required for use in buildings. Therefore, lumber must be dried (or seasoned). The term *seasoning* implies a controlled rate of drying. Controlled drying minimizes the separation of fibers that occurs during the drying process.

The moisture content (MC) in a piece of wood is the weight of water in the wood divided by its oven-dry (completely dry) weight, expressed as a percentage:

$$\text{MC} = \frac{\text{weight of water in wood}}{\text{weight of oven-dry wood}} \times 100$$

Let us assume that a piece of wood weighs 6 lb. Now let us dry it until all of its water has evaporated and then weigh it again. If the weight of this dry piece is 4.5 lb (implying that the weight of water in the wood was 1.5 lb), the moisture content in the wood before drying, as obtained from the preceding equation, is

$$\text{MC} = \frac{1.5}{4.5} \times 100 = 33.3\%$$

FIGURE 13.23 Free water and bound water in wood cells. (Photo courtesy of Southern Forest Products Association)

The moisture content in the tree before it is cut varies with the species and may be as high as 200%. Some of this water is contained in cell walls, which are in their fully swollen state, and some of the water is contained in cell cavities. During the drying process, the cavity water, referred to as *free water,* evaporates first.

After all the free water has evaporated, the cell wall water, called the *bound water,* begins to evaporate, Figure 13.23. The stage at which all the free water has evaporated and the bound water has just begun to evaporate is referred to as the *fiber saturation point* (FSP). At this point, the cell walls are in their fully swollen state, just as they were when there was free water in the cells.

GREEN VERSUS DRY LUMBER

The average moisture content in lumber at the FSP is approximately 30%. It is only when the moisture content in wood falls below FSP (i.e., when the bound water begins to evaporate) that the water leaves the cell walls and the wood begins to shrink. No shrinkage of wood occurs if its moisture content is greater than its fiber saturation point. Dimensional changes in lumber (shrinkage and swelling) are discussed later in this section.

Lumber is typically seasoned in the mill to a moisture content of 19% or less—a value that distinguishes *dry* lumber from *green* lumber. Thus, a piece of lumber whose moisture content is less than or equal to 19% is referred to as *dry lumber*, and that whose moisture content is 20% or greater is referred to as *green lumber*, Figure 13.24.

The value of 19% has been selected because lumber with a moisture content of 19% or less is not susceptible to fungal decay. Additionally, the strength properties of lumber do not improve significantly below 19% moisture content. Note that the strength and stiffness of lumber increase as the moisture content decreases beyond the fiber saturation point. Seasoning also helps to reduce transportation costs because a drier piece of lumber weighs less.

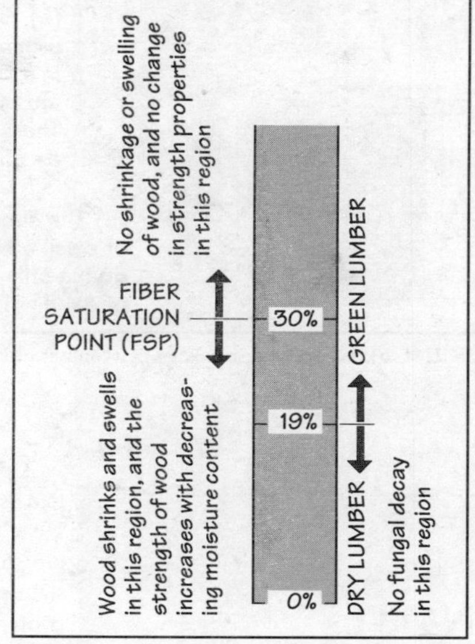

FIGURE 13.24 Definitions of dry lumber and green lumber as a function of lumber's moisture content.

AIR SEASONING VERSUS KILN SEASONING

Two processes are used to season lumber: *air seasoning* and *kiln seasoning*. In air seasoning, lumber pieces dry naturally. They are simply stacked one over the other in piles in such a way that air can circulate freely around them. The piles are kept under a roof to prevent the direct action of rain, snow, or sun.

Air seasoning is a slow process that can take several months, depending on the thicknesses and species of lumber. Kiln seasoning is much faster. The high temperature in the kiln also kills any fungus that may be present in the living tree. In kiln seasoning, the lumber is stacked in a chamber through which warm air with a controlled amount of humidity (steam) is circulated. Figure 13.25 shows a typical seasoning kiln.

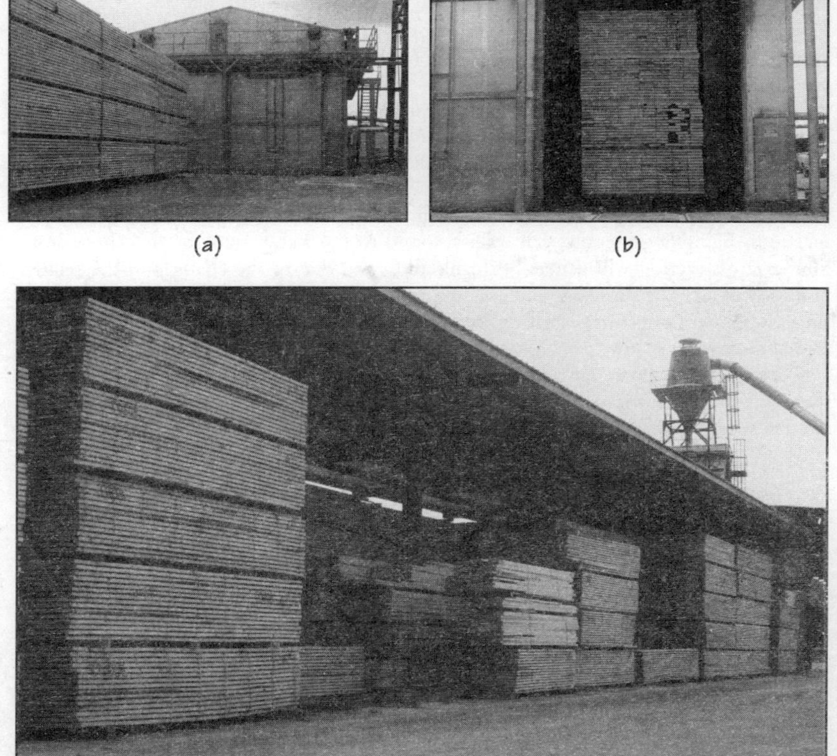

FIGURE 13.25 **(a)** A stack of unseasoned sawed lumber stacked outside the kiln, ready to be moved into the seasoning kiln seen in the background. **(b)** A view of a seasoning kiln with its doors open. **(c)** Seasoned lumber stored in the yard.

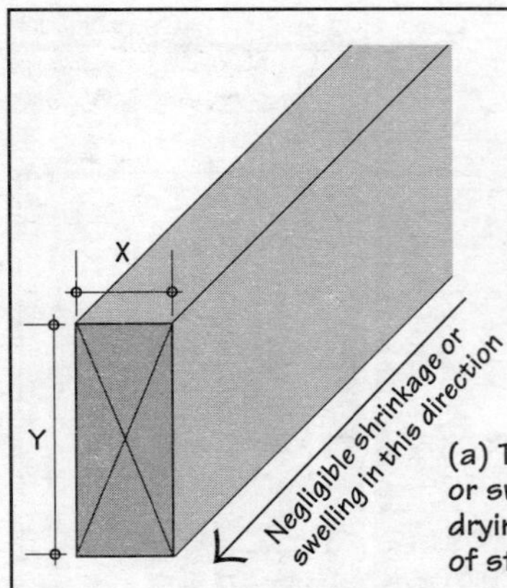

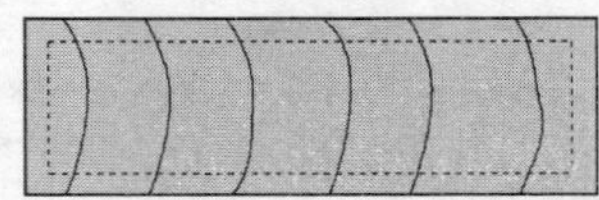

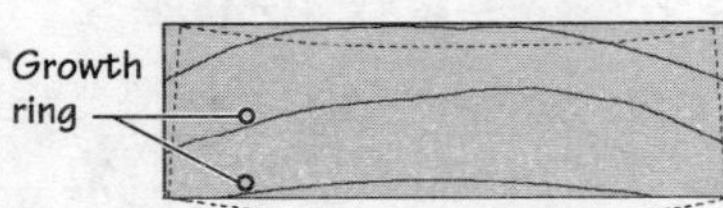

(b) Shrinkage and swelling of wood is primarily in the direction of growth rings. In a radially-sawn lumber, growth rings are of the same approximate size. Therefore, it shrinks uniformly on drying as shown by dashed lines.

(c) In a flat-sawn lumber, the growth rings are of different sizes. Therefore, a flat-sawn lumber member tends to cup on drying as shown by dashed lines.

(a) The dimensions X and Y in a lumber member will shrink or swell with moisture changes. Thus, a stud will shrink on drying only in its cross-sectional dimensions. The length of stud will not be greatly affected.

FIGURE 13.26 Drying shrinkage of a lumber member.

A piece of lumber (air or kiln) seasoned to a moisture content of 19% or less is identified as *surfaced dry* (S-DRY). If the moisture content at the time of surfacing is more than 19%, the lumber is identified as *surfaced green* (S-GRN). Some mills season lumber to a moisture content of 15% or less. The identification in that case is MC 15 or KD 15; KD stands for *kiln dried*. As we will see in Section 13.11, the identification of its moisture content is an important component of the lumber grade.

Lumbers marked S-DRY, MC 15, or KD 15 are considered to be equivalent for most structural purposes. However, for nonstructural uses, such as for wall and ceiling paneling and flooring, where greater dimensional stability is needed, a lower moisture content is desirable. Additionally, it is generally immaterial whether lumber is kiln or air dried.

NOTE

S-DRY, MC 15, and KD 15 are considered equivalent moisture-content designations for framing lumber. However, sawn lumber used in glulam beams and wood I-joists is generally seasoned to a moisture content of less than or equal to 15%.

Shrinkage and Swelling of Lumber

As stated previously, wood begins to shrink only when the water from cell walls begins to evaporate, that is, when the moisture content in wood is less than the fiber saturation point. The drying shrinkage of cell walls affects the dimensions of lumber's cross section only, not its length. Consequently, a vertical lumber member, such as a stud, shortens negligibly along its height on drying. Similarly, the shrinkage of a floor joist is negligible along its length. In other words, wood shrinks only across the grain, not along the grain, Figure 13.26(a).

Cell walls not only shrink on drying but also expand when their moisture content increases. Being hygroscopic, cell walls respond to the water vapor in air. Therefore, a lumber cross section will generally shrink and swell due to the changes in the relative humidity of air. The shrinkage and swelling are much larger along the direction of growth rings in lumber (tangential direction) than perpendicular to the direction of growth rings (radial direction).

Consequently, quarter-sawn lumber shrinks uniformly, Figure 13.26(b) and tangentially (flat-) sawn lumber will generally cup, bow, or twist on drying, Figure 13.26(c). This is another reason why radially sawn lumber is preferred for wood flooring.

EXPAND YOUR KNOWLEDGE

Moisture Content of Wood and Relative Humidity of Air

A piece of lumber swells and shrinks with changes in the relative humidity (RH) of the surrounding air and gradually reaches an equilibrium moisture content (EMC). Although the relationship of EMC with RH varies somewhat with the species, the following represents a good average.

Based on this relationship, the average moisture content in framing lumber lies between 5% and 14%, depending on how arid or humid the location is.

RH	EMC	RH	EMC
0%	0%	75%	14%
25%	4.5%	90%	20%
50%	8.5%	100%	FSP approx. 30%

Source: Bruce Hadley. *Understanding Wood*. Newtown, CT: Taunton Press, 1980, p. 69.

13.6 LUMBER SURFACING

Saw blades used for mass manufacturing are coarse-edged. Therefore, lumber obtained after sawing is quite rough. Because rough surfaces can cause injuries to framers, lumber used in conventional wood light framing is surfaced smooth.

Surfacing is done by high-speed planing machines—a replacement for the traditional hand-held carpenter's plane. Surfacing smooths lumber, rounds off the edges, and makes it more square, removing some of the distortions that have occurred during the seasoning process. That is why seasoning usually precedes surfacing. Lumber that has been surfaced before seasoning is identified as S-GRN (surfaced green).

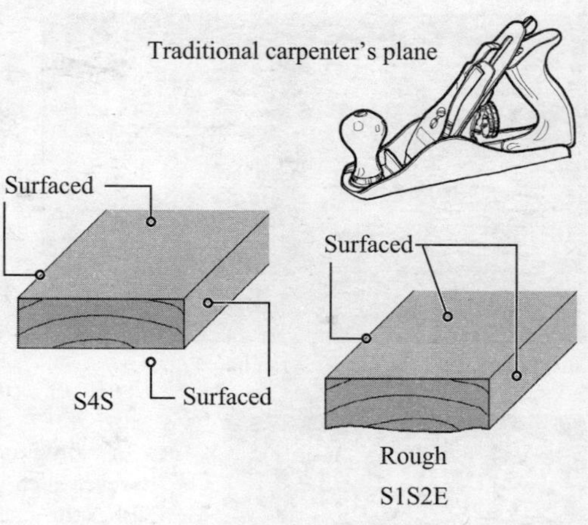

Framing lumber is surfaced on its longitudinal faces, not the ends, because the ends are usually resawn during construction. A piece surfaced on one wide face only is referred to as *S1S* (surfaced on one side); if surfaced on both wide faces, it is called *S2S* (surfaced on two sides); if surfaced on all four faces, it is called *S4S*.

A piece that is surfaced on one wide face and one narrow face (edge) is called *S1S1E* (surfaced on one side and one edge). Other designations are *S2S1E* and *S1S2E*. Lumber used for structural framing is typically surfaced S4S, fascia boards, as S1S2E, and so on.

In construction drawings, rough or surfaced lumber is graphically depicted by diagonal lines through its cross section. Two crossed diagonal lines indicate a continuous member, Figure 13.27(a), and one diagonal line indicates an interrupted (discontinuous) member, such as blocking or shims, Figure 13.27(b). Surfaced lumber is used for framing members, which are typically covered over by finish materials such as gypsum wallboard or plywood subflooring. Rough lumber is used for exposed lumber components (unprotected by gypsum board, etc.), particularly for beams and columns of large cross sections.

A *worked (finished) lumber* member is represented by growth ring symbols in the cross section, Figure 13.27(c). Worked lumber is lumber that has been dressed by additional machining to obtain the required profile. Worked lumber is used for interior finish members, such as door and window frames, trims and moldings, finish flooring, paneling, and so on. Figure 13.28 shows a typical section through the head of a door, illustrating the use of lumber's graphic symbols.

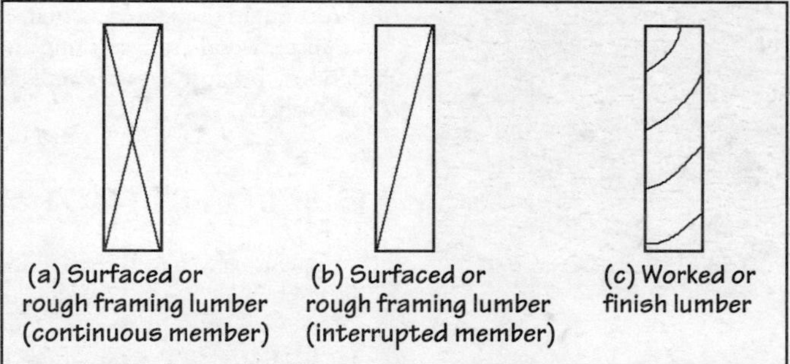

FIGURE 13.27 Graphic symbols for framing lumber and worked (finished) lumber.

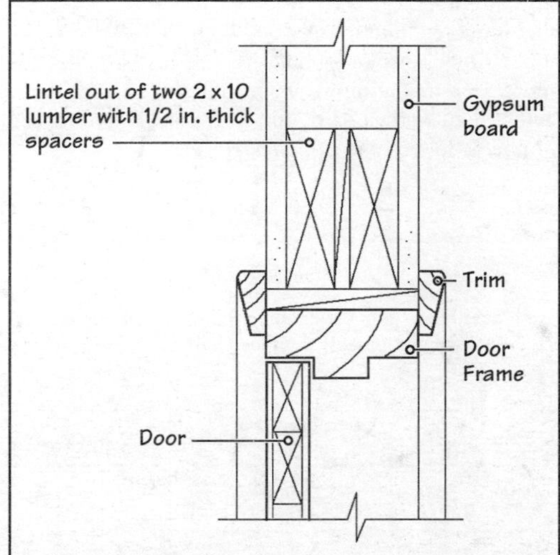

FIGURE 13.28 A section through the head of a door illustrating the use of graphic symbols for lumber members.

FINGER-JOINTED LUMBER

Two lumber pieces can be glued together to produce a longer member using complementary finger joints on either side, Figure 13.29. By finger jointing, defects in a wood member are cut away and the good pieces are joined to obtain higher-grade lumber.

Because of high-strength and water-resistant glues, a finger-jointed member is as strong as a single-length (unjointed) member. Jointed and unjointed members can be used interchangeably.

FIGURE 13.29 Finger jointing of lumber pieces.

13.7 NOMINAL AND ACTUAL DIMENSIONS OF LUMBER

In North America, a softwood lumber's cross section is specified by its nominal dimensions rather than by its actual dimensions. Thus, when we say that a lumber piece is 2 × 4 (two by four), we refer to its nominal dimensions and imply that its cross section measures 2 in. × 4 in. nominally. The actual dimensions of a 2 × 4 (surfaced dry) are $1\frac{1}{2}$ in. × $3\frac{1}{2}$ in.

In stating the nominal dimensions, the inch label is not used. For the actual dimensions, however, inch labels must be used. Thus, the size with and without the inch labels distinguishes the actual size from the nominal size.

Nominal dimensions correspond roughly to the wood in the log before it is sawn, seasoned, and surfaced. The difference between the two represents the amount of wood lost due to shrinkage, sawing, and surfacing. The relationship between the nominal and actual dimensions of S-DRY lumber is shown in Table 13.3. Thus, a 2 × 10 (two by ten) dry lumber cross section actually measures $1\frac{1}{2}$ in. × $9\frac{1}{4}$ in.

Although the stated actual dimensions are reasonably accurate for construction purposes, wood cross sections do not have the same dimensional precision as steel cross sections, because wood swells and shrinks with changes in relative humidity in the atmosphere.

TABLE 13.3 NOMINAL AND ACTUAL SIZES OF S-DRY LUMBER

Nominal size	Actual size
1	$\frac{3}{4}$ in.
$1\frac{1}{2}$	$1\frac{1}{4}$ in.
2	$1\frac{1}{2}$ in.
3	$2\frac{1}{2}$ in.
4	$3\frac{1}{2}$ in.
5	$4\frac{1}{2}$ in.
6	$5\frac{1}{2}$ in.
8	$7\frac{1}{4}$ in.
10	$9\frac{1}{4}$ in.
12	$11\frac{1}{4}$ in.
14	$13\frac{1}{4}$ in.

13.8 BOARD FOOT MEASURE

In the United States, softwood lumber is sold by volume, similar to concrete.* The obvious volumetric measure is the cubic foot (or cubic yard; in the SI system, it is a cubic meter). However, a volumetric measure that is unique to lumber is the *board foot* (bd ft). One board foot is the amount of lumber contained in a 1-in.-thick board that measures 1 ft × 1 ft on its face. In other words, 1 ft^3 = 12 bd ft.

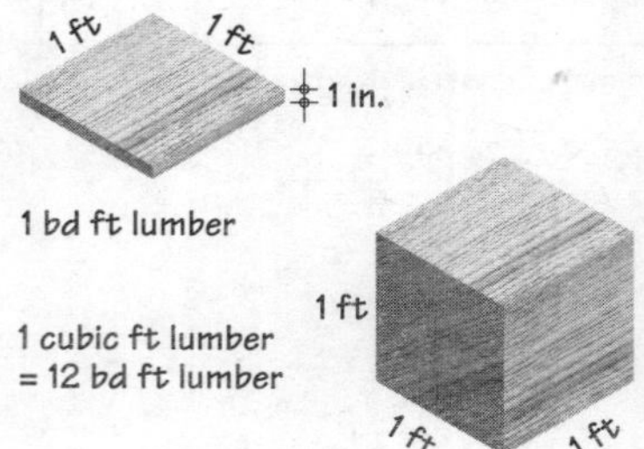

Thus, a 2 × 12 piece of lumber that is 20 ft long contains 40 bd ft. One hundred such pieces contain 4,000 bd ft. In calculating board feet, we use the nominal dimensions, not the actual dimensions, of lumber.

The price of lumber is quoted in terms of 1,000 bd ft (MBF) and varies according to the lumber's cross-sectional dimensions and grade. Larger cross sections, longer lengths, and higher grades cost more. In a home-improvement building material store, lumber is typically sold per piece rather than by the board foot. In Europe and other countries, where the SI system prevails, lumber is sold by the cubic meter.

Example

Determine how much a builder will pay a lumber wholesaler for 900 pieces of 2 × 12 lumber, each 16 ft long. The price of lumber is $800 per MBF.

Solution

$$\text{Total volume of lumber} = \frac{(16 \times 2 \times 12)(900)}{(12 \times 12)} = 2{,}400 \text{ ft}^3 = 2{,}400(12)\text{bd ft} = 28.8 \text{ MBF}$$

$$\text{Total price of lumber} = 28.8(800) = \$23{,}040$$

*Concrete is sold per cubic yard of green (wet) concrete.

PRACTICE QUIZ

Each question has only one correct answer. Select the choice that best answers the question.

18. Structural lumber is typically
- **a.** flat-sawn.
- **b.** quarter-sawn.
- **c.** either (a) or (b), depending on the type of structural member.

19. Flat-sawn lumber is dimensionally more stable than quarter-sawn lumber.
- **a.** True
- **b.** False

20. Which of the following terms means the same as *quarter-sawn*?
- **a.** *Tangentially sawn*
- **b.** *Radial-sawn*

21. The weight of a piece of lumber is 2 lb. The weight of the same piece after it is fully dried is 1.8 lb. What was the original moisture content in the wood?
- **a.** 7%
- **b.** 9%
- **c.** 11%
- **d.** 15%
- **e.** None of the above

22. Fungal decay in wood occurs when the moisture content in wood is
- **a.** greater than or equal to 15%.
- **b.** greater than or equal to 20%.
- **c.** greater than or equal to 25%.
- **d.** between 15% and 25%.
- **e.** none of the above.

23. The moisture content in a piece of lumber has been measured to be 15%. According to the wood industry, this piece of lumber will be classified as
- **a.** dry.
- **b.** green.
- **c.** wet.
- **d.** moist.
- **e.** arid.

24. From the user's view point, the difference between MC 15 and KD 15 is generally ignored.
- **a.** True
- **b.** False

25. The term *S-DRY* implies that the lumber has been
- **a.** stored in a dry climate.
- **b.** sawed when its moisture content was ≤25%.
- **c.** surfaced when its moisture content was ≤25%.
- **d.** sawed when its moisture content was ≤19%.
- **e.** surfaced when its moisture content was ≤19%.

26. The term *S2S* implies that the lumber has been
- **a.** sawed twice.
- **b.** sawed from two sides.
- **c.** surfaced on two sides.
- **d.** none of the above.

27. A 10-ft-long stud was installed in position when the wood's moisture content was 19%. During a long dry spell, the wood's moisture content became 8%. The new stud length will be
- **a.** approximately 9 ft 10 in.
- **b.** approximately 9 ft $10\frac{1}{2}$ in.
- **c.** approximately 9 ft 11 in.
- **d.** approximately the same as the initial stud length of 10 ft.
- **e.** none of the above.

28. A piece of lumber has been specified as 2 in. × 4 in. in cross section. This refers to its
- **a.** actual cross-sectional dimensions.
- **b.** nominal cross-sectional dimensions.

29. The actual dimensions of 4 × 8 lumber are
- **a.** $3\frac{1}{2}$ in. × $7\frac{1}{2}$ in.
- **b.** $3\frac{1}{4}$ in. × $7\frac{1}{2}$ in.
- **c.** $3\frac{1}{4}$ in. × $7\frac{1}{4}$ in.
- **d.** $3\frac{1}{2}$ in. × $7\frac{1}{4}$ in.
- **e.** none of the above.

30. The actual dimensions of 2 × 12 lumber are
- **a.** $1\frac{1}{2}$ in. × 11 in.
- **b.** $1\frac{1}{2}$ in. × $11\frac{1}{2}$ in.
- **c.** $1\frac{1}{2}$ in. × $11\frac{3}{4}$ in.
- **d.** $1\frac{3}{4}$ in. × $11\frac{3}{4}$ in.
- **e.** none of the above.

31. In calculating the board foot measure, we must use the actual dimensions of lumber, not its nominal dimensions.
- **a.** True
- **b.** False

32. A retailer of building materials purchased 1,000 pieces of 2 × 12 lumber, each 10 ft long. How many board feet did the retailer purchase?
- **a.** 1,000 bd ft
- **b.** 10 MBF
- **c.** 20,000 bd ft
- **d.** 100,000 bd ft
- **e.** None of the above

13.9 SOFTWOOD LUMBER CLASSIFICATION

A lumber piece is classified into one of the following three categories, Figure 13.30:

- Board lumber, or simply boards
- Dimension lumber
- Timbers

The distinction between them is based on the thickness of the member—thickness being the smaller of the two dimensions of its cross section. Thus, the thickness of a 4 × 10 member is 4 in. (nominal).

Board lumber includes lumber whose thickness is less than 2 in. nominal. Thus, 1 × 4, 1 × 12, $1\frac{1}{2}$ × 6, and so on, are boards. Usually, 16 in. (nominal) is the maximum available width in board lumber. Board lumber is used in nonstructural applications such as sheathing, fencing, and shelving.

Dimension lumber includes lumber whose (nominal) thickness is 2 to 4 in. The width of dimension lumber varies from 2 to 14 in. in steps of 2 in.—that

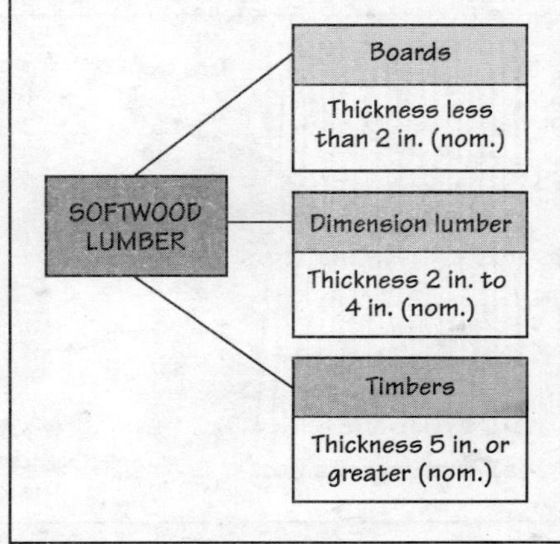

FIGURE 13.30 Size-based softwood lumber classification.

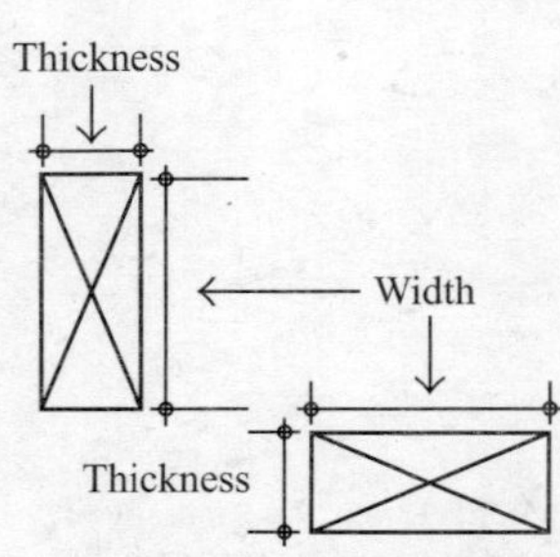

The smaller cross-sectional dimension of a lumber member is its thickness, and the larger dimension is its width.

is, 2 in., 4 in., 6 in., 8 in., and so on.* Thus, a 2 × 10, 3 × 6, 4 × 4, 4 × 8, 4 × 12, and so on, are all in the dimension lumber category.

Timbers includes lumber whose nominal thickness is 5 in. or greater. Thus, 5 × 8, 6 × 6, and 8 × 10 are timbers. Dimension lumber and timbers are graded based on strength.

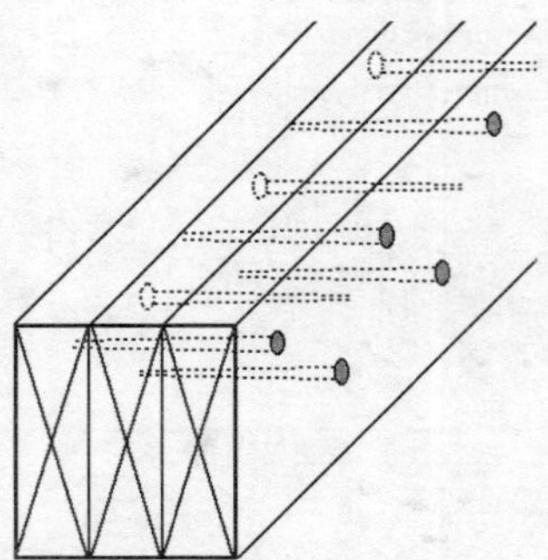

Two or three 2-by lumber members may be nailed, screwed, or bolted together to obtain a thicker member.

DIMENSION LUMBER AND TIMBERS

Most of the lumber used in structural framing of wood buildings is dimension lumber. Out of this, the 2 × (referred to as *two-by* or *2-by*) lumber is most commonly used. In fact, a whole house or an apartment block can be framed, for all practical purposes, using 2-by lumber only, such as 2 × 4, 2 × 6, 2 × 8, 2 × 10, and 2 × 12. If a member thicker than 2 in. is required, two or three 2-by members can be nailed, screwed, or bolted together.

Dimension lumber is available in lengths of 8 ft, 10 ft, 12 ft, and so on, in steps of 2 ft up to 28 ft. Most lumber yards do not regularly stock lumber longer than 20 ft. Local availability should be checked before specifying a certain length.

Two-by lumber is typically available as S4S and in surfaced dry condition. Three- or 4-in.-thick dimension lumber is available in dry or green condition. Timbers (5 in. and thicker) are generally shipped in green condition and are manufactured in S4S or rough condition.

13.10 LUMBER'S STRENGTH AND APPEARANCE

As a naturally grown material, wood does not have the same degree of uniformity as steel or concrete. Although two steel beams of the same length and cross section have identical load-carrying capacities, this is not so with lumber beams. Thus, two lumber beams of the same length and cross-sectional dimensions, even obtained from the same log, can have vastly different load-carrying capacities.

As previously discussed, the strength of a piece of lumber is affected by its species and is a function of specific gravity. Strength is also affected by its growth and manufacturing characteristics. These include slope of grain, knots, checks, shakes, splits, and wane.

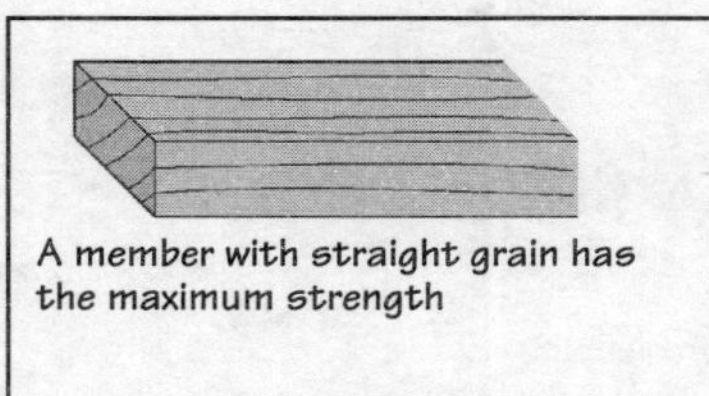

A member with straight grain has the maximum strength

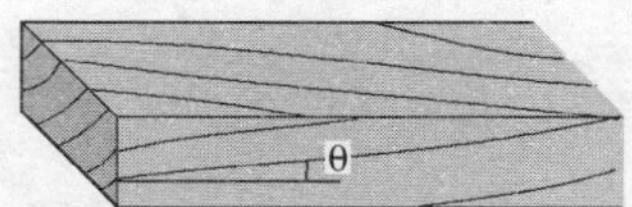

As the slope of the grain increases, (angle θ) the strength of the member decreases

FIGURE 13.31 The slope of the grain affects the lumber's strength. It is most easily observed by examining the longitudinal grain of the lumber in comparison with the edge of the member.

SLOPE OF GRAIN

An important factor that affects the strength of lumber is the slope of the grain. A piece of lumber whose grain runs parallel to the long axis of the member will have maximum strength for that species. As the slope of the grain with respect to the axis increases, the strength of the member decreases, Figure 13.31.

KNOTS

A knot occurs where a branch emerges from the tree trunk, Figure 13.32(a). The growth rings in the trunk of a tree are interlaced with the growth rings of its branches. Within the trunk, the growth rings of a branch form a cone whose end points toward the center of the trunk and whose diameter gradually increases toward the bark. Hence, the grain of wood in the vicinity of a branch is not straight, but highly disturbed from straightness (yielding cross-graining), Figure 13.32(b). Consequently, the presence of a knot in a member reduces its strength. The more numerous the knots in a member, the lower its strength.

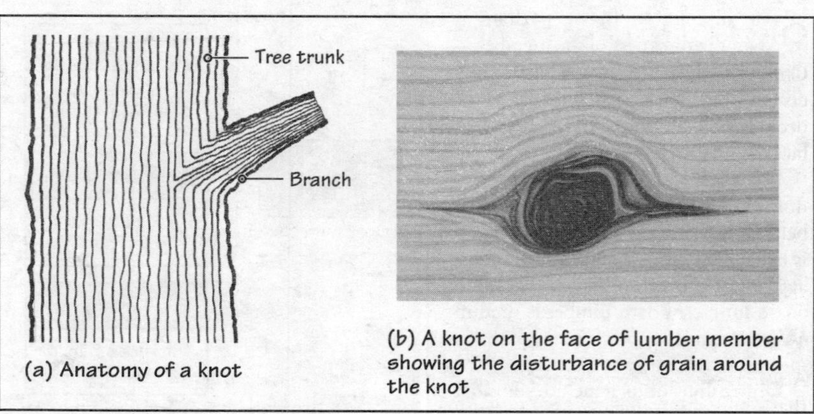

FIGURE 13.32 **(a)** Anatomy of a knot. **(b)** Grain orientation in the vicinity of a knot.

*A width greater than 12 in. may not be available in most lumber yards.

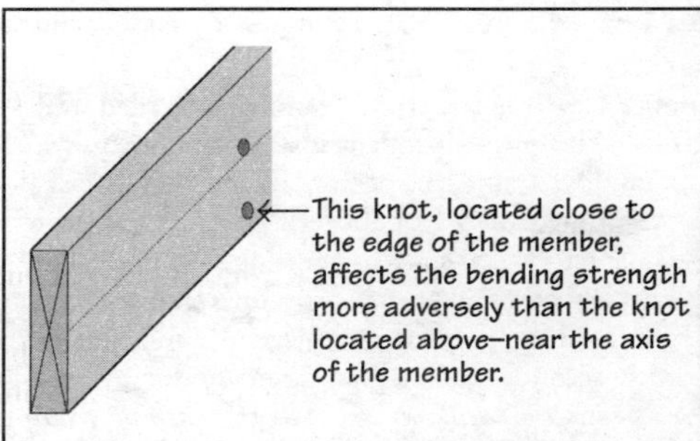

This knot, located close to the edge of the member, affects the bending strength more adversely than the knot located above—near the axis of the member.

FIGURE 13.33 Effect of the location of a knot on the bending strength of a member.

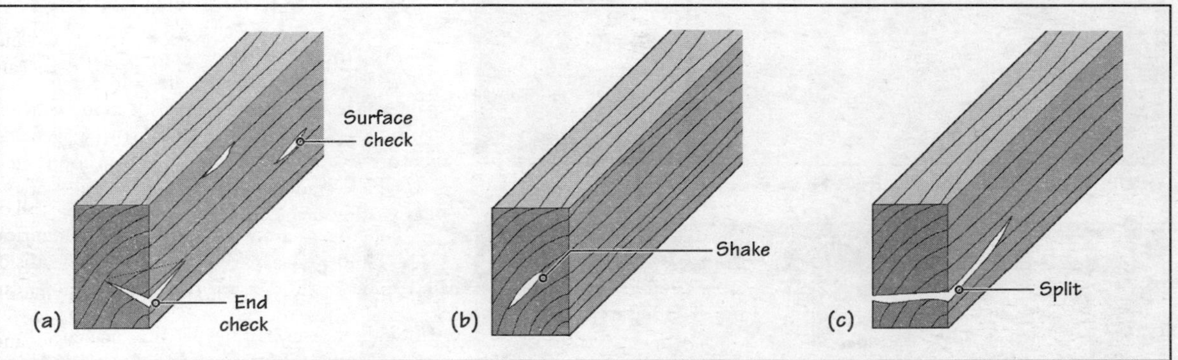

FIGURE 13.34 Check, shake, and split in lumber.

The adverse effect of a knot on the member's strength is more pronounced if the knot is a *loose knot.* A loose knot results when the branch dies during the tree's growth and the successive growth rings of the trunk encircle this dead branch. The death of the branch may be biologically caused—for example, by the overshadowing of the branch by the upper parts of the tree—or mechanically caused—such as by the effect of strong wind or a hail storm.

A loose knot reduces the strength of wood substantially, and if rotting occurs in the knot, a *knot hole* results. A knot that is not loose, but is tightly intergrown with the adjoining tissue, is referred to as an *encased knot.*

Apart from the number, size, and type of knots (encased or loose), the location of the knot within the member also affects its strength. A knot located on or near the longitudinal axis of the member does not affect the strength as adversely as a knot located near the edges of the member, Figure 13.33. Remember that in a member subjected to bending (such as a beam, floor joist, rafter, etc.), the bending stresses are maximum at the extreme edges of the member and zero at the center (neutral axis).

CHECKS, SHAKES, AND SPLITS

Checks, shakes, and splits are separations of wood fibers. A *check* is separation of wood fibers along the rays (perpendicular to growth rings). It is caused by the drying of wood and occurs at the ends of the member and also on its faces, Figure 13.34(a). It results from the fact that the surfaces of wood dry faster than its interior.

A *shake* is separation of wood fibers along the growth rings, Figure 13.34(b). It occurs during the growth of the tree and is not due to drying. A *split* occurs at the ends of a member and is a complete separation of wood fibers through the entire end, Figure 13.34(c). It is believed to be caused by a weakness that occurred during the growth of the tree and was aggravated during drying.

WANE

A *wane* is the absence of wood or the presence of bark at the corner or the edge of a piece that results from the sawing process, Figure 13.35. Like knots, checks, splits, and shakes, wanes also reduce the strength of the member. Therefore, their sizes and numbers are taken into account in the grading of lumber.

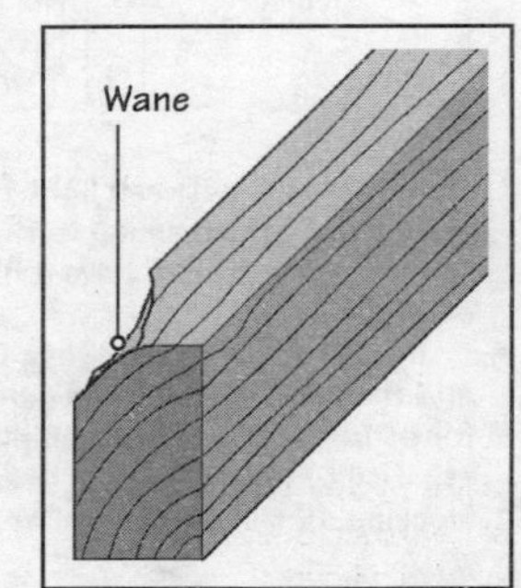

FIGURE 13.35 A wane on a piece of lumber.

13.11 LUMBER GRADING

In North America, each piece of lumber used for structural framing is graded and so stamped, Figure 13.36. The grading is done by an independent inspection agency, which employs trained inspectors. Each piece is visually examined on all of its four surfaces for growth characteristics, such as the slope of grain, knots, checks, and shakes. An inspector's trained eyes can determine the grade of a piece of lumber within a few seconds. Such a grading method is referred to as *visual grading*, as opposed to *machine grading*, described later.

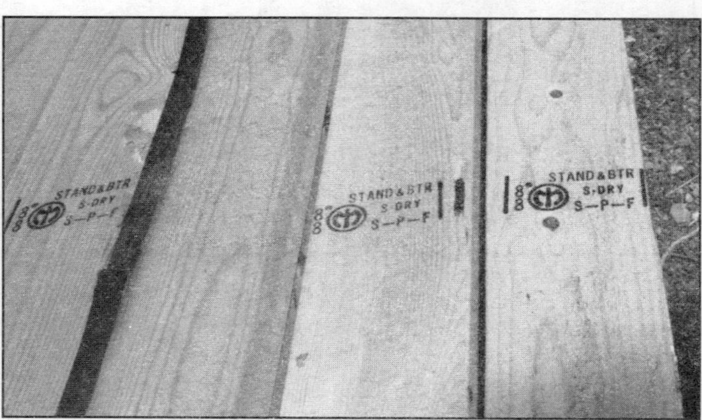

FIGURE 13.36 Grade stamps on lumber members.

In visual grading, the inspection agency grades lumber according to the grading rules prescribed by a grading rules–writing agency. In North America, there are seven *grading rules–writing agencies* (six in the United States and one in Canada), which also function as inspection agencies, Table 13.4. In addition to rules-writing agencies, there are several nonrules-writing inspection agencies, which grade lumber as per the rules of the rules-writing agencies.

TABLE 13.4 GRADING RULES–WRITING AGENCIES

U.S. Agencies
Northeastern Lumber Manufacturers Association, Inc. (NELMA)
Northern Softwood Lumber Bureau (NSLB)
Redwood Inspection Service (RIS)
Southern Pine Inspection Bureau (SPIB)
West Coast Lumber Inspection Bureau (WCLIB)
Western Wood Products Association (WWPA)
Canadian Agency
National Lumber Grades Authority (NLGA)

VISUAL GRADING—DIMENSION LUMBER GRADES AND GRADE STAMPS

Although the seven rules-writing agencies have different grading rules for boards and timbers, their grading rules for dimension lumber are identical and are referred to as the *National Grading Rules (NGR) for Dimension Lumber*. According to these rules, dimension lumber can carry any one of the several grades shown in Figure 13.37.

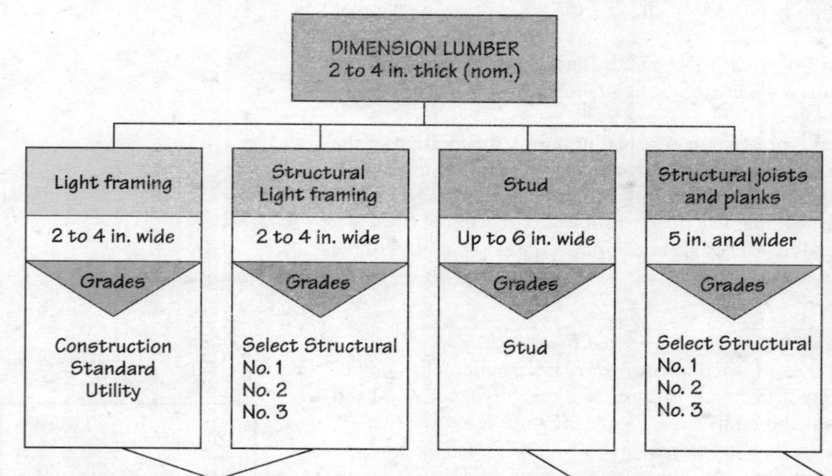

FIGURE 13.37 Various grades for dimension lumber.

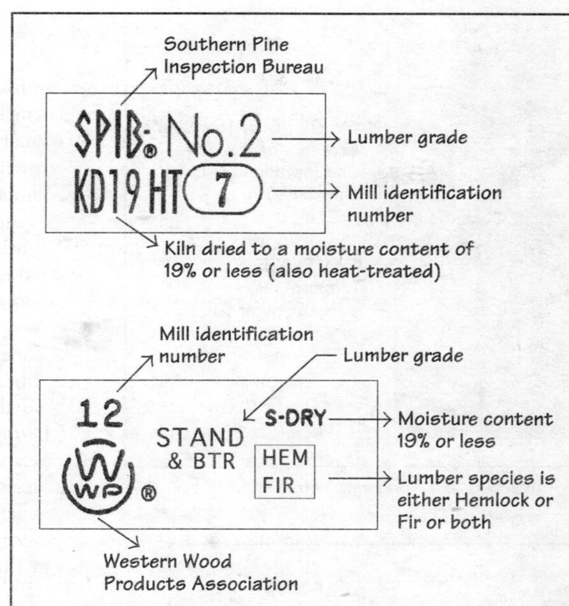

Southern pine grade stamp does not specify lumber species, since Southern pine is a mixture of four species: Longleaf pine, Shortleaf pine, Slash pine and Loblolly pine. These species are mixed together because they are almost equivalent in appearance and performance.

Grade stamps of other agencies include species information. However, two or more species may be grouped together. For example, the WWPA grade stamp given below shows that two species–Hemlock and Fir–have been combined together. Similarly, a species designation "S-P-F" means that Spruce, Pine and Fir have been grouped together.

To distinguish Canadian S-P-F designation from U.S. designation, S-P-F from the U.S. is called S-P-F (south) or S-P-F^S. Similarly, D-Fir (North) indicates that it is Canadian Douglas Fir.

Heat Treatment (HT) of Lumber is intended to eliminate pests and to make it suitable for packaging industry, which utilizes a great deal of dimension lumber. Heat treatment is not required for lumber used in construction. Heat treatment standards require that lumber be subjected to a kiln temperature of 133°F (56°C) for a minimum of 30 minutes. Kiln drying of lumber for construction purposes is generally more stringent.

FIGURE 13.38 Typical grade stamps. A comprehensive grade stamp list of North American lumber is given at the end of this chapter.

Two typical grade stamps for dimension lumber are shown in Figure 13.38. Note that a grade stamp identifies the following items:

- Species of lumber
- Moisture content at the time of surfacing
- The mill that produced the piece
- The inspection (grading) agency
- The structural grade

Examples of grade stamps of various agencies for dimension lumber available in North America are given in the Principles in Practice section at the end of this chapter.

Visual Grading—Allowable Stresses from the Grade Stamp

Through extensive structural testing of solid lumber members, the allowable stress values and the modulus of elasticity of lumber of various species and grades have been compiled by the wood industry. Table 13.5 shows these values for only one species group—spruce-pine-fir. The purpose of this table is simply to illustrate the nature of the data available for the structural design of wood frame buildings.

TABLE 13.5 BASE ALLOWABLE STRESSES OF VISUALLY GRADED DIMENSION LUMBER

Species and grade	Size classification	Bending F_b	Tension parallel to grain F_t (psi)	Shear parallel to grain F_v (psi)	Compression perpendicular to grain $F_{c\perp}$ (psi)	Compression parallel to grain $F_{c\parallel}$ (psi)	Modulus of elasticity E (ksi)
Spruce-Pine-Fir (South)							
Select Structural	2 in.–4 in. thick	1,300	575	135	335	1,200	1,300
No. 1		875	400	135	335	1,050	1,200
No. 2		795	350	135	335	1,000	1,100
No. 3	2 in. and wider	450	200	135	335	575	1,000
Stud		600	275	135	335	625	1,000
Construction	2 in.–4 in. thick	875	400	135	335	1,200	1,000
Standard		500	225	135	335	1,000	900
Utility	2 in.–4 in. wide	225	100	135	335	650	900

Base values: The values given here are base values. Design values are a function of base values and several other factors, such as lumber's moisture content, whether the member is a single-use member (such as a beam) or a repetitive-use member, type of load, and so on.
Source: American Forest & Paper Association (AF&PA) and American Wood Council (AWC): 2005 *National Design Specifications (NDS) for Wood Construction*. Courtesy American Wood Council/American Forest & Paper Association, Washington, DC, with permission.

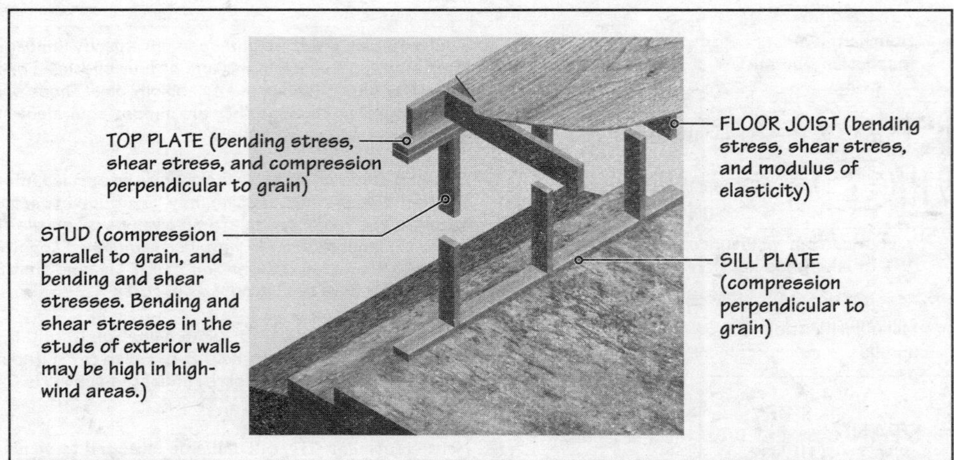

FIGURE 13.39 Structural properties that are critical for the design of various members of a wood frame building.

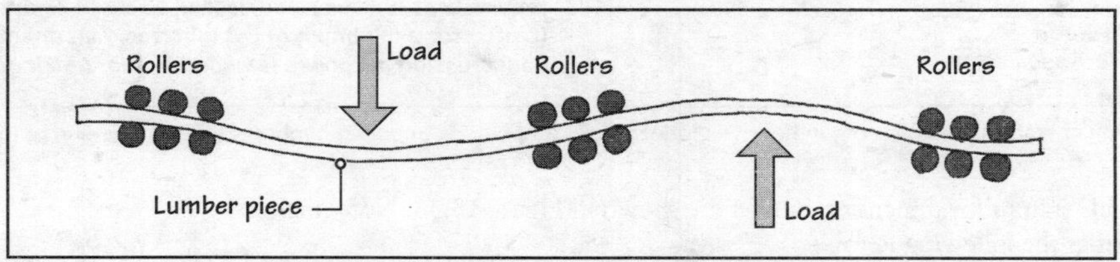

FIGURE 13.40 Outline of the equipment used for machine stress-rated lumber. Load-deflection information is fed into a computer, which automatically determines the values of the modulus of elasticity and allowable bending strength of the lumber.

Figure 13.39 shows the types of stresses that are critical for the structural design of various members of a conventional wood frame building.

Machine Stress-Rated Lumber

Lumber grading discussed previously is called *visual grading,* because the grading is done by visual examination of the lumber piece by a trained inspector. Whereas the lumber used in the structural frame of a typical wood building is visually graded, the lumber used in more demanding structural applications, such as wood trusses, is machine graded. Machine-graded lumber is referred to as *machine stress-rated (MSR) lumber,* or simply *machine-rated lumber.*

In MSR lumber, the stiffness (modulus of elasticity) of a piece of lumber is determined by a nondestructive testing procedure in which the piece is subjected to a given load and its deflection is noted, Figure 13.40. Because various structural properties of lumber are related to its stiffness, these can be obtained from the already-established statistical relationships.

Although a certain amount of visual inspection of lumber is necessary even in MSR lumber, machine grading is heavily automated, using high-speed equipment to obtain stiffness and other properties, including the affixing of the grade stamp.

An MSR grade stamp is similar to the grade stamp of visually graded lumber, except that in place of the grade mark in visually graded lumber, the MSR stamp gives E and F_b values. Thus, in the grade stamp of Figure 13.41, the lumber has the following properties: allowable bending stress (F_b) = 2,400 psi and modulus of elasticity (E) = 2.0×10^6 psi. Other structural properties of this piece, such as the allowable compressive stress, shear stress, and so on, are obtained from the tables prepared by the lumber industry.

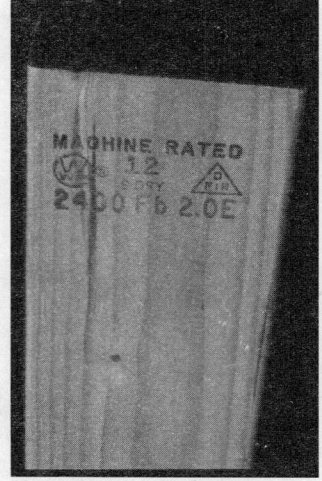

FIGURE 13.41 A typical grade stamp on a machine stress-rated (MSR) lumber.

13.12 DURABILITY OF WOOD

Like any other material, wood is subject to deterioration. Mechanical and chemical deterioration are the two most common ways in which a building material deteriorates. Mechanical deterioration of wood is due to physical wear, such as the abrasion caused by foot traffic or wheeling equipment on a floor. Other causes of mechanical wear, which occur primarily in wood exposed to the exterior climate, are (a) erosion of the material by wind and water,* (b) material fatigue caused by its repeated expansion and contraction, and (c) deterioration caused by freeze-thaw cycles.

Chemical deterioration is due to the effect of chemicals (acids and alkalies) in the environment and the chemical breakdown of wood's constituents by the ultraviolet rays of the sun. Generally, mechanical and chemical deterioration of wood is of little importance, because the major cause of wood's deterioration is biological, referred to as *biodeterioration.*

BIODETERIORATION

Biodeterioration is caused by living organisms that use wood as food. Two such groups of organisms consume wood:

- Fungi
- Insects

Deterioration of wood caused by fungi is called *fungal decay* or simply the *decay of wood* (also called *rotting*). Land-based insects that consume wood are *termites,* and water-based wood-eating insects are called *marine borers.* In addition to termites and marine borers, there are carpenter ants, which, though not using wood as food, can nest inside it, thereby weakening the wood to some degree.

13.13 FUNGAL DECAY

Unlike other plants, fungi cannot manufacture their own food but live off the food produced by other plants. Because wood is mainly cellulose and lignin, it is an excellent source of food for these parasites.

Not all forms of fungi invade the cell structure of wood. Those that do are the ones that cause structural damage to wood (decay). As the decay-causing fungus consumes this food, the strength of wood decreases, which is usually accompanied by a change in the external appearance of wood.

For instance, a type of fungus that consumes cellulose (white in color) makes wood turn more brown, and the fungus that consumes lignin (brown in color) makes wood turn whiter. *Brown rot* and *white rot* are the terms used for the two forms of decay.

Because fungus is a plant life, four factors are necessary for its survival and growth:

- Oxygen
- Mild temperature
- Water
- Food

When any one of these four elements is removed, fungal decay cannot take place. Thus, wood completely immersed in water at all times does not decay due to the absence of oxygen. Similarly, fungal decay does not generally occur in extremely cold climates. The most favorable temperature range for fungal growth and survival is 70°F to 85°F (20°C to 30°C), as it is with most other plants. Thus, in a seasoning kiln, where high-temperature air is used to dry the wood, most types of fungi perish.

PREVENTING FUNGAL DECAY

Because it is not possible to control temperature and the supply of oxygen in building assemblies, the only methods available to prevent fungal decay are (a) to control the supply of water in wood and (b) to make wood toxic to fungi.

It has been found that if the moisture content in wood is less than 20%, fungal decay does not occur. That is why 19% moisture content has been fixed as the dividing line between dry and green lumber. Specifying dry lumber and keeping it dry (below 19% moisture content) is the best insurance against decay. This fact highlights the importance of ventilating attics and crawl spaces in wood structures, because excessive moisture buildup in confined spaces can raise the moisture content of wood above 20%.

Although the lumber that is not exposed to rain or groundwater may be kept dry, this is not true with lumber exposed to a water source. Therefore, exterior lumber, lumber buried

*The erosion due to wind is caused by dust and grit in the air. Thus, in terms of the weatherability of materials, wind behaves as a low-intensity sand blasting and rain serves as water blasting.

in the ground, or lumber close to or in contact with the ground will decay. Two options are available to prevent the decay of exposed lumber:

- Use of naturally decay-resistant species
- Use of preservative-treated lumber

Whereas the heartwood of a tree is generally more decay resistant than its sapwood, the heartwood of a few species, such as cedar, redwood, cypress, and black locust, have a high degree of decay resistance. These species may be used in decay-prone situations. Preservative-treated lumber is treated with chemicals that are toxic to fungi and insects.

13.14 TERMITE CONTROL

Most termite damage in buildings is caused by subterranean termites. Subterranean termites live in underground colonies. Due to their distaste for light, they normally cannot reach above-ground wood. However, they can build tubular mud tunnels over foundation walls and piers to reach the above-ground wood.

Because they can obtain water from the ground, termites do not have to depend on the water in the wood. Thus, although wood with a moisture content of less than 20% is not subject to fungal decay, it is vulnerable to termites if the termites can reach the wood.

Subterranean termites are more commonly found in warm climates; the termite hazard is usually higher in warm, damp climates than in warm, dry climates. The hazard is relatively low in colder regions. Figure 13.42 shows the termite-infestation probability map of the United States.

NOTE

Termites and Building Materials

Cellulose-based materials are food for termites, but termites also attack plastics. They are known to chew through cable shields, plastic laminates, and plastic foam insulation. The use of below-ground plastic foam insulation is, therefore, not recommended in termite-infested regions.

Reducing (or Preventing) Termite Damage

A multipart strategy is required to reduce or prevent termite damage in wood buildings. For wood members not buried in the ground, all or a combination of the following strategies are generally recommended:

- Maintain distance between wood and ground
- Provide a soil barrier, that is, a chemical soil treatment
- Use naturally decay-resistant or preservative-treated wood
- Use a termite shield
- Inspection and remediation

The best line of defense against termites is to provide sufficient distance between the ground and the wood member, supplemented by frequent inspection to ensure that no mud tunnels exist. Figure 13.43 shows some of the commonly recommended construction details for preventing termite attack.

In most wood frame buildings, preservative-treated lumber is specified only for the sill plate when it is anchored to the concrete or masonry foundation. All other framing lumber is

NOTE

Importance of a Termite Barrier

Experts estimate that termites can go through a space as small as $\frac{1}{24}$ in. Therefore, a concrete slab-on-grade should be without visible cracks. Hollow concrete masonry foundations must be fully grouted, and the termite shield must be carefully soldered at all joints.

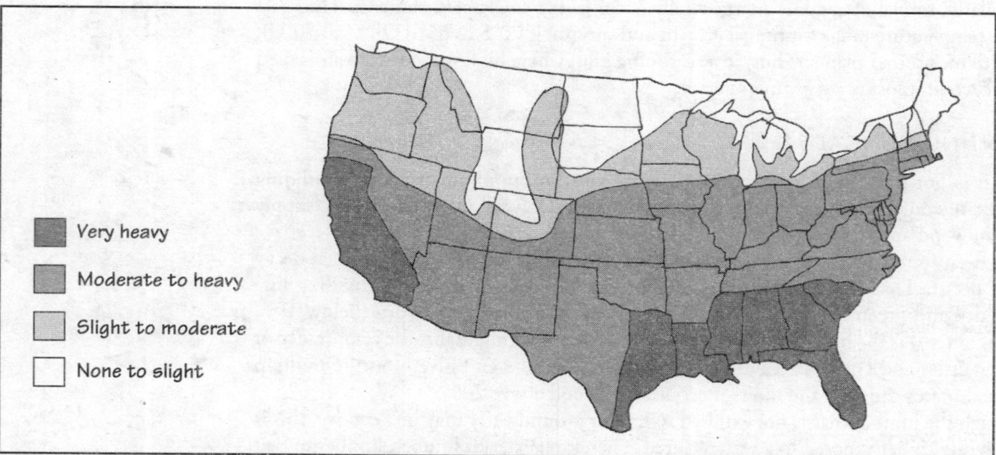

FIGURE 13.42 Termite infestation probability map of the United States. This map should be regarded only as approximate and must be verified with its original source and the information obtained from the building code of the jurisdiction.
Source: United States Forest Service publication: URL: http//www.treesearch.fs.fed.us/pubs/26674.

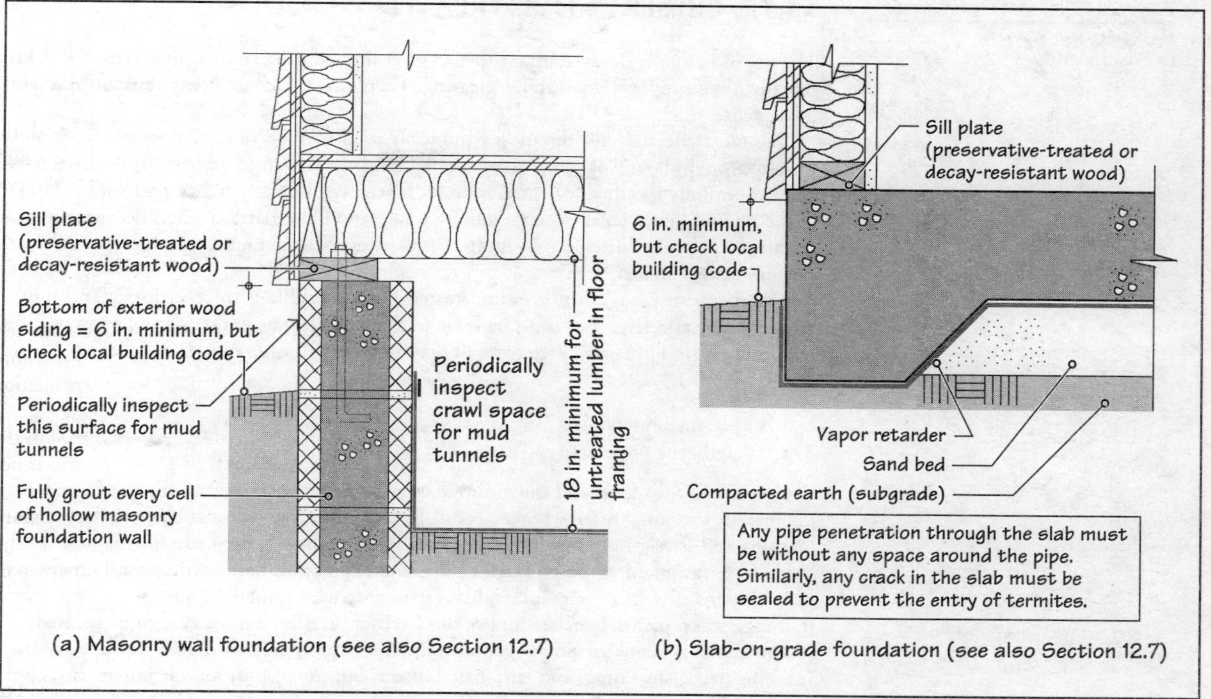

FIGURE 13.43 The use of preservative-treated or decay-resistant lumber in a conventional wood frame building.

usually untreated lumber. Wood floor joists over a crawl space are required to be of preservative-treated or decay-resistant species only if the distance between the bottom of the joists and the ground is less than 18 in. Wood beams less than 12 in. above the ground must satisfy the same criterion.

Treating the soil around foundation walls and the soil below concrete slabs-on-grade or basement slabs provides additional protection against termites. Although the method of application of the chemicals (e.g., aldrin or dieldrin—0.5% in water emulsion) is fairly simple, it is recommended that the application be done by a specialist agency. Booster applications may be needed in some locations every few years.

In locations subjected to heavy or very heavy termite hazard (based on local experience), the use of a termite shield is recommended. A termite shield is generally made of galvanized sheet steel profiled with a drip-edge-type bend. Termites are unable to maneuver around this bend in building mud tunnels. All wood must be placed above termite shields, Figure 13.44.

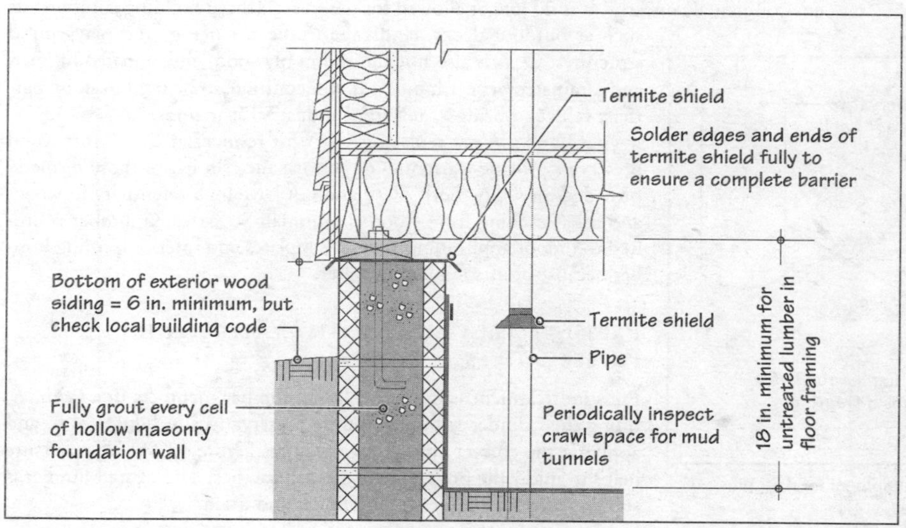

FIGURE 13.44 The use of a termite shield is recommended as an additional protection in regions with heavy infestation of termites.

13.15 PRESERVATIVE-TREATED WOOD

The use of naturally decay-resistant species gives limited protection against termites. Additionally, their supply is small and decreasing. Therefore, the use of decay-resistant species is infrequent.

The most effective and the most commonly used method of achieving termite protection is through the use of preservative-treated lumber. Therefore, preservative-treated lumber is commonly specified for outdoor decks, fences, verandahs, porches, and lumber buried in the ground (fence poles, wood retaining walls, wood foundations, etc.). Because the preservative is pressure injected into lumber, preservative-treated lumber is also referred to as *pressure-treated lumber.*

The preservatives used in pressure-treated lumber enable it to resist insect and fungal attack. To be effective, they must be toxic to these organisms, nondamaging to wood, and relatively safe to humans. Three types of preservatives are commonly used for the purpose:

- Creosote
- Oil-borne preservatives
- Waterborne preservatives

Creosote is the oldest and the most effective preservative against all forms of wood-eating insects, including marine borers. A distillate of coal tar, creosote is black to deep brown in color and is relatively insoluble in water. It is commonly used in railroad ties, utility poles, highway guardrail posts, marine bulkheads, and piles. Because frequent human contact with creosote is not recommended, creosote-treated lumber is not used in structural framing, decks, patios, benches, and so on. Creosote-treated lumber cannot be painted.

Pentachlorophenol (or simply penta) is the most commonly used oil-borne preservative. It is effective against fungi and land-based insects but not against marine borers. It is commonly used for bridge timbers and utility poles.

WATERBORNE PRESERVATIVES

Wood treated with waterborne preservatives has several applications. Common applications of waterborne preservative–treated wood are on outdoor decks, fences, gazebos, playground equipment, structural framing, and highway noise barriers. Commonly used preservatives are

- Chromated copper arsenate (CCA)
- Alkaline copper quat (ACQ)
- Copper azole (CA)
- Sodium borates (SBX)

CCA is a waterborne mixture of the compounds of copper, chromium, and arsenic, giving the name *chromated copper arsenate.* Copper is toxic to fungi, and arsenic is an insecticide. Chromium is added to lock the preservative into wood.

CCA was the most widely used waterborne preservative until 2003. However, due to environmental concerns, it is no longer allowed for use in residential building components such as outdoor decks, landscape timbers, fencing, and playground structures. CCA is also not allowed in plywood, glulaminated lumber, and laminated veneer lumber. It will continue to be used in noise barriers, vehicular bridges, and some industrial structures.

ACQ and CA are replacing CCA for residential uses. Most wood preservers provide warranties of performance similar to those formerly provided for CCA. Both ACQ and CA have high retentivity in wood, and their leaching from wood is minimal. SBX-treated lumber is limited to indoor applications such as sill plates and interior framing lumber because of its solubility in water.

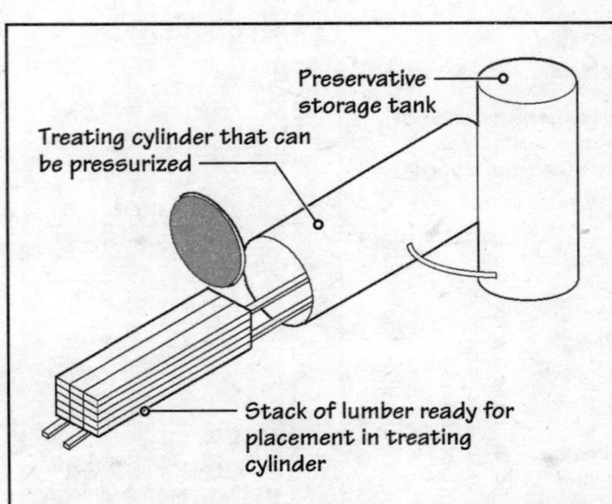

FIGURE 13.45 Outline of the equipment used for pressure treating lumber with a preservative.

PRESSURE TREATMENT WITH WATERBORNE PRESERVATIVES

Pressure treatment is done by placing lumber members in a cylinder, filling the cylinder with waterborne preservatives in liquid form, and applying the correct amount of pressure, Figure 13.45. The pressure helps to inject the preservative into cell cavities. The treated lumber is subsequently taken out of the cylinder and dried.

A piece of lumber kiln dried (to less than or equal to 19% moisture content) after treatment carries the stamp KDAT, meaning *kiln dried after treatment.* The amount of preservative injected into wood is a function of the hazard to which it would be subjected, as shown in Table 13.6.

Not all wood species can be effectively pressure treated; more permeable ones are better for treatment. Some of the commonly treated species are shown in Table 13.7. Of these, some species need incisions (slits along the grain of wood) prior to treatment to help push the preservative into the required depth of the member.

Because of the presence of copper in CCA, ACQ, and CA, pressure-treated lumber is generally greenish in color. However, a coloring agent can be added to the preservative to change the color of treated lumber to one that approaches the color of untreated lumber. Treated lumber may be stained, painted, or sealed.

Treated lumber carries a treater's stamp. It is either an imprint (ink stamp) on the lumber or, more commonly, a plastic tag stapled to an end of the member, referred to as an *end tag* in the industry, Figure 13.46. Note that treated lumber carries two stamps: One stamp refers to the mill that originally produced that member and its grade, and the other stamp is the treater's stamp (or end tag). Figure 13.47 shows the information that is provided on treated lumber's end tag or stamp.

TABLE 13.6 RETENTION OF WATERBORNE PRESERVATIVES AND LOCATION OF TREATED LUMBER OR PLYWOOD IN A BUILDING

Location of lumber or plywood	Retention (lb/ft³)
Aboveground	0.25
Ground contact	0.40
Permanent wood foundation	0.60
Structural poles	0.60
Fresh water immersion	0.60
Salt water immersion [a]	2.50

[a]Marine borers are usually found in salt water or brackish water.

TABLE 13.7 MAJOR PRESSURE-TREATABLE SOFTWOOD SPECIES

Southern pine	Douglas fir
Ponderosa pine	Hem-fir
Red pine	Western larch
Jack pine	Redwood
Sugar pine	
White pine	

FIGURE 13.46 A pressure-treated lumber pile with end tags.

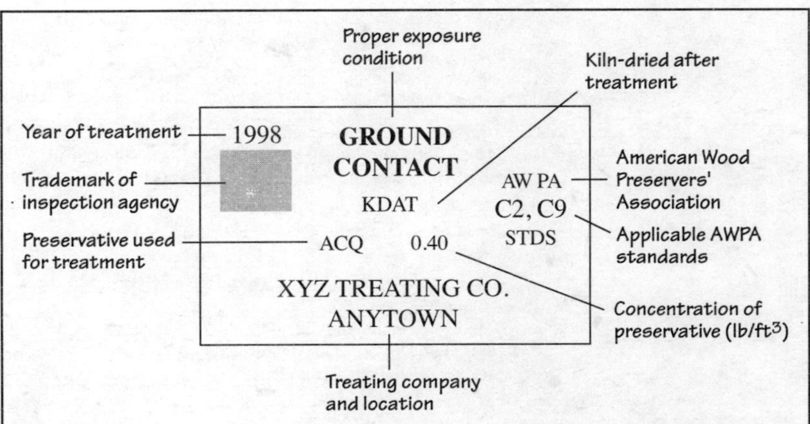

FIGURE 13.47 Information provided on an ink stamp or an end tag on pressure-treated lumber.

FOCUS ON SUSTAINABILITY

Sustainability Features of Lumber

Wood is biodegradable and the most rapidly renewable building material. It is the only material that is theoretically inexhaustible and uses the most renewable form of energy for its production—solar energy. The only nonrenewable energy used in wood's production is in sawing, planing, and dressing of wood to its finished form. Additionally, although the production of other materials creates environmental pollution from gaseous and liquid emissions, the production of wood (the growth of trees) produces oxygen that purifies the environment. The use of wood as a building material is, therefore, a sustainable practice.

However, because forests are essential for the survival of life on earth, excessive and uncontrolled harvesting of trees to obtain wood is damaging to the environment. The wood industry has, therefore, introduced third-party certification of lumber and engineered wood products. The certification of wood is a means of ensuring that the forests are managed for sustainability through

- Maintenance of biodiversity of plant and animal species in forests
- Conservation of soil and water quality in forests
- Controlled harvesting of trees for socioeconomic benefits
- Sustained improvement in forest management practices

Certification of wood is voluntary, and a number of agencies conduct wood certification under the rules established by the Forest Stewardship Council (FSC). Certified wood is more expensive. However, it is expected that with increasing use of certified wood, a greater number of commercial forests will participate in the certification of their products, bridging the cost gap between certified and uncertified wood.

FSC is a global umbrella organization with a small permanent secretariat in Oaxaca, Mexico, that accredits wood-certifying agencies in various countries. Wood certified by an FSC-accredited certifier (e.g., the SmartWood and Scientific Certification systems) carries the FSC logo.

The FSC stamp is an index of sustainable forest-management practices. It does not include other issues of sustainability, such as embodied energy and emissions of volatile organic compounds from engineered wood products.

In addition to special handling precautions, treated wood has special disposal restrictions. For instance, treated wood must not be burned in stoves or fireplaces because of the toxic chemicals that it releases. The fasteners (nails and screws) and other steel components (joist hangers, etc.) used with treated wood must be of galvanized steel or stainless steel because of the corrosive effect of the preservative.

13.16 FIRE-RETARDANT-TREATED WOOD

Fire-retardant-treated wood (FRTW) is lumber or plywood that has been pressure treated with chemicals that retard the development of fire. Although not recognized as a noncombustible material (such as steel, concrete, or masonry), FRTW is permitted by building codes in non-load-bearing situations, where only noncombustible materials are allowed. FRTW is required to have a flame spread rating of 25 or less (see Chapter 7 for the definition of flame spread rating). Other uses of fire-retardant treatment are in wood roofing shingles and shakes. Each FRTW piece must carry the Underwriters Laboratory identification.

PRACTICE QUIZ

Each question has only one correct answer. Select the choice that best answers the question.

33. Which of the following lumber cross sections is that of a dimension lumber?
a. 4 × 8 **b.** 6 × 8
c. 1 × 8 **d.** 1 × 12
e. 6 × 12

34. Which of the following is not a lumber grade?
a. No. 1 **b.** No. 2
c. No. 3 **d.** No. 4
e. Construction

35. The basis for distinguishing lumber as *board, dimension lumber,* and *timber* is the
a. species of lumber.
b. strength of lumber.
c. cross-sectional dimensions of lumber.
d. length of lumber.
e. durability of lumber.

36. A lumber grade stamp must include information on the cross-sectional dimensions of lumber.
a. True **b.** False

37. A lumber grade stamp must include the mill identification number.
a. True **b.** False

38. Which of the following grades will not apply to a 2 × 8 lumber?
a. No. 1 **b.** No. 2
c. No. 3 **d.** Construction

39. Which of the following grades can only be used as blocking?
a. No. 1 **b.** No. 2
c. Construction **d.** Utility
e. Standard

40. Lumber used for structural framing members, such as studs, floor joists, and rafters, is graded using ____.
a. only visual inspection.
b. only machine grading.
c. both visual and machine grading must be used to arrive at lumber's grade.

41. Preservative-treated lumber is effective
a. only against termite attack. **b.** only against fungal attack.
c. against both fungi and termites.

42. A termite shield is generally made of
a. asphalt-treated felt. **b.** galvanized sheet steel.
c. lead sheet. **d.** kraft paper.

43. Wood certified by the Forest Stewardship Council (FSC) is available
a. only in the United States.
b. in the United States and Canada.
c. in the United States, Canada, and Mexico.
d. in (c) and several other countries.

PRINCIPLES IN PRACTICE

Typical Grade Stamps of Visually Graded Lumber

U.S. Grading Agencies

Northeastern Lumber Manufacturers Association (NeLMA), Inc. Cumberland Center, Maine—a rules-writing agency and inspection agency

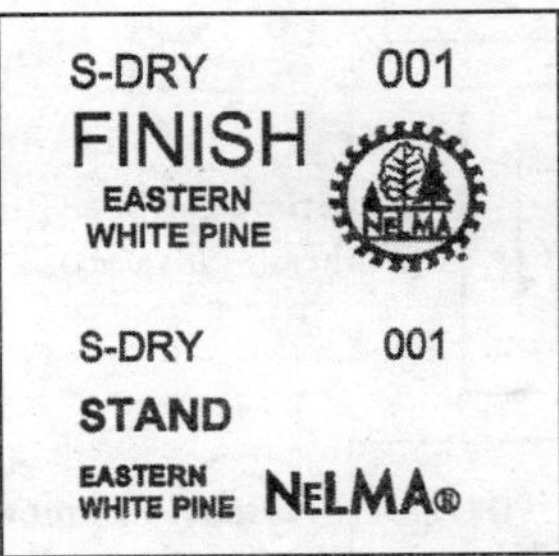

Northern Softwood Lumber Bureau (NSLB), Cumberland Center, Maine—a rules-writing agency and inspection agency

Pacific Lumber Inspection Bureau (PLIB), Inc., Bellevue, Washington—an inspection agency

Redwood Inspection Service (RIS), Novato, California—a rules-writing agency and inspection agency

Renewable Resource Associates (RRA), Inc., Atlanta, Georgia—an inspection agency

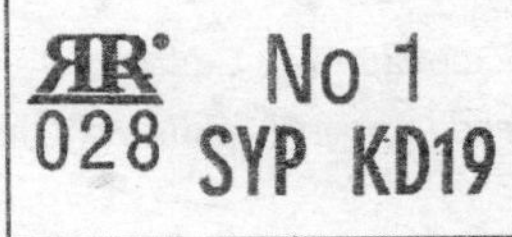

Southern Pine Inspection Bureau (SPIB), Pensacola, Florida—a rules-writing agency

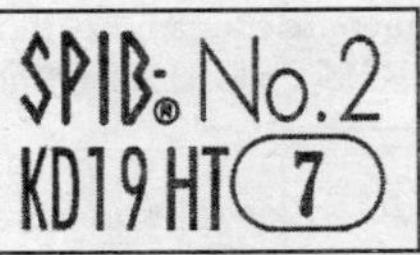

Timber Products Inspection (TPI), Conyers, Georgia—an inspection agency

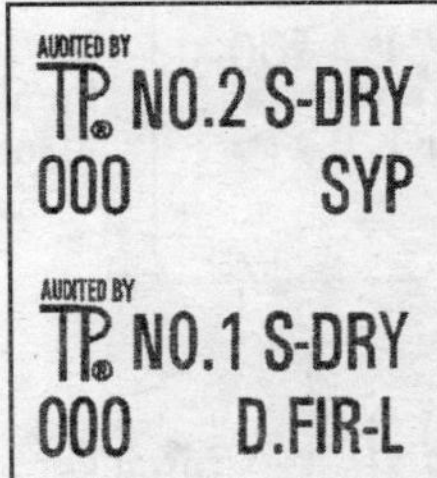

West Coast Lumber Inspection Bureau (WCLIB), Portland, Oregon—a rules-writing agency and inspection agency

Western Wood Products Association. Portland, Oregon—a rules-writing agency and inspection agency

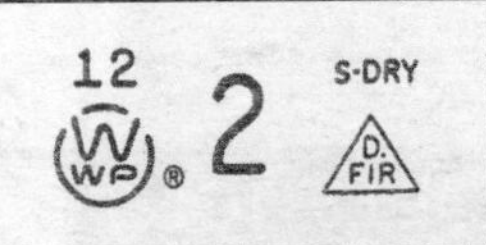

(Continued)

PRINCIPLES IN PRACTICE

Typical Grade Stamps of Visually Graded Lumber (*Continued*)

Canadian Inspection Agencies

The following Canadian inspection agencies have been certified by the American Lumber Standards Committee. There is only one rules-writing agency in Canada—the National Lumber Grades Authority (NLGA). NLGA's rules for dimension lumber are the same as those of the U.S. National Grading Rules (NGR) for Dimension Lumber.

A.F.P.A.® 00
S-P-F NLGA 1
KD-HT

Alberta Forest Products Association (AFPA), Edmonton, Alberta

CL®A S-P-F 100 No. 2 S-GRN.
CL®A 100 SPRUCE-PINE-FIR NO. 1 S-DRY

Canadian Lumbermen's Association (CLA), Ottawa, Ontario

CFPA® 00
S-P-F S-DRY
CONST

Central Forest Products Association (CFPA), Hudson Bay, Saskatchewan

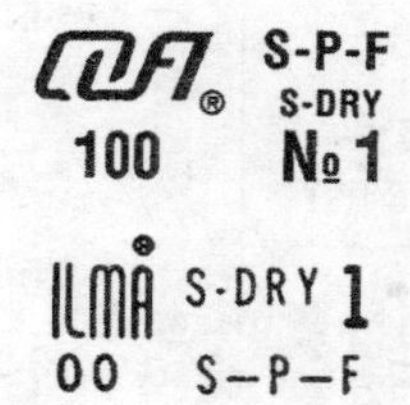

Council of Forest Industries (COFI), Vancouver, British Columbia

MacDonald Inspection (MI), Coquitlam, British Columbia

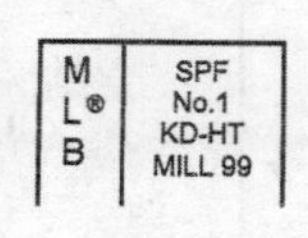

Maritime Lumber Bureau (MLB), Amherst, Nova Scotia

O.L.M.A.® 01-1
CONST. S-DRY
SPRUCE - PINE - FIR

Ontario Lumber Manufacturers Association (OLMA), Toronto, Ontario

Pacific Lumber Inspection Bureau (PLIB), Bellevue, Washington , U.S.A., British Columbia Division: Vancouver, British Columbia

Quebec Lumber Manufacturers Association, (QFIC), Ste-Foy, Quebec

Source: These grade stamps have been adapted from the American Lumber Standards Committee's (ALSC) website (www.alsc.org), with permission. Note that this is an incomplete list. For a complete list refer to the above website.

REVIEW QUESTIONS

1. Describe the essential differences between softwood and hardwood trees. Give at least three commonly used species of each.
2. What is the difference between heartwood and sapwood? Explain.
3. What does the term *SPF* mean? What is the difference between SPF^s and SPF? Explain.
4. Using sketches and notes, explain why wood is stronger along the grain than across the grain.
5. Using sketches and notes, explain how to distinguish graphically between framing lumber and finished lumber in cross-sectional drawings.
6. In visually grading of lumber, what do graders typically look for in the piece of lumber? Explain.
7. What information is typically provided on the grade stamp of visually graded lumber? Explain with the help of an example of a typical grade stamp.
8. With the help of a sketch, explain how machine-rated lumber is graded. What information does the grade of machine-rated lumber provide?
9. Explain how fungal decay of lumber can be prevented.
10. With the help of a sketch, explain what a termite shield is and where in a building it is typically provided.

CHAPTER 14

Materials for Wood Construction–II (Engineered Wood Products, Fasteners, and Connectors)

CHAPTER OUTLINE

There is a growing number of wood products that are produced using techniques that extend beyond sawing and planing. One of the earliest examples—plywood—originated from a laminating process that can be traced back to ancient Egypt and China. However, it was not until the invention of waterproof adhesives in the 1930s that plywood became the first commercially produced *manufactured wood product.*

Manufactured wood products are made by bonding together lumber members, wood veneers, wood strands, wood particles, and other forms of wood fibers to produce a composite material. Their manufacturing process eliminates defects and weak points in wood (or spreads them over the product), giving a stiffer, stronger, and more homogeneous material than lumber. The process also utilizes wood materials that would have otherwise become a waste.

Two types of manufactured wood products are used in construction:

- Engineered wood products
- Industrial wood products

Engineered wood products are those that are engineered for structural applications. They include glue-laminated wood, structural composite lumber, wood I-joists, plywood, oriented strandboard, and wood trusses.

Industrial wood products include particle board, medium-density fiberboard (MDF), and high-density fiberboard (HDF), also called *hardboard.* Particle board is commonly used for cabinet work, furniture, heavy-duty shipping containers, and so on. MDF is used as a replacement for solid lumber. It has a smooth surface that allows precise machining to form complex and intricate moldings. Hardboard is commonly used as floor underlayment (Section 14.8).

This chapter begins with a discussion of various engineered wood products. It then deals with fasteners and connectors used in wood frame construction.

14.1 GLULAM MEMBERS

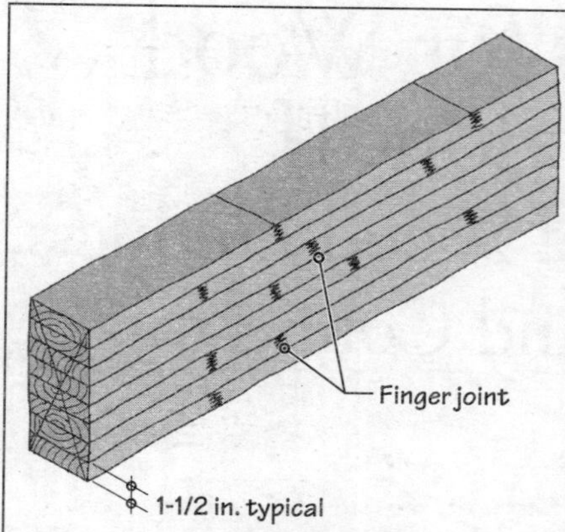

FIGURE 14.1 Anatomy of a glulam beam. A glulam beam is typically made by gluing together $1\frac{1}{2}$-in.-thick laminations of solid sawn lumber.

Sawn lumber has several limitations where large cross-sectional wood members are required to span long distances between structural supports. Some of these limitations are:

- Large sawn lumber cross sections can be obtained only from trees with large girths. Such trees are generally protected from harvesting. Additionally, trees with excessively long life spans make lumber farming uneconomical.
- Large sawn lumber cross sections cannot be dried to an acceptable moisture content.
- Because sawn lumber is naturally grown, there is little control over its structural properties.

Glue-laminated wood (referred to as *glulam*) is the response to these limitations. It is made from individual lengths of dimension lumber that are glued together to form large cross sections. The individual lengths are joined horizontally (face laminated) as well as vertically (end jointed), Figure 14.1. The ends are generally finger jointed.

Laminating is an effective way of using short lengths of high-grade sawn lumber obtained by eliminating pieces that are of low grade due to knots, shakes, and splits. Because glulam is generally made from dimension lumber, it uses lumber that has been dried to a fairly low moisture content, generally less than 15%. Additionally, laminating allows the use of smaller trees harvested from younger forests.

The adhesive used between laminations is of high strength and is fully water resistant. Consequently, a glulam member is stronger and stiffer than a sawn lumber of the same dimensions. The use of water-resistant adhesive means that a glulam member can be used in externally exposed conditions without delaminating along adhesive lines.

FIGURE 14.2 A 12-in.-wide, 7-ft-deep glulam beam over a long span (82 ft) in Gunter Primary School, Gunter, Texas. Architect: SAI Architects, Keller, Texas. (Photo courtesy of APA—The Engineered Wood Association)

USES AND SIZES OF GLULAM MEMBERS

The most common use of glulam members is for long-span beams, Figure 14.2. Other uses are for heavy columns, Figure 14.3, and heavy trusses, Figure 14.4. In some situations, glulam members are specified because their large cross sections provide greater fire resistance. A fire resistance rating of up to 1 h is obtainable with unprotected glulam members.

Glulam manufacturers produce beams of standard cross sections that are cut to the required lengths. Because 2-by (i.e., $1\frac{1}{2}$-in.-thick) lumber is used in producing glulam beams, its depth is generally in multiples of $1\frac{1}{2}$ in., such as $7\frac{1}{2}$ in., 9 in., or $10\frac{1}{2}$ in.

The width of glulam beams varies depending on the width of the dimension lumber used. Standard beam widths vary from nearly 3 in. to nearly 14 in. Wider beams can be custom produced. Theoretically, a glulam member can be made to any cross-sectional size and length by gluing pieces of dimension lumber side by side and face to face, Figure 14.5. However, limitations of transportation to the construction site and installation of the member must be considered in choosing a large glulam member.

Because glulam is made from small wood sections under controlled conditions, it can also be curved along its length. Thus, arches and other contours can be easily obtained in glulam, Figure 14.6. In curved glulam members, laminations are generally $\frac{3}{4}$ in. thick instead of the standard thickness of $1\frac{1}{2}$ in.

FIGURE 14.3 Glulam columns used in Beaverton Public Library, Beaverton, Oregon. Architect: Thomas Hacker and Associates, Portland, Oregon. (Photo courtesy of APA—The Engineered Wood Association)

FIGURE 14.4 Glulam trusses used in the gymnasium of Thunder Mountain Middle School, Enumclaw, Washington. Architect: BLRB Architects, Tacoma, Washington. (Photo courtesy of APA—The Engineered Wood Association)

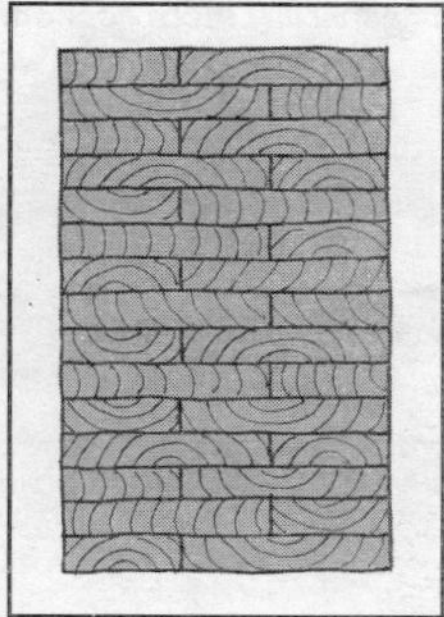

FIGURE 14.5 Large cross sections of glulam members are made by gluing laminations side-by-side and face-to-face.

FIGURE 14.6 Seshan Golf Course bridge near Shanghai, China. (Photo courtesy of APA—The Engineered Wood Association)

BALANCED AND UNBALANCED GLULAM BEAMS

Balanced glulam beams are symmetrical in lumber quality above and below the beam's mid-depth. In an unbalanced beam, the quality of the lumber used in upper laminations is different from that of the lumber used in lower laminations to account for the fact that the lower half of the beam will be in tension and the upper half of the beam will be in compression. Unbalanced beams are, therefore, stamped with *TOP* to ensure their correct placement, Figure 14.7. They can be used only as single-span beams, Figure 14.8.

Balanced glulam beams are more versatile because they can be used for any span condition. However, they are generally used where the beams are continuous over two or more supports, because a continuous beam is subjected to tension and compression on both faces of the beam at different locations. Similarly, a beam with an overhang (cantilever) will also require the use of a balanced glulam beam.

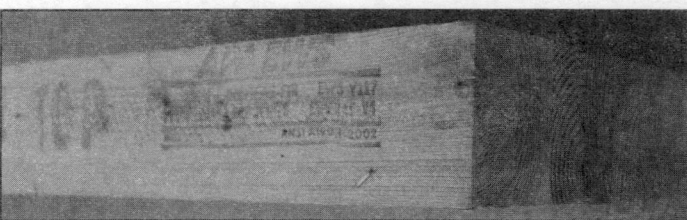

FIGURE 14.7 An unbalanced glulam beam showing the beam's top face.

NOTE

APA, PFS, and TECO

Three U.S. organizations provide third-party certification and quality control of glulam and other engineered wood (EW) products: APA—The Engineered Wood Association, PFS Corporation, and TECO Corporation.

APA originated as the certifier of plywood products. At that time, APA was the acronym for *American Plywood Association*. With the introduction of other engineered wood products, it has changed its name to "APA—The Engineered Wood Association" to better describe its new role.

PFS and TECO are sister organizations. TECO deals with the quality certification of wood panels—plywood and oriented strandboard—and PFS deals with the other engineered wood products.

Of these three organizations, APA has by far the largest market share at the present time.

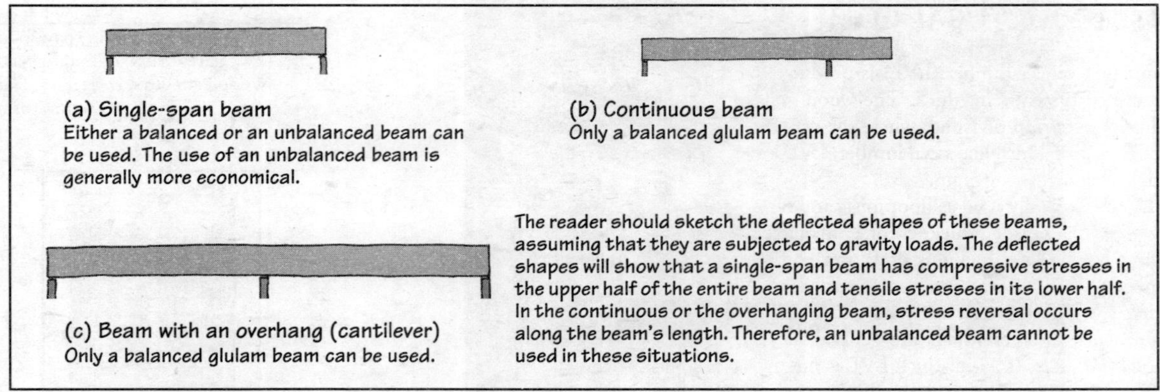

(a) Single-span beam
Either a balanced or an unbalanced beam can be used. The use of an unbalanced beam is generally more economical.
(b) Continuous beam
Only a balanced glulam beam can be used.
(c) Beam with an overhang (cantilever)
Only a balanced glulam beam can be used.
The reader should sketch the deflected shapes of these beams, assuming that they are subjected to gravity loads. The deflected shapes will show that a single-span beam has compressive stresses in the upper half of the entire beam and tensile stresses in its lower half. In the continuous or the overhanging beam, stress reversal occurs along the beam's length. Therefore, an unbalanced beam cannot be used in these situations.

FIGURE 14.8 Single-span bean, continuous beam, and cantilevered beam.

SPECIFYING GLULAM MEMBERS—GLULAM MEMBER GRADES

Most glulam producers in the United States conform to the quality certification mechanism developed by APA—The Engineered Wood Association, which requires each glulam member to be grade stamped. Among other information, the grade stamp on a glulam member provides the structural properties of the beam and its appearance. The appearance of a glulam member is classified into one of the following four grades:

- *Framing Appearance Grade*—the lowest appearance grade in which the glue smears and squeeze-out, knots, and splits are acceptable. Framing grade glulam members should not be exposed to view.
- *Industrial Appearance Grade*—commonly specified in situations where the appearance of the member is relatively unimportant, such as in a warehouse, or if the member is to be covered with a finish material. Some manufacturers make industrial grade glulam beams called *headers*. These are beams whose thickness is consistent with the commonly used widths of wood frame walls. Thus, glulam headers are made to widths of $3\frac{1}{2}$ in., $5\frac{1}{2}$ in., and $7\frac{1}{4}$ in. The $3\frac{1}{2}$-in.-thick header can be used over an opening in a wood frame wall with 2×4 studs. A $5\frac{1}{2}$-in.-thick header is used in a wall with 2×6 studs.
- *Architectural Appearance Grade*—an appearance grade that is intermediate between those of industrial and premium grades.
- *Premium Appearance Grade*—a grade commonly specified where the appearance of the beam is important.

A typical APA grade stamp on a glulam member is shown in Figure 14.9.

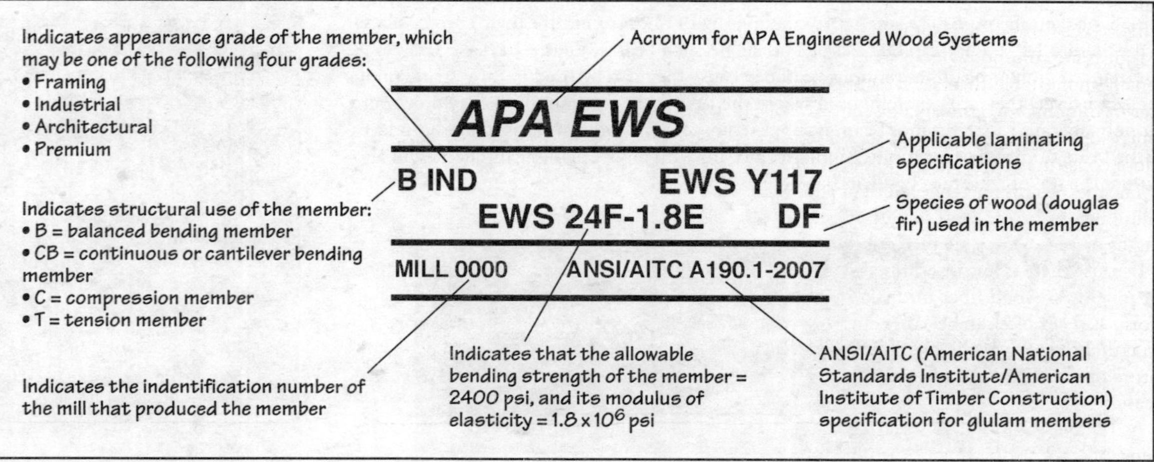

FIGURE 14.9 A typical grade stamp on a glulam member.

14.2 STRUCTURAL COMPOSITE LUMBER—LVL AND PSL

Laminated veneer lumber (LVL) is produced by gluing together dried wood veneers that are approximately $\frac{1}{8}$ in. thick. The wood grain in all veneers runs in the same direction, unlike the grain in plywood, where the veneers are cross-grained between laminations (see Section 14.5). Thus, like sawn lumber, LVL is stronger along the grain and weaker across the grain.

LVL is generally used as floor joists and rafters. Therefore, it is usually made to a finished thickness of $1\frac{3}{4}$ in. and to depths of up to 18 in., Figure 14.10. The thickness of $1\frac{3}{4}$ in. (compared to $1\frac{1}{2}$ in. for sawn lumber) provides a greater fastening surface for floor or roof sheathing. Other thicknesses, such as $3\frac{1}{2}$ in. and $5\frac{1}{4}$ in., are also available. The $3\frac{1}{2}$-in.-thick LVL members are generally used as headers in 2 × 4 wood stud walls. If $3\frac{1}{2}$-in.-thick LVL is not available, two $1\frac{3}{4}$-in.-thick LVL members can be nailed together to make a $3\frac{1}{2}$-in.-thick member.

Because LVL is made from veneers that are dried to a moisture content of approximately 12% (the equilibrium moisture content in a typical building interior), it is less likely to warp or split compared to sawn lumber. Because the defects in veneers are cut away, LVL is stronger than sawn lumber of the same dimensions. Like glulam, LVL allows the use of smaller trees.

LVL members are produced by gluing veneers to form large billets, approximately 2 ft × 4 ft in cross section and nearly 80 ft long. These billets are then sawn to yield the desired (smaller) LVL members.

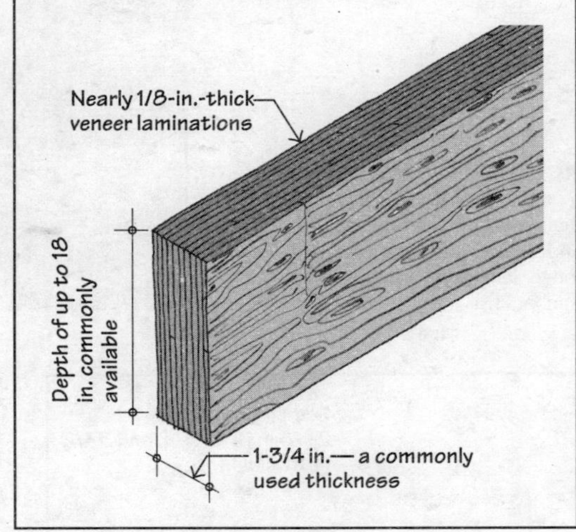

FIGURE 14.10 A typical LVL member.

PARALLEL STRAND LUMBER

A variation of LVL is called *parallel strand lumber* (PSL). PSL is made by gluing together narrow strands of veneer in place of wide veneers. The strands are made by chopping the veneers into strips about $\frac{1}{2}$ in. wide and 8 ft long. Like LVL, PSL is first made into large billets, which are then sawn to the required dimensions. Both LVL and PSL are proprietary to each manufacturer. Manufacturers provide structural data and other recommendations that must be adhered to in the use of LVL and PSL members. LVL and PSL together are referred to as *structural composite lumber* (SCL).

14.3 WOOD I-JOISTS

Wood I-joists (so named because of their I-shaped cross section) are made by gluing wood flanges to a wood web, Figure 14.11. The flanges are made of either sawn lumber or LVL. LVL has gained greater acceptance because I-joist flanges must be jointless, which is more easily done with LVL members because they are available in much longer lengths than sawn lumber.

FIGURE 14.11 Typical wood I-joists.

The web in a wood I-joist is either of plywood or oriented-strandboard (OSB) panel. OSB is more commonly used because of the absence of core voids in OSB (Section 14.7), which gives greater shear strength than a plywood web. Joints in the web are permissible, either in the form of butt joints or scarf joints.

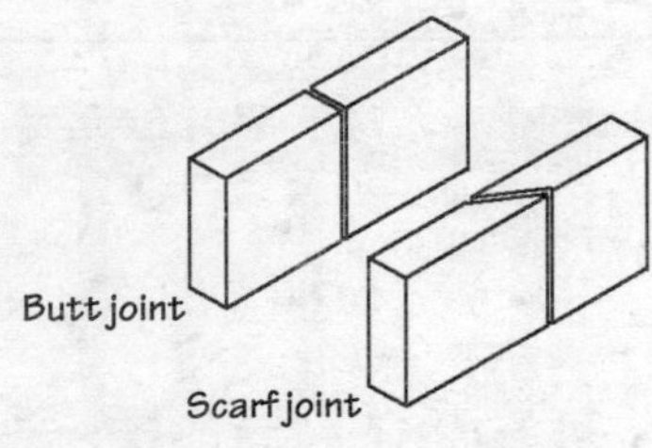

I-joists are commonly used as floor joists, Figure 14.12, and roof rafters. They are dimensionally more stable than sawn lumber due to their plywood or OSB webs. (As we will see later in this chapter, atmospheric moisture changes do not affect plywood or OSB member dimensions as much as those of solid lumber.) That is why I-joist manufacturers claim that a floor constructed of I-joists is less prone to squeaking than one made of sawn lumber joists.

The most common flange width of I-joists is $1\frac{3}{4}$ in., with total depth ranging from 10 in. to 18 in., Figure 14.13. Most manufacturers supply the joists to their distributors and dealers in lengths of 60 ft, which are then cut to the lengths desired for the project.

Because there is less wood fiber in an I-joist, it is lighter than a corresponding sawn lumber joist. Less wood fiber means a more resource-efficient material. Another advantage of I-joists is that holes can be easily cut in webs to carry utility lines without compromising the structural capacity of the member. However, because of the proprietary and preengineered nature of the joists, manufacturers provide strict guidelines concerning the size, shape, and location of holes.

A disadvantage of I-joists is their thin web, which makes them relatively unstable. Hence, they must be braced during construction, that is, until the installation of floor or roof sheathing.

FIGURE 14.12 I-joists are commonly used as floor joists because of their greater dimensional stability and stiffness and their availability in longer lengths than solid sawn lumber joists.

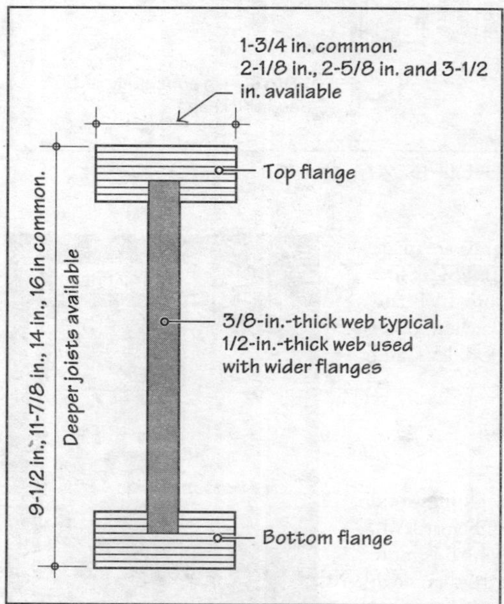

FIGURE 14.13 Commonly used dimensions of an I-joist. Manufacturers' literature must be consulted for exact dimensions.

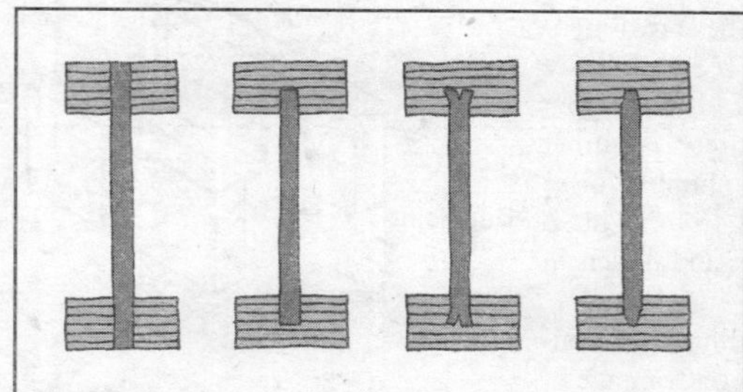

FIGURE 14.14 A few commonly used web-flange connections in I-joists.

SPECIFYING I-JOISTS

Like the structural composite lumber industry, the wood I-joist industry is also proprietary. Each I-joist manufacturer formulates its own design by testing its products. Thus, the connection between the web and the flanges is proprietary, and so is the type of adhesive. Some of the commonly used flange-web connections are shown in Figure 14.14.

I-joist manufacturers provide ready-to-use span tables and other data to help design and construction professionals select the right types of joists for a project. All important construction details are also provided.

14.4 WOOD TRUSSES

A truss consists of individual members that are joined together to form an array of interconnected triangular frames. Because a triangle is a naturally rigid geometric shape that resists being distorted when loaded from any direction, a truss is more rigid than a beam with the same amount of material. In fact, a truss is one of the most efficient means of carrying loads between supports. It is able to carry a greater load over a given span using less material than a rectangular beam or an I-beam. Figure 14.15 shows a typical wood truss, along with the important truss vocabulary.

Being made of individual triangles, truss members need not be continuous. Because the members are small in size, sawn lumber (generally 2 × 4 lumber) is most commonly used for trusses meant for residential or light commercial buildings. LVL, PSL, or glulam are also used by some manufacturers as members in trusses for heavy commercial or industrial applications.

In a sawn lumber truss, the individual members are connected together by metal nail plates, Figure 14.16(a). A nail plate is generally a 16-, 18-, or 20-gauge galvanized steel plate with the nails punched out of the sheet so that the nails and the plate are of one piece, Figure 14.16(b). The nails are generally $\frac{3}{8}$ in. long, with nearly eight nails per square inch. Nail plates are machine applied in manufacturing plants so that trusses are fully fabricated before they are transported to the construction site.

Several shapes, some of which are shown in Figure 14.17, are common for wood trusses. The shape and size of a truss are limited by manufacturing and transportation capabilities. Depending on its overall shape, a wood truss is one of the following:

- Roof truss
- Floor truss

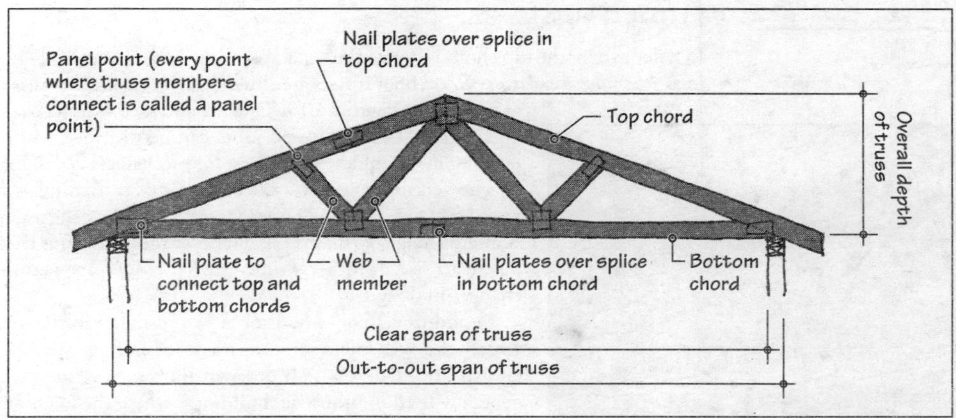

FIGURE 14.15 Important terms related to trusses.

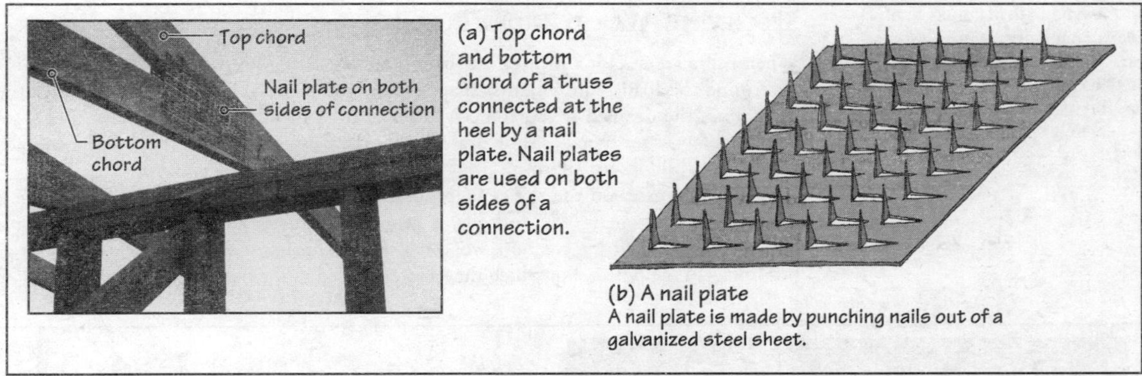

Top chord
Nail plate on both sides of connection
Bottom chord
(a) Top chord and bottom chord of a truss connected at the heel by a nail plate. Nail plates are used on both sides of a connection.
(b) A nail plate
A nail plate is made by punching nails out of a galvanized steel sheet.

FIGURE 14.16 Galvanized steel nail plate.

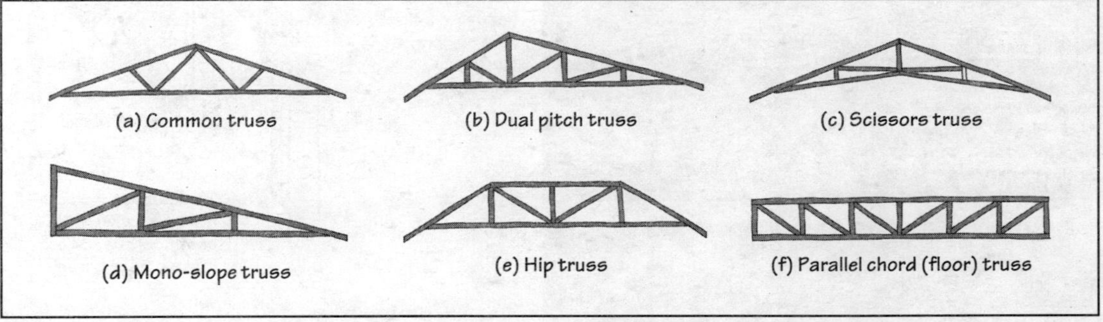

FIGURE 14.17 Commonly used shapes of wood trusses for residential and light commercial structures.

Roof Trusses

In a roof truss, the top chord is pitched. The bottom chord is generally horizontal, but it can be sloping, as in a scissors truss, Figure 14.17(c). Roof trusses are generally designed to bear on exterior walls, with the bottom chord providing the nailing surface for the ceiling. In other words, most interior walls of a building do not carry any roof loads. This is in contrast to the stick-built rafter and ceiling joist roof, in which the ceiling joists are generally supported on interior as well as exterior walls.

The use of roof trusses, therefore, gives greater freedom in the layout of interior walls and their future reconfiguration. Wood roof trusses for residential and light commercial buildings are generally spaced 24 in. on center, Figure 14.18.

FLOOR TRUSSES

In a floor truss, the top chord is horizontal—parallel to the bottom chord. Thus, it is referred to as a *parallel chord truss.* Wood floor trusses are also called *trussed joists* because they function as floor joists, Figure 14.19. They are also used as lintel beams over openings, Figure 14.20. Trussed joists can be provided with a rectangular opening in the middle of the joist for mechanical ducts, Figure 14.21.

The versatility and cost effectiveness of trussed joists make them a popular choice in wood buildings with long spans (greater than 20 ft). A disadvantage of trussed joists (also of roof trusses) is that they cannot be cut to size. In other words, their use does not permit last-minute changes in design.

In addition to all-wood joists, wood-metal joists are available, in which the web members are made of hollow steel pipes and the chords are made of LVL or sawn lumber, Figure 14.22. They are generally used in industrial buildings, where the loads are greater and the spans are longer. The use of steel in webs increases the strength and stiffness of the joists, and the wood chords provide a nailable surface for roof and floor sheathing and the ceiling.

FIGURE 14.18 Mono-slope roof trusses. Two mono-slope trusses laid end-to-end and field connected make a common roof truss. Mono-slope trusses are generally used where a common truss is too large to transport to the site. Wood trusses are generally spaced 24 in. on center.

SPECIFYING WOOD TRUSSES

When using trusses for a project, all that an architect, engineer, or builder has to do is to determine the load requirements, shape, span, and slope. The truss manufacturer then determines the detailed truss geometry, sizes of truss members, wood species, and so on. The truss fabricator prepares shop drawings for the trusses and sends them to the engineer, architect, or builder for approval.

Installing trusses on the site is as important as truss design or fabrication. A crane or other lifting equipment (e.g., fork lift) is generally required, and installers must be skilled and experienced. During erection, wood trusses should almost always be held in an upright position—in the position in which they are to be used in the building, Figure 14.23. When

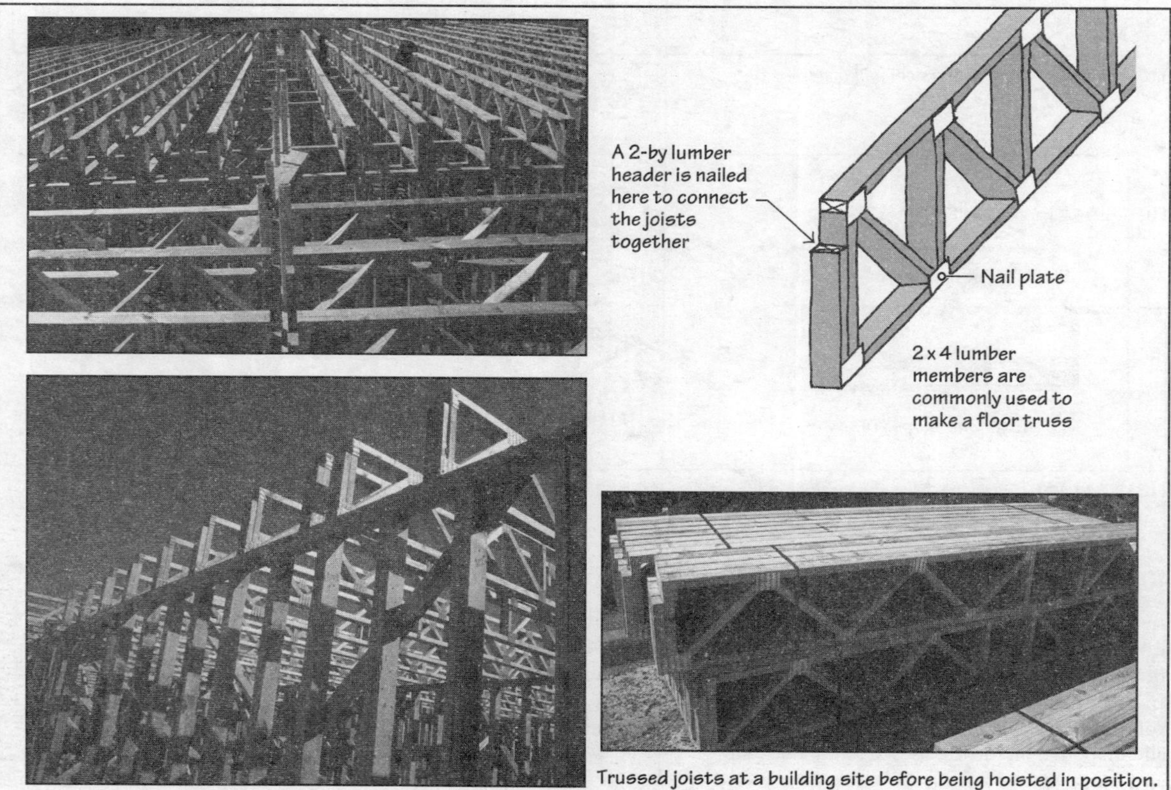

FIGURE 14.19 Floor trusses (trussed joists).

FIGURE 14.20 Trussed lintel over an opening.

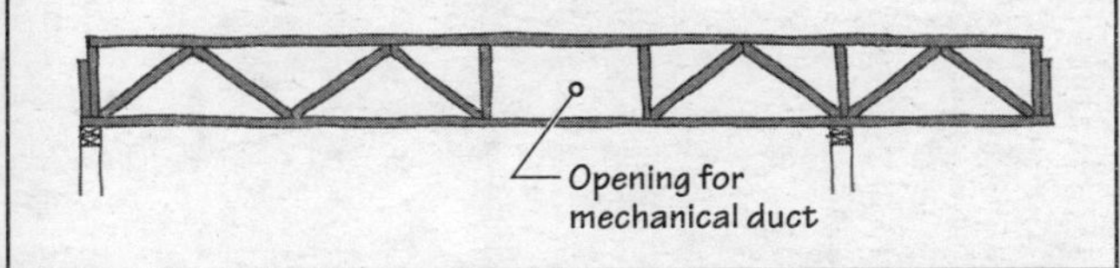

FIGURE 14.21 A floor truss with a rectangular opening to accommodate a mechanical duct.

FIGURE 14.22 Wood-steel trussed joists.

FIGURE 14.23 Because wood trusses are made from thin (2-by) members and connected together with (sheet steel) nail plates, they must be handled carefully during hoisting or placing in position so that the nail-plate connected joints do not open up. That is why, as shown in this figure, a roof truss is being carried in vertical alignment by a fork lift.

held horizontally, the lateral bending of truss members during lifting may overstress the connections, causing nail plates to loosen or pop out.

PRACTICE QUIZ

Each question has only one correct answer. Select the choice that best answers the question.

1. The most common use of glulam members is in
   a. structural panels. **b.** long-span beams.
   c. short-span beams. **d.** none of the above.
2. Unprotected glulam members with a large cross section are considered to provide a fire resistance rating of up to
   a. $\frac{1}{2}$ h. **b.** 1 h.
   c. 2 h. **d.** 3 h.
   e. 4 h.
3. A glulam member
   a. must be made using long, continuous lengths of high-grade sawn lumber.
   b. is generally stronger and stiffer than sawn lumber of the same dimensions.
   c. requires complete protection from exterior elements due to its water-soluble adhesive.
   d. none of the above.
   e. all of the above.
4. Unbalanced beams can be used only as single-span beams.
   a. True **b.** False
5. Balanced glulam beams are mandated for
   a. continuous beams. **b.** single-span beams.
   c. beams with an overhang. **d.** (a) and (c).
   e. (b) and (c).
6. Laminated veneer lumber (LVL) is
   a. produced by gluing together wood veneers that are approximately $\frac{1}{8}$ in. thick.
   b. generally used as floor joists.
   c. glued with all veneers running in the same direction.
   d. stronger along the grain and weaker across the grain.
   e. all of the above.
7. Wood I-joists are made of LVL or solid sawn lumber flanges and webs.
   a. True **b.** False
8. Compared to sawn lumber joists, wood I-joists are
   a. less expensive. **b.** slightly heavier.
   c. dimensionally more stable. **d.** all of the above.
9. Characteristics of a truss include
   a. greater spanning capability than a sawn lumber beam having the same amount of material.
   b. individual members joined together to form an array of interconnected rectangular frames.
   c. individual members joined together to form an array of interconnected triangular frames.
   d. (a) and (b).
   e. (a) and (c).
10. Nail plates, used to join members of a truss, are generally used only on one face of the truss.
    a. True **b.** False
11. Wood trusses are generally spaced at
    a. 12 in. on center. **b.** 24 in. on center.
    c. 48 in. on center. **d.** 60 in. on center.

14.5 WOOD PANELS

Wood panels are an important part of wood frame construction. They are used structurally—as floor sheathing, roof sheathing and wall sheathing—and nonstructurally—as exterior siding and interior paneling. Wood panels are divided into two types:

- *Veneered panels*—consisting of plywood panels.
- *Nonveneered panels*—consisting of oriented strandboard (OSB) and particle board panels. Particle board panels are generally used in shelving and furniture making and are not discussed here.

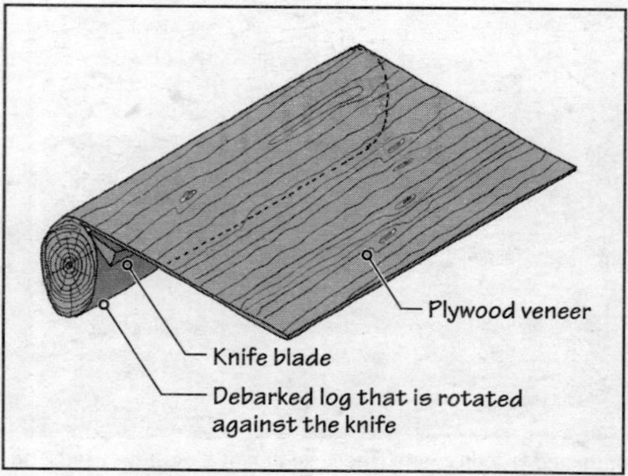

FIGURE 14.24 Rotary slicing of a log—a commonly used method of making plywood veneers.

14.6 PLYWOOD PANELS

Plywood panels are made by gluing wood veneers under heat and pressure. Veneers are generally produced by a machine that holds a debarked log at two ends in a lathe and rotates the log against a stationary knife blade extending throughout the length of the log, Figure 14.24. This operation peels the log off, giving a continuous veneer about $\frac{1}{8}$ in. thick, in much the same way that one would unroll a paper towel. The veneer so obtained is subsequently cut to desired sizes.

To make a plywood panel, the defects in veneers, such as knot holes and splits, are cut away or repaired where necessary. The veneers are then dried and glued together so that the grain direction in each veneer is oriented at a right angle to the grain direction of the adjacent veneer, Figure 14.25.

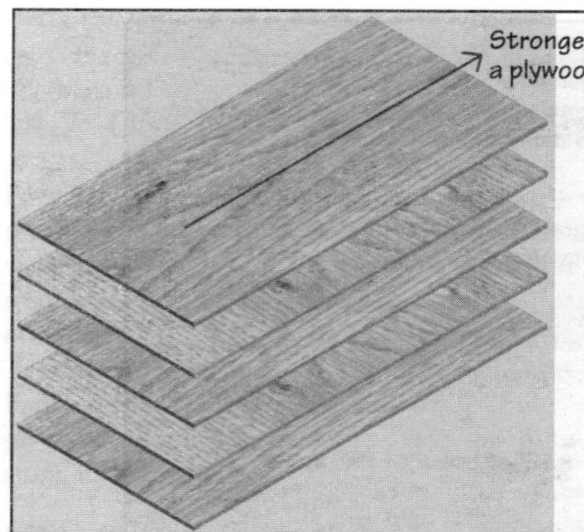

The grain directions in alternate veneer layers are at right angles to each other. Generally, this arrangement implies an odd number of veneers in a panel. However, an even number of veneers is also used in which the two innermost veneers are oriented in the same direction so that the top and bottom veneers (referred to as *face veneers*) are always in the same direction.

For example, if a panel is made of six veneers, of which the two innermost veneers are in the same direction, it is referred to as a five- layer, six-veneer panel. The panel shown here is a five-layer, five-veneer panel.

Because bending stresses maximize at the top and bottom fibers in a panel and because wood is stronger along the grain than across the grain, a plywood panel is stronger in the direction of the grain of the face veneers.

FIGURE 14.25 Orientation of veneer grains in a plywood panel.

Because wood is stronger along the grain than across the grain, cross-graining tends to equalize the strength of a plywood panel in its two principal directions. It also makes a plywood panel dimensionally more stable because shrinkage and swelling of wood are high across the grain and negligible along the grain. For the same reason, a plywood panel is less likely to split than sawn lumber. Therefore, it can be nailed near its edge without splitting.

The most commonly used plywood panel size is 4 ft × 8 ft, with thickness varying from $\frac{1}{4}$ in. to 1 in. Longer plywood panels are manufactured for siding and industrial use. The panel dimension of 4 ft × 8 ft is its nominal dimension. Its actual dimension is $\frac{1}{8}$ in. smaller in length as well as in width. In other words, the actual dimensions of a 4-ft × 8-ft panel are $47\frac{7}{8}$ in. × $95\frac{7}{8}$ in. This allows a panel to be installed with a $\frac{1}{8}$-in. space all around for moisture expansion, Figure 14.26.

The grain direction of face veneers of a panel is oriented along the panel's length (along the 8-ft dimension in a 4-ft × 8-ft panel), making the panel stronger in its long direction. In structural applications, the long direction of a plywood panel should, therefore, be perpendicular to the supporting joists or rafters.

Veneer Grades of Softwood Plywood

Plywood veneers used in building construction are generally obtained from softwoods. Softwood veneers are graded in five grades—A, B, C-plugged, C, and D—based on the type and size of defects such as knots and splits, Figure 14.27. Veneer grade A is the highest

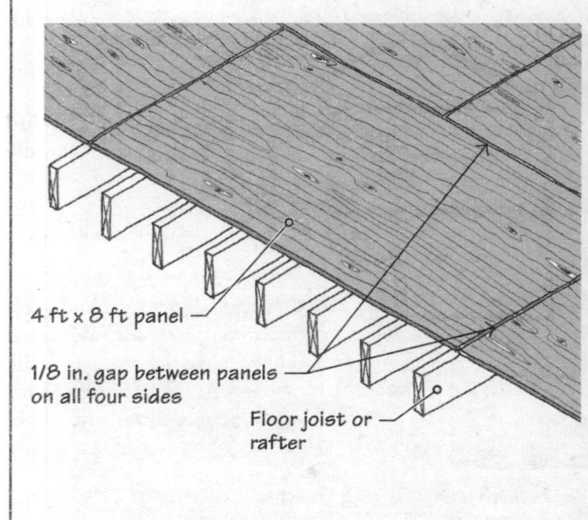

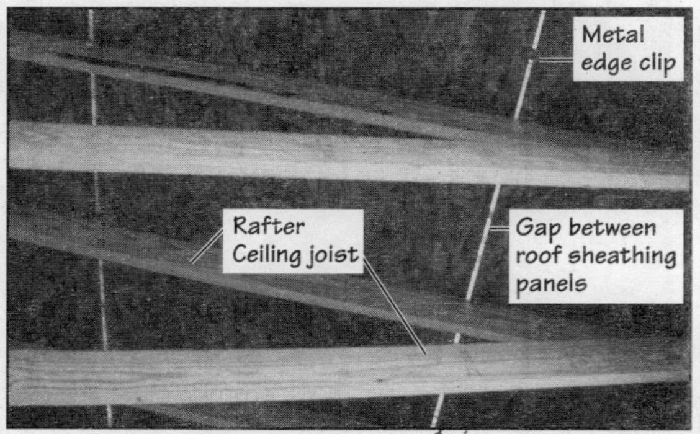

1/8-in. gap between roof sheathing panels. The white line in this photo represents daylight coming through the gap. Note that roof sheathing requires (metal) edge clips between adjacent panels to add to their continuity.

FIGURE 14.26 Plywood panels must be oriented with their long direction perpendicular to the supporting members. Additionally, a gap of nearly $\frac{1}{8}$ in. must be left all around panels to accommodate moisture expansion.

FIGURE 14.29 Because of their lower cost and greater racking resistance, OSB panels are generally used to sheath the entire envelope of a wood frame building (see also Chapter 15, Section 15.9).

for prolonged periods. Additionally, OSB panels cannot be treated with preservatives, whereas preservative-treated plywood is available.

14.8 SPECIFYING WOOD PANELS—PANEL RATINGS

Wood panels can be divided into two categories based on their end use:

- Performance-rated engineered wood panels
- Sanded and touch-sanded plywood panels

PERFORMANCE-RATED ENGINEERED WOOD PANELS

Performance-rated engineered wood panels are meant for structural applications (e.g., wall sheathing, floor sheathing, roof sheathing) as well as for nonstructural use in exterior siding. The panels are preengineered and rated accordingly. The rating provides the user with the panel's structural capacity and other performance data, such as its intended end use and durability. This simplifies the specification of the panel and relieves the user of any further investigation.

Performance rating means that as long as the panel meets the specified requirements for end use (exposure durability and span rating), it does not matter which material—plywood or OSB—it is made of. The rating is achieved through a third-party certification process and stamped accordingly. Typical ratings (grade stamps) of panels are given in Figure 14.30. The following performance properties are of particular importance:

Intended End Use Performance-rated panels can be used in one of the following three situations:

- Sheathing (over studs, floor joists, or rafters)
- Combination floor sheathing
- Exterior siding

Combination floor sheathing is a special type of floor sheathing that replaces a two-layer covering over floor joists. If regular (square-edged) sheathing is used over floor joists, it requires an additional layer of panels over the sheathing panels. The additional panel layer is called *underlayment* because it functions as an underlayer for the floor finish and is typically $\frac{1}{4}$-in.-thick sanded plywood or hardboard.

Combination floor sheathing panels have been developed to act as sheathing as well as underlayment and are generally provided with a tongue-and-groove (T&G) profile along the long (8-ft) edges, Figure 14.31, although square-edged panels are also available.

Whereas sheathing and combination sheathing panels may be of plywood or OSB, only plywood is used in exterior siding panels for aesthetic and durability reasons.

Exposure Durability Engineered wood panels are produced in two exposure-durability classifications:

- Exterior
- Exposure 1

NOTE

Sturd-I-Floor as Combination Floor Sheathing

APA-rated Sturd-I-Floor is a commonly specified floor sheathing, ranging in thickness from $\frac{3}{4}$ in. to $1\frac{1}{8}$ in. It does not require any underlayment and is commonly used in situations where carpet is the required floor finish. According to APA, the panel surface of Stud-I-Floor has extra resistance to punch-through damage.

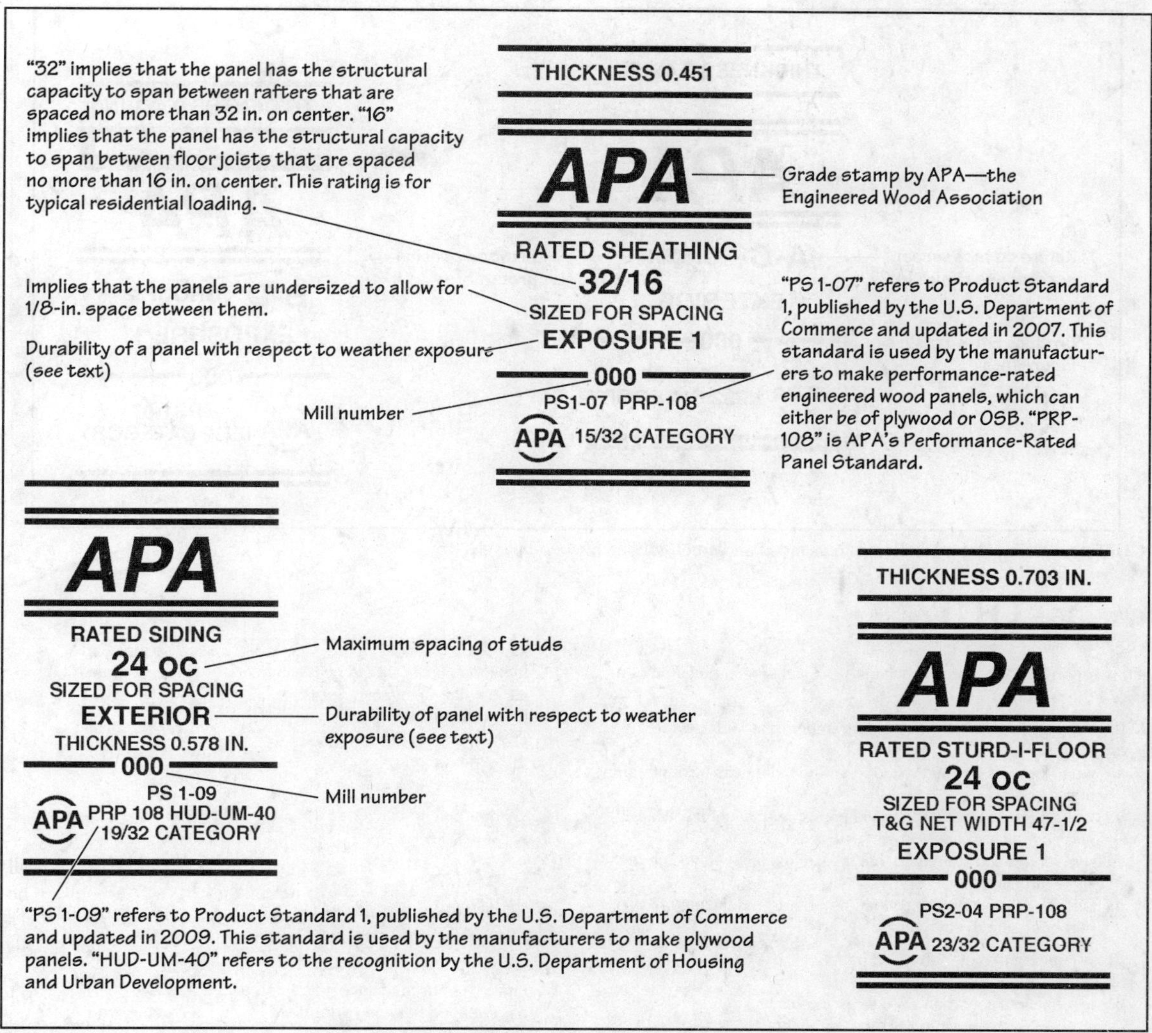

"32" implies that the panel has the structural capacity to span between rafters that are spaced no more than 32 in. on center. "16" implies that the panel has the structural capacity to span between floor joists that are spaced no more than 16 in. on center. This rating is for typical residential loading.
THICKNESS 0.451
APA
Grade stamp by APA—the Engineered Wood Association
RATED SHEATHING
32/16
Implies that the panels are undersized to allow for 1/8-in. space between them.
SIZED FOR SPACING
Durability of a panel with respect to weather exposure (see text)
EXPOSURE 1
000
Mill number
PS1-07 PRP-108
"PS 1-07" refers to Product Standard 1, published by the U.S. Department of Commerce and updated in 2007. This standard is used by the manufacturers to make performance-rated engineered wood panels, which can either be of plywood or OSB. "PRP-108" is APA's Performance-Rated Panel Standard.
APA 15/32 CATEGORY
APA
RATED SIDING
24 oc
Maximum spacing of studs
SIZED FOR SPACING
EXTERIOR
Durability of panel with respect to weather exposure (see text)
THICKNESS 0.578 IN.
000
Mill number
PS 1-09
APA PRP 108 HUD-UM-40
19/32 CATEGORY
"PS 1-09" refers to Product Standard 1, published by the U.S. Department of Commerce and updated in 2009. This standard is used by the manufacturers to make plywood panels. "HUD-UM-40" refers to the recognition by the U.S. Department of Housing and Urban Development.
THICKNESS 0.703 IN.
APA
RATED STURD-I-FLOOR
24 oc
SIZED FOR SPACING
T&G NET WIDTH 47-1/2
EXPOSURE 1
000
PS2-04 PRP-108
APA 23/32 CATEGORY

FIGURE 14.30 Typical APA grade stamps on performance-rated engineered wood panels. TECO, another grading agency, uses similar stamps.

Exterior-rated panels are designed for permanent exposure to the weather—to withstand the effect of rain, humidity, and sunshine. *Exposure 1* panels are meant for use in protected situations, that is, where the panels are to be covered with an exterior facing material. However, Exposure 1 panels are designed to withstand the effect of weather for several days due to construction delays. Both Exterior and Exposure 1 panels use the same waterproof glue. The difference lies in the panels' composition.

Note that the exposure-durability classification of panels relates to the moisture resistance of the glue bond, not the fungal resistance of the panel. Thus, an Exterior-rated panel, if not used correctly, is subject to fungal attack.

Span Rating and Overall Thickness As stated previously, performance-rated panels are preengineered, which means that architects, engineers, and builders do not need to calculate the load capacity of the panel. This information is provided by the manufacturer and is stamped on the panel, along with the panel's end use, exposure classification, and thickness.

1/8-in. gap

The tongue in a Sturd-I-Floor brand panel has a thin end so that it squashes when the panels expand.

FIGURE 14.31 Tongue-and-groove joint between Sturd-I-Floor panels.

Sanded and Touch-Sanded Plywood Panels

Sanded and touch-sanded panels are rated for performance as well as exterior appearance. The exterior appearance refers to the grade of face veneers of the panels. A few typical grade stamps of such panels are shown in Figure 14.32. Note that the performance-rated panels are rated only for performance, not for appearance.

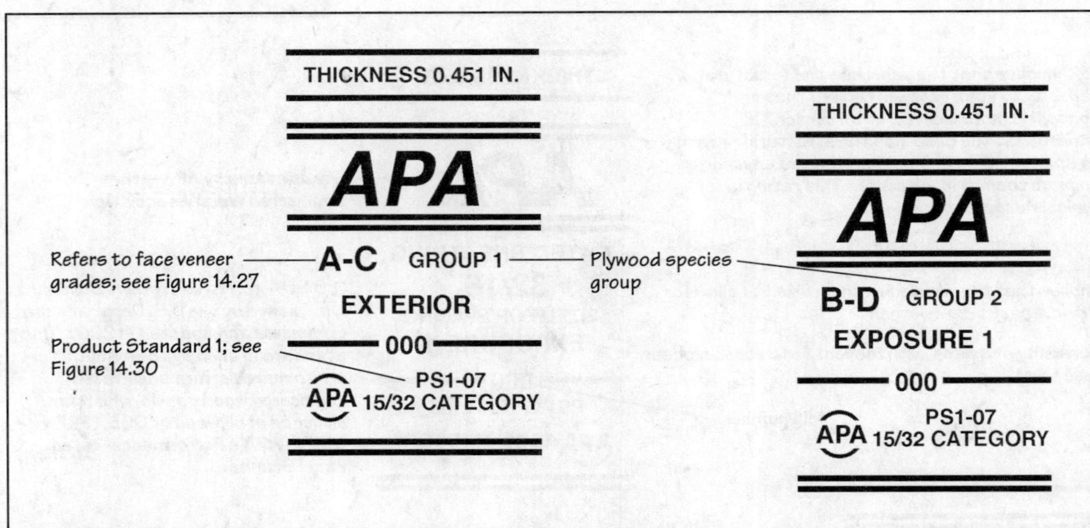

FIGURE 14.32 Typical grade stamps on sanded and touch-sanded plywood panels.

PRACTICE QUIZ

Each question has only one correct answer. Select the choice that best answers the question.

12. Plywood panels are made by gluing under heat and pressure layers of
- **a.** wood veneers so that the wood grain in all layers is in the same direction.
- **b.** wood veneers so that the wood grains in adjacent layers are perpendicular to each other.
- **c.** shredded wood strands so that the strands in all layers are in the same direction.
- **d.** shredded wood strands so that the strands in adjacent layers are perpendicular to each other.

13. Cross-graining makes a plywood panel
- **a.** dimensionally more stable.
- **b.** less likely to split than solid sawn lumber.
- **c.** stronger in the direction of face veneer grains.
- **d.** all of the above.
- **e.** (a) and (c).

14. The most commonly used plywood panel (nominal) size is 4 ft × 8 ft.
- **a.** True
- **b.** False

15. Plywood panels should be placed with their longer dimension
- **a.** parallel to rafters or joists.
- **b.** perpendicular to rafters or joists.
- **c.** diagonal to rafters or joists.
- **d.** either (a) or (b).
- **e.** either (a) or (c).

16. With respect to veneer quality, softwood plywood is graded in grades
- **a.** A to C.
- **b.** B to D.
- **c.** A to D.
- **d.** A to E.
- **e.** B to F.

17. An OSB panel generally has greater shear strength than a plywood panel of the same size and thickness.
- **a.** True
- **b.** False

18. Grade stamps on engineered wood panels specify
- **a.** intended use and exposure.
- **b.** allowable spans.
- **c.** mill number.
- **d.** thickness.
- **e.** all of the above.

19. Exterior siding panels may be made of OSB or plywood.
- **a.** True
- **b.** False

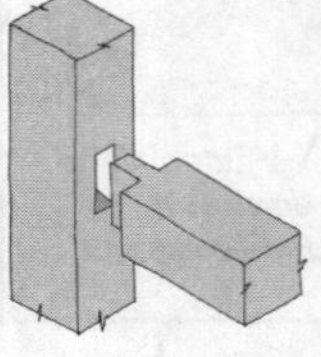

Mortise-and-tenon joint

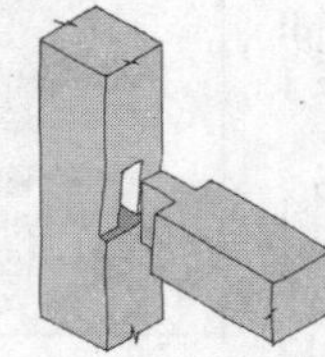

Housed mortise-and-tenon joint

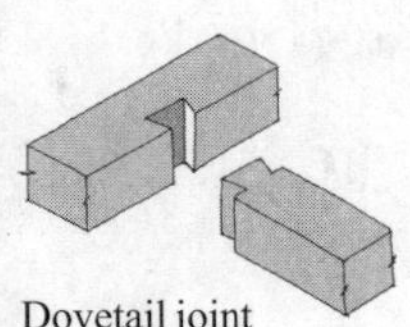

Dovetail joint

14.9 FASTENERS FOR CONNECTING WOOD MEMBERS

The traditional method of connecting wood members was through different types of interlocking joints, such as a mortise-and-tenon joint, housed mortise-and-tenon joint, and dovetail joint. Except for fine furniture and to some extent in timber frame construction (see the Expand Your Knowledge section at the end of Chapter 15), these *joinery methods,* as they are referred to, have been replaced by simpler and more effective methods that often yield stronger connections. Most joints in contemporary wood construction are made by simply nailing the members together or by nailing them through sheet metal connectors. In some joints, adhesives are used in addition to nails, whereas in others, screws and bolts are necessary.

TYPES OF NAILS

A nail is generally made of low or medium carbon steel wire that is heat treated to increase its stiffness. Where increased impact resistance is needed, such as for masonry nails, steel with a higher carbon content is used. Nails made in this way without any further treatment for corrosion are called *brite nails.* In exterior siding and decks where greater corrosion resistance is needed, hot-dip galvanized nails are used. (Stainless steel nails can provide even higher corrosion resistance but are expensive.)

For increased holding power, nails are phosphate or vinyl coated. Vinyl-coated nails produce heat due to friction when the nail is driven, melting the vinyl, which increases the bond between the wood and the nail. They have a thinner shank and are easier to drive into wood and are, therefore, called *sinker nails.* For most structural connections, however, brite (ungalvanized, uncoated) steel nails are used.

A nail has three basic parts: (a) the tip, (b) the shank, and (c) the head, Figure 14.33. The tip of most nails is diamond shaped. Therefore, the nail type is distinguished by the type of its head and the type of shank. Some of the commonly used nail types in wood frame construction are shown in Figure 14.34.

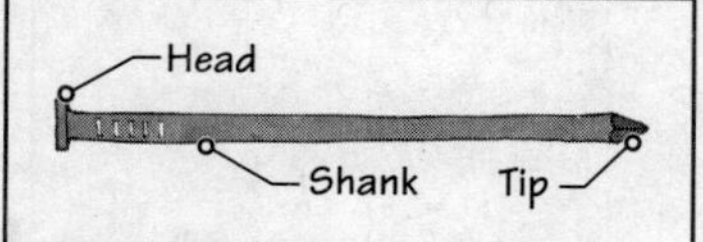

FIGURE 14.33 Parts of a nail.

For framing connections, common nails are most frequently used. Box nails are similar to common nails but have a thinner shank, which reduces wood splitting. They are generally used for attaching wood shingles. Casing nails and finish nails are used for finish carpentry. Deformed shank nails are used for attaching wood flooring or gypsum wallboard.

NAIL SIZES

The length of common nails in the United States is specified by a *penny* (abbreviated as *d*) designation. For example, a 10d nail (called *a 10 penny nail*) is 3 in. long. The penny designation originated in England centuries ago when 1 poundweight of 10d nails cost 10 pence, 1 poundweight of 12d nails cost 12 pence, and so on.

Common nails are available in lengths ranging from 2d to 60d. A 2d nail is 1 in. long, and a 60d nail is 6 in. long. From 2d to 10d, the increase in length is $\frac{1}{4}$ in. per penny. Therefore, a 10d nail is 3 in. long. The next two nail sizes are 12d and 16d, with lengths of $3\frac{1}{4}$ in. and $3\frac{1}{2}$ in., respectively. A 20d nail is 4 in. long, Figure 14.35. Most commonly used nail sizes in wood frame construction are 6d, 8d, 10d, and 16d, highlighted in Figure 14.35.

NAILED CONNECTIONS

Nails work best when they are subjected to shear—that is, when the load is perpendicular to the length of the nails—or when they are in compression. Nails are particularly weak in withdrawal. Withdrawal resistance is needed when the load is parallel to the length of the nails, trying to pull the connected members apart, Figure 14.36. Three types of nailed connections are used in wood frame construction, Figure 14.37:

- Face-nailed connection
- End-nailed connection
- Toe-nailed connection

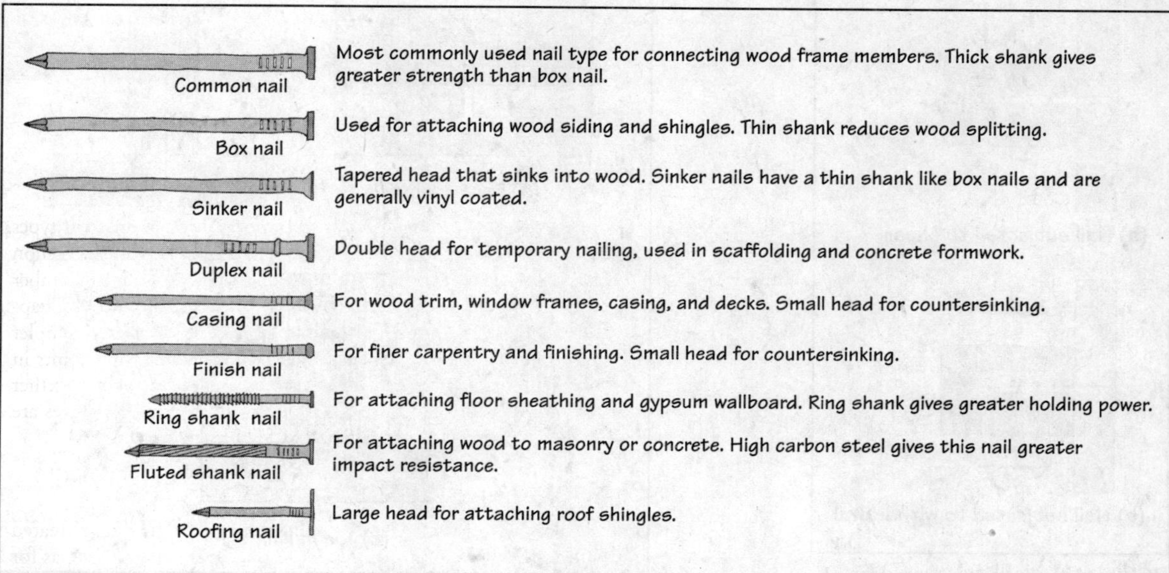

Common nail
Most commonly used nail type for connecting wood frame members. Thick shank gives greater strength than box nail.
Box nail
Used for attaching wood siding and shingles. Thin shank reduces wood splitting.
Sinker nail
Tapered head that sinks into wood. Sinker nails have a thin shank like box nails and are generally vinyl coated.
Duplex nail
Double head for temporary nailing, used in scaffolding and concrete formwork.
Casing nail
For wood trim, window frames, casing, and decks. Small head for countersinking.
Finish nail
For finer carpentry and finishing. Small head for countersinking.
Ring shank nail
For attaching floor sheathing and gypsum wallboard. Ring shank gives greater holding power.
Fluted shank nail
For attaching wood to masonry or concrete. High carbon steel gives this nail greater impact resistance.
Roofing nail
Large head for attaching roof shingles.

FIGURE 14.34 Commonly used types of nails in wood frame construction.

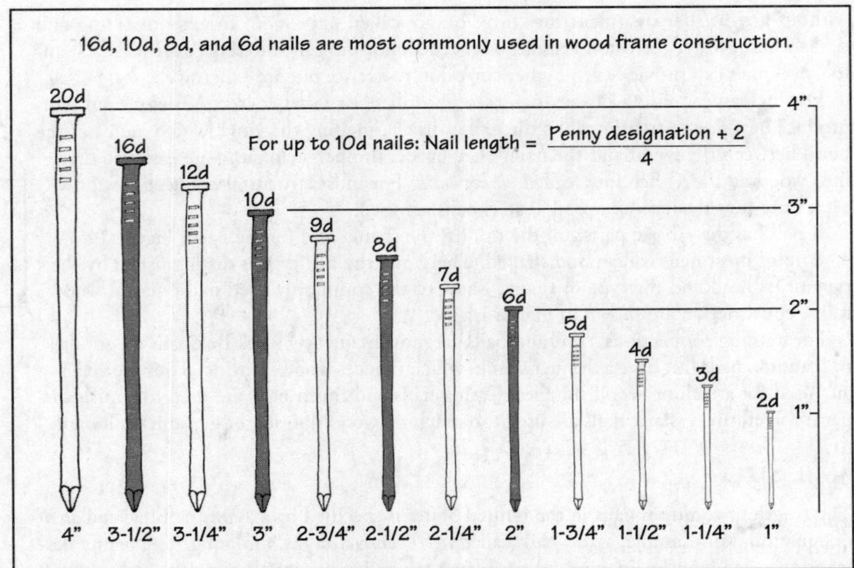

FIGURE 14.35 Standard lengths of common nails.

A face-nailed connection is the strongest of the three because it has the highest withdrawal resistance. The withdrawal resistance of a nail is a function of the nail's orientation with respect to the grain of wood in the holding member (the member that contains the tip of the nail). If the axis of the nail is parallel to the grain of wood in the holding member, the withdrawal resistance is extremely small. In practice, it is assumed equal to zero. Withdrawal resistance is highest if the nail's axis is perpendicular to the grain of wood in the holding member, such as in a face-nailed connection.

End nailing, in which the nails are parallel to the grain in the holding member, is the weakest connection. End nailing is acceptable only in situations where the member is not subjected to withdrawal. Toe nailing is stronger than end nailing, but it is used where access for end nailing is unavailable. A fourth type of connection, referred to as *blind nailing,* is used in finished wood flooring.

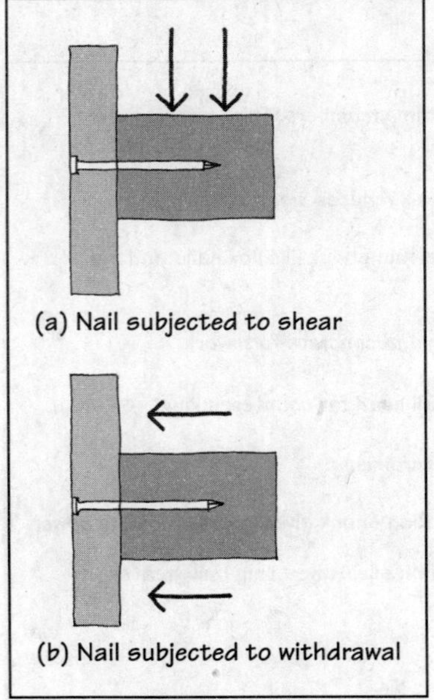

FIGURE 14.36 Nails subjected to **(a)** shear and **(b)** withdrawal.

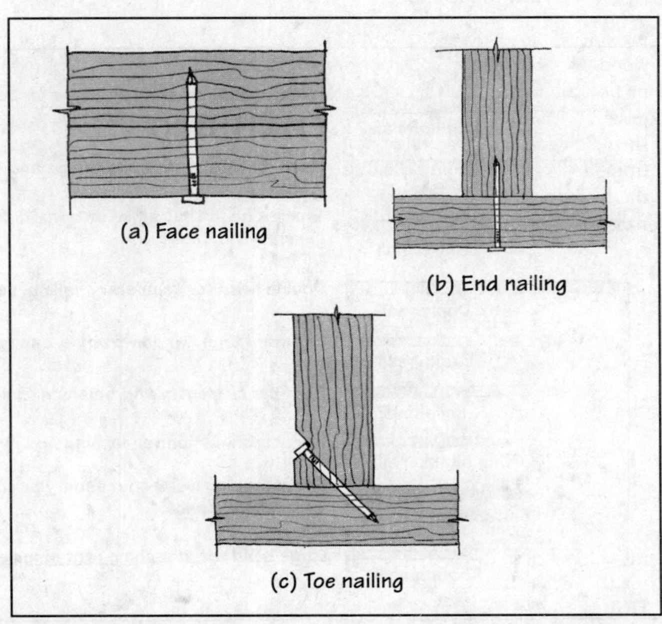

FIGURE 14.37 Types of nailed connections used in wood frame construction.

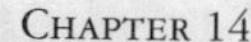

FIGURE 14.38 A typical nailing gun.

POWER-DRIVEN NAILING AND STAPLING

Although manual hammering is still used to drive nails, power-driven nailing (through pneumatic or electric nailing guns) has become the preferred method of making connections, Figure 14.38. In the power-driven system, nailing is achieved by the pull of a trigger, which not only speeds the process substantially but is also less tiring for the worker.

Nails used in nailing guns are thinner and smaller than the corresponding pennyweight sizes of hand-driven nails. Manufacturers provide the equivalence between the nails used in guns and the nails used in manual hammering. Power stapling is also a recognized method of fastening wood members and is generally used as an alternative to 8d or 6d nails.

NAIL POPPING

Nail popping—nails sticking out of the wood members—is a problem primarily in floor sheathing that is nailed to the joists. It occurs as the floor joists dry and shrink in size, which pulls them away from the sheathing, Figure 14.39. Because the load on sheathing is downward, the nails pop out through the sheathing.

NOTE

The Importance of the Lowly Nail

Though seemingly insignificant, nails are an extremely important component in contemporary wood construction. It is the nails that hold the thousands of individual wood members together. It is estimated that approximately 35 nails are used in 1 square foot of a typical wood frame dwelling. Thus, in an average modern U.S. home (of about 2,500 ft^2), approximately 90,000 nails may be used. Before the introduction of nail guns, driving such a large quantity of nails was not only time-consuming but also labor intensive. The use of nail guns has dramatically increased construction efficiency by increasing the speed and reducing the cost of making connections.

With such a large quantity of nails required, it is obvious that without the discovery of the mass production of nails, which made nails affordable and available in large quantities, contemporary wood construction would not have been possible [14.1]. Until the dawn of the nineteenth century, nails were generally hand made by forging a piece of iron. They were, therefore, expensive and had limited availability. A legend has it that nails were so important and expensive at that time that old, dilapidated homes were burned to retrieve them. After sifting nails from the ashes, a blacksmith would straighten them for reuse.

Because they were hand-made, nail making was a cottage industry, and many people, including some well-to-do ones, made nails for personal use and (or) sale using their fireplaces. The third U.S. President (1801–1809), Thomas Jefferson, had a nail-making shop at his Monticello farm [14.2].

Nail-making machines at the time made nails by shearing them from steel plates. Sheared nails, referred to as *cut nails*, were tapered at the tip on one face; the other face had a constant thickness, representing the thickness of the plate from which they were sheared. The tips of hand-forged nails, by contrast, were tapered all around.

The making of cut nails, though much faster than hand forging, was still quite labor intensive compared with modern nail making because different machines were used for shearing, heading, and tipping. The introduction of improved nail-making machines around 1850, which cut and headed the nails using a single machine, increased their availability and affordability. Since then, nail-making technique has advanced considerably. Today, most nails are made from wires, though cut nails are still made because of their greater holding power compared with that of round wire nails.

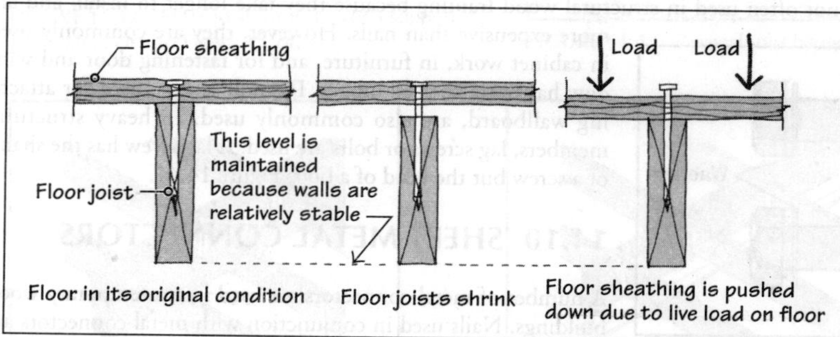

FIGURE 14.39 Mechanics of nail popping.

PRACTICE QUIZ

Each question has only one correct answer. Select the choice that best answers each question.

20. Several different types of nails are used in wood frame construction. They are typically distinguished from each other by the
- **a.** nail head and nail shank.
- **b.** nail head and nail tip.
- **c.** nail tip and nail shank.
- **d.** nail head and type of threads.
- **e.** none of the above.

21. The nails most commonly used in wood frame construction are
- **a.** 4d, 6d, 9d, and 12d.
- **b.** 6d, 12d, 18d, and 24d.
- **c.** 6d, 10d, 16d, and 20d.
- **d.** 6d, 8d, 10d, and 16d.
- **e.** 6d, 8d, 10d, and 12d.

22. Nails work best when they are subjected to
- **a.** shear.
- **b.** tension.

23. Which of the following nailed connections is the strongest?
- **a.** Face nailing
- **b.** Toe nailing
- **c.** End nailing

24. Nail popping is primarily a problem in floor sheathing and is caused by the shrinkage of floor sheathing.
- **a.** True
- **b.** False

25. Joist hangers are steel connectors
- **a.** that connect floor joists with a supporting beam.
- **b.** that connect floor joists with the supporting wall.
- **c.** that connect floor joists with floor sheathing.
- **d.** all of the above.

26. The adhesives used in making engineered wood products contain
- **a.** urethane.
- **b.** styrene.
- **c.** formaldehyde.
- **d.** lignin.
- **e.** none of the above.

27. Gaseous emissions produced by engineered wood products decrease over time.
- **a.** True
- **b.** False

REVIEW QUESTIONS

1. Using sketches and notes, explain the difference between a balanced and an unbalanced glulam beam and the situations in which they are specified.
2. Using sketches and notes, explain the composition of a wood I-joist. What are the advantages and disadvantages of wood I-joists compared with solid lumber?
3. Using a sketch and notes, explain the following terms related to a wood roof truss: (a) *top chord*, (b) *bottom chord*, (c) *web member*, (d) *panel point*, (e) *nail plate*.
4. Using a sketch, explain the difference between a roof truss and a floor truss. In which situations will you consider the use of wood floor trusses?
5. Which type of nail is commonly used for connections between framing members in wood light-frame structures? How do we specify the size of nails?
6. Discuss the relative advantages and disadvantages of OSB and plywood panels.
7. With the help of an example, explain what type of information is provided in the grade stamp of a typical performance-rated engineered wood panel.
8. Using sketches, illustrate the following connections in wood frame members: (a) face-nailed connection, (b) end-nailed connection, and (c) toe-nailed connection.
9. With the help of a three-dimensional sketch, explain the connection of a floor joist with a supporting beam using a joist hanger.

CHAPTER 15

Wood Light-Frame Construction–I

CHAPTER OUTLINE

Wood light-frame (WLF) construction is the dominant system for contemporary residential and light commercial construction in the United States and several other countries. It is examined in this chapter and the next. The framing aspects of WLF construction are considered in this chapter, and the interior and exterior finishes in the next chapter.

This chapter begins with a brief overview of the evolution of WLF, followed by a discussion of the organizational principles of this construction system. Then the essential characteristics and components of the system, including the floor, wall, and roof framing and sheathing, are investigated. At the end of this chapter, two Principles in Practice features are presented. The first provides an example of the WLF construction process and sequence, and the second describes how loads are resisted by a typical WLF structure.

Concluding this chapter is an Expand Your Knowledge section that provides a brief account of the wood construction that existed in the United States before the discovery of (and its evolution into) the contemporary WLF construction.

15.1 EVOLUTION OF WOOD LIGHT-FRAME CONSTRUCTION

As the European settlers colonized America and built on the new land, they tried to re-create the ambience and spirit of their original countries. Early American buildings and construction systems were, therefore, derived mainly from European practices. Wood light-frame construction, on the other hand, was invented in the United States and is uniquely American.

WLF construction was first used around 1830 in the Chicago area, which was then developing as a major urban center in the United States. There is some controversy about who should be credited with the development of this new system. Two persons—Augustine Taylor and George Washington Snow—both carpenters from the Chicago area, are said to have invented the system independent of each other.

The WLF system has evolved over nearly two centuries and has adapted marvelously to several technical innovations in buildings, such as insulation, electricity, piped water supply, and sewage disposal, which did not exist during the time of Taylor or Snow. The system's adaptability and the ease with which it could be used led to its success and popularity. So successful has the system been that it is now the dominant system for contemporary residential and light commercial construction in the United States and several other countries.

Balloon Frame

Prior to the invention of the WLF system, most buildings in the United States were built of thick masonry walls, heavy timber posts, beams, and thick wood plank floors and roofs. The joints between heavy timber posts and beams consisted of mortise-and-tenon, or dovetail, joints (see the Expand Your Knowledge section at the end of this chapter). Because no nails or screws were used to secure the joints, expert craftsmanship was needed to provide tight joints.

When the WLF system was invented to replace the heavy construction system just described, it was called the *balloon frame*. It consisted of thin, closely spaced vertical wood members (called *studs*) and similar floor and roof framing members (called *joists* and *rafters*).

The connections between members were made with simple nails. Measuring, hammering, and sawing were, therefore, the only skills needed for the connections, making sophisticated carpentry and joinery skills unnecessary. In fact, the entire structural frame of a WLF building could be erected by a relatively unskilled crew of two or three people.

An additional advantage of the WLF system lay in its lightweight members, which needed little effort to hoist and install. Because the members were small in cross section, the total amount of wood consumed in a balloon frame building was only a fraction of that consumed in a traditionally built building of the same size. In fact, many historians believe that various U.S. cities could not have developed as rapidly as they did without the invention of the balloon frame.

As with most other inventions, the birth of the WLF system may be attributed to necessity. The necessity arose from an abundance of wood in the new country, where skilled carpenters were scarce. Immigrants at that time consisted largely of poor, unskilled workers unable to find work in Europe. This was in direct contrast to the wealth of skilled carpenters in Europe, where forests were gradually becoming depleted.

The term *balloon frame* was coined out of contempt by the critics of the system at the time, who considered it too insubstantial to withstand the forces of nature. Although time proved them wrong, the term continued to be used. In the balloon frame, the studs run the full height of the building, from the sill plate at the bottom to the top plate under the rafters. The intermediate floor joists (in a two-story building) rest on a 1×6 (or 1×4) ledger—called a *ribband*—notched into the studs, Figure 15.1.

In addition to ribbands, fire-stops are provided at floor lines in a balloon frame, Figures 15.1 and 15.2. Without the fire-stops, the air space between the studs is continuous from the top to the bottom. This creates a fire risk because a fire occurring in the stud cavity of one floor can spread to another floor. The fire-stops separate the cavities and deprive the fire of the oxygen required for its growth.

The continuity of studs was recognized as a major limitation of the balloon frame. This limitation became more acute as the greater use of wood reduced the availability of long, straight members and made them more expensive. The balloon frame was, therefore, modified over time into the *platform frame*, in which the individual studs are only one story high. Contemporary WLF construction consists mainly of the platform frame, the subject of discussion in this chapter.

NOTE

Balloon Frame and Nails

As stated, the balloon frame was invented in the United States in 1830. However, as with all innovations, wide acceptance of the balloon frame had to wait for the right environment to emerge. That happened with the mass production of nails in the mid-nineteenth century (see the margin note in Section 14.9 entitled "The Importance of the Lowly Nail").

The balloon frame was indeed much lighter and simpler to construct than the wood construction system that preceded it (see the Expand Your Knowledge section at the end of this chapter). However, the balloon frame construction was heavily dependent on nailed connections. Without the availability and affordability of nails, the balloon frame construction would not have gone much further [15.1]. The convergence of the two innovations—mass production of nails and invention of the balloon frame—revolutionized wood construction. As nail manufacturing advanced, making nails inexpensive, balloon frame construction increased rapidly throughout the United States and other countries.

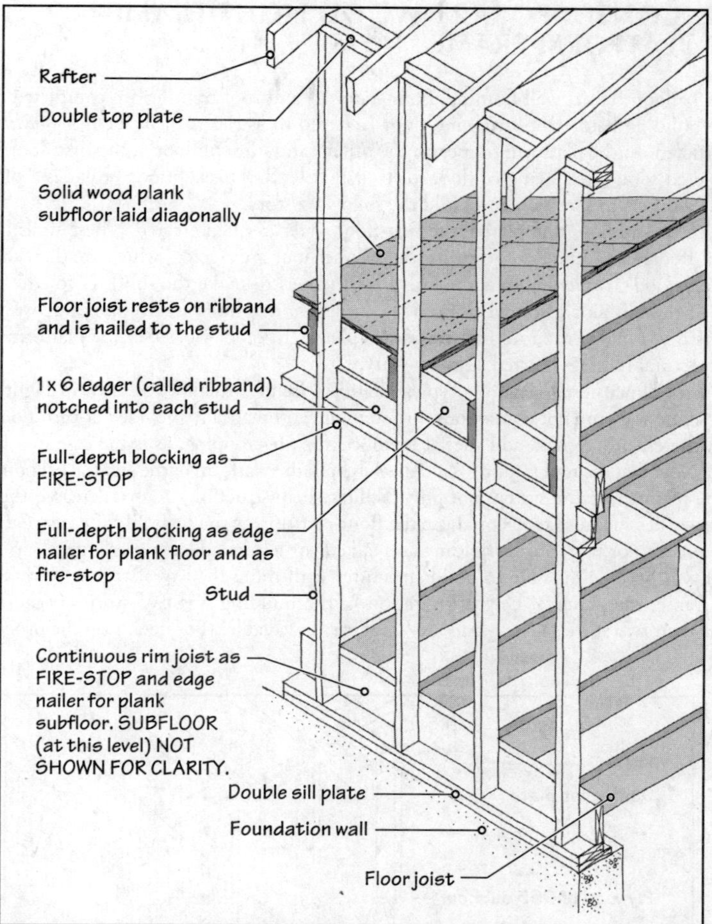

FIGURE 15.1 A typical balloon frame construction—a precursor of the contemporary platform frame construction.

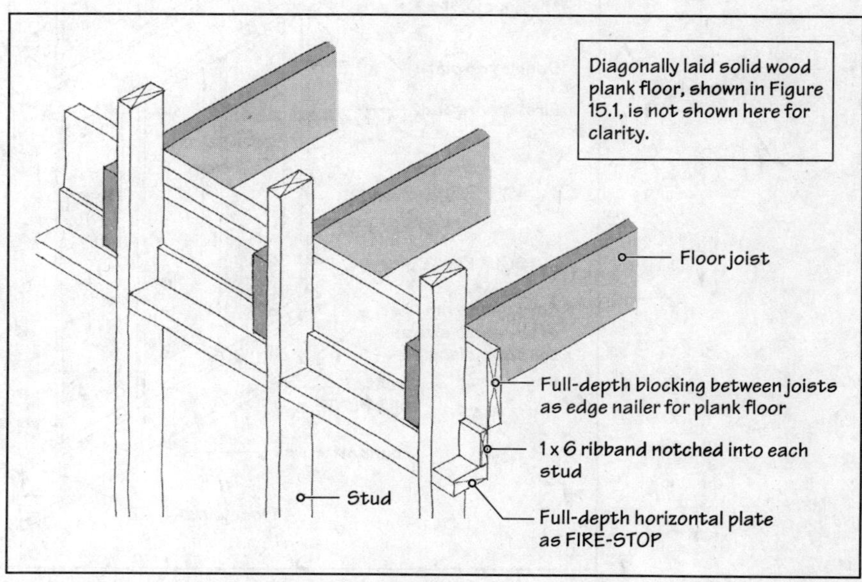

FIGURE 15.2 Detail of balloon frame construction at the intermediate floor level.

15.2 CONTEMPORARY WOOD LIGHT-FRAME—THE PLATFORM FRAME

In the balloon frame, wall framing, floor framing, and roof framing are completed for the entire building before the structural floor (referred to as the *subfloor*, or *floor sheathing*) is constructed. In the platform frame, on the other hand, the subfloor at the first floor level is completed soon after laying of floor joists at that level. This subfloor provides a platform for the workers to stand on and build the following story.

With the subfloor completed, the first-floor walls are then erected. These walls bear on the subfloor. Later, the floor joists for the second floor are erected, which are then sheathed with plywood or oriented strandboard (OSB) panels to make the subfloor for the second story. If there is an additional story in the building, it is built exactly the same way as the lower story. Finally, the roof framing is completed. In other words, in the platform frame, the structural frame is erected story by story.

A three-dimensional view of a typical platform frame structure is shown in Figure 15.3. Because the walls are not continuous, fire-stops are automatically provided at each floor level. Thus, there is no need to add special members for fire-stopping, as is the case in a balloon frame. Note that the roof framing is identical in both the platform frame and the balloon frame.

The platform frame system not only facilitated construction, it also increased the safety of the workers because they could use the floor platform as the scaffold for constructing the subsequent floor. Exterior scaffolding and tall construction ladders were unnecessary. Additionally, it became possible to build structures with more than two stories. With the balloon frame, the practical height limitation of the building was two stories because studs longer than two stories were—and are—not easily available. (For details on the process and

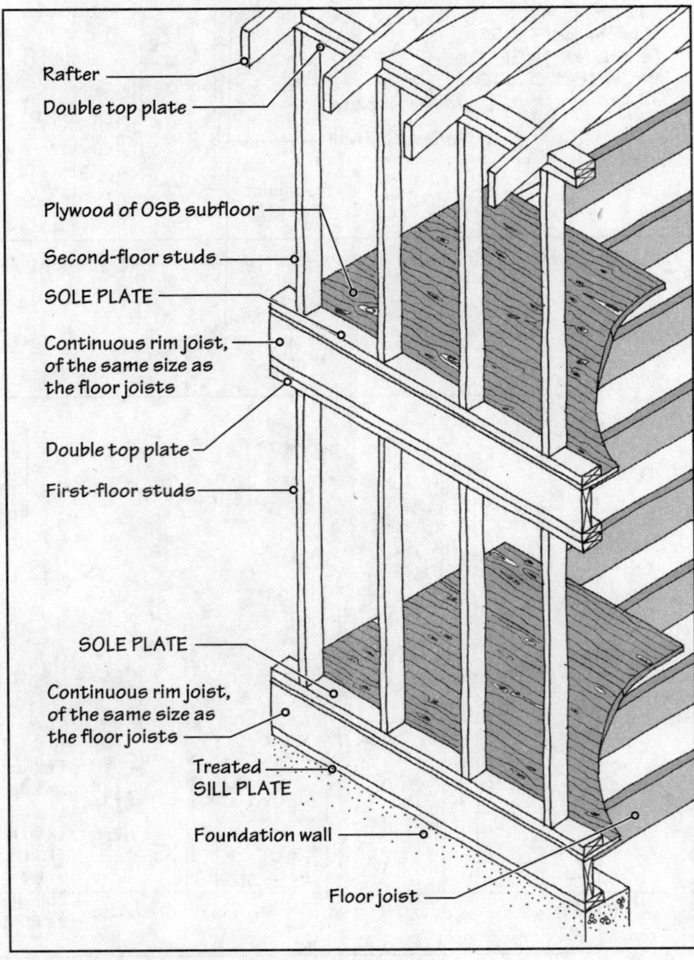

FIGURE 15.3 A typical platform frame construction.

sequence of the construction of a typical WLF building, see the first Principles in Practice section at the end of this chapter.

DIMENSIONAL STABILITY OF BALLOON AND PLATFORM FRAMES

The continuity of studs in a balloon frame, however, gives it some benefits over the platform frame. For instance, the balloon frame is dimensionally more stable because it has a much smaller overall cross-grain lumber dimension than the platform frame.

Figure 15.4 shows the cross sections through a typical platform frame and a typical balloon frame. Assuming that 2 × 12 solid lumber floor joists are used in both frames, the total cross-grain lumber in the platform frame is 33 in., compared to 6 in. in a two-story balloon frame. Because lumber shrinks negligibly along the grain but significantly across the grain (Section 13.5), the shrinkage and swelling in a balloon frame are much smaller than in a platform frame.

Excessive shrinkage and swelling result in nail popping and cracking of brittle finishes applied to the frame, such as, gypsum board and brick veneer. The detailing of the platform frame to accommodate shrinkage and swelling effects is covered in Section 15.11.

STRUCTURAL STABILITY OF BALLOON AND PLATFORM FRAMES

An additional advantage of the balloon frame is its greater structural stability. Continuous studs provide a more positive connection between the first and second floors of a balloon frame, giving greater strength and stability in resisting lateral loads.

In the platform frame, the connection between the two floors is provided by nailing the bottom plate of walls to the underlying floor frame. Additional connection is provided through the exterior sheathing. The connection so obtained between the floors is, however,

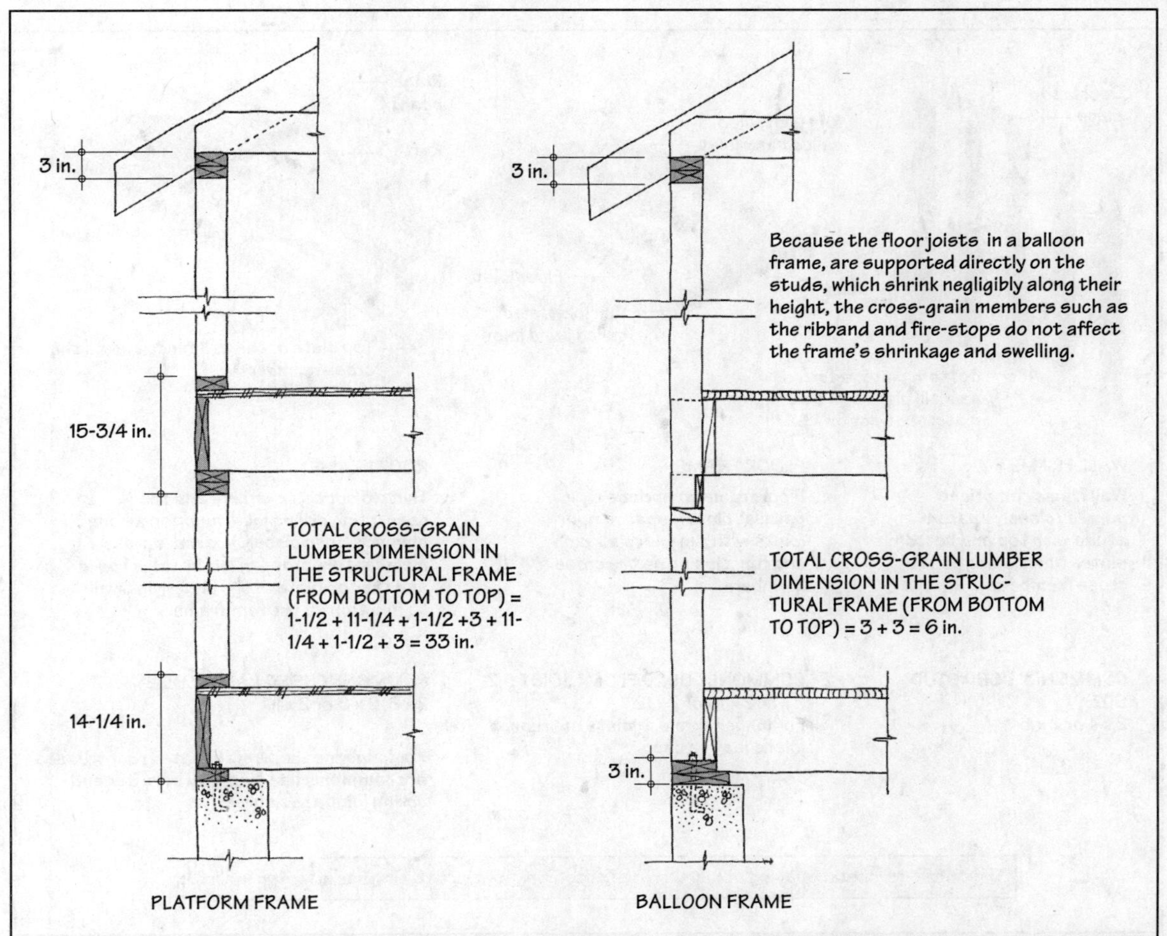

3 in.
15-3/4 in.
14-1/4 in.
TOTAL CROSS-GRAIN LUMBER DIMENSION IN THE STRUCTURAL FRAME (FROM BOTTOM TO TOP) = 1-1/2 + 11-1/4 + 1-1/2 +3 + 11-1/4 + 1-1/2 + 3 = 33 in.
PLATFORM FRAME
3 in.
Because the floor joists in a balloon frame, are supported directly on the studs, which shrink negligibly along their height, the cross-grain members such as the ribband and fire-stops do not affect the frame's shrinkage and swelling.
TOTAL CROSS-GRAIN LUMBER DIMENSION IN THE STRUCTURAL FRAME (FROM BOTTOM TO TOP) = 3 + 3 = 6 in.
3 in.
BALLOON FRAME

FIGURE 15.4 Cross sections through the structural frames of typical platform and balloon frames, highlighting the cross-grain lumber members in each.

not as strong as that provided by continuous studs. Therefore, in high-wind or seismic regions, a platform frame requires metal straps and connectors to improve the connectivity between floors. See the second Principles in Practice section at the end of this chapter.

15.3 FRAME CONFIGURATION AND SPACING OF MEMBERS

A common feature of the three principal structural elements—the walls, floors, and roof—of a platform frame building is that each of these elements is made of several parallel, closely spaced, repetitive members joined at each end to a continuous cross member that runs perpendicular to the parallel members, Figure 15.5. Cross members connect the parallel members together to form a *frame* and also play an important structural role.

In a wall assembly, the cross members are the top and bottom plates. In a floor assembly, the cross member at each end is a continuous member of the same size as the joists, called the *rim joists* or *band joist*. In a rafter-and-ceiling joist assembly, the top plates of the supporting walls on either side function as the cross members. The triangular rafter-and-ceiling joist assembly constitutes repetitive members. In a trussed roof, the trusses are the repetitive members.

Most framing members in a WLF building are 2-by members (the actual thickness is $1\frac{1}{2}$ in.). Generally, 2×4 or 2×6 members are used for walls; 2×8, 2×10, or 2×12 members are used for floor joists; and 2×6, 2×8, or 2×10 members are used for rafters.

The center-to-center spacing of members is either 12 in., 16 in., 19.2 in., or 24 in., as determined by the loads. These dimensions are also based on the size of the panels used as *sheathing* over the framing members—plywood, OSB, or gypsum board panels (typically 4 ft $\times$ 8 ft). Although 19.2-in. spacing is not commonly used, it is recognized as a possibility, because, like 12-in., 16-in., and 24-in. spacings, it also divides the 8-ft (96-in.) dimension by a whole number.

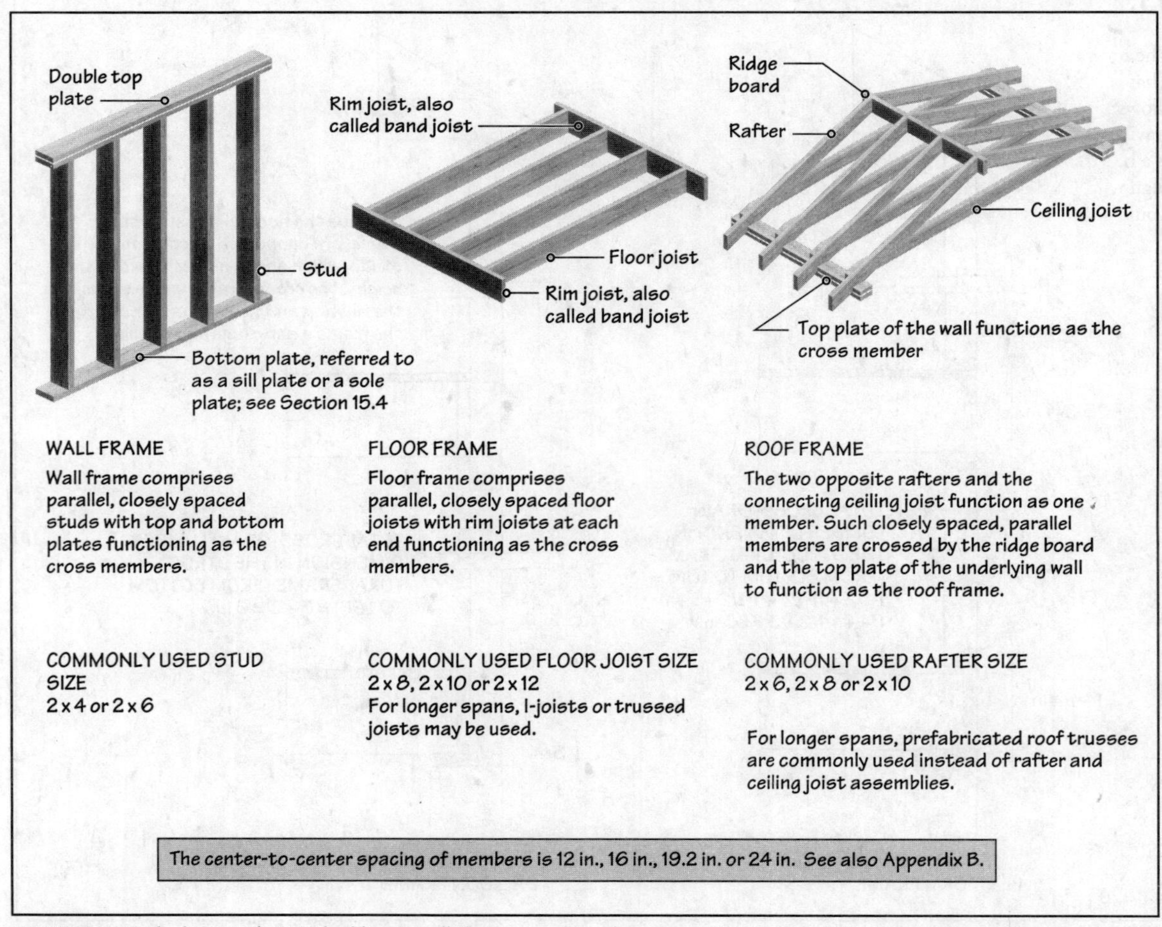

Double top plate
Stud
Bottom plate, referred to as a sill plate or a sole plate; see Section 15.4
Rim joist, also called band joist
Floor joist
Rim joist, also called band joist
Ridge board
Rafter
Ceiling joist
Top plate of the wall functions as the cross member
WALL FRAME
Wall frame comprises parallel, closely spaced studs with top and bottom plates functioning as the cross members.
FLOOR FRAME
Floor frame comprises parallel, closely spaced floor joists with rim joists at each end functioning as the cross members.
ROOF FRAME
The two opposite rafters and the connecting ceiling joist function as one member. Such closely spaced, parallel members are crossed by the ridge board and the top plate of the underlying wall to function as the roof frame.
COMMONLY USED STUD SIZE
2 x 4 or 2 x 6
COMMONLY USED FLOOR JOIST SIZE
2 x 8, 2 x 10 or 2 x 12
For longer spans, I-joists or trussed joists may be used.
COMMONLY USED RAFTER SIZE
2 x 6, 2 x 8 or 2 x 10
For longer spans, prefabricated roof trusses are commonly used instead of rafter and ceiling joist assemblies.
The center-to-center spacing of members is 12 in., 16 in., 19.2 in. or 24 in. See also Appendix B.

FIGURE 15.5 Each element of a WLF building (a wall, floor, or roof) is a frame consisting of several closely spaced, repetitive members that are connected at each end by a cross member.

PRACTICE QUIZ

Each question has only one correct answer. Select the choice that best answers the question.

1. Roughly how old is WLF construction?
   a. 1,000 years
   b. 800 years
   c. 600 years
   d. 400 years
   e. 200 years

2. When WLF was first discovered, it was called a
   a. balloon frame.
   b. rigid frame.
   c. skeleton frame.
   d. platform frame.
   e. portal frame.

3. WLF construction originated in
   a. the United Kingdom.
   b. France.
   c. Australia.
   d. Norway.
   e. the United States.

4. In the early version of WLF construction, the studs were continuous from the foundation to the roof.
   a. True
   b. False

5. In time, the early version of WLF construction was modified to what is now called a
   a. balloon frame.
   b. rigid frame.
   c. skeleton frame.
   d. platform frame.
   e. portal frame.

6. The type of WLF construction that has the largest amount of cross-grain lumber is a
   a. balloon frame.
   b. rigid frame.
   c. skeleton frame.
   d. platform frame.
   e. portal frame.

7. The framing members in a WLF building are generally
   a. 1-by solid lumber members.
   b. 2-by solid lumber members.
   c. 3-by solid lumber members.
   d. 4-by solid lumber members.
   e. none of the above.

8. The center-to-center spacings of framing members in a WLF building are generally
   a. 12 in., 16 in., or 24 in.
   b. 12 in., 24 in., or 36 in.
   c. 2 ft, 3 ft, or 4 ft.
   d. 4 ft, 6 ft, or 8 ft.
   e. virtually any spacing may be used, depending on the architectural and structural considerations.

9. The center-to-center spacing of framing members in a WLF building is mainly based on
   a. age-old practice.
   b. structural considerations.
   c. the size of sheathing panels.
   d. insulation requirements.
   e. (b) and (c).

15.4 ESSENTIALS OF WALL FRAMING

The top plate in a wall assembly consists of two members, each of the same size as the studs. That is why it is referred to as a *double top plate.* Doubling the top plate makes it stronger and allows the floor joists or rafters to be placed anywhere on the top plate, Figure 15.6(a). The gravity loads from the joists or rafters are transferred to the double top plate, which functions as a beam supported on the studs. If a single top plate is used, the floor joists and rafters must align with the underlying studs, Figure 15.6(b). In this case, the loads from the joists or rafters would go directly into the studs, and the top plate does not have a beam action.

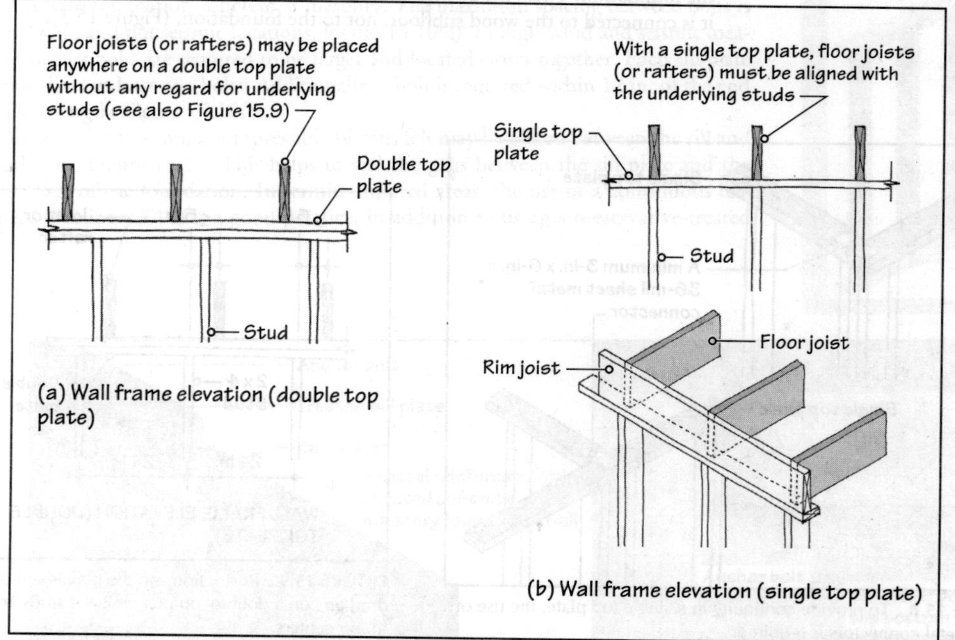

Floor joists (or rafters) may be placed anywhere on a double top plate without any regard for underlying studs (see also Figure 15.9)
Double top plate
Stud
(a) Wall frame elevation (double top plate)
With a single top plate, floor joists (or rafters) must be aligned with the underlying studs
Single top plate
Stud
Floor joist
Rim joist
(b) Wall frame elevation (single top plate)

FIGURE 15.6 If a double top plate is used in a wall frame, the floor joists or roof rafters need not align with the underlying studs; see Figure 15.9 for exceptions.

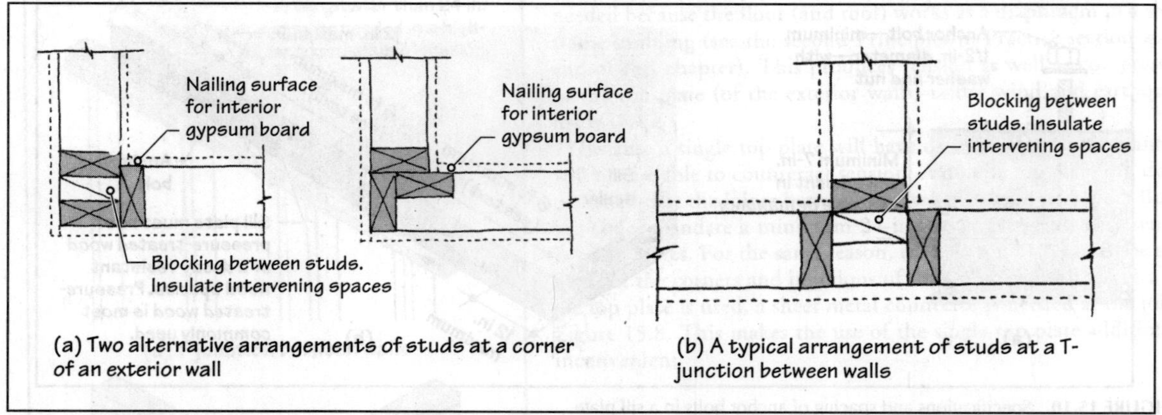

FIGURE 15.12 Typical arrangements of studs (shown in plan) at a corner of an exterior wall and a T-junction between walls.

Number of Studs at a Wall Corner or a T-Junction Between Walls

An exterior wall must function as a shear wall (see the second Principles in Practice section at the end of this chapter). A shear wall is subjected to tension at one end and compression at the other end, caused by the tendency of the wall to overturn under the action of wind or earthquake loads. The compression caused by the lateral loads adds to the compression due to the gravity loads. Therefore, the corner of an exterior wall must be stronger than the field of the wall, requiring a minimum of three studs at the corner.

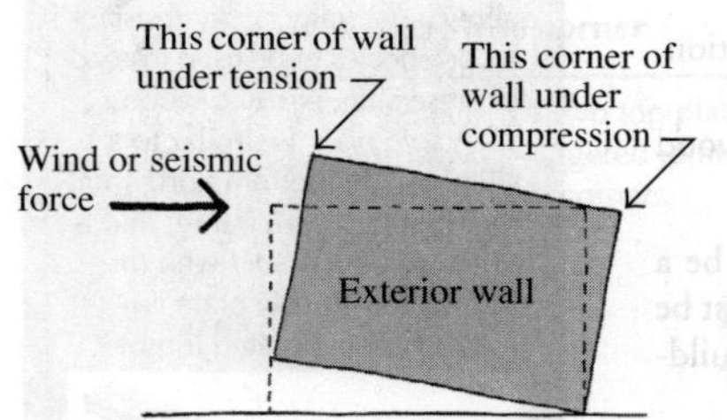

In addition to the structural consideration, the provision of three studs at a corner is required to obtain an adequate nailing surface for interior gypsum board and exterior sheathing. Two alternative details are typically used for a three-stud corner, Figure 15.12. This figure also shows a typical detail of studs at a wall T-junction, which also requires three studs.

15.5 FRAMING AROUND WALL OPENINGS

A typical framing around openings in a wall (doors or windows) is shown in Figure 15.13. A wall opening requires the use of *jack studs* on both sides of the opening. A jack stud is a partial-height stud that supports the lintel beam, generally referred to as a *header* (or *lintel header* in some literature). One jack stud on both sides of the opening gives $1\frac{1}{2}$-in. bearing for the header. This is adequate for a small opening in a one- or two-story building. If the opening size is large or if the number of floors in the building exceeds two, two or three jack studs may be needed.

Header

A header is typically made of two or three 2-by lumber members, depending on the thickness of the wall. The members are face nailed to form a beam. If the wall is framed of 2×4 members, two 2-by lumber members are required, with a $\frac{1}{2}$-in.-thick filler, Figure 15.14. The filler is usually a plywood or an OSB sheet. For exterior walls, a rigid plastic foam insulation is used. In a 2×6 wall, three 2-by lumber members are required, with two filler sheets. For large openings, trussed headers (Figure 14.20) or glulam headers are used.

Precut Length of Studs and Oversized Headers

Lumber is usually available in lengths of 8 ft, 10 ft, 12 ft, and so on, up to a maximum length of 26 ft. (Some lumber dealers may stock lumber only up to a length of 22 ft.) A

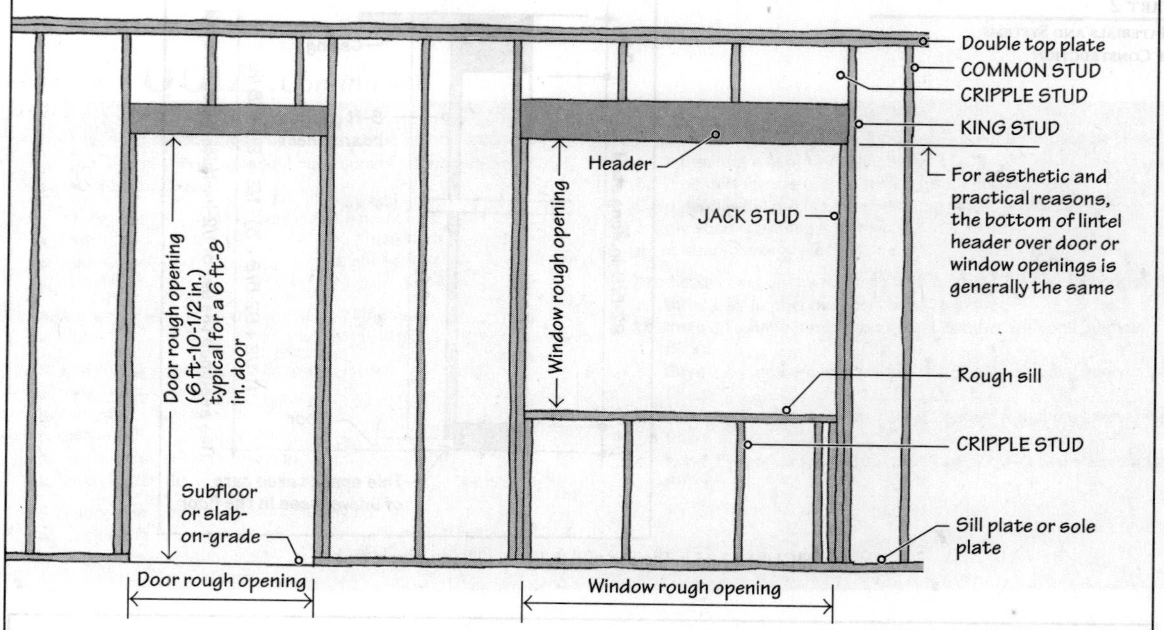

Double top plate
COMMON STUD
CRIPPLE STUD
KING STUD
Header
For aesthetic and practical reasons, the bottom of lintel header over door or window openings is generally the same
JACK STUD
Door rough opening (6 ft-10-1/2 in.) typical for a 6 ft-8 in. door
Window rough opening
Rough sill
CRIPPLE STUD
Subfloor or slab-on-grade
Sill plate or sole plate
Door rough opening
Window rough opening

Number of Jack Studs and King Studs in Load-bearing Wall Openings

Although one jack stud has been indicated in this figure, two or three jack studs may be required if the opening is subjected to a greater dead load, which is a function of the number of floors above the opening and the roof snow load. Generally, for a load-bearing wall, the total number of jack studs on both sides of the opening should at least be equal to the number of common studs removed from the wall by the opening.

Similarly, although one king stud has been shown, in high-wind regions the use of two or three kings studs may be needed in a wall.

FIGURE 15.13 Wall framing around openings.

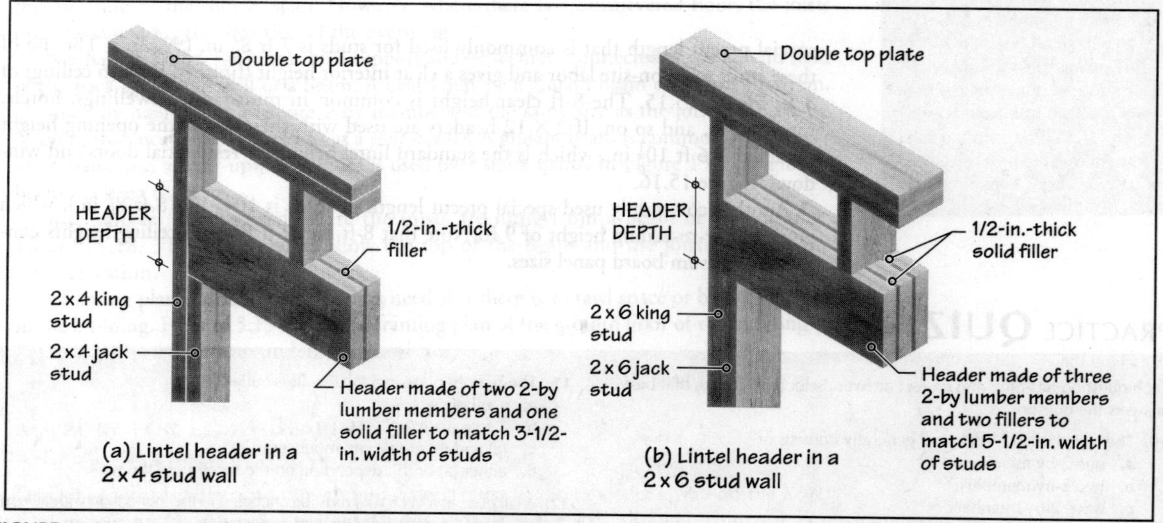

FIGURE 15.14 Anatomy of a header made with 2-by lumber members.

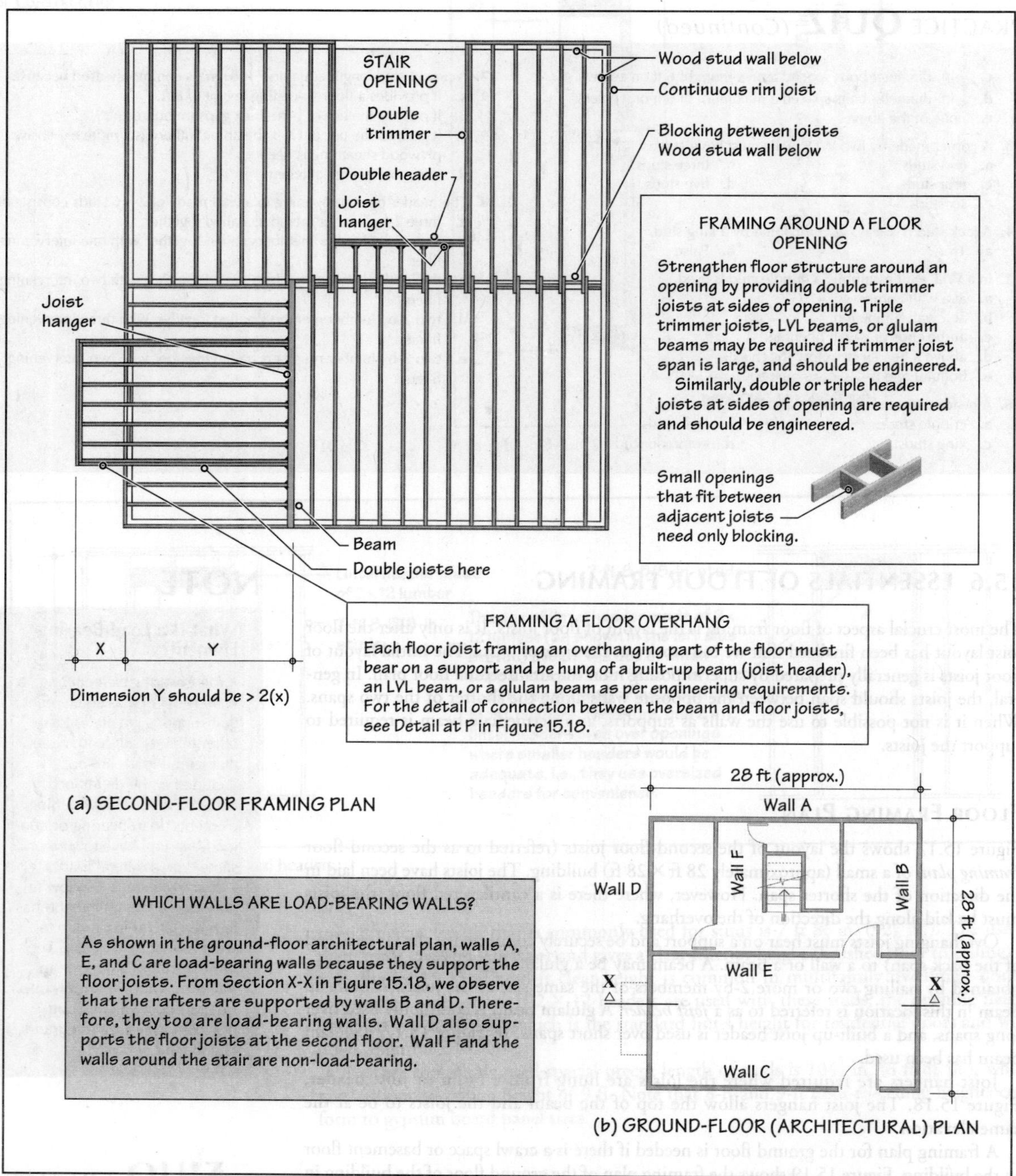

STAIR OPENING
Double trimmer
Double header
Joist hanger
Joist hanger
Wood stud wall below
Continuous rim joist
Blocking between joists
Wood stud wall below
FRAMING AROUND A FLOOR OPENING
Strengthen floor structure around an opening by providing double trimmer joists at sides of opening. Triple trimmer joists, LVL beams, or glulam beams may be required if trimmer joist span is large, and should be engineered.
Similarly, double or triple header joists at sides of opening are required and should be engineered.
Small openings that fit between adjacent joists need only blocking.
Beam
Double joists here
X
Y
Dimension Y should be > 2(x)
FRAMING A FLOOR OVERHANG
Each floor joist framing an overhanging part of the floor must bear on a support and be hung of a built-up beam (joist header), an LVL beam, or a glulam beam as per engineering requirements. For the detail of connection between the beam and floor joists, see Detail at P in Figure 15.18.
(a) SECOND-FLOOR FRAMING PLAN
28 ft (approx.)
Wall A
Wall D
Wall F
Wall B
Wall E
Wall C
28 ft (approx.)
X
X
(b) GROUND-FLOOR (ARCHITECTURAL) PLAN
WHICH WALLS ARE LOAD-BEARING WALLS?
As shown in the ground-floor architectural plan, walls A, E, and C are load-bearing walls because they support the floor joists. From Section X-X in Figure 15.18, we observe that the rafters are supported by walls B and D. Therefore, they too are load-bearing walls. Wall D also supports the floor joists at the second floor. Wall F and the walls around the stair are non-load-bearing.

FIGURE 15.17 Second-floor framing plan of a building.

Enhancement of Floor Framing

Carefully examine the layout of Figure 15.17. Observe that the joists along the two opposite ends of the cantilevered floor must be doubled. Similarly, the joists framing around an opening (in this case, a stair opening) are also required to be doubled.

Doubling of floor joists (or decreased spacing between joists) may also be needed in areas of the floor that carry excessive load, such as a floor topped with concrete and ceramic tiles, Figure 15.22.

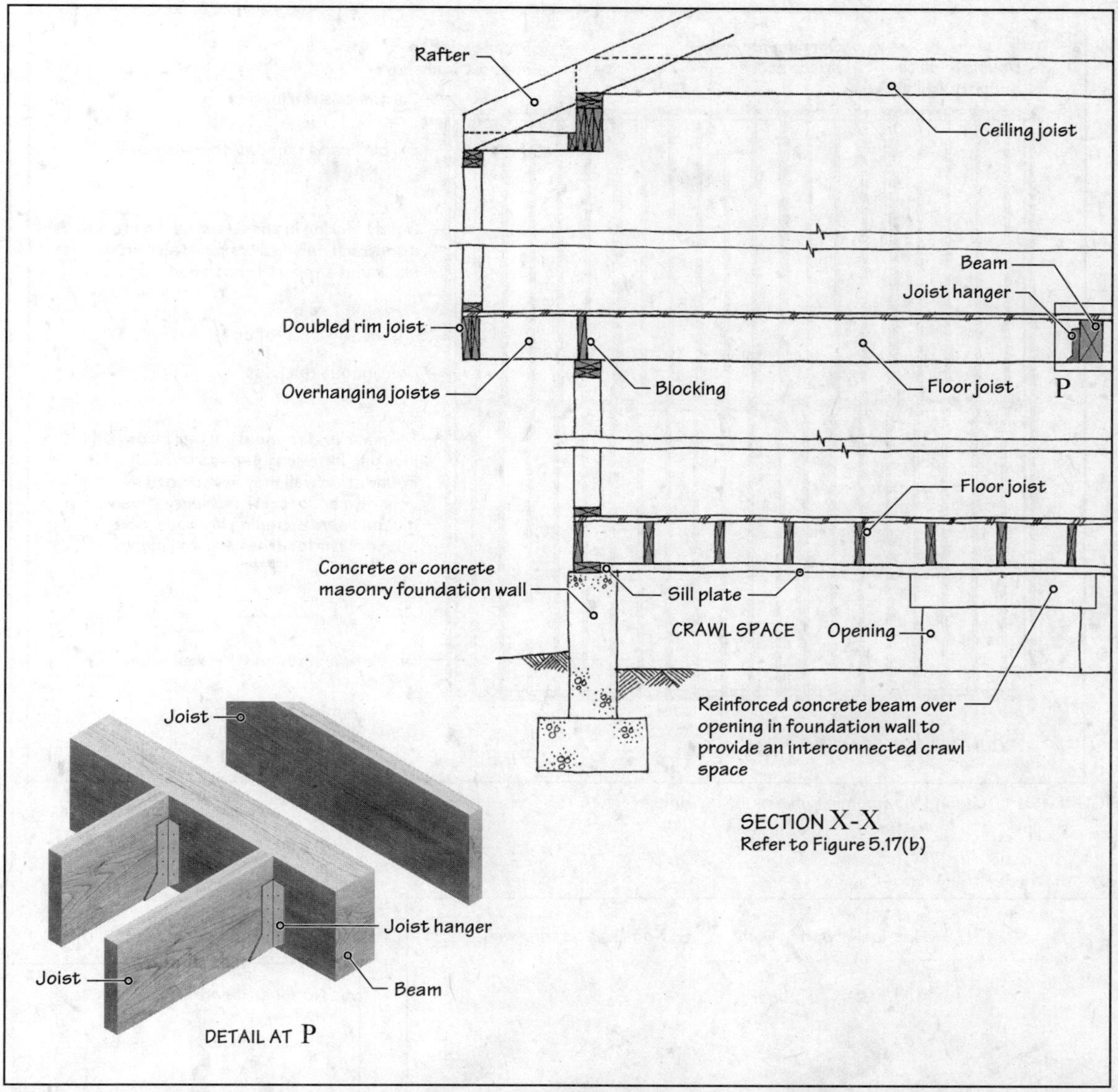

FIGURE 15.18 Section through the building in Figure 15.17. Detail at P shows the connection between the overhanging floor joists at the second floor and the supporting beam.

Role of Rim (or Band) Joists

Because floor joists are slender members, they are prone to lateral buckling (Section 4.9). Rim joists provide lateral restraint to the floor joists, reducing their tendency to buckle as well as connecting the ends of the joists.

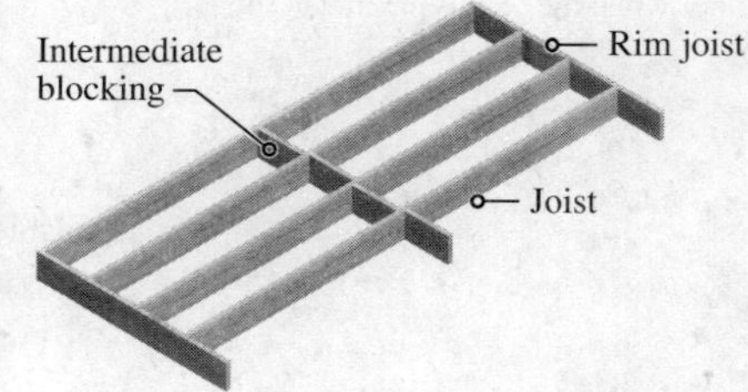

For the same reasons, floor joists that have a long span may require intermediate full-depth *blocking* or *diagonal bridging,* Figure 15.23. Generally, intermediate blocking or bridging is required if the floor joist depth is greater than 12 in. (nominal). However, it is prudent to provide rows of blocking or bridging for all floor joists generally at 8 ft on centers.

Diagonal bridging is generally provided through 1 × 3 lumber with bevel-cut ends to fit tight at the joists. Full-depth blocking has the advantage of not requiring bevel-cut ends. It also stiffens the floor and distributes the load between adjacent joists. However, full-depth blocking obstructs pipes running

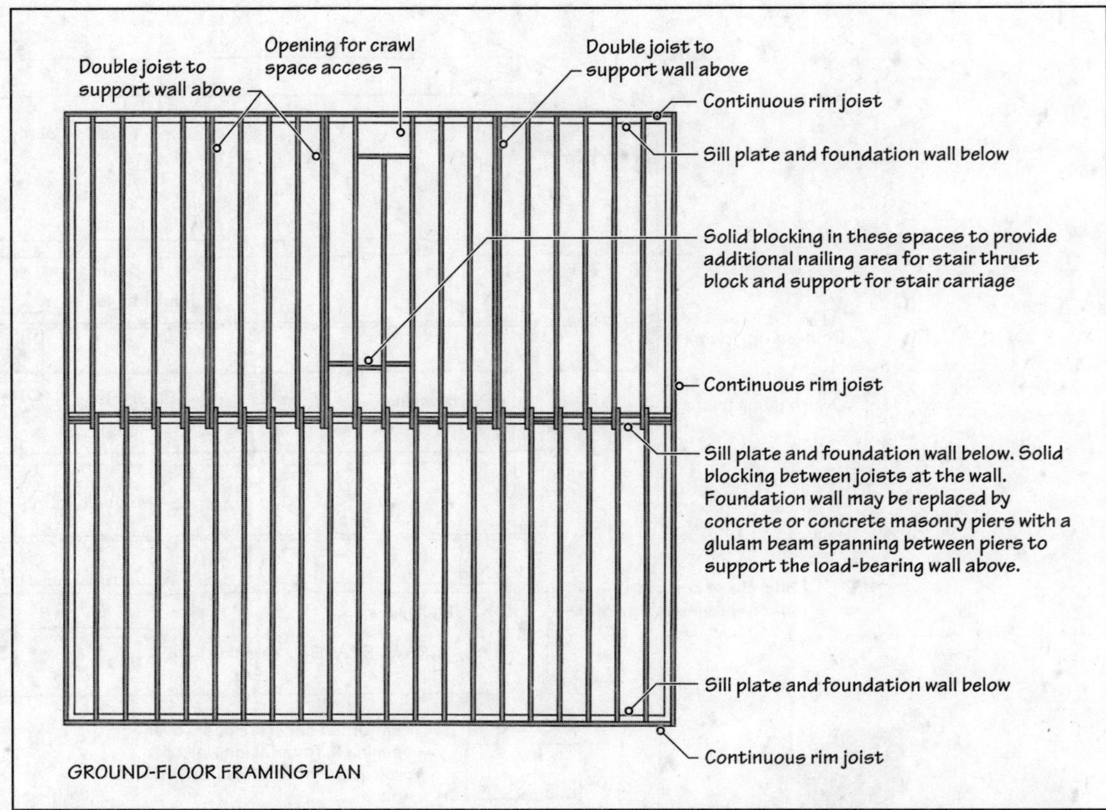

FIGURE 15.19 Ground-floor framing plan of the building in Figure 15.17 (assuming a crawl space below).

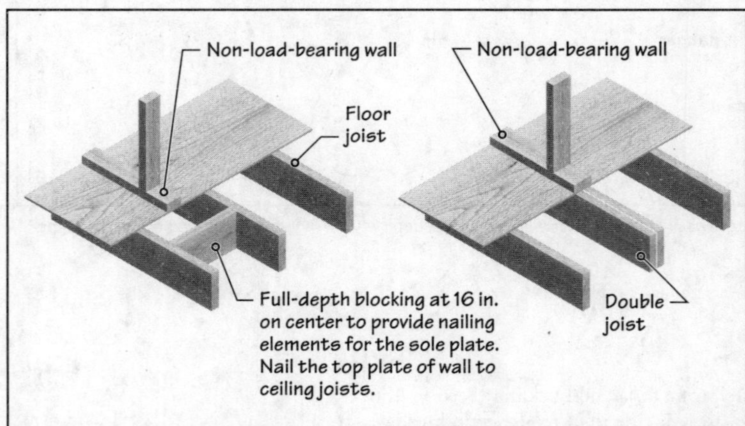

FIGURE 15.20 Two alternative ways of supporting a non-load-bearing wall on a wood floor when the wall is parallel to the joists.

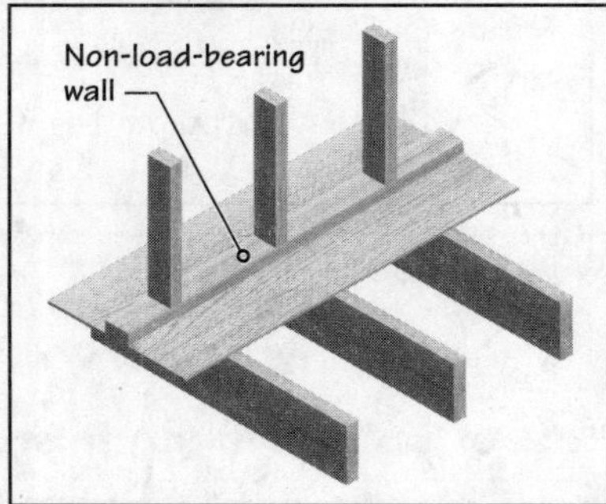

FIGURE 15.21 Support for a non-load-bearing wall on a wood floor when the wall is perpendicular to the joists.

in the space between adjacent joists. Diagonal bridging consisting of metal straps is also available from several manufacturers. It is easier to install than lumber bridging.

Structural Continuity in Floor Joists and Rim Joists

Floor joists should have continuity, provided either through overlapped ends nailed together at the supporting wall, Figure 15.24(a), or through scabs nailed to butt-jointed

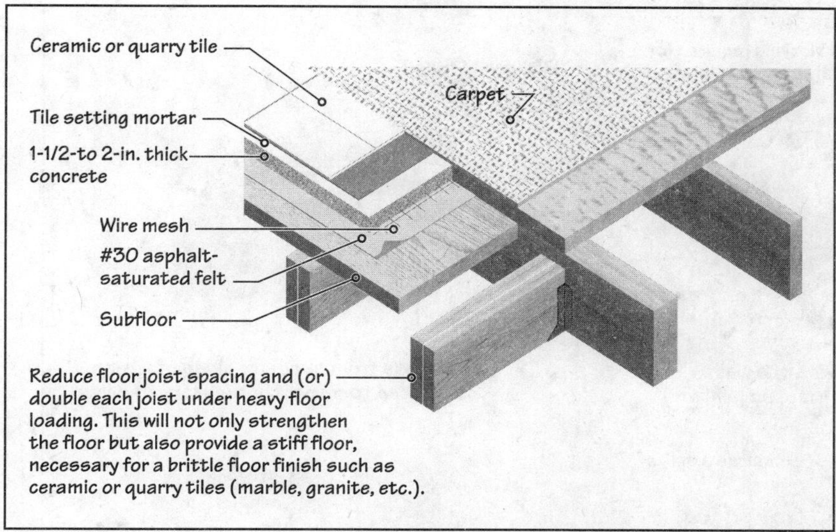

Ceramic or quarry tile

Tile setting mortar
1-1/2-to 2-in. thick concrete

Reduce floor joist spacing and (or) double each joist under heavy floor loading. This will not only strengthen the floor but also provide a stiff floor, necessary for a brittle floor finish such as ceramic or quarry tiles (marble, granite, etc.).

FIGURE 15.22 Strengthening of the floor frame under a heavily loaded part of the floor.

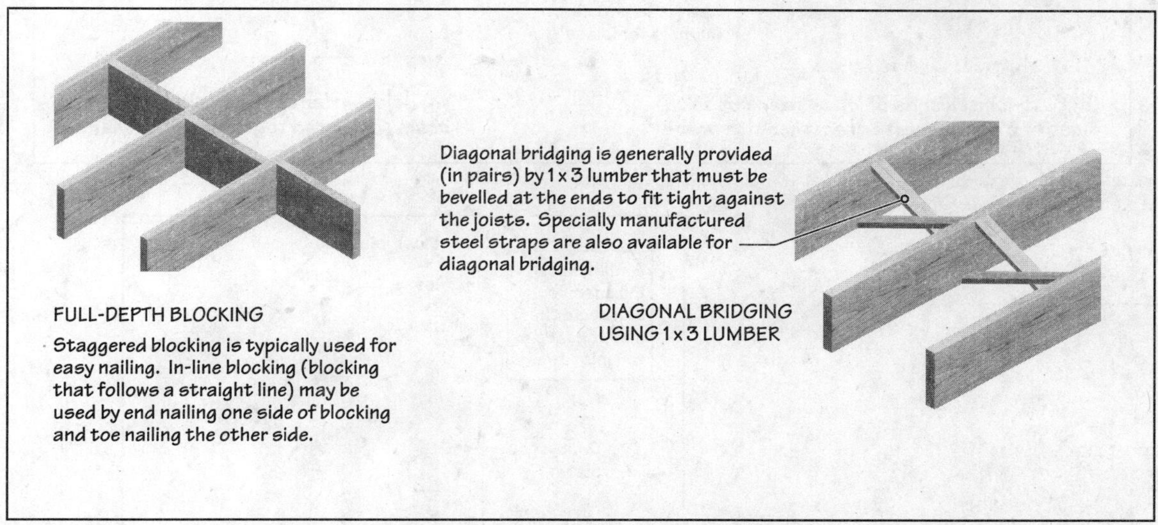

Diagonal bridging is generally provided (in pairs) by 1 x 3 lumber that must be bevelled at the ends to fit tight against the joists. Specially manufactured steel straps are also available for diagonal bridging.
FULL-DEPTH BLOCKING
Staggered blocking is typically used for easy nailing. In-line blocking (blocking that follows a straight line) may be used by end nailing one side of blocking and toe nailing the other side.
DIAGONAL BRIDGING USING 1 x 3 LUMBER

FIGURE 15.23 Two commonly used means of providing intermediate lateral restraint in floor joists.

ends, Figure 15.24(b). Similar details are used with joists that bear on an underlying beam, Figure 15.24(c). For joists that are connected to a supporting beam with joist hangers, the continuity between joists is provided through the connections, Figure 15.24(d).

A rim joist should also be continuous. Continuity in a rim joist adds to the continuity of the wall's top plate and is typically provided by nailing a 2-by splice member at the joint, Figure 15.25.

15.7 ROOF TYPES AND ROOF SLOPE

The roof of a WLF building is generally sloped. The more commonly used roof shapes are gable, hip, and shed roofs. Often these roof types are used in combination, Figure 15.26. Figure 15.26 also provides an introduction to some of the roof vocabulary.

The slope of a roof is not expressed in degrees but as a rise-to-run ratio. In this expression, the run is generally kept at a constant value of 12. Thus, a roof slope is expressed as 2:12, $2\frac{1}{2}$:12, 5:12, 6:12, and so on. The greater the rise, the greater the roof slope. Figure 15.27

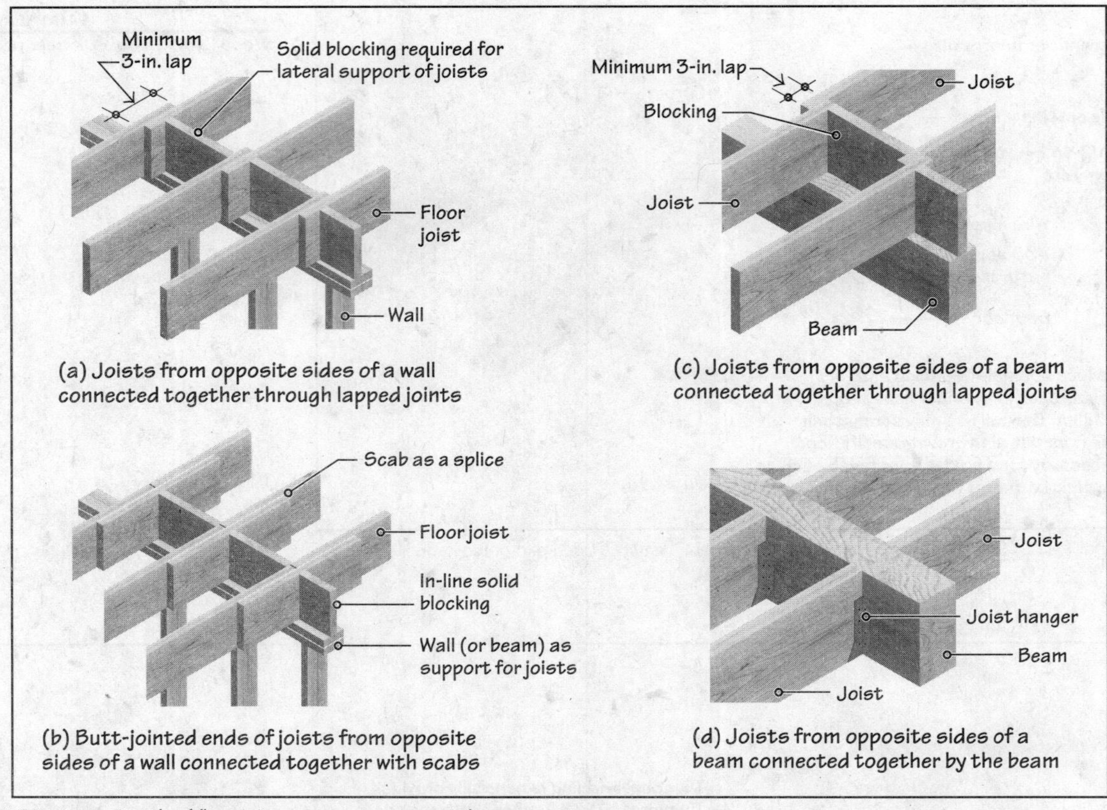

FIGURE 15.24 Details of floor joists resting on an intermediate support.

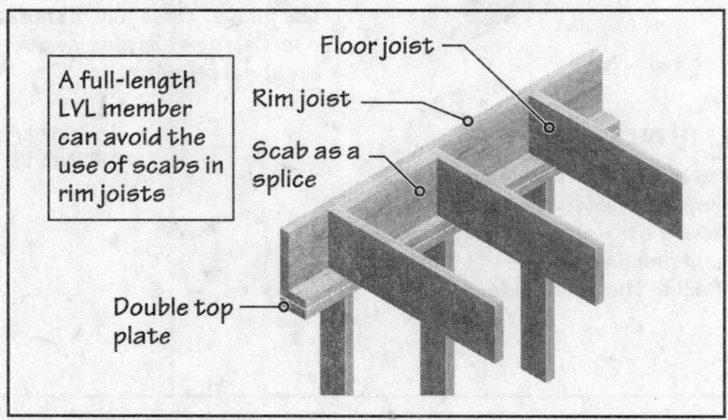

FIGURE 15.25 Detail of a splice-connected rim joist.

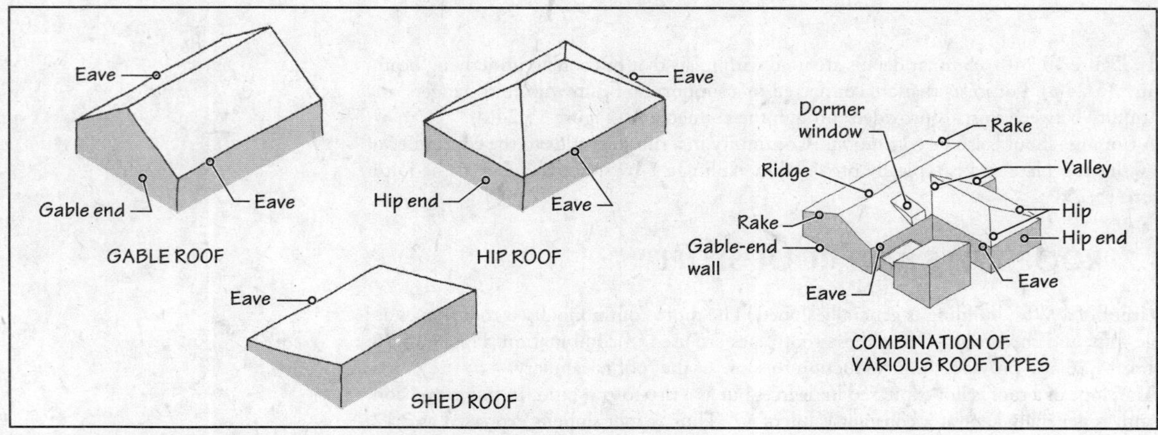

FIGURE 15.26 Commonly used roof shapes and roof-related terminology.

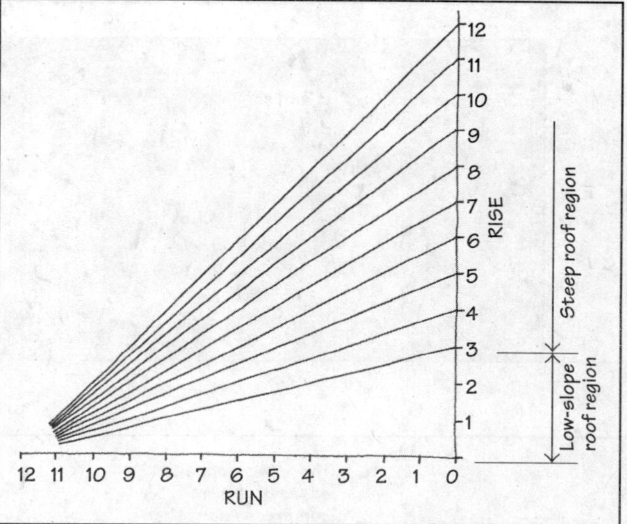

FIGURE 15.27 Roof slope.

gives a visual picture of various roof slopes. As we will observe in Chapter 33, the roofing industry divides roofs into two types based on roof slope:

- Low-slope roof
- Steep roof

A low-slope roof is a roof whose slope is less than 3:12. A steep roof has a slope of greater than or equal to 3:12. Using a 3:12 roof slope as the dividing line is based on structural considerations, as explained later in this chapter.

An additional basis for the dividing line is roof cover considerations. Several asphalt shingle manufacturers do not warrant the use of their product on a roof with a slope of less than 3:12.

PRACTICE QUIZ

Each question has only one correct answer. Select the choice that best answers the question.

19. In a WLF building, a floor framing plan indicates
- **a.** the layout of studs and their spacing.
- **b.** the layout of studs and exterior wall sheathing.
- **c.** the layout of floor joists and blocking.
- **d.** the layout of floor joists, floor beams, and blocking.
- **e.** the layout of floor beams, studs, and wall sheathing.

20. A non-load-bearing wall in a building is one that supports the loads from the roof but not from any floor.
- **a.** True
- **b.** False

21. The primary purpose of full-depth blocking in a floor frame is to
- **a.** reduce compressive stresses in floor joists.
- **b.** prevent buckling of floor joists.
- **c.** reduce tensile stresses in floor joists.
- **d.** give additional nailing surface for a gypsum board ceiling.
- **e.** give additional nailing surface for floor sheathing.

22. Blocking between floor joists is required where
- **a.** the joists bear on a wall.
- **b.** the joists bear on a beam.
- **c.** the joists are hung on a beam using joist hangers.
- **d.** all of the above.
- **e.** both (a) and (b).

23. A floor in a WLF building that supports a non-load-bearing wall generally requires no additional strengthening or stiffening when
- **a.** the wall runs parallel to floor joists.
- **b.** the wall runs perpendicular to floor joists.

24. A roof with gable ends on both sides in a rectangular building has
- **a.** eaves on all four sides.
- **b.** eaves on three sides.
- **c.** eaves on two sides.
- **d.** eave on one side.
- **e.** no eaves.

25. A rectangular roof with hip ends on two sides has
- **a.** eaves on all four sides.
- **b.** eaves on three sides.
- **c.** eaves on two sides.
- **d.** eave on one side.
- **e.** no eaves.

26. In the building industry, the slope of a roof is generally expressed
- **a.** in degrees.
- **b.** in radians.
- **c.** as a rise-to-run ratio.
- **d.** as a run-to-rise ratio.
- **e.** none of the above.

15.8 ESSENTIALS OF ROOF FRAMING

The roof framing of a typical WLF building consists either of trusses or of rafter-and-ceiling joist assemblies. Theoretically, there is little difference between them, except that a truss is a shop-fabricated, multitriangle frame and a rafter-and-ceiling-joist assembly is a site-fabricated, single-triangle frame.

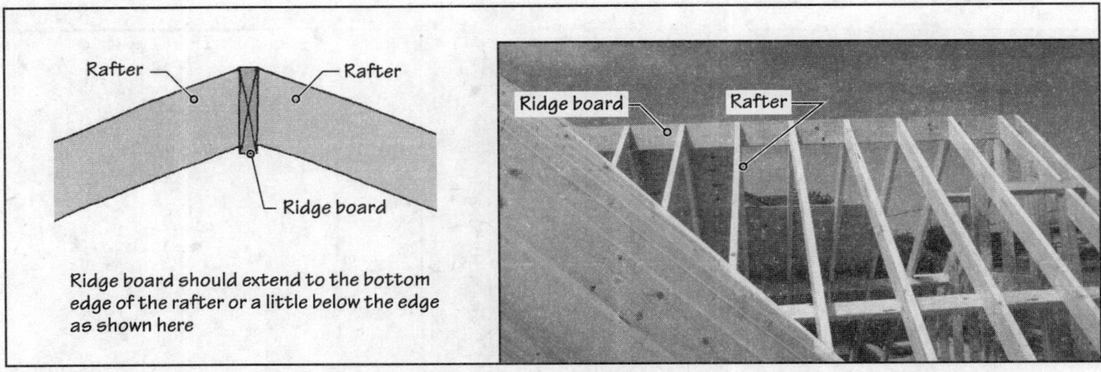

FIGURE 15.28 Ridge board and rafters.

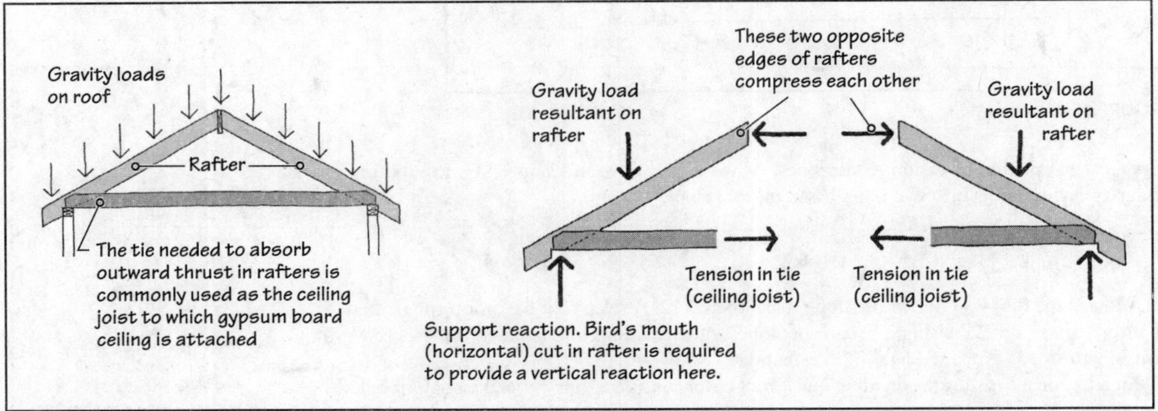

FIGURE 15.29 Forces created by gravity loads in a rafter couple and ceiling joist assembly.

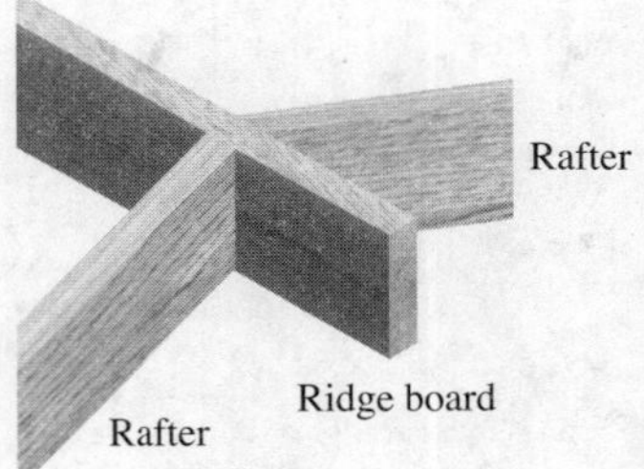

In a rafter-joist assembly, the rafters rest on the walls at one end and are connected to the opposite rafters at the ridge. The connection between the two rafters is made through a continuous *ridge board* that runs perpendicular to the rafters.

A ridge board is generally a 2-by member that must extend from the top edge to the bottom edge of the rafter (or a little below), Figure 15.28. It has no structural function except to align the rafter ends in a straight line at the top. That is why the codes only require a 1-by member, although a 2-by member is generally used as the ridge board. Many framers use an LVL member to ensure a continuous, straight ridge.

Each rafter pair (connected at the ridge) must be tied together at the bottom to resist the outward thrust created by the gravity loads on the roof. In a typical WLF roof, this tie also functions as the ceiling joist, Figure 15.29. The rafter pair and the ceiling joist thus make a stable triangular frame, which rests on two opposite supports and delivers a vertical load at the supports with no lateral thrust.

To deliver the load vertically at the support, each rafter is cut to have a horizontal bearing on the supporting walls. Each rafter is, therefore, specially notched near the bottom. This notch is referred to as a *bird's mouth,* Figure 15.30. The horizontal part of the bird's mouth should equal the width of the supporting wall (including the thickness of sheathing) to ensure that the wall sheathing engages the top plate fully. If the sheathing is applied to the wall after the rafters are in place, it must be cut around the rafters.

Support for Ceiling Joists

Because the distance between the supporting walls is usually too large to allow the use of one-piece members as ceiling joists, the ceiling joists are generally two-piece members connected together at an intermediate support. The intermediate support is necessary because the ceiling joists are subjected not only to tension but also to bending due to gravity loads from the attic. The intermediate support is usually an underlying wall, Figure 15.31. If a wall is not available, a beam should be introduced. Alternatively, prefabricated roof trusses may be used, which do not need intermediate supports.

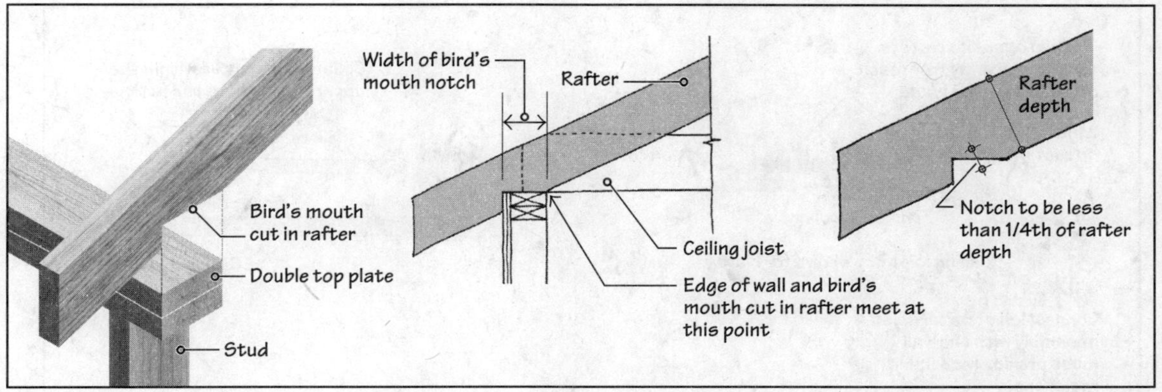

FIGURE 15.30 Bird's mouth cut in a rafter.

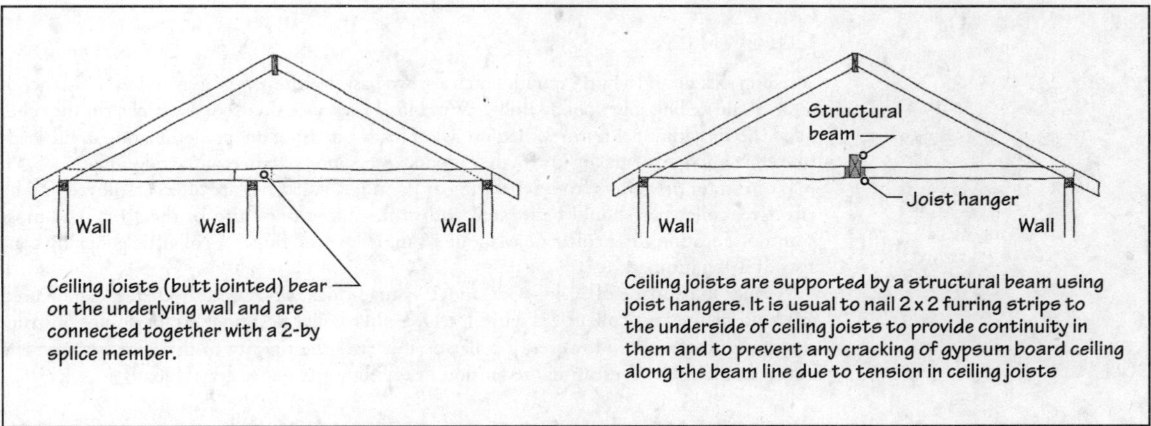

FIGURE 15.31 Ceiling joist support.

Purlins and Purlin Braces as Intermediate Rafter Supports

Apart from providing support to ceiling joists, an intermediate wall may also be used to provide additional support to rafters. This is done by using a purlin to support the rafters. The purlin is supported on closely spaced purlin braces, Figure 15.32. The use of purlins and braces not only reduces the size of rafters, it also increases the stiffness of the roof. This is particularly helpful in roofs supporting large gravity loads due to snow or heavy roofing materials.

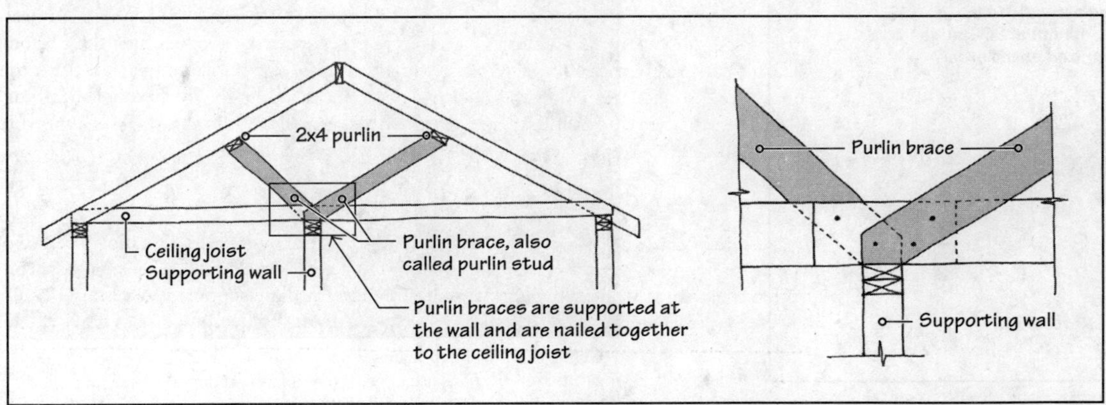

FIGURE 15.32 Purlins and purlin braces.

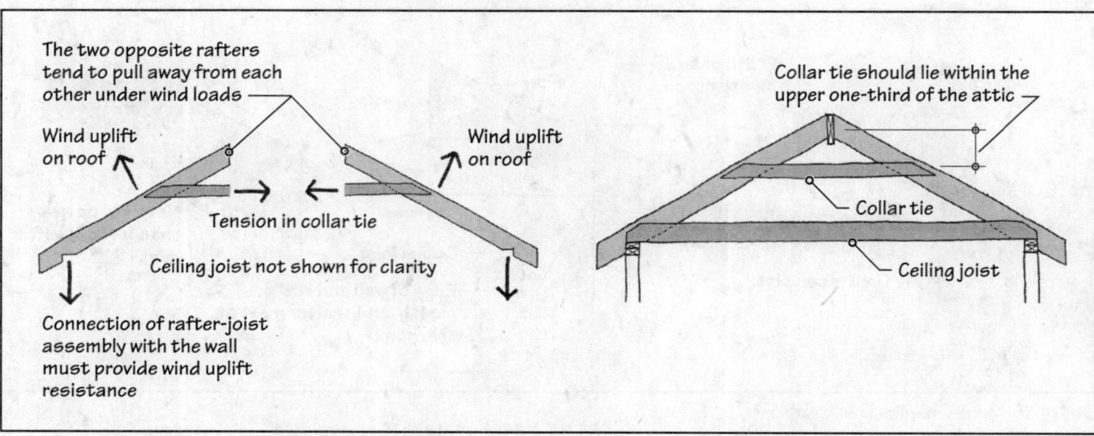

FIGURE 15.33 The use of collar ties for wind-uplift resistance in a roof.

Collar Ties

A ceiling joist tie and a bird's mouth notch are two basic features of the gravity-load resistance of a rafter-and-ceiling-joist roof assembly. Wind-load resistance in a roof assembly, on the other hand, needs additional features. Because wind loads on a roof are predominantly uplift loads (the exact opposite of gravity loads), they produce separation of rafters at the ridge, Figure 15.33.

To restrain the rafters from separating at the ridge, collar ties are often employed. To be effective, collar ties should be located within the upper one-third of the attic. The most common location for a collar tie is about 18 in. below the ridge. A collar tie generally consists of 2 × 4 lumber.

A *ridge strap,* made of $1\frac{1}{4}$-in.-wide and 20-gauge-thick galvanized sheet steel can be used as an alternative to a collar tie, Figure 15.34. Unlike collar ties, ridge straps do not obstruct the space in the attic. However, a collar tie increases the rigidity of the joint between the rafters at the ridge, decreasing the tension in ceiling joists under gravity loads.

Prefabricated Trusses as an Alternative to a Rafter-Joist Assembly

Wood trusses are an alternative to a stick-built rafter-joist assembly. Their use is particularly attractive for single-family dwellings in regions where prefabrication is more economical because of the climate's severity and/or the higher labor costs.

Because of the length limitations of sawn lumber, a rafter-joist assembly has limited span capabilities. When the width of the building is large (typically greater than 35 ft), the use of trusses is a more viable solution for roof framing, Figure 15.35. Wood trusses are,

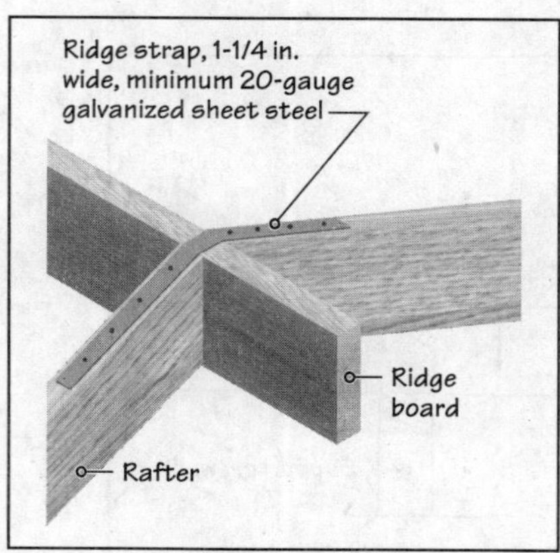

FIGURE 15.34 Ridge straps as alternatives to collar ties.

FIGURE 15.35 Prefabricated wood trusses are commonly used in buildings where roof spans are large. Prefabrication reduces on-site labor and hence makes trusses a popular alternative in regions with unfavorable weather conditions or high labor costs.

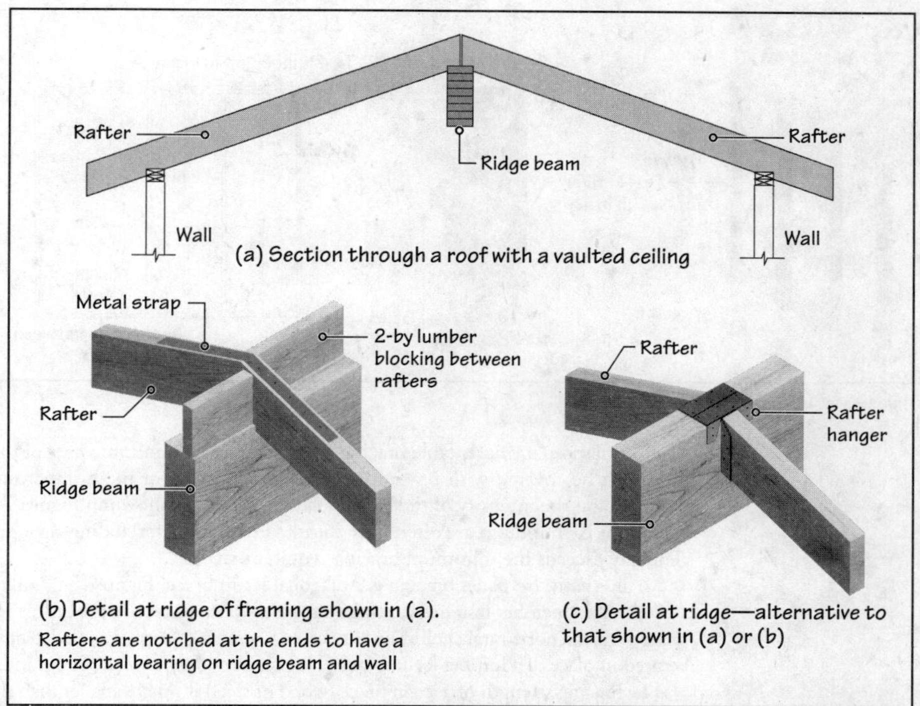

FIGURE 15.36 If the rafters are supported on a (structural) ridge beam, ceiling joists are not required to absorb the lateral thrust. Such a roof system is called a *vaulted* (or *cathedral*) *ceiling*.

therefore, a norm in wood frame commercial and multifamily residential buildings. Truss manufacturers provide all the necessary engineering and details that should be used with their trusses.

15.9 VAULTED CEILINGS

The rafter-ceiling-joist assembly described previously works only if the roof slope is greater than 3:12. If the slope is less than 3:12, the tension in ceiling joists becomes too large to be handled by conventional 2-by members. In such a case, the ridge board must be replaced by a *ridge beam*. A ridge beam is a structural member, unlike a ridge board. The rafters bear on the ridge beam at the top and on the supporting walls at the eave.

Such an assembly, consisting of two opposite rafters and the ridge beam, creates no outward thrust in the rafters under gravity loads. Consequently, there is no need for ceiling joist ties. The gypsum board ceiling in such a roof may be attached directly to the rafters and ridge beam, and/or the ridge beam may remain exposed. Such a ceiling is referred to as a *cathedral ceiling* or a *vaulted ceiling*, Figure 15.36. A vaulted ceiling may also be used with a roof slope greater than 3:12.

15.10 SHEATHING APPLIED TO A FRAME

The wall, floor, and roof frames must be covered with a sheathing material. Sheathing (generally a panel material) serves both structural and nonstructural functions.

WALL SHEATHING AND LATERAL BRACING OF THE FRAME

Wall sheathing furnishes a base over which exterior wall finishes are applied. It also provides a nailing base for an air-weather retarder. Structurally, wall sheathing integrates the studs into a composite wall system. It also provides bracing to the frame against lateral loads.

The most commonly used material for exterior wall sheathing (and bracing) is OSB. However, plywood, gypsum sheathing, and several other panel materials (including rigid

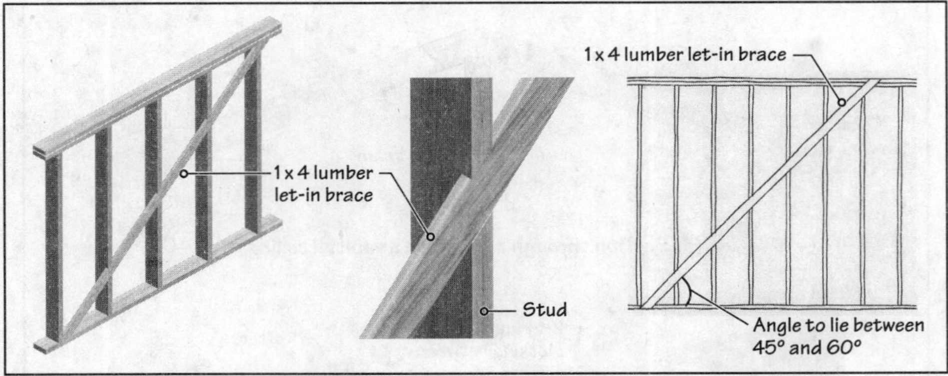

FIGURE 15.37 Lumber let-in lumber brace.

foam insulation) are acceptable bracing alternatives. The minimum area of the exterior wall that must be covered with bracing material is a function of the number of stories of the building and the intensity of the lateral load. However, as shown in Figure 14.29, the walls of most WLF buildings are often fully sheathed with OSB (excluding the openings), which generally exceeds the minimum bracing requirement.

An alternative to panel bracing is a diagonal let-in brace, Figure 15.37. This consists of a 1×4 lumber member fastened to studs that are notched to receive the brace. The angle of the brace with the horizontal should lie between 45° and 60°. A steel angle let-in brace may also be used in place of a lumber let-in brace. This requires only a small slit cut into the studs.

The bracing strength of a let-in brace (wood or steel) is much smaller than that of plywood or OSB panel bracing. Therefore, the use of a let-in brace is generally limited to situations where the lateral loads are small.

WLF buildings that have a relatively large footprint may require bracing in the interior walls in addition to the bracing provided in exterior walls. Interior-wall bracing can be provided by sheathing the interior walls on both sides with plywood or OSB and covering them with gypsum board. Alternatively, 2-by lumber (blocking) members placed diagonally through the wall may be used to obtain an interior braced wall, Figure 15.38.

Floor Sheathing

Floor sheathing (also called a *subfloor*) is a structural element because it transfers floor dead and live loads to the joists. Floor sheathing also provides diaphragm action in resisting lateral loads. The most commonly used material for a subfloor is OSB or plywood panels, generally used in 4-ft $\times$ 8-ft dimensions. Panels must be laid with the 8-ft dimensions perpendicular to the joists, Figure 15.39. Two types of subfloor panels are used, depending on the profile of the long (8-ft) edges:

- All four panel edges have a straight profile.
- Two long edges have a tongue-and-groove profile, whereas the other two edges are straight (Section 14.8).

If carpet is used as the floor finish, straight-edge subfloor panels require blocking along the long edges. Resilient floor finishes (such as vinyl or linoleum tiles or sheets) generally

FIGURE 15.38 Diagonal bracing of an interior wall with 2-by lumber members.

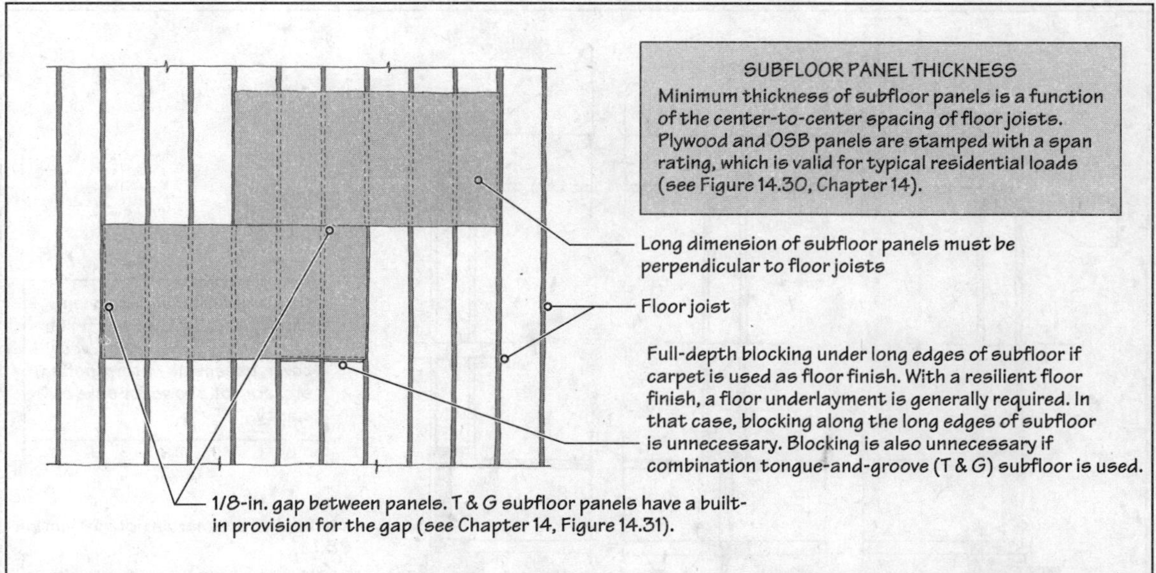

SUBFLOOR PANEL THICKNESS
Minimum thickness of subfloor panels is a function of the center-to-center spacing of floor joists. Plywood and OSB panels are stamped with a span rating, which is valid for typical residential loads (see Figure 14.30, Chapter 14).
Long dimension of subfloor panels must be perpendicular to floor joists
Floor joist
Full-depth blocking under long edges of subfloor if carpet is used as floor finish. With a resilient floor finish, a floor underlayment is generally required. In that case, blocking along the long edges of subfloor is unnecessary. Blocking is also unnecessary if combination tongue-and-groove (T & G) subfloor is used.
1/8-in. gap between panels. T & G subfloor panels have a built-in provision for the gap (see Chapter 14, Figure 14.31).

FIGURE 15.39 Subfloor requirements.

require a smooth underlayment over an OSB or plywood subfloor (Chapter 36). When a floor underlayment is used, blocking of the long edges of the subfloor is generally not required. Blocking is also not required if tongue-and-groove panels are used for the subfloor.

Roof Sheathing

Like floor sheathing, roof sheathing is also a structural component, and OSB or plywood panels are most commonly used. The thickness of panels is based on their span rating. Roof sheathing panels do not require blocking, but metal edge clips (generally H-shaped) are used along the long (8-ft) edges to join adjacent panels, Figure 15.40. The clips automatically leave a gap of $\frac{1}{8}$ in. between panels and allow the edges of panels to expand.

15.11 EQUALIZING CROSS-GRAIN LUMBER DIMENSIONS

In Section 15.2, reference was made to the moisture-related dimensional instability of cross-grain lumber. In a WLF building, cross-grain lumber is present in floor and wall frames. In the walls, only the top and bottom plates constitute cross-grain lumber. In the floors, floor joists and supporting beams constitute cross-grain lumber.

It is usually not possible to reduce cross-grain lumber dimensions in a building. All that can be done is to equalize cross-grain lumber dimensions throughout the structural frame so that the entire frame moves as

NOTE

WLF and Building Codes

As stated in Chapter 7, the building codes recognize WLF construction as Type V construction—the least fire-resistive type. Type V construction is further sub-classified into Type V(A) and Type V(B). In Type V(A), all critical assemblies (such as exterior walls, floors, roofs, and interior partitions) are required to be 1-h rated. A 1-h rating is typically achieved by covering the wood frame with $\frac{5}{8}$-in.-thick type X gypsum board on both sides (Chapter 16).

Type V(B) construction, commonly used for individual homes, is a nonrated WLF construction requiring no fire-rated assembly. Thus, a typical Type V(B) assembly is covered with $\frac{1}{2}$-in.-thick type R gypsum board on the interior, whose fire rating is much less than 1 h.

The use of Type V(A) is permitted for all occupancies. With some exceptions, Type V(B) is also permitted for all occupancies. However, because of the combustibility of a wood frame, the allowable areas and heights permitted by the codes for Type V buildings are fairly limited. For instance, the maximum permissible height for Type V construction used in a single-family dwelling is three stories. If automatic sprinklers are used throughout the building, the building height may be increased by one story.

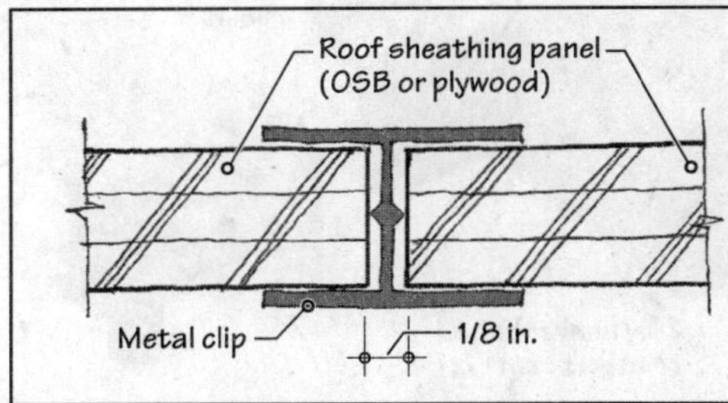

FIGURE 15.40 Clip between adjacent roof sheathing panels.

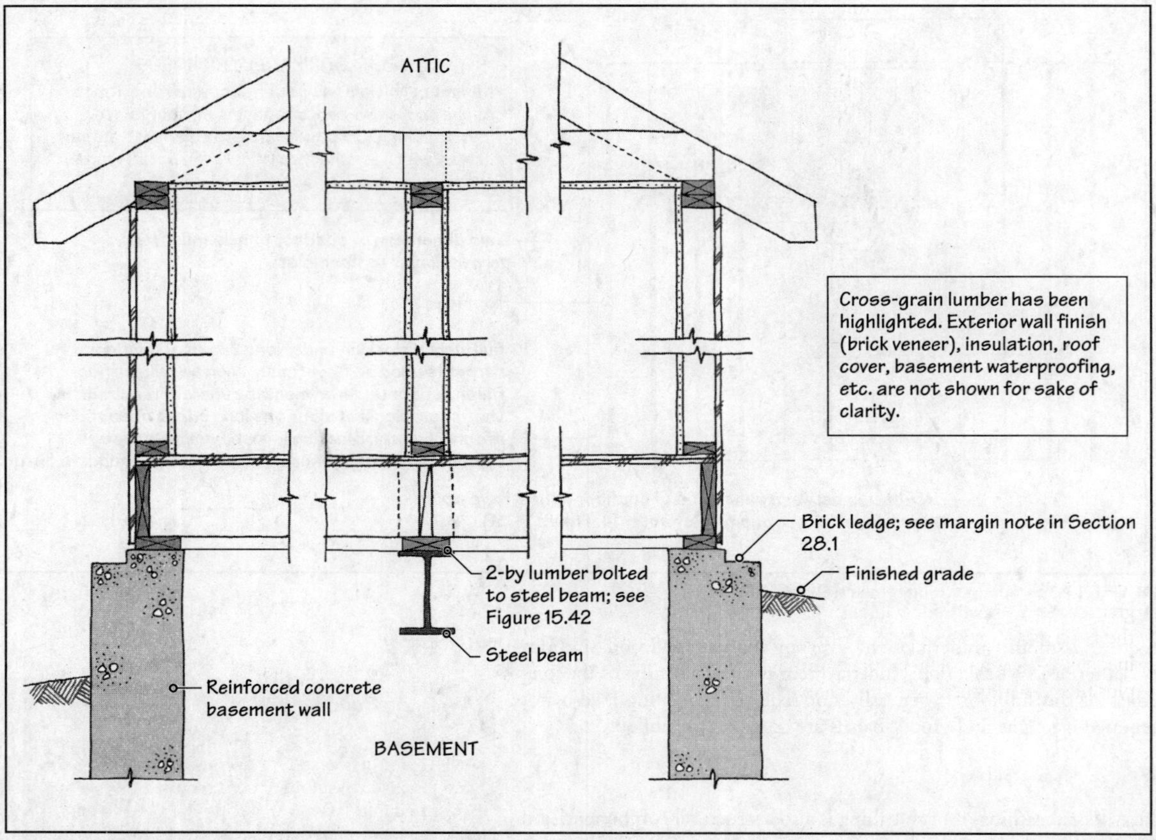

FIGURE 15.41 Equalizing cross-grain lumber dimensions in a WLF building using a steel beam.

a unit. Unequal cross-grain lumber dimensions lead to differential swelling and shrinkage, creating bending of framing members, which can crack brittle interior (gypsum) finishes. It may also lead to squeaking floors.

Consider the building in Figure 15.41, where a steel beam has been used in the basement to support the overlying floor and wall. This is a commonly used detail because the steel beam is unaffected by moisture changes. The provision of a 2-by member directly above the steel beam matches the sill plate over the foundation wall, equalizing the cross-grain lumber dimension. It also provides a nailing surface for the joists, Figure 15.42.

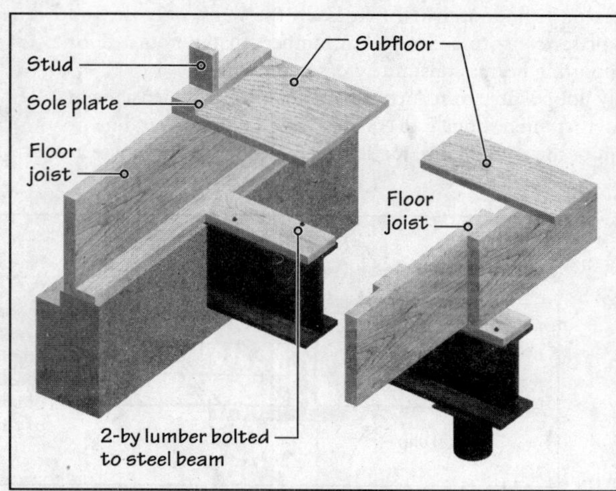

FIGURE 15.42 Floor-level detail of the structure in Figure 15.41.

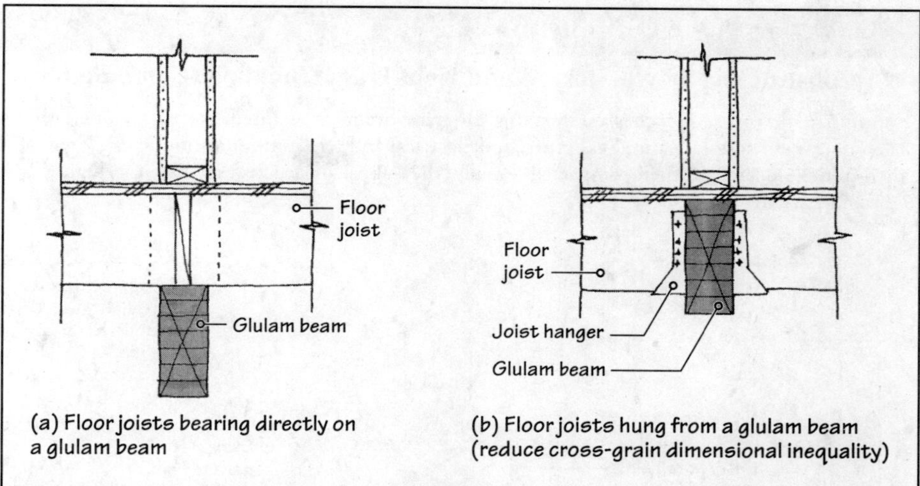

FIGURE 15.43 Use of a glulam beam and its effect on equalizing cross-grain lumber dimensions.

If the steel beam is replaced by a timber or glulam beam, as shown in Figure 15.43(a), cross-grain lumber dimensions will become unequal, creating dimensional instability. However, if the floor joists are hung from a timber or glulam beam using joist hangers, the dimensional inequality is substantially reduced. If the depth of the beam equals exactly the depth of joists plus the sill plate, there is no inequality of cross-grain lumber dimensions, Figure 15.43(b).

PRACTICE QUIZ

Each question has only one correct answer. Select the choice that best answers the question.

27. Because a ridge board is a structural member, its size depends on roof loads.
a. True **b.** False

28. A bird's mouth cut in rafters is made at the ridge.
a. True **b.** False

29. In a roof with a ridge beam,
a. ceiling joists are spaced at 12 in. on center maximum.
b. ceiling joists are spaced at 16 in. on center maximum.
c. ceiling joists are spaced at 24 in. on center maximum.
d. none of the above.

30. When a collar tie is provided in a pitched WLF roof, it should be located
a. in the upper one-fourth of the attic.
b. in the upper one-third of the attic.
c. in the lower one-fourth of the attic.
d. in the lower one-third of the attic.
e. in the center of the attic.

31. The primary purpose of collar ties is to
a. increase the wind uplift resistance of roof framing.
b. increase the gravity load resistance of roof framing.
c. increase the wind uplift resistance of floor framing.
d. increase the gravity load resistance of floor framing.

32. If the cross-grain lumber dimensions in a WLF building are the same throughout the entire building, it is not subjected to any swelling or shrinkage.
a. True **b.** False

33. As per the building codes, the type of construction that WLF represents is
a. Type I construction. **b.** Type II construction.
c. Type III construction. **d.** Type IV construction.
e. Type V construction.

34. The maximum number of floors allowed by building codes for a single-family dwelling built using wood light-frame construction is
a. one. **b.** two.
c. three. **d.** five.

PRINCIPLES IN PRACTICE

Constructing a Two-Story Wood Light-Frame Building

The following text and illustrations show the process and sequence of construction of a two-story wood light-frame (WLF) building. Some builders may follow a slightly different process or sequence.

Several foundation systems can be used in WLF construction, depending on the soil conditions and climate (Chapter 12). In this example, a concrete slab-on-ground (also called *slab-on-grade*) foundation has been used.

STEP 1: FOUNDATION PREPARATION

If the foundation for a WLF building consists of a slab-on-grade, as shown in Figure 1, the ground must first be prepared to receive the slab. This includes clearing the site of vegetation and undesirable topsoil and grading the site

(Continued)

PRINCIPLES IN PRACTICE

Constructing a Two-Story Wood Light-Frame Building (*Continued*)

to provide appropriate drainage. Trenches are then dug for grade beams as required, as well as underground utilities (sewage, wastewater, water supply, natural gas pipes, and electrical and telecommunication lines). Trenches are backfilled after the utilities are laid and then compacted. Finally, the slab is constructed, as shown in Figure 2 (see also Sections 22.11 and 22.12).

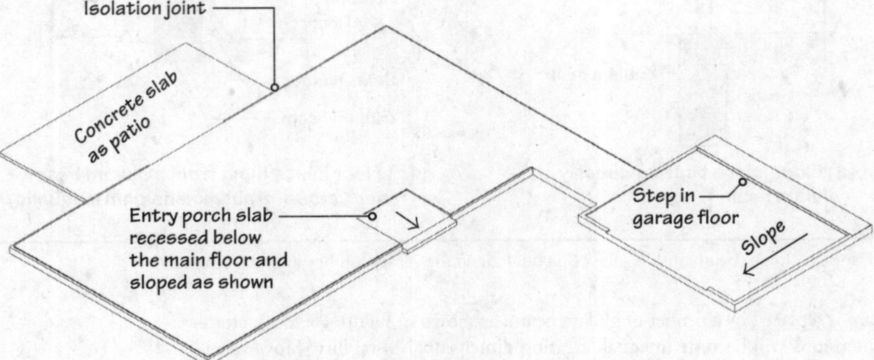

FIGURE 1 Diagram of a concrete slab-on-grade (also called a *concrete slab-on-ground*) foundation.

FIGURE 2 A typical completed slab-on-grade ready to receive the WLF superstructure. Note that the utilities extend above the slab (referred to as *stubbed out* of the slab).

STEP 2: FRAMING THE FIRST-FLOOR WALLS

Before the wall framing for the ground floor begins, chalk lines are placed (snapped) on the slab to locate the walls per the architectural floor plan, Figures 3 and 4.

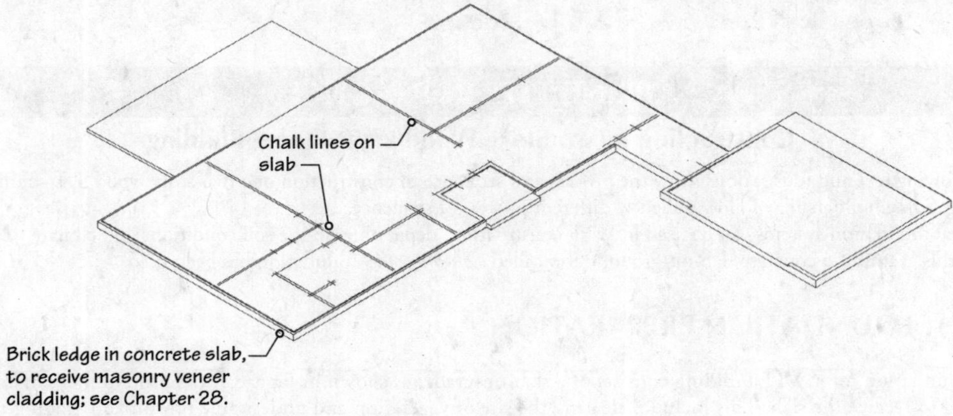

FIGURE 3 Slab-on-grade with chalk lines.

PRINCIPLES IN PRACTICE

With the chalk lines in place, framing the ground floor walls can begin, Figures 5, 6, and 7. Temporary braces stabilize the walls until the frame is complete and the exterior walls are sheathed. When double-height spaces are required, continuous or double studs are used to frame the high segments of the wall.

FIGURE 4 Chalk lines on the slab indicate the location of the walls. Note that plumbing pipes align and are located within the thickness of the wall.

FIGURE 5 Wall assembly under construction.

FIGURE 6 Walls are typically framed on the slab or floor platform and subsequently tilted up into position.

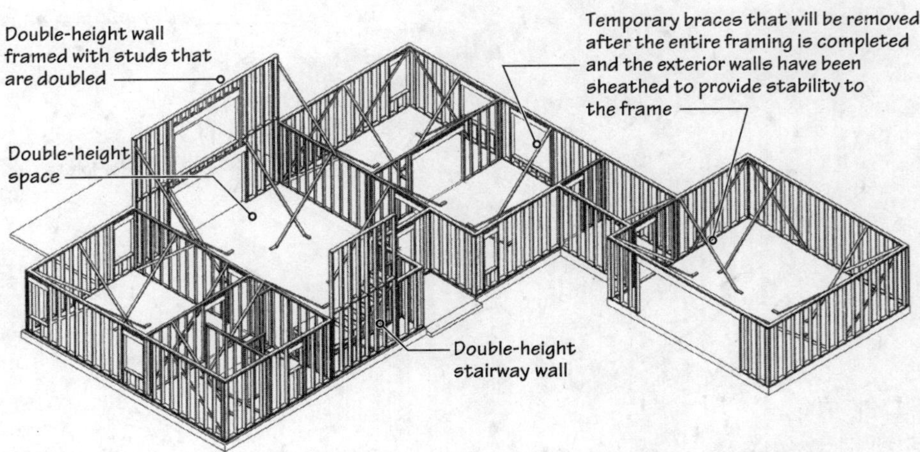

FIGURE 7 Ground-floor wall framing is completed.

(Continued)

PRINCIPLES IN PRACTICE

Constructing a Two-Story Wood Light-Frame Building (*Continued*)

STEP 3: FRAMING THE SECOND FLOOR

Upon completion of the first-floor walls, joists can be installed for the second-story floor, Figures 8 and 9. Unless the plan dictates otherwise, the floor joists are oriented to span the short direction. At locations such as stair openings or step-down floors, floor joists may be doubled or tripled, or a beam may be used to carry the greater load.

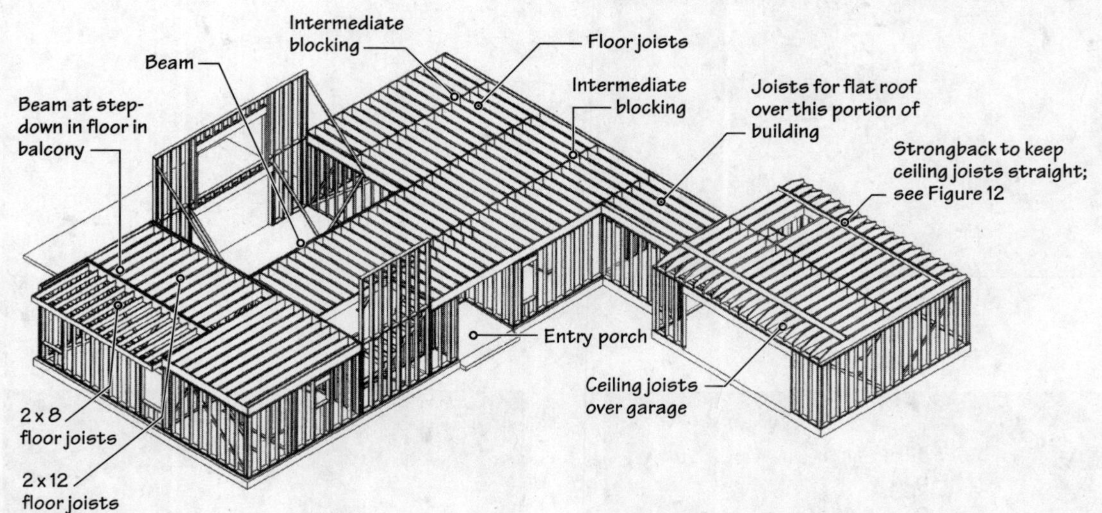

FIGURE 8 Floor framing at the second floor.

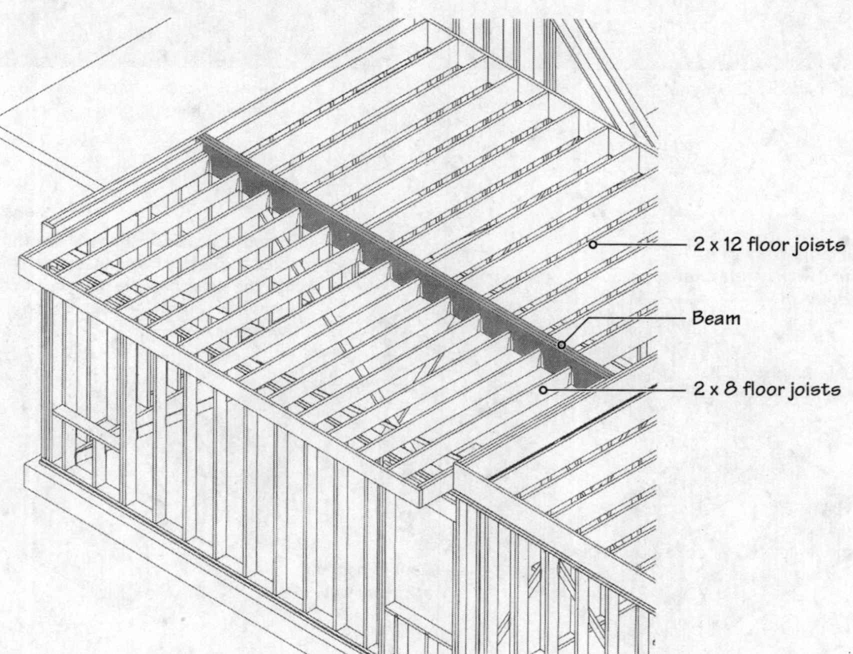

FIGURE 9 Detail of the floor framing at a step down in the floor (see Figure 8). Note that 2 × 12 floor joists are tripled to form a built-up beam. The 2 × 8 floor joists supported by the beam are cantilevered beyond the supporting wall below. This is similar to the example in Figures 15.17 and 15.18.

STEP 4: FLOOR SHEATHING AT THE SECOND FLOOR

Next, the sheathing is applied to the floor joists to form the second-story floor assembly, Figure 10.

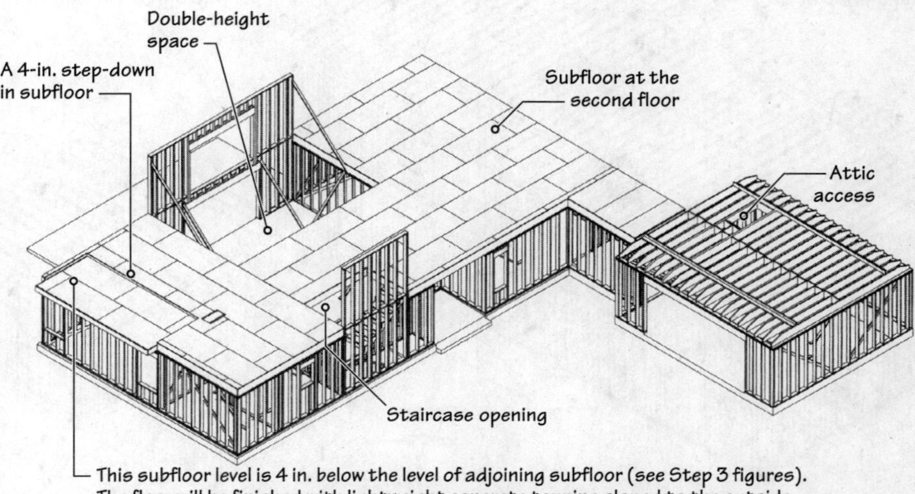

FIGURE 10 The plywood and OSB sheathing are staggered, glued, and nailed to the floor joists.

STEP 5: SECOND-FLOOR WALL FRAMING

The walls for the second story are constructed on the subfloor surface, Figure 11, in a manner similar to that shown in Step 2.

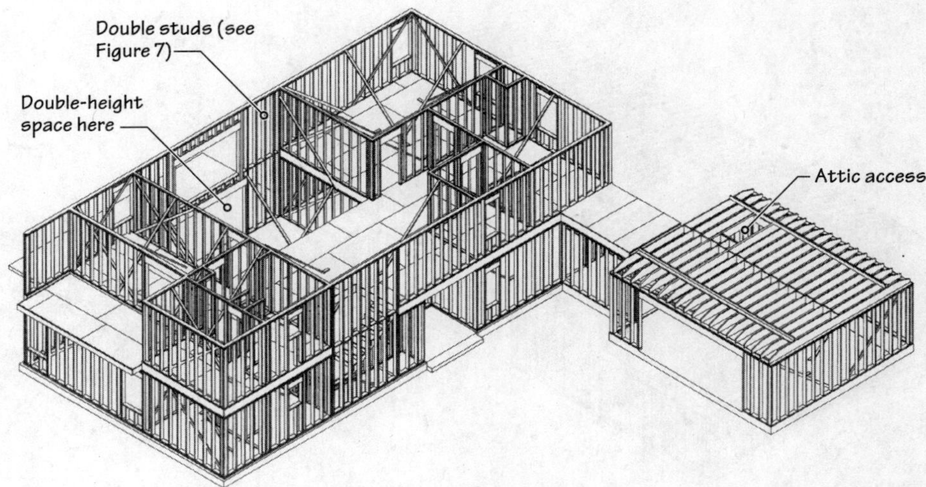

FIGURE 11 Walls are erected, framed, and braced.

STEP 6: CEILING JOIST AND ROOF FRAMING

In a stick-framed roof, ceiling joists span across the walls, Figure 12. Rafters are then set in place to complete the roof framing. When roof trusses are used, they replace the function of ceiling joists and rafters in one unit.

(Continued)

PRINCIPLES IN PRACTICE

Constructing a Two-Story Wood Light-Frame Building (*Continued*)

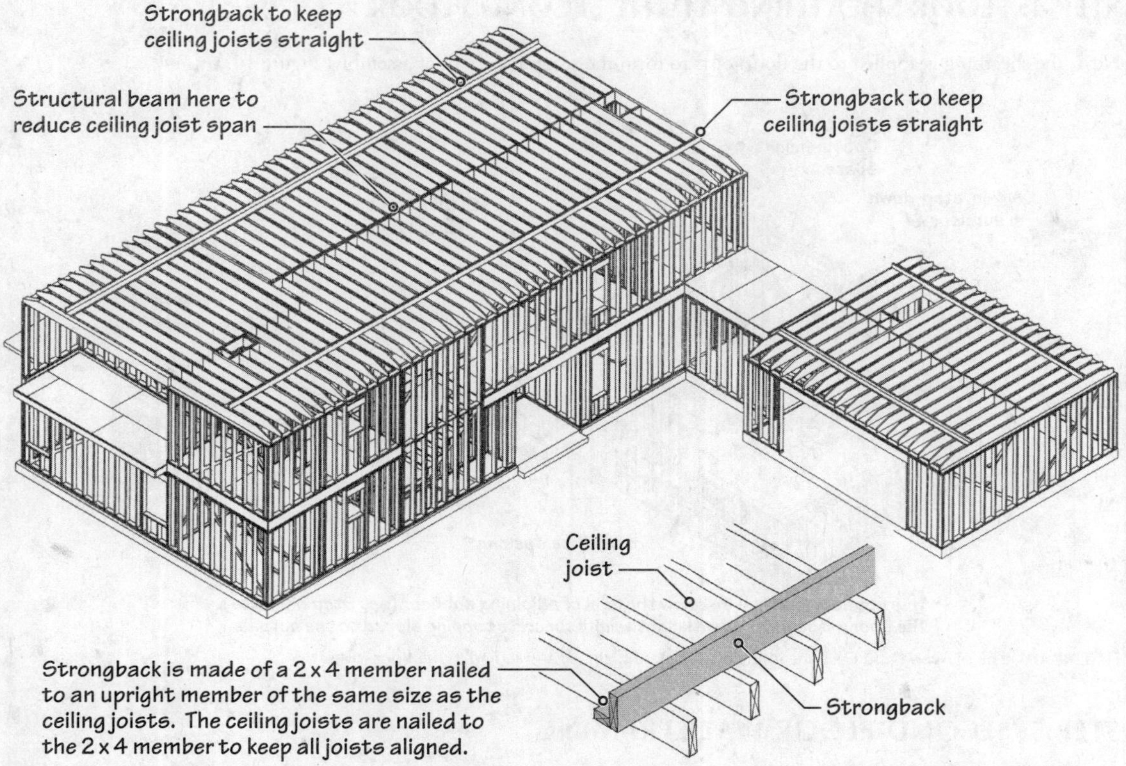

Strongback to keep ceiling joists straight

Structural beam here to reduce ceiling joist span

Strongback to keep ceiling joists straight

Ceiling joist

Strongback is made of a 2 x 4 member nailed to an upright member of the same size as the ceiling joists. The ceiling joists are nailed to the 2 x 4 member to keep all joists aligned.

Strongback

FIGURE 12 Ceiling joist framing. Note the use of strongbacks and of the structural beam (resting on two opposite walls) to reduce the ceiling joist span.

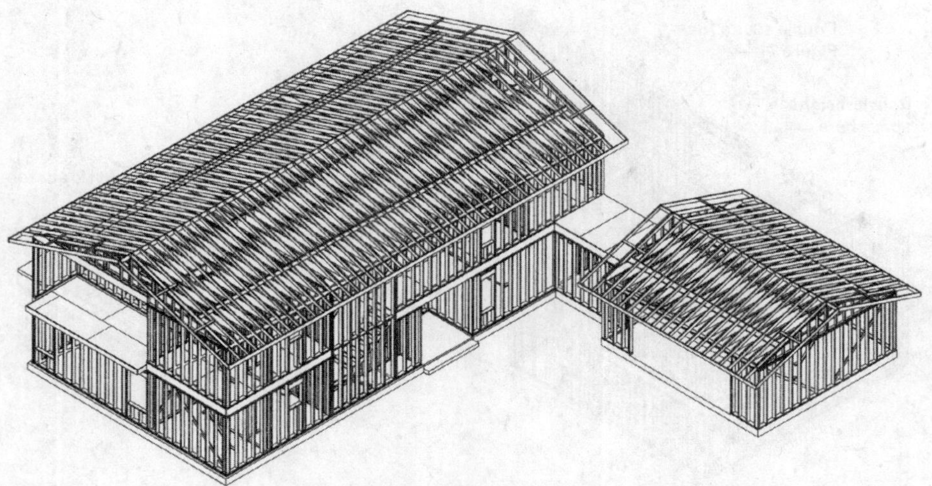

FIGURE 13 Roof framing is completed. Note the conditions at the gable end and eave overhangs.

STEP 7: WALL AND ROOF SHEATHING

With the framing complete, Figure 13, the building is ready to receive wall sheathing, Figure 14, and then roof sheathing, Figure 15, which greatly enhances the structural rigidity. Note that some framers may apply wall sheathing prior to roof framing. After the walls and roof have been sheathed, the structural frame is stable. Temporary interior wall supports can now be removed, allowing work on interior finishes.

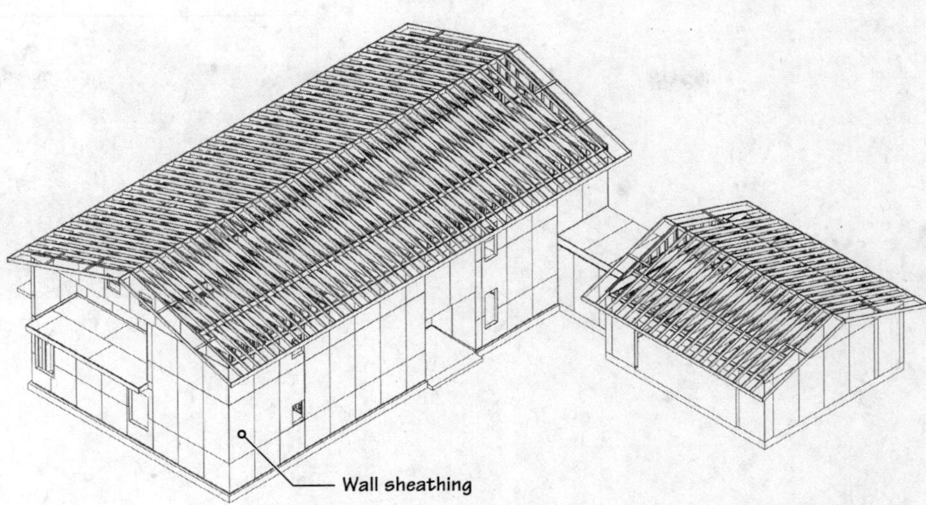

FIGURE 14 Sheathing is installed on all exterior walls.

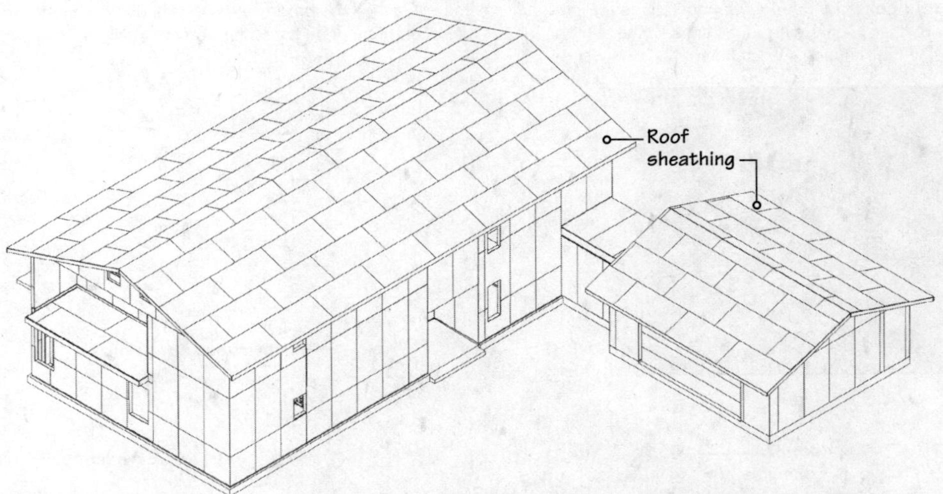

FIGURE 15 Roof sheathing has been installed in a staggered pattern. After the walls and roof have been sheathed, the structural frame is stable. Temporary interior wall supports can now be removed, allowing work on interior finishes.

STEP 8: EXTERIOR AND INTERIOR FINISHES

Roofing, including underlayment, Figure 16, and finish cover, Figure 17 (discussed in Chapter 34), are generally applied first to dry-in the interior spaces and allow inside work to progress. Interior work, including rough plumbing and electrical and wall insulation, is completed before drywall and interior finishes. Exterior work includes windows and doors, moisture/air retarder, and finish surface treatment, Figure 18.

(Continued)

PRINCIPLES IN PRACTICE

Constructing a Two-Story Wood Light-Frame Building (*Continued*)

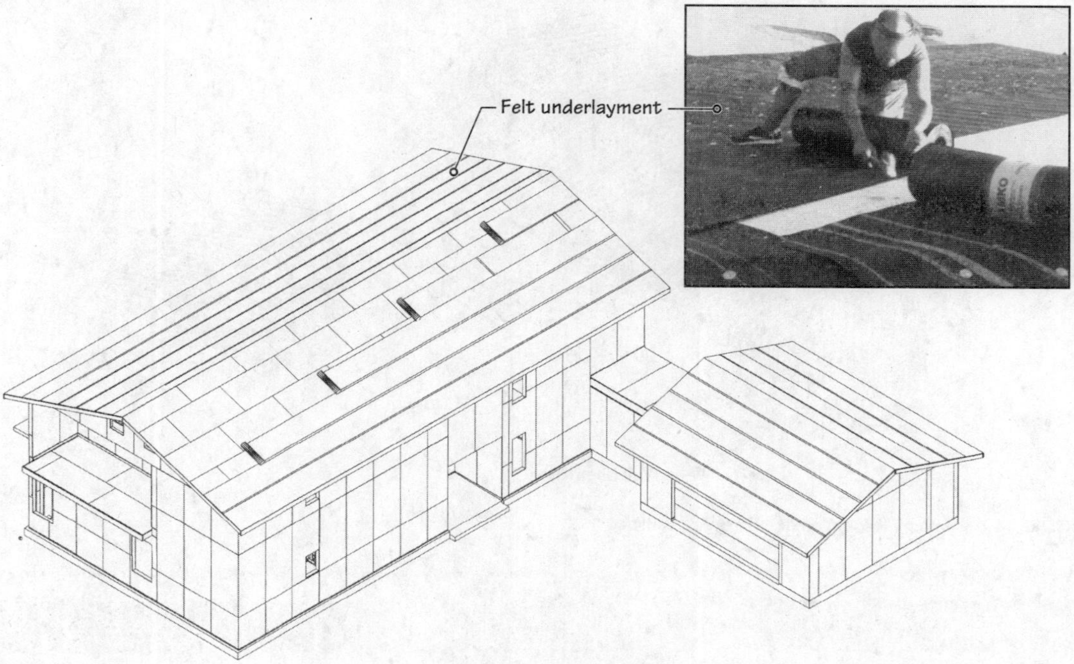

FIGURE 16 Application of roof felt underlayment. An asphalt-treated felt (called *roof underlayment*) is laid over the roof sheathing (deck) before it is covered with roof shingles. The felt functions as the second line of defense against water penetration, the first line of defense being the shingles; see Chapter 34.

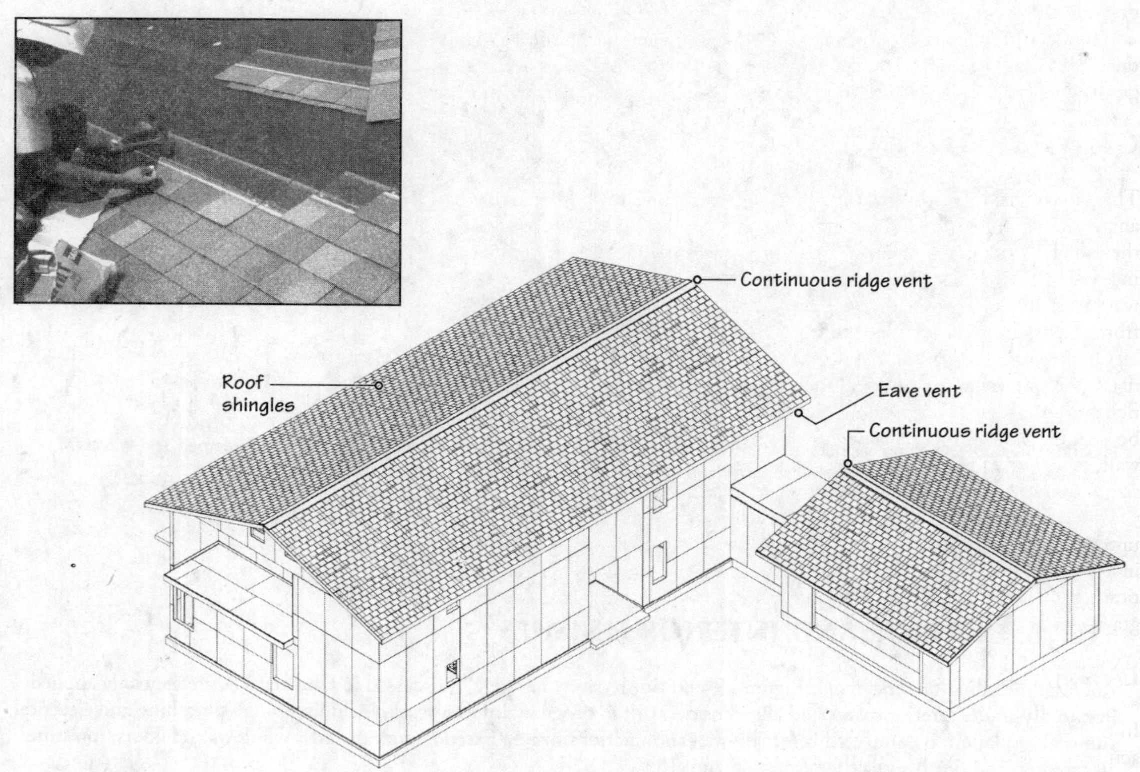

FIGURE 17 Roof shingles are now installed as roofing material. Several alternative roofing materials are available; see Chapter 34.

PRINCIPLES IN PRACTICE

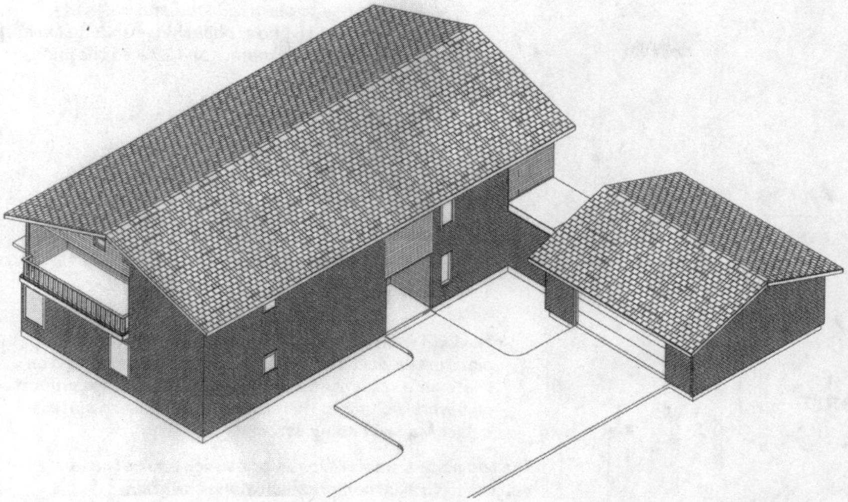

FIGURE 18 Exterior wall finishes (siding, brick veneer, stucco, etc.) are now applied (see Chapters 16, 28, and 29).

PRINCIPLES IN PRACTICE

How a WLF Building Resists Loads

All loads on a building, regardless of the type, must terminate in the ground, where they are finally absorbed. In other words, the ground serves as an infinitely large sink for all loads. In proceeding from its point of origin in the building down to the ground, a load will generally traverse through several intermediate components.

The path that a particular load takes through various components helps us understand the role of each component in resisting the loads. Additionally, it helps to determine the magnitude and types of stresses to which each component is subjected. Both gravity and lateral loads must be considered, and the stresses caused by each load are added to determine a component's size.

GRAVITY LOAD PATH

The path of gravity loads through a WLF building is relatively simple to trace and is shown in Figure 1. In tracing the gravity load path, we proceed from the top and work our way down. Rafters transmit roof loads to the supporting walls. As these loads travel down, they meet the load transmitted to the wall from the floor, where they add up. Further down into the wall, the load from the next floor is added. Finally the load reaches the foundation.

Observe that the load on the wall increases as it proceeds toward the foundation. This implies that the wall should be made stronger at each lower floor. For the sake of simplicity, however, wall strength at upper floors may be kept the same as that on the lowest floor. For instance, if the first-floor walls are required to consist of 2×6 studs at 16 in. on center, the same stud size and spacing can also be used in the upper floors.

The load paths shown in Figure 1 indicate that a load-bearing wall at an upper floor must be aligned with that at the lower floor. Although a small offset in load-bearing walls can be managed structurally, it is advisable to plan for vertical alignment of all load-bearing walls. Figure 2 shows the details of how the gravity loads acting on an intermediate floor are transmitted to the foundation.

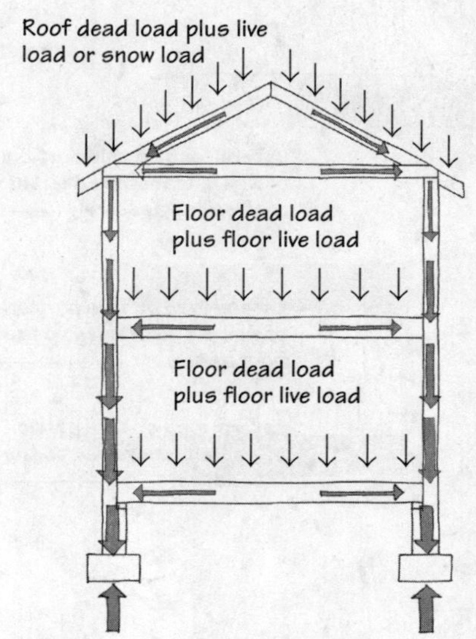

FIGURE 1 Journey of gravity loads through a WLF building.

LATERAL LOAD PATH

In tracing the lateral loads through the structure, let us consider the loads acting on one of the walls (say, the gable-end wall) of a two-story WLF building, Figure 3. In resisting these loads, the wall behaves as a vertical slab

(Continued)

PRINCIPLES IN PRACTICE

How a WLF Building Resists Loads (*Continued*)

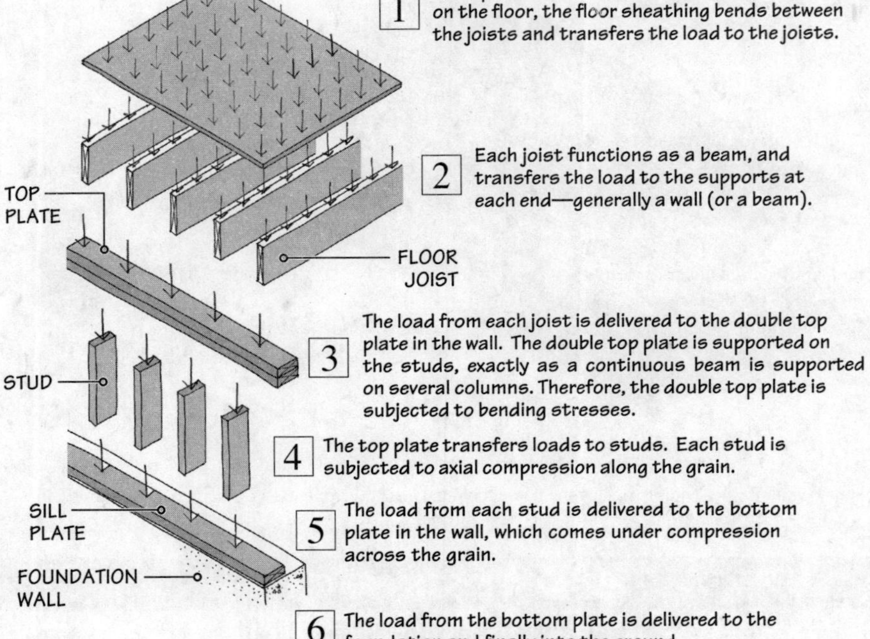

1 In response to the dead load and live load on the floor, the floor sheathing bends between the joists and transfers the load to the joists.
2 Each joist functions as a beam, and transfers the load to the supports at each end—generally a wall (or a beam).
TOP PLATE
FLOOR JOIST
3 The load from each joist is delivered to the double top plate in the wall. The double top plate is supported on the studs, exactly as a continuous beam is supported on several columns. Therefore, the double top plate is subjected to bending stresses.
STUD
4 The top plate transfers loads to studs. Each stud is subjected to axial compression along the grain.
SILL PLATE
5 The load from each stud is delivered to the bottom plate in the wall, which comes under compression across the grain.
FOUNDATION WALL
6 The load from the bottom plate is delivered to the foundation and finally into the ground.

FIGURE 2 Details of the path of gravity loads acting on an intermediate floor of a WLF building.

supported by three elements—the foundation at the bottom, the intermediate floor, and the roof-ceiling assembly. The loads transmitted to each of these three elements are proportional to their respective tributary areas.

Thus, if the floor-to-floor height and the floor-to-roof height are each 10 ft, the lateral loads on the lowest 5 ft of the wall are transmitted to the foundation. The loads on the next 10 ft of the wall are transmitted to the (second) floor, whereas the loads on the remaining (uppermost) portion of the wall are transmitted to the roof.

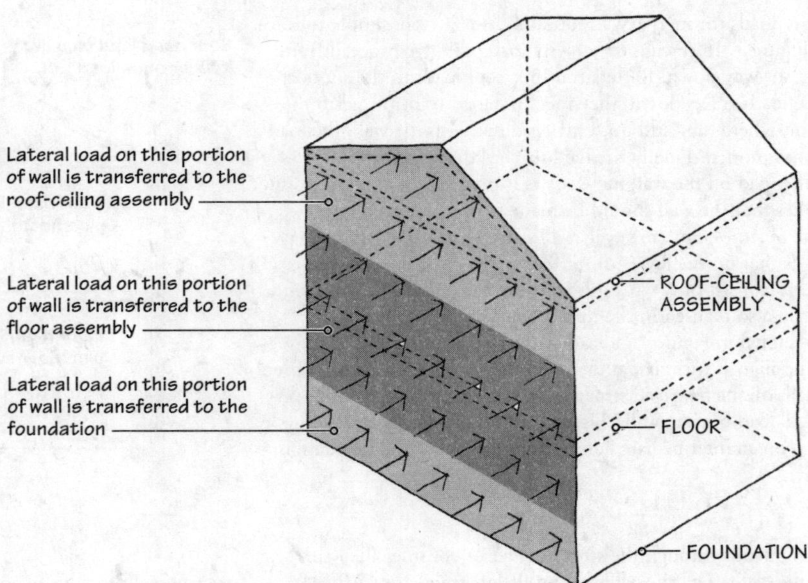

FIGURE 3 Journey of lateral loads acting on one of the walls of a WLF building. Note that the lateral loads assumed on the (gable-end) wall are shown as push-in loads, but they could just as well be suction loads.

PRINCIPLES IN PRACTICE

In a WLF building, the transfer of lateral loads from the wall to the floor or the roof occurs through the studs, which function as vertical beams. Therefore, in high-wind regions, the size and spacing of studs in exterior walls may be dictated by the wind loads. In regions where the wind loads are small, gravity loads dictate the size and spacing of studs.

HORIZONTAL DIAPHRAGMS

The load transmitted from the wall to the foundation is absorbed into the ground, where all loads must eventually reach. Therefore, this load will not be traced any further.

The load transmitted to the (second) floor lies in the plane of the floor, Figure 4. In resisting this load, the floor behaves as a deep beam, with side walls functioning as its supports. One of the characteristics of a deep beam is that the tensile and compressive stresses created in it are concentrated mainly at its extreme edges.

The action of the floor in transferring the lateral loads to its supports (side walls) through deep beam action is called *diaphragm action,* and the tensile and compressive stresses in it are called *diaphragm tension* and *diaphragm compression,* respectively.

In practice, we assume that diaphragm tension and compression are resisted by the double top plate supporting the floor. The presence of diaphragm tension requires that the top plate, as well as the adjacent rim joist, be continuous.

Just as a floor works as a diaphragm in transferring lateral loads to side walls, a roof works in the same way—although as a folded diaphragm.

SHEAR WALLS

The loads acting on a side wall are those that have been transferred to it from the floor and roof diaphragms. In resisting these loads, the wall is supported only at the foundations. Therefore, it behaves as a vertical cantilever carrying in-plane loads. This makes the wall *rack,* or deform angularly, Figure 5.

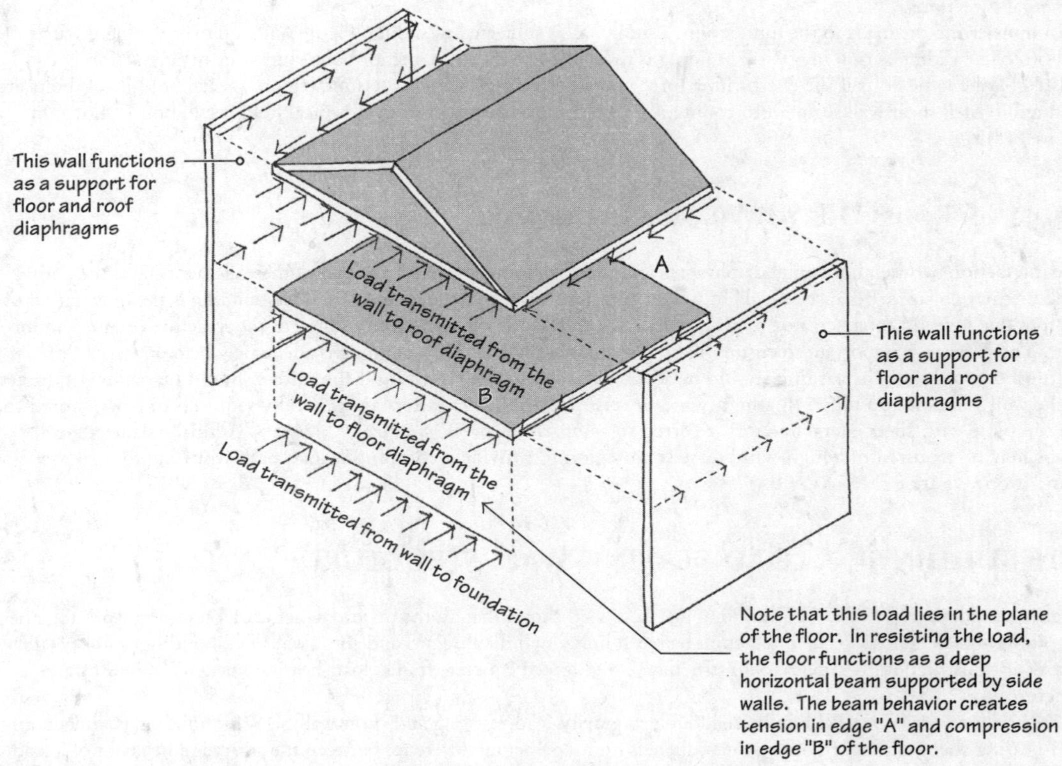
This wall functions as a support for floor and roof diaphragms
A
Load transmitted from the wall to roof diaphragm
B
Load transmitted from the wall to floor diaphragm
Load transmitted from wall to foundation
This wall functions as a support for floor and roof diaphragms
Note that this load lies in the plane of the floor. In resisting the load, the floor functions as a deep horizontal beam supported by side walls. The beam behavior creates tension in edge "A" and compression in edge "B" of the floor.

FIGURE 4 Load transfer from the (gable-end) wall to floor and roof diaphragms.

(Continued)

PRINCIPLES IN PRACTICE

How a WLF Building Resists Loads (*Continued*)

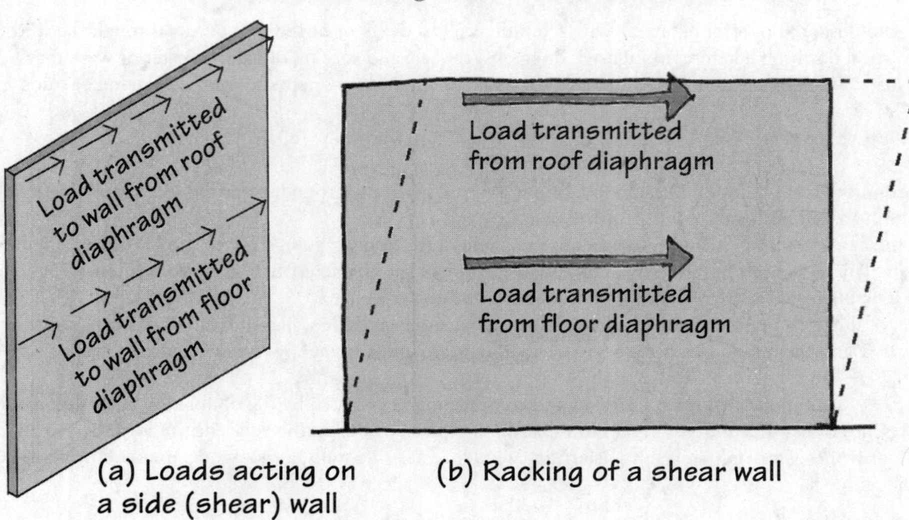

FIGURE 5 Racking in a WLF wall due to in-plane loads.

A WLF wall must, therefore, be braced against racking, which is achieved through the use of plywood, OSB sheathing, or diagonal braces (Section 15.10). Because the racking and the consequent angular deformation are caused by shear forces, a wall braced to resist racking is referred to as a *shear wall.* In a rectilinear box-type building with four walls, all walls must be designed as shear walls to account for lateral loads from either one of the two principal directions.

In transferring the loads to the foundation, a shear wall is subjected to sliding, Figure 6(a) and overturning, Figures 6(b) and 6(c). The sliding action in a WLF building is resisted by foundation anchor bolts. The overturning action is also resisted by the same bolts if the overturning force is small. In a high-wind or seismic region, special hold-down bolts are required in each shear wall segment to counteract excessive overturning forces. A typical hold-down bolt is shown in Figure 6(d).

Uplift on Roofs Due to Wind or Earthquake

The interaction between horizontal diaphragms and shear walls means that if any one of these elements fails, the entire structure may collapse. Because of its light weight, a frequent cause of the failure of a WLF building is the blowing off of the roof due to uplift forces created by wind. Once the roof blows off, complete collapse of the structure becomes imminent. Therefore, it is important to ensure that the roof assembly in a WLF building is adequately held down.

In other words, roof sheathing should be securely nailed to the rafters, and the rafters should be securely fastened to the wall's double top plate. In a high-wind or seismic region, the rafters are generally required to be anchored to the top plate with sheet-metal anchors, referred to as *hurricane* or *seismic straps,* Figure 7. Additional sheet steel straps may be required in a high-wind or seismic region to provide a continuous load path from the roof to the foundation, Figure 8.

WLF BUILDING—A LOAD-BEARING WALL STRUCTURE

The structural behavior of a WLF building just described shows that all the major elements of a building work together in countering lateral loads. The interaction between horizontal diaphragms and shear walls in a building is not peculiar to a WLF construction. It applies to all structures—reinforced concrete frame, load-bearing masonry, or steel frame structure.

Because most walls in a WLF building double as gravity-load-resisting and shear walls, a WLF building is, in fact, a load-bearing-wall building. In other words, there is no fundamental difference between the structural behavior of a load-bearing masonry or concrete building and that of a contemporary WLF building. The reason for referring to it as a wood *light-frame* building is that the walls consist of a frame instead of solid concrete or masonry walls.

PRINCIPLES IN PRACTICE

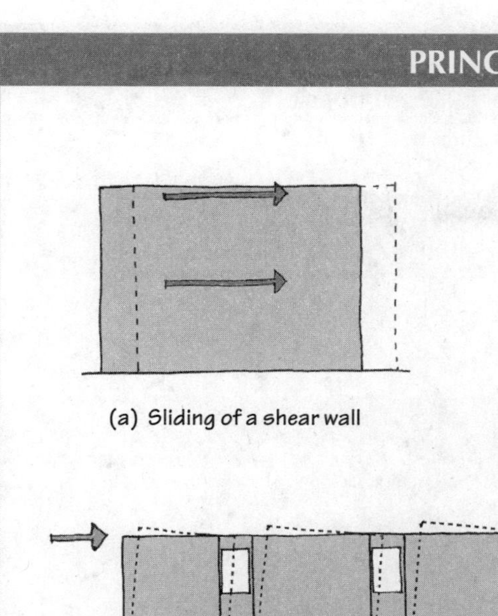

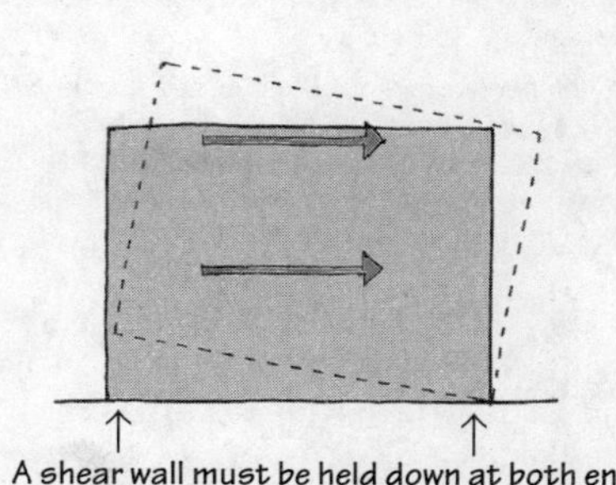

A shear wall must be held down at both ends
(b) Overturning of a shear wall

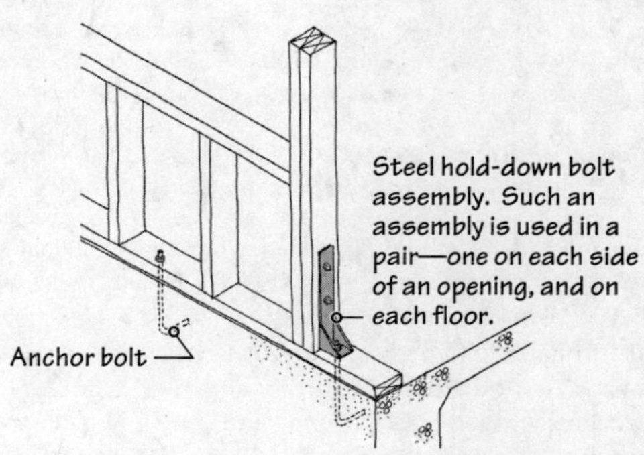

Steel hold-down bolt assembly. Such an assembly is used in a pair—one on each side of an opening, and on each floor.

FIGURE 6 Sliding and overturning of a shear wall.

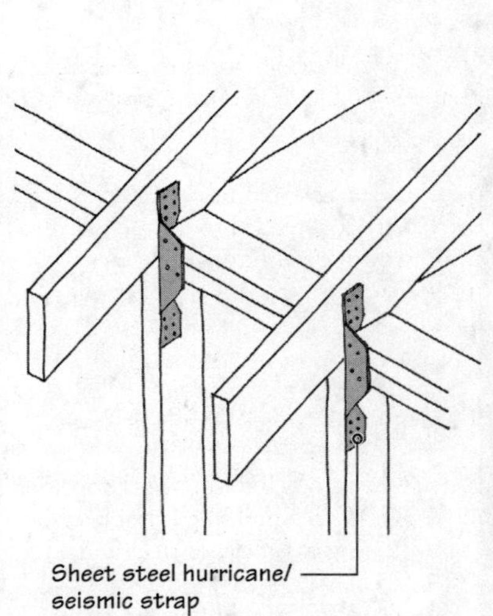

FIGURE 7 Commonly used hurricane/seismic straps for a roof.

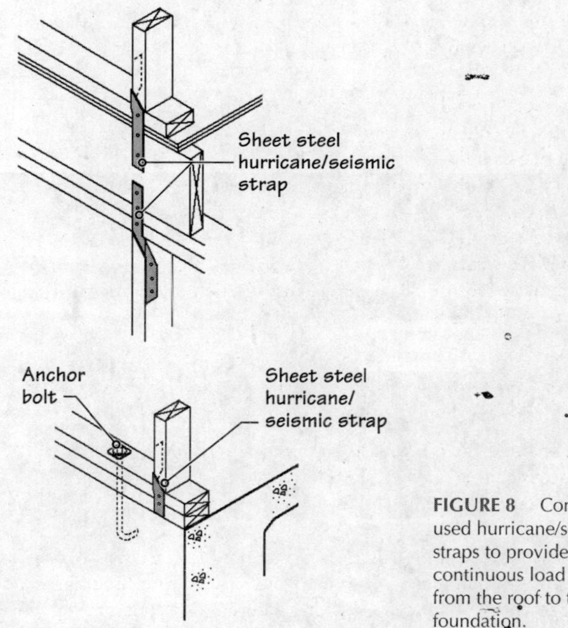

FIGURE 8 Commonly used hurricane/seismic straps to provide a continuous load path from the roof to the foundation.

PRACTICE **QUIZ**

Each question has only one correct answer. Select the choice that best answers the question.

35. In referring to a floor or roof of a building as a *horizontal diaphragm*, the refererence is with respect to

a. gravity loads. **b.** lateral loads.
c. both gravity and lateral loads. **d.** none of the above.

36. Hold-down bolts, when provided in a WLF wall building, are used in

a. floors. **b.** walls.
c. the roof.

37. A wood light-frame building is essentially a

a. load-bearing-wall structure. **b.** frame structure.

EXPAND YOUR KNOWLEDGE

Wood Construction Prior to Balloon Frame

Ignoring some primitive precedents such as the tepee—a cone of wood poles covered with hides—or a pit dwelling roofed with wood beams and thatch or hides, wood construction in the United States developed from the early log construction to the present-day platform frame construction. As discussed in Section 15.1, platform frame construction evolved quickly and seamlessly from its predecessor, balloon frame construction. The wood construction that existed immediately prior to the balloon frame was, however, vastly different from it.

The purpose of this section is to provide a brief background on pre–balloon frame wood construction techniques (using heavy timber members) to help the reader visualize the radical change brought about by balloon frame construction. In addition, because pre–balloon frame construction has recently seen a renaissance, this section also gives a brief account of the pre–balloon frame's contemporary incarnation.

Types of Heavy Timber Frame

Prior to the invention of the balloon frame, that is, until the mid-nineteenth century, the structural members of wood frame buildings in the United States ranged in thickness from 6 in. to 10 in., and even larger in some instances. Compare these dimensions with those of balloon frame members, which were nominally 2 in. thick. In other words, the construction system that existed before the balloon frame used heavy sections of wood. We refer to this construction system as *heavy timber construction* or *timber frame construction*. (In Section 13.9, *timbers* are defined by contemporary construction standards as wood members with both nominal cross-sectional dimensions greater than or equal to 5 in.)

It is because of the substantial difference between the cross-sectional dimensions of timber frame members and those of the balloon frame that the balloon frame came to be known as *wood light-frame construction*. Two types of timber frame systems, differing in terms of their lateral-force resisting mechanisms, were common:

- Timber frame with exterior masonry walls
- Braced timber frame

Timber Frame with Exterior Masonry Walls

This system consisted of load-bearing exterior walls of brick or stone with an interior structure consisting of (heavy) timber beams and columns. The floor framing consisted of thick lumber planks or tongue-and-groove lumber decking, and roof framing was either flat or pitched and also consisted of heavy timber.

This version of timber framing came to be known as *mill construction* because the factories of that era (producing textiles, shoes, and other consumer items) built with this system could obtain insurance coverage against fire. Mill construction was recognized as slow-burning, providing greater fire resistance than the alternative available at the time, which consisted of exterior masonry walls with interior framing comprising wrought iron (or cast iron) beams and columns. (See Sections 7.2 and 7.5, which highlight the greater fire resistance of heavy timber members compared with that of unprotected iron or steel. Spray-on fire protection, currently used on steel frames, did not exist at the time.)

Mill construction was used for nonfactory occupancies as well—commercial, institutional, and storage buildings. In this construction system, the exterior walls provide both gravity load resistance and lateral load resistance. With all the lateral load resistance provided by masonry walls, the interior columns need to resist only gravity loads. An example of historic mill construction, used for a commercial building completed in 1901, is shown in Figure 1.

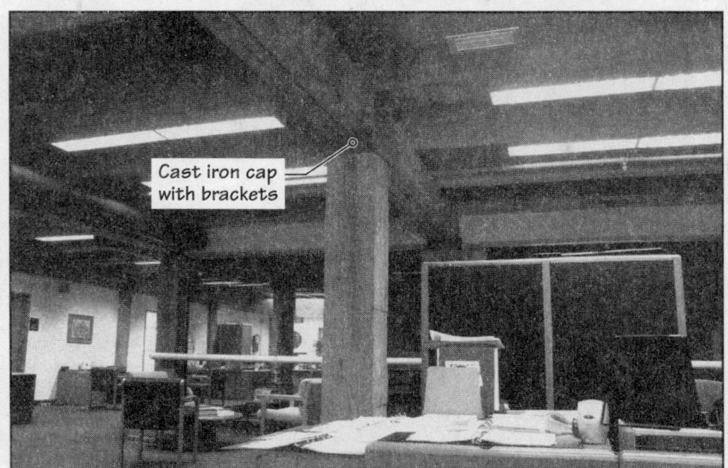

FIGURE 1 The interior of the second floor of the historic Texas School Book Depository Building in Dallas, Texas, illustrates the typical mill construction used for a commercial building. This seven-story building was built in 1901, with heavy timber beams and columns (approximately 15 in. × 15 in.) and brick walls about 30 in. thick at the ground floor. The sixth floor of the building was used to assassinate President John F. Kennedy in 1963. The building is being preserved for its historic importance, and the lower floors of the building are currently used by the Dallas County Administration.

(Continued)

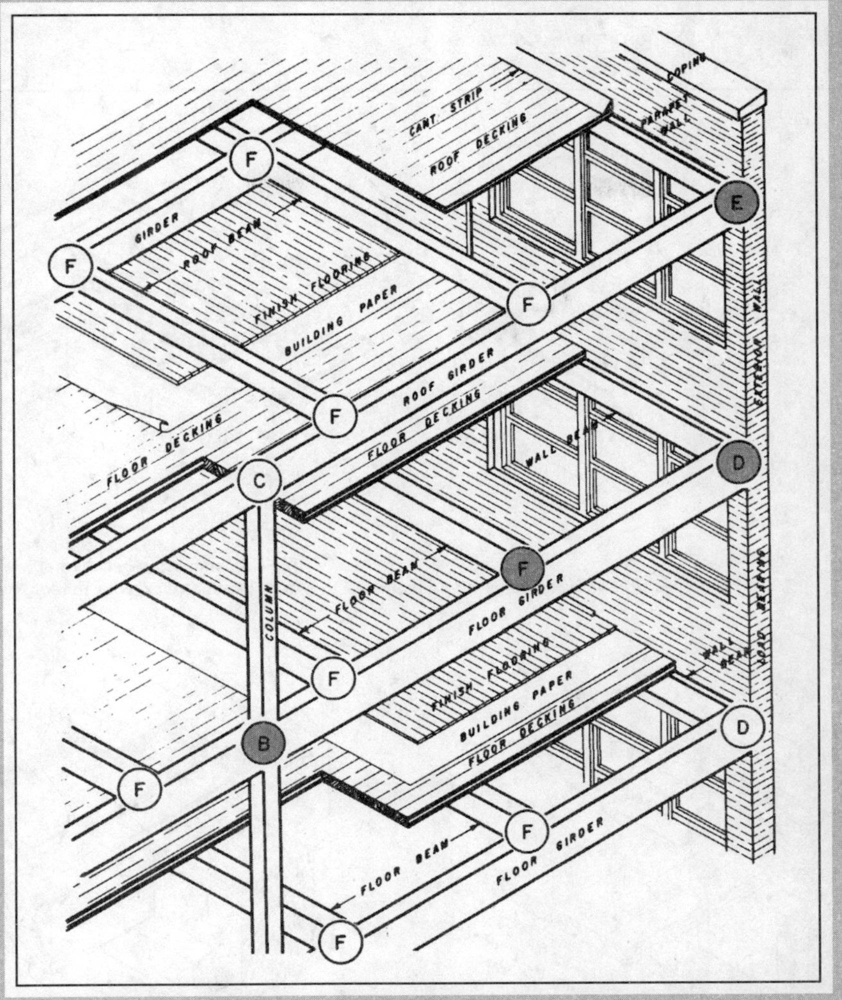

FIGURE 2 The typical arrangement of structural members in Type IV heavy timber (HT) construction is almost identical to that of the (historic) mill construction. Figure 3 shows a few details of this system. For a more comprehensive discussion of HT construction, refer to Publication WCD5, *Heavy Timber Construction,* by the American Wood Council. The publication can be viewed in PDF format at http:// www.awc.org/pdf/wcd5.pdf. (Illustration courtesy of the American Wood Council, Leesburg, Virginia)

A construction system quite similar to historic mill construction is also used today and is referred to by the International Building Code (IBC) as *Type IV, heavy timber (HT) construction.* Although a *timber* is defined as a member with minimum nominal lumber dimensions of 5 × 5, heavy timber (Type IV) construction is further qualified by the IBC as one in which the minimum nominal lumber dimensions are 6 × 10 for floor beams, 6 × 8 for roof beams, 8 × 8 for columns of a multistory building, 6 × 8 for columns of a single-story building, and so on.

As shown in Table 7.1, Type IV construction does not allow concealed spaces in the structural frame (such as the spaces between studs or floor joists in WLF construction), that is, the timber members are exposed. Exterior load-bearing walls, when provided, must be of noncombustible materials with a minimum 2-h fire rating.

The salient features of contemporary Type IV construction are shown in Figure 2, and a few representative details are given in Figure 3. The difference between historic mill construction and Type IV construction is that the connections between connecting beams, between beams and columns, and between beams and walls use hardware made of steel and are more sophisticated than the cast iron or wrought iron hardware used in mill construction. Another difference is that in contemporary Type IV construction, engineered lumber, such as glulam, is more commonly used than solid sawn lumber.

Braced Timber Frame

Braced timber frame construction is similar to Type IV or mill construction but without the exterior load-bearing walls. In the absence of walls that provided lateral load resistance in mill construction, a braced timber frame depended on inclined (knee) braces and close-fitting joints between the members of the frame. Because knee braces could not provide the same degree of lateral resistance as masonry walls, a braced timber frame was generally used for one- or two-story buildings—individual residences, barns, and small commercial buildings.

(Continued)

EXPAND YOUR KNOWLEDGE

Wood Construction Prior to Balloon Frame (*Continued*)

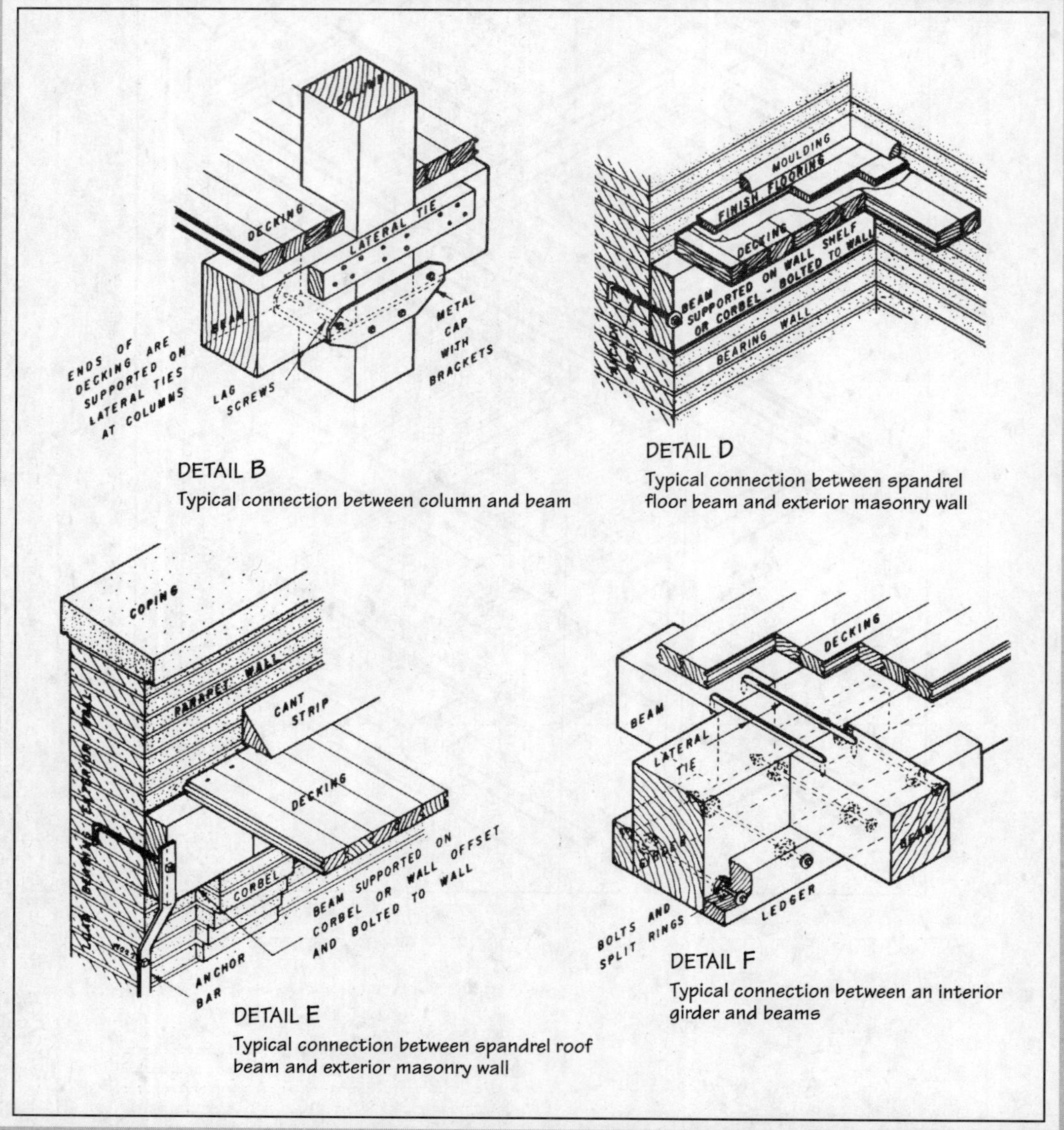

FIGURE 3 Typical details used in Type IV HT construction, similar to historic mill construction. The details are keyed to Figure 2. For a more comprehensive discussion of HT construction, refer to Publication WCD5, *Heavy Timber Construction*, by the American Wood Council. The publication can be viewed in PDF format at http:// www.awc.org/pdf/wcd5.pdf. (Illustrations courtesy of the American Wood Council, Leesburg, Virginia)

A typical braced timber frame used at the time is shown in Figure 4. The framing varied because several variations of the system existed, depending on the scale of the building, its occupancy, and regional preferences. The primary structural components of the building consisted of semirigid wood frames called *bents*.

Most dwellings were small. Therefore, each bent was typically 20 to 25 ft wide, spaced about 12 ft to 15 ft apart. Thus, the structure shown in Figure 4, consisting of four bents (three bays), is about 40 ft long. A fireplace, built of brick or stone, forming the main feature of the dwelling, was located between the two interior bents.

NOTE

What Is a Bent?

A bent is an assembly of linear structural members, all of which lie in a vertical plane that is perpendicular to the ridge. The term *bent* is not in common use in contemporary literature because it has been replaced by the term *frame*. In other words, the terms *two-dimensional frame* and *bent* may be used interchangeably.

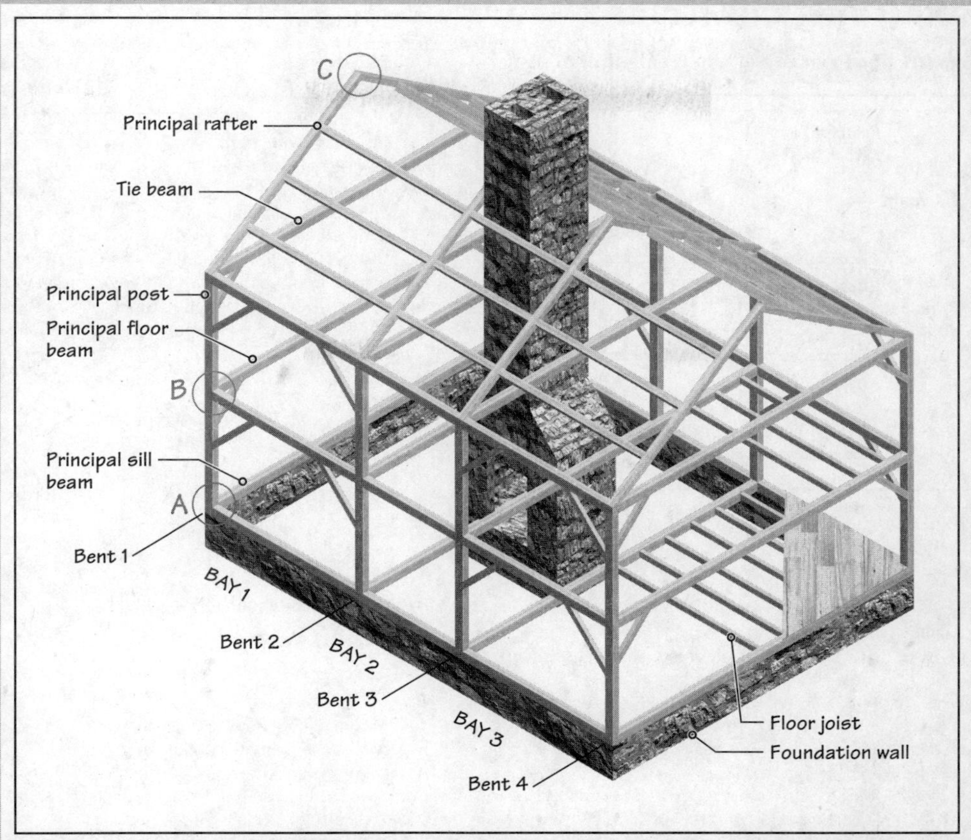

FIGURE 4 A typical (historic) braced timber frame.

Each bent consisted of two inclined rafters, a beam at the intermediate floor, a (sill) beam at the foundation level, and knee braces for lateral load resistance, Figure 5(a). The members of each bent were called *principal members*—for example, *principal post, principal rafter, principal floor beam,* and *principal sill beam*. The beam that resists the lateral thrust in the rafters (i.e., ties the rafters together) was appropriately called the *tie beam* or *crossbeam*.

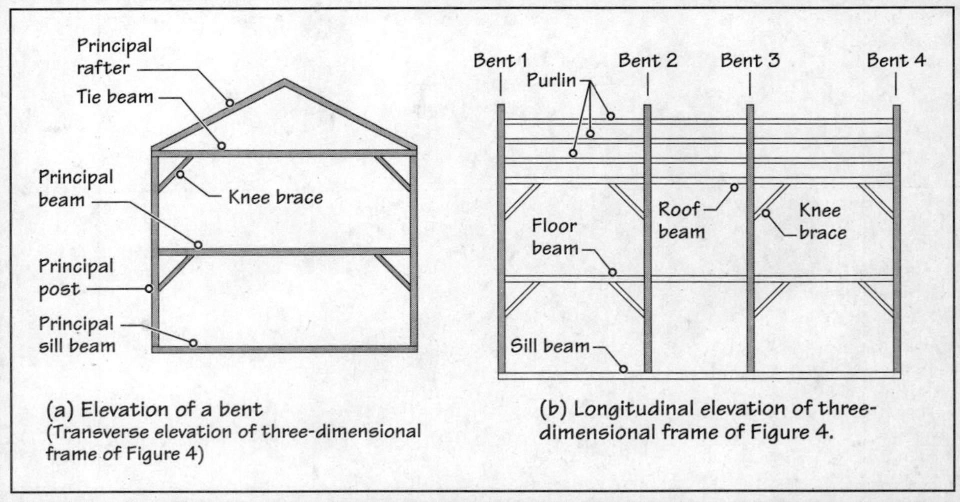

FIGURE 5 Transverse and longitudinal elevations of the three-dimensional timber frame shown in Figure 4 with commonly used nomenclature for various members.

(Continued)

Wood Construction Prior to Balloon Frame (*Continued*)

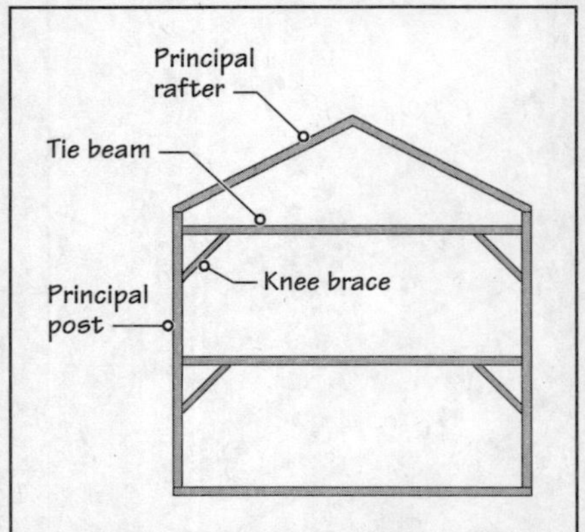

FIGURE 6 Elevation of a bent with a tie beam placed below the eave to simplify the joints between tie beam, rafter, and principal post.

A bent cannot stand up on its own. Two or more bents (generally identical in configuration) must be joined along the length of the building to make a stable three-dimensional structure. These joints are provided by a longitudinal sill beam at the foundation level, and by floor and roof beams at the floor and roof levels, respectively, Figure 5(b). Floor joists placed between principal sill beams and principal floor beams further add to the stability of the structure (see Figure 4).

At the roof level, various alternatives were used to connect the bents using a ridge beam and (or) purlins. The framing shown in Figure 4 and 5(b) uses *purlins* that span between the principal rafters. The purlins support wood decking that runs parallel to the rafters. The lattice formed by the purlins and decking adds to the structural stability at the roof. The exclusion of a ridge beam simplified the joinery at the ridge because the joint only needed to accommodate the two opposite rafters. For the same reason, the tie beam was placed slightly below the eave in many cases, Figure 6.

Joints Between Framing Members

The profile of each joint between the members of the timber frame is particularly important because no metal connection hardware or fasteners were available. Not only was the framing geometry modified to obtain a suitable joint detail, but the cross-sectional dimensions of framing members were often dictated by joint detail rather than member spans or loads.

The most versatile joint profile between the members is the *mortise-and-tenon joint*, in which the mortise (slot) in one member accepts the tenon (tongue) from the opposite member. Most joints in a braced timber frame are some version of this joint.

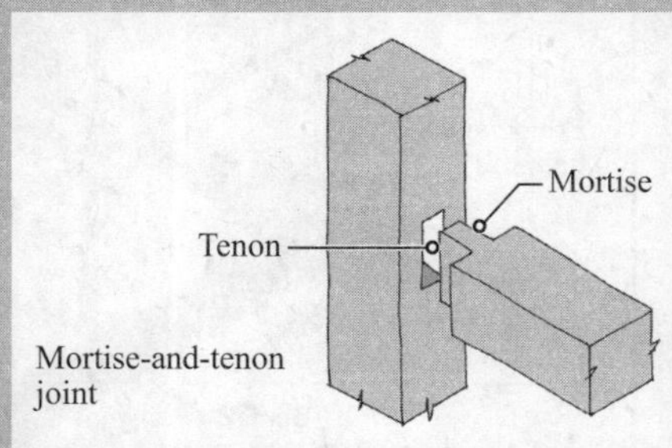

The joints in a timber frame must be tight and close-fitting so that they lock the members of the frame together. Expert carpenters and joiners were needed to make precise joints. Wooden pegs were used to increase the tensile strength of the joint, and wedges were used for joint tightness.

Both pegs and wedges helped maintain joint tightness even after the building's construction. As the (heavy) wood sections changed from their original "green" condition to equilibrium moisture content, the joints lost some tightness, a problem that was overcome by hammering the pegs and wedges in. A few examples of joint details in a braced timber frame are shown in Figures 7 to 9. These details are keyed to the frame of Figure 4.

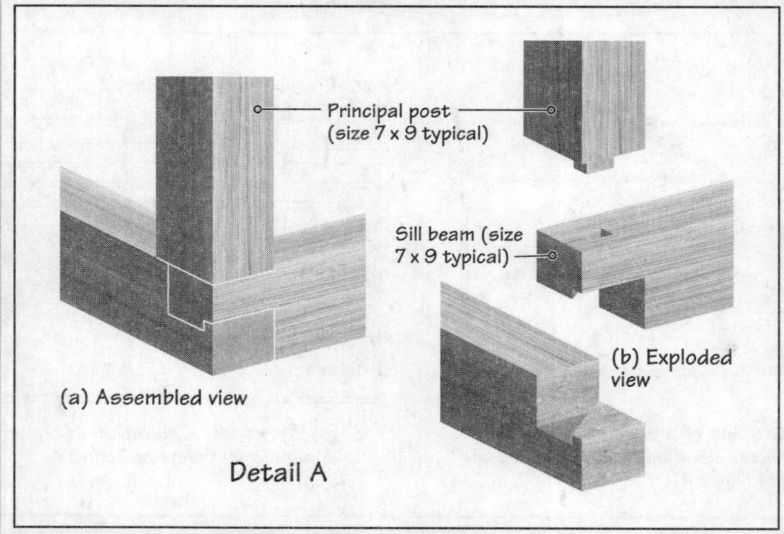

FIGURE 7 Detail of the joint between (corner) principal post and sill beams from two directions. The joint between the post and sill beam is a mortise-and-tenon joint; the joint between the sill beams is a complex form of lap joint. (Keyed to Figure 4.)

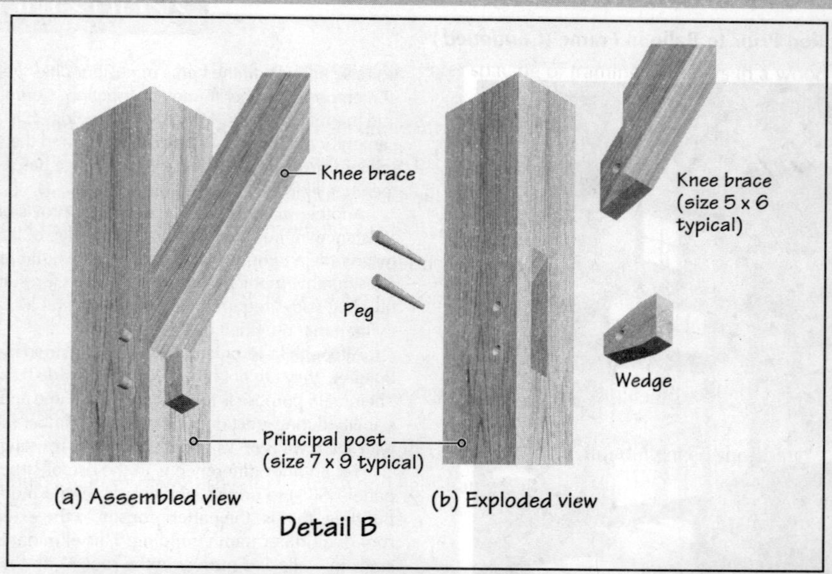

FIGURE 8 Detail of the (mortise-and-tenon) joint between a principal post and a knee brace. Notice the use of pegs and wedges in the joint. (Keyed to Figure 4.)

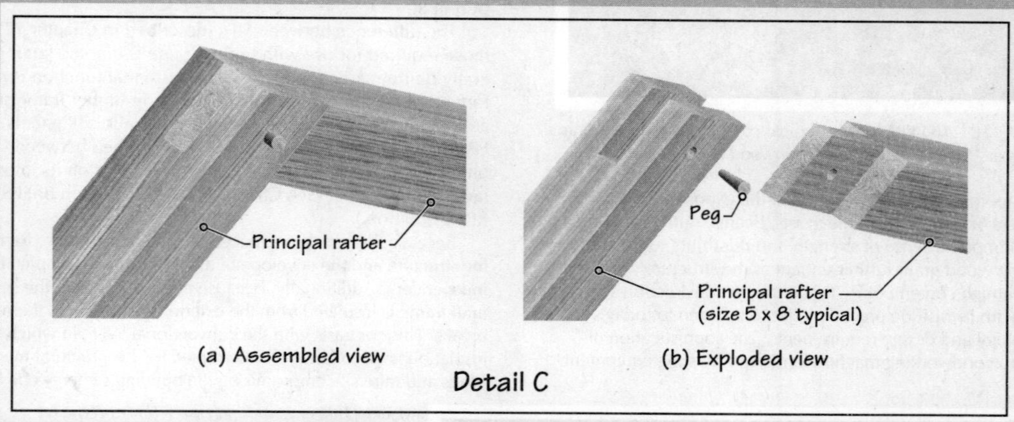

FIGURE 9 Detail of the (open mortise-and-tenon) joint between the opposite principal rafters; the joint between common rafters is similar. (Keyed to Figure 4.)

Erection of a Braced Timber Frame

Each bent was generally assembled flat on the ground, raised to the vertical orientation, and lifted into position to bear on the masonry foundation wall. Generally, when all bents were completed, a day was fixed for tilting and lifting the bents to the required locations. On this day, the entire community would assemble to help in the effort [15.2]. As the bents were raised, they were supported by temporary braces, which allowed the longitudinal framing members to be connected to them.

Frame Infill and Finishes

In order to provide a support surface for finish materials, infill systems were attached to the frame to form walls, floors, and (in some cases) roofs. For the walls, studs were used as infill members. On the interior surface, the studs were covered with *wattle and daub* [15.3]. Wattle consisted of small horizontal and vertical pieces (scantlings) joined together in a basket-weave pattern. This served as what is now called *lathe*.

The wattle was plastered with clay reinforced with horsehair. (The term *daub* was then used for plaster.) In later times, the space between the studs was filled with brick masonry or sand-lime plaster over wattle. The exterior finish on the studs was generally horizontal wood boards, called *clapboard*, which is similar to contemporary wood lap siding. The floors were finished with wood boards pegged to joists.

The Rennaissance of Braced Timber Frame

The historic braced timber frame construction, described in the previous section, has recently seen a revival in North America by builders who have revived the craft, lost to us for about two centuries. Figures 10 and 11 illustrate the use of timber frame in

(Continued)

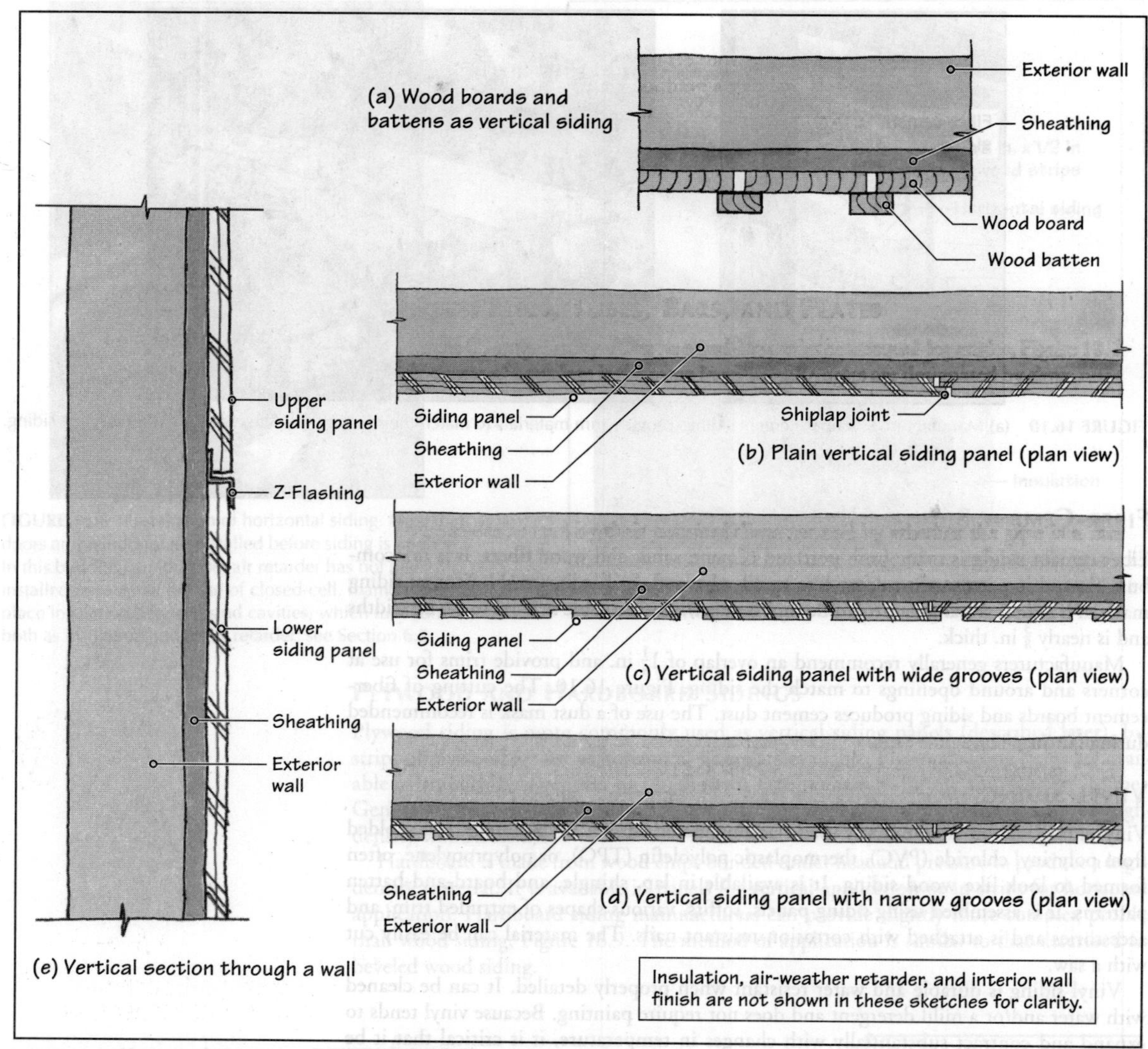

FIGURE 16.11 Vertical plywood or hardboard siding details.

one more commonly used. The projection may be wide (2 to 3 ft) or narrow (a few inches) and may be finished with a horizontal soffit, Figure 16.12.

If the slope of the roof is large, an eave with a horizontal soffit cannot project too far beyond the wall, or the bottom of the soffit will interfere with the door or window head, Figure 16.13. A sloping soffit, Figure 16.14, does not have this limitation. A sloping soffit has another advantage over a horizontal soffit in that the transition from rake to eave is seamless. With a horizontal soffit, the transition requires additional detailing, Figure 16.15.

Ridge Ventilation

In Chapter 6 (Section 6.8), the importance of attic ventilation was discussed. As stated there, attic ventilation consists of intake and exhaust ventilation in the roof. Although intake ventilation is provided through soffit vents, exhaust ventilation can be provided in many ways (Figure 6.12).

One of the means of providing exhaust ventilation is to install a continuous ridge ventilator. Several proprietary ridge ventilators are available that can be installed over felt underlayment and covered over by roof shingles. The sheathing at the ridge is set back from the ridge board (by 1 to 2 in. on either side) to provide a clear space between rafters for ventilation, Figure 16.16.

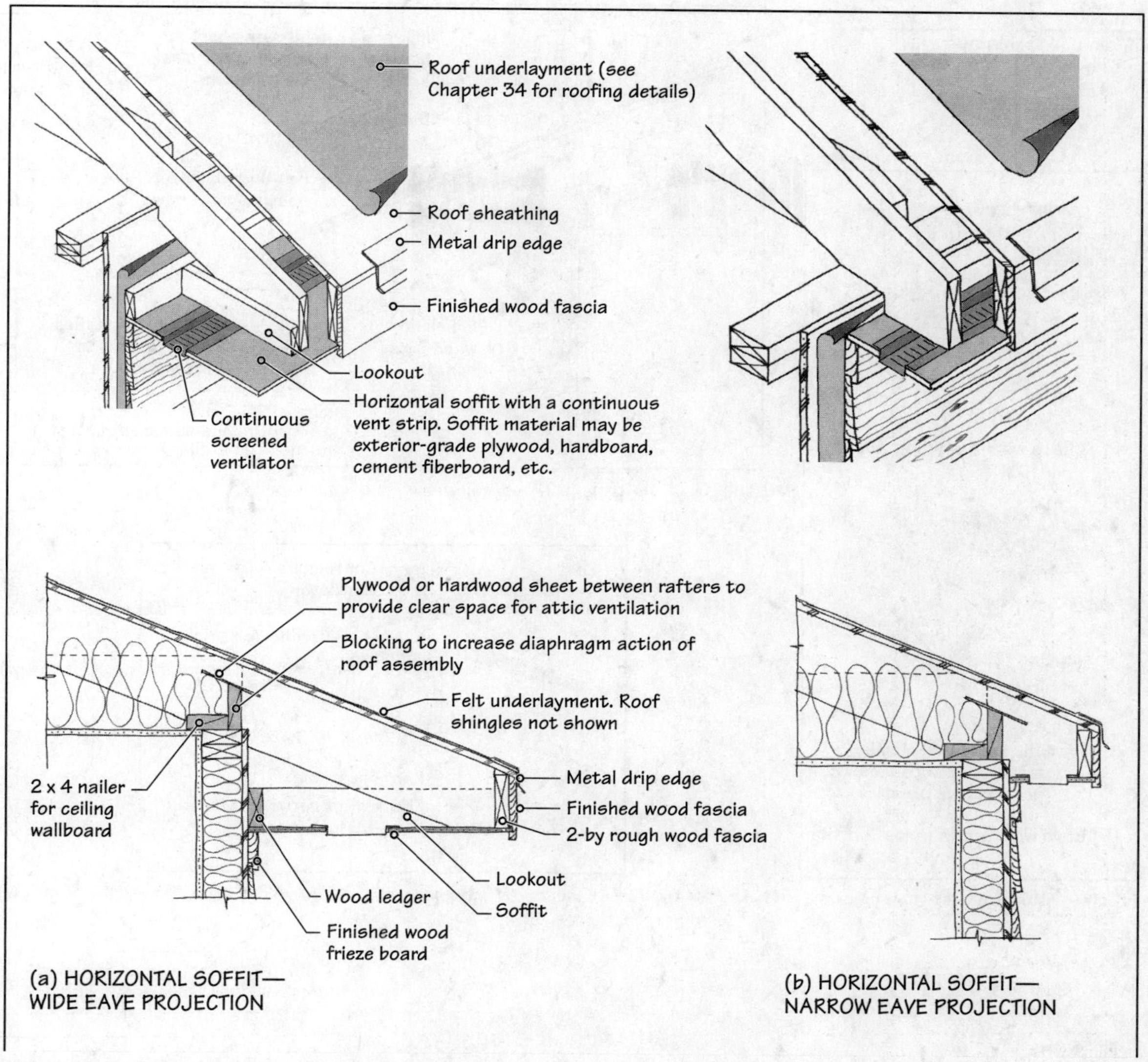

FIGURE 16.12 Details of wide and narrow eave projections.

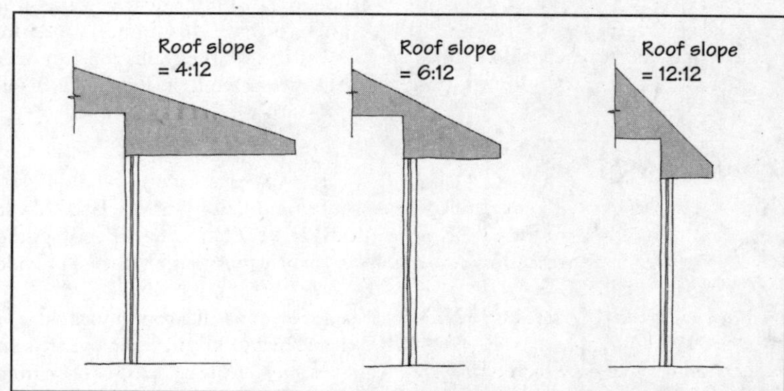

FIGURE 16.13 Effect of roof slope on an eave projection with a horizontal soffit.

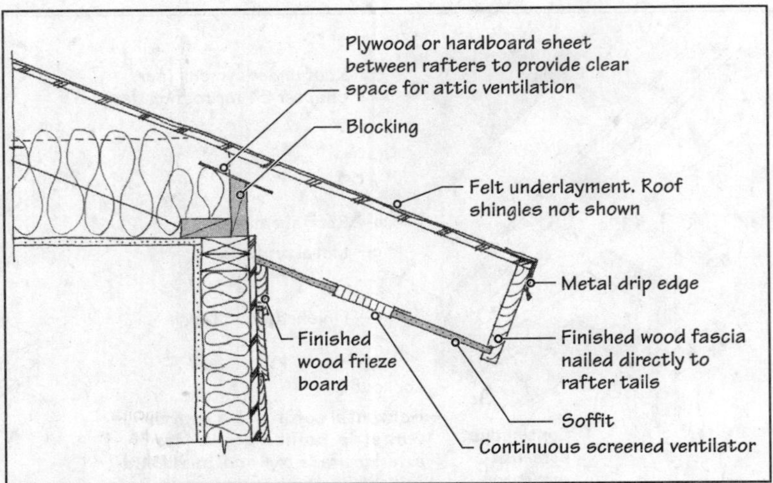

FIGURE 16.14 Details of an eave projection with a sloping soffit.

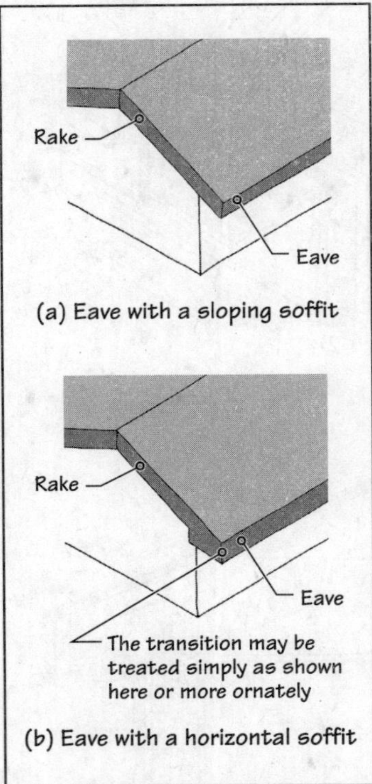

FIGURE 16.15 Eave-rake transitions.

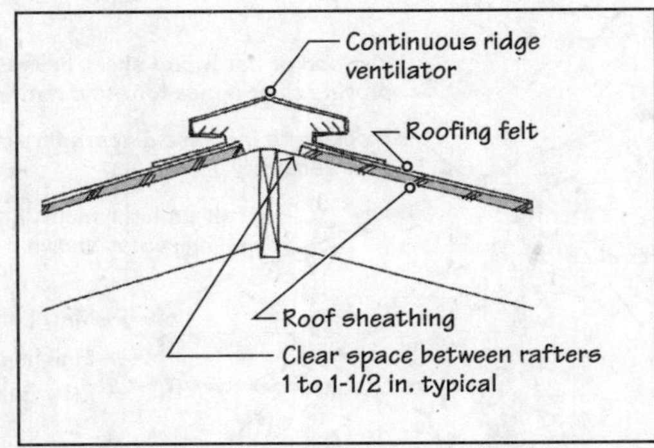

FIGURE 16.16 One of the several ways of detailing for ridge ventilation.

PRACTICE QUIZ

Each question has only one correct answer. Select the choice that best answers the question.

1. Horizontal beveled wood siding is generally made by slicing a
   - a. solid lumber piece.
   - b. LVL member.
   - c. glulam member.
   - d. OSB member.
   - e. plywood member.

2. Horizontal siding is provided on
   - a. the walls of a wood light-frame building.
   - b. the eave of a wood light-frame building.
   - c. the rake of a wood light-frame building.
   - d. the roof of a wood light-frame building.

3. Horizontal wood siding material is generally finished smooth on the front and back faces.
   - a. True
   - b. False

4. Each course of beveled horizontal wood siding is nailed to the wall through
   - a. one line of nails near the top of the siding.
   - b. one line of nails near the bottom of the siding.
   - c. one line of nails in the center of the siding.
   - d. two lines of nails, one at the bottom and one at the top of the siding.

5. Vertical joints between adjacent horizontal siding boards are generally
   - a. lapped $\frac{1}{2}$ in.
   - b. lapped at least 1 in.
   - c. lapped at least $1\frac{1}{2}$ in.
   - d. lapped at least 2 in.
   - e. none of the above.

6. To produce a drainage layer with horizontal siding, we generally use
   - a. 2-in.-wide horizontal strips nailed to sheathing.
   - b. $1\frac{1}{2}$-in.-wide horizontal strips nailed to sheathing.
   - c. 2-in.-wide horizontal strips nailed to studs before installing sheathing.
   - d. $1\frac{1}{2}$-in.-wide horizontal strips nailed to studs before installing sheathing.
   - e. none of the above.

7. Fiber-cement siding is
   - a. noncombustible but prone to fungal decay.
   - b. combustible and prone to fungal decay.
   - c. noncombustible and not prone to fungal decay.
   - d. none of the above.

8. Vertical siding for a wood light-frame building is generally made from small strips of plywood.
   - a. True
   - b. False

9. An eave is generally horizontal and the rake is inclined.
   - a. True
   - b. False

10. The transition from a projecting eave to a rake needs special detailing when the eave is provided with a
    - a. horizontal soffit.
    - b. sloping soffit.
    - c. no soffit.

16.5 GYPSUM BOARD

Gypsum board is the most commonly used interior wall and ceiling finish in virtually all types of buildings. It provides a strong, fire-protective cover over framing members and can be finished with paint, wallpaper, wood paneling, and so on. It is more economical and much easier to install and finish than the lath-and-plaster alternative that preceded it. Additionally, being a dry form of construction, it leaves less construction mess than lath and plaster.

Gypsum board is known by several names that are used synonymously—*drywall, wallboard, gypsum wallboard* (GWB), and *plasterboard.* The term *sheetrock* is not a generic name but is proprietary to a gypsum board manufacturer.

WHY GYPSUM IS A WONDROUS MATERIAL WITH RESPECT TO FIRE

Gypsum is a rocklike mineral found extensively on the earth's surface. Its chemical name is calcium sulfate dihydrate ($CaSO_4.2H_2O$), which indicates that every molecule of gypsum contains two molecules of water (H_2O) chemically combined with one molecule of calcium sulfate ($CaSO_4$). The consequence of gypsum's chemical formulation is that 100 lb of gypsum rock contains 21 lb of water, that is, 21% of gypsum rock is water. (The prefix *di* in dihydrate means "two".)

It is the water contained in gypsum that gives it its fire-protective property. Under the action of heat, the water in gypsum converts to steam. Steam not only helps to absorb the heat, but also provides a cooling effect. Because the temperature of water cannot exceed its boiling point (212°F), regardless of how long the fire acts on it, the temperature of gypsum rock cannot equal that of fire unless all the water in gypsum has converted to steam. Once all the water has turned into steam, gypsum begins to disintegrate under the effect of fire. Therefore, the presence of water in gypsum slows the disintegration process.

The temperature gradient of the wall after 2-h fire exposure, when a 6-in.-thick solid gypsum wall is exposed to fire in the standard test and the temperature of the wall at various points within the wall, is shown in Figure 16.17. Observe that after 2 h, the temperature of the unexposed face of the wall is only 130°F, and the temperature of the wall 2 in. away from the fire side is the boiling point of water—212°F.

ANATOMY AND SIZES OF GYPSUM BOARD PANELS

Gypsum board consists of a gypsum core, which is sandwiched between specially treated paper faces, Figure 16.18. The paper adds strength to the board and provides a smooth surface for finishing and painting.

The most commonly used size of gypsum board is a 4 ft × 8 ft panel. Some of the other commonly used sizes are 4 ft × 9 ft, 4 ft × 10 ft, and 4 ft × 12 ft. Boards of width 4 ft 6 in. are also available and are used horizontally on walls in rooms with 9-ft-high ceilings. Longer

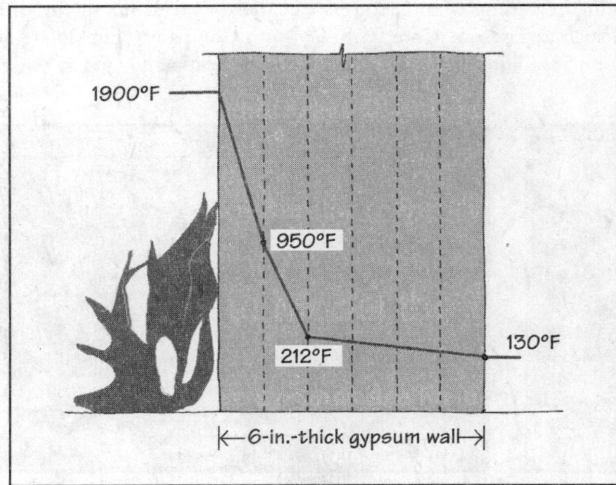

FIGURE 16.17 Temperature gradient along a 6-in.-thick gypsum wall after 2-h exposure to fire under the ASTM E119 test—the test used to determine the fire rating of assemblies (see the Expand Your Knowledge section in Chapter 7).

NOTE

Plaster of Paris

When gypsum is heated to remove 75% of its water, the resulting product is calcium sulfate hemihydrate (*hemi* means "half"), referred to as *plaster of paris*. The chemical notation of plaster of paris is $CaSO_4 \cdot \frac{1}{2} H_2O$. (Ca is the symbol for calcium and SO_4 is the symbol for sulfate, so $CaSO_4$ stands for calcium sulfate.) When water is added to plaster of paris, it returns to its original hard, rocklike form—gypsum—containing calcium sulfate and two molecules of water.

Because plaster of paris reverts to gypsum with the addition of water and its subsequent setting, plaster of paris expands on setting. This is unlike portland cement, which shrinks on setting due to the evaporation of water. Thus, plaster of paris is used for casting in molds because it accurately assumes the shape of the mold.

If water is completely removed from gypsum, the resulting product, which is pure calcium sulfate, is called *anhydrite*. Anhydrite has a longer setting time than plaster of paris and is marketed as Keene's cement, used as a finish coat in interior plaster.

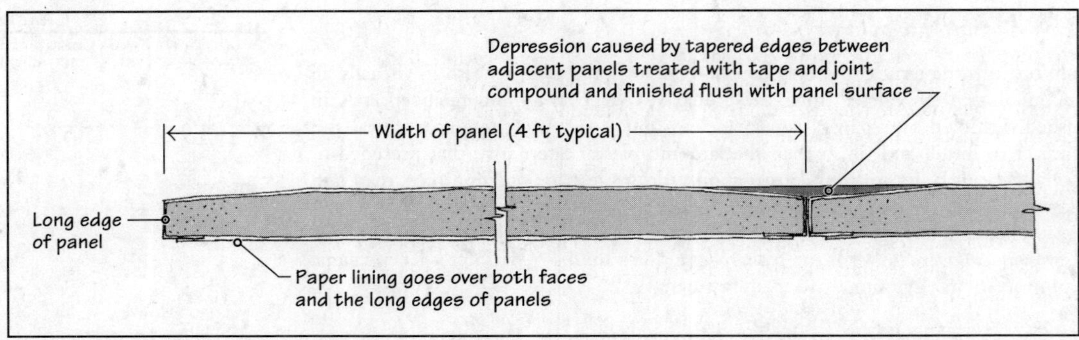

FIGURE 16.18 Section through a gypsum board panel.

boards are more cumbersome to install. They are heavier to carry and more difficult to maneuver in stair enclosures and other interior spaces. However, fewer joint seams in longer boards reduce cutting of boards and finishing of seams.

The most frequently used board thickness for residential buildings is $\frac{1}{2}$ in. For commercial buildings, the corresponding thickness is $\frac{5}{8}$ in. Other available thicknesses are $\frac{3}{8}$ in. and $\frac{1}{4}$ in.

The long edges of a gypsum board are tapered, as shown in Figure 16.18. The taper allows the joints to be taped over and filled with a joint compound so that a finished wall or ceiling surface appears jointless. The boards that do not require finishing (e.g., prefinished boards and gypsum sheathing panels) have square (nontapered) long edges.

The manufacturing process of gypsum boards yields an endlessly long board wrapped with paper, which is cut to the desired lengths. Therefore, the short (4-ft-wide) edges where the cut is made do not have paper wrapping and have smooth, square edges (not tapered).

Types of Gypsum Board Panels

Several different types of gypsum board panels, with different properties, are available. The more commonly used types are as follows:

- *Type R (regular) Boards:* Type R boards are the standard boards, typically used in $\frac{1}{2}$-in. thickness.
- *Type X Boards:* Type X boards are more fire-resistive than Type R boards. The core in Type X boards contains noncombustible fibers (e.g., fiberglass) mixed with gypsum. The fibers provide greater strength so that the boards, if subjected to fire, maintain their structural integrity for a longer time than Type R boards.

  By definition, a Type X gypsum board, when used in $\frac{5}{8}$-in. thickness on both sides of a wood or metal stud, gives a minimum 1-h fire-rating. A $\frac{5}{8}$-in.-thick board that does not give a 1-h fire rating is not a Type X board. A $\frac{1}{2}$-in.-thick Type X panel gives a 45-min fire rating. Manufacturers call their Type X boards by different brand names, such as Firecode Core (U.S. Gypsum Company), Firebloc (American Gypsum Company), Fire-shield (National Gypsum Company), and so on, Figure 16.19.

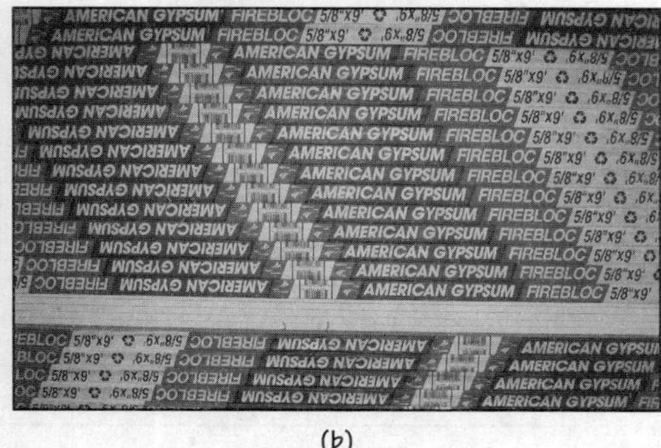

FIGURE 16.19 Type X gypsum panels: (a) called Firecode by U. S. Gypsum Company and (b) called Firebloc by American Gypsum Company.

- *Mold-and-moisture-resistant (MMR) Boards:* These boards employ several different proprietary technologies to resist mold and moisture absorption. Some combine a moisture-resistant core with paper that is treated on one side to be more resistant to mold. Another uses a moisture-resistant core sandwiched between glass mats on both faces. Mold resistance is determined by an ASTM test that rates panels on a scale from 1 to 10, 1 being the least resistant and 10 being the most resistant panel. The test is conducted over a 4-week period, giving only a short-term rating.

  MMR boards are specified for bathroom walls and ceilings and for any other areas where greater resistance to mold is required, such as the walls and ceilings of hospitals and nursing homes.
- *Flexible Boards:* These boards are more flexible than other gypsum boards and are meant for use on curved walls or ceilings. They are $\frac{1}{4}$ in. thick and have heavier paper facing to resist cracking. They are generally used in two layers or more to give the required gypsum thickness.
- *High-impact Boards:* These are boards that have a thick fiberglass mesh embedded in the core near the back to provide greater impact and penetration strength compared with other boards. They are available as Type R or X.
- *Prefinished Boards:* These boards are covered on one face with vinyl instead of paper and do not require any finishing (tape and bed, paint, or wallpaper). Generally, both long and short panel edges are square (not tapered). Manufacturers provide special nails to match the color and texture of the finish in addition to joint-cover accessories. The vapor impermeability of vinyl covering should be taken into account before specifying the use of prefinished boards in warm, humid climates.
- *Gypsum Wall Sheathing Boards:* Although gypsum wall sheathing in wood frame buildings is used much less since the introduction of OSB boards, it is used in situations where the fire rating of the exterior walls is a concern. Gypsum wall sheathing boards have a specially treated core and water-resistant paper facings. The long edges of boards are profiled to shed water, Figure 16.20. The boards are available as Type R or X. A particularly popular gypsum sheathing for commercial construction is made from gypsum core embedded in fiberglass mats on both faces.

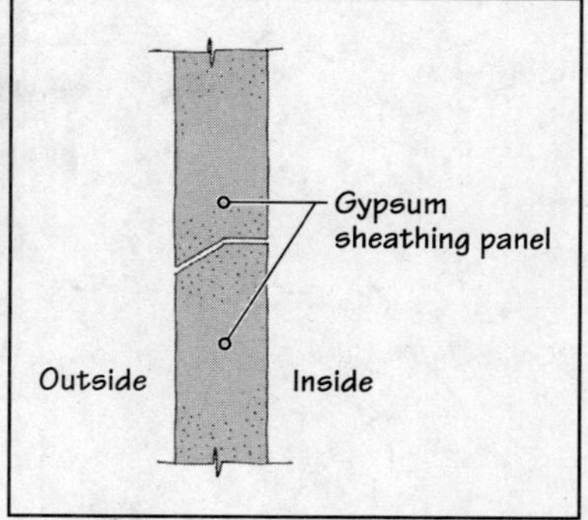

FIGURE 16.20 Vertical cross section through horizontally applied, paper-faced gypsum sheathing panels.

NOTE

Portland Cement Boards

While MMR boards are specified for bathroom walls and ceilings because of the greater relative humidity in bathrooms, they cannot be used on surfaces that receive direct water, such as the glazed ceramic tile walls surrounding a shower or bathtub. On such surfaces, the tiles should not be backed by gypsum boards, but by portland cement boards because of their greater water resistance. A portland cement board is made of portland cement, sand, and glass fibers or cellulose fibers. It is stronger than gypsum board of the same thickness but more difficult to install. Portland cement boards are also used as underlayment for ceramic tile flooring.

LIMITATIONS OF GYPSUM BOARDS

Gypsum is a water-soluble material. Therefore, it is basically an interior-use material. Gypsum boards (including MMR boards) are not recommended for use in locations with direct exposure to water or continuous interior high humidity, such as in swimming pool enclosures, steam rooms, or gang shower rooms. However, gypsum boards are available for use in protected exterior locations such as for eave soffits (soffit panels) or as wall sheathing.

16.6 INSTALLING AND FINISHING INTERIOR DRYWALL

Gypsum boards are fastened to framing members using nails or screws, Figure 16.21. Screw application is preferred over nails because screws eliminate nail popping caused by wood shrinkage. On metal studs, only screws can be used. However, modern pneumatic nailers make nailing gypsum boards much faster than fastening with screws. The nail or screw heads are slightly depressed into the panel surface so that they can be filled with joint compound and leveled with the panel surface.

Joint compound (also referred to as *mud* in the drywall trade) has a plaster-type consistency and is available in pails or cartons in a ready-to-use formulation. Drywall finishers use a pan to hold the mud in one hand and a knife in the other to apply the mud as needed, Figure 16.22.

Gypsum boards must be fastened to every framing member. On walls, Type R boards may be applied either horizontally or vertically. However, because Type X boards are used in fire-rated assemblies, they must be installed as per the instructions of the manufacturer of the rated assembly. Fastener spacing depends on the spacing between framing members and the required fire rating of the assembly.

Ceiling boards are attached first so that the wall boards provide edge support to the ceiling boards, Figure 16.23. Joints between boards must occur on framing members. The

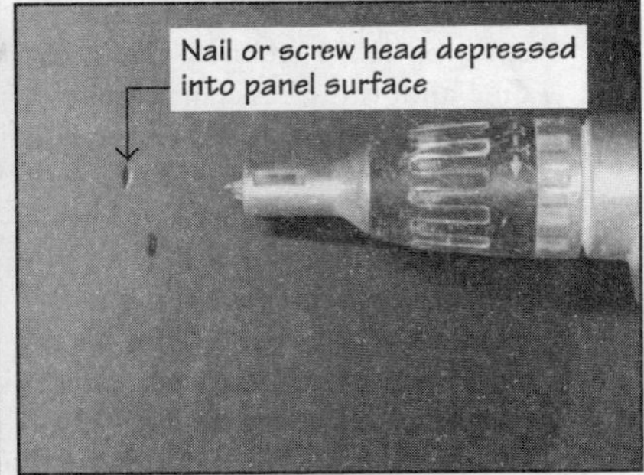

FIGURE 16.21 Gypsum panels may be installed using either nails or screws on wood framing members. On metal studs, only screws can be used. The nails and screws are slightly recessed into the panel surface so that they can be covered over by joint compound and finished smooth with panel surface. (Photos courtesy of Tommy Eaton)

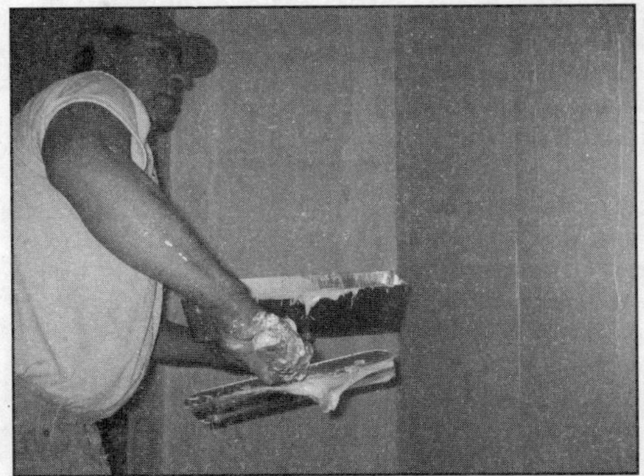

FIGURE 16.22 A drywall worker uses a pan containing the joint compound (mud) and a knife to apply it. (Photo courtesy of Tommy Eaton)

joints are covered with a tape that reinforces the joint against movement and cracking of the joint compound. Two types of tape are used—paper tape and fiberglass scrim tape, Figure 16.24. The paper tape is adhered by spreading a thin coat of joint compound, into which the tape is lightly pressed. Fiberglass tape is self-adhering and does not need an undercoat of joint compound.

Taping of the joint is followed by an overcoat of joint compound, Figure 16.25. After the coating has dried, it is feathered smooth with a sander, and another coat of joint compound and sanding is done until the joint is ready to receive paint.

Inside corners of rooms are taped with paper tape. The paper tape has a built-in crease, along which the tape is folded, Figure 16.26. The folded tape is installed in the corner, covered with joint compound, and finished smooth in the same way as on a flat joint.

Outside corners require L-shaped metal or vinyl trims, which are nailed into the studs through the drywall panel. The trim is covered with three layers of joint compound coatings and finished flush with the surface of the boards, Figure 16.27. When all joints and corners are finished and the interior is ready to receive the paint, it looks somewhat as shown in Figure 16.28.

FIGURE 16.23 Gypsum board panels for ceilings are typically attached prior to wall panels.

FIGURE 16.24 Paper tape and self-adhering fiberglass scrim joint tape. (Photo courtesy of Tommy Eaton)

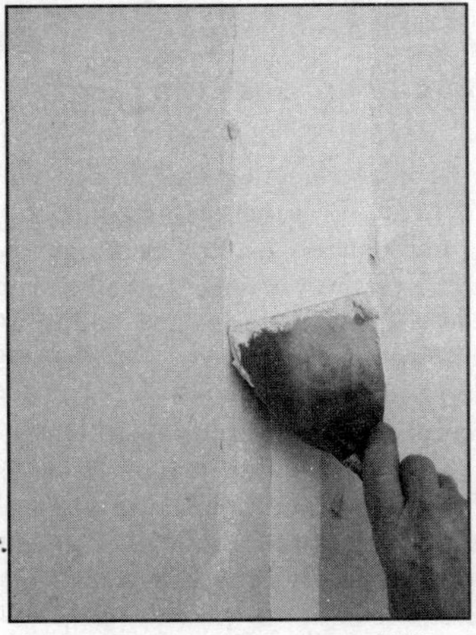

FIGURE 16.25 Joint compound filling over joint tape. (Photo courtesy of Tommy Eaton)

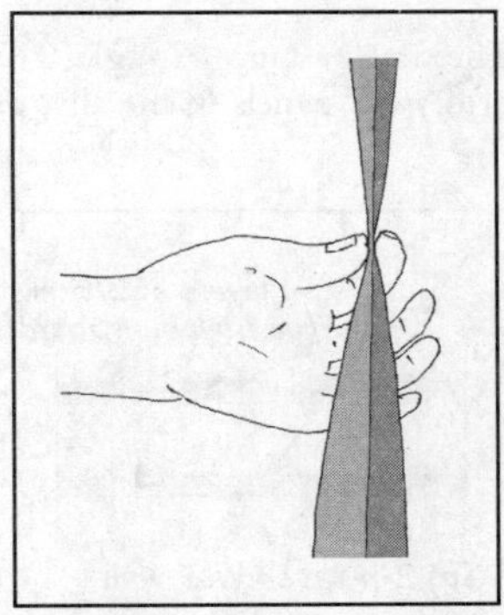

FIGURE 16.26 Paper tape is folded along the crease before installation in an inside corner.

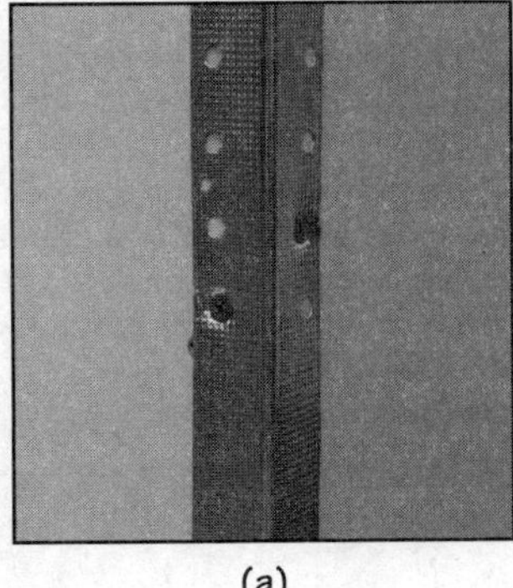

(a)

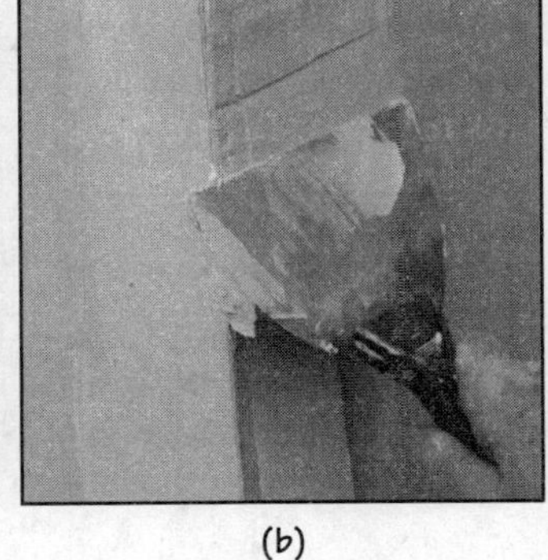

(b)

FIGURE 16.27 An outside corner is reinforced with metal or vinyl trim (shown in **(a)**) before being covered by joint compound (shown in **(b)**). (Photos courtesy of Tommy Eaton)

FIGURE 16.28 A room interior ready to receive paint. (Photo courtesy of Tommy Eaton)

MULTILAYER DRYWALL

Although one-layer application is the norm, sometimes two or more layers of drywall panels may be needed on one or both faces of the assembly to obtain a higher fire rating or higher sound insulation. In such an assembly, each layer may be nailed or screwed. The screw (or nail) size and spacing must be as per the assembly tested for fire rating or sound rating.

With prefinished boards, in which two-layer application is common, the base layer is fastened with nails or screws and the face layer is adhered to the base layer using manufacturer-provided adhesives. Temporary nailing is generally required on the edges of face panels to hold them in place until the adhesive has attained strength.

16.7 FIRE-RESISTANCE RATINGS OF WLF ASSEMBLIES

The fire-resistance rating of a WLF wall or ceiling-floor assembly depends on the type of drywall panels. Generally, one $\frac{5}{8}$-in.-thick, Type X drywall panel on each side of a wood stud wall gives a 1-h fire-resistance rating. If two layers of the same drywall panels are used, a 2-h fire-resistance rating is obtained. Three such layers give a 3-h rating, Figure 16.29.

However, the fire-resistance rating also depends on the spacing of studs, the spacing and type of nails or screws, and whether or not wall cavities are insulated. Drywall manufacturers provide the fire-rating data for various assemblies in which their product can be used. Another standard source for this information is the *Fire Resistance Design Manual* published by the Gypsum Association, which is updated every few years. This publication gives the specifications and sketches of 1-h, 2-h, and 3-h rated assemblies. These ratings have been established based on measurements in standard fire tests. Because the ratings are in integer hours, an assembly whose measured fire-resistance rating is 1 h 59 min is listed as a 1-h assembly. To be listed as a 2-h assembly, its fire rating must lie between 2 h and 2 h 59 min.

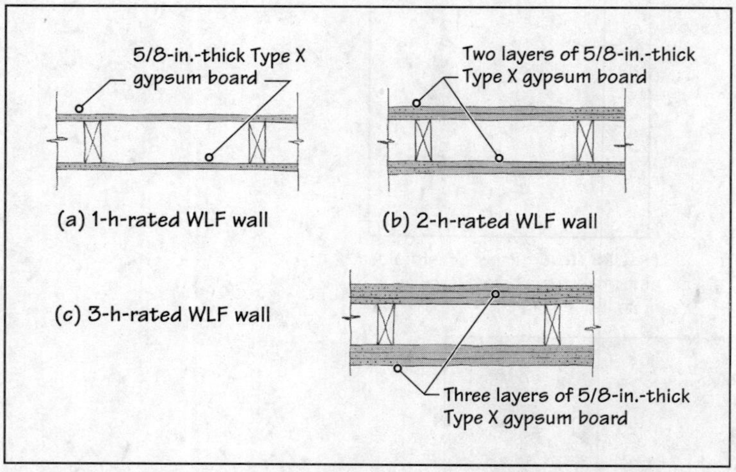

FIGURE 16.29 Fire-rated wood frame assemblies are generally obtained by using one or more layers of $\frac{5}{8}$-in.-thick Type X gypsum board. Although the type and thickness of gypsum board are the primary factors that determine the fire rating of a wood frame assembly, a few other factors are also important. These include the type and spacing of nails or screws, the spacing of framing members (studs or ceiling joists), the provision of insulation within framing members, and whether the panel is applied horizontally or vertically.

PRACTICE QUIZ

Each question has only one correct answer. Select the choice that best answers the question.

11. Which of the following terms for gypsum board is proprietary to a manufacturer?
- **a.** *Wallboard*
- **b.** *Sheetrock*
- **c.** *Plasterboard*
- **d.** *Drywall*

12. Gypsum board is one of the best materials for fire protection because
- **a.** it is noncombustible.
- **b.** it is heat-treated during its manufacture.
- **c.** each molecule of gypsum contains two molecules of water that retard the progress of fire.
- **d.** each molecule of gypsum contains two molecules of carbon dioxide that retard the progress of fire.

13. The most commonly used size for gypsum boards is
- **a.** 4 ft × 8 ft.
- **b.** 4 ft × 8 ft 6 in.
- **c.** 4 ft 6 in. × 8 ft.
- **d.** 3 ft × 6 ft.
- **e.** 4 ft × 4 ft.

14. Of Types R and X gypsum board panels, both of the same thickness, Type R is more fire-resistive.
- **a.** True
- **b.** False

15. One sheet of Type X gypsum board panel, $\frac{5}{8}$in. thick, used on each side of a stud wall gives a fire rating of
- **a.** 30 min.
- **b.** 45 min.
- **c.** 60 min.
- **d.** 75 min.
- **e.** 90 min.

16. Moisture-resistant gypsum board panels are generally
- **a.** green in color.
- **b.** pink in color.
- **c.** white in color.
- **d.** yellow in color.
- **e.** gray in color.

17. Gypsum board panels used on curved walls are generally
- **a.** $\frac{5}{8}$ in. thick.
- **b.** $\frac{1}{2}$ in. thick.
- **c.** $\frac{3}{8}$ in. thick.
- **d.** $\frac{1}{4}$ in. thick.
- **e.** $\frac{1}{8}$ in. thick.

PRACTICE QUIZ *(Continued)*

18. High-impact gypsum board panels generally have a
  a. steel lining at the back side.
  b. aluminum foil lining at the back side.
  c. fiberglass mesh at the back side.
  d. plastic sheet at the back side.
  e. none of the above.

19. Gypsum sheathing is generally used on the
  a. exterior side of an exterior stud wall.
  b. interior side of an exterior stud wall.
  c. exterior side of an exterior masonry wall.
  d. interior side of an exterior masonry wall.

20. It is better to install ceiling gypsum board panels before installing the wall panels in a room.
  a. True   **b.** False

21. The most important reason for using more than one layer of gypsum board panels on one or both faces of a wall is to obtain a stronger wall assembly.
  a. True   **b.** False

REVIEW QUESTIONS

1. List various exterior wall finishes used in wood light-frame buildings. Which of these finishes are unique to (wood or steel) light-frame buildings.
2. Draw a detail section at the foundation level through a wood light-frame building whose exterior wall is finished with beveled wood siding. Assume a crawl space under the ground floor.
3. List various types of sidings commonly used in buildings.
4. Explain why gypsum board is an excellent material for providing fire resistance in building assemblies.
5. Explain the difference between a Type R gypsum board and a Type X gypsum board.
6. What is an MMR gypsum board? Where would you specify it?
7. What are the important differences between portland cement boards and gypsum boards?
8. Using a sketch and notes, explain how you will detail a 2-h fire-rated wood light-frame wall assembly.

CHAPTER 17

Structural Insulated Panel Construction

CHAPTER OUTLINE

The structural insulated panel system of construction described in this chapter is an alternative to the conventional WLF system. It is a panelized system that reduces on-site construction time and allows the use of less skilled labor. Panelization, however, necessitates the use of hoisting equipment such as cranes, fork lifts, booms, and so on during construction, Figure 17.1, whose use in WLF construction is generally limited.

Because insulation is part of the panels, the insulation subtrade is eliminated, which further increases construction efficiency. Several other advantages of the system, as well as its limitations, are described at the end of this chapter. Because the structural insulated panel is a relatively new system that is still evolving, its current market share is small compared with that of the well-established conventional WLF system that dominates the residential and light commercial construction market in the United States.

In addition to all structural insulated panel construction, structural insulated panels are used as wall and roofing panels in timber frame buildings (see the Expand Your Knowledge section in Chapter 15) and low-rise steel frame buildings.

17.1 BASICS OF THE STRUCTURAL INSULATED PANEL (SIP) SYSTEM

The structural insulated panel (SIP) system consists of panels of sandwich composition. Each panel comprises two facing boards bonded to a core consisting of rigid plastic foam insulation, Figure 17.2. The core generally consists of expanded polystyrene (EPS). (Extruded polystyrene and polyisocyanurate cores may also be used, but they are more expensive.) The material used for the facings may be oriented strandboard (OSB) or plywood. OSB is commonly used because of its lower cost and its availability in much larger sizes. The most common thickness of OSB facing is $\frac{1}{2}$ in.

Panels are manufactured by applying structural-grade adhesive on both faces of the core and then laminating the OSB to it. The assembly is kept under pressure for the required

FIGURE 17.1 Two examples of typical structural insulated panel buildings under construction. (Photos courtesy of the Structural Insulated Panel Association [SIPA])

duration. Because insulation is included in the panels, SIPs are used only in the building envelope. Therefore, in a SIP building, the interior walls are generally framed with 2-by lumber and the intermediate floors are also framed with 2-by lumber or engineered wood members. The use of SIPs for an intermediate floor is, in fact, discouraged because of its lower structure-borne sound insulation. The foam in SIPs functions as an efficient acoustical bridge. SIPs may, however, be used in a floor that is required to be insulated if its lower sound-insulation value is of no concern, such as a floor over a crawl space or basement.

Although there are many similarities between the characteristics of the SIP systems of various manufacturers, SIP is a proprietary system that differs somewhat from manufacturer to manufacturer. The discussion and illustrations provided here are generic. Most SIP manufacturers have details that have been carefully developed and approved by the code authorities and should be strictly followed.

In considering the use of a SIP system, the general procedure is to send the architectural drawings of the project to the selected SIP manufacturer. Based on the analysis of the drawings, the manufacturer determines the size, shape, configuration, and thickness of the panels and prepares the necessary shop drawings for their fabrication.

Typical panel thickness for walls: 4-1/2 or 6-1/2 in.

Expanded polystyrene (EPS) core

Oriented strandboard (OSB) facing on both sides of expanded polystyrene (EPS) core; 1/2-in.-thick OSB typical

FIGURE 17.2 Composition of a typical structural insulated panel.

After the approval of the shop drawings by the architect and the constructor, the manufacturer fabricates the panels. Each panel is uniquely identified as to its location in the building. All panels required for the project are packaged and sent to the construction site for assembly.

Because almost no cutting is required on site, the panels are shipped to the site when needed, eliminating the need for elaborate site storage facilities. The plant typically packages the panels in reverse order of their use, increasing the builder's productivity. The entire fabrication work, that is, the analysis of architectural drawings, preparation of shop drawings, and fabrication of panels, is generally computerized.

Structural Behavior of a SIP Panel

Structurally, a SIP panel functions as a composite panel in which the core integrates the two facings. Thus, under an axial compressive load, even if the load is initially delivered to one facing, both facings combine to resist the load because the core transfers the load to the other, initially unloaded facing through shear mechanism.

Similar behavior comes into play when the panel is subjected to bending. Under bending, one facing is subjected to compressive stresses and the other facing is under tensile stresses. The core resists the shear stresses generated by bending. The behavior of a SIP panel is, therefore, identical to that of an I-section, where the facings function as the flanges of the I-section and the foam core functions as the section's web, Figure 17.3.

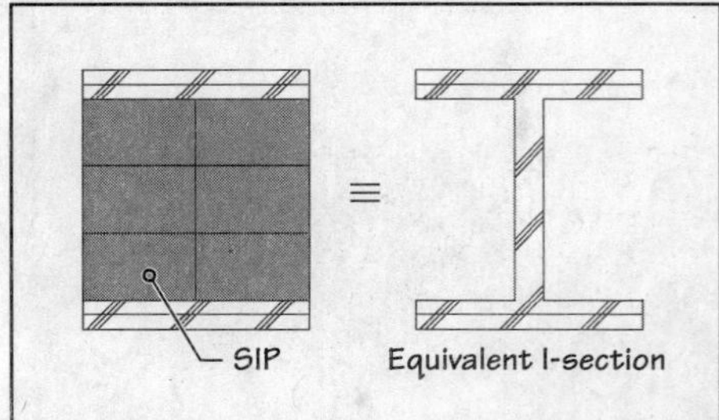

SIP
Equivalent I-section

FIGURE 17.3 In carrying both axial and lateral loads, a SIP behaves as an I-section where the core functions as the web of the section and the facings as its flanges.

The core also helps to prevent the buckling of thin facings under compressive stresses, created either by the bending of the panels (in a floor or roof) or under compressive loads (in a wall). In other words, all three elements of a SIP are stressed under axial or lateral loads.

The facing-core composite behavior also comes into action in providing racking resistance to the structure where the facings and the core act together as shear walls. In a conventional wood frame building, only the exterior sheathing and the studs provide racking resistance.

Panel Sizes

Panels are produced in various thicknesses, depending on whether they are used in walls, floors, or roofs. Wall panels are generally $4\frac{1}{2}$ in. or $6\frac{1}{2}$ in. thick. A $4\frac{1}{2}$-in.-thick SIP consists of a $3\frac{1}{2}$-in.-thick core, matching a wood light-frame wall made of 2×4 studs. A $6\frac{1}{2}$-in.-thick panel (with a $5\frac{1}{2}$-in.-thick core) corresponds to a wood frame wall comprising 2×6 studs. Panel thicknesses of $8\frac{1}{4}$ in., $10\frac{1}{4}$ in., and $12\frac{1}{4}$ in. are used for floors or roofs.

The surface dimensions (length and width) of panels vary, depending on whether lifting and placing the panels in position at the site is done manually or using hoisting equipment. Panels of up to 8 ft × 24 ft can be produced by fabricators.

17.2 SIP WALL ASSEMBLIES

In general, SIP wall panels are either 4 ft or 8 ft wide. They may be continuous over the entire height of the building or they may be only one floor tall, requiring floor-by-floor installation. The floor-by-floor installation of panels is similar to the construction of walls in a conventional WLF building, where the upper floor walls are installed over the subfloor at that level.

The details of the connection of wall panels to the foundation, between adjacent panels, at wall corners, and to the floors and roofs are critical. Figure 17.4 provides an

FIGURE 17.4 General layout of exterior walls in a SIP structure.

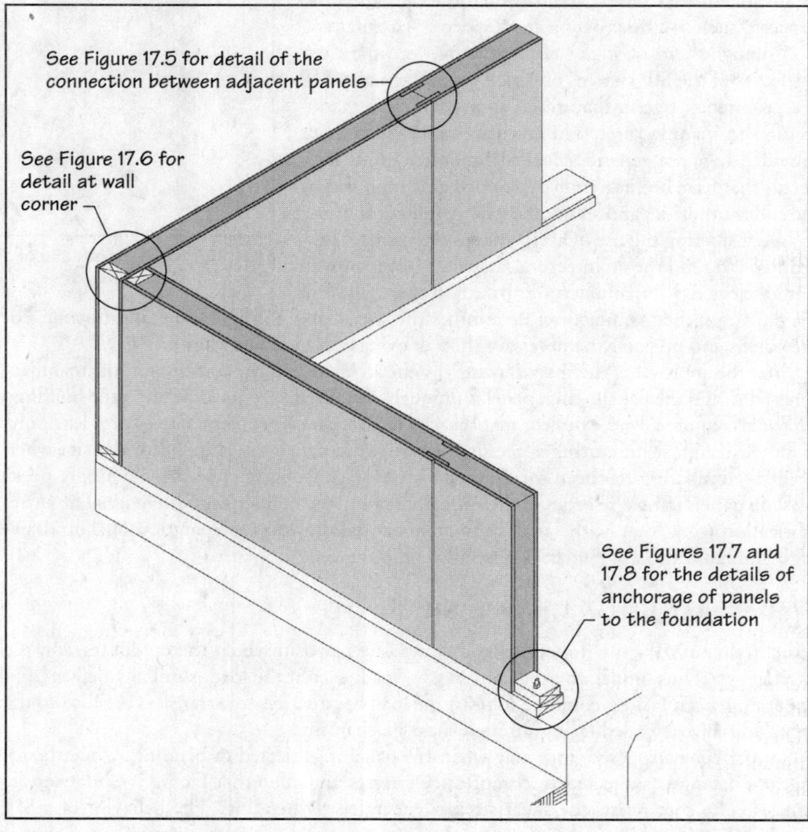

See Figure 17.5 for detail of the connection between adjacent panels
See Figure 17.6 for detail at wall corner
See Figures 17.7 and 17.8 for the details of anchorage of panels to the foundation

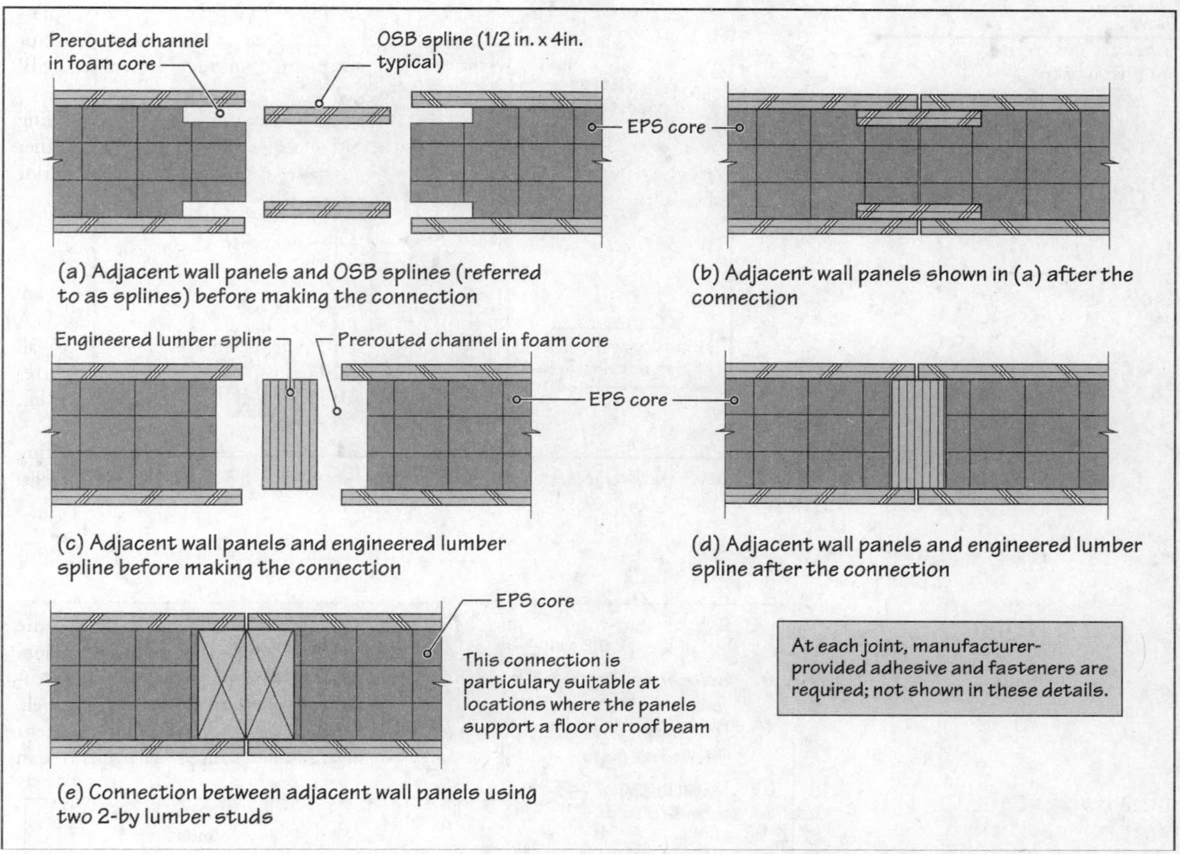

FIGURE 17.5 Three commonly used alternative ways of connecting adjacent wall panels. These details are keyed to Figure 17.4.

overall view of a typical panel layout. The adjacent wall panels are connected together with splines, Figure 17.5.

The splines may consist of $\frac{1}{2}$-in.-thick OSB strips (referred to as *surface splines*), engineered wood studs, or double 2-by studs, depending on the requirement of the connection. Nails, screws, and adhesives are used in a connection per the manufacturer's details.

The use of connecting splines requires the foam core to be routed and grooved to accept the splines. This operation is performed at the panel-fabrication plant so that the panels arrive at the site already routed. Chases for electrical wiring (generally $1\frac{1}{2}$ in. in diameter) are also routed before the panels are shipped to the site. Chases for other utility pipes are not included in SIPs because these pipes typically occur in the interior walls.

Figure 17.6 shows the connection between panels at a wall corner. Figures 17.7 and 17.8 show the details of the anchorage of wall panels to foundations. A two-sill-plate assembly at the foundation is commonly recommended by the manufacturers. The lower sill plate consists of preservative-treated lumber. The upper sill plate functions as a connecting spline. The details of Figures 17.7 and 17.8 ensure that the OSB facings bear directly over the lower sill plate so that the gravity load on the wall is transferred through the facings to the foundations in bearing.

There are various ways in which windows and doors can be detailed in a SIP wall. A commonly used method is shown in Figure 17.9. As stated previously, the interior partitions in a SIP building are framed by using conventional 2-by studs. Figure 17.10 shows the detail of the junction between an exterior SIP panel and an interior wall.

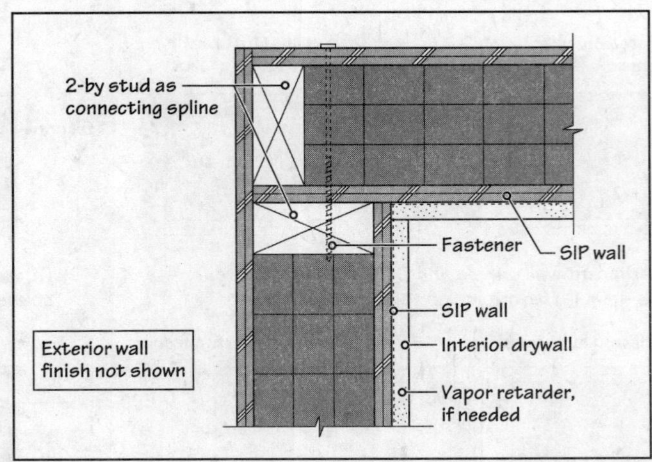

FIGURE 17.6 Typical connection at the corner of two structural insulated wall panels. This detail is keyed to Figure 17.4.

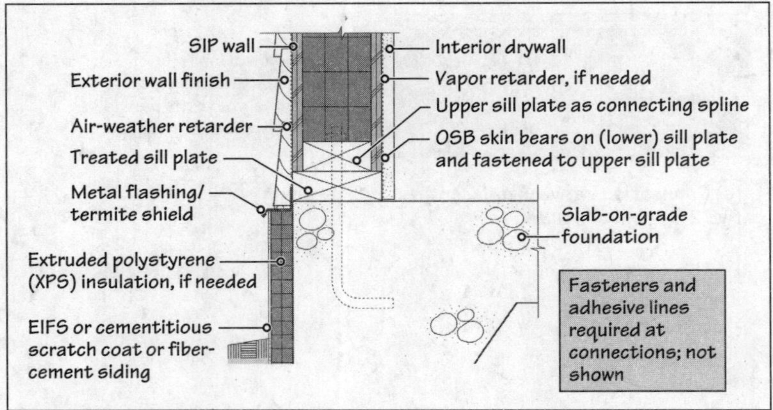

FIGURE 17.7 Typical detail of the anchorage of structural insulated wall panels to a slab-on-grade foundation. This detail is keyed to Figure 17.4.

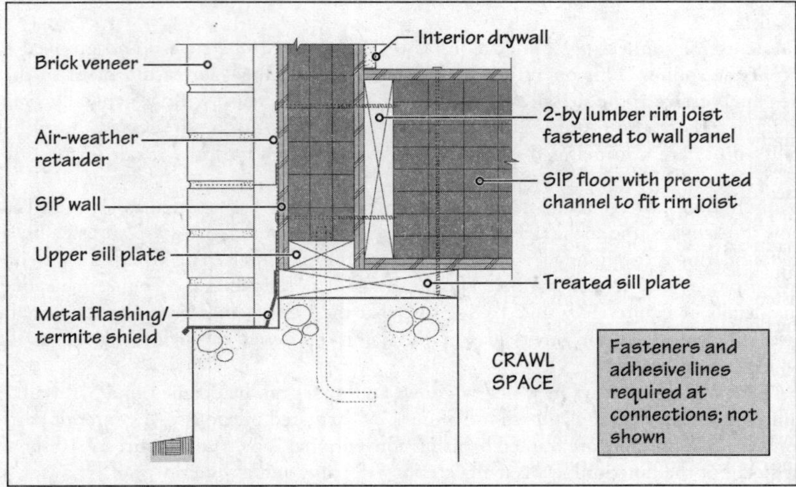

FIGURE 17.8 Typical detail of the anchorage of structural insulated wall panels to a foundation with a crawl space.

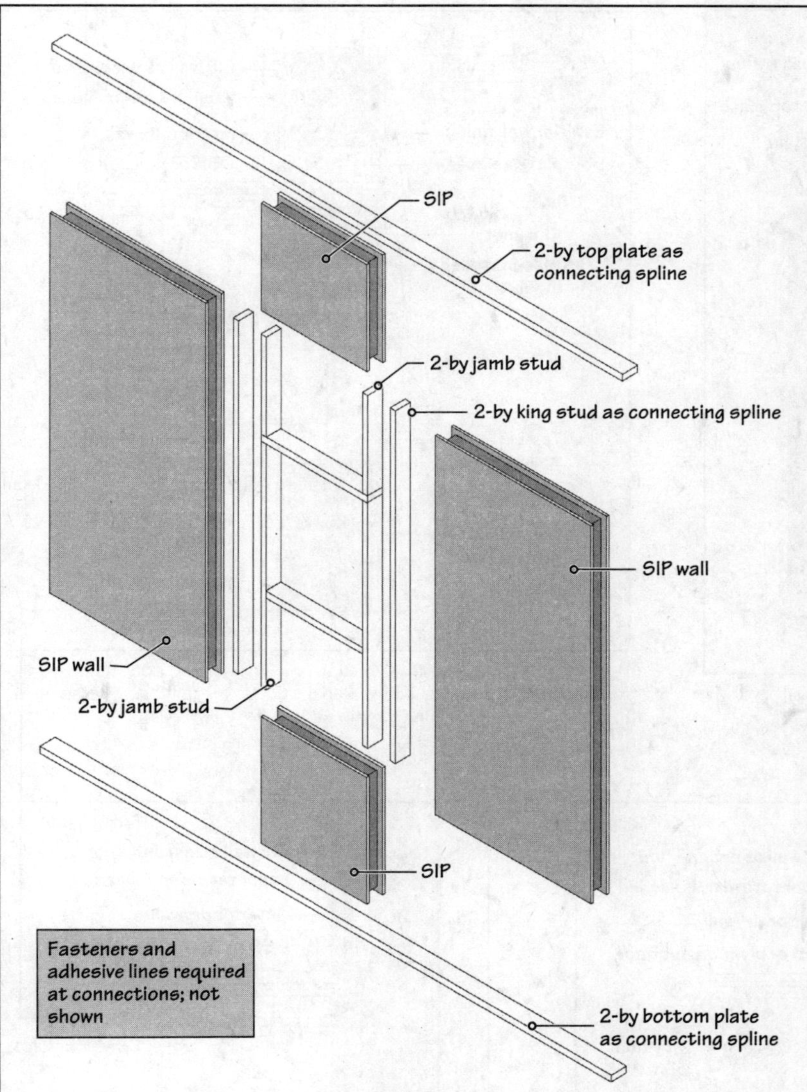

FIGURE 17.9 A typical detail of framing around a window. Framing around a door is similar.

17.3 SIP FLOOR ASSEMBLIES

As stated earlier, an intermediate floor in a SIP building is framed with 2-by lumber joists, engineered wood members (e.g., wood I-joists), or trussed joists. The floor joists may be placed on the wall panel and fastened to the top plate spline using a detail similar to that used in a conventional wood frame building, Figure 17.11.

The detail of Figure 17.11 creates a thermal bridge at the rim joist. Therefore, some SIP manufacturers recommend the detail shown in Figure 17.12, in which the floor joists are hung off the top plate spline using joist hangers. The joist hangers are top-supported, that is, fastened to the top plate in the SIPs. Where the SIPs are continuous, that is, two floors high, the floor joists are supported on joist hangers that are fastened to a ledger beam, Figure 17.13.

17.4 SIP ROOF ASSEMBLIES

Two alternative roof assemblies can be used in a SIP building. One alternative is to use any one of the various assemblies that are used in a conventional wood frame building—that is, a stick-frame rafter-ceiling assembly or wood roof truss assembly. In these assemblies, insulation is placed in the attic.

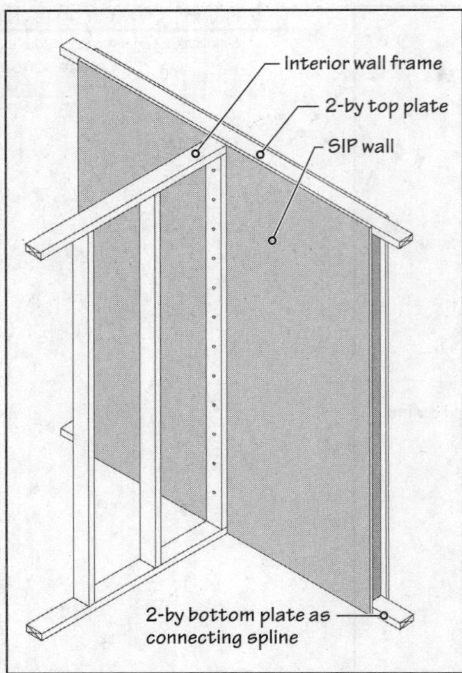

FIGURE 17.10 Connection of an interior wall to a SIP wall.

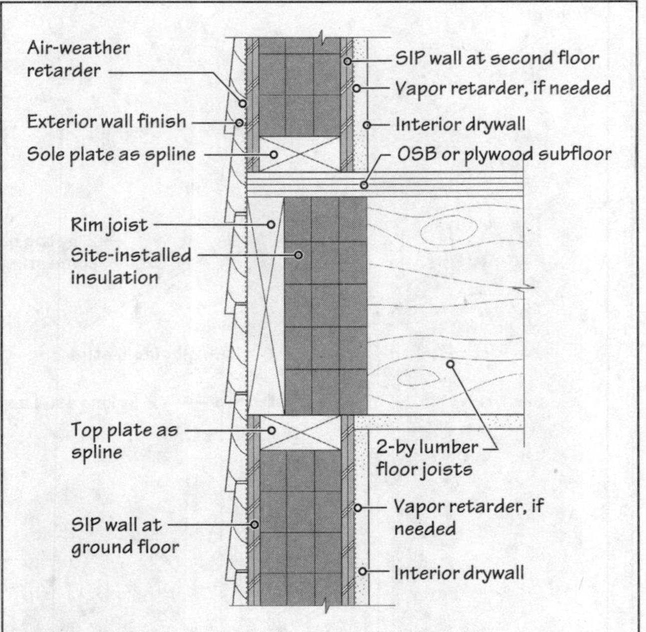

FIGURE 17.11 Detail at the junction of a SIP floor and a SIP wall. The detail is similar to that used in a conventional wood frame building. Site-installed insulation is needed behind the rim joist.

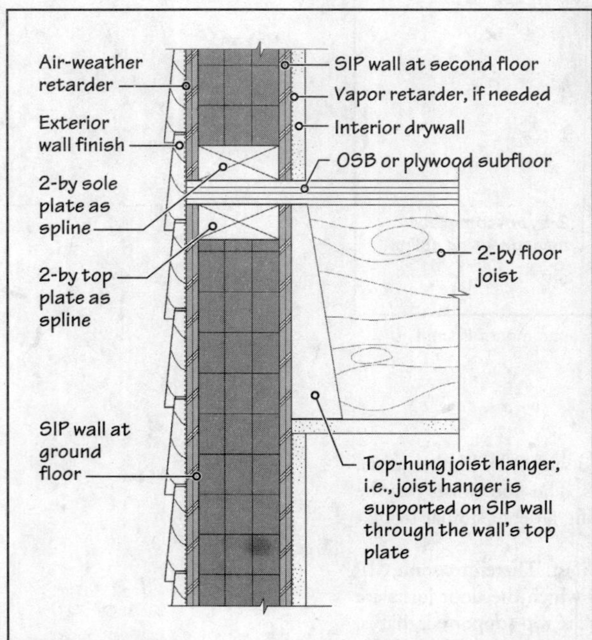

FIGURE 17.12 Detail at the intersection of a floor and a SIP wall—an alternative to the detail in Figure 17.11. In this detail, site-installed insulation is not needed.

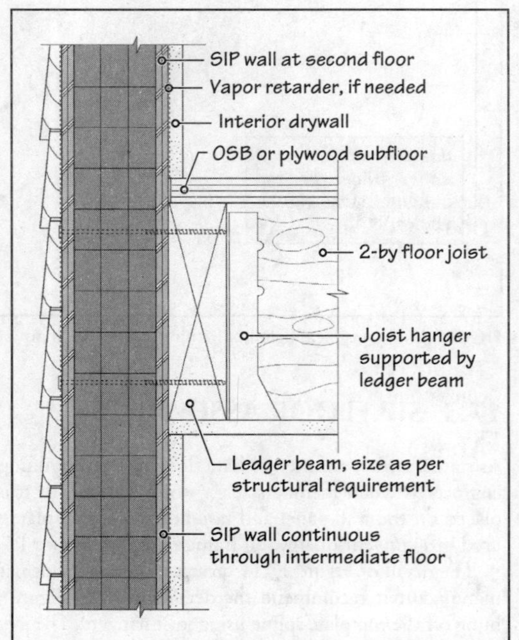

FIGURE 17.13 Detail at the intersection of a floor and a SIP wall where the SIP wall is continuous through the floor.

The second alternative is to use a SIP roof, where the roof SIPs span from the ridge beam to the exterior wall, creating a cathedral ceiling. Typical details of such an alternative are shown in Figure 17.14.

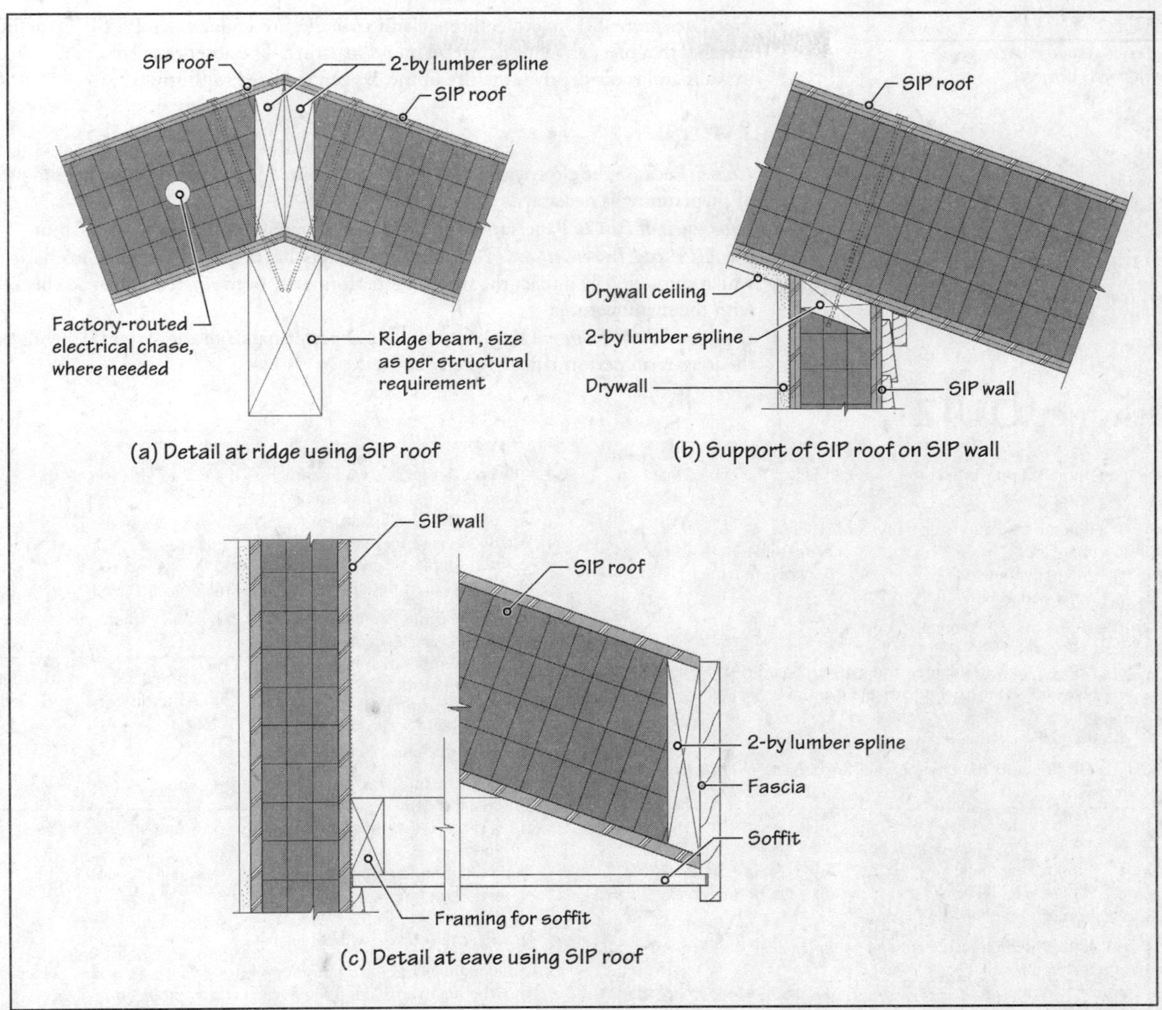

FIGURE 17.14 Details at the ridge and eave using roof and wall SIPs.

17.5 ADVANTAGES AND LIMITATIONS OF SIPS

The SIP system of construction has several advantages and limitations compared with the conventional WLF system.

ADVANTAGES

- *Panelization and Deletion of Insulation Subtrade:* As stated at the beginning of this chapter, panelization and deletion of the insulation subtrade in a SIP structure increase the speed and efficiency of construction.
- *Air-Weather Retarders:* Because the panels are relatively airtight, air-weather retarders are needed only for their water-resistive properties.
- *Continuous Nailable Surface:* Because OSB facings provide a continuous nailable surface, interior drywall and exterior wall finishes can be applied without having to locate the studs.
- *Pneumatic Nailers:* Although several details require the use of screws, pneumatic nailers are used extensively.
- *On-site Waste:* Very little on-site waste is produced in a SIPs structure because most members arrive at the site precut to size.
- *Energy Efficiency:* One of the major advantages of SIP is their energy efficiency. A SIP structure has virtually no thermal bridges in the envelope. The use of foam core allows no air movement within the insulation, unlike that in fiberglass insulation.

- *Fire Endurance:* SIP manufacturers claim that the fire endurance of a SIP structure matches that of a conventional wood frame structure. The absence of concealed spaces in walls and roofs deprives the fire of the oxygen required for burning.

LIMITATIONS

- *Chases:* Because the electrical chases must be prerouted in the panels, a greater amount of preplanning is necessary.
- *Squareness of Panels:* Panels must be absolutely square for a successful installation.
- *Termites and Insect Attack:* Termites and insects boring through foam insulation, which can seriously impact the structural performance of the system, must be checked with the manufacturer.
- *Long-term Performance:* Long-term structural performance of the system depends on the long-term performance of the adhesives.

PRACTICE **QUIZ**

Each question has only one correct answer. Select the choice that best answers the question.

1. In a SIP, the insulation generally consists of
a. fiberglass. **b.** extruded polystyrene.
c. polyisocyanurate. **d.** polyurethane.
e. expanded polystyrene.

2. The facing in a SIP generally consists of
a. OSB on both faces.
b. OSB on the exterior face and gypsum board on the interior face.
c. plywood on the exterior face and gypsum board on the interior face.
d. plywood on the exterior face and OSB on the interior face.
e. none of the above.

3. In a SIP structure, the interior walls are constructed of 4 ft × 8 ft SIP panels.
a. True **b.** False

4. In a SIP building, exterior wall panels are typically
a. 6 in. or 7 in. thick. **b.** $5\frac{1}{2}$ in. or 6 in. thick.
c. 5 in. or 6 in. thick. **d.** 4 in. or 6 in. thick.
e. $4\frac{1}{2}$ in. or $6\frac{1}{2}$ in. thick.

5. Adjacent structural insulated wall panels in a SIP building are connected through
a. 2-by lumber studs. **b.** engineered lumber splines.
c. OSB splines. **d.** any one of the above.
e. special sheet metal connectors.

6. A corner between structural insulated wall panels in a SIP building requires a minimum of
a. four 2-by lumber studs.
b. three 2-by lumber studs.
c. two 2-by lumber studs.
d. one 2-by lumber stud.
e. 2-by lumber studs are not required.

7. The connection of a wall panel to a concrete foundation, as shown in this text, requires the use of
a. one treated sill plate.
b. two treated sill plates.
c. two sill plates, one of which is treated.
d. a sill plate is not needed in a SIP structure.

8. In a SIP structure, an intermediate floor may be framed with any one of the following except
a. 2-by lumber members.
b. wood I-joists.
c. wood trussed joists.
d. SIP panels.
e. all of the above (without exception).

9. In a SIP structure, the roof may be framed with any one of the following except
a. a rafter-and-ceiling-joist assembly consisting of 2-by lumber members.
b. a rafter-and-ceiling-joist assembly consisting of wood I-joists.
c. wood roof trusses.
d. SIP panels.
e. all of the above (without exception).

10. A SIP manufacturer generally supplies standard-size panels, which must be cut to size at the site as required by project drawings.
a. True **b.** False

11. Similar to the conventional WLF structure, the interior gypsum board wall finish must be nailed to the studs in a SIP structure.
a. True **b.** False

12. The maximum size of available SIP panels is
a. 4 ft × 8 ft. **b.** 4 ft × 10 ft.
c. 8 ft × 8 ft. **d.** 8 ft × 16 ft.
e. 8 ft × 24 ft.

REVIEW QUESTIONS

1. Using a sketch and notes, explain the anatomy of a SIP. In what thickness are the panels commonly used?
2. Using sketches and notes, show two alternative ways in which two adjacent SIP wall panels are joined.
3. Using a sketch and notes, explain the support of SIP walls on a slab-on-grade foundation.
4. Explain why SIPs are not recommended for an intermediate floor but can be used for roofs.
5. List the two most important advantages and the two most important disadvantages of SIP construction in comparison with WLF construction.

CHAPTER 18 Material Steel and Steel Components

CHAPTER OUTLINE

Because of its extensive use in industry, construction, and weaponry, iron is by far the most important of all metals. That is why metals are generally divided into two categories:

- Ferrous metals—metals that contain iron (the Latin term for iron is *ferrum*)
- Nonferrous metals—metals that do not contain iron (e.g., aluminum and copper)

Steel is the most important ferrous metal. Its high strength in relation to its weight makes it the material of choice for skyscrapers and long-span structures, such as sports stadiums and bridges. Its malleability and weldability allow it to be shaped, bent, and made into different types of components. These characteristics provide the versatility that architects and engineers have exploited in creating a wide range of highly expressive structures, as exemplified by Figures 18.1 and 18.2.

The history of development of steel has been long and arduous because the fuel technology necessary to produce temperatures high enough to melt iron ore on a large scale was not available until the early eighteenth century. The melting point of iron is 2,800°F (1,540°C). By contrast, copper's melting point is approximately 2,000°F (1,100°C)—a temperature that could be obtained in a wood or charcoal fire. That is why the use of copper and, later, bronze (an alloy of copper and tin) preceded that of steel by several centuries.

Wrought Iron—The Earliest Form of Iron

The earliest predecessor of steel was wrought iron. It was produced by heating iron ore in a charcoal fire. Though insufficient to melt the ore, a charcoal fire was sufficient to extract the oxygen from the ore, which consists of iron oxide mixed with some sand (silicon dioxide). In this reaction, carbon dioxide and carbon monoxide are driven off, and iron oxide is

18.1 MAKING OF MODERN STEEL

Extracting carbon from cast iron to obtain the so-called perfect iron-carbon alloy was a major metallurgical problem of the nineteenth century. It was finally resolved by the development of the basic oxygen furnace (BOF) by Henry Bessemer in 1855. In the BOF, the excess carbon in molten pig iron is converted to carbon dioxide by blowing oxygen through it, leaving behind that perfect iron-carbon alloy—steel. After the discovery of the BOF, the electric arc furnace was developed, which obtains the same end result as the BOF.

MAKING STEEL FROM VIRGIN IRON ORE—THE INTEGRATED MILL METHOD

The traditional method of making steel is by using an *integrated mill*, so called because it converts the iron ore into a finished steel product (e.g., steel I-beams) by using a number of intermediate processes—all integrated into one large complex. The most important part of an integrated mill is the blast furnace, a huge steel cylinder whose interior is lined with refractory bricks, Figure 18.3.

The raw materials—iron ore, coke, and limestone—are charged into the furnace from the top in alternate layers so that the three materials are mixed together. Limestone acts as a flux. A flux is a material that removes impurities and reduces the melting point of the main raw material—iron ore. Hot air from the stove is fed into the furnace from the bottom of the stack, which ignites the coke. As the iron ore and limestone melt, they travel down the stack and settle at the bottom of the furnace. The molten material consists of two parts: molten iron and molten slag.

The *slag* is essentially molten limestone, but it also contains molten sand and other elements present in iron ore. Being lighter, molten slag floats over molten iron. Therefore, both slag and molten iron are drained separately from the bottom of the furnace.

Molten iron from the blast furnace is transported to BOF. Here oxygen is blown into molten iron, which converts the carbon (in iron) to carbon dioxide. At this stage, small quantities of a few other metals, such as copper, nickel, chromium, and so on, are added to the furnace to obtain the steel with the desired chemistry. The molten steel goes through continuous casting that yields long, rectangular shaped members, called *billets*, *blooms*, or *slabs*, distinguished by their cross-sectional dimensions.

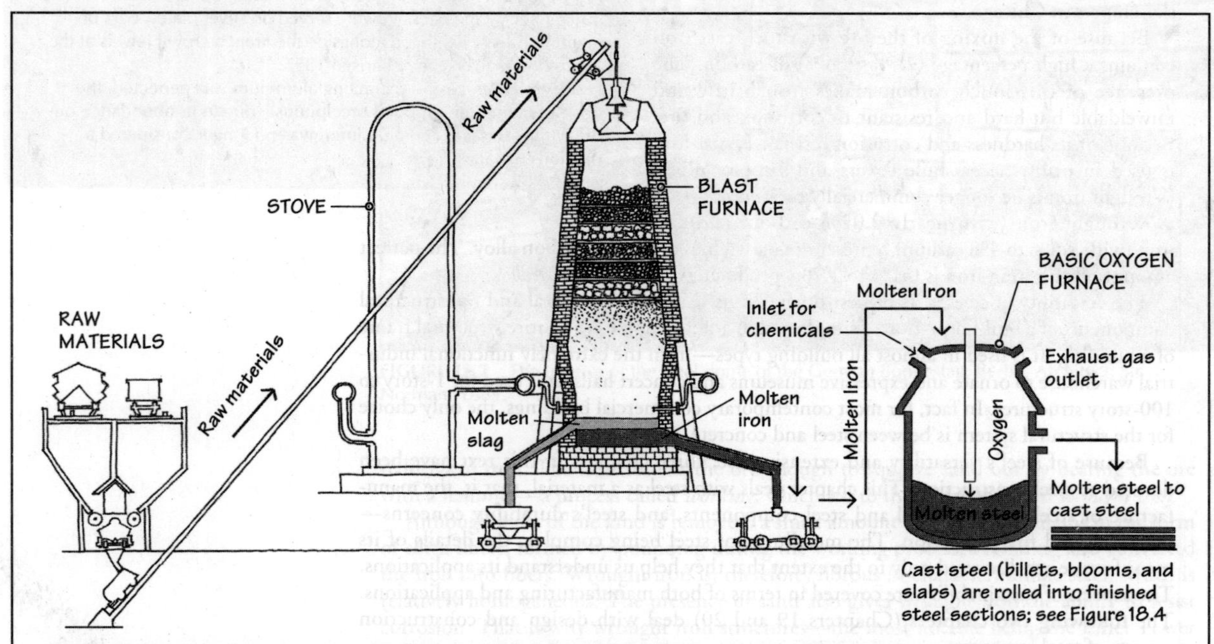

FIGURE 18.3 This illustration shows the process of manufacturing steel in an integrated mill. The blast furnace and the basic oxygen furnace (BOF) are the two (of the three) major components of an integrated mill. Raw materials (iron ore, limestone, and coke) are fed into the blast furnace, which yields molten iron. Molten iron is used as a charge for the BOF. In the BOF, oxygen is blown into molten iron, which converts the carbon in iron to carbon dioxide, leaving behind molten steel. Molten steel is cast into billets, blooms, and slabs for rolling into finished steel sections. A rolling mill is the third major component of an integrated mill.

Billets, blooms and slabs are reheated and rolled into the finished steel cross sections. Billets are of smaller cross section than blooms and are used for making I-sections, C-sections, angles, bars, and rods. Blooms are used for heavier I-sections. From slabs, steel plates and sheets are obtained, Figure 18.4.

SUSTAINABLE STEEL MANUFACTURING—THE MINI-MILL METHOD

A method increasingly used for making steel is one that recycles steel-based scrap from automobiles, transport vehicles, machinery, and so on, instead of using the virgin ore. This requires a much smaller outlay compared to the integrated mill, and the related setup is called the *mini mill.* The steel obtained from a mini mill is more economical and also more sustainable than that obtained from an integrated mill. Consequently, the number of integrated mills has declined over the years. However, they continue to be used because of the increase in steel demand and also because some specialty steel products can only be made in an integrated mill.

In the mini-mill method, steel scrap is charged from the top into a large furnace called an *electric arc furnace* (EAF). An EAF has three graphite electrodes that produce a high-temperature arc that converts the scrap into molten steel—three electrodes because of the three-phase electric power, Figure 18.5. Calcium oxide (lime) and magnesium oxide are fed into the EAF from a side door as needed. These materials form a foamy slag on melting. The foam insulates the furnace lining from the high temperature of the arc.

The molten steel from the EAF is transported to a ladle (a refractory-lined vessel) where the steel chemistry is modified by adding the required chemicals. From the ladle, molten steel goes through continuous casting to billets, blooms, or slabs and is rolled into the desired cross sections in exactly the same way as in an integrated mill. Some mini mills cast billets into dog-bone cross-sectional profiles from which I-section or H-section members are made. These billets are called *near-net shape billets* or *beam blanks,* Figure 18.6. Rectangular shape billets, Figure 18.7, are used for making angles, bars, and rods.

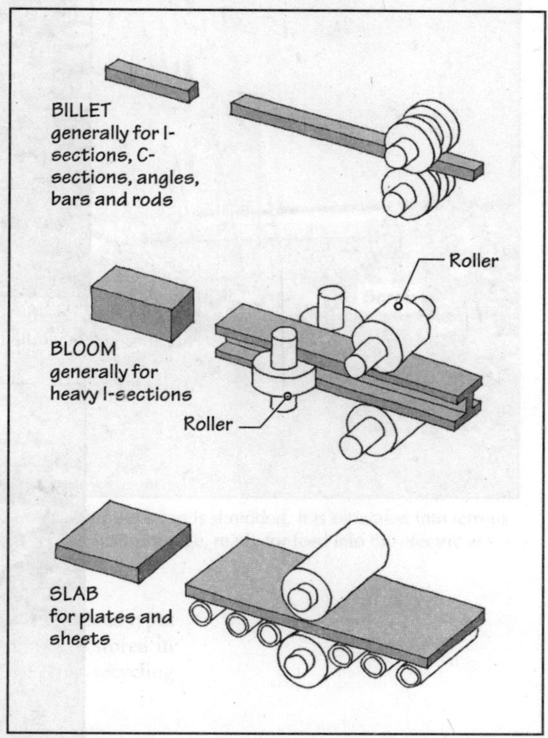

FIGURE 18.4 Process of hot rolling to obtain finished steel sections. In this process, billets, blooms, and slabs, obtained from the basic oxygen furnace, are reheated to virtual softening and rolled to the required cross sections as shown.

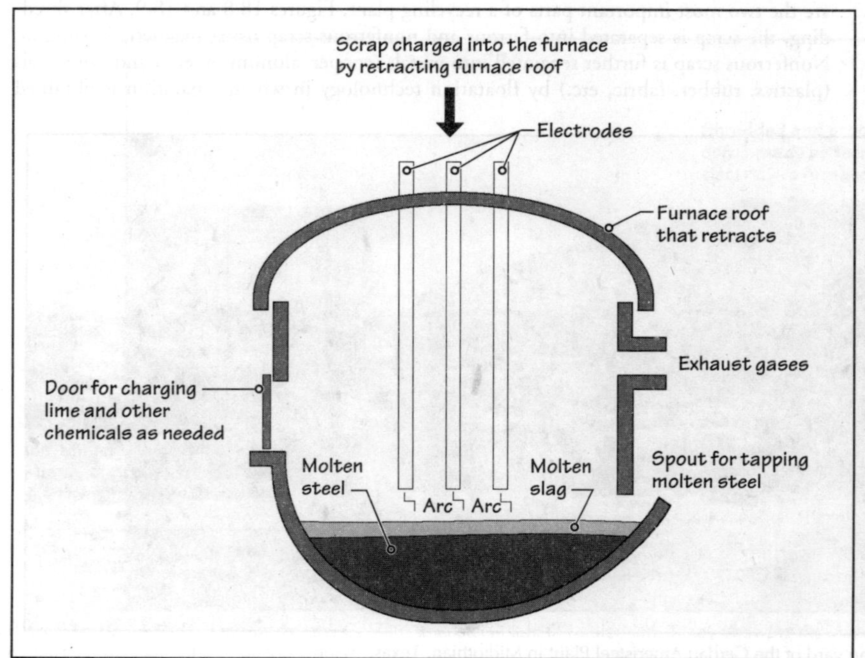

FIGURE 18.5 An electric arc furnace (EAF) converts scrap steel into molten steel. This requires the use of lime and other chemicals, which are fed into the furnace as and when needed through a side door. The furnace has a retractable roof to allow the charging of steel scrap. Molten steel from the EAF is sent to a vessel (called a *ladle*) to modify its chemistry as needed. Molten steel from the ladle is cast into billets, blooms, and slabs, and subsequently reheated and rolled into finished cross sections.

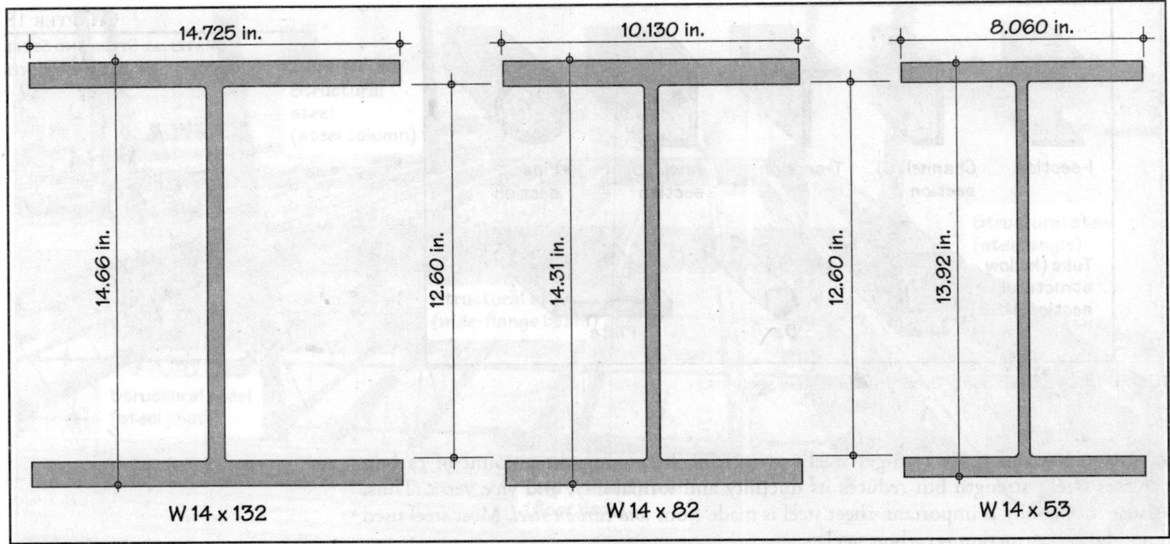

FIGURE 18.16 Most wide-flange sections that share a common nominal depth designation have the same interior flange-to-flange dimension. This illustration shows three sections, all with a nominal depth of 14 in. Their overall depths are nearly 14 in., and several other dimensions are different from each other. However, the interior flange-to-flange dimensions are 12.60 in. for all sections. Several other sections with a nominal depth of 14 in. are available (see Table 18.1).

Another difference between W- and S-shapes is that, for the same depth, the flanges in a W-shape are wider than those in an S-shape. That is why a W-shape is called a *wide-flange section.* Because of the parallel flange surfaces and wide flanges, W-shapes are the ones most often used for beams and columns. S-shapes (also called *American standard* shapes) are used in cranes and rails.

NOTE

Nominal Depths of W-Shapes

W44, W40, W36

W33, W30, W27, W24, W21, W18

Each W-shape is designated by two numbers: the first number refers to the nominal depth of the section (in inches), and the second number gives its weight for a 1-ft length (pounds per foot). For example, W21 × 68 means that the nominal depth of the section is 21 in. and its nominal weight is 68 lb/ft. The nominal depth is the section's approximate (not exact) depth.

W-shapes are available in a large range of sizes and weights, from W40 × 503 to W4 × 13. A few manufacturers also make 44-in.-deep W-shapes. Several W-shapes share a common nominal depth but have different flange widths, flange thicknesses, and web thicknesses. Sections such as W14 × 132, W14 × 82, and W14 × 53 have different weights but the same nominal depth.

The interior flange-to-flange dimensions of most W-shapes sharing the same nominal depth designation are equal. For example, as shown in Figure 18.16, W14 × 132, W14 × 82, and W14 × 53 have two dimensional facts in common: (a) their interior flange-to-flange dimension is the same (12.60 in.), and (b) their overall depths are approximately 14 in. All other dimensions of these sections are different from each other. Generally, W-shapes with squarish proportions (depth and flange width nearly equal), such as W14 × 132, are used as columns. Tall, narrow sections, such as W14 × 53, are used as beams.

An HP-shape is so called because it is generally used as a bearing pile (see Sections 11.6 and 12.3) and is squarish in proportion; that is, its overall depth and flange width are nearly equal, resembling the letter *H.* The thickness of the web and flanges are equal in an HP-shape. An M-shape is one that cannot be classified as W, S, or HP; the letter *M* is an acronym for *miscellaneous.*

Dimensional information and engineering properties of various structural steel shapes are available in the *Steel Construction Manual* published by the American Institute of Steel Construction (AISC). Table 18.1 gives an indication of some of the information available in this publication.

C-SHAPES

Steel channels (C-shapes) are similar in profile to S-shapes, that is, their inner flange surfaces are inclined at an angle of 2:12, Figure 18.17(a). They are designated by two numbers after the letter C. The first number gives the overall depth of the section, and the second

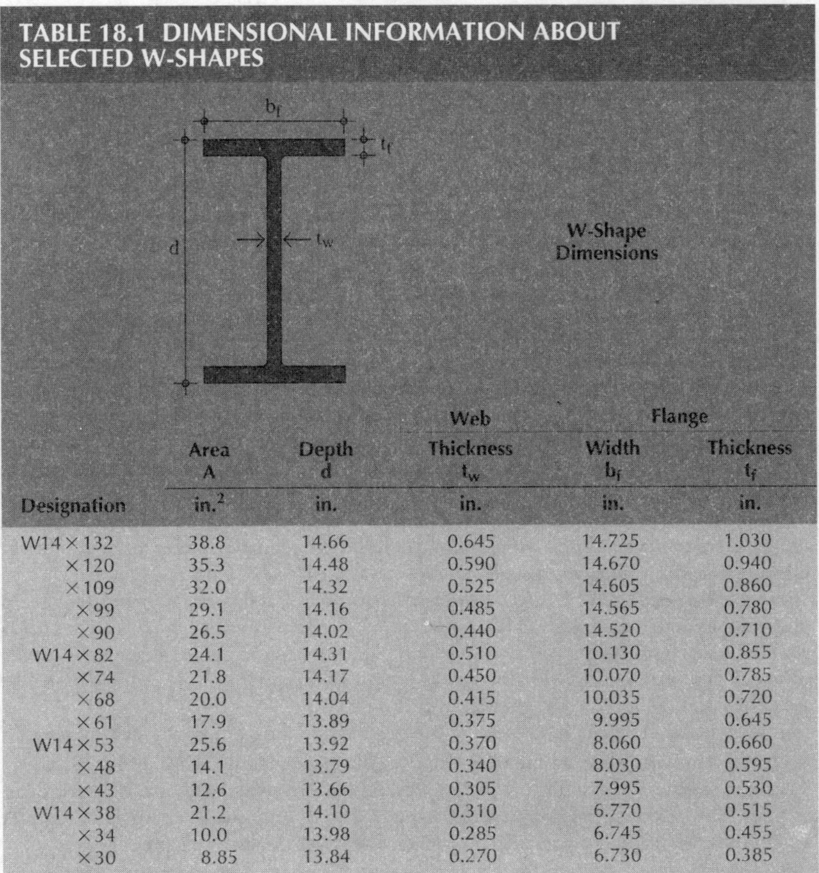

TABLE 18.1 DIMENSIONAL INFORMATION ABOUT SELECTED W-SHAPES

Designation	Area A in.^{2}	Depth d in.	Web Thickness t_w in.	Flange Width b_f in.	Flange Thickness t_f in.
W14×132	38.8	14.66	0.645	14.725	1.030
×120	35.3	14.48	0.590	14.670	0.940
×109	32.0	14.32	0.525	14.605	0.860
×99	29.1	14.16	0.485	14.565	0.780
×90	26.5	14.02	0.440	14.520	0.710
W14×82	24.1	14.31	0.510	10.130	0.855
×74	21.8	14.17	0.450	10.070	0.785
×68	20.0	14.04	0.415	10.035	0.720
×61	17.9	13.89	0.375	9.995	0.645
W14×53	25.6	13.92	0.370	8.060	0.660
×48	14.1	13.79	0.340	8.030	0.595
×43	12.6	13.66	0.305	7.995	0.530
W14×38	21.2	14.10	0.310	6.770	0.515
×34	10.0	13.98	0.285	6.745	0.455
×30	8.85	13.84	0.270	6.730	0.385

Source: This table has been adapted from *Steel Construction Manual* (13th ed.), published by the American Institute of Steel Construction (AISC), with permission.

The table is an abridged version. A complete version includes additional dimensions and information about the engineering properties of each section, such as the moment of inertia, radius of gyration, and so on.

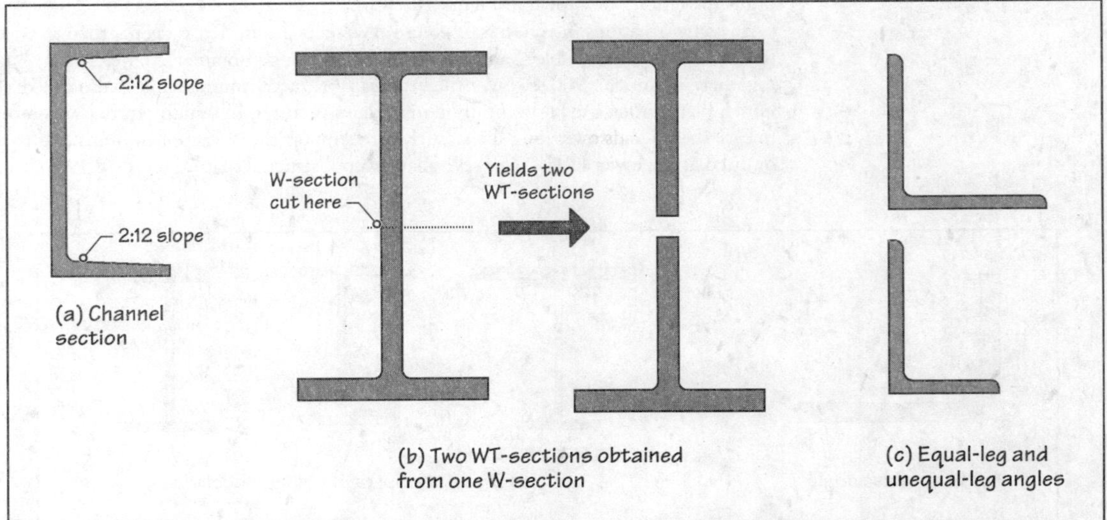

FIGURE 18.17 **(a)** Channel section profile. **(b)** WT-sections obtained from a W-section. **(c)** Equal-leg and unequal-leg angles.

number gives the weight of a 1-ft length of the section. Thus, C8 × 11.5 means that the channel is 8 in. deep and weighs 11.5 lb/ft.

Miscellaneous channels do not have a standard slope on the inner flange surfaces. They are designated in the same way as C-shapes, for example, MC12 × 50.

T-SHAPES

A T-shape section is made by splitting a W-shape, M-shape, or S-shape into two equal parts, Figure 18.17(b). It is, therefore, called WT, MT, or ST, depending on its origin. For example, two WT6 × 29 sections are obtained from one W12 × 58.

L-SHAPES, PIPES, TUBES, BARS, AND PLATES

Steel angles (L-shapes) may either be equal-leg angles or unequal-leg angles, Figure 18.17(c). The thickness of both legs is the same in an angle. Angles are designated by three numbers. The first two numbers give the length of each leg, and the third number gives the thickness of the legs. L4 × 4 × $\frac{1}{2}$ is an example of an equal-leg angle with legs equal to 4 in. each; the thickness of each leg is $\frac{1}{2}$ in. In an unequal-leg angle, the longer leg is mentioned first, as in L4 × 3 × $\frac{1}{4}$. Angles have various uses, such as for masonry lintels and members of steel trusses.

Pipes are designated by their nominal diameter and by whether the pipe is a *standard weight, extra strong,* or *double-extra strong.* These three designations refer to the pipe's wall thickness. Pipes are generally used as columns or as members of a truss.

A tube is referred to as a *hollow structural section* (HSS) and is made by bending a steel plate and welding it seamlessly. That is why the edges of a tube are rounded (Figure 18.12). An HSS may be square, rectangular, or round. Square or round HSSs are generally used as columns, and rectangular HSSs are used as beams. Like pipe trusses, HSS member trusses are fairly common for long-span structures.

A rectangular or square HSS member is designated by three dimensions, all expressed in fractional numbers. Thus, an HSS 12 × 8 × $\frac{1}{2}$ is a rectangular HSS measuring 12 in. × 8 in. on the outside; the wall thickness is $\frac{1}{2}$ in. A round HSS is designated by the outside diameter and wall thickness, both dimensions expressed to three decimal places. Thus, an HSS 10.000 × 0.500 has an outer diameter of 10 in. and a wall thickness of $\frac{1}{2}$ in.

BENT-PLATE SECTIONS AND BUILT-UP SECTIONS

When size and other requirements dictate the use of nonstandard shapes and (or) cross-sectional dimensions, they can be obtained by (a) bending a steel plate to the required profile, called *bent-plate sections,* or (b) welding two or more standard sections together, called *built-up sections.* The most commonly used bent-plate profile is an angle. Bent-plate angles have a rounded corner, whereas a standard-size (hot-rolled) angle has a sharp corner, Figure 18.18(a). Bent-plate angles are used where an angle with odd dimensions (not available as a standard angle) is mandated by the detail (see Figure 28.33). A few examples of built-up sections are shown in Figure 18.18(b).

In some situations, it is necessary to use a W-shape beam that is deeper than the deepest (44 in.) W-shape available. Such a built-up beam can be obtained by welding steel plates together, Figure 18.19(a). Because the beam is fabricated from plates, it is also called a *plate girder.* Plate girders are fairly common in bridge structures. In buildings, they are employed to carry heavy loads over long spans, such as to support the weight of columns that must be omitted at the lower levels to achieve a larger unobstructed space, Figure 18.19(b).

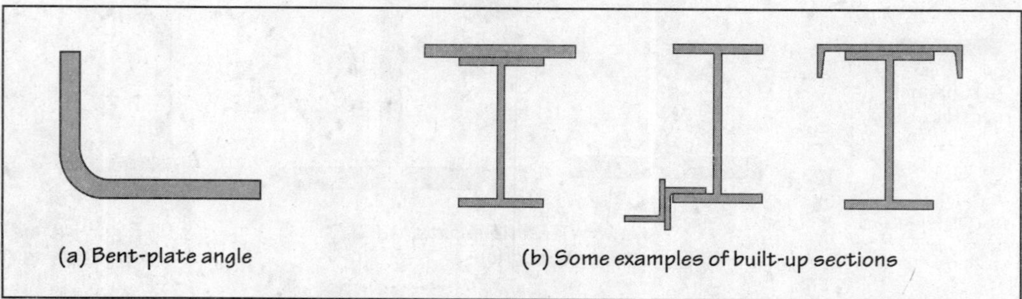

FIGURE 18.18 Bent-plate angle and built-up sections.

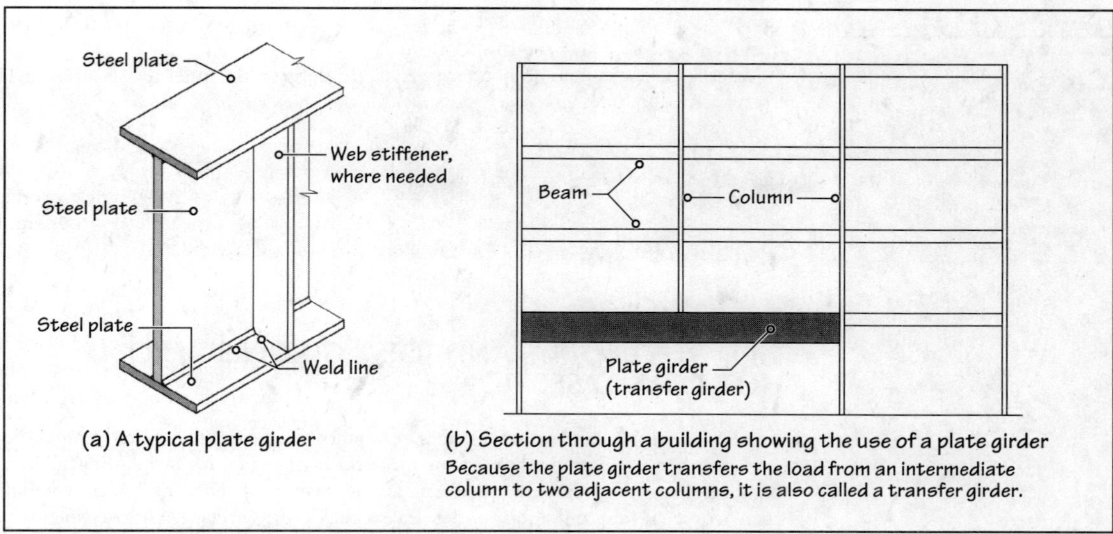

FIGURE 18.19 **(a)** Anatomy of a typical plate girder. **(b)** One of the several uses of a plate girder.

PRACTICE QUIZ

Each question has only one correct answer. Select the choice that best answers the question.

1. Which of the following ferrous metals has the largest amount of carbon?
   - **a.** Steel
   - **b.** Wrought iron
   - **c.** Cast iron
2. Which of the following ferrous metals is least corrosion resistant?
   - **a.** Steel
   - **b.** Wrought iron
   - **c.** Cast iron
3. An integrated mill produces steel using
   - **a.** steel scrap.
   - **b.** iron ore.
   - **c.** 25% to 50% iron ore plus 25% to 50% scrap.
   - **d.** none of the above.
4. The mini-mill method of steel manufacturing predates the integrated-mill method.
   - **a.** True
   - **b.** False
5. An electric arc furnace uses
   - **a.** two electrodes to melt iron ore.
   - **b.** two electrodes to melt steel scrap.
   - **c.** three electrodes to melt iron ore.
   - **d.** three electrodes to melt steel scrap.
   - **e.** none of the above.
6. Slag is
   - **a.** a waste product generated from the manufacture of steel.
   - **b.** the ore used for making steel.
   - **c.** the ore used for making aluminum.
   - **d.** a ferrous metal that has been replaced by steel.
   - **e.** none of the above.
7. When steel structures first appeared, they were called
   - **a.** skeleton frame.
   - **b.** skeleton cage.
   - **c.** steel skeleton.
   - **d.** all of the above.
   - **e.** none of the above.
8. Which of the following statements is true?
   - **a.** All structural steel sections are made from billets or blooms that are rolled to shape at room temperature.
   - **b.** All structural steel sections are made by preheating steel sheets and bending them to the required cross sections.
   - **c.** All structural steel sections are made by bending steel sheets to the required cross sections at room temperature.
   - **d.** All structural steel sections are made by preheating billets or blooms and rolling them to shape.
   - **e.** none of the above.
9. Steel decks are made from cold-formed steel.
   - **a.** True
   - **b.** False
10. The steel type most commonly used for framing members (columns and beams) of a steel building is
    - **a.** steel type A36 with a yield strength of 36 ksi.
    - **b.** steel type A36 with a yield strength of 360 ksi.
    - **c.** steel type A992 with a yield strength of 50 ksi.
    - **d.** steel type A992 with a yield strength of 40 ksi.
    - **e.** none of the above.
11. A W-shape section is so called because
    - **a.** it is an I-section with wide flanges.
    - **b.** it is an I-section with a wide web.
    - **c.** its cross-sectional shape resembles the letter *W*.
    - **d.** it is a rectangular section whose width is larger than its depth.
12. W-shape, S-shape, HP-shape, and M-shape are all structural steel I-sections.
    - **a.** True
    - **b.** False
13. Which of the following steel shapes is most commonly used for the structural frame of buildings?
    - **a.** HP-shape
    - **b.** S-shape
    - **c.** M-shape
    - **d.** W-shape
    - **e.** C-shape
14. The designation W14 × 38 means that the
    - **a.** width of the section is 14 in. and the depth is 38 in.
    - **b.** depth of the section is 14 in. and the width is 38 in.
    - **c.** depth of the section is 14 in. and its weight is 38 lb/ft.
    - **d.** nominal depth of the section is 14 in. and its weight is 38 lb/ft.
    - **e.** none of the above.
15. To obtain the structural engineering properties of standard steel sections, the following publication is generally used.
    - **a.** *Steel Fabricators' Manual*
    - **b.** *Steel Construction Manual*
    - **c.** ASTM standard on steel sections
    - **d.** Local manufacturers' literature
    - **e.** *Steel Erectors' Manual*

(*Continued*)

PRACTICE QUIZ (Continued)

16. In a steel angle, both legs must have the same thickness but may be of different lengths.
    a. True
    b. False

17. The acronym HSS stands for
    a. heat-strengthened steel.
    b. high-strength steel.
    c. high-strength section.
    d. hollow steel section.
    e. none of the above.

18. HSS members are made by
    a. bending and welding heated steel plates.
    b. cold rolling of steel blooms.
    c. hot rolling of steel blooms.
    d. casting molten steel to shape.
    e. any one of the above, depending on the manufacturer.

18.5 STEEL JOISTS AND JOIST GIRDERS—FROM HOT-ROLLED SECTIONS

In addition to the standard steel shapes and built-up sections, two types of prefabricated steel members are commonly used for roof and floor structures in buildings. These are trusslike, open-web members, called *joists* and *joist girders*. There is no fundamental difference between a joist and a joist girder, except that a joist girder is a heavier member and spans from column to column, whereas a joist is a lighter member that spans between the girders, Figure 18.20.

Joists and joist girders are fabricated by manufacturers according to the specifications prepared by the Steel Joist Institute (SJI). They are generally recommended for use in light, uniformly loaded roofs and floors.

STEEL JOISTS

The Steel Joist Institute classifies joists into three categories:

- K-series joists (joist depth ranges from 8 in. to 30 in.)
- LH-series joists (joist depth ranges from 18 in. to 48 in.)
- DLH-series (joist depth ranges from 52 in. to 72 in.)

K-series joists are more frequently used because they fall within the range of loads and spans that are common in buildings. Manufacturers producing K-series joists must comply with SJI's specifications, which, for the most part, are performance-oriented, giving a measure of freedom to manufacturers to select the members' sizes and shapes. However, a joist bearing the standard SJI designation must have a minimum specified load-carrying capacity.

FIGURE 18.20 A steel frame structure showing the use of steel joists and steel joist girders. Observe that a joist girder spans from column to column. A joist, on the other hand, spans from joist girder to joist girder.

A joist bears on a girder only through its top chord; the bottom chord of the joist is free (see Figures 18.21(e) and (f)). Both the top and bottom chords of a joist girder, on the other hand, are connected to the columns at both ends—a connection that provides lateral stability to the structure (i.e., resistance to lateral loads).

Figure 18.21 shows a typical elevation, section, and other important details of a K-series joist. Each joist bears a designation, consisting of two numbers sandwiching the letter *K*. The first number gives the depth of joists in inches, and the second number gives the relative stiffness of the joist. The designation system is further explained in Figure 18.21.

The load-carrying capacities of various K-series joists are tabulated as shown in Table 18.2. Thus, in practice, the joists are not structurally designed but are simply selected from SJI's load tables based on the dead load and live load that they are required to support.

LH- and DLH-series joists are designated similarly to K-series joists, for example, 24LH10 and 56DLH16. The

NOTE

Joist Series' Nomenclature

Until 1986, SJI had three series of joists: H-series, LH-series, and DLH-series. H-series joists (ranging from 8 in. to 30 in. deep) were most frequently used because their depths and load-carrying capacities were within the range commonly required. LH-series and DLH-series joists, referred to as *long-span* and *deep long-span* joists, respectively, were intended for spans and loads not covered by H-series joists.

In 1986, SJI introduced K-series joists, which replaced the earlier H-series joists primarily because of the savings resulting from the improved design concepts.

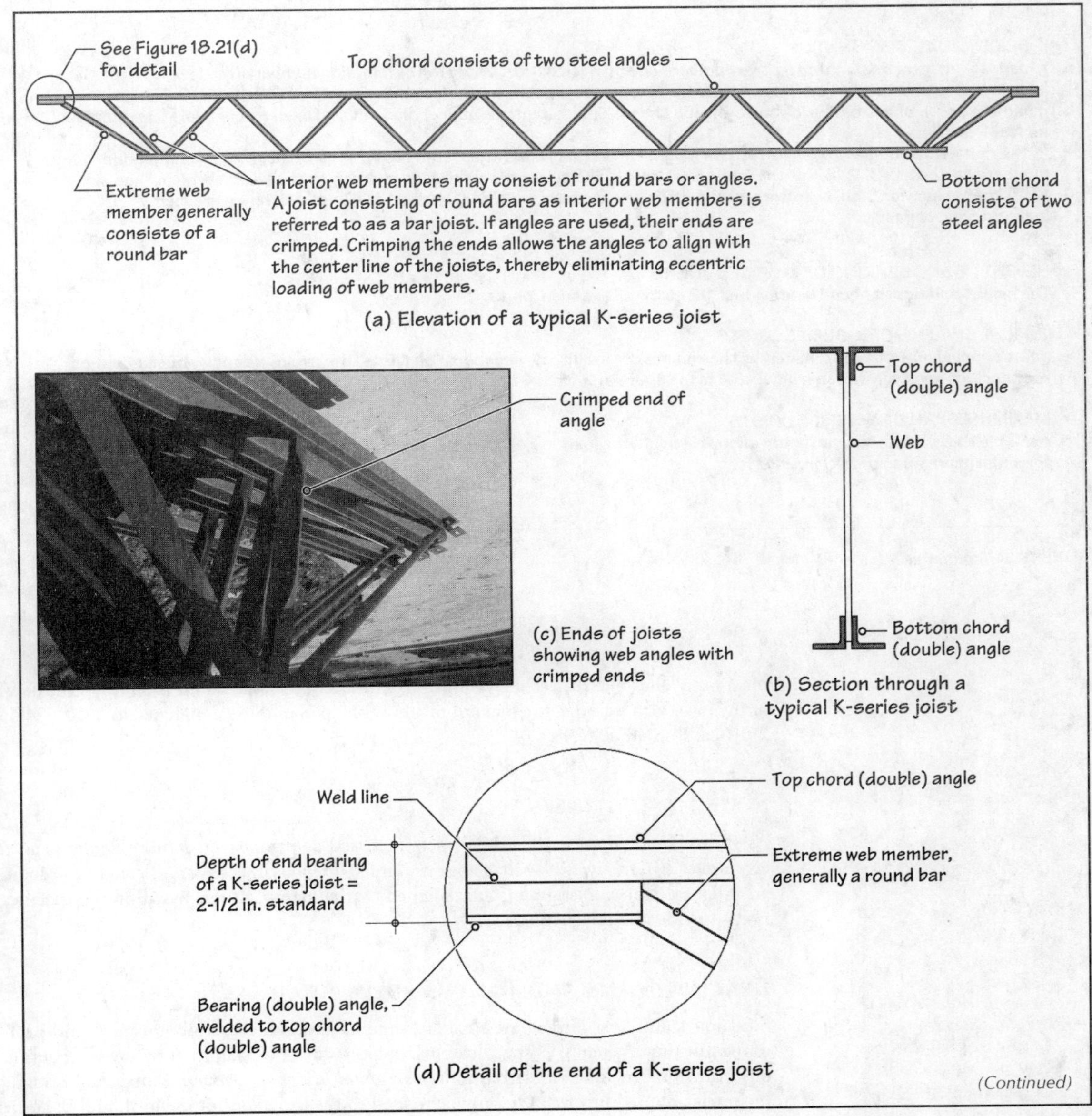
See Figure 18.21(d) for detail
Top chord consists of two steel angles
Extreme web member generally consists of a round bar
Interior web members may consist of round bars or angles. A joist consisting of round bars as interior web members is referred to as a bar joist. If angles are used, their ends are crimped. Crimping the ends allows the angles to align with the center line of the joists, thereby eliminating eccentric loading of web members.
Bottom chord consists of two steel angles
(a) Elevation of a typical K-series joist
Crimped end of angle
(c) Ends of joists showing web angles with crimped ends
Top chord (double) angle
Web
Bottom chord (double) angle
(b) Section through a typical K-series joist
Weld line
Top chord (double) angle
Depth of end bearing of a K-series joist = 2-1/2 in. standard
Extreme web member, generally a round bar
Bearing (double) angle, welded to top chord (double) angle
(d) Detail of the end of a K-series joist
(Continued)

FIGURE 18.21 Essential details of K-series joists.

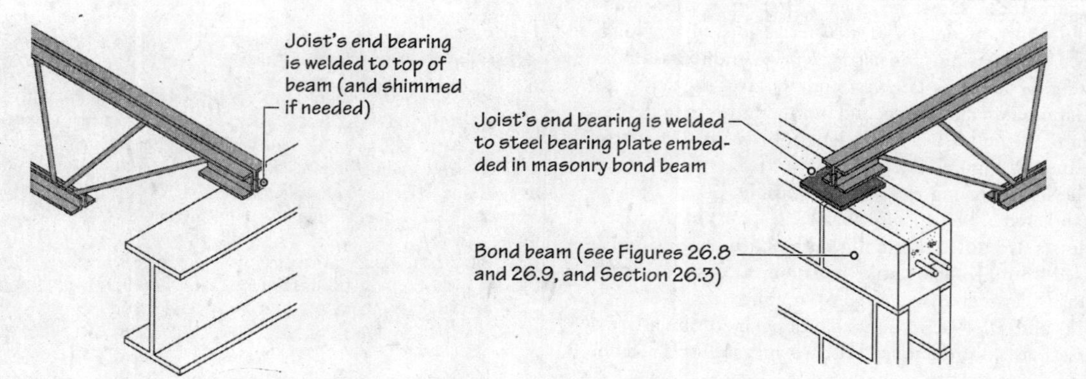

Joist's end bearing is welded to top of beam (and shimmed if needed)
Joist's end bearing is welded to steel bearing plate embedded in masonry bond beam
Bond beam (see Figures 26.8 and 26.9, and Section 26.3)

(e) Support detail of a K-series joist bearing on a wide-flange beam. Bearing of joist on a joist girder is similar.

(f) Support detail of a K-series joist bearing on a masonry wall. Detail of joist bearing on a reinforced concrete wall is similar.

K-SERIES JOIST DESIGNATION
K-series joists are designated by two numbers one on each side of letter K. The left-side number gives the depth of the joist in inches and varies from 8 to 30 in 2-in. increments, i.e., the left side numbers can be 8, 10, 12, ... 30. Thus, the minimum depth of a K-series joist is 8 in. and the maximum depth is 30 in. For example, the depth of a joist designated as 16K4, is 16 in.

The right-side number gives the relative weight of the joist and varies from 1 to 12 in steps of 1, i.e., the right-side numbers can be 1, 2, 3, ... 12. For the same joist depth, the larger the right-side number, the heavier the joist. Thus, a 16K5 joist is heavier (and, therefore, being of the same depth, is stiffer) than a 16K4 joist and has a greater load-carrying capacity.

HEIGHT OF END BEARING OF K-SERIES JOISTS
The height and depth of end bearing is 2-1/2 in. for all K-series joists.

TYPICAL SPACING OF K-SERIES JOISTS
Joist spacing must be determined by the engineer or architect, depending on the load-carrying capacity of the floor or roof deck. It generally varies from 2 to 4 ft for floors and from 4 to 6 ft for roofs.

MAXIMUM SPAN OF K-SERIES JOISTS
As Table 18.2 shows, the maximum allowable span of a joist is 24 times its depth. Thus, a 10-in.-deep joist can be used for a maximum span of 240 in. = 20 ft.

FIGURE 18.21 (*continued*) Essential details of K-series joists.

maximum allowable span of a given joist (regardless of the joist series) is 24 times the depth of the joist. For example, the maximum allowable span of joists designated as 24K4, 24K5, 24K6, and so on, is 48 ft.

Steel Joist Girders

As shown in Figure 18.18, joist girders are used as primary structural elements as an economical alternative to W-shape beams. Unlike the joists, which are selected from standard SJI load tables, joist girders are designed by the manufacturers based on the specifications provided by the engineer.

Bracing Joists and Joist Girders Against Instability

Steel joists and joist girders are slender elements and are, therefore, unstable and prone to overturning. As per SJI's specifications, the joists must be stabilized by rows of continuous horizontal members, referred to as *horizontal bridging members.* Horizontal bridging members are used in rows: One row is welded to the top chord of the joist, and the other row is welded to the bottom chord, Figure 18.22.

TABLE 18.2 STANDARD LOAD TABLE OF SELECTED K-SERIES JOISTS

Joist Designation Depth (in.) Approx. weight (lb/ft)	8K1 8 5.1	10K1 10 5.0	12K1 12 5.0	12K3 12 5.7	12K5 12 7.1	14K1 14 5.2	14K3 14 6.0	14K4 14 6.7	14K6 14 7.7
Span (ft)									
8	550 **550**								
9	550 **550**								
10	550 **480**	550 550							
11	532 **377**	550 **542**							
12	444 **288**	550 **455**	550 **550**	550 **550**	550 **550**				
13	377 **225**	479 **363**	550 **510**	550 **510**	550 **510**				
14	324 **179**	412 **289**	500 **425**	550 **463**	550 **463**	550 **550**	550 **550**	550 **550**	550 **550**
15	281 **145**	358 **234**	434 **344**	543 **428**	550 **434**	511 **475**	550 **507**	550 **507**	550 **507**
16	246 **119**	313 **192**	380 **282**	476 **351**	550 **396**	448 **390**	550 **467**	550 **467**	550 **467**
17		277 **159**	336 **234**	420 **291**	550 **366**	395 **324**	495 **404**	550 **443**	550 **443**
18		246 **134**	299 **197**	374 **245**	507 **317**	352 **272**	441 **339**	530 **397**	550 **408**
19		221 **113**	268 **167**	335 **207**	454 **269**	315 **230**	395 **287**	475 **336**	550 **383**
20		199 **97**	241 **142**	302 **177**	409 **230**	284 **197**	356 **246**	428 **287**	525 **347**
21			218 **123**	273 **153**	370 **198**	257 **170**	322 **212**	388 **248**	475 **299**
22			199 **106**	249 **132**	337 **172**	234 **147**	293 **184**	353 **215**	432 **259**
23			181 **93**	227 **116**	308 **150**	214 **128**	268 **160**	322 **188**	395 **226**
24			166 **81**	208 **101**	282 **132**	196 **113**	245 **141**	295 **165**	362 **199**
25						180 **100**	226 **124**	272 **145**	334 **175**
26						166 **88**	209 **110**	251 **129**	308 **156**
27						154 **79**	193 **98**	233 **115**	285 **139**
28						143 **70**	180 **88**	216 **103**	265 **124**

Note: The normal-weight numbers represent the total safe, uniform load-carrying capacity of joists in pounds per foot. Dead loads, including the self-load of joists, must be deducted to determine the joist's live-load capacity. Boldface numbers represent the live load (in pounds per foot) that will produce a deflection in the joist equal to the joist span divided by 360.

Source: This table has been adapted from *Standard Specifications Load Tables & Weight Tables for Steel Joists and Joist Girders*, published by the Steel Joist Institute (SJI), with permission. For a complete version of the table, refer to this publication

Generally, the bridging members are $1\frac{1}{4}$ in. $\times$ $1\frac{1}{4}$ in. angles, but they must meet the stiffness and strength requirements of SJI's specifications. An alternative to horizontal bridging is *diagonal bridging,* as shown in Figure 18.23. Regardless of which bridging alternative is used, it is important that the end spans of the bridging are securely connected to the wall or spandrel beam. The spacing between the rows of bridging is also specified by

FLOOR DECKS

A steel floor deck functions as a working platform as well as permanent formwork for the concrete fill, so that the deck and the fill form a concrete floor slab. Two types of floor decks are used:

- A *form deck* functions as permanent formwork only. Thus, the concrete slab must be reinforced with conventional reinforcement, Figure 18.27.
- In a *composite deck*, the deck also functions as steel reinforcement for the concrete slab, reducing the need for conventional concrete reinforcing bars.

The surface of a composite floor deck is designed so that the deck bonds integrally with the overlying concrete fill. Protrusions and depressions are, therefore, stamped on the deck surface, Figure 18.28(a). Alternatively, the deck may be profiled with dovetail flutes, as shown in Figure 18.28(b), to provide the composite action.

Because a composite deck provides permanent reinforcement, the deck must be protected from corrosion. Two alternatives are generally used. The better alternative is a hot-dip galvanized finish. Another alternative is a phosphatized/painted finish, in which the exposed deck surface is painted and the unexposed surface is phosphatized. The minimum concrete cover above the deck is 2 in. A composite deck is not recommended for use in parking garages in cold regions where salt spread is used on the roads.

Because a form deck supports only the weight of wet concrete (no floor live load) and does not function as concrete reinforcement, it is usually much shallower and of a thinner gauge than a composite deck. A composite deck, on the other hand, must support the dead load of wet concrete, and when the concrete has hardened, the deck and the concrete together must support the entire load on the deck. A comparison of the typical profiles of composite and form decks is shown in Figure 18.29.

The anchorage of a floor deck to supporting joists or beams is similar to that of a roof deck, that is, through puddle welds. The connection of the longitudinal (side) edges of adjacent deck panels is also similar.

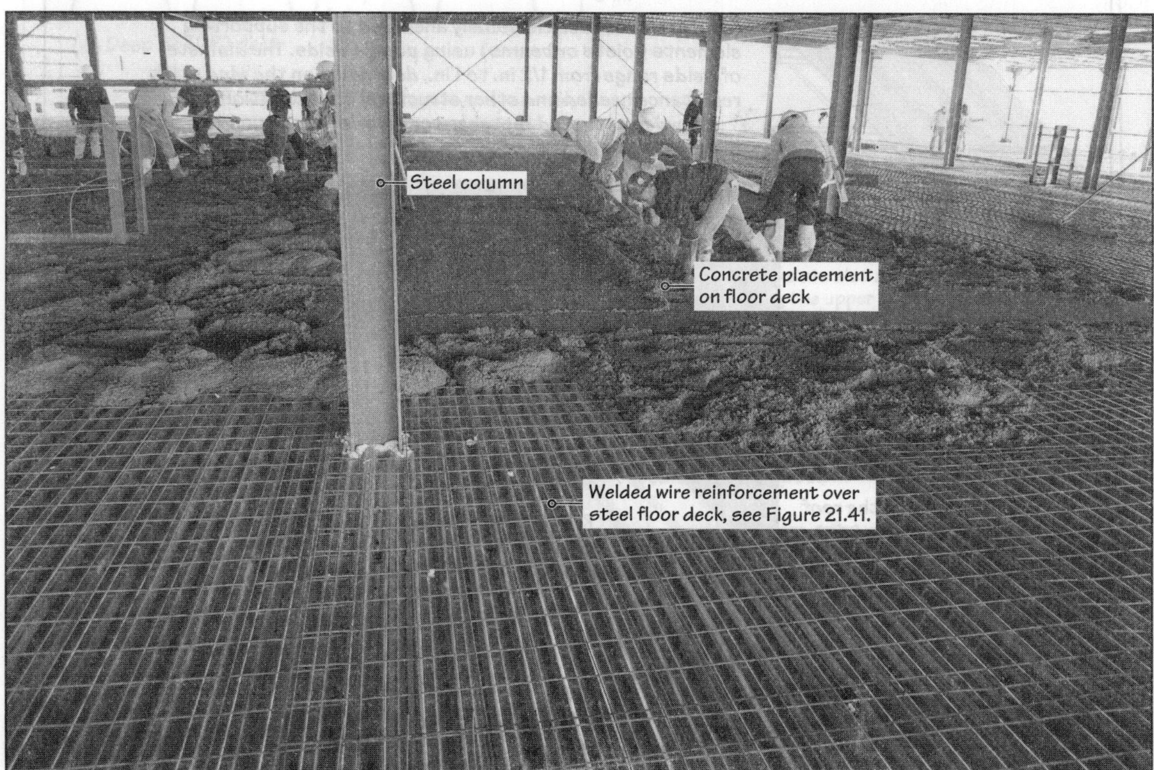

FIGURE 18.27 This illustration shows the placement of concrete on a floor deck (a form deck in this case). Observe the steel reinforcing bars placed over the deck.

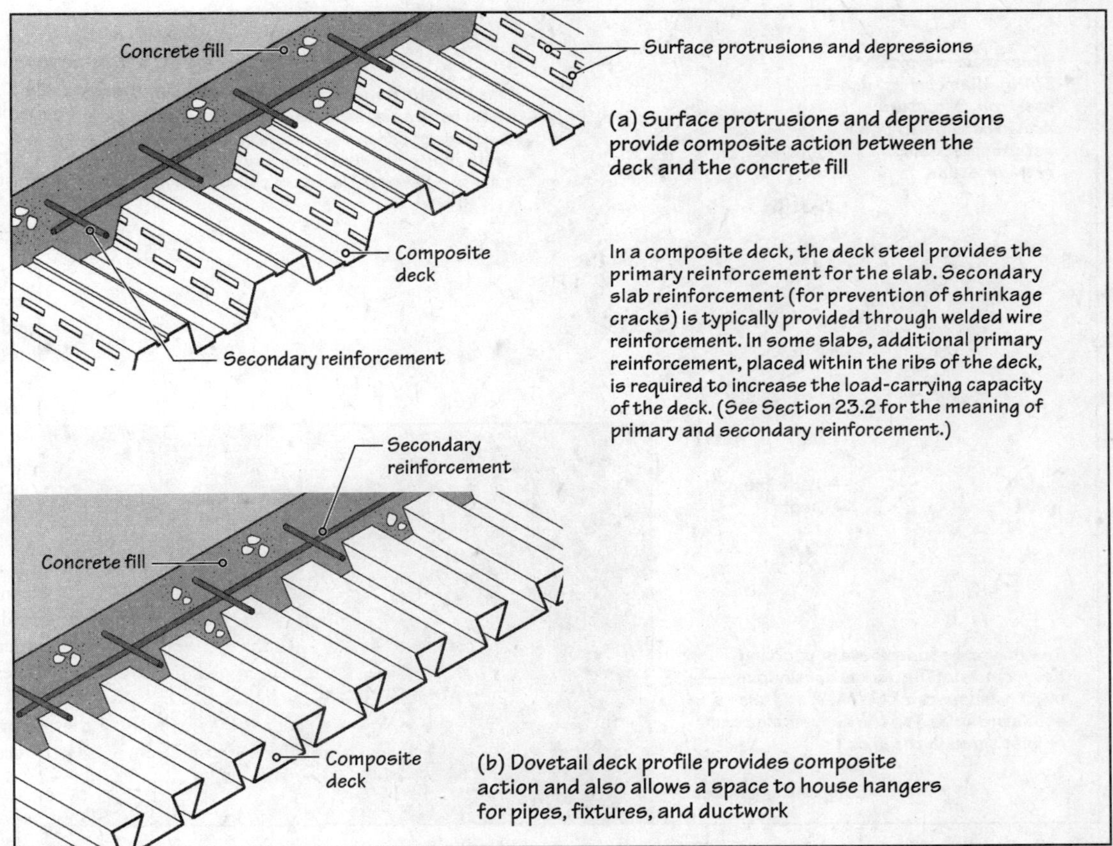

Concrete fill
Surface protrusions and depressions
(a) Surface protrusions and depressions provide composite action between the deck and the concrete fill
Composite deck
In a composite deck, the deck steel provides the primary reinforcement for the slab. Secondary slab reinforcement (for prevention of shrinkage cracks) is typically provided through welded wire reinforcement. In some slabs, additional primary reinforcement, placed within the ribs of the deck, is required to increase the load-carrying capacity of the deck. (See Section 23.2 for the meaning of primary and secondary reinforcement.)
Secondary reinforcement
Secondary reinforcement
Concrete fill
Composite deck
(b) Dovetail deck profile provides composite action and also allows a space to house hangers for pipes, fixtures, and ductwork

FIGURE 18.28 Two commonly used composite deck types.

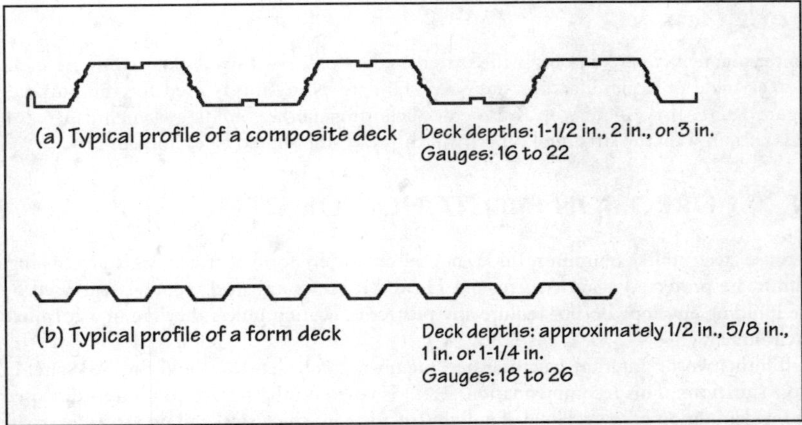

FIGURE 18.29 Comparison between the general profiles, depths, and gauges of composite decks and form decks.

Composite Action Between Floor Slab and Supporting Members

Both a composite deck and a form deck can be made to act compositely with the supporting beams by using shear studs, Figure 18.30. Shear studs prevent slippage of the deck under bending of the underlying beam. They are similar to nails that connect a plywood deck with supporting wood joists.

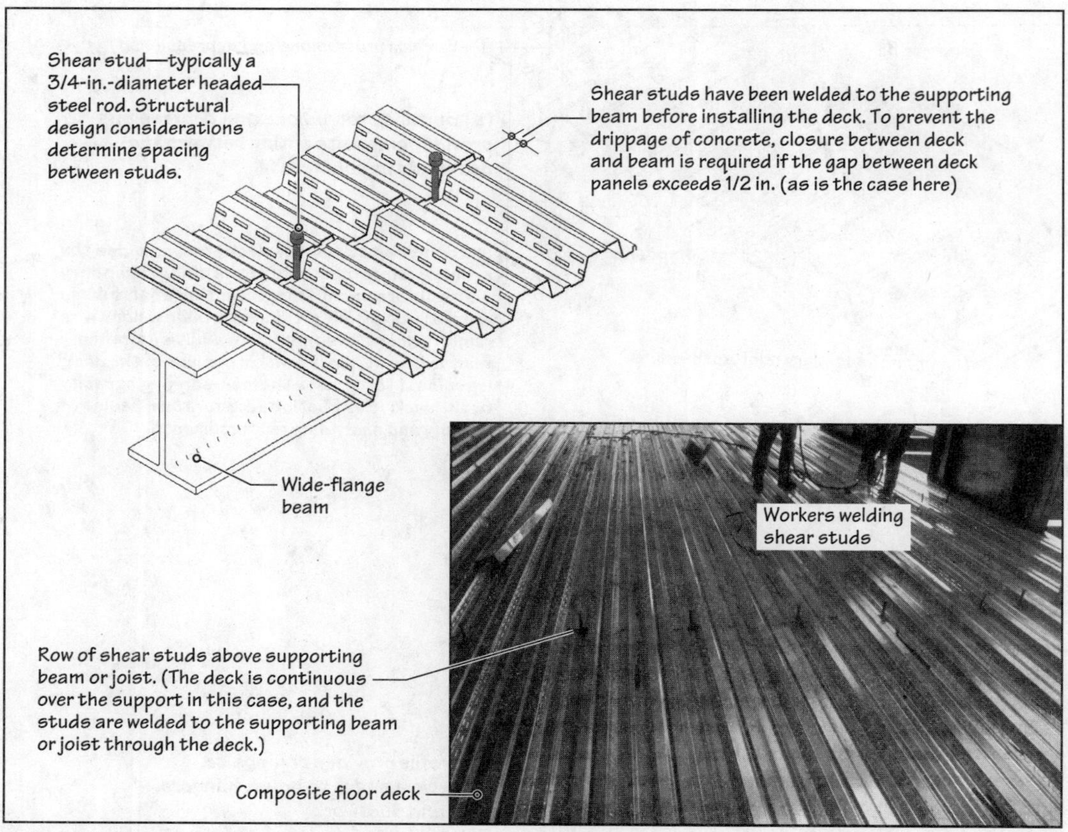

Shear stud—typically a 3/4-in.-diameter headed steel rod. Structural design considerations determine spacing between studs.
Shear studs have been welded to the supporting beam before installing the deck. To prevent the drippage of concrete, closure between deck and beam is required if the gap between deck panels exceeds 1/2 in. (as is the case here)
Wide-flange beam
Workers welding shear studs
Row of shear studs above supporting beam or joist. (The deck is continuous over the support in this case, and the studs are welded to the supporting beam or joist through the deck.)
Composite floor deck

FIGURE 18.30 Use of shear studs for composite action between floor slab and beam.

Pour Stops at Spandrel Beams and Edges of Floor Openings

To terminate a concrete pour at the extreme edges of a floor deck, pour stops are used. A steel angle or a manufacturer-supplied standard pour stop is used for the purpose, Figure 18.31. If the pour stop is used for anchoring another building element, such as a glass curtain wall, the structural capacity of the pour stop should be examined.

18.7 CORROSION PROTECTION OF STEEL

Because steel (unlike aluminum) does not automatically form a protective oxide coating, it must be protected against corrosion. However, structural steel members enclosed by the building envelope do not require any protective coating unless they are in a corrosive environment.

In other words, interior structural steel members can be left bare (mill-finished state) in most situations. This recommendation [18.2] is based on the surface conditions disclosed by the demolition of several long-standing structures in which steel had no protective coating. These structures suffered no corrosion or the corrosion was localized to small areas that were affected by persistent water leakage or condensation.

Although bare steel is acceptable, almost all structural steel members generally receive a prime coat in the fabricator's shop before being delivered to the construction site. The prime coat provides temporary protection until the steel is wrapped by the building envelope. Structural steel that is permanently exposed to the atmosphere must, however, receive further protection. Additionally, concealed steel components that may be subjected to wetting, such as steel anchors buried in masonry or concrete, should also be protected.

Several protective coatings are available for steel to suit different environmental conditions, aesthetic requirements, and budgets. These include acrylics, epoxies, polyurethanes,

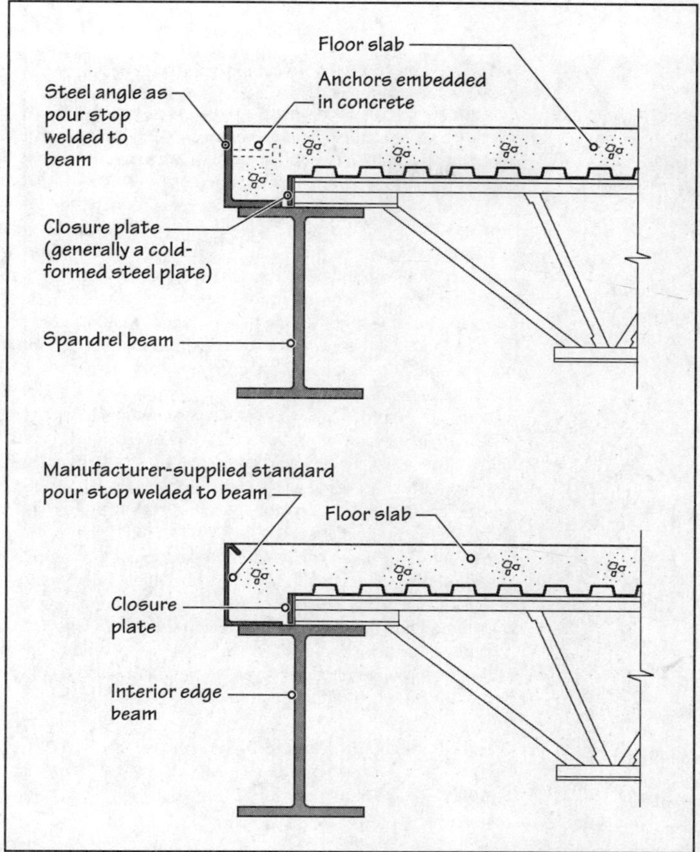

FIGURE 18.31 The use of pour stops at spandrel beams and interior mezzanine floor beams.

and zinc coating. For exposed structural steel members, polyurethane coatings are the hardest, toughest, and most versatile. The prime coat is generally applied in the shop, followed by one or two field-applied finish coats. A well-applied polyurethane coating provides more than 20 years' protection.

For cold-formed and light structural steel members, zinc coating (referred to as *galvanizing*) is a cost-effective solution. Of the various types of galvanizing, hot-dip galvanizing is by far the most reliable and the most widely used.

HOT-DIP GALVANIZING

In the hot-dip galvanizing process, the steel member is immersed in a kettle of molten zinc, which is at a temperature of about 850°F (475°C). This results in the formation of a zinc-iron alloy coating that is metallurgically bonded to steel. After the member is removed from the kettle, the excess zinc is drained and shaken off. The galvanized member is then cooled in air or quenched in water. As a result of its immersion, the entire member is coated with zinc, including the protrusions, recesses, and corners.

Kettles up to 50 ft long are available. If a member is longer than 50 ft, it can be galvanized in two halves. After one half has been galvanized, it is removed; then the other half is immersed in the kettle.

Hot-dip galvanizing should be done after the member has been completely fabricated. No cutting, grinding, or process that eradicates the coating should be done after galvanizing.

The thickness of the coating influences the degree of protection. The thicker the coating, the greater the protection. The thickness of a zinc coating is specified as Gxx, where G stands for galvanizing and xx indicates the average coating mass in ounces per square foot (oz/ft^2) of the steel member. Thus, a G60 coating means that the member is coated on all of its surfaces with an average of 0.60 oz/ft^2 of zinc. If the member is a sheet, G60 means the presence of 0.30 oz/ft^2 of zinc on each surface of the sheet. A G90 coating indicates 0.90 oz/ft^2 of zinc coating on all surfaces.

NOTE

Corrosion—an Electrochemical Process

Corrosion is not simply a chemical process; it is an electrochemical process that results in the eating away of the surface of a metal when it is exposed to weather. It occurs when electrons flow through an electrolyte. (An electrolyte is a liquid that allows the flow of electrons through it.)

Water is a necessary ingredient for corrosion. Acidic water or water containing salt is a better electrolyte. That is why coastal regions are more corrosive than inland areas and deserts are less corrosive than humid regions.

EXPAND YOUR KNOWLEDGE

Corrosion of Metals

Metals, by virtue of their atomic structure, are chemically reactive materials. They react readily with atmospheric oxygen to form metal oxides, which are chemically stable. That is why metals are generally obtained on the earth's surface in the form of their oxides, that is, iron oxide (iron ore), aluminum oxide (bauxite ore), and so on.

Atmospheric Corrosion

Because of their reactivity, metals have a tendency to return to their original (oxide) state. The oxidation reaction obviously requires atmospheric oxygen, but it also requires the presence of water. The reaction is accelerated by the presence of atmospheric impurities because they, in combination with water, can form weak acids.

For example, sulfur released from automobile and industrial exhausts dissolves into rainwater and converts to weak sulfuric acid. Sulfuric acid exists in most contemporary urban environments. It was also present in the environment before the introduction of automobiles due to the burning of coal, which contains a substantial amount of sulfur.

Water that contains salt also accelerates the oxidation of metals. Corrosion of metals is more rapid in coastal areas than in inland areas. Salt spread on roads in snow-prone regions accelerates the corrosion of steel bridges and of steel reinforcement embedded in floor slabs of parking garages. The oxidation of metals, referred to as *corrosion*, begins at the surface of the metal, and if the reaction is not interrupted, it proceeds inward and can corrode the entire metal over time.

The corrosion of steel results in the formation of iron oxide—a brown-colored material called *rust*. Because rust is a loose and flaky material, it separates from the steel member and exposes the member to further corrosion. The volume of rust is greater than that of the original iron. That is why the corrosion of reinforcing steel not only reduces the amount of steel, but also tends to spall the concrete.

By contrast with iron oxide, aluminum oxide, formed by the oxidation of aluminum, clings securely to aluminum. Being a stable compound, aluminum oxide forms a protective layer and prevents further oxidation of aluminum. When an aluminum surface is scratched and aluminum is exposed, a new oxide layer is immediately formed over the scratch, maintaining the protection.

Galvanic Corrosion

The type of corrosion just described is called *atmospheric corrosion* because it is caused by a metal's exposure to the atmosphere. An additional type of corrosion of interest to design and construction professionals is *galvanic corrosion*. Galvanic corrosion, also called *bimetallic corrosion*, is caused by the physical contact between two dissimilar metals. Galvanic corrosion is not chemically different from atmospheric corrosion; that is, both galvanic and atmospheric corrosion processes produce the same end product. The difference is that the contact between the two dissimilar metals changes the metals' respective corrosion rates. One metal corrodes more rapidly than its atmospheric corrosion rate, and the other metal corrodes more slowly. The change in their corrosion rates depends on the type of metals in contact.

The metal whose corrosion rate declines is called a *more noble* metal, and the one whose corrosion rate increases is called a *less noble* metal. The relative nobility of metals is given by the *galvanic series*, produced by Luigi Galvani in the late eighteenth century. The galvanic series for a few selected metals is given below. Of various metals, gold and platinum are the most noble, and magnesium and zinc are the least noble.

The change in the corrosion rates of metals that are close to each other in the galvanic series is smaller. The farther apart the metals are in the series, the greater the change in their respective corrosion rates. Thus, if zinc and copper, which are far apart in the series, are placed in contact, the corrosion rate of zinc will be accelerated and that of copper will decline substantially.

Galvanic corrosion is easily controlled by separating the metals from each other either by a plastic separator or by coating one or both metals. Coatings also inhibit atmospheric corrosion.

Galvanic Series of Selected Metals

Less noble	Magnesium
	Aluminum
	Zinc
	Steel
	Stainless steel
	Tin
	Lead
	Copper
	Bronze
	Brass
	Titanium
	Silver
More noble	Platinum
	Gold

In the SI system, the same zinc coating is specified as Zyy, where yy indicates the average coating mass in grams per square meter (g/m^2). Because 1 $oz/ft^2 = 305\ g/m^2$, a G60 coating is equal to Z180, and a G90 coating is equal to Z275.

Three levels of hot-dip galvanizing are commonly specified: G30 for a mildly corrosive environment, G60 for a more corrosive environment, and G90 or G120 for a highly corrosive environment.

18.8 FIRE PROTECTION OF STEEL

The fire endurance of steel is poor, and its inherent incombustibility gives a false sense of security. Exposed (unprotected) steel, unless it is very thick, cannot withstand long exposure to fire. Steel's yield strength and modulus of elasticity drop to nearly 60% of their original values at about 1,100°F (600°C)—a temperature that is well below the temperature used in a standard fire test.

Steel has a high coefficient of thermal expansion. When subjected to fire, an unrestrained steel member expands, pushing on the members to which it is connected. A restrained member develops high internal stresses. In either case, the consequences can be disastrous. It is important, therefore, that structural steel be protected against fire. Unprotected steel buildings may be used, but building codes severely limit the allowable areas and heights of such structures.

FUNDAMENTALS OF FIRE PROTECTION OF STEEL

There are basically two ways to protect steel against fire:

- Insulate the steel component with a noncombustible thermal insulation. Spray-applied fire protection and intumescent paints fall in this category.
- Encase the steel component with a noncombustible material with high thermal capacity, such as concrete, gypsum board, or water (see Section 5.8). These materials retard the buildup of temperature, producing the same end result as thermal insulation.

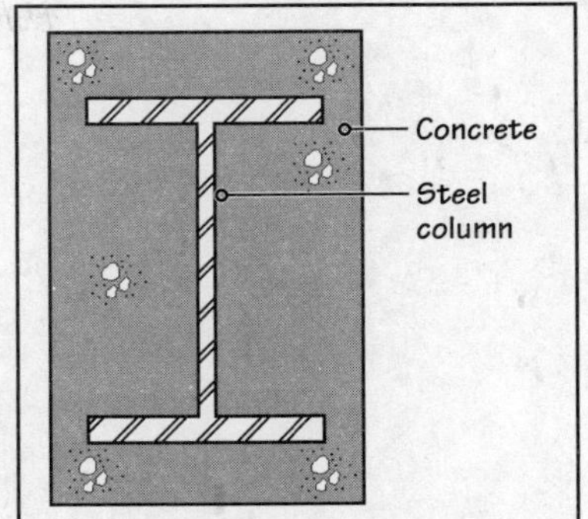

FIGURE 18.32 Encasing a steel member in concrete (or masonry) is one of the oldest methods of protecting steel against fire. The use of this method is relatively uncommon in contemporary construction because of the availability of more efficient and economical alternatives.

ENCASEMENT OF STEEL IN CONCRETE

Figure 18.32 shows a steel member encased in concrete. Concrete encasement is generally used in columns and floor systems that consist of reinforced concrete slabs supported by steel beams so that the concrete-encased steel beams function monolithically with the slab. The disadvantages of protecting steel with concrete encasement are the high cost of formwork and the large dead load it poses on the structure. It has, therefore, been replaced by more efficient techniques, which are lightweight and more economical, such as spray-applied fire protection and gypsum board encasement.

GYPSUM BOARD ENCASEMENT

Figure 18.33 shows steel members wrapped with gypsum board, in which gypsum boards are mechanically fastened to steel sections, usually screw attached to light-gauge steel studs, Figure 18.33. Almost any degree of fire rating can be obtained by varying the number of board layers. (See Chapter 16 for a discussion of gypsum boards.)

SPRAY-APPLIED FIRE PROTECTION

Spray-applied fire protection is by far the most commonly used method because of its lower cost and greater convenience. It has two disadvantages: (a) the finished profile of the steel component on which it is sprayed becomes jagged and uneven, and (b) sprayed material may come lose from the member with a decrease of adhesion over time. Two types of materials are used for spray-applied (or spray-on) protection:

- Mineral fiber and binder, usually fiberglass and portland cement
- Cementitious mixture consisting of portland cement mixed with a lightweight aggregate such as expanded perlite or vermiculite

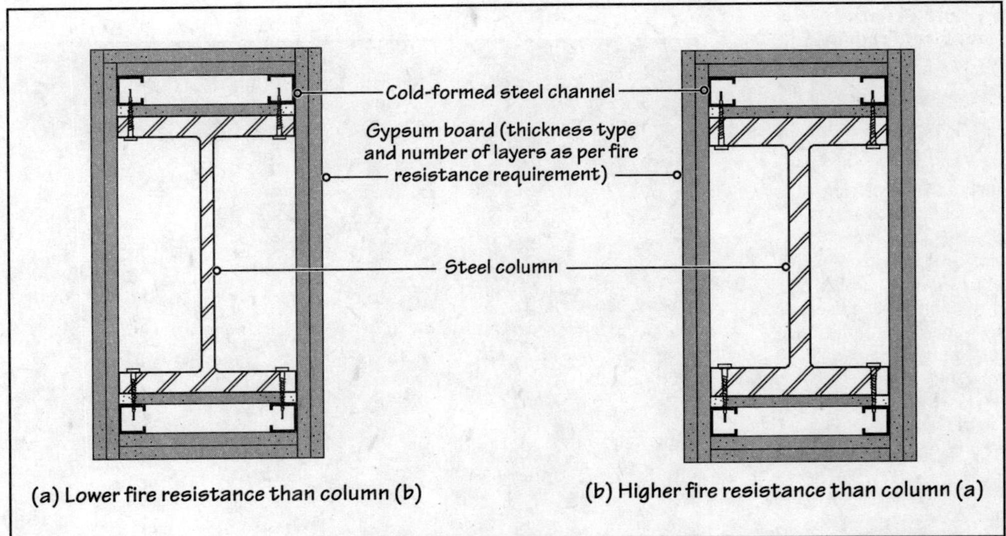

FIGURE 18.33 Details of a steel column covered with gypsum board layers. Details **(a)** and **(b)** are identical except that the column in detail **(b)** is heavier than that of the column in detail **(a)**. Because the thickness of steel also affects the member's fire resistance, in addition to the thickness and type of gypsum board, the fire resistance rating of column **(b)** is higher than that of column **(a)**.

FIGURE 18.34 Spray-on fire protection of a steel roof assembly. Observe that both joists as well as the roof deck are protected.

Figure 18.34 shows spray-applied fire protection on a steel roof assembly. Note that the fire protection extends over steel joists as well as the roof deck. Figure 18.35 shows fire protection on a steel floor assembly. Note that the deck is not covered with protection because of the presence of concrete fill over the deck, which provides sufficient thermal capacity to obviate the need for insulation through spray-on protection. Floor decks may need spray-on protection if the required fire rating is high. Roof decks almost always require spray-on protection due to the absence of concrete fill on roof decks.

Virtually any amount of fire resistance can be obtained by varying the thickness of spray-on protection. For a given fire rating, the thickness of spray-on protection depends on the spray-on material and the thickness of the steel. The heavier the member, the thinner the required protection. However, a $\frac{5}{8}$-in. thickness per 1 h of fire rating is a good rough estimate of the required thickness of protection. Thus, a 1-h rating is obtained by applying about $\frac{5}{8}$-in.-thick material, a 2-h rating by applying $1\frac{1}{4}$-in.-thick material, and so on.

The fire rating is not the only criterion that spray-on protection must meet. It must also meet the minimum requirements for damageability resistance, adhesion to steel, deflection (so that the protection does not delaminate when the steel member deflects under loads), air-erosion resistance, and corrosion resistance (spray-on protection also protects steel from corrosion).

NOTE

Use of Unprotected Structural Steel Framing Members

Unprotected steel in structural framing of building is routinely used. Unprotected steel-framed buildings, defined as Type II(B) construction (Chapter 7), are allowed by building codes for all occupancy groups except I-2 occupancy group (hospitals, nursing homes, etc.).

However, if protected steel framing is used for the same building, the built-up area allowed by the building codes is greater, depending on the type of construction used. Protected steel construction types may be Type I(A), Type I(B), or Type II(A). The allowable built-up area may also be increased through the use of automatic sprinklers (see Principles in Practice in Chapter 2).

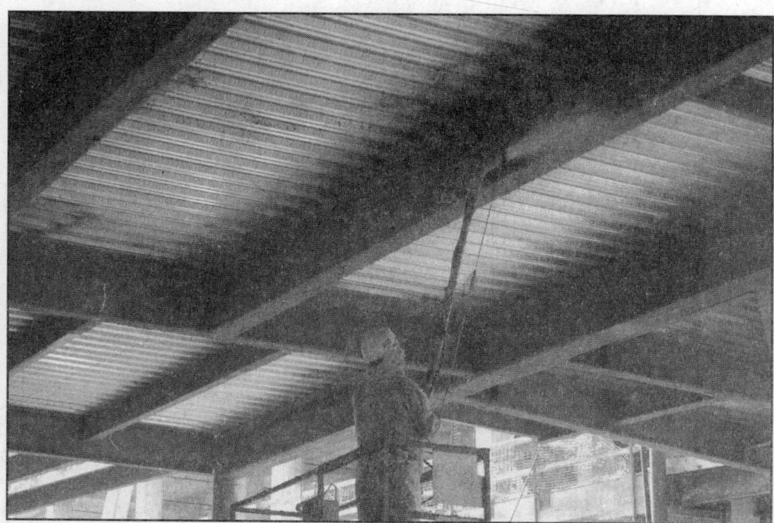

FIGURE 18.35 Spray-on fire protection of a steel floor assembly. Observe that only the floor beams are protected. Fire protection is generally not necessary for the floor deck because of the concrete topping.

INTUMESCENT PAINTS

An alternative to gypsum board encasement is intumescent paint on steel members. Intumescent paint is typically 20 to 50 mil (0.5 to 1.3 mm) thick. When exposed to the heat of fire, the paint intumesces, or swells, yielding an insulating char cover on steel that is 2 in. to 4 in. thick. It is this char layer that protects steel from fire. However, the char layer is damaged by long-term exposure to fire; hence, this technique can be used only for low levels of fire rating, generally not exceeding 2 h.

Intumescent paint is a thin film that does not alter the overall profile of steel sections. It gives the structure the look of exposed steel and is commonly used in situations where the expression of exposed steel is required from an architectural and interior design perspective, Figure 18.36.

Intumescent paints are available in different colors. Thus, the painted structure can be color coordinated with the overall character of the interior. Other advantages of intumescent paint protection are its abrasion resistance and its relatively smooth surface (which does not easily collect dust). It is, therefore, ideal for areas where a dust-free environment is required, such as laboratories and hospitals. If a washable top coat is provided on intumescent paint, the structure can be washed with water, which is an advantage in food-processing plants.

Intumescent paint protection, however, is far costlier than conventional spray-on fireproofing. It is, therefore, not practical for buildings where economy is a major consideration. In buildings with large areas of steel deck supported on steel beams and columns, intumescent paint may be used for beams and columns, but the deck may be protected by a more economical alternative.

FIGURE 18.36 A steel truss painted with intumescent paint providing a 1-h fire rating, Hospital for Sick Children, Toronto, Ontario. (Photo courtesy of AD Fire Protection Systems, Scarborough, Ontario. Architect: Zeidler Roberts Partnership. Photo by Lenscape Inc.)

SUSPENDED CEILINGS

Suspended ceilings consisting of gypsum lath and plaster, gypsum boards, or acoustical tiles are also used to provide fire protection to otherwise unprotected steel beams or trusses in roof-ceiling or floor-ceiling assemblies. The ceiling grid may be directly attached to, or hung from, the bottom flanges of beams or the bottom chords of trusses.

The effectiveness of such a system depends on how well the ceiling continues to perform as a barrier. Often the maintenance is not adequate, resulting in a defective ceiling tile or gypsum board not being replaced properly, so that hot gases enter the plenum space to attack the unprotected steel.

FOCUS ON SUSTAINABILITY

Sustainability Features of Steel

Steel can be easily recycled by melting old steel and reforming it, as discussed in Section 18.1. Large quantities of new steel are being made from steel recovered from old cars, washing machines, refrigerators, structural members from old buildings, and so on. The steel industry claims that along with aluminum, steel is the most recycled building material. In 2008, approximately 83% of steel was recycled.* In the United States, all structural steel sections, such as the W-shape, C-shape, and L-shape, are produced by the electric arc furnace process, which utilizes mostly recycled steel.

As mentioned earlier, steel does not corrode if used in interior locations. It is, therefore, a durable material even without any corrosion protection. Another sustainable feature of steel is that it lends itself readily to deconstruction, particularly if mechanical fasteners are used for component assembly. In a well-designed assembly, steel components can be disassembled, refabricated, and reassembled at a new location (and for a new use) without the need to recycle.

Steel structures, being lighter in weight than concrete and masonry structures, require smaller foundations, conserving the use of concrete and excavations.

Although the manufacture of steel produces pollution, there are no pollution issues with steel during its service. Unlike engineered wood components, there are no outgassing issues with steel. However, steel members protected by fibrous fire protection may dispel stray fibers in the air.

**Modern Steel Construction*, February 2010.

PRACTICE QUIZ

Each question has only one correct answer. Select the choice that best answers the question.

19. Which of the following does not represent a steel joist series?
- **a.** H-series
- **b.** J-series
- **c.** K-series
- **d.** LH-series
- **e.** DLH-series

20. In the steel joist designation 24LH10, the number 24 gives the
- **a.** length of the joist in feet.
- **b.** depth of the joist in inches.
- **c.** weight of the joist in pounds per foot.
- **d.** number of web members.
- **e.** none of the above.

21. A K-series steel joist that is 20 in. deep cannot be used to span more than
- **a.** 20 ft.
- **b.** 25 ft.
- **c.** 30 ft.
- **d.** 35 ft.
- **e.** 40 ft.

22. Steel joists are braced against instability by
- **a.** horizontal bridging members at the bottom of the joists.
- **b.** horizontal bridging members at the top of the joists.
- **c.** (a) or (b).
- **d.** (a) and (b).
- **e.** none of the above.

23. Diagonal bridging may be used as an alternative to horizontal bridging to stabilize steel joists.
- **a.** True
- **b.** False

24. Steel joists and joist girders are best utilized in a building that has concentrated loads.
- **a.** True
- **b.** False

25. Steel roof decks are either form decks or composite decks.
- **a.** True
- **b.** False

26. Steel floor decks are either form decks or composite decks.
- **a.** True
- **b.** False

27. The thickness of sheet steel in a roof or floor deck is generally specified in terms of
- **a.** inches.
- **b.** millimeters.
- **c.** gauge number.
- **d.** any one of the above.

28. Which of the following decks requires structural concrete topping?
- **a.** Form deck
- **b.** Composite deck
- **c.** Floor deck
- **d.** all of the above
- **e.** none of the above

29. Roof decks are anchored to supporting members by continuous welds along the support lines.
- **a.** True
- **b.** False

30. A pour stop is generally used with
- **a.** roof decks.
- **b.** floor decks.
- **c.** (a) and (b).
- **d.** none of the above.

31. Two dissimilar metals are generally separated from each other by a separator, which helps to prevent
- **a.** atmospheric corrosion.
- **b.** galvanic corrosion.

32. In a galvanized steel plate that carries the designation G60, the number 60 means that
- **a.** the total mass of zinc coating on both surfaces of the plate is 60 oz/ft^2.
- **b.** the total mass of zinc coating on each surface of the plate is 60 oz/ft^2.
- **c.** the total mass of zinc coating on both surfaces of the plate is 60 g/ft^2.
- **d.** the total mass of zinc coating on each surface of the plate is 60 g/ft^2.
- **e.** none of the above.

33. Spray-on fire protection materials for steel contain
- **a.** portland cement with mineral fibers or lightweight aggregate.
- **b.** polymer-based cement with mineral fibers or lightweight aggregate.
- **c.** portland cement with normal-weight aggregate.
- **d.** polymer-based cement with normal-weight aggregate.
- **e.** any one of the above, depending on the manufacturer.

34. Spray-on fire protection of steel gives a much cleaner appearance than gypsum board encasement of steel.
- **a.** True
- **b.** False

35. Fire-protection of structural steel framing members is mandated by building codes for all buildings.
- **a.** True
- **b.** False

REVIEW QUESTIONS

1. Discuss the basic differences (in terms of chemical composition and important properties) between wrought iron, cast iron, and steel. Where would you recommend the use of wrought iron?
2. Discuss the two methods used in manufacturing steel, giving their relative advantages and disadvantages.
3. Discuss the difference between hot rolling and cold forming of steel and their uses in making steel components.
4. Sketch the commonly used hot-rolled sections used in building construction.
5. Using sketches, show the important difference between the W-shapes used for beams and those used for columns.
6. Using sketches and notes, explain:
   a. what a plate girder is and where it is used.
   b. what a built-up steel section is and where it is used.
7. Sketch the elevation of a typical K-series joist.
8. Sketch the end of a K-series joist in three dimensions. What is the standard end-bearing depth of a K-series joist?
9. Using sketches and notes, explain how joists and joist girders are stabilized against overturning.
10. Discuss the two basic approaches used for fire protection of steel.
11. In using spray-on fire protection for floor and roof assemblies, we generally apply the protection on roof decks and supporting beams, but in the case of the floor decks, we generally apply it only on the supporting beams, leaving the floor deck without spray-on protection. Explain the reason.

CHAPTER 19 Structural Steel Construction

CHAPTER OUTLINE

The previous chapter covered various aspects of steel as a construction material—for example, fundamentals of steel manufacturing, types of steel components, and the methods used for protecting steel from corrosion and fire. This chapter deals with the application of that information to the design and construction of steel-frame buildings. It begins with the basics of structural framing, that is, location, and spacing of the components of (a) the building's vertical support subsystem (columns and [or] walls) and (b) the building's horizontal support subsystem (floors and roof).

The structural frame of a steel building is constructed by assembling prefabricated steel components at the site. This is in contrast to the buildings whose structural system consists of load-bearing masonry or cast-in-place reinforced concrete. In these buildings, almost the entire structure is site-constructed, with virtually no prefabrication. Therefore, issues related to component fabrication, component erection, and component assembly are important in a steel structure and are discussed in this chapter.

A unique feature of steel structures (as distinct from wood, cast-in-place concrete, and masonry structures) is that the assembly details—details of the connections between steel components—are not prepared by the architect or the structural engineer, but by the steel detailer. Steel detailing is, therefore, a specialized expertise. As explained in the chapter, steel detailing may be done either by an independent detailing company or by an in-house outfit of a steel fabricator.

19.1 PRELIMINARY LAYOUT OF FRAMING MEMBERS

Preparing the *preliminary framing layout* of a steel-frame building (and also of wood or concrete buildings) is not simply a structural engineering exercise. It involves careful integration of structural considerations with several nonstructural considerations, such as HVAC, the building envelope, fire resistance, interior finishes, aesthetics, and cost. Because of its integrative nature, preparation of the preliminary framing layout is at the heart of an architect's expertise and is generally undertaken during the design development phase (see Chapter 1).

For a small or midsized building, an experienced architect would generally prepare the layout without the consultants' help. For a complex or large building, coordinating this activity with the consultants is usually necessary. The final framing layout is generally prepared by the structural engineer.

FRAMING PLANS AND THE STRUCTURAL GRID

The framing layout of a structure is usually prepared in plan view; therefore, it is referred to as a *framing plan.* Generally, a roof framing plan is different from a floor framing plan because of the difference in floor and roof loads. In some cases, framing plans may also differ from floor to floor. For example, the ground floor of a building may have large areas requiring larger column spacing in comparison with that at the upper floors. This and other differences will require framing plans to differ from floor to floor.

In preparing the framing plans, establishing the *structural grid* is the first step. A majority of buildings are either rectangular in plan or rectangle-based, such as an L-shape, T-shape, and so on. For such buildings, the structural grid is orthogonal, that is, the grid consists of two sets of lines perpendicular to each other. Additionally, several buildings, though rectilinear in plan, have parts that break from rectangularity, such as triangles, trapeziums, and so on. For these buildings, the grid is partially orthogonal and partially nonorthogonal. Though uncommon, the grid in some buildings may be entirely nonorthogonal, such as a triangular grid.

A structural grid determines the column locations, which are generally placed at the intersection of the grid lines. Therefore, the grid lines are also called *column lines.* The grid also defines the orientation of the components constituting the floors and roof. As discussed in Chapter 18, a floor or roof in a steel-frame building consists of (a) beams or beamlike members, such as girders, joist girders, and joists, and (b) floor and roof deck panels. In most buildings, the deck panels are the *tertiary components* of the floor or roof, while the beams or beamlike members form the *primary* and *secondary components.*

The primary components span along one direction of the grid and support the secondary components that span along the other grid direction. The secondary elements support the tertiary elements—the deck panels—which span perpendicular to the secondary elements. Thus, the load from a roof or floor is transferred from the deck panels to secondary components, and then to the primary components, and finally to the columns. Thus, an important feature of the framing of a floor or roof of a steel building is that all components are *one-way components,* that is, they span along one direction only. This is in contrast to cast-in-place concrete slabs, which may either be *one-way* or *two-way slabs.*

For the primary and secondary elements in most buildings, the choice is between the use of W-sections or steel joists, as indicated by the following four alternatives:

- W-sections for both primary and secondary elements, Figure 19.1. In this arrangement, the primary elements are generally called *girders* and the secondary elements are called *beams.* Some designers may refer to the primary elements as *beams* and the secondary elements as *intermediate beams.*

NOTE

Structural Grid and Structural Bays

The space enclosed by adjacent column lines is called a *structural bay* or simply a *bay.* In a building, the number of bays in one direction may be different from those in the other direction. Thus, in Figure 19.5, the number of bays in the horizontal direction is four and the number of bays in the vertical direction is three.

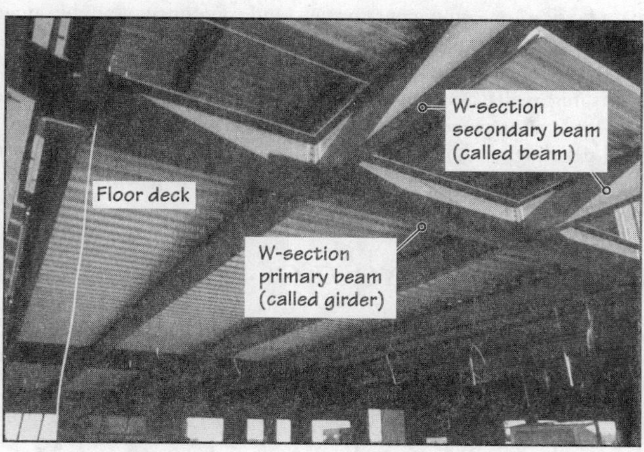

FIGURE 19.1 A floor structure with W-sections for primary and secondary framing elements. Note the hierarchy in the load transfer mechanism.

The floor load is transferred from the deck panels (tertiary elements) to secondary elements (secondary beams). From the secondary beams, the load is transferred to the primary elements (primary beams) and then to columns.

Instead of using the terms *primary beam* and *secondary beam,* the terms *girder* and *beam,* respectively, are more common. The term *girder* is used for a beam that supports other beams.

FIGURE 19.2 A floor structure with W-sections as primary framing elements and steel joists as secondary elements. Note the hierarchy in the load transfer mechanism. The floor load is transferred from tertiary elements (deck panels) to joists, from joists to W-section beams, and then to columns. The photograph was taken during the installation of floor deck on the second floor of the building, making the ground-floor structure visible through second-floor framing elements.

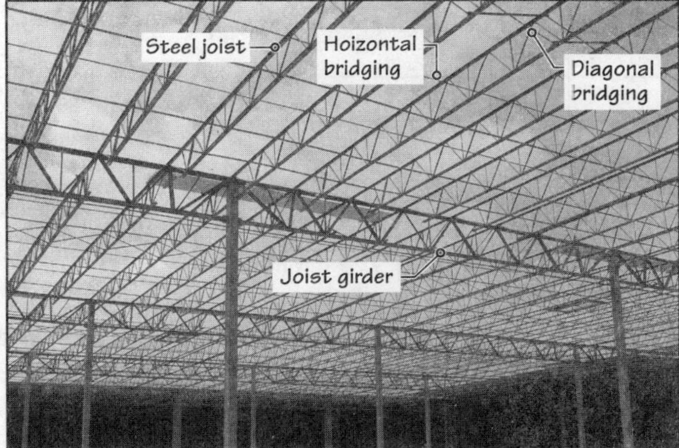

FIGURE 19.3 A roof structure with steel joist girders as primary framing elements and steel joists as secondary elements. Note that the roof load is transferred from deck panels to joists, from the joists to joist girders, and then to columns.

See Figure 18.22 for an explanation of horizontal bridging elements. Diagonal bridging is generally required at the center of joists.

- W-sections for primary elements and steel joists for secondary elements, Figure 19.2.
- Steel joist girders or trusses for primary elements and steel joists for secondary elements, Figure 19.3.
- In some buildings, such as those with load-bearing walls, the secondary supporting elements may be omitted and the roof deck may be supported directly on the primary elements, Figure 19.4.

LATERAL LOAD RESISTANCE

The structural framing of a building must consider both gravity and lateral loads. Therefore, simultaneously with the preparation of the floor and roof framing plans, the architect (in consultation with the structural engineer) should determine how to provide lateral load resistance to the structure. In a steel-frame building, the choice is between the use of the following alternatives or their combinations:

- Rigid frames
- Braced frame
- Reinforced concrete shear walls
- Steel shear walls

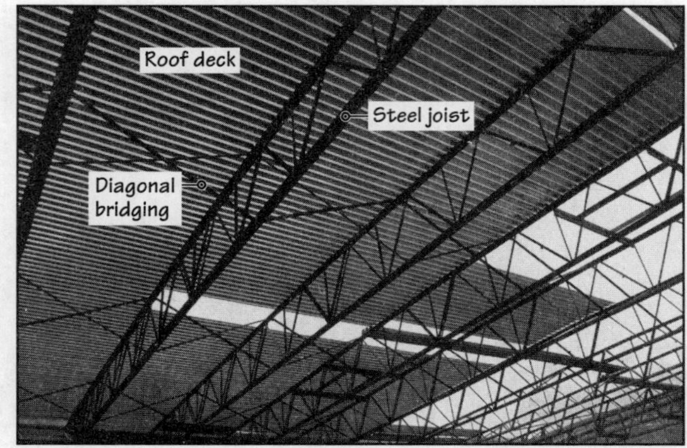

FIGURE 19.4 In this roof structure, the roof deck panels transfer the load to steel joists, which transfer the load to a masonry wall. Thus, the joists function as primary elements and the deck panels as secondary elements. Observe the daylight filtering through the deck. This is because the deck is a perforated (acoustical) deck; see Figure 18.25(g).

The concepts related to these alternatives are explained in detail in the Principles in Practice section at the end of this chapter. Reading (even browsing) through this section should be helpful to the reader. The concept of framing is presented in the following examples.

EXAMPLES OF FRAMING LAYOUTS

A typical floor-framing plan of a low-rise rigid-frame rectangular building, using W-sections for both girders and intermediate beams, is shown in Figure 19.5. A version of the same

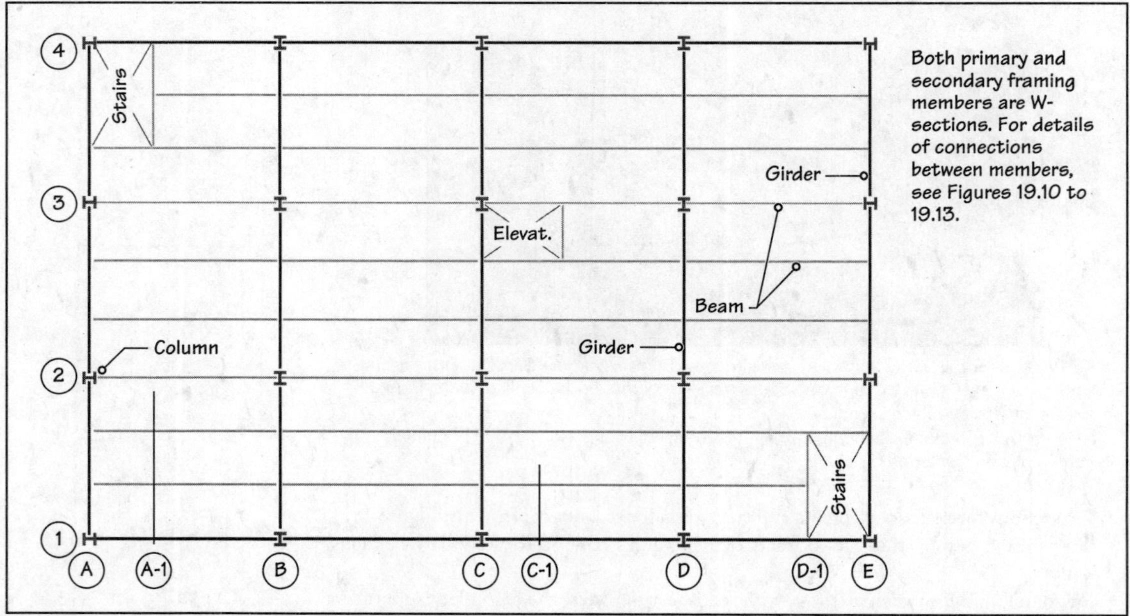

FIGURE 19.5 A typical floor framing plan for a steel rigid frame building using W-sections for primary and secondary framing members. Observe that column grid lines are identified by A, B, C, and so on along one direction and 1, 2, 3, and so on in the other direction.

Rigid frames have been identified by black lines. Because rigid frames provide lateral stability (resistance to lateral loads), they need to be provided in both major directions of the building. Thus, in this building, (three-bay) rigid frames have been provided along grid lines A, B, C, D, and E, and (four-bay) rigid frames along grid lines 1 and 4.

A rigid frame means that the connections between the beams and columns are rigid connections. Thus, in this building, the connections between beams and columns along grid lines A, B, C, D, E, 1, and 4 are rigid connections. All other beam-column or beam-beam connections are simple (shear) connections.

In a rigid frame building (such as this one), every column line does not need to be a rigid frame. The number of rigid frame lines along a given direction depends on the magnitude of lateral load on that facade—a function of building length and lateral load intensity. Thus, because this building is longer along grid lines A, B, C, and so on, every column line is a rigid frame line. In the other direction, only two out of four column lines are rigid frame lines because of the smaller length of the building in that direction. In a low-rise building, rigid frames along grid lines B, C, and D (or B and D) may be replaced by simple frames.

The columns along grid lines A and E have been turned through 90° to increase the stiffness of the frames along grid lines 1 to 4.

To fully appreciate the essence of framing layout of a building, the reader should read Principles in Practice at the end of this chapter.

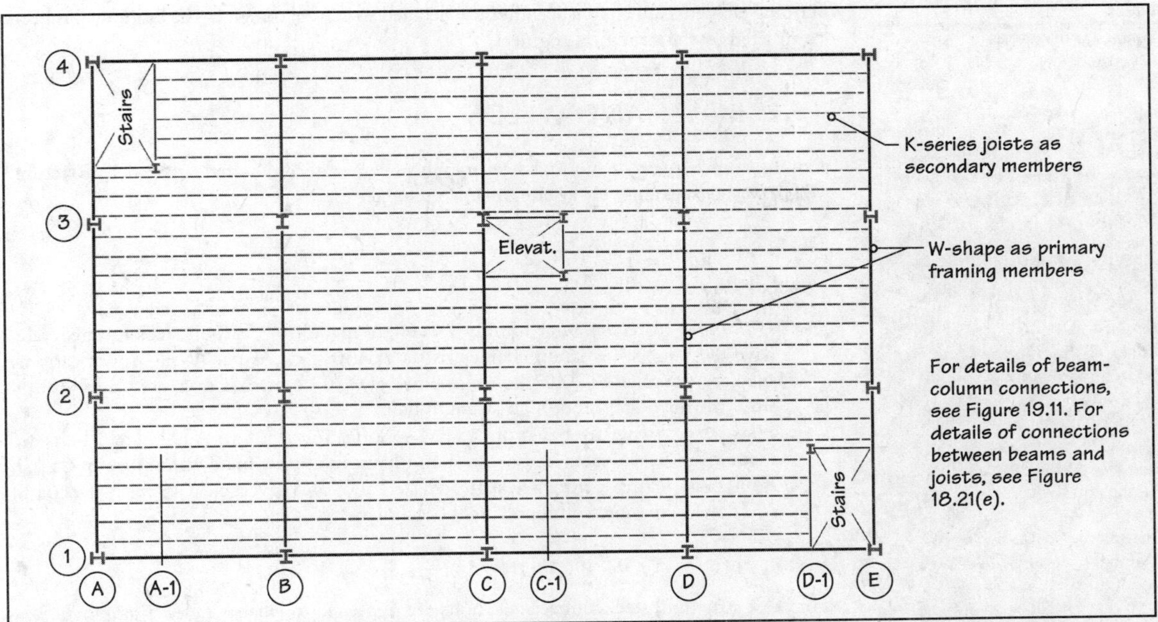

FIGURE 19.6 Floor framing plan of the rigid frame building in Figure 19.5 with steel joists as secondary framing members instead of W-section beams. Rigid frame lines are the same as in Figure 19.5, that is, along grid lines A, B, C, D, and E in one direction and along grid lines 1 and 4 in the other direction.

framing plan using steel joists as secondary elements is shown in Figure 19.6. Figure 19.7 shows the framing plan of the same building using braced frames instead of rigid frames.

The preliminary depth of floor- and roof-framing members should be established at this stage, in addition to the type of floor and roof decks to be used. This information helps to provide the approximate floor-to-floor height of the building. The exact dimensions of the

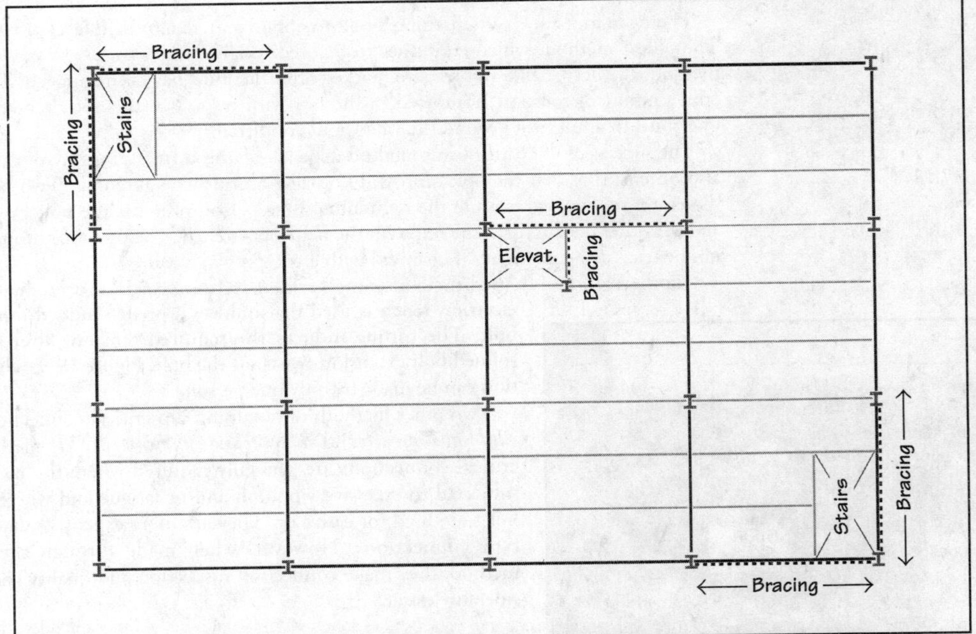

FIGURE 19.7 Framing plan of a typical braced frame building. Braced frames have been identified by dashed lines. Because braced frames provide lateral stability (resistance to lateral loads), they need to be provided in both major directions of the building. Thus, in this building, there are four exterior braced bays and two interior braced bays. In many buildings, exterior braced bays are avoided; see Principles in Practice at the end of this chapter. All beam-column and beam-beam connections are simple connections.

framing elements are established after the detailed structural design of the building has been completed by the structural engineer.

19.2 BOLTS AND WELDS

Connections between structural steel members can either be bolted or welded. Riveting, which was used extensively at one time, is no longer used.

BOLTS

Steel bolts are of two types:

- *Unfinished* (or *common* or *ordinary*) *bolts* are made from carbon steel and generally have the same stress-strain characteristics as A36 steel. As per ASTM specifications, they are classified as A307 bolts. The use of A307 bolts has decreased significantly since the introduction of high-strength bolts.
- *High-strength bolts* are based on ASTM specification A325 or A490. A325 bolts are made from heat-treated carbon steel and have an approximate yield stress of 85 ksi. A490 bolts are made from a heat-treated steel alloy and have a yield strength of 120 ksi. A325 bolts are more commonly used because they cost less.

SNUG-TIGHT CONNECTION

For most structural steel connections, bolts are tightened to what is called a *snug-tight condition.* This condition is obtained when the connected members have been bolted together using a spud wrench with the full force of a person, Figure 19.8. The load transfer between members in a snug-tight connection (also called a *bearing connection*) occurs through shear in the bolts.

SLIP-CRITICAL BOLTED CONNECTION

Another commonly used bolted connection is a *slip-critical connection.* The bolts in a slip-critical connection are tightened to a (high) tensile stress so that shear resistance is provided through friction between the connected surfaces, not through bearing, as in a snug-tight connection. In a slip-critical connection, the bolts are first brought to a snug-tight condition. Subsequently, they are tightened further until the bolt shank is under a predetermined level of tensile stress.

There are four ways of tightening a bolt to obtain a predetermined level of tensile stress. The oldest method, which continues to be used today, is the *turn-of-nut method.* In this method, the bolt shank and nut are marked after the joint has been snugged. Then a specific amount of rotation is induced in the nut with respect to the bolt, generally from a one-third to a full turn beyond the snug-tight condition.

The success of the turn-of-nut method depends on the correct snugging of the joint and also on ensuring that the bolt head will not turn on further tightening. Two persons are, therefore, required to execute the tightening. One person prevents the bolt from turning, and the other person turns the nut with the help of a wrench. The turn-of-nut method cannot be used if the members are painted with a compressible paint.

Another commonly used method is the *twist-off bolt method.* In this method, a special electric wrench is used that induces a predetermined torque in the bolt. The torque induces the required tension, after which the splined bolt extension twists off the bolt, Figure 19.9. This connection can be made by only one person.

Two other methods of obtaining slip-critical connections are the *calibrated wrench* and *direct-tension indicator* (DTI) methods. Slip-critical connections are generally required where the members are subjected to excessive vibration and/or fatigue and where oversized holes are used for easier fit. They are more expensive than bearing-type connections. However, when made through the twist-off method, they make connection inspection and quality control considerably easier.

WELDS

Welding is a process by which connected steel parts are brought to a plastic or fluid state through heating of the parts, allowing them

NOTE

Spacing of Framing Members

The spacing of columns is generally a function of the architectural plan. If the dimensions of bays are unequal, the primary horizontal framing members should generally be placed along the shorter dimension, and the secondary members should be placed along the longer dimension (Figure 19.5). This strategy reduces the difference between the depths of members and also yields an economical framing system.

If the secondary members are W-shape beams, typical spacing between them is 8 ft to 12 ft. If joists are used as secondary members, their typical spacing is 2 ft to 4 ft for floors and 4 ft to 6 ft for roofs.

Approximate Depths of Horizontal Framing Members

Depth of joist = span/20; depth of joist girder = span/14

Depth of W-section beam = span/22

Depth of W-section girder = span/16

See also Appendix B.

FIGURE 19.8 A typical spud wrench for making snug-tight bolted connections.

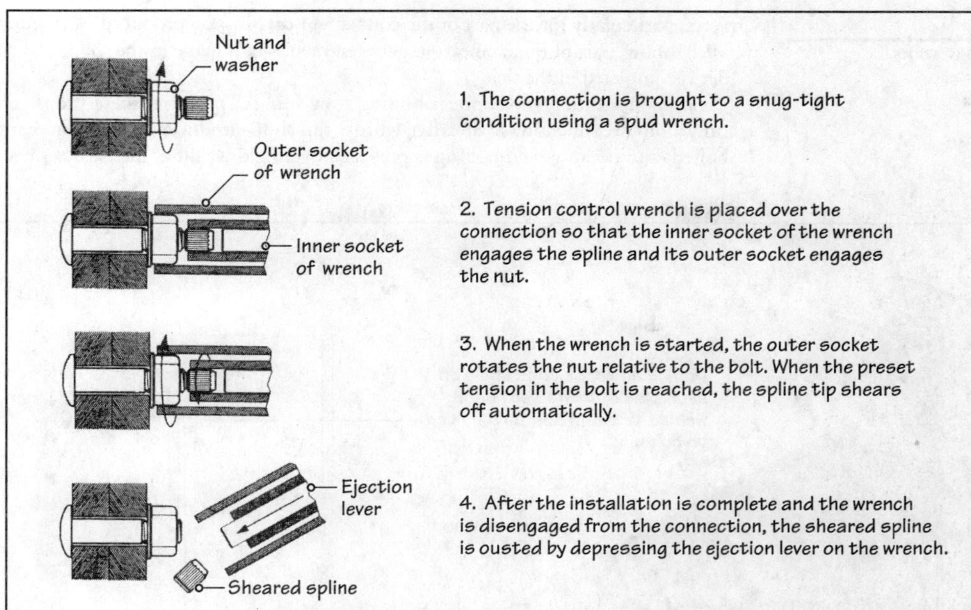

Nut and
washer
1. The connection is brought to a snug-tight condition using a spud wrench.
Outer socket
of wrench
Inner socket
of wrench
2. Tension control wrench is placed over the connection so that the inner socket of the wrench engages the spline and its outer socket engages the nut.
3. When the wrench is started, the outer socket rotates the nut relative to the bolt. When the preset tension in the bolt is reached, the spline tip shears off automatically.
Ejection
lever
Sheared spline
4. After the installation is complete and the wrench is disengaged from the connection, the sheared spline is ousted by depressing the ejection lever on the wrench.

FIGURE 19.9 Steps in making a slip-critical connection using the twist-off bolt method. (Illustration courtesy of Nucor Fasteners with permission)

to fuse together, generally with the addition of another molten metal. Welding may be done using either gas welding or arc welding. In arc welding, used more commonly today, two terminals of a high-voltage electric circuit are brought close together, creating a sustained spark across the space between the terminals. A temperature of 6,000°F to 10,000°F (3,300°C to 5,500°C) is produced in the arc.

One of the terminals in arc welding is the electrode, which is in the welder's hand, and the other terminal is provided by the two metals to be welded together. In the welding process, the electrode also melts and becomes part of the connection between the two metals being fused.

WELDING VERSUS BOLTING

Welding has a much larger range of applicability than bolting. For example, hollow structural steel sections (round or rectangular) cannot be easily bolted together, but they can be easily welded.

Welded connections also eliminate the need for bolts and connection gusset plates, which can amount to a substantial saving of steel in some structures. Welding also creates continuity between the connected members, which is more difficult to obtain through bolting. However, welding requires a much greater level of skill than bolting. Welding should preferably be carried out under the controlled conditions of a shop. Good welding practice requires that the members to be welded must be dry and free from dirt and grease. Therefore, field welding is generally avoided as much as possible.

Bolting, on the other hand, is rapid, involves less-skilled labor, and can be accomplished without any surface cleaning. It is more suited to field conditions than welding because weather conditions have relatively little effect on bolting.

19.3 CONNECTIONS BETWEEN FRAMING MEMBERS

Structural steel connections are so varied and numerous that they cannot all be discussed here. Therefore, only the more commonly used connection types are illustrated:

- Column-to-beam connections
- Beam-to-beam connections
- Column-to-column connections (i.e., column splice)

Each of these connections can be detailed in several ways. One of the major reasons for variations in the details for the same connection is the substitution of welds for bolts, or vice versa, particularly for the part of the connection that is shop produced. As Figures 19.10 to 19.13 show, part of the connection between members is made in the shop, and the remainder is completed in the field.

Some fabrication shops prefer bolting to welding. Therefore, these shops generally use fully bolted connections. In other words, the shop-produced part of the connection is bolted, and because field bolting is preferred over field welding, the part of the connection

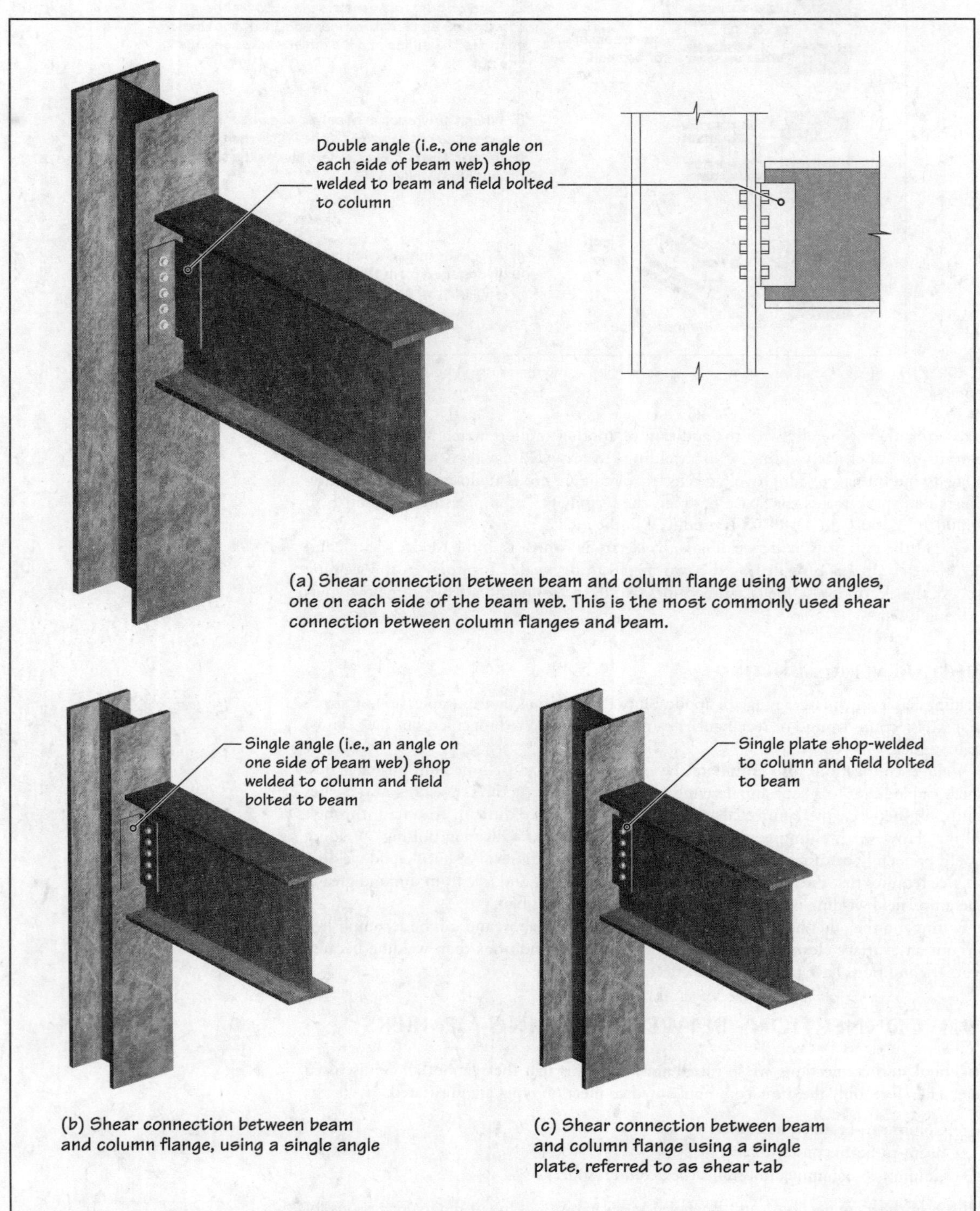

Double angle (i.e., one angle on each side of beam web) shop welded to beam and field bolted to column
(a) Shear connection between beam and column flange using two angles, one on each side of the beam web. This is the most commonly used shear connection between column flanges and beam.
Single angle (i.e., an angle on one side of beam web) shop welded to column and field bolted to beam
Single plate shop-welded to column and field bolted to beam
(b) Shear connection between beam and column flange, using a single angle
(c) Shear connection between beam and column flange using a single plate, referred to as shear tab

FIGURE 19.10 Typical column-beam shear (Type II) connections.

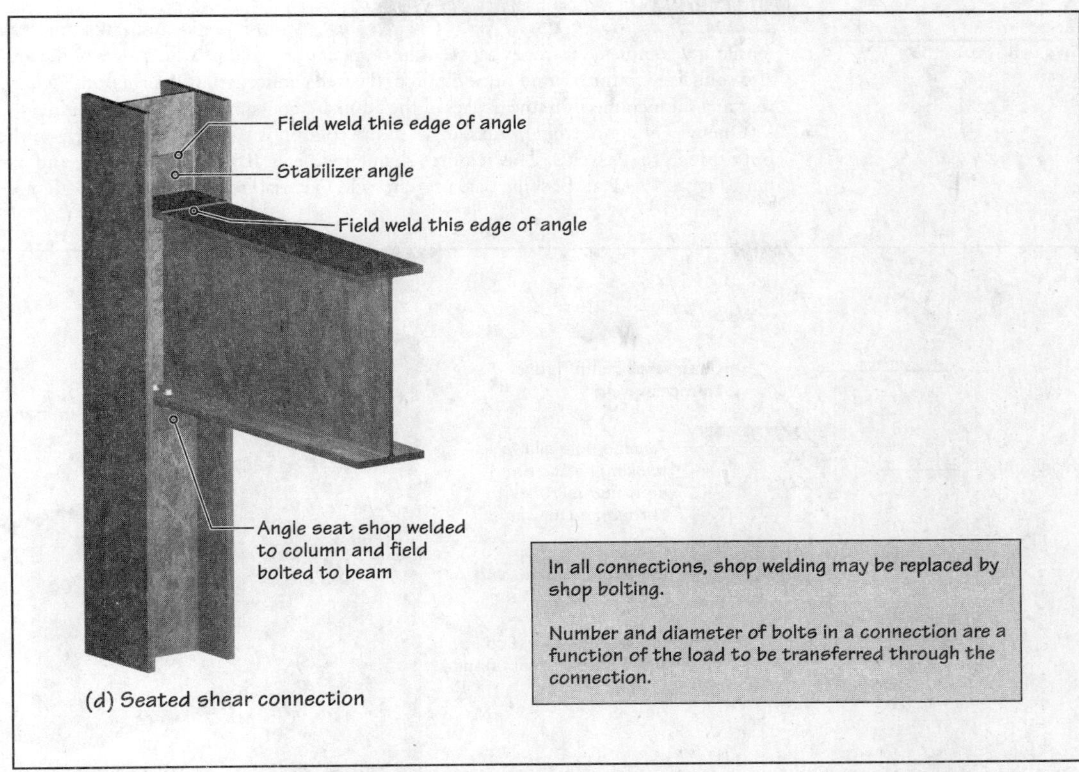
Field weld this edge of angle
Stabilizer angle
Field weld this edge of angle
Angle seat shop welded to column and field bolted to beam
In all connections, shop welding may be replaced by shop bolting.
Number and diameter of bolts in a connection are a function of the load to be transferred through the connection.
(d) Seated shear connection

FIGURE 19.10 (*Continued*) Typical column-beam shear (Type II) connections.

completed in the field is also bolted. Fabrication shops that are better equipped for welding prefer shop-welded and field-bolted connections.

In addition to variations caused by substituting bolting for welding (and vice versa), the ease of erection governs the choice of connection details.

NOTE

Interchangeability of Terms Used for Steel Member Connections

With steel connections, the terms *simple connection, shear connection,* and *AISC Type II connection* are interchangeable. The terms *moment connection, rigid connection,* and *AISC Type I* are also interchangeable.

COLUMN-BEAM CONNECTIONS

The American Institute of Steel Construction (AISC) divides column-beam connections as follows:

- Rigid connection, also called *moment connection* or *AISC Type I connection*
- Simple connection, also called *shear connection* or *AISC Type II connection*
- Semirigid connection, also called *AISC Type III connection*

In practice, however, most connections are either simple or rigid connections. Therefore, only these two types are illustrated.

Column-Beam Simple (AISC Type II) Connections A simple connection is a flexible connection that allows the ends of the beam to rotate under the loads. (Although simple connections used in practice have some end restraint, i.e., moment resistance, the restraint is small and is, therefore, neglected.)

The most frequently used column-beam simple connection uses two angles that are shop welded to a beam and field bolted to a column flange (or web), as required, Figure 19.10(a). Where a single-angle connection is acceptable, it may be shop welded to the column and field bolted to the beam, Figure 19.10(b). A single angle may be replaced by a single plate, Figure 19.10(c).

A seated shear connection is well suited for connecting the beam with a column web, particularly if there is a beam on each side of the web, Figure 19.10(d). A seated shear connection consists of one angle at the bottom of the beam and the other at the top.

The top angle simply provides stability to the connection. Hence, it is called a *stabilizer angle.* To ensure that the stabilizer angle does not introduce moment resistance in the connection, it is generally field welded to both column and beam at the toes only (the edges away from the angle heel).

Column-Beam Rigid (AISC Type I) Connections Because a rigid connection must transfer tensile and compressive stresses between the beam and the column, the flanges of the beam are also connected to the column (in addition to the web connection). Observe from Figure 19.10 that in a shear connection, the flanges of the beam are not connected to the column.

One way of connecting beam flanges to the column is to weld them directly to the column using groove welds. This requires chamfered beam flanges, access holes, and backing bars, Figure 19.11(a). Backing bars prevent weld material from dripping down. To prevent

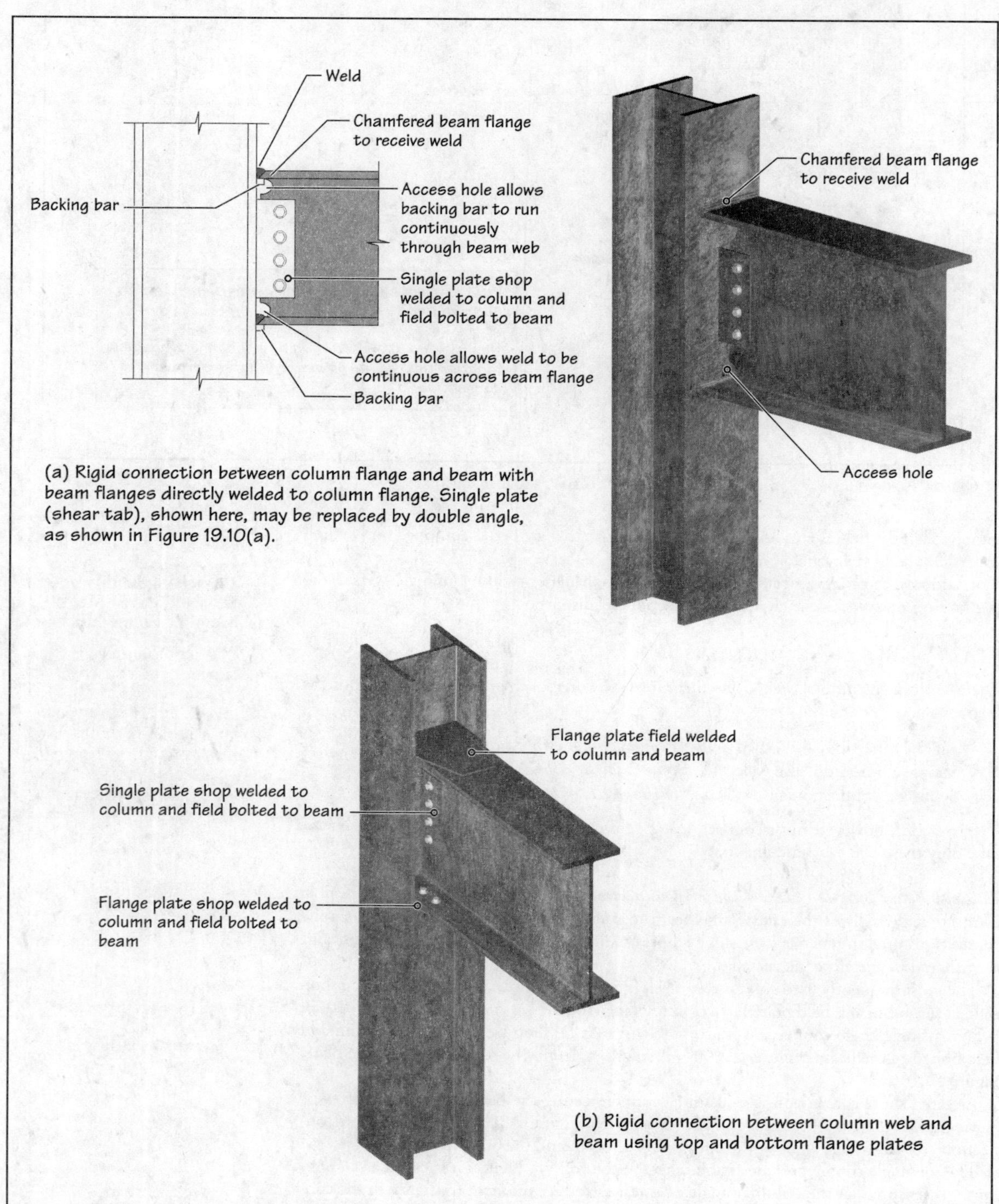

FIGURE 19.11 Typical column-beam rigid connections.

weld material from flowing off the sides, runoff tabs are used, which are removed after the connection is made. Backing bars are, however, left in place. (In seismic zones, the backing bar at the bottom flange must be removed.)

Figure 19.11(b) shows another alternative to column-beam rigid connections, in which steel plates have been used to connect beam flanges to the column instead of direct welding the beam flanges to the column. These plates are referred to as *flange plates.*

BEAM-BEAM (OR GIRDER-BEAM) CONNECTIONS

Beam-to-beam connections, where one beam supports another beam, are fairly common in steel floors and roofs. These connections are generally simple connections and are similar to simple column-to-beam connections, Figure 19.12.

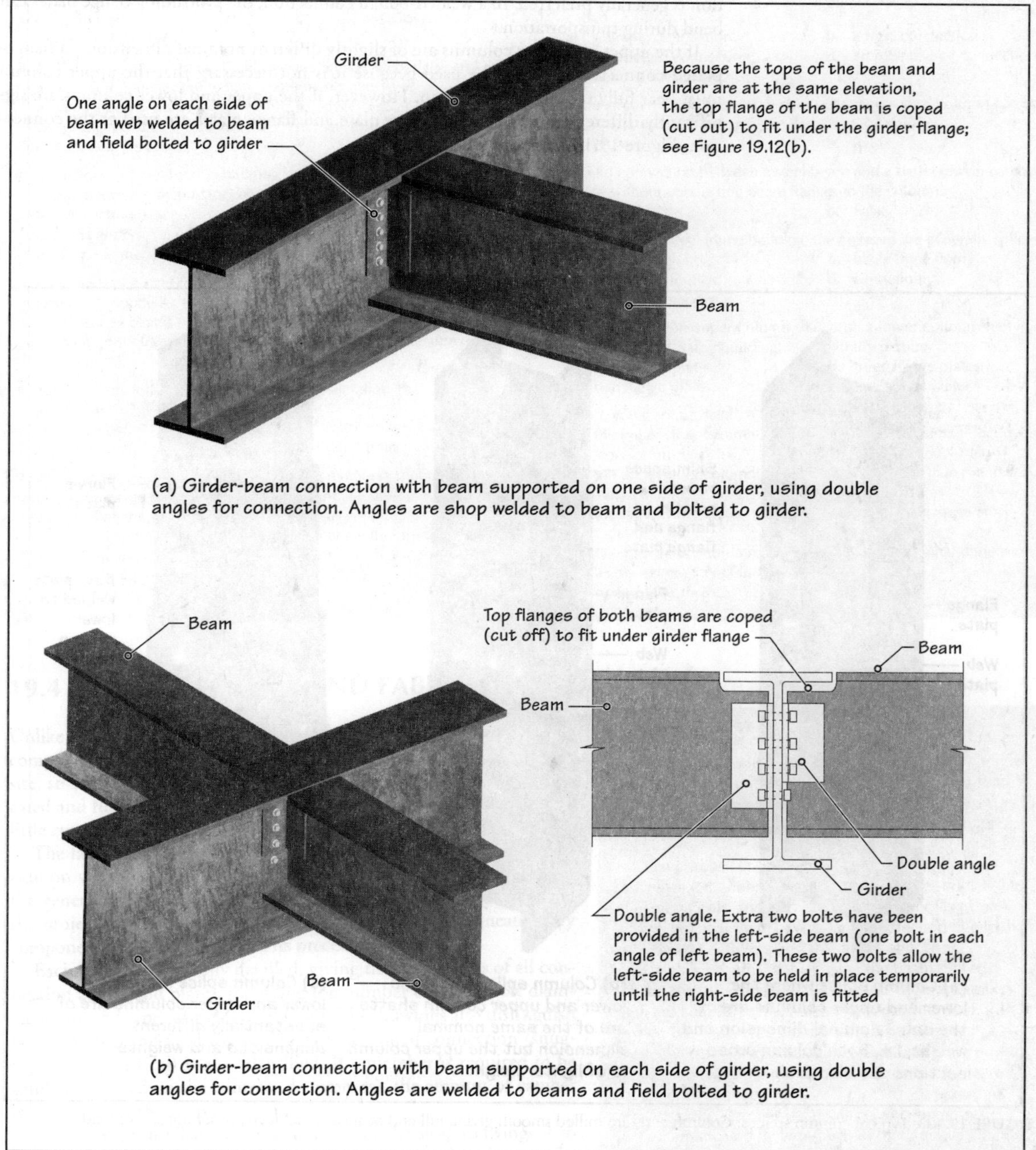
Girder
One angle on each side of beam web welded to beam and field bolted to girder
Because the tops of the beam and girder are at the same elevation, the top flange of the beam is coped (cut out) to fit under the girder flange; see Figure 19.12(b).
Beam
(a) Girder-beam connection with beam supported on one side of girder, using double angles for connection. Angles are shop welded to beam and bolted to girder.
Beam
Top flanges of both beams are coped (cut off) to fit under girder flange
Beam
Beam
Double angle
Girder
Double angle. Extra two bolts have been provided in the left-side beam (one bolt in each angle of left beam). These two bolts allow the left-side beam to be held in place temporarily until the right-side beam is fitted
Beam
Girder
(b) Girder-beam connection with beam supported on each side of girder, using double angles for connection. Angles are welded to beams and field bolted to girder.

FIGURE 19.12 Typical girder-beam (shear) connections.

STEEL DETAILING

For most buildings, the design and detailing of connections between components (such as those shown in Figures 19.10 to 19.13) are prepared by the fabricator based on the framing plans and other information provided in the project's structural drawings. Leaving connection detailing to the fabricator allows the fabricator to use the details that are most economical and best suited for erection, scheduling, and the peculiarities of the site. Where details are aesthetically significant, the fabricator must conform to the architectural and structural requirements of the project.

Some fabricators have an in-house detailing staff. Most fabricators, however, subcontract the work to an independent steel detailer. Structural steel detailing is both a science and an art because the details must not only be able to withstand the forces (as required by the engineer's drawings) but also be economical, be easy to erect, and suit the fabrication shop's capabilities and resources.

ORDERING MATERIALS FOR FABRICATION

After the shop drawings are completed, they are reviewed first by the general contractor and then by the project architect and structural engineer. Because the modifications expected from the review process are generally minor, the fabricator orders the structural steel sections from the rolling mills while the review of shop drawings is in progress.

Most rolling mills need sufficient lead time to deliver the material, particularly the heavier structural steel sections. This is because a rolling mill rolls a certain section (and a steel grade) for a few days; then the rollers are changed to produce a different section (Figure 18.4). Thus, a given section is rolled at intervals that may range from weeks to months. Therefore, a rolling mill generally rolls a batch containing different sections on order. To hold various grades, sections, and lengths in the mill's stock waiting to receive an order is too uneconomical for most mills.

A fabricator may order heavy sections (for columns and beams) cut to the required lengths by the mill. On the other hand, the fabricator may opt to cut the sections to lengths from standard mill lengths in the shop if that is more economical. Standard lengths are generally 30 ft, 35 ft, 40 ft, and so on, up to 60 ft. Length greater than 60 ft may be available as a special order. Smaller sections, such as plates and angles, are always cut by the fabricator.

It is important to realize that structural sections with exactly the same specified dimensions and weight may not, in reality, be of the same dimensions. Wear of the rollers, thermal distortion of hot cross sections and their subsequent cooling, and several other factors cause the dimensions of the sections with the same specifications to be slightly different. Steel detailers and fabricators recognize this fact and provide the required connection and erection tolerances.

FABRICATION

Most large steel-fabrication shops are semiautomated, that is, computer assisted, which makes the fabrication process not only precise but also quick. Figure 19.14 shows a typical

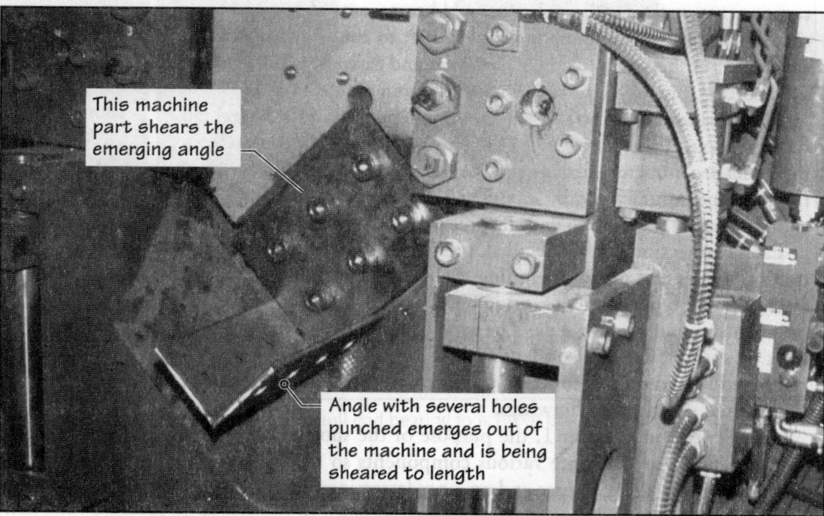
This machine part shears the emerging angle
Angle with several holes punched emerges out of the machine and is being sheared to length

FIGURE 19.14 A computer-aided plate and angle punching machine cuts plates and angles to the required size and drills holes of the required diameters and spacings without any manual labor. Photo taken at the Irwin Steel Fabrication Plant, Justin, Texas.

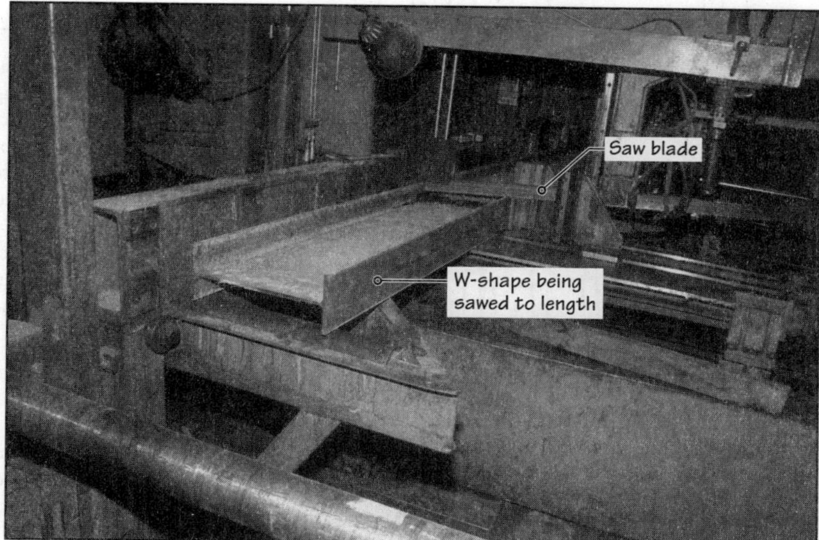

FIGURE 19.15 A computer-aided band saw cuts heavy steel members to size. The machine is versatile enough to make skew cuts instead of a right-angle cut, as shown here. Photo taken at the Irwin Steel Fabrication Plant, Justin, Texas.

angle-shearing machine that shears the angles to size and drills holes of the required diameters and spacings without any manual labor. Whereas lighter sections, such as plates and angles, are sheared, heavier steel sections, such as beams and columns, are sawed to desired length by semi-automated or fully automated saws, Figure 19.15.

Welding in most fabrication shops is carried out manually. For long, continuous lines of welds or repetitive welding, an automated welding system is used.

Most structural steel members do not require any surface preparation, such as blast-cleaning or a prime coat. However, if the steel has developed loose mill scale because of its extended exposure to weather, it should be blast-cleaned, particularly if it is to be treated with spray-on fire protection or intumescent paint or galvanized. Steel components that are to be finished with paint generally require a shop coat of primer.

As the components are being fabricated, they are identified by marks that help further fabrication and erection. Unless dispatched to the construction site immediately on fabrication, the fabricated components are stored in the fabricator's shed. The storage is planned in a way that allows easy retrieval for delivery to the site in predetermined batches.

19.5 STEEL ERECTION

The erection of the structural steel frame at the site may be performed by the fabricator or by a separate erection company. In most cases, the general contractor will seek separate bids for fabrication and erection. If the fabricator, selected on the basis of a competitive bid for fabrication, also submits a competitive bid for erection, the fabrication and erection may be done by the same organization. If not, the erection contract may be given to a different company. Fabrication and erection by the same entity is obviously preferred.

Steel frame erection begins with the erection of columns, Figure 19.16, using a crane. For a small building, a mobile crane on rubber tires is generally adequate. For a large building, a tower crane is needed in addition to one or more mobile cranes, Figure 19.17. Generally, a tower crane is built over the foundations of an elevator shaft. Climbing cranes, secured to the building's columns, can also be used. In that case, a climbing crane climbs up the columns as the building's height increases. These columns are specially engineered and provided with special attachment points.

Structural steel columns are anchored to the foundations through anchor bolts. The anchor bolts are set into column footings before footing concrete is placed. The bolts are supplied by the fabricator but embedded into the footings by the general contractor per the fabricator's details.

A steel base plate is shop welded to the column bottom with predrilled holes to penetrate the anchor bolts. Columns are leveled to ensure that they are plumb, which can be done in a number of ways. One method uses leveling bolts under the base plate, Figure 19.18(a).

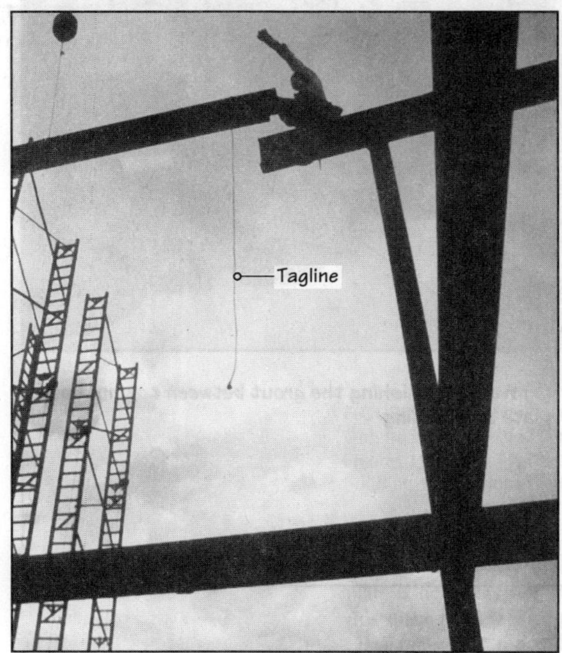

FIGURE 19.20 A steel erector (with his body harness tied to the beam on which he sits) maneuvers the new beam into position, initially using the rope attached at the end of the beam and later by hand. The rope is referred to as a *tagline* in erection vernacular.

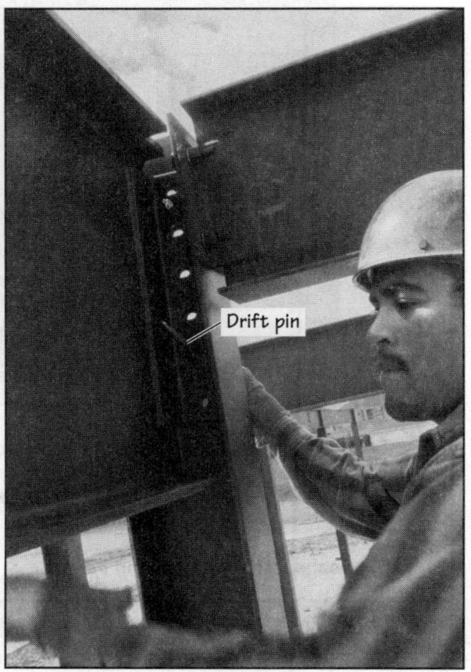

FIGURE 19.21 Generally, members at a connection are forced to align. In this image, a drift pin has been used to make the alignment. In some connections, crowbars or hammers may have to be used.

The initial connection between members is made using some of the bolts in the connection. After most of the members have been placed in position, the remaining bolts are installed and the connection is completed by hand tightening the bolts. Finally, an impact wrench is used in case of slip-critical connections.

The placement of beams for a tier is followed by floor and roof decking. The installation of decking is done by the general contractor or the erector, depending on the contract.

PRACTICE QUIZ

Each question has only one correct answer. Select the choice that best answers the question.

16. The fabrication of structural steel components is done based on
- **a.** architectural construction drawings.
- **b.** structural construction drawings.
- **c.** both (a) and (b).
- **d.** shop drawings.
- **e.** none of the above.

17. Details of connections between structural steel members are prepared by the
- **a.** architect.
- **b.** structural engineer.
- **c.** both (a) and (b).
- **d.** general contractor.
- **e.** none of the above.

18. The erection of structural steel members is generally done by the
- **a.** general contractor.
- **b.** structural engineer.
- **c.** American Institute of Steel Construction.
- **d.** steel detailer.
- **e.** none of the above.

19. The installation of a floor or roof deck is done by the
- **a.** general contractor.
- **b.** erector.
- **c.** manufacturer of the deck.
- **d.** (a), (b), or (c).
- **e.** (a) or (b).

20. Camber refers to
- **a.** downward deflection of a beam.
- **b.** upward deflection of a beam to counter downward deflection by a live load.
- **c.** upward deflection of a beam to counter downward deflection by a dead load.
- **d.** upward deflection of a beam to counter downward deflection by a dead load plus a live load.
- **e.** none of the above.

21. In connecting a steel column to concrete foundations, anchor bolts are used. These bolts are embedded into the foundation by
- **a.** the general contractor.
- **b.** the fabricator.
- **c.** the erector.
- **d.** a specialized bolt-embedding outfit.
- **e.** none of the above.

22. A steel column may or may not need a base plate at the bottom.
- **a.** True
- **b.** False

23. To level a column so that it is plumb,
- **a.** steel shims are used.
- **b.** high-strength plastic shims are used.
- **c.** leveling bolts are used.
- **d.** any one of the above.
- **e.** none of the above.

EXPAND YOUR KNOWLEDGE

A Building Separation Joint in a Steel-Frame Building

As discussed in Section 9.2, continuous building separation joints are required in many buildings. A preferred method is to use two columns and two beams at the joint. Often for the sake of economy, however, a single column with two beams is used. Figure 19.22 explains the construction of such a joint in a building with W-shape columns and beams. (See Chapter 23 for a similar detail in a reinforced concrete building.)

Details of a building separation joint for a single-story building, consisting of joists and joist girders, are shown in Figure 19.23.

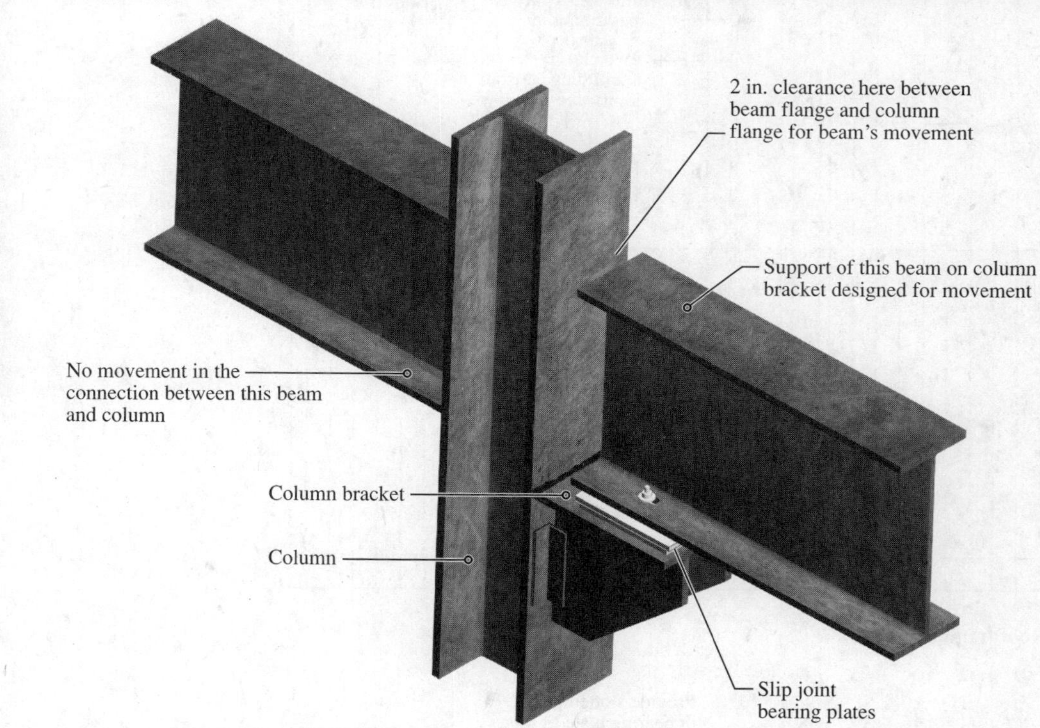

(a) SLIP-JOINT (STEEL BEARING PLATE ASSEMBLY)
This assembly consists of two steel bearing plates. Each bearing plate (shown below) is factory epoxied to a Teflon sheet. The lower bearing plate is welded to column bracket and consists of two round holes. The upper bearing plate consists of slotted holes. The slotted holes allow the beam to slide over the lower bearing plate. The upper bearing plate is slightly larger than the lower bearing plate, which produces a gap at the end between plates to allow it to be filled with an elastomeric sealant. The sealant protects Teflon sheet from being damaged by dust and other desposits.

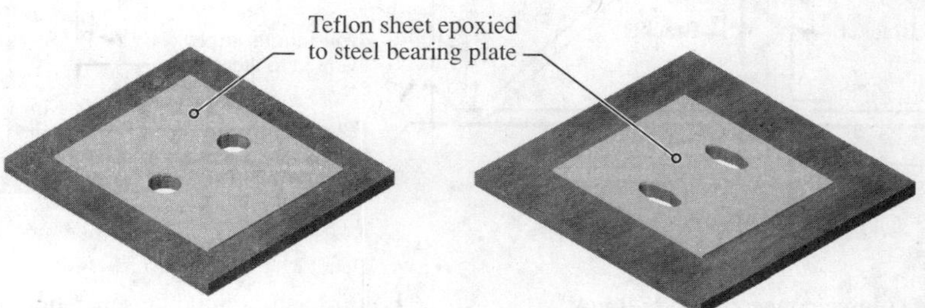

(b) STEEL BEARING PLATES
Although the plates have been shown separated from each other, they are placed over each other with Teflon surfaces in contact. (For properties of Teflon sheet, see Expand Your Knowledge, "Building a Separation Joint in a Sitecast Concrete Building" in Chapter 23)

FIGURE 19.22 Details of the connection between W-shape beams and a column at a building separation joint using only one column (see Section 9.2 for an explanation of a building separation joint).

(Continued)

EXPAND YOUR KNOWLEDGE

A Building Separation Joint in a Steel-Frame Building (*Continued*)

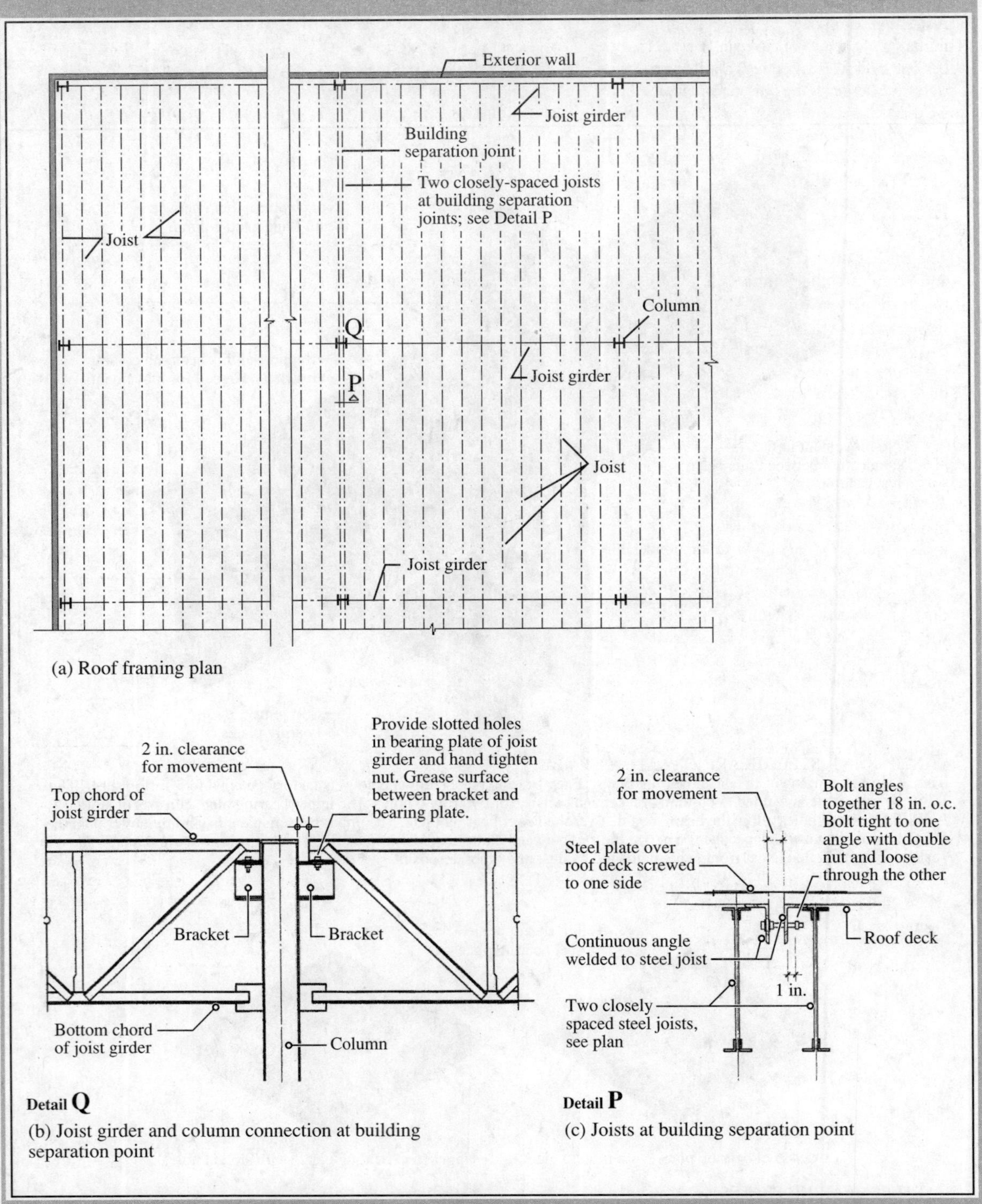

FIGURE 19.23 Details of a building separation joint in a single-story building with steel joists and joist girders (see Section 9.2 for an explanation of a building separation joint).

PRINCIPLES IN PRACTICE

Fundamentals of Frame Construction

Although frame construction in timber is much older, the real use of frame construction originated with the discovery of steel. A steel frame can be stabilized without braces. In a timber frame, on the other hand, some form of bracing is essential to obtain frame stability (see Principles in Practice at the end of Chapter 15). Frame stability without the use of braces is possible in steel because steel-frame members can be fused together at the joints through welding, giving continuity across the joints, that is, rigid connections. The only other material in which connection rigidity is obtained is cast-in-place concrete. Therefore, when we think of contemporary frame construction, we generally think either of *steel-frame construction* or *concrete frame construction.*

The fundamentals of frame construction described here, though focused on steel-frame construction, apply to concrete frame construction as well.

POST-AND-BEAM CONSTRUCTION

The earliest predecessor of the frame construction is the *post-and-beam* (also called *post-and-lintel*) *construction.* A post-and-beam assembly consists of columns (posts) and beams (lintels) with little or no connection between the posts and the beams. It was among the earliest construction systems devised by humankind, Figure 1. Ancient Egyptians, Greeks, and Romans used post-and-beam construction extensively to produce architectural masterpieces that stand until this day (see Figure 4.1).

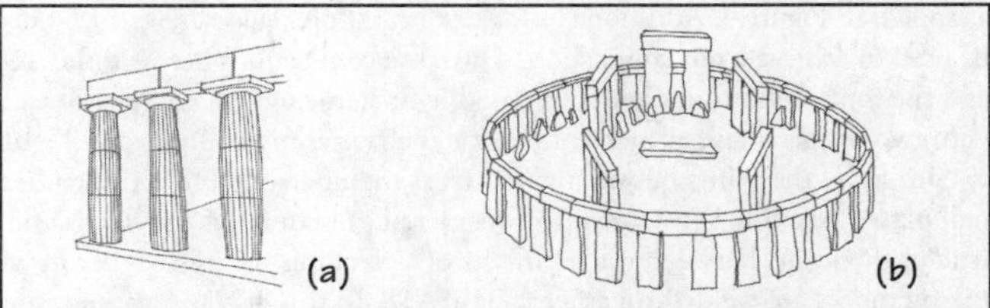

FIGURE 1 **(a)** A typical post-and-beam structure used by ancient Egyptians, Greeks, and Romans (Figure 4.1). **(b)** Stonehenge (United Kingdom, approximately 2500 to 1700 BC) is generally cited as the earliest surviving example of the post-and-beam structure.

A post-and-beam assembly works well for gravity loads. Under gravity loads, the beam is subjected to bending. Because there is no connection between the posts and the beam, the bending in the beam is not transferred to the posts, Figure 2. Consequently, the posts are under pure compression—no bending.

Under lateral loads (wind and earthquake), the post-and-beam assembly is theoretically unstable. When the lateral loads are within the plane of the post-and-beam assembly, the beam tends to slide over the column, and the column tends to overturn or slide along its base, Figure 3(a). If the lateral loads are perpendicular to the plane of the assembly, the entire assembly tends to overturn, Figure 3(b).

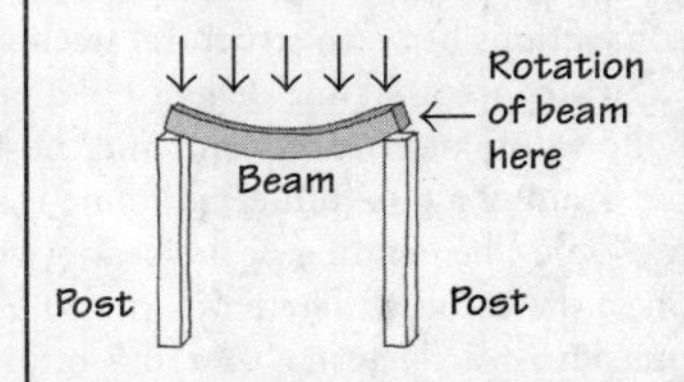

FIGURE 2 Under gravity loads, the beam in a post-and-beam structure is subjected to bending, which causes it to rotate at the joint. Because of the absence of a connection between the post and the beam, the rotation of the beam at the joint is not transferred to the post. The post is, therefore, subjected to compression only.

Early post-and-beam buildings used large, hefty stones. Stability of the building against overturning and sliding was achieved through large dead load of members and the friction between their surfaces.

A frame is similar to a post-and-beam assembly, except that the members of the frame (the columns and beams) are connected together. The connection between a column and a beam can be a *pin connection,* giving a pin-connected frame, or it can be a *rigid connection,* giving a rigid frame.

A pin-connected column-beam frame can be one of the following two types:

- Pin-connected cantilevered frame
- Pin-connected braced frame

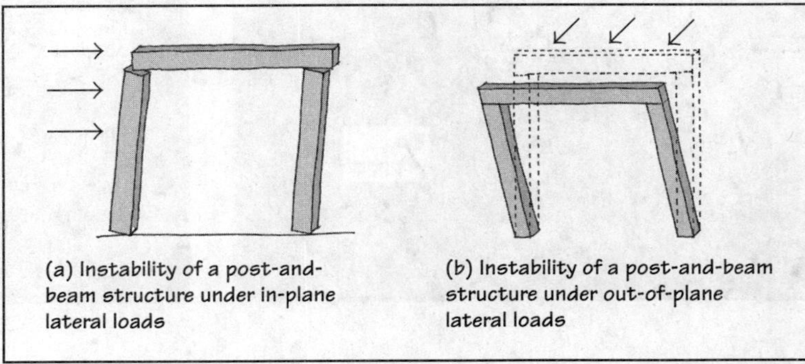

(a) Instability of a post-and-beam structure under in-plane lateral loads
(b) Instability of a post-and-beam structure under out-of-plane lateral loads

FIGURE 3 Under the action of both in-plane and out-of-plane lateral loads, a post-and-beam structure is unstable.

(Continued)

Fundamentals of Frame Construction (*Continued*)

PIN-CONNECTED CANTILEVERED FRAME

A pin connection is defined as a connection that allows the members to rotate at the connection independently of each other. In other words, the rotation of one member at the connection is not transferred to the other member. The simplest pin-connected frame consists of a single-bay frame (i.e., one beam and two columns) in which the beam is connected to each column through a single pin (a nail, screw, or bolt), Figure 4.

This connection allows the beam to bend and rotate under gravity loads without transferring the bending to the column. The connection, however, transfers the gravity load on the beam through shear created in the pin. A pin connection is, therefore, also referred to as a *shear connection.* Another term used for this connection is a *simple connection* (Section 19.3).

In practice, a shear (pin or simple) connection need not have lapped members, as shown in Figure 4. Additionally, welds or multiple nails, screws, or bolts may be used to achieve a pin connection. Thus, the connections between the studs and the top and bottom plates of a wood light-frame, which are typically joined with two or three nails at each connection, behave as pin connections, Figure 5.

Similarly, the joints between wood truss members, connected together with nail plates (Section 14.4), are also considered pin connections. The same is true of the joints between the members of a steel joist or joist girder in which the members are welded together (Figures 18.20 to 18.22). Although such joints are not true (100%) pin connections, they provide adequate rotation of members at a joint to be considered pin connections.

In buildings, a joint detailed as a *true* pin connection is used for both structural and ornamental reasons, Figure 6. True pin connections are more common in civil engineering structures, such as waterway, highway, or railroad bridges. Details of commonly used shear connections between structural steel beams and columns were illustrated in Section 19.3.

In a conceptual line diagram of a pin-connected frame, a pin connection is represented by a hollow dot, and the columns and the beams are represented by single heavy lines, Figure 7. If the beam of a pin-connected frame is subjected to gravity loads, it will bend as shown in Figure 7(a). The bending of the beam (which causes the beam to rotate at the joint) is not passed on to the column. Therefore, the columns in a pin-connected frame come under pure compression, exactly as in a post-and-beam structure.

One way to ensure the stability of the frame in Figure 7 against lateral loads is to anchor the columns rigidly into the ground so that they are cantilevered from the ground like utility poles. Under the action of a lateral load, both columns will deflect horizontally, as shown in Figure 7(b)—a

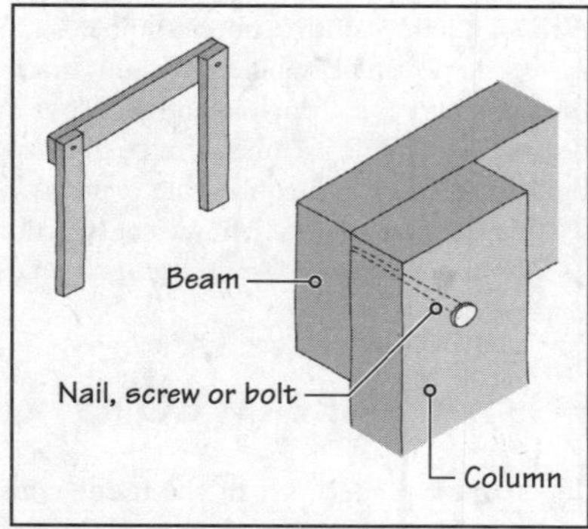

FIGURE 4 A single-bay, pin-connected frame in which the beam and column are connected together with a single nail, screw, or bolt. A pin connection is also referred to as a *shear connection* because the gravity load from the beam is transferred to the column, creating shear in the nail, screw, or bolt.

FIGURE 5 A connection between a stud and the top (or bottom) plate of a wood light-frame, typically using two or three nails, is considered a pin connection.

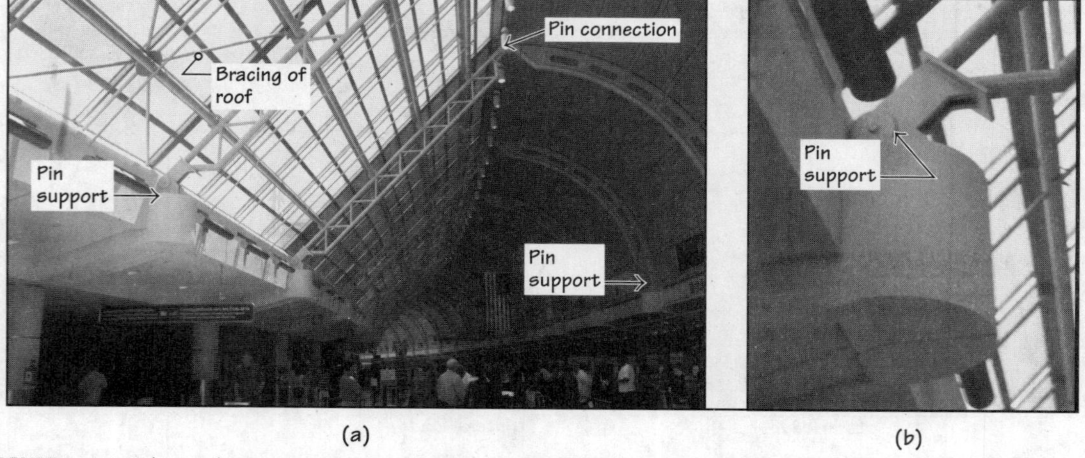

FIGURE 6 **(a)** A glass roof at Toronto Airport supported by three-pin frames. Each frame consists of two members, a linear trussed member and a curved hollow-web member. The members are joined at the top with a pin connection. **(b)** A pin connection has also been used to support each member at its support (base). All three pins in each frame have been detailed as true pins.

PRINCIPLES IN PRACTICE

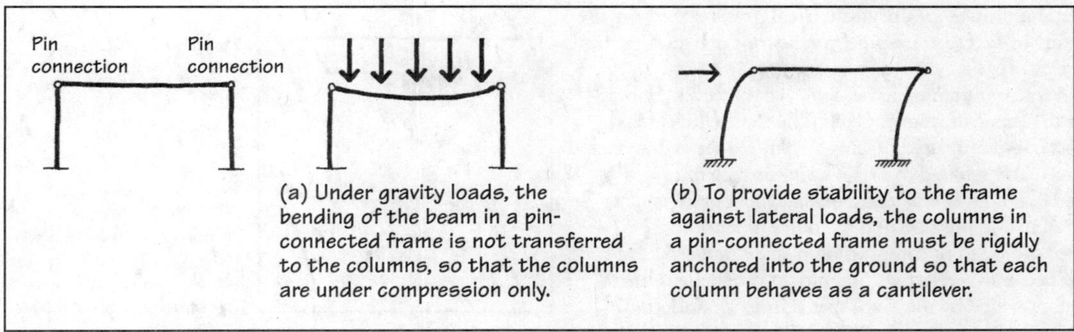

FIGURE 7 In a conceptual line diagram of a pin-connected frame, a pin is represented by a circular dot, and the frame members are represented by single heavy lines.

tendency called *racking*. Excessive racking of the cantilevered frame in Figure 7 is one of its major limitations. Racking can be reduced only by increasing the size and stiffness of columns—a solution that becomes increasingly inefficient as the column height increases.

Note that because of the pin connection between the beam and the columns, the bending in the columns of the cantilevered frame is not transferred to the beam. Thus, in transferring the lateral load from one column to the other, the beam in Figure 7 comes under compression.

Another limitation of the cantilevered frame in Figure 7 is the need for rigid column bases. The rigidity at the base of each column can be achieved by anchoring the column to the foundation, that is, by using a concrete footing. The footing must be heavy and sufficiently large so that the bending in the columns will not cause the footing to rotate.

A shallow, isolated column footing is generally not rigid enough to resist rotation, particularly if the footing is small and the soil is compressible. Such a footing simply rotates with the column, Figure 8(b), representing a *pinned column base*. A deep, large, and isolated column footing, Figure 8(c), or a mat footing in which several columns share a common footing (Section 12.1), is treated as a *rigid* (or *fixed*) *column base*.

NOTE

Racking of a Frame

The term *racking* refers to lateral deflection (angular deformation) of a frame under lateral loads and is an important design consideration. Building codes limit the amount of racking in buildings. Excessive racking causes occupant discomfort, damage to nonstructural components such as doors and windows, and fatigue in structural components, which may lead to structural failure over a period of time.

PIN-CONNECTED BRACED FRAME

In the conceptual line diagram, a pinned column base is represented in the same way as a pin connection, that is, by a circular dot. Thus, a single-bay, pin-connected frame in which the columns have pinned bases is a four-pin frame, Figure 9. Such a frame is obviously unstable, as it will simply fold under lateral loads. A four-pin frame is also

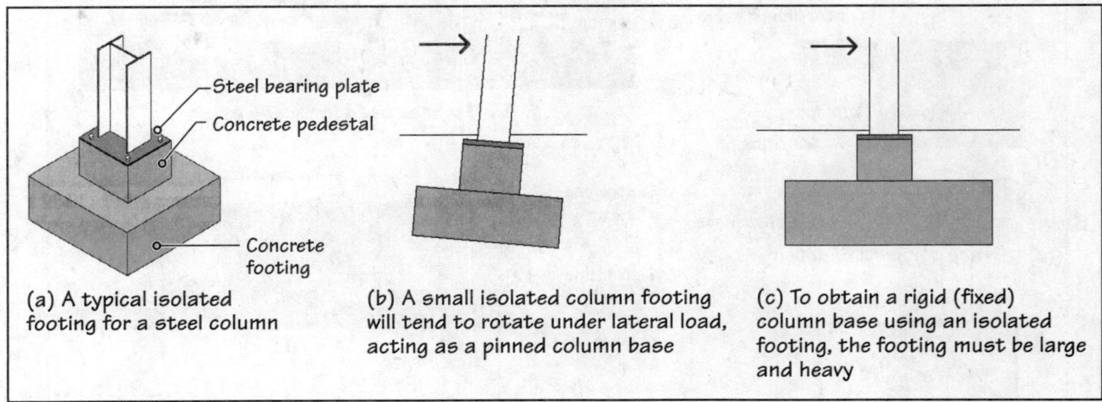

FIGURE 8 **(a)** A typical isolated footing for a steel column. **(b)** A small column footing is believed to provide a pinned column base because it tends to rotate under the action of a lateral load. **(c)** To obtain cantilever behavior from a column with an isolated footing, the footing must be large and heavy.

(Continued)

PRINCIPLES IN PRACTICE

Fundamentals of Frame Construction (*Continued*)

unstable under an unsymmetrical gravity load on the beam. In fact, a four-pin frame is not a structure, but a *mechanism.*

A four-pin frame can be stabilized, however, by bracing the frame in its own plane. The most efficient and most commonly used bracing system is a set of diagonal braces, also referred to as *X-bracing* or *cross bracing,* Figure 10. Under lateral loads, one diagonal brace is subjected to compression and the other to tension. The brace subjected to compression buckles under the load and becomes ineffective. Stability of the frame is, therefore, provided by the brace that is in tension. Because steel is strong in tension, X-bracing members need not be heavy. Therefore, X-bracing generally consists of steel rods, hollow pipes, plates, or angles. In high-rise structures, heavier sections such as wide-flange or hollow structural steel members are used.

Other bracing alternatives are the *K-brace* and the *eccentric K-brace,* Figure 11. K-bracing and eccentric K-bracing are generally used in multistory buildings, where substantial openings (doors or windows) are required in braced bays. The difference between the K-brace and the eccentric K-brace lies in their details. The braces in an eccentric K-brace do

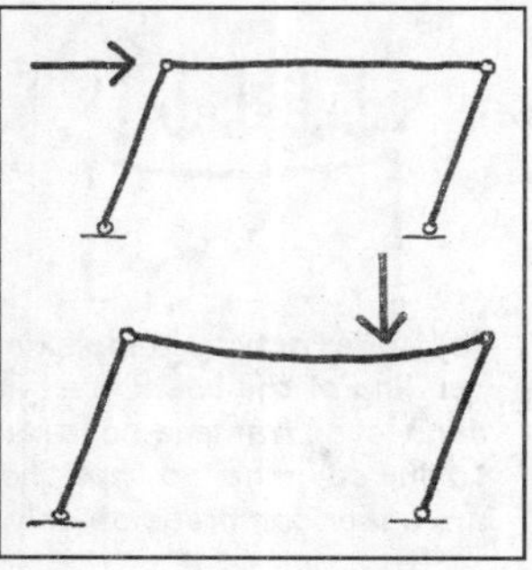

FIGURE 9 A four-pin frame is not a structure but a mechanism because it is unstable under both lateral loads and asymmetrical gravity loads.

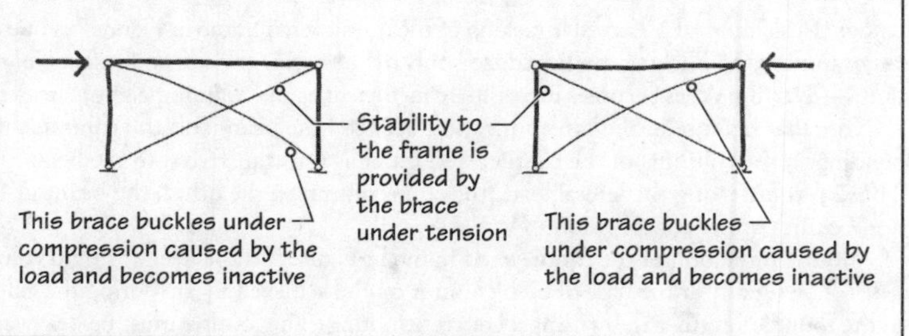

FIGURE 10 A four-pin frame can be stabilized by using X-bracing. Note that only one of the two X-braces is active under a lateral load.

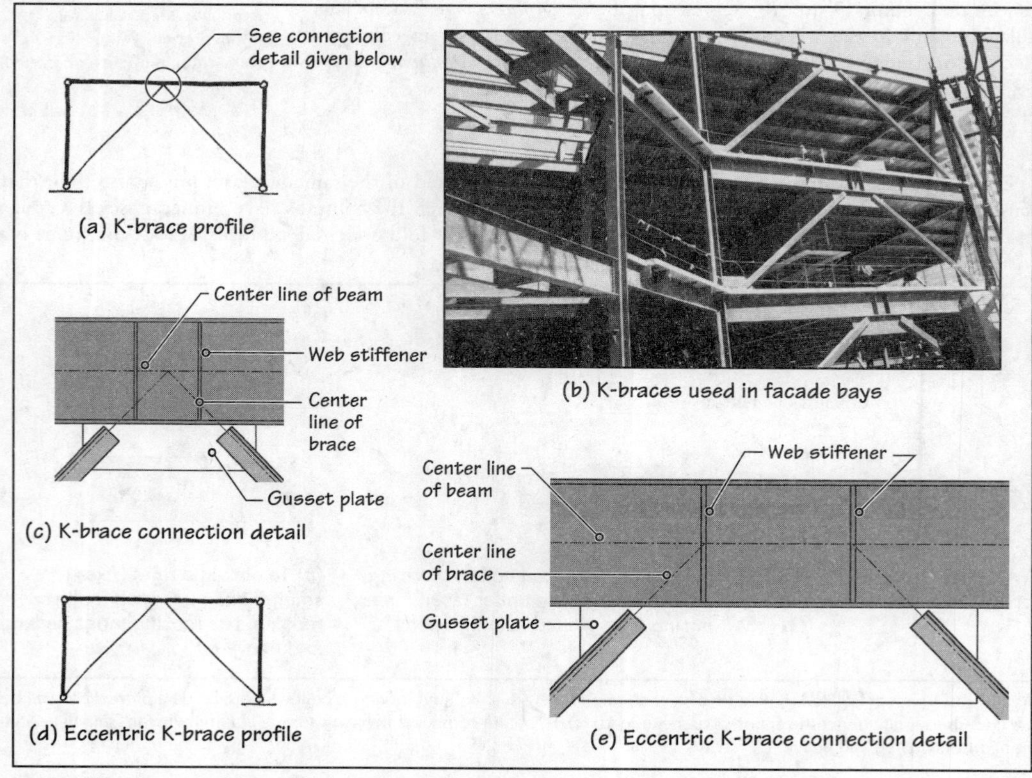

FIGURE 11 K-braces and eccentric K-braces.

PRINCIPLES IN PRACTICE

not meet at a point, leaving a small length of beam between the connections. An eccentric K-brace is generally recommended for use in seismic areas due to its ability to provide ductility of the frame.

Because a K-brace (or an eccentric K-brace) is connected to the center of the overlying beam, it must carry gravity loads on the beam, which halves the beam's span, reducing the beam size. Gravity load support implies that K-braces are heavier than X-braces. Thus, unlike X-braces, in which only one brace is active at a time, both K-braces are designed to resist tension as well as compression.

In low-rise buildings, one diagonal brace may be used, which must be heavy enough to resist compression as well as tension, Figure 12(a). A particularly interesting bracing alternative for a pin-connected frame is a *panel brace,* in which the entire space within the frame is filled with a panel, Figure 12(b). In concrete or steel structures, the panel brace consists of a masonry or concrete infill wall, referred to as an *infill brace* or *shear wall.*

(a) Single brace

(b) Panel brace, also called shear wall brace

FIGURE 12 **(a)** If a single brace is used, it must be sufficiently heavy against buckling under compression. **(b)** Shear wall bracing.

RIGID FRAME

In a rigid frame the connections between the beam and the columns are rigid. Unlike a pin connection, a *rigid connection* retains a 90° angle between the connected members on deformation under the loads. Rigidity implies that the bending of one member is transferred through the joint to the other member. Therefore, a rigid connection is also called a *moment-resisting connection* or simply a *moment connection.*

Because both the beam and the columns bend together, all three members of a rigid frame function collectively in resisting a load. Consequently, when the beam of a rigid frame is subjected to gravity loads, its deformation is smaller than that of the beam in a pin-connected frame. In the conceptual line diagram, a moment connection is represented by a small diagonal thickening of the joint, as shown in Figure 13.

Rigid connection

Rigid connection

Tension tie

FIGURE 13 Unlike a pin connection, a rigid connection retains its 90° angle between the connected members on deformation. Consequently, a rigid frame resists gravity loads by creating bending in all three frame members. Because bending is structurally inefficient, a rigid frame is generally less economical than a comparable braced frame.

Note that a rigid frame under gravity loads experiences an outward thrust in column bases similar to that of an arch.

In a single-line diagram of the frame, a rigid connection is denoted by diagonal thickening of the joint.

RIGID FRAME VERSUS BRACED FRAME

A rigid frame does not require bracing for lateral stability, providing an unobstructed space. However, a rigid connection is more difficult to make than a shear connection. In a wood rigid frame, joint rigidity may be obtained by securely fastening a plywood or OSB gusset plate on both sides of the joint, Figure 14. Compare this with simply nailing the members together to obtain a pin connection.

As illustrated later, the difference between a rigid connection and a shear connection is even more pronounced in a steel-frame structure. Simplicity of connections is one reason why braced steel frames are preferred over rigid steel frames.

Braced frames are generally less expensive and allow faster erection. Rigid frame structures are used only where braced frames are architecturally unacceptable because of the obstruction caused by the braces. Some of the other advantages of a braced frame over a rigid frame are as follows:

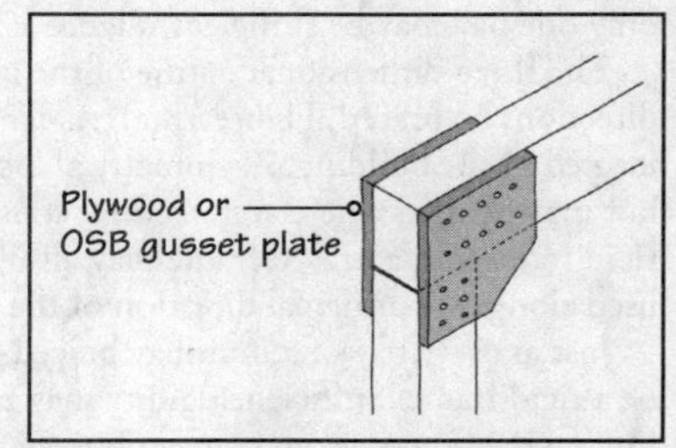

FIGURE 14 A rigid connection between two wood frame members can be obtained by nailing a plywood or OSB gusset plate on both sides of the joint. A rigid connection between two steel members is more complex (see Figure 19.11).

- The columns in a rigid frame are subjected to bending under lateral loads (Figure 13). In a braced frame, the columns are subjected to axial stresses only. Because bending is an inefficient structural action, the columns in a rigid frame are heavier than the columns in a braced frame. This difference becomes progressively more pronounced as the height of the structure increases.

(Continued)

PRINCIPLES IN PRACTICE

Fundamentals of Frame Construction (*Continued*)

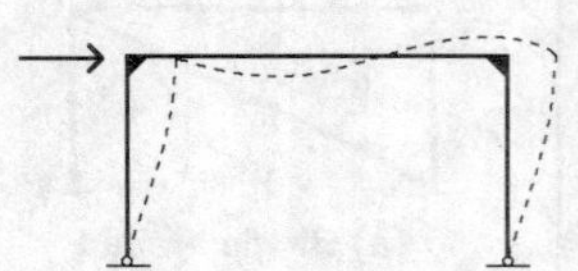

(a) Racking of a rigid frame with pinned bases under lateral loads.

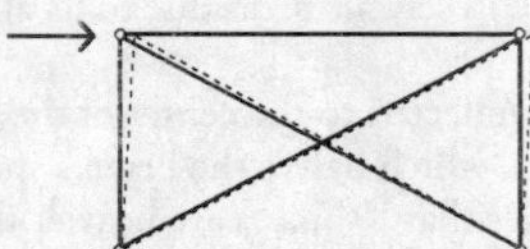

(b) Racking of a braced frame under lateral loads is due to the elongation of the brace, and shortening and settlement of column footing under compression.

The racking of a braced frame is much less than that of a rigid frame even when the rigid frame is supported on fixed column bases. A braced frame is, therefore, much stiffer than a corresponding rigid frame.

A K-braced frame is stiffer than a corresponding X-braced frame because of its shorter diagonals.

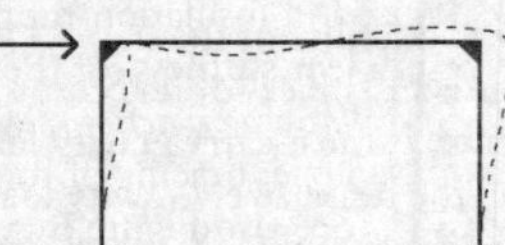

(c) Racking of a rigid frame with fixed bases under lateral loads is smaller than that of a rigid frame with pinned bases.

FIGURE 15 Comparison of the racking of a rigid frame and a braced frame.

- The racking of a rigid frame is larger than that of a comparable braced frame, Figures 15(a) and 15(b). In fact the term *rigid* is a misnomer because a braced frame is much stiffer than a corresponding rigid frame. The racking of a rigid frame can be decreased somewhat by using fixed (rigid) column bases, but the increase in stiffness so obtained does not match the inherent stiffness of a braced frame, Figure 15(c). Racking of a rigid frame also occurs under asymmetrical gravity loads, Figure 16.
- Another limitation of a rigid frame is its arching action, which makes the columns spread out under gravity loads, requiring a heavier footing or a tie that connects the column footings (Figure 13).

Because of its several limitations, a rigid-frame solution for a building that is more than six or seven floors in height becomes too uneconomical to justify its use.

(a) Rigid frame with pinned bases

(b) Rigid frame with fixed bases

FIGURE 16 Racking of the frame is produced even under asymmetrical gravity loads, which is more pronounced in a rigid frame than in a braced frame.

MULTIBAY SINGLE-STORY FRAMES

The fundamentals discussed with respect to single-bay frames also apply to multibay frames. However, it is not necessary to brace every bay of a multibay frame. Sufficient frame stiffness can be achieved by bracing a few selected bays in a single-story, multibay frame building. For example, in a four-bay, single-story frame, either two end bays or two central bays may be braced, Figures 17(a) and (b). Where lateral loads are small, bracing of only one bay may be sufficient, Figure 17(c).

The three-dimensional nature of the building requires that the frames in both principal directions be braced, Figure 18(a). As far as possible, braced bays should be symmetrically located in the building. Asymmetrical locations of braced bays are structurally inefficient but may be used where architectural considerations dictate their use. A combination of a rigid frame and a braced frame may also be used. In a common braced and rigid-frame combination, the rigid frame is used along one principal direction of the building and the braced frame is used along the other direction, Figure 18(b).

Just as every bay of a multibay braced frame structure need not be braced, every bay of a multibay rigid frame need not be a rigid frame. Sufficient rigidity may be obtained by using rigid column-beam connections for a few selected bays and leaving the others as pin connections, Figures 19(a) and (b).

STIFFNESS OF HORIZONTAL DIAPHRAGMS (FLOORS AND ROOF)

Resistance to lateral loads in a building requires that various components of the building function interactively. This implies that the horizontal diaphragms (floors and roof) of the building should be sufficiently stiff so that they can transfer lateral loads from the vertical elements on one face of the building to the vertical elements on other faces.

Generally, reinforced concrete floors, concrete-filled steel deck floors, and insulation-topped steel roof decks have sufficient stiffness for the purpose (referred to as providing *diaphragm action*). Where diaphragm action is not available, bracing of roofs may be needed (see Figure 18). (See also Figure 6, where the large glazed roof is braced by diagonal braces.)

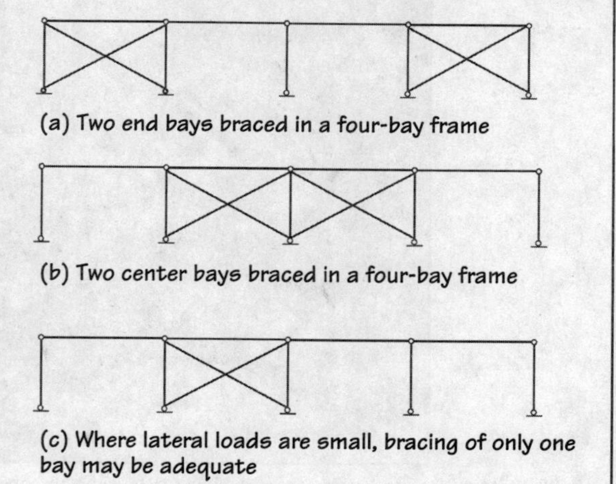

FIGURE 17 Some of the ways of bracing a four-bay, single-story frame.

MULTISTORY STEEL FRAMES

Although the principles of framing a multistory building are similar to those of framing a multibay, single-story building, a much larger number of solutions exist as the building height increases. Complete coverage of the framing solutions for high-rise buildings is outside the scope of this text. The coverage provided here is limited to buildings up to 30 to 40 stories in height.

For a low-rise multistory building, a steel rigid frame may be used. Note once again that a rigid frame solution becomes progressively uneconomical with increasing building height due to the inordinate increase in column and beam sizes. However, because of the absence of braced bays, it provides greater adaptability in organizing the interior spaces and exterior elevations. A typical steel rigid frame structure is shown in Figure 20.

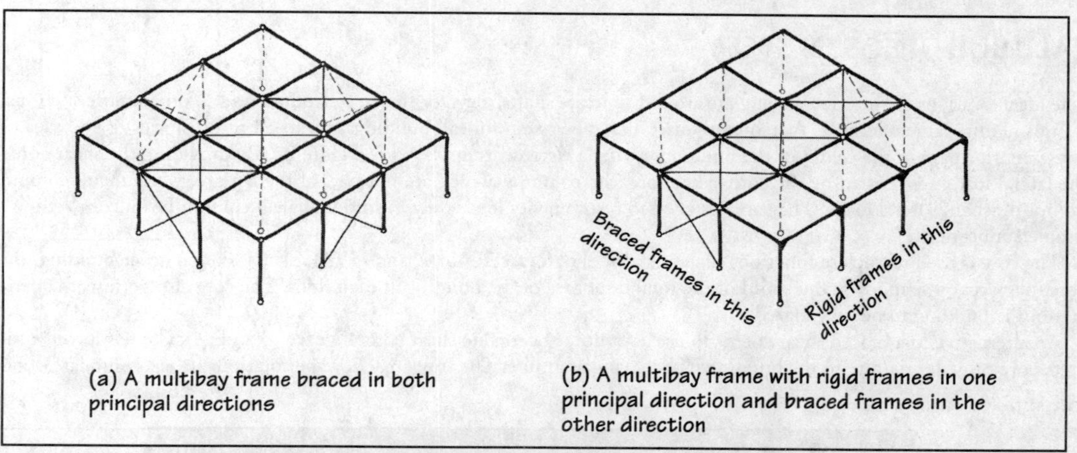

Braced frames in this direction
Rigid frames in this direction
(a) A multibay frame braced in both principal directions
(b) A multibay frame with rigid frames in one principal direction and braced frames in the other direction

FIGURE 18 **(a)** Braced frames in both principal directions. **(b)** Braced frames in one direction and rigid frames in the other direction.

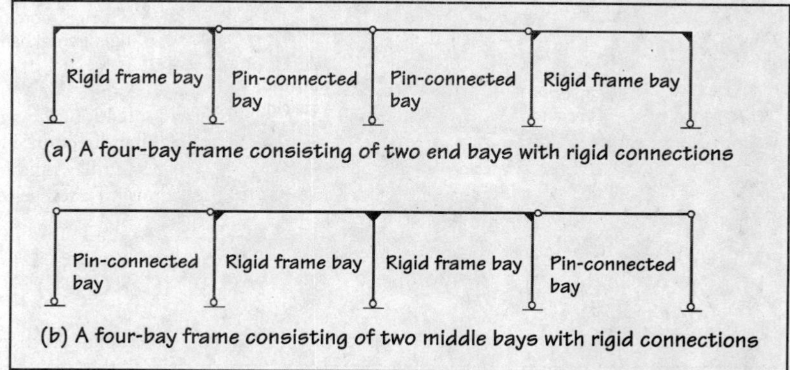

FIGURE 19 It is possible to obtain sufficient rigidity by using rigid connections in a few bays of a multibay frame.

(Continued)

Fundamentals of Frame Construction (*Continued*)

FIGURE 20 A typical steel rigid frame building. Observe the absence of braced bays, which gives greater freedom in organizing the interior spaces and building's facade.

A braced frame structure is more commonly used for both low-rise and high-rise buildings. The bracing may be incorporated in the exterior bays, where it is generally most efficient. In a large building, a few interior bays may also need to be braced, Figure 21. In a braced steel frame building, all column-beam connections are simple connections.

TALL BUILDINGS IN STEEL

The ideal locations for interior bracing are around staircase shafts, elevator shafts, restrooms, and so on, because such spaces do not require large openings. A frequently used bracing system for tall buildings (up to 30 or 40 stories) is to brace only a few interior bays in the center of the building, giving a rigid central *shear core*, Figure 22. By providing the shear core, all the lateral loads are resisted by the core. Therefore, the columns, which are not part of the core (e.g., perimeter columns), do not resist any lateral loads. They are designed to resist gravity loads only. Additionally, all column-beam connections are simple connections.

The core is used to enclose functions such as stairs, elevators, restrooms, and HVAC shafts. For an office building, the core may occupy as much as one-third of the total floor area of the building at each floor. The core also permits the perimeter of the building to be fully glazed.

An alternative to a braced shear core is to use a reinforced concrete shear core, Figures 23 and 24. The reinforced concrete core provides rigidity to the building by behaving as a tube. The tube may have perforations to accommodate doors

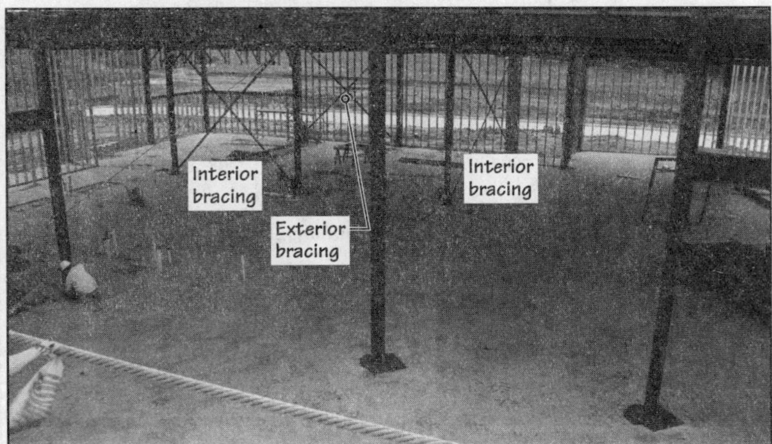

FIGURE 21 A typical braced frame building. Observe the provision of braced bays in the interior as well as the exterior of the building.

PRINCIPLES IN PRACTICE

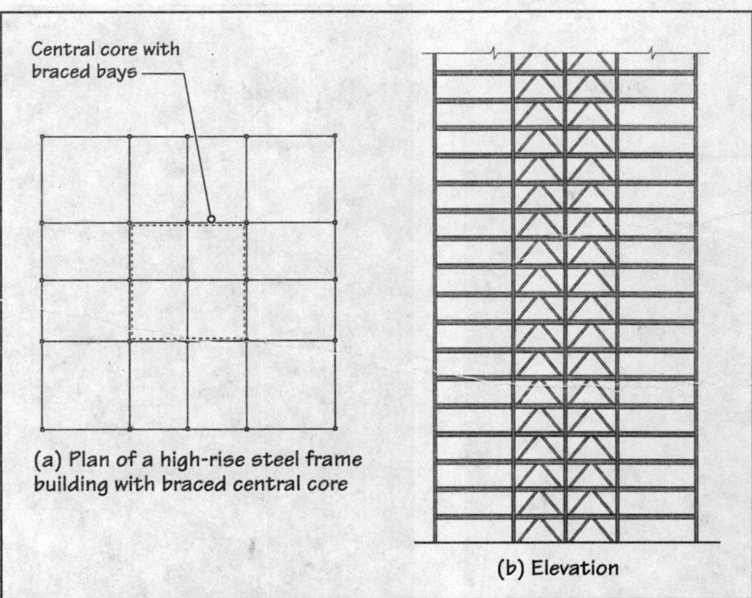

FIGURE 22 Plan and elevation of a high-rise steel-frame building with a braced steel-frame shear core.

and other openings as long as the perforations are not too large. A reinforced concrete core has an advantage over a braced steel-frame core because of its inherent high fire resistance, which is generally mandated for stair and elevator shafts.

A particularly elegant building consisting of a reinforced concrete core is the 12-story office building shown in Figure 25. In this building there are no perimeter columns, and the floors, consisting of steel wide-flange beams, have been suspended from the top of the core with steel cables.

In a high-rise building, which is much longer in one direction than in the other direction, a combination of a shear core and braced end shear walls is often used, Figure 26.

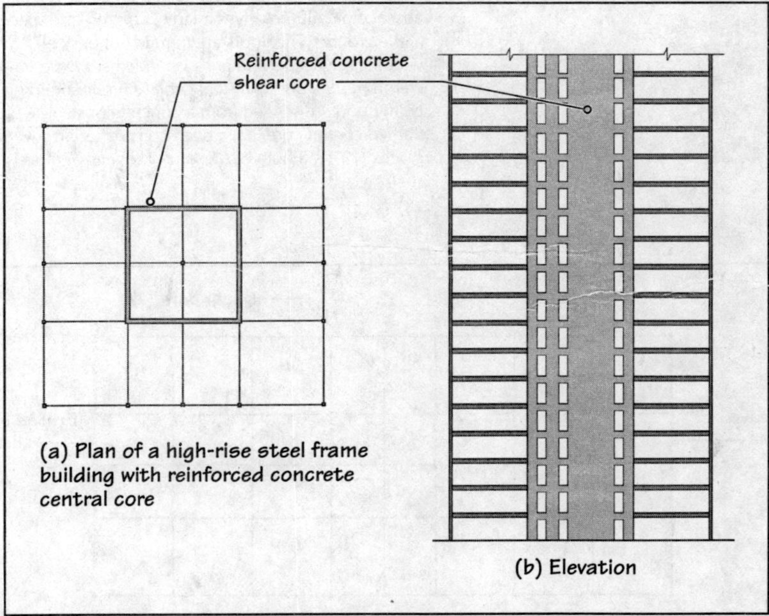

FIGURE 23 Plan and elevation of a high-rise steel-frame building with a reinforced concrete shear core.

(Continued)

PRINCIPLES IN PRACTICE

Fundamentals of Frame Construction (*Continued*)

FIGURE 24 A steel-frame building with a reinforced concrete central core under construction in Hong Kong.

FIGURE 25 The 12-story West Coast Transmission Company Building, Vancouver, Canada, consisting of a reinforced concrete shear core. Note that there are no columns in the building, and the only support system is a reinforced concrete central core. The absence of columns gives a large, column-free space at the ground. The floors are framed out of steel wide-flange beams and are suspended at the perimeter from the core with steel cables. Similar buildings (with floors suspended from a concrete or steel-framed central core) have been constructed in several places [19.1]. (Photo byAlexander Ulyanov/Emporis)

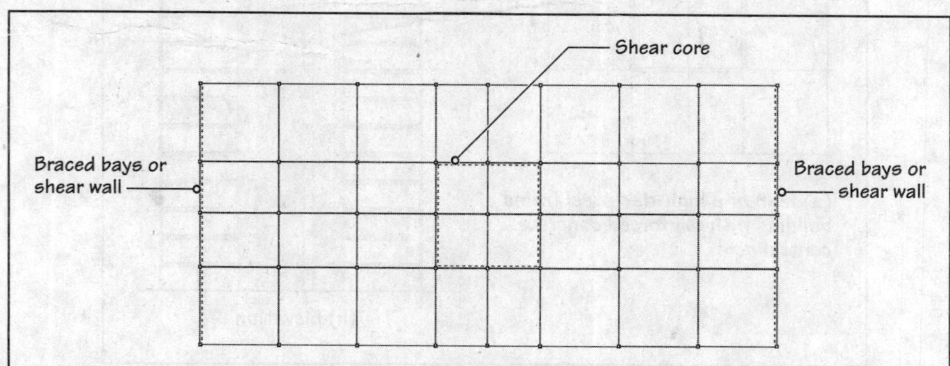

FIGURE 26 A commonly used method of framing a tall building that is much longer in one direction than the other is to use a central core and two end shear walls (or braced bays).

PRINCIPLES IN PRACTICE

LONG-SPAN STEEL BUILDINGS

Just as steel is the only structural material available for very tall buildings (skyscrapers), it is also the only material suitable for long-span structures (spans greater than 100 ft). Long-span structures are typically required for sports facilities, convention centers, atriums, and other assembly spaces.

For spans of up to about 200 ft, the use of steel joists provides an economical solution, Figure 27. For longer spans, two-way trusses, two-way W-shape grid structures, and steel space frames are often used, Figures 28 to 30.

FIGURE 27 The use of steel trusses in the University of Houston Athletic Center (truss span is approximately 200 ft; truss depth is approximately 10 ft at midspan). (Photo courtesy of Nucor-Vulcraft Group)

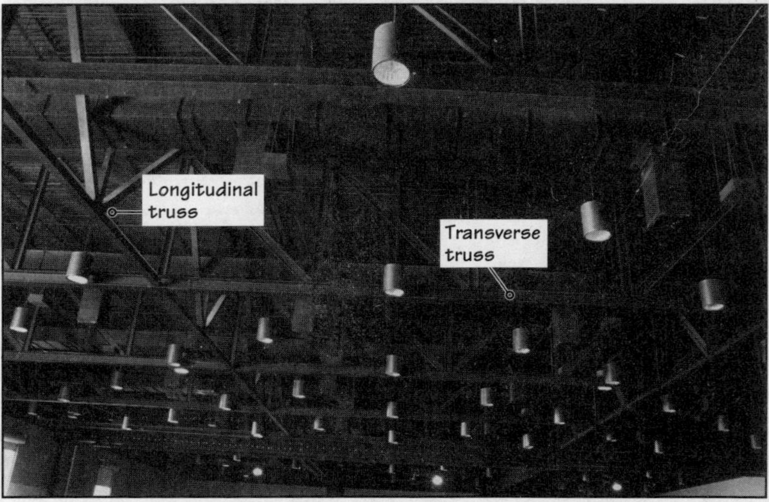

FIGURE 28 The use of two-way steel trusses in the main hall of the Convention Center, Arlington, Texas (truss spans are 300 ft x 400 ft).

(Continued)

PRINCIPLES IN PRACTICE

Fundamentals of Frame Construction (*Continued*)

FIGURE 29 The roof of a large atrium space consisting of a two-way grid of W-shape beams, King Saud University, Riyadh, Saudi Arabia.

FIGURE 30 Arched steel space frame used in the atrium of William Beaumont Hospital, Royal Oak, Michigan. (Photo courtesy of Novum Structures, Inc.)

PRACTICE **QUIZ**

Each question has only one correct answer. Select the choice that best answers the question.

24. A braced-frame steel structure is generally more economical than a comparable rigid frame structure.

a. True **b.** False

25. The difference between diagonal bracing and K-bracing between steel columns is that

a. K-bracing provides less spatial interruption than diagonal bracing.

b. K-bracing members are heavier than diagonal bracing members.

c. K-bracing members also carry some gravity loads, while diagonal framing members do not.

d. all of the above.

e. none of the above.

26. Eccentric K-bracing is generally recommended for use

a. because it is more economical than other forms of bracing.

b. because it increases frame stability more than other forms of bracing.

c. in hurricane-prone regions.

d. in seismic regions.

e. all of the above.

27. The term *racking* refers to the effect produced on the structure mainly by

a. dead loads. **b.** floor live loads.
c. roof live load. **d.** lateral loads.
e. none of the above.

28. A steel-frame building may consist of either a rigid frame or a braced frame. A combination of the two systems (braced- and rigid-frame systems) cannot be used in the same building.

a. True **b.** False

29. Which of the two—a rigid frame or a braced frame—provides greater structural stiffness, that is, greater lateral stability?

a. Rigid frame **b.** Braced frame

30. In a braced-frame steel structure, the braced bays

a. should be provided only on the exterior facade of the building.
b. should be provided only in the interior of the building.
c. should be provided on both the exterior facade and the interior of the building.
d. may be provided on the exterior facade or the interior of the building.
e. may be provided on the exterior facade, the interior of the building, or both.

31. A multistory steel-frame structure must be provided with a central shear core.

a. True **b.** False

32. The central shear core in a steel-frame building

a. should consist of reinforced concrete walls in both directions of the building.
b. should consist of steel-braced frames along both directions of the building.
c. either (a) or (b).
d. should consist of reinforced concrete walls in one direction of the building and steel-braced frames in the other direction.

REVIEW QUESTIONS

1. Explain what the term *framing plan* means. What information does a framing plan include?
2. In a typical structural steel floor or roof framing, there are three types of elements, called *primary*, *secondary* and *tertiary elements*. Explain what these are and what material(s) they consist of.
3. Explain the pros and cons of welding versus bolting in steel structures.
4. With the help of sketches and notes, explain the difference between a simple connection and a rigid connection between a wide-flange column and a wide-flange beam.
5. With the help of sketches and notes, explain the connection of a column of an upper floor with that of a lower floor.
6. Explain the roles of the general contractor, fabricator, and erector of a structural steel building.
7. Using sketches and notes, explain how the joists and joist girders are stabilized against overturning.

CHAPTER 20

Cold-Formed (Light-Gauge) Steel Construction

CHAPTER OUTLINE

As stated in Section 18.3, cold-formed steel (also called *light-gauge steel*) members are made by bending steel sheets into required profiles at room temperature. An important use of cold-formed steel is in (corrugated) floor and roof decks, commonly used in buildings framed with structural steel. Steel decks have been covered in Chapter 18. This chapter deals with the use of cold-formed steel in other building assemblies, identified in the following two paragraphs.

Cold-Formed Steel in Non-Load-Bearing Wall Assemblies

An extensive use of cold-formed steel is in interior partitions and exterior non-load-bearing walls in Type I and Type II construction, in which the building codes mandate the use of noncombustible materials (see Chapter 7 for the definition of Type I and Type II construction). In both applications (interior partitions and exterior non-load-bearing walls), cold-formed steel members are required to resist lateral loads only. (The reader is advised to review the difference between load-bearing and non-load-bearing elements covered in the margin note in Section 15.6.)

Cold-Formed Steel in Load-Bearing Assemblies

While the use of cold-formed steel in non-load-bearing walls has existed for a long time, its use in the load-bearing arena is much more recent. In this application, cold-formed steel is used in all load-bearing and non-load-bearing assemblies of the building—walls, floors, and roofs—in the same way as wood light-frame (discussed in Chapter 15).

One region that has seen an increasing use of cold-formed steel in load-bearing applications is the western United States, where cold-formed steel's higher strength, greater

Thickness	18 mil	27 mil	30 mil	33 mil	43 mil	54 mil	68 mil	97 mil
Gauge and color mark	25	22	20 DW	20 STR	18	16	14	12
				White	Yellow	Green	Orange	Red
Strength	33 ksi		33 ksi and 50 ksi			50 ksi		
Protection	G40		G40 and G60			G60		
	Used for interior drywall framing		Used for interior drywall framing, exterior non-load-bearing walls and low-rise, load-bearing applications			Commonly used in mid-rise, load-bearing applications		

FIGURE 20.1 Available thickness, yield strength, and corrosion protection of cold-formed steel members. Cold-formed steel members 33 mil and thicker (commonly used in load-bearing applications) are identified with a color stripe, as per the ASTM-specified color code, shown in the third row here. No such color code exists for 18-mil-, 27-mil-, and 30-mil-thick CFS members, commonly used for interior partitions.

ductility, and lower dead load give it an advantage over the wood frame in resisting seismic forces. Other areas of growth are warm, humid areas that are infested with termites and/or susceptible to mold growth. Durability against termites is the primary reason for the popularity of cold-formed steel in the structural framing of homes in Hawaii, Florida, Georgia, and parts of Texas.

20.1 COLD-FORMED STEEL (CFS) SPECIFICATIONS

Because they are made from thin sheets of steel, cold-formed steel (CFS) members must be protected from corrosion. Except for steel roof decks, which may be paint finished, CFS members are virtually always hot-dip galvanized. The thickness of the zinc coating (the degree of corrosion protection) depends on the member's requirement. G40 is used for interior partitions and G60 for load-bearing members and exterior non-load-bearing walls. In coastal regions, G90 is generally specified. (See Section 18.7 for the meaning of G40, G60, etc.)

CFS Sheet Strength

CFS members are made from sheet steel of various yield strengths. The most commonly used yield strengths are 33 ksi and 50 ksi. Steel with 33 ksi yield strength is used in all applications—interior partitions, exterior non-load-bearing walls, and load-bearing members. Steel with 50 ksi yield strength is used in applications where 33 ksi steel is inadequate, such as in load-bearing members under high gravity loads and non-load-bearing members under high lateral loads.

CFS Sheet Thickness

The thickness of sheet steel used for CFS members varies from 18 mil to 97 mil, Figure 20.1. These values represent bare steel thickness and do not include the thickness of the zinc coating. The Steel Stud Manufacturers Association (SSMA) recommends specifying sheet thickness in mils rather than traditional gauge numbers. However, gauge number specification is commonly used in practice. (The higher the gauge number, the thinner the sheet, and 1 mil $= \frac{1}{1{,}000}$ in.)

To facilitate identification, manufacturers label the strapped bundle of identical CFS members and mark the ends of members (33 mil or thicker) with a color stripe as per the color code specified in ASTM Standard C955, Figure 20.2 (see also Figure 20.1).

20.2 CFS FRAMING MEMBERS

Four types of CFS framing members are in common use:

- Studs and joists (symbolized by *S* in the SSMA designation)—C-shaped
- Tracks (symbolized by *T* in the SSMA designation)—U-shaped
- Bridging channels (symbolized by *U* in the SSMA designation)
- Furring channels (symbolized by *F* in the SSMA designation)

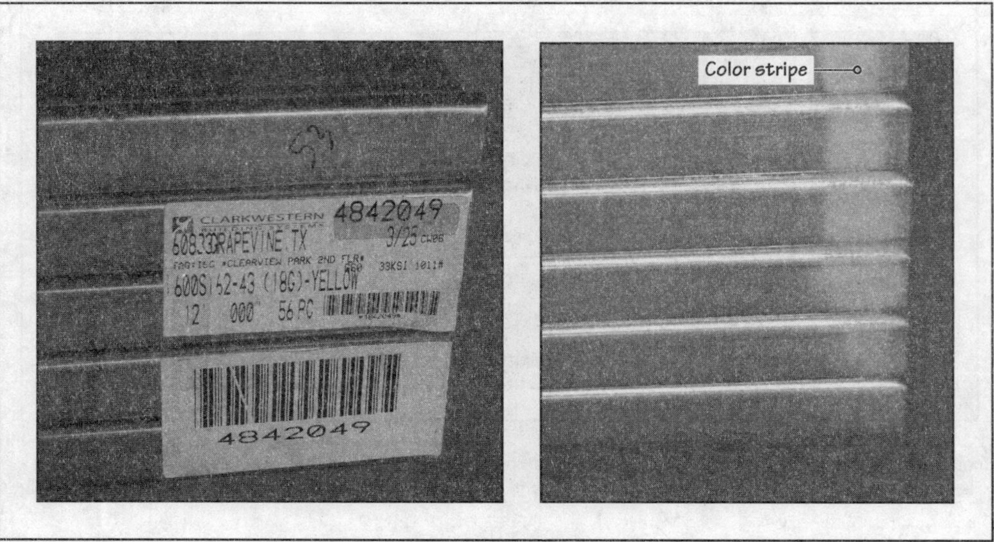

FIGURE 20.2 Cold-formed steel members carry a label on the strapped bundle that identifies their cross-sectional dimensions and sheet thickness. Additionally, members that are 33 mil and thicker, carry a color stripe that identifies the member's gauge number (sheet thickness) as per ASTM Standard C955

The word *STUF* may be invoked to recall the symbols—*S* for stud, *T* for track, *U* for channel, and *F* for furring section. The SSMA uses a standard designation system to identify its products. The designation system requires four types of information: (a) member depth, (b) member type (S, T, U, or F), (c) flange width, and (d) sheet thickness. Thus, as shown in Figure 20.3, a member identified as "600 S 162-33" means that it is a stud with a depth of 6.00 in., flange width of 1.62 (actually, $1.625 = 1\frac{5}{8}$-in.), and sheet steel thickness of 33 mil.

CFS Studs

CFS studs are available in several depths, flange widths, and gauge numbers, Figure 20.4. Interior partitions commonly use $3\frac{5}{8}$-in.-deep studs (25 or 22 gauge). Deeper and heavier studs may be required for tall partitions, load-bearing walls, and exterior non-load-bearing walls. Studs are generally punched with holes at 4 ft on center. Punched holes allow the passage of electrical conduits and other utility lines through the wall. They also allow the use of bridging channels.

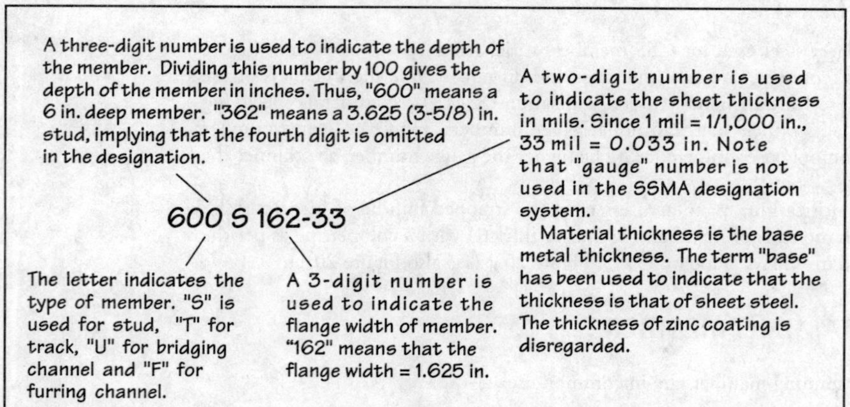

A three-digit number is used to indicate the depth of the member. Dividing this number by 100 gives the depth of the member in inches. Thus, "600" means a 6 in. deep member. "362" means a 3.625 (3-5/8) in. stud, implying that the fourth digit is omitted in the designation.

A two-digit number is used to indicate the sheet thickness in mils. Since 1 mil = 1/1,000 in., 33 mil = 0.033 in. Note that "gauge" number is not used in the SSMA designation system.

Material thickness is the base metal thickness. The term "base" has been used to indicate that the thickness is that of sheet steel. The thickness of zinc coating is disregarded.

600 S 162-33

The letter indicates the type of member. "S" is used for stud, "T" for track, "U" for bridging channel and "F" for furring channel.

A 3-digit number is used to indicate the flange width of member. "162" means that the flange width = 1.625 in.

Because both studs and joists have the same cross-sectional shape (C-shape), the symbol "S" is also used for joists.

FIGURE 20.3 Designation system for identifying cold-formed steel members established by the Steel Stud Manufacturers Association, SSMA).

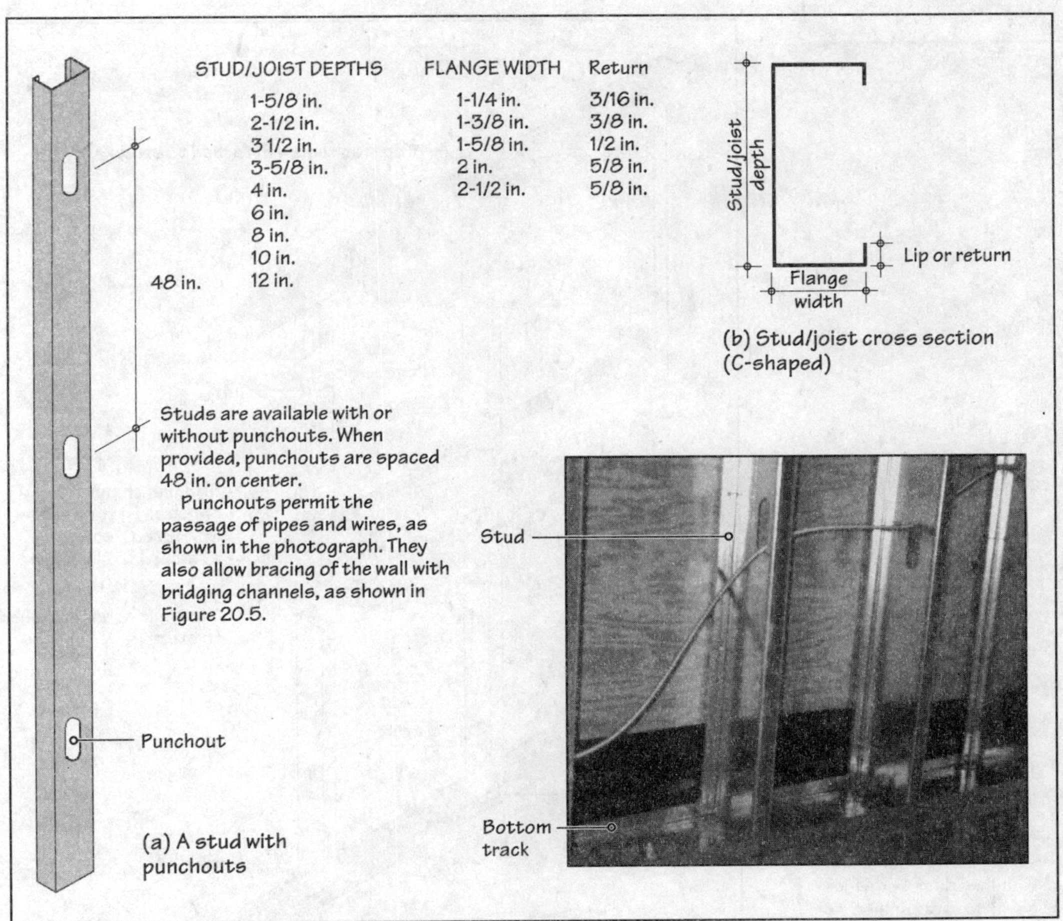

FIGURE 20.4 Cross-sectional profile and commonly used sizes (thickness and depth) of CFS studs and joists.

CFS Tracks

CFS tracks match the studs so that the studs fit within the tracks, Figure 20.5. The tracks run at the top and bottom of the studs, simulating the top and bottom plates in WLF construction. Studs and tracks used in a wall assembly must be of the same gauge number.

CFS Bridging Channels

Because of their thin walls and open cross section, CFS studs are prone to twisting under the loads, referred to as *torsional buckling*. Bridging channels help prevent such twisting. They are U-shaped and fit within the holes in punched studs, as shown in Figure 20.5. (A tubular section is a closed section and is less prone to twisting.)

Bridging channels are generally required for non-load-bearing walls. They cannot be used to resist axial buckling in load-bearing walls because they do not extend the full width of studs (see Figure 20.20).

Bridging channels are unnecessary where gypsum board (drywall) extends over the full height of studs. However, most contractors provide one or two lines of bridging channels to prevent the twisting of studs from construction-related impacts before the installation of gypsum board.

CFS Furring Channels

Two types of furring channels are in common use: (a) hat channels and (b) resilient channels, Figure 20.6. A hat channel is generally used on the interior face of masonry covered with gypsum board (see Chapter 28). A resilient channel is used to increase the airborne sound insulation of a wall or floor (see Chapter 8).

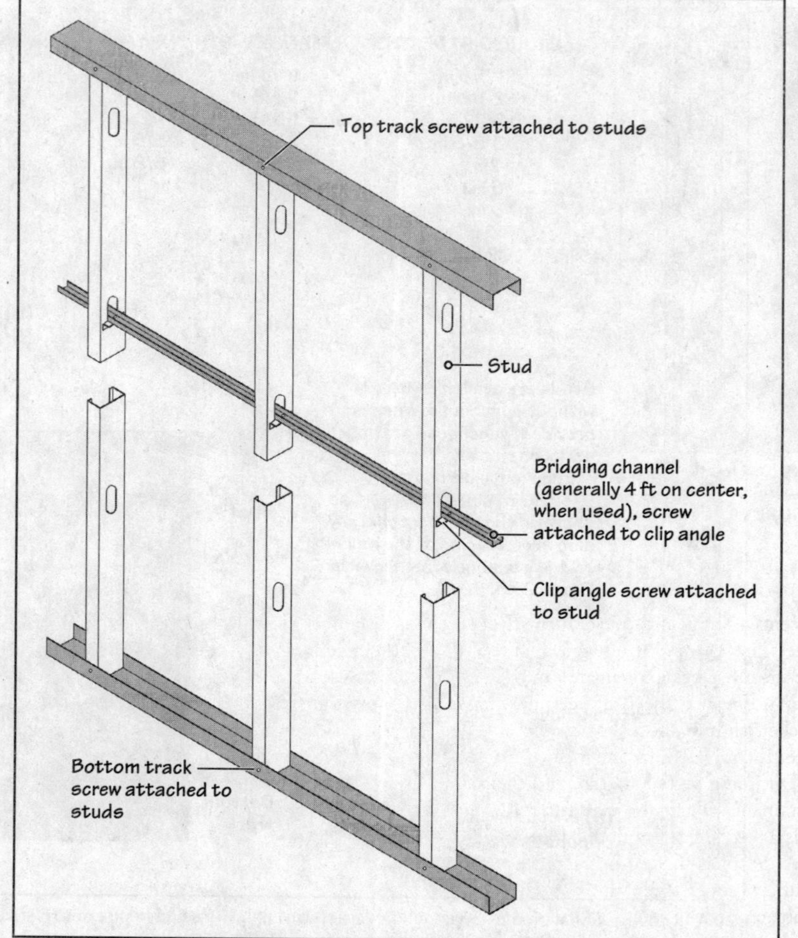

FIGURE 20.5 Typical framing of a wall using CFS members. Note that bridging channels can only be used with punched studs. They help prevent the twisting of studs under lateral loads, but they do not provide adequate resistance against the buckling of studs under axial loads (in a load-bearing wall).

CFS MEMBER IDENTIFICATION MARK

The cross-sectional size and material specifications of CFS studs and tracks are printed 4 ft on center on every piece. A typical identification mark is

600T125-43-33K-G60.

As explained in Figure 20.3, this mark indicates that the member is a "track" whose depth = 6.00 in., flange width = 1.25 in., sheet thickness = 43 mil, steel strength = 33 ksi, and corrosion protection is G60. In addition to the printed mark on each member, manufacturers attach a label to the bundle of CFS members that are identical in length, material specifications, and cross-sectional dimensions (see Figure 20.2).

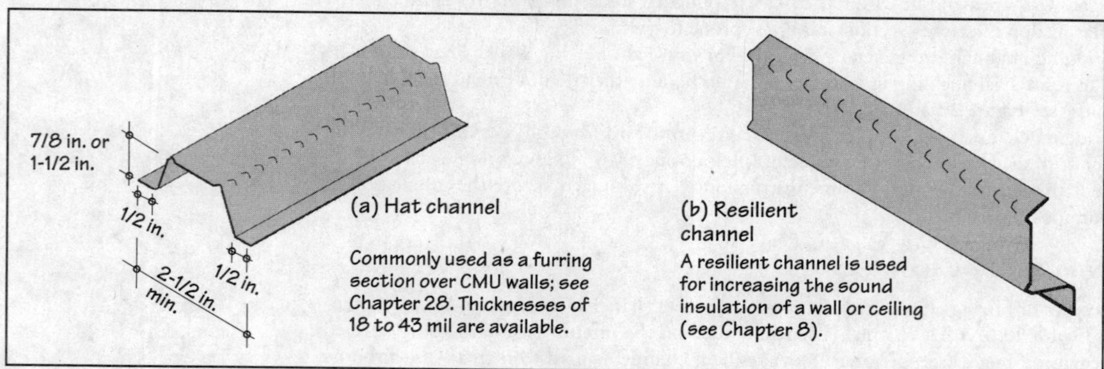

FIGURE 20.6 Two commonly used CFS furring channels.

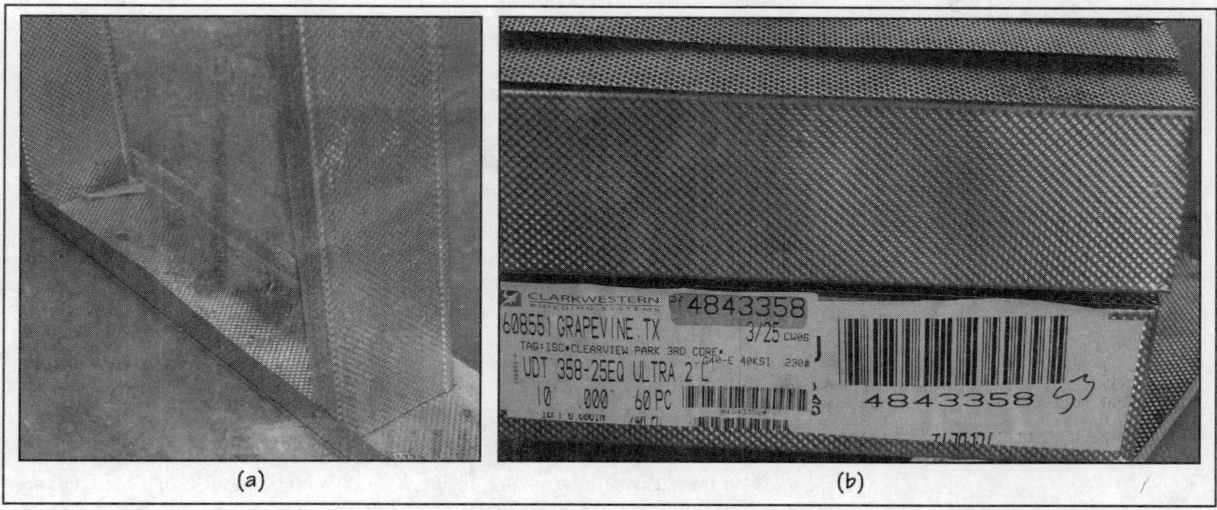

FIGURE 20.7 (**a**) Track and studs formed from dimpled sheet steel. (**b**) A close-up of the dimpled surface.

20.3 DIMPLED STUDS AND TRACKS

Recently, studs and tracks formed out of dimpled sheet steel (in place of plain, smooth sheets) have been introduced, Figure 20.7. Dimpling of a sheet increases its strength, as if the sheet has been corrugated in both directions. Consequently, a CFS stud or track made from a dimpled sheet of a certain thickness is equivalent in strength to one made from a thicker smooth sheet. For example, a CFS member formed from a 15-mil-thick dimpled sheet is equivalent to one formed from an 18-mil-thick smooth sheet.

Dimpling saves material, and the manufacturers claim that dimpled products are more sustainable. They also claim that dimpled studs provide greater sound insulation and fire resistance. The strength of steel used in dimpled products is 40 ksi.

A dimpled surface is created by a patented technology called UltraSteel, invented by Hadley Industries in the United Kingdom. The technology has been licensed to a few North American manufacturers. The use of dimpled steel has become quite common for interior drywall framing.

PRACTICE **QUIZ**

Each question has only one correct answer. Select the choice that best answers the question.

1. A load-bearing CFS frame structure is generally lighter than a comparable wood light-frame structure.
   - **a.** True
   - **b.** False
2. Bridging channels are used to prevent the buckling of studs in load-bearing walls.
   - **a.** True
   - **b.** False
3. The amount of corrosion protection required in CFS members is generally
   - **a.** G30.
   - **b.** G40.
   - **c.** G50.
   - **d.** G60.
   - **e.** varies with the application.
4. The yield strength of CFS used for interior partitions is generally
   - **a.** 33 ksi.
   - **b.** 40 ksi.
   - **c.** 50 ksi.
   - **d.** either (a) or (b).
   - **e.** either (a) or (c).
5. The yield strength of CFS used in non-load-bearing applications is generally
   - **a.** 33 ksi.
   - **b.** 40 ksi.
   - **c.** 50 ksi.
   - **d.** either (a) or (b).
   - **e.** either (a) or (c).
6. Punchouts in CFS studs are generally spaced at
   - **a.** 48 in. on center.
   - **b.** 40 in. on center.
   - **c.** 24 in. on center.
   - **d.** 16 in. on center.
   - **e.** none of the above.
7. CFS studs and tracks are U-shaped.
   - **a.** True
   - **b.** False
8. The identification mark on each CFS member is painted
   - **a.** 2 ft on center.
   - **b.** 3 ft on center.
   - **c.** 4 ft on center.
   - **d.** 6 ft on center.
   - **e.** none of the above.
9. In the identification mark on a CFS member "362S125-30 33K G40," the letter *S* implies that
   - **a.** it is an S-shaped member.
   - **b.** it is a straight, unbent member.
   - **c.** it can only be used for a single-story building.
   - **d.** it is made from steel with a strength of 362 ksi.
   - **e.** none of the above.
10. In the identification mark on a CFS member "362S125-30 33K G40," the number 30 implies that
    - **a.** it is a 30-ft-long member.
    - **b.** it is a 30-in.-long member.
    - **c.** the thickness of steel used for the member is gauge 30.
    - **d.** the thickness of steel used for the member is 30 mil.
    - **e.** none of the above.

(Continued)

11. The use of bridging channels requires
    a. punched studs.
    b. unpunched studs.
    c. punched tracks.
    d. unpunched tracks.
    e. none of the above.

12. Dimpling of sheet steel to form CFS members increases the strength of members but reduces corrosion protection.
    a. True
    b. False

13. Building codes mandate the use of CFS framing for interior partitions in buildings of
    a. Type I construction.
    b. Type II construction.
    c. Types I and II construction.
    d. Types I, II, and III construction.
    e. none of the above.

14. If a masonry wall must be covered with drywall on its interior face, the framing generally used for the purpose consists of CFS
    a. studs.
    b. track.
    c. studs and tracks.
    d. bridging channel.
    e. hat channel.

20.4 CFS FRAMING AND GYPSUM BOARD INTERIOR PARTITIONS

Gypsum board interior partitions with CFS framing are constructed by first fastening the top and bottom tracks to the structure, Figure 20.8(a). With both tracks in place, studs are inserted between them and screw fastened to the tracks. Finally, gypsum board is screw attached to the studs, Figure 20.8(b).

Size, Gauge, and Spacing of Studs

The structural design of CFS framing for an interior gypsum board partition is fairly elementary. The only gravity load borne by CFS framing is self-weight and the weight of gypsum board, which is fairly small. Therefore, the structural design of studs—size, spacing, and gauge—is a function of the lateral load. Building codes mandate that interior partitions be designed for a minimum lateral load of 5 psf, resulting from incidental impacts, vibration, and slamming of doors. (In seismic zones, partitions are subjected to a greater lateral load, which must be calculated and designed for.)

Because steel is a strong material, the deflection of studs under the lateral load (not steel strength) governs their design. Gypsum board is a brittle material with a limited capacity to withstand deflection. Excessive deflection leads to cracking of gypsum board at taped joints. The maximum permissible deflection for gypsum board partitions is

Partition height divided by 240, generally expressed as (L/240),

where L = partition height. Many designers prefer limiting the deflection to (L/360), which gives a stiffer partition.

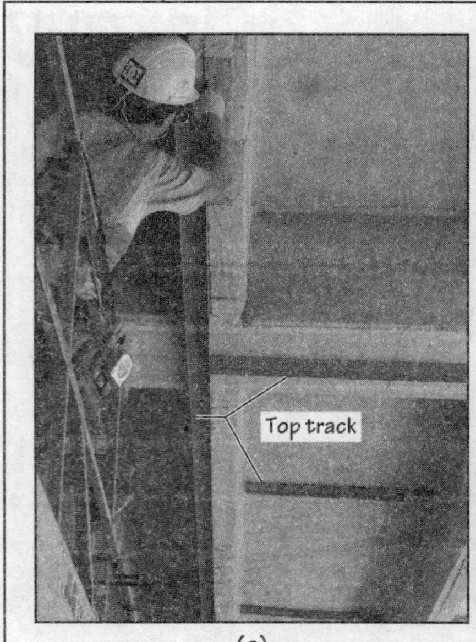

(a)

(b)

FIGURE 20.8 **(a)** A worker fastens the top track for an interior partition to the underside of the floor slab. **(b)** Gypsum board being fastened to CFS framing of an interior partition.

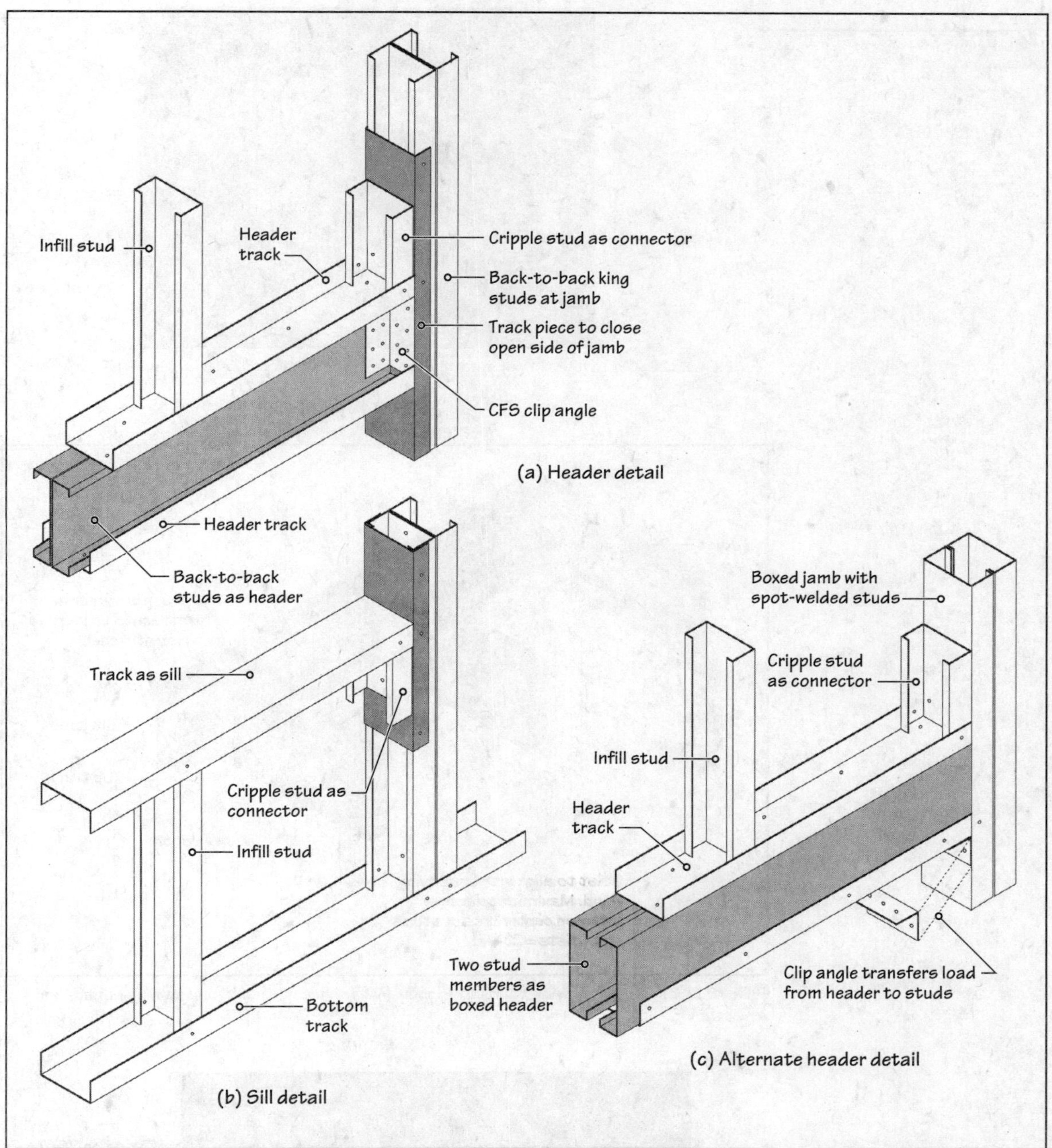

FIGURE 20.21 Details of a window header and sill in a load-bearing wall. The door header detail is similar.

Once the floor joists are in place, the plywood or OSB subfloor for the next floor is laid in exactly the same way as in a WLF building. After the completion of the subfloor, the walls for the next floor are erected.

CFS Roof Framing

Figure 20.25 shows the framing of a typical CFS roof, and Figure 20.26 shows the framing detail at an eave. Because assemblies are prefabricated, roof framing is in the form of trusses. Being lighter than corresponding wood trusses, cold-formed steel trusses can generally be hoisted into position without special lifting and hoisting equipment. Additionally, the joints between members of a cold-formed steel truss are more rigid, so that the trusses are easier to handle and more forgiving during their placement.

FIGURE 20.22 Layout of floor joists in a typical CFS structure. (Photo courtesy of Dietrich Metal Framing)

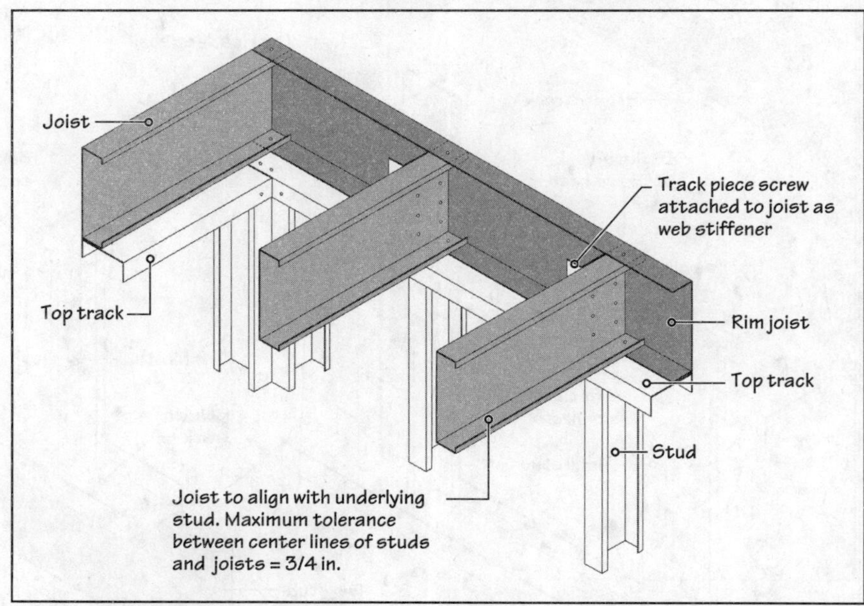

FIGURE 20.23 Detail showing floor joist supports, web stiffening, and in-line framing of studs and joists.

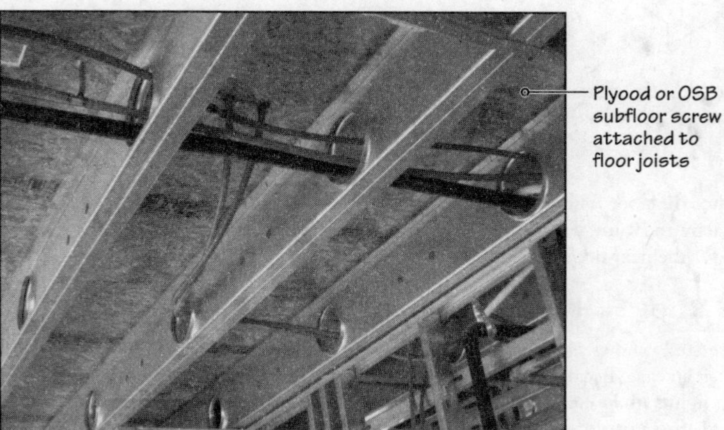

FIGURE 20.24 CFS floor joists with extruded holes allow utility lines to pass through. (Photo courtesy of Dietrich Metal Framing)

FIGURE 20.25 Use of CFS roof trusses. (Photo courtesy of American Residential Steel Technologies)

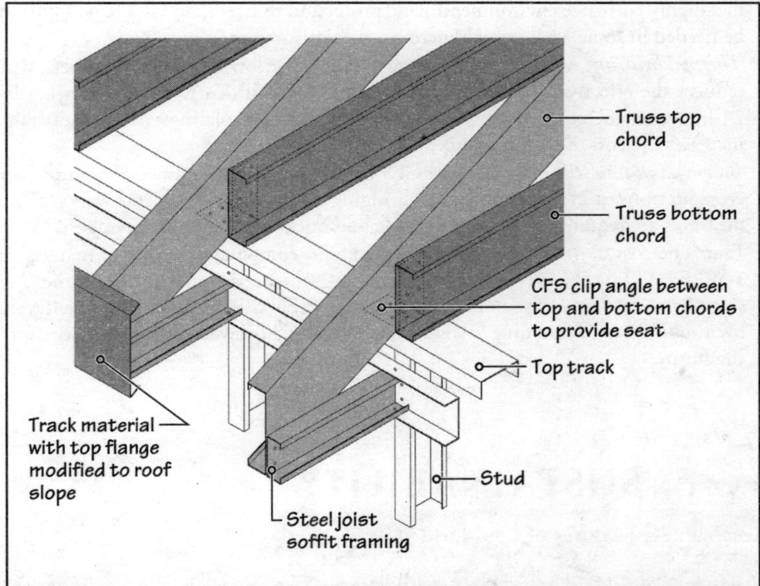

FIGURE 20.26 A typical eave detail in a CFS roof.

20.8 ADVANTAGES AND LIMITATIONS OF CFS CONSTRUCTION

Because CFS and WLF systems are similar, they share several attributes:

- Both are dry systems of construction.
- Both systems provide spaces between framing members to allow infill insulation and passage of utility pipes.
- Both systems are lightweight, which eliminates the need for heavy construction and hoisting equipment.
- Both systems require simple, inexpensive tools for connecting the members. Welding is generally unnecessary in CFS structures.
- Both systems accept the same interior and exterior finishes.

In addition to the shared characteristics, a CFS building has several advantages and disadvantages vis-à-vis a WLF building.

Advantages

- *Lighter weight:* CFS members are much lighter than corresponding WLF members, reducing workers' fatigue. Lighter weight is an advantage in seismic zones.

- *Higher strength:* For the same overall cross-sectional dimensions, CFS members are stronger and more ductile than WLF members. Higher strength and greater ductility are the reasons for the increased use of CFS for residential framing in seismic and high-wind areas.
- *Dimensional stability:* Steel is not subjected to dimensional instability—shrinkage, swelling, twisting, and warpage caused by moisture changes. CFS members are straight and true.
- *Uniform quality:* Unlike wood, steel is a manufactured material with uniform and consistent properties.
- *Durability:* Steel is not subject to fungal decay or termite attack.
- *Noncombustibility:* A major advantage of steel compared to wood is steel's noncombustibility.
- *Waste recycling:* Prefabrication of assemblies substantially reduces on-site labor and waste recycling.

NOTE

Dust Marking

If a CFS frame structure is not properly detailed, a phenomenon known as *ghosting*, or *dust marking*, may occur due to the temperature differential between the framing members and the rest of the wall. This phenomenon, which generally occurs in cold climates, creates dust deposits on the interior drywall, highlighting the framing members.

LIMITATIONS

- *Corrosion:* Although CFS members are galvanized, they can be subjected to corrosion in a highly corrosive environment. Zinc protection thicker than G60 (e.g., G90) may be needed in some areas, which increases the cost substantially.
- *Thermal bridging:* A major disadvantage of CFS is the low R-value of members, which reduces the effectiveness of infill insulation. Although this is offset somewhat by a 24-in. center-to-center spacing of studs and the use of insulating exterior sheathing, it increases the cost of the building.
- *Bracing and web stiffening:* Because CFS members are thin, lightweight, and of open cross sections, a greater number of bridging and bracing lines and web-stiffening measures are required to prevent twisting and buckling of framing members.
- *Cost:* The cost of materials for a CFS building is comparable to that of materials for a WLF building. However, CFS framing has a higher labor cost due to the use of screw guns instead of the pneumatic nailing used in a WLF building. The development of improved fastening systems for CFS may, however, change the situation in the future.

FOCUS ON SUSTAINABILITY

Sustainability Features of Cold-formed Steel

Absence of Mold and Toxic Preservatives

Because steel is an inorganic material, CFS framing does not allow the growth of mold. Additionally, CFS is termite and decay resistant. This implies that, unlike a WLF building, a CFS building does not require any toxic preservatives (termicide) in its structural frame that might eventually leach into the soil or end up in landfills and from there go into the environment—air and water.

Additionally, because of its termite resistance, the structural frame of a CFS building will remain intact even if the (preconstruction) chemical barrier around the building's foundations is inadequate or compromised.

Recyclability

CFS members are manufactured from coils of galvanized sheet steel that have a large percentage of recycled content (see Chapter 18 for a discussion of the overall recycling of steel). Because screw connections are used, virtually all CFS members can be easily disassembled after the building's useful life. They may be reused, and, if not, they are 100% recyclable.

According to the Steel Recycling Institute,* a 2,000-ft^2 house constructed of WLF requires wood from 40 to 50 mature trees (covering about 1 acre of forest land). A house of the same size made with CFS requires steel that is equivalent to six scrapped cars.

Thermal Performance

Because steel is a highly conducting material, its thermal performance is poor. Thermal bridging reduces the efficacy of fibrous insulation placed in framing cavities. The use of insulating sheathing, however, can reduce thermal bridging substantially.

*http://www.recycle-steel.org

PRACTICE QUIZ

Each question has only one correct answer. Select the choice that best answers the question.

15. CFS framing for interior gypsum board partitions is designed for a minimum lateral load of
- **a.** 30 psf.
- **b.** 25 psf.
- **c.** 20 psf.
- **d.** 15 psf.
- **e.** none of the above.

16. CFS framing for exterior non-load-bearing walls must be designed to a maximum permissible deflection
- **a.** of L/240.
- **b.** of L/360.
- **c.** as needed, depending on the exterior sheathing applied to the CFS framing.
- **d.** as needed, depending on the type of exterior cladding.
- **e.** as needed to resist wind loads.

17. The size, spacing, and gauge of CFS studs for a gypsum board partition are governed by the
- **a.** shear strength of CFS members.
- **b.** axial compressive strength of CFS members.
- **c.** bending strength of CFS members.
- **d.** deflection of CFS members.
- **e.** none of the above.

18. The CFS industry provides two types of tables from which the size, spacing, and gauge of studs may be selected for an interior partition—composite design tables and noncomposite design tables. Of the two, noncomposite design tables are more conservative.
- **a.** True
- **b.** False

19. Two types of CFS buildings have been identified in this text: (a) CFS buildings as Type V(B) construction and (b) CFS buildings as Type II(A) construction. The distinction between the two is primarily based on the
- **a.** type of sheathing material used in wall, floor, and roof assemblies.
- **b.** size of CFS studs.
- **c.** corrosion protection of CFS members.
- **d.** extent of prefabrication used in construction.
- **e.** none of the above.

20. Like WLF walls, CFS walls are assembled on site on the floor and subsequently tilted up and moved into position.
- **a.** True
- **b.** False

21. The anchorage of the bottom track of a load-bearing CFS wall to the concrete footing
- **a.** requires 6-in.-long stud material to strengthen the connection.
- **b.** requires 6-in.-long additional track material to strengthen the connection.
- **c.** requires a 6-in.-long cold-formed furring section to strengthen the connection.
- **d.** requires no additional strengthening.

22. The corner of a load-bearing CFS wall requires a minimum of
- **a.** five studs.
- **b.** four studs.
- **c.** three studs.
- **d.** two studs.

23. The top track in a load-bearing CFS wall must be doubled in the same way as the top plate in a WLF wall.
- **a.** True
- **b.** False

24. *In-line framing* in a CFS structure means that
- **a.** the studs must be continuous across an intermediate floor.
- **b.** the floor joists must align with the studs in the underlying supporting wall.
- **c.** the floor joists from two opposite directions of an underlying supporting wall must be in line, not staggered.
- **d.** the floor-to-floor height in a two- or three-story structure must be the same.

25. The benefit of using load-bearing CFS framing instead of load-bearing WLF in seismic zones is the
- **a.** lighter weight of CFS framing.
- **b.** ductility of steel.
- **c.** thermal bridging.
- **d.** (a) and (b).
- **e.** (b) and (c).

REVIEW QUESTIONS

1. In which regions of the United States is CFS gaining popularity in load-bearing applications, and why? Explain.
2. Explain what the following CFS member designation means: 362S125-30-33K-G40.
3. Using sketches, explain the difference between the cross-sectional shapes of studs and tracks.
4. Sketch a typical hat section and explain where it is commonly used.
5. Explain what dimpled sheet steel is and its advantages over smooth sheet steel for use in studs and tracks.
6. Explain the difference between composite design tables and noncomposite design tables for sizing steel studs for an interior partition. Explain in which situation you will use (or not use) each table.
7. List all possible applications of CFS in contemporary buildings.
8. What level(s) of corrosion protection is(are) required for CFS members? Explain which level you will specify for various applications.
9. Explain the essential difference between (a) a CFS building as Type V(B) construction and (b) a CFS building as Type II(A) construction. Explain in which situation(s) each type is commonly used.

CHAPTER 21

Lime, Portland Cement, and Concrete

CHAPTER OUTLINE

This chapter begins with a discussion of lime, which, until the mid-nineteenth century, was the only cement available for use in masonry mortar, plaster, and concrete. Today, the use of lime is limited to modest quantities in mortar, stucco, whitewash, and soil stabilization.

Portland cement, whose discovery led to the replacement of lime in most applications, is covered next. Without portland cement, modern architecture, as we know it, would not be possible. In fact, it is virtually impossible to construct a building today without using portland cement in some form.

Portland cement is the main ingredient in concrete. Therefore, a discussion of concrete follows that of portland cement. Concrete is covered here as a material. Its applications, that is, concrete construction systems, are discussed in the following two chapters.

Because concrete is generally used in conjunction with steel reinforcement (as reinforced concrete), we also cover steel reinforcement here.

21.1 INTRODUCTION TO LIME

Lime is a cement (binder) and has been used for thousands of years in masonry mortars to bind the stone and brick units in walls. Until the discovery of portland cement, lime-sand mortar was the only masonry mortar available. Egyptians used lime-sand mortar in the construction of pyramids, and virtually all historic stone buildings in Europe used lime-sand mortar. In lime-sand mortar, lime is the binder and sand is the filler.

Lime is made from limestone, one of the most abundant rocks in the earth's crust. Chemically, limestone is calcium carbonate. Lime is produced by simply heating limestone—a process known as *calcining*.

Calcining is typically done by first crushing limestone rock into approximately 2-in. particles and then heating them in a kiln to a temperature of approximately 1,800°F (1,000°C). Calcining expels the carbon dioxide from calcium carbonate, leaving calcium oxide. (It is an endothermic reaction, implying that heat is required for the reaction.)

Calcium oxide is called *quicklime*. Quicklime is a caustic substance that corrodes metals and causes severe damage to human skin. However, it reacts readily with water to form calcium hydroxide. Calcium hydroxide is called *hydrated lime*, or *slaked lime*, because it contains water that is chemically combined with calcium oxide. It is a relatively benign material and is the one that is commonly used in building construction.

NOTE

A Note on Quicklime

Historical records indicate that British soldiers used quicklime to incapacitate their French enemies in the war of 1217 by throwing it in their faces—one of the earliest records of chemical warfare.

The chemical reaction of quicklime with water (hydration reaction) is exothermic, implying that heat is produced during the reaction. The amount of heat produced in the hydration of 1 lb of quicklime is nearly 450 Btu. Because 1 Btu is the amount of heat required to raise the temperature of 1 lb of water by 1°F (Chapter 5), 450 Btu will raise the temperature of 3.2 lb of water from 70°F to 212°F (21°C to 100°C)—the boiling point of water. In other words, if 1 lb of quicklime is mixed with 3.2 lb of water initially at room temperature, that water will begin to boil.

Wood box carts carrying quicklime have ignited through accidental penetration of rainwater through leaks in the carts. Criminals misused quicklime to dispose of the corpses of their victims by dumping them in a pit, covering them with quicklime, and backfilling with wet soil.

Both quicklime and hydrated lime are white in color. Hydrated lime is generally used in powder form. The amount of water used to hydrate quicklime is much greater than that required for the hydration reaction. Therefore, after completion of the hydration reaction, the quicklime and water mixture is dried to evaporate the excess water. The dried hydrated lime is then pulverized to powder form. In the United States, hydrated lime is sold in 50-lb bags, Figure 21.1.

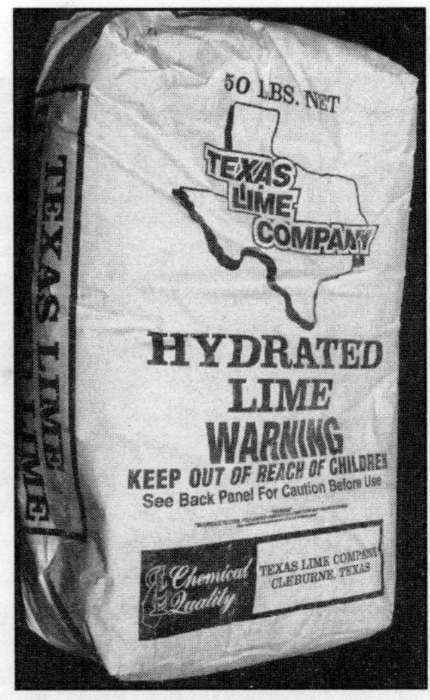

FIGURE 21.1 A typical lime bag.

Carbonation of Lime

The hardening property of lime results from its ability to react with atmospheric carbon dioxide to form calcium carbonate—the raw material used in making lime. Thus, the chemical reaction of lime with carbon dioxide (carbonation reaction) returns lime to its parent material—limestone—which is hard and strong. Stated differently, the calcination of limestone results in quicklime; the hydration of quicklime produces hydrated lime; and the carbonation of hydrated lime changes it to limestone. These three processes constitute a cycle, Figure 21.2.

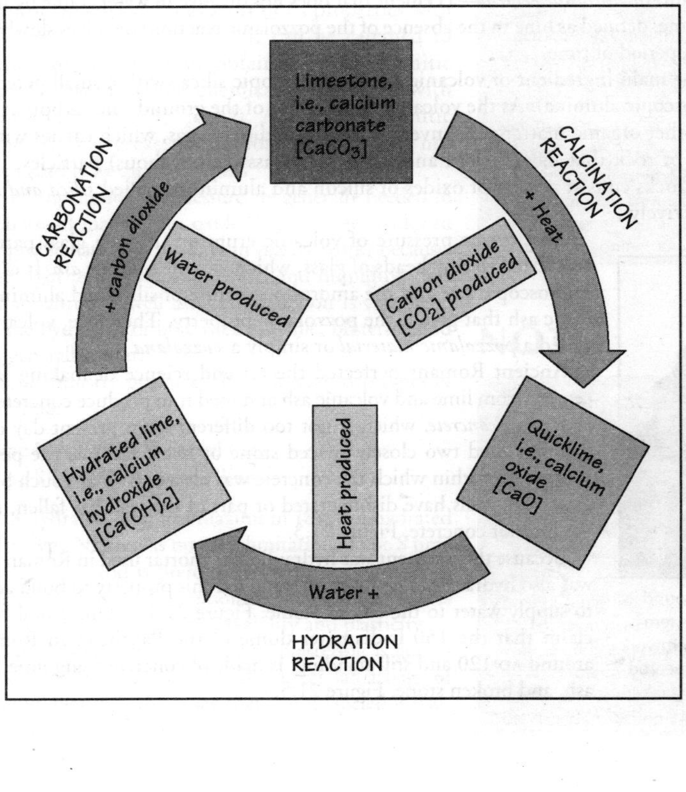

FIGURE 21.2 Limestone cycle—calcination, hydration, and carbonation.

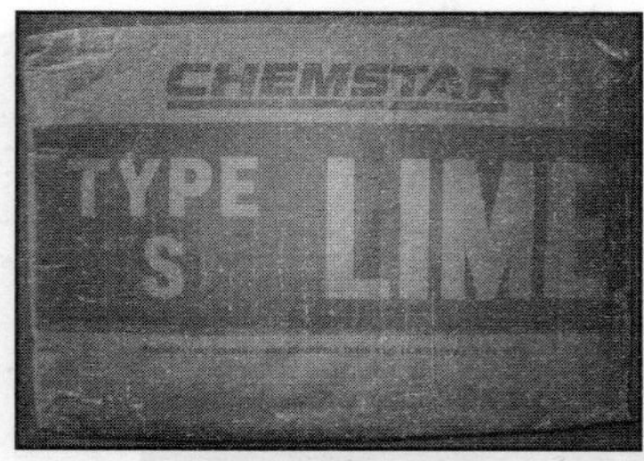

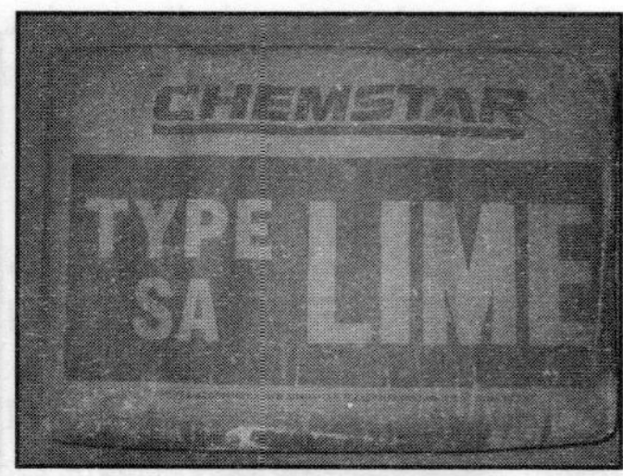

FIGURE 21.6 Type S hydrated lime and Type SA hydrated lime bags.

Type S lime can be produced from either dolomitic or high-calcium limestone. In the United States, dolomitic limestone is preferred because it gives greater workability to lime—a quality that masons and plasterers require. Lime is also produced with air-entraining agents mixed with it. Air entrainment increases the resistance of lime mortar against the freeze-thaw effect but decreases its strength (Section 21.4). Air-entrained lime is labeled by manufacturers as NA lime or SA lime.

USES OF LIME

Lime is used in construction as well as nonconstruction activities. Some of the nonconstruction uses of lime are in water purification, water softening, sewage sludge treatment, manufacturing of plastics and paints, and so on. Lime is used in municipal water-treatment plants because of its disinfecting property. It removes hazardous compounds, such as lead, from water and also softens hard water. In construction activities, lime is principally used in

- Masonry mortar (Chapter 24)
- Plaster and stucco (Chapter 29)
- Soil stabilization

PRACTICE QUIZ

Each question has only one correct answer. Select the choice that best answers the question.

1. Lime is made by
   - **a.** heating crushed limestone and then pulverizing it.
   - **b.** mixing crushed limestone with water and then pulverizing the mixture.
   - **c.** mixing crushed limestone with sulfuric acid and then pulverizing the mixture.
   - **d.** mixing crushed limestone with nitric acid and then pulverizing the mixture.
2. Quicklime reacts readily with water.
   - **a.** True
   - **b.** False
3. Hydrated lime is obtained by treating
   - **a.** slaked lime with water.
   - **b.** quicklime with water.
   - **c.** limestone with water.
   - **d.** slaked lime with portland cement.
   - **e.** quicklime with portland cement.
4. Quicklime is most commonly used in building construction.
   - **a.** True
   - **b.** False
5. Hydrated lime–water paste sets and hardens due to the
   - **a.** chemical reaction of water with lime.
   - **b.** chemical reaction of the oxygen in air with lime.
   - **c.** chemical reaction of the carbon dioxide in air with lime.
   - **d.** chemical reaction of the nitrogen in air with lime.
   - **e.** chemical reaction of the water in lime with the carbon dioxide in air.
6. Which of the following materials is a pozzolana?
   - **a.** Portland cement
   - **b.** Lime
   - **c.** Fly ash
   - **d.** Sand
   - **e.** Concrete
7. One of the important cementitious ingredients in making Roman concrete was
   - **a.** clay.
   - **b.** volcanic ash.
   - **c.** sand.
   - **d.** portland cement.
   - **e.** none of the above.
8. Hydrated lime is available in two types, which are referred to as
   - **a.** Types N and S lime.
   - **b.** Types I and II lime.
   - **c.** Types X and Y lime.
   - **d.** Types A and B lime.
   - **e.** Types 1 and 2 lime.
9. A pozzolanic material must be composed mainly of
   - **a.** microscopic silica.
   - **b.** microscopic and amorphous silica.
   - **c.** microscopic and crystalline silica.
   - **d.** silica and alumina.
   - **e.** amorphous silica and alumina.

21.3 PORTLAND CEMENT

Portland cement, a hydraulic cement, was patented in 1824 by Joseph Aspidin, a British stonemason. Aspidin burned finely ground limestone and clay in a kiln and found that the resulting product was a hydraulic cement, which set and gained strength much more quickly than lime.

Aspidin called his new product *portland cement* because the concrete that he made from it resembled, in color and strength, a highly sought-after natural limestone quarried from the Isle of Portland, off the British coast. Note that portland cement is no longer a brand name but a generic one; therefore, a large number of manufacturers produce portland cement.

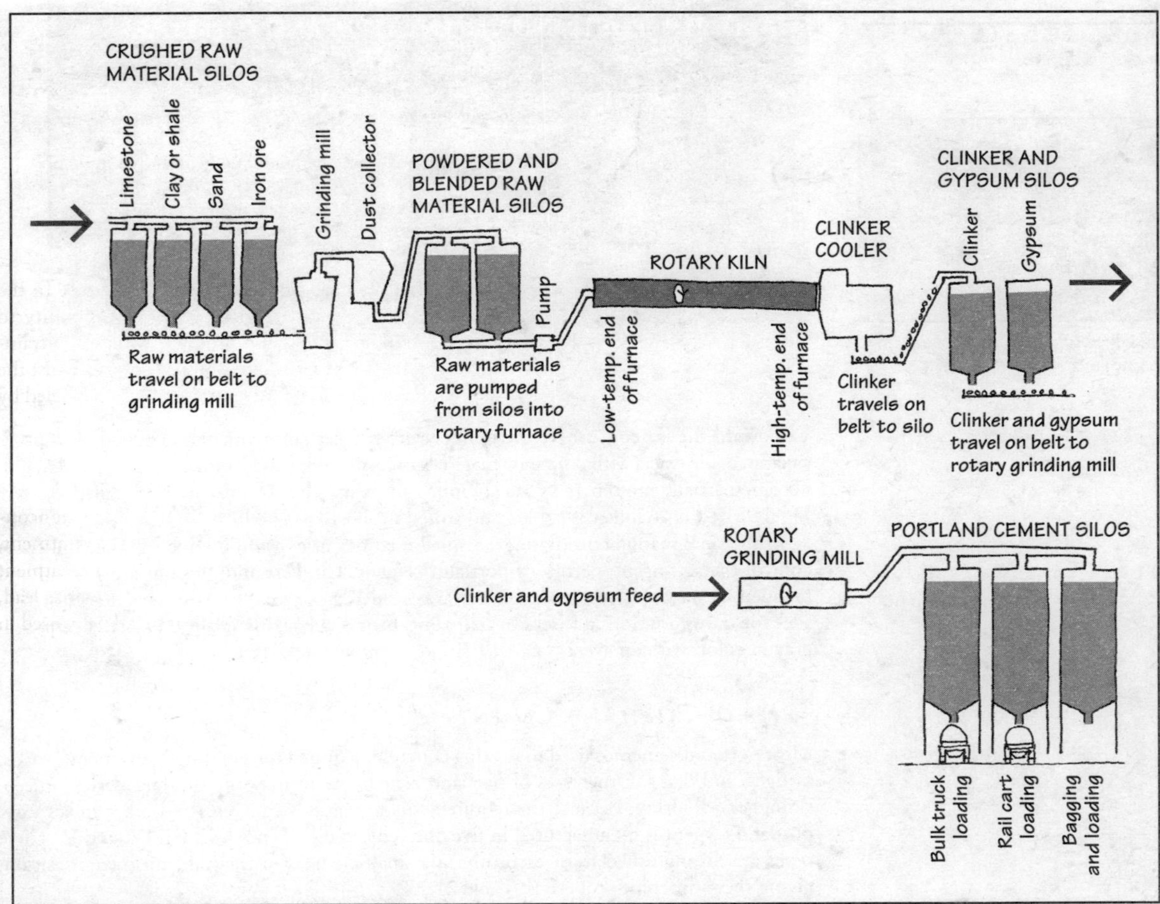

FIGURE 21.7 An outline of the portland cement manufacturing process.

Modern portland cement is made from almost the same basic raw materials and virtually by the same basic process that Aspidin used, that is, by heating a mixture of limestone and clay. Limestone comprises approximately two-thirds of the raw material required for making portland cement.

Another major raw material required is clay or shale. (Shale is a highly compressed form of clay.) Clay consists of the oxides of silicon, aluminum, and iron—called *silica, alumina,* and *iron oxide,* respectively. Some clays can provide all the silica, alumina, and iron oxide required in the manufacture of portland cement. Other clays may be supplemented by sand (to provide silica), bauxite (to provide alumina), and iron ore (to provide iron oxide). Generally, therefore, one or two raw materials, in addition to limestone and clay, are needed in portland cement's manufacture.

Manufacturing Process of Portland Cement

The manufacture of portland cement begins with quarrying limestone, which is later crushed to approximately $\frac{5}{8}$-in.-size particles. Crushed limestone and other raw materials (clay, sand, and iron ore) are stored in silos for processing as needed. Figure 21.7 is a conceptual diagram of the entire manufacturing process.

The raw materials are mixed together in required proportions and ground to a fine powder in a grinding mill. Powerful fans draw the powder from the grinding mill into a dust collector, and from there it is taken to storage silos. From the silos, the powder goes into a rotating, or rotary, kiln.

The rotary kiln—a huge steel cylinder lined inside with refractory bricks—is at the heart of the manufacturing process. It is considered to be the heaviest rotating industrial equipment. It is slightly sloped to allow the material to tumble forward by gravity. The kiln temperature increases as the material moves forward. In the initial section of the kiln, where the temperature is around 1,800°F (1,000°C), calcium carbonate is converted to calcium oxide, similar to the manufacturing of lime. The discharge of carbon dioxide is, therefore, a major environmental pollutant in the neighborhood of a portland cement plant.

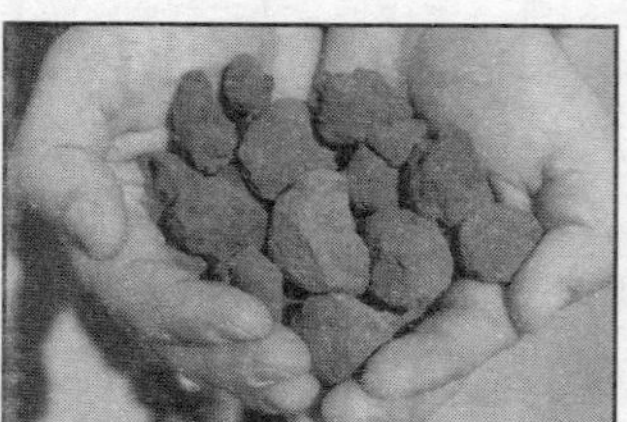

FIGURE 21.8 Clinker size and shape.

TABLE 21.1 PORTLAND CEMENT TYPES AND THEIR USES

Portland cement type	Description and use
I	**General-purpose** portland cement, used where there is no special requirement
II	**Moderate-sulfate-resistant and moderate-heat-of-hydration** portland cement
III	**High-early-strength** portland cement, used in precast elements and where high early strength is required, such as in cold weather
IV	**Low-heat-of-hydration** portland cement, used in massive civil engineering structures
V	**Sulfate-resistant** portland cement, used where high sulfate resistance is required

Toward the far end, where the temperature reaches approximately 3,400°F (1,900°C), calcium oxide reacts with other raw materials to form complex calcium compounds. Materials become partially molten and coalesce into approximately 1-in.-size nodules called *clinkers,* Figure 21.8. The clinker is cooled and stored in silos.

In the final manufacturing stage, a small quantity of gypsum is added to the clinkers to control the setting properties of portland cement. Clinkers and gypsum are ground and blended into a very fine powder, which is considered portland cement. It is then stored in silos for transportation in trucks or railroad cars or to a bagging facility. Portland cement is gray in color, with an average particle size of approximately 45 μm.

Types of Portland Cement

Most portland cement is used in making concrete, required for buildings, pavements, roads, bridges, and dams. Other uses of portland cement are in masonry mortar, plaster, stucco, siding, and flooring. Because the requirements for various concretes and other mixes vary, portland cement is manufactured in five different types—Types I, II, III, IV, and V. These types are distinguished from each other by small changes in their chemical composition, giving them different properties, Table 21.1.

Type I is a *general-purpose* portland cement. Type III portland cement is *high-early-strength* portland cement. It develops strength at a faster rate than other types in the initial stage, but its final strength is the same as that of the other types. As described in Section 21.9, portland cement develops its ultimate strength over several months.

Type III portland cement is generally used in making precast concrete elements, such as hollow-core slabs, double-tees, and concrete pipes. Economic considerations require that the formwork for precast elements be used as frequently as possible. Generally, precast concrete elements are cast one morning and stripped off the form the following morning, giving a 24-h turnover cycle (see Chapter 23).

Type IV is *low-heat-of-hydration* portland cement. It is meant for use in massive concrete structures (dam walls or bridge piers), where the temperature rise due to heat generated from hydration must be minimized. Due to its limited use, Type IV cement is produced by manufacturers only on special request.

Type V is *sulfate-resistant* portland cement. Portland cement can be affected adversely by the presence of sulfur. In fact, concrete made out of Type I portland cement will decompose into small fragments (spall) after just a few years in the presence of high-sulfur-containing environments. Some soils and groundwater have a high sulfur content. The use of Type V portland cement is recommended in such environments.

Type II is *moderate-sulfate-resistant* and *moderate-heat-of-hydration* portland cement. It combines the properties of Types IV and V to a moderate degree. Therefore, most manufacturers make portland cement that meets the requirements of both Types I and II and label the product as Type I/II portland cement, Figure 21.9. Approximately 90% of all portland cement used is Type I (or Type I/II) followed by Type III (approximately 5%).

The particle size of portland cement affects its performance. The finer the particles, the greater the surface area available for water to coat cement particles. This increases the rate of hydration, leading to higher early strength of portland cement paste. Type III portland cement, therefore, has finer particles than the other types.

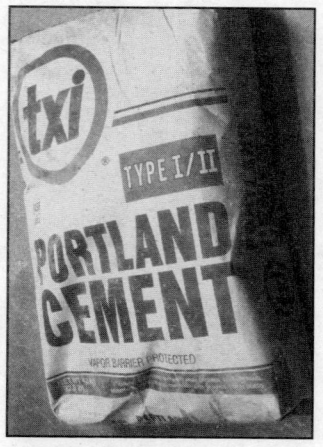

FIGURE 21.9 Type I/II portland cement. A portland cement bag (1 ft^3) generally weighs 94 lb.

EXPAND YOUR KNOWLEDGE

Composition of Portland Cement

Complex calcium compounds are formed during the heating of raw materials in a rotary kiln. These compounds impart different properties to portland cement, and their relative proportions determine the type of portland cement. These compounds are

Tricalcium silicate (C_3S)—called *alite*—approximately 50%
Dicalcium silicate (C_2S)—called *belite*—approximately 25%
Tricalcium aluminate (C_3A)—called *celite*—5 to 12%
Tetracalcium-alumino-ferite (C_4AF)—called *iron*—up to 8%

Alite hydrates rapidly. It is responsible for initial set and early strength (from 1 to 7 days). Belite hydrates slowly and contributes to later strength gain (28+ days).

Celite liberates a great deal of heat in the initial few days and contributes somewhat to early strength. Too much celite can reduce the sulfate resistance of portland cement.

Iron has little impact on portland cement properties but is needed as a flux to reduce energy consumption during manufacturing. A flux reduces the temperature at which the raw materials begin to melt and coalesce. The characteristic gray color of portland cement is due to the presence of iron and a small amount of manganese. White portland cement is made by reducing the iron and manganese contents.

Gypsum (calcium sulfate) is added to increase the setting time of portland cement. *Setting* is stiffening (early hardening) of the mix, that is, when the mix becomes semisolid. The mix should set neither too rapidly nor too slowly. A rapid set does not give enough time for placing and finishing the mix. A slow set delays finishing operations such as troweling and floating. Set control is, therefore, important. The amount of gypsum is carefully controlled.

Portland cement has a small quantity of alkalis—sodium and potassium oxides—which can react adversely with some forms of aggregates used in concrete.

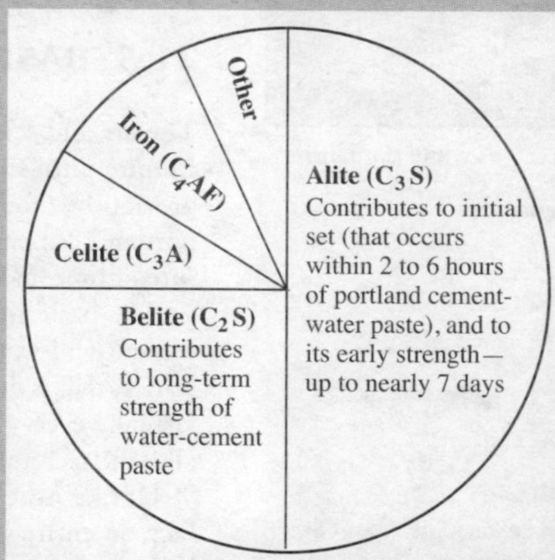

21.4 AIR-ENTRAINED AND WHITE PORTLAND CEMENTS

The portland cement types previously described are the basic types. These types are also available with an integrally combined air-entraining agent. Air entrainment in concrete, mortar, or plaster increases the durability of these materials against freezing and thawing.

The process of placing concrete in forms requires a much greater amount of water in a mix (called the *water of convenience*) than that required for the chemical reaction between portland cement and water (see Section 21.9). An inadequate amount of water in concrete does not allow concrete to be fluid enough to flow into the form and fill it completely. This results in honeycombed concrete. The excess water in a concrete mix evaporates as it stiffens, leaving voids. These voids are in addition to the voids between concrete ingredients—portland cement and aggregate—and those caused by lack of (or poor) consolidation. These voids contain *entrapped air*.

Because of the entrapped air, concrete is porous and tends to absorb rainwater. During freezing weather, the absorbed water turns into ice and expands. If a concrete is critically saturated and undergoes cycles of freezing and thawing, it will spall unless air voids are present to dissipate the pressure caused by the increased volume. To reduce freeze-thaw damage, tiny particles of air are introduced in a concrete mix, referred to as *air entrainment*. As the absorbed water in concrete expands on freezing, the entrained air relieves the pressure. Air-entrained concrete is commonly specified in exposed concrete elements subjected to freeze-thaw cycles.

Air entrainment can be achieved in two ways. The preferred way is to add an air-entraining chemical to the mix, referred to as an *air-entraining admixture*. The alternative is to use an air-entrained portland cement. Air-entrained portland cement has an air-entraining chemical as its integral part. All five basic types of portland cement are available as air-entrained portland cement. They are identified as Types IA, IIA, IIIA, IVA, and VA. With an air-entraining admixture, normal portland cement (Type I, II, III, IV, or V) is used.

WHITE PORTLAND CEMENT

The more commonly used portland cement is gray in color. For aesthetic reasons, white portland cement is preferred, particularly for terrazzo flooring, stucco, and architectural concrete. Colored aggregates in a terrazzo floor show up much better against a background of

NOTE

Entrapped Air versus Entrained Air in Concrete

Entrapped air in concrete does not provide freeze-thaw resistance because it is not uniformly distributed in concrete and the air particle size is too large to be effective (generally 1 mm or larger). Additionally, entrapped air voids are irregular in shape and are generally interconnected.

Entrained air voids, on the other hand, are much smaller (10 to 100 μm in size), almost spherical in shape, and uniformly distributed in concrete, but not interconnected.

A typical (non-air-entrained) concrete has an air content of approximately 1.5%. An air-entrained concrete requires about 6% air in order to provide high freeze-thaw resistance.

white portland cement. Additionally, when pigments are added to white portland cement, the resulting color is better than that obtained with the use of gray portland cement.

White portland cement is made by reducing the iron and manganese contents. This complicates the manufacturing process and increases the energy cost. Consequently, white portland cement is approximately twice the cost of normal (gray) portland cement. However, because portland cement is only one part of a concrete or stucco mix, the final product may only be 50% more expensive.

Like the gray cement, white portland cement is available in bags, Figure 21.10, or in bulk. It is generally produced as Type I or Type III. These types have the same properties as the gray Type I and Type III portland cement, respectively.

FIGURE 21.10 A white portland cement bag.

NOTE

What Is Concrete?

Concrete consists of

- Aggregate (coarse and fine aggregate) as matrix or filler
- Portland cement–water paste as an adhesive

Any other material used in concrete is called a *concrete admixture* (see Sections 21.10 and 21.11).

21.5 BASIC INGREDIENTS OF CONCRETE

The art and science of making concrete virtually disappeared after the fall of the Roman Empire until the discovery of portland cement approximately 1,400 years later. As stated previously, modern concrete is similar to Roman concrete, except that in modern concrete, portland cement is used as the glue to bind the stone aggregate together instead of a mixture of lime and volcanic ash.

The basic ingredients of modern concrete are portland cement paste (i.e., portland cement and water) and aggregate. Together, these materials provide a hard, rocklike substance, that is durable, fire resistant, and relatively inexpensive. It is used universally because virtually every region of the earth has the raw materials necessary to produce it. Additionally, the technology associated with its production and use is fairly simple.

Unlike other structural materials, concrete can be formed to any desired shape. In fact, an entire structure, regardless of its shape or size, can be formed monolithically—as if it were one object (*mono* means "one", and *lithos* means "piece of stone"). Some pioneers of the modern movement in architecture (e.g., Le Corbusier and Frank Lloyd Wright) and several structural engineers (e.g., Robert Maillart and Pier Nervi, and, more recently, Santiago Calatrava) have exploited the sculptural property of concrete, Figure 21.11.

SIZE OF THE AGGREGATE

Aggregate in concrete occupies 65% to 80% of the volume of concrete, the remainder being portland cement. The aggregate must consist of different-sized particles, referred to as *size gradation.* Proper gradation ensures that smaller particles fit within the voids created by larger particles, so that the entire mass of concrete is relatively dense, Figure 21.12(a).

FIGURE 21.11 **(a)** Chapel at Ronchamp, France, by architect Le Corbusier. Le Corbusier was among the pioneers in exploiting the sculptural properties of concrete in building structures. (Photo by Lazar Mihai-Bogdan/Shutterstock)

FIGURE 21.11 **(b)** Concrete structural frame work in the City of Arts and Sciences, Valencia, Spain, by architect-engineer Santiago Calatrava, illustrating the sculptural properties of concrete.

NOTE

Gravel

Gravel consists of naturally occurring particles of rock 0.2 in. to 3 in. size (see Figure 11.1). Because gravel is formed by erosion and weathering of rocks, it is round in shape, in contrast to crushed stone. The term *pea gravel* refers to gravel that is approximately $\frac{1}{4}$ in. to $\frac{3}{8}$ in. size.

A well-graded aggregate (implying that it consists of particles of various sizes) not only gives a stronger concrete, but also reduces the amount of portland cement necessary to wrap the particles and fill the spaces between them. An aggregate that consists of particles of only one or two sizes has a higher percentage of voids and, therefore, requires a much larger amount of portland cement, Figure 21.12(b). Because portland cement is far more expensive than aggregates, this gives an uneconomical concrete.

In general, therefore, the aggregate in a concrete mix consists of several sizes. However, the concrete industry divides the aggregate into two size groups:

- Fine aggregate
- Coarse aggregate

Fine aggregate is generally sand, but more precisely it is that material of which 95% passes through a No. 4 sieve. A No. 4 sieve consists of a wire mesh with wires spaced at $\frac{1}{4}$ in. on center. Because the wires have a certain standard thickness, the largest particle size of fine aggregate that can pass through a No. 4 sieve is slightly smaller than $\frac{1}{4}$ in.

Fine aggregate needs to be graded from a No. 4 sieve down to a No. 100 sieve. Recommended grading is shown in Table 21.2.

Coarse aggregate is that aggregate of which 95% is retained on a No. 4 sieve. It consists of either crushed stone or gravel. Gravel has several advantages over crushed stone, but crushed stone is commonly used because it is more economical.

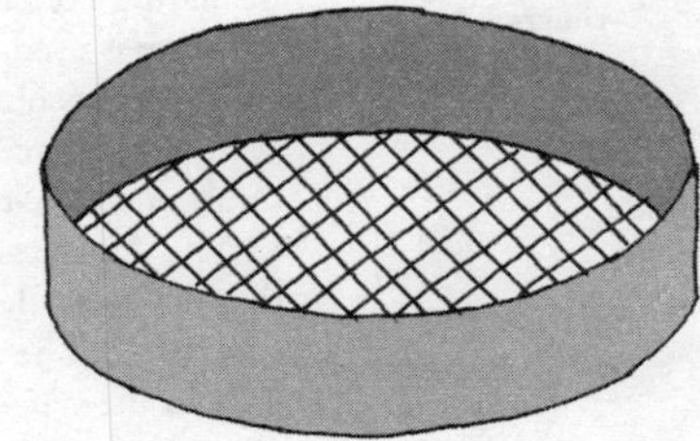

A typical sieve used for grading of aggregates in concrete laboratories. The sieves used in a concrete laboratory are the same as those used in a geotechnical laboratory. A No. 4 sieve has wires spaced at 1/4 in. (0.25 in.) on center, a No. 10 sieve has wires at 1/10 in. (0.1 in.) on center, and so on. The clear opening between wires in a No. 4 sieve = 4.75 mm and that of a No. 10 sieve = 2.00 mm (see Figures 11.1 and 11.4).

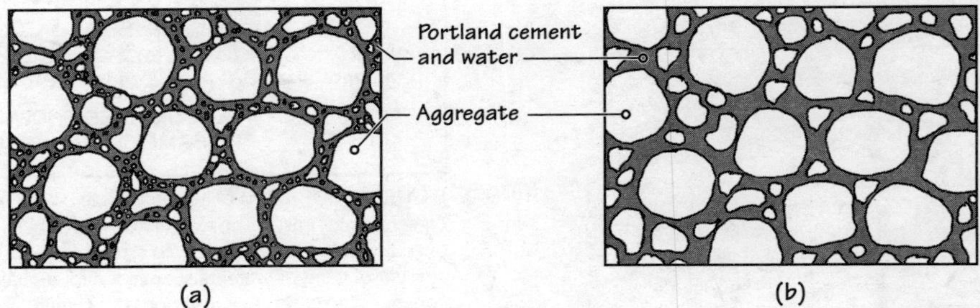

FIGURE 21.12 **(a)** Well-graded aggregate. **(b)** Poorly graded aggregate. The colored areas in these illustrations represent portland cement and water paste. Observe that poorly graded aggregate requires a greater amount of portland cement.

TABLE 21.2 RECOMMENDED GRADING OF FINE AGGREGATE FOR CONCRETE SLABS

Sieve size	Percent of aggregate passing through a sieve
$\frac{3}{8}$ in.	100
No. 4	95–100
No. 8	80–90
No. 16	50–75
No. 30	30–50
No. 50	10–20
No. 100	2–5

Although the minimum size of coarse aggregate is limited by the No. 4 sieve, its maximum size is a function of the smallest dimension of the building element and the spaces between steel reinforcement. For example, the American Concrete Institute (ACI) Code requires that for a concrete wall, the maximum coarse aggregate size should be smaller than one-fifth of the thickness of the wall. Obviously, if the size of coarse aggregate is too large for the space in which it is placed, it will be difficult to consolidate, giving nonhomogeneous and inconsistent concrete.

Similarly, if the steel reinforcement is closely spaced, the maximum coarse aggregate size must be small. The objective is that the concrete should flow easily around the reinforcement. In most concrete used in buildings, the maximum coarse aggregate is $\frac{3}{4}$ in. Concrete used in thick walls and heavy foundations may have aggregate up to $1\frac{1}{2}$ in. in size; it may be up to 6 in. in dam walls.

The size of coarse aggregate has a bearing on the concrete's cost. Larger particles have a smaller surface area for a given volume. Therefore, concrete made with large particles requires less portland cement, giving a more economical concrete, Figure 21.13.

Concrete
- Structural concrete (compressive strength ≥ 2,000 psi)
  - Normal-weight structural concrete (130 – 155 pcf)
  - Lightweight structural concrete (85 – 115 pcf)
- Insulating concrete (compressive strength 150 – 500 psi, density = 30 – 50 pcf (generally used as roof insulation))

NORMAL-WEIGHT AND LIGHTWEIGHT COARSE AGGREGATES

Concrete may either be structural or nonstructural. Nonstructural concrete is generally insulating concrete, which is used as roof insulation (see Chapters 5 and 33). Structural concrete is used in a structural frame—beams, slabs, columns, and walls.

The most commonly used structural concrete has a compressive strength of 4,000 psi. However, any concrete with a compressive strength of 2,000 psi or above is considered to be structural concrete. Structural concrete may either be normal-weight or lightweight concrete.

The American Concrete Institute Code defines lightweight structural concrete as one that, in its hardened state, weighs 85 to 115 lb/ft^3. Normal-weight concrete weighs 130 to 155 lb/ft^3. The difference between normal-weight and lightweight structural concretes is only in the type of coarse aggregates used. The fine aggregate in both is the same.

NOTE

Cost of Concrete and Coarse Aggregate Size

In a concrete mix, portland cement paste coats the surface of aggregates. If the surface area of aggregates is small, the amount of cement paste needed is small. Large aggregates have a small surface area for a given volume. Therefore, everything else being the same, a mix consisting of larger aggregates is more economical.

That the surface area of smaller particles contained in a given volume is greater than that of larger particles can be explained by considering 2-in. and 1-in. cubes. The surface area of a 2-in. cube is 24 in.^{2}, and its volume is 8 in.^{3}.

Now consider 1-in. cubes. To obtain a volume of 8 in.^{3} using 1-in. cubes, eight cubes are needed. The surface area of these eight cubes is 8 × 6 = 48 in.^{2}—double that of a 2-in. cube. Thus, the use of half-size particles has doubled the surface area, although the space occupied by the particles has not changed.

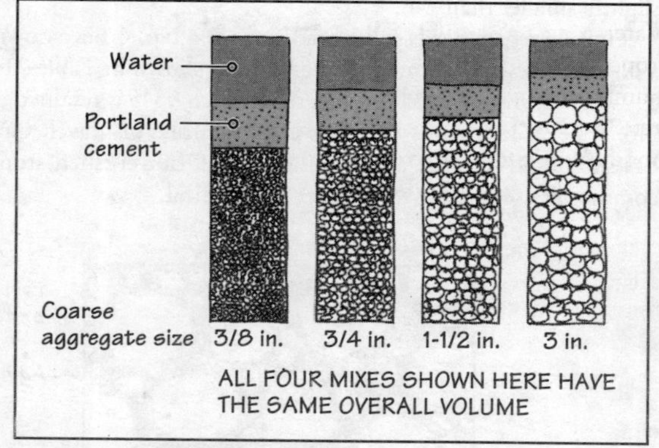

FIGURE 21.13 Effect of aggregate size on the amount of cement-water paste. The four mixes shown here have the same overall volume. Note that as the size of aggregate increases, the amount of cement-water paste needed for a given volume of concrete decreases. Because portland cement is the major cost-contributing ingredient in a mix, concrete with larger aggregate is generally more economical.

FIGURE 21.14 Two commonly used aggregates in concrete—crushed limestone as normal-weight aggregate and expanded shale as lightweight aggregate. The quantities shown here are such that both aggregate piles have the same approximate weight. Expanded shale, being fired clay, is generally brown in color, similar to that of a fired clay brick.

Normal-weight concrete is obtained using normal-weight aggregate. These aggregates include crushed limestone, granite, quartz, and so on. Lightweight structural concrete is obtained using lightweight aggregate.

One commonly used lightweight aggregate is expanded shale, obtained by heating crushed shale to a high temperature. This converts any water present in a shale particle into steam, which—in trying to exit—expands the particle, reducing its density. Because it is a baked material and because shale is a compressed form of clay, the color of expanded shale is similar to that of brick. Figure 21.14 portrays the difference between normal-weight limestone aggregate and lightweight expanded shale aggregate.

Lightweight structural concrete costs more than normal-weight concrete because of the higher cost of producing the aggregate. However, it can yield savings in the overall cost of the structure because it reduces dead loads. This is particularly attractive in a high-rise structure, where any reduction in the dead load of floor slabs reduces the size of supporting beams. This, in turn, reduces the size of columns, which leads to lighter foundations.

Lightweight structural concrete has, however, limitations as to strength. High-strength concrete is difficult to obtain using lightweight aggregates. However, the commonly used concrete strength of 4,000 psi is easily obtained using lightweight aggregates. Therefore, in a high-rise structure, lightweight concrete is used in floor beams and slabs, and normal-weight concrete is used in columns. The reason is that, whereas the columns have to support the load of all the upper floors, the floor beams and slab are required to support only the load of one floor.

Quality of Water

Water is an important component of concrete. Portland cement derives its cementing property from its reaction with water. Water used in concrete must be clean. A rule of thumb in the concrete industry is that if the water is fit for drinking, it is fit for use in concrete. Because seawater contains salts, its use leads to corrosion of reinforcing steel. It also modifies the portland cement–water reaction and is, therefore, not appropriate for use.

NOTE

Carbonate versus Silicious Aggregates

Concrete aggregates are generally obtained from limestone, granite, quartz, sandstone, and so on. Limestone (calcium carbonate) aggregate yields concrete that is slightly more fire resistant than the concrete obtained from aggregates produced from granite, quartz, sandstone, and so on.

In general, carbonate-containing aggregates give a concrete that is more fire resistant than that obtained from silicious aggregates (see Table 23.1).

NOTE

Wash Water

Concrete mixers are washed after delivery and before being recharged with the next batch. Ready-mix plants are increasingly using the wash water in the mix—a more sustainable alternative.

Practice QUIZ

Each question has only one correct answer. Select the choice that best answers the question.

10. The main ingredient used in the manufacture of portland cement is
    a. marble. b. granite.
    c. limestone. d. sand.
    e. iron ore.

11. ASTM specifications classify portland cement into five basic types, referred to as
    a. Types A, B, C, D, and E.
    b. Types I, II, III, IV, and V.
    c. Types PC1, PC2, PC3, PC4, and PC5.
    d. Types X, Y, Z, P, and Q.
    e. none of the above.

12. The most commonly specified type of portland cement is
    a. Type C. b. Type III.
    c. Type PC1. d. Type I/II.
    e. Type II/III.

13. Which type of portland cement is commonly used in making precast concrete members?
    a. Type C b. Type III
    c. Type PC1 d. Type I/II
    e. Type II/III

14. White portland cement is generally weaker than normal (gray-colored) portland cement.
    a. True b. False

(Continued)

PRACTICE QUIZ (Continued)

15. The aggregate for concrete is generally subdivided into coarse aggregate and fine aggregate. The distinction between the two types of aggregate is based on the
    a. aggregate's strength.
    b. aggregate's surface texture.
    c. aggregate's density.
    d. aggregate's particle size.
    e. aggregate's parent rock.
16. In a No. 4 sieve, the dimensions of voids are approximately
    a. 4 in. × 4 in.
    b. $\frac{1}{4}$ in. × $\frac{1}{4}$ in.
    c. 2 in. × 2 in., giving an area of approximately 4.0 in.^{2}.
    d. 1 in × 4 in.
    e. none of the above.
17. A well-graded coarse aggregate is one in which the size of all aggregate particles is approximately the same.
    a. True
    b. False
18. Everything else being the same, a concrete with a larger coarse aggregate requires less portland cement and water paste to give the same concrete strength.
    a. True
    b. False
19. Lightweight structural concrete is heavier than insulating concrete.
    a. True
    b. False

21.6 IMPORTANT PROPERTIES OF CONCRETE

Properties of concrete that are of interest to architects, engineers, and builders may be divided into the following two categories:

- Fresh (plastic-state) concrete properties
- Hardened concrete properties

FRESH CONCRETE PROPERTIES—WORKABILITY (SLUMP) OF CONCRETE

A good concrete mix should be able to retain its homogeneity (nonsegregation of particles) during all operations involved in transporting the concrete from the mixer to the form and placing, compacting, and finishing the concrete. Additionally, it should be possible to perform these operations with relative ease and in a reasonable amount of time. In other words, a good concrete should be easy to pump (when required), place, and compact, and it should set within a reasonable amount of time so that finishing can be done without delay.

Whether a concrete meets these requirements or not can often be judged visually by an experienced concrete technician. However, a measure commonly used to describe them collectively is called *workability*. Workability of concrete is the ease with which a concrete can be placed and compacted in the form with minimum loss of consistency and homogeneity. Concrete that is not workable is referred to as a *harsh* concrete. To obtain a workable concrete,

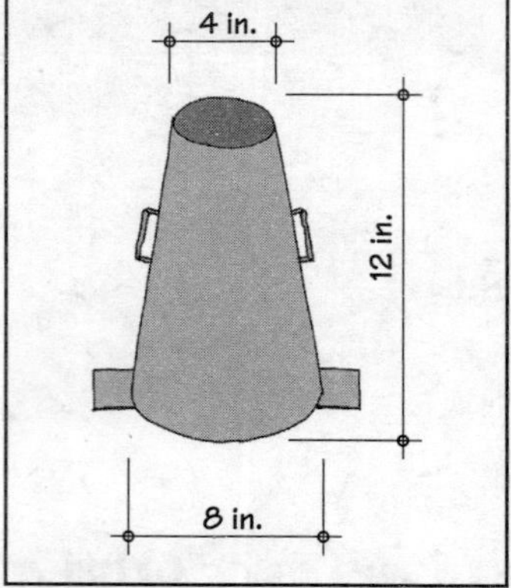

FIGURE 21.15 A slump cone is open at the top and bottom.

- The aggregates should be well graded.
- Everything else being the same, large aggregates reduce workability. Adequate fine material is important in providing a workable concrete. That is why the addition of fly ash or silica fume, for example, in a mix improves its workability. Because the particle size of portland cement is small, increasing the amount of portland cement also increases workability. However, excessive portland cement in a mix makes it too gluey and tacky, reducing workability.
- The aggregates should not be too angular in shape. Gravel, because of its naturally round particles, gives greater workability than crushed stone. However, crushed stone, if properly graded, yields good workability. Similarly, sand obtained from river beds or mined from sand pits gives greater workability than that obtained by crushing and pulverizing stone.
- An adequate amount of water is necessary for workability.

A commonly used measure of workability is *slump*. Although slump does not evaluate all aspects of workability, it is its best available measure. The slump of concrete is measured using a truncated cone. The cone is open at the top and bottom and has handles and foot tabs. It measures 4 in. at the top and 8 in. at the bottom and is 12 in. high, Figure 21.15.

Slump measurements are performed by a qualified concrete technician at the construction site immediately after the concrete is received. To measure slump, the inside surface of the slump cone is dampened with water, and any old concrete sticking to its surface is removed.

The cone is placed on a rigid, flat, smooth, and nonabsorbent surface. It is held firmly in place by standing on foot tabs, Figure 21.16(a). Concrete is placed in the cone in three layers,

(a) A technician filling a slump cone and rodding the concrete inside the cone with a steel rod.

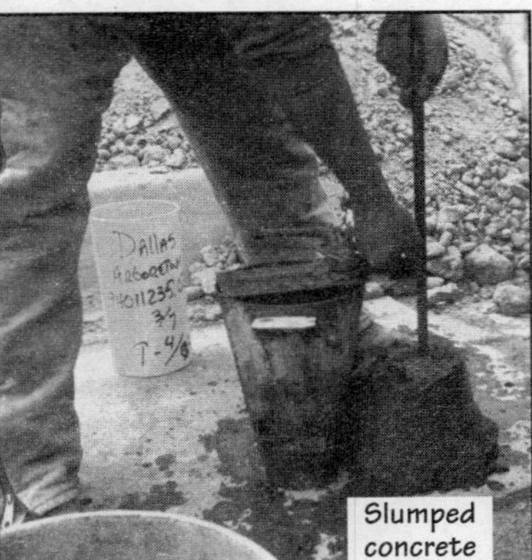

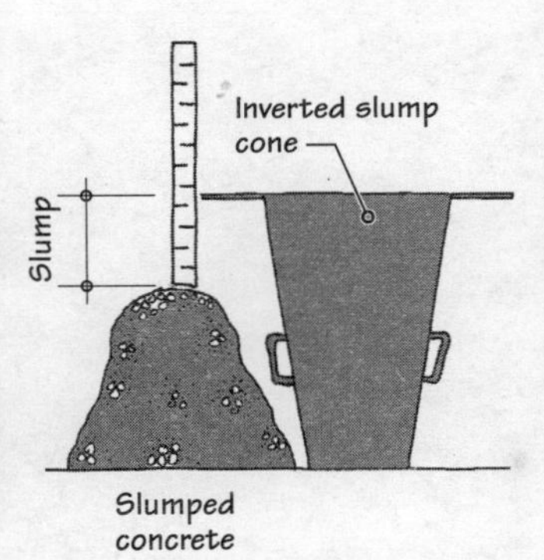

(b) After the cone is full, the technician lifts the slump cone and inverts it to measure concrete's slump using a trowel to line with the cone's top.

FIGURE 21.16 Measurement of the slump of concrete.

each 4 in. high (as per ASTM C143). Each layer is tamped 25 times through its depth by a $\frac{5}{8}$-in.-diameter smooth steel rod—a process also referred to as *rodding.*

After the last layer has been fully rodded, the excess concrete is struck off so that the concrete is flush with the top of the cone. The cone is then removed with a slow upward motion. As soon as the cone is removed, the concrete settles. The slump of concrete is a measure of settlement in inches—the difference between the 12-in. height of the cone and the top of the slumped concrete, Figure 21.16(b).

Workable concrete will slump uniformly, retaining its overall conical profile. A poorly graded concrete will slump nonuniformly.

Different slumps are required for different applications of concrete. A low value of slump gives a concrete that will not completely flow in the form and will require careful compaction.

In some situations, a high-slump concrete is desired, such as in the grout used in masonry walls. In most situations (unless a water-reducing admixture has been used; see Section 21.10), a high value of slump indicates the use of excess water. This, as we will see later, reduces concrete's strength. Generally, for concrete walls, columns, and beams, a slump of 4 to 5 in. is required. For slabs, a slump of 3 to 4 in. is generally adequate.

NOTE

Approximate Required Slump Values

Walls and columns	4–5 in.
Beams	4–5 in.
Slabs	3–4 in.

Fresh Concrete Properties—Particle Segregation and Bleeding

If concrete is not carefully placed in the forms or if it is compacted excessively, larger aggregates settle down and smaller ones rise to the top. The segregation of particles gives a nonhomogeneous concrete, reducing its strength. It is important, therefore, that concrete be placed in position carefully and not thrown from a large distance.

Segregation of aggregates is generally accompanied by *bleeding.* Bleeding is the rising of water to the top. This is similar to the rise of water that occurs in a heap of wet sand on a beach when tapped a few times. Bleeding occurs naturally when the mix is compacted or floated. Therefore, a certain amount of bleeding is inherent during placing, compacting, and finishing. Excessive bleeding should be avoided.

Hardened Concrete Properties—Compressive Strength of Concrete

The two most important properties of hardened concrete are durability and compressive strength. The first step in determining concrete's strength is to cast test cylinders from the concrete received at the job site. The casting of cylinders and the slump test are generally performed at the same time.

FIGURE 21.17 Casting of concrete test cylinders.

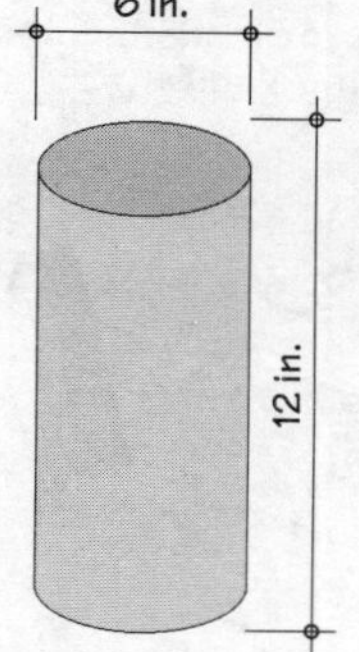

Dimensions of a concrete test cylinder

A concrete cylinder is 6 in. in diameter and 12 in. high. Building codes and standards specify the number of test cylinders that must be cast for a certain amount of concrete used. A concrete test cylinder is made by placing concrete in a mold in three layers, similar to the slump test. Each layer is rodded 25 times, and finally the excess concrete is struck off, leaving it flush with the top of the mold, Figure 21.17.

The molds for cylinders are generally of plastic, steel, or coated cardboard. Plastic and steel molds are slightly oiled before being filled with concrete. Cardboard molds have an integral waterproofing on the inside surface. The outside of a mold is identified with an indelible marker as to the job site and the date of casting. After the molds have been cast, their tops are covered with a plastic sheet to retain the water in the concrete. They are kept at the construction site for 24 h after casting, care being taken to ensure that they remain undisturbed for this period.

After 24 h, they are carefully transported to a concrete testing laboratory, where the molds are stripped off the cylinders. At the time of removal of the molds, the identifying information is transferred to the cylinders, which are then placed in a moist chamber. The air in a moist chamber is fully saturated at approximately 73°F (20°C). The cylinders remain in the moist chamber until they are tested to failure. Some research laboratories that do not have a moist chamber use a water tank in which the cylinders are kept immersed.

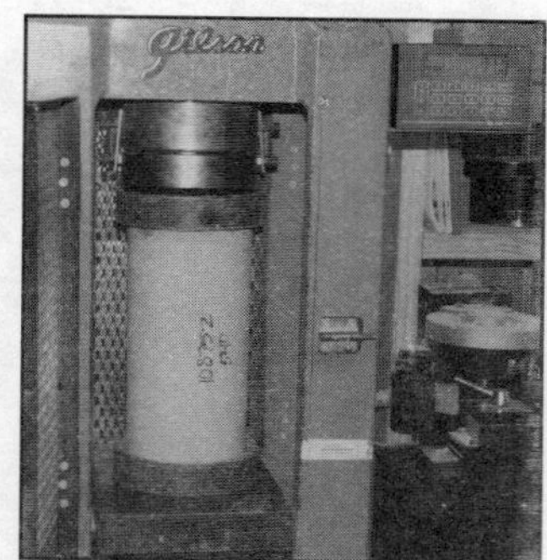

FIGURE 21.18 A concrete test cylinder in position in the testing machine, ready to be compressed to failure (see also Figure 4.6).

As described later, concrete develops its strength over a long period of time. The typical concrete strength used for design purposes is its 28-day strength. Cylinders are, therefore, generally tested 28 days after casting. To determine the strength of concrete, the cylinders are removed from the moist chamber (or water tank) and crushed to failure, Figure 21.18. The compressive strength of concrete is simply the failure load divided by the cross-sectional area of the cylinder (see Chapter 4).

Because the top and bottom surfaces of a concrete cylinder are rough, they are capped with a smooth-surfaced material before the cylinder is placed in the test machine. Smooth ends give a more reliable test result. Rough ends lead to local crushing of protruding concrete particles, falsifying the result. Rough ends also misalign the cylinder; the cylinder must be in a vertical position in the machine. The capping material commonly used with cylinders is a 6-in.-diameter neoprene pad set in a steel containment ring. The containment ring prevents the neoprene pad from expanding laterally.

Concrete is the only building material whose quality is verified after it has been used in the building. Cylinder and slump tests verify that the concrete supplied by the concrete manufacturer has the strength and slump specified for the project.

NOTE

Field Curing of Concrete Cylinders

To obtain the compressive strength of concrete that is representative of the field conditions, cylinders may be field cured.

21.7 MAKING CONCRETE

Concrete is made by mixing coarse and fine aggregates, portland cement, and water. A small amount of concrete for a do-it-yourself job may be made by mixing various ingredients with a shovel, adding water, and mixing the ingredients further until all materials have blended. Alternatively, bags of dry concrete mix (consisting of premixed aggregates and portland cement) may be obtained from a building material store. A dry concrete mix needs only the addition of water.

If a slightly larger quantity of concrete is required, an on-site mobile concrete mixer can be used, Figure 21.19. However, where even a small degree of control on the quality of concrete is required, the concrete should be obtained from a ready-mix plant. A ready-mix plant is a concrete-manufacturing facility. Approximately 95% of all concrete used in contemporary building construction is obtained from such a facility. Most cities in the United States have at least one ready-mix plant within the city or close by.

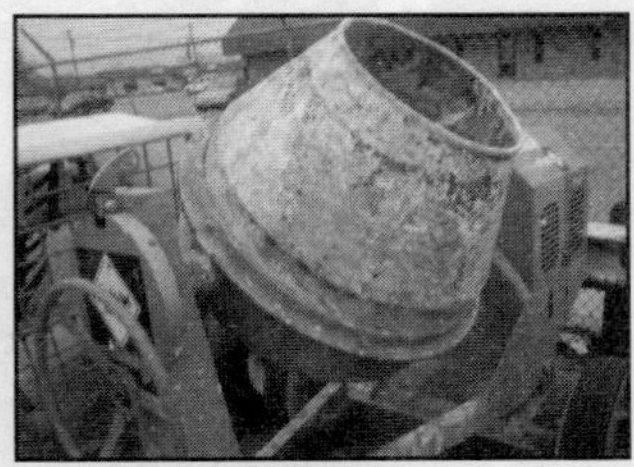

FIGURE 21.19 A transportable mixer for on-site mixing of small quantities of concrete.

An important part of concrete manufacturing is *mix design*. Mix design involves determining the correct amounts of various ingredients to give a concrete that has the required durability, strength, workability, and any other property specified by the architect or engineer. Mix design is fairly complex and is covered in detail in texts devoted entirely to the subject. One reason for its complexity is the enormous variability in the quality of aggregates, even when they are obtained from the same source. A typical ready-mix plant has technical personnel with expertise on mix design.

Figure 21.20 shows an overview of a typical ready-mix plant. The most prominent and visible part of the plant is a set of cylindrical silos. The silos contain coarse aggregate, fine aggregate, and portland cement. The aggregates are initially stockpiled in an open yard within the plant compound, Figure 21.21. From the stockpiles, the aggregates are conveyed into the silos as needed. The space directly under the silos is designed to hold a concrete mixer truck. Correct amounts of various materials are fed by gravity from the silos into the mixer, Figure 21.22.

Weighing the materials and charging the truck mixer are fully automated and are controlled from a central location that overlooks the charging operation, Figure 21.23. The first material fed into the mixer is usually water, followed by aggregates and portland cement. A small amount of water may be added at the construction site if needed. During travel, the mixer drum is set to rotate, so that when the concrete arrives at the site, it is almost ready for discharge (with some site remixing, as needed). A typical concrete mixer truck holds 10 yd^3 of fresh concrete.

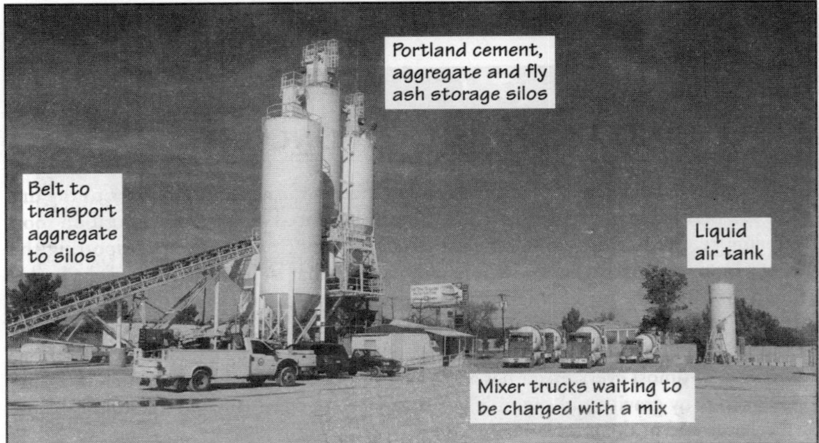

FIGURE 21.20 An overview of a ready-mix concrete plant. Liquid air may be needed in the concrete mix in warm weather to control its temperature. Ice may be used as an alternative, depending on availability and cost.

FIGURE 21.31 Troweling of concrete with a power-driven troweling machine, which is available in two different types, as shown. Each machine has metal blades that rotate over the concrete surface.

PRACTICE QUIZ

Each question has only one correct answer. Select the choice that best answers the question.

20. The slump of concrete is measured
- **a.** when the concrete is fresh.
- **b.** 28 days after casting the concrete in a test specimen.
- **c.** 14 days after casting the concrete in a test specimen.
- **d.** 7 days after casting the concrete in a test specimen.
- **e.** 1 day after casting the concrete in a test specimen.

21. The slump of concrete is a measure of the
- **a.** compressive strength of concrete.
- **b.** tensile strength of concrete.
- **c.** modulus of elasticity of concrete.
- **d.** workability of concrete.

22. The test specimen used for determining the compressive strength of concrete in the United States is a
- **a.** 6-in. diameter, 6-in.-high cylinder.
- **b.** 6-in. diameter, 12-in.-high cylinder.
- **c.** 6-in. diameter, 18-in.-high cylinder.
- **d.** 6-in. diameter, 12-in.-high cone.

23. The compressive strength of concrete is generally measured
- **a.** when the concrete is fresh.
- **b.** 28 days after casting the concrete in a test specimen.
- **c.** 14 days after casting the concrete in a test specimen.
- **d.** 7 days after casting the concrete in a test specimen.
- **e.** 1 day after casting the concrete in a test specimen.

24. In a high-rise building, the concrete may be brought from the ground to the location of placement by
- **a.** chutes.
- **b.** buckets.
- **c.** pumping.
- **d.** (a) or (b).
- **e.** (b) or (c).

25. After concrete has been placed in the form, it should be compacted as much as is economically practical.
- **a.** True
- **b.** False

26. The consolidation and compaction of concrete on a construction site are generally done
- **a.** manually using a 2-in.-diameter, 3-ft-long steel rod.
- **b.** manually using a 1-in.-diameter, 3-ft-long steel rod.
- **c.** using a power-driven rod vibrator.
- **d.** using a power-driven plate vibrator.

27. The finishing operations of a concrete surface usually involve striking, floating, and troweling. The sequence in which these three operations are generally performed is
- **a.** striking, floating, and troweling.
- **b.** floating, striking, and troweling.
- **c.** striking, troweling, and floating.
- **d.** troweling, floating, and striking.

21.9 PORTLAND CEMENT AND WATER REACTION

Concrete and other mixes made from portland cement gain their strength due to the reaction of portland cement with water, referred to as the *hydration of portland cement.* The amount of water required for complete hydration is about 40% of the weight of portland cement. In other words, for complete hydration, the water-cement ratio (referred to as the *w-c ratio*) should be 0.40. Often, however, a larger quantity of water is needed to provide the requisite workability of concrete.

To obtain a normal-weight concrete with a slump of 4 in. or so (used for beams and columns), a w-c ratio of 0.55 to 0.6 is usually needed, which exceeds that needed for complete hydration. In some concretes, the w-c ratio required may be as high as 0.7. The excess water eventually evaporates, leaving air voids in hardened concrete, which reduce concrete's strength.

Experiments have indicated that the strength of concrete is inversely proportional to the w-c ratio, Figure 21.32. Therefore, a concrete should contain the minimum amount of water that gives it the required workability.

DURATION OF HYDRATION REACTION—THE CURING OF CONCRETE

The hydration reaction begins as soon as water and portland cement come into contact, but the rate at which this reaction proceeds is extremely slow. It takes up to 6 months or longer

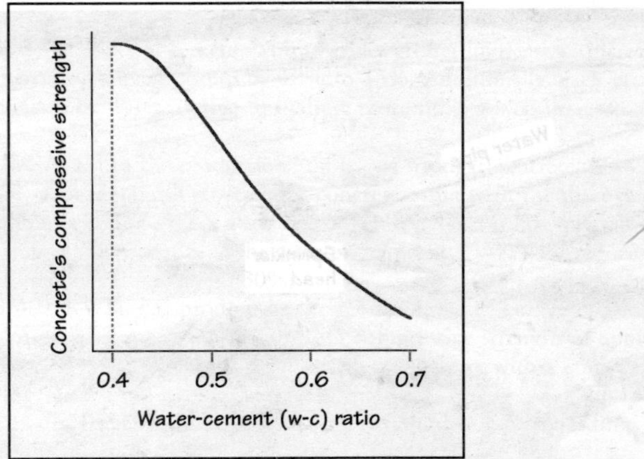

FIGURE 21.32 Strength of concrete as a function of the w-c ratio. For a given concrete (type of aggregate and amount of portland cement), concrete's strength increases as the amount of water used is reduced.

for concrete to gain its full strength, Figure 21.33. However, approximately 80% of concrete strength develops in 28 days.

Approximately two-thirds of the 28-day strength is obtained in the first 7 days and approximately half in the first 3 days. This is true only if sufficient water and favorable temperature are available for the hydration reaction to continue. That is why concrete test cylinders are kept in a moist chamber until tested. Providing moisture to concrete continuously for hydration is called *curing of concrete.*

Curing not only increases the strength of concrete, but also reduces its permeability to water. A well-cured concrete is denser and, hence, stronger and more durable. On construction sites, curing is begun as soon as the concrete has fully set (solidified), which is generally 12 to 24 h after placing the concrete. Curing in the initial stages of hardening is extremely important and should continue as long as possible, not less than 7 days. The following methods of field curing are common:

- *Keeping concrete wet with water:* This is the most effective curing method, which includes covering the exposed surfaces of concrete with water-absorptive materials, such as burlap cloth, rags, or blankets, and sprinkling water on them, Figure 21.34. If it is possible to continuously sprinkle water on concrete, cover materials are not necessary.

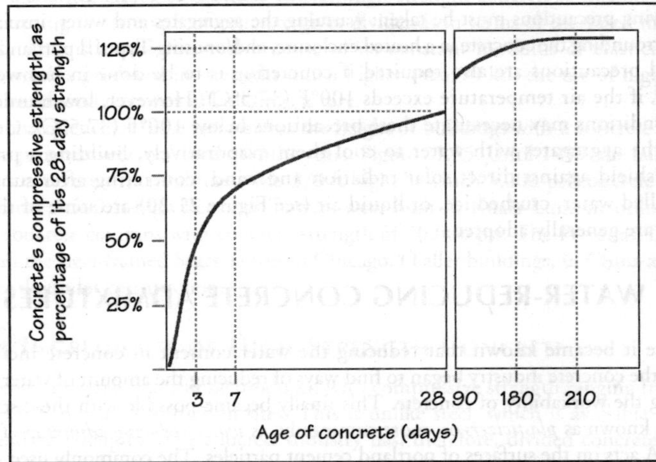

FIGURE 21.33 Compressive strength of concrete as a function of its age. Observe that concrete keeps gaining strength well beyond 28 days. Since we generally use concrete's 28-day strength as the design strength, the additional strength adds to the safety of a concrete structure.

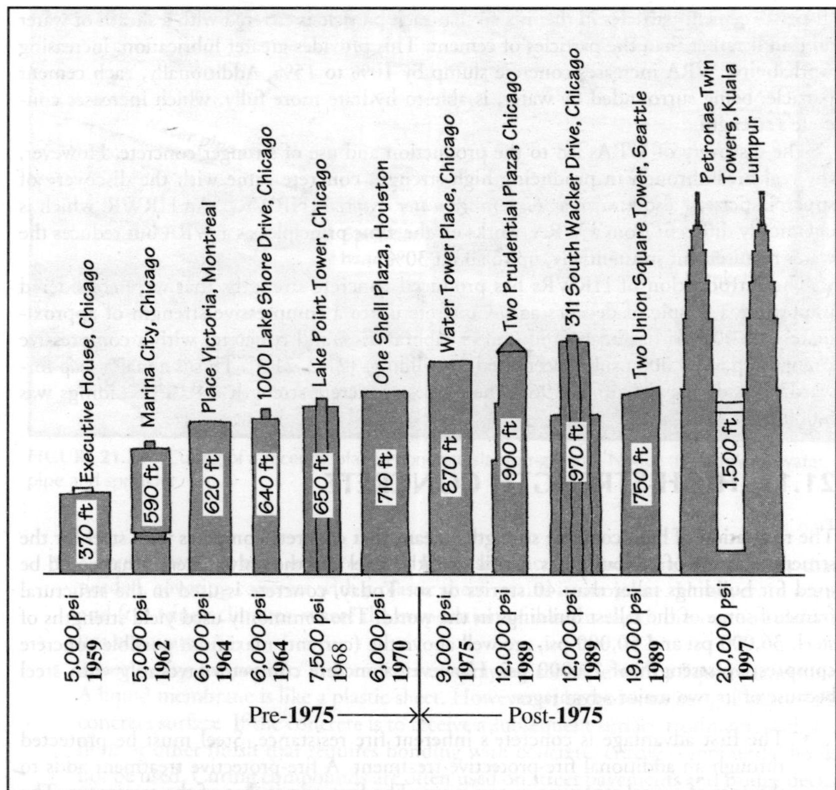

FIGURE 21.35 Progressive increase in concrete's strength in the structural frame of buildings.

This division is totally arbitrary because no one factor affects concrete strength abruptly. Generally, conventional concrete may be produced without the use of admixtures. High-strength concrete requires the use of admixtures. A *concrete admixture* is defined as a material other than portland cement, coarse and fine aggregates, and water. WRAs and HRWRs are not the only materials required for the production of high-strength concrete. High-strength concrete also requires the use of other admixtures, such as fly ash and silica fume.

FLY ASH—A SUSTAINABLE CONCRETE INGREDIENT

As mentioned previously, fly ash is a waste product produced by coal-fired power stations. It is used in both high-strength and conventional concretes. Because fly ash particles are microscopic in size, they densify concrete, filling in voids between portland cement particles. The densification of concrete increases its strength and durability. Small particles of fly ash also increase concrete's workability and reduce the permeability and creep of concrete (Section 9.6).

However, the main benefit of fly ash is in its pozzolanic property. During the manufacture of portland cement, a certain amount of lime (calcium hydroxide) does not convert to complex portland compounds—alite, belite, celite, and so on. This lime is referred to as *free lime.* Fly ash reacts with free lime, converting it into a hydraulic cement, and adds to concrete's strength.

SILICA FUME—A SUSTAINABLE CONCRETE INGREDIENT

Silica fume is a waste product from the silicon industry. It is an extremely fine and amorphous silica (sand) with average particle size of 0.1 μm [21.4]. In other words, more than 100 silica fume particles occupy the same space as 1 particle of portland cement. Silica fume fills the voids that other concrete materials cannot. It is an extremely lightweight material and is available in bags, Figure 21.36. A large silica fume bag weighs less than 10 lb. Silica fume is an expensive concrete admixture. Therefore, it is not commonly used in conventional concrete. It has excellent pozzolanic properties. Like fly ash, it reacts with free lime to convert it into a hydraulic cement.

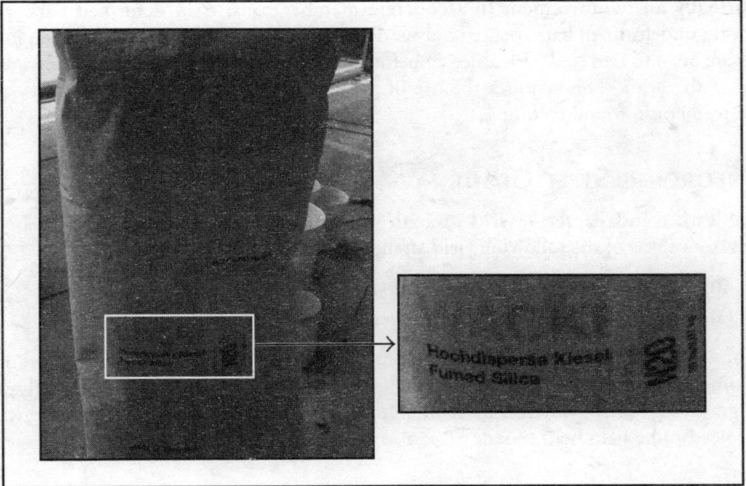

FIGURE 21.36 A silica fume bag.

HIGH-STRENGTH CONCRETE AND PORTLAND CEMENT QUANTITY

Because portland cement is the binding agent, its contribution to concrete's strength is obvious. In general, the greater the amount of portland cement, the greater the concrete's strength. However, increasing the amount of portland cement in a mix yields exponentially smaller increases in its strength. In general, more than 900 lb (approximately 10 bags) of portland cement per cubic yard does not increase concrete's strength significantly.

DISADVANTAGES OF HIGH-STRENGTH CONCRETE

The cost of concrete increases with increasing strength. At 20,000 psi or so, the mix is not only very expensive but requires a great deal of quality control at the site, which increases the cost further. Remember, concrete's strength depends on the quality of aggregates and admixtures, grading of aggregates, and the quantity of portland cement. It also depends on placing, compacting, and curing. An ultra-high-strength concrete requires that all of these factors be very tightly controlled.

21.12 STEEL REINFORCEMENT

Concrete is much weaker in tension than in compression. Its tensile strength is approximately 10% of its compressive strength. Therefore, concrete is generally used in conjunction with steel reinforcement, which provides the tensile strength in a concrete member. The use of plain concrete—concrete without steel reinforcement—is limited to pavements and some slabs-on-ground.

Steel is the ideal material to complement concrete because the thermal expansion of both materials is the same. In other words, when heated or cooled, both steel and concrete expand or contract equally. Consequently, no stress is caused by differential expansion or contraction. Composite materials that expand differentially are subjected to such stresses.

Steel also bonds well with concrete. In a composite material, the bond between two materials is necessary for it to function as a single material. The bond between steel and concrete is due to the chemistry of the two materials, which produces a *chemical bond* between them. Additionally, as water from concrete evaporates, it shrinks and grips the steel bars, making a *mechanical bond.*

The mechanical bond is enhanced by using reinforcing bars, or *rebar,* that have surface deformations, Figure 21.37. Because a mechanical bond is a function of the area of contact between the two materials, surface deformations increase that area, thereby increasing the bond. For the same reason, rebar that have a light, firm layer of rust bonds better with concrete. Rust that is produced by leaving rebar outdoors on a construction site for a few days or weeks is not objectionable as long as the rust is not loose or flaky. Loose and flaky rust should be scraped using burlap or a piece of cloth. Excessively rusted rebar should not be used.

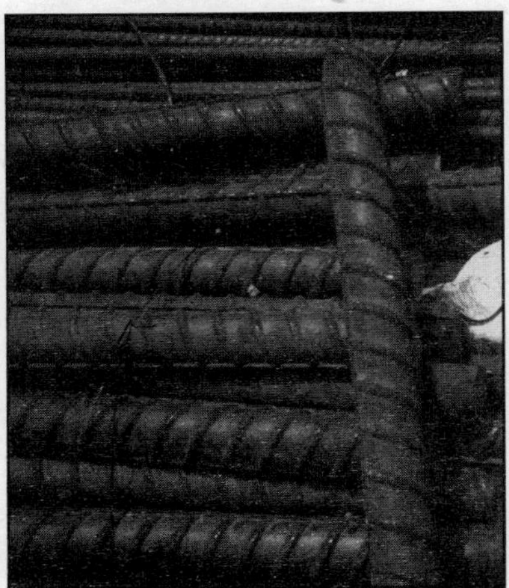

FIGURE 21.37 Deformations in steel bars.

Virtually all reinforcement in a concrete member consists of deformed bars. Plain (smooth, undeformed) bars are used only as dowels at expansion joints. At expansion joints, rebar are oiled (or covered with a sleeve) before being embedded in concrete to allow movement at the joint. This requires the use of plain bars. Plain rebar are also used as spiral reinforcement in round columns.

REINFORCING STEEL GRADE

As we learned in Chapter 4, yield strength is an important steel property. Rebar are hot rolled from steels of the following yield strengths:

- 40,000 psi—referred to as *grade 40 steel*
- 60,000 psi—referred to as *grade 60 steel*

The cost of both steel grades is approximately the same. Grade 60 steel is one-and-a-half times stronger than grade 40 steel. Therefore, it is by far the most frequently used steel grade. Grade 40, which is more ductile, that is, more easily bendable, is used where rebar need to be field bent. Grade 50 is also available but is rarely used.

DIAMETER AND LENGTH OF BARS

Eleven different diameters of rebar are available, from $\frac{3}{8}$ in. to $2\frac{1}{4}$ in., Table 21.3. The diameter of a bar is generally stated in terms of a number. Thus, a No. 3 bar (commonly stated as a #3 bar) has a diameter of $\frac{3}{8}$ in. A #4 bar has a diameter of $\frac{1}{2}$ in., a #5 bar has a diameter of $\frac{5}{8}$ in., and so on, up to a #8 bar, which has a diameter of 1 in.

According to this numbering system, a #9 bar should have a diameter of $\frac{9}{8}$ in. (1.125 in.). However, its actual diameter is 1.128 in., providing a cross-sectional area of 1.0 in.^{2}. In structural calculations, the cross-sectional area of a bar is more important than the bar diameter. Therefore, rationalizing area is more important than the diameter. That is why a #18 bar has a diameter of 2.257 in., which gives an area of 4.0 in.^{2}. Note that $\frac{18}{8} = 2.25$.

Deformed rebar are available from #3 up. A $\frac{1}{4}$-in.-diameter bar is available only as a plain bar, but it is not used as concrete reinforcement except in welded wire reinforcement, as described later. The diameter of a deformed bar is its nominal diameter. The nominal diameter is the diameter that gives the same cross-sectional area as a plain bar.

Bar lengths of 20 ft and 40 ft are standard stock sizes. Lengths up to 60 ft are generally available as special orders. The 60-ft length is based more on transportation limitations than on manufacturing.

BAR IDENTIFICATION MARKINGS

Each rebar in the United States carries an identification marking, Figure 21.38. The marking consists of four pieces of information in the following order:

- Producing mill's identification number or symbol.
- Type of steel. The most commonly used steel is new billet steel. Its identification mark is *N*. Mark *S* is used for special billet steel. Mark *R* is used for rail steel. Some mills close to old, unused rail lines produce this steel. Mark *A* stands for axle steel.
- Bar size.

TABLE 21.3 STANDARD BAR DIAMETERS AND AREAS

Bar size and designation	Diameter (in.)	Cross-sectional area (in.^{2})
#3	0.375	0.11
#4	0.500	0.20
#5	0.625	0.31
#6	0.750	0.44
#7	0.875	0.60
#8	1.000	0.79
#9	1.128	1.00
#10	1.270	1.27
#11	1.410	1.56
#14	1.693	2.25
#18	2.257	4.00

- Steel grade. Each bar carries at least two vertical lines. These are simply deformation lines. An additional vertical line on a bar is the grade line, implying that it is grade 60 bar. No grade line means that it is grade 40 or 50 bar. Bars that have diagonal deformations have the grade stamped on them during manufacturing if they are of grade 60. No grade stamp means a grade 40 or 50 bar.

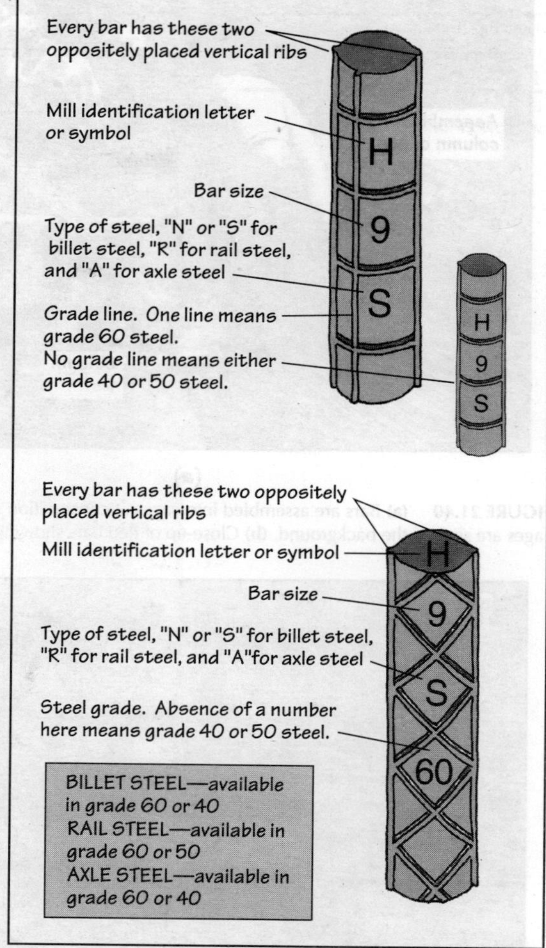

FIGURE 21.38 Two commonly used identification markings on reinforcing bars.

Bending and Tying of Bars

In a concrete member (beam, slab, or column), some rebar need to be straight, and others need to be bent to required profiles. The cutting and bending of rebar is done in a rebar fabrication plant. Rebar arrive at a construction site prebent and in the required lengths, Figure 21.39. Rebar bundles arrive at the construction site with identification tags that identify the structural members to which the bars belong.

After the bars are received at the site, they are assembled in beam-and-column cages. The cage assembly requires tying the bars together, Figure 21.40. Tie wires are approximately $\frac{1}{20}$-in.-diameter steel wires, which can be easily bent with the hands.

Epoxy-Coated Bars

Most reinforcing bars in concrete members are black bars, that is, they do not have any surface coating. However, concrete members that are exposed to corrosive atmospheres require coated bars. Fusion-bonded, epoxy-coated bars are commonly used in concrete bridges and parking garages exposed to deicing salts, which gradually penetrate through concrete to corrode the bars. Epoxy-coated bars may also be used in other corrosive atmospheres.

21.13 WELDED WIRE REINFORCEMENT (WWR)

Welded wire reinforcement (WWR) is a prefabricated reinforcing steel available in rolls or mats, Figure 21.41. Rolls come in widths of 5 to 7 ft. Mats come in various dimensions. They are commonly used in ground-supported slabs (or pavement) and steel deck–supported slabs, where reinforcement requirements are marginal, that is, much smaller than those needed for reinforced concrete suspended slabs.

FIGURE 21.39 Bars generally arrive at a construction site bent to shape and cut to required lengths. Tags attached to the bundles identify the building component (beam, column, or slab) to which the bars belong.

REVIEW QUESTIONS

1. Describe the difference between quicklime and slaked lime.
2. Describe how lime-sand mortar hardens and functions as a cement.
3. List various types of portland cement and where they are typically specified.
4. What is air-entrained portland cement? Where is it typically specified?
5. What is lightweight structural concrete? How does it differ from normal-weight concrete? What are the advantages and disadvantages of using lightweight structural concrete?
6. With the help of a sketch, explain how the slump of concrete is determined.
7. Using sketches, explain how the compressive strength of concrete is determined.
8. Discuss the importance of the water-cement ratio in concrete.
9. Explain what curing of concrete is, and why it is necessary.
10. What is a concrete admixture, and why is it used? Discuss the commonly used concrete admixtures.

CHAPTER 22 Concrete Construction–I (Formwork, Reinforcement, and Slabs-on-Ground)

CHAPTER OUTLINE

As stated in Chapter 21, concrete is weak in tension. On average, the tensile strength of concrete is 10% of its compressive strength. The low tensile strength of concrete is offset by combining it with steel, which is strong in tension. In this concrete-steel combination, steel is fully embedded in the concrete so that both concrete and steel function integrally.

Plain concrete (i.e., concrete without steel reinforcement) may be used only in situations where tensile stresses are minimal, such as in pavements, ground-supported slabs, and lightly loaded wall footings on stable, uniformly compacted soils. Even in these situations, a nominal amount of steel is desirable to resist the tensile stresses caused by the shrinkage of concrete or unforeseen bending. There are two ways in which steel is used to strengthen concrete.

Reinforced Concrete

In one option, steel reinforcing bars (rebars), discussed in Chapter 21, are used. In most concrete members subjected to bending (e.g., beams and slabs), reinforcing bars are positioned in locations where tensile stresses are likely to be produced. Because steel is also much stronger than concrete in compression, reinforcing bars are also used to provide compressive strength (e.g., in concrete columns, walls, and some beams).

A concrete member containing reinforcing bars is called a *reinforced concrete member.* If an entire structure consists of reinforced concrete members, which is fairly common, the structure is referred to as a *reinforced concrete structure.*

PRESTRESSED CONCRETE

The alternative to the use of reinforcing bars is to introduce a permanent initial compression in a concrete member, giving a *prestressed concrete member.* The compression is introduced in regions of the member where tension is expected to be produced by the loads. The magnitude of compression produced by prestressing is controlled to ensure that it reduces or cancels the tension created by the loads.

Prestressing of a member is obtained using prestressing cables called *strands,* which are made of high-strength steel wires. The most commonly used strands have a yield strength of 270 ksi, which is 4.5 times stronger than the most commonly used reinforcing bars (Grade 60 steel; see Chapter 21). Prestressing can be accomplished in two ways:

- Pretensioning
- Posttensioning

In pretensioning, the strands are tensioned between two fixed abutments using hydraulic jacks, Figure 22.1. Subsequently, concrete is cast in the form enclosing the tensioned strands and allowed to cure, creating a bond between the strands and concrete. When the concrete has developed sufficient strength, the strands are cut free at the ends of the member. Because the cut strands tend to relax and attempt to return to their original untensioned state (but are unable to do so because of their bond with concrete), they compress the member.

If the strands are placed in the center of the member's cross section, a uniform compressive stress is produced over the entire cross section. Generally, however, the strands are placed eccentrically so that the prestress developed in the member is the reverse of the stresses that are expected to be produced by the loads.

For example, for a single-span beam supported at two ends, the strands are placed near the bottom of the beam, Figure 22.1(b), so that after the strands are released, compressive stresses are created at the bottom of the beam and an upward curvature (camber) is developed, Figure 22.1(c). When loads are applied, the beam deflects downward and becomes straight.

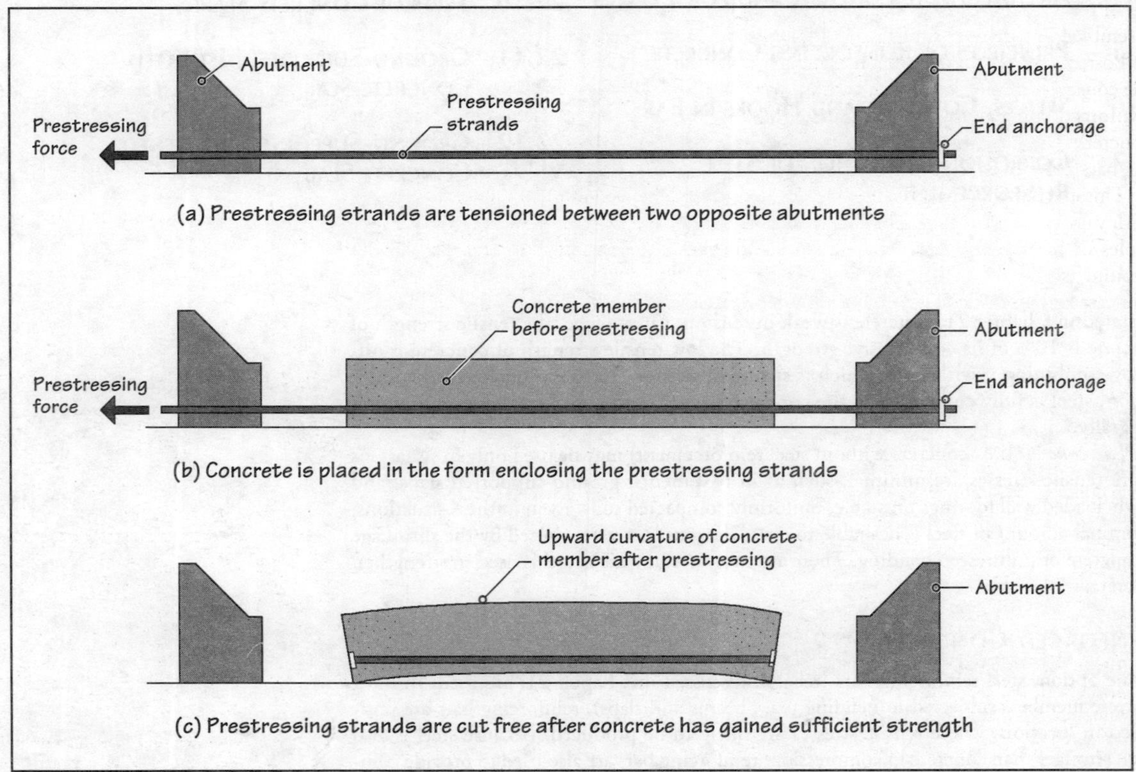

FIGURE 22.1 Three important stages in making a pretensioned (precast, prestressed) concrete member.

The strands shown in Figure 22.1 are straight. A more effective use of the strands is to profile them so that their eccentricity changes along the length of the member in response to the stresses created by the loads. In posttensioned concrete members, the profiling of strands (called *draped strands*) is in the shape of a smooth curve (see Figure 23.19). Due to the fabrication process, draped strands in pretensioned concrete members, however, must have abrupt changes in their profiles.

Posttensioning differs from pretensioning in that in posttensioning, the strands are tensioned after the concrete has been placed. The strands for posttensioning are encased in sleeves that are placed in the form before the concrete is placed. After the concrete has gained sufficient strength, the strands are tensioned. Because the strands are free to move within the sleeves, they can be tensioned without simultaneously tensioning the concrete. After tensioning of the strands to the extent required, they are anchored in that position by mechanical wedges at the member's ends, which restrains them from returning to their original (untensioned) state. The prestress created in the member by posttensioning is identical to that created by pretensioning.

STRANDS AND TENDONS

The combination of strands, sleeves, end anchorages, and so on, is referred to as *tendons.* The tendons are laid within the concrete forms in the same way as the reinforcing bars. A tendon may consist of one strand (see Figure 22.42) or multiple strands. (Because a strand is the most important component of a tendon, the strand is sometimes referred to as a tendon.)

ADVANTAGES AND DISADVANTAGES OF PRESTRESSING

There are several benefits of prestressing. The two most important ones are reduced cracking in concrete and the reduction in the dimensions of structural members, resulting in smaller dead loads and overall economy. The disadvantages of prestressing are the higher cost of tendons compared with reinforcing bars and the need for greater quality control and a more skilled labor force.

SITE-CAST CONCRETE AND PRECAST CONCRETE

Pretensioning is generally used for precast concrete members and is accomplished at the precaster's plant. Therefore, a pretensioned member is also called a *precast, prestressed member.* Precast, prestressed concrete members are transported to the construction site and assembled in a manner similar to that of structural steel members.

Posttensioning, on the other hand, is generally used for site-cast members and is done at the construction site. Thus, site-cast (also called *cast-in-place* or *CIP*) *concrete* may either be reinforced concrete or posttensioned concrete. (As illustrated in Chapter 23, reinforced concrete structures can also be posttensioned, in which prestressing tendons and reinforcing bars are used in the same component).

This and the following chapter are devoted to concrete construction. This chapter deals with various aspects of site-cast concrete construction, such as formwork and shoring, principles of reinforcing, and concrete surface finishes. Construction of concrete slabs-on-ground is also covered in this chapter. Other issues of concrete construction, such as site-cast and precast concrete framing systems, are discussed in Chapter 23.

EXPAND YOUR KNOWLEDGE

Some Important Terms

Reinforced Concrete and Nonprestressed Concrete

In concrete literature, the term *reinforced concrete* is also referred to as *nonprestressed concrete*. Similarly, steel reinforcement (steel rebars) is referred to as *nonprestressing reinforcement*. This recognizes that rebars and prestressing strands are two different types of concrete reinforcement. It also recognizes the growing use of prestressed concrete in buildings, so that referring to steel rebars as *conventional reinforcement* is becoming less common.

Site-Cast Concrete, Cast-in-Place Concrete, and Precast Concrete

Although the terms *site-cast concrete* and *cast-in-place concrete* are generally used interchangeably, there is a subtle difference between them. The terms *cast-in-place*, *cast-in-situation*, and *cast-in-situ* are synonymous; where the member is cast in the location where it is supposed to remain. This distinguishes it from a *site-precast* concrete member, which, although cast on site, is not cast in situ. For example, concrete tilt-up walls are site precast. Concrete tilt-up wall construction is covered in Chapter 26.

22.1 VERSATILITY OF REINFORCED CONCRETE

Reinforced concrete can be used for any structure, small, big, low-rise, or high-rise—in buildings, roads, bridges, tunnels, retaining walls, dams, and so on. In one form or another, reinforced concrete is used in almost every contemporary building, regardless of the geographical location.

The materials for reinforced concrete (crushed stone, sand, and water) are relatively inexpensive and locally available. In most locations, the steel required for reinforcing bars and the portland cement are also locally produced. Additionally, the skill and training required for quality control in reinforced-concrete construction can be imparted to a local labor force far more easily than that required for steel construction (steel erection, welding, bolting, fire-retardant treatment, etc.).

The integral fire resistance of reinforced concrete and the various types of finishes that can be achieved add to its versatility, making it the most popular construction system. In many developing countries, reinforced concrete is used in almost every element of an architecturally significant building—structural frame, exterior and interior walls, basements, foundations, water-containing structures, pavements, street furniture, and so on, Figure 22.2.

The versatility of reinforced concrete is also due to the fact that the two structural materials (concrete and steel) are complementary, which allows variation in the size of a reinforced-concrete member for a given load and span. Thus, the size of a beam or column, providing the same strength, can be varied (to some extent) by varying the relative amounts of the two materials—a flexibility not available in steel or wood construction.

For instance, by increasing the amount of steel in a beam, it is possible to reduce the amount of concrete needed, or vice versa. The reduction in the amount of concrete results in a smaller beam size. (Note that the amount of steel even in a highly reinforced concrete

NOTE

Concrete Construction and Building Codes

Because concrete is inherently fire enduring, concrete members can easily achieve the fire ratings required for Type I(A) construction—the most fire-resistive construction. For most occupancies, the building codes allow an unlimited area and an unlimited height if the building is built with Type I(A) construction. Unlike steel, concrete elements do not require add-on fire protection to meet the fire-rating requirements of a given type of construction.

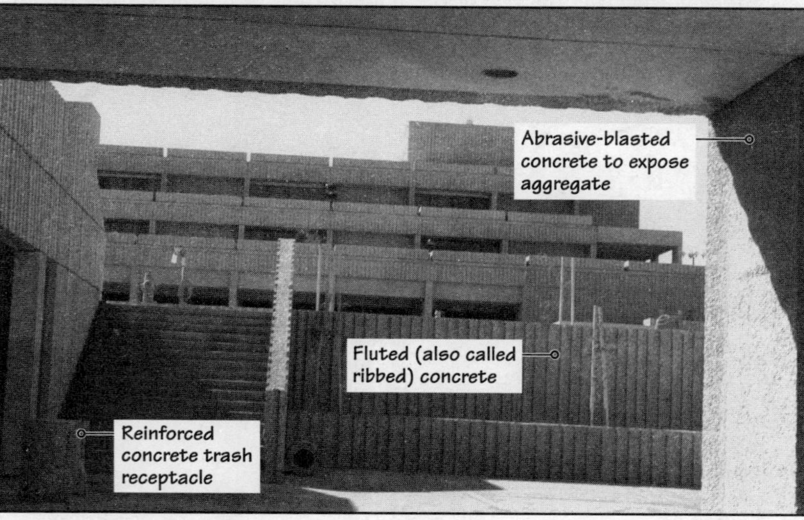

FIGURE 22.2 Almost all buildings on the campus of King Fahd University of Petroleum and Minerals in Dhahran, Saudi Arabia, are constructed of reinforced concrete. It has been used in the walls, structural frame, roof and floor slabs, stairs, water tower, roads and pavements, fountains, street furniture, and so on, highlighting the versatility of this material.

beam seldom exceeds 2% of the beam's cross section. Therefore, a change in the amount of steel alters the beam's strength but does not impact the physical dimensions of the beam.)

22.2 CONCRETE FORMWORK AND SHORES

In reinforced-concrete construction, concrete and reinforcement must be contained in molds, referred to as *formwork*. The formwork for elevated slabs and beams must be supported on vertical supports, referred to as *shores*. Both formwork and shores are temporary structural elements that are removed after the reinforced-concrete members are able to support themselves.

For a reinforced-concrete slab-on-ground, the formwork is simple and generally consists of 2-by solid lumber (or plywood) edge forms braced by stakes, Figure 22.3. Shores are not required in this case because the concrete is ground supported.

For an elevated concrete slab, the formwork and shores are more complex. Figure 22.4 shows the formwork and shores for an elevated concrete slab without beams. The formwork for a complex floor system is more complicated. That is why the formwork in a typical reinforced-concrete building is generally the single most expensive item.

On average, the concrete costs nearly 25% to 30% of the cost of the structure (including placing and finishing of concrete), and steel reinforcement costs approximately 20% to

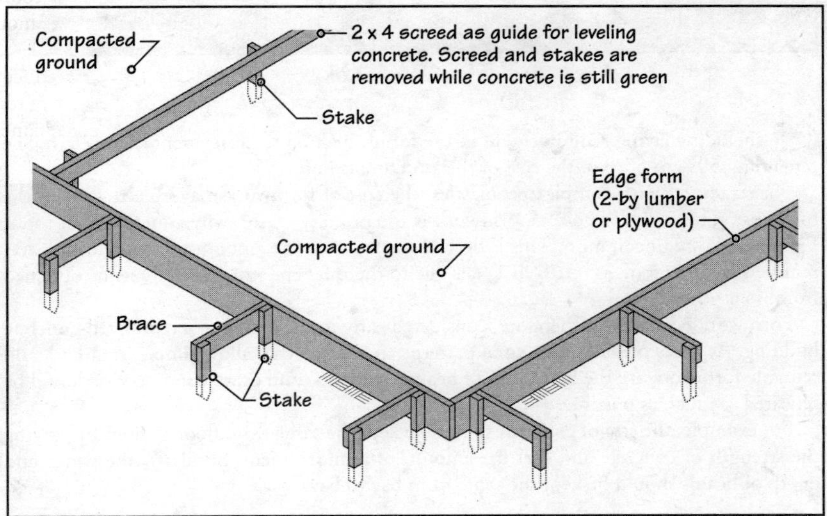

FIGURE 22.3 Formwork for a ground-supported concrete slab.

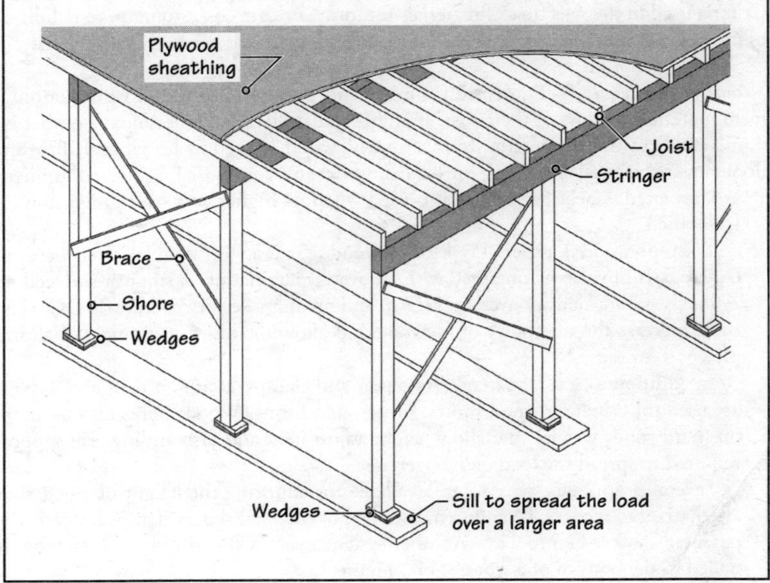

FIGURE 22.4 Formwork and shores for an elevated concrete slab. In this illustration, the shores and formwork are constructed of dimension lumber and plywood. As shown in Figure 22.11, steel and aluminum are also used as formwork and shoring materials.

FIGURE 22.5 Reinforced-concrete shell roof in the Oceonographic Museum in the City of Arts and Sciences, Valencia, Spain, by architect Santiago Calatrava.

Observe the large roof span constructed of extremely thin concrete. The amount of reinforcement required is also small. Concrete shell structures are highly efficient in the use of materials because of their geometry, which induces mainly compression and very little bending stress in the structure. However, their geometry makes formwork extremely expensive. The overall cost of concrete shell roofs is quite high.

25% (including laying reinforcement in the forms, making reinforcement cages, etc.). The remaining 50% to 55% of the cost of the structure is in formwork.

For structures with complex geometries, the cost of formwork may exceed 75% of the total cost. The excessive cost of formwork is the primary reason why concrete shell roofs, Figure 22.5, are uncommon. This is despite the fact that the amount of concrete and steel required in these roofs is relatively small due to the inherent structural efficiency obtained from their geometry.

Formwork economy is, therefore, considered early in the design of a reinforced-concrete building. As far as possible, structural patterns and shapes that allow simple, standard, and reusable formwork are the best. Sizes of beams, columns, and other components should be repeated as much as possible.

For example, the size of columns should be kept the same from floor to floor by varying the strength of concrete and (or) the amount of reinforcement. Similarly, the width and depth of beams should be kept the same from bay to bay.

MATERIALS USED FOR FORMWORK AND SHORES

Criteria used in the selection of materials for formwork are cost, strength, reusability, durability, ease of assembly, and weight. The following materials generally satisfy these criteria.

Wood and Plywood Dimension lumber and plywood are the most commonly used form materials because of their cost, strength, nailability, and reusability. Lumber is commonly used in the supporting frame, and plywood is used as horizontal sheathing for floors and as vertical sheathing for beams, walls, and columns. Engineered lumber, such as laminated veneer lumber, is also used when its higher cost over dimension lumber is justified.

Form plywood is a special type of plywood that can withstand repeated wetting from concrete. It can also be obtained with an overlay lamination of smooth-surfaced materials, such as high-density overlay (HDO) and medium-density overlay (MDO). Laminations increase the durability of plywood and allow for easier stripping and cleaning of plywood for reuse.

In addition to several patented fasteners and clamps, common nails and duplex nails are used for fastening wood shores, braces, and forms. Wood shores may be supported on (hardwood) wedges that allow height adjustment and dismantling. Sill supports are required to spread the load on a larger area.

Several proprietary systems are available for adjusting the height of wood shores as alternatives to wedges. One system uses two overlapped shores clamped together with a patented device, Figure 22.6. Another system uses a screw-based steel assembly that is nailed to the bottom of a wood shore, Figure 22.7.

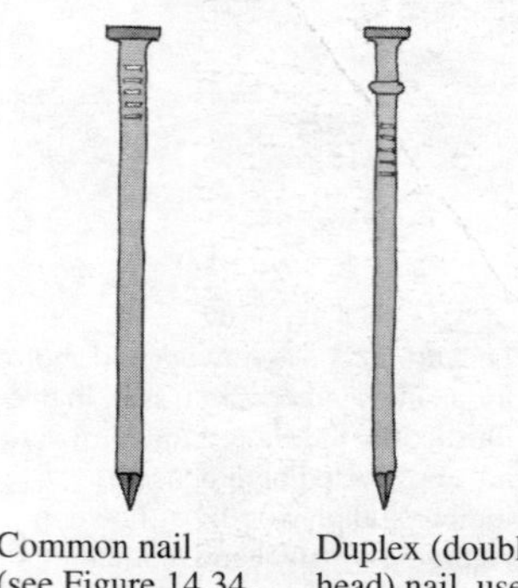

Common nail (see Figure 14.34, Chapter 14)

Duplex (double-head) nail, used for temporary nailing

FIGURE 22.6 Shores for a reinforced-concrete floor slab in which two dimension lumber members are lapped and clamped together with a patented device. The lapped and clamped assembly provides adjustment in the height of a shore.

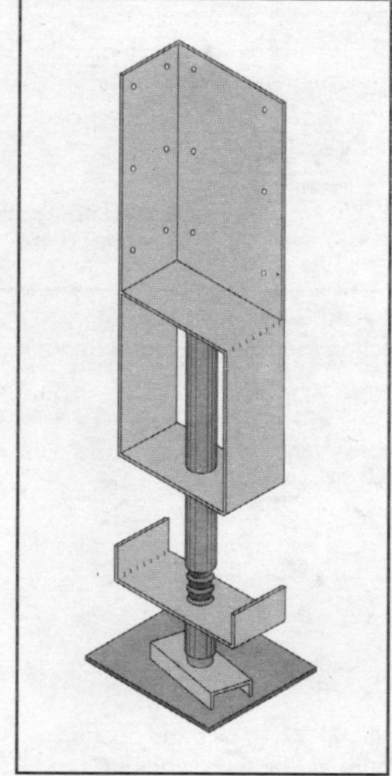

FIGURE 22.7 A screw-based metal assembly that can be nailed to one end of a lumber shore to give it height adjustability.

Steel Steel is also a commonly used form and shore material because of its high strength. Steel angles, channels, and prefabricated steel joists are used in the supporting frame for slab forms and shores. The use of steel joists can substantially reduce the congestion of shores that typically occurs with lumber-based formwork.

Steel pipe shores can be used as individual shores, but they are more commonly used as scaffold-type framed shores. They are generally provided with screw-type adjustment at the bottom and an adjustable head at the top, Figure 22.8.

Aluminum as Form Material Aluminum alloys (which do not react with fresh concrete) are being increasingly used as form material because of their lighter weight and high

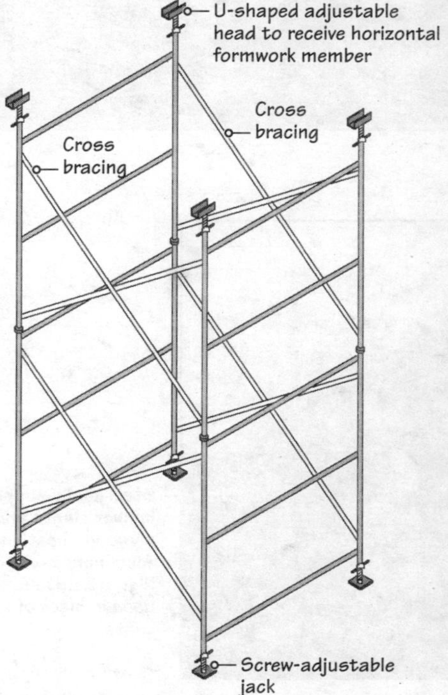

FIGURE 22.8 Steel pipe scaffold shores are available in panel frames in widths of 2 to 5 ft and heights of 4 to 6 ft. Panel frames can be assembled vertically and braced diagonally to make a stable shore frame.

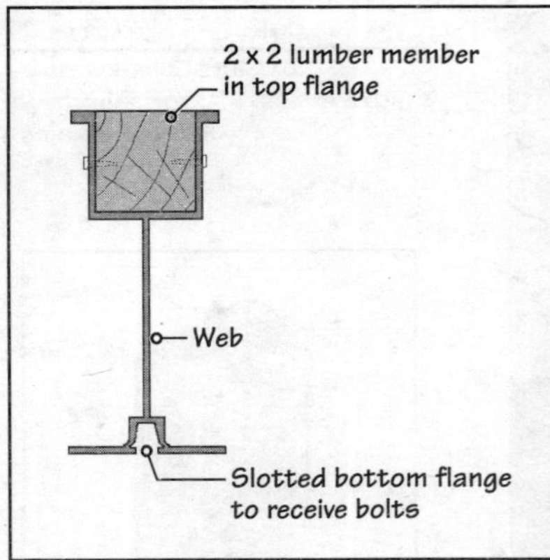

FIGURE 22.9 A typical extruded aluminum section used as a joist for formwork.

strength-to-weight ratio. Generally, aluminum angles are used in trusses or in extruded one-piece joists with U-shaped flanges to provide nailing and bolting capabilities, Figure 22.9.

Glass Fiber–Reinforced Plastic (GFRP) as Form Material Because it can be molded into shape and has a smooth surface, GFRP is commonly used for pans and domes required for one-way or two-way joist floor systems (see Chapter 23).

PREFABRICATED PANEL FORMS

Because the major cost of formwork in a multistory building is in the floors, it is fairly common to prefabricate the formwork and shores for each bay of the floor as a complete unit and "fly" it into position with a crane, Figure 22.10. The form units, referred to as *flying forms*, are generally complete, freestanding units with their own adjustable shores designed for easy removal and hoisting to the next level, Figure 22.11.

The term *flying form* is typically used with floor slab forms, but other large prefabricated panel forms (called *gang forms*) are also used for walls and other reinforced concrete elements.

FIGURE 22.10 Formwork for an entire bay between columns can be assembled as a large panel and "flown" into position by a crane. Referred to as *flying forms*, these panels can be demounted with the help of jacked shores and lifted to the next floor for reuse. Flying forms are particularly convenient in mid- and high-rise buildings with identical dimensions of floor slab components. In this figure, the right-hand panel is being lowered into position by the crane.

FIGURE 22.11 Details of the panelized flying form in Figure 22.10. The panel consists of steel joists that are supported on two aluminum trusses, one at each end. The vertical members of trusses extend down to form shores, which are provided with adjustable jacks. The steel joists support lumber planks, which in turn support plywood sheathing.

PRACTICE QUIZ

Each question has only one correct answer. Select the choice that best answers the question.

1. According to building codes, plain concrete (concrete without steel reinforcement) is not permitted in new construction.
   a. True b. False
2. The yield strength of prestressing steel is
   a. 40 ksi. b. 50 ksi.
   c. 60 ksi. d. 100 ksi.
   e. none of the above.
3. The term *prestressed concrete* refers to both pretensioned and posttensioned concrete.
   a. True b. False
4. Which of the following statements is true?
   a. The term *strands* includes tendons.
   b. The term *tendons* includes strands.
   c. The term *tendons* refers to reinforcing bars that have been bent to required shapes.
   d. The term *strands* refers to reinforcing bars that have been bent to required shapes.
   e. none of the above.
5. Which of the following statements is true?
   a. Pretensioning is generally done at the construction site and yields a precast, prestressed concrete member.
   b. Pretensioning is generally done at the precaster's plant and yields a precast, prestressed concrete member.
   c. Posttensioning is generally done at the construction site and yields a precast, posttensioned concrete member.
   d. Posttensioning is generally done at the precaster's plant and yields a precast, posttensioned concrete member.
6. In a typical sitecast reinforced-concrete structure, the cost of formwork and shores is
   a. approximately equal to the combined cost of concrete and steel.
   b. approximately 75% of the combined cost of concrete and steel.
   c. approximately 50% of the combined cost of concrete and steel.
   d. approximately 25% of the combined cost of concrete and steel.
   e. none of the above.
7. Concrete shell roofs are not commonly used because
   a. they require a relatively large amount of concrete per square foot of floor area.
   b. they require a relatively large amount of steel reinforcement per square foot.
   c. they require prestressing, which substantially raises their cost.
   d. all of the above.
   e. none of the above.
8. In site-cast concrete construction, shores are used
   a. in ground-supported concrete slabs.
   b. in elevated concrete slabs.
   c. as formwork for concrete walls.
   d. as formwork for concrete columns.
   e. none of the above.
9. Height adjustability in shores made of dimension lumber is provided through
   a. lapped members clamped together with specially made clamps.
   b. a screw-based metal assembly.
   c. (a) or (b).
   d. height adjustability is not provided in shores.
10. Scaffold-type shores are generally made from
   a. dimension lumber. b. plywood.
   c. steel pipes. d. steel plates.
   e. all of the above.
11. High-density overlay (HDO) on form plywood
   a. improves the strippability of forms.
   b. allows easier cleaning of forms.
   c. increases reusability of forms.
   d. all of the above.
   e. none of the above.
12. Aluminum is commonly used as form and shore material because
   a. it is chemically compatible with wet concrete.
   b. it is an easily nailable material.
   c. its strength-to-weight ratio is low, which reduces the weight of forms and shores.
   d. its strength-to-weight ratio is high, which reduces the weight of forms and shores.
   e. all of the above.
13. The term *flying form* refers to
   a. a large formwork assembly for floor slabs.
   b. a large formwork assembly for roof slabs.
   c. a large formwork assembly for floor and roof slabs.
   d. a large formwork assembly for concrete walls.
   e. a large formwork assembly for columns.

22.3 FORMWORK REMOVAL AND RESHORING

To speed construction, forms and shores are removed as soon as possible. Another advantage of early removal of formwork is that if repair or patching of a concrete surface is needed, it can be accomplished while the concrete is in the early stages of curing, favorable to a good bond.

However, premature removal of forms may not only be dangerous but may also affect the quality of the finish. The time between the end of concreting and the removal of forms (referred to as the *stripping time*) is a function of the ambient temperature, strength of concrete, and the type of component. Higher ambient temperature and higher strength of concrete require less stripping time. Forms from vertical surfaces of members, such as column forms, the sides of beams, and slabs, can be removed earlier. After their removal, concrete surfaces are generally wrapped with blankets or burlap and moist cured.

Forms from the horizontal surfaces can be removed only when the concrete has gained sufficient strength to support itself, which is generally assumed to be 70% of the specified strength of concrete. Stripping time is typically specified by the project's structural engineer.

RESHORING

After the formwork from the horizontal surfaces of the structure (beams and slabs) is stripped and moved to the next higher level, the surfaces are supported by shores, a process

FIGURE 22.12 In this building under construction, the concreting for the top floor has been completed and the forms for the next floor are being installed. The two underlying floors have been reshored, while the reshores from the ground floor have been removed to allow interior construction. As far as possible, reshores must align in the same vertical position from floor to floor.

called *reshoring*. Reshoring facilitates maximum reuse of formwork by taking advantage of the partial strength of the lower floors to support the weight of wet concrete on the floors above. Other objectives of reshoring are

- To prevent deflection and creep of concrete (concrete that has not gained sufficient strength creeps and deflects more)
- To reduce hairline cracks in concrete

In a multistory building, where floors are generally cast 7 to 10 days apart, one shored story and two reshored stories are generally required below the story on which wet concrete is placed. This requirement is relaxed if there is a longer gap between the construction of the floors, Figure 22.12. Reshores are generally placed in the same vertical position from floor to floor.

Form Release Agents

Before concrete is placed in the forms, their interior surfaces are coated with release agents to allow the forms to be stripped without damage to both the forms and the concrete. Release agents lubricate the surfaces and, in the case of wood and plywood forms, they also seal the surfaces against water absorption, which increases form life. Various proprietary release agents are available, which are either oil- or wax-based coatings. Better-quality release agents allow several reuses of forms between coatings.

Responsibility for Formwork Design

The structural safety of formwork and shores is obviously vital. Because construction site safety is in the realm of the general contractor (see Chapter 1), the structural design of formwork is also the general contractor's responsibility. Thus, the general contractor is free to choose the forming and shoring systems that will yield an economical structure within the requirement of the finished product specified by the architect/engineer.

22.4 ARCHITECTURAL CONCRETE AND FORM LINERS

In the early days of its development, concrete was considered only an engineering material, suitable for gutters, sewers, pavements, roads, and structural frames of buildings. It was considered unsuitable for building facades because of its dull gray, unappealing appearance, which deteriorates with time because of shrinkage cracks, attracting dust and moss.

Le Corbusier was among the pioneers who, apart from exploiting its sculptural properties, made concrete count as an architectural material—suitable for building facades and interiors. In most of Le Corbusier's buildings, *smooth, off-the-form concrete* was used; that is, concrete was left exposed (in its natural state), without any additional surface treatment.

An acceptable surface quality straight from the forms can be obtained only if the formwork surface is good, the concrete has an optimum water-cement ratio, and the concrete is vibrated sufficiently to avoid honeycombing. Excessive vibration can cause bleeding or separate the ingredients.

SMOOTH, OFF-THE-FORM CONCRETE

Although smooth, off-the-form concrete provides the most economical surface finish, it is generally not recommended for contemporary facades or interiors because

- Color variations are difficult to avoid between batches of concrete, particularly with gray portland cement. They are relatively easier to control with (more expensive) white portland cement and are less noticeable with aesthetic (relief) joints on the surface.
- Smooth forms should be free from surface imperfections. Additionally, with smooth forms, the vibration of forms is recommended unless care is taken to ensure that internal vibrators do not damage form surfaces.
- Repair of smooth-surface concrete after removal of the formwork is generally noticeable.

TEXTURED CONCRETE SURFACES

Because of the concerns with smooth, off-the-form concrete, the use of textured concrete is more common for building facades. (Interior application of textured concrete is relatively uncommon because of its propensity to collect dust.) Textured concrete surfaces can be obtained by (a) surface treatment after formwork removal, (b) using form liners, or (c) both.

The following are commonly used surface treatments on concrete after removal of the forms:

- *Exposed aggregate finish by water washing:* This is achieved by applying chemical retarders on formwork, which retards the setting of concrete near the surface. After removal of the form, the surface layer of concrete (cement-and-fine-aggregate paste) is carefully removed by pressure washing with water to expose the coarse aggregates. Retarders that expose coarse aggregates to various depths (generally up to one-third of its size) are available.
- *Abrasive blasting:* Depending on the type of abrasive and the amount of pressure applied, abrasive blasting can be light, medium, or heavy. The difference between the three types is easily noticeable (see Figure 28.39). Blasting removes the natural gloss of aggregates and provides a mat (nonglossy) appearance compared with that obtained from chemical retarders.
- *Acid etching:* Acid etching involves removing surface cement paste with dilute hydrochloric acid, which reveals the sand and a small percentage of coarse aggregate. In this application, acid-resistant sand and coarse aggregate are recommended (such as quartz or granite). Carbonate aggregates, such as marble, limestone, and dolomite, are not acid resistant (see Chapter 25).

ARCHITECTURAL CONCRETE THROUGH FORM LINERS

A large variety of textures and patterns can be created on concrete by lining the formwork with the required pattern. The use of form liners is generally more economical than modifying the concrete surface after removing the forms. A few patterns obtained through the use of form liners are shown in Figure 22.13. Form liners, which are generally made from polymeric materials, are available for single or repeat applications (see Figure 28.47). They are attached to the formwork or casting beds (for tilt-up concrete walls or precast concrete panels).

22.5 PRINCIPLES OF REINFORCING CONCRETE

A reinforced-concrete member must obviously have an adequate amount of steel. In addition, the reinforcement must be placed in the correct location. Sometimes a seemingly trivial error in the location of reinforcement can have serious consequences. For example,

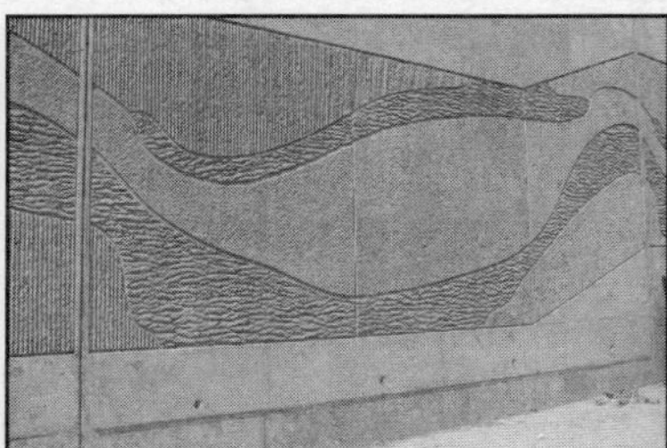

FIGURE 22.13 Some of the several patterns and textures that can be obtained in architectural concrete through the use of form liners. (Photos courtesy of The Greenstreak Group)

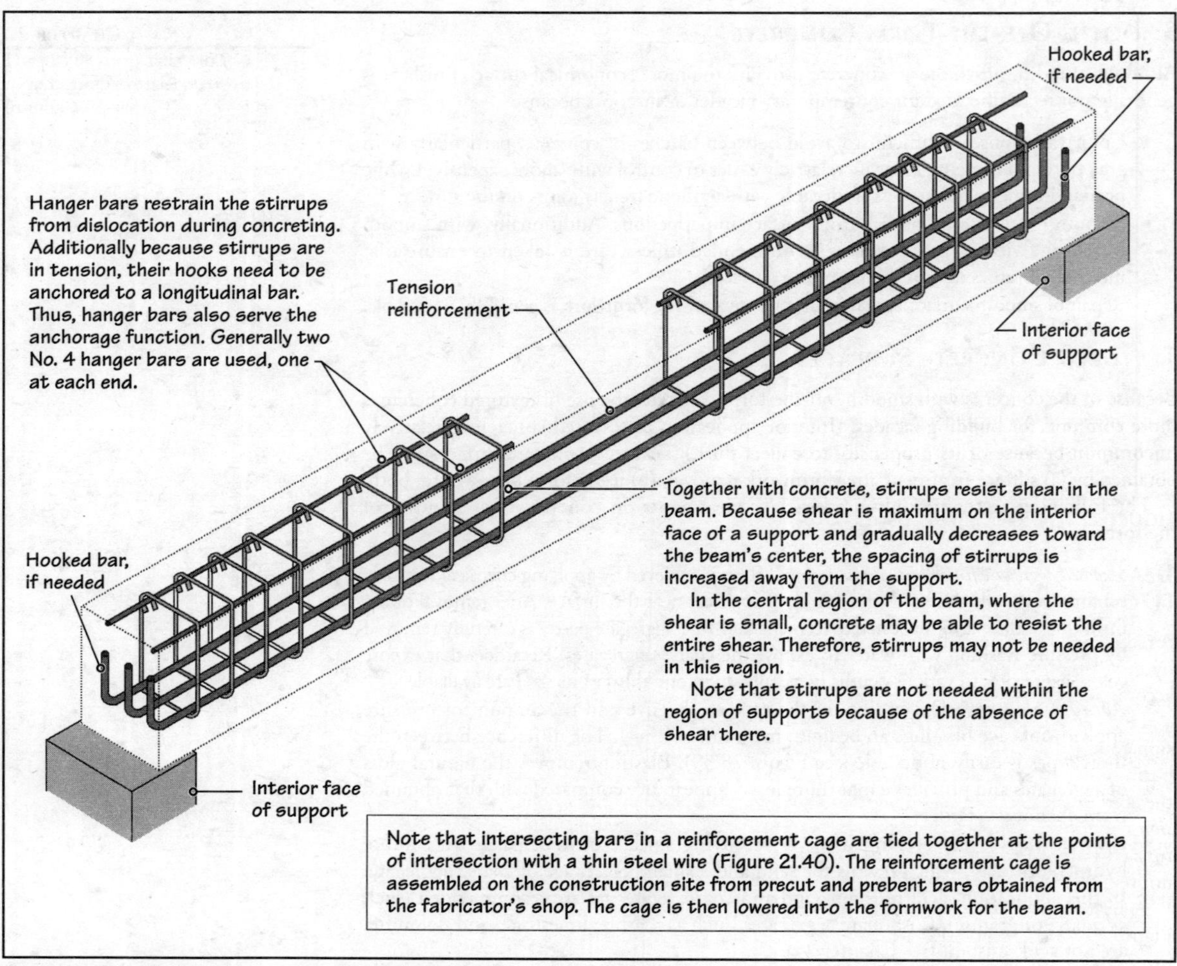

Hooked bar, if needed
Hanger bars restrain the stirrups from dislocation during concreting. Additionally because stirrups are in tension, their hooks need to be anchored to a longitudinal bar. Thus, hanger bars also serve the anchorage function. Generally two No. 4 hanger bars are used, one at each end.
Tension reinforcement
Interior face of support
Together with concrete, stirrups resist shear in the beam. Because shear is maximum on the interior face of a support and gradually decreases toward the beam's center, the spacing of stirrups is increased away from the support.
In the central region of the beam, where the shear is small, concrete may be able to resist the entire shear. Therefore, stirrups may not be needed in this region.
Note that stirrups are not needed within the region of supports because of the absence of shear there.
Hooked bar, if needed
Interior face of support
Note that intersecting bars in a reinforcement cage are tied together at the points of intersection with a thin steel wire (Figure 21.40). The reinforcement cage is assembled on the construction site from precut and prebent bars obtained from the fabricator's shop. The cage is then lowered into the formwork for the beam.

FIGURE 22.14 Reinforcement cage for a simply supported, reinforced-concrete beam.

because an overhanging portion of a concrete slab is subjected to tension at its top surface (and compression at the bottom), the reinforcement must be located near the top surface. If the reinforcement is inadvertently located at the bottom of the slab (as is generally the case with slabs without overhangs), the slab may collapse when the formwork is removed. Such mistakes, though not common, have occurred in buildings with poor inspection protocols.

In addition to the amount and location, there are several other issues related to reinforcement. Consider a typical reinforcement cage for a simply supported beam, Figure 22.14.

NOTE

Simply Supported Beams and Continuous Beams

A simply supported beam is a beam that bears on two end supports. When the beam bends under loads, the ends of the beam rotate freely; that is, the supports do not present any restraint to end rotation.

A beam whose ends are framed into a column is unable to rotate freely at the ends, such as the continuous beam of Figure 22.18.

Tension Reinforcement, Stirrups, and Hanger Bars

Under gravity loads, a *simply supported beam* is subjected to tensile stresses at the bottom and compressive stresses at the top. Reinforcement is, therefore, required at the beam's bottom, referred to as *tension reinforcement.* No bars are required at the top of the beam unless it also has compression reinforcement.

In addition to tension reinforcement, stirrups are generally required in a beam to resist shear. Also referred to as *shear reinforcement,* stirrups are typically made from No. 3 or No. 4 bars. In response to the decreasing shear force in a beam away from the supports, stirrup spacing is increased from the face of the support toward the beam's center. No change is made in the diameter of stirrup bars.

Stirrups generally consist of a loop with two vertical legs. Because only the vertical legs of a stirrup resist shear, stirrups can be open-loop stirrups, Figure 22.15(a). Open-loop stirrups reduce reinforcement congestion and allow easier placement of concrete. Closed-loop stirrups are more commonly used, Figures 22.15(b) and 22.15(c), and are mandated in seismic zones, where stress reversal may occur, and in beams with compression reinforcement.

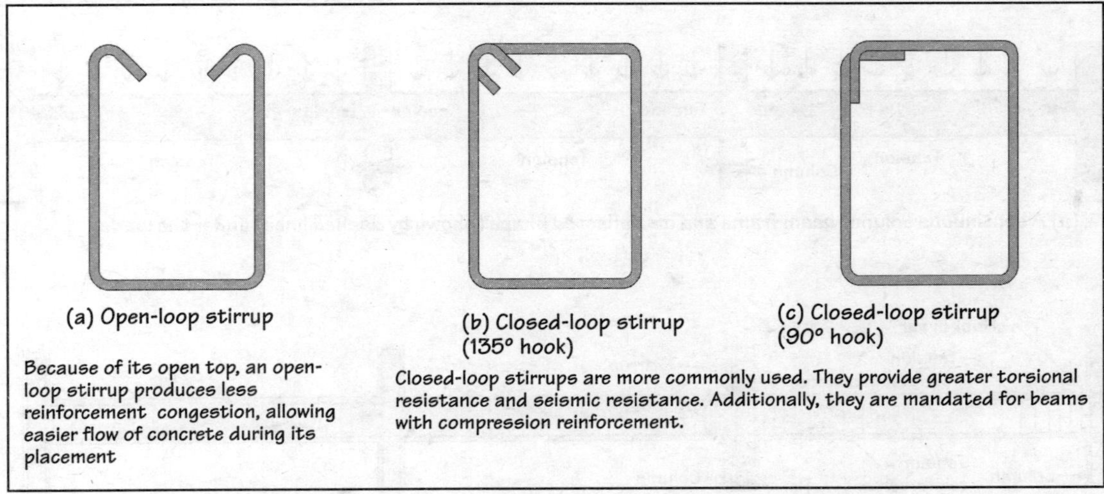

(a) Open-loop stirrup
(b) Closed-loop stirrup (135° hook)
(c) Closed-loop stirrup (90° hook)
Because of its open top, an open-loop stirrup produces less reinforcement congestion, allowing easier flow of concrete during its placement
Closed-loop stirrups are more commonly used. They provide greater torsional resistance and seismic resistance. Additionally, they are mandated for beams with compression reinforcement.

FIGURE 22.15 Commonly used stirrup shapes.

Beams with Compression Reinforcement (Doubly Reinforced Beams)

A reduction in the size of a reinforced-concrete beam for a given span and load can be obtained by increasing the amount of tension reinforcement. If the tension reinforcement cannot be accommodated in a single layer (Figure 22.14), it may be provided in two layers, Figure 22.16.

However, the American Concrete Institute (ACI) Code does not allow an increase in tension reinforcement beyond a certain percentage. This is to ensure a ductile failure of the beam under an overload. For further reduction in beam size, additional steel may be used in the compression region, that is, at the top of a simply supported beam, referred to as *compression reinforcement.* When compression reinforcement is used in a simply supported beam, it runs continuously from support to support, like the tension reinforcement. Hanger bars are not required in this beam type, and stirrups are required throughout the beam's length.

A beam provided with compression reinforcement is called a *doubly reinforced beam,* Figure 22.17. A beam with with tension reinforcement only, such as that shown in Figure 22.16, is called a *singly reinforced beam.*

Continuity in Reinforced-Concrete Members

A single-span, simply supported beam (Figure 22.14) is seldom used in site-cast reinforced-concrete construction. Instead, reinforced-concrete members are generally continuous, which creates a stress pattern different from that of a simply supported beam. For instance, in the three-span beam-column frame of Figure 22.18(a), tension is cre-

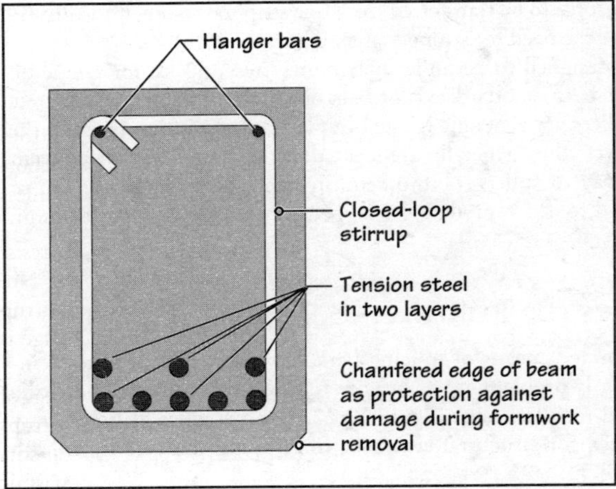

FIGURE 22.16 A concrete beam with tension steel in two layers.

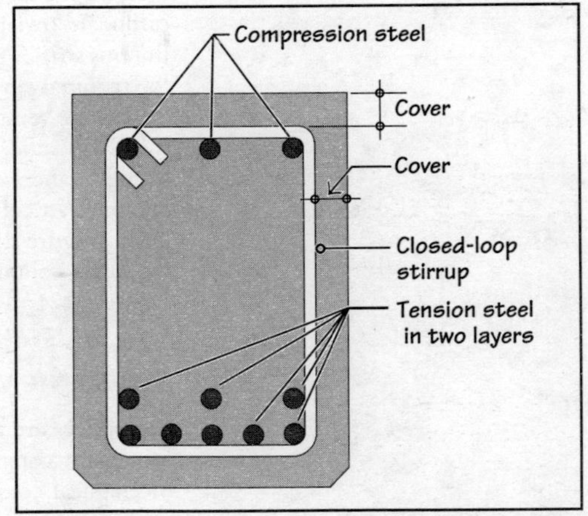

FIGURE 22.17 A section through a doubly reinforced concrete beam. The beam shown in Figure 22.16 (and also in Figure 22.14) is a singly reinforced beam.

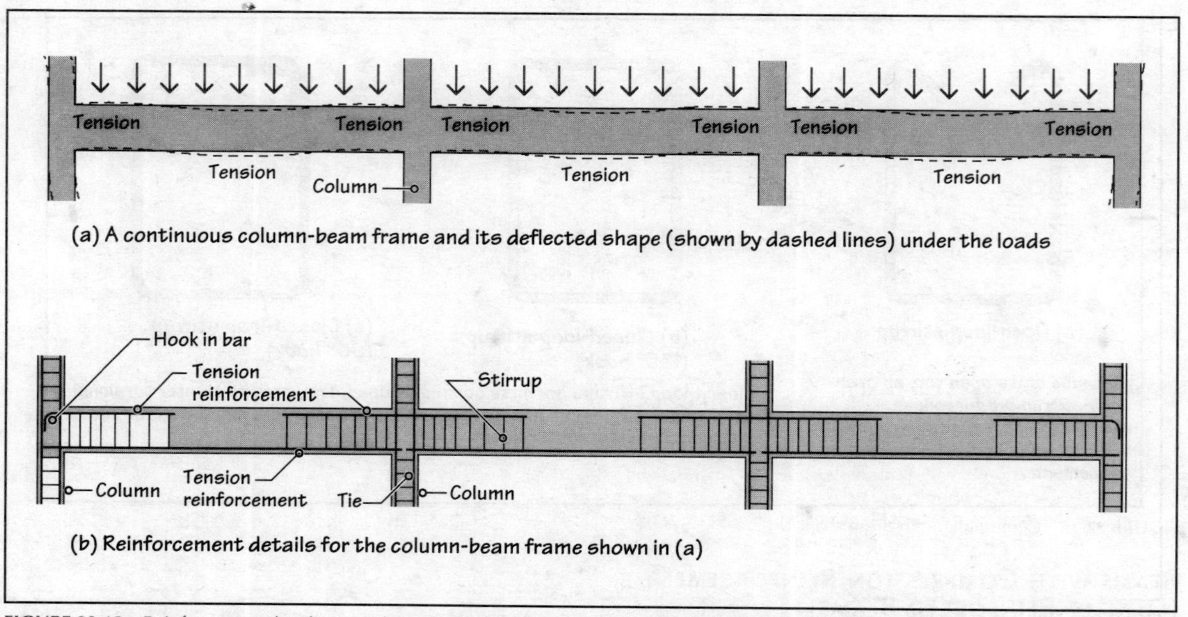

FIGURE 22.18 Reinforcement details in a column-beam frame in response to the stresses created by the loads. Note that the longitudinal bars at both the top and bottom of the beam are tension reinforcing bars.

ated at the bottom of the beam in the middle region and at the top of the beam near the supports. The reinforcement in such a beam is, therefore, more complex, Figure 22.18(b). Note that both top and bottom longitudinal bars in the beam are tension-reinforcing bars.

22.6 SPLICES, COUPLERS, AND HOOKS IN BARS

Reinforced concrete can function as a structural material only if there is a perfect bond (adhesion) between the concrete and the reinforcing bars. This bond allows two lengths of reinforcing bars to function as one continuous bar through lap splices. Because reinforcing bars are available in finite lengths (generally up to a maximum of 60 ft), lap splicing allows reinforcement to extend continuously in long beams, large slabs, and multistory columns.

Because concrete works as glue, the stress in a bar is transferred to concrete and from there to the lapped adjacent bar, Figure 22.19(a). This is similar to joining two strips of cardboard by lapping them and placing glue in the lap to make them function as one continuous strip. The greater the force to be transferred from one strip to the other, the greater the required lap length and/or the need for stronger glue, Figure 22.19(b).

Because reinforcement is designed to be under maximum stress (60 ksi for Grade 60 steel), lap splice lengths have been standardized for bars of different diameters, concrete strength, and steel grade. Splices are generally located in places of minimum stress in the member. In columns, however, they generally are located at the floor level for convenience, Figure 22.20. One way to splice column reinforcement is to bend the top of the lower column bars inward to accommodate the lapped bars from the upper column, Figure 22.21.

MECHANICAL COUPLERS FOR REINFORCING BARS

Lap splices are the most common means of splicing bars. However, they produce reinforcement congestion, which is problematic in heavily reinforced columns. Therefore, mechanical couplers are used to splice bars, Figure 22.22. Various types of proprietary couplers that have been tested for structural adequacy in gripping the bars from both ends are available.

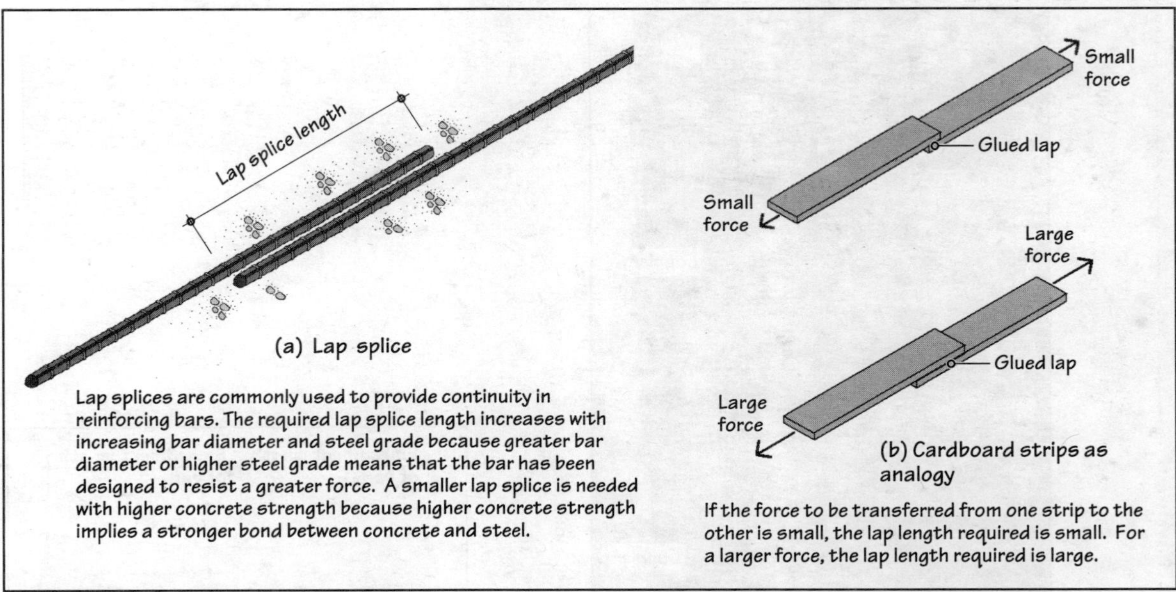

Lap splice length
(a) Lap splice
Lap splices are commonly used to provide continuity in reinforcing bars. The required lap splice length increases with increasing bar diameter and steel grade because greater bar diameter or higher steel grade means that the bar has been designed to resist a greater force. A smaller lap splice is needed with higher concrete strength because higher concrete strength implies a stronger bond between concrete and steel.
Small force
Glued lap
Small force
Large force
Glued lap
Large force
(b) Cardboard strips as analogy
If the force to be transferred from one strip to the other is small, the lap length required is small. For a larger force, the lap length required is large.

FIGURE 22.19 Principles of lap splicing of reinforcing bars.

FIGURE 22.20 Lap splicing of column bars at the floor level. The lapped bars are generally placed adjacent to each other and tied together. A maximum spacing of 6 in. between lapped bars is, however, permitted by the American Concrete Institute (ACI) Code.

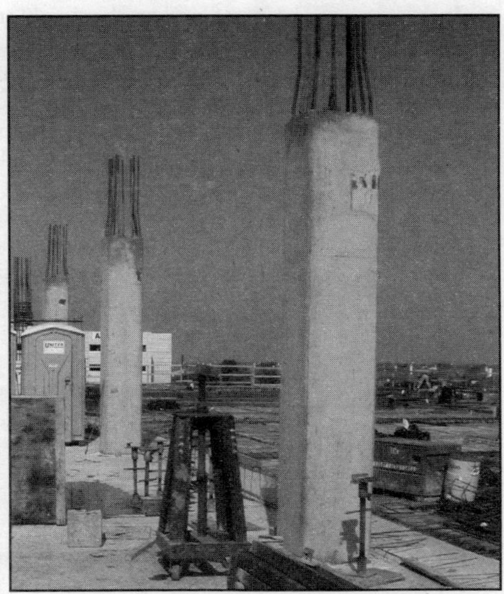

FIGURE 22.21 Column bars bent inward to receive bars for the upper floor.

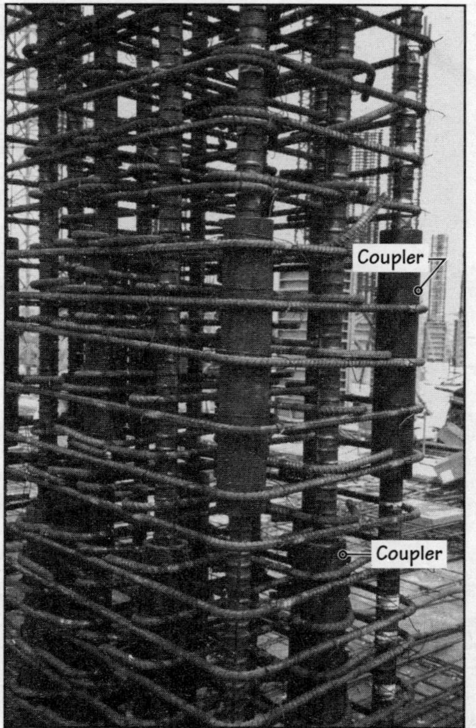

FIGURE 22.22 Reinforcement cage of a heavily reinforced concrete column showing the use of couplers for splicing of bars. Note that the couplers have been staggered in adjacent bars to reduce congestion.

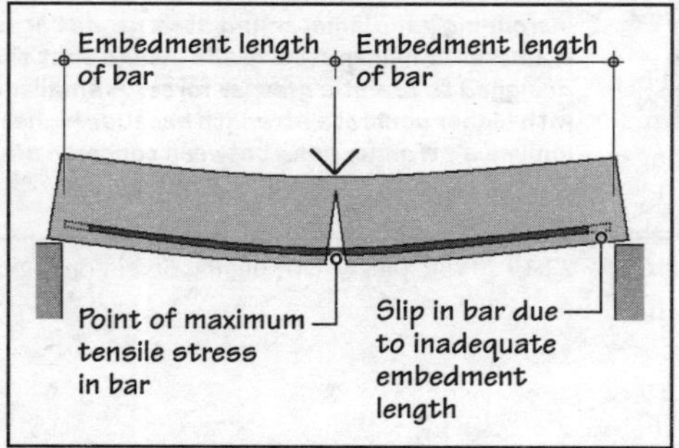

FIGURE 22.23 Bond failure of a bar due to inadequate embedment length.

Use of Hooked Longitudinal Bars

For a reinforcing bar to develop the design tensile stress (60 ksi for Grade 60 steel), the bar must be embedded in concrete for a sufficient length. If the embedment length of the bar is inadequate, the bar will slip due to loss of anchorage, which may cause the member to fail.

Consider the simply supported beam in Figure 22.23. If the length of a bar from the point of maximum tensile stress (the beam's center in this case) to the end is inadequate, the bar will lose its anchorage, causing it to slip. To increase anchorage in such a case, hooks are provided at the ends of bars. If sufficient embedment length can be provided through straight bars, hooks are not needed. Thus, the provision of a hook is tantamount to increasing the embedment length of a bar. For bars used in a beam or slab, 90° or 180° hooks are used, Figure 22.24.

Hooks are generally required at the junction of a beam meeting the last column in a multicolumn bay because sufficient space is not available there for straight bars (Figure 22.18(b)). Note that hooks are effective only for bars in tension, not for bars in compression.

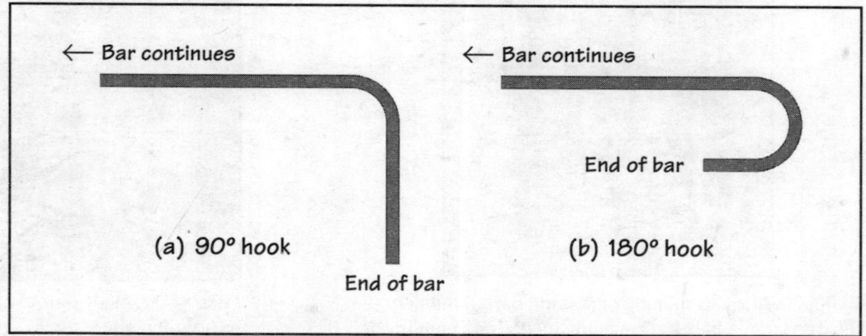

FIGURE 22.24 Note the 90° and 180° hooks in bars. Hook dimensions (diameter of the bend and extension of the bar from the bend to the end of the bar) have been standardized by the American Concrete Institute (ACI) Code so that both hooks have the same anchorage capacity and can be used interchangeably.

TABLE 22.1 MINIMUM CONCRETE COVER FOR REINFORCEMENT

Exposure of concrete	Minimum cover (in.)
Cast against and permanently exposed to earth	3.0
Exposed to earth or weather	2.0
Not exposed to earth or weather	
Beams and columns	1.5
Slabs and walls	0.75

Source: Building Code Requirements for Structural Concrete, ACI 318-05.

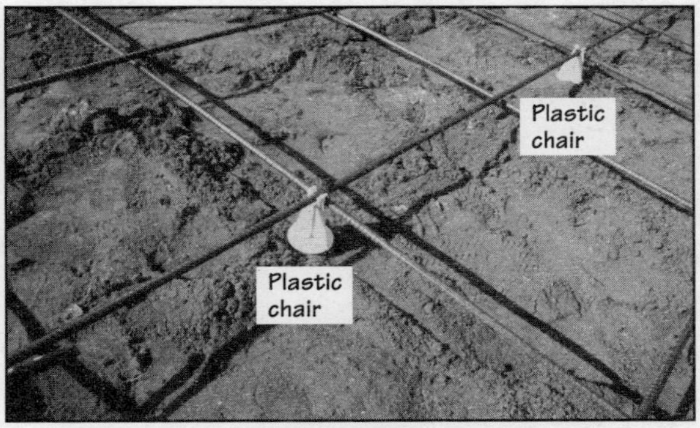

(a)

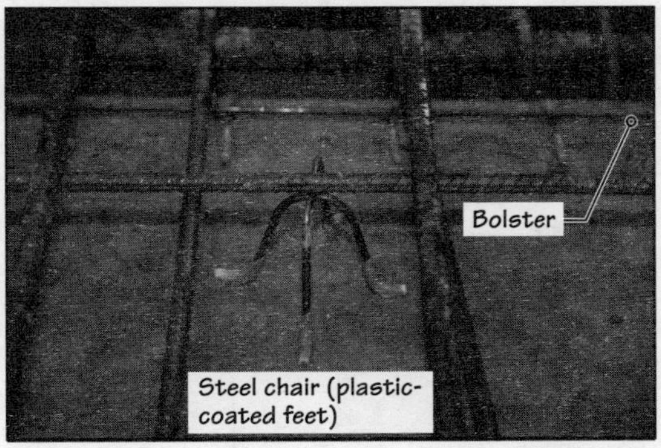

(b)

FIGURE 22.25 **(a)** Plastic chairs as bar support. Note that the bars from one direction are supported on chairs and the bars from the other direction are tied to the supported bars. **(b)** Chairs and bolsters as bar supports. A bolster is an assembly of several steel chairs joined together with a steel bar. Note that the steel chairs and bolsters have plastic-coated feet.

22.7 CORROSION PROTECTION OF STEEL REINFORCEMENT

To protect steel reinforcement from corrosion, a minimum amount of concrete cover is required. For reinforcement in footings, the minimum required cover is 3 in. For interior beams, the minimum cover required is 1.5 in., and exterior beams require a 2-in. cover, Table 22.1.

The cover is measured from the exposed concrete surface to the nearest edge of reinforcement. Therefore, in a beam, the cover is measured from the outer edge of stirrups (Figure 22.17), and in a column, from the outer edge of ties (Figure 22.27(b)).

To ensure the required cover in a beam or slab, bar supports are used. These are available in the form of either individual *chairs* or bolsters. Chairs may be of molded plastic, Figure 22.25(a), or of steel wire, Figure 22.25(b). Steel chairs have plastic-coated feet to prevent corrosion. A bolster is made from steel wire and several chairs, connected together in a long length. Both chairs and bolsters are available in various heights.

PRACTICE **QUIZ**

Each question has only one correct answer. Select the choice that best answers the question.

14. Reshoring concrete structures is necessary where excessive deflection of floor slabs or foundation settlement is anticipated on removal of shores.
a. True **b.** False

15. The structural design of formwork design is the responsibility of the
a. general contractor.
b. structural engineer.
c. structural engineer, but the design must be approved by the building official.
d. general contractor, but the design must be approved by the structural engineer.
e. none of the above.

16. Form liners are used to
a. ensure easy strippability of forms.
b. allow acid etching of the concrete surface after stripping the forms.
c. allow acid etching of concrete form before placing concrete.
d. produce textured concrete surfaces.
e. allow water washing of concrete after stripping the forms.

17. The primary purpose of steel reinforcement is to increase the tensile strength of concrete elements. However, steel reinforcement is also used to increase the shear strength and compressive strength of concrete elements.
a. True **b.** False

18. Stirrups are used in a concrete beam
a. to increase the beam's compressive strength.
b. to provide tensile strength in the beam since concrete's tensile strength is negligible.
c. to increase the tensile strength of the beam beyond that provided by concrete.
d. to provide shear strength in the beam because concrete's shear strength is negligible.
e. to increase the shear strength of the beam beyond that provided by concrete.

19. Stirrups are generally made of
a. No. 3 or No. 4 bars.
b. No. 4, No. 5, or No. 6 bars.
c. No. 6 or No. 7 bars.
d. bars of any diameter, as needed.

20. A doubly reinforced beam is one in which
a. reinforcing bars are provided in bundles of two bars.
b. reinforcing bars are provided at the top and bottom of the beam.
c. reinforcing bars are provided in two layers at the bottom of the beam.
d. reinforcing bars are provided in two layers at the top of the beam.
e. none of the above.

(Continued)

PRACTICE **QUIZ** *(Continued)*

21. Mechanical couplers are used
  a. to lap splice reinforcing bars.
  b. as an economical alternative to lap splicing of reinforcing bars.
  c. where the lap-splice length of reinforcing bars is excessive.
  d. where lap splicing of reinforcing bars will produce excessive congestion.
  e. none of the above.

22. Hooks are used
  a. to lap splice reinforcing bars.
  b. where the lap-splice length is excessive.
  c. where lap splicing of bars will produce excessive congestion.
  d. all of the above.
  e. none of the above.

23. Hooks in bars are produced by turning the bars through
  a. 30° or 60°.
  b. 45° or 90°.
  c. 90° or 180°.
  d. 180° or 270°.

24. Steel reinforcement in concrete elements that are permanently exposed to the earth, such as in footings, must have a minimum concrete cover of
  a. 5 in.
  b. 4 in.
  c. 3 in.
  d. 2 in.
  e. 1 in.

25. The chairs used to support reinforcing steel bars are generally
  a. plastic chairs.
  b. steel-encased plastic chairs.
  c. steel chairs with plastic-coated feet.
  d. (a) and (b).
  e. (a) and (c).

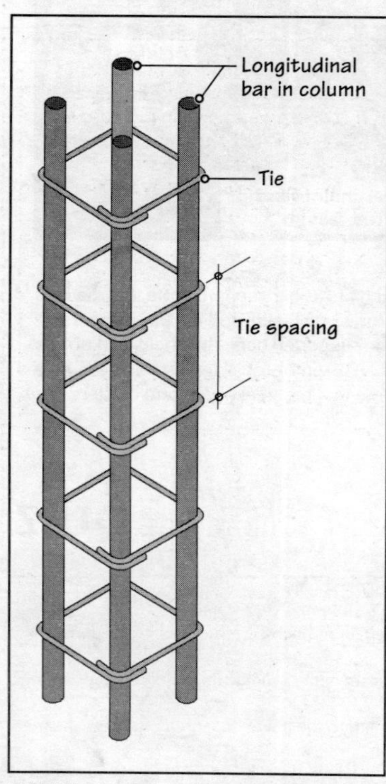

FIGURE 22.26 Reinforcement cage for a column consists of longitudinal bars and ties (transverse reinforcement). The minimum number of longitudinal bars in a square or rectangular column is four (as shown here). In a column with a larger number of bars, the tie pattern is more complex, as shown in Figure 22.27.

22.8 REINFORCEMENT AND FORMWORK FOR COLUMNS

The reinforcement in a column is provided primarily to resist compressive forces. However, because columns are also subjected to bending, the reinforcement must be located on the column periphery. Another reason for the peripheral location is that the reinforcement cage of a column confines the concrete within the cage, even after the concrete has been crushed to failure, resulting in a more gradual ductile column failure.

A column reinforcement cage consists of longitudinal reinforcement (vertical bars) and transverse reinforcement (ties), Figure 22.26. The total area of the longitudinal reinforcement for a column is based on structural requirements but must lie between 1% and 8% of the gross area of the column. Generally, however, it is difficult to accommodate a reinforcement area of more than 4% unless the bars are welded together at splices or coupled with mechanical couplers.

TRANSVERSE REINFORCEMENT IN COLUMNS—INDIVIDUAL TIES

Ties are similar to stirrups but have two structural functions. One is to prevent the buckling of the longitudinal reinforcement (which consists of slender bars under compression). If the bars are not prevented from buckling, they will burst out of the concrete. The other function is to provide shear resistance to columns subjected to bending. Ties also help in the formation of a reinforcement cage. Their center-to-center spacing is dictated by simple rules given in the American Concrete Institute (ACI) Code.

Ties are generally No. 3 or No. 4 bars. Each longitudinal column bar must be enclosed by a tie corner (see Figure 22.27(a) for an exception). Several tie shapes may be needed in the same column, Figure 22.27. In round columns, round ties may be used, Figure 22.28.

TRANSVERSE REINFORCEMENT IN COLUMNS—CONTINUOUS SPIRAL

In round columns, a continuous helical spiral may be used in place of individual ties. A column with transverse spiral reinforcement is more robust because of its greater ductility. That is why a column with spiral reinforcement is preferred in seismic zones. Continuous transverse spiral reinforcement may also be provided in square columns by using a circular reinforcement cage in place of a square cage.

COLUMN FORMS

Column forms for rectangular or square columns are generally made of wood or steel. A typical column form consists of four panels braced by a steel angle frame on all sides. The angle frame is hinged at one end and clamped at the opposite end so that the form can be stripped by simply unclamping and rotating the unclamped side, Figure 22.29.

Round column forms may be of steel plate or waterproof fiberboard, Figure 22.30. Steel forms are reusable, and are generally semicircular and of short lengths that are bolted together to obtain the correct height. Fiber forms are one piece and meant for one-time use.

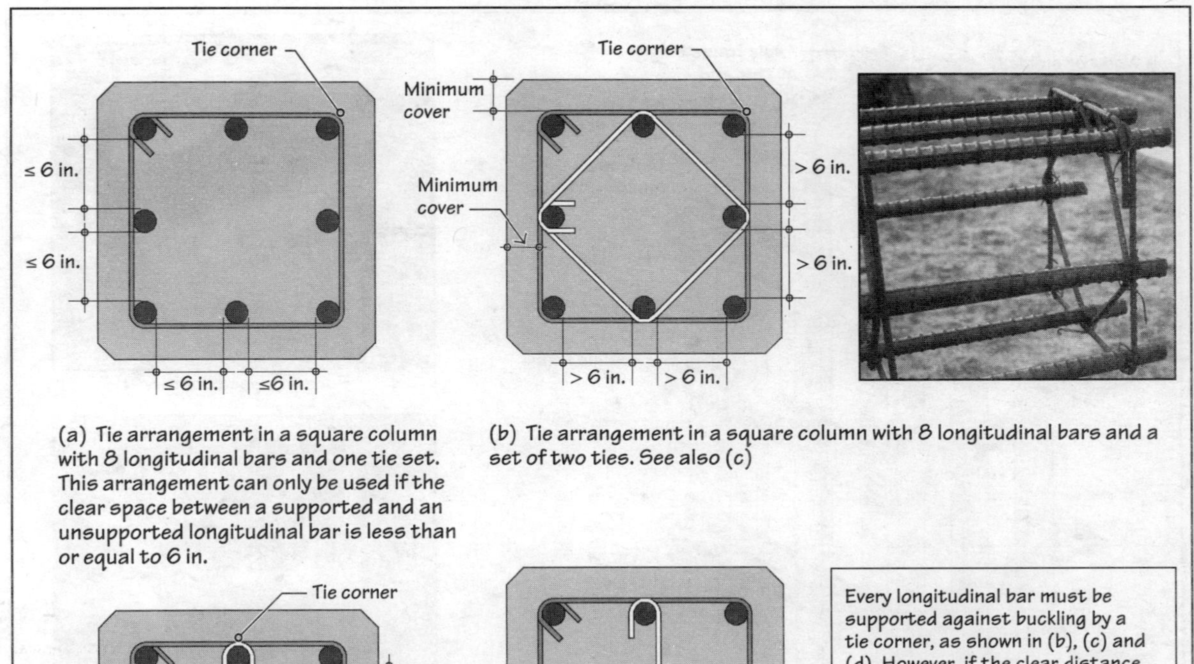

(a) Tie arrangement in a square column with 8 longitudinal bars and one tie set. This arrangement can only be used if the clear space between a supported and an unsupported longitudinal bar is less than or equal to 6 in.

(b) Tie arrangement in a square column with 8 longitudinal bars and a set of two ties. See also (c)

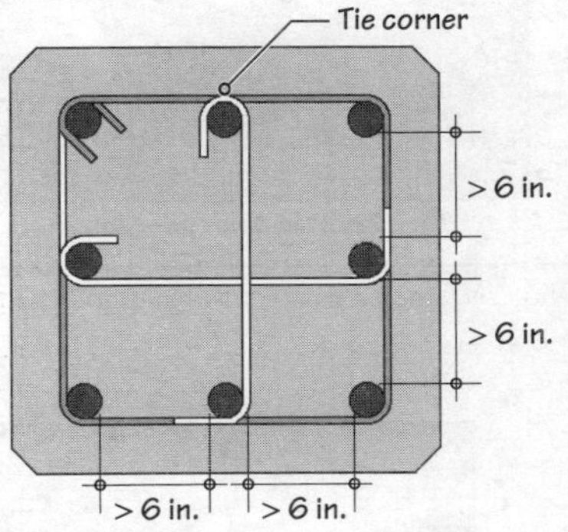

(c) Tie arrangement in a square column with 8 longitudinal bars that uses a set of three ties—an alternative to the arrangement shown in (b)

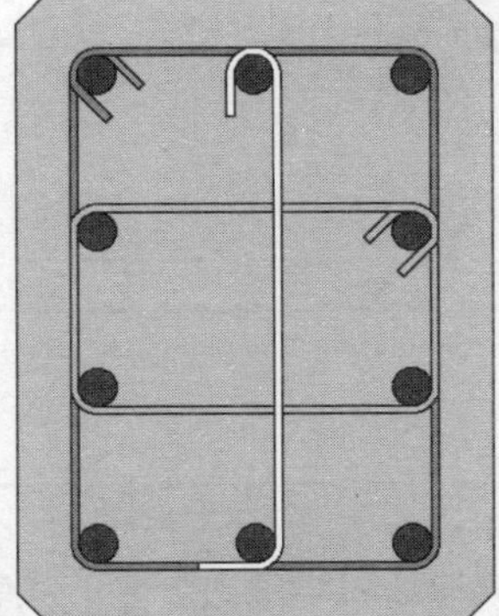

(d) Tie arrangement in a rectangular column with 10 longitudinal bars and a set of three ties

Every longitudinal bar must be supported against buckling by a tie corner, as shown in (b), (c) and (d). However, if the clear distance between longitudinal bars is ≤ 6 in., an alternate bar may be left unsupported, as shown in (a).

A set of ties may consist of one tie shape, as shown in (a), a set of two ties, as shown in (b), a set of three ties, as shown in (c) and (d), etc.

The center-to-center spacing of the tie set is governed by empirical code requirements.

FIGURE 22.27 Tie arrangements in a typical rectangular or square reinforced-concrete column.

FIGURE 22.28 Reinforcement cage for a round column with circular ties.

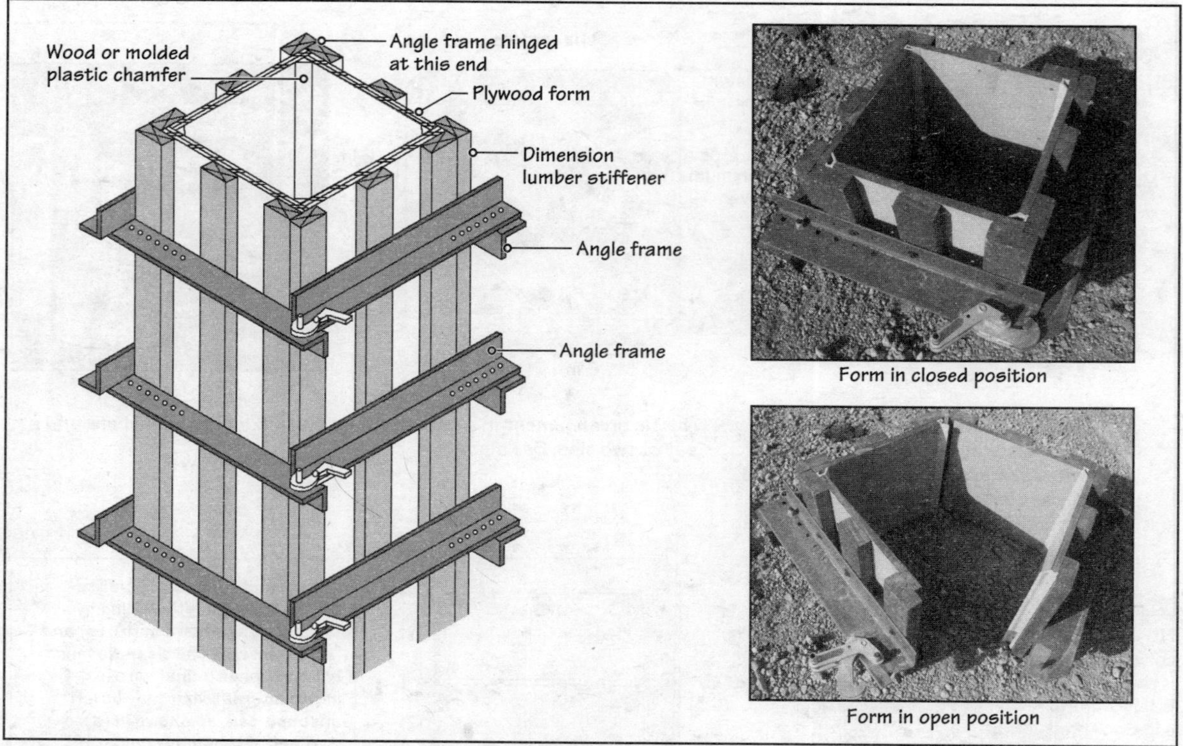

FIGURE 22.29 A typical rectangular or square column form. The photographs show a short section of a hinged column form in a formwork fabricator's yard.

(a) Reusable formwork for a concrete column made of 2-ft-long semicircular steel plates with end stiffeners that are bolted together at the stiffeners.

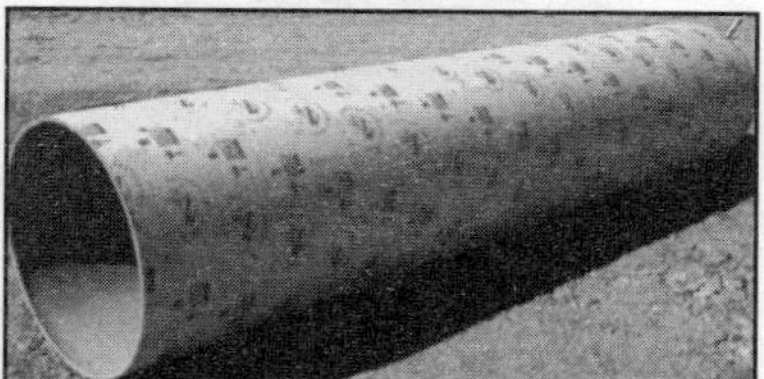

(b) One-time use fiberboard form for a round concrete column.

FIGURE 22.30 Two types of forms for round columns.

22.9 REINFORCEMENT AND FORMWORK FOR WALLS

Concrete walls are commonly used in buildings as retaining walls, basement walls, shear walls, or load-bearing walls. The essential difference between the reinforcement in a wall and a column is that ties are not required in a wall unless the wall supports a large gravity load.

The reinforcement in a wall consists of a mesh formed by vertical and horizontal bars provided in one central layer. In a wall 10 in. or thicker, two layers of reinforcement are required close to each face of the wall, but separated from the face by minimum required cover, Figure 22.31.

FORMWORK FOR WALLS

The formwork for each face of a wall is generally made of plywood panels (generally $\frac{3}{4}$ in. thick) supported by horizontal members (generally 2 × 4 lumber) called *walers*, Figure 22.32. The walers are supported by vertical lumber called *stiffbacks*. The stiffbacks are supported by diagonal braces, Figure 22.33. The formwork for each face is separated by *form ties*—specially shaped steel wires that separate and hold the formwork from two opposite faces and resist the pressure of wet concrete.

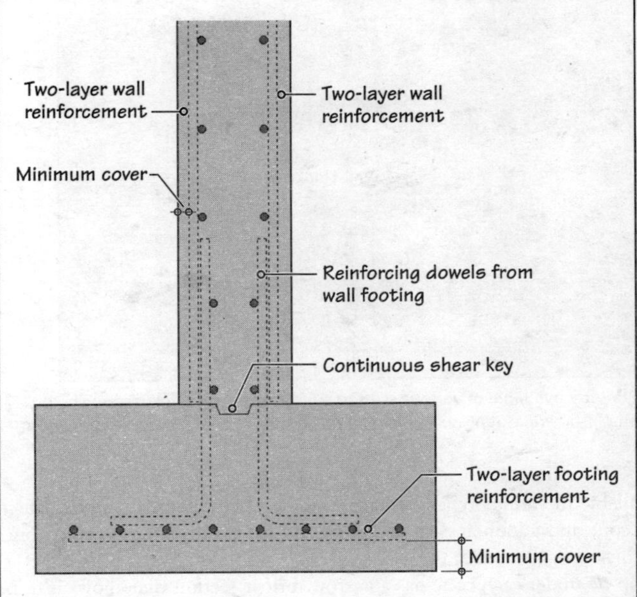

FIGURE 22.31 A section at the foundation level through a reinforced-concrete wall with two-layer reinforcement. Note that the footing of the wall is constructed first, with reinforcing dowels projecting out of the footing. (This is similar to construction of the footing for a column; see Expand Your Knowledge in Section 22.11.)

FIGURE 22.32 Workers equipped with safety harness erect the formwork for a reinforced-concrete wall. (See Figure 22.33 for details.)

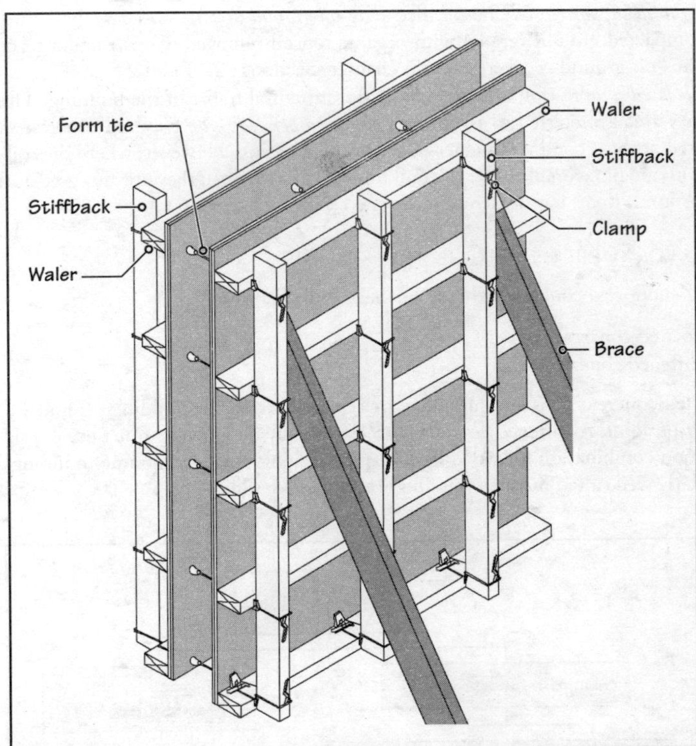

FIGURE 22.33 Isometric view of the formwork shown in Figure 22.32. Note that the formwork is made from plywood and dimension lumber. The entire assembly is fastened together with removable clamps. Form ties between opposite plywood panels separate the panels and hold them against the pressure of wet concrete.

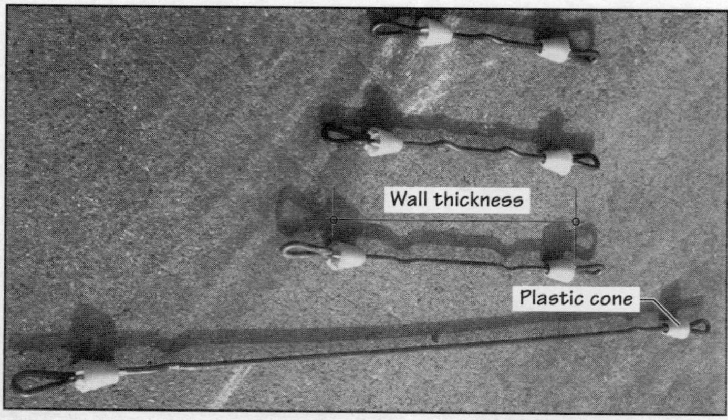

FIGURE 22.34 Snap ties are available in various sizes to suit various wall thicknesses. The out-to-out dimension between plastic cones represents the wall thickness. Several other form tie designs are available.

Form ties are available in various types. One commonly used type is a *snap tie*, which consists of a plastic cone and a loop at each end. The end-to-end distance between cones represents the wall thickness, Figure 22.34.

The wire of a snap tie under each cone has a narrow, brittle section that allows it to be snapped there by a plier-assisted twist of the end loop. The snapping is done after the formwork has been stripped. After removal of the end loops and the plastic cones, the pockets left by the cones are filled with a sealant.

22.10 TYPES OF CONCRETE SLABS

As shown in Figure 22.35, site-cast concrete slabs can be of two types:

- *Ground-supported slabs* (also called *slabs-on-ground* or *slabs-on-grade*) bear directly on compacted ground (grade) with organic topsoil removed. Construction of concrete slabs-on-ground is covered in this chapter (Sections 22.11 and 22.12).
- *Elevated slabs* rest on and are part of the structural frame of the building. Therefore, they are also referred to as *framed slabs* or *suspended slabs*. Elevated concrete slabs are used at the second and higher floors of the building. However, where the soil conditions are unfavorable (or if the building has a basement), they are also used at the first (ground) floor. Construction of elevated slabs is covered in Chapter 23.

NOTE

Classification of Ground-Supported Slabs as Type I, II, or III

A classification of ground-supported concrete slabs similar to that mentioned in Figure 22.35 is given by the Building Research Advisory Board (BRAB). In this classification, isolated slabs are called *Type I* or *Type II*, respectively, depending on whether they are unreinforced or reinforced with nominal reinforcement. A ground-supported stiffened slab is called *Type III* [22.1].

GROUND-SUPPORTED CONCRETE SLABS

Ground-supported concrete slabs are further subdivided into

- Isolated concrete slabs
- Stiffened concrete slabs

Isolated concrete slabs are those that are separated from the building's foundation with an isolation joint. A stiffened concrete slab is generally designed to function as a slab-and-foundation combination for wood light-frame (or light-gauge steel frame) buildings and is particularly well suited for such buildings on expansive soils.

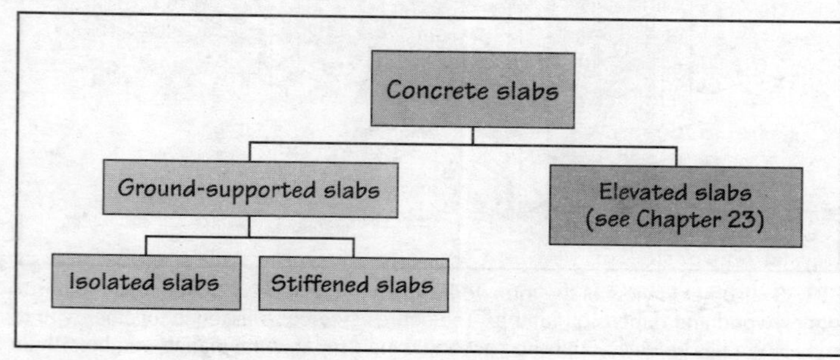

FIGURE 22.35 Types of concrete slabs.

22.11 GROUND-SUPPORTED, ISOLATED CONCRETE SLAB

An isolated, ground-supported concrete slab is commonly used on stable, undisturbed soils (with all organic material removed) or on select fills compacted to provide a uniformly strong base. As long as the bearing capacity of the supporting ground is not too low, it is generally not a critical issue because concrete slabs are fairly rigid. Therefore, concentrated loads on the slab (from foot traffic, moving furniture, forklifts, and other items) are spread over a large area. Consequently, the stress on the ground due to the loads on the slab is relatively small.

VAPOR RETARDER

A vapor retarder (generally a 6-mil-thick polyethylene sheet with lapped joints) is placed immediately below an isolated slab. It prevents the passage of subsoil water and water vapor through the slab into the building's interior. In the absence of a vapor retarder, the vapor from below the slab may condense under an object covering the slab, such as a rug, furniture legs, or boxes.

A vapor retarder is particularly important if the interior of the building is heated or cooled and/or if the slab is covered with a vapor-impermeable floor finish such as vinyl tiles, linoleum, or thin-set terrazzo. A vapor retarder is not required under a slab if the passage of water vapor does not affect the building's interior, such as a garage or an open shed. It is also not required under driveways, pavements, or patios, for example.

A secondary advantage of a vapor retarder is that it works as a slip membrane between the slab and the subbase so that the slab moves more freely as it shrinks. In the absence of a slip membrane, the shrinkage of the slab is restrained, which induces tensile stresses in the slab. These stresses are in addition to the tensile stresses caused by shrinkage. Because the reinforcement in an isolated slab-on-ground is provided to resist shrinkage stresses, a vapor retarder reduces reinforcement requirements.

A disadvantage of using a vapor retarder is that it aggravates the differential shrinkage between the top and bottom of the slab. Water is lost by evaporation from the top of the slab but remains relatively intact at the bottom, leading to dishlike curling of the slab at the edges, Figure 22.36. Curling can be reduced by using concrete with a low water-cement ratio and proper curing to ensure the availability of sufficient moisture at the top of the slab.

SUBBASE AS THE DRAINAGE LAYER

A subbase—a layer of granular material placed immediately above the top of compacted ground (below the vapor retarder, if used)—is generally required. Although a properly placed and consolidated concrete slab is resistant to infiltration of subsoil water, a subbase provides additional protection by providing a capillary break. The subbase also equalizes slab support by filling any uneven spots in the ground.

The required subbase thickness is generally 4 in. A thicker subbase does not yield greater benefit. Subbase material can be crushed stone, gravel, or sand. Sand is preferred in slabs provided with a vapor retarder because, unlike crushed stone or gravel, it does not puncture the vapor retarder.

CONCRETE STRENGTH AND THICKNESS

A concrete strength of 3,000 psi for residential slabs and 4,000 psi for commercial and light-industrial slabs is generally recommended. A higher strength is needed only for more heavily loaded slabs or where greater wear resistance is needed. The thickness of an isolated reinforced concrete slab is also a function of the load it must carry, with a typical minimum of 4 in.

AMOUNT OF STEEL REINFORCEMENT

A concrete slab supported on stable, compacted soil may not require steel reinforcement. However, a small (i.e., nominal) amount of reinforcement is generally provided in an isolated slab, Figure 22.37. Its purpose is to reduce the size of shrinkage cracks in the slab. (See [22.2] for determining the amount of reinforcement.)

NOTE

The Terms *Grade* and *Ground*

The terms *grade* and *ground* are generally used interchangeably. Thus, *slab-on-ground* and *slab-on-grade* imply the same thing. However, *slab-on-ground* is more commonly used in the literature published by the American Concrete Institute (ACI), Portland Cement Association (PCA), and Post-Tensioning Institute (PTI).

By contrast, terms such as *natural grade, existing grade, finished grade, compacted grade, above grade, below grade, grade beams, grading lines* (contour lines), and so on, are in common use.

NOTE

Fibrous Concrete Reinforcement in Slabs-on-Ground

Because the purpose of reinforcement in an isolated concrete slab-on-ground is to control shrinkage cracks, synthetic (polypropylene, nylon, and polyester) fibers are being increasingly used as replacement for steel reinforcement. They are mixed with other concrete ingredients.

Synthetic fibers are generally 0.5 to 1.5 in. (12 to 38 mm) long and approximately 0.04 in. (1 mm) thick. Shorter fibers are effective during the early stages of concrete hardening, and the longer fibers increase the tensile strength of hardened concrete. Fiber-reinforced concrete is generally more economical than hand-tied steel bars.

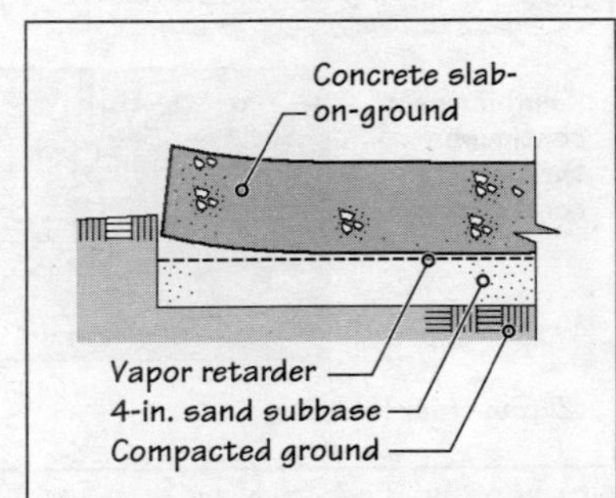

FIGURE 22.36 Curling of a concrete slab-on-ground due to differential drying shrinkage.

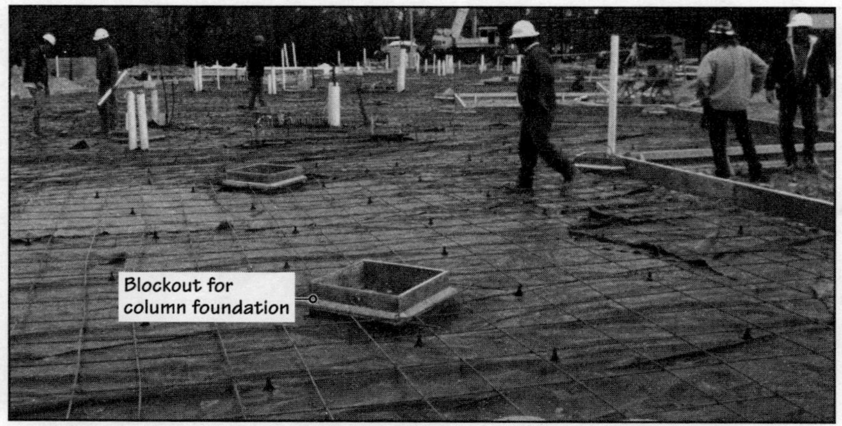

FIGURE 22.37 An isolated slab being inspected before concreting. Observe the vapor retarder and the reinforcing bars supported on chairs. Also observe the blockouts for columns, which isolate the slab from the building's structure.

Note that reinforcement does not prevent the slab from cracking, but it reduces the width of each individual crack, keeping the overall crack width unaffected. Note also that reinforcement does not increase the load-carrying capacity of the slab, which is primarily a function of slab thickness. Because the size of individual cracks is reduced, control joints can be spaced farther apart in a reinforced slab than in an unreinforced slab. It follows, therefore, that if control joints in a slab are unacceptable, the slab must be provided with a greater amount of reinforcement.

Location of Steel Reinforcement in an Isolated Slab

Because the purpose of reinforcement in an isolated, ground-supported concrete slab is to control shrinkage cracks, it should be located as close to the top of the slab as possible, keeping in view the requirements for cover and surface finish.

Joints in an Isolated Concrete Slab-on-Ground

An isolated concrete slab-on-ground requires the following types of joints:

- Control joints
- Isolation joints
- Construction joints

Control Joints

Control joints accommodate the shrinkage of concrete. Their purpose is to provide weakness in the slab at predetermined locations to force the slab to crack there. In the absence of control joints, the slab will crack in a random, haphazard pattern. Control joints are generally provided by sawing the slab at intervals to a depth of 0.25 times the thickness of the slab, Figure 22.38. The width of a saw-cut joint is approximately $\frac{1}{8}$ in.

Because concrete begins to shrink and crack as it hardens, the slab must be cut as soon as it is hard enough to provide a clean, unraveled joint (usually within 48 h of concrete placement). Control joints are spaced at 24 times the thickness of slab in both directions, but not more than 15 ft apart. (In exterior concrete walkways and pavements, control joints may be provided through saw-cut joints or tooled joints. A tooled joint is a shallow groove made while the concrete is green.)

The crack that develops under a control joint follows a zigzag pattern along the boundaries of coarse aggregate particles. The zigzag crack, shown in Figure 22.38, provides shear interlock between adjacent sections of the slab, referred to as *aggregate interlock*.

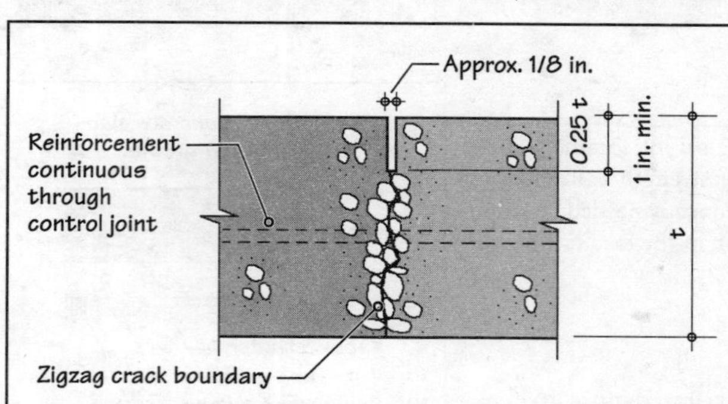

FIGURE 22.38 Section through a control joint in a slab. The slab generally cracks through its entire thickness under a control joint. Shear transfer between adjacent sections of the slab at a control joint is provided by the zigzag crack along aggregate boundaries.

Isolation Joints

Unlike control joints, *isolation joints* in a concrete slab extend the entire thickness of the slab. They are typically $\frac{1}{2}$ in. wide and are provided to ensure that the slab is isolated from the building's structural components so that their movement (creep, foundation settlement, etc.) is not transferred to the slab. The joint space is generally filled with asphalt-saturated fiberboard and covered with a sealant. Figure 22.39 shows the location of control joints and isolation joints in a typical isolated interior slab.

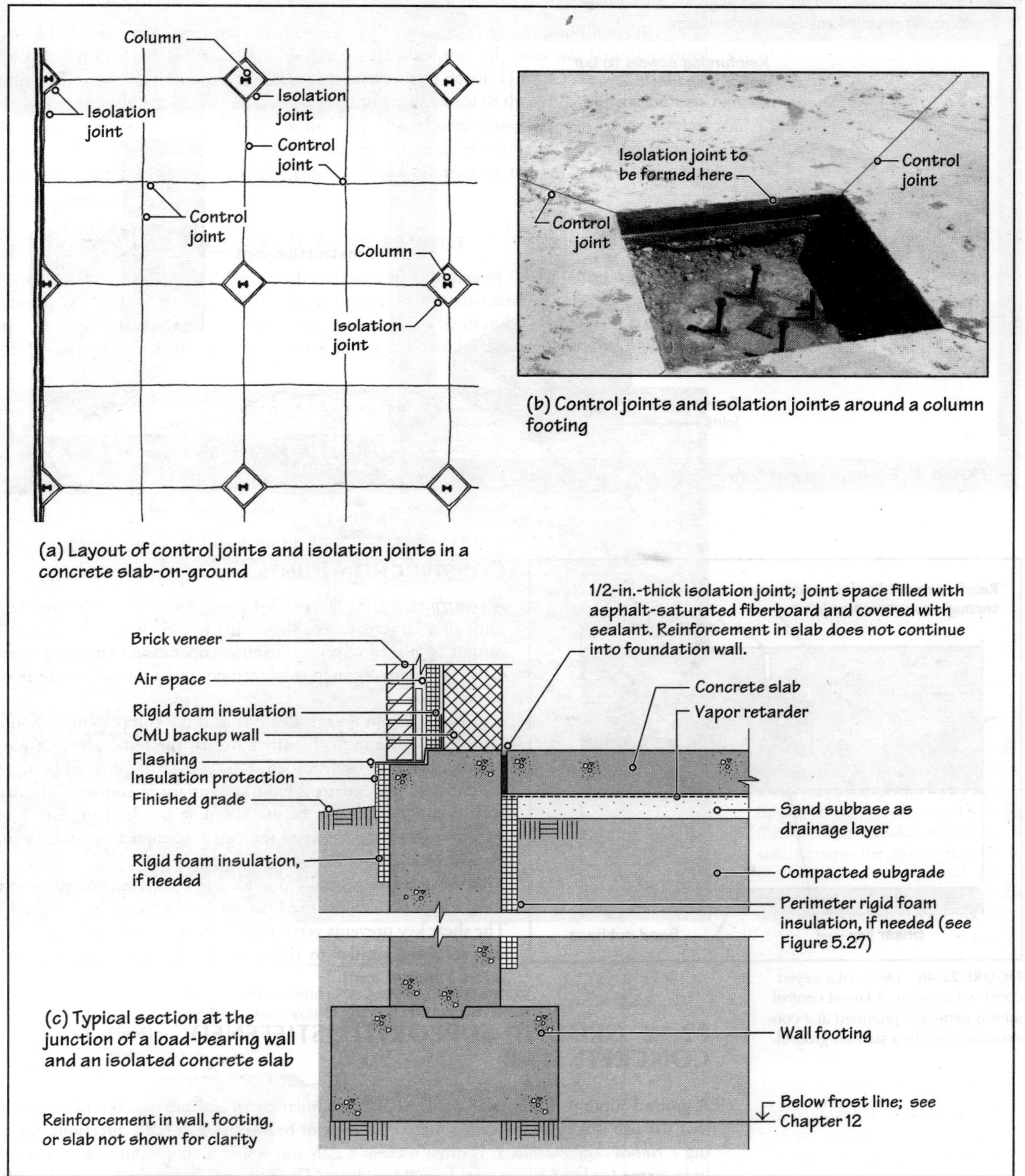

FIGURE 22.39 Isolation joints and control joints in an interior isolated concrete slab-on-ground.

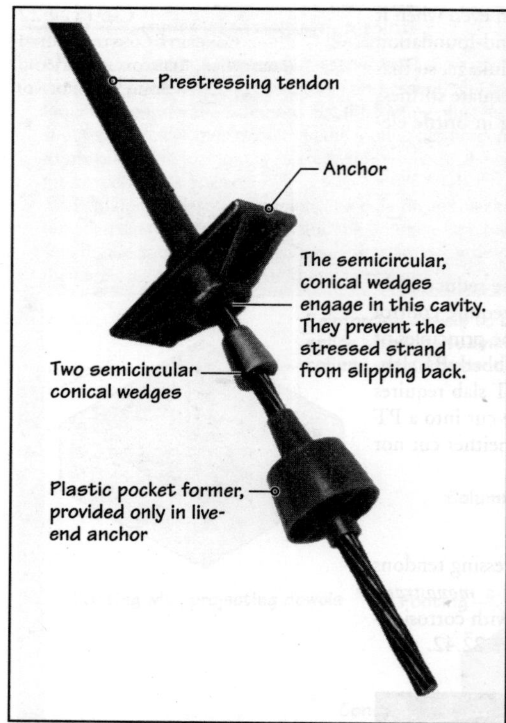

FIGURE 22.43 Anatomy of a live-end anchor. A dead-end anchor is similar; see Figure 22.48. (Photo courtesy of Jack W. Graves, Jr)

Each tendon has two anchor assemblies, one at each end. One end has a fixed anchor assembly called a *dead end*. The opposite end has the stressing-end assembly, called a *live end*. The tendon is stressed from the live end. Dead-end and live-end anchors are almost identical. The anchor contains a cavity to accommodate two semicircular conical steel wedges that prevent the stressed tendon from slipping back, Figure 22.43.

The only difference between a live end and a dead end is that the live-end anchor is provided with a plastic pocket former, which gets embedded in concrete. The pocket former is removed before stressing of the tendons and grouting of the pocket, as described later.

A PT ribbed slab-on-ground is typically 4 to 5 in. thick. The ribs are typically 10 to 12 in. wide and 24 to 36 in. deep, spaced a maximum of 10 to 15 ft on center. Figure 22.44 shows the plan and section of one such slab, and Figure 22.45 shows the details of dead-end and live-end anchors.

Stressing of Tendons

A concrete strength of 3,000 psi is generally used for a PT slab. The concrete is placed, vibrated, finished, and cured in the same way as for any other slab. After the concrete has hardened (generally 24 to 48 h after its placement), the edge forms from the slab are removed. This allows easy removal of pocket formers while the concrete is still green.

The tendons are stressed after the concrete has gained sufficient strength (approximately 2,000 psi). However, stressing should not be delayed excessively to forestall the development of shrinkage cracks in the slab. Generally, an interval of 3 to 10 days is used between concrete placement and stressing, depending on factors

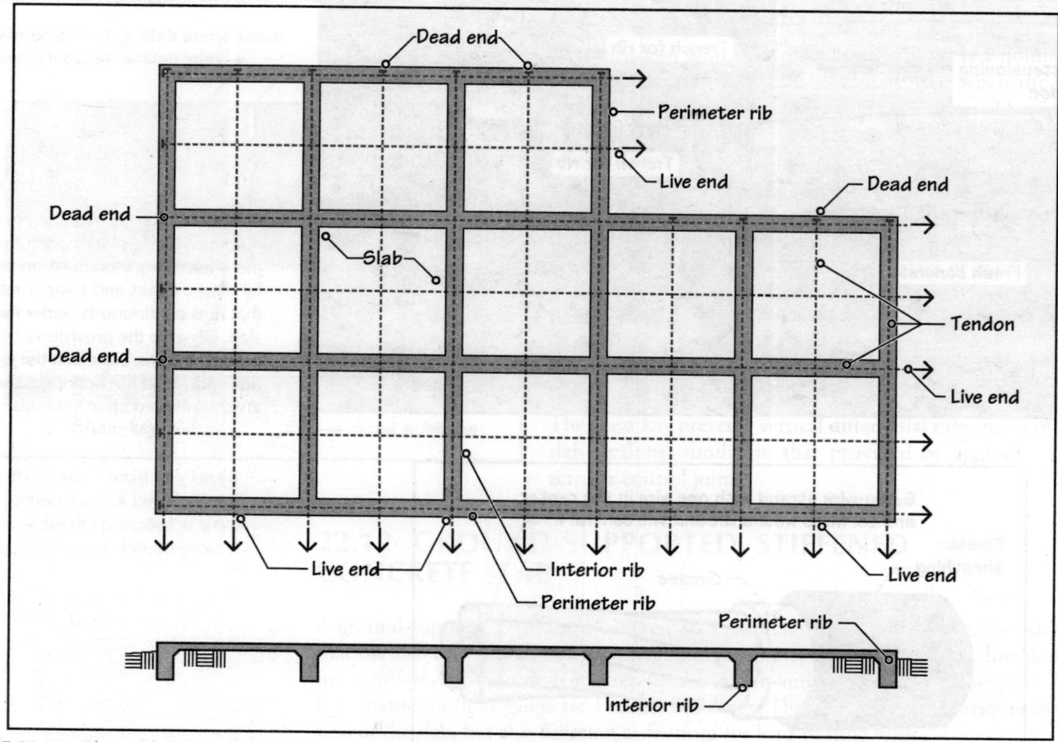

FIGURE 22.44 Plan and section of a typical PT slab-foundation combination showing the layout of ribs and prestressing tendons.

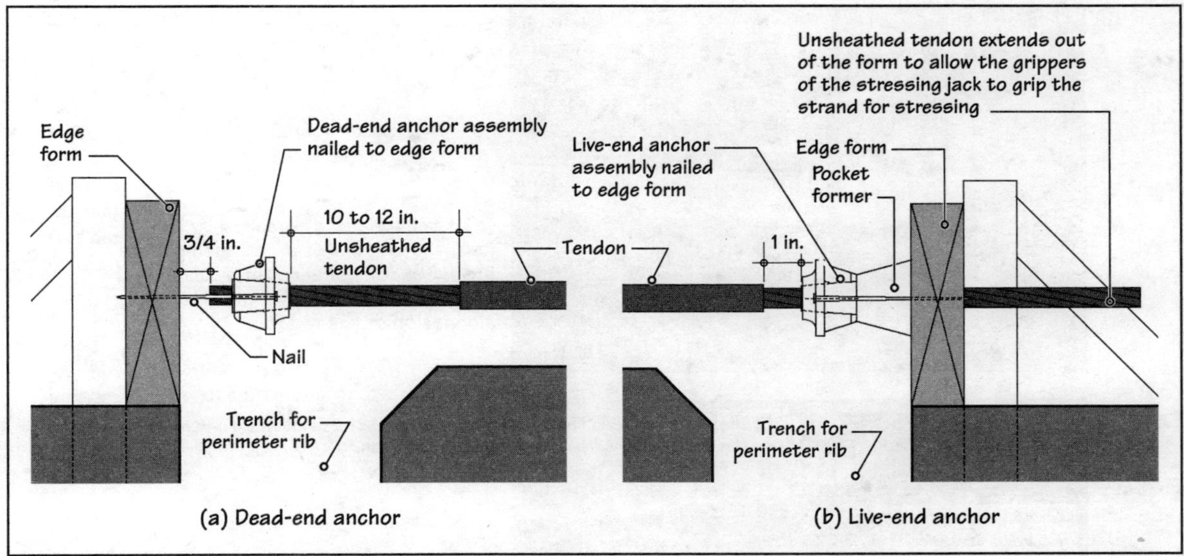

FIGURE 22.45 Details of a dead-end and a live-end anchor.

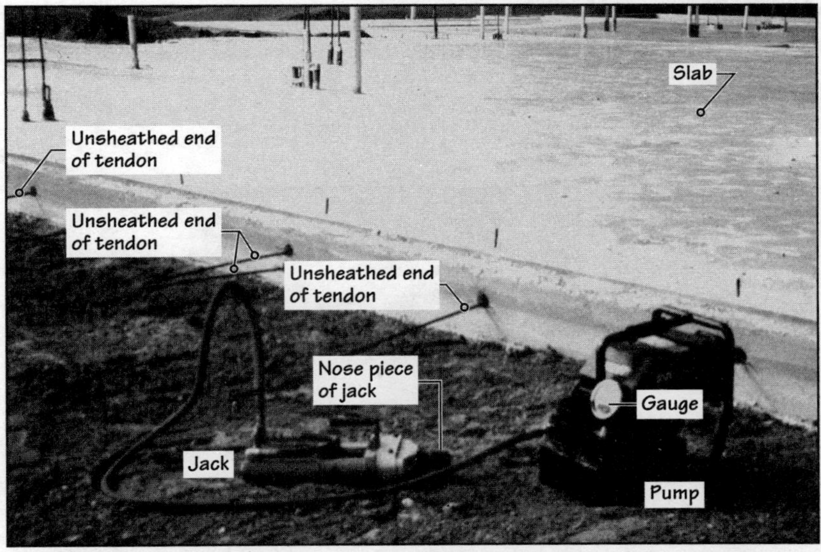

FIGURE 22.46 A PT concrete slab with tendons projecting at the live ends ready for stressing. Seen in the foreground is prestressing equipment consisting of a jack that is connected to a hydraulic pump. The nose piece of the jack pulls the strand. (Photo courtesy of Jack W. Graves, Jr)

such as ambient temperature, water-cement ratio of concrete, and admixtures used in concrete.

The prestressing equipment consists of a jack connected to a hydraulic pump, Figure 22.46. The effective compressive prestress introduced in the slab is 50 to 100 psi, which is determined by the gauge readings on the pump and verified by measuring the elongation in tendons after stressing, Figure 22.47(a). Therefore, before stressing begins, each tendon is reference marked with paint at the live end at a constant distance from the slab edge, Figure 22.47(b).

After the tendon has been tensioned by the required force, the excess length of tendon is either saw-cut or flame-cut, and the pocket left by the pocket former is patched with non-shrink grout, Figure 22.48.

NOTE

Ribbed Slab versus Constant-Thickness Slab

A constant-thickness PT slab-on-ground is often used in place of a ribbed PT slab-on-ground. Referred to as a *California slab*, it is 7.5 in. or thicker (without any beams). Another version of a constant-thickness slab is a 5-in.-thick slab with perimeter beams and no intermediate beams.

Source: *Engineering News-Record* (http://www.enr.com), September 20, 2004, p. 8.

(a) A tendon being stressed by the jack. The amount of prestressing force in the tendon is indicated by the gauge reading and verified by a predetermined elongation of the tendon.

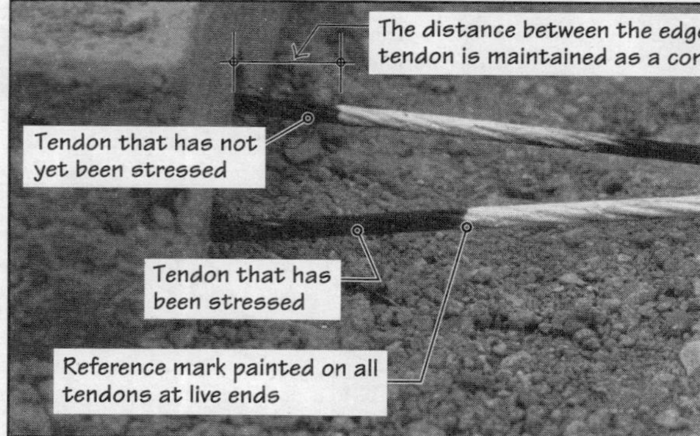

(b) Stressed and unstressed tendons.

FIGURE 22.47 Procedure used for verifying the elongation in tendons. (Photo courtesy of Jack W. Graves, Jr.)

(a) After stressing, the tendon is cut with a saw (or a torch).

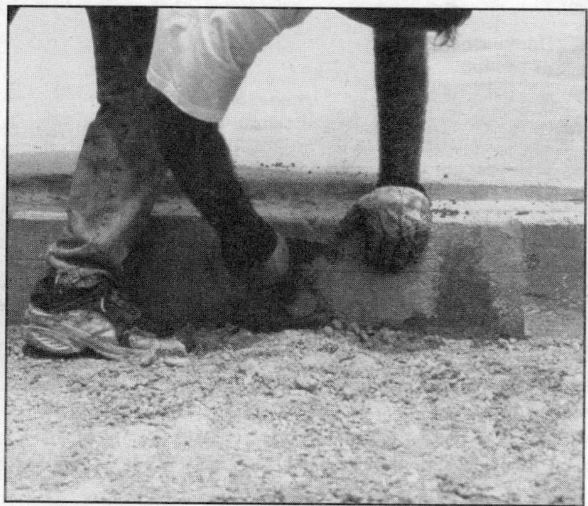

(b) After the tendon is cut, the pocket is filled with a nonshrink grout to prevent corrosion of the tendon.

FIGURE 22.48 Cutting of a tendon and patching of the pocket with grout. (Photo courtesy of Jack W. Graves, Jr)

POSTTENSIONED SLAB-ON-GROUND TO ELIMINATE CONTROL JOINTS

Although posttensioned, ground-supported slabs are generally used as a slab-foundation combination, they are also used in situations where they do not function as foundations, such as where control joints are unacceptable and the cracking of the slab is to be minimized. In such a slab, posttensioning of the slab is done in two stages. The first stage of posttensioning is done within 36 to 48 h of concrete placement to prevent any initial cracking of the slab, followed by the second stage a few days later.

PRACTICE QUIZ

Each question has only one correct answer. Select the choice that best answers the question.

26. The most important reason for providing ties in a concrete column is to
- **a.** prevent the buckling of longitudinal column bars.
- **b.** increase the compressive strength of a column.
- **c.** increase the shear strength of a column.
- **d.** allow the formwork to be tied to the longitudinal bars in the column.
- **e.** Ties are generally not provided in a concrete column.

27. The most important reason for providing ties in a concrete wall is to
- **a.** prevent the buckling of longitudinal bars in the wall.
- **b.** increase the compressive strength of the wall.
- **c.** increase the shear strength of the wall.
- **d.** allow the formwork to be tied to the longitudinal bars in the wall.
- **e.** Ties are generally not provided in a concrete wall.

28. Formwork for a concrete wall generally consists of
- **a.** plywood, form ties, and walers fastened together with duplex nails.
- **b.** plywood, form ties, walers, and stiffbacks fastened together with duplex nails.
- **c.** plywood, form ties, walers, and stiffbacks fastened together with clamps.
- **d.** plywood and stiffbacks fastened together with clamps.

29. A vapor retarder is required under
- **a.** all concrete slabs-on-ground.
- **b.** all exterior concrete slabs-on-ground.
- **c.** all interior concrete slabs-on-ground.
- **d.** under some interior concrete slabs-on-ground.

30. The strength of concrete in a residential concrete slab-on-ground is generally
- **a.** 1,000 psi.
- **b.** 2,000 psi.
- **c.** 3,000 psi.
- **d.** 4,000 psi.
- **e.** 5,000 psi.

31. An isolated concrete slab-on-ground is generally
- **a.** used on expansive soils.
- **b.** provided with perimeter beams.
- **c.** provided with perimeter and intermediate beams.
- **d.** unreinforced.
- **e.** none of the above.

32. A construction joint in a concrete slab-on-ground is required if the concrete in the slab cannot be placed in one continuous operation.
- **a.** True
- **b.** False

33. The tendons used for prestressing a PT concrete slab-on-ground generally consist of
- **a.** a single $\frac{1}{2}$-in.-diameter, nine-wire strand.
- **b.** a single $\frac{1}{2}$-in.-diameter, seven-wire strand.
- **c.** two-strand tendons, each strand made of seven wires.
- **d.** three-strand tendons, each strand made of seven wires.
- **e.** none of the above.

34. In a PT concrete slab-on-ground, the tendons are stressed from the
- **a.** dead ends.
- **b.** live ends.
- **c.** upper ends.
- **d.** lower ends.
- **e.** any one of the above.

35. Pocket formers in a PT concrete slab-on-ground are used at
- **a.** dead ends.
- **b.** live ends.
- **c.** upper ends.
- **d.** lower ends.
- **e.** any one of the above.

36. A PT concrete slab-on-ground must have perimeter and intermediate ribs.
- **a.** True
- **b.** False

REVIEW QUESTIONS

1. Describe the differences between pretensioning and posttensioning of concrete. Is there any difference in the type of steel used in each case?
2. Describe the difference between (a) reinforcing steel and prestressing steel, (b) formwork and shores, (c) formwork and flying forms, and (d) shoring and reshoring.
3. Concrete reinforcing bars come in limited lengths. Explain how we ensure their continuity in a long structural element, such as a column in a high-rise building or a long, continuous beam.
4. Explain what concrete cover is and the functions it serves.
5. What is the function of stirrups? Sketch commonly used open-loop and closed-loop stirrups.
6. Using sketches and notes, explain the typical layout of reinforcement in a simply supported, reinforced-concrete beam.
7. List various types of reinforced-concrete slabs as described in this text, and the essential differences between them.
8. Sketch a typical (a) control joint in a slab-on-ground, (b) construction joint in a slab-on-ground, (c) isolation joint in a slab-on-ground, and (d) prestressing strand.
9. Sketch in section a typical ribbed concrete slab-on-ground.

CHAPTER 23

Concrete Construction–II (Site-Cast and Precast Concrete Framing Systems)

CHAPTER OUTLINE

This chapter is a continuation of Chapter 22. It begins with a discussion of commonly used types of elevated site-cast concrete slabs. (Slabs-on-ground were covered in Chapter 22.)

This is followed by coverage of precast, prestressed concrete elements such as hollow-core slabs, solid planks, double-tee units, and inverted-tee beams. Construction systems related to these elements are also discussed.

NOTE

Preliminary Thickness of One-Way and Two-Way Slabs

Estimate the thickness of a one-way slab by dividing the slab span by 24. Thus, if the slab span is 12 ft, use a slab approximately 6 in. thick.

Two-way slab thickness is span/36. Use the longer of the two spans. Thus, if a slab panel measures 16 ft × 21 ft, the slab thickness is approximately 21/36 = 7/12 ft, or a 7-in.-thick slab.

Also see Appendix B, and consult a structural engineer for exact sizes.

23.1 TYPES OF ELEVATED CONCRETE FLOOR SYSTEMS

Site-cast reinforced-concrete framing systems consist of horizontal elements (elevated floor/roof slabs and beams) and vertical elements (columns and walls). Approximately 80% to 95% of the cost of materials and formwork of a concrete structural frame is in the horizontal framing elements of the frame. Consequently, the choice of the elevated floor system is the most important item in a concrete structure.

Elevated concrete floor systems can be classified as (a) beam-supported floors and (b) beamless floors. They are further divided into several types, Figure 23.1.

23.2 BEAM-SUPPORTED CONCRETE FLOORS

A reinforced-concrete floor slab with beams on all four sides can either be a *one-way slab* or a *two-way slab*. They are also called *one-way solid slabs* or *two-way solid slabs* to distinguish them from one-way joist slabs or two-way joist slabs, which are not completely solid.

The transfer of loads from a solid slab to the four supporting beams is represented approximately by 45° lines originating from the slab corners, Figure 23.2(a). If the slab panel is a square, each supporting beam receives the same amount of load, Figure 23.2(b). If the slab panel is rectangular, one pair of beams carries a greater load than the other pair.

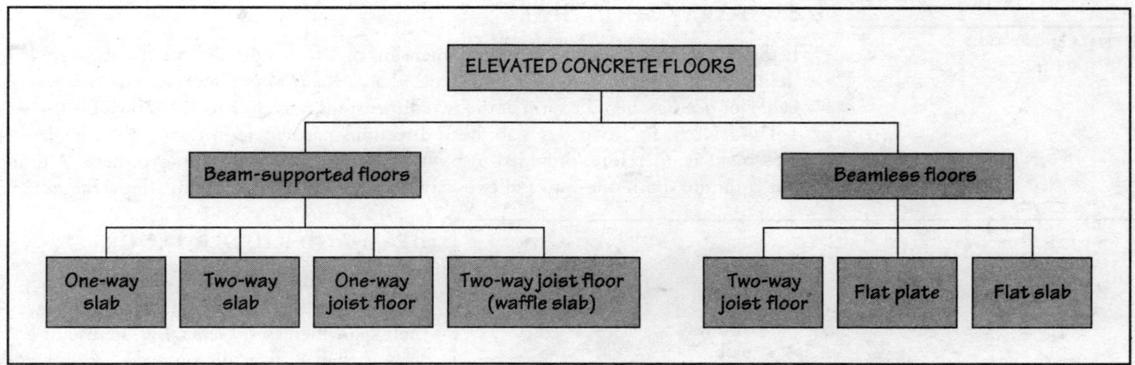

FIGURE 23.1 Types of elevated concrete floors.

ONE-WAY SOLID SLAB

If the ratio of the long dimension to the short dimension of a four-side-supported slab panel is greater than or equal to 2.0, most of the load on the slab is transferred to the long pair of beams, that is, the load path is along the short dimension of the slab panel, Figure 23.2(c). The load path along the long dimension of the slab is negligible.

Because the load is effectively transferred along one direction in Figure 23.2(c), the slab behaves as a one-way slab. The reinforcement in a one-way slab is placed along the short direction, referred to as the *primary reinforcement* to distinguish it from the nominal reinforcement placed along the perpendicular direction, called the *secondary reinforcement.* The purpose of secondary reinforcement is to resist stresses caused by concrete shrinkage and thermal expansion and contraction of the slab.

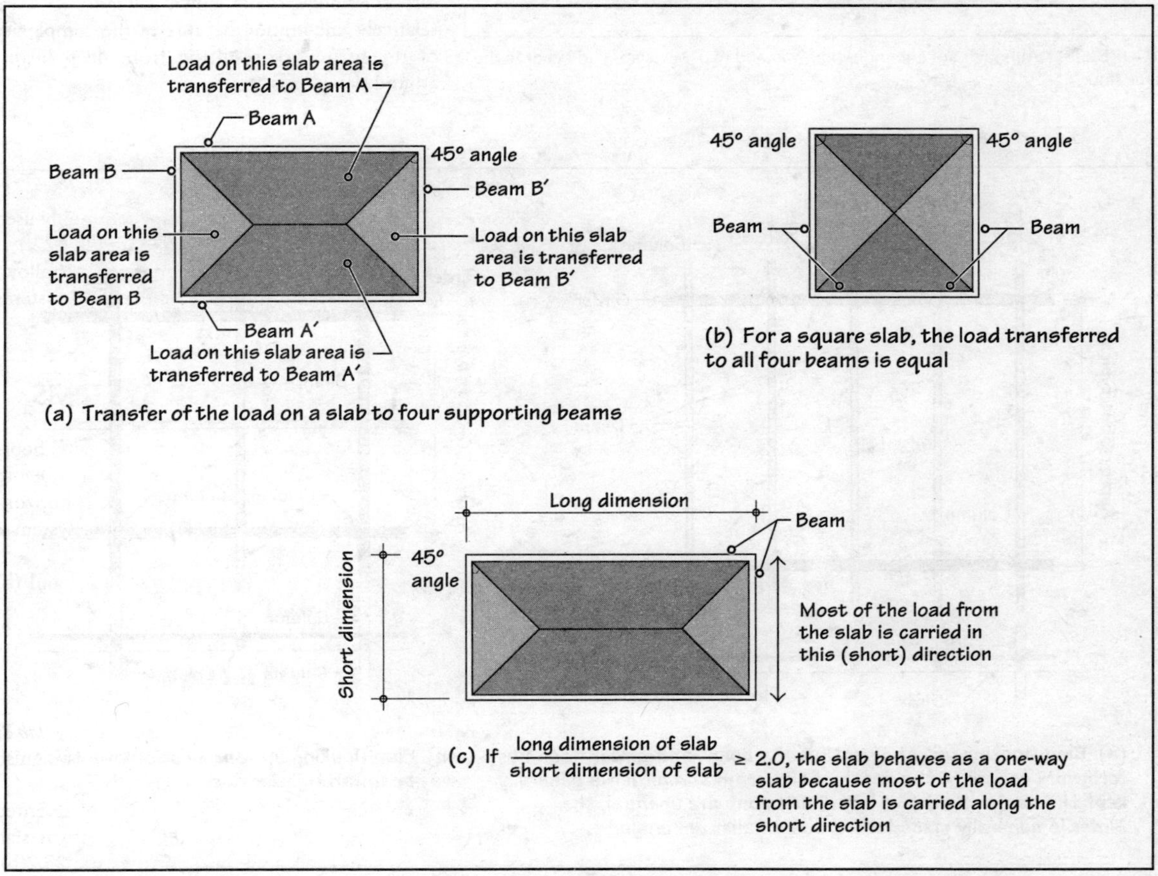

FIGURE 23.2 Distinction between one-way and two-way concrete slabs supported by beams on all four sides.

TWO-WAY SOLID SLAB

If the ratio of the long to the short dimension of a four-side-supported slab panel is less than 2.0, the slab is considered to behave as a *two-way slab.* However, real two-way slab behavior occurs when the ratio of the two dimensions is as close to 1.0 as possible (between 1.0 and 1.25). In a two-way slab, both directions participate in carrying the load. Reinforcement is, therefore, provided in both directions as primary reinforcement. Although not common, both one-way and two-way slabs may occur in the same floor, Figure 23.3.

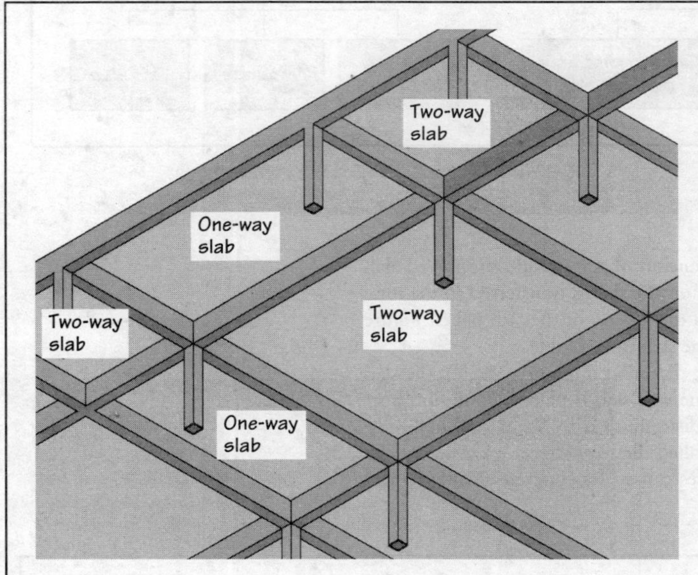

FIGURE 23.3 Although not common, one-way and two-way slabs can occur in the same floor.

BEAM-AND-GIRDER FLOORS

One-way and two-way solid slabs become increasingly thick and hence uneconomical as their span increases. Generally, the use of a slab thicker than 8 in. is discouraged because it creates a large dead load on the floor. For a one-way slab, an 8-in. slab thickness is reached with a span of approximately 16 ft. For a square two-way slab, a span of approximately 24 ft requires an 8-in.-thick slab.

Because 16-ft and 24-ft dimensions are relatively small for column spacing, one-way and two-way slabs are generally used in a *beam-and-girder floor,* Figure 23.4(a), or in a *two-way beam-and-girder floor,* Figure 23.4(b).

BAND BEAM FLOOR

A concrete floor that cannot be constructed with a flat form deck becomes uneconomical. Therefore, the floor systems shown in Figure 23.4 are relatively uncommon because of the complexity of the formwork resulting from deep beams around slab panels.

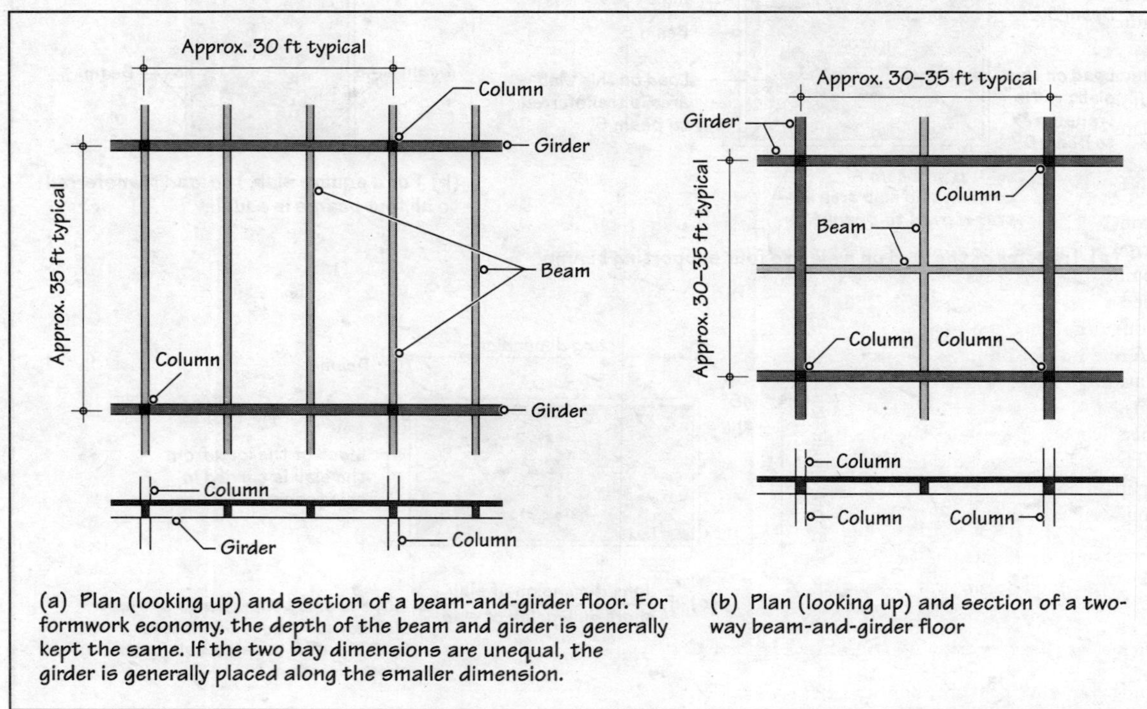

FIGURE 23.4 Plan and section through beam-and-girder floors.

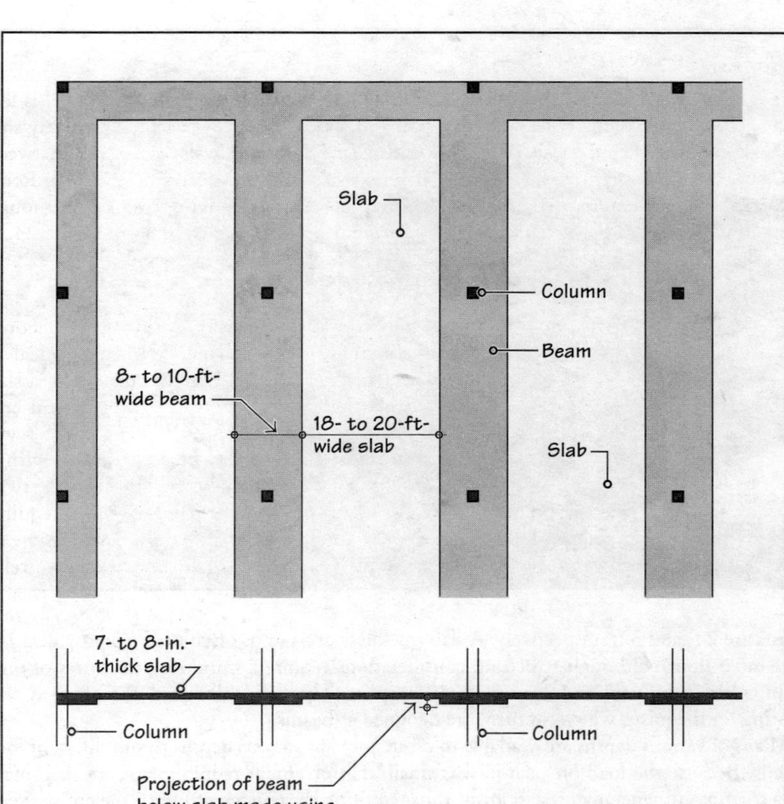

FIGURE 23.5 Plan (looking up) and section through a typical banded slab, which yields a more economical formwork than the beam-and-girder system in Figure 23.4. The projection of the beam below the bottom of the slab is generally obtained by using dimension lumber, such as 2 × 8, 2 × 10, and so on. This illustration also shows typical spans and member thickness (also see Appendix B).

A one-way slab floor with wide and shallow, continuous beams, referred to as *band beams* (in contrast with the conventional *narrow beams*), gives more economical formwork, Figure 23.5. Because the beams are wide, the slab span is reduced, reducing the slab thickness. Additionally, because the beams are shallow, the floor-to-floor height is smaller, reducing the height of columns, interior partitions, and exterior cladding. A smaller floor-to-floor height also reduces the overall height of the building, which reduces the magnitude of lateral loads on the building.

One-Way Joist Floor

A concrete floor that results from extremely economical formwork consists of closely spaced, narrow ribs in one direction supported on beams in the other direction, Figure 23.6. Because the ribs are narrow and closely spaced, the floor resembles a wood joist floor. It is, therefore, called a *joist floor* or a *ribbed floor*, but it is more commonly known as a *one-way joist floor* to distinguish it from the *two-way joist floor* described later.

A one-way joist floor is constructed with U-shaped pans as formwork placed over a flatform deck. The gap between the pans represents the width of the joists, which can be adjusted by placing the pans closer together or farther apart, Figure 23.7. The pans are generally made of steel or glass fiber–reinforced plastic (GFRP) and can be used repeatedly.

The vertical section through a pan tapers downward for easy stripping and has supporting lips at both ends. Pan widths and heights have been standardized to give two categories of one-way joist floors:

- Standard-module one-way joist floor
- Wide-module one-way joist floor

Standard-Module One-Way Joist Floor

Standard-module pans are 20 in. and 30 in. wide, Figure 23.8. These dimensions have been standardized so that, with 4-in.- and 6-in.-wide joists, the center-to-center spacings between

FIGURE 23.6 A one-way joist floor. Observe that the bottom of the beams and joists is at the same level to allow the entire slab to be formed on a flat-form deck.

joists are 2 ft and 3 ft, respectively. A slab thickness of $3\frac{1}{2}$ in. is often used (to provide a 1-h fire-rated floor), although structural considerations require a minimum thickness of only 2 in. with 20-in. pans and $2\frac{1}{2}$ in. with 30-in. pans. The slab is designed as a one-way slab resting on the joists, which, in turn, are designed as beams.

Pans of various depths are available to create joists of various depths to suit different joist spans. Because the load on each joist is small, reinforcement requirements are also small. No stirrups are generally used in joists, thus requiring the concrete to resist the entire shear. The increase in shear near-joist supports is resisted, if needed, by increasing the joist width, Figure 23.9. Special pans with closed ends on one side are available to produce the widening of joists.

A 4- to 5-in.-wide load-distribution rib is often provided in the center of joists.

WIDE-MODULE ONE-WAY JOIST FLOOR

Wide-module pans are available in 53-in. and 66-in. widths, Figure 23.10. They are generally used with 5-ft and 6-ft center-to-center joist spacings, giving joist widths of 7 in. and 6 in., respectively. However, almost any joist width can be created to suit the load and span conditions. Joist widths of 8 and 9 in. are also common.

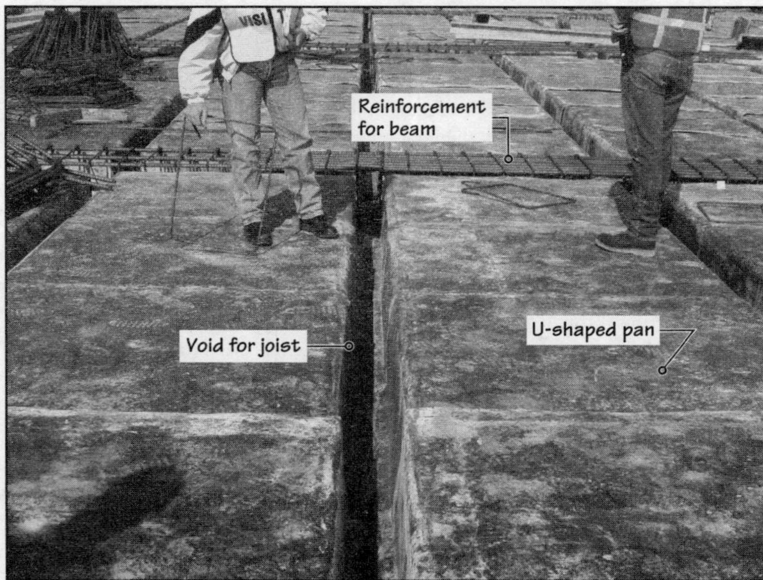

FIGURE 23.7 Formwork for a one-way joist floor showing U-shaped pans. The space between the pans represents the width of the joists. Notice that reinforcement for beams has already been laid. Reinforcement for joists and the slab is yet to be completed (also see Figure 23.11).

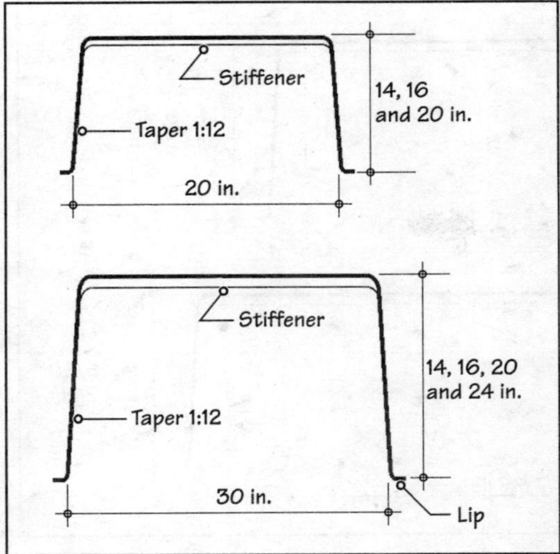

FIGURE 23.8 U-shaped standard-module pans (consult the manufacturer for available sizes). Note that the pans have open ends on both sides, except the pans used adjacent to beam or the distribution rib, which have closed ends (see Figure 23.9).

Because a wide-module floor has a larger spacing between joists, the slab thickness required is also larger. Therefore, a wide-module floor is generally used where a minimum fire rating of 1 h (required slab thickness $= 3\frac{1}{2}$ in.) or 2 h (required slab thickness $= 4\frac{1}{2}$ in.) is needed.

Figure 23.11 shows the overall view of a wide-module one-way joist floor with rebars in place. As shown in this figure, the joists in a wide-module floor are generally designed with shear reinforcement (stirrups). Therefore, the widening of joist ends at beams is unnecessary. Special end caps are available to close the open ends of pans near the beams, Figure 23.12.

NOTE

Preliminary Depth of a One-Way Joist Floor

Estimate the depth of a one-way joist floor by dividing the joist span by 18. Thus, if the span of joists is 36 ft, use 24-in.-deep joists, which includes the thickness of the slab. Using a $4\frac{1}{2}$-in.-thick slab and 20-in.-deep pans for a wide-module floor gives a floor depth of $24\frac{1}{2}$ in.

Where the spacing of columns in either direction is unequal, the supporting beams should preferably be oriented along the shorter direction and the joists in the longer direction.

Also see Appendix B, and consult a structural engineer for exact sizes.

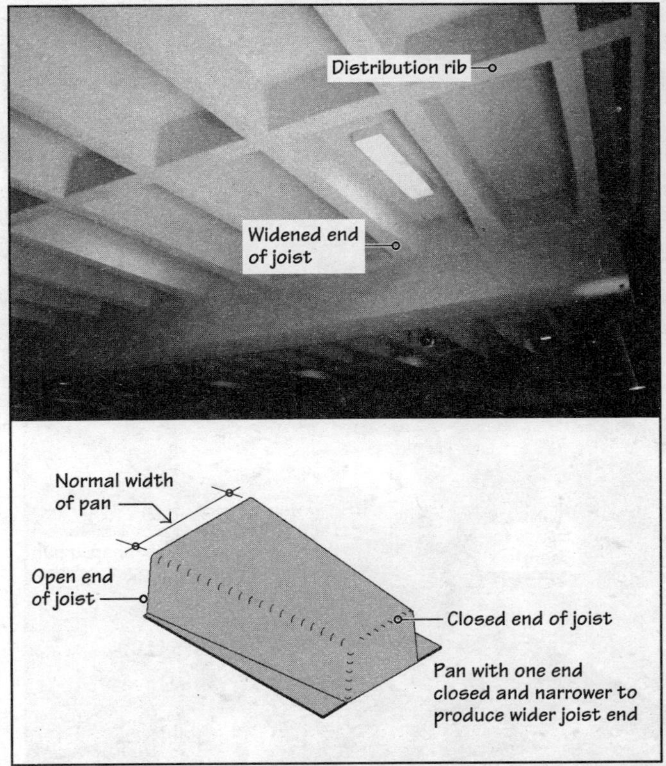

FIGURE 23.9 A one-way joist floor with standard pans. Observe that the joist width has been increased near the beams to accommodate greater shear in the joists at that location. The beams in this slab are deeper than the joists. This is unusual because it increases the cost of formwork compared with the slab with the same depth of joists and beams, as shown in Figure 23.6. A distribution rib has been provided in the center of the slab.

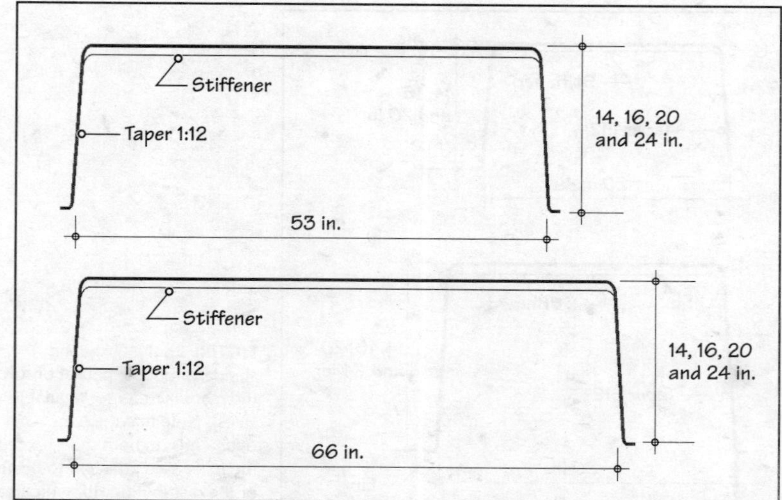

FIGURE 23.10 Wide-module pan dimensions. Consult manufacturers for available sizes.

FIGURE 23.11 An overall view of a wide-module one-way joist floor formwork with rebars in place—ready to receive concrete. Observe the use of shear reinforcement (stirrups) in the joists.

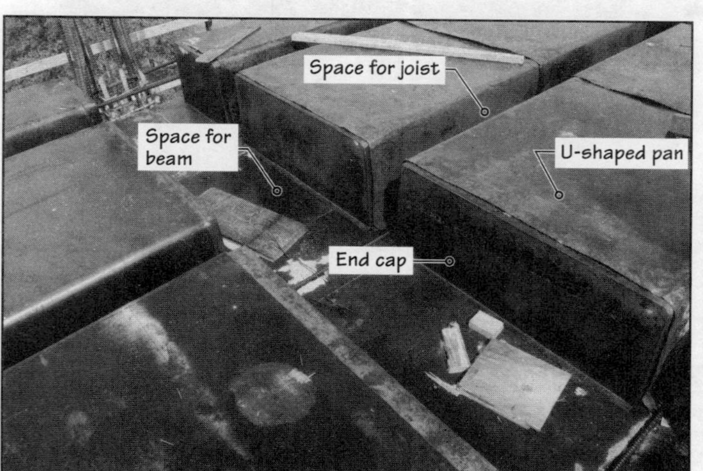

FIGURE 23.12 Wide-module pans and end caps. Because both ends of the pans are open, end caps are needed to close the gap to form a void for the beam.

Two-Way Joist Floor (Waffle Slab)

A two-way joist floor, also called a *waffle slab,* consists of joists in both directions, Figure 23.13. For the same depth of joists, a waffle slab yields a stiffer floor than a one-way joist floor. It is, therefore, used where the column-to-column spacing lies between 35 and 50 ft. A waffle slab is best suited for square or almost square column-to-column bays. When left exposed to the floor below, the waffle slab provides a highly articulated ceiling.

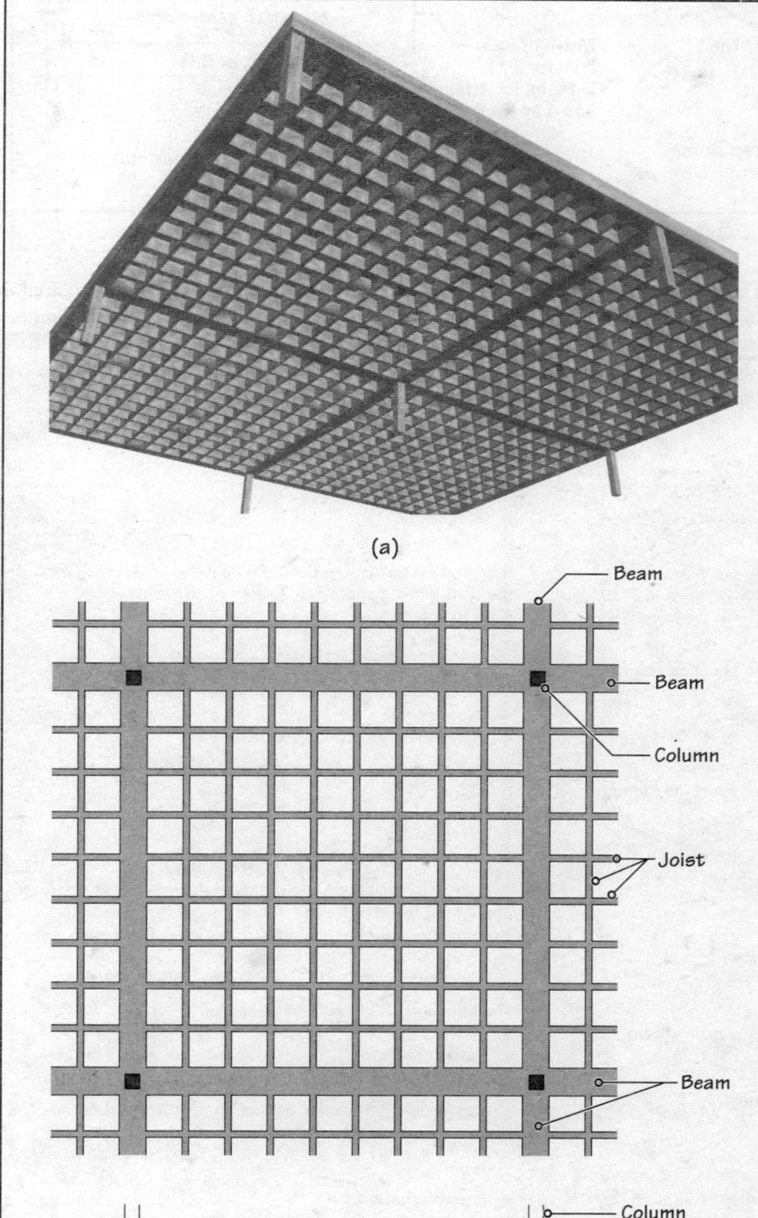

FIGURE 23.13 Two-way joist floor (waffle slab) supported on beams on all sides. **(a)** Isometric from below. **(b)** Plan (looking up) and section through the slab. Note that a waffle slab can also be constructed without beams, as shown in Figure 23.15.

NOTE

Preliminary Depth of a Two-Way Joist Floor

Estimate the depth of a waffle slab by dividing the longer span by 22. Thus, if the two-way joist floor span is 44 ft, use a 2-ft (i.e., 24-in.)-deep floor that includes the thickness of the slab. Using a $4\frac{1}{2}$-in.-thick slab for a 2-h rated floor and 20-in.-deep domes gives a floor depth of $24\frac{1}{2}$ in.

Also see Appendix B, and consult a structural engineer for exact sizes.

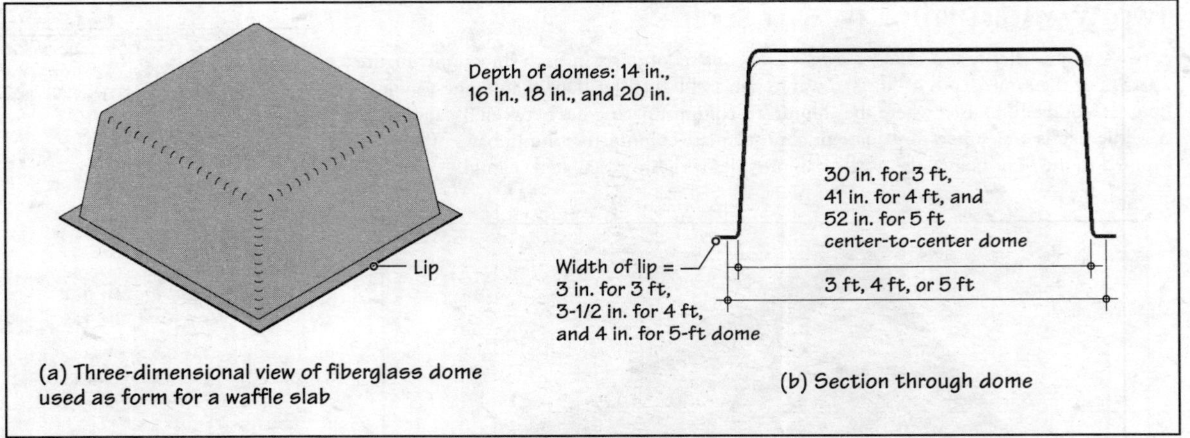

FIGURE 23.14 Standard GFRP dome sizes for a waffle slab. Consult manufacturers for available sizes.

The formwork for the slab consists of glass fiber–reinforced plastic domes placed on a flat form deck with the lips butting each other, Figure 23.14. Like the pans, the domes can withstand repeated use. Dome dimensions have been standardized to produce 3-ft, 4-ft, and 5-ft center-to-center distances between domes in a variety of depths. The domes have a wide supporting lip on all sides and are laid on a flat-form deck.

PRACTICE **QUIZ**

Each question has only one correct answer. Select the choice that best answers the question.

1. A reinforced concrete, four-side-supported slab panel is
   - **a.** a one-way slab.
   - **b.** a two-way slab.
   - **c.** a four-way slab.
   - **d.** (a) or (b).
   - **e.** (b) or (c).
2. A reinforced-concrete, four-side-supported slab panel measuring 15 ft × 22 ft is
   - **a.** a one-way slab.
   - **b.** a two-way slab.
   - **c.** a four-way slab.
   - **d.** (a) or (b).
   - **e.** none of the above.
3. A reinforced-concrete, four-side-supported slab panel measuring 15 ft × 32 ft is
   - **a.** a one-way slab.
   - **b.** a two-way slab.
   - **c.** a four-way slab.
   - **d.** (a) or (b).
   - **e.** none of the above.
4. In a band beam reinforced-concrete floor, the beams are
   - **a.** narrow and deep.
   - **b.** wide and shallow.
   - **c.** located at a spacing of 8 to 10 ft on center.
   - **d.** unnecessary.
5. A band beam floor is generally more economical than a beam-and-girder floor because
   - **a.** it requires a smaller quantity of concrete.
   - **b.** it requires a smaller quantity of reinforcement.
   - **c.** its formwork is more economical.
   - **d.** none of the above.
6. A standard-module one-way joist floor is constructed using pans that are
   - **a.** 20 in. wide.
   - **b.** 30 in. wide.
   - **c.** 40 in. wide.
   - **d.** (a) or (b).
   - **e.** (b) or (c).
7. Based on structural considerations only, the thickness of the slab in a standard-module one-way joist floor need not exceed
   - **a.** 2 in. to $2\frac{1}{2}$ in.
   - **b.** 3 in. to $3\frac{1}{2}$ in.
   - **c.** 4 in. to 5 in.
   - **d.** 5 in. to 6 in.
8. Pans used as formwork for one-way joist floors have a
   - **a.** U-shaped profile.
   - **b.** Z-shaped profile.
   - **c.** dome profile.
   - **d.** any one of the above.
   - **e.** none of the above.
9. A wide-module one-way joist floor is constructed using pans that are
   - **a.** 66 in. wide.
   - **b.** 60 in. wide.
   - **c.** 53 in. wide.
   - **d.** (a) or (b).
   - **e.** (a) or (c).
10. In a one-way joist floor, the supporting beams run in one direction and the joists run in the other direction. If the spans in the two directions are unequal, the beams should preferably span along the
    - **a.** shorter direction.
    - **b.** longer direction.
11. The formwork used for a waffle slab has a
    - **a.** U-shaped profile.
    - **b.** Z-shaped profile.
    - **c.** dome profile.
    - **d.** any one of the above.
    - **e.** none of the above.
12. Using standard formwork components for a waffle slab, the center-to-center distance between waffles (voids) is
    - **a.** 3 ft, 4 ft, or 5 ft.
    - **b.** 2 ft, 4 ft, or 6 ft.
    - **c.** 4 ft, 5 ft, or 6 ft.
    - **d.** 5 ft, 6 ft, and 7 ft.
    - **e.** none of the above.

23.3 BEAMLESS CONCRETE FLOORS

A waffle slab is more commonly constructed as a beamless slab, Figure 23.15. In a beamless waffle slab, a few domes on all sides of a column are omitted so that the thickness of the slab at the columns is the same as the depth of the joists. The thickening of the slab at the columns provides shear resistance (against the slab punching through the columns).

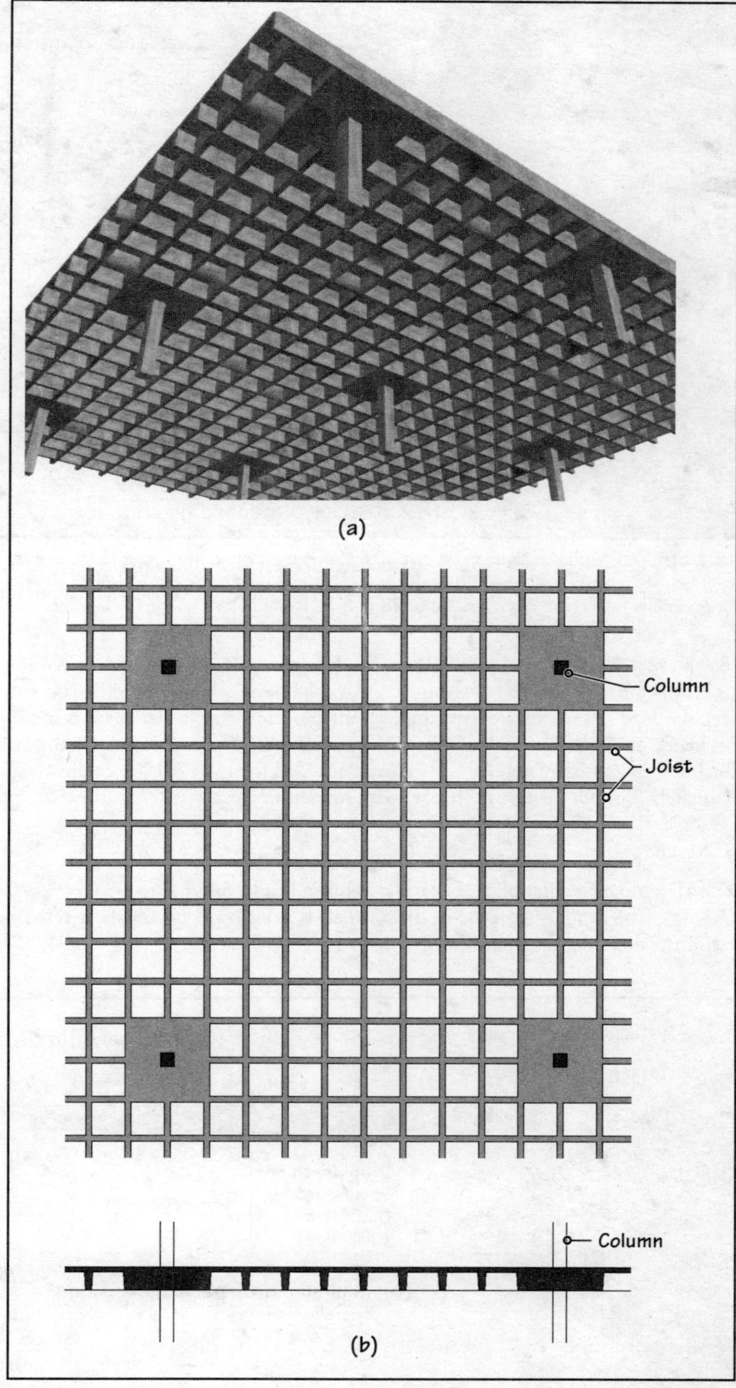

FIGURE 23.15 A beamless waffle slab, i.e., a waffle slab supported on column heads (capitals). Note the similarity with a flat slab of Figures 23.17. **(a)** Isometric from below. **(b)** Plan (looking up) and section through the slab. Note that a waffle slab can also be constructed with beams, as shown in Figure 23.13. A beamless waffle slab may be provided with spandrel beams, if needed.

Flat Plate

A flat plate consists of a solid slab supported directly on columns, Figure 23.16. A flat plate is similar to a two-way banded slab, except that the beam bands in both directions are concealed within the thickness of the slab. Therefore, the spans that can be achieved economically with a flat-plate floor are smaller than those obtained from one-way or two-way joist floors.

Flat-plate slabs are suitable for occupancies with relatively light live loads, such as hotels, apartments, and hospitals, where small column-to-column spacing does not pose a major design constraint. Additionally, a drop ceiling is not required in these occupancies and HVAC ducts can be run within the corridors, where a lower ceiling height is acceptable.

FIGURE 23.16 A flat-plate slab in an office building under construction. Observe the use of perimeter beams, which may be required with heavy exterior wall cladding. Flat-plate slabs without perimeter beams are common. Columns may be rectangular, square, or round.

NOTE

Preliminary Depths of Flat-Plate and Flat-Slab Floors

Estimate the depth of a flat-plate floor by dividing the longer clear distance between columns by 30. Thus, if the distance between columns is 20 ft, use an 8-in.-thick slab.

For a flat-slab floor, estimate the thickness of the slab by dividing the longer clear distance between columns by 35. Thus, if the distance between columns is 24 ft, use an 8-in.-thick slab.

Also see Appendix B, and consult a structural engineer for exact sizes.

A flat-plate slab results in a low floor-to-floor height, and its formwork is economical. Because the beams are concealed within the slab thickness, columns need not be arranged on a regular grid—a major architectural advantage. However, a flat plate is a two-way system; hence, the column spacing in both directions should be approximately the same. A slab thickness of approximately 6 in. is generally needed for 15-ft × 15-ft column bays and approximately 8 in. for 20-ft × 20-ft bays with residential loads.

FLAT SLAB

A *flat slab* is similar to a flat plate, but it has column heads, referred to as *drop panels,* Figure 23.17(a). The primary purpose of drop panels is to provide greater shear resistance at the columns, where the shear maximizes.

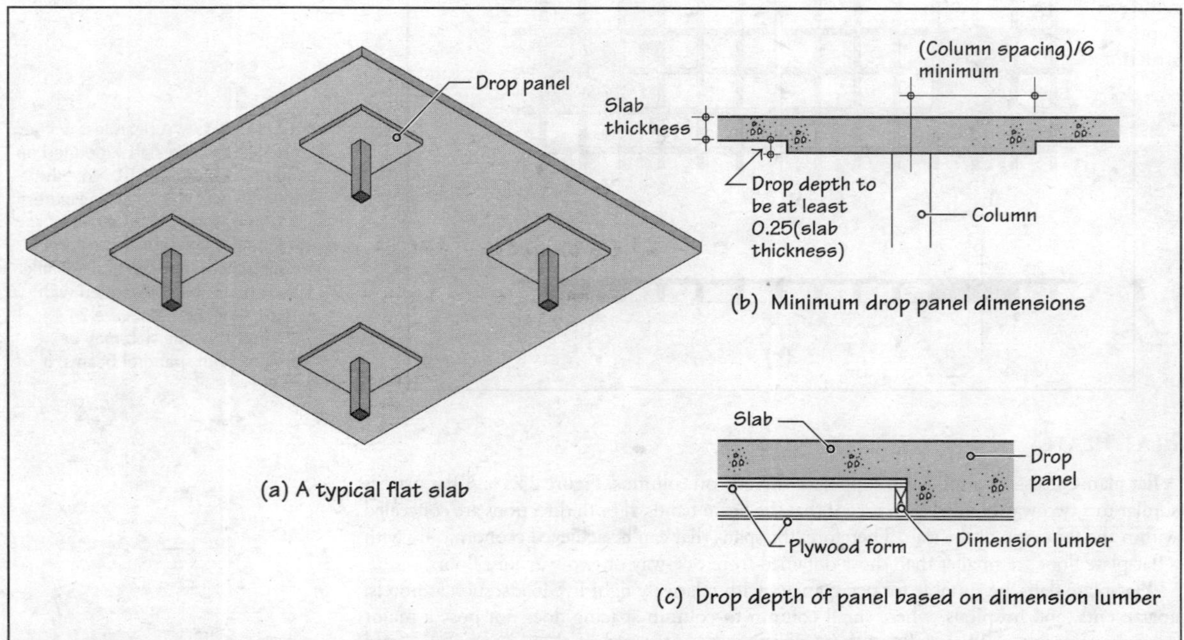

FIGURE 23.17 Typical details of a flat slab and minimum code requirements for drop panel dimensions.

Structurally, the drop panel must extend a minimum of one-sixth of the slab span in each direction, and its drop below the slab must at least be 25% of the slab thickness, Figure 23.17(b). For formwork economy, the drop depth is also based on lumber dimensions, Figure 23.17(c). With round columns, however, manufacturers supply column forms that have built-in drop panels and column capitals, Figure 23.18.

A flat slab is generally used where the live loads are relatively high, such as in parking garages or storage or industrial facilities.

FIGURE 23.18 A flat slab with (round) drop panels and column capitals under construction.

23.4 POSTTENSIONED ELEVATED CONCRETE FLOORS

The reinforced-concrete elevated-slab systems described previously can also be posttensioned. The posttensioning of slabs reduces slab and beam dimensions, reducing the dead load of the floor. This is particularly helpful in seismic areas, where a lower dead load imparts a lower seismic load on the structure. Smaller slab and beam dimensions also result in a lower floor-to-floor height, reducing the cost of exterior cladding. Column and foundation sizes are also reduced.

Because building codes place a limit on the maximum amount of prestress that can be introduced, an elevated, posttensioned concrete slab has a significant amount of conventional reinforcement as well, Figure 23.19. However, the amount of conventional reinforcement in a posttensioned slab is lower than that required if the slab were not posttensioned, reducing reinforcement congestion in long-span and/or heavily loaded members.

Generally, the prestressing (posttensioning) tendons in a slab are distributed in a *banded arrangement* rather than being individually distributed throughout the span in a *basket-weave arrangement*. Generally, the tendons are grouped together along the support lines in one direction, and bands of tendons are placed at equal spacings in the transverse direction, Figure 23.20.

Posttensioned (PT) concrete floors are being increasingly favored for all types of structures—parking garages, high-rise office buildings, hospitals, and condominiums. Figure 23.21 shows a one-way joist floor with PT supporting beams. Posttensioning reduces the depth of beams, which may be necessary in some situations to equalize the depth of the beams with those of the joists. The process of posttensioning in an elevated structure is similar to that of a PT slab-on-ground, covered in Chapter 22.

NOTE

Two Most Commonly Used Concrete Floor Systems

Although various site-cast concrete floor systems have been described in this text, the two most economical floor systems are the *flat-plate floor* and the *one-way joist floor*. For typical live loads of 50 psf, the one-way flat-plate floor is economical for spans of up to approximately 30 ft. For spans between 30 ft and 50 ft, a one-way joist floor is economical [23.1].

FIGURE 23.19 A cage for a concrete beam consisting of conventional reinforcement and posttensioning tendons (red color) ready to be lowered into the beam form. Observe the drape (generally parabolic) provided in posttensioning tendons to balance the stresses created by gravity loads on the beam. The steel frames shown are used to make the cage. They will be removed after the cage has been placed in the form. Reinforcement for the slab is yet to be completed. Figure 23.20 shows the same beam after its reinforcement has been lowered and the reinforcement and posttensioning tendons for the slab are in place.

FIGURE 23.20 Layout of reinforcement and posttensioning tendons (red color) in the slab of a building. The reinforcement and posttensioning tendons for the beam are shown in Figure 23.19. Note the banded arrangement of tendons in the slab.

FIGURE 23.21 A wide-module, one-way joist floor with posttensioned (PT) supporting beams. Note the use of (conventional) reinforcement in the slabs and joists. The beams have both conventional reinforcement and posttensioning tendons (red color). PT beams are commonly used in a one-way joist floor to reduce beam size so that joists and beams have the same depth.

23.5 INTRODUCTION TO PRECAST CONCRETE

As stated in the introduction to Chapter 22, precast concrete members (except concrete tilt-up walls) are fabricated in a precast plant and transported to the construction site for assembly. Precasting, which is generally done in covered or sheltered spaces, is particularly helpful in climates that limit the use of site-cast concrete.

Additionally, because precasting is done at the ground level, the cost of formwork and shoring is considerably reduced. Formwork cost reduction is also achieved through the use of standard-size elements cast in permanent forms, which are reused several times more than the formwork used for site-cast members. Precasting also allows greater quality control over the strength of concrete and surface finishes. Most surface finishes are more easily obtained in a precast plant than at the site—often several floors above ground.

Precast elements used as horizontal framing members are generally prestressed (i.e., pretensioned). To ensure rapid reuse of forms and prestressing equipment, Type III (high-early-strength) concrete is generally used with a relatively higher-strength concrete (5,000 to 6,000 psi). To accelerate the setting and hardening of concrete, members are steam cured. Consequently, most precast concrete members are made on a 24-h cycle.

Precast concrete also has many disadvantages. Its main disadvantage is the cost of transportation. Although precast members are generally lighter than corresponding site-cast members (because of prestressing), they are still fairly heavy. Transportation also limits the length and width of precast members.

Another disadvantage of precasting is the need for heavier hoisting equipment at the construction site and additional safety measures that must be observed during erection. Erection and assembly at the site also introduce the need for a more skilled workforce compared with site-cast concrete construction. Architecturally, the most limiting factor in the use of precast concrete is the difficulty in sculpting concrete at a large scale, which is more easily realized with site-cast concrete. Precast elements are generally straight, with standard profiles.

ARCHITECTURAL PRECAST AND STRUCTURAL PRECAST CONCRETE

Precast concrete members are classified as (a) *architectural precast concrete* and (b) *structural precast concrete. Architectural precast* refers to concrete elements that are used as nonstructural cladding elements. Their most common use is in precast concrete curtain walls (see Chapter 28).

Structural precast concrete, which is covered in this chapter, includes all elements of a building's structural frame (floor/roof slabs, columns, and walls). Although an entire room or assembly of rooms can be precast, most structural precast concrete is used in standard elements that are assembled on site to form spaces.

STRUCTURAL PRECAST—TOTAL PRECAST AND MIXED PRECAST CONSTRUCTION

A building can be constructed of all precast concrete members in which all structural components—columns, load-bearing walls, and floor and roof slabs—are of precast concrete. This system is referred to as total *precast concrete construction.* In mixed precast construction, some elements of the building consist of precast concrete members, while the others are of cast-in-place concrete, steel, or masonry.

Mixed precast construction is far more common than total precast. It combines the benefits of both precast and conventional construction. In mixed precast, precast concrete is used only in floor and roof slabs. Using precast concrete floor/roof slabs yields significant savings because a large percentage of the cost of materials and formwork in a concrete structure is embedded in floor and roof slabs (see Section 23.1).

PRACTICE **QUIZ**

Each question has only one correct answer. Select the choice that best answers the question.

13. A flat slab floor consists of
a. a slab of constant thickness.
b. a slab with beams on all four sides of the slab panel.
c. a slab with drop panels at each column.
d. none of the above.

14. A flat plate floor consists of
a. a slab of constant thickness.
b. a slab with beams on all four sides of the slab panel.
c. a slab with drop panels at each column.
d. none of the above.

15. Commonly used beamless concrete floors are
a. one-way and two-way joist floors.
b. flat-plate and flat-slab floors.
c. beam-and-girder floors.
d. all of the above.

16. In a posttensioned elevated concrete floor, prestressing tendons are combined with conventional steel reinforcement.
a. True **b.** False

17. The primary reason for using posttensioned concrete floors is to
a. reduce or prevent the cracking of concrete.
b. improve safety against failure.
c. reduce floor depth.
d. reduce or prevent corrosion of reinforcement.
e. reduce formwork cost.

18. The strength of concrete used in precast concrete members is generally
a. 1 to 2 ksi. **b.** 2 to 3 ksi.
c. 3 to 4 ksi. **d.** 4 to 5 ksi.
e. 5 to 6 ksi.

19. Portland cement used in precast concrete members is generally
a. Type V. **b.** Type IV.
c. Type III. **d.** Type II.
e. Type I.

20. *Architectural precast concrete* refers to concrete members used in building interiors.
a. True **b.** False

NOTE

Approximate Spanning Capability of Hollow-Core Slabs

To determine the approximate thickness of a hollow-core slab floor, divide the span by 40. Thus, if the distance between supporting members is 25 ft, the thickness of a hollow-core slab is $(25 \times 12)/40 = 7.5$ in. Hence, use an 8-in. slab. A 12-in. slab has an approximate spanning capability of 40 ft.

Due to lighter loads on roofs, the slab thickness for roofs may be determined by dividing the span by 50.

Also see Appendix B, and consult a structural engineer for exact sizes.

23.6 STRUCTURAL PRECAST CONCRETE MEMBERS

Structural precast concrete members may be classified as

- Horizontal-spanning elements: These consist of (a) hollow-core slabs, (b) solid planks, (c) double-tee units, and (d) inverted-tee beams.
- Vertical-spanning members: These consist of (a) columns and (b) walls.

Horizontal-spanning elements are almost always prestressed, and vertical spanning elements are generally not prestressed. A brief description of horizontal-spanning members and their use in mixed precast construction follows.

HOLLOW-CORE SLABS AND SOLID PLANKS

Hollow-core slabs are precast, prestressed concrete slabs that contain voids in their central region. The voids reduce the dead load of the slab by 40% to 50% compared with a site-cast concrete slab for the same span.

Hollow-core slabs are produced in thicknesses of 4 in., 6 in., 8 in., 10 in., 12 in., and 14 in., with the 8-in.-thick slab being most commonly used. The width of the slabs is generally 4 ft. Narrow-width slabs are produced in the plant by cutting standard 4-ft slabs along their length to fit a space that is not a multiple of 4 ft. In addition to 4-ft-wide slabs, some plants make slabs that are 3 ft 4 in. and 8 ft wide.

Figure 23.22 shows a typical section through an 8-in.-thick, 4-ft-wide hollow-core slab. The number of voids and their cross-sectional profiles are manufacturer specific. Manufacturers provide the load capacity of their slabs, along with suggested details for connecting the slabs with supporting members.

Hollow-core slabs are generally used in mixed precast buildings, where the supporting structure may be of steel, masonry or site-cast concrete. Figure 23.23 shows the use of hollow-core slabs in a steel-frame building. Their use with a site-cast concrete frame is

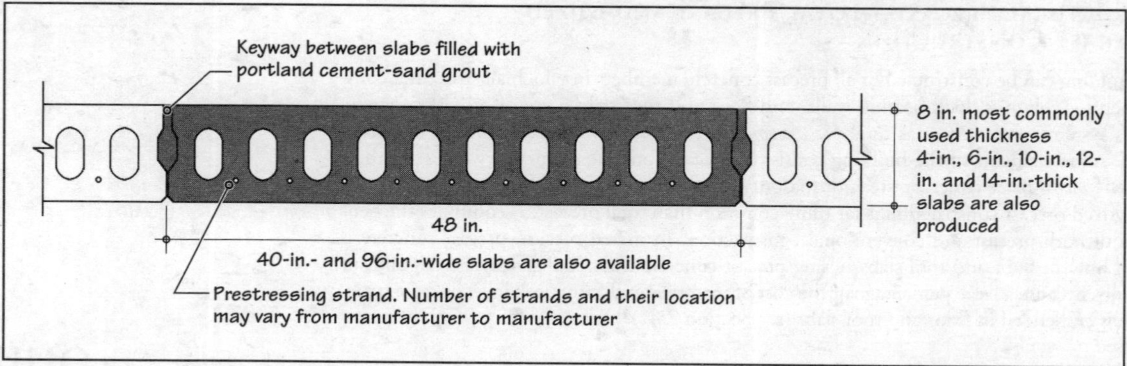

FIGURE 23.22 A typical section through a 4-ft-wide, 8-in.-thick hollow-core slab. The dimensions of voids, their number, and their profile are manufacturer specific.

FIGURE 23.23 Mixed precast concrete construction consisting of a steel-frame structure with hollow-core slabs.

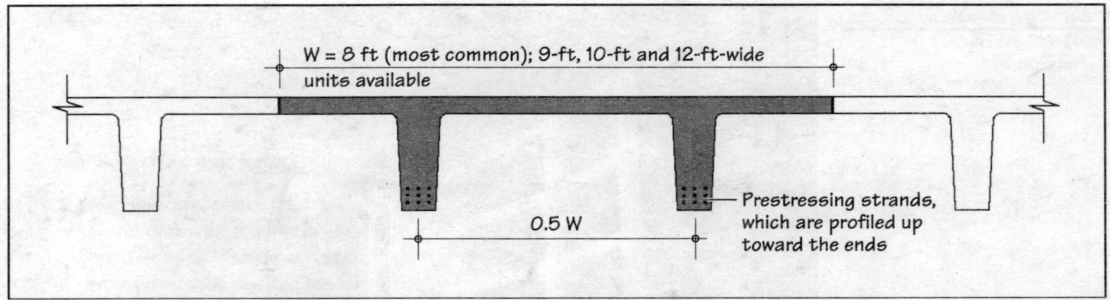

FIGURE 23.24 Typical section through a double-tee floor.

shown in Chapter 12 (Figure 12.34); Chapter 26 (Sections 26.5 and 26.6) shows their use in load-bearing masonry structures.

Connections between hollow-core slabs and supporting members are made using site-cast concrete fill and reinforcing steel (see Sections 26.5 and 26.6). In addition to the concrete fill used for connections, a site-cast concrete topping is generally used over the slabs. The topping provides structural integration of slab units and increases the floor's fire resistance and sound insulation. It also functions as a leveling bed, particularly with units with uneven camber. In many buildings, however, topping is omitted for the sake of economy. Topping, when used, is generally 2 in. to $2\frac{1}{2}$ in. thick and reinforced with welded wire reinforcement (WWR).

Hollow-core slabs are produced through a long-line extrusion process. The machine, housed in a precasting enclosure, lays concrete on a casting bed over prestressing tendons, which have already been pretensioned to the required level of stress. It forms continuous voids as it travels along the length of the bed. The concrete used is high-strength, zero-slump concrete.

After casting, the slab is steam cured and cut to customer-specified lengths. The cut lengths of slabs are removed from the bed the morning after casting, and the process just described is repeated on a 24-h basis.

Solid precast concrete planks are similar to hollow-core planks and are generally used where superimposed loads are relatively light.

DOUBLE-TEE UNITS

A typical section through a double-tee unit is shown in Figure 23.24. The width of units is generally 8 ft, with a center-to-center distance of 4 ft between beams. Double-tee units with widths of 9 ft, 10 ft, and 12 ft are also available from some manufacturers. Overall depths of double-tee units vary from 16 in. to 36 in. to give different spanning capabilities.

Double-tee units are used where the spans are large and cannot be provided economically with site-cast concrete construction or hollow-core slabs. They are commonly used for hotel and bank lobbies. As with hollow-core slabs, a topping of concrete and WWR may be used on double-tees for structural integration and leveling.

Another common use of double-tees is in multistory parking garages, where a minimum distance of 60 ft between columns is generally necessary. In garages, double-tees are generally supported on site-cast or precast inverted T-beams. Double-tee units are cast on a long line of steel formwork. Several units are made in one casting operation by dividing the bed with separators.

INVERTED-TEE BEAMS

Precast inverted-tee beams are generally used as supporting members for hollow-core slabs or double-tee units. Like hollow-core slabs and double-tee units, they are prestressed. They are commonly used in total precast construction (see Figure 23.28).

23.7 TOTAL PRECAST CONCRETE CONSTRUCTION

Figure 23.25 shows a total precast building under construction, whose structural frame comprises double-tee units, inverted-tee beams, columns, and walls. The construction process of a total precast building has much in common with structural steel construction; that is, members are brought into position using a crane, as seen in Figure 23.26, and are connected together either through welds or bolts. Steel embeds are included in precast members to facilitate member bolting and/or welding.

Figure 23.27(a) shows the top of a lower-floor column with threaded bolts projecting out, and Figure 23.27(b) shows the bottom of an upper-floor column with an embedded

NOTE

Approximate Spanning Capability of Double-Tee Units

To determine the approximate depth of a double-tee unit floor, divide the span by 28. Thus, if the span of units is 60 ft, the approximate depth of the floor is $(60 \times 12)/28 = 26$ in.

Standard available depths of double-tee units are generally 14 in., 18 in., 24 in., and 30 in. The manufacturer should be consulted for more precise information.

Also see Appendix B, and consult a structural engineer for exact sizes.

NOTE

Single-Tee Units

Single-tee units were used formerly (generally for long spans), but they are not commonly used today because of erection instability problems.

FIGURE 23.25 A total precast concrete parking garage under construction.

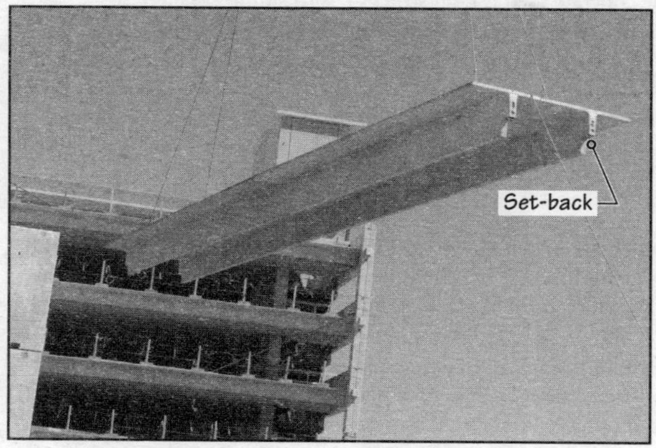

FIGURE 23.26 A double-tee floor unit being flown into position in the total precast building in Figure 23.25. The setbacks in double-tee stems reduces floor height.

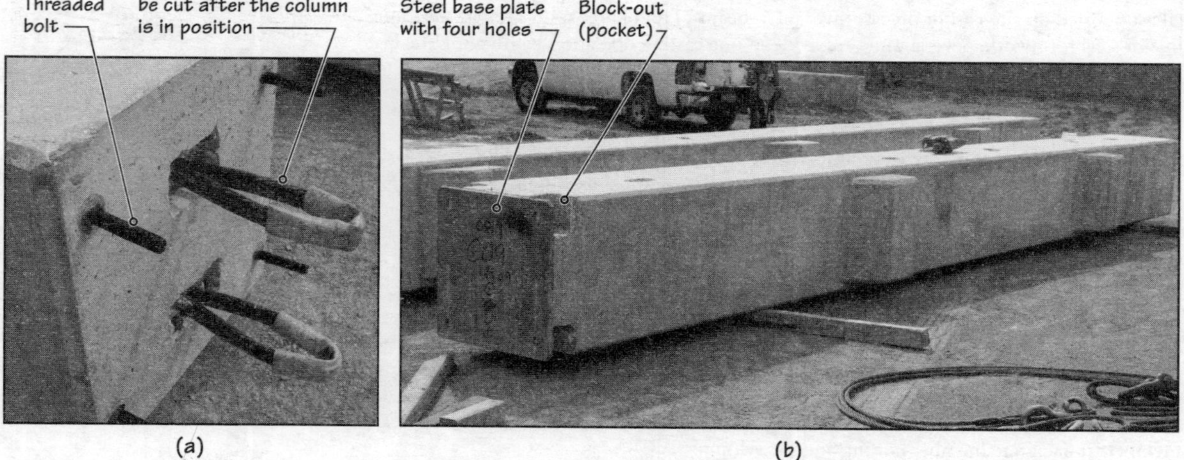

FIGURE 23.27 **(a)** Top of the lower-floor column with threaded bolts. **(b)** Bottom of the upper-floor column (two stories high). The block-outs (pockets) in the columns allow a bolted connection between the lower- and upper-floor columns.

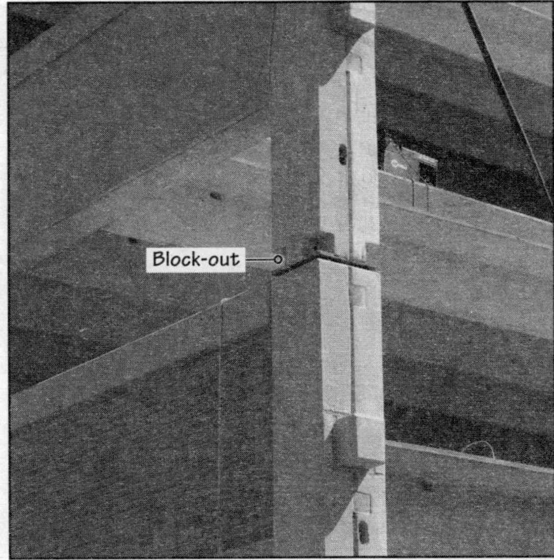

FIGURE 23.28 Close-up of the bolted connection between the lower- and upper-floor columns. The block-outs (pockets) will be filled with concrete and finished smooth to match the column surface.

FIGURE 23.29 Projections in precast concrete walls to support **(a)** inverted-tee beams and **(b)** double-tee floor slabs.

base plate containing holes to engage the bolts. The block-outs in the column above the holes are filled with concrete after the connection has been made. Figure 23.28 shows a close-up of the connection between the lower- and upper-floor columns.

Walls in a total precast building carry gravity loads and also function as shear walls. Where needed, the walls are provided with projecting brackets to receive inverted-tee beams, Figure 23.29(a) or double-tee units, Figure 23.29(b).

23.8 FIRE RESISTANCE OF CONCRETE MEMBERS

The fire resistance of a concrete member is a function of three variables:

- Thickness of the member
- Type of coarse aggregate
- Cover over reinforcement and prestressing tendons

TABLE 23.1 FIRE RESISTANCE OF CONCRETE MEMBERS

Member	Aggregate type	Fire resistance 1 h	2 h	3 h
Minimum thickness (in.) of floor slabs or walls	Lightweight	2.5	3.6	4.4
	Carbonate	3.2	4.6	5.7
	Siliceous	3.5	5.0	6.2
Minimum thickness (in.) of columns	Lightweight	8	9	10.5
	Carbonate	8	10	11
	Siliceous	8	10	12

A thicker member has greater fire resistance. For the same member thickness, a lightweight aggregate gives greater fire resistance than a normal-weight aggregate. Among the normal-weight aggregates, carbonate aggregate (i.e., limestone) gives greater fire resistance than noncarbonate (i.e., siliceous) aggregates.

Local building codes prescribe the minimum thickness of members to achieve a given fire resistance, Table 23.1. In addition to meeting minimum thickness requirements, the member must satisfy minimum cover requirements. The minimum cover required for corrosion protection (Table 19.1) is adequate to satisfy the fire-resistance requirements in most situations. However, the local building code must be referenced for precise information.

PRACTICE QUIZ

Each question has only one correct answer. Select the choice that best answers the question.

21. In mixed precast construction, precast concrete members are generally used for
- **a.** shear walls.
- **b.** load-bearing walls.
- **c.** columns.
- **d.** floors.
- **e.** all of the above.

22. In total precast construction, precast concrete members are generally used for
- **a.** shear walls.
- **b.** load-bearing walls.
- **c.** columns.
- **d.** floors.
- **e.** all of the above.

23. A 12-in.-thick hollow-core floor slab will generally span up to approximately
- **a.** 40 ft.
- **b.** 30 ft.
- **c.** 25 ft.
- **d.** 20 ft.
- **e.** none of the above.

24. When a concrete topping is used over hollow-core slabs, its thickness is approximately
- **a.** 1 in.
- **b.** $1\frac{1}{2}$ in.
- **c.** 2 in.
- **d.** 3 in.
- **e.** 4 in.

25. The purpose of concrete topping on hollow-core slabs is to
- **a.** structurally integrate the slabs.
- **b.** level the top of slabs.
- **c.** increase the sound insulation of the floor.
- **d.** increase the fire rating of the floor.
- **e.** all of the above.

26. The most commonly used width of a double-tee floor unit is
- **a.** 8 ft.
- **b.** 6 ft.
- **c.** 4 ft.
- **d.** none of the above.

27. Double-tee floors are commonly used
- **a.** in residential occupancies.
- **b.** where spans lie between 30 ft and 40 ft.
- **c.** where spans lie between 40 ft and 50 ft.
- **d.** where spans are in excess of 50 ft.

28. The fire-resistance rating of concrete members is a function of
- **a.** member thickness.
- **b.** type of coarse aggregate in concrete.
- **c.** concrete cover for corrosion protection.
- **d.** all of the above.

EXPAND YOUR KNOWLEDGE

Building a Separation Joint in a Site-Cast Concrete Building

As discussed in Chapter 9, continuous building separation joints are required in many buildings. A preferred method is to use two columns and two beams at the joint. Often for the sake of economy, however, a single column with two beams is used.

The following text explains the construction of such a joint in a site-cast, reinforced-concrete building. (See Chapter 19 for similar detail in a steel-frame building.)

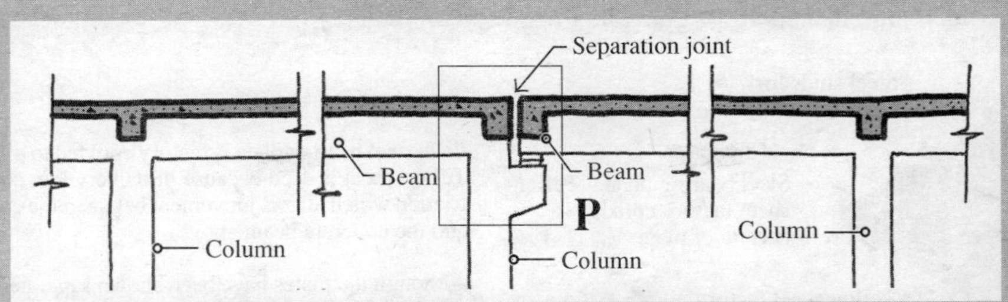

FIGURE 1 Section through a concrete floor at a building separation joint using one column at the joint (also see Chapter 9).

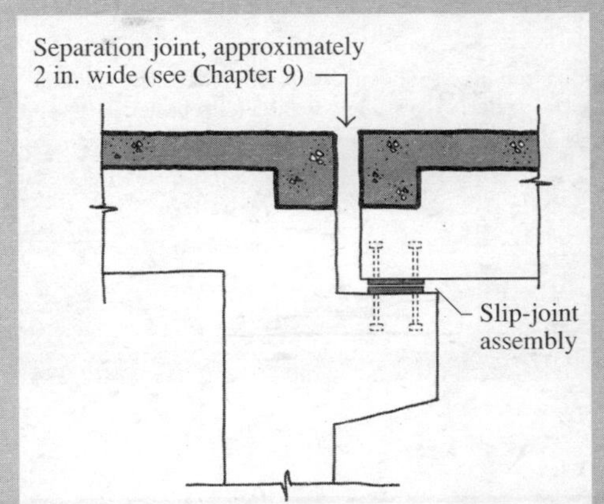

FIGURE 2 Detail P.

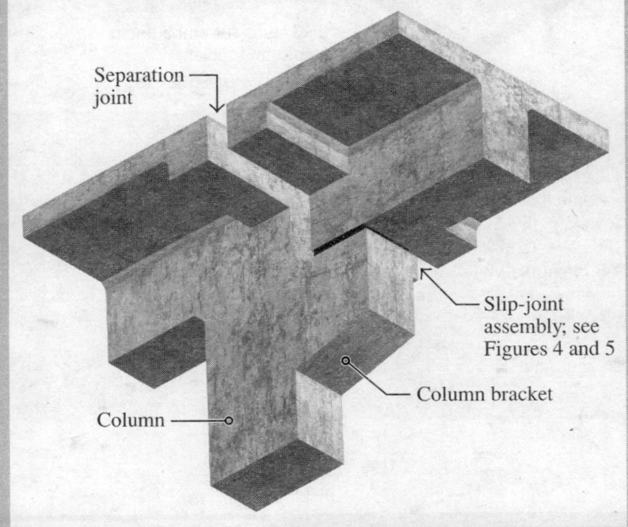

FIGURE 3 Three-dimensional view of Detail P.

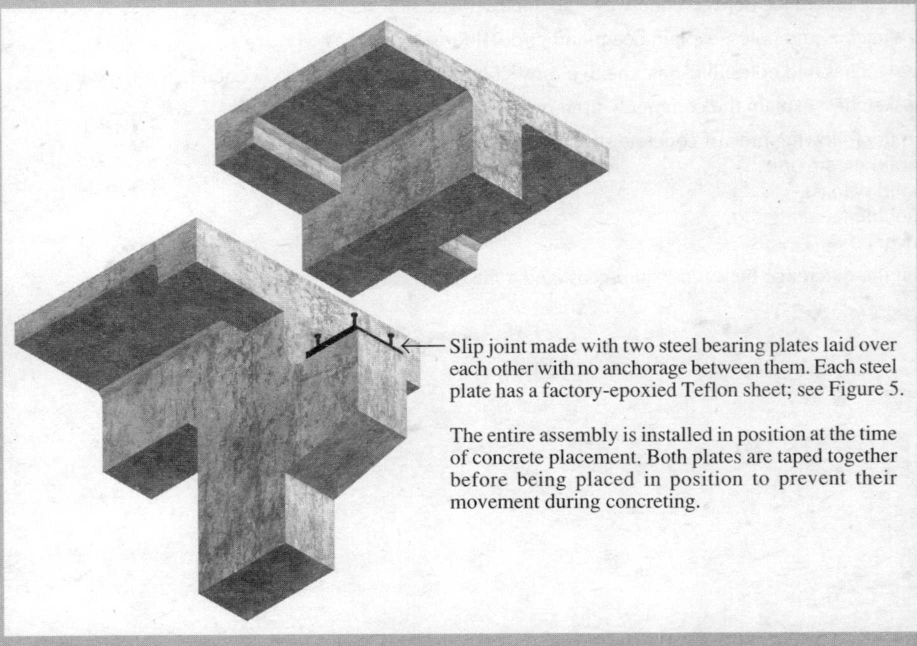

Slip joint made with two steel bearing plates laid over each other with no anchorage between them. Each steel plate has a factory-epoxied Teflon sheet; see Figure 5.
The entire assembly is installed in position at the time of concrete placement. Both plates are taped together before being placed in position to prevent their movement during concreting.

FIGURE 4 Detail of the slip joint at P.

(Continued)

Building a Separation Joint in a Site-Cast Concrete Building (*Continued*)

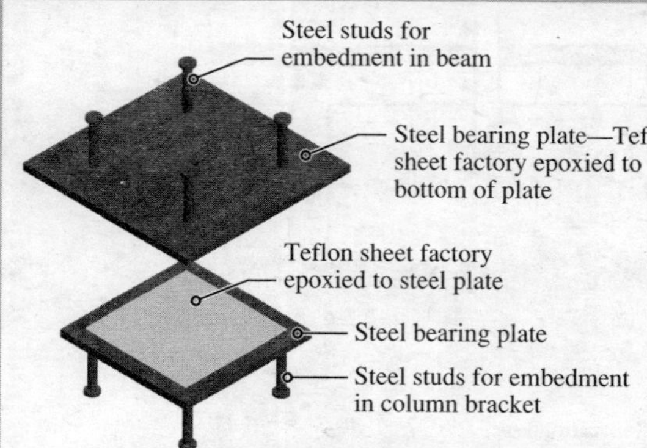

Each steel bearing plate is factory epoxied to a Teflon sheet. Teflon sheet is used because of its very low coefficient of friction which allows movement between the column bracket and the concrete beam above.

Although the plates have been shown separated from each other, they are taped together before being placed in position. The upper plate is slightly larger than the lower plate to provide protection to the Teflon sheet and to allow the space between the plates to be filled with an elastomeric sealant (also see Figure 19.22).

FIGURE 5 Bearing plates in a slip-joint assembly.

Coefficient of Friction

Assume that two rectangular blocks are laid, one over the other. A weight, P, is placed on the upper block, which is then made to slide over the lower block with the help of a horizontal force, F. If F is the horizontal force that is just able to slide the upper block, then the *coefficient of friction*, ϕ, between the surfaces of the blocks is given by

$$\phi = \frac{F}{P}$$

If the force required to slide is small, ϕ is small. The value of ϕ between two Teflon sheets is nearly 0.05 (as reported by a separation joint assembly manufacturer, Conserve Inc., Georgetown, South Carolina).

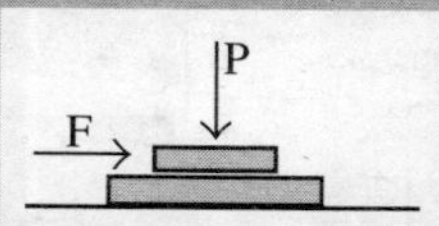

REVIEW QUESTIONS

1. Explain the difference between one-way slab and two-way slab actions.
2. Using sketches and notes, explain beam-and-girder floor systems.
3. Using sketches and notes, illustrate the two most commonly used reinforced-concrete floor systems.
4. Using sketches, explain the commonly used beamless concrete floor systems.
5. Sketch the following precast concrete structural members:
   a. Hollow-core units
   b. Solid planks
   c. Double-tee units
   d. Inverted-tee beams
6. Explain the difference between total precast and a mixed precast construction.

CHAPTER 24 Masonry Materials–I (Mortar and Brick)

CHAPTER OUTLINE

Masonry is one of the oldest building materials. Sun-dried clay (adobe) bricks were used as early as 8,000 BC. The origin of stone masonry is generally traced by historians to the early Egyptian and Mesopotamian civilizations, which existed around 3,000 BC. Indeed, until steel and portland cement were discovered in the mid-nineteenth century, stone was the only building material available for the construction of large building structures and bridges.

The history of architecture is replete with examples of magnificent buildings in which dressed (and partially dressed) stones were used in almost every building element—walls, columns, beams, arches, roofs (vaults and domes), and floors. In some buildings, the stones were so large that historians have differing theories as to the hoisting apparatus used by the builders at the time.

The history of architecture is also a testament to the durability and aesthetics of stone. It is these properties that render stone a matchless material even for present-day buildings, particularly those that require durable and maintenance-free facades, Figure 24.1. In fact, together with glass, stone is one of the most-used material for the exterior facades of contemporary skyscrapers.

Whereas stone is uniquely suited for cladding high-rise and significant buildings, other masonry materials, such as brick and block, are more economical facade alternatives and are widely used. If we lump all masonry materials together, we observe that the use of masonry on contemporary building facades exceeds that of all other materials combined.

FIGURE 24.1 (a) Split-face (cleft) Italian travertine used as cladding material in the J. Paul Getty Center, Los Angeles, California (a complex of several buildings). The building in the background is clad with metal panels. Architect: Richard Meyer. (Photo courtesy of Dr. Jay Henry)

FIGURE 24.1 (b) Smooth-surfaced limestone cladding, Morton Meyerson Symphony Center, Dallas, Texas. Architect: I. M. Pei and Partners.

UNIT MASONRY

Because bricks are generally made from clay, brick masonry is also referred to as *clay masonry.* Block masonry is called *concrete masonry* because the blocks are made from concrete. Because masonry is laid unit by unit (e.g., brick by brick or block by block), it is also referred to as *unit masonry.* Bricks and blocks are, therefore, called *clay masonry units* and *concrete masonry units* (CMU), respectively.

Masonry units are bonded together with mortar to yield a composite building component—generally a wall. Thus, mortar is the common ingredient in all masonry construction. Mortarless masonry, although possible, is uncommon and has relatively few applications; it is not discussed in this text. Because mortar is common to virtually all masonry, we discuss it first, followed by a discussion of bricks. Other masonry materials, including CMU, stone, and glass masonry, are discussed in Chapter 25. Masonry construction systems are discussed in Chapters 26 and 28.

24.1 MASONRY MORTAR

Mortar consists of a binder (cementitious material), a filler, and water. Portland cement and hydrated lime comprise the binder, and the filler is sand. When these three elements are mixed together with the required quantity of water, mortar results.

The primary function of masonry mortar is to bond the masonry units into an integral masonry wall. Because mortar, in its plastic state, is pliable, it molds itself to the surface profile of the units being mortared. This not only helps seal the wall against water and air infiltration, it also provides a cushion between the units.

The mortar cushion also compensates for size variations between individual units. Another important role of mortar is to provide surface character to masonry through shadow lines at mortar joints and color intervention between the units.

WORKABILITY AND WATER RETENTIVITY OF MORTAR: THE ROLE OF LIME

Although portland cement is the primary cementitious material in mortar, lime imparts several useful and important properties to the plastic as well as the hardened mortar. In the plastic (wet) state of mortar, lime improves its workability and water retentivity. A mortar comprising portland cement only (without lime) is coarse and, hence, less workable. In the hardened state, lime improves the water resistance of the wall. A wall built with portland cement and lime mortar is more watertight than a wall built with only portland cement mortar.

In Chapter 21, we observed that the workability of concrete is a quantifiable property. This is not the case with mortar. The workability of mortar is difficult to quantify because it is a function of several interdependent factors. However, a mason with even a limited amount of experience and training can easily distinguish between a workable and a nonworkable mortar.

A workable mortar is cohesive and spreads easily on the units using a trowel. Because of its cohesiveness, it clings to the vertical surfaces of the units and the trowel without sliding down. It extrudes easily so that excess mortar in the joints can be troweled off without the mortar dropping off or smearing the units. A lay (and rather crude) explanation of the difference between a workable and a nonworkable mortar is the difference between spreading a creamy (more workable) and a crunchy (less workable) peanut butter on a piece of toasted bread.

Another important property of plastic mortar is its water retentivity. This is the ability of mortar to retain water without letting it bleed out. A mortar with good water retentivity remains soft and plastic for a long period of time and allows only a limited amount of water to be absorbed by the units.

Water retentivity and workability are directly related to each other. Extremely fine sand particles, air-entraining agents, and lime increase the workability and water retentivity of mortar. While a certain amount of water absorption by the units is necessary for the bond between the mortar and the units, excessive water retentivity is to be avoided because it reduces the strength of the bond.

WATERTIGHTNESS OF A MASONRY WALL: THE ROLE OF LIME

Lime also improves the elasticity of hardened mortar. In other words, a lime-based mortar is able to flex somewhat in its hardened state. This reduces the cracks caused by the bending of a wall under lateral loads. Lime also provides an *autogenous healing* property to mortar. Autogenous healing refers to the self-sealing of small cracks produced either within the mortar or at the interface between the mortar and the units. The cracks may result from either the bending stresses in masonry or the drying shrinkage of portland cement in mortar.

NOTE

Mortarless Masonry Walls

A common contemporary use of mortarless masonry is in segmental retaining walls. Mortarless masonry is extremely uncommon in a conventional wall.

If a mortarless masonry wall (also called *dry-stacked masonry*) is used in a conventional wall, the units may need grinding to provide a smooth bed surface. To provide bending resistance in a mortarless masonry wall, the voids in the units may be grouted and reinforced, and/or both surfaces of the wall may be coated with a surface-bonding material, such as plaster.

NOTE

Lime for Masonry Mortar

The lime recommended for use in masonry mortar is Type S hydrated lime (see Chapter 21).

NOTE

Autogenous Healing

Because a lime-based mortar is somewhat elastic, it is less liable to crack. However, if cracks are formed, the carbonation of lime fills them. In the carbonation process, lime absorbs atmospheric carbon dioxide (see Chapter 21), which increases its mass and volume. The increase helps to fill the voids caused by drying shrinkage of portland cement and by flexural cracking of masonry—a process known as *autogenous healing*.

The term *autogenous healing* has been borrowed from medicine and refers to the natural healing process of human bones. When a bone in the body fractures, a surgeon simply aligns the fractured parts in their original position and sustains them in that position using a cast. The fractured pieces fuse together automatically in due course. Although the healing of a bone fracture is complex, it is primarily due to the calcium in human bones—similar to the calcium in lime.

MORTAR STRENGTH: THE ROLES OF PORTLAND CEMENT AND LIME

Two strength properties of mortar are generally of interest:

- Compressive strength
- Flexural tensile bond strength

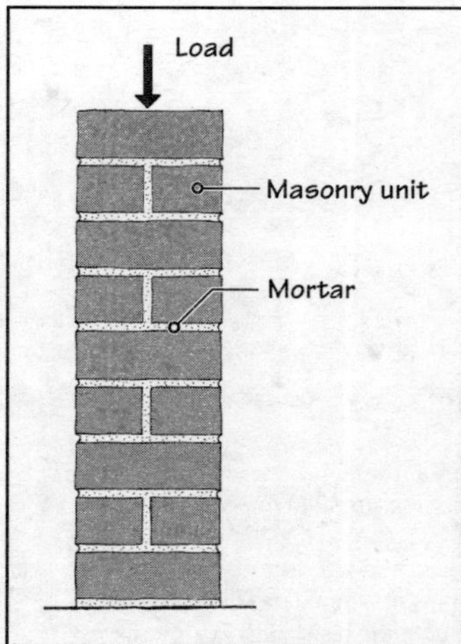

FIGURE 24.2 Because a masonry wall is composed of masonry units and mortar, the compressive strength of the wall is a function of the strength of the units and the strength of the mortar.

Although several factors affect the compressive strength of mortar, the most important factor is the mortar's cementitious content. As we will observe later, the total amount of cementitious content (portland cement plus lime) in various types of mortar is roughly constant with respect to the amount of sand. The relative proportions of portland cement and lime are, however, different.

Increasing the amount of portland cement with respect to lime increases the mortar's compressive strength. Conversely, increasing the amount of lime with respect to portland cement decreases the mortar's compressive strength. Because mortar is an integral part of a masonry wall, its strength affects the compressive strength of the wall, Figure 24.2.

Unlike the compressive strength of mortar, which is a property only of the mortar, bond strength is a property of the masonry wall. It is a measure of the bond between the masonry units and the mortar. It comes into play when an unreinforced masonry wall is subjected to bending (flexure), Figure 24.3.

The bond between a masonry unit and the mortar is both a chemical and a mechanical bond. Therefore, the bond strength of masonry is a function of several factors, such as the types of units, surface roughness of units, workmanship (such as the pressure applied between the units at the time of mortaring), curing conditions (air temperature, wind and humidity), and so on.

Another important factor that affects the bond between the mortar and the units is the amount of water in mortar. The amount of water in mortar must be sufficiently high so that the mortar can flow and be sucked into the minute crevices in masonry units. This develops and improves the bond between the units and the mortar. Everything else being the same, increasing the amount of portland cement in mortar with respect to lime increases the bond strength of masonry.

The bond strength of masonry is pertinent primarily in an unreinforced masonry wall, such as masonry veneer, or in an unreinforced masonry backup wall. In a vertically reinforced masonry wall, Figure 24.4, steel reinforcement resists flexural tension. Therefore, from a purely structural viewpoint, the bond strength is not relevant in a reinforced masonry wall (Figure 24.4).

NOTE

Flexural Tensile Bond Strength

A wall subjected to bending (flexure) experiences tension as well as compression. These stresses are referred to as *flexural tension* and *flexural compression*, respectively. Because unreinforced masonry is relatively stronger in compression than in tension, the *flexural tensile strength* of masonry is more critical than the *flexural compressive strength*.

Because flexural tensile strength (of an unreinforced masonry wall) is due to the bond between the units and the mortar, it is called *flexural tensile bond strength*, or simply *bond strength* (Figure 24.3). As stated in the text, bond strength is not relevant in a masonry wall provided with steel reinforcing bars to resist bending (Figure 24.4).

REQUIRED STRENGTH OF MASONRY MORTAR

The foregoing discussion indicates that the strength of a masonry wall is directly related to the strength of the mortar. It also indicates that an increase in lime (with respect to portland cement) decreases the strength of mortar and, hence the strength of a masonry wall, but it increases the wall's watertightness.

EXPAND YOUR KNOWLEDGE

A Historical Note on Cementitious Materials in Mortar

Before the discovery of portland cement, masonry mortar consisted of lime (as the cementitious material) and sand. After the discovery of portland cement, masons discarded the use of lime and made the mortar using only portland cement and sand. The change from lime mortar to portland cement mortar was instinctive and spontaneous because portland cement gives a much stronger mortar than lime. Additionally, portland cement sets and hardens far more rapidly than lime, allowing masons to lay more masonry units per day than using lime mortar, increasing the masons' productivity.

Remember from Chapter 21 that the setting and hardening of lime are due to its carbonation reaction with carbon dioxide in air. Because air contains only about 0.04% carbon dioxide, the setting and hardening of lime are slow. The setting and hardening of portland cement, on the other hand, are due to portland cement's reaction with water, whose quantity is generally far greater than that needed for the hydration reaction.

A few decades later, however, it was discovered that the walls made with portland cement mortar leaked far more than those made with lime mortar. The investigation of the walls' permeability to water has led to our understanding of the importance and benefits of using lime in contemporary masonry mortar.

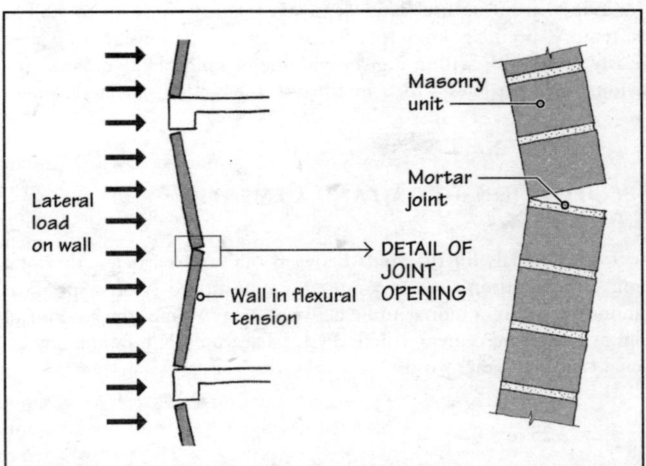

FIGURE 24.3 A wall is subjected to flexural tension when it is subjected to lateral loads (wind or earthquake). The flexural tension can lead to the wall's cracking. Because the bond between the units and the mortar is much weaker than the tensile strength of the units, the wall generally cracks at the mortar joints, that is, the mortar joints open up.

This illustration shows the opening up of horizontal mortar joints caused by the vertical bending of the wall. If the wall bends horizontally, vertical joints will tend to open.

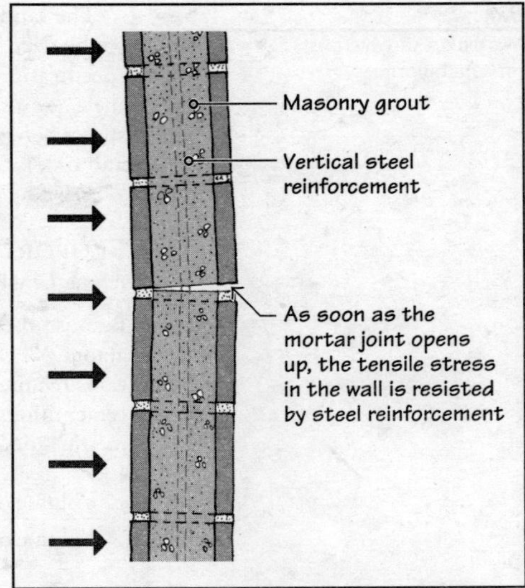

FIGURE 24.4 Flexural tensile stress in a reinforced (and grouted) masonry wall is resisted by the bond between the masonry and mortar only until the mortar joints have opened. This is the case with relatively small lateral loads. As the load increases and a mortar joint opens, the entire tensile stress is now borne by the steel reinforcement. The bond between the masonry and the mortar is now irrelevant. For a three-dimensional illustration of a (vertically) reinforced masonry wall, see Figure 25.7.

A wall's watertightness is just as important as its strength. In fact, both watertightness and strength are somewhat interrelated. Water penetrating a wall corrodes steel reinforcement, ties, and other embedded accessories, which may ultimately reduce the wall's strength. The masonry industry, therefore, favors a relatively low-strength mortar for most masonry walls.

Another reason for choosing a low-strength mortar is to ensure that if the cracking of a masonry wall occurs, it should occur in mortar joints, not in the units, because it is much easier to repair a broken mortar bed than a broken masonry unit. Because a low-strength mortar is more workable, it provides better workmanship and full coverage of joints and, hence, a more watertight wall. In fact, a general recommendation for masonry walls is:

Specify the weakest mortar that will give the required performance.

However, there are structures in which the strength of masonry is more important than its watertightness, such as the walls located in high-wind or seismic zones and heavily loaded interior walls. In these structures, a high-strength mortar is recommended.

NOTE

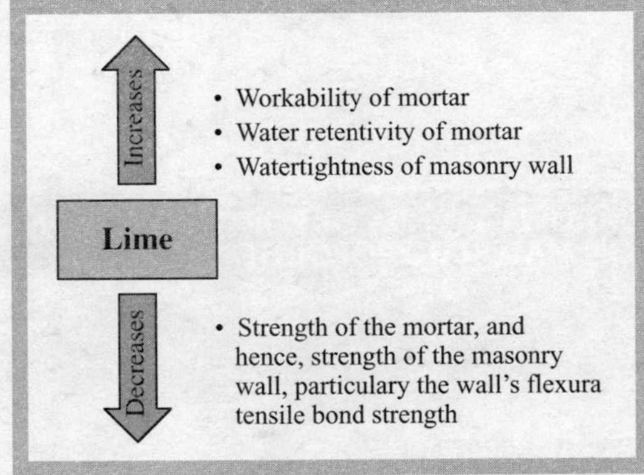

24.2 MORTAR MATERIALS AND SPECIFICATIONS

Mortar is prepared at the construction site in a small mixer (see Figure 21.19). Hand mixing with a shovel is appropriate only for a very small job. The amount of water needed in a mortar mix is not controlled by specifications but left to the discretion of the mason to obtain the required workability. Water for mortar must be clean, potable, and free of impurities.

Materials for Masonry Mortar

Portland cement used in mortar is the same as that used in making concrete (see Chapter 21). Generally, Type I/II portland cement is used. Lime for mortar is Type S lime (see Chapter 21).

The sand used in mortar is referred to as *mason's sand.* It may be manufactured sand (pulverized stone) or mined from natural sand deposits. It must conform to ASTM specifications, which require specific grading (particle size variation). This grading gives the mason's sand nearly 30% voids within a given volume of sand. The voids are the spaces between individual sand particles, which are filled by the binder (portland cement and lime).

PROPORTION SPECIFICATION—PORTLAND CEMENT AND LIME (PCL) MORTAR

Because the cementitious materials fill the voids between the sand particles, the total amount of cementitious compounds in a masonry mortar is almost fixed. Mortar specifications require the amount of sand in a mortar to lie between $2\frac{1}{4}$ to 3 times the amount of cementitious compounds—portland cement + lime (PCL). The sand : PCL ratio is a volumetric ratio, not a weight ratio. In other words:

$$\frac{\text{Volume of sand (ft}^3\text{)}}{\text{Volume of PCL (ft}^3\text{)}} = 2.25 \text{ to } 3.0$$

The range allows the masons to adjust the mix for workability and water retentivity of the mortar.

Varying the relative proportions of portland cement and lime in a mortar mix provides different types of mortar—referred to as *Types M, S, N,* and *O,* Table 24.1. Different mortar types have different recommended applications.

Mortar Type M has the maximum amount of portland cement and hence the least amount of lime. It gives the strongest mortar. Because of its low lime content, Type M mortar is the least workable, is less elastic in its hardened state, exhibits more shrinkage, and, therefore, results in a wall with greater potential for leakage. It is not recommended for use in walls, except below-grade (foundation) walls. It may be used in interior masonry floors for its greater strength and abrasion resistance. (Exterior masonry paving is generally mortarless.)

Type O mortar has the least amount of portland cement and the maximum amount of lime, and hence is the weakest mortar. Because of its low strength (and hence low durability), Type O mortar is not generally recommended for use in contemporary exterior walls. The most common use for type O mortar is for tuck-pointing repairs of old historic masonry.

TABLE 24.1 TYPES OF PORTLAND CEMENT PLUS LIME (PCL) MORTARS BASED ON MATERIAL PROPORTIONS

		Materials to Be Proportioned by Volume			
	Mortar type	Portland cement (PC)	Lime (L)	Sand	Recommended applications
Strongest Mortar	M	1	$\frac{1}{4}$	As per ASTM specifications, the amount of sand for all types of mortar is $2\frac{1}{4}$ to 3 times the amount of cementitious materials (PCL). Generally, however, the amount of sand used is 3 times (PCL). Thus, in Type S mortar, if PC = 1 ft^3 and L = 0.5 ft^3, the amount of sand is 3(1 + 0.5) = 4.5 ft^3	Interior masonry floors, foundation walls, parapet walls, and so on
	S	1	Over $\frac{1}{4}$ to $\frac{1}{2}$		Load-bearing and non-load-bearing walls in seismic or high-wind regions, chimneys, and so on
	N	1	Over $\frac{1}{2}$ to $1\frac{1}{4}$		General-purpose mortar for exterior load-bearing and non-load-bearing walls
Weakest Mortar	O	1	Over $1\frac{1}{4}$ to $2\frac{1}{2}$		Interior non-load-bearing walls in nonseismic and low-wind regions

The amount of water should be as much as that required for workability by the mason. This recommendation is unlike that in concrete, where, for strength reasons, the minimum amount of water is recommended to produce the required workability and consistency.

Type S mortar is commonly recommended for exterior (load-bearing or non-load-bearing) walls in seismic or high-wind regions. It is preferred over Type N mortar because of its higher bond strength, although it is less water resistant. Type N mortar is the general-purpose mortar, used in most non-load-bearing exterior walls, such as masonry veneer walls. It may also be used in load-bearing walls. However, where greater strength and durability are needed, Type S mortar should be considered.

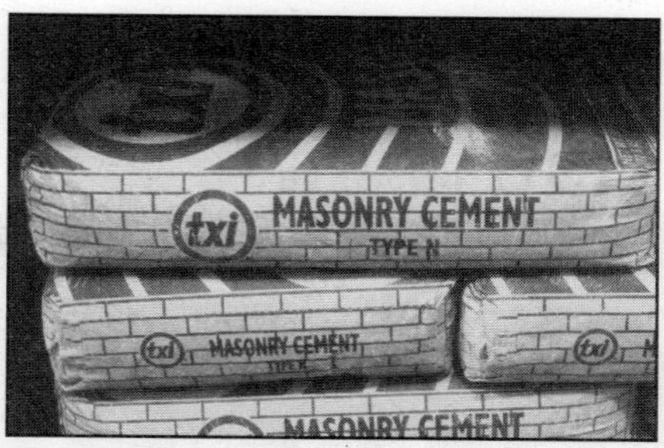

FIGURE 24.5 A Type N masonry cement bag.

Proportion Specification Using Masonry Cement or Mortar Cement

Although PCL mortar is commonly used, a preblended cementitious mix for mortar (in one bag) is available. It is sold as *masonry cement* and is available as Types M, S, and N, Figure 24.5. Thus, to produce a Type S mortar, the mason simply mixes 1 ft^3 of Type S masonry mortar with $2\frac{1}{4}$ to 3 ft^3 of sand. There is no need to use separate portland cement and lime bags.

The use of one-bag masonry cement precludes the need to measure and mix the cementitious materials on site, thus simplifying the preparation of mortar. It also results in a mortar that is more consistent in quality and appearance than a PCL mortar. In masonry cement, cementitious materials are factory blended under a stricter quality-control environment than a site-mixed PCL mortar.

Masonry cement may contain cementitious materials other than portland cement and may contain pulverized limestone in place of lime. To improve the workability of mortar, some manufacturers use air-entraining agents in their masonry cement. This reduces the bond between the units and the mortar. Codes recognize these facts by reducing the allowable bond strength of masonry built with masonry cement mortar.

Another preblended cementitious mix for mortar is called *mortar cement.* Like masonry cement, it is also a one-bag mix, but it differs from masonry cement in that it gives the same bond strength as PCL mortar. Masonry codes treat mortars made with mortar cement or PCL as being equivalent. Like masonry cement, mortar cement bags are available in Types M, S, and N.

Property Specification of Mortar

An alternative way to specify mortar is by its properties, similar to the way concrete is specified. The most distinguishing mortar property is its compressive strength, determined by crushing 2-in. × 2-in. cubes after 28 days.

By testing the cubes, the amounts of cementitious materials can be established in the laboratory to give the required mortar strength. The proportions so established are used in producing the mortar at the construction site. The sand content in a mix obtained from property specification is generally greater than that used in the proportion specification ($2\frac{1}{4}$ to 3 times the cementitious materials).

Mortar produced in the laboratory using the property specification is referred to as *Type M* mortar if its 28-day compressive strength is at least 2,500 psi, as *Type S* mortar if its compressive strength is at least 1,800 psi, and so on, as shown in Table 24.2.

NOTE

A Note on the Strength of Mortar

The compressive strengths of mortar given in Table 24.2 are 28-day strengths, obtained from tests of 2-in. mortar cubes prepared in the laboratory. The actual compressive strength of the corresponding mortars in the wall is much higher than the tabular values because of the small ($\frac{3}{8}$-in.) thickness of mortar and the confinement provided by the units.

EXPAND YOUR KNOWLEDGE

Letter Designations (M, S, N, and O) for Mortar Types

The letters *M, S, N,* and *O* are the alternate letters in the word *masonwork:*

MaSoNwOrK

Although the letter *K* does not represent any current ASTM mortar type, it is used in preservation work (as a more appropriate substitute for Type O mortar) to simulate historic masonry, where only lime mortars were used. Type K mortar has 1 part portland cement and $2\frac{1}{2}$ to 4 parts lime.

Types M, S, N, and O have been preferred over Types I, II, III, and IV (or Types 1, 2, 3, and 4) because no type is better than the other. Each type has its own unique set of applications.

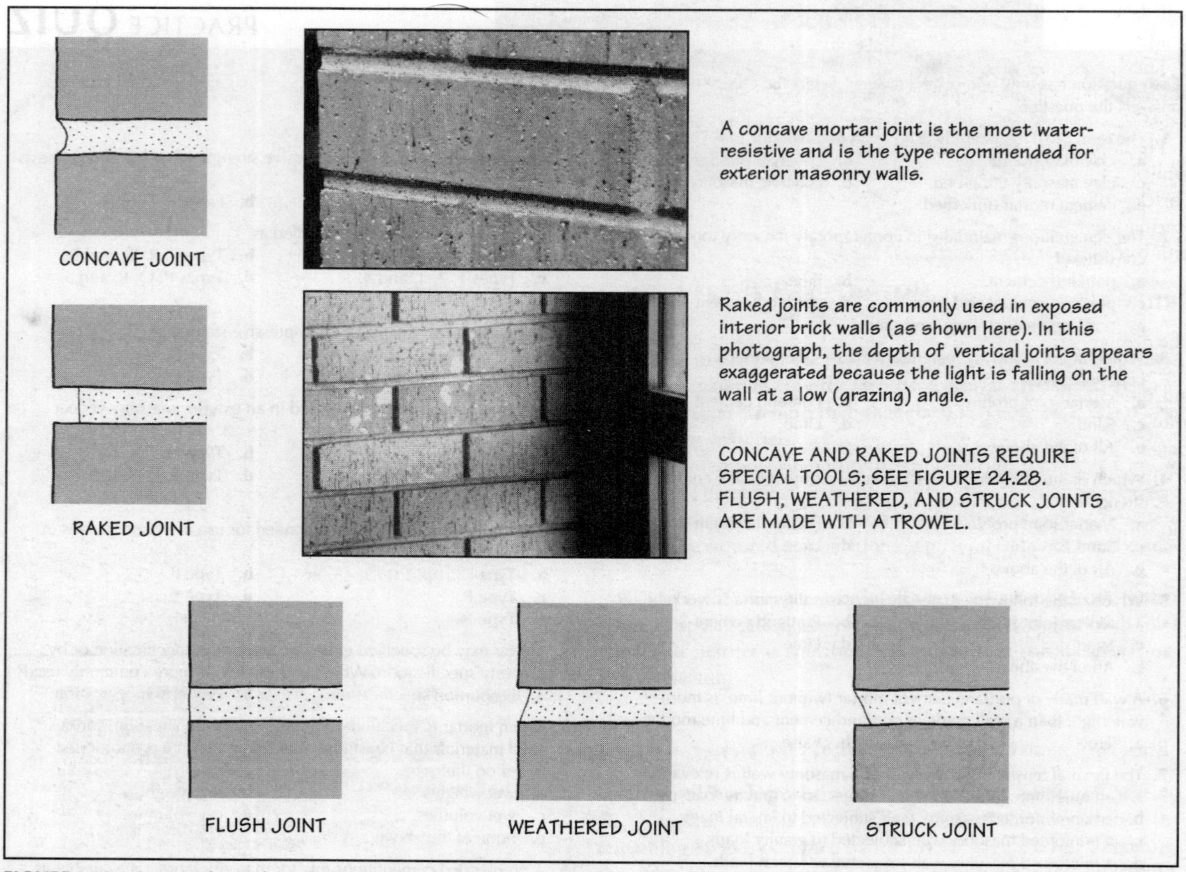

A concave mortar joint is the most water-resistive and is the type recommended for exterior masonry walls.

CONCAVE JOINT

Raked joints are commonly used in exposed interior brick walls (as shown here). In this photograph, the depth of vertical joints appears exaggerated because the light is falling on the wall at a low (grazing) angle.

CONCAVE AND RAKED JOINTS REQUIRE SPECIAL TOOLS; SEE FIGURE 24.28. FLUSH, WEATHERED, AND STRUCK JOINTS ARE MADE WITH A TROWEL.

RAKED JOINT

FLUSH JOINT

WEATHERED JOINT

STRUCK JOINT

FIGURE 24.8 Commonly used mortar joint profiles.

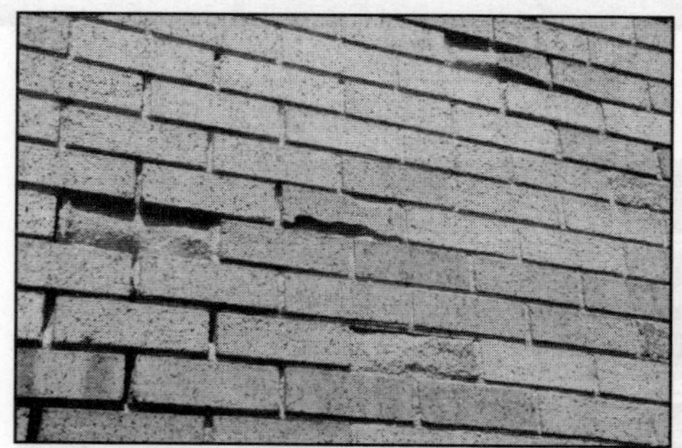

FIGURE 24.9 Spalling of bricks due to freeze-thaw action. In freeze-thaw action, the water absorbed by a wall expands on freezing, producing compressive stresses in the wall. Over repeated freeze-thaw cycles, a wall (made of a low-strength material) spalls, that is, crumbles and falls off.

The most watertight joint profile is the concave joint. It is obtained by tooling the joint with a cylindrical tool that compresses the mortar against itself and the units (Figure 24.28). Because most water penetrates a masonry wall through the unit-to-mortar interface (relatively little penetrates through the units or the mortar), compressing the mortar improves the seal between the units and the mortar. A concave joint is also more resistant to freeze-thaw damage, Figure 24.9.

The compression and the resulting seal provided by a concave joint are not provided by other joint profiles. That is why a tooled concave joint profile is generally recommended for all exterior masonry.

For interior masonry, raked joints and flush joints are common. Raked joints highlight the units by providing deeper shadows within the mortar. This effect is dramatized if the light falls at a low angle to the wall. The disadvantage of raked joints is that they collect dust. Flush joints are used for interior masonry where dust is a concern. Between weathered and struck joints, the weathered joint gives greater water resistance by directing the water away from the joint.

24.4 MANUFACTURE OF BRICKS

In North America, brick (clay) masonry is generally used either as an exterior or an interior wall cladding material, for example, brick veneer (see Chapter 28). The high strength of bricks makes them durable against freezing and thawing and against slow erosion by rainwater and wind—important requirements for an exterior finish. Several brick manufacturers make their bricks with a compressive strength of 6,000 to 10,000 psi. Clay masonry is also suitable as an exterior cladding due to its mass, which provides high fire resistance and sound-insulating properties (see Chapter 8). Finally, constructing a brick wall requires a high level of craft. This gives brick facades an aesthetic character not available in other facade materials, such as site-cast concrete, precast concrete, or insulated metal panels.

The appearance of a brick facade conveys an image of permanence and stability; brick facades are used frequently in significant civic buildings, schools, and college campuses. In fact, as a facade material, brick is by far the most widely used material in contemporary buildings.

The high strength of bricks should make them a logical choice for load-bearing wall applications. However, this is not the case. The reason is that it is more difficult to insulate a brick wall or reinforce it with steel bars than a concrete masonry wall. Therefore, in load-bearing masonry construction, the load-bearing component is generally concrete masonry, with brick as a facade material (see Chapter 28).

BRICK MANUFACTURING

The primary raw materials for making bricks are clay and shale. The two main constituents of clay and shale are the oxides of silicon and aluminum. Some minor components are iron and other metal oxides, which are particularly responsible for giving brick its red-brown color. White and light-colored bricks are made by using clay that is naturally deficient in metal oxides and removing whatever metal oxides are present in it. White bricks are generally more expensive than the normal (red-brown) bricks.

Although modern technology has substantially changed the details of brick manufacturing, it is conceptually simple and consists of the following six operations:

- Mining clay from the ground
- Grinding and sieving clay to a fine powder
- Mixing water with sieved clay
- Forming wet clay into the desired brick shape (green bricks)
- Drying green bricks
- Firing dried bricks in a kiln

Brick shapes can be formed by one of the following two methods:

- Extrusion of wet clay through a die—*extruded bricks*
- Molding wet clay—*molded bricks*

NOTE

Difference Between Clay and Shale

Clay is available at the surface of the earth. Therefore, clay and *surface clay* are synonymous terms. *Shale* is also clay, but it is available deep in the ground. Because of the pressure of the overlying material, shale has hardened to a high compressive strength, almost equaling that of stone.

EXTRUDED BRICKS

The extrusion of wet clay is done by forcing it through a die, which yields a column of clay that slides over a moving belt, Figure 24.10. The process is similar to the extrusion of toothpaste through a tube. The die consists of conical rods, which create *core holes* in the clay column. The cross-sectional dimensions of the clay column determine the length and width of the brick.

As the clay column moves forward, a wire cutter, consisting of a number of parallel wires, cuts it into individual bricks, Figure 24.11. The spacing between wires is the brick

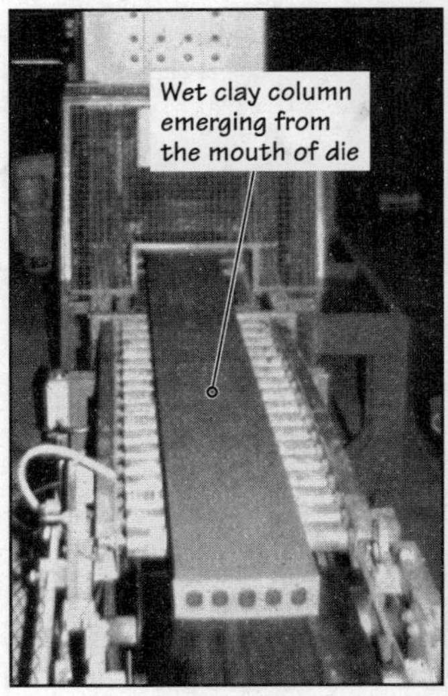

(a)

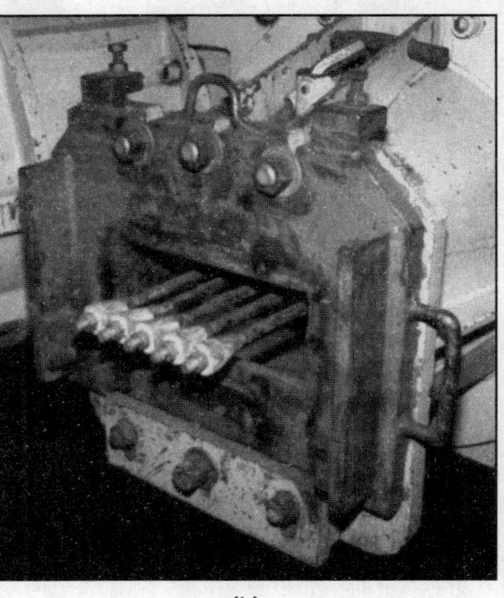

(b)

FIGURE 24.10 Manufacture of extruded (clay) bricks. **(a)** A wet clay column emerging from the die. **(b)** Die mouth with conical rods. (Photos courtesy of Acme Brick Company, Fort Worth, Texas)

FIGURE 24.11 A rotary wire cutter cuts the wet clay column into several bricks in one pass. (Photo courtesy of Acme Brick Company, Fort Worth, Texas)

FIGURE 24.12 Some manufacturers make bricks with special surface textures that are applied to the wet clay column with rotating textured drums. (Photo courtesy of Acme Brick Company, Fort Worth, Texas)

height. If any surface texture is to be applied to bricks, it is applied before the cutting operation, Figure 24.12. If no surface texture is applied, the brick has a smooth surface finish resulting from the pressure applied by the steel die as the clay passes through it. That is why the smooth texture is referred to as *die skin texture.* A few other commonly used surface finishes are shown in Figure 24.13.

After the bricks have been cut, they are transported to a drying chamber, in which the temperature ranges from 100°F (38°C) to 400°F (150°C). Bricks must be dried before being fired in the kiln to prevent cracking of green bricks. The heat used in the drying chamber is the exhaust heat from the kiln where the bricks are fired.

The kiln used in modern brick manufacturing plants is a long, tunnel kiln, Figure 24.14. Cars of dried bricks move slowly through the kiln on a rail. The kiln temperature varies from nearly 400°F at the entry point of the dried bricks to a maximum of about 2,000°F (1,100°C). At nearly 2,000°F (1,100°C), clay particles soften (virtually melt) and fuse together, creating a hard, strong brick. Fired bricks are removed from the kiln, sorted, strapped in cubes, and stored in the yard until delivery to the construction site, Figure 24.15.

Core Holes in Extruded Bricks

Extruded bricks have through-and-through core holes. The primary reason for core holes is that they lead to more uniform drying and firing of bricks. Although the core holes reduce

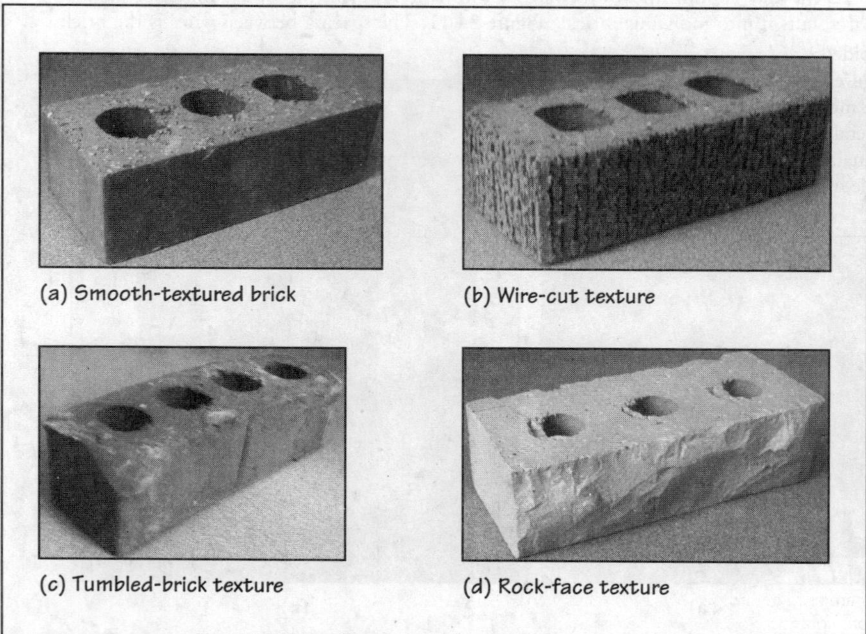

FIGURE 24.13 A few of the several textures available on extruded bricks.

FIGURE 24.14 A typical tunnel kiln. (Photo courtesy of Acme Brick Company, Fort Worth, Texas)

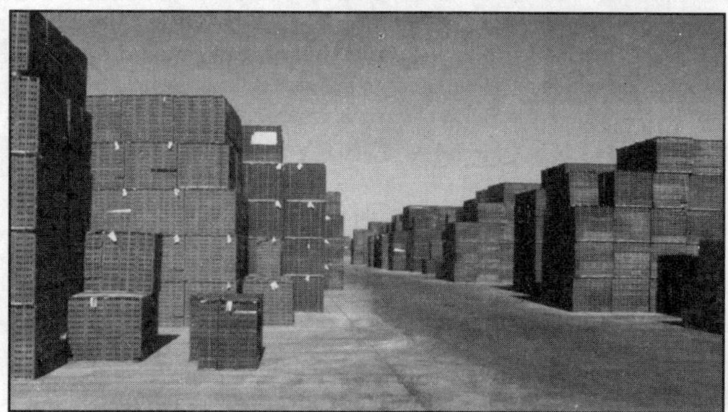

FIGURE 24.15 A typical storage yard at a brick manufacturing plant.

the bearing area of a brick and, hence, the compressive strength of masonry, the improved unit strength resulting from uniform firing compensates for the bearing area reduction. Core holes also improve the bond between the bricks and the mortar. The number and sizes of core holes are manufacturer specific, depending on the size of the brick.

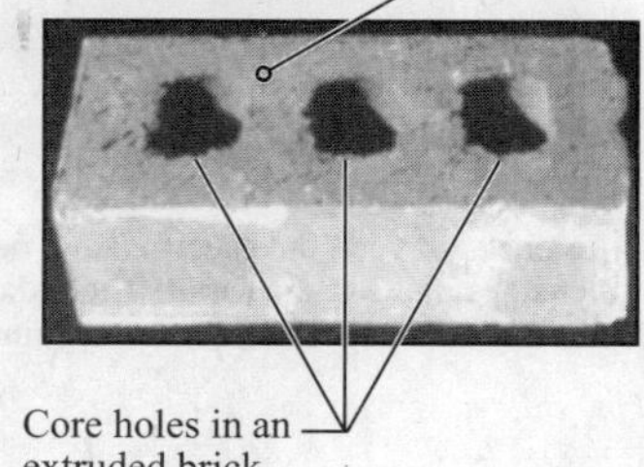

MOLDED BRICKS AND FROGS

Molded bricks are made by force dropping individual lumps of wet clay into brick molds, Figure 24.16. The excess clay is scraped off the molds, and green bricks are removed from the mold immediately thereafter. Modern brick manufacturing is highly mechanized, and several bricks can be molded in one single pass rather than by the historic molding of individual bricks, Figure 24.17. In both extruded and molded-brick manufacturing, there is little, if any, manual handling.

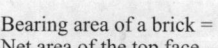

FIGURE 24.16 Individual lumps of wet clay are used in forming molded bricks. (Photo courtesy of Acme Brick Company, Elgin, Texas)

FIGURE 24.17 Several hundred bricks are molded in one single pass in the molding machine. (Photo courtesy of Acme Brick Company, Elgin, Texas)

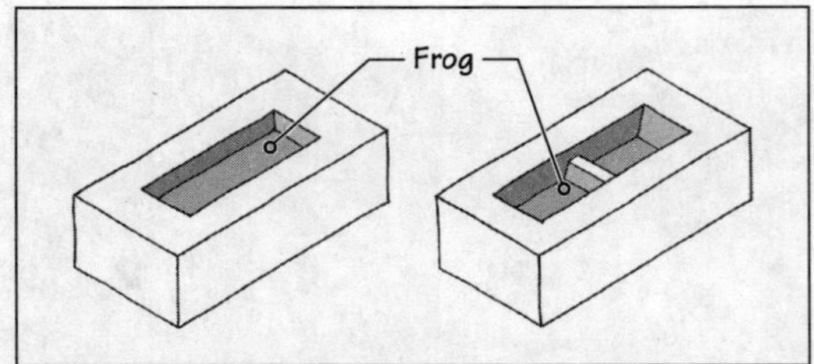

FIGURE 24.18 Two of the several shapes of frogs used in molded bricks.

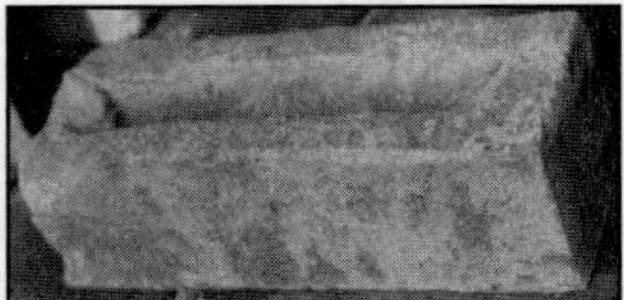

A typical molded brick.

Because the clay-water mix is pushed under pressure through a die for extruded bricks, it is a fairly stiff mix. By contrast, the clay-water mix for molded bricks must contain more water. As the water evaporates, it leaves voids in bricks. Consequently, a molded brick is generally weaker and softer than an extruded brick. However, molded bricks are preferred by some designers and owners because they do not have the machine-like (more precise) appearance of extruded bricks.

Whereas an extruded brick has through-and-through core holes, a molded brick contains a depression, referred to as a *frog*. The frog can take many shapes, depending on the manufacturer. Two such frogs are shown in Figure 24.18. In laying the molded bricks in a wall, the frog faces downward, so that any water entering the wall will not be held by the frog recess.

24.5 DIMENSIONS OF MASONRY UNITS

A masonry unit (both clay brick and concrete block) has three types of dimensions:

- Specified dimensions
- Nominal dimensions
- Actual dimensions

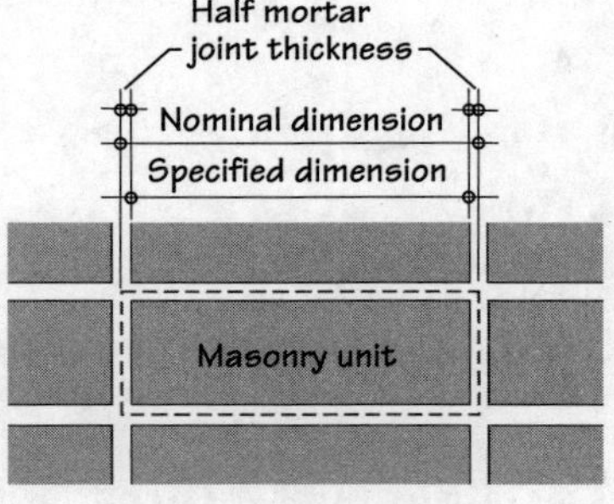

Difference between the specified and nominal dimensions of a masonry unit. The red dashed lines represent the center lines of mortar joints.

The *specified dimension* of a masonry unit is the finished dimension that the specifier has requested and the manufacturer desires to achieve. However, because the manufacturing process is not perfect, the *actual dimensions* of a unit are different from the specified dimensions. The difference between the specified and actual dimensions of a masonry unit must lie within the *dimensional tolerance* established by the industry for that product.

The *nominal dimension* of a unit is the specified dimension plus one mortar joint thickness. The nominal dimensions of a unit refers to the space occupied by one unit (and the associated mortar) in the wall. Thus,

Nominal dimension = specified dimension + one mortar joint dimension

Because the standard mortar joint thickness is $\frac{3}{8}$ in., the nominal dimension of a masonry unit is $\frac{3}{8}$ in. greater than the specified dimension. Thus, if the specified length of a masonry unit is $7\frac{5}{8}$ in., its nominal length is 8 in.

In practice, the nominal dimensions of units are generally given, and the inch labels are not used. Inch marks are used with the specified dimensions. Thus, if

$$\text{A unit's nominal dimensions} = 4 \times 2\tfrac{2}{3} \times 8$$
$$\text{Its specified dimensions} = 3\tfrac{5}{8}\text{ in.} \times 2\tfrac{1}{4}\text{ in.} \times 7\tfrac{5}{8}\text{ in.}$$

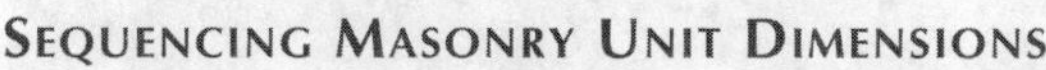

SEQUENCING MASONRY UNIT DIMENSIONS

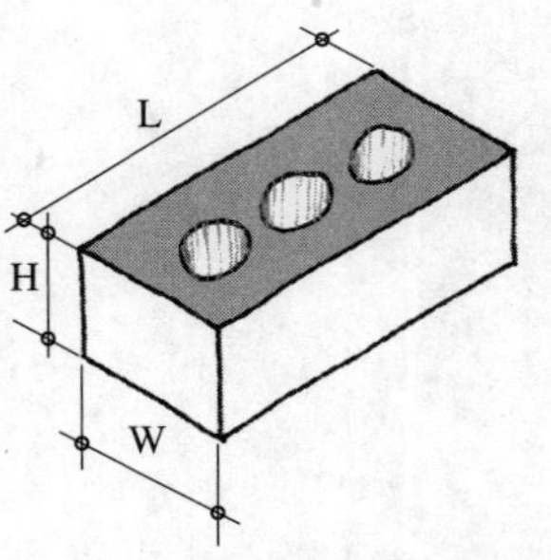

Masonry unit dimensions must be stated in the following sequence: W × H × L.

Observe the sequence in which the unit's dimensions have been stated. The width (W, i.e., *through-the-wall* dimension) is stated first, followed by the face dimensions—height (H) × length (L). Specifying the unit's dimensions using a sequence other than W × H × L may confuse the subcontractor, manufacturer, or supplier.

HOLLOW VERSUS SOLID MASONRY UNITS

Both bricks and concrete blocks generally contain voids. In the masonry industry, a *solid* masonry unit is less than 25% hollow (≥75% solid cross-sectional bearing area). A *hollow* masonry unit is 25% or more hollow.

Most bricks are solid units because their total core area is less than 25%. A brick without cores must be specified as *100% solid.* A 100% solid brick is generally used for paving or as coping at the top of walls. Because brick-manufacturing machinery is set up to make cored or frogged bricks, 100% solid bricks are special bricks.

PRACTICE QUIZ

Each question has only one correct answer. Select the choice that best answers the question.

17. The most commonly used mortar joint thickness is
- **a.** $\frac{3}{4}$ in.
- **b.** $\frac{5}{8}$ in.
- **c.** $\frac{1}{2}$ in.
- **d.** $\frac{3}{8}$ in.
- **e.** $\frac{1}{4}$ in.

18. Which of the following mortar joint profiles should be specified for an exterior wall?
- **a.** Concave joint
- **b.** Raked joint
- **c.** Flush joint
- **d.** Weathered joint
- **e.** Struck joint

19. Bricks that have core holes are
- **a.** extruded bricks.
- **b.** molded bricks.
- **c.** (a) and (b).

20. Bricks that have frogs are
- **a.** extruded bricks.
- **b.** molded bricks.
- **c.** (a) and (b).

21. The primary reason for providing core holes in bricks is to
- **a.** reduce the weight of bricks.
- **b.** provide uniform drying and firing of bricks.
- **c.** increase the shear resistance of masonry walls.
- **d.** none of the above.

22. The nominal dimension of a masonry unit whose actual dimension equals $7\frac{5}{8}$ in. is
- **a.** 7 in.
- **b.** $7\frac{1}{4}$ in.
- **c.** $7\frac{1}{2}$ in.
- **d.** $7\frac{3}{4}$ in.
- **e.** none of the above.

23. A masonry unit that has no voids (e.g., no core holes) is referred to as a
- **a.** solid unit.
- **b.** 100% solid unit.
- **c.** (a) or (b).

24. A masonry unit that has 20% voids (e.g., 20% core hole area) is referred to as a
- **a.** solid unit.
- **b.** hollow unit.
- **c.** partially hollow unit.
- **d.** none of the above.

25. In specifying the dimensions of a masonry unit, which dimension is stated first?
- **a.** Length, L, of the unit
- **b.** Height, H, of the unit
- **c.** Width, W, of the unit

26. In specifying the dimensions of a masonry unit, which dimension is stated last?
- **a.** Length, L, of the unit
- **b.** Height, H, of the unit
- **c.** Width, W, of the unit

24.6 TYPES OF CLAY BRICKS

Bricks are available in various sizes. The most commonly used brick is the (extruded) *modular brick,* which measures $3\frac{5}{8}$ in. × $2\frac{1}{4}$ in. × $7\frac{5}{8}$ in. Its nominal dimensions are $4 \times 2\frac{2}{3} \times 8$. Some of the other commonly used brick sizes are shown in Figure 24.19.

TYPES OF CLAY BRICKS BASED ON USE

Bricks used in different applications must obviously be different. The bricks used as face veneer must have lower dimensional tolerances and be more weather resistant than the bricks used in

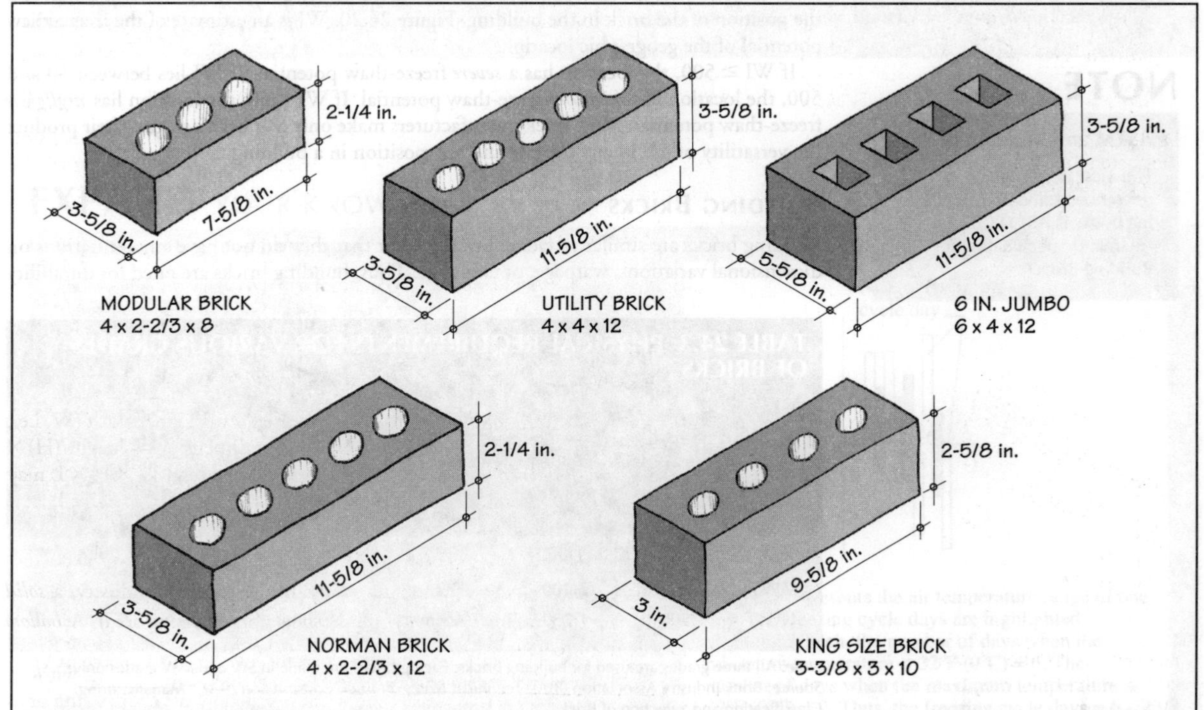

FIGURE 24.19 Commonly used brick types.

and strength only; that is, they carry a grade—Grade SW, MW, or NW. The specifications for Grade SW and MW in both facing bricks and building bricks are identical. Building bricks have an additional grade NW (Table 24.3).

Building bricks are generally used in brick walls that are covered with another facing material. However, they may be used in face veneers if their warpage, chippage, and dimensional variations are acceptable.

NOTE

ASTM and Paving Bricks

The most frequently referenced standard for paving bricks is ASTM C902, "Specification for Pedestrian and Light Traffic Paving Bricks," or ASTM C1272, "Specification for Heavy Vehicular Paving Bricks."

PAVING BRICKS

Paving bricks are graded for freeze-thaw as Class SX, MX, or NX—from most freeze-thaw resistant to least freeze-thaw resistant. They are also graded for abrasion resistance as Types I, II, and III—from highest abrasion resistance to lowest abrasion resistance.

24.7 BOND PATTERNS IN MASONRY WALLS

Bricks can be assembled in a wall in several patterns, referred to as *bond patterns* or simply as *bonds.* The purpose of a bond is functional as well as aesthetic. Functionally, the bond is meant to stagger the units so that the load on one unit is shared by an increasing number of underlying units. A one-*wythe* masonry wall (a wall whose thickness equals the width, W, of one unit), built from whole (uncut) units, can have two types of bonds, Figure 24.21:

- Running bond
- Stack bond

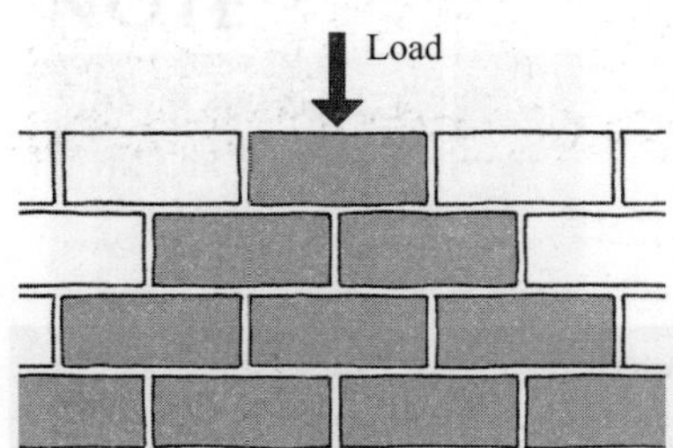

The staggering of masonry units and gravity load distribution. Also see Expand Your Knowledge in Section 25.1.

A stack bond is commonly used at a sharply curved corner, Figure 24.22(a). It is also used for aesthetic reasons, Figure 24.22(b). The (horizontal) bending strength of a stack-bond wall is lower than that of a running-bond wall. Therefore, a stack-bond wall must be provided with horizontal reinforcement.

The bricks used in a stack-bond wall should have as much dimensional uniformity as possible so that the mortar joints have the least variation in width, Figure 24.23.

BRICK ORIENTATIONS ON WALL ELEVATION

In Figure 24.21, the bricks have only one orientation. The orientation refers to the exposed face of the brick. The exposed face of the brick in Figure 24.21 is called the *stretcher face* or simply the *stretcher.* Being a six-faced figure, a brick can have six different face orientations on a wall elevation, Figure 24.24:

- Stretcher
- Header
- Rowlock
- Soldier
- Shiner
- Sailor

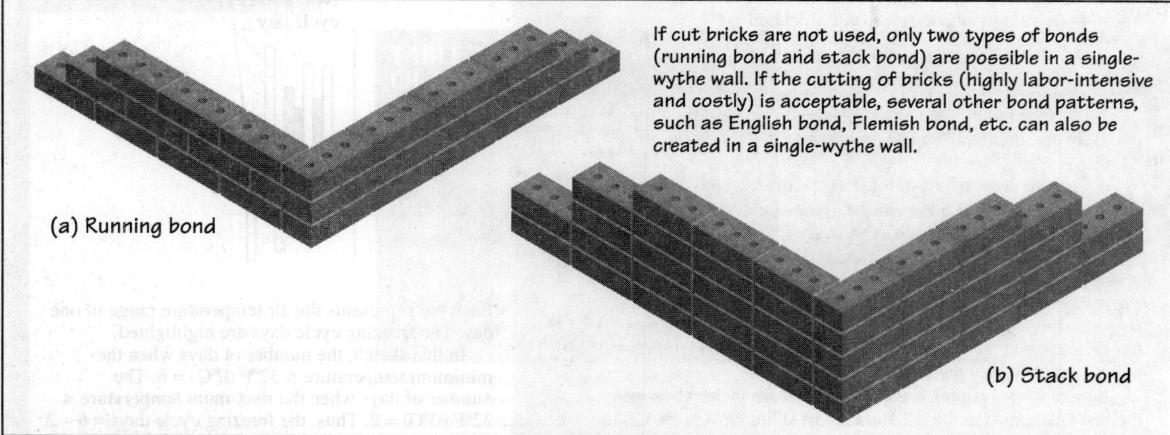

If cut bricks are not used, only two types of bonds (running bond and stack bond) are possible in a single-wythe wall. If the cutting of bricks (highly labor-intensive and costly) is acceptable, several other bond patterns, such as English bond, Flemish bond, etc. can also be created in a single-wythe wall.
(a) Running bond
(b) Stack bond

FIGURE 24.21 Single-wythe walls in **(a)** running bond and **(b)** stack bond.

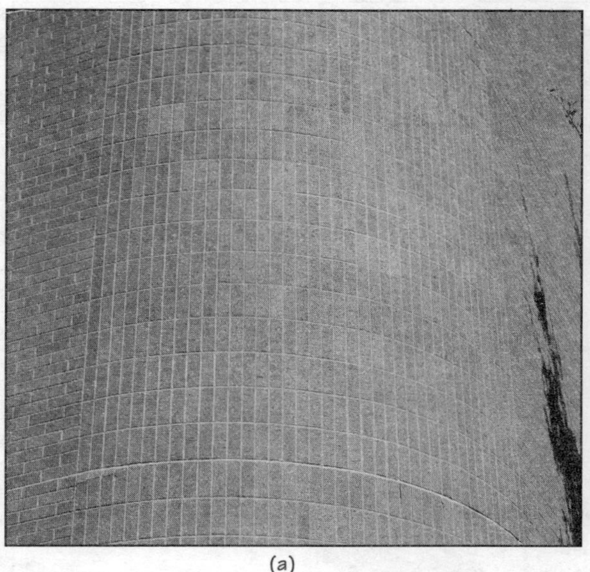

(a)

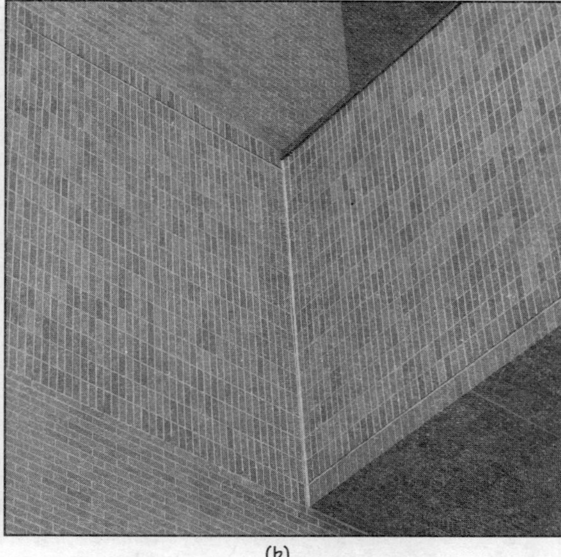

(b)

FIGURE 24.22 A stack-bonded brick wall used **(a)** to achieve a sharp curve and **(b)** for aesthetic reasons.

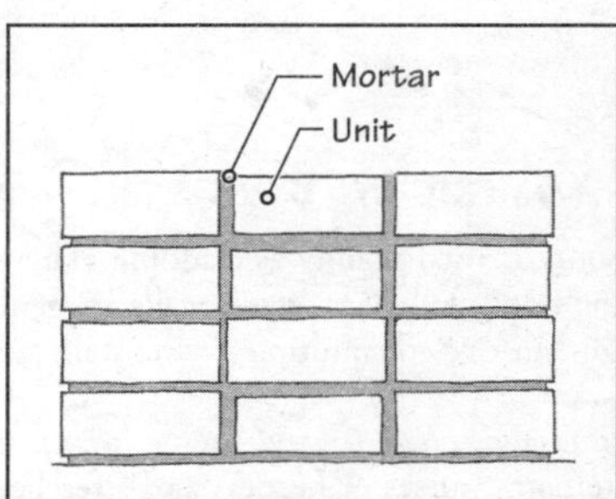

FIGURE 24.23 A stack-bonded wall with large variations in masonry unit sizes is visually unacceptable.

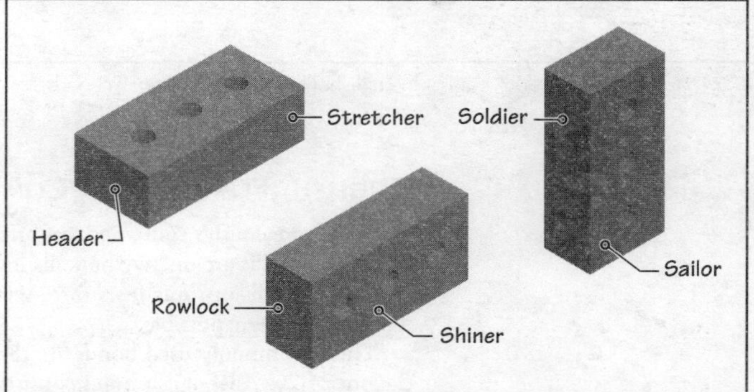

FIGURE 24.24 Terms used to distinguish between the six orientations of a brick unit in a wall elevation.

When a course of masonry consists of stretchers only (as in the running-bond and stack-bond walls of Figure 24.21), it is called a *stretcher course.* If it consists of headers only, it is called a *header course,* and so on. Single-wythe walls are generally made of all stretcher courses (Figure 24.21) and are often terminated at the top with one or more soldier courses to cover the core holes, Figure 24.25.

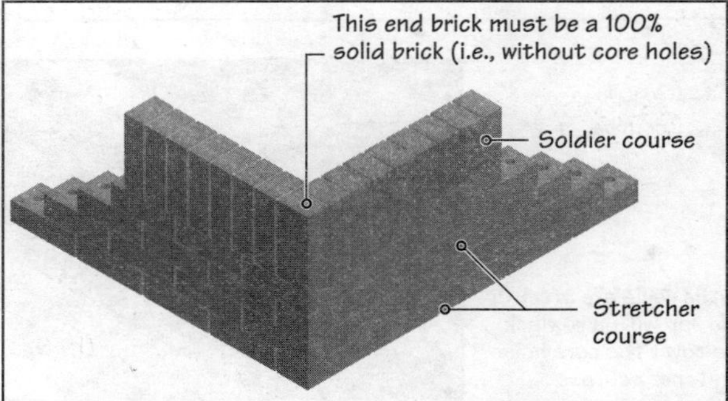

FIGURE 24.25 A single-wythe wall is often terminated with a soldier course at the top to cover the core holes in stretcher courses.

NOTE

The Terms: Masonry Wythe and Masonry Course

A masonry *wythe* is a vertical section of masonry that is one unit thick. The walls shown in Figure 24.21 are both one-wythe walls. Thus, if a wall is built of modular bricks and its thickness is 4 in. nominal, it is a one-wythe wall. If the wall is 8 in. (nominal) thick, it is a two-wythe wall. A 12-in.-thick wall is a three-wythe wall, and so on. The walls shown in Figure 24.26 are two-wythe walls.

A masonry *course* is a horizontal layer of masonry units. The English and Flemish bond walls of Figures 24.26(a) and (b) consist of four courses. The wall of Figure 24.26(c) consists of 14 courses.

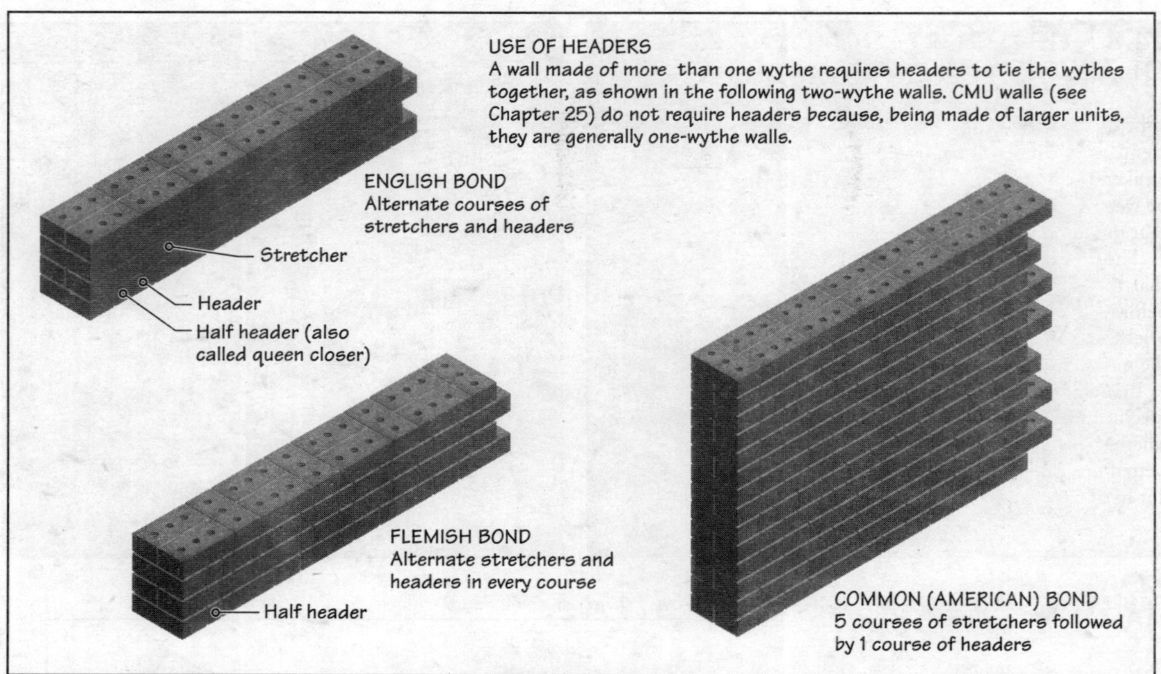
USE OF HEADERS
A wall made of more than one wythe requires headers to tie the wythes together, as shown in the following two-wythe walls. CMU walls (see Chapter 25) do not require headers because, being made of larger units, they are generally one-wythe walls.
ENGLISH BOND
Alternate courses of stretchers and headers
Stretcher
Header
Half header (also called queen closer)
FLEMISH BOND
Alternate stretchers and headers in every course
Half header
COMMON (AMERICAN) BOND
5 courses of stretchers followed by 1 course of headers

FIGURE 24.26 Two-wythe brick walls—in English, Flemish, and Common bonds.

English, Flemish, and Common (American) Bonds

As stated previously, contemporary brick masonry is used mainly as cladding. Therefore, most brick walls are one-wythe walls in running bond. However, fence walls are generally two-wythe walls or sometimes three-wythe walls. In two- or multiple-wythe walls, several bond patterns are possible.

Three commonly used bonds are the *English bond, Flemish bond,* and *American bond,* Figure 24.26. An English bond consists of alternate courses of headers and stretchers. In Flemish bond, there are alternate headers and stretchers in each course. In common (American) bond, there is a header course after every five courses of stretchers.

Just as a single-wythe wall is terminated with a soldier course to cover the core holes, a two-wythe wall is terminated with a rowlock course, Figure 24.27. Alternatively, a precast concrete or stone coping may be used as termination.

The bonds shown in Figure 24.26 use headers to tie the two wythes of masonry. If a header-tied masonry wall is used in a heated or cooled building where the interior and exterior faces of the wall have a large temperature difference, the headers can break. Therefore, in contemporary masonry, English, Flemish, or American bonds, are normally used in fence walls in which the opposite faces are not exposed to temperature differentials.

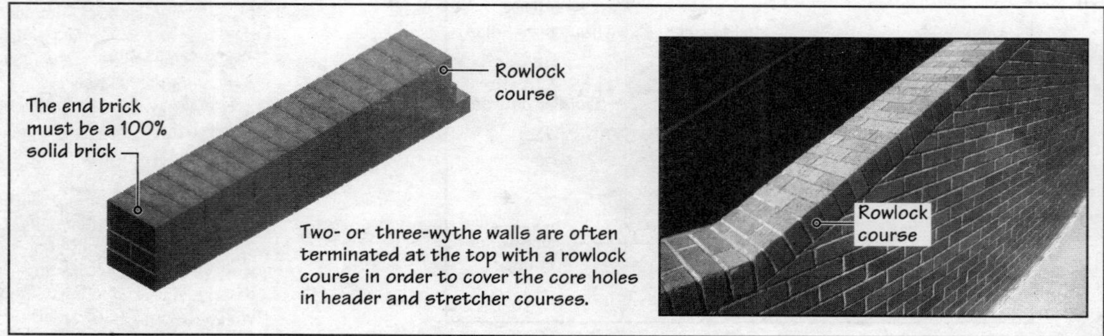
The end brick must be a 100% solid brick
Rowlock course
Two- or three-wythe walls are often terminated at the top with a rowlock course in order to cover the core holes in header and stretcher courses.
Rowlock course

FIGURE 24.27 Rowlock course as the terminal course in multiwythe brick walls.

24.8 THE IMPORTANCE OF THE INITIAL RATE OF ABSORPTION (IRA) OF BRICKS

If bricks are too porous, they can suck too much water out of the mortar, leaving insufficient water for the hydration reaction (between portland cement and water). This can weaken the mortar and, hence, the wall. Bricks that are highly absorptive must, therefore, be wetted with water prior to being laid in the wall. The wetting of bricks should be such that the interior of the brick is moist but the exterior surfaces of the brick are dry.

The preferred way of wetting absorptive bricks is to let water run on the pile of bricks so that the bricks become wet. Depending on the weather, this may be done the evening before (or a few hours before) the bricks are laid. This ensures that the interiors of the bricks are wet but the exteriors are dry. Bricks that are wet on the surface tend to float on the mortar bed and may also bleed water out of the mortar.

Bricks that are not highly absorptive do not need wetting and can be laid dry. Brick specifications use a measure referred to as the *initial rate of absorption* (IRA) to determine whether the bricks require wetting or not. If IRA > 30 g/min/30 in.^{2}, the bricks should be wetted prior to laying; otherwise, not. (Note that in colder regions, wetting the bricks may not be appropriate.)

24.9 THE CRAFT AND ART OF BRICK MASONRY CONSTRUCTION

The construction of a brick wall can appear deceptively simple to a lay observer, particularly in the hands of an expert mason. Compared to wood framing or concrete placing, however, the construction of brick walls requires much greater skill and craft. Additionally, it is slow and labor intensive. To speed construction, several masons work simultaneously on building a wall. This requires coordination and quality control to ensure uniform workmanship.

The first step in constructing a brick wall is to lay the corners. The string line is then stretched between the corners and the bricks are laid to the string line, maintaining a level and plumb wall. In addition to the string line, the primary tools used by a mason are a trowel and a level. Some masons use a story pole to give uniformity in course heights.

The mortar joints are tooled after the mortar has become sufficiently hard but is still pliable (referred to as *thumbprint hard*). Figure 24.28 shows a typical sequence of steps involved in constructing a brick wall.

REPOINTING OF MORTAR JOINTS

Mortar may deteriorate due to freeze-thaw action and/or erode from water. This leads to receding joint depth, resulting in a leaky wall. The recessed joints need to be refilled with mortar—a process referred to as *repointing.*

Repointing of mortar joints used to be far more common when lime mortar was used. It is relatively rare in contemporary masonry because of the use of some portland cement in mortar. However, there are situations, particularly with Type O mortar, in which the joints in some masonry walls may need repointing. Repointing involves raking the existing joints and filling them with mortar.

NOTE

Retempering Mortar

Retempering mortar (adding of water to unused mortar that has somewhat dried) is acceptable provided that the mortar has not stiffened excessively. In general, mortar should be prepared immediately prior to its use. If needed, the plastic life of unused mortar can be extended somewhat by covering it with a plastic sheet.

Retempering reduces the strength of mortar, but if it is done within $2\frac{1}{2}$ h of initial preparation, the results are acceptable. Therefore, mortar prepared more than $2\frac{1}{2}$ h (depending on the weather) prior to its use should be discarded. Retempering colored mortar may result in undesirable color variation.

EXPAND YOUR KNOWLEDGE

Initial Rate of Absorption (IRA) and Its Measurement

The IRA is obtained by immersing a dry brick in water to a depth of $\frac{1}{8}$ in. The immersion is done for only 1 min, after which the brick is removed. The brick is weighed before and after immersion. If the weight of the dry brick is W_d and that of the wet brick is W_w, then IRA is given by

$$IRA = \frac{W_w - W_d}{A_n}(30 \text{ in.}^2) \quad \text{where } A_n = \text{the immersed area of the brick}$$

If IRA > 30 g/min/30 in.^{2} of the brick's immersed surface area, the bricks should be wetted; otherwise, not. A surface area of 30 in.^{2} has been chosen because it gives a convenient figure of 1 g of water absorption per square inch of a brick's surface.

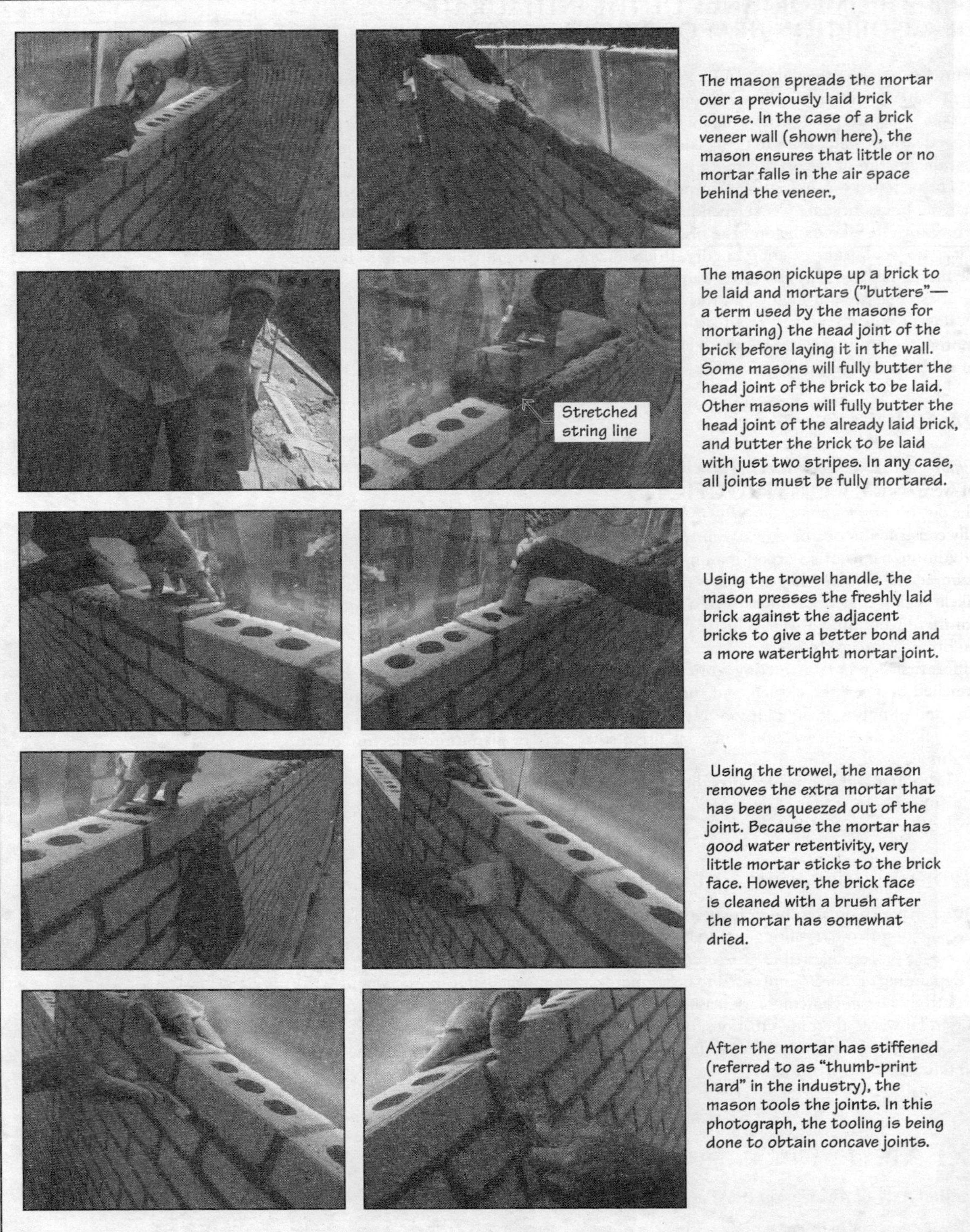

The mason spreads the mortar over a previously laid brick course. In the case of a brick veneer wall (shown here), the mason ensures that little or no mortar falls in the air space behind the veneer.,
The mason pickups up a brick to be laid and mortars ("butters"—a term used by the masons for mortaring) the head joint of the brick before laying it in the wall. Some masons will fully butter the head joint of the brick to be laid. Other masons will fully butter the head joint of the already laid brick, and butter the brick to be laid with just two stripes. In any case, all joints must be fully mortared.
Stretched string line
Using the trowel handle, the mason presses the freshly laid brick against the adjacent bricks to give a better bond and a more watertight mortar joint.
Using the trowel, the mason removes the extra mortar that has been squeezed out of the joint. Because the mortar has good water retentivity, very little mortar sticks to the brick face. However, the brick face is cleaned with a brush after the mortar has somewhat dried.
After the mortar has stiffened (referred to as "thumb-print hard" in the industry), the mason tools the joints. In this photograph, the tooling is being done to obtain concave joints.

FIGURE 24.28 Steps involved in constructing a single-wythe brick wall.

The left-side image shows a concave joint tool and the right-side image shows a raked joint tool.

FIGURE 24.28 (*Continued*) Steps involved in constructing a single-wythe brick wall.

Raking involves grinding into the joint to obtain adequate depth (typically, twice the mortar joint thickness) and filling the joint with the new mortar, Figure 24.29. Before filling of the joint, the joint is wetted to improve the bond between the existing and new mortar.

24.10 EFFLORESCENCE IN BRICK WALLS

Efflorescence is the deposit of a white substance on a masonry wall. This substance consists of water-soluble salts, present in masonry units, in the mortar, or both, which migrate to the outside of the units as the water evaporates, Figure 24.30. Efflorescence does not generally create any structural or sanitation problem, but it is unsightly.

Although efflorescence may occur in both concrete and clay masonry, it is more common and more obvious in brick masonry walls because of the dark color of bricks. The likelihood of efflorescence can be reduced by selecting masonry units that have been tested for the absence of soluble salts (per ASTM C67).

Efflorescence may also originate from mortar due to soluble salts in water or sand. Minor efflorescence can be removed by washing the wall, allowing it to dry, and then brushing off

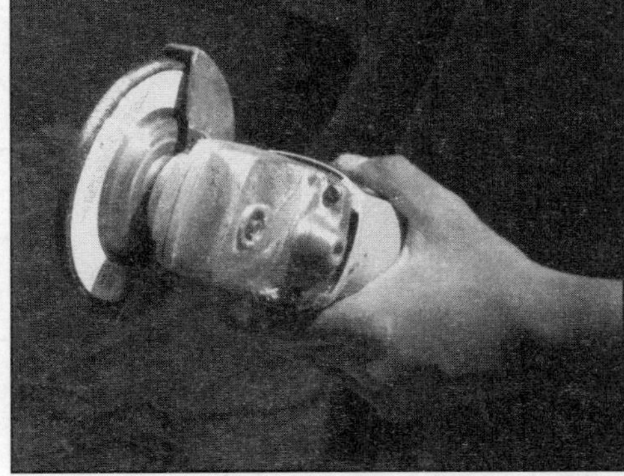

FIGURE 24.29 Grinding of mortar joints in preparation for repointing.

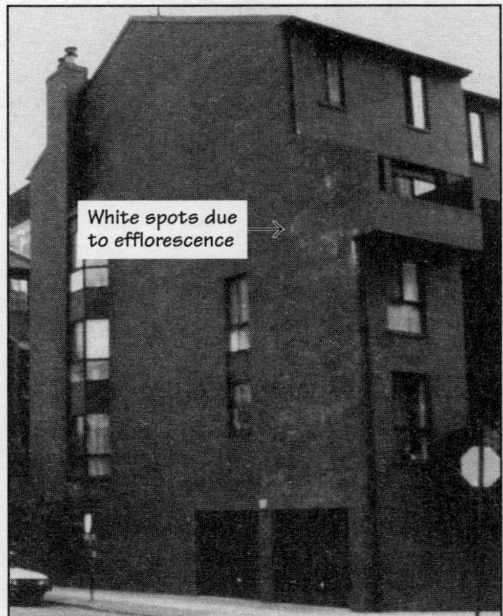

FIGURE 24.30 Efflorescence (white spots) on a brick wall.

the dry salt. Repeating this cycle a few times should leach out all the salts. Efflorescence that appears several years after the wall was built could be due to unclean water penetrating the wall, such as from a lawn sprinkler system, or leakage from a roof.

24.11 EXPANSION CONTROL IN BRICK WALLS

As described in Chapter 9, brick walls expand irreversibly due to the absorption of moisture. They also expand or contract due to temperature changes and foundation settlements. Therefore, brick masonry is divided into segments, with continuous vertical joints between the segments. Because expansion of brick walls is more dominant than contraction, the vertical joints are detailed as expansion joints.

Most brick masonry is used as veneer in the exterior cladding of buildings, covered in Chapter 28. Expansion control of brick veneer is covered in that chapter.

PRACTICE QUIZ

Each question has only one correct answer. Select the choice that best answers the question.

27. Per ASTM standards, facing bricks are classified as Types FBS, FBX, and FBA. Which of these types has the most stringent requirements for dimensional tolerance, chippage, and warpage?
a. FBS **b.** FBX
c. FBA

28. The freeze-thaw resistance of bricks has been found to be a function of the
a. compressive strength of bricks.
b. boiling-water absorption of bricks.
c. saturation coefficient (SC) of bricks.
d. all of the above.
e. (a) and (b).

29. The weathering index (WI) of a location is a function of the
a. average annual freezing-cycle days.
b. average annual winter rainfall of the location.
c. average annual air temperature of the location.
d. all of the above.
e. (a) and (b).

30. In a running-bond wall
a. each course consists of headers.
b. each course consists of stretchers.
c. each course consists of alternate headers and stretchers.
d. the courses alternate between headers and stretchers.
e. none of the above.

31. In an English-bond wall,
a. each course consists of headers.
b. each course consists of stretchers.
c. each course consists of alternate headers and stretchers.
d. the courses alternate between headers and stretchers.
e. none of the above.

32. In a Flemish-bond wall,
a. each course consists of headers.
b. each course consists of stretchers.
c. each course consists of alternate headers and stretchers.
d. the courses alternate between headers and stretchers.
e. none of the above.

33. IRA is a measure of the tensile strength of a brick.
a. True **b.** False

34. Bricks that are highly absorptive may need to be wetted before being laid in the wall.
a. True **b.** False

35. Retempering of mortar means
a. adding portland cement to mortar.
b. adding lime to mortar.
c. adding sand to mortar.
d. adding water to mortar.

36. Retempering of mortar, if needed, should be generally done within
a. $5\frac{1}{2}$ h of the initial preparation of mortar.
b. 4 h of the initial preparation of mortar.
c. $2\frac{1}{2}$ h of the initial preparation of mortar.
d. 1 h of the initial preparation of mortar.
e. none of the above.

37. Efflorescence in masonry refers to
a. excessive chippage of masonry units.
b. excessive warpage of masonry units.
c. white spots in masonry walls.
d. yellow spots in masonry walls.
e. none of the above.

REVIEW QUESTIONS

1. Describe the importance of lime in mortar.
2. Describe the importance of portland cement in mortar.
3. Give the ASTM classification of mortar types and their respective applications.
4. Explain why the masonry industry recommends the use of the weakest mortar that will give the required performance.
5. Using sketches and notes, describe various commonly used mortar joint profiles. Which profile gives the most watertight wall and why?
6. Explain the difference between the actual, nominal, and specified dimensions of masonry units.
7. Bricks may carry a grade of SW, MW, or NW. Which factors determine the grade?
8. Explain what the weathering index (WI) means, and how it relates to building construction.
9. Explain what efflorescence is and how it can be mitigated.

CHAPTER 25

Masonry Materials–II (Concrete Masonry Units, Natural Stone, and Glass Masonry Units)

CHAPTER OUTLINE

This chapter is a continuation of the previous chapter and deals with masonry materials not covered there—concrete masonry units, natural stone, and glass masonry units. Because joint reinforcement and grout are an integral part of most concrete masonry walls, these are also discussed in this chapter.

Steel reinforcement, used in reinforced masonry, is the same as that used in reinforced concrete construction and was covered in Chapter 21. Masonry construction systems are discussed in Chapters 26 and 28.

25.1 CONCRETE MASONRY UNITS—SIZES AND SHAPES

Concrete masonry units (CMUs), also called *concrete blocks,* are more versatile and more complex than clay bricks. A much larger number of surface finishes and colors can be obtained in CMU than in brick.

Additionally, a CMU is a much larger unit than a brick. A typical CMU occupies 12 times the space of a modular brick. Thus, laying one CMU is equivalent to laying 12 bricks, which reduces the labor cost in erecting a wall. Because of its much larger size, a CMU contains large voids.

As shown later in this chapter, the voids in a CMU wall are continuous through its height. This arrangement allows the voids to be filled with grout (which is like concrete) and reinforced vertically—a requirement for load-bearing walls and walls required to resist

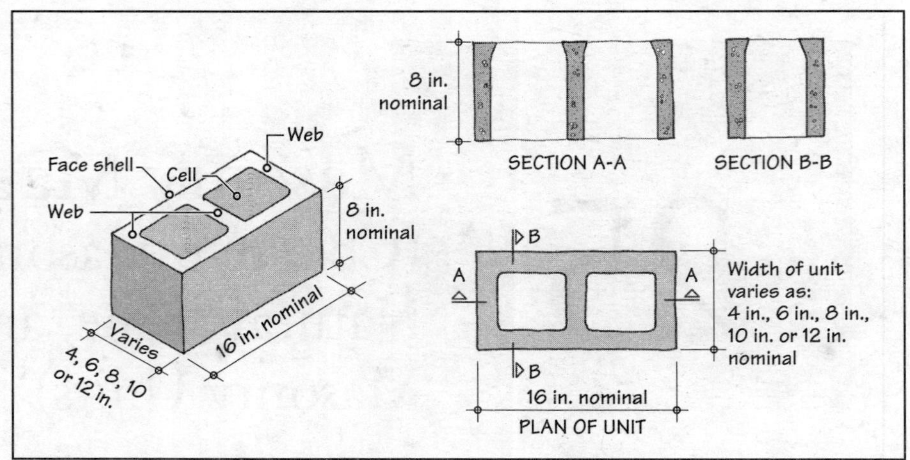

FIGURE 25.1 Sizes of CMUs.

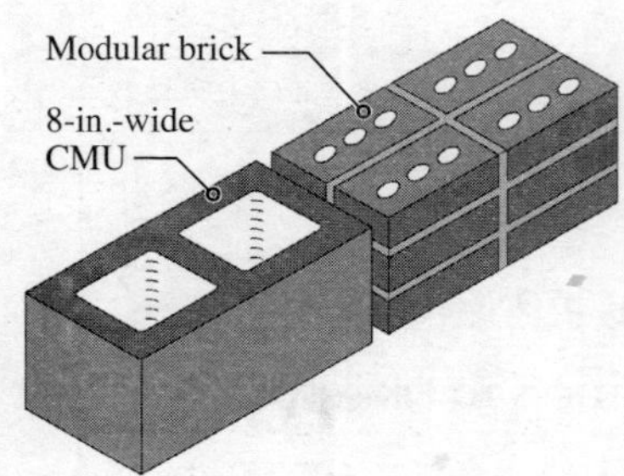

Relative sizes of CMU and bricks.
Volume of 1 CMU = volume of 12 modular bricks

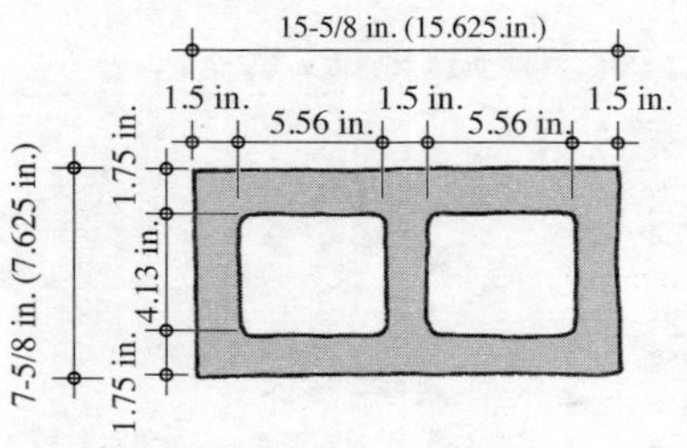

Assuming that the average thickness of face shells and webs is 1.75 in. and 1.5 in., respectively, the cells in a typical 8-in. unit are 4.13 in. × 5.56 in. = 23 in.^{2}

lateral loads. Unlike clay bricks, which are generally used in face veneers, CMUs are used in both structural and face veneer applications.

SIZES OF CMUS

CMUs are available in several different sizes. The shape of a typical unit is shown in Figure 25.1. The length and height of all units are generally the same: length = 16 in. nominal ($15\frac{5}{8}$ in. specified) and height = 8 in. nominal ($7\frac{5}{8}$ in. specified). The width varies from 4 in. to 12 in. nominal in steps of 2 in., that is, width = 4, 6, 8, 10, 12, or 14 in. Because the length and height of a CMU are generally fixed, a CMU's size is generally specified by its width. An 8-in.-wide unit is most commonly used.

The face shells and the webs of a unit are generally flared at the top of the unit as laid in the wall. Thus, a unit has a top and a bottom. The taper helps the mason in lifting and placing the unit in a wall. ASTM specifications provide a minimum thickness of face shells and webs for CMUs of different sizes. For example, an 8-in. CMU must have a minimum face shell thickness of $1\frac{1}{4}$ in. at the thinnest point and a 1-in. minimum thickness for the webs.

The voids in a CMU are called *cells.* Most manufacturers make a two-cell CMU. The longitudinal walls of a CMU are called *face shells,* and the transverse walls are called *webs.* Thus, a two-cell unit has three webs. In an 8-in. unit, the cells are approximately 23 in.^{2}—a space that is large enough for reinforcing and grout (see Section 25.5 for a discussion of grout).

SCORED, BULLNOSE, AND SASH UNITS, AND UNITS WITH PROJECTING FACE SHELLS

Various shapes of CMUs are available to serve different functions in a wall. A typical manufacturer may make 20 or more different shapes and several different finishes. When all sizes, shapes, surface finishes, and colors are considered, a CMU manufacturer may have hundreds of different products.

Although the unit shape of Figure 25.1 constitutes the bulk of a CMU wall, a half-unit is required at the end of a wall, Figure 25.2, and is generally provided as a standard unit by the manufacturers. Where an odd-dimension unit is needed, a full-size unit may be sawed at the site. The sawing of units is time and labor intensive and raises the cost of construction.

Figure 25.3 shows a bullnose unit and a scored unit. Bullnose units are used where a sharp wall corner is to be avoided. When scored units are used, the wall has the appearance of a stack-bond wall but the strength of a running-bond wall, Figure 25.4 (see also Figure 25.23).

The units shown in Figure 25.5 (top) have projecting face shells at one or both ends. They are generally used interchangeably with the flush-end units of Figure 25.1. However, as described later, a CMU is generally mortared on face shells only. The projecting face shells make it more convenient to mortar the head joints. Units with projecting face shells are also used in control joints (see Figure 25.25 and Figure 9.18). Control joints can also be made with sash units, Figure 25.5 (bottom).

Cutting CMU, as shown here, is
to be avoided as much as
possible to reduce construction
cost and increase efficiency.

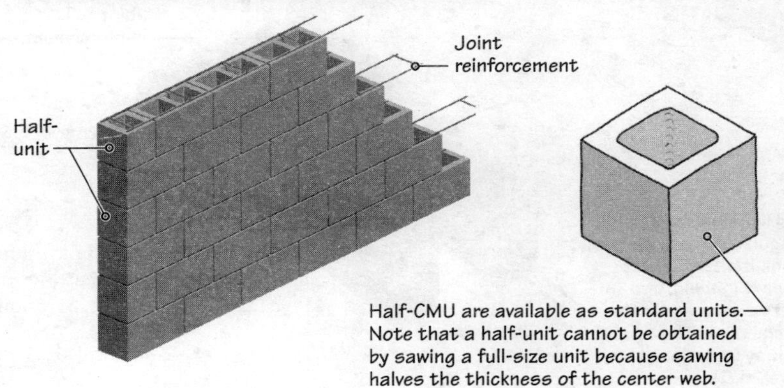

Note that a half-unit cannot be obtained
by sawing a full-size unit because sawing

FIGURE 25.2 Half-units and their use in a wall.

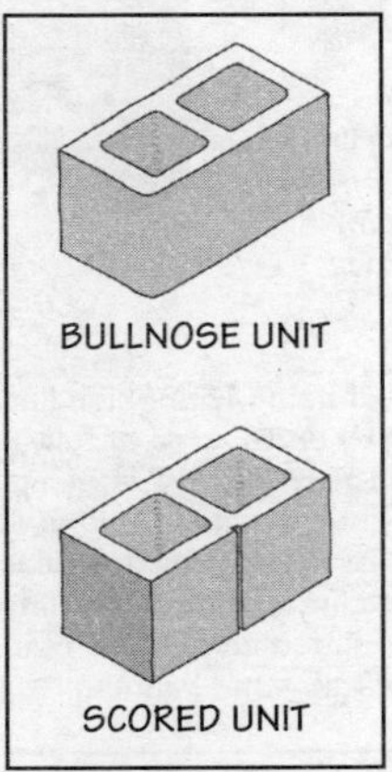

FIGURE 25.3 Bullnose and scored units.

LINTEL UNIT

A lintel unit is U-shaped, Figure 25.6. When grouted and reinforced horizontally, several lintel units in the same course function like a concrete beam that can be used to span an opening. For a large opening requiring a deeper lintel, some manufacturers make a 16-in.-high (two-course-high) unit.

A 16-in.-deep lintel can also be constructed by using an 8-in.-high lintel unit and fully grouting the lintel course and the course immediately above it. Similarly, a 24-in.-deep

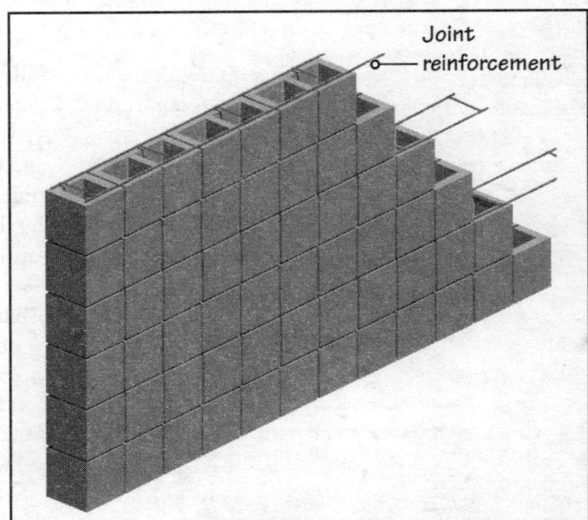

FIGURE 25.4 The use of scored units in a wall.

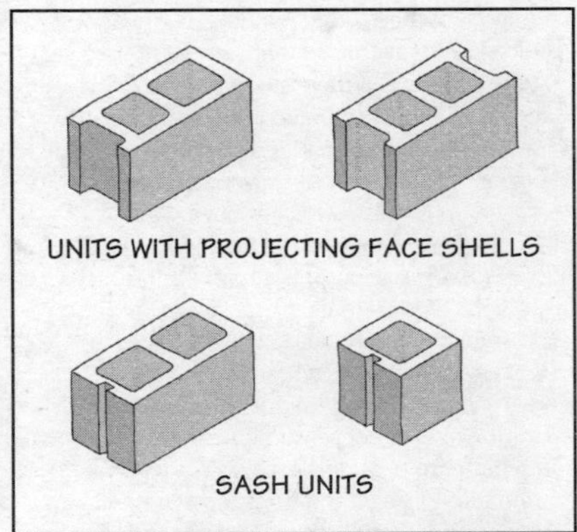

FIGURE 25.5 CMUs with projecting face shells and sash units. CMUs with projecting face shells on one or both ends can be substituted for the plain-end units (Figure 25.1) and can also be used in control joints (see Figure 25.25).

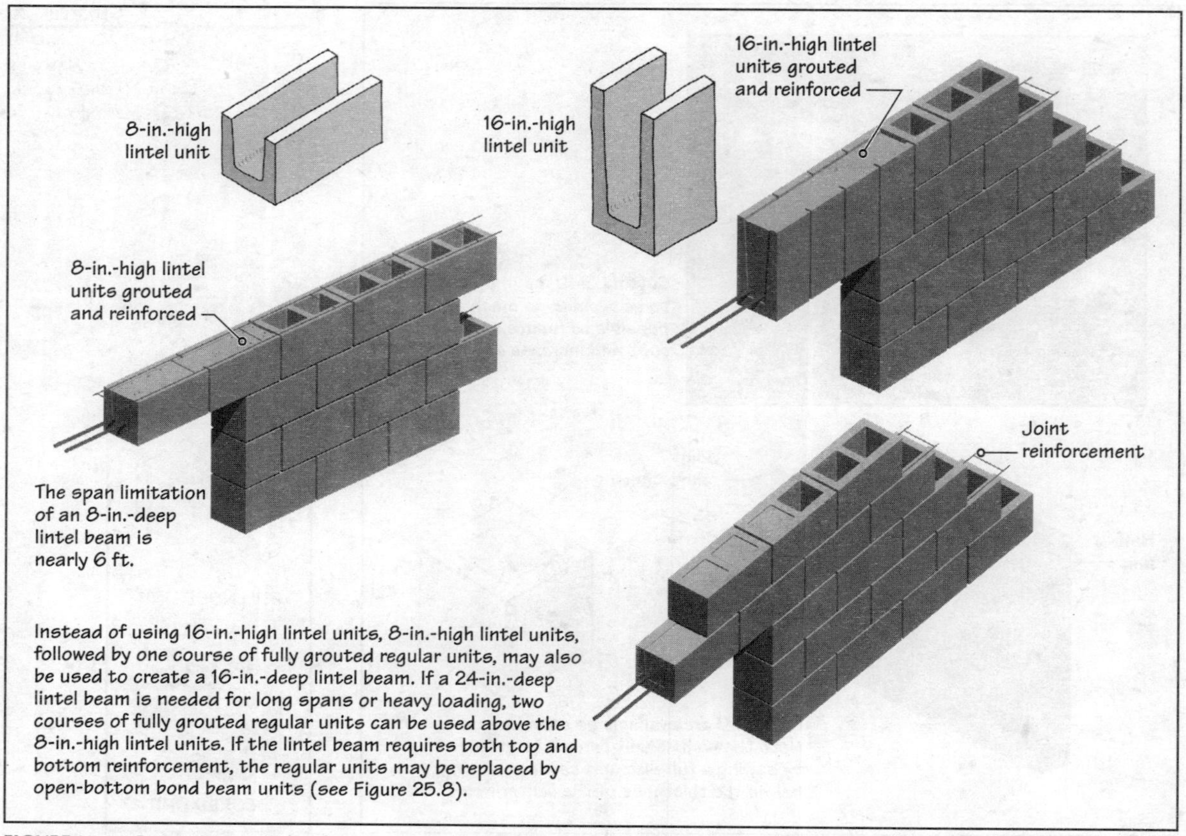

8-in.-high lintel unit

16-in.-high lintel unit

16-in.-high lintel units grouted and reinforced

8-in.-high lintel units grouted and reinforced

Joint reinforcement

The span limitation of an 8-in.-deep lintel beam is nearly 6 ft.

Instead of using 16-in.-high lintel units, 8-in.-high lintel units, followed by one course of fully grouted regular units, may also be used to create a 16-in.-deep lintel beam. If a 24-in.-deep lintel beam is needed for long spans or heavy loading, two courses of fully grouted regular units can be used above the 8-in.-high lintel units. If the lintel beam requires both top and bottom reinforcement, the regular units may be replaced by open-bottom bond beam units (see Figure 25.8).

FIGURE 25.6 Concrete masonry lintels.

lintel can be obtained by grouting two courses of masonry above the standard 8-in.-high lintel.

Lintels in a CMU wall may also be made of steel, precast concrete, or site-cast concrete. Concrete masonry lintels have the advantage of maintaining the bond pattern, color, and surface texture of the wall.

EXPAND YOUR KNOWLEDGE

Arching Action in Masonry Walls

Masonry walls are generally built in running bond, English bond, Flemish bond, and so on. In all of these bond patterns, the head joints are staggered so that a concentrated gravity load, as it travels down toward the foundation, is distributed over an increasingly larger number of masonry units, Figure 1. In other words, a vertical load placed on a wall is directed downward at a 45° angle on either side of the load, Figure 2.

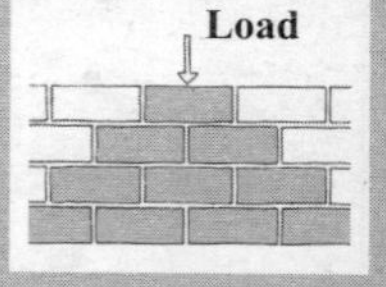

Load

FIGURE 1

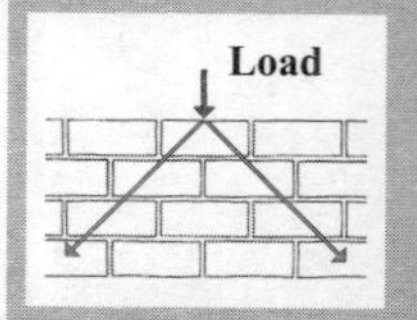

Load

FIGURE 2

The 45° flaring of the load in masonry walls is referred to as the *arching action* because the phenomenon is similar to the behavior of an arch, which transfers the gravity loads placed on it to side abutments. Arching action occurs in all masonry walls in which the head joints in one course are (sufficiently) staggered from those in the upper and lower courses. Arching action is altogether absent in a dry-stacked, stack-bond wall. In a mortared stack-bond wall, some arching action occurs due to vertical shear transfer from one unit to the other through mortar joints, but it is ignored.

It is because of the arching action that sizable holes can be created in an existing masonry wall without causing the wall to collapse. The practical consequence of arching action is that a lintel beam (or simply a lintel) spanning an opening may not receive all of the load placed directly above it, Figure 3. If the height of masonry above the lintel is more than half the width

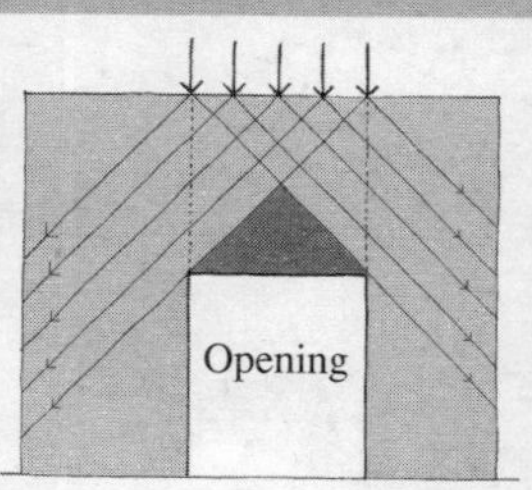

FIGURE 3 A load placed on the top of a masonry wall is directed downward, flaring at 45° from the vertical. The lintel over the opening, therefore, receives the load from the triangular (highlighted) area only.

Arching Action in Masonry Walls (*Continued*)

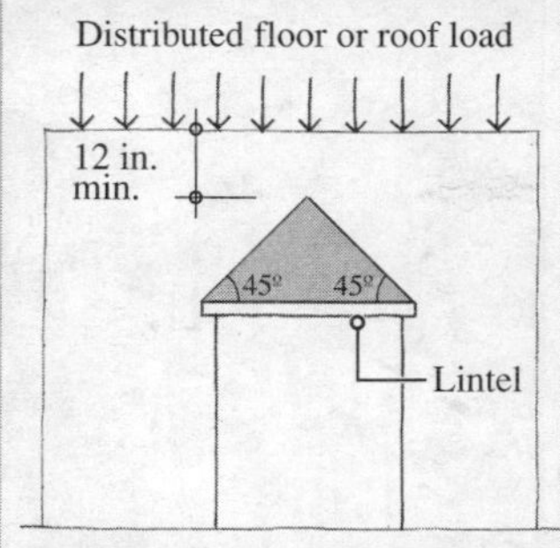

FIGURE 4 The lintel is designed to carry this triangular load of masonry plus the self-load of the lintel. The distributed superimposed (floor or roof) load on the top of the wall is not carried by the lintel. However, a concentrated load on the top of the wall, lying directly above the opening, must be considered a distributed load on the lintel.

of the opening (to enable a 45° triangle to be formed within the wall), the load carried by the lintel is only the load of the masonry within the 45° triangle, Figure 4.

In practice, a 12-in. depth of masonry above the apex of a 45° triangle is needed to assume a triangular load on the lintel. In other words, if the height of masonry above the lintel is half the lintel span plus 12 in., arching action is assumed and the lintel is designed for triangular loading plus the self-load of the lintel. An additional requirement for arching action is that abutments on either side of the opening are sufficiently thick to absorb the lateral thrust caused by the arching action.

If the wall in Figure 4 carries a distributed superimposed load (due to an overlying floor or roof), it is not carried by the lintel but by the wall on either side of the opening. Because this condition occurs commonly, the lintels over openings in masonry walls need only a small amount of steel reinforcement.

However, a concentrated load directly above the opening should be considered a concentrated load on the lintel.

In the absence of arching action, the lintel must be designed for the rectangular load of the wall plus any distributed and concentrated superimposed load above the lintel opening, Figure 5.

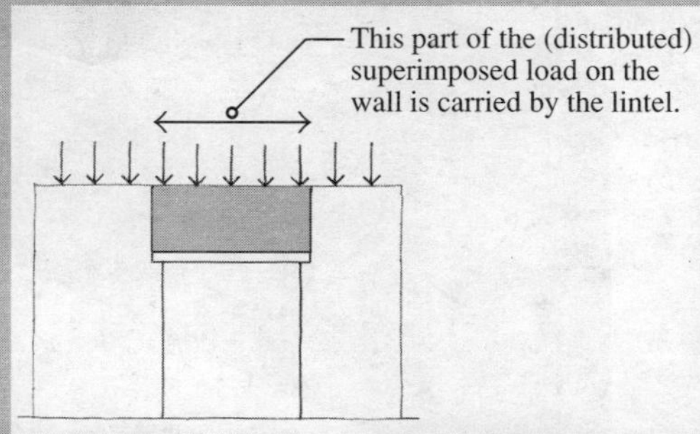

FIGURE 5 In the absence of arching action, the lintel must be designed to carry this rectangular load of masonry plus the self-load of the lintel and any (distributed and concentrated) superimposed wall load directly above the opening.

A-Unit, H-Unit, and Bond Beam Unit

An A-unit is used in a reinforced concrete masonry wall, Figure 25.7. The use of an A-unit eliminates the need to place (string) the unit over the reinforcement or electrical conduit. An H-unit is used where every cell is reinforced. Because the cells in a concrete masonry wall are 8 in. on center, reinforcing each cell implies that the reinforcement in the wall is 8 in. on centers. A concrete masonry wall with reinforcement at 8 in. on centers is a fully grouted wall (no ungrouted cells). Note that a reinforced concrete masonry wall can have reinforcement only at 8 in., 16 in., 24 in., 32 in., and so on, on centers.

Some other commonly used CMUs are bond beam units and pilaster units, Figure 25.8. The use of bond beam units is illustrated later in this chapter (Section 25.4).

Acoustical Units

An acoustical unit is commonly used where a concrete masonry wall with sound-absorptive properties is needed. It has fiberglass embedded behind open slits.

NOTE

Pilasters, Columns, and Piers

Pilasters, columns, and piers are similar elements. However, a *column* refers to an independent, isolated, vertical load-bearing masonry or nonmasonry member. The term *pier* is generally used for a masonry column that is short in height, generally used as foundation for wood frame buildings with an underlying crawl space (see Chapter 12, e.g., Figure 12.24).

A *pilaster* is a column formed by thickening a small area of a masonry wall, which may project on one or both sides of the wall. Like a reinforced concrete column, a reinforced masonry column requires ties. A pilaster does not require ties unless it is provided with compression reinforcement.

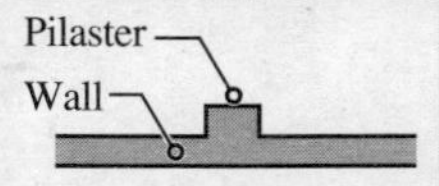

Surface Textures—Split-Face Unit and Ribbed Unit

CMUs can be provided with surface textures other than the (standard) smooth texture. A fairly rough texture is obtained in a split-face unit, Figure 25.9. A split-face unit mimics a rough, stonelike texture and is produced by fracturing a fully hardened double CMU with a guillotine, which produces two split-face units. Split-face units are available with rough texture on one face or on two adjacent faces (corner unit). Half-units are also available.

Another highly textured unit is the ribbed unit, which has several vertical ribs. A ribbed unit is also available as a split-face unit so that the ribs are not smooth. When ribbed units

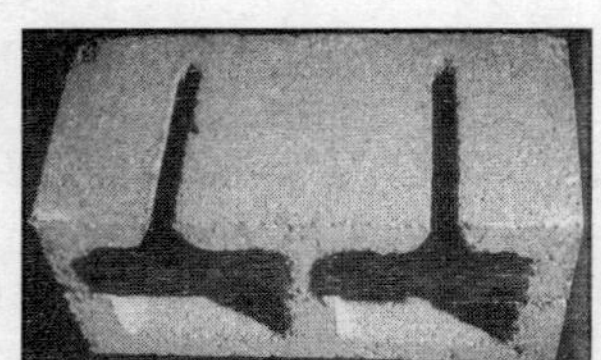

Acoustical unit

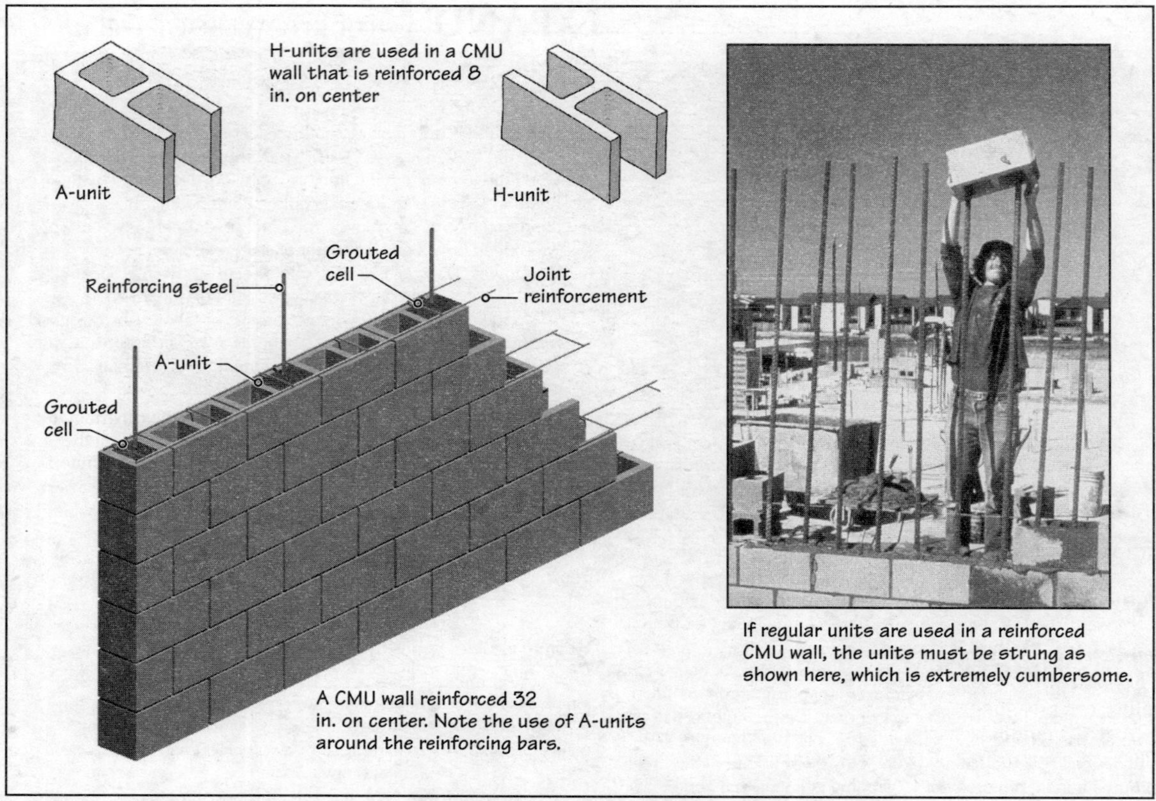

FIGURE 25.7 The use of A-units and H-units in a reinforced CMU wall. (Photo courtesy of Brick Industry Association)

are laid in a running-bond pattern, the ribs align to produce continuous vertical lines, Figure 25.10.

Surface Finishes—Burnished Unit and Glazed Unit

The exposed surface of a CMU can be ground smooth to reveal the aggregate. Variations in aggregate color, size, and type and the use of integral pigments can produce a surface that resembles a smooth-surfaced granite. Ground-face CMUs are called

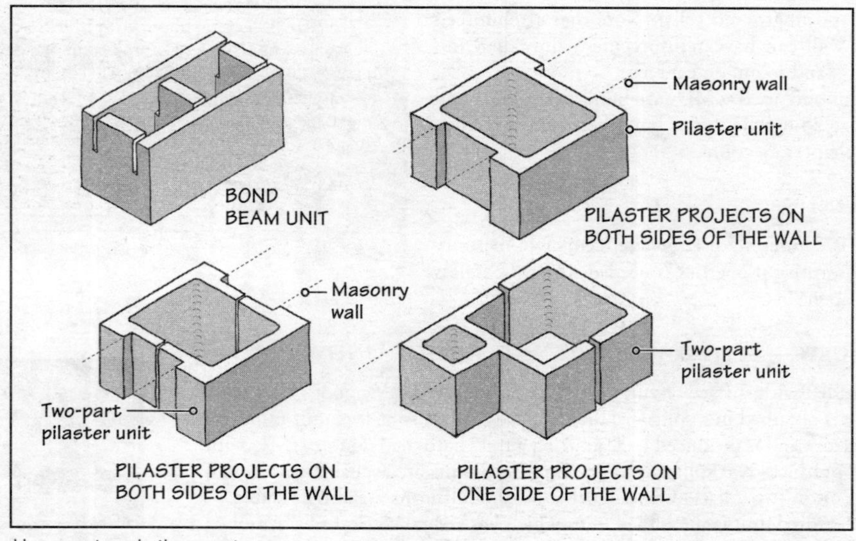

FIGURE 25.8 Bond beam unit and pilaster units.

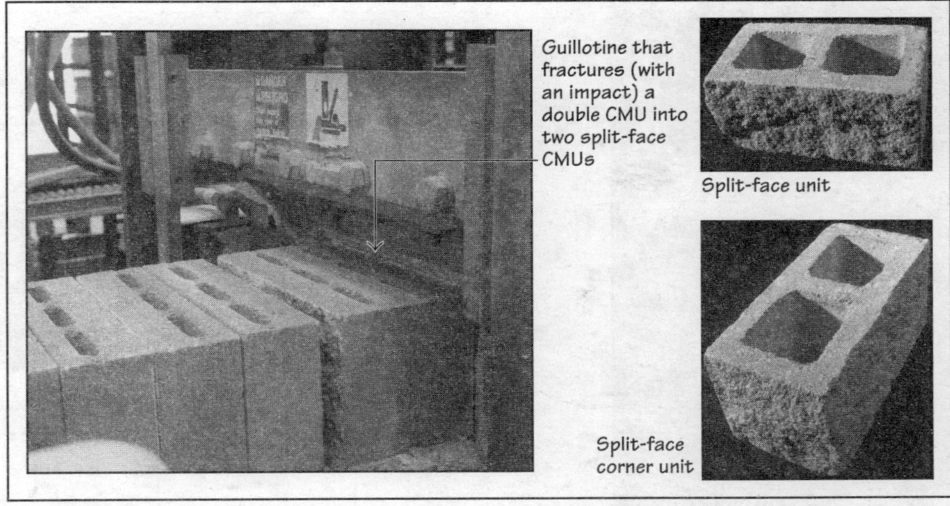

FIGURE 25.9 Split-face units are made by impact fracturing a double CMU through the middle into two units.

burnished units and are often used in interiors where no additional finish is required. They can also be used in an exterior wall, often as accent bands in a wall made with split-face units, Figure 25.11.

A glazed unit has a facing of a glazing material bonded to one or more faces of the unit. The facing, approximately $\frac{1}{10}$ in. thick, is applied after the block has been made. The facing extends a little beyond the CMU face on all sides, Figure 25.12.

A glazed CMU surface is impervious to moisture and dust collection and is easy to clean, so it is sanitary. Glazed CMUs, which are available in various colors, are often used in hospital interiors, public kitchens, and so on—wherever surface cleanliness is important. The suitability of glazed CMUs on exteriors must be verified with the manufacturer to ensure color fastness and resistance to delamination by freeze-thaw action.

FIGURE 25.10 Ribbed units in Crawford Housing, New Haven, Connecticut. Architect: Paul Rudolph. (Photo courtesy of Dr. Jay Henry)

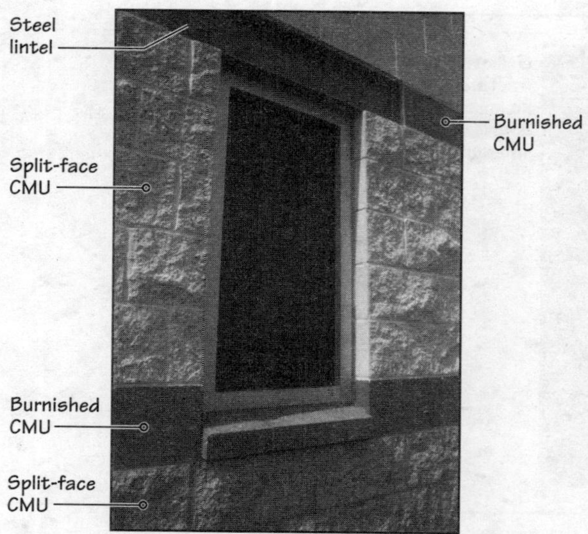

FIGURE 25.11 Use of burnished units as accent bands in a split-face CMU wall.

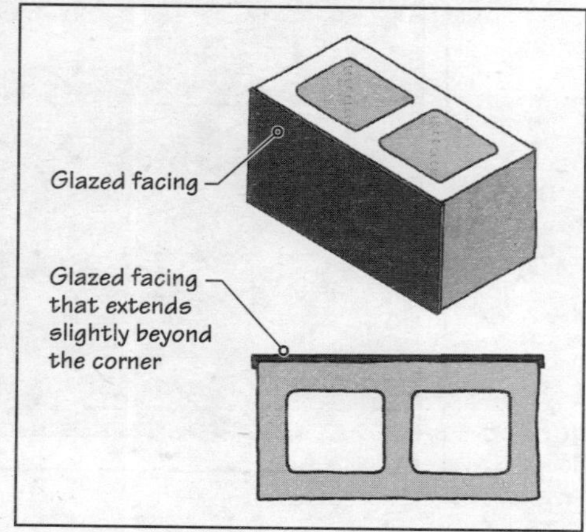

FIGURE 25.12 A glazed concrete masonry unit.

25.2 CONCRETE MASONRY UNITS—MANUFACTURING AND SPECIFICATIONS

CMUs are made from concrete. Therefore, a CMU-manufacturing plant has concrete-making equipment in addition to the equipment required to manufacture the units. Briefly, CMU manufacturing involves casting fairly dry (zero-slump) concrete mix in molds of the desired shape and compacting the concrete.

Because of the zero-slump concrete, a fresh-cast (green) CMU has adequate strength to be removed from the mold immediately. A CMU casting and compacting machine will generally produce three to four units in one pass, Figure 25.13. The green units travel on rollers to be stacked on a multistack cart. As the cart, generally mounted on wheels, gets filled, it is moved to a curing chamber, where the units undergo accelerated curing by warm (approximately 150°F), supersaturated air, Figure 25.14.

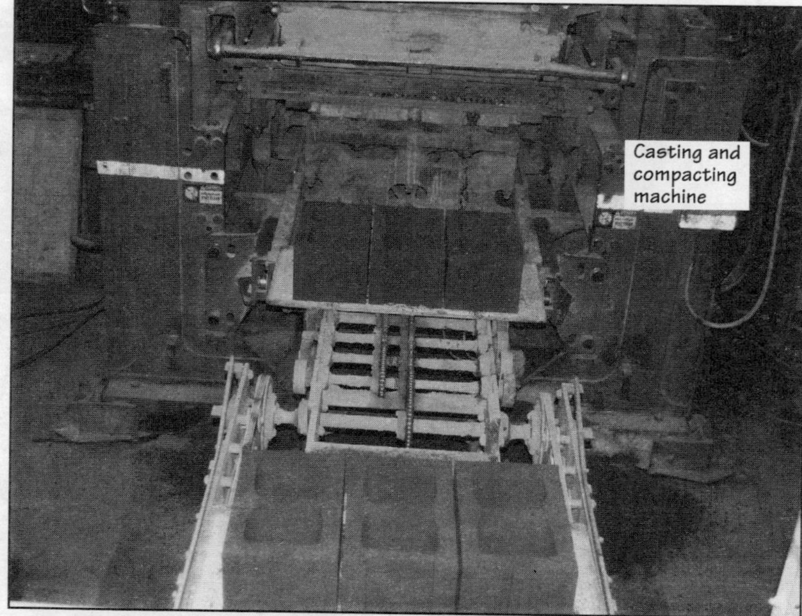

FIGURE 25.13 Fresh-cast (green) CMUs travel on rollers to be stacked on a cart that delivers them to the curing chamber (Figure 25.14).

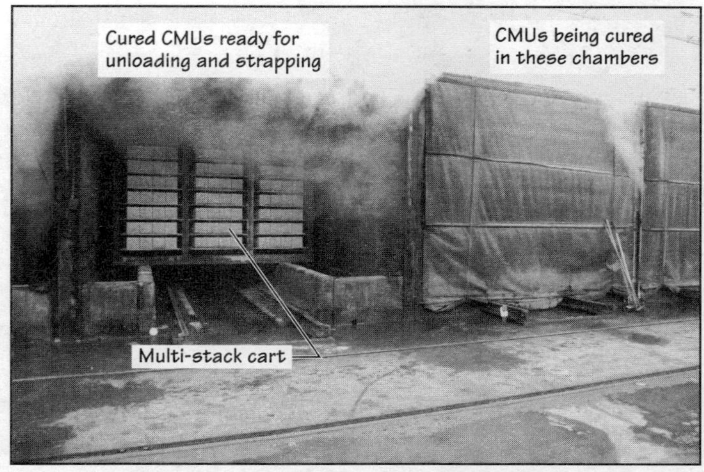

FIGURE 25.14 Curing chambers for accelerated curing of CMUs. Observe the warm saturated-air clouds. (The photograph was taken on a cold day; on a really warm day, clouds may not be visible.)

The units are cured (typically for 10 hours) and then unloaded from the multistack cart to wooden palettes, strapped into cubes, and stored in the yard for shipping to the construction site. Because the manufacturers do not want a large inventory in their storage yards, it is possible that a unit may be sent to the construction site within a day or two of being manufactured.

SPECIFICATIONS FOR CMUs

Virtually all shapes of CMUs can be made from lightweight or normal-weight concrete. A lightweight CMU is made from lightweight coarse aggregate, and normal-weight units are made from normal-weight coarse aggregate. The weight classification is based on the following concrete densities:

- *Lightweight CMU*—dry concrete density < 105 lb/ft^3
- *Normal-weight CMU*—dry concrete density ≥ 125 lb/ft^3

A *medium-weight* unit is also made by some manufacturers, with a dry concrete density of 105 to 125 lb/ft^3. A typical-8 in. (two-cell) lightweight unit weighs approximately 28 lb, and a normal-weight unit weighs approximately 38 lb. Lightweight units are preferred because they reduce the dead load on a structure, and they are easier to work with. The masons may charge less for laying them. However, because lightweight aggregate costs more (see Chapter 21), lightweight units are generally more expensive than normal-weight units.

Lightweight units have the necessary strength for most situations. Where high unit strength is needed, normal-weight units should be considered. In any case, the minimum compressive strength of a CMU used in load-bearing applications is 1,900 psi (based on the net bearing area of the unit).

PRACTICE QUIZ

Each question has only one correct answer. Select the choice that best answers the question.

1. The voids in CMUs are called
   - **a.** voids.
   - **b.** cores.
   - **c.** cells.
   - **d.** frogs.
   - **e.** none of the above.
2. The nominal length of a typical CMU is
   - **a.** 18 in.
   - **b.** 16 in.
   - **c.** 12 in.
   - **d.** 8 in.
   - **e.** variable.
3. The nominal height of a typical CMU is
   - **a.** 18 in.
   - **b.** 16 in.
   - **c.** 12 in.
   - **d.** 8 in.
   - **e.** variable.
4. The nominal width (through-wall thickness) of a typical CMU is
   - **a.** 18 in.
   - **b.** 16 in.
   - **c.** 12 in.
   - **d.** 8 in.
   - **e.** variable.
5. A typical CMU has
   - **a.** one web.
   - **b.** two webs.
   - **c.** three webs.
   - **d.** four webs.
   - **e.** five webs.
6. A typical CMU has
   - **a.** one face shell.
   - **b.** two face shells.
   - **c.** three face shells.
   - **d.** four face shells.
   - **e.** five face shells.

(Continued)

7. Which of the following CMU shapes is used to simulate a stack-bond wall, although the wall is, in fact, made with a running-bond pattern?
   a. Lintel
   b. Bullnose unit
   c. Sash unit
   d. Unit with projecting face shells
   e. Scored unit
8. Which of the following CMU shapes has a rounded corner?
   a. Lintel
   b. Bullnose unit
   c. Sash unit
   d. Unit with projecting face shells
   e. Scored unit
9. CMUs with projecting face shells can be used on both sides of a control joint in a wall. The other unit that is commonly used in the same situation is a
   a. scored unit. b. bullnose unit.
   c. sash unit. d. lintel unit.
10. Which of the following CMU shapes has a U-shaped profile in its vertical cross section?
    a. Lintel unit
    b. Bullnose unit
    c. Sash unit
    d. Unit with projecting face shells
    e. Scored unit
11. Arching action occurs in a masonry wall with
    a. a running-bond pattern.
    b. a stack-bond pattern.
    c. both (a) and (b).
    d. neither (a) nor (b).
12. A lintel over an opening in all masonry walls is designed to carry the load from triangle-shaped superimposed masonry.
    a. True b. False
13. If a CMU wall is vertically reinforced at 16 in. on centers, which of the following units will you recommend for use in the wall?
    a. A-unit b. H-unit
    c. Standard unit with three webs d. Sash unit
    e. Bond beam unit
14. If a CMU wall is vertically reinforced at 8 in. on centers, which of the following units will you recommend for use in the wall?
    a. A-unit b. H-unit
    c. Standard unit with three webs d. Sash unit
    e. Bond beam unit
15. If a CMU wall is vertically reinforced at 24 in. on centers, which of the following units will you recommend for use in the wall?
    a. A-unit b. H-unit
    c. Standard unit with three webs d. Both (a) and (b)
    e. Both (a) and (c)
16. The center-to-center spacing of vertical reinforcement in a CMU wall cannot be less than 8 in.
    a. True b. False
17. Which of the following CMUs has slotted webs that are broken before being laid in the wall?
    a. A-unit b. H-unit
    c. Standard unit with three webs d. Sash unit
    e. Bond beam unit
18. The terms *pilaster* and *pier* are synonymous.
    a. True b. False
19. A split-face CMU is made by treating a standard CMU with an oxyacetylene torch to produce a very smooth surface on one or more faces of the unit.
    a. True b. False
20. A burnished CMU is made by treating a standard CMU with an oxyacetylene torch to produce a very smooth surface on one or more faces of the unit.
    a. True b. False

25.3 CONSTRUCTION OF A CMU WALL

The construction of a CMU wall differs from that of a brick wall in the following ways:

- Because a CMU is much heavier than a brick, it requires two hands to lay instead of the one-hand operation used in laying bricks, Figure 25.15. While it takes greater effort and time to lay a CMU than a brick because of a CMU's much larger size, the construction of a CMU wall is much faster.
- Unlike bricks, whose bed and head joints are fully mortared, a CMU is generally mortared only on its exterior periphery. The masonry industry refers to it as *face-shell mortaring* of the units, Figure 25.16. In face-shell mortaring, the webs of a CMU are not mortared. The exceptions are stack-bond walls and partially grouted CMU walls. Mortaring the webs on both sides of a grouted cell in a partially grouted wall prevents the grout from flowing into ungrouted cells. The first course of a CMU wall is laid over a full bed of mortar regardless of the type of wall.
- A typical CMU wall contains joint reinforcement for shrinkage control, which is not required in brick walls other than a stack-bonded brick wall.
- A reinforced CMU wall is generally *partially grouted* because only cells that contain reinforcement need the grout. Other cells are left ungrouted unless fire-resistance and/or sound-insulation requirements mandate a *fully grouted* wall. A reinforced brick wall, on the other hand, must be fully grouted in all cases—the grout being provided in the space between two masonry wythes, Figure 25.17.

The images in Figure 25.18 highlight some of the similarities and differences between the construction of brick and CMU walls.

NOTE

Why Only Face-Shell Mortaring of CMUs?

1. When a wall bends, the tensile and compressive stresses are concentrated primarily at the front and back faces of the wall. Mortaring the webs, therefore, does not significantly increase a wall's bending strength.
2. Because CMUs are laid in running bond, the webs of a CMU in the upper course do not fully align with the webs of CMUs in the lower course. Therefore, if the mortar is placed on the webs, it will not serve any purpose. Additionally, not mortaring the webs speeds construction.

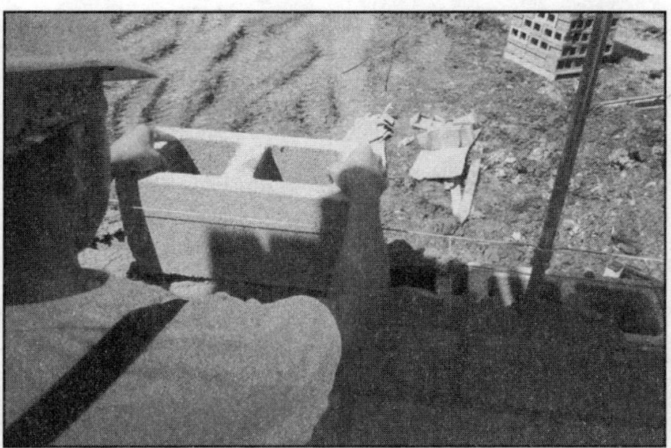

FIGURE 25.15 The laying of a CMU is a two-hand operation versus the one hand used in laying bricks.

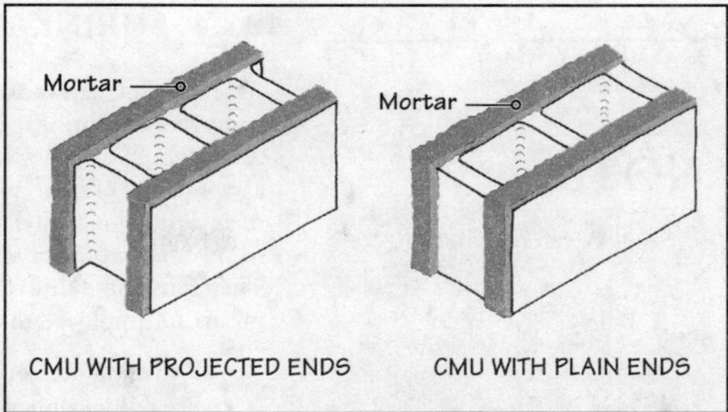

FIGURE 25.16 In a CMU wall, only face shells are generally mortared; that is, the webs are not mortared, and the head joints are mortared to the depth of the face shells.

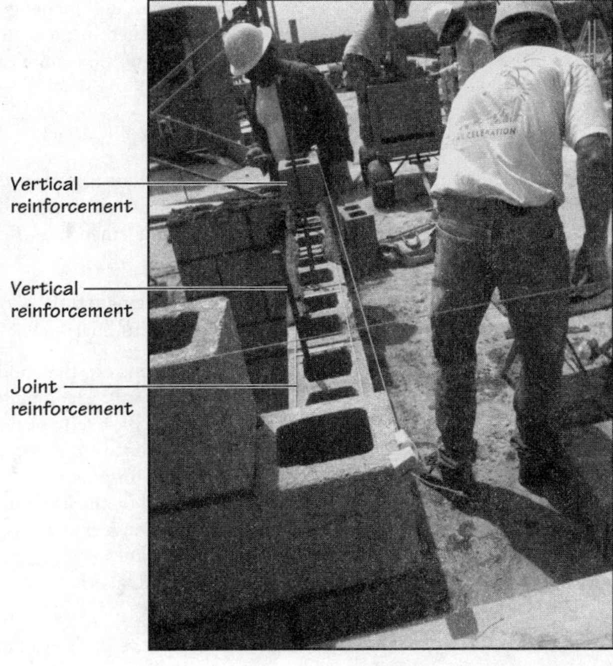

(a)

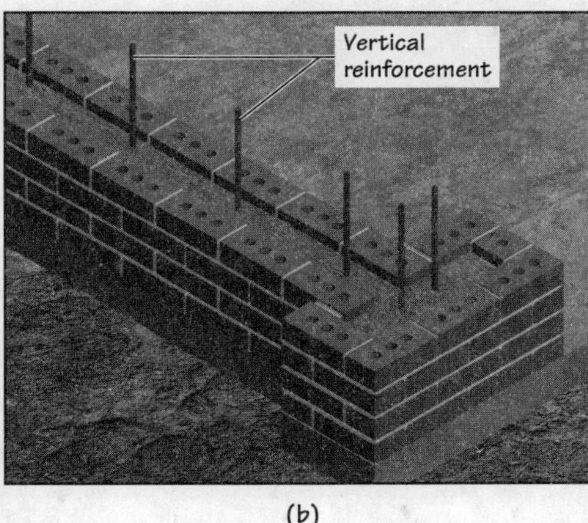

(b)

FIGURE 25.17 (a) A CMU wall is called a *reinforced CMU wall* if it is provided with vertical steel reinforcement. Generally, only CMU cells with reinforcement are grouted. The other cells are left ungrouted, unless the grouting of unreinforced cells is required for greater fire resistance and (or) greater sound insulation of the wall. Note that in this photograph the CMU wall is reinforced 24 in. on center. Observe that the first CMU course is laid in full mortar bed. (b) A reinforced brick masonry wall must be fully grouted in the space between the two wythes of the wall.

(a)

(b)

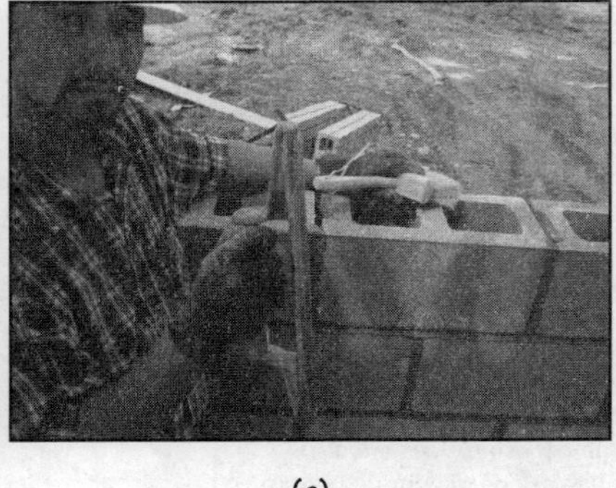

(c)

FIGURE 25.18 (a) The unit is aligned, and pressed downward and sideways using the trowel. (b) The mortar is brushed off the face of the walls when it has dried sufficiently. (c) The concave joint tool used with CMUs is longer than the one used with brick masonry (see Figure 24.28).

25.4 SHRINKAGE CONTROL IN CMU WALLS

As stated in Chapter 9, clay bricks expand during service, whereas CMUs shrink. As with concrete, the shrinkage of CMU continues for several months after their manufacture. Although ASTM specifications require that the manufacturers limit the maximum shrinkage potential of their units to the specified value, it is not economical to let CMUs stabilize their dimensions fully before selling them.

In other words, a CMU that arrives on the construction site may still be shrinking. Therefore, some shrinkage of units is to be expected after they are placed in the wall. Two means are employed to control the shrinkage of CMU walls:

- Providing horizontal reinforcement in the wall
- Providing control (shrinkage) joints so that a long CMU wall is divided into smaller segments

The purpose of horizontal reinforcement is to resist tensile stresses in the wall caused by the shrinkage of units. Note that horizontal reinforcement cannot prevent the formation of shrinkage cracks. Nor can it reduce the total width of cracks in a given length of a wall. However, reinforcement distributes the cracks in the wall by increasing their number and reducing the width of individual cracks—the total width of cracks remaining unchanged. Thus, instead of one or two large cracks, numerous small cracks are formed in the wall if horizontal reinforcement is provided. Small cracks are more water resistant, heal more easily, and can be sealed with a coating.

NOTE

Steel Wire Designation

A cold-drawn steel wire is designated by the cross-sectional area of the wire (see also Section 21.13). Thus, a W1.7 wire has a cross-sectional area of 0.017 in.^{2} (diameter = 0.147 in., or nearly $\frac{1}{7}$ in.). It is also referred to as *No. 9 gauge wire*.

A W2.1 wire has a cross-sectional area of 0.021 in.^{2} It is also referred to as *No. 8 gauge wire*. Some other commonly used wires are

W1.1—gauge No. 11, area = 0.011 in.^{2}

W2.8—$\frac{3}{16}$-in. diameter, no gauge designation, area = 0.028 in.^{2}

W4.9—$\frac{1}{4}$-in. diameter, no gauge designation, area = 0.049 in.^{2}

HORIZONTAL REINFORCEMENT—JOINT REINFORCEMENT

Horizontal reinforcement is generally provided by steel wire reinforcement placed in the mortar joints, referred to as *joint reinforcement.* Joint reinforcement consists of two parallel longitudinal wires welded to cross wires. Two types of cross-wire arrangement are used, Figure 25.19.

- In *ladder-type joint reinforcement,* cross wires are perpendicular to the longitudinal wires.
- In *truss-type joint reinforcement,* cross wires are diagonal to the longitudinal wires.

As we will see in Chapter 28, both ladder- and truss-type reinforcement can have additional wires welded to it to tie a masonry veneer wall to a CMU backup wall. The longitudinal wires in joint reinforcement are embedded in the mortar. This is achieved by placing the joint reinforcement on the masonry bed and then placing the mortar over it, Figure 25.20. Because of face-shell mortaring, the cross wires are not covered by mortar.

The tensile stresses caused by the shrinkage of a CMU wall are resisted by the longitudinal wires. The cross wires simply hold the longitudinal wires together. The maximum permissible diameter of the longitudinal wires is half the mortar joint thickness (i.e., $\frac{3}{16}$ in.), but the smaller-diameter W1.7 wire (No. 9 gauge wire) is most commonly used.

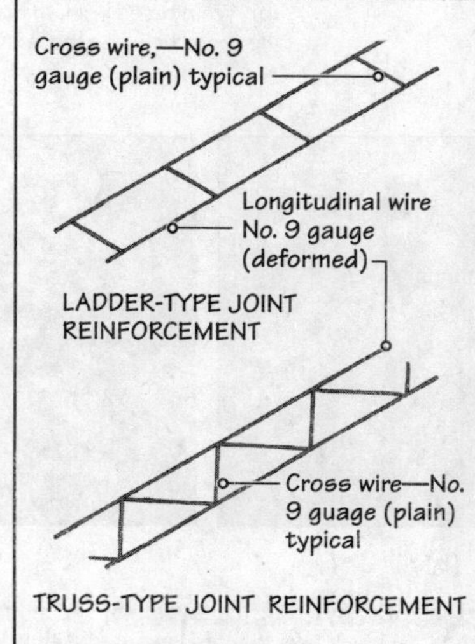

FIGURE 25.19 Types of joint reinforcement commonly used. Although 9-gauge wires are most common, other sizes are also available. The maximum permissible diameter of wire in joint reinforcement is half the mortar thickness.

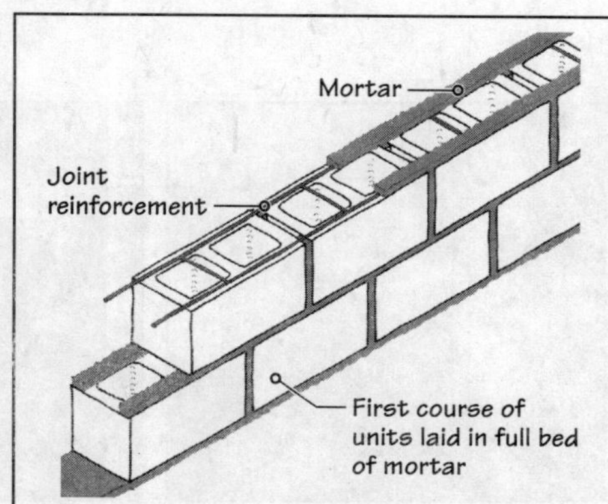

FIGURE 25.20 To embed joint reinforcement in mortar, standard practice is to place the joint reinforcement on the bed surface of the units first and then lay the mortar over it (see also Figure 25.17).

The longitudinal wires are slightly deformed (textured) to improve their bond with mortar. The cross wires (also No. 9 gauge) are generally plain, spaced no more than 16 in. on centers. Hot-dip galvanizing, in which the reinforcement is dipped in a hot bath of zinc, is required for corrosion protection. The thickness of galvanizing (zinc) film required depends on the exposure of the wall to corrosion. Where corrosion protection is critical, stainless steel joint reinforcement is recommended.

The fabrication of joint reinforcement (welding of cross wires with longitudinal wires and cutting to required lengths) should precede galvanizing. The alternative, in which the reinforcement is fabricated with galvanized wires, leaves the welded spots without any protection. If the latter is the mode of fabrication, the welded spots and sheared ends should be coated with zinc.

Joint reinforcement is generally available in 10-ft lengths. The distance between longitudinal wires has been standardized for CMU walls of different thickness. The standard requires that the longitudinal wires be spaced as far apart as possible. However, a minimum $\frac{5}{8}$-in. cover must be provided by the mortar, Figure 25.21. Therefore, for use in an 8-in. CMU wall, the out-to-out distance between wires is $(7\frac{5}{8} - \frac{5}{8} - \frac{5}{8}) = 6\frac{3}{8}$ in.

The vertical spacing of joint reinforcement in a wall is a function of the diameter of longitudinal wires and the thickness of the wall. With No. 9 gauge wires, the recommended spacing is 16 in. on center. This amounts to placing the joint reinforcement every other course of masonry. If $\frac{3}{16}$-in. longitudinal wires are used, the vertical spacing of joint reinforcement can be increased.

More stringent horizontal reinforcement requirements (wire diameter and vertical spacing of joint reinforcement) than those given here for shrinkage crack control may be needed to resist high-wind and seismic loads.

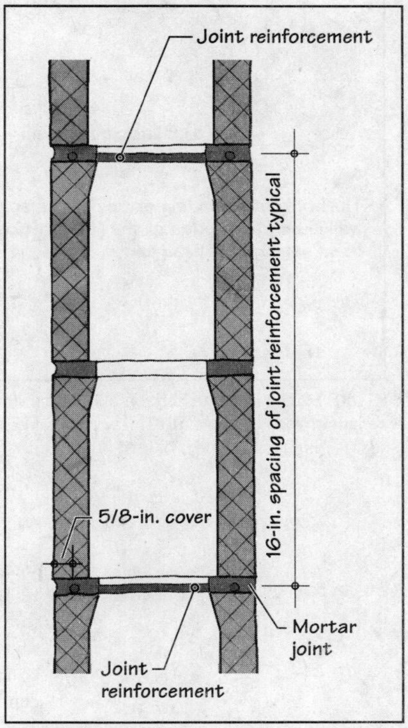

FIGURE 25.21 The width of joint reinforcement for CMU walls is standardized to provide $\frac{5}{8}$-in. mortar cover to longitudinal wires.

Horizontal Reinforcement—Bond Beams

Horizontal reinforcement in CMU walls may also be provided through bond beams. A bond beam is a CMU beam formed by grouting the bond beam units. The slotted webs of bond beam units are broken by the mason before being laid, providing a continuous space for horizontal reinforcement and grouting.

Because bond beam units are open at the top and the bottom, a plastic mesh fabric is placed below the units to prevent the grout from leaking into the lower cells. For shrinkage crack control, bond beams are generally spaced 4 ft on center vertically with only one No. 4 reinforcing bar ($\frac{1}{2}$-in. diameter) in each bond beam, Figure 25.22.

Bond beams are more cumbersome to provide than joint reinforcement. However, the reinforcement in a bond beam is better protected, and a bond beam can be reinforced more heavily to resist high-wind and seismic forces. As we will observe in Chapter 26, bond beams are required for reasons other than crack control in load-bearing masonry structures.

Horizontal Reinforcement and Stack-Bond Walls

A stack-bond wall is much weaker in horizontal bending than a running-bond wall because the continuous head joints open easily, presenting little resistance to horizontal bending, Figure 25.23(a). In a running-bond wall, the masonry units have to overcome the shear (sliding) resistance at the bed joints before the head joints can open, Figure 25.23(b).

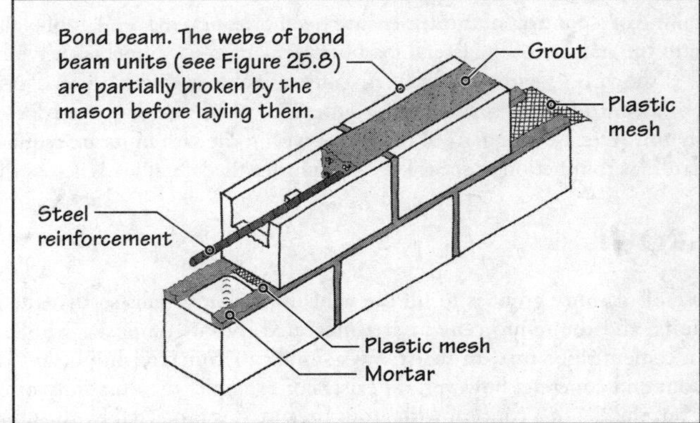

FIGURE 25.22 The use of bond beams for shrinkage crack control.

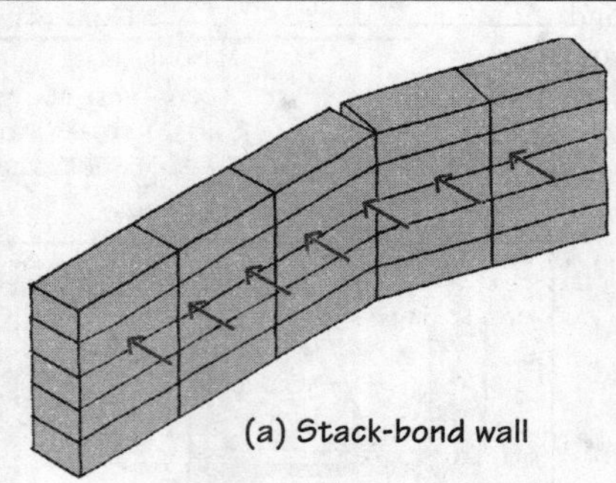

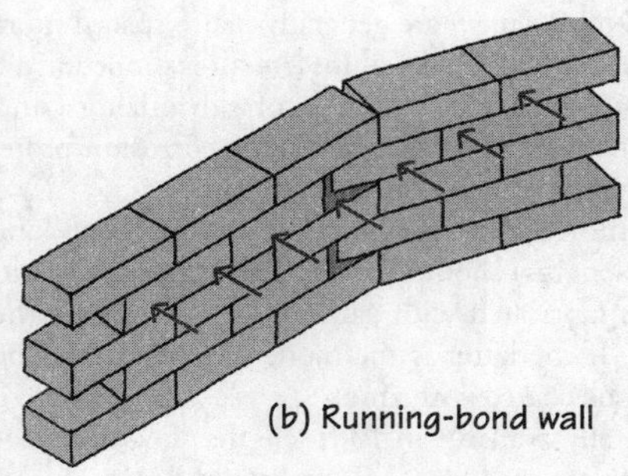

(b) Running-bond wall
When a running-bond wall bends horizontally, the units must slide over the underlying units. The resistance of the units to sliding increases the bending strength of the wall. Note that this phenomenon is absent in a stack-bond wall.
The vertical bending strength of both a stack-bond and a running-bond wall is the same (see Figure 24.3). It can be increased by providing vertical reinforcement in walls (see Figure 24.4).

FIGURE 25.23 **(a)** The horizontal bending strength of a stack-bond wall is low because it is only a function of the flexural tensile bond strength of vertical joints. **(b)** The horizontal bending strength of a running-bond wall is greater than that of a stack-bond wall because of the shear (sliding) resistance provided by the units.

Horizontal reinforcement substantially increases the horizontal bending strength of a stack-bond wall. Tests have shown that if a CMU stack-bond wall is provided with joint reinforcement at 16 in. on centers, its bending strength is 20% greater than that of an unreinforced running-bond wall. If the running-bond wall is also reinforced horizontally in the same way as a stack-bond wall, both the stack-bond wall and the running-bond wall are equal in horizontal bending strength. Horizontal reinforcement in bond beams (4 ft on center) achieves the same result as joint reinforcement laid 16 in. on center.

Control Joints

In addition to horizontal reinforcement, crack control in CMU walls requires the provision of continuous vertical control joints. The width of a control joint is usually the same as the thickness of a mortar joint—$\frac{3}{8}$ in. The recommendations for control joint spacing are as follows:

- The length-to-height ratio of a wall between control joints should be less than or equal to 1.5. In any case, the distance between control joints should not exceed 25 ft. Thus, if the height of a wall (from floor to floor or from floor to roof) is 10 ft, the maximum distance between control joints is 15 ft. However, if the wall height is 20 ft, the maximum distance between control joints is 25 ft.
- In addition to meeting the first provision, several other provisions should also be met, as shown in Figure 25.24.

The detailing of a control joint must ensure free horizontal movement of wall segments in addition to the transfer of the lateral load between adjacent segments. A typical control joint detail is shown in Figure 25.25(a). An alternative detail, which uses a sash unit on both sides of a control joint, is shown in Figure 25.25(b). A cross-shaped PVC gasket is used in sash unit recesses. The parts of the gasket within the sash units are required to have adequate hardness to function as shear key—to transfer the lateral loads across the joint.

25.5 GROUT

The purpose of masonry grout is to fill the voids in masonry walls so that the grout, the masonry units, and the reinforcement are integrated into a composite whole. Grout is, therefore, a cementitious mix, in many ways similar to concrete. Important differences between grout and concrete, however, do exist. For example, the durability of concrete is important, but this is not a concern with grout because it is protected by the units. A small amount of lime is permitted in grout but not in concrete.

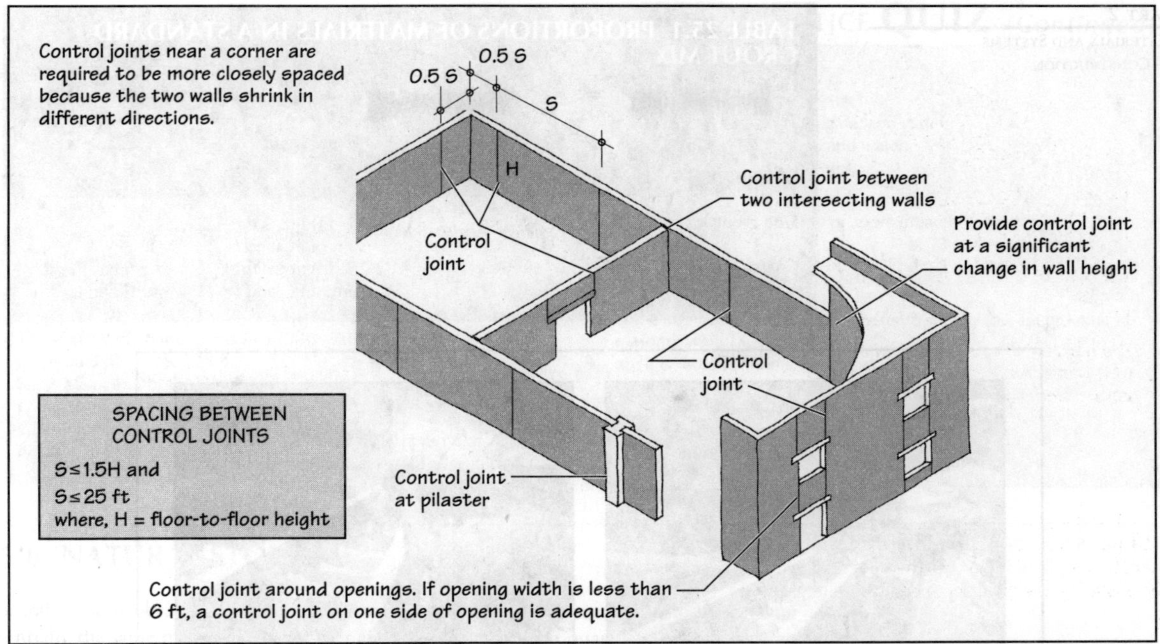

FIGURE 25.24 Recommendations for control joint spacing in CMU walls.

Because grout is placed in masonry voids, it contains more water than concrete. In other words, grout is a "soupy" mix to allow it to flow down the voids, which are relatively small in size. The required slump of grout is 8 to 11 in., depending on the absorption characteristics of masonry units, void dimensions, and the ambient temperature and humidity conditions at the time of grout placement.

Types of Grout

Void dimensions also affect the type of aggregate used in grout. Grout is classified as *fine grout* and *coarse grout.* In fine grout, the aggregate consists of sand only. In coarse grout, the aggregate consists of sand and coarse aggregate. The maximum size of coarse aggregate is generally limited to $\frac{3}{8}$ in.

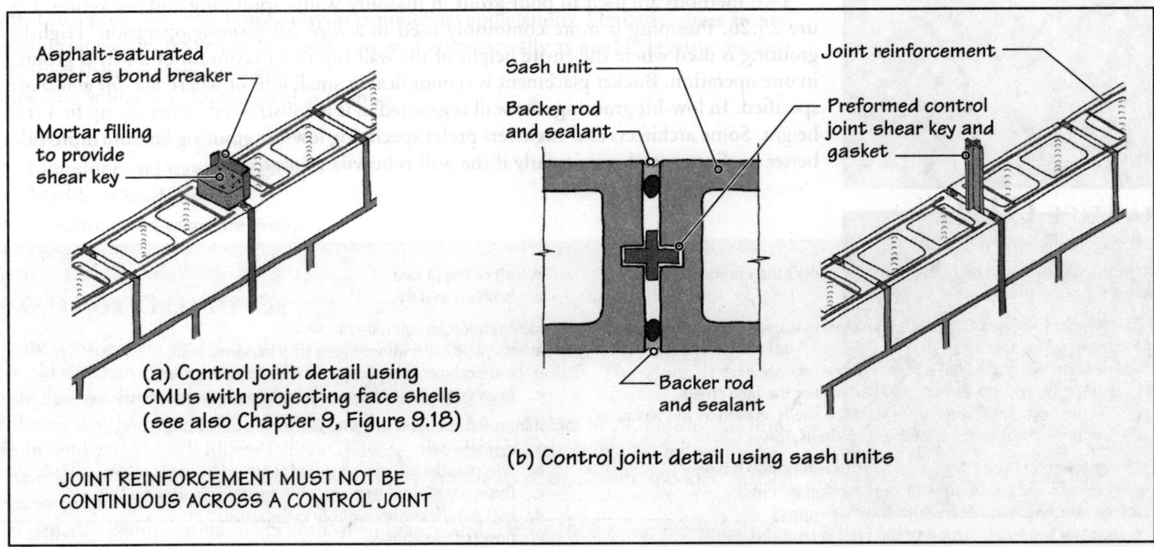

FIGURE 25.25 Two alternative details for control joints in CMU walls.

EXPAND YOUR KNOWLEDGE

Geological Classification of Rocks

Geologists classify the rocks on the earth's surface into

- Igneous rock
- Sedimentary rock
- Metamorphic rock

As the earth's crust began to cool from its original molten state to its present solid state, the first rock to form was igneous rock. In its molten form, the earth's surface was essentially the same as the one that is present deep inside the earth today. This molten rock, which is forced out during a volcanic eruption, is referred to as *lava* or *magma*.

The texture of *igneous rock* depends mainly on how slowly or rapidly the molten mass cooled. Slow cooling allowed the molecules to arrange themselves in crystals. If the molten mass cooled rapidly, the molecules did not have the time to arrange themselves in crystals. Thus, a rapidly cooled volcanic rock is noncrystalline (nongranular) and glassy—brittle and relatively weak.

The igneous rock that exists deeper inside the earth's crust cooled rather slowly because of the insulating effect of the overburden. The coarse-grained crystalline structure of granite—an igneous rock—is the result of the slow cooling process that took place over thousands (millions) of years. The crystalline structure of granite gives it the hardness and strength for which it is well known.

Sedimentary rock began to form as the igneous rock disintegrated due to erosion by water. As the eroded particles were carried by water, many marine and other life forms were also carried with it. These substances were deposited in seas, lakes, valleys, and deltas and were compressed by the pressure of overlying material and bonded together, forming a different kind of rock, called *sedimentary rock*.

Limestone, travertine, and sandstone are sedimentary rocks. The difference between them is the chemical composition of the sediment beds and the material that cemented the beds. Limestone and travertine consist primarily of calcium carbonate. Sandstone consists of silicon dioxide.

Metamorphic rocks originate from either igneous or sedimentary rock that was altered (morphed) in its chemical structure by the action of heat and pressure. Marble, quartzite, and slate are metamorphic rocks. Marble is morphosed from limestone; quartzite, from sandstone; and slate, from shale. Because of the action of pressure and heat, a metamorphic rock is generally stronger than the rock from which it originated.

Chemical Classification of Rocks

In terms of their chemistry, rocks may be classified as

- Siliceous rocks
- Calcareous rocks
- Argillaceous rocks

Siliceous rocks (such as sandstone, quartzite, and granite) are rich in silica (silicon dioxide). Calcareous rocks (such as limestone, marble, and travertine) are rich in calcite (calcium oxide). Argillaceous rocks (such as slate) are rich in alumina (aluminum oxide). Calcareous rocks are generally more prone to disintegration by acids but are slightly more fire resistive than silicious rocks.

granules determine the strength, color, and surface appearance of granite. Quartz is the strongest and most durable of the three minerals. Therefore, a greater amount of quartz gives a stronger and harder granite that is more difficult to process—sawing, profiling, and grinding.

Being silicon dioxide, quartz is chemically the same as sand and, hence, is white to light pink in color. A granite low in quartz is generally darker in color. Commercial black granite is extremely low in quartz, approaching 0%, and is really not granite, but a stone called *basalt* by geologists. Black granite (basalt), which has a handsome finish, is commonly used as tabletops or countertops. Being relatively weaker, it is not favored for use where high strength or abrasion resistance is necessary, such as on floors and stair treads.

Both feldspar and mica present weakness in granite, particularly mica, because it decomposes more easily. Quartzite, a stone that is almost 100% quartz, is extremely strong. It is commonly used as an aggregate for ultra-high-strength concretes.

LIMESTONE

Limestone is a *sedimentary rock,* consisting primarily of the carbonates of calcium and magnesium, with small amounts of clay, sand, and organic material, such as seashells and other fossils. Limestone consisting of approximately 95% calcium carbonate and 5% impurities is called *calcite limestone.* That consisting of 60% to 80% calcium carbonate and 20% to 40% magnesium carbonate is called *dolomitic limestone.* Dolomitic limestone is generally stronger than calcite limestone.

Unlike granite, limestone is generally nongranular, with a relatively uniform surface appearance. It is softer than both marble and granite and hence easier to quarry, saw, and shape. It ranges in color from white to gray and does not take a polish.

Calcium carbonate reacts with acids. Acid in some foods (citric acid in lemons, limes, oranges, etc., and acetic acid in vinegar and pickles) reacts with limestone. That is why limestone is not recommended for use as kitchen, dining, or bar tabletops. Limestone facades in areas with serious acid rain problems have deteriorated. However, because acid rain is a relatively recent problem (related primarily to emissions from the use of petroleum and coal), the deterioration of limestone is also a recent problem.

Several historic buildings with limestone facades have performed quite well in the absence of reactive atmospheres, including the Empire State Building and Rockefeller Center in New York City, the Pentagon in Washington, D.C., and the Chicago Tribune Tower in Chicago, Illinois. Limestone is also commonly used as concrete aggregate.

TRAVERTINE

Travertine is a *sedimentary rock* obtained from the sediments of limestone dissolved in springwater. Springwater (particularly from hot springs) running over limestone deposits dissolved the limestone, which subsequently sedimented (i.e., were deposited) in a nearby location. Sometimes the sediment trapped water, which eventually evaporated, leaving voids in the rock. Travertine is, therefore, a porous stone, and travertine slabs are pitted with voids.

Travertine is closely related to limestone, but because it did not morph (change) like marble, it is a softer stone. Most travertine varieties, therefore, do not take a polish. Denser varieties, which take a good polish, are referred to as *travertine marble*.

Because of its porous structure, travertine does not have the durability required for use in thin slabs for exterior cladding. However, architects, who like the surface texture created by the pits, have used travertine in building exteriors as (thick) masonry walls. When travertine is used as a flooring material, its pitted surface can collect dirt, requiring greater maintenance unless the pits are filled with a portland cement (or an epoxy) and ground smooth.

MARBLE

Geologically, marble is different from limestone because it is a *metamorphic rock*. Chemically, however, marble is similar to limestone. In fact, marble is limestone, which under centuries of high pressure and heat in the earth's crust morphed from a sedimentary rock to a metamorphic rock.

Because of the pressure and heat, marble is stronger and denser than the original limestone but weaker than granite. Like granite, it takes a good polish. As stated previously, denser varieties of limestone that accept polish are sold (and used) commercially as marble.

Marbles vary in surface appearance, which can be without patterns or with patterns—veiny, or mottled, or both, Figure 25.28. The presence of veins often indicates the presence of faults, weakness, or cracks in marble. The use of veiny and variegated marble is generally discouraged for exterior applications. Marbles vary greatly in color—from white to black, pink, and so on. Being chemically identical to limestone, marble is also vulnerable to acid attack.

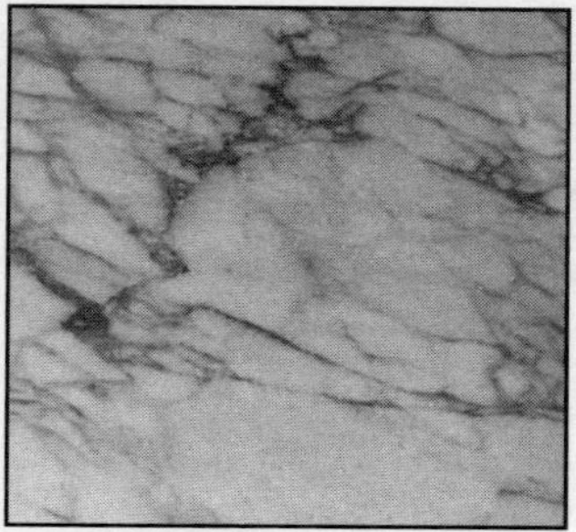

FIGURE 25.28 Veiny surface of marble.

SANDSTONE

Sandstone is a *sedimentary rock* formed by layers of sand (quartz) particles with oxides of calcium, silicon, and iron as cementing agents. If the cementing agent consists primarily of the oxide of silicon, sandstone is light in color and strong. If cemented by iron oxide, the sandstone is brown or red in color and softer. Sandstone that has a large amount of calcium oxide as cement is relatively more vulnerable to disintegration.

SLATE

Slate is a *metamorphic rock* formed by the morphosis of clay and mica sediments. Slate has a nongranular, smooth texture and is characterized by distinct cleavage planes that permit easy splitting in slabs as thin as $\frac{1}{4}$ in. That is why slate is used as roofing and flooring material. It is available in various colors, ranging from black to pink and light green.

25.7 FROM BLOCKS TO FINISHED STONE

Natural stone is procured by stone fabricators from quarries in the form of large blocks. Because of the way they are extracted from the rocks, blocks are irregular in size, Figure 25.29. The blocks are converted into slabs and other cross-sectional profiles in stone-fabrication plants. The conversion is done by sawing the blocks—a process similar to sawing lumber, except that water is used continuously during the sawing process to keep the saw blades cool, Figure 25.30.

The saw shown in Figure 25.30 uses one reciprocating blade that moves both horizontally and vertically so that only one slab is sliced at a time. In more sophisticated machines, several saw blades are ganged together to make multiple cuts in one pass.

FIGURE 25.29 Blocks of stone outside a stone-fabricating plant. Photo taken at Texas Quarries, Austin, Texas.

FIGURE 25.30 Conversion from stone blocks to slabs through sawing. Note that the sawing operation needs a great deal of water to cool the blade. Photo taken at Texas Quarries, Austin, Texas.

Cross-sectional profiles for round columns, column caps, balusters, wall copings, window sills, and so on, are obtained by machining the sawn material to the desired shape. Different types of power-driven tools or tool attachments are available for different profiles. Although most stone fabrication is automated, complex ornamental work requires working with hand tools such as chisels, picks, and hammers, Figure 25.31.

FIGURE 25.31 While a great deal of stonework is done through machines, some detailing requires working with hand tools. (Photo courtesy of Texas Quarries, Austin, Texas)

FINISHES ON STONE SLABS AND PANELS

The surfaces of stone slabs and panels can be finished in several ways. The finish not only affects the surface appearance of the stone but also its durability. A honed or polished finish is generally more durable because it facilitates the drainage of water from the surface. The following are some of the commonly used finishes on stone slabs and panels:

FIGURE 25.32 Honing (grinding) of stone slabs using water to cool the grinders. Photo taken at Texas Quarries, Austin, Texas.

- *Sawn finish:* If the stone is not finished beyond sawing, the surface is called a *sawn finish.* A sawn finish has visible saw marks.
- *Honed finish:* When a sawn finish is ground smooth with an abrasive material, a honed finish is obtained. Honing requires repeated grinding with increasingly fine abrasives. Water is used continuously during the honing process to control dust, Figure 25.32.
- *Polished finish:* Conceptually, there is no difference between a honed and a polished finish. A honed finish is smooth but with a matt appearance. A polished finish is obtained by grinding the stone surface beyond the honed finish with finer abrasives and finally buffing it with felt until the surface develops a sheen. A polished finish brings out the color of stone to its fullest extent by reflecting light like a mirror. The difference in color of a rough and a polished finish on the same stone is easily noticeable, often significant.

  A clear penetrating sealer that adds to the sheen is generally applied to the surface of a polished stone. The sealer increases the durability of stone by sealing the pores, adding resistance to chemical attack and the formation of stains. Only dense stones, such as granite, marble, and dense varieties of limestone, travertine, and slate, can develop a polish.

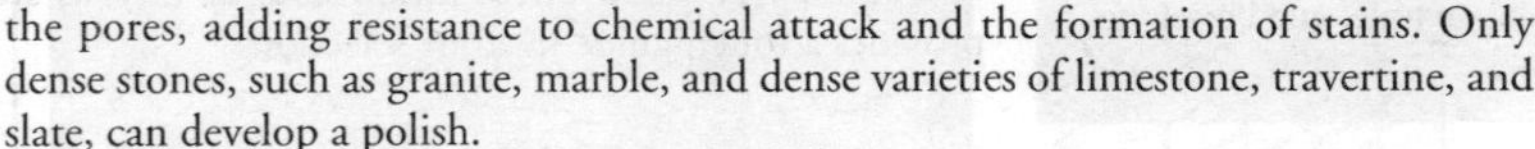

- *Flame-cut finish:* A flame-cut finish, also referred to as a *thermal finish,* is a rough finish obtained by torching the stone surface with a natural gas or oxyacetylene torch. Before torching, the stone is thoroughly wetted. The heat from the torch expands the absorbed water into steam, which breaks loose the surface particles in the stone, leaving behind a rough surface. Generally, a flame-cut finish is used only on granite because other stones are too porous to break only at the surface.

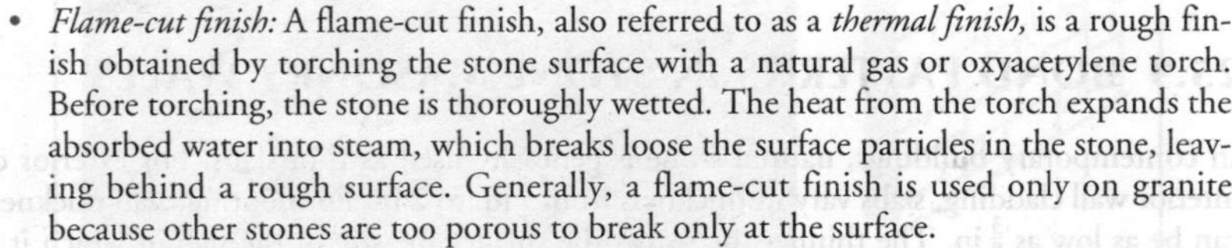

  The roughness of a flame-cut finish makes it ideal for floors, particularly those that are subject to frequent wetting. Often, a flame-cut finish is used in the treads of granite-topped stairs and a polished finish is used for the risers, which gives the desired contrast between the treads and the risers.
- *Bush-hammered finish:* A bush-hammered finish is also a rough finish and is obtained by hammering off the surface of stone with picks.
- *Split-face (cleft) finish:* Like a CMU, stone can also be split through one of its faces, yielding two split-face slabs. Splitting is easier in slate, which has natural cleavage planes, in which case the stone is referred to as *cleft-finished.* In the stone industry, however, the terms *cleft* and *split-face* are used interchangeably.
- *Sandblasted finish:* Although not as commonly used as other finishes, sandblasting stone yields a rough surface.

25.8 STONE SELECTION

The selection of stone for a particular use is a function of several factors. Budget and aesthetics (color, pattern, and surface appearance) are the two most important factors to be considered for stone used in building interiors. For exterior use, the performance history of a stone in the local environment (durability) is obviously another important factor. Where the material has no track record, the physical properties of the stones being evaluated should be compared. Generally, the following properties are of importance:

- Density
- Water absorption
- Compressive strength

FOCUS ON SUSTAINABILITY

Sustainability Features of Masonry

The most important sustainability feature of masonry is its durability. Masonry structures can stand and remain serviceable for a very long time, requiring little maintenance. Their durability is a result of several factors. First, they degrade slowly because they are resistant to physical, biological, and chemical deterioration. Second, they are noncombustible. The following are additional sustainability features of masonry:

Local production: Clay masonry units and CMUs are generally produced in most communities.

Reusability: Masonry materials can be easily reused. Salvaged bricks and stones are available in most communities. Some owners and architects prefer to use salvaged masonry. Interlocking concrete masonry pavers or units used in retaining walls do not require any mortar and can be disassembled and reused

Recyclability: Masonry materials can also be easily recycled. Crushed and pulverized bricks (referred to as *grog*) can be mixed with clay and used to manufacture new bricks. Stone can be used as aggregate for concrete, and crushed CMUs can be used as underbed for concrete footings and slabs-on-grade.

Absence of VOC emissions: Masonry materials do not emit any volatile organic compounds (VOCs). Although radon gas has been associated with certain soils, brick manufacturers ensure the absence of radon in the clay used in the manufacture of bricks.

Thermal mass: All masonry materials provide thermal mass, which may help conserve energy in some climates (see Chapter 5).

Embodied energy: The embodied energy in stone and CMUs is lower than that in clay brick, which in turn has much lower embodied energy than metal or glass.

Negative sustainability aspects of masonry include soil erosion and habitat loss caused by the mining of clay and stone. However, like the other industries, the masonry industry is becoming increasingly aware of its responsibilities to the environment. Brick manufacturers are increasingly converting abandoned pits into lakes for use by the community.

PRACTICE QUIZ

Each question has only one correct answer. Select the choice that best answers the question.

31. Geologically, the earliest rock on the earth's surface was
- a. igneous rock.
- b. sedimentary rock.
- c. metamorphic rock.
- d. none of the above.

32. Marble is
- a. an igneous rock.
- b. a sedimentary rock.
- c. a metamorphic rock.
- d. none of the above.

33. Granite is
- a. an igneous rock.
- b. a sedimentary rock.
- c. a metamorphic rock.
- d. none of the above.

34. Slate is
- a. an igneous rock.
- b. a sedimentary rock.
- c. a metamorphic rock.
- d. none of the above.

35. Which of the following stones is generally more vulnerable to disintegration by an acidic atmosphere?
- a. Granite
- b. Marble

36. In general terms, which of the following stones is considered most durable?
- a. Marble
- b. Granite
- c. Limestone
- d. Sandstone
- e. Slate

37. The bond patterns that are used in brick masonry, such as running bond, English bond, and Flemish bond, are also used in stone masonry.
- a. True
- b. False

38. Dressed stones are required to be used in
- a. rubble masonry.
- b. ashlar masonry.
- c. CMU masonry.
- d. CSMU masonry.

REVIEW QUESTIONS

1. Explain the differences between lightweight and normal-weight CMUs, including their uses.
2. Explain why CMUs are generally face-shell mortared. In which situations is full-bed mortaring of CMUs used?
3. Explain the purpose of joint reinforcement in CMU walls. Sketch a typical joint reinforcement used in CMU walls.
4. With the help of sketches, explain why a running-bond wall has higher horizontal bending strength than a stack-bond wall, whereas the vertical bending strength of both walls is equal.
5. What measures are used to respond to the shrinkage of CMU walls?
6. Explain the differences between fine grout and coarse grout and where each is used.
7. Explain why granite is the most durable of all building stones. Where would you recommend the use of granite?
8. Explain the essential differences between marble and granite and their uses in buildings.
9. Explain what arching action in masonry is and its significance in building construction.

CHAPTER 26

Masonry and Concrete Bearing Wall Construction

CHAPTER OUTLINE

The use of masonry in contemporary buildings is primarily in the walls. Minor uses occur in interior flooring or exterior paving. Although masonry roofs (vaults, domes, and shells) were frequently constructed in the past, they are relatively uncommon in modern buildings.

Masonry walls may either be load-bearing or non-load-bearing walls. Non-load-bearing masonry walls include retaining walls, exterior infill and cladding, and interior partitions, Figure 26.1. Masonry cladding functions as an exterior weather-resistant cover and is similar to the exterior siding in a wood light-frame building. Masonry cladding is discussed in Chapter 28.

Masonry infill occupies the space between the concrete or steel frame members. It may be used to resist the racking of the frame—that is, to serve as a shear wall. In that case, the infill must completely fill the space between the columns of the frame so that the lateral load is transferred from the columns to the infill and vice versa, Figure 26.2(a).

If the infill is not used as a shear wall, adequate separation between the columns and the infill is required so that the lateral load is not transferred from the columns to the infill, Figure 26.2(b).

ADVANTAGES OF A MASONRY BEARING WALL BUILDING

A load-bearing wall (by definition) is a wall that supports gravity loads in addition to its self-load. Load-bearing walls may also act as shear walls. (However, as shown in Figure 26.1, some shear walls may be non-load-bearing walls, i.e., they may not support any gravity load other than their self-loads.)

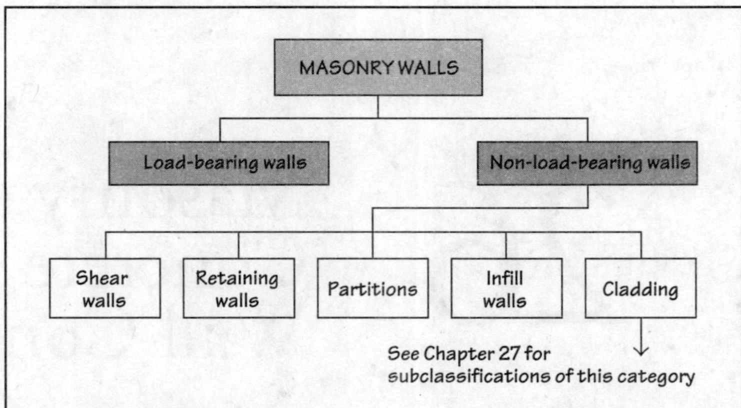

FIGURE 26.1 Various applications of masonry walls.

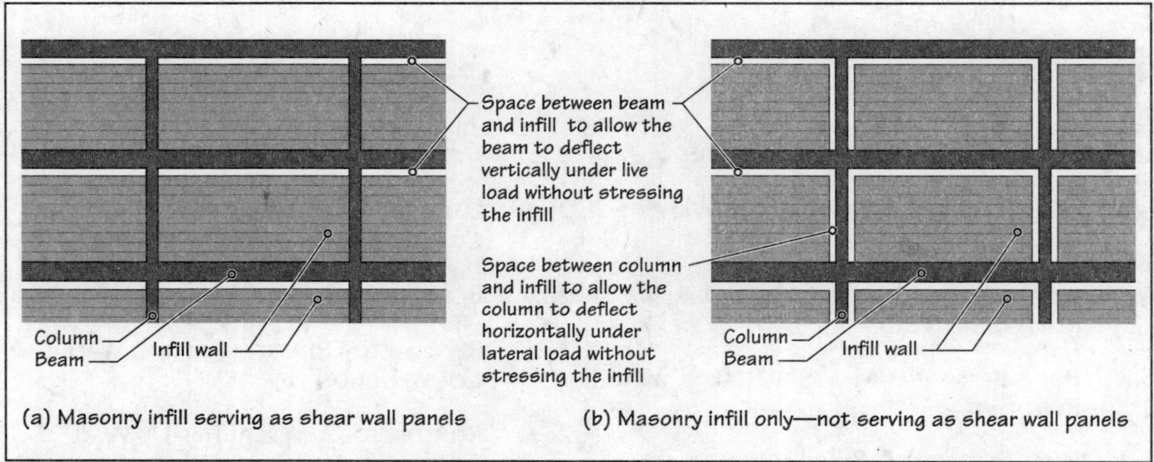

FIGURE 26.2 Masonry infill between the members of a structural frame may or may not be used to resist shear (racking of the frame).

A major advantage of a load-bearing masonry structure, also called a *masonry bearing wall structure,* is that the walls that are required for load-bearing purposes may also function as shear walls in addition to enclosing and dividing spaces. This contrasts with a steel or concrete frame structure, where the frame serves the structural function only. Space-enclosing and dividing functions must, therefore, be performed by nonstructural walls. The separation of the structural and nonstructural elements in a frame structure requires careful detailing of the connections between them. Such details are unnecessary in a load-bearing masonry structure.

Constructed of an inherently fire-resistive, mold-free, sound-insulating, relatively inexpensive, and durable material, load-bearing masonry has a favorable life-cycle cost index. That is why masonry experts claim that load-bearing masonry is a more sustainable alternative to a concrete or steel-frame structure in many applications.

BEARING WALL STRUCTURES IN REINFORCED MASONRY AND REINFORCED CONCRETE

Although masonry has been used historically without any steel reinforcement, in contemporary times the use of plain (unreinforced) masonry is less common. Reinforced masonry has several advantages over plain masonry at little additional cost. As discussed in Section 26.2, reinforcement increases a masonry wall's flexural and shear strengths. Additionally, because steel is a ductile material, a reinforced-masonry wall is better able to absorb structural deformations by the yielding of steel—an important requirement for structures located in seismic regions. Plain masonry's ability to absorb deformations is limited due to the brittle nature of masonry.

Poor performance of unreinforced-masonry structures has been demonstrated again and again through studies of postearthquake and posthurricane building damage all over the world. However, it must be added that low-rise, plain masonry bearing wall structures have excellent performance records in locations that are not subjected to earthquakes or extreme winds.

This chapter begins with an introduction to traditional load-bearing masonry, followed by a discussion of its contemporary alternative. Because there is a great deal of similarity between bearing wall construction in reinforced masonry and reinforced concrete, reinforced-concrete bearing wall construction is also covered in this chapter.

Reinforced-concrete bearing wall structures can be constructed of site-cast concrete or precast concrete. The precast-concrete bearing wall system is typically a concrete tilt-up wall system, which is also discussed in this chapter.

26.1 TRADITIONAL MASONRY BEARING WALL CONSTRUCTION

The masonry bearing wall system is one of the oldest construction systems. Until the invention of the frame structure, the masonry bearing wall system was the only system available for constructing major buildings. Although several large and tall masonry structures were built, their design was based on arbitrary rules formulated on an intuitive rather than a scientific basis. Gothic cathedrals symbolize the daring extremes to which such intuitive understanding was extended.

Over time, this intuitive understanding was embedded in the building codes as empirical rules for the design of masonry bearing walls. These rules required the thickness of exterior masonry walls to increase progressively toward the lower floors. The increase in wall thickness was necessary to accommodate greater gravity loads at the lower floors.

Another important reason for progressively increasing wall thickness at the base was to ensure stability against overturning by wind loads. (Very little was known at the time about seismic loads.) The code-required thickness was based on the assumption that all wind loads on the building were resisted by exterior walls only. In providing this resistance, each exterior wall functioned as a freestanding wall, behaving as a vertical cantilever fixed in the ground—like a flagpole, Figure 26.3(a).

This assumption made building code requirement highly conservative, necessitating extremely thick walls at the base of a tall building. That is why the exterior brick walls of

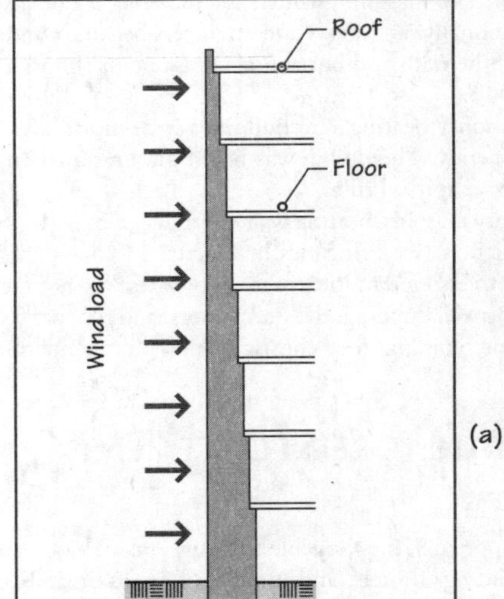

FIGURE 26.3 **(a)** In resisting wind loads, the exterior walls in historic load-bearing masonry buildings were designed as freestanding cantilevers. Therefore, the thickness of walls at lower floors increased substantially with an increase in the height of the building.
(b) Monadnock Building, Chicago, completed in 1891, preserved as the last high-rise traditional bearing-wall structure. The walls at the ground floor are approximately 6 ft thick. Architects: Burnham and Root. (Photo courtesy of Dr. Jay Henry)

the 215-ft-high, 16-story Monadnock Building (completed in 1891 in Chicago) are approximately 6 ft thick at the ground level, Figure 26.3(b).

Because the lower floors of a building are the most usable and profitable, it was apparent that high-rise masonry structures were economically unviable because of the unduly large percentage of floor area occupied by the walls. In fact, the Monadnock Building was the last high-rise masonry bearing wall structure in North America until the mid-twentieth century, when the system was revived using a more efficient structural design methodology.

In general, the interior structure of most historical masonry bearing wall buildings consisted of columns and beams. The columns were of cast iron and the beams were of wrought iron. In later structures, steel was substituted for both cast iron and wrought iron. In several masonry buildings, (heavy) timber columns and beams were used in the interior structure (Figure 1, Expand Your Knowledge, Chapter 15).

BEARING WALL SYSTEM—ITS ABANDONMENT AND REVIVAL

The use of the masonry bearing wall system for high-rise buildings went out of favor as the skeleton (iron and, later, steel) frame became popular. Steel is much stronger than masonry. Therefore, the columns in a frame structure were much smaller than those in masonry bearing walls. Because the walls in a frame structure are nonstructural, their thickness was considerably reduced. This conserved valuable floor space. A secondary advantage of the skeleton frame was the speed with which it could be erected compared to thick masonry walls.

The beginning of the twentieth century saw major advances in the science of structural engineering, particularly in the area of design and analysis of frames and continuous beams. Advances were also made in the science of materials related to steel and reinforced concrete. Because masonry does not lend itself to frame construction, it was largely displaced from the realm of structural materials.

CONTEMPORARY BEARING WALL CONSTRUCTION

The situation began to change with reconstruction work in Europe after World War II, which required the construction of large-scale and relatively inexpensive housing. This was also the time when the structural design theories of reinforced concrete became fairly sophisticated. The extension of these theories to masonry, which, as a material, is not vastly different from concrete, was logical. Additionally, architects and engineers began to understand the interactive structural behavior of the walls and horizontal components (floors and roof) of a building.

During the 1950s, several high-rise masonry bearing wall buildings were built, initially in Europe and subsequently in North America. The design was based on a rational engineering approach rather than on the earlier empirical rules.

The structural behavior of a contemporary masonry bearing wall building is similar to that of conventional wood light-frame construction (see Principles in Practice in Chapter 15). However, the construction details required to respond to the structural behavior are not identical. The reason is that in a masonry bearing wall building, the walls, floors, and roof are solid components, whereas in a wood light-frame building, they consist of linear frame members with intervening cavity spaces.

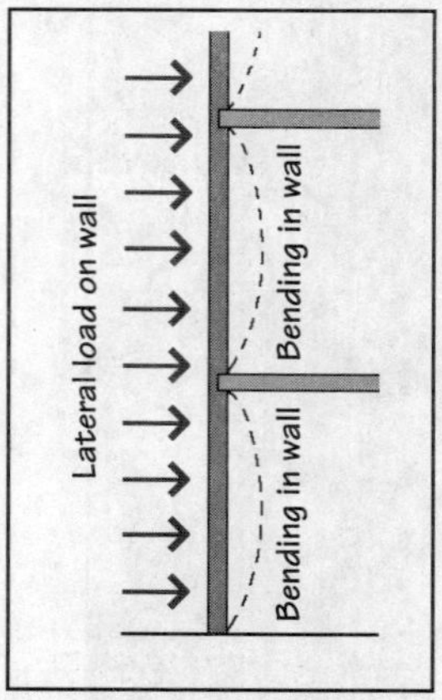

26.2 IMPORTANCE OF VERTICAL REINFORCEMENT IN MASONRY WALLS

Using the contemporary rational design approach, it is possible to design masonry bearing wall residential structures (apartment buildings, hotels, and motels) of approximately 20 stories using 8- to 10-in.-thick walls. In fact, an 8-in.-thick CMU wall is the industry's standard for masonry bearing wall structures. For long-span buildings with relatively high walls, such as those used in gymnasiums, warehouses, and large assembly spaces, 10- or 12-in.-thick CMU walls may be needed. CMU walls thicker than 12 in. are rare.

For a one- or two-story building in a low-wind or nonseismic region, it may be possible to design masonry walls without any vertical reinforcement in some situations. In a taller building or a building in a high-wind or seismic region, vertical reinforcement is generally always needed. (Masonry walls without vertical reinforcement are referred to as *plain-masonry* walls, and those with vertical reinforcement are called *reinforced-masonry* walls.)

The benefits of reinforced masonry over plain masonry are as follows:

- Reinforcement increases the strength of a wall against bending caused by (a) eccentric gravity loads on the wall and (b) lateral loads perpendicular to the wall, Figure 26.4.

Lateral load on wall
Bending in wall
Bending in wall

FIGURE 26.4 Bending in a wall caused by lateral loads or eccentric gravity loads on the wall. Vertical wall reinforcement increases the wall's bending strength.

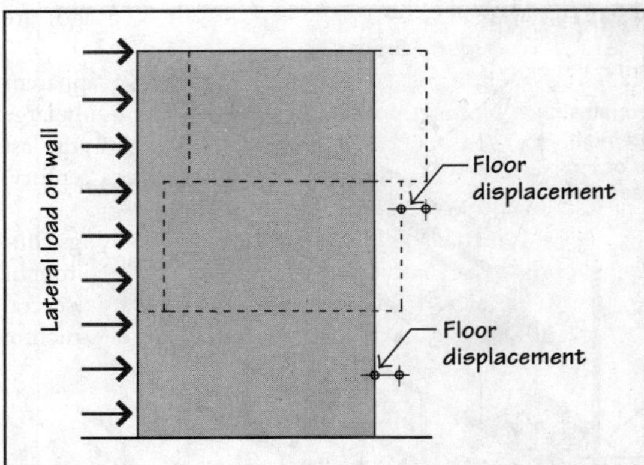

FIGURE 26.5 Vertical reinforcement helps stitch floors together, reducing the horizontal displacement between floors (story drift) caused by lateral loads on the building.

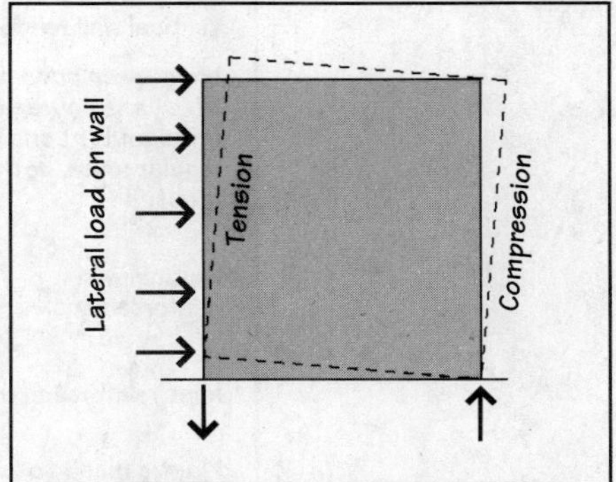

FIGURE 26.6 Vertical reinforcement helps resist tension caused by overturning of the wall by lateral loads. Because lateral loads can act on either side of a wall, both ends of a wall require reinforcing.

- Reinforcement increases the sliding resistance of the wall to in-plane lateral loads (as in a shear wall), reducing horizontal displacement between floors, Figure 26.5. In other words, reinforcement stitches the floors together. Additionally, it provides positive anchorage between the wall and the foundation.
- Reinforcement helps to resist the tension caused by overturning of the wall, Figure 26.6 (and functions like hold-down bolts in a wood light-frame building (see Principles in Practice in Chapter 15). Therefore, both ends of a masonry wall are typically reinforced. Reinforcement is also provided near both ends of a door or window opening. Reinforcement at the ends of an opening also helps to absorb the impact from slamming of a door or window.

26.3 BOND BEAMS IN A MASONRY BEARING WALL BUILDING

In most masonry bearing wall structures, the floors consist of precast concrete hollow-core slabs, metal deck, or wood deck. Being individual elements, they do not have the continuity to resist diaphragm tension (see Principles in Practice in Chapter 15). Therefore, they are anchored to masonry beams called *bond beams.* Because a bond beam is reinforced and continuous, it is able to resist the diaphragm tension.

A bond beam is provided at each floor level and roof level in all exterior walls and at all interior bearing walls and shear walls, Figure 26.7. A bond beam is also provided at the top of the parapet. A parapet-level bond beam does not have a structural function but instead facilitates the anchorage of coping on the wall.

Floor- and roof-level bond beams are similar to the bond beams used for shrinkage crack control in CMU walls (see Figure 25.22). However, floor- and roof-level bond beams are generally more heavily reinforced than those used for crack control.

Generally, a bond beam is only one unit (8 in.) deep but, if needed, its depth can be increased to 16 in. by grouting the regular concrete masonry units below (or above) the bond beam units, Figure 26.8. The reinforcement in a bond beam must be continuous to be able to effectively absorb diaphragm tension. It must, therefore, be adequately lap spliced between the ends of bars, particularly at wall corners and intersections.

Because a bond beam is a grouted element, it provides a suitable base on which the roof or floor trusses and joists can be supported and anchored. Some architects and engineers prefer to use site-cast reinforced-concrete bond beams, Figure 26.9, which are more robust than masonry bond beams. However, because a site-cast concrete bond beam requires formwork, its construction is more cumbersome than that of a masonry bond beam.

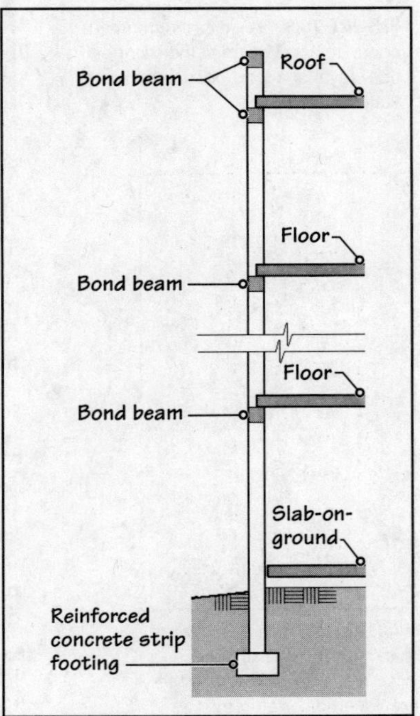

FIGURE 26.7 Locations of bond beams in a masonry bearing wall structure.

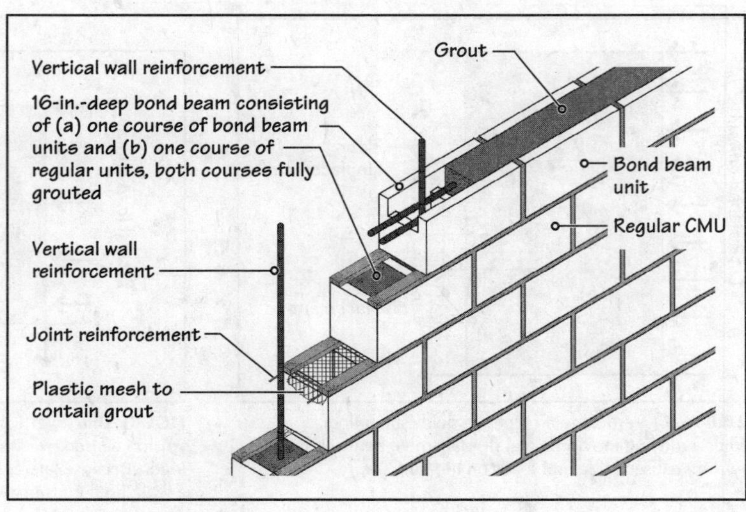

FIGURE 26.8 Detail of a 16-in.-deep bond beam in a CMU wall (see also Chapter 25, Figure 25.22).

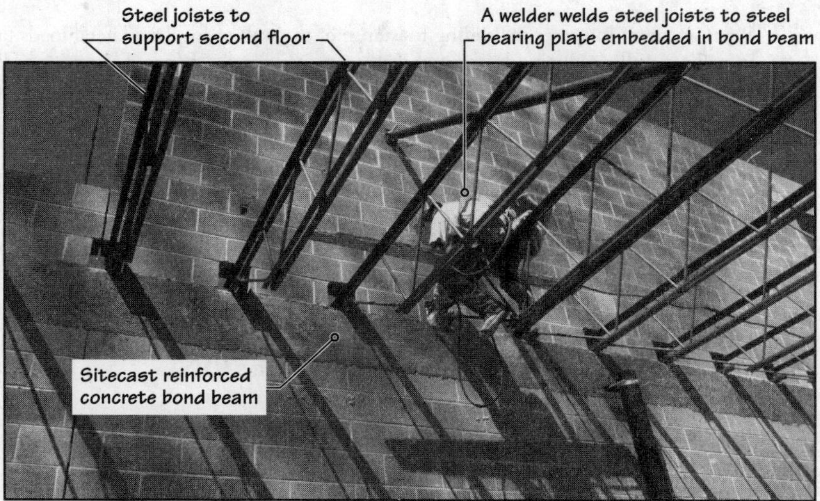

FIGURE 26.9 A sitecast reinforced-concrete bond beam at the second-floor level of a load-bearing masonry wall building.

Vertical Reinforcement and Bond Beam Combination—A Reinforced Masonry Frame

The combination of bond beams and vertically reinforced (and grouted) CMU cells forms a reinforced-masonry frame, Figure 26.10. The frame ties the intervening plain masonry into an integral mass, increasing the wall's resistance to lateral loads. The system is identical to a reinforced-concrete frame with plain-masonry infill that is a common form of construction in many countries.

26.4 WALL LAYOUT IN A BEARING WALL BUILDING

For a bearing wall building to be structurally efficient, a sufficient number and adequate lengths of walls must be provided in both principal directions of a building, yielding a *cellular-type* plan. The simplest cellular-type plan is a single-story rectangular (box-type) structure, commonly used for long-span bearing wall buildings (gymnasiums, lecture halls, and other large assembly or storage spaces), Figure 26.11.

For a multifloor bearing wall structure, a cellular-type structure is best achieved when the walls at an upper floor are at the same location as the ones on a lower floor, resulting in repetitive floor plans, Figure 26.12. Repetitive floor plans are not only structurally more efficient but also more easily constructible. They allow greater prefabrication and familiarity with the construction process, giving economy.

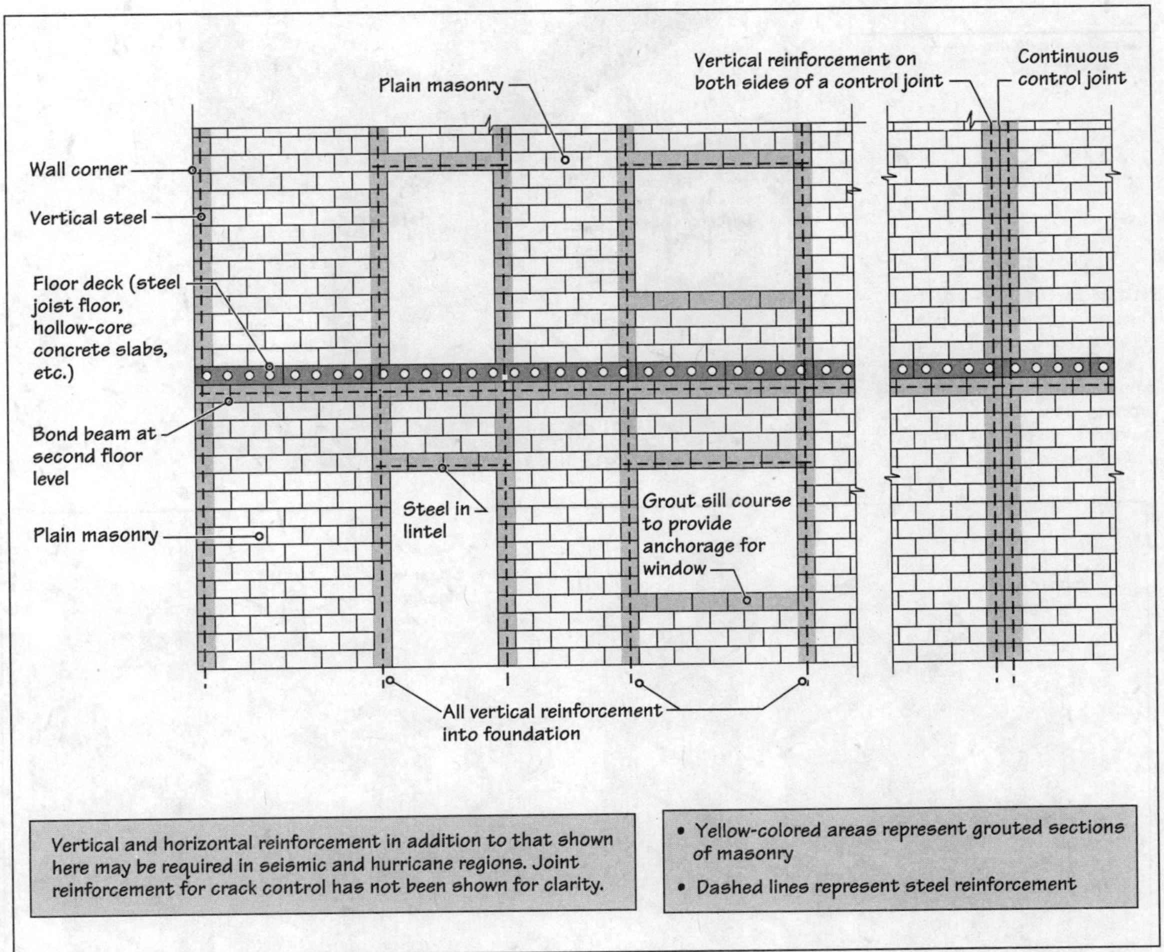

FIGURE 26.10 The combination of (reinforced) bond beams and vertically reinforced cells in a masonry wall (both yellow highlighted) is similar to a reinforced-concrete beam-and-column structural frame with intervening plain masonry infill.

FIGURE 26.11 (a) A gymnasium with 12-in.-thick CMU bearing walls (in a box-type floor plan) and a steel joist roof. The joists span nearly 100 ft between walls and are approximately 60 in. deep.

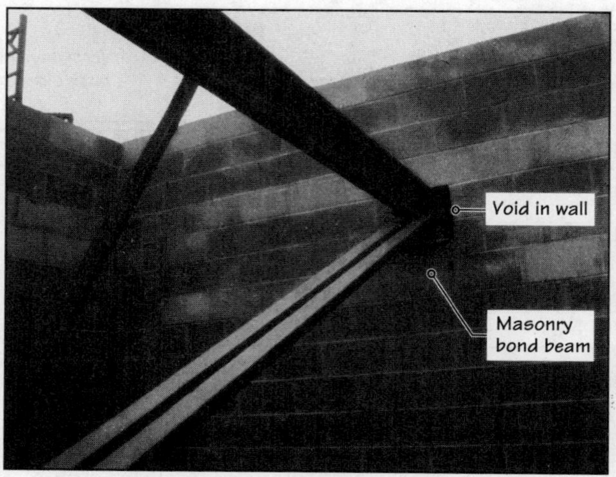
Void in wall
Masonry bond beam

FIGURE 26.11 (b) Support for joists of the building in Figure 26.11(a). Voids, which contain steel bearing plates, are left in the wall at support locations during the construction of the wall. A continuous masonry bond beam is provided under the voids.

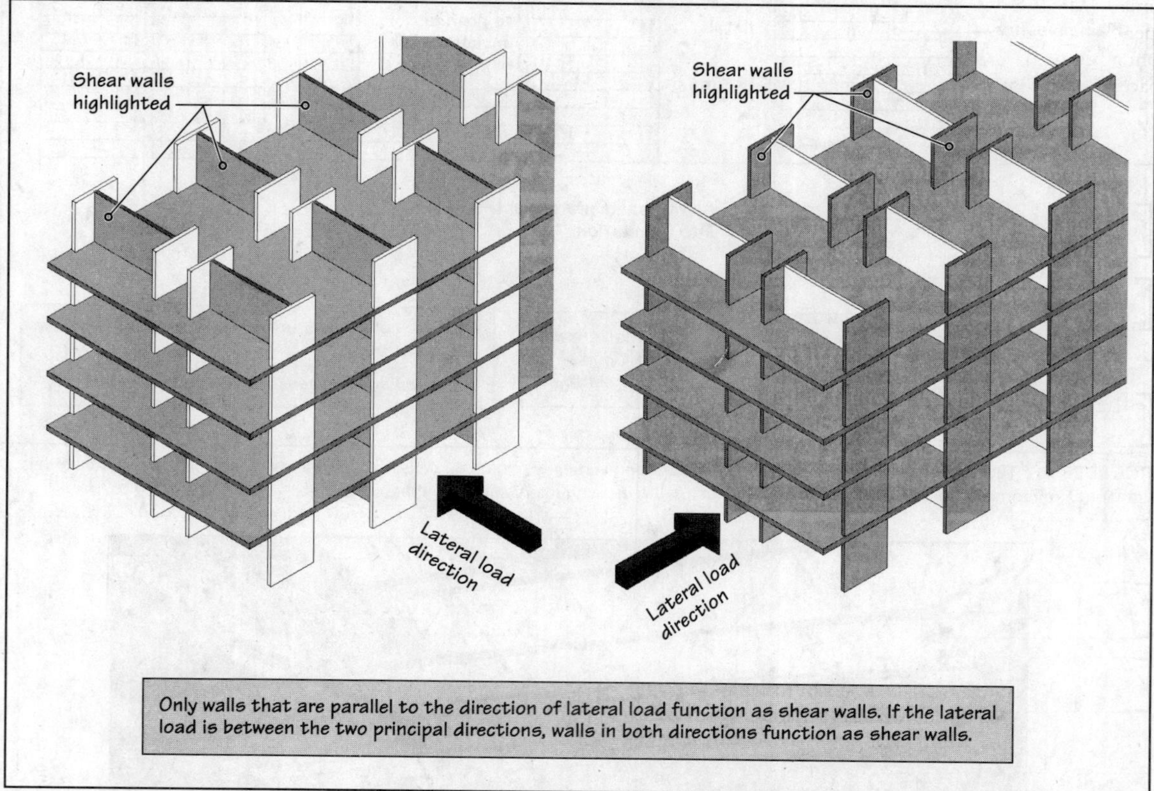
Shear walls highlighted
Lateral load direction
Shear walls highlighted
Lateral load direction
Only walls that are parallel to the direction of lateral load function as shear walls. If the lateral load is between the two principal directions, walls in both directions function as shear walls.

FIGURE 26.12 A bearing wall structure should have lateral load–resisting (shear) walls in both principal directions. Each figure shows the walls that function as shear walls when subjected to lateral load in one principal direction.

A multifloor bearing wall structure is, therefore, ideal for occupancies in which economy is a major consideration, and a cellular-type, repetitive floor layout is intrinsic to the occupancy. Such occupancies are generally residential—apartment buildings, hotels, motels, student dormitories, correction facilities, hospital wards, and so on.

Repetitiveness of floor layout requires that all load-bearing or shear walls are continuous to the foundation. This is particularly important in high-rise bearing wall structures, where there is a temptation to provide larger uninterrupted spaces at the first-floor or basement level.

The discontinuity of walls at foundations, referred to as a *soft story,* is structurally feasible but requires a heavy transfer structure, Figure 26.13. The transfer structure carries the superimposed gravity loads and provides the required lateral load resistance. A soft story is particularly problematic in seismic zones.

Another desirable attribute of a bearing wall structure is symmetry in wall layout. An asymmetrical wall layout leads to rotation of the building (torsion) under lateral loads. The greater the asymmetry, the greater the torsion created, which increases the cost of the structure.

Role of Floor and Roof Decks in the Wall Layout

Floor and roof decks can either be one-way or two-way decks (see Chapter 23). In a one-way deck, the deck spans between two opposite walls; that is, the gravity load on the deck is carried in one direction across the two opposite walls. In a two-way deck, the load is transferred to all four walls. Two-way decks are generally of site-cast concrete.

One-way decks, which are more commonly used, may be of steel, precast concrete, or wood. The use of one-way decks yields the following two types of bearing wall plans:

- Cross-bearing wall plan
- Longitudinal-bearing wall plan

In a cross-bearing wall plan, the bearing walls are transverse to the main axis of the building. Figure 26.14 compares the suitability of a few cross-bearing wall plans. A similar analysis for longitudinal-bearing wall plans is presented in Figure 26.15. Larger openings are generally possible with a cross-bearing wall plan. It is, therefore, more commonly used. A longitudinal-bearing wall plan provides larger, unobstructed interior spaces. It is used where interior walls are architecturally undesirable.

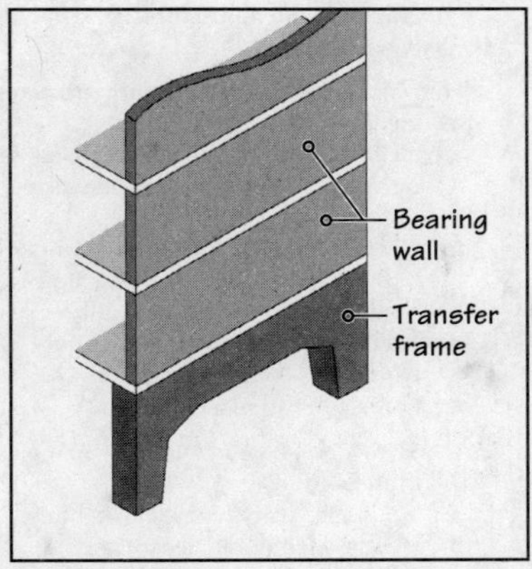

FIGURE 26.13 Transfer frame at the first floor or basement level of a multistory bearing wall structure. The frame must have adequate stiffness to preclude the existence of a soft story.

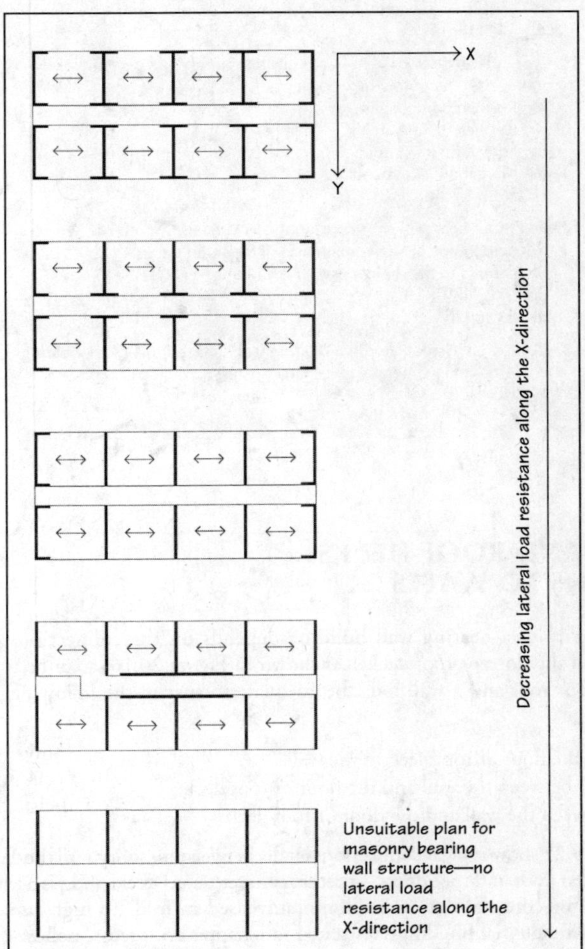

FIGURE 26.14 Relative suitability of various cross-bearing wall plans [26.1].

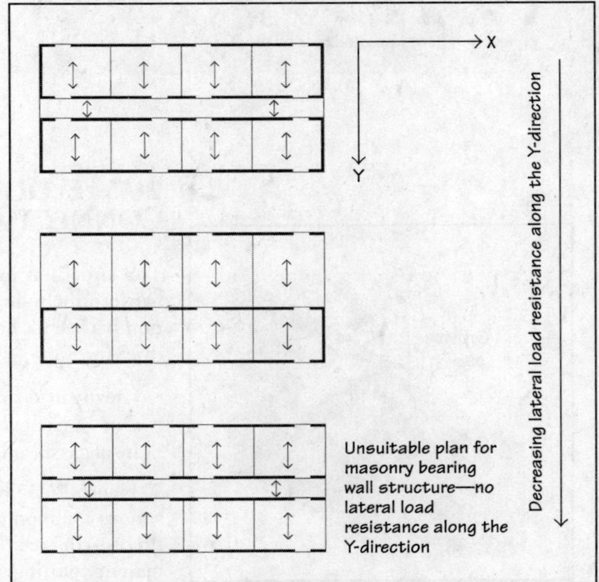

FIGURE 26.15 Relative suitability of various longitudinal-bearing wall layouts [26.1]. Note that a longitudinal-bearing wall plan provides larger, unobstructed interior spaces compared with a cross-bearing wall plan.

PRACTICE QUIZ

Each question has only one correct answer. Select the choice that best answers the question.

1. Historical load-bearing masonry structures were designed with the assumption that
   a. all wind loads on the building were resisted by exterior walls only, and in doing so, they interacted with each other structurally.
   b. all wind loads on the building were resisted by exterior walls only, with no structural interaction between them.
   c. both interior and exterior walls resisted wind loads.
   d. interior and exterior walls and floor and roof diaphragms resisted wind loads collectively.
   e. none of the above.
2. Use of the load-bearing masonry wall system became extinct for a while and was revived
   a. around the early nineteenth century.
   b. around the mid-nineteenth century.
   c. around the late nineteenth century.
   d. after World War I.
   e. after World War II.
3. The structural behavior of a contemporary load-bearing masonry wall building is similar to that of a
   a. conventional wood light-frame building.
   b. site-cast reinforced-concrete frame building.
   c. precast-concrete frame building.
   d. steel-frame building.
   e. none of the above.
4. Referring to a masonry wall as a *reinforced-masonry wall* implies that the wall contains
   a. horizontal reinforcing bars.
   b. joint reinforcement.
   c. vertical reinforcing bars.
   d. bond beams.
   e. none of the above.
5. Masonry walls without joint reinforcement are called *plain masonry walls*.
   a. True b. False
6. A bond beam in a masonry bearing wall building is required for structural reasons
   a. above all openings in exterior walls.
   b. at each floor level.
   c. at each floor level and roof level.
   d. at each floor level, roof level, and top of the parapet.
   e. none of the above.
7. A bond beam in a masonry bearing wall building must be a
   a. steel, reinforced-concrete, reinforced-masonry, or wood beam.
   b. steel, reinforced-concrete, or reinforced-masonry beam.
   c. steel or reinforced-concrete beam.
   d. reinforced-concrete or reinforced-masonry beam.
   e. none of the above.
8. A bond beam in a masonry bearing wall building
   a. is preferably used along the shorter span.
   b. is preferably used along the longer span.
   c. may be used along the shorter or longer span.
   d. is embedded in the walls.
   e. none of the above.
9. A typical bond beam is provided with
   a. horizontal reinforcement.
   b. stirrups.
   c. horizontal reinforcement and stirrups.
   d. none of the above.
10. A load-bearing wall structure works best when the floor plan of the building has
    a. walls that are distributed almost uniformly in both principal directions.
    b. walls at an upper floor that align with the walls at a lower floor.
    c. walls that are continuous up from the foundations.
    d. all of the above.
11. A cellular-type floor plan in a multistory building is generally inherent in the following occupancies:
    a. Residential occupancies
    b. Business occupancies
    c. Educational occupancies
    d. Mercantile occupancies
    e. None of the above.
12. In a cross-bearing wall structure, the load-bearing walls are
    a. perpendicular to the main axis of the building.
    b. parallel to the main axis of the building.
    c. (a) or (b).
    d. (a) and (b).
13. Compared to a longitudinal-bearing wall floor plan, a cross–bearing wall floor plan generally gives larger exterior wall openings.
    a. True b. False

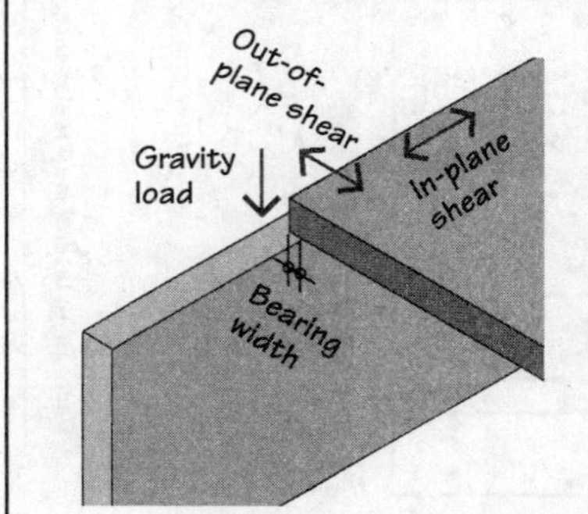

FIGURE 26.16 Connection forces between a floor (or roof) deck and a bearing wall.

26.5 FLOOR AND ROOF DECKS—CONNECTIONS TO WALLS

The structural integrity of a bearing wall building depends on the connections between the walls and the floor or roof decks. As shown in Figure 26.16, a connection between a floor or roof and a wall must be adequate to sustain the following three forces:

- Gravity load from the floor or roof deck to the wall
- Out-of-plane shear between the wall and the floor or roof deck
- In-plane shear between the wall and the floor or roof deck

Figures 26.17 to 26.21 show typical connection details between masonry walls and various commonly used floor and roof decks—precast concrete, steel deck, and wood. Precast-concrete hollow-core slabs are most commonly used in mid- to high-rise masonry bearing wall residential buildings. As stated in Chapter 23, precast hollow-core slabs are produced in various widths (widths of 4 ft and 8 ft are more common). Steel and wood decks are common in low-rise buildings.

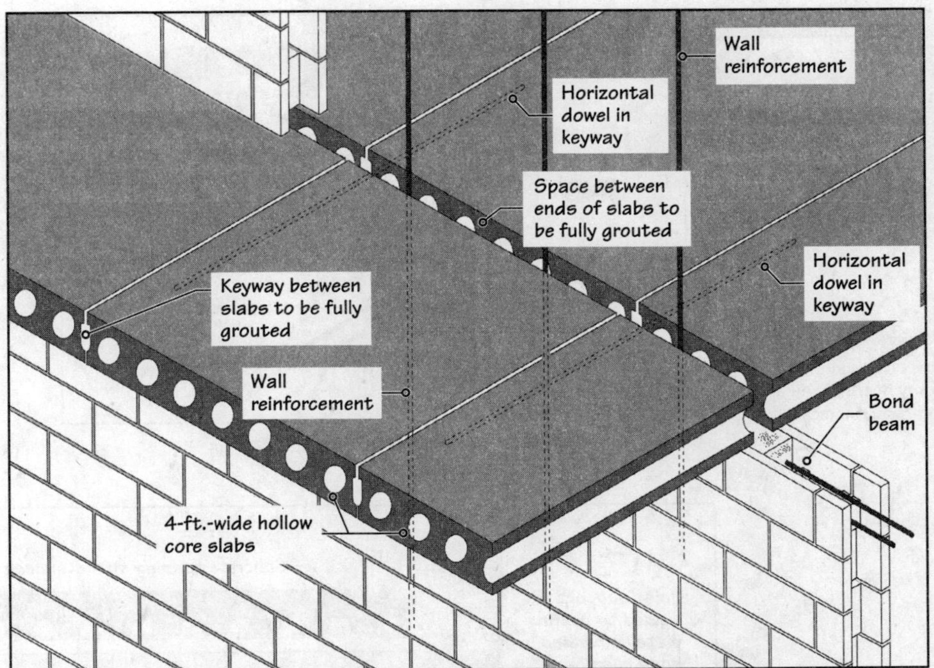

FIGURE 26.17 Schematic connection detail between an interior concrete masonry bearing wall and 4-ft-wide hollow-core precast concrete slabs. Detail is similar if the hollow-core slabs are 8 ft wide; see Figure 26.23(c).

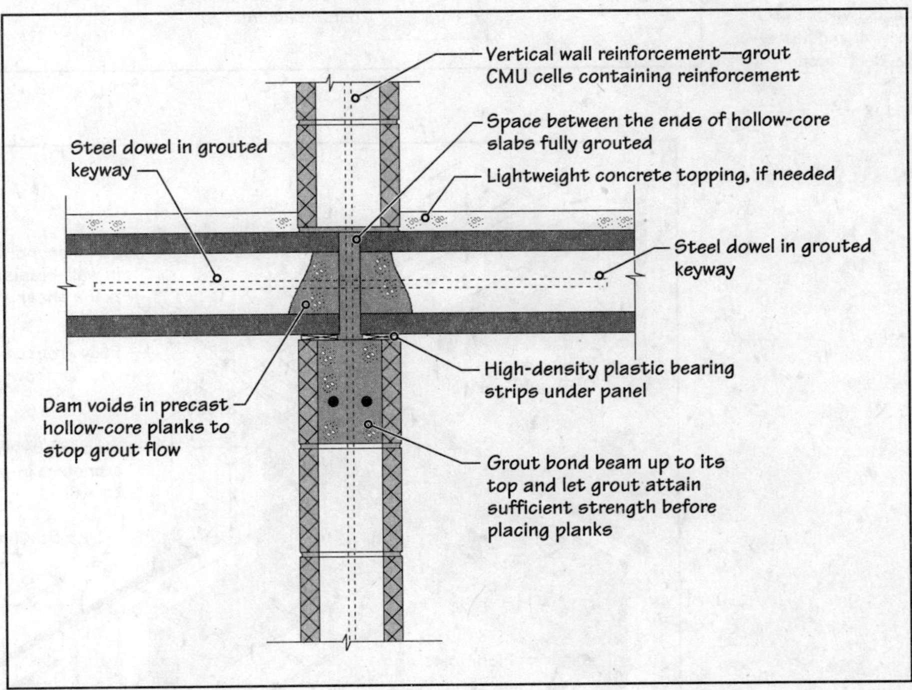

FIGURE 26.18 Detail of Figure 26.17 shown in section.

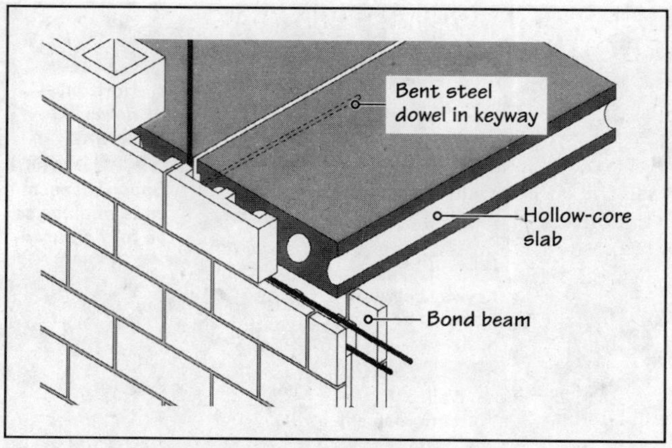

Bent steel dowel in keyway
Hollow-core slab
Bond beam

FIGURE 26.19 Schematic connection detail between an exterior concrete masonry bearing wall and hollow-core precast concrete slabs.

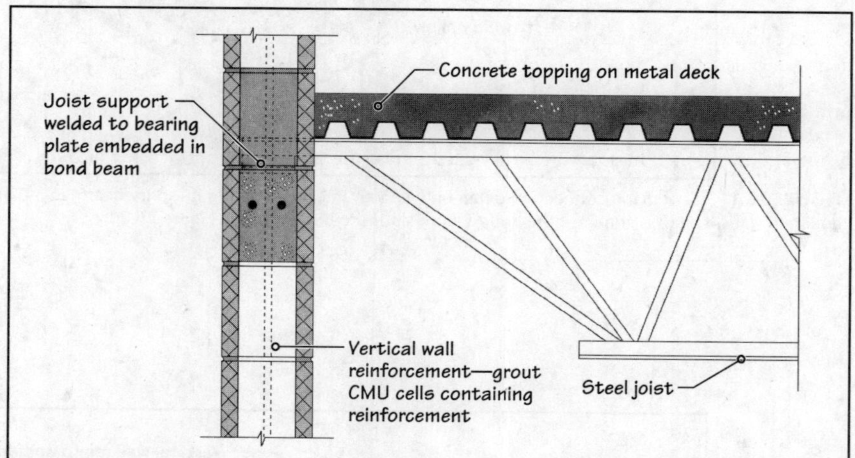

FIGURE 26.20 A schematic connection detail between a load-bearing masonry wall and a steel joist with a metal deck floor.

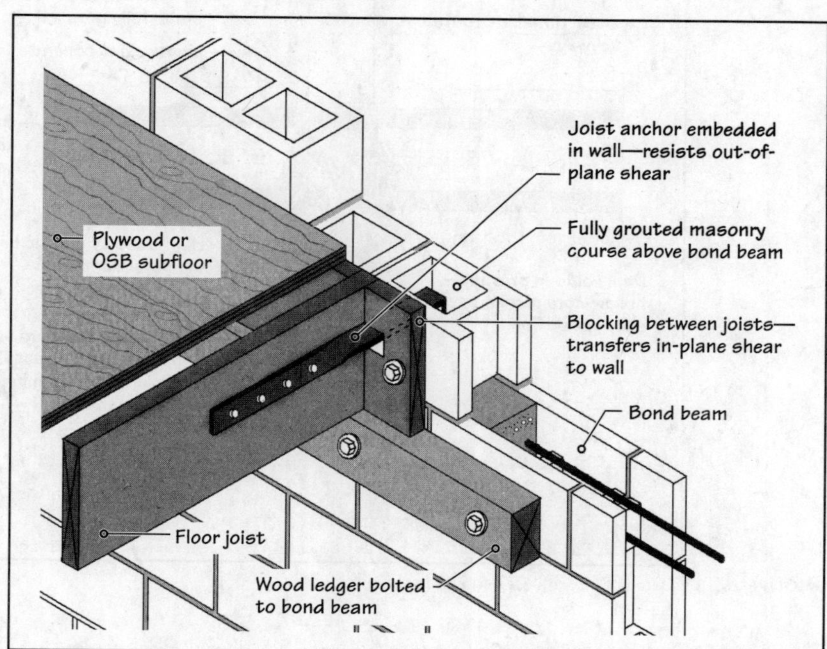

FIGURE 26.21 A schematic connection detail between a load-bearing masonry wall and a wood light-frame floor.

26.6 LIMITATIONS OF MASONRY BEARING WALL CONSTRUCTION

As stated in Section 25.1, cutting masonry units increases the cost of construction and also slows it down, so it is generally avoided in masonry structures. Consequently, the length of a masonry walls is based on modular dimensions. In CMU walls, the module is 8 in. Therefore, the lengths of CMU walls and opening widths are multiples of 8 in.

With a mortar joint thickness of $\frac{3}{8}$ in., the clear size of an opening is $\frac{3}{8}$ in. greater than that given by the multiple of 8 in. In other words, the rough opening width is 3 ft $4\frac{3}{8}$ in. ($40\frac{3}{8}$ in.), not 3 ft 4 in. (40 in.), Figure 26.22(a).

Similarly, the width of the masonry wall between openings is $\frac{3}{8}$ in. less than that given by the multiples of 8 in. Thus, the length of masonry between openings is 1 ft $3\frac{5}{8}$ in. ($15\frac{5}{8}$ in.), not 1 ft 4 in. (16 in.).

The 8-in. module also applies to height dimensions. Thus, the floor-to-ceiling heights in a CMU masonry wall building are generally 8 ft, 8 ft 8 in., 9 ft 4 in., and so on. Similarly, a rough door-opening height of 7 ft $4\frac{3}{8}$ in., shown in Figure 26.22(b), is commonly used. This dimension is (slightly) excessive for a standard 7-ft-0-in.-high door.

Where standard-height doors are desired, a starter course of saw-cut masonry units may be used. (However, note that in most commercial construction, doors and windows are custom fabricated to opening dimensions. Therefore, standard door and window dimensions generally do not apply to commercial construction.)

NOTE

The dimension of a masonry opening is $\frac{3}{8}$ in. greater than the multiples of a masonry unit module. The dimension of the masonry between openings is $\frac{3}{8}$ in. less than the multiples of the masonry unit module.

WATER-RESISTIVE EXTERIOR FINISH REQUIRED

Because contemporary masonry bearing walls are relatively thin, they require some form of weather-resistive exterior coating (clear coating or paint), plaster (stucco or EIFS), or some

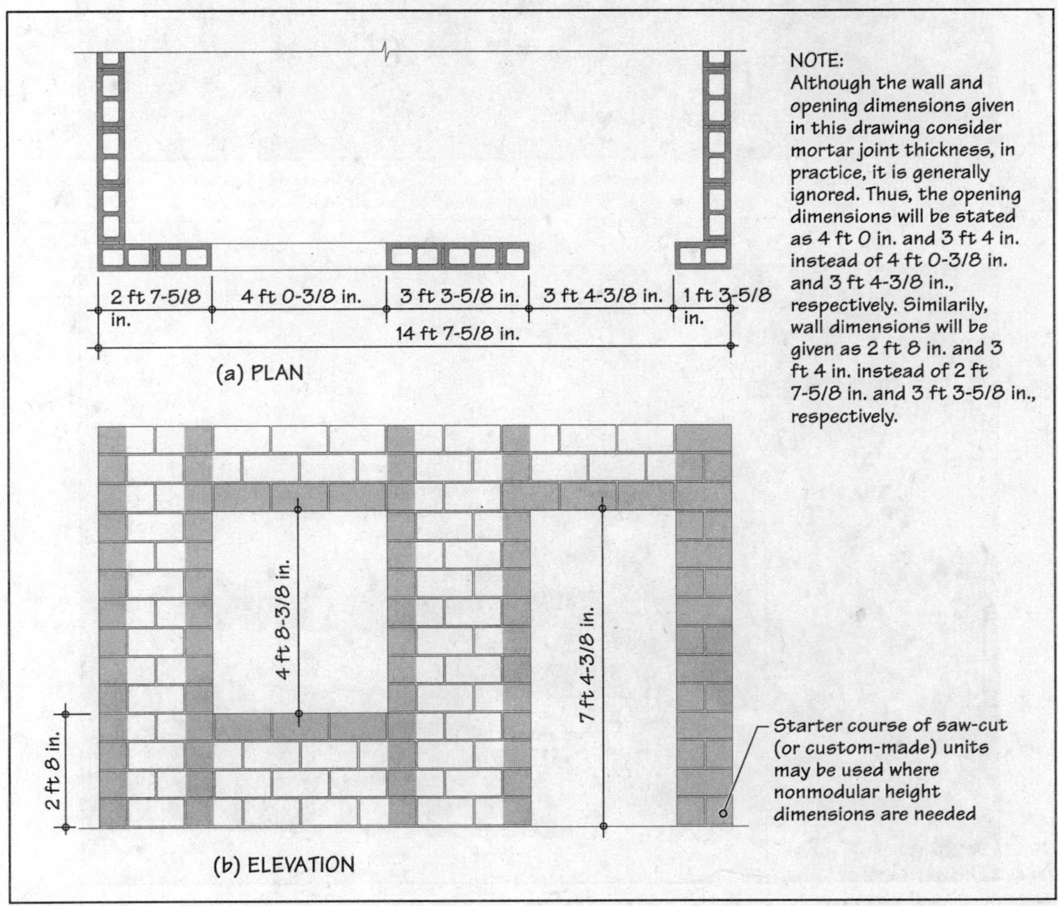
NOTE:
Although the wall and opening dimensions given in this drawing consider mortar joint thickness, in practice, it is generally ignored. Thus, the opening dimensions will be stated as 4 ft 0 in. and 3 ft 4 in. instead of 4 ft 0-3/8 in. and 3 ft 4-3/8 in., respectively. Similarily, wall dimensions will be given as 2 ft 8 in. and 3 ft 4 in. instead of 2 ft 7-5/8 in. and 3 ft 3-5/8 in., respectively.
2 ft 7-5/8 in.
4 ft 0-3/8 in.
3 ft 3-5/8 in.
3 ft 4-3/8 in.
1 ft 3-5/8 in.
14 ft 7-5/8 in.
(a) PLAN
4 ft 8-3/8 in.
7 ft 4-3/8 in.
2 ft 8 in.
Starter course of saw-cut (or custom-made) units may be used where nonmodular height dimensions are needed
(b) ELEVATION

FIGURE 26.22 Effect of modularity of CMUs on the dimensions of walls and openings.

form of cladding to prevent water from entering the interior (see Chapters 28 and 29 for a discussion of exterior cladding). Exterior walls in traditional masonry structures did not require such cladding because their thickness made them inherently water resistive.

COORDINATION BETWEEN TRADES DURING CONSTRUCTION

Electrical conduit and junction boxes and other utility lines need to be built into walls as the construction progresses. Door frames are generally erected in position, and the walls are built around them. Therefore, greater coordination between various trades is needed during the construction of a masonry building than is required for a steel- or concrete-frame building. Where precast-concrete floor slabs are used, detailed shop drawings are required to indicate the location of openings for vertical continuity of service lines.

Figures 26.23(a) to (e) illustrate the highlights of the construction process of a typical high-rise masonry bearing wall building, and Figures 26.23(f) shows the finished building.

FIGURE 26.23 (a) This and the following five images show the construction of an eight-story load-bearing masonry hotel building. The building, with a cross-bearing wall plan (see Figure 26.14), is in its early stages of construction.

FIGURE 26.23 (b) An 8-ft-wide precast-concrete hollow-core slab being flown into position. Observe the vertical reinforcement projecting from the lower-level walls.

FIGURE 26.23 (c) Keyway between 8-ft-wide hollow-core slabs. Additional connection slots have been provided in the center of each slab. Observe that the floor plan of the building is a cross-bearing wall plan (see Figure 26.14).

FIGURE 26.23 (d) An 8-ft-wide hollow-core slab has a factory-provided slot in the center in order to achieve a connection between the walls and the hollow-core slabs at 4 ft on center.

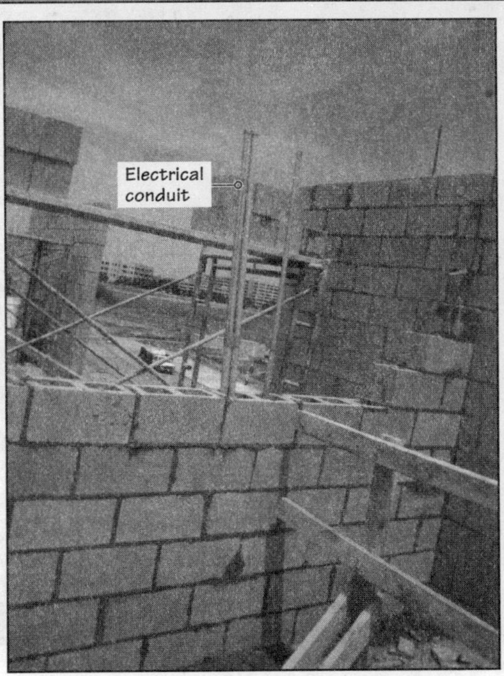

FIGURE 26.23 (e) Electrical conduits and other utility lines must be built into walls during construction.

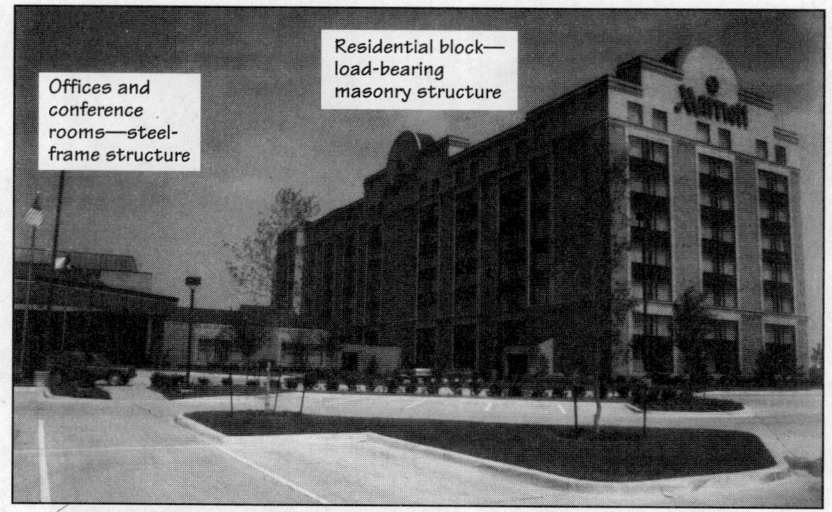

FIGURE 26.23 (f) The eight-story hotel residential block with a masonry bearing wall structure (shown under construction in (a) to (e)). The office and conference areas in this building are in steel frame because of their large, unobstructed spaces.

26.7 BEARING WALL AND COLUMN-BEAM SYSTEM

Although masonry bearing wall construction is most suited for mid- to high-rise residential occupancies, as previously described, it is also well suited to low-rise building types. Chief among them are school and college buildings, offices, shopping centers, and religious buildings.

However, such buildings generally consist of a hybrid construction system, in which all exterior walls are masonry bearing walls and the interior structure is a combination of masonry bearing walls, drywall partitions, and a column-beam frame. The lateral load resistance is generally designed into the masonry walls.

The column-beam frame may consist of cast-in-place concrete or steel. Concrete is better suited because both concrete and concrete masonry have similar properties, causing few differential-movement problems. However, a steel frame is generally more economical and is, therefore, widely used.

Differential-movement problems between steel columns and masonry walls must be considered; they are generally more easily managed in low-rise buildings. Figure 26.24 shows a typical building that combines load-bearing masonry with a steel frame.

FIGURE 26.24 A school building with exterior load-bearing masonry walls and interior steel frame under construction.

26.8 REINFORCED-CONCRETE BEARING WALL CONSTRUCTION

Site-cast reinforced-concrete bearing walls can be used in place of masonry bearing walls. When used in conjunction with site-cast reinforced-concrete floor slabs, they provide a robust structure because of the inherent continuity between the vertical and horizontal elements of the building.

In other words, the joints between the walls and the floor slabs of a site-cast reinforced-concrete bearing wall structure are tougher than those in masonry wall and precast-concrete hollow-core slab structures. Site-cast-concrete bearing wall structures are, therefore, better able to resist lateral loads. Additionally, site-cast reinforced-concrete walls can seamlessly integrate with a site-cast reinforced-concrete column-beam frame in the case of a hybrid (bearing wall and frame) system.

Load-Bearing Reinforced-Concrete Masonry Walls with Tunnel Forms

In general, a site-cast-concrete bearing wall structure can be designed with thinner walls than a similar masonry bearing wall structure. As stated in Chapter 22, a site-cast-concrete structure is economical only if the cost of formwork is controlled. A type of formwork that has gained favor in constructing high- and mid-rise reinforced-concrete bearing wall structures is the *tunnel form,* Figure 26.25. A tunnel form consists of prefabricated and collapsible steel forms (of an inverted L-shape) that allow the walls and the overlying floor slabs to be cast simultaneously.

Tunnel forms are generally removed the day following the casting of concrete for a given floor, cleaned and prepared for reuse. Depending on the time required to lay the reinforcement and utility conduits for the walls and floors, a three- to four-day cycle for constructing a floor is typical.

Walls and slabs of virtually any thickness can be constructed by increasing or decreasing the distance between tunnel forms. However, the system is feasible only with a modular and repetitive floor plan, such as that obtained in residential structures. Figure 26.26 shows a mid-rise reinforced-concrete bearing wall hotel building constructed using tunnel forms, and Figure 26.27 shows a corresponding high-rise apartment building.

A reinforced-concrete bearing wall building has the same advantages as a masonry bearing wall building, that is, good sound insulation, high fire resistance, and a mold-free structure. The disadvantages are also the same, the primary disadvantage being the inflexiblity in spatial organization. Like their masonry counterpart, exterior site-cast concrete walls generally require a water-resistive cladding.

FIGURE 26.25 Prefabricated, reusable, and collapsible steel tunnel forms.

REVIEW QUESTIONS

1. With the help of sketches and notes, explain why the walls in a traditional masonry bearing wall structure had to be extremely thick.
2. Explain the importance of vertical reinforcement in load-bearing masonry walls.
3. With the help of sketches and notes, explain why it is desirable to reinforce the ends, corners, and jambs around openings in the walls of load-bearing masonry buildings.
4. What is the purpose of bond beams, and where are they required in a typical load-bearing masonry structure?
5. List the various load-bearing wall systems described in this chapter.
6. With the help of a sketch, explain what the tunnel form is and its benefits and limitations.
7. Explain why the loads on the slab-on-ground in concrete tilt-up wall buildings are generally higher during the construction of the building than after the building is complete.
8. With the help of a sketch and notes, explain what a closure strip is.

CHAPTER 27

Exterior Wall Cladding–I (Principles of Rainwater Infiltration Control)

CHAPTER OUTLINE

Exterior walls are one of the major determinants of the appearance of a building. They convey images such as strength or solidity (brick- or stone-clad walls), lightness or openness (glass-metal curtain walls), or a sense of movement or activity (bright, glistening metal curtain walls). In terms of the building's performance, the importance of exterior walls is even greater. Together with the roof, they constitute the building's envelope, providing the separation between inside and outside and serving to maintain an acceptable interior environment.

The exterior walls can serve the envelope function only if they are able to perform well under a range of environmental conditions. They must

(a) Prevent water infiltration from rain and snow
(b) Control heat loss and heat gain
(c) Control air leakage and water vapor transmission
(d) Resist fire
(e) Control sound transmission
(f) Accommodate movement due to thermal, moisture, and other causes

Factors (b) to (f) are covered in Chapters 5–9. Water infiltration is, however, the most critical of all concerns. Its importance is universal because no building can be considered a shelter unless its envelope is water resistant. Additionally, a water-resistant envelope has been the fundamental requirement of buildings throughout the ages—well before other environmental factors emerged as design concerns.

This chapter focuses on the principles of water-infiltration control in exterior-wall assemblies. The construction and detailing of a variety of these assemblies is covered in Chapters 28 (Exterior Wall Cladding-II), 29 (Exterior Wall Cladding-III) and 32 (Exterior Wall Cladding-IV). The principles of water-infiltration control in basement walls are covered in Chapter 12, and roofs are discussed in Chapters 33 and 34.

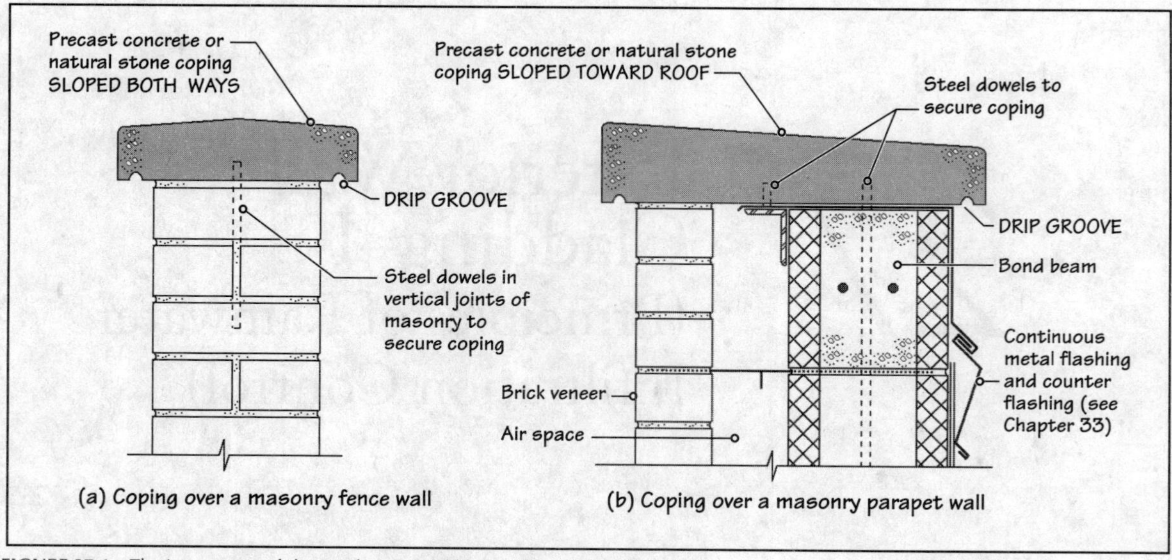

FIGURE 27.1 The importance of sloping the exposed horizontal surfaces of a wall.

27.1 RAINWATER INFILTRATION CONTROL—GENERAL PRINCIPLES

Three major forces that affect water infiltration through exterior walls are gravity, the capillary effect, and wind. Because these forces frequently act on an assembly simultaneously, multiple strategies must be used in the same assembly to counter water infiltration.

GRAVITY-INDUCED INFILTRATION

Gravity generally affects water penetration through an exposed horizontal surface of a wall and can be countered by providing a nominal slope in the surface. The purpose of slope (usually 1:12) is to drain water away from vulnerable parts of a building. Thus, a roof overhang protects the upper portion of a wall, and the coping on the top of a masonry fence wall is sloped on both sides of the wall, Figure 27.1.

The coping over a roof parapet is generally sloped toward the roof so that the water drains on the roof. In exterior window sills and thresholds under entrance doors, the slope is directed away from the building.

CAPILLARY-INDUCED INFILTRATION

Water molecules are subjected to forces of attraction between them, creating an intermolecular cohesive bond between them. In addition to the cohesive bond, water molecules experience forces of attraction from the molecules of the surface with which they are in contact. These forces create an adhesive bond. For most building surfaces, the adhesive bond is stronger than the cohesive bond. It is because of the stronger adhesive bond that a thin film of water is able to travel horizontally along the soffit instead of dropping vertically by gravity. A drip mechanism, which consists of either a continuous groove or a vertical projection, counteracts this phenomenon, as shown in Figures 27.1 and 27.2.

Cohesive and adhesive forces also create the capillary effect, which is responsible for suction forces that occur in tiny spaces between two surfaces (surfaces in very close proximity to each other). In this case, the water is sucked between the surfaces.

Water can also be sucked into the tiny pores of porous materials such as brick, CMU, or stone. Coating the surface of a porous material with a sealer helps close the pores. Because the sealer degrades rapidly due to ultraviolet radiation, abrasion, and other physical processes, the use of sealers is not a durable means of waterproofing a porous surface. To be effective, seals on exterior walls need frequent reapplication.

Capillary suction can also occur in a joint between two nonporous materials if the joint is narrow. In narrow joints, capillary suction can be prevented by introducing a larger space,

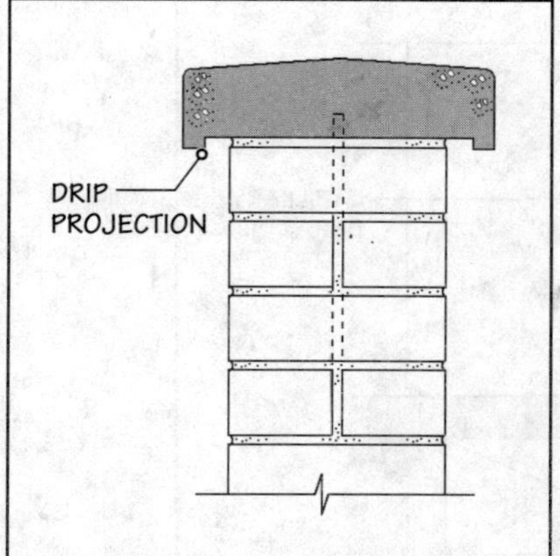

FIGURE 27.2 Detail of Figure 27.1(a) redrawn with a drip projection in the coping instead of a drip groove.

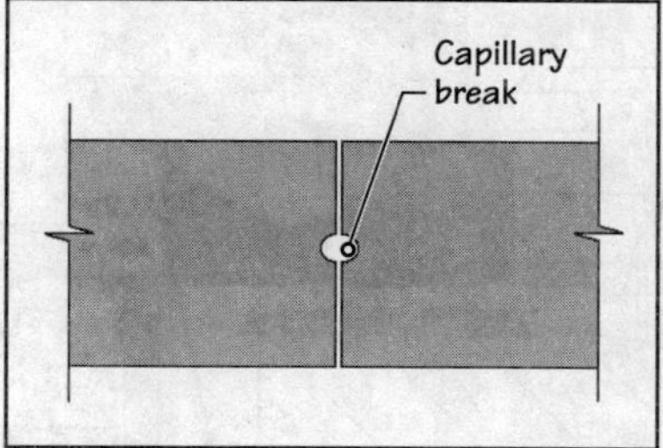

FIGURE 27.3 A narrow space can be made more water-resistant by providing a capillary break.

nearly $\frac{1}{4}$ in. or wider, immediately beyond the capillary space. This space is referred to as the *capillary break,* Figure 27.3.

Wind-Induced Infiltration

Providing an *overlap* between members meeting at a joint is another commonsense measure. The force of wind can impart kinetic energy to water, causing the water to travel horizontally as well as vertically in the joint. Therefore, the overlap must be sufficiently large to present a reliable baffle against wind-driven rain, Figure 27.4. A lapped horizontal joint performs better than a lapped vertical joint, because water has to work against the force of gravity to get across the overlap in a horizontal joint. Capillary breaks can also be included in a lapped joint.

A horizontal butt joint can be made infiltration resistant by incorporating L- or Z-flashing, Figure 27.5. L- or Z-flashing consists of an impervious membrane that provides a barrier to water and directs it to the exterior surface. Various types of sheet materials—galvanized steel, stainless steel, copper, lead, and polyvinyl chloride (PVC)—are used as flashing materials. Other uses of flashing are shown in Figures 27.6 and 27.8.

The Importance of Joint Sealants

Filling joints with sealants (see Chapter 9) is another commonly used strategy to keep water from penetrating through joints between building components. Because sealants degrade

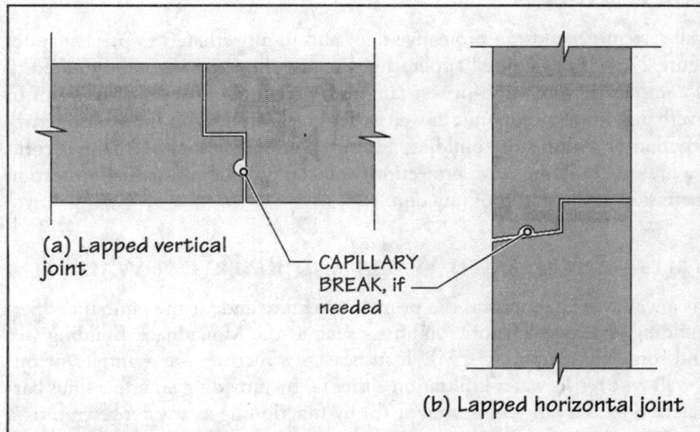

FIGURE 27.4 Lapped vertical and horizontal joints.

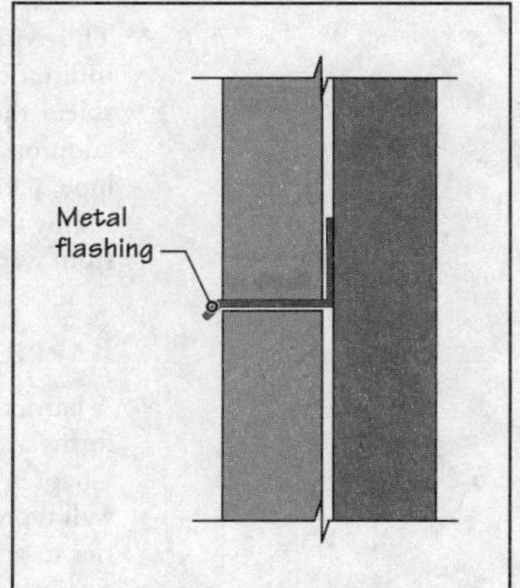

FIGURE 27.5 L-flashing at a horizontal butt joint. If the projecting part of flashing is turned down at a 90° angle instead of a 45° angle, it is referred to as *Z-flashing*.

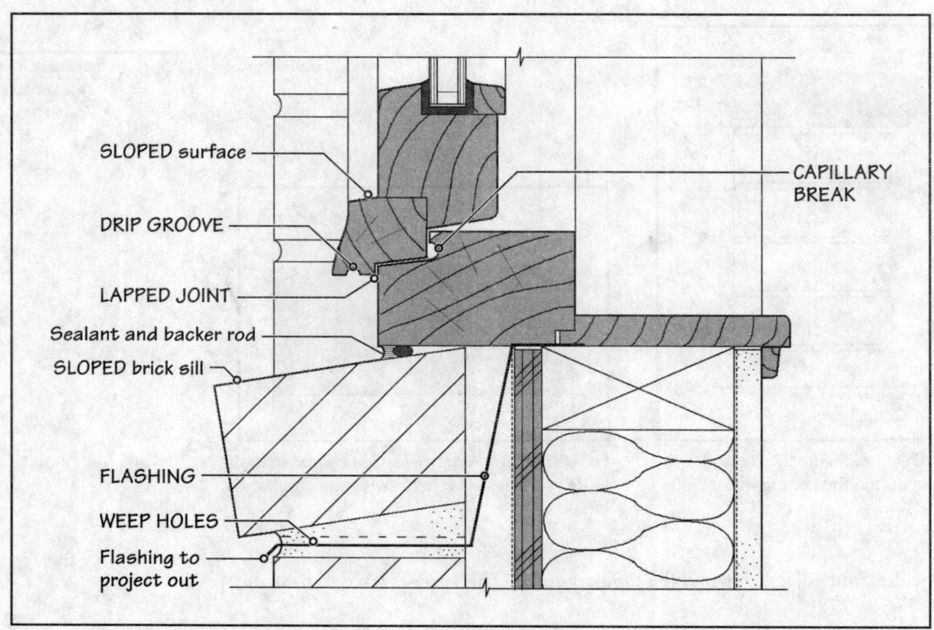

FIGURE 27.6 A section at a window sill showing various rainwater leakage control strategies used in one detail.

over time from exposure to environmental factors, particularly solar radiation and water, they should not be relied upon as the only water-resisting element. A certain amount of redundancy must be built into the detail in case the sealant fails. Usually all or most of the strategies previously discussed are used in a single detail, as shown in a section at the sill level of a typical wood window, Figure 27.6.

27.2 RAINWATER INFILTRATION CONTROL AND EXTERIOR WALLS

Water-resistant exterior walls may be divided into three types:

- Walls with overhangs
- Barrier walls
- Drainage walls

Walls with Overhangs

This type of wall system depends on protective floor and roof overhangs to prevent water infiltration, Figure 27.7. It has limited application because the overhangs are required to shield the wall completely from wind-driven rain and, therefore, must be quite deep. In addition, the overhangs impose economic as well as design constraints on the exterior envelope, particularly that of a multistory building. However, a shallow roof overhang is commonly used in a low-rise building (eave projection) and can provide substantial protection from water penetration at the wall-roof junction.

Barrier Walls—Face-Sealed Walls and Reservoir Walls

A barrier wall is a wall that functions as the primary structure and, at the same time, performs as the building envelope. Historic buildings such as the Monadnack Building (see Chapter 26) and some contemporary CMU load-bearing structures are examples of this wall type. This wall type resists water infiltration either (a) by providing an impervious barrier to water, referred to as a *face-sealed wall*, or (b) by functioning as a water reservoir—a *reservoir wall*.

An example of a face-sealed wall is a CMU or concrete wall system with an applied water-resistant veneer, such as EIFS or stucco (see Chapter 29). On the other hand, an

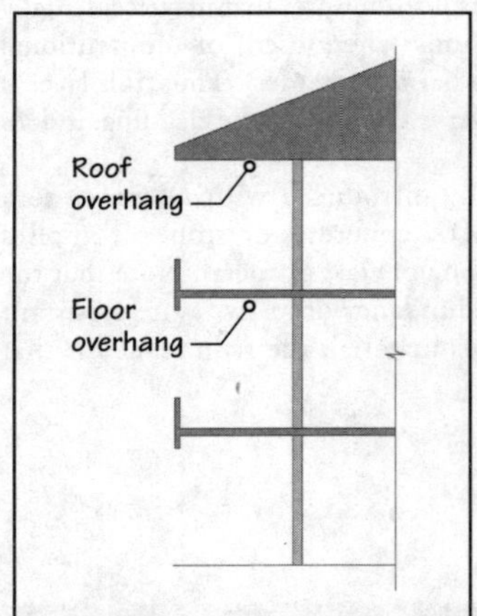

FIGURE 27.7 Deep roof and floor overhangs can substantially reduce water leakage from wind-driven rain through walls.

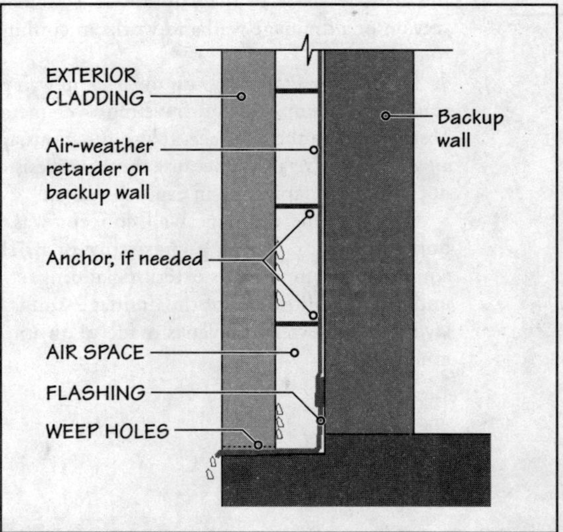

FIGURE 27.8 A section through a drainage wall showing its important water infiltration control features. (Note that the air space in a drainage wall may also have insulation, which is not part of water infiltration control.)

example of a barrier wall that acts as a reservoir is a porous masonry or concrete wall that is so thick that any absorbed water is unable to migrate to the wall's interior surface. The thicker the wall, the larger the reservoir. In this type of wall, only the outer layer of the wall gets wet, even during a long rainy spell; given a long dry spell, the entire wall will dry again.

It is very important that a barrier wall does not develop through-wall cracks that may be caused by weathering, expansion or contraction, foundation settlement, and so on. If cracks develop, water infiltration will result. Careful detailing and attention to water resistance in materials are, therefore, essential to the long-term performance of a barrier wall.

Drainage Wall

A relatively more water-resistant exterior wall is a drainage wall. It is a wall that consists of an exterior cladding, an inner backup wall, and an intervening air space between the cladding and the backup.

In a drainage wall, the exterior cladding is the first defense against water infiltration. Therefore, it is made as water resistant as possible. However, some water will inevitably penetrate through the cladding. Any water that leaks through the cladding collects in the air space and is drained out through small openings at the bottom of the cladding called *weep holes,* Figure 27.8.

Important Features of a Drainage Wall

A drainage wall must have sufficient built-in redundancy to remain infiltration-free for the entire life of the building and includes the following features:

- Exterior cladding
- Air space
- Flashing
- Water-resistant backup
- Weep holes

Flashing in a drainage wall is a continuous waterproof membrane providing a continuous barrier that originates at the backup wall and penetrates through the air space and cladding. The purpose of flashing is to collect and channel any water that gets into the air space back to the exterior. It is attached to the backup wall at the bottom of a section of a drainage wall and works in conjunction with weep holes, which allow water to escape.

The *water-resistant layer* on the backup wall protects it from water that may accidentally reach the backup wall by traveling over metal anchors (where used) or unintentional obstructions in the air space. It also functions as an air barrier (retarder). Thus, this layer is an air-weather retarder. Because the water-resistant layer is shielded by the cladding, it does not degrade as rapidly as an exposed surface.

Although the drainage wall concept was initially introduced with reference to a porous exterior cladding consisting of brick, CMU, concrete, or stone, it applies equally to an impervious exterior cladding, such as one of glass or metal. Note that the underlying philosophy of the drainage wall is the redundancy of its two water-resistant layers. If the exterior layer is made of an impervious material, the redundancy is that much greater.

EXPAND YOUR KNOWLEDGE

Exterior-Wall Cladding Systems: Curtain Wall, Veneer Wall, and Infill Wall

The primary exterior-wall types in use today are curtain walls, veneer walls, and infill walls.

Curtain Walls

As the name suggests, a *curtain wall* forms a curtain on the exterior face of a building. It is used with a frame structure, covers the building's structural frame, and is directly suspended from it. In a curtain wall, the wind loads are transferred directly from the curtain wall to the building's structural frame.

A 2-in. minimum separation is generally required between the curtain wall and the building's structural frame to accommodate unintended dimensional irregularities in the frame.

A typical example of a curtain wall is a *glass-aluminum curtain wall* (see Chapter 32). *Opaque curtain walls* consist of precast concrete panels, glass fiber–reinforced panels, metal panels, and so on. A backup wall may be used in an opaque curtain wall to provide an interior wall finish and to incorporate insulation, electrical, and other utility conduits within the wall. Because the wind load is resisted by the curtain wall, the backup wall in this assembly (if provided) does not experience any wind loads.

Veneer Walls

Another exterior wall cladding, which is similar to an opaque curtain wall is a *veneer wall*. A veneer wall can be used with both a frame structure and a bearing wall structure. A backup wall is always required in a veneer wall. The wind loads that act on the veneer wall are transferred to the backup wall. The backup wall then transfers them to the building's structural frame. A veneer wall may be of two types:

- Anchored veneer
- Adhered veneer

In an *anchored veneer*, the anchors connect the veneer to the backup; therefore, they participate in transferring the wind loads from the veneer to the backup wall. An anchored veneer wall is typically designed as a drainage wall with a minimum 2-in. air space between the veneer and the backup wall.

A veneer may also be applied without anchors, in which case it is adhered to the backup wall. Such a veneer is called an *adhered veneer*. An adhered veneer wall is a face-sealed (barrier) wall, unlike the anchored veneer. Stucco and EIFS wall assemblies are examples of face-sealed barrier walls.

Infill Walls

Like a curtain wall and a veneer wall, an infill exterior wall is a non-load-bearing wall assembly. It occupies the space between the building's columns and beams, leaving the structural frame exposed to the outside. An infill wall can be used only with a frame structure. It can be designed as either an anchored infill wall or an adhered infill wall.

Before energy-use concerns became critical, most exterior-wall assemblies were designed as infill walls, Figure 1. In these buildings, the steel or concrete structural frame, having an extremely low R-value, contributed to substantial thermal short-circuiting. In contemporary buildings, the use of curtain walls or veneer walls is the norm because they cover the building's structural frame and, therefore, can be designed with greater thermal efficiency.

Exterior Cladding

The term *exterior-wall cladding* (or simply *cladding*) is a general term that is used for all exterior wall finishes.

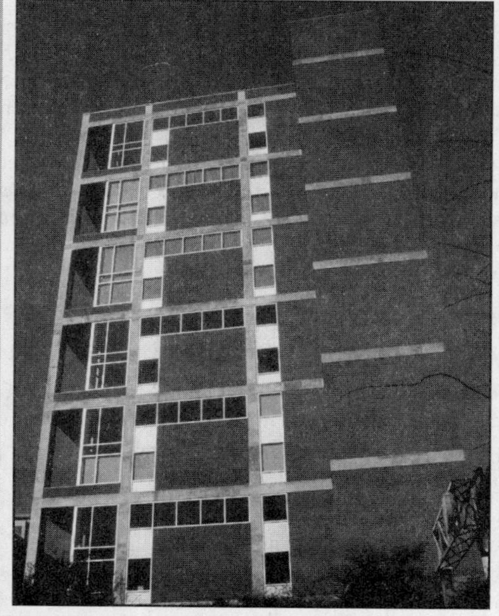

FIGURE 1 A typical example of an exposed structural frame and exterior infill walls.

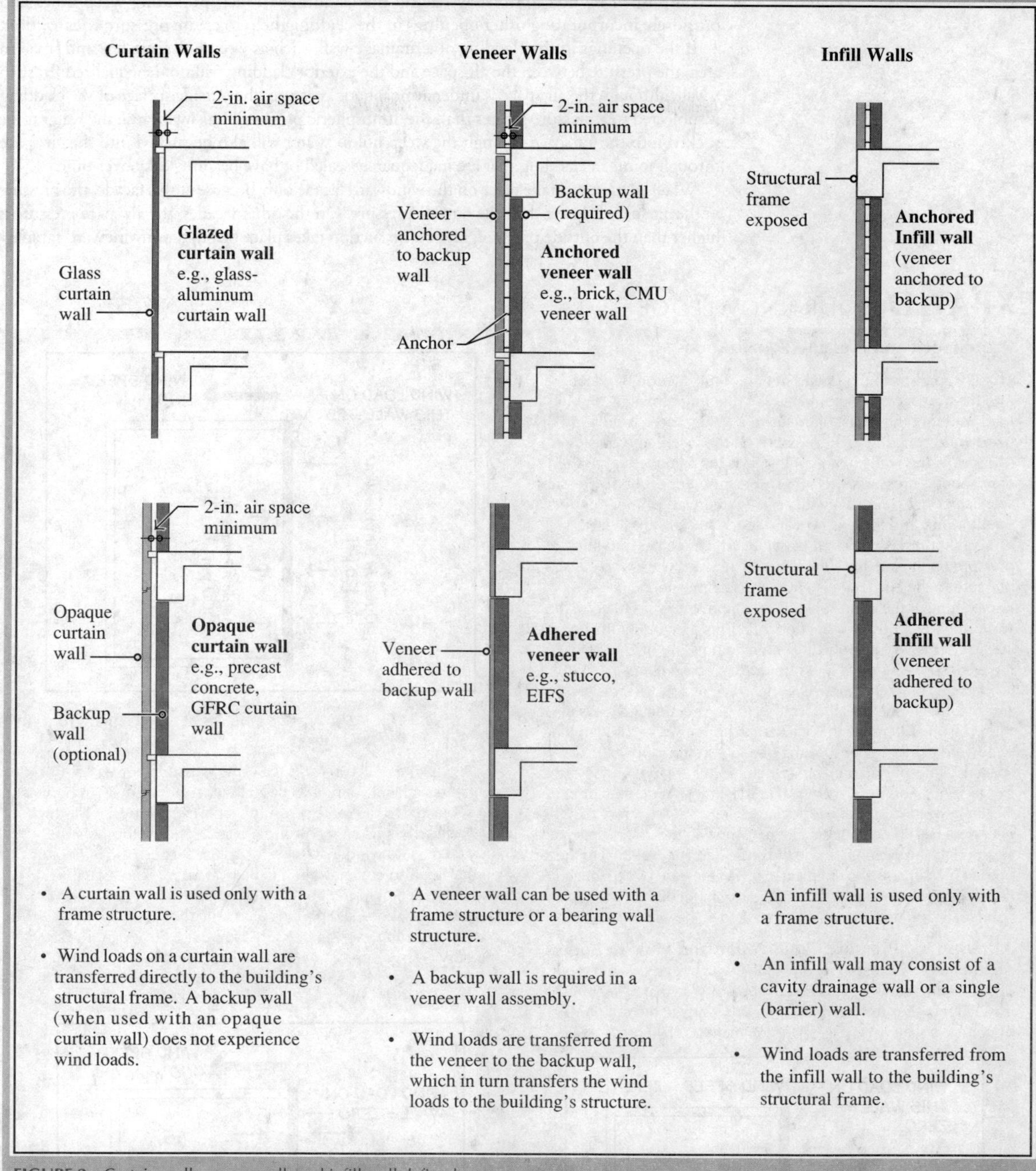

FIGURE 2 Curtain wall, veneer wall, and infill wall defined.

27.3 RAIN-SCREEN EXTERIOR CLADDING

A rain-screen cladding system includes the basic principles of the drainage wall system, but it also addresses the issue of water penetration due to unequal distribution of air pressure on the exterior of the wall and in the air space between the cladding and the backup wall. The equalization of pressure is required when gaps and breaches in exterior cladding (such as weep holes, cracks, and apertures between masonry units and mortar joints) become sources of water penetration by suction under conditions of wind-driven rain.

In order to reduce the suction of water in a drainage wall, the pressure between the air space and the outside should be equalized as much as possible. Pressure equalization is accomplished by not completely sealing the cladding. In fact, providing weep holes and purposely incorporating other openings in the cladding help to create pressure equalization.

If the openings in the cladding of a drainage wall, such as weep holes, are few and small in area, the pressure between the air space and the exterior cladding will not be equalized. In such a wall, although the air space is under atmospheric pressure, the outside surface of the cladding is subjected to a pressure greater than the atmospheric pressure. This will cause the water to be sucked into the air space through the weep holes. Water will also be sucked into the air space through joints in cladding that are inadequately sealed or have become leaky over time.

Water suction will take place on the windward facade only because at this facade, the pressure in the air space is lower than the outside pressure. On the other facades, the air space pressure is higher than the outside pressure. Hence, no suction takes place through nonwindward facades.

EXPAND YOUR KNOWLEDGE

Rain-Screen and Pressure Equalization

To fully appreciate the phenomenon of the suction of water into the air space in a drainage wall, the fundamentals of wind loads on buildings must be reviewed. When there is no wind (zero wind speed), the air pressure on the inside and outside surfaces of the walls of an enclosure is the same, equal to the atmospheric pressure—2,100 psf. Because atmospheric pressure works on both sides of a wall equally, the wall is under perfect equilibrium, and there is no wind load on the wall, Figure 1.

Equilibrium is disturbed by wind, which creates additional pressure on the outside surface of a windward wall, whereas the pressure in the enclosure is equal to the atmospheric pressure. For example if the wind speed is 100 mph, the wind load on the wall is nearly 25 psf (see Section 3.5). This means that the outside surface of the windward wall is under a pressure of 2,125 psf, but the enclosure is under a pressure of 2,100 psf, Figure 2.

It is this difference between the inside and outside air pressures that we refer to as the *wind load* (see Section 3.5). It is also this pressure difference that causes suction of water through the windward cladding because the pressure in the (drainage) air space is atmospheric pressure, whereas the pressure on the exterior face of cladding is greater than the atmospheric pressure.

If the windward wall has large openings and the rest of the enclosure has no openings at all, the wind will move into the enclosure and the air pressure inside the enclosure will be equal to the outside pressure, as shown in Figure 3. In the case of perfect equalization of pressures, the windward wall will not be subjected to any wind load, and no suction will be created at the windward wall.

Atmospheric Pressure Equalization and Human Bodies

Atmospheric air pressure is due to the weight of a column of air on the earth's surface. This equals the pressure that a 29.92-in. (say, 30 in. = 2.5 ft) column of mercury exerts at the column's bottom. It equals 14.7 psi, or approximately 2,100 psf.

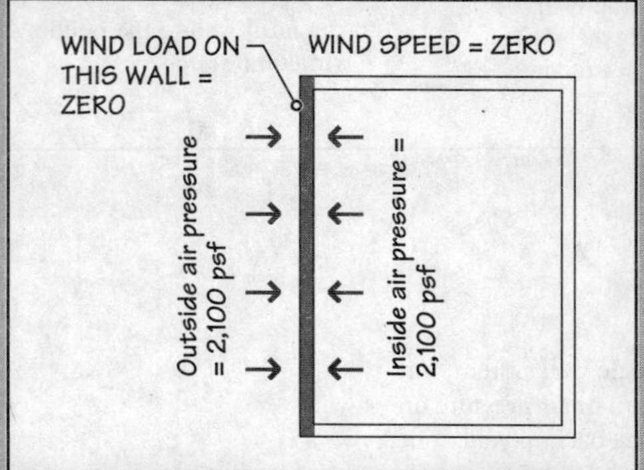

FIGURE 1 Outside and inside air pressures on a wall under zero wind speed.

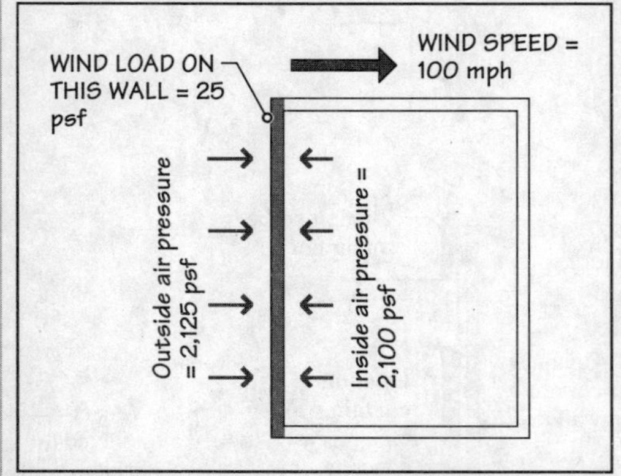

FIGURE 2 Outside and inside air pressures on a windward wall of a fully enclosed building.

A pressure of 2,100 psf is the same pressure that a 14-ft-thick concrete slab will exert at its base. In other words, human bodies are subjected to the same pressure that is exerted by a 14-ft-thick concrete slab at its base. We are not squashed by this pressure (i.e., there is no wind load on our bodies) because the interior of our bodies also contains air, and there is air pressure equalization between the inside and outside of our bodies. (Because concrete weighs about 150 pounds per cubic foot [pcf], a 14-ft-high concrete column exerts a pressure of about 2,100 psf at its base. Mercury weighs approximately 850 pcf. Therefore, only a [30-in.] 2.5-ft-high mercury column exerts the same pressure.)

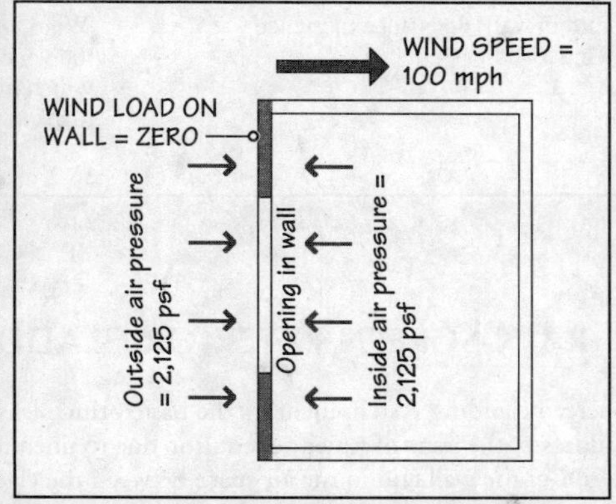

FIGURE 3 Outside and inside air pressures in an enclosure that has a large opening on the windward wall.

Wind Load on the Cladding of a Pressure-Equalized Wall

Because there is no pressure differential between the inside and outside faces of cladding in a pressure-equalized drainage wall, the cladding is not subjected to any wind load. In such a wall, all of the wind load acts on the backup, and the cladding functions as a nonstructural element whose primary environmental function is to control water penetration by allowing free movement of air for pressure equalization. Therefore, the exterior cladding in a pressure-equalized wall is referred to as a *screen*—or, more commonly, as a *rain screen.*

Practical Limitations of a Rain-Screen Wall

Theoretically, the exterior cladding in a pressure-equalized drainage wall should not be subjected to any wind load. In practice, however, it is not possible to achieve a completely load-free cladding. The reason is that pressure equalization does not occur instantaneously. It takes some time for pressure equalization to take place, depending on how rapidly the outside air can move into the intervening air space. If the area of openings in the wall is large, pressure equalization will be rapid. In the case of a small opening area, pressure equalization will be slow.

In other words, there is a time lag between a change in the outside pressure and the corresponding change in air space pressure. During this period (when pressure equalization is taking place), the cladding is subjected to wind loads. If there is no further change in the outside pressure, the air space and the outside will continue to be under equal pressures.

However, wind in a storm or hurricane occurs in gusts, so exterior air pressure changes almost continuously. This disturbs pressure equalization in a drainage wall. Therefore, the cladding is constantly subjected to wind loads, although the magnitude of the wind load on the cladding of a pressure-equalized wall is smaller than that on a non-pressure-equalized wall.

Airtight Backup Wall

Once the pressure in the air space is equalized with respect to the pressure on the wall's facade, no air movement will occur in the space (because any movement of air implies the existence of a pressure differential). The consequence of this fact is that the rain-screen principle works only if the backup is completely airtight. If the backup has openings, air will move from the air space into the enclosure, implying the existence of a pressure differential (hence, the absence of pressure equalization).

If the air is able to move into the enclosure through the backup wall, water may be sucked into the enclosure. Therefore, in a pressure-equalized wall, the backup wall must be made airtight. This is usually achieved by sealing all the joints in the backup wall and/or providing an air barrier. The air barrier is usually placed on the outside face of the backup wall so that it can also function as the damp-proofing layer.

Compartmentalization of the Air Space

Another cause of air movement in an intervening air space is shown in Figure 27.9. Under the action of wind, the windward facade is under positive pressure and the side facades are under negative pressure. If the air space is continuous from one facade to the other, air will move through the space as shown, negating attempts at pressure equalization.

Therefore, the air space of a pressure-equalized drainage wall must be closed at wall corners. This is usually achieved by using a continuous vertical closure at the corners of the air space, Figure 27.10. A closed-cell compressible filler, such as neoprene sponge, may be used as the closure material.

In addition to providing vertical closures at wall corners, the air space should be closed by horizontal closures. In fact, the air space should be subdivided into independent compartments. The compartmentalization is in response to the variation of wind pressure over the building's facade. Wind pressure is greater at upper floors of a building than at lower floors, and is greater at edges and corners of the building than in the middle (see Section 3.5). If the air space is not compartmentalized, the pressure inequality over the facade will generate air movement inside the space.

Each compartment should have openings in the cladding at the top and bottom so that pressure equalization can take place. The openings should preferably be far enough away

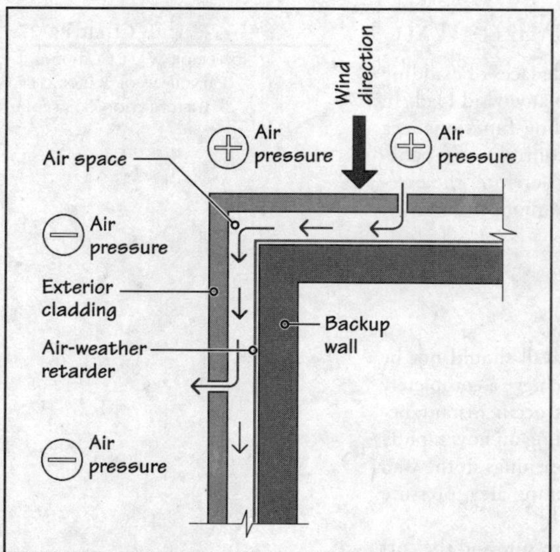

FIGURE 27.9 Because wind produces positive pressure on the windward wall and negative pressure on other walls of a building, the air in a continuous air space will circulate and move in and out through weep holes and other openings (joints, cracks, etc.) in the cladding. This makes it impossible to achieve pressure equalization between the outside and the air space.

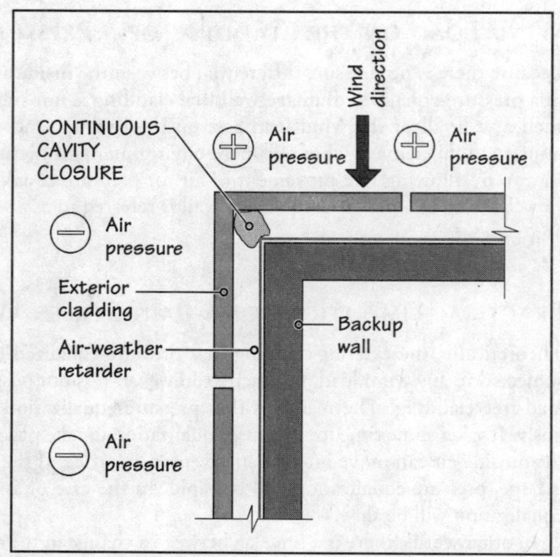

FIGURE 27.10 To prevent air movement within the air space, the air space should be divided into small sections with continuous cavity closures, particularly at corners and floor levels.

from each other to prevent airflow short-circuiting in the compartment. Each compartment should also be relatively small and should be independently drained by weep holes. Although preferable, it is not necessary that air space closures form absolutely airtight compartments. In summary, an ideal rain-screen wall should consist of the following three water-infiltration-control features:

- Voids in cladding for pressure equalization in the air space
- Compartmentalization of the air space
- Airtight backup wall

PRACTICE QUIZ

Each question has only one correct answer. Select the choice that best answers the question.

1. The coping on a roof parapet is generally
   - a. sloped toward the interior (roof).
   - b. sloped toward the outside.
   - c. sloped on both sides.
   - d. either (a), (b), or (c).
   - e. dead flat.
2. The coping on a masonry fence wall is generally
   - a. sloped toward the interior.
   - b. sloped toward the outside.
   - c. sloped on both sides.
   - d. either (a), (b), or (c).
   - e. dead flat.
3. A drip mechanism at the underside of a projecting horizontal surface counters the effect of
   - a. gravity.
   - b. adhesive forces between water molecules and the building surface.
   - c. cohesive forces between water molecules.
   - d. cohesive forces between water molecules and the building surface.
   - e. kinetic energy.
4. A capillary break is a
   - a. vertical barrier between two abutting surfaces of different materials.
   - b. horizontal barrier between two abutting surfaces of different materials.
   - c. wider space in an otherwise narrow joint.
   - d. narrow space in an otherwise wide joint.
   - e. none of the above.
5. Z-flashing is effective in
   - a. a horizontal butt joint in cladding.
   - b. a vertical butt joint in cladding.
   - c. a horizontal lapped joint in cladding.
   - d. a vertical lapped joint in cladding.
   - e. all of the above.
6. In terms of water infiltration, the exterior masonry walls in historic buildings functioned as
   - a. waterproof walls.
   - b. drainage walls.
   - c. face-sealed barrier walls.
   - d. dam walls.
   - e. none of the above.
7. The most important feature of an (exterior) drainage wall is
   - a. exterior cladding.
   - b. waterproofing of the backup wall.
   - c. flashing.
   - d. weep holes.
   - e. all of the above.
8. Metal anchors are needed to transfer lateral loads to the backup wall in
   - a. some veneer walls.
   - b. all veneer walls.

9. A curtain wall is separated from the structural frame of the building by a minimum of
a. 1 in. **b.** 2 in.
c. 3 in. **d.** 4 in.
e. none of the above.

10. A curtain wall is typically used in buildings whose structural system consists of a
a. load-bearing masonry wall structure.
b. load-bearing reinforced-concrete wall structure or load-bearing masonry wall structure.
c. reinforced-concrete frame structure or steel-frame structure.
d. steel-frame structure or load-bearing masonry wall structure.
e. all of the above.

11. A backup wall is required in a
a. curtain wall. **b.** veneer wall.
c. infill wall. **d.** all of the above.

12. Suction that pulls wind-driven rain into the air space of a drainage wall takes place on the windward face of a wall.
a. True **b.** False

13. In an ideal pressure-equalized drainage wall,
a. the cladding resists all of the wind load.
b. the backup resists all of the wind load.
c. the backup is provided with openings to equalize air pressure in the air space.
d. the joints in the cladding are sealed to control water infiltration.
e. all of the above.

14. An ideal pressure-equalized drainage wall is also referred to as a
a. pressure-screen wall. **b.** wind-screen wall.
c. no-pressure wall. **d.** rain-screen wall.
e. all of the above.

15. For a pressure-equalized drainage wall to function effectively, the air space must be continuous from floor to floor.
a. True **b.** False

REVIEW QUESTIONS

1. Using a sketch and notes, explain what a capillary break is and where it is commonly used.
2. Explain the differences between a barrier wall and a drainage wall.
3. Using sketches and notes, explain the important features of a drainage-type exterior wall.
4. Using sketches and notes, explain the difference between an anchored veneer and an adhered veneer.
5. Using sketches and notes, explain the salient features of a pressure-equalized drainage wall.
6. Explain why the air space in a pressure-equalized drainage wall must be compartmentalized.

CHAPTER 28

Exterior Wall Cladding–II (Masonry, Precast Concrete, and GFRC)

CHAPTER OUTLINE

This is the first of the two chapters on exterior wall finishes; it includes masonry veneer, precast concrete, glass fiber–reinforced concrete (GFRC), and prefabricated masonry panels. Other exterior wall finishes—stucco, exterior insulation and finish systems (EIFS), stone cladding, and insulated metal panel walls—are discussed in the next chapter (Chapter 29).

ANCHORED MASONRY VENEER AND ADHERED MASONRY VENEER

Masonry veneer may either be (a) anchored masonry veneer or (b) adhered masonry veneer. In this chapter, only anchored masonry veneer is covered. Adhered masonry veneer is discussed in the following chapter because its detailing and construction process are similar to those of stucco.

28.1 ANCHORED MASONRY VENEER ASSEMBLY—GENERAL CONSIDERATIONS

Among all contemporary exterior wall cladding systems, masonry (brick, CMU, and stone) veneer is most widely used, and within the three masonry veneer systems, *brick veneer* is by far the most popular system. Its popularity lies in its durability, fire resistance, and aesthetic appeal. Additionally, the system requires almost no maintenance and can be used for buildings of all heights and complexity—from high-rise to low-rise and from simple rectilinear

FIGURE 28.1 Because of its versatility, durability, economic advantages and aesthetic appeal, brick veneer is among the most commonly used facades for all types of buildings, particularly for mid-rise and high-rise buildings, as this New York City street shows. Seen in the background in this photo is the glass-clad Bloomberg Tower by Cesar Pelli and Associates with Schuman, Lichtenstein, Claman and Effron. (Photo by Marshall Gerometta of Emporis)

facades to intricate ones. A cursory survey of building facades in North American cities confirms this assertion, Figure 28.1.

A brick veneer generally consists of a single-wythe brick wall (generally 4 in. nominal thickness). The backup wall used with brick veneer may be load-bearing or non-load-bearing and may consist of one of the following:

- Wood or cold-formed steel stud
- Concrete masonry
- Reinforced concrete

A wood stud (or steel stud) load-bearing backup wall, Figure 28.2, is typically used in low-rise residential construction. Concrete masonry, non-load-bearing steel stud, and reinforced-concrete backup walls are generally used in commercial construction. In fact, brick veneer with concrete masonry backup is the wall assembly of choice for many building types, such as schools, university campus buildings, and offices.

The discussion presented here refers to brick veneer. It can, however, be extended to include other (CMU and stone) masonry veneers with little or no change.

ANCHORS

In a brick veneer assembly, the veneer is connected to the backup wall with steel anchors, which transfer the lateral load from the veneer to the backup wall. In this load transfer, the

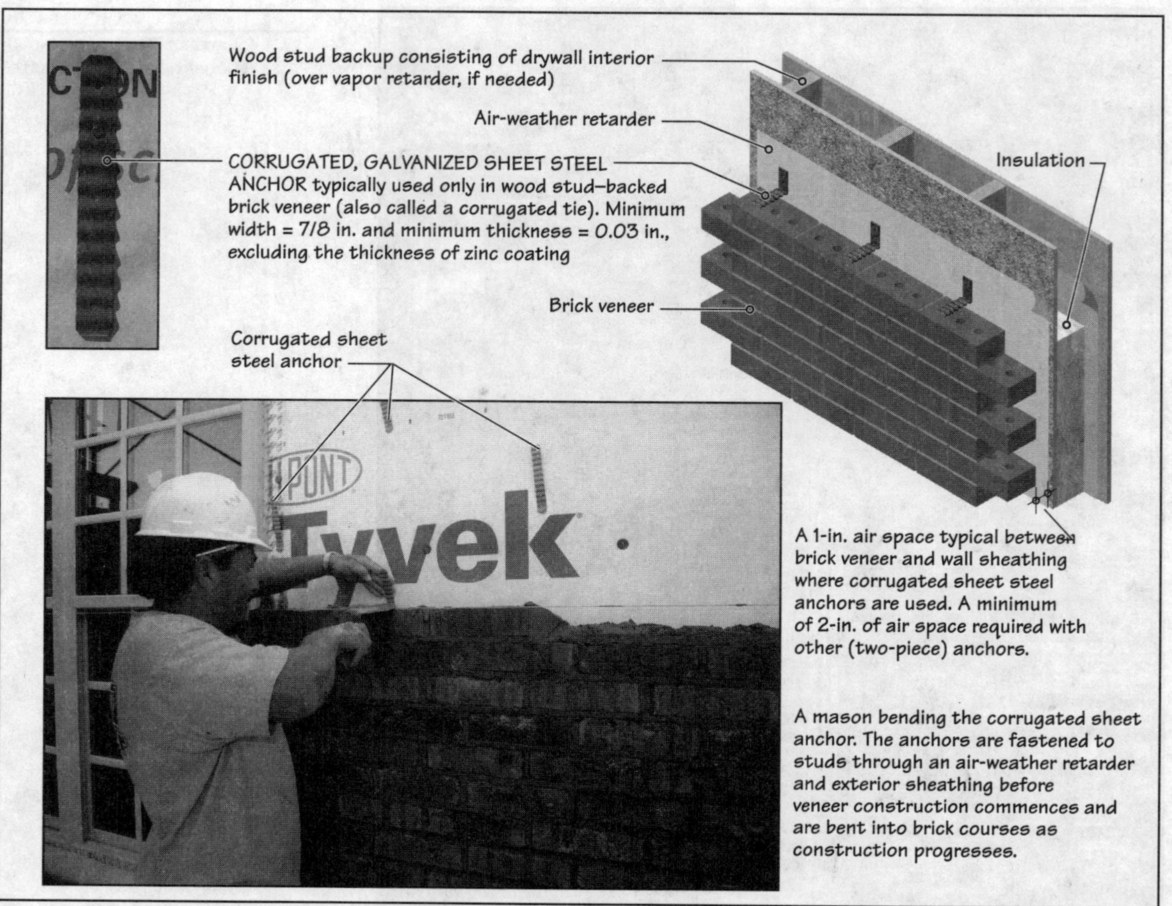
Wood stud backup consisting of drywall interior finish (over vapor retarder, if needed)
Air-weather retarder
CORRUGATED, GALVANIZED SHEET STEEL ANCHOR typically used only in wood stud–backed brick veneer (also called a corrugated tie). Minimum width = 7/8 in. and minimum thickness = 0.03 in., excluding the thickness of zinc coating
Insulation
Brick veneer
Corrugated sheet steel anchor
Tyvek
A 1-in. air space typical between brick veneer and wall sheathing where corrugated sheet steel anchors are used. A minimum of 2-in. of air space required with other (two-piece) anchors.
A mason bending the corrugated sheet anchor. The anchors are fastened to studs through an air-weather retarder and exterior sheathing before veneer construction commences and are bent into brick courses as construction progresses.

FIGURE 28.2 Brick veneer with a wood stud backup wall.

anchors are subjected to either axial compression or tension, depending on whether the wall is subjected to inward or outward pressure.

The anchors must, therefore, have sufficient rigidity and allow little or no movement in the plane perpendicular to the wall. However, because the veneer and the backup will usually expand or contract at different rates in their own planes, the design of anchors must accommodate upward-downward and side-to-side movement, Figure 28.3.

Anchors for a brick veneer wall assembly are, therefore, made of two pieces that engage each other. One piece is secured to the backup, and the other is embedded in the horizontal mortar joints of the veneer. An adjustable two-piece anchor should allow the veneer to move with respect to the backup in the plane of the wall but not perpendicular to it.

An exception to this requirement is a one-piece sheet steel corrugated anchor (Figure 28.2). The corrugations in the anchor enhance the bond between the anchor and the mortar, increasing the anchor's pullout strength. But the corrugations weaken the anchor in compression by making it more prone to buckling. A one-piece corrugated anchor is recommended for use only in low-rise, wood light-frame buildings in low-wind and low-seismic-risk locations.

Galvanized steel is commonly used for anchors, but stainless steel is recommended where durability is an important consideration and/or where the environment is unusually corrosive.

The spacing of anchors should be calculated based on the lateral load and the strength of the anchor. However, the maximum spacing for a one-piece corrugated anchor or an adjustable two-piece wire anchor (wire size W1.7) is limited by the codes to one anchor for every 2.67 ft^2 [28.1]. Additionally, the anchors should not be spaced more than 32 in. on center horizontally and not more than 18 in. vertically, Figure 28.4.

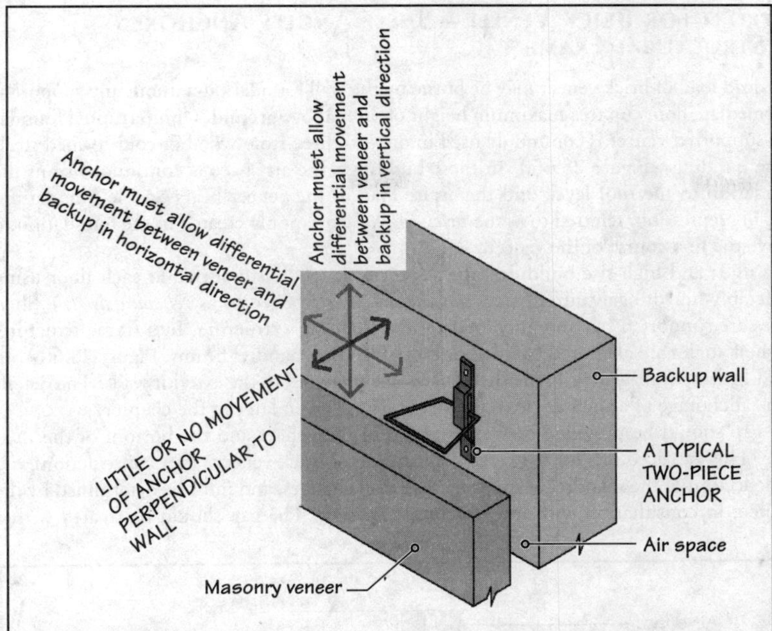

FIGURE 28.3 Adjustability requirements in various directions of a two-piece anchor.

Air Space

An air space of 2 in. (clear) is recommended for brick veneer. Thus, if there is $1\frac{1}{2}$-in.-thick (rigid) insulation between the backup wall and the veneer, the backup and the veneer must be spaced $3\frac{1}{2}$ in. apart so that the air space is 2 in. clear, Figure 28.5.

In a narrower air space, there is a possibility that if the mortar squeezes out into the air space during brick laying, it may bridge over and make a permanent contact with the backup wall. A 2-in. air space reduces this possibility. In a wood-stud-backed wall assembly with one-piece corrugated anchors, however, a 1-in. air space is commonly used (Figure 28.2).

The maximum distance between the veneer and the backup wall is limited by the masonry code to $4\frac{1}{2}$ in. unless the anchors are specifically engineered to withstand the compressive forces imposed on them by the lateral load. With a large gap between the veneer and the backup wall, the anchors are more prone to buckling failure.

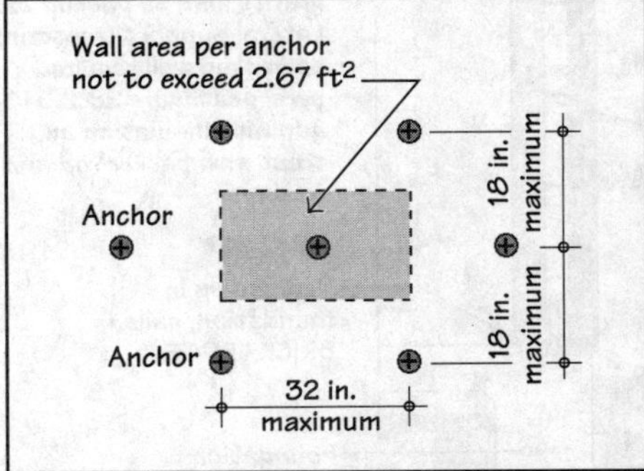

FIGURE 28.4 Anchor spacing should be determined based on the lateral load and the strength of anchors. However, for a one-piece corrugated sheet anchor or a two-piece adjustable wire anchor (wire size W1.7), the maximum spacing allowed is as shown. W1.7 means that the cross-sectional area of wire = 0.017 in.^{2} (see Section 21.13). Anchors should preferably be staggered, as shown.

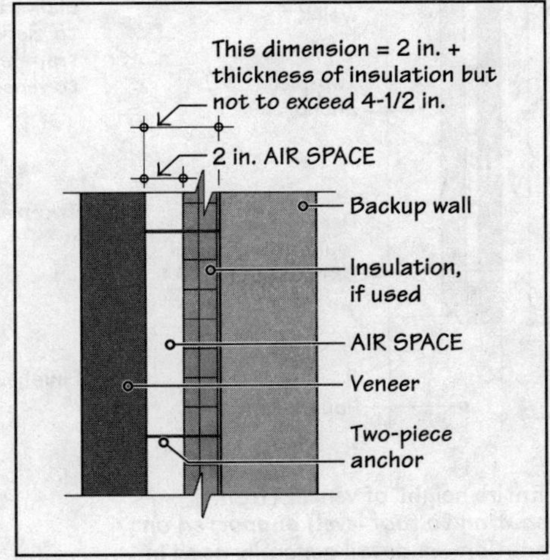

FIGURE 28.5 A minimum air space of 2 in. is required except in low-rise buildings, where a 1-in. air space is common. The gap between the veneer and the backup wall should not exceed $4\frac{1}{2}$ in.

SUPPORT FOR BRICK VENEER—SHELF ANGLES ANCHORED TO STRUCTURAL FRAME

The dead load of brick veneer may be borne by the wall foundation without any support at intermediate floors up to a maximum height of 30 ft above ground. Uninterrupted foundation-supported veneer is commonly used in one- to three-story wood or cold-formed steel-frame buildings, Figure 28.6(a). In these buildings, the air space is continuous from the foundation to the roof level, and the entire load of the veneer bears on the foundation. A $1\frac{1}{2}$-in. depression, referred to as the *brick ledge*, is commonly created in the foundation to receive the first course of the veneer.

In mid- and high-rise buildings, the veneer is generally supported at each floor using (preferably hot-dip galvanized) steel *shelf angles* (also referred to as *relieving angles*). Shelf angles are supported by, and anchored to, the building's structure. In a frame structure, the shelf angles are anchored (welded or bolted) to the spandrel beams, Figure 28.6(b). In a load-bearing wall structure, the shelf angles are anchored to the exterior walls. The details of the anchorage of a shelf angle to the structure are given later in the chapter.

A gap should be provided between the top of the veneer and the bottom of the shelf angle. This gap accounts for the vertical expansion of brick veneer (after construction) and the deflection of the spandrel beam under live load changes, and must be determined by the architect in consultation with the structural engineer. The gap should be treated with a

NOTE

Brick Ledge

A brick ledge is the depression in the concrete foundation at the base of brick veneer and is generally $1\frac{1}{2}$ in. deep, because it is formed by 2-by lumber used as a block-out when placing concrete (Figure 28.6). The depression further prevents water intrusion into the backup wall.

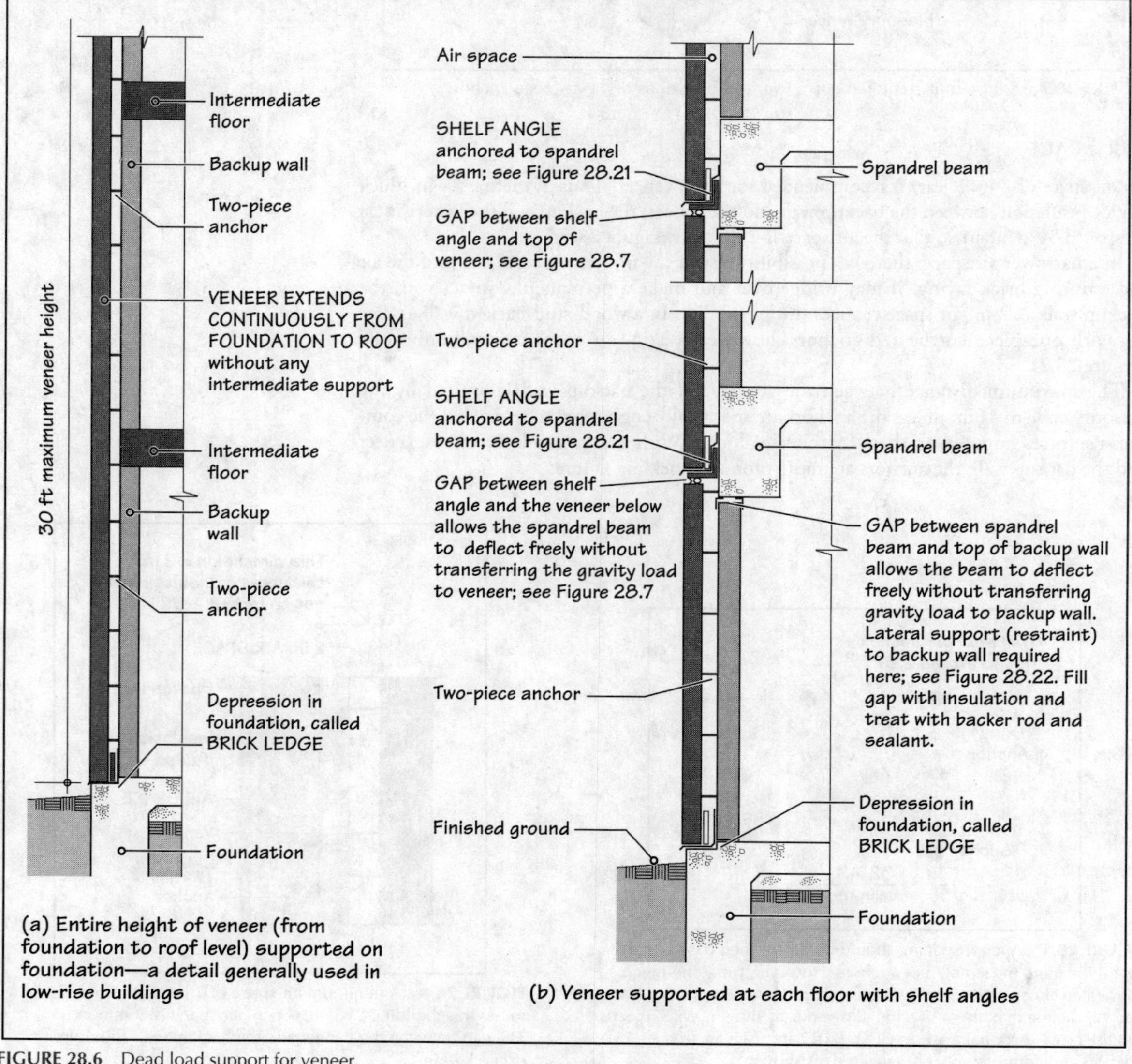

FIGURE 28.6 Dead load support for veneer.

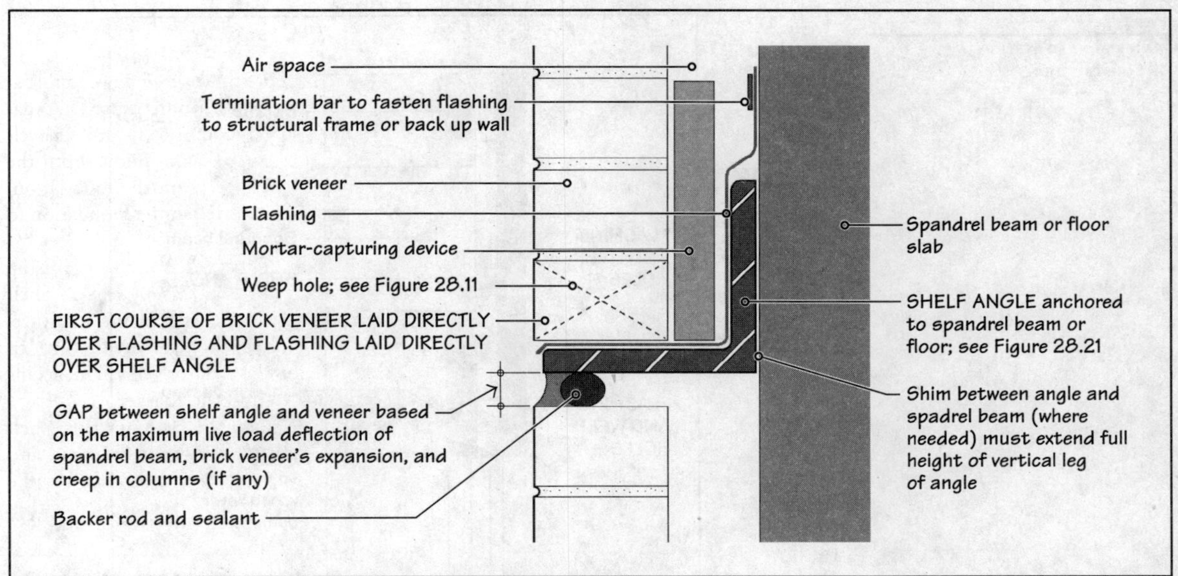

FIGURE 28.7 Schematic detail at a typical shelf angle.

backer rod and sealant, Figure 28.7. The veneer may project beyond the shelf angle, but the projection should not exceed one-third of the thickness of the veneer.

Shelf angles must not be continuous. A maximum length of about 20 ft is used for shelf angles, with nearly a $\frac{3}{8}$-in. gap between adjacent lengths to provide for thermal movement. The gap should ideally be at the same location as the vertical expansion joints in the veneer.

Lintel Angles—Loose-Laid

Whether the veneer is supported entirely on the foundation or at each floor, additional dead load support for the veneer is needed over wall openings. The lintels generally used over an opening in brick veneer are of steel (preferably hot-dip galvanized) angles, Figure 28.8. Unlike the shelf angles, lintel angles are not anchored to the building's structural frame but are simply placed (loose) on the veneer, Figure 28.9.

To allow the lintel to move horizontally relative to the brick veneer, no mortar should be placed between the lintel bearing and the brick veneer. Flashing and weep holes must be provided over lintels in exactly the same way as on the shelf angles.

Locations of Flashings and End Dams

As shown in Figure 28.8, flashings must be provided at all interruptions in the brick veneer:

- At the foundation level
- Over a shelf angle
- Over a lintel angle
- Under a window sill

Joints between flashings must be sealed, and all flashings must be accompanied by weep holes. The flashing should preferably project out of the veneer face to ensure that the water will drain to the outside of the veneer (Figure 28.7). Where the flashing terminates, it must be turned up (equal to the height of one brick) to form a dam to prevent water from entering the air space, Figure 28.10.

Flashing Materials

Flashing material must be impervious to water and resistant to puncture, tearing, and abrasion. Additionally, flashing must be flexible so that it can be bent to the required profile. Durability is also important because replacing failed flashing is cumbersome and expensive. Therefore, metal flashing must be corrosion resistant. Resistance to ultraviolet

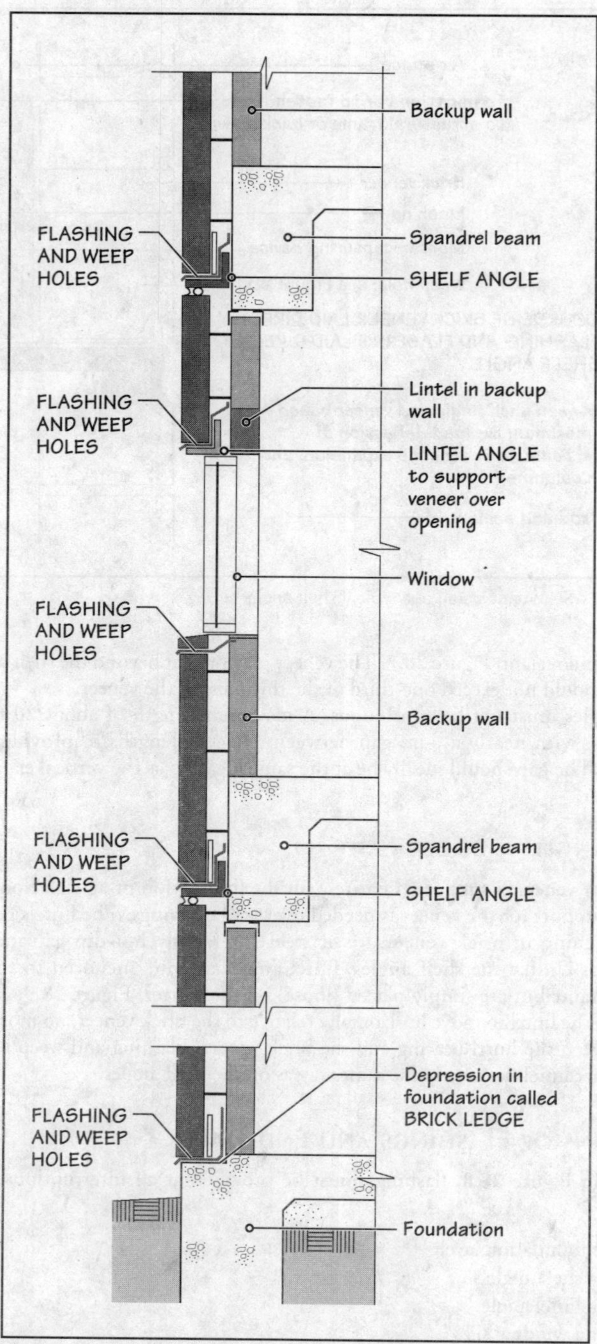

FIGURE 28.8 A schematic section showing the locations of shelf angles, lintel angles, and flashings in a typical brick veneer assembly.

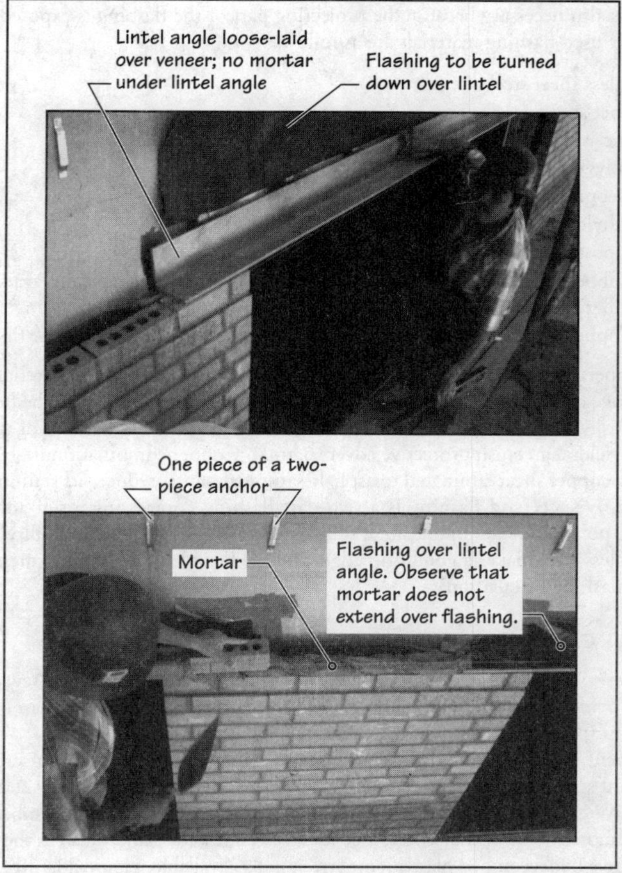

FIGURE 28.9 Lintel angles in a brick veneer wall. Note that the first course of veneer over flashing is laid without any mortar bed between the bricks and flashing; see also Figure 28.7.

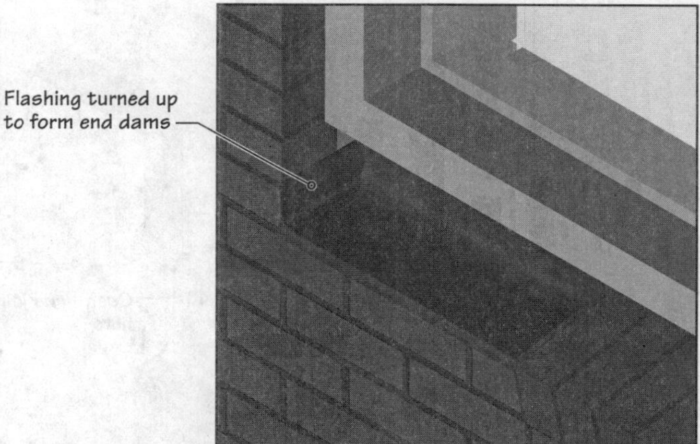

FIGURE 28.10 End dams in flashing, revealed by removing a few bricks of the window sill (in the rowlock course). Weep holes in the brick sill (over the flashing) are not shown for clarity.

radiation is also necessary because the projecting part of the flashing is exposed to the sun. Commonly used flashing materials are as follows:

- Stainless sheet steel
- Copper sheet
- Plastics such as
  - Polyvinyl chloride (PVC)
  - Neoprene
  - Ethylene propylene diene monomer (EPDM)
- Composite flashing consisting of
  - Rubberized asphalt with cross-laminated polyethylene, typically available as self-adhering and self-healing flashing
  - Copper sheet laminated on both sides to asphalt-saturated paper or fiberglass felt

Termination bar
Flexible, self adhering (peel-and-stick) and self-healing flashing
Anchor stainless steel flashing to angle here
Stainless steel flashing
Shelf angle or lintel angle

Copper and stainless steel are among the most durable flashing materials. Copper's advantage over stainless steel is its greater flexibility, which allows it to be bent to shape more easily. However, copper will stain light-colored masonry because of its corrosion, which yields a greenish protective cover (patina). Copper combination flashing, consisting of a copper sheet laminated to asphalt-saturated paper, reduces its staining potential.

A fairly successful flashing is a two-part flashing comprising a self-adhering, self-healing polymeric membrane and a stainless steel drip edge. The durability and rigidity of stainless steel makes a good drip edge, and the flexible, self-adhering membrane simplifies flashing installation.

CONSTRUCTION AND SPACING OF WEEP HOLES

Weep holes must be provided immediately above the flashing. There are several different ways to provide weep holes. The simplest and most effective weep hole is an open vertical mortar joint (open-head joint) in the veneer, Figure 28.11.

To prevent insects and debris from lodging in the open-head joint, joint screens may be used. A joint screen is an L-shaped sheet metal or plastic element, Figure 28.12(a). Its vertical leg has louvered openings to let the water out, and the horizontal leg is embedded in the horizontal mortar joint of the veneer. The joint screen has the same width as the head joints. An alternative honeycombed plastic joint screen is also available, Figure 28.12(b).

Instead of the open-head joint, wicks or plastic tubes ($\frac{3}{8}$-in. diameter) may be used in a mortared-head joint. Wicks, which consist of cotton ropes, are embedded in head joints,

NOTE

Weep Hole Spacing

A center-to-center spacing of 24 in. is generally used for weep holes when open-head joints are used. A spacing of 16 in. is used with weep holes consisting of wicks or tubes.

FIGURE 28.11 Open-head joints as weep holes.

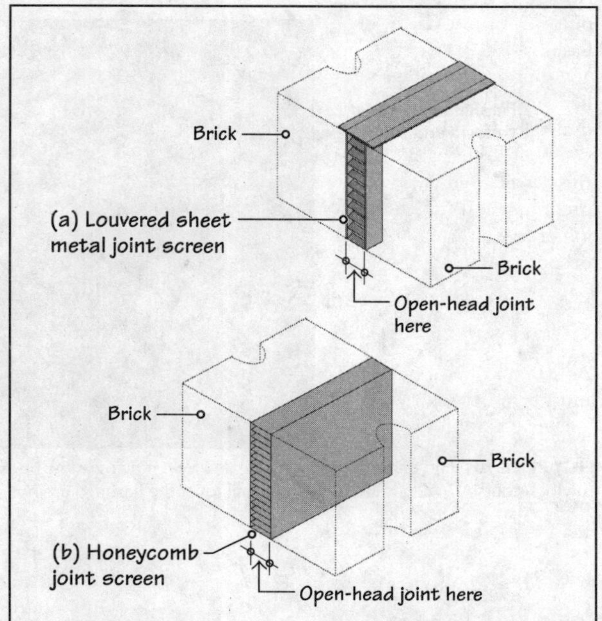

FIGURE 28.12 Two alternative screens used with weep holes consisting of open-head joints.

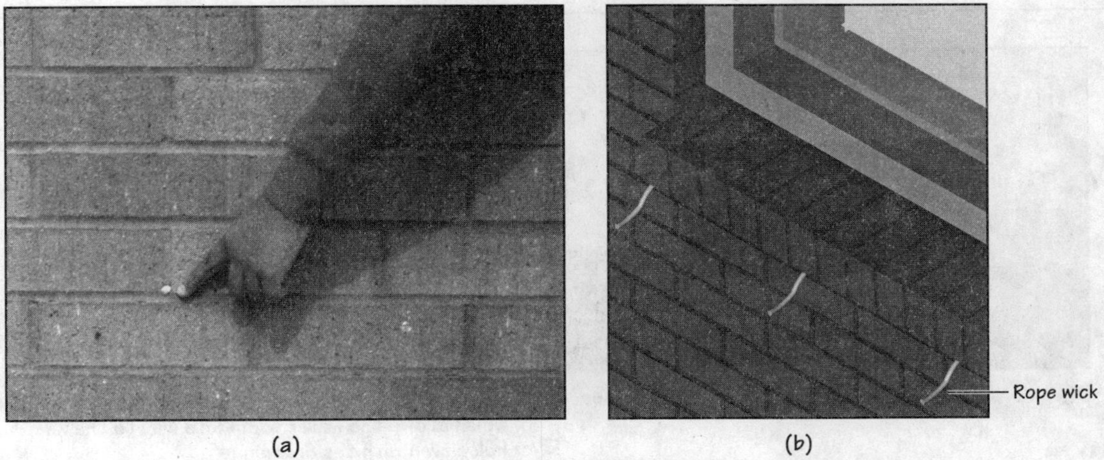

FIGURE 28.13 **(a)** Wicks as weep holes. **(b)** The wick ropes will be trimmed similar to those shown in (a).

Figure 28.13. They absorb water from the air space by capillary action and drain it to the outside. Their drainage efficiency is low.

Plastic tubes are better than wicks, but they do not function as well as open-head joints. They are placed in head joints with a rope inside each tube. The ropes are pulled out after the veneer has been constructed. This ensures that the air spaces of the tubes are not clogged by mortar droppings.

A sufficient number of weep holes must be provided for the drainage of the air space. Generally, a weep hole spacing of 24 in. is used with open-head joints; 16-in. spacing is used with wicks or tubes.

Mortar Droppings in the Air Space—Mortar-Capturing Device

For the air space to function as an effective drainage layer, it is important to minimize mortar droppings in the air space. Excessive buildup of mortar in the air space bridges the space. Additionally, the weep holes function well only if they are not clogged by mortar droppings. Poor bricklaying practice can result in substantial accumulations of mortar on the flashing. Care in bricklaying to reduce mortar droppings is therefore essential.

Additional measures must also be incorporated to keep the air space unclogged. An earlier practice was to use a 2-in.-thick bed of pea gravel over the flashing. This provides a drainage bed that allows the water to percolate to the weep holes.

A better alternative is to use a *mortar-capturing device* in the air space immediately above the flashing. This device consists of a mesh made of polymeric strands, which trap the droppings and suspend them permanently above the weep holes, Figure 28.14. The use of a mortar-capturing device allows water in the air space to percolate freely through mortar droppings to reach the weep holes.

Continuous Vertical Expansion Joints in Brick Veneer

As stated in Section 9.8, brick walls expand after construction. Therefore, a brick veneer must be provided with continuous vertical expansion joints at intervals, Figure 28.15.

The maximum recommended spacing for vertical expansion joints is 25 ft in the field of the wall and not more than 10 ft from the wall's corner [28.2]. The joints are detailed so that sealant and backer rods replace mortar joints for the entire length of the continuous vertical expansion joint, allowing the bricks on both sides of the joint to move while maintaining a waterproof seal, Figure 28.16. The width of the expansion joint is $\frac{1}{2}$ in. (minimum) to match the width of the mortar joints.

With vertical expansion joints and the gaps under shelf angles (which function as horizontal expansion joints), a brick veneer essentially consists of individual brick panels that can expand and contract horizontally and vertically without stressing the backup wall or the building's structure.

(a) Mortar-capturing device, as installed in air space

(b) Mortar-capturing device with mortar droppings, which allows the water to find its way to the weep holes even with the droppings

(c) Image illustrates the relative ineffectiveness of a bed of pea gravel in an air space that was earlier used as a mortar-capturing device

FIGURE 28.14 Mortar-capturing device. (Photos courtesy of Mortar Net USA Ltd., producers of The Mortar Net™)

FIGURE 28.15 Continuous vertical expansion joints in brick veneer. Also note the continuous horizontal joints under the shelf angles.

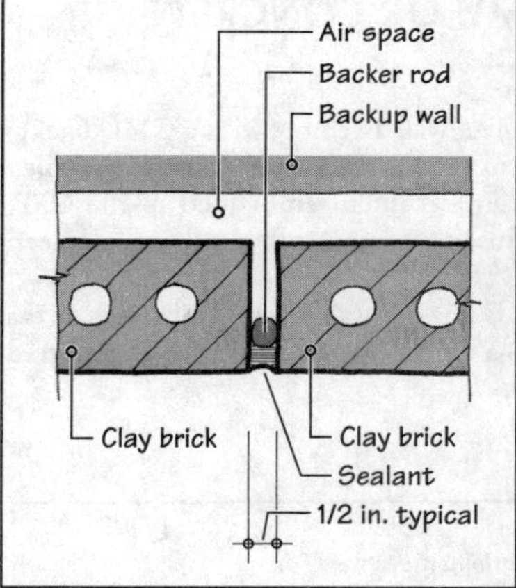

FIGURE 28.16 Detail plan of a vertical expansion joint in brick veneer. This illustration is the same as Figure 9.17.

MORTAR TYPE AND MORTAR JOINT PROFILE

Type N mortar is generally specified in all-brick veneer except in seismic zones, where Type S mortar may be used (see Section 24.2). A concave joint profile yields veneer with more water resistance (see Section 24.3).

PRACTICE QUIZ

Each question has only one correct answer. Select the choice that best answers the question.

1. The backup wall in a brick veneer wall assembly consists of a
a. reinforced-concrete wall. **b.** CMU wall.
c. wood or steel stud wall. **d.** all of the above.
e. (b) and (c) only.

2. In a brick veneer wall assembly, the wind loads are transferred directly from the veneer to the building's structure.
a. True **b.** False

3. The anchors used to anchor the brick veneer to the backup wall are generally two-piece anchors to allow differential movement between the veneer and the backup
a. in all three principal directions.
b. perpendicular to the plane of the veneer.
c. within the plane of the veneer.
d. none of the above.

4. The anchors in a brick veneer wall assembly provide
a. gravity load support to both veneer and backup.
b. lateral load support to both veneer and backup.
c. gravity load support to the veneer.
d. lateral load support to the veneer.

5. The minimum required width of the air space between a brick veneer and a CMU backup wall is
a. 1 in. **b.** $1\frac{1}{2}$ in.
c. 2 in. **d.** 3 in.
e. none of the above.

6. The minimum width of the air space generally used between a brick veneer and a wood stud backup wall is
a. 1 in. **b.** $1\frac{1}{2}$ in.
c. 2 in. **d.** $2\frac{1}{2}$ in.
e. 3 in.

7. A steel angle used to support the weight of a brick veneer over an opening is called a
a. lintel angle. **b.** shelf angle.
c. relieving angle. **d.** all of the above.
e. (a) or (b).

8. A shelf angle must be anchored to the building's structural frame.
a. True **b.** False

9. A lintel angle must be anchored to the building's structural frame.
a. True **b.** False

10. In a multistory building, shelf angles are typically used at
a. each floor level.
b. each floor level and at midheight between floors.
c. at the foundation level.
d. all of the above.
e. (a) and (c).

11. For a brick veneer that bears on the foundation and continues to the top of the building without any intermediate support, the maximum veneer height is limited to
a. 40 ft. **b.** 35 ft.
c. 30 ft. **d.** 25 ft.
e. 20 ft.

12. A shelf angle in a brick veneer assembly must provide
a. gravity load support to the veneer.
b. lateral load support to the veneer.
c. gravity load support to both veneer and backup.
d. lateral load support to both veneer and backup.

13. In a brick veneer assembly, flashing is required
a. at the foundation level. **b.** over a lintel angle.
c. over a shelf angle. **d.** under a window sill.
e. all of the above.

14. In a brick veneer assembly, weep holes are required at
a. each floor level.
b. each alternate floor level.
c. immediately above the flashing.
d. immediately below the flashing.
e. 2 in. above a flashing.

15. The most efficient weep hole in a brick veneer consists of a
a. wick. **b.** plastic tube.
c. open-head joint. **d.** None of the above.

16. A mortar-capturing device in a brick veneer assembly is used
a. at each floor level.
b. at each alternate floor level.
c. immediately above a flashing.
d. immediately below a flashing.
e. 2 in. above a flashing.

17. In a brick veneer assembly, vertical expansion joints should be provided at a maximum distance of
a. 40 ft in the field of the wall and 40 ft from a wall's corner.
b. 30 ft in the field of the wall and 30 ft from a wall's corner.
c. 30 ft in the field of the wall and 20 ft from a wall's corner.
d. 25 ft in the field of the wall and 10 ft from a wall's corner.

28.2 BRICK VENEER WITH A CMU OR CONCRETE BACKUP WALL

Figure 28.17 shows an overall view of a brick veneer wall assembly with a CMU backup wall. The steel anchors that connect the veneer to the CMU backup wall are two-piece wire anchors. One piece is part of the joint reinforcement embedded in the CMU walls, Figure 28.18(a). The other piece fits into this piece and is embedded in the veneer's bed joint.

Several other types of anchors used with a CMU backup wall are available, such as that shown in Figure 28.18(b). Figure 28.18(c) shows a typical anchor used with a reinforced-concrete member (spandrel beam or backup wall).

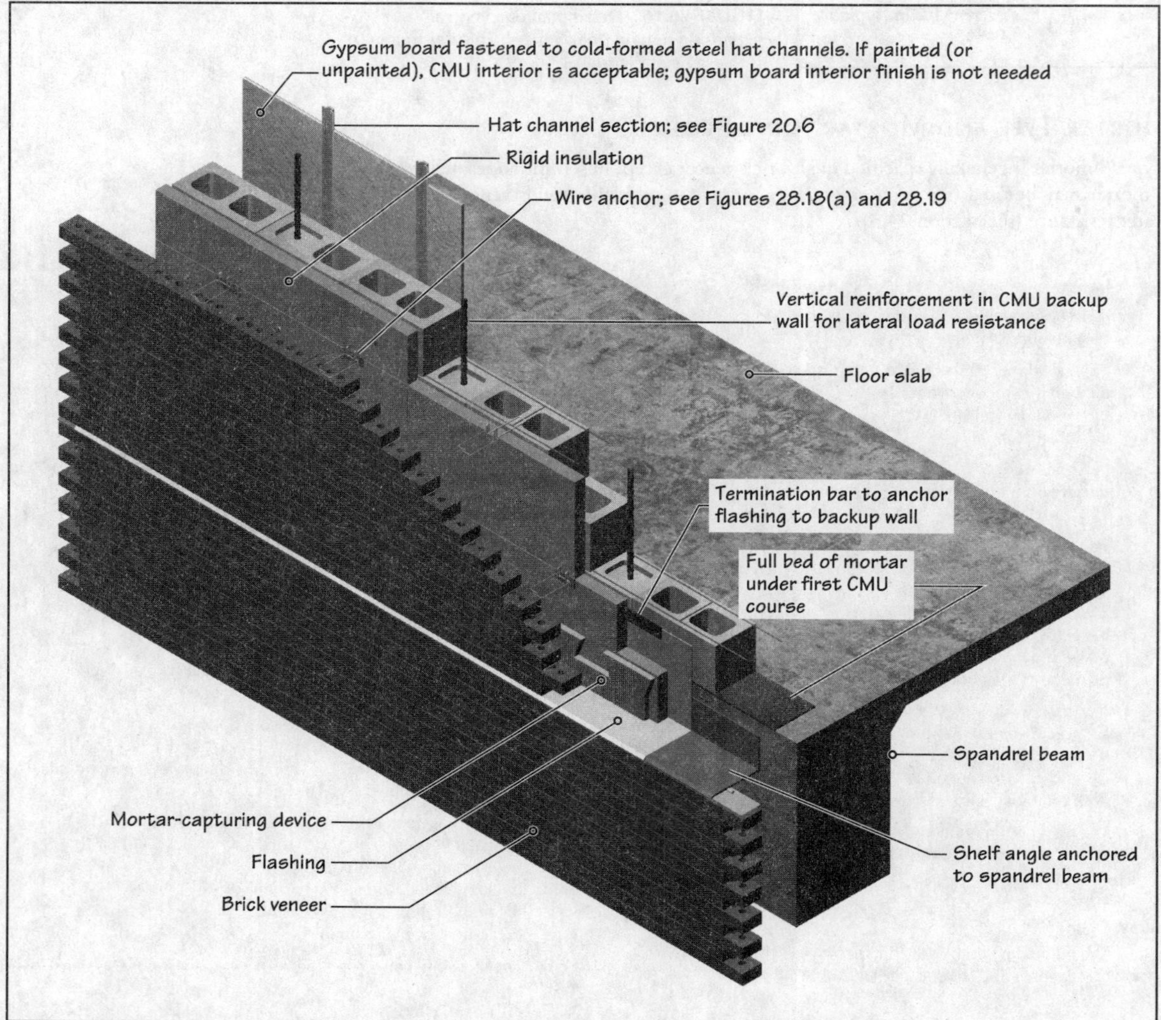

FIGURE 28.17 Brick veneer with CMU backup wall.

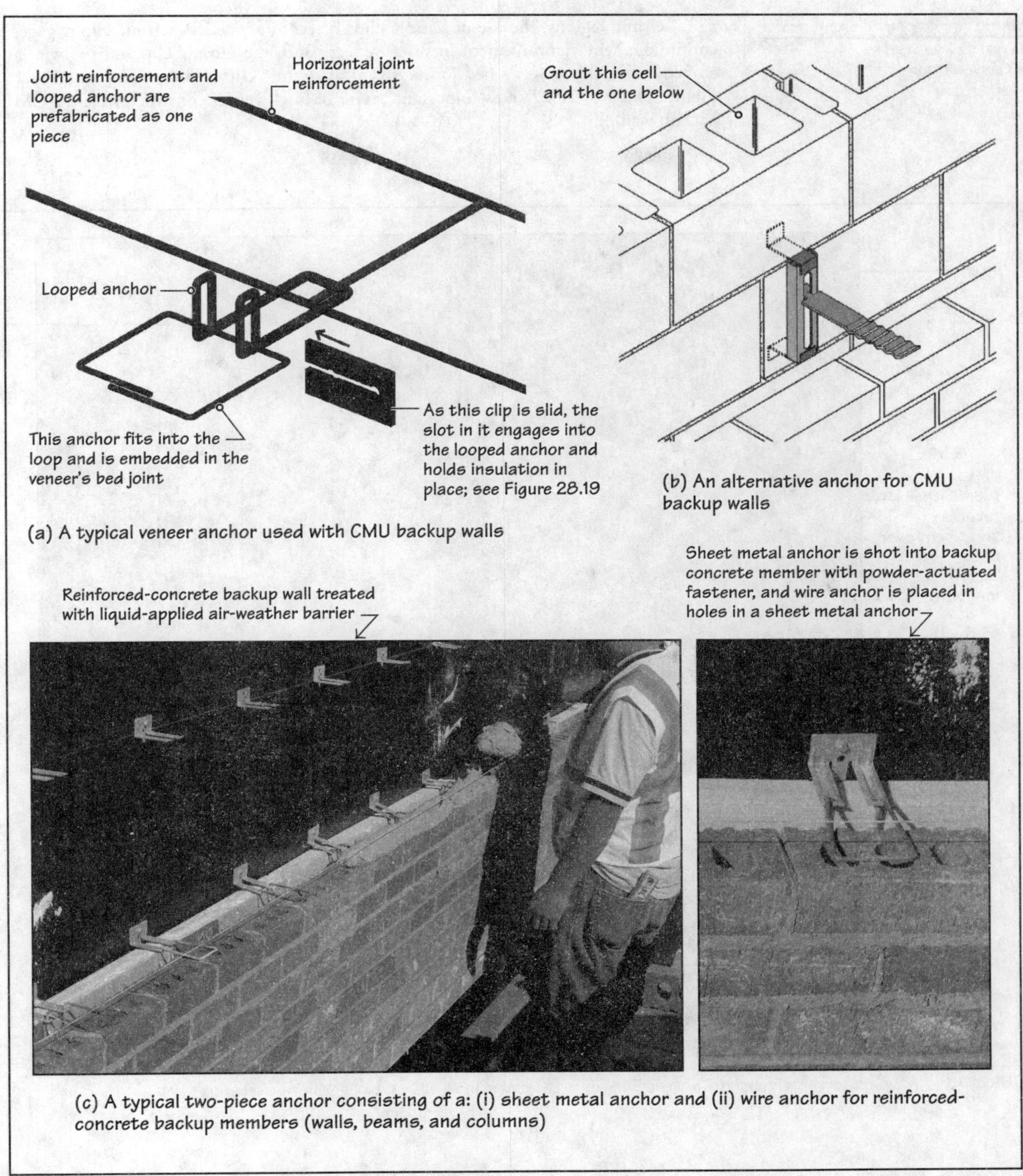

Joint reinforcement and looped anchor are prefabricated as one piece
Horizontal joint reinforcement
Looped anchor
This anchor fits into the loop and is embedded in the veneer's bed joint
As this clip is slid, the slot in it engages into the looped anchor and holds insulation in place; see Figure 28.19
(a) A typical veneer anchor used with CMU backup walls
Grout this cell and the one below
(b) An alternative anchor for CMU backup walls
Reinforced-concrete backup wall treated with liquid-applied air-weather barrier
Sheet metal anchor is shot into backup concrete member with powder-actuated fastener, and wire anchor is placed in holes in a sheet metal anchor
(c) A typical two-piece anchor consisting of a: (i) sheet metal anchor and (ii) wire anchor for reinforced-concrete backup members (walls, beams, and columns)

FIGURE 28.18 Typical anchors used with CMU and concrete backup walls.

In seismic regions, the use of seismic clips is recommended. A seismic clip engages a continuous wire reinforcement in brick veneer. Both the seismic clip and the wire are embedded in the veneer's bed joint. A typical seismic clip is shown in Figure 28.19. Figures 28.20 to 28.22 show important details of brick veneer construction and a CMU backup wall.

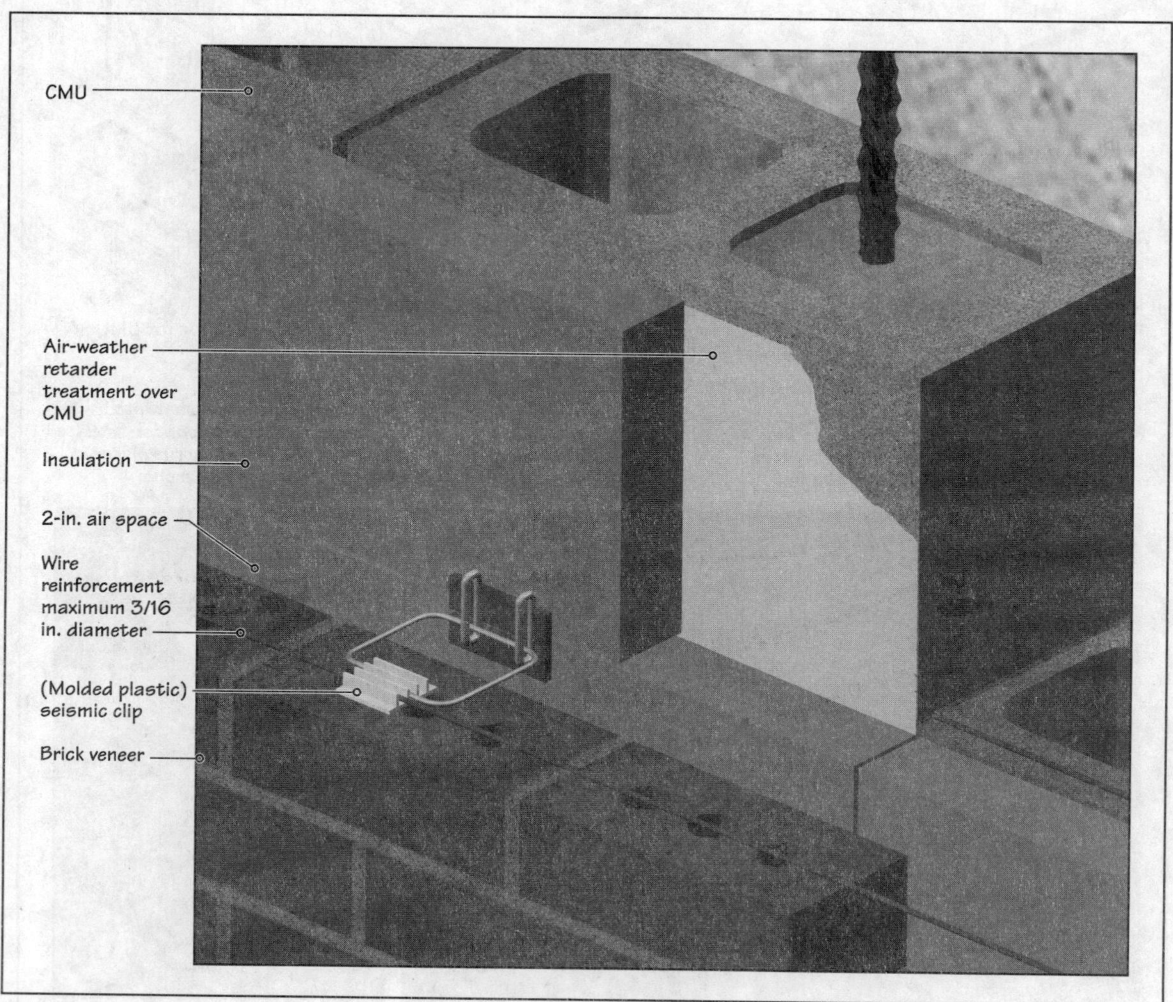

FIGURE 28.19 A typical seismic clip in brick veneer. The clip is embedded in the mortar along with wire reinforcement.

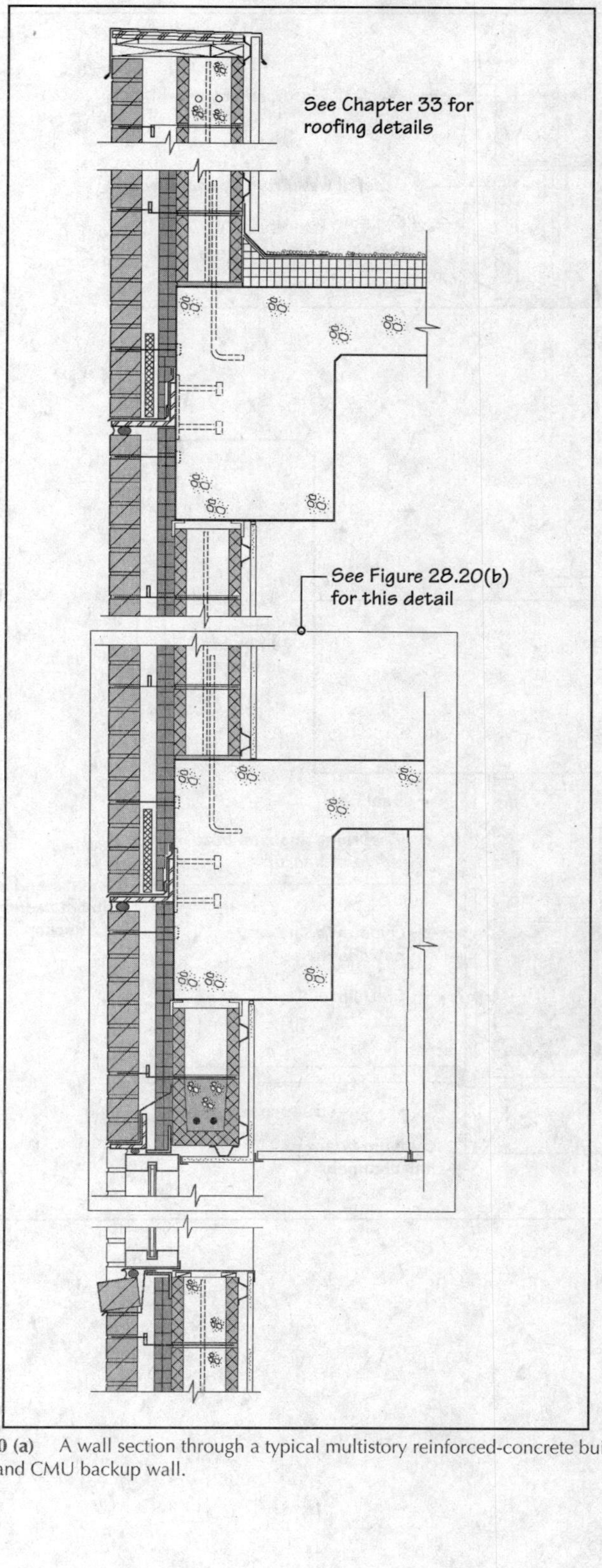

See Chapter 33 for roofing details

See Figure 28.20(b) for this detail

FIGURE 28.20 (a) A wall section through a typical multistory reinforced-concrete building with brick veneer and CMU backup wall.

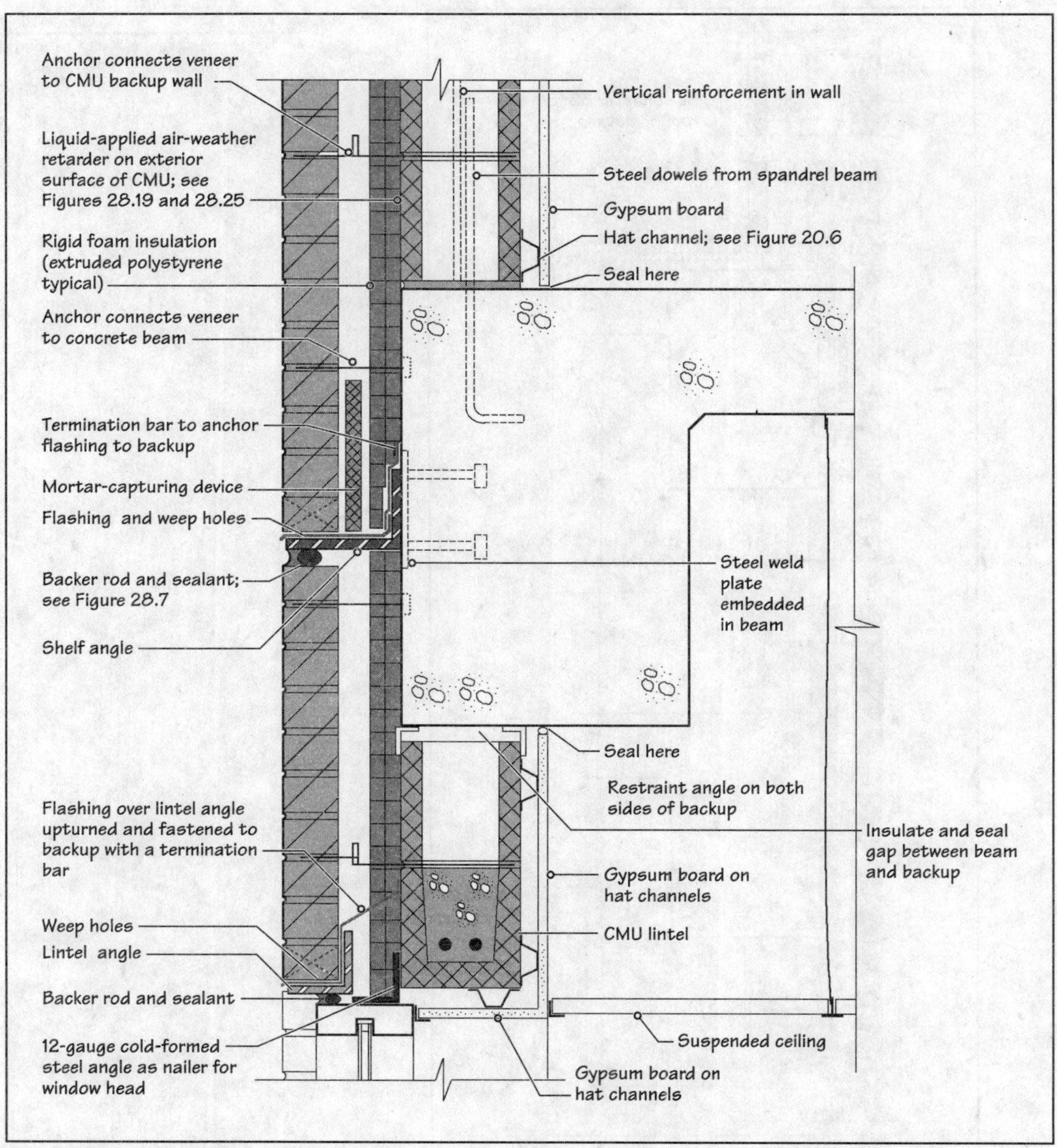
Anchor connects veneer to CMU backup wall
Liquid-applied air-weather retarder on exterior surface of CMU; see Figures 28.19 and 28.25
Rigid foam insulation (extruded polystyrene typical)
Anchor connects veneer to concrete beam
Termination bar to anchor flashing to backup
Mortar-capturing device
Flashing and weep holes
Backer rod and sealant; see Figure 28.7
Shelf angle
Flashing over lintel angle upturned and fastened to backup with a termination bar
Weep holes
Lintel angle
Backer rod and sealant
12-gauge cold-formed steel angle as nailer for window head
Vertical reinforcement in wall
Steel dowels from spandrel beam
Gypsum board
Hat channel; see Figure 20.6
Seal here
Steel weld plate embedded in beam
Seal here
Restraint angle on both sides of backup
Insulate and seal gap between beam and backup
Gypsum board on hat channels
CMU lintel
Suspended ceiling
Gypsum board on hat channels

FIGURE 28.20 (b) This refers to Figure 28.20(a).

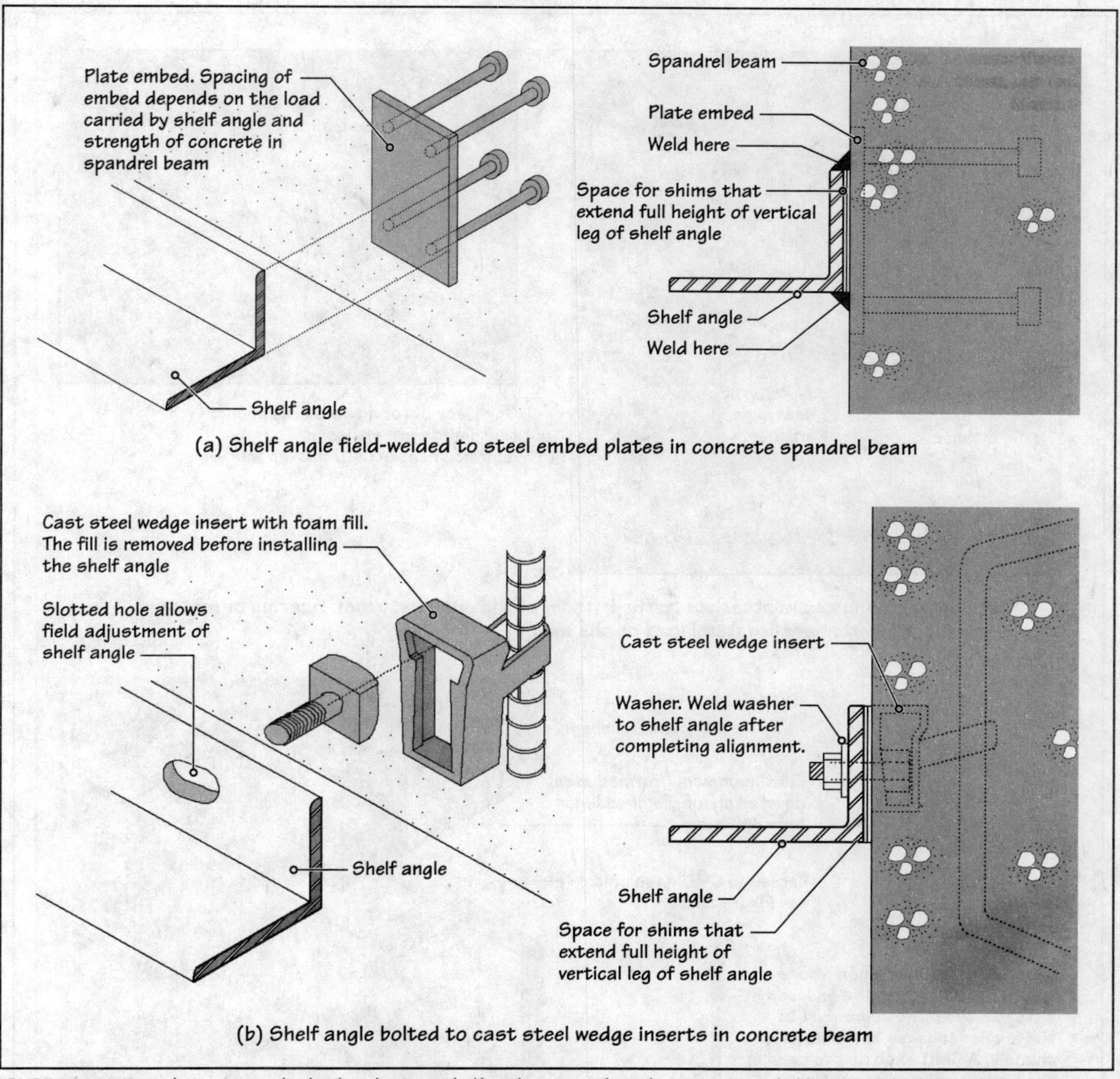

FIGURE 28.21 . Two alternative methods of anchoring a shelf angle to a reinforced-concrete spandrel beam. Method (a) is more commonly used. In this method, the shelf angle for a given floor must be in place before the construction of the veneer below it begins. Method (b) allows the shelf angle to be installed after the veneer below the shelf angle has been constructed, and is needed where the veneer consists of large natural stone or cast-stone panels, requiring a mechanical lifting device to bring them into position from above.

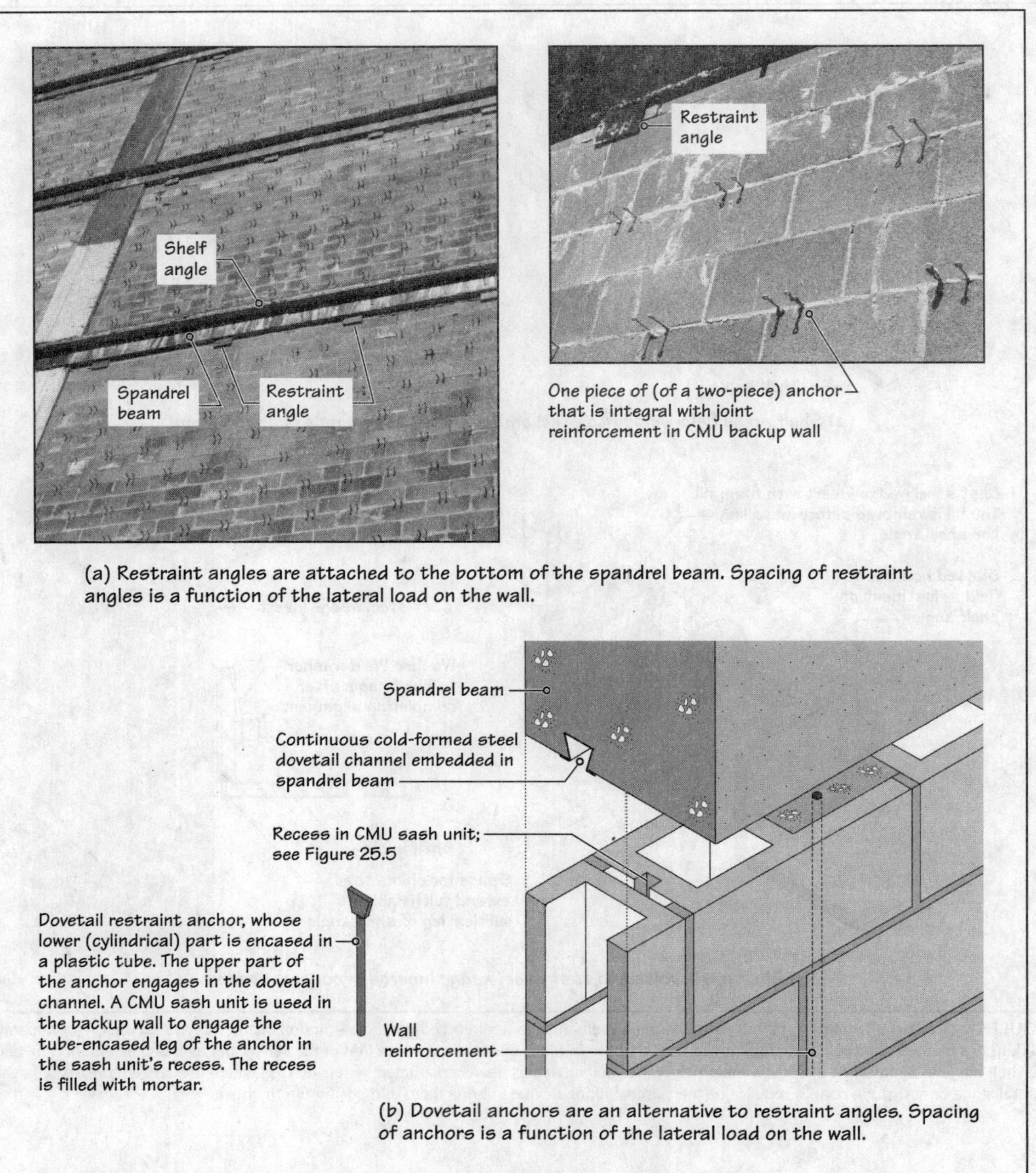

(a) Restraint angles are attached to the bottom of the spandrel beam. Spacing of restraint angles is a function of the lateral load on the wall.

(b) Dovetail anchors are an alternative to restraint angles. Spacing of anchors is a function of the lateral load on the wall.

FIGURE 28.22 Two alternative methods of providing lateral load restraint at the top of a CMU backup wall.

The shelf angle and lintel angles can be combined into one by increasing the depth of the spandrel beam down to the level of the window head, Figure 28.23—a strategy that is commonly used with ribbon windows and continuous brick veneer spandrels, Figure 28.24.

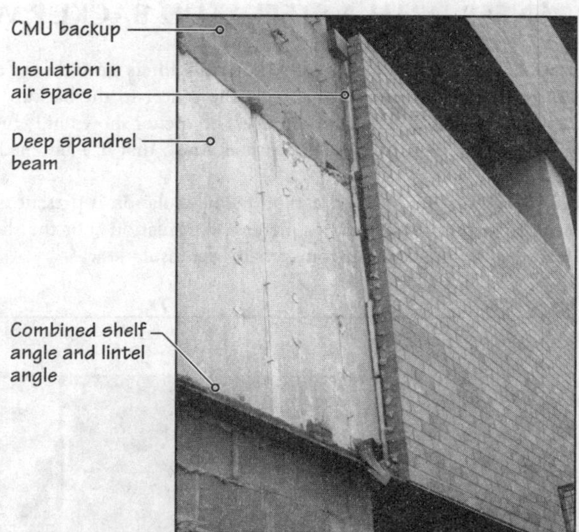

FIGURE 28.23 With a deep spandrel beam, which extends from the window head in the lower floor to the upper-floor level, the shelf angle and lintel angle are combined into one angle.

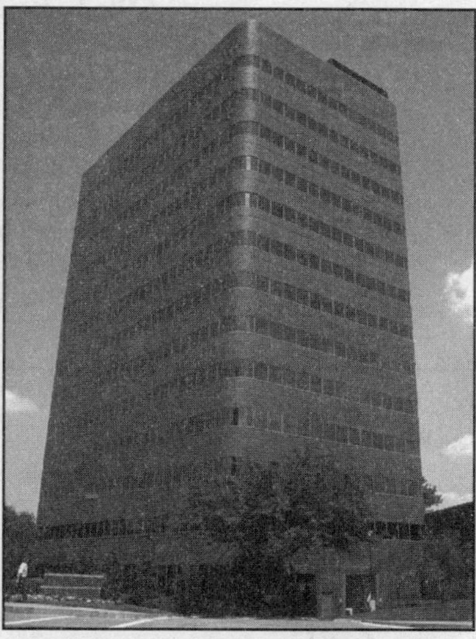

FIGURE 28.24 A building with brick veneer spandrels and ribbon windows.

Air and Water Resistance of CMU Backup Walls

It is important that the exterior face of CMU backup walls is suitably treated to resist the flow of air and water while allowing water vapor to pass through. Several proprietary liquid-applied air-weather barriers are available, which can be brushed on, sprayed on, or mop-applied, Figure 28.25.

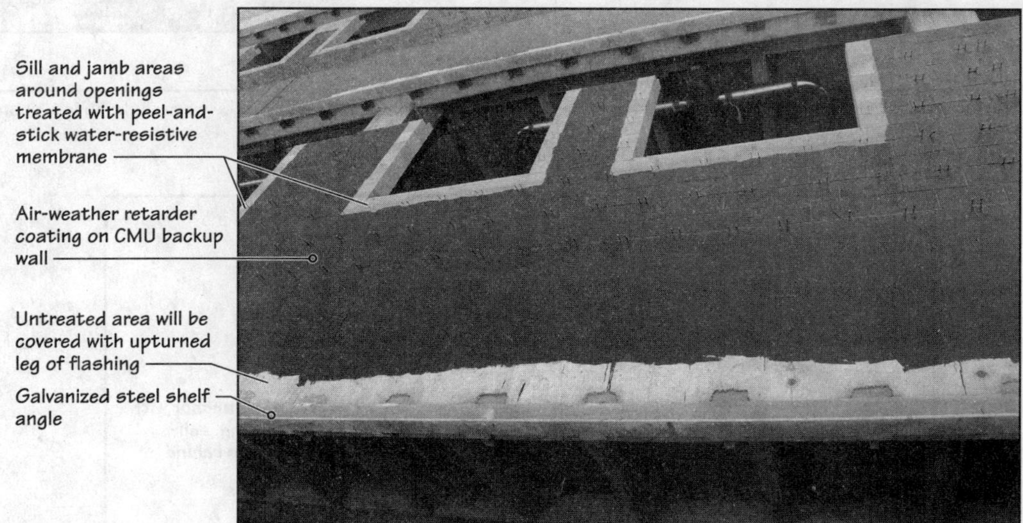

FIGURE 28.25 The exterior surface of a CMU backup wall (and the underlying spandrel beam) has been treated with a liquid air-weather retarder (blue color) before the brick veneer is constructed.

28.3 BRICK VENEER WITH A STEEL STUD BACKUP WALL

The construction of brick veneer with a steel stud backup wall differs from that of a CMU-backed wall mainly in the anchors used for connecting the veneer to the backup. Various types of anchors are available to suit different conditions. The anchor shown in Figure 28.26 is used if the air space does not contain rigid foam insulation so that it is fastened to steel studs through exterior sheathing.

The anchor shown in Figure 28.27 is used if rigid foam insulation is present in the air space. The sharp ends of the prong-type anchor pierce the insulation (not the sheathing) and transfer lateral load to the studs without compressing the insulation.

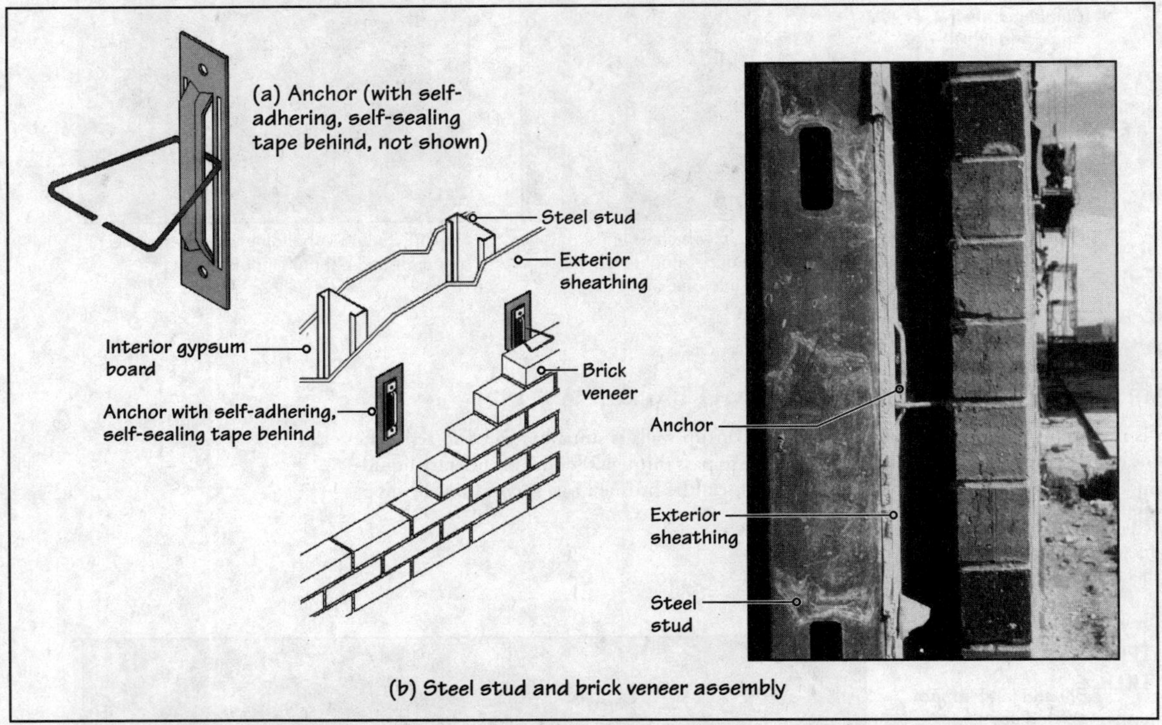

FIGURE 28.26 A typical anchor used in a steel stud and brick veneer assembly with insulation provided between studs; no insulation is provided outside of the exterior sheathing.

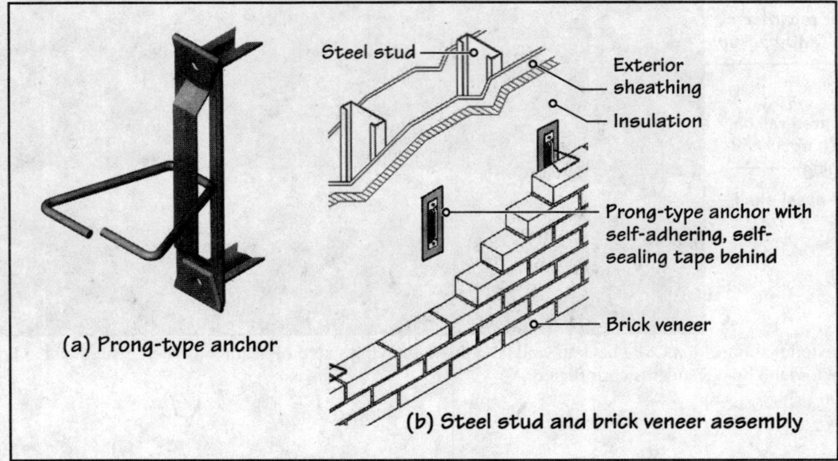

FIGURE 28.27 A typical steel stud and brick veneer assembly with insulation provided outside of the exterior sheathing (insulation may or may not be provided between studs); this requires a prong-type anchor. (Illustration courtesy of Hohmann and Barnard, Inc.)

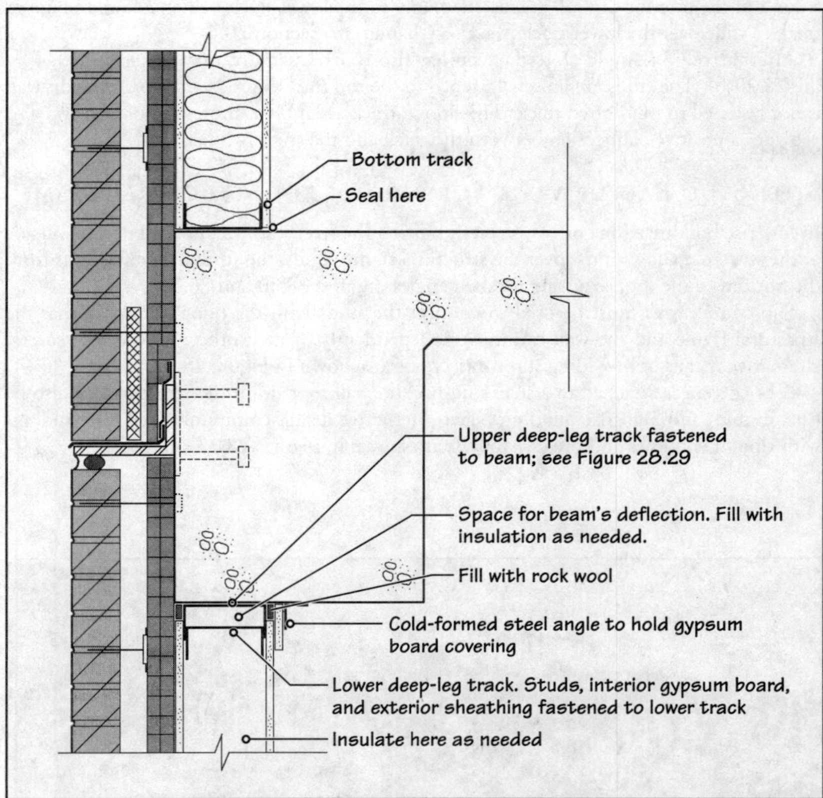

FIGURE 28.28 Detail of a typical steel stud and brick veneer assembly in a reinforced-concrete structure. The shelf angle, mortar-capturing device, flashing, and other details have not been labeled on the illustration for the sake of clarity. See also Figure 20.10.

STEEL STUD BACKUP WALL AS INFILL WITHIN THE STRUCTURAL FRAME

Figure 28.28 shows a detailed section of brick veneer applied to a steel stud backup wall with a reinforced-concrete structural frame. The (vertical) deflection of the spandrel beam is accommodated by providing a two-track assembly consisting of two nested deep-leg tracks, Figure 28.29(a). The upper track of this slip assembly is fastened to the beam. The

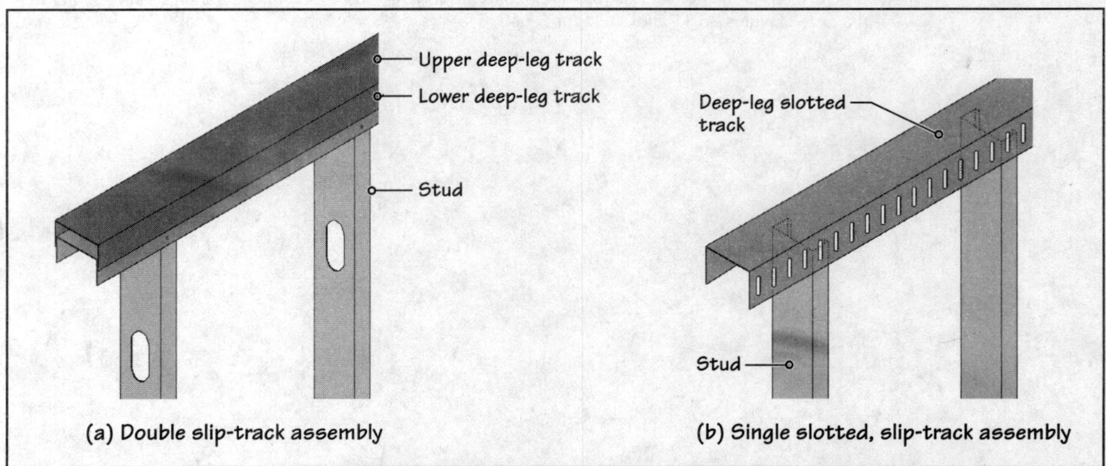

FIGURE 28.29 Two alternative methods of providing a slip-track assembly in a steel stud backup wall (refers to Figure 28.28).

studs and the interior drywall are fastened only to the lower track, which allows the upper track to slide over the lower track (see also Chapter 20, Section 20.5).

Alternatively, a single slotted, deep-leg slip-track assembly may be used. See Figure 28.29(b). The studs are loosely fastened to the top track through the slots. The drywall is not fastened to the slotted track. The slotted track assembly is more economical and also provides a positive connection between the track and the studs.

STEEL STUD BACKUP WALL FORWARD OF THE STRUCTURAL FRAME

In low-rise buildings (one or two stories), putting the steel stud backup wall on the outside of the structure allows it to cover the structural frame. Thus, the studs are continuous from the bottom to the top, requiring no shelf angles, Figures 28.30 and 28.31.

Slip connections must be used to connect the studs with the floor or roof so that the structural frame and the wall can move independently of each other. Two of the several alternative means of providing slip connections are shown in Figure 28.32.

A brick veneer attached to a steel stud backup wall forward of the structure can also be used in mid- and high-rise buildings. Two alternative details commonly used for buildings with ribbon windows and brick spandrels are shown in Figures 28.33 and 28.34.

FIGURE 28.30 In a one- or two-story building with a steel stud backup wall and a steel-frame structure, the studs are generally continuous from the foundation to the roof. The vertical continuity of studs across a floor and (or) roof increases their lateral load resistance. Refer to Figures 28.31 and 28.32 for details.

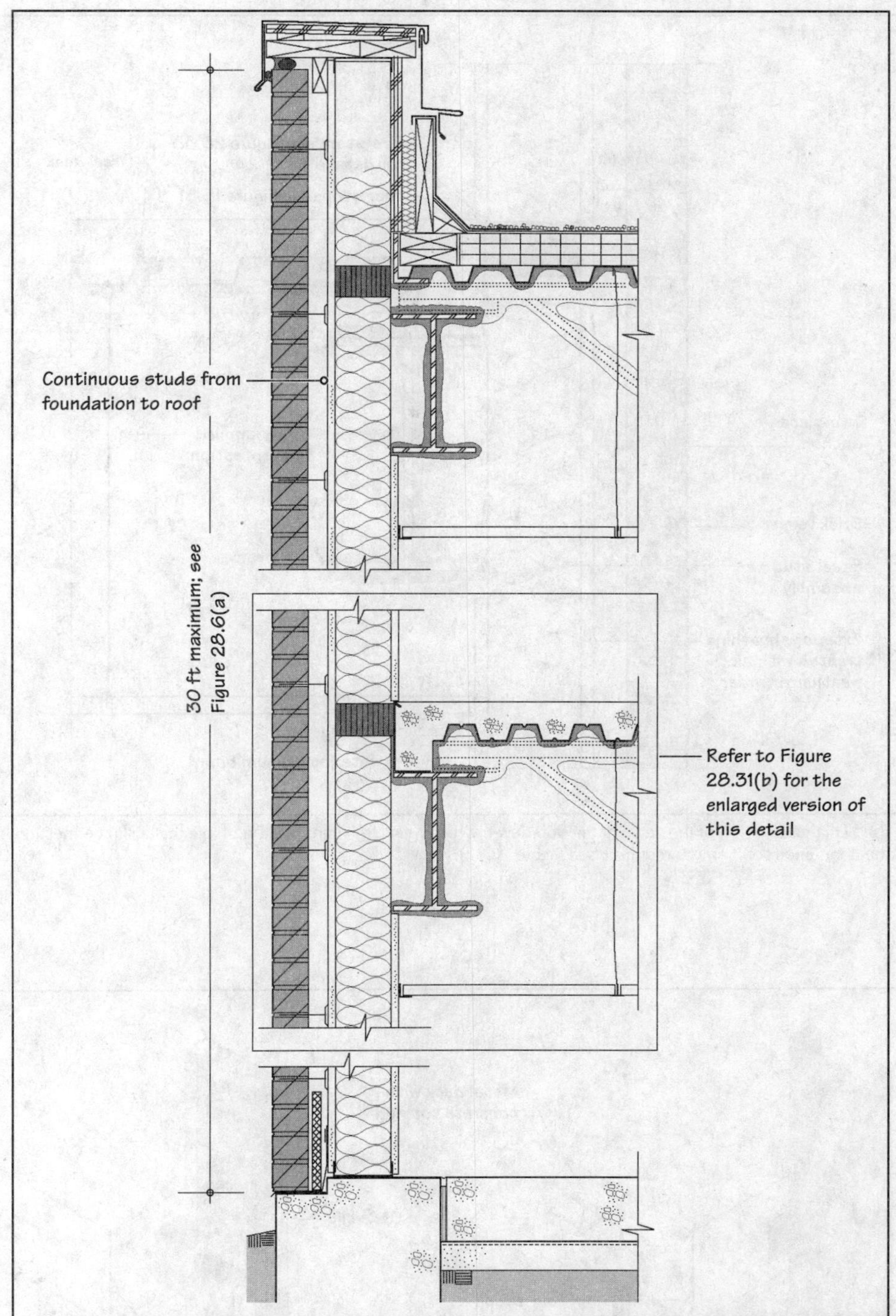

FIGURE 28.31 (a) A typical section through a low-rise (one- or two-story) steel-frame building with a brick veneer and steel stud backup wall assembly. Steel studs in the backup framing are continuous from the foundation to the roof (refer to the details in Figures 28.31(b) and 28.32).

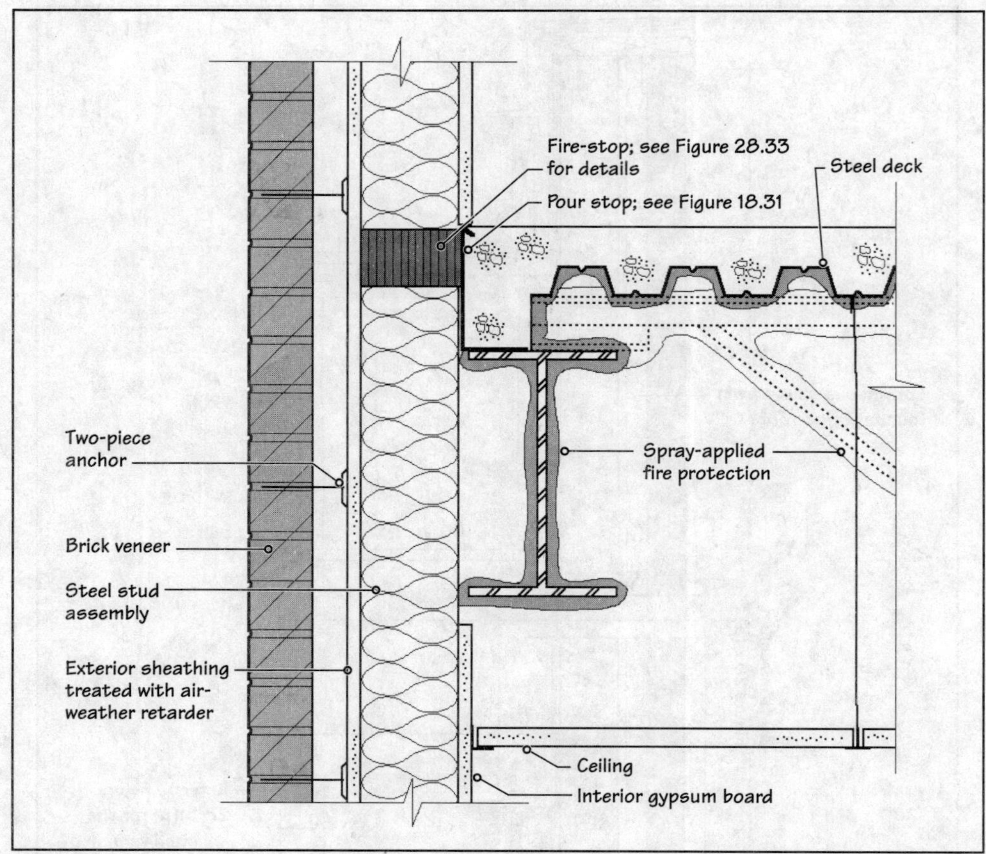

FIGURE 28.31 (b) Enlarged version of the detail at the floor level, which refers to Figure 28.31(a). Note that a slip connection between the steel studs and the floor (and roof) frame is required (see Figure 28.32).

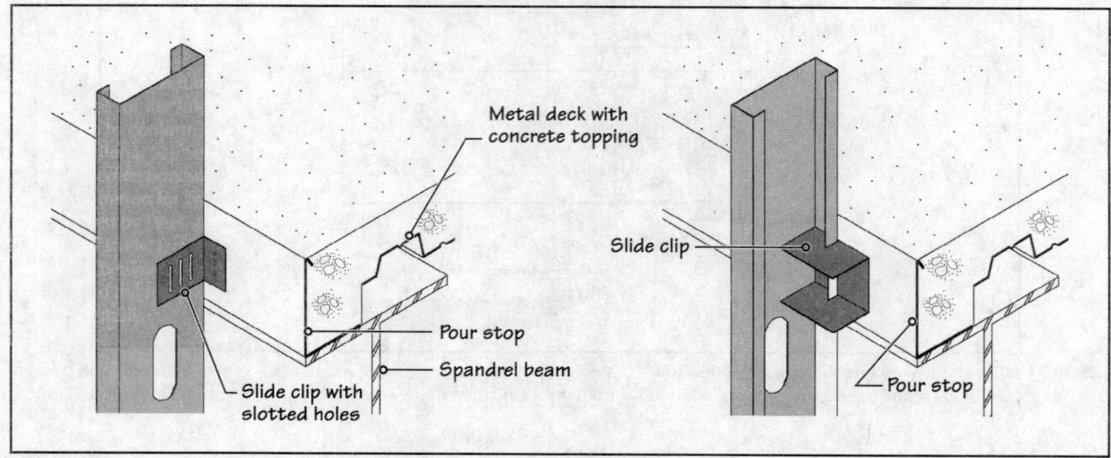

FIGURE 28.32 Two alternative methods of providing a slip connection between a steel stud backup wall and the floor/roof structure; this refers to Figure 28.31.

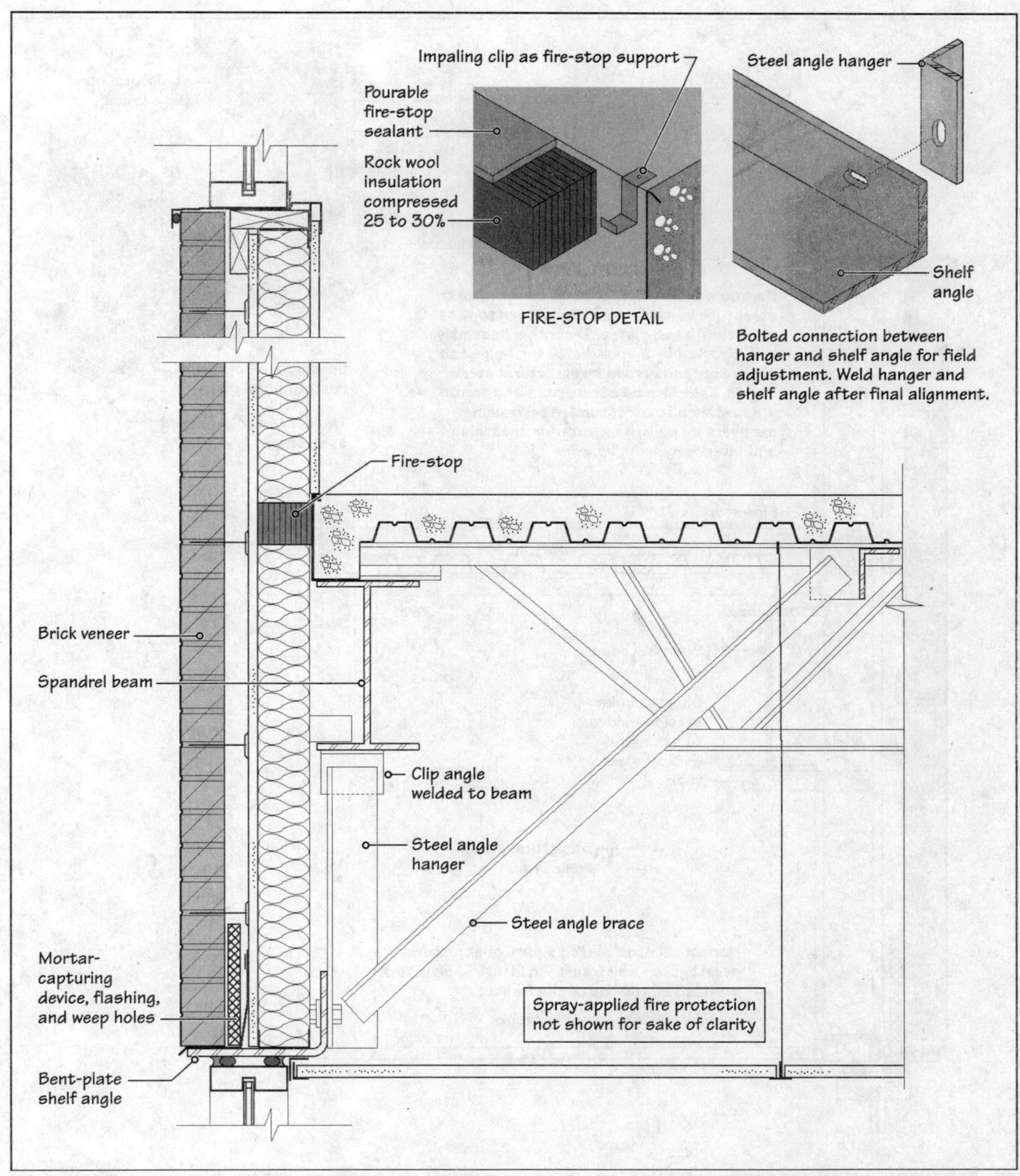

FIGURE 28.33 Brick spandrel with a steel stud backup wall. Steel studs and brick veneer bear on a bent-plate shelf angle. (For a discussion of bent-plate angles, see Chapter 18, Section 18.4.)

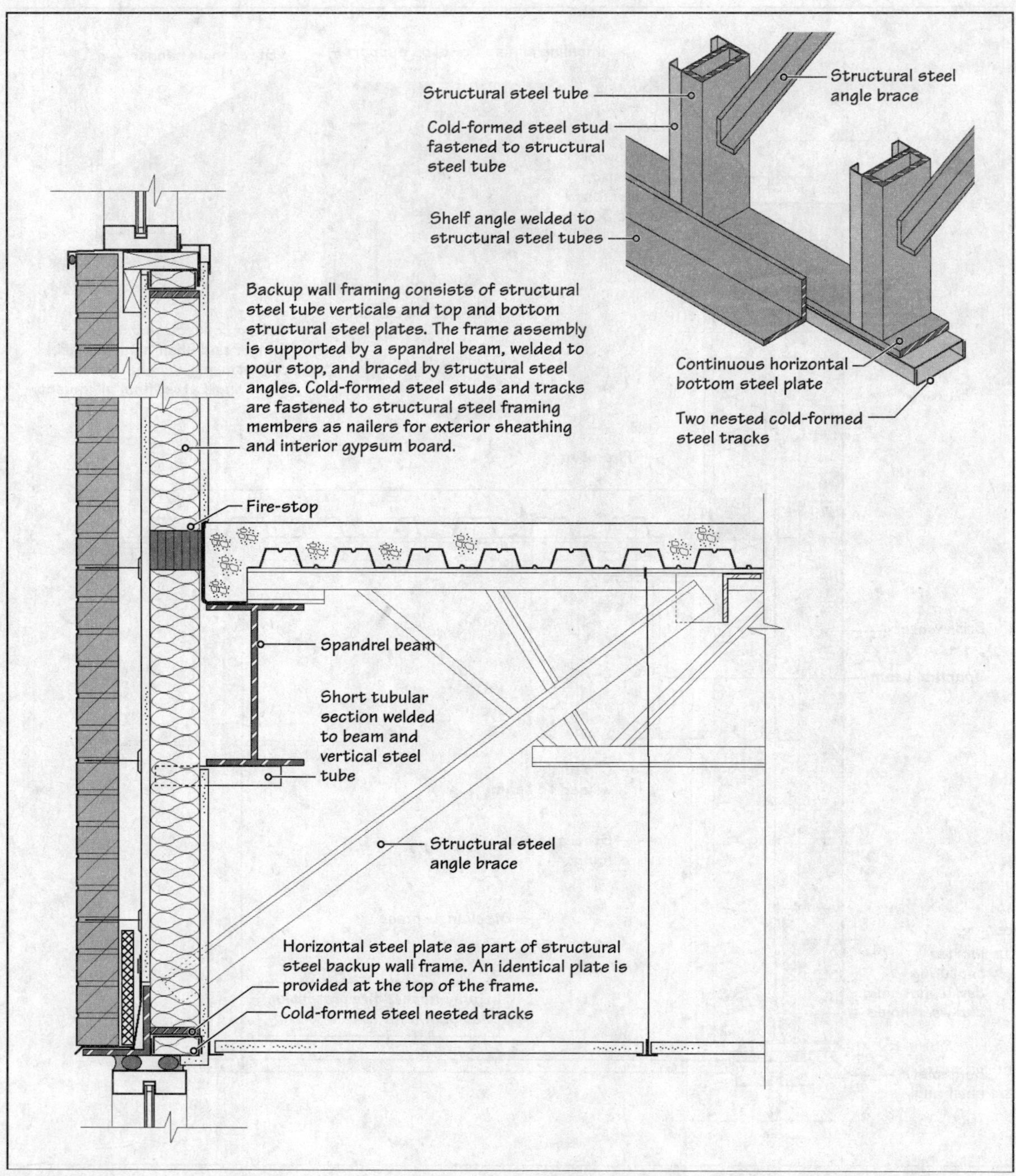
Structural steel tube
Structural steel angle brace
Cold-formed steel stud fastened to structural steel tube
Shelf angle welded to structural steel tubes
Backup wall framing consists of structural steel tube verticals and top and bottom structural steel plates. The frame assembly is supported by a spandrel beam, welded to pour stop, and braced by structural steel angles. Cold-formed steel studs and tracks are fastened to structural steel framing members as nailers for exterior sheathing and interior gypsum board.
Continuous horizontal bottom steel plate
Two nested cold-formed steel tracks
Fire-stop
Spandrel beam
Short tubular section welded to beam and vertical steel tube
Structural steel angle brace
Horizontal steel plate as part of structural steel backup wall frame. An identical plate is provided at the top of the frame.
Cold-formed steel nested tracks

FIGURE 28.34 (a) Brick spandrel with structural steel backup wall framing. The shelf angle is hung from the structural steel frame and supports only the brick veneer (and insulation outside of the exterior sheathing, where provided); also see Figure 28.34(b).

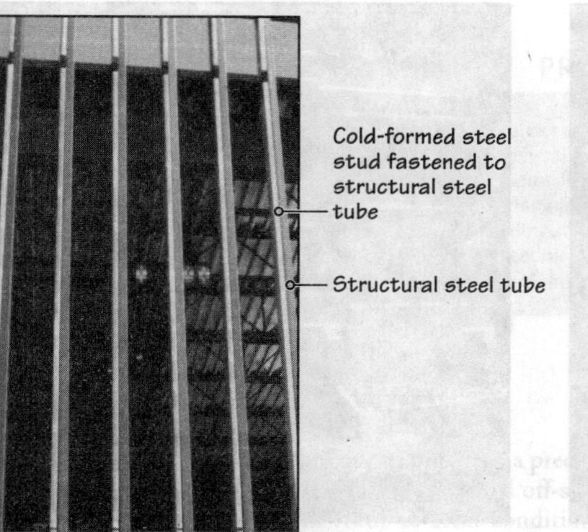

FIGURE 28.34 (b) Structural steel framing for a brick veneer panel. The framing is supported by the floor and roof beams. The shelf angle is hung from the vertical structural steel frame and supports only the brick veneer.

AIR AND WATER RESISTANCE OF A STEEL STUD BACKUP WALL

Weather resistance of exterior sheathing on steel studs is necessary, and may be accomplished by using either (a) a wrap-type air-weather retarder or (b) a liquid-applied air-weather retarder. The latter alternative is more commonly used for commercial buildings.

28.4 CMU BACKUP VERSUS STEEL STUD BACKUP

A major benefit of a steel stud backup wall in a brick veneer assembly (compared with a CMU backup wall) is its lighter weight. For a high-rise building, the lighter wall not only reduces the size of spandrel beams but also that of the columns and footings, yielding economy in the building's structure. However, this benefit is accompanied by several concerns.

Steel studs can deflect considerably before the bending stress in them exceeds their ultimate capacity. Brick veneer, on the other hand, deflects by a very small amount before the mortar joints open. Open mortar joints weaken the wall and increase the probability of leakage, corroding the anchors.

Thus, the steel stud backup and brick veneer assembly performs well only if the stud wall is sufficiently stiff. To obtain the necessary stiffness, the deflection of studs must be controlled to a fairly small value. In fact, the design of a steel stud backup wall to resist the lateral loads is governed not by the strength of studs but by their deflection.

The Brick Industry Association (BIA) recommends that the lateral load deflection of steel studs, when used as backup for brick veneer assembly, should not exceed

$$\frac{\text{stud span}}{600}$$

where the stud span is the unsupported height of studs. For example, if the height of studs (e.g., from the top of the floor to the bottom of the spandrel beam in Figure 28.7) is 10 ft, the deflection of studs under the lateral load must be less than

$$\frac{(10 \times 12)}{600} = 0.5 \text{ in.}$$

Increasing the stiffness of studs increases the cost of the assembly. Another concern with steel stud backup is that the veneer is anchored to the backup only through screws that engage the threads within a cold-formed stud sheet. Over a period of time, condensation

NOTE

Deflection of a Steel Stud Wall Assembly

The design criterion of deflection not exceeding span divided by 600, suggested by BIA, is the minimum requirement. For critical buildings, a more stringent deflection criterion, such as span divided by 720 or span divided by 900, is recommended by some experts.

Steel stud manufacturers generally provide tables for the selection of their studs to conform to the deflection criteria.

(a)

(b)

FIGURE 28.38 **(a)** A precast concrete panel being unloaded from the delivery truck for hoisting into position by a crane. **(b)** The same panel being hoisted into its final position.

In a PC curtain wall project, an important role for the architect is to work out the aesthetic expression of the panels (shapes, size, exterior finishes, etc.). This should be done in consultation with the precast plant. A great deal of coordination between the architect, engineer of record, general contractor, precast plant, and erection subcontractor is necessary for a successful PC curtain wall project.

Because of the sculptability of concrete and the assortment of possible finishes of the concrete surface (smooth, abrasive-blasted, acid-etched, etc.), PC curtain walls lend themselves to a variety of facade treatments. The use of reveals (aesthetic joints), moldings, and colored concrete further adds to the design variations, Figure 28.39.

Panel Shapes and Sizes

Another key design decision is the size, shape, and function of each panel. These include window wall panels, spandrel panels, spandrel plus infill panels, and so on, Figure 28.40. The panels are generally one floor high, but those spanning two floors may be used.

PC wall panels are generally made as large as possible, limited only by the erection capacity of the crane, the transportation limitations, and the gravity load delivered by the panel to the structural frame. For structural reasons, the panels generally extend from column to column. Smaller panel sizes mean a larger number of panels, requiring a greater number of

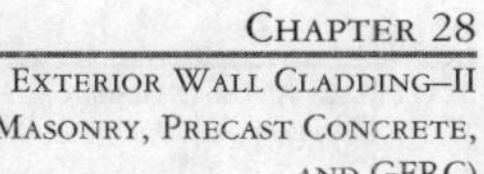

FIGURE 28.39 (a) Lightly abrasive-blasted panels with a great deal of surface detailing.

FIGURE 28.39 (b) The part of this panel on the left side of the reveal is lightly abrasive-blasted, and the right side part is medium abrasive-blasted.

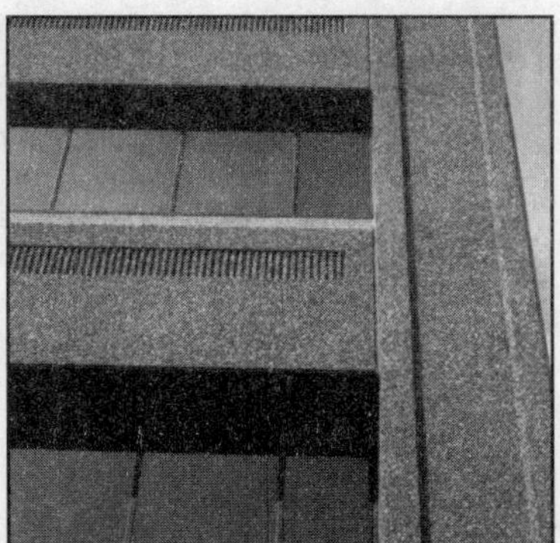

FIGURE 28.39 (c) Panel with an exposed aggregate finish.

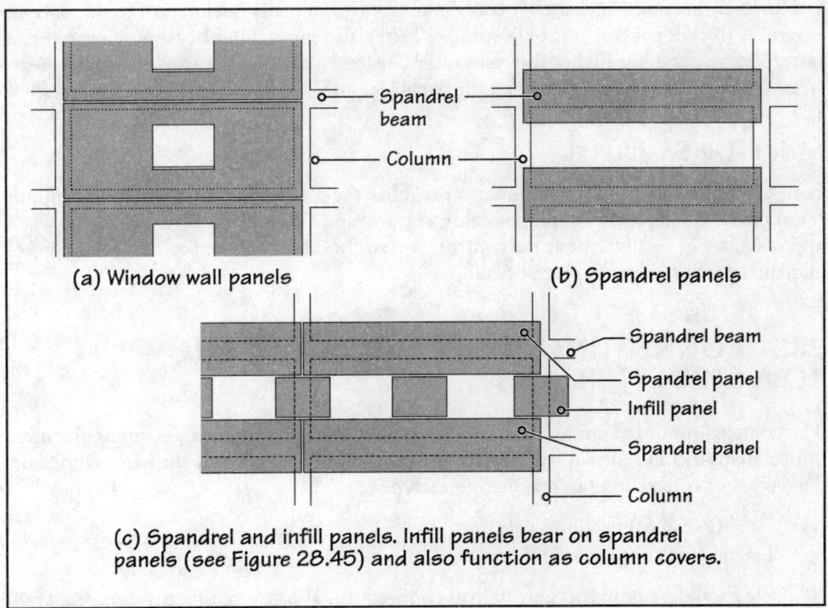

FIGURE 28.40 A few commonly used shapes of precast concrete wall panels. Panels should preferably extend from column to column so that their dead load (which is delivered through two supports only) is transferred to the beam as close to the columns as possible (see Section 28.7).

support connections, longer erection time, and, hence, a higher cost. If the scale of large panels is visually unacceptable, false joints can be incorporated in the panels, as shown in preceding images.

CONCRETE STRENGTH

PC curtain wall panels are removed from the form as soon as possible to allow rapid turnover and reuse of the formwork. This implies that the 28-day concrete strength should be reasonably high so that when the panel is removed from the form, it can resist the stresses to which it may be subjected during the removal and handling processes.

The required concrete strength is also a function of the curtain wall's exposure, durability requirements, shape, and size of the panels. Flat panels may require higher strength (or greater thickness) compared to ribbed or profiled panels. Therefore, the strength of concrete must be established in consultation with the precast plant supplying the panels.

The most commonly used 28-day strength of concrete for PC curtain wall panels is 5,000 psi. This relatively high strength gives greater durability, greater resistance to rainwater penetration, and improved in-service performance. In other words, the panels are better able to resist stresses caused by the loads, building movement, and volume changes induced by thermal, creep, and shrinkage effects.

For aesthetic and economic reasons, a panel may use two mixes—a face (architectural) mix and a backup (structural) mix. In this case, the two mixes should have nearly equal expansion and contraction coefficients to prevent undue bowing and warping of the panel. In other words, the strength, slump, and water-cement ratios of the two mixes should be nearly the same.

Because panels are generally fabricated face down on flat formwork, the face mix is placed first, followed by the backup mix. The thickness of the face mix is a function of the aggregate size but should not be less than 1 in. The precaster's experience should be relied upon in determining the thicknesses and properties of the two mixes.

PANEL THICKNESS

The thickness of panels is generally governed by the handling (erection) stresses rather than the stresses caused by in-service loads. A concrete cover on both sides and the two-way reinforcement in a panel generally give a total of about 4 in. of thickness. Add to this the thickness that will be lost due to surface treatment, such as abrasive blasting and acid etching, and the total thickness of a PC wall panel cannot be less than 5 in.

However, because panel size is generally maximized, a panel thickness of less than 6 in. is rare. A thicker panel is not only stronger but is also more durable, is more resistant to water leakage, and has higher fire resistance. Greater thickness also provides greater heat-storage capacity (Chapter 5), making the panel less susceptible to heat-induced stresses.

MOCK-UP SAMPLE(S)

For PC curtain wall projects, the architect requires the precast plant to prepare and submit for approval a sample or samples of color, texture, and finish. The *mock-up* panels, when approved, are generally kept at the construction site and become the basis for judgment of all panels produced by the precast plant.

28.7 CONNECTING THE PC CURTAIN WALL TO A STRUCTURE

The connections of PC curtain wall panels to the building's structure are among the most critical items in a PC curtain wall project and are typically designed by the panel fabricator. Two types of connections are required in each panel:

- Gravity load connections
- Lateral load connection

There should be only two gravity load connections, also referred to as *bearing supports*, per panel located as close to the columns as possible. The lateral load connections, also referred to as *tiebacks*, may be as many as needed by structural considerations, generally two or more per panel, Figure 28.41.

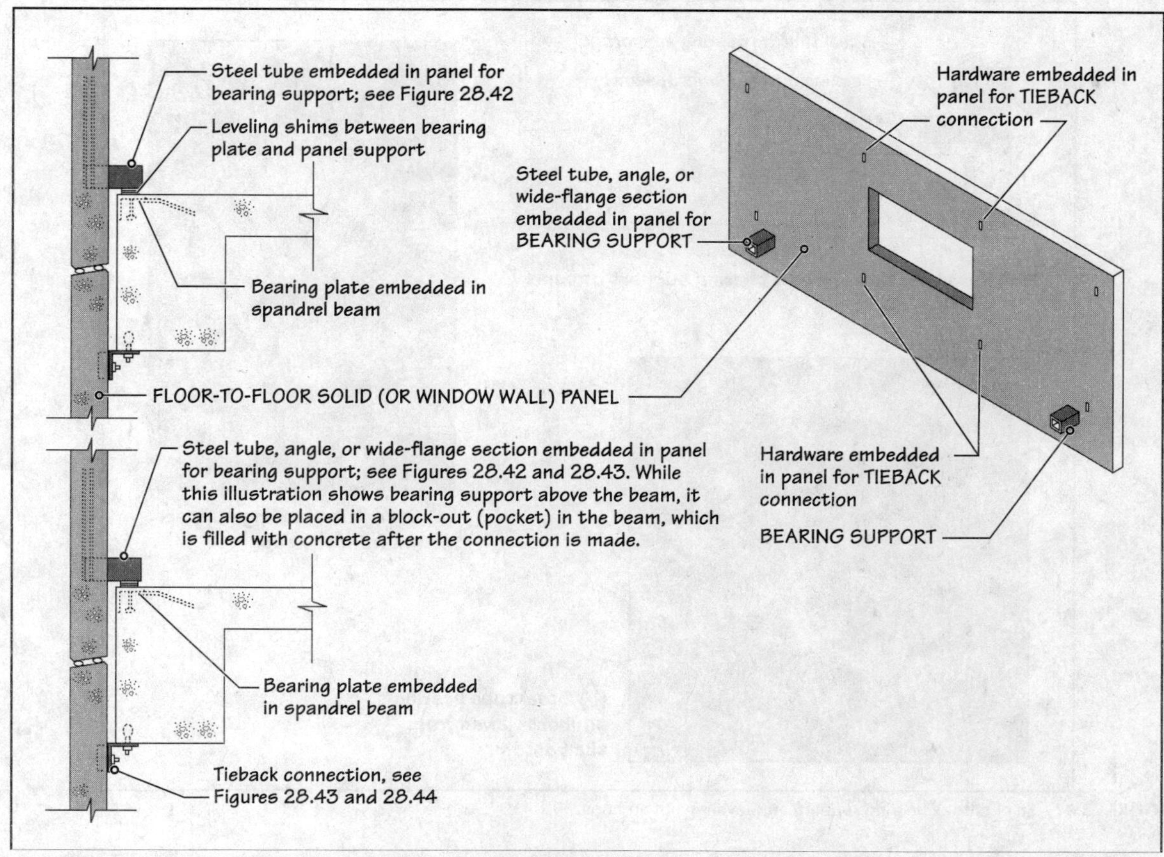

FIGURE 28.41 Support connections for a typical floor-to-floor curtain wall panel.

Bearing Supports

A commonly used bearing support for floor-to-floor panels is provided by a section of steel tube, part of which is embedded in the panel and part of which projects out of the panel. The projecting part rests on the (steel angle) bearing plate embedded at the edge of the spandrel beam. Dimensional irregularities, both in the panel and in the structure, require the use of leveling shims (or bolts) under bearing supports during erection.

After the panels have been leveled, the bearing supports are welded to the bearing plate. The bearing support system is designed to allow the panel to move within its own plane so that the panel is not subjected to stresses induced by temperature, shrinkage, and creep effects.

In place of leveling shims in a bearing support, a leveling bolt is often used, Figure 28.42. The choice between the shims and the bolt is generally left to the preference of the precast manufacturer and the erector. Alternatives to the use of steel tube for bearing supports are steel angles or a wide-flange (I-) section, Figure 28.43.

Tiebacks

A lateral load connection (tieback) is designed to resist horizontal forces on the panel from wind and/or earthquake and due to the eccentricity of panel bearing. Therefore, it must be able to resist tension and compression perpendicular to the plane of the panel.

A tieback is designed to allow movement within the plane of the panel. The connection must, however, permit adjustment in all three principal directions during erection. A typical tieback is shown in Figure 28.44; see also Figure 28.43.

NOTE

The connection system of a PC curtain wall resembles that of a brick veneer connection system and is common to all types of curtain walls. The shelf angles in brick veneer provide the gravity load connection, and the anchors between the veneer and the backup wall provide the lateral load connection.

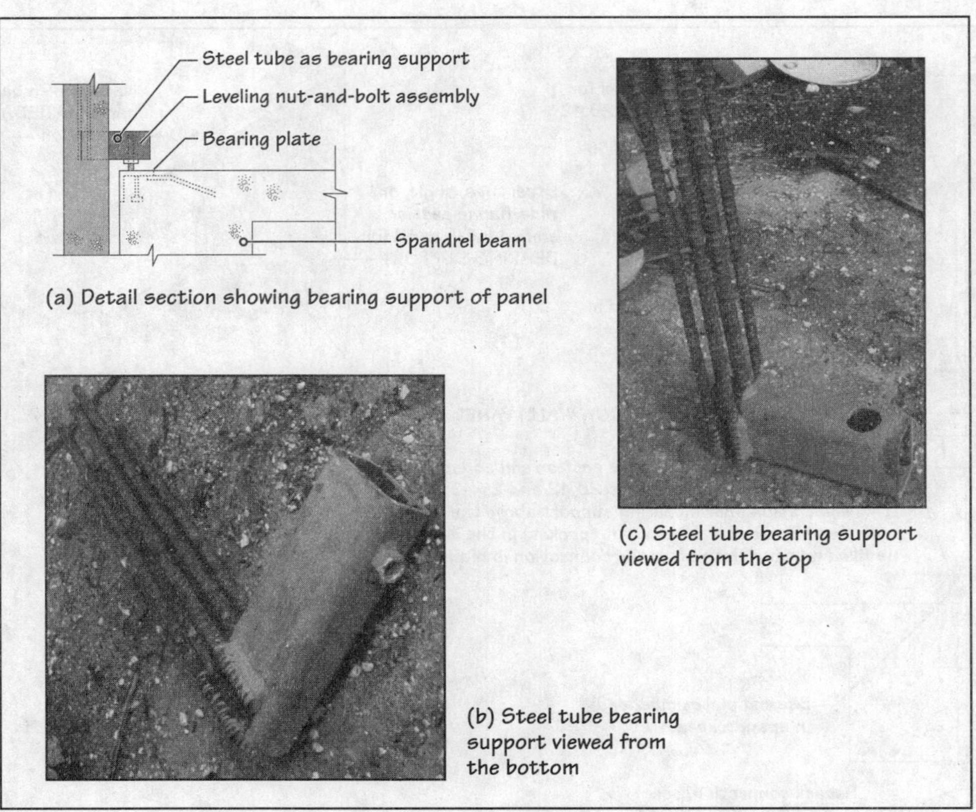

FIGURE 28.42 Steel tube as bearing support, and leveling nut and bolt.

Support Systems for Spandrel Panels and Infill Panels

Support systems for a curtain wall consisting of spandrel and infill panels are shown in Figure 28.45; see also Figure 28.40(c).

Panels and Steel Frame Structure

PC wall panels, which create eccentric loading on the spandrel beams, create torsion in the beams. Due to the lower torsional resistance of wide-flange steel beams, PC panels used with a steel-frame structure are generally designed to span from column to column and made to bear directly on them. Tiebacks, however, are connected to the spandrel beams.

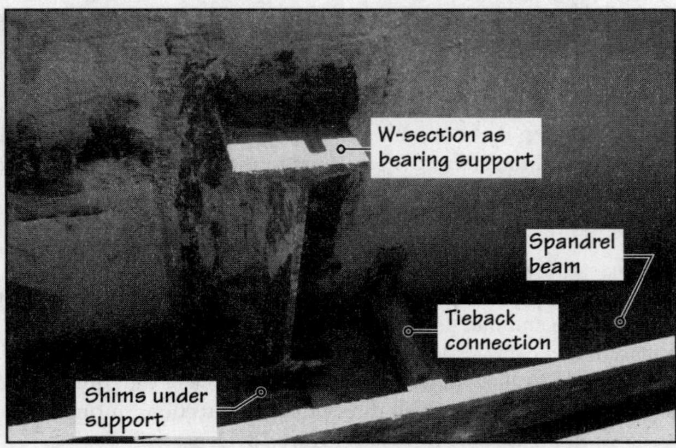

FIGURE 28.43 Wide-flange section used as bearing support in a PC panel.

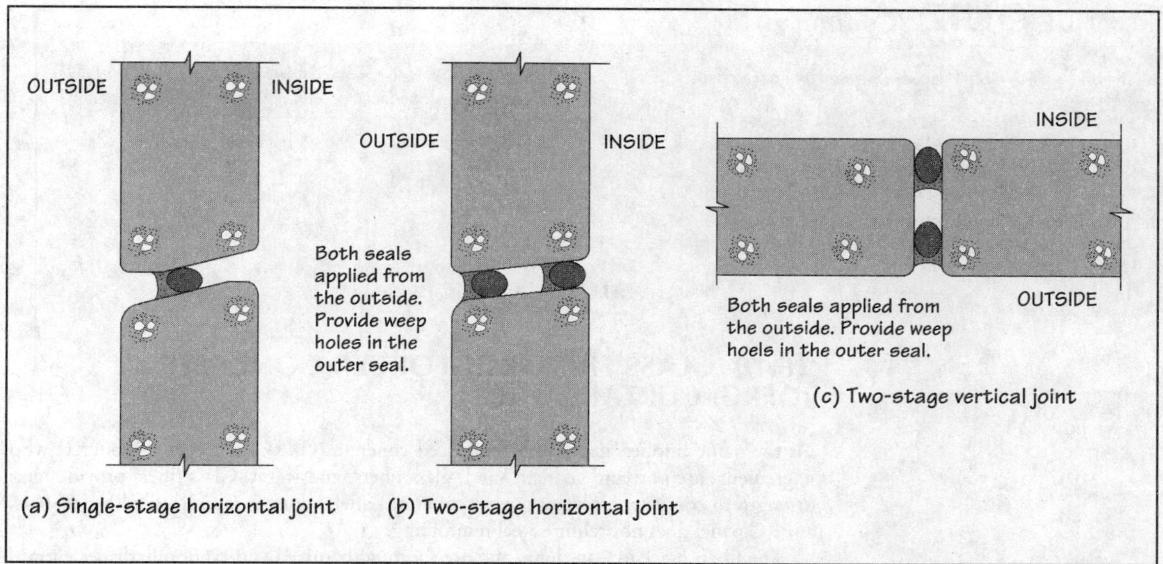
OUTSIDE
INSIDE
OUTSIDE
INSIDE
Both seals applied from the outside. Provide weep holes in the outer seal.
INSIDE
OUTSIDE
Both seals applied from the outside. Provide weep hoels in the outer seal.
(a) Single-stage horizontal joint
(b) Two-stage horizontal joint
(c) Two-stage vertical joint

FIGURE 28.50 Treatment of joints between PC curtain wall panels.

Both inner and outer seals should be applied from the outside to avoid discontinuities at the spandrel beams and floor slabs. This requires a deep-stem roller to push the backer rod deep into the joint and a long-nozzle sealant gun.

In addition to sealed joints, PC curtain walls should be provided with stainless steel channels and weep tubes to collect and drain any condensed water that may collect on the back face of the panels (see Figure 28.49).

Insulated, Sandwiched Panels

PC curtain wall panels with rigid plastic foam insulation sandwiched between two layers of concrete may be used in cold regions. A sandwich panel consists of an outer layer of concrete and an inner layer of concrete, in which both layers are connected together with ties through an intermediate layer of rigid plastic foam insulation.

The outer layer is a nonstructural layer, whereas the inner layer is designed to carry the entire load and transfer it to the structural frame. The panel is fabricated by first casting the concrete for the nonstructural layer. This is followed by embedding the ties and placing the insulation boards over the ties. The insulation is provided with holes so that the ties project above the insulation and are embedded in the concrete that is cast above the insulation.

PRACTICE QUIZ

Each question has only one correct answer. Select the choice that best answers the question.

22. The strength of concrete used in precast concrete curtain walls is generally
- **a.** greater than or equal to 3,000 psi.
- **b.** greater than or equal to 4,000 psi.
- **c.** greater than or equal to 5,000 psi.
- **d.** none of the above.

23. Precast concrete curtain wall panels are generally
- **a.** cast on the construction site.
- **b.** fabricated by a ready-mix concrete plant and transported to the construction site.
- **c.** (a) or (b).
- **d.** none of the above.

24. The number of bearing supports in a precast concrete curtain wall panel must be
- **a.** one.
- **b.** two.
- **c.** three.
- **d.** four.
- **e.** between three and six.

25. The structural design of precast concrete curtain wall panels is generally the responsibility of the
- **a.** structural engineer, who designs the structural frame of the building in which the panels are to be used.
- **b.** structural engineer retained by the panel fabricator.
- **c.** structural engineer retained by the general contractor of the building.
- **d.** structural engineer specially retained by the owner.
- **e.** none of the above.

26. During the erection of precast concrete curtain wall panels, the panels are leveled. The leveling of the panels is provided in the
- **a.** panels' bearing supports.
- **b.** panels' tieback connections.
- **c.** (a) or (b).
- **d.** (a) and (b).

(Continued)

27. The bricks used in brick-faced concrete curtain wall panels are generally
    a. of the same thickness as those used in brick veneer construction.
    b. thicker than those used in brick veneer construction.
    c. thinner than those used in brick veneer construction.
    d. any one of the above, depending on the building.
28. A bond break such as a polyethylene sheet membrane is generally used between the concrete and bricks in a brick-faced precast concrete curtain wall panel.
    a. True   b. False
29. The connection between the natural stone facing and the backup concrete in a stone-faced precast concrete curtain wall panel is obtained by
    a. the roughness of the stone surface that is in contact with the concrete.
    b. nonmoving steel dowels.
    c. flexible steel dowels.
    d. any one of the above.

28.10 GLASS FIBER–REINFORCED CONCRETE (GFRC) CURTAIN WALL

As the name implies, glass fiber–reinforced concrete (GFRC) is a type of concrete whose ingredients are portland cement, sand, glass fibers, and water. Glass fibers provide tensile strength to concrete. Unlike a precast concrete panel, which is reinforced with steel bars, a GFRC panel does not require steel reinforcing.

The fibers are 1 to 2 in. long, and are thoroughly mixed and randomly dispersed in the mix. The random and uniform distribution of fibers not only provides tensile strength, but also gives toughness and impact resistance to a GFRC panel. Because normal glass fibers are adversely affected by wet portland cement (due to the presence of alkalies in portland cement), the fibers used in a GFRC mix are alkali-resistant (AR) glass fibers.

GFRC Skin and Cold-Formed Steel Frame

A GFRC curtain wall panel consists of three main components:

- GFRC skin
- Cold-formed steel backup frame
- Anchors that connect the skin to the steel backup frame

Of these three components, only the skin is made of GFRC, which is generally $\frac{1}{2}$- to $\frac{3}{4}$-in.-thick. The skin is anchored to a frame consisting of cold-formed (galvanized) steel members, Figure 28.51. In this combination, the GFRC skin transfers the loads to the frame, which, in turn, transfers the loads to the building's structure. The size and spacing of backup frame members depend on the overall size of the panel and the loads to which it is subjected.

Flex Anchors

The skin is hung 2 in. (or more) away from the face of the frame using bar anchors. The gap between the skin and the frame is essential because it allows differential movement between the skin and the frame, particularly during the period when the fresh concrete in the skin shrinks as water evaporates.

FIGURE 28.51 (a) The front face of a large GFRC panel after light abrasive blasting. It is being lifted for storage in the fabricator's shed.

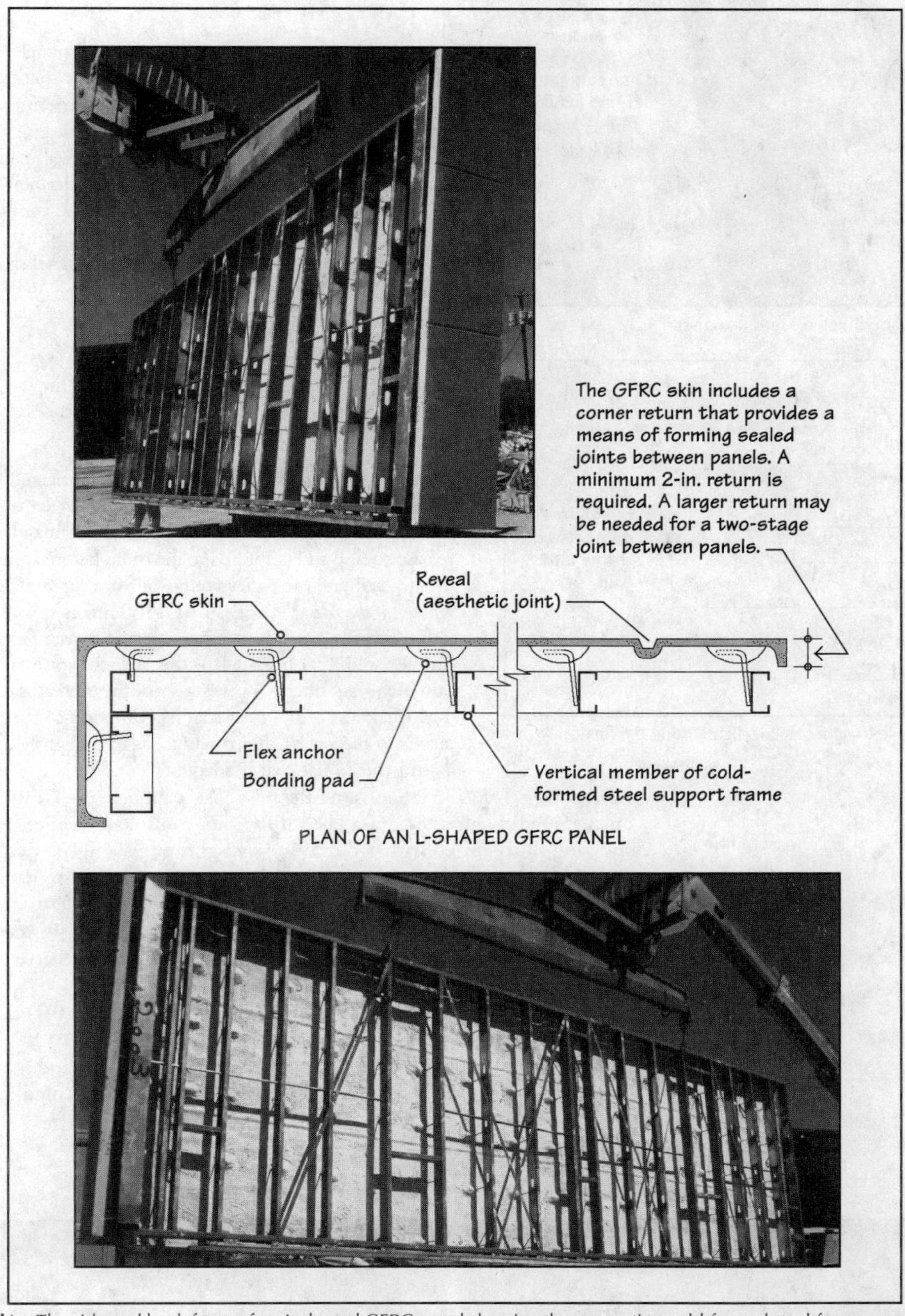
The GFRC skin includes a corner return that provides a means of forming sealed joints between panels. A minimum 2-in. return is required. A larger return may be needed for a two-stage joint between panels.
Reveal (aesthetic joint)
GFRC skin
Flex anchor
Bonding pad
Vertical member of cold-formed steel support frame
PLAN OF AN L-SHAPED GFRC PANEL

FIGURE 28.51 (b) The side and back faces of an L-shaped GFRC panel showing the supporting cold-formed steel frame.

The anchors are generally $\frac{3}{8}$-in.-diameter steel bars bent into an L shape. To provide corrosion resistance, cadmium-plated steel is generally used for the anchors. One end of an anchor is welded to the frame, and the other end is embedded in the skin. The skin is thickened around the anchor embedment. The thickened portion of the skin is referred to as a *bonding pad*, Figure 28.52.

The purpose of the anchors is to transfer both gravity and lateral loads from the skin to the frame. In doing so, the anchors must be fairly rigid in the plane perpendicular to the panel; that is, they should be able to transfer the loads without any deformation in the anchors.

However, the skin will experience in-plane dimensional changes due to moisture and temperature effects. The anchor must, therefore, have sufficient flexibility to allow these changes to occur without excessively stressing the skin. One way of achieving this goal is to flare the anchor away from the frame and weld it to the frame member at the far end, Figure 28.53. The term *flex anchor* underscores the importance of anchor flexibility.

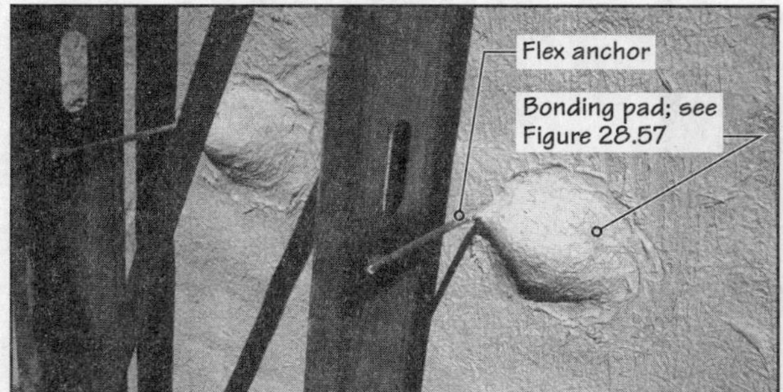

FIGURE 28.52 Bonding pads and flex anchors.

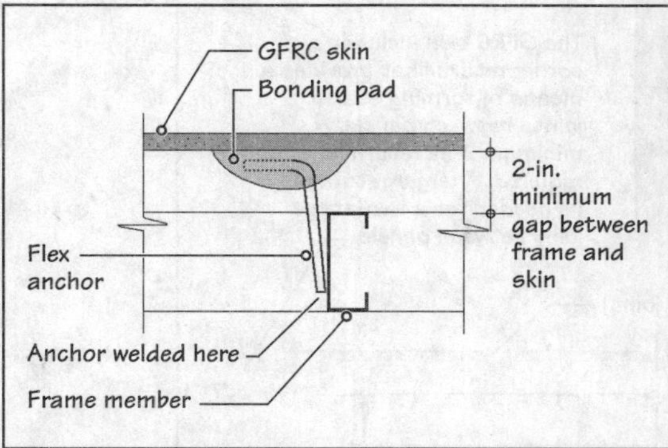

FIGURE 28.53 Flex anchor detail highlighting the flexibility of a flex anchor.

Panel Shapes

Like the precast concrete curtain wall panels, GFRC curtain wall panels can be formed into different shapes, depending on the building's facade. The most commonly used panels are floor-to-floor panels (Figure 28.51), window wall panels, and spandrel panels, similar to PC curtain wall panels. A spandrel panel is essentially a solid panel of smaller height than a floor-to-floor panel. Because the panels can be configured in several ways for a given facade, the architect must work with the GFRC fabricator before finalizing the panel shapes.

28.11 FABRICATION OF GFRC PANELS

Figures 28.54 to 28.57 show the six-step process of fabricating a GFRC panel.

- *Preparing the Mold:* The mold must be fabricated to the required shape, Figure 28.54(a). The mold generally consists of plywood, but other materials, such as steel or plastics, may be used. A form-release compound is applied to the mold before applying the GFRC mix to facilitate the panel's removal from the mold.
- *Applying the Mist Coat:* Before GFRC mix is sprayed on the panel, a thin cement-sand slurry coat, referred to as a *mist coat,* is sprayed on the mold, Figure 28.54(b). Because the mist coat does not contain any glass fibers, it gives a smooth, even surface to the panel. The thickness of the mist coat is a function of the finish the panel face is to receive. If the panel face is to be lightly abrasive-blasted, a $\frac{1}{8}$-in.-thick mist coat is adequate.
- *Applying the GFRC Mix:* Soon after the mist coat is applied, the GFRC mix is sprayed on the mold, Figure 28.55(a). The GFRC mix consists of (white) portland cement and sand slurry mixed with about 5% (by weight) of glass fibers. Because air may be trapped in the mix during spraying, the mix is consolidated after completion of the spray application by rolling, tamping, or troweling, Figure 28.55(b).
- *Frame Placement:* After the GFRC spray is completed, a cold-formed steel support frame is placed against the skin, leaving the required distance between the skin and the frame, Figure 28.56.
- *Bonding Pads:* Finally, bonding pads are formed at each anchor using the same mix as that used for the skin, Figure 28.57.
- *Removing the Panel from the Mold and Curing:* The panel (skin and frame) is generally removed from the mold 24 h after casting and subsequently cured for a number of

FIGURE 28.54 **(a)** The mold of a roof cornice GFRC panel. **(b)** Application of a mist coat on the mold.

(a)
(b)

FIGURE 28.55 **(a)** Spraying of a GFRC mix over the mist coat. **(b)** Consolidating the GFRC mix using a roller around the bends, edges, and corners.

days. Because the panel has not yet gained sufficient strength, special care is needed during its removal.

Surface Finishes on GFRC Panels

The standard finish on a GFRC panel is a light abrasive-blasted finish to remove the smoothness of the surface obtained from the mist coat. However, a GFRC panel can also be given an exposed aggregate finish, which results in a surface similar to that of a precast concrete panel.

To obtain an exposed aggregate finish, the mist coat is replaced by a concrete layer, referred to as a *face mix*. The thickness of the face mix is about $\frac{3}{8}$ in., and the aggregate used in the face mix is between $\frac{3}{16}$ in. and $\frac{1}{4}$ in. The exposure of aggregate is obtained by abrasive blasting or acid etching.

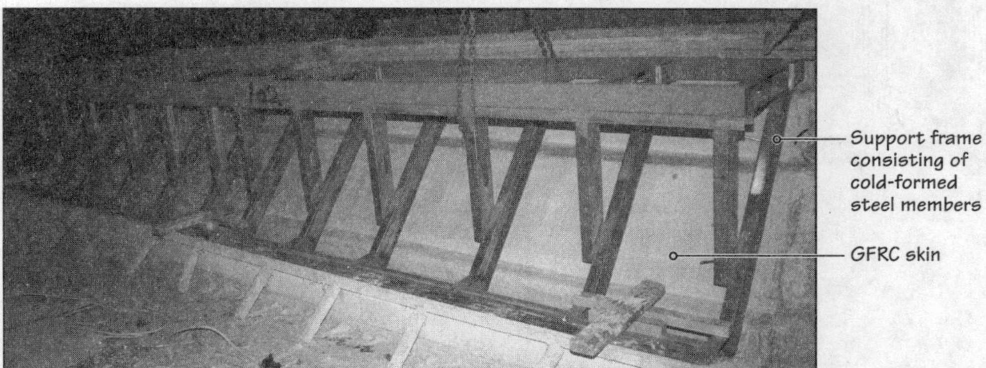

FIGURE 28.56 After spraying and consolidating the mix for the GFRC skin on the mold, a cold-formed steel supporting frame is placed against the mold and bonding pads are formed.

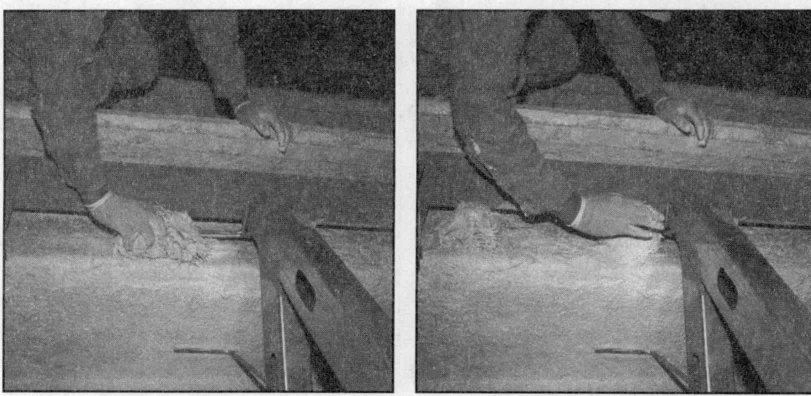

FIGURE 28.57 Making a bonding pad (also see Figures 28.52 and 28.53).

28.12 DETAILING A GFRC CURTAIN WALL

A GFRC panel is connected to the building's structure in a system similar to that of a precast concrete panel. In other words, a GFRC panel requires two bearing connections and two or more tieback connections. However, because GFRC panels are much lighter than the precast concrete panels, their connection hardware is lighter.

As with precast concrete curtain walls, the structural design of panels and their support connections is generally the responsibility of the panel fabricator. Figures 28.58 and 28.59 show two alternative schematics for the connection of a GFRC spandrel panel to the building's structure.

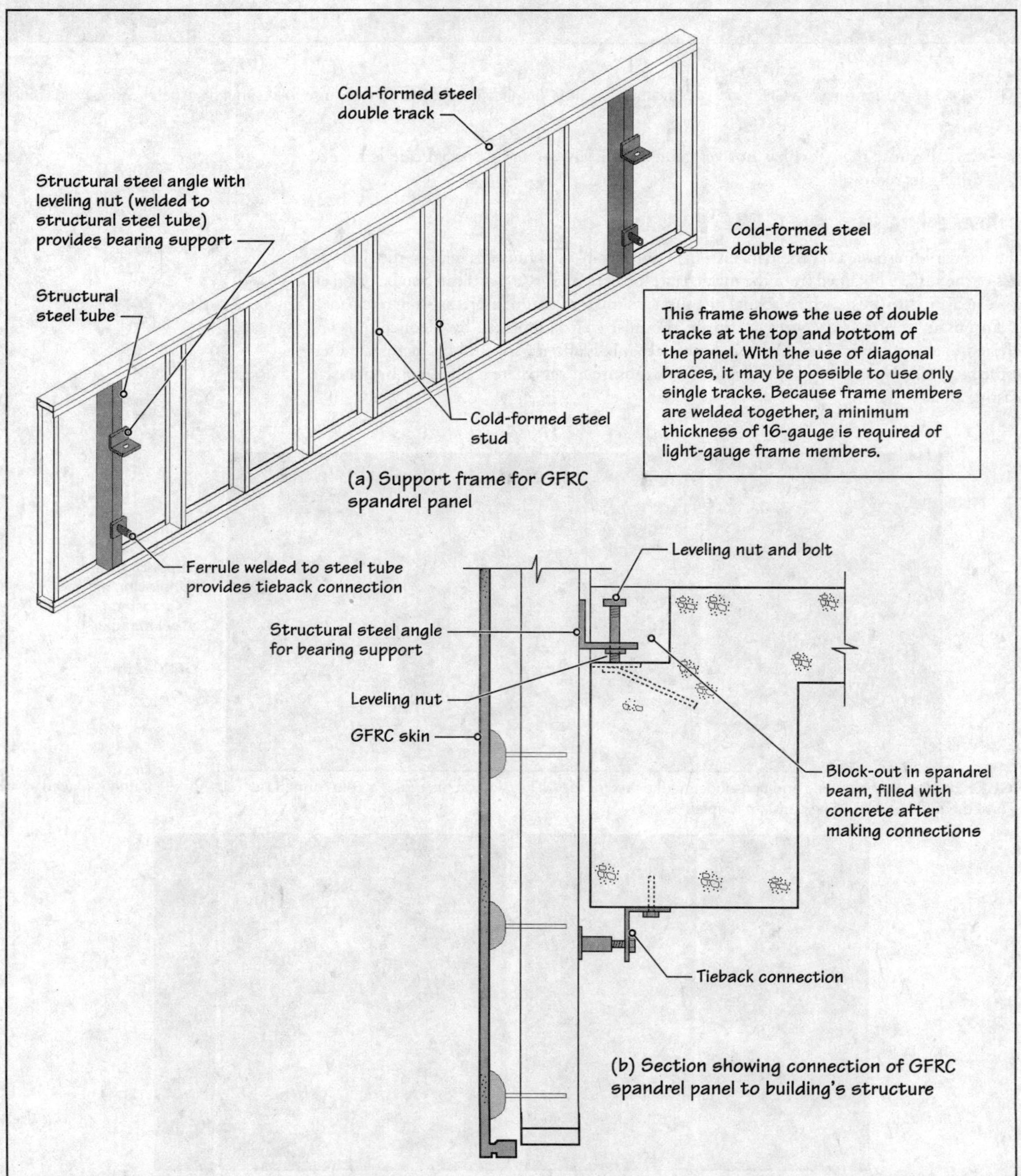

FIGURE 28.58 Schematic details of bearing and tieback connections in a GFRC spandrel panel (also see Figures 28.59 and 28.60).

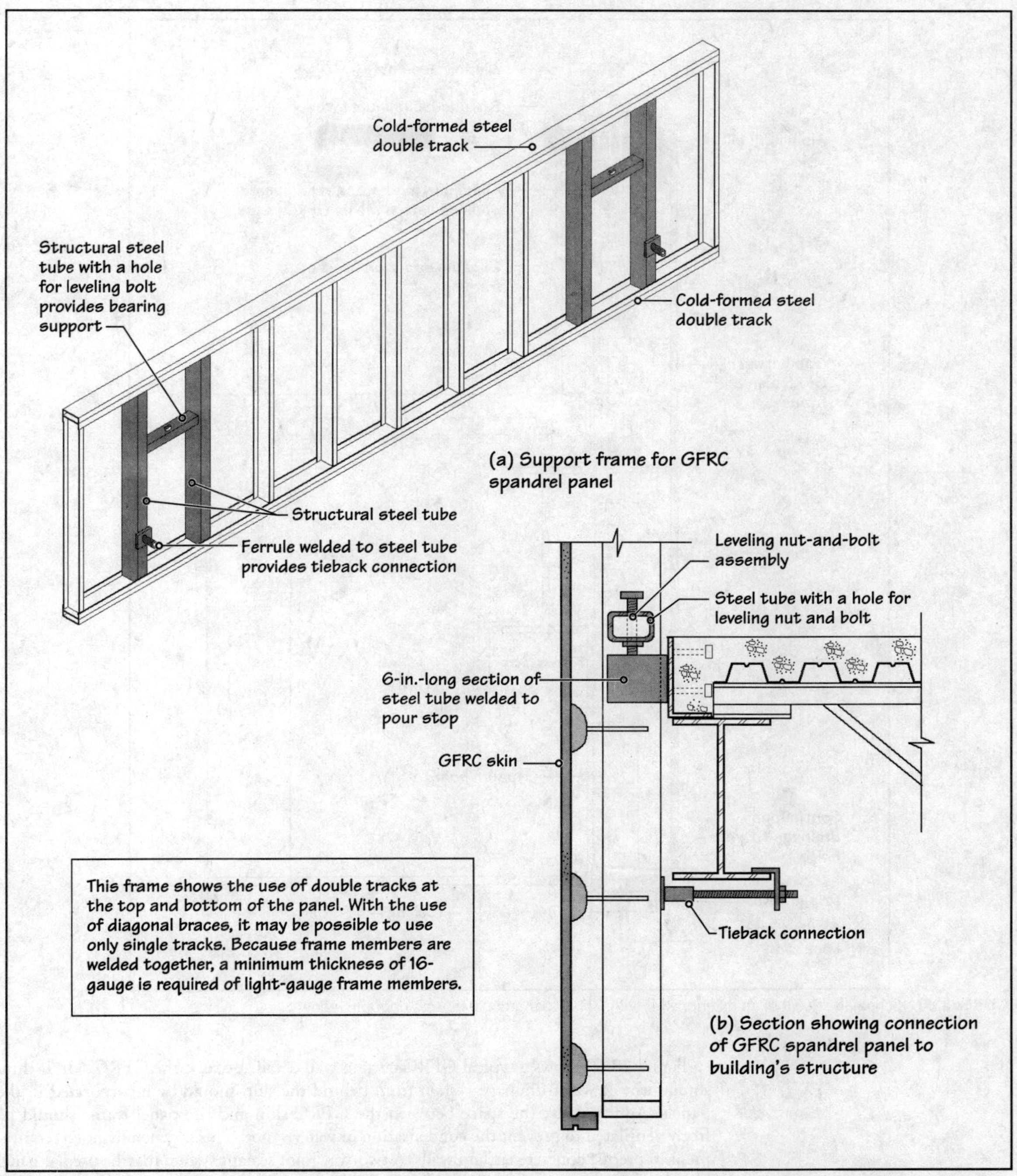

FIGURE 28.59 Schematic details of bearing supports and tieback connections in a spandrel GFRC panel (also see Figures 28.58 and 28.60).

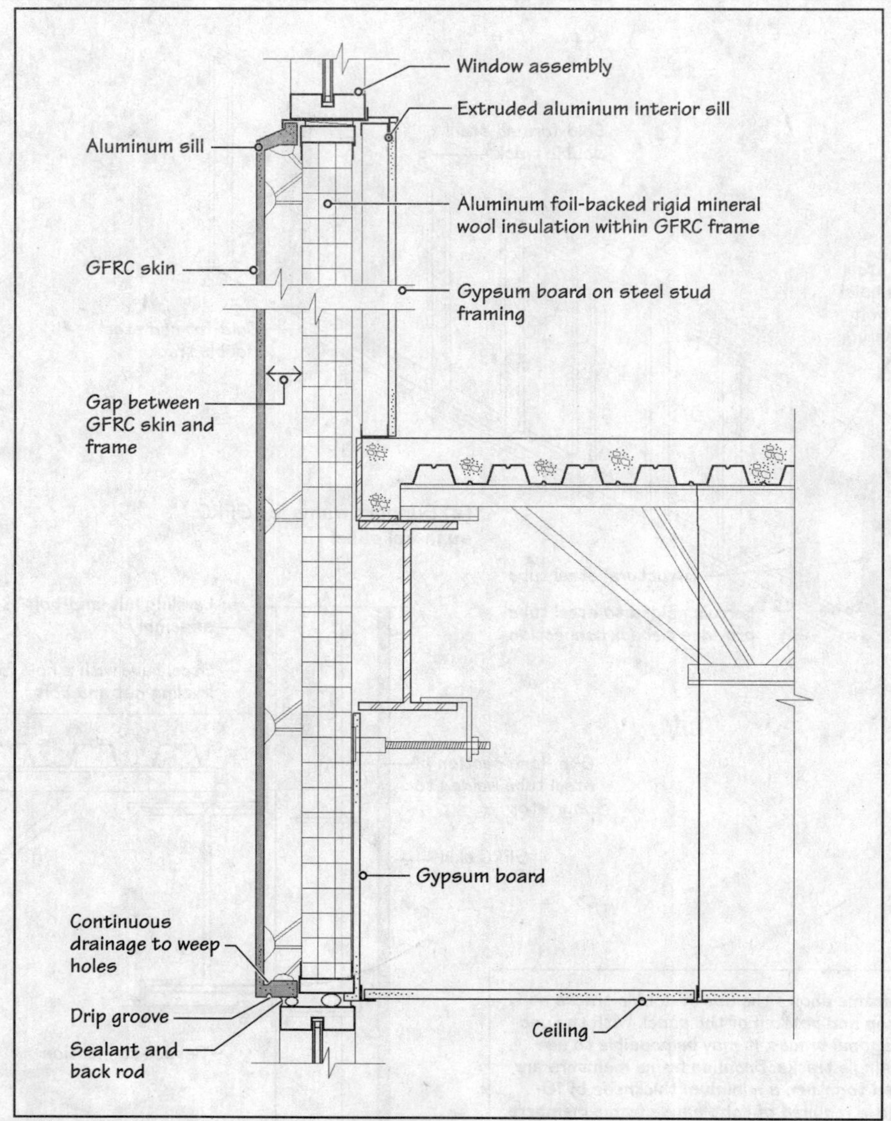

FIGURE 28.60 Schematic detail of an exterior wall with GFRC spandrel panels and ribbon windows.

Figure 28.60 shows a typical GFRC curtain wall detail. Because the GFRC skin is thin, some form of water drainage system from behind the skin should be incorporated in the panels. Additionally, the space between the GFRC skin and the panel frame should be freely ventilated to prevent the condensation of water vapor. This is generally not a requirement in precast concrete curtain walls. A two-way joint sealant system may be used at panel junctions similar to that for PC curtain walls (Figure 28.50).

PRACTICE **QUIZ**

Each question has only one correct answer. Select the choice that best answers the question.

30. The term *GFRC* is an acronym for
- **a.** glass fiber–reinforced concrete.
- **b.** glass fiber–reinforced cement.
- **c.** glass fiber–restrained concrete.
- **d.** glass fiber–restrained cement.
- **e.** none of the above.

31. A GFRC curtain wall panel consists of
- **a.** a GFRC skin.
- **b.** a cold-formed steel frame.
- **c.** anchors.
- **d.** all of the above.
- **e.** (a) and (b).

32. The GFRC skin is typically
- **a.** 2 to 3 in. thick.
- **b.** $1\frac{1}{2}$ to $2\frac{1}{2}$ in. thick.
- **c.** 1 to 2 in. thick.
- **d.** $\frac{1}{2}$ to $\frac{3}{4}$ in. thick.
- **e.** none of the above.

33. The glass fibers used in the GFRC skin are referred to as *AR fibers*. The term *AR* is an acronym for
a. alkali resistant.
b. acid resistant.
c. alumina resistant.
d. aluminum reinforced.
e. none of the above.

34. The anchors used in GFRC panels are typically
a. two-piece anchors.
b. one-piece anchors.
c. either (a) or (b), depending on the thickness of the GFRC skin.
d. either (a) or (b), depending on the lateral loads to which the panel is subjected.
e. none of the above.

35. The GFRC skin is obtained by
a. casting the GFRC mix in a mold similar to casting a precast concrete member.
b. applying the mix with a trowel.
c. spraying the mix over a mold.
d. any of the above, depending on the panel fabricator.

36. GFRC curtain wall panels are generally lighter than the corresponding precast concrete curtain wall panels.
a. True
b. False

37. In prefabricated brick curtain wall panels, the bricks used are
a. generally of the same thickness as those used in brick veneer construction.
b. generally thicker than those used in brick veneer construction.
c. generally thinner than those used in brick veneer construction.

38. Prefabricated brick curtain wall panels are generally fabricated
a. in a brick-manufacturing plant.
b. in a masonry contractor's fabrication yard.
c. in a precast concrete fabricator's plant.
d. at the construction site.

REVIEW QUESTIONS

1. Using sketches and notes, explain the adjustability requirements of anchors used in a brick veneer wall assembly.
2. Using sketches and notes, explain the functions of a shelf angle and a lintel angle in a brick veneer assembly.
3. With the help of sketches and notes, explain various ways in which weep holes can be provided in a brick veneer assembly.
4. Discuss the pros and cons of using a CMU backup wall versus a steel stud backup wall in a brick veneer wall assembly.
5. With the help of sketches and notes, explain the support system of a precast concrete curtain wall panel.
6. Using sketches, explain the commonly used shapes of precast concrete wall panels.
7. With the help of a sketch, explain the two-stage joint seal in precast concrete wall panels. What are its advantages over a single-stage joint seal?
8. Using a sketch, explain the purpose of a (a) bonding pad and (b) flex anchor in a GFRC panel.

CHAPTER 29

Exterior Wall Cladding–III (Stucco, Adhered Veneer, EIFS, Natural Stone, and Insulated Metal Panels)

CHAPTER OUTLINE

This chapter continues the discussion of exterior-wall cladding systems. Topics discussed in this chapter are portland cement plaster (stucco), exterior insulation and finish systems (EIFS), stone cladding, and insulated metal wall panels.

29.1 PORTLAND CEMENT PLASTER (STUCCO) BASICS

Plaster has been used for centuries as an exterior and interior wall and ceiling finish. Apart from rendering the wall and ceiling surfaces smooth and paintable, plastering makes them more resistant to water and air infiltration and increases sound insulation and resistance to fire.

A plaster mix is similar to a masonry mortar mix and consists of cementitious material(s), sand, and water. In some plasters, a fibrous admixture is also used. Prior to the discovery of portland cement and gypsum, lime was the only cementitious material available for plaster. In contemporary construction, gypsum and portland cement are the primary cementitious materials. Because gypsum is not a hydraulic cement (i.e., it will dissolve in water), gypsum plaster is suitable only for interior applications not subjected to high humidity levels or wetting.

As noted in Chapter 16, gypsum plaster has largely been replaced by (prefabricated) gypsum boards. Therefore, we limit the discussion to portland cement plaster. Unlike gypsum plaster, portland cement plaster can be used on both interior and exterior surfaces.

The predominant use of portland cement plaster in contemporary buildings is as an exterior wall finish—the topic of discussion in this chapter. Its use as an interior finish is limited to situations where high humidity levels or wetting of the plastered surfaces occur, such as in saunas, public shower rooms, and commercial kitchens. In most parts of the United States, exterior portland cement plaster is referred to as *stucco.*

Although stucco can be applied on adobe walls, it is more commonly used on

- Cold-formed steel stud walls
- Wood stud walls
- Masonry walls
- Concrete walls

Because stucco is a portland cement–based material, the application of stucco requires appropriate temperature conditions. Generally, stucco should be applied if the ambient air temperature is at least 40°F (5°C) and rising.

Mix Composition for Stucco Coats

Stucco is typically applied sequentially in two coats over a wall, called the *base coat* and the *finish coat.* In some situations, the base coat is applied in two layers, referred to as the *scratch coat* and the *brown coat.* The ingredients of the base coat are portland cement, lime, sand, and water.

Portland cement is the glue that bonds all constituents of the mix, which eventually cure into a strong and rigid surface. Lime imparts plasticity and cohesiveness to the mix. Plasticity implies that the mix can be spread easily, and cohesiveness implies that the mix will hold and not sag on a vertical surface during application.

Generally, portland cement and lime for the base coat are factory blended into one bag, Figure 29.1(a). The bag ingredients are site mixed with sand and water in a mixer. Like the base coat, the finish coat is factory blended into one bag, Figure 29.1(b). Color is integral to the finish coat mix. The finish coat does not require the addition of sand; hence, water is the only additional material needed to prepare the finish coat for application. Although a one-bag finish coat helps to provide consistent quality from batch to batch, it is necessary that it be applied continuously, with interruptions only at the control joints or expansion joints.

Two types of factory-blended mixes for the finish coat are available:

- Portland cement–based mix
- Acrylic polymer–based mix

The use of an acrylic polymer–based finish coat is more common because it is more flexible, reduces cracking of the stucco surface, and provides consistent and relatively nonfading colors. However, a portland cement–based finish coat is more breathable (vapor-permeable) than a polymer–based finish so that any moisture trapped within or behind the stucco dries out faster.

(a)

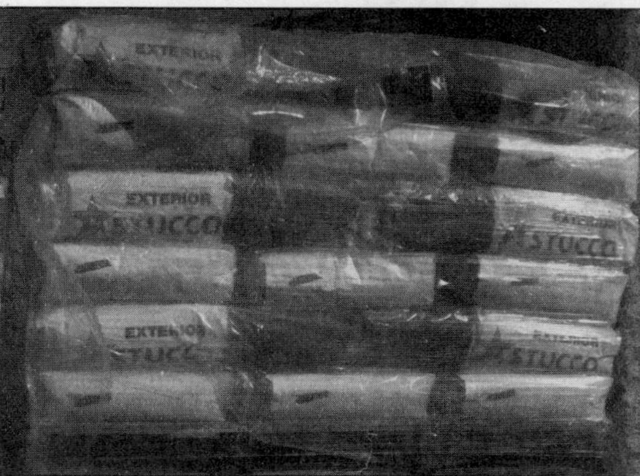

(b)

FIGURE 29.1 **(a)** Bags containing base coat material (portland cement and lime blended together). The bags are emptied in a mixer, to which sand and water are added to obtain the stucco base coat mix. **(b)** Bags containing finish coat material, which requires only the addition of water to obtain the finish coat mix.

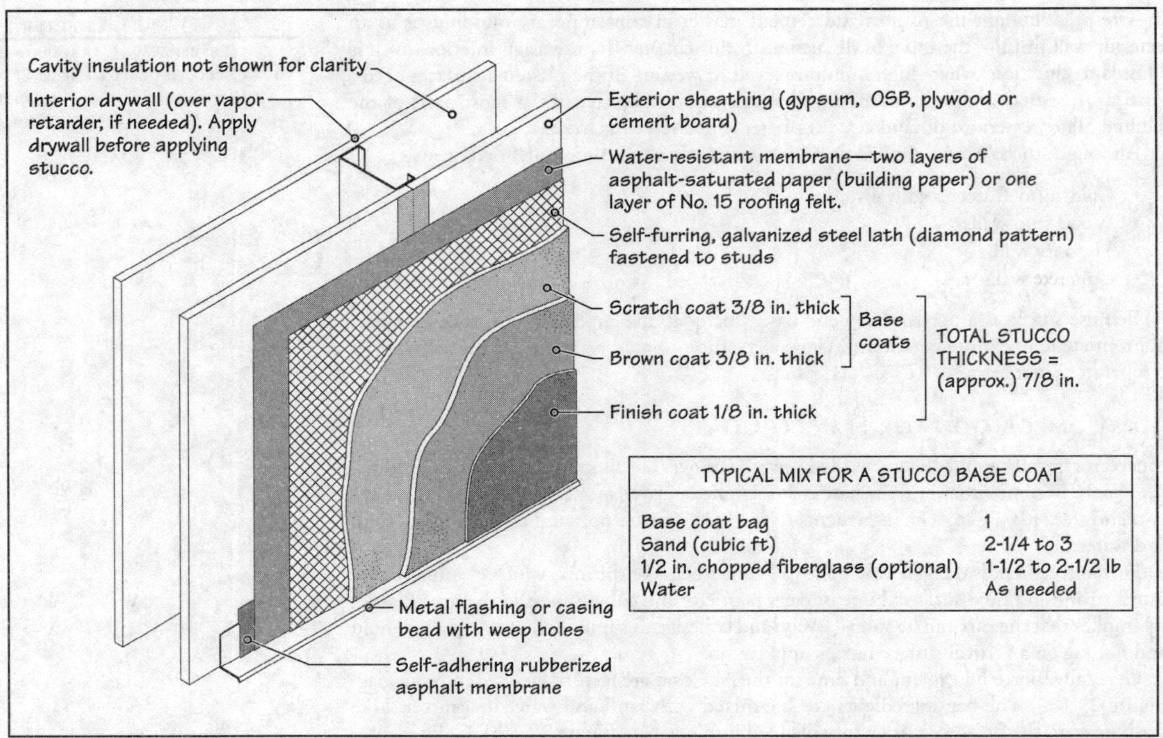

TYPICAL MIX FOR A STUCCO BASE COAT	
Base coat bag	1
Sand (cubic ft)	2-1/4 to 3
1/2 in. chopped fiberglass (optional)	1-1/2 to 2-1/2 lb
Water	As needed

FIGURE 29.2 Anatomy of a steel- or wood-stud wall with a stucco finish.

29.2 STUCCO ON STEEL- OR WOOD-STUD WALLS

The anatomies of wood and cold-formed steel frame walls with a stucco finish are essentially identical, as shown in Figure 29.2. Both wall assemblies require an exterior (gypsum, plywood, OSB, or cement board) sheathing, a water-resistant membrane, and a self-furring metal base.

The water-resistant membrane is the second line of defense against water intrusion. The first line of defense is the stucco finish itself. In most projects, two layers of asphalt-saturated building paper or one layer of No. 15 asphalt-saturated roofing felt (see Chapter 33) is used as the water-resistant membrane. It is applied horizontally with laps between sheets.

The metal base generally consists of self-furring galvanized steel lath (diamond pattern), Figure 29.3. The lath is fabricated from steel sheets, which are slit at regular intervals and

NOTE

Building Paper or Roofing Felt Backing for Stucco—The Drainage Plane

The building paper or the roofing felt absorbs a certain amount of water during the application of stucco. As the paper (or felt) and stucco dry, the paper (or felt) wrinkles and the stucco shrinks, leaving small vertical drainage channels behind the stucco. Although the drainage channels are small (unlike an air space in a drainage wall), they help keep the stucco wall dry. The multitude of drainage channels are referred to as a *drainage plane*.

A proprietary air-retarder-type backing for stucco is also available that has a specially made crinkled surface, providing a drainage plane similar to the one that develops in a paper- or felt-backed stucco.

FIGURE 29.3 A worker fastening self-furring metal lath (diamond pattern) over No. 15 roofing felt.

then stretched. For this reason, the lath is also known as *expanded metal lath.* The lath sheets are finally hot-dip galvanized, as needed.

The lath provides a mechanical key to which the stucco bonds. The self-furring character of the lath is obtained by incorporating dimples or other means in the lath during the stretching process that hold it about $\frac{1}{4}$ in. away from the substrate. Thus, when the stucco coat is applied, the lath is embedded in it, becoming an integral part of the stucco, much like the reinforcing bars in a reinforced-concrete slab.

Self-furring lath is also available with a continuous backing of building paper integral with the lath. This combination increases work efficiency, particularly in large stucco projects.

In a wood-stud wall, the lath is fastened to the studs with large-head nails. In a cold-formed steel-stud wall, the lath is anchored to the studs with self-drilling, self-tapping screws with large (nearly $\frac{1}{2}$-in.-diameter) heads. Because the lath is anchored to the studs, the stucco is structurally engaged to the studs.

APPLICATION OF STUCCO

Figure 29.4 shows the exterior of a building covered with scaffolding in readiness for stucco application. Stucco applied on (wood or steel) stud walls generally consists of two coats (a scratch coat and a base coat), each approximately $\frac{3}{8}$-in.-thick, and an approximately $\frac{1}{8}$-in.-thick finish coat, giving a total stucco thickness of approximately $\frac{7}{8}$ in.

The ingredients for all coats are mixed in a mixer, which is connected to a pump, Figure 29.5. The mix is delivered to the point of application through a long pipe terminating in a nozzle that sprays the mix, Figure 29.6.

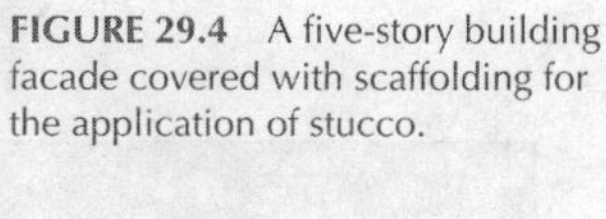

FIGURE 29.4 A five-story building facade covered with scaffolding for the application of stucco.

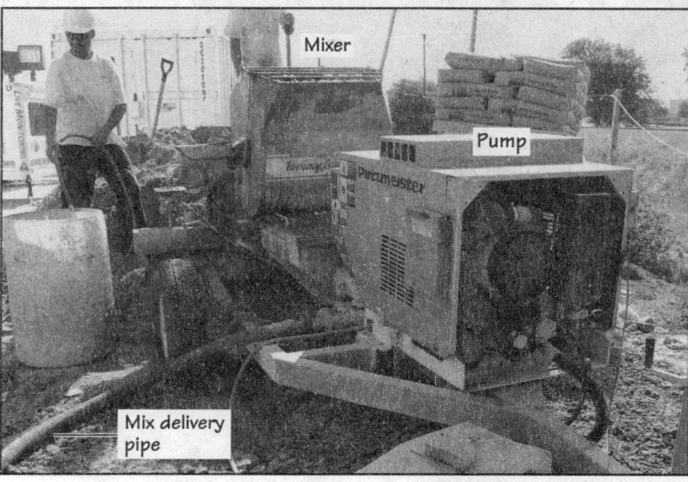

FIGURE 29.5 A stucco mixer and pump. The pump receives the mix from the mixer and delivers it through a pipe to the point of application.

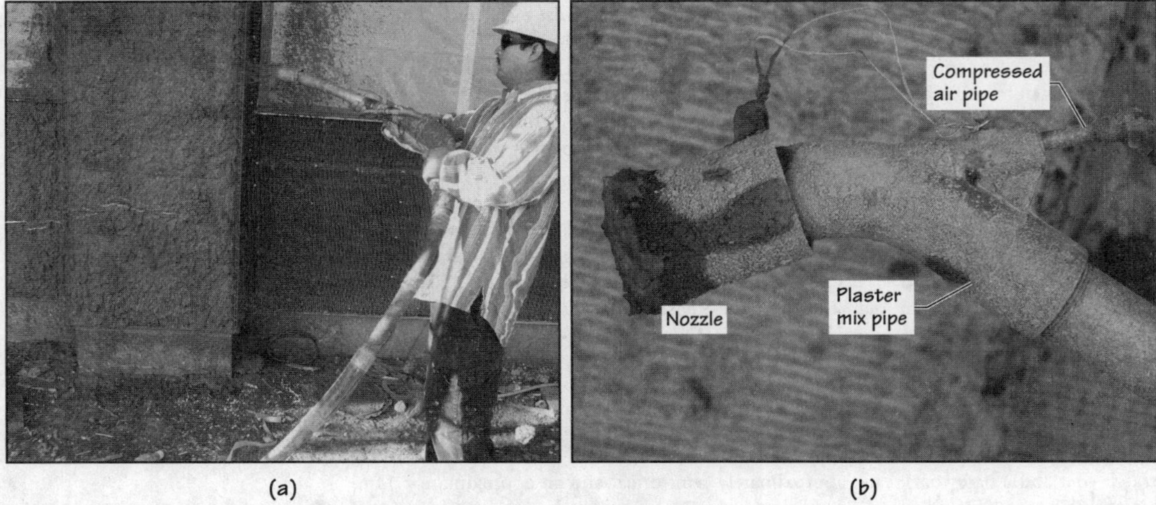

FIGURE 29.6 **(a)** Spraying of the stucco base coat over metal lath. **(b)** The pipe used for spraying the stucco mix terminates in a nozzle. The spray action for the mix is provided by a separate compressed-air pipe that also terminates in the same nozzle as the plaster mix pipe.

For the scratch coat, the sprayed-on material is troweled with sufficient pressure to squeeze it through the lath so that the lath is embedded completely in the scratch coat, Figure 29.7(a). Generally, one plasterer sprays the mix, and then another plasterer hand trowels the sprayed material. After the troweling operation, the material is scratched (hence the name *scratch coat*), Figure 29.7(b). The scratched surface provides a mechanical key for the following brown coat. On a wall (vertical surface), the scratches are horizontal.

The mix for the brown coat is also spray-applied in the same way as the scratch coat. However, the sprayed-on material is brought to an even plane using a wood or metal float, which also densifies the applied material, Figure 29.8(a). After floating, the surface is troweled with a steel trowel, which smoothes the surface further and prepares it for the finish coat, Figure 29.8(b).

The time interval between the scratch and brown coats depends on the ambient temperature and humidity conditions. It is important that the scratch coat develop sufficient strength and rigidity to carry the weight of the brown coat without damaging the grooves. At the same time, the scratch coat should not set and dry so much that its bond with the subsequent coat is compromised.

A 24- to 48-h interval between the scratch and brown coats is commonly used in most climates. If a larger interval is used, the scratch coat should be moist cured until the application of the brown coat.

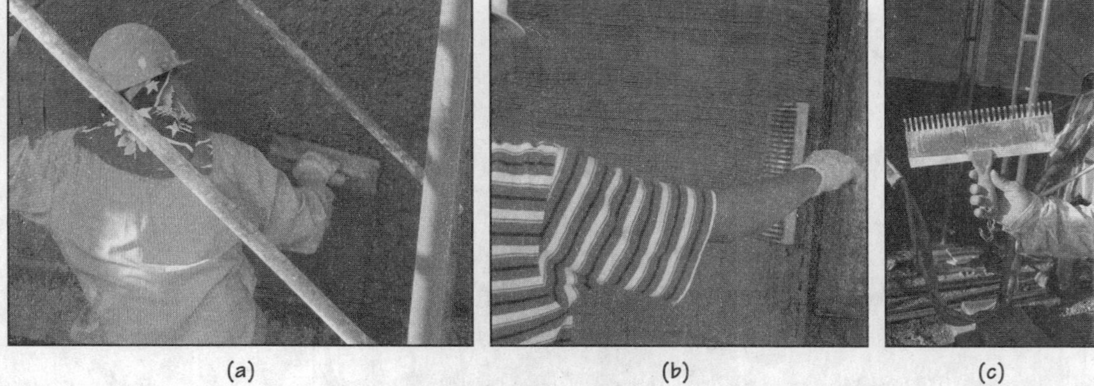

FIGURE 29.7 Application of a scratch coat. **(a)** After spraying the mix, one plasterer spreads the mix with a trowel, forcing it into the lath. **(b)** Another plasterer follows behind the first plasterer and scratches the surface with a scratching tool. **(c)** A close-up of scratching tool.

(a) Floating the brown coat

(b) Troweling the brown coat

FIGURE 29.8 Application of the brown coat. The sprayed mix is first floated and then troweled smooth.

The finish coat is obviously the most important component of the stucco finish because it provides the required color and texture. It can be applied by hand or sprayed. The sprayed finish may be left as is or textured using a sponge float or other suitable means. A longer time interval (generally, 7 to 10 days) between the brown coat and the finish coat is required to allow the base (scratch and brown) coats to stabilize from shrinkage.

Some finish coat manufacturers require two applications of their material. The first application is that of a primer, which prepares the base coat surface to receive the finish coat and improves adhesion between the base coat and the finish coat. The finish coat develops the final stucco color and texture.

Control Joints and Expansion Joints

Because of the presence of portland cement, shrinkage is an inherent feature of a stucco surface, which leads to its cracking. Although the cracking of stucco cannot be fully eliminated, it can be controlled by providing closely spaced control joints.

ASTM standards require that the area between control joints in stucco applied on a stud-wall assembly not exceed 144 ft^2, with neither dimension of this area exceeding 18 ft. As far as possible, control joints should also be provided around openings in walls, Figure 29.9.

In addition to control joints, expansion joints are needed in a stucco finish. Although control joints are a means of controlling the shrinkage of the stucco finish, expansion joints respond to large movements in the building structure. Thus, expansion joints should be provided at each floor level in the exterior wall in a multistory building to absorb the movement in the spandrel beam, Figure 29.10. Expansion joints are also needed where there is a major change in building elevation or where a stucco-finished wall abuts a wall made of a different material.

NOTE

Control Joint Locations on Stucco Applied on Wood- or Steel-Stud Walls

- The maximum area between control joints is 144 ft^2.
- The maximum length or width of an area between control joints is 18 ft.
- The length-to-width ratio of an area between control joints should be between 1 and 2.5.

FIGURE 29.9 Control joints on a stucco facade.

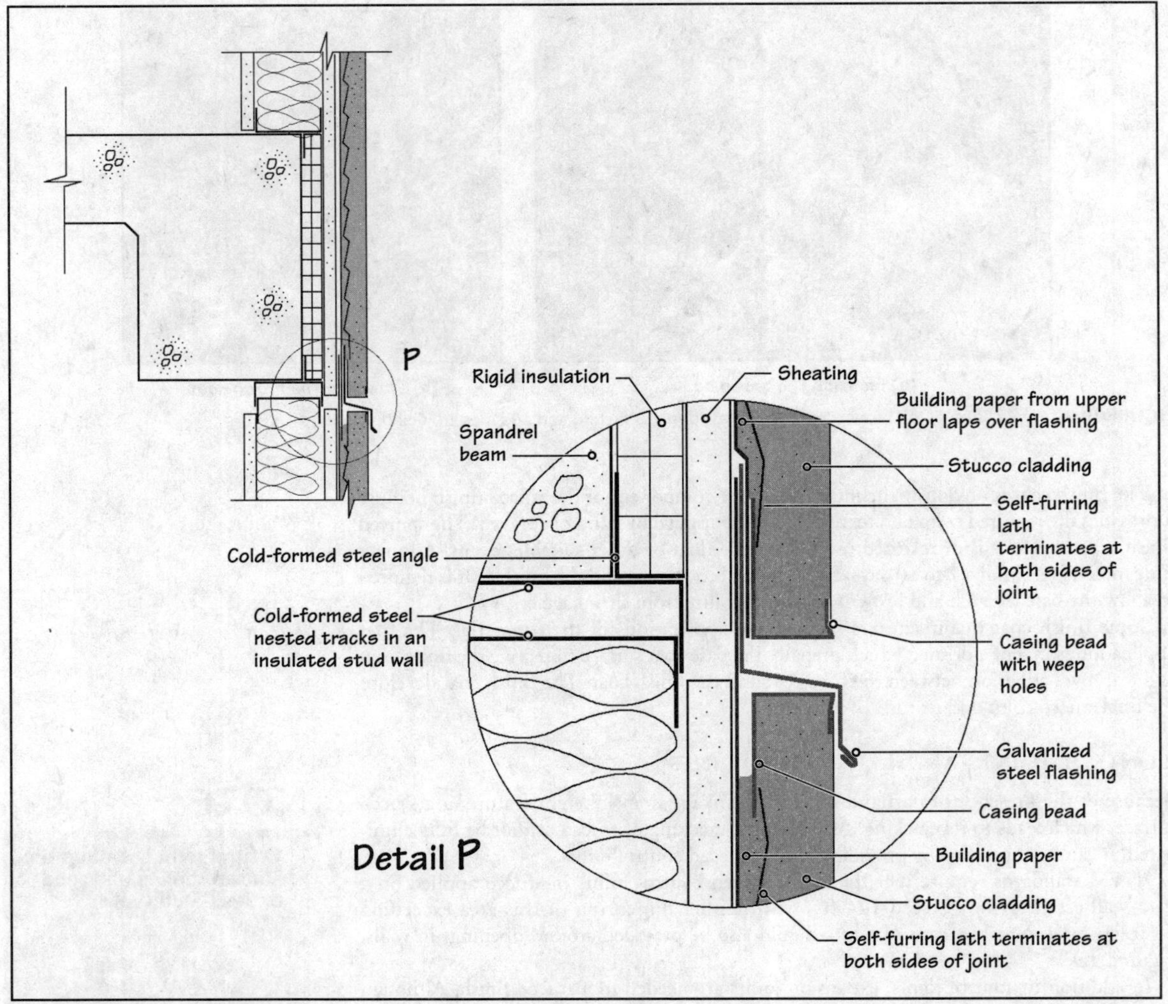

FIGURE 29.10 A typical horizontal expansion joint detail in a stucco-clad exterior stud wall. For a vertical expansion joint, a two-piece expansion joint accessory may be used; see Figure 29.12(c).

If the movement of the spandrel beam is not transferred to the wall, the wall needs only control joints. This is shown in Figure 29.11, which illustrates the detailing of a typical control joint and the termination of stucco at the foundation level.

Control joints and expansion joints are constructed using accessories, Figure 29.12. The flanges of joint accessories consist of expanded metal or a slotted material so that they can integrate with the lath and be fastened to the studs in the same way as the lath.

A control joint accessory consists of a single piece, whereas an expansion joint is a two-piece accessory. In the case of an expansion joint, the lath must terminate on both sides of the joint. The lath should, however, be continuous under a control joint.

Apart from the control and expansion joint accessories, several other accessories and trims, such as casing beads, corner beads, and flashing, are required in a typical stucco project. A casing bead is used at the terminal edge of a stucco surface. Therefore, it is also referred to as a *stucco stop.*

An exterior corner bead provides a straight vertical or horizontal intersection between two surfaces. It also guards against chipping at corners from impact and establishes the thickness of the stucco finish. An interior corner bead is generally not needed with stucco.

The materials used for joint accessories and trims are zinc, galvanized steel, and PVC. In a highly corrosive environment, the use of zinc or PVC should be considered, depending on the local experience. Where accessories abut, the joints at the ends of two lengths or their intersections should be sealed with exterior-grade sealant.

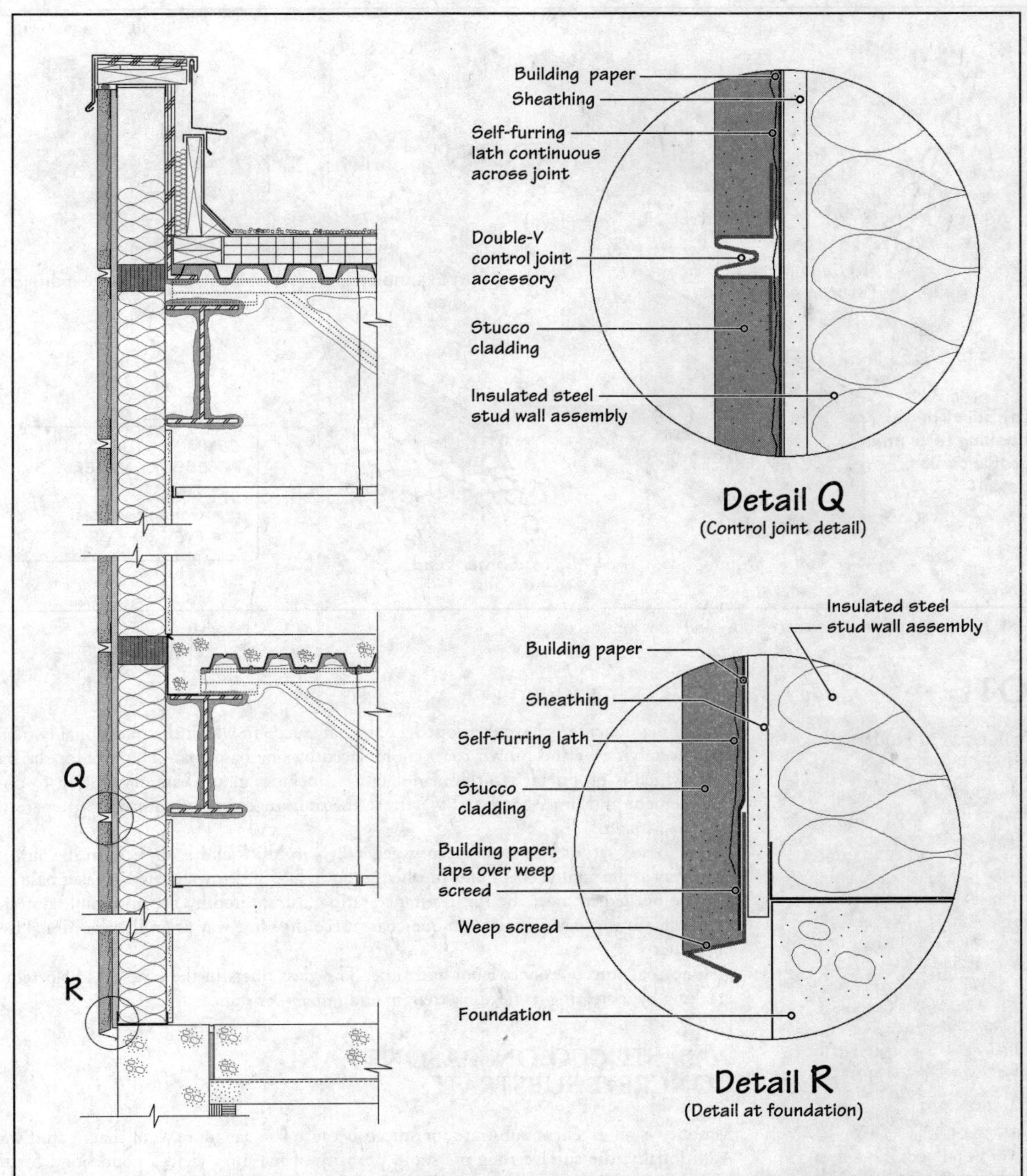

FIGURE 29.11 A typical wall section and corresponding details of a low-rise, stucco-clad steel-frame building.

Rigidity of the Stud Wall Assembly

Because a fully cured stucco surface is relatively thin ($\frac{7}{8}$ in.) and brittle, it is important that the backup wall assembly be sufficiently rigid. A flexible assembly will aggravate cracking, leading to rapid deterioration of the wall from water penetration, freeze-thaw damage, and so on.

Building codes and standards require that the deflection of wood- and steel-stud wall assemblies clad with stucco be controlled to a maximum of span divided by 360 (L/360). More stringent deflection-design criterion (e.g., maximum deflection not to exceed L/420 or even L/480) should be considered for a better-performing stucco wall.

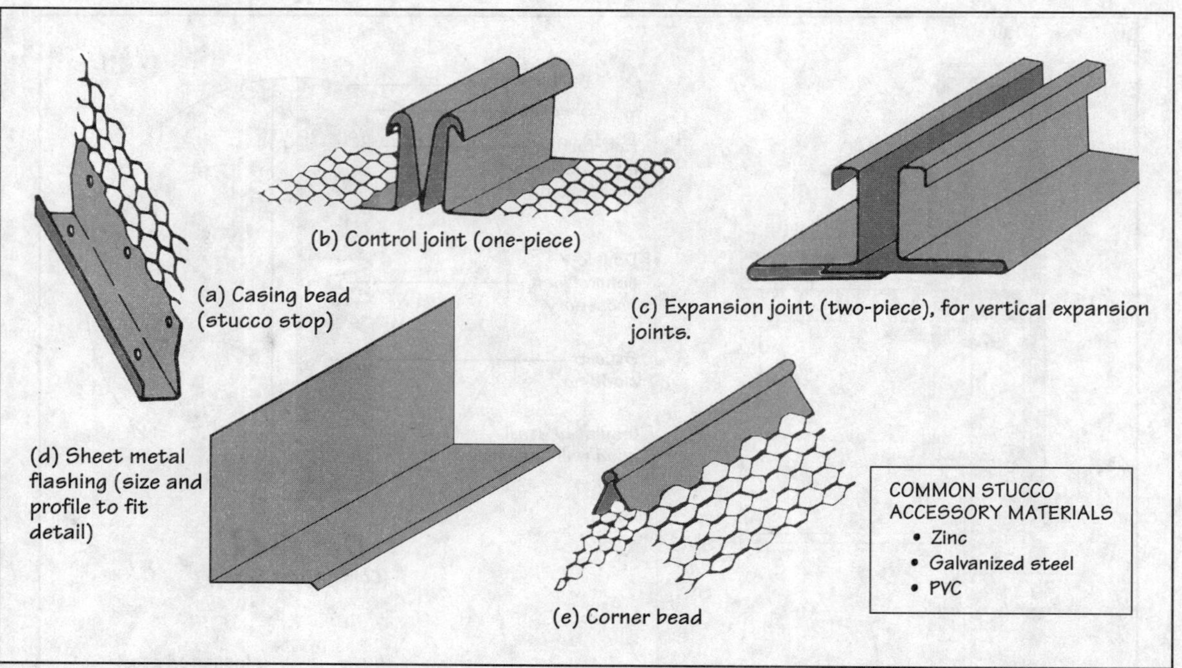

FIGURE 29.12 Commonly used accessories and trims for stucco.

NOTE

Deflection of Stud-Wall Assemblies

The deflection of stud walls that receive stucco finish should not exceed span divided by 360, that is, L/360 caused by lateral loads. Here, L is the vertical span of studs. For steel studs, L is generally the distance between the bottom track and the top track in the assembly.

Steel-stud manufacturers provide tables for the selection of their studs to conform to the deflection criteria. See Section 28.4 for the corresponding deflection criterion for brick veneer backed by steel studs.

For wood studs, L refers to the distance between the bottom plate and the top plate in the wall assembly.

One-Coat Stucco

Several manufacturers have developed a one-coat stucco in which the traditional two base coats (scratch coat and brown coat) are replaced by one base coat. The mix for the base coat, which is proprietary to the manufacturer, includes glass fibers in addition to portland cement and lime. Sand and water are the only ingredients required to prepare the base coat material.

The base coat for one-coat stucco is generally $\frac{1}{2}$ in. thick and is applied on the lath the same way as the scratch coat but is finished smooth, like the brown coat. A thicker base coat may be needed to meet the fire-resistance rating, as required by the applicable building code. The finish coat is applied on one-coat stucco the same way as on the traditional two-coat stucco.

One-coat stucco reduces labor and time. The glass fibers in the base coat help reduce cracking by increasing its flexural strength and impact resistance.

29.3 STUCCO ON MASONRY AND CONCRETE SUBSTRATES

Masonry is an excellent substrate for stucco because it is far more rigid than a stud wall. Additionally, the surface roughness and porosity of masonry yield a good bond for the stucco finish. Therefore, stucco applied on a masonry wall does not need lath. (Remember that a self-furring lath is required on a stud wall to develop a surface to which stucco will bond. The bond between stucco and concrete masonry is particularly strong because both are portland cement–based materials.)

Stucco applied on a masonry wall usually consists of two coats (a base coat to smooth any irregularities on the wall's surface and a finish coat) with a total thickness of $\frac{5}{8}$ to $\frac{3}{4}$ in., Figure 29.13. To retain the natural roughness of masonry, the mortar joints in masonry are left flush, that is, they are not tooled. Obviously, the masonry surface must be clean and free from defects that may compromise the bond between stucco and masonry.

Because masonry is porous, it may suck water from the mix, leaving insufficient water in stucco. Therefore, a masonry surface may need prewetting before applying the base coat.

A concrete wall is neither as absorptive nor as rough as a masonry wall. Therefore, light sand blasting followed by application of a liquid bonding coat is generally required on a concrete wall, as recommended by the stucco manufacturer. The total thickness of stucco on a concrete wall is nearly the same as on a masonry wall.

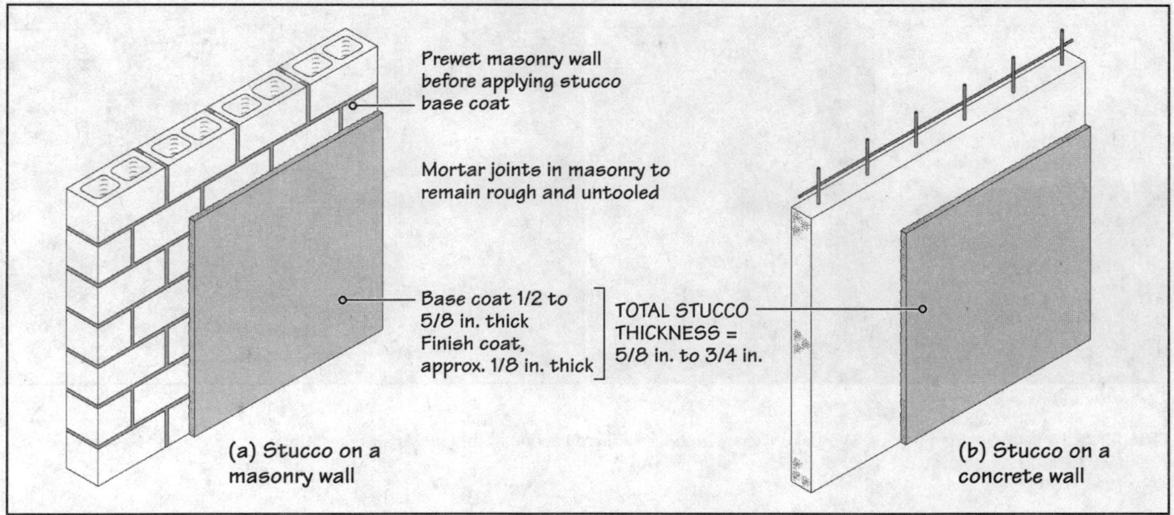

FIGURE 29.13 Anatomy of a stucco-clad **(a)** masonry wall and **(b)** concrete wall.

JOINTS IN STUCCO-FINISHED MASONRY AND CONCRETE WALLS

Because masonry and concrete walls are more rigid than wood- or steel-stud walls, the control joints in stucco applied on them can be spaced farther apart. The recommended maximum area of a stucco panel between control joints on masonry or concrete substrates is 250 ft^2.

As far as possible, control joints and expansion joints in stucco should be in the same locations as the corresponding joints in substrate masonry or concrete. The control joints and other accessories are fastened to masonry or concrete using masonry or concrete nails.

NOTE

Control Joint Locations on Stucco Applied on Masonry or Concrete Walls

- Maximum area between control joints = 250 ft^2.
- Other requirements are the same as for stud walls.

29.4 LIMITATIONS AND ADVANTAGES OF STUCCO

A stucco-clad masonry or concrete wall is a barrier wall (Chapter 27). It does not have any means of draining rainwater out of the assembly should rainwater permeate through the stucco. A stucco-clad steel- or wood-stud framed wall, on the other hand, is not a barrier wall.

Although the anatomy of a stucco-clad framed wall does not qualify it to be called a drainage wall, water penetration tests have indicated that if water intrudes through stucco cladding under a long spell of wind-driven rain, it is able to drain out from over the water-resistant barrier behind stucco (i.e., through the *drainage plane*). However, good workmanship, rigidity of the substrate, and use of control joints, expansion joints, and sealants are essential for a well-performing stucco assembly, particularly in wet climates.

Additionally, because stucco is a portland cement–based material, it is a breathable surface. Should water permeate stucco, it will begin to evaporate as soon as the rain stops and the wall begins to dry. (With an acrylic polymer–based finish coat, the evaporation of water may be slower.)

Although its relative thinness and light weight may present some challenges in keeping a stucco-clad wall dry in wet regions, these properties are advantages in seismic zones. Remember, from Section 3.7, that the earthquake load on a building component is directly related to its dead load. Consequently, a stucco-finished wall is subjected to a smaller earthquake load than a wall clad with heavier finishes, such as masonry veneer or concrete curtain wall.

Additionally, because stucco is an adhered finish (in contrast with anchored masonry veneer), it is better able to resist the vibrations caused by an earthquake. A stucco finish is often recommended for use in relatively dry and seismically active regions.

Another advantage of stucco is the clarity of form that results from its use. The variety of colors that are available add to the aesthetic quality of a stucco-finished building. Architects have exploited this attribute of stucco, as shown in the two images of Figure 29.14.

(a)

(b)

FIGURE 29.14 **(a)** Solana Campus, Westlake, Texas, Architect: Ricard Legoretta. **(b)** Fine Arts Center, College of Santa Fe, New Mexico, Architect: Ricardo Legoretta.

PRACTICE QUIZ

Each question has only one correct answer. Select the choice that best answers the question.

1. In addition to water, a stucco base coat mix consists of
   - **a.** gypsum, lime, and sand.
   - **b.** gypsum, portland cement, and sand.
   - **c.** portland cement, lime, and sand.
   - **d.** lime, glass fibers, and sand.
2. On wood-stud and cold-formed steel-stud wall assemblies, stucco is typically applied in three coats. Beginning from the first to the last, the coats are generally called
   - **a.** brown coat, scratch coat, and finish coat.
   - **b.** scratch coat, brown coat, and finish coat.
   - **c.** prime coat, scratch coat, and finish coat.
   - **d.** prime coat, brown coat, and finish coat.
   - **e.** prime coat, base coat, and finish coat.
3. On wood-stud and cold-formed steel-stud wall assemblies, stucco is typically applied in three coats. The total typical thickness of the three coats is
   - **a.** $1\frac{1}{2}$ in.
   - **b.** $1\frac{1}{4}$ in.
   - **c.** $1\frac{1}{8}$ in.
   - **d.** 1 in.
   - **e.** $\frac{7}{8}$ in.
4. The required color of a stucco surface is
   - **a.** obtained by spray painting the surface with the required color after the finish coat has fully cured.
   - **b.** obtained by spray painting the surface with the required color immediately following application of the finish coat.
   - **c.** obtained by using colored portland cement and lime.
   - **d.** integral to the mix of all coats used in stucco.
   - **e.** integral to the mix of the finish coat.
5. Metal lath is required when stucco is applied on
   - **a.** stud-wall assemblies.
   - **b.** concrete walls.
   - **c.** masonry walls.
   - **d.** all of the above.
   - **e.** none of the above.
6. Metal lath, when used in stucco, is installed on the wall
   - **a.** before the application of the first stucco coat.
   - **b.** sandwiched between the first and second stucco coats.
   - **c.** directly under the finish coat.
   - **d.** between the second and finished stucco coats.
7. Control joints on a stucco surface are provided primarily to
   - **a.** control thermal expansion and contraction of the metal lath.
   - **b.** control thermal expansion and contraction of the stucco surface.
   - **c.** control differential movement between the stucco surface and the substrate.
   - **d.** control drying shrinkage of stucco.
8. Control joints on a stucco surface are formed by using accessories, which consist of
   - **a.** galvanized steel.
   - **b.** zinc.
   - **c.** PVC.
   - **d.** any one of the above.
   - **e.** none of the above.
9. Control joints are required to be more closely spaced on a stucco surface backed by a stud wall than on a stucco surface backed by a masonry wall.
   - **a.** True
   - **b.** False
10. The total thickness of stucco on a concrete or concrete masonry wall is generally
    - **a.** the same as that on a stud-backed wall.
    - **b.** greater than that on a stud-backed wall.
    - **c.** smaller than that on a stud-backed wall.
11. When control joints are provided in a stucco-clad wall, expansion joints are unnecessary.
    - **a.** True
    - **b.** False

29.5 ADHERED MASONRY VENEER

As mentioned in the introduction to Chapter 28, masonry veneer may also be adhered to the backup wall in place of being anchored to it. In an adhered masonry veneer, the masonry units are relatively thin. The most commonly used adhered masonry veneer units consist of artificial (manufactured) stone with an average thickness of about $1\frac{1}{2}$ in., although natural stone and thin brick units are also used. Thin brick units range from $\frac{1}{2}$ to $\frac{3}{4}$ in. thick.

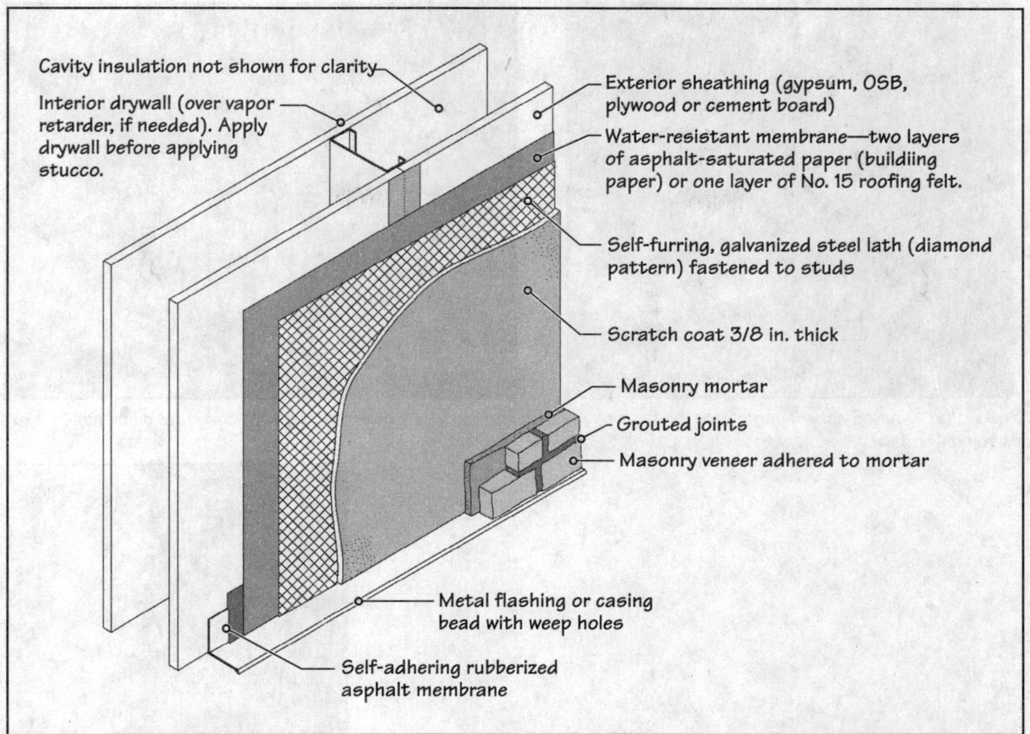

FIGURE 29.15 Anatomy of adhered veneer on a (wood or steel) stud wall.

Apart from its lower cost, a major advantage of manufactured stone veneer is that it is light weight, about 10 psf, compared with anchored brick veneer, which is about 30 psf. Because of its light weight, manufactured stone veneer units adhere to the backup immediately. Additionally, the units are of consistent quality and are available in various shapes, colors, and profiles, including corner units.

Veneer Anatomy and the Construction Process

The construction process and the details of an adhered masonry veneer are similar to those of stucco. Figure 29.15 shows the anatomy of adhered masonry veneer on a stud wall, which is similar to that of Figure 29.2. A $\frac{3}{8}$-in-thick scratch coat on metal furring lath is the backup used for the veneer. After approximately 24 h of completion of the scratch coat, veneer units are adhered to the scratch-coated wall using Type S or Type N masonry mortar (see Chapter 24). Generally, $\frac{3}{8}$-in-thick mortar is adequate, but larger thickness may be needed if the units are irregular.

Although not essential, the addition of an acrylic-polymer bonding agent to the mortar is recommended. The bonding agent reduces the curing time of the mortar and increases its bond strength. Thus, typically, the mason applies the mortar to the back of the unit and presses and taps the mortared unit against the backup at the desired location. The unit adheres to the backup almost instantaneously, Figure 29.16. Some masons call it *lick-and-stick* masonry. The joints between the units are grouted (pointed) with mortar as soon as the bond between the units and the backup is sufficiently strong, generally after 24 h. Control joints and expansion joints in adhered masonry veneer should be treated with a backer rod and a sealant in place of grout. Figure 29.17 shows the wall of Figure 29.16 in the finished state.

Adhered Masonry Veneer with Portland Cement Board Backup

An alternative to the detail shown in Figure 29.15 is that of Figure 29.18. In this detail, a $\frac{1}{2}$-in.-thick portland cement board, applied over exterior sheathing and an air-weather retarder, replaces metal furring lath and the scratch coat. Other details are essentially identical to those

FIGURE 29.16 **(a)** A mason applying mortar to the back of a manufactured stone (corner) unit. **(b)** Masonry unit being adhered to a scratch-coated wall surface.

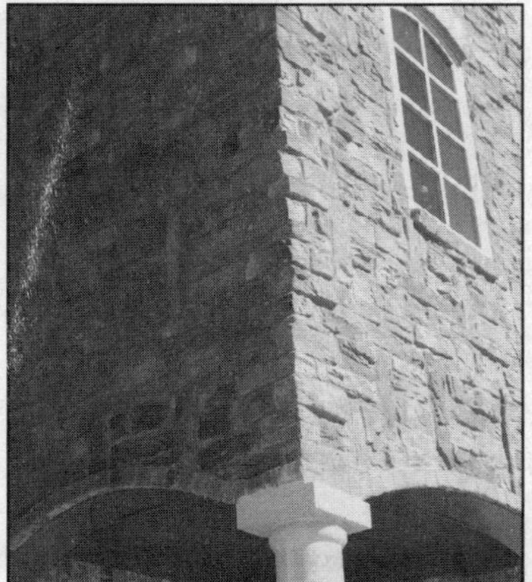

FIGURE 29.17 Adhered stone veneer wall (of Figure 29.16) in the finished state.

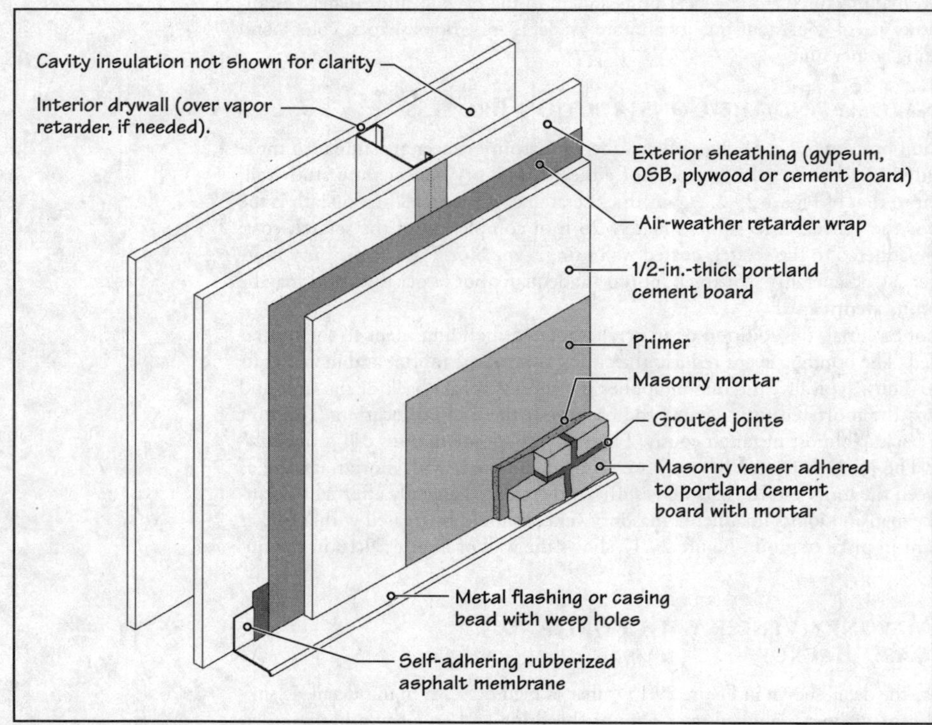

FIGURE 29.18 Anatomy of an adhered masonry veneer using portland cement board as backup in place of a scratch-coated surface.

previously given. Because the cement board surface is not as rough as the scratch coat, manufacturers of this system have developed a surface primer that is applied over the cement board to improve the adhesion between the mortar and the board, which also improves the water resistance of the system.

29.6 EXTERIOR INSULATION AND FINISH SYSTEM (EIFS) BASICS

The exterior insulation and finish system (abbreviated EIFS and generally pronounced as "eefs") consists of a layer of a rigid polystyrene foam insulation, a fiberglass-reinforcing mesh, a polymer-based base coat, and a polymer-based finish coat. The reinforcing mesh is embedded in the base coat.

Because, in its finished appearance, an EIFS-clad facade looks like a stucco facade, it is also referred to as *synthetic stucco* to distinguish it from portland cement stucco—the *conventional stucco*.

The description of the EIFS just given is that of a basic system. More sophisticated (and, hence, more complex) systems are available (see Section 29.8). The important fact to recognize is that an EIFS assembly is a system because it comprises several chemically complex materials. Compatibility between system parts and also among the substrate, trims and joint accessories, flashings, sealants, and so on, must be ensured. Therefore, all materials for the assembly must be obtained from one manufacturer and applied per the manufacturer's instructions. The manufacturer's warranty for the system is generally contingent on the use of its approved material distributor, certified applicator, and recommended construction details.

The EIFS Industry Members Association (EIMA), a trade organization representing EIFS manufacturers, classifies an EIFS into two categories:

- Polymer-based (PB) EIFS, also called *soft-coat* EIFS
- Polymer-modified (PM) EIFS, also called *hard-coat* EIFS

In the PB system, the insulation consists of expanded polystyrene (EPS, i.e., molded and beaded) boards, which are adhered to the substrate, Figure 29.19. The total thickness of the PB EIFS lamina (mesh, base coat, and finish coat) is approximately $\frac{1}{8}$ in.

The PM system uses extruded polystyrene (XPS) or polyisocyanurate (iso) boards, which are anchored to the substrate using steel screws and plastic caps. The base coat on a PM system consists of a polymer-modified portland cement and is at least $\frac{1}{4}$ in. thick.

The PB system is far more commonly used and is the one described here in further detail. Although most manufacturers' systems are similar, there are differences between them. Therefore, the manufacturer's literature should be consulted for more authoritative information.

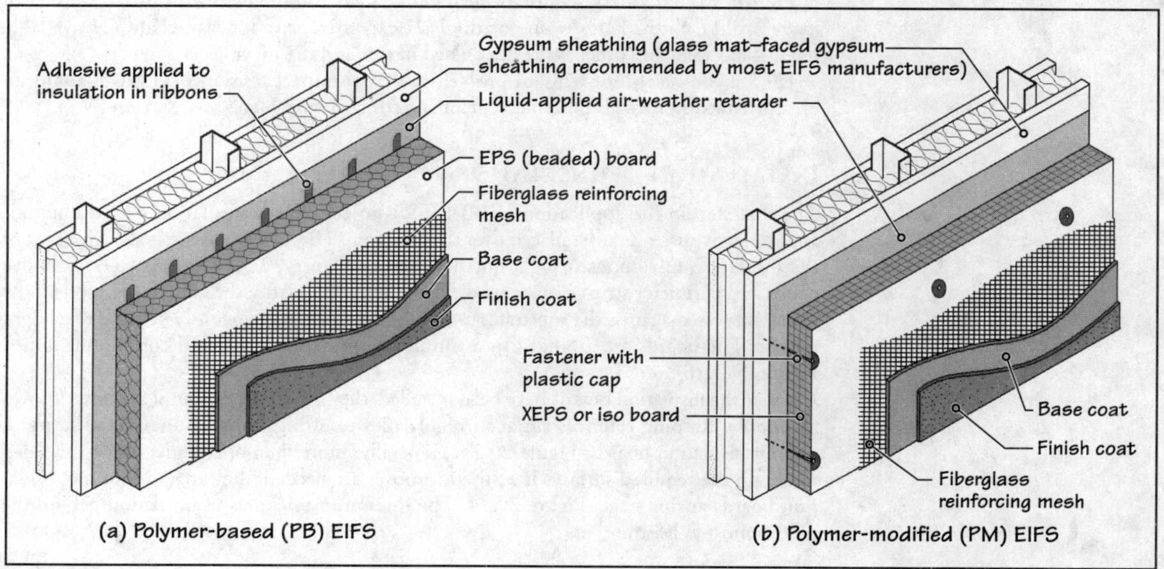

FIGURE 29.19 Difference between PB and PM EIFS.

(a)

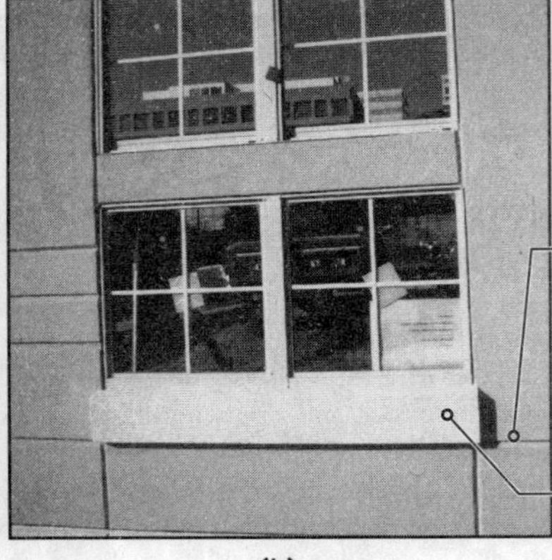

(b)

FIGURE 29.20 **(a)** To make a selection, three facade alternatives (mock-ups) were requested by the architect from the EIFS subcontractor for use in a high-rise condominium project; see also Figure 29.16(b). **(b)** A close-up of one of the mock-ups shown in Figure 29.16(a).

29.7 APPLICATION OF POLYMER-BASED EIFS

EIFS is a versatile cladding and is used for all types of projects—low-rise, mid-rise, and high-rise buildings—in commercial as well as residential projects. It can be used over wood frame, cold-formed steel-stud frame, concrete, or masonry walls.

The use of rigid foam insulation in EIFS has two major advantages. First, it is energy efficient because placing insulation on the exterior of a wall assembly results in a higher effective R-value than placing it within or toward the interior of the assembly (see Section 5.9).

Second, the foam insulation allows a great deal of detail work to be easily incorporated on the facade. The foam can be molded to shape, it can be cut out to provide grooves for surface relief, and its thickness can be varied to give accent bands. The surface detail remains virtually unchanged after the application of the base and finish coats because of their relatively small total thickness.

Third, a large variety of colors are available in an EIFS finish coat. Most EIFS-clad facades are richly detailed and variously colored. Few other cladding materials lend themselves to such ornateness and bright, relatively fade-resistant colors as EIFS. Therefore, a mock-up panel of EIFS cladding is all the more important in most projects, Figure 29.20.

Fourth, EIFS does not require control joints, such as those used with stucco, adding to its versatility. Fifth, EIFS is one of the least expensive exterior wall claddings available today—less than any other system discussed in this and the previous chapter.

EIFS is, however, not without problems. Its low impact resistance and its inability to breathe (in contrast with portland cement stucco) create problems (see Section 29.8).

INSTALLATION OF INSULATION

The first step in the application of PB EIFS is to adhere the insulation over the liquid-applied air-weather retarder placed over the substrate. The adhesive is applied to the back of the insulation in ribbons using a notched trowel, Figure 29.21. (Alternatively, adhesive daubs and perimeter strips may be used, Figure 29.22(a).) After the adhesive is applied, the board is pressed against the substrate, to which it clings instantaneously because of its light weight. The boards are arranged in a running-bond pattern to avoid continuous joints, Figure 29.22(b).

Once the insulation boards have been installed, they are sanded smooth, a process known as *rasping*. Rasping removes surface irregularities resulting from an uneven substrate or uneven insulation boards, Figure 29.23. Generally, more than one sander grade is needed to obtain the required surface. If aesthetic grooves are needed, they are cut into the insulation boards at this stage, Figure 29.24. The minimum thickness of insulation behind the groove must at least be $\frac{3}{4}$ in.

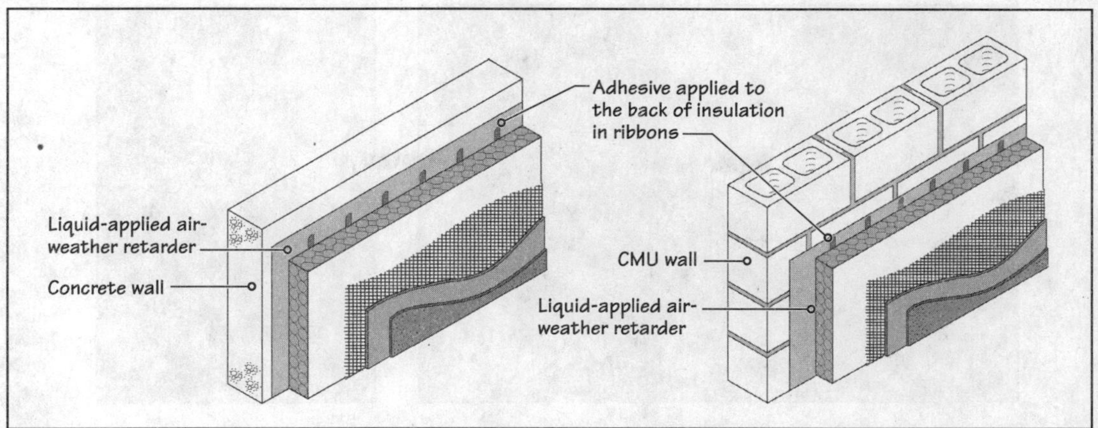

FIGURE 29.21 EIFS on a concrete or CMU wall. Manufacturer-provided adhesive is applied to the back of an insulation board, which is then pressed against the concrete or CMU wall.

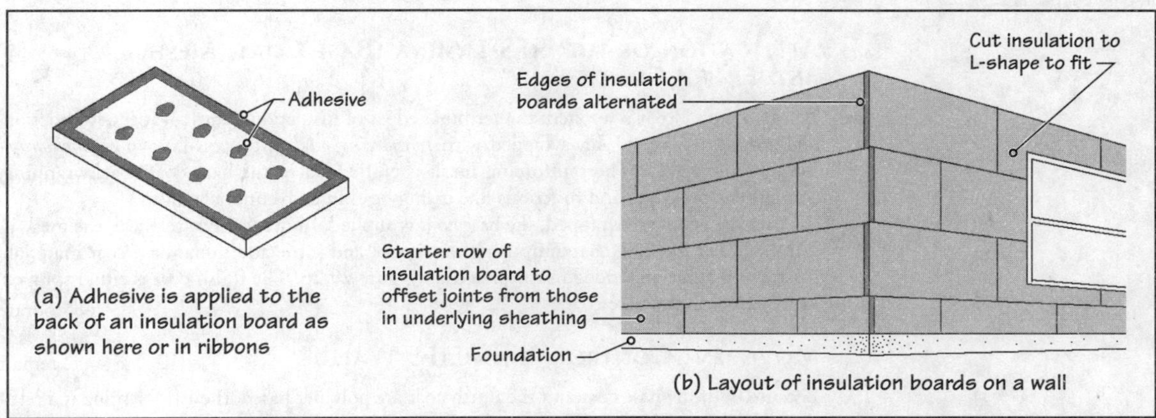

FIGURE 29.22 **(a)** An alternative method of applying adhesive to the back of an insulation board using daubs of adhesive (instead of ribbons of adhesive). **(b)** Layout of insulation boards on a wall.

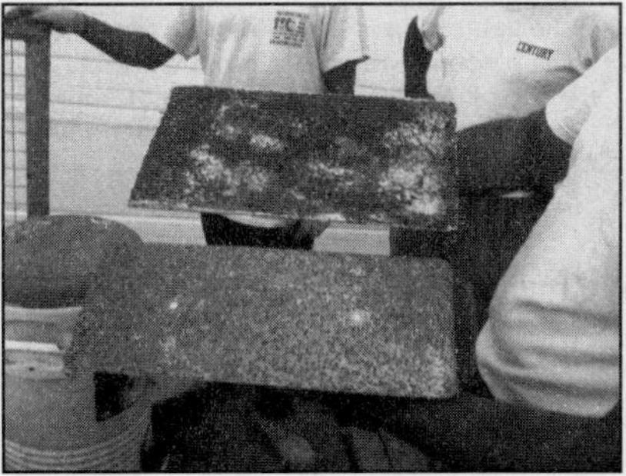

FIGURE 29.23 After the insulation has been adhered to the substrate, it is sanded smooth, a process called *rasping*. The right-hand lower photo shows two rasps with different roughness grades.

(a)

(b)

FIGURE 29.24 **(a)** Cutting of reveals (aesthetic grooves) in the insulation after it has been adhered to the substrate with the help of a groove-making tool and a level. **(b)** A close-up of the groove-making tool that makes a V-shaped groove. The tool can be adapted to make grooves with different profiles.

Application of the EIFS Lamina (Base Coat, Mesh, and Finish Coat)

Before the base coat is applied, the terminal edges of insulation boards—at the foundation and around openings—are wrapped with *reinforcing mesh,* a process known as *backwrapping,* Figure 29.25. The reinforcing mesh is embedded in the base coat. Backwrapping strengthens the edges and functions like casing bead in conventional stucco.

After the edges are wrapped, the base coat is applied. Immediately afterward, the mesh is unrolled over the base coat and, using the trowel and some additional base coat material, the mesh is fully embedded in the base coat, Figure 29.26. The finish coat is either sprayed or troweled on the base coat.

Movement Control in an EIFS Wall

Because both the base coat and the finish coat are polymer based, the EIFS lamina is a relatively flexible membrane. Therefore, an EIFS-clad wall does not require any control joints.

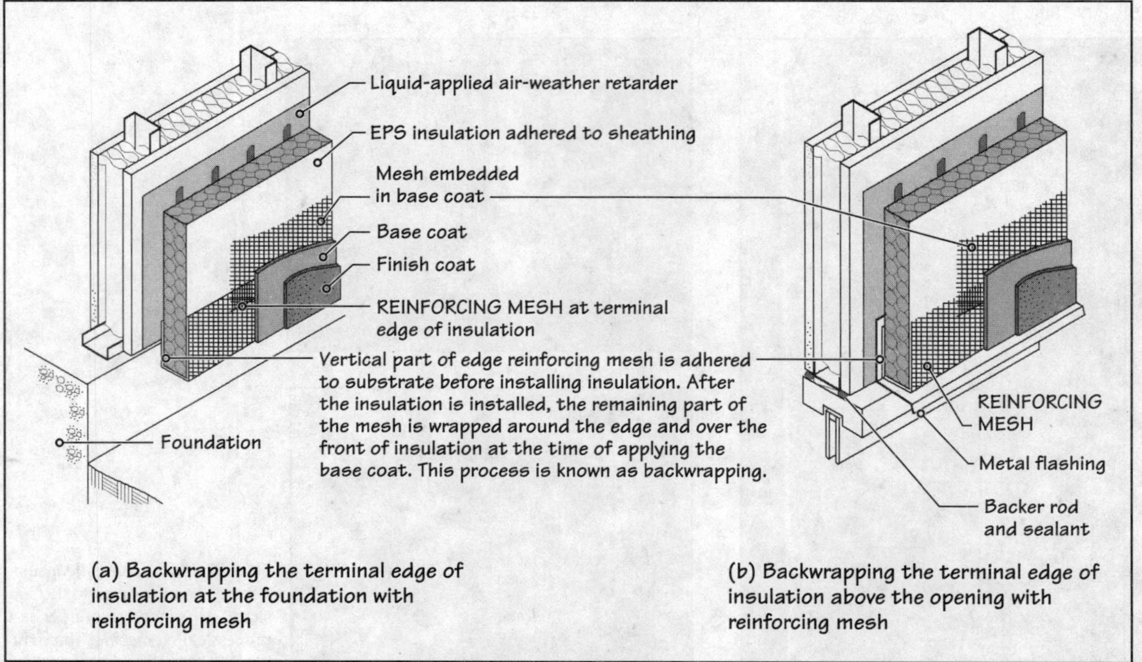
Liquid-applied air-weather retarder
EPS insulation adhered to sheathing
Mesh embedded in base coat
Base coat
Finish coat
REINFORCING MESH at terminal edge of insulation
Vertical part of edge reinforcing mesh is adhered to substrate before installing insulation. After the insulation is installed, the remaining part of the mesh is wrapped around the edge and over the front of insulation at the time of applying the base coat. This process is known as backwrapping.
Foundation
REINFORCING MESH
Metal flashing
Backer rod and sealant
(a) Backwrapping the terminal edge of insulation at the foundation with reinforcing mesh
(b) Backwrapping the terminal edge of insulation above the opening with reinforcing mesh

FIGURE 29.25 Backwrapping of insulation, that is, wrapping the terminal edges of insulation with reinforcing mesh.

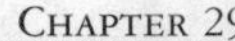

FIGURE 29.26 In applying the base coat, the base coat adhesive is first troweled on the insulation. Then the mesh is unrolled over the adhesive and an additional layer of base coat adhesive is applied. This embeds the mesh fully in the base coat.

Expansion joints are, however, required where large movements in the structure or substrate are expected. These locations are (a) at the spandrel beam level, (b) where an EIFS-clad wall abuts a wall of a different material, (c) at a major change in wall elevation, and (d) at a building expansion joint. Figure 29.27 shows a typical expansion joint detail in an EIFS wall.

29.8 IMPACT-RESISTANT AND DRAINABLE EIFS

As stated earlier, there are two major concerns in the standard EIFS cladding system described in the previous section. The first concern is its low resistance to impact. This is due to the low compressive strength of EPS insulation and the thinness of the EIFS lamina. Damage by hailstorms has been reported on EIFS cladding, particularly on horizontal surfaces, such as aesthetic bands and window sills.

The second concern is that an EIFS-clad wall is a barrier wall. Being polymer based, it does not breathe as well as a portland cement stucco wall. Any water that permeates through an EIFS cladding will generally take a long time to evaporate. This leads to an overall deterioration of the wall assembly and may result in the growth of mold in some walls. Therefore, good workmanship and quality control during application and good detailing with respect to flashing and sealants, particularly around openings and terminations, are

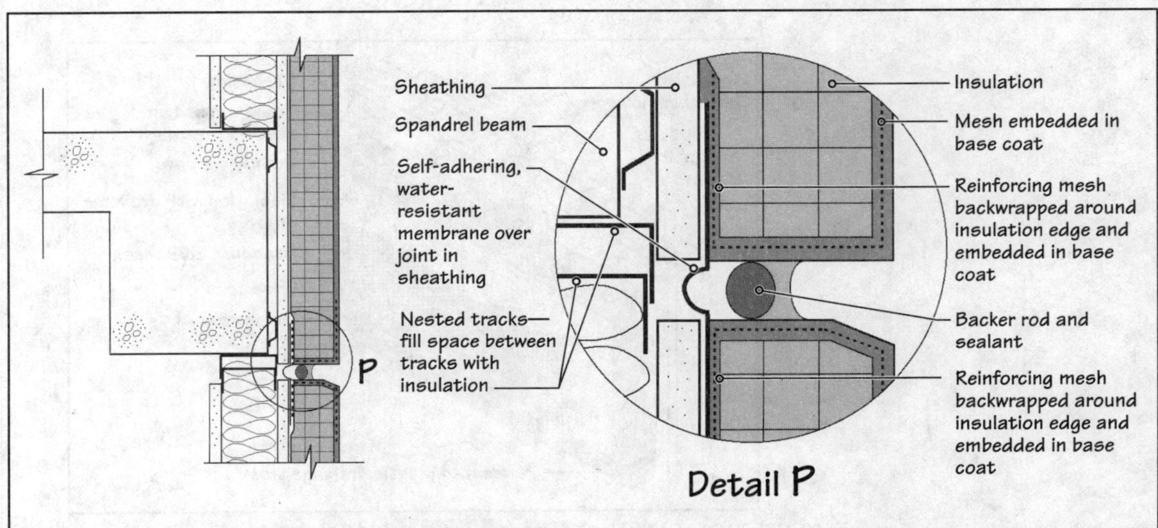

FIGURE 29.27 A typical expansion joint detail in an EIFS-clad wall.

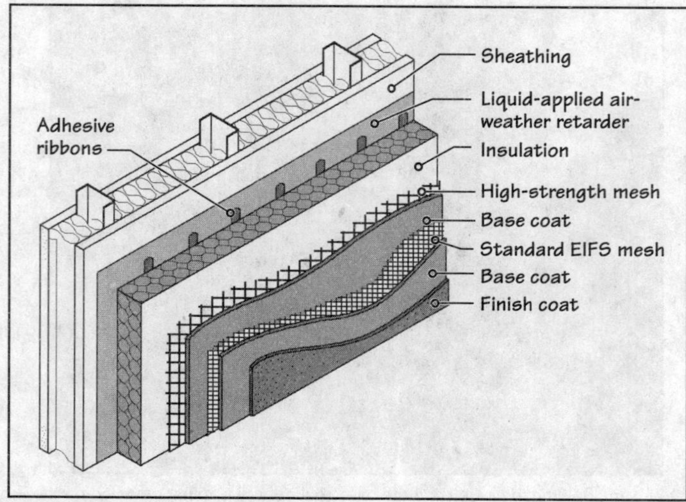

FIGURE 29.28 Anatomy of an impact-resistant EIFS assembly.

extremely important in an EIFS-clad wall. To address these concerns, EIFS manufacturers have introduced an impact-resistant EIFS and a drainable EIFS.

IMPACT-RESISTANT EIFS

An impact-resistant EIFS consists of two layers of base coat. The first layer is embedded in a thicker reinforcing mesh, and the second layer has the same mesh as the standard EIFS, Figure 29.28. Although the impact-resistant EIFS has higher compressive strength than the standard EIFS, it is not equivalent to the compressive strength of most other cladding systems. Therefore, in many projects, the first few floors of the building are clad with a high-compressive-strength cladding, such as masonry veneer, concrete, or stucco, and the higher floors are clad with EIFS.

DRAINABLE EIFS

Two types of drainable EIFS wall systems are available. The system with insulation adhered to the substrate has vertical drainage grooves at the back of the insulation, Figure 29.29.

The other system, in which the insulation is mechanically anchored to the substrate, has a thin drainage mat and an air-weather retarder behind the insulation, Figure 29.30. The drainage mat is typically made of interwoven plastic strands. As an alternative to the mat and building paper, one manufacturer uses a layer of synthetic material with drainage lines.

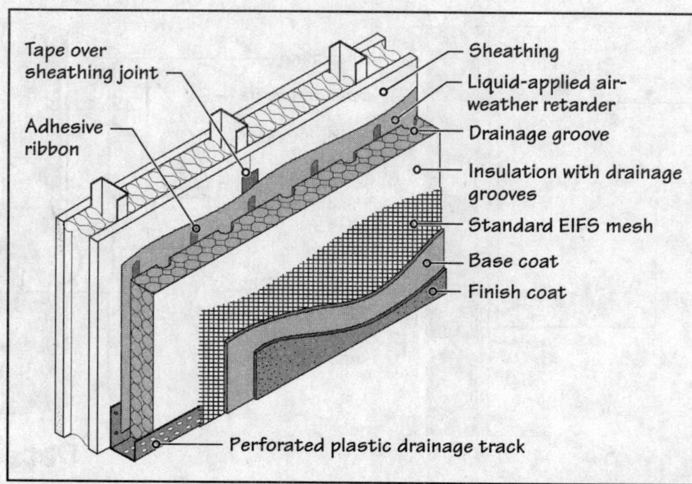

FIGURE 29.29 Drainable EIFS system in which the insulation is adhered to the substrate and the drainage is provided by vertical grooves in the insulation.

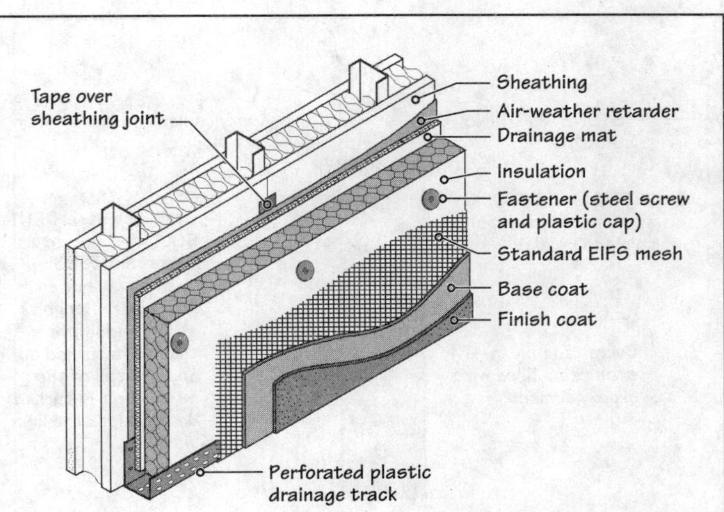

FIGURE 29.30 Drainable EIFS system in which the insulation is mechanically fastened to the wall framing members and the drainage is provided by a drainage mat made of interwoven plastic strands.

Both systems require a perforated plastic track at the bottom to allow water to weep at the base. (The bottom track is similar to a casing bead in a portland cement stucco wall.) In a multistory building, the tracks are provided at each floor level.

PRACTICE QUIZ

Each question has only one correct answer. Select the choice that best answers the question.

12. In an adhered masonry veneer, the most commonly used masonry unit is
- **a.** thin bricks.
- **b.** regular bricks.
- **c.** concrete masonry units (CMUs).
- **d.** natural stone.
- **e.** manufactured stone.

13. In an adhered masonry veneer, masonry units are adhered to the backup surface using
- **a.** masonry mortar.
- **b.** epoxy resin.
- **c.** portland cement slurry.
- **d.** a polymer-acrylic bonding agent.
- **e.** any one of the above.

14. Masonry units in an adhered masonry veneer on a stud wall are applied directly on
- **a.** gypsum sheathing.
- **b.** gypsum sheathing treated with a liquid-applied air-weather retarder.
- **c.** a stucco scratch coat over metal furring lath.
- **d.** a stucco brown coat.
- **e.** any one of the above.

15. Masonry units in an adhered masonry veneer
- **a.** require pointing of joints before adhering the units.
- **b.** require pointing of joints after adhering the units.
- **c.** require pointing of joints while the units are being adhered.
- **d.** do not require pointing of joints.

16. The term *EIFS* is an acronym for
- **a.** exterior insulation and finished stucco.
- **b.** exterior insulation and finish system.
- **c.** externally insulated finish system.
- **d.** envelope insulation with finished stucco.

17. EIFS is classified into two categories: polymer-based (PB) EIFS and polymer-modified (PM) EIFS. Which of these two is more commonly used?
- **a.** PB EIFS
- **b.** PM EIFS

18. In PB EIFS, the insulation commonly used is
- **a.** fiberglass.
- **b.** polyisocyanurate.
- **c.** extruded polystyrene.
- **d.** expanded polystyrene.
- **e.** none of the above.

19. In PB EIFS, the EIFS lamina consists of
- **a.** base coat, insulation, and finish coat.
- **b.** base coat, mesh, and finish coat.
- **c.** insulation, mesh, and finish coat.
- **d.** insulation, lath, base coat, and finish coat.

20. The thickness of the lamina in PB EIFS is approximately
- **a.** 1 in.
- **b.** $\frac{1}{2}$ in.
- **c.** $\frac{1}{4}$ in.
- **d.** $\frac{1}{8}$ in.
- **e.** $\frac{1}{16}$ in.

21. Backwrapping in an EIFS-clad wall refers to
- **a.** wrapping the edges of the openings in the wall with a water-resistant tape.
- **b.** wrapping the terminal edges of EIFS insulation with EIFS mesh.
- **c.** wrapping the joints between EIFS insulation with EIFS mesh.
- **d.** wrapping the joints between EIFS insulation with a water-resistant tape.

22. The requirements for control joints in an EIFS wall are
- **a.** generally the same as those in a stucco-clad wall.
- **b.** more stringent than those in a stucco-clad wall.
- **c.** less stringent than those in a stucco-clad wall.
- **d.** none; control joints are not required in an EIFS-clad wall.

23. Impact-resistant EIFS generally has
- **a.** one base coat and one finish coat.
- **b.** one base coat and two finish coats.
- **c.** two base coats and two finish coats.
- **d.** two base coats and one finish coat.
- **e.** three base coats and one finish coat.

24. Expansion joints are required in an EIFS-clad wall.
- **a.** True
- **b.** False

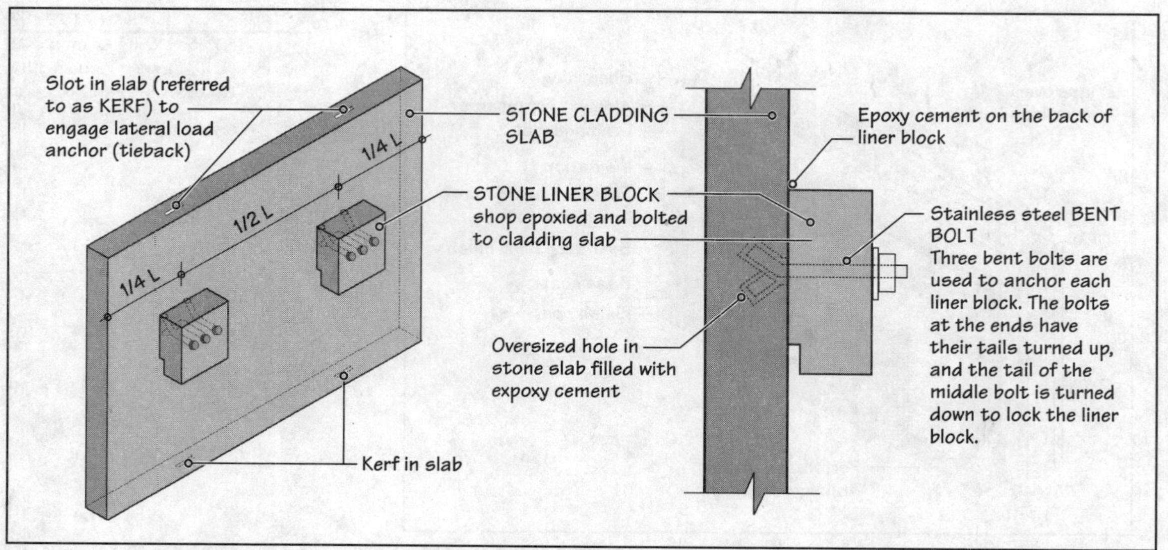

FIGURE 29.31 A stone cladding slab with liner blocks at quarter points.

29.9 EXTERIOR CLADDING WITH DIMENSION STONE

Granite, marble, and limestone are the three stones commonly used for exterior cladding. The minimum recommended thickness for exterior granite cladding slabs is $1\frac{1}{4}$ in. (3 cm) with a panel size of 20 ft^2 or less. The corresponding thickness for marble and limestone is 2 in. (5 cm). Greater thickness is used for larger slab sizes or for greater durability.

Stone cladding can either be field installed, slab by slab, at the construction site or prefabricated into curtain wall panels. Field installation can be done by one of the following two methods:

- Standard-set installation
- Vertical channel support installation

NOTE

Quarter-Point Supports

Quarter-point dead-load supports for slabs (as shown in Figure 29.31) give uniform center-to-center spacing between supports and also produce least bending stresses in slabs.

29.10 FIELD INSTALLATION OF STONE—STANDARD-SET METHOD

In the standard-set method, each stone slab is directly anchored to the backup wall with its own dead-load and lateral-load supports. Two dead-load supports are required for each slab, which are provided by stone liner blocks. The liner blocks are bolted to the slab at a stone-fabrication plant with stainless steel bent bolts set in epoxy resin (typically at quarter points), Figure 29.31.

In installing stone slabs, each liner block is made to bear on a J-shaped shelf angle clip that is anchored to a CMU or reinforced-concrete backup wall, Figures 29.32 and 29.33. A setting pad that functions as a cushion and a shim is typically used under each liner block.

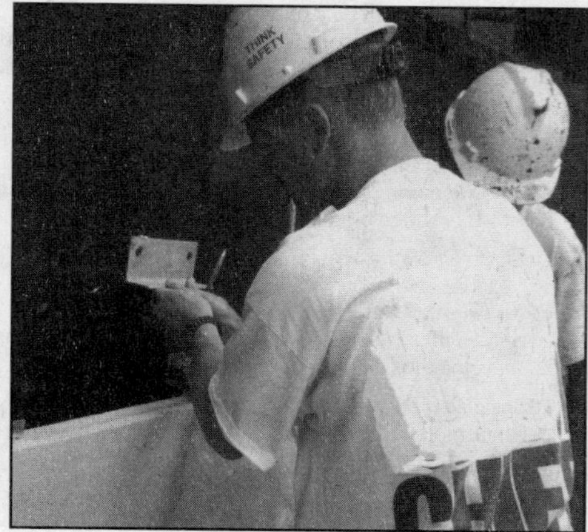

FIGURE 29.32 Anchorage of a J-shaped shelf angle clip to a CMU backup wall whose surface has been treated with a liquid-applied air-weather retarder.

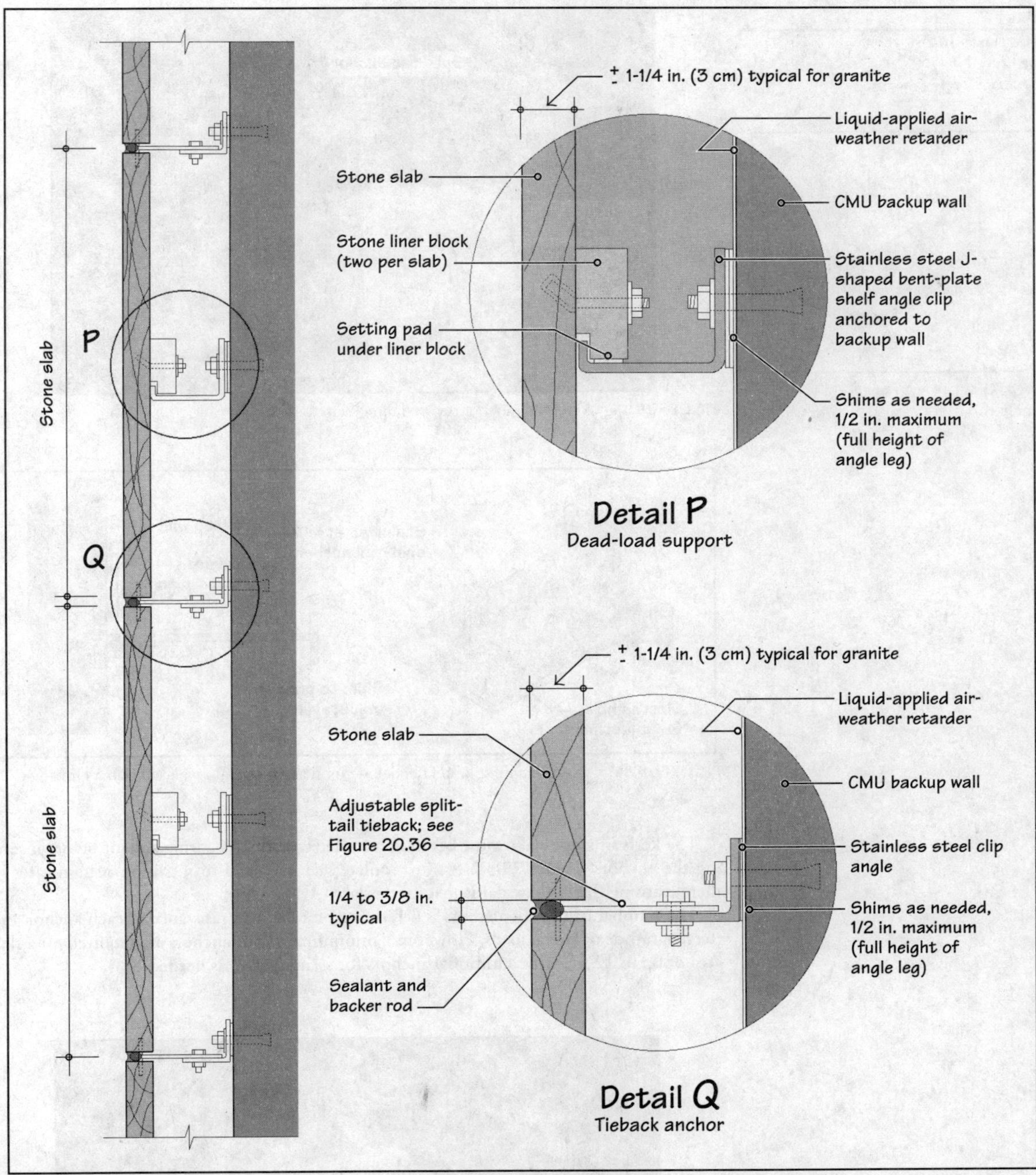

FIGURE 29.33 A section through stone cladding with concrete masonry backup wall using the standard-set method.

Tieback anchors provide lateral-load supports and consist of split-tail anchors, whose tails engage in the slots (kerfs) provided in the cladding slabs, Figure 29.34. Split-tail anchors may be secured to the backup wall directly, as shown in Figure 29.35. Alternatively, they may be secured through a clip angle anchored to the backup wall, Figure 29.36. Securing the split-tail anchors through an angle provides greater adjustability in their location.

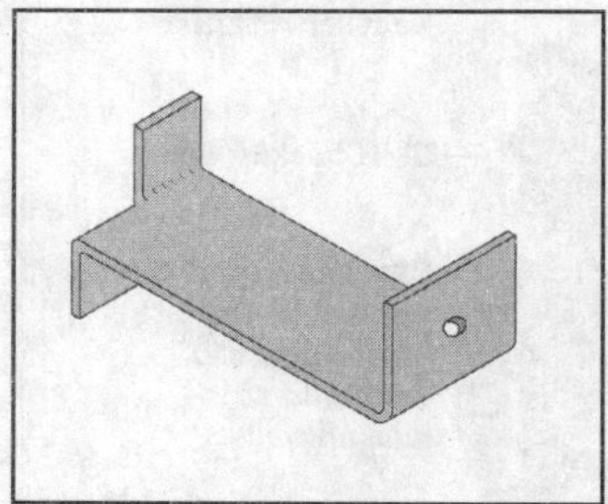

FIGURE 29.34 A typical split-tail anchor.

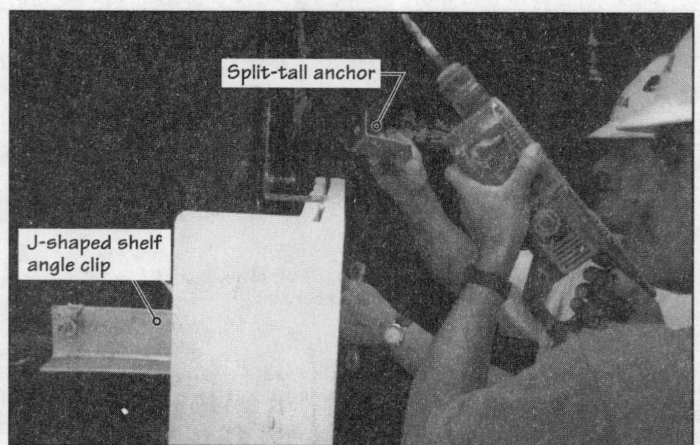

FIGURE 29.35 A split-tail anchor secured directly to the backup wall.

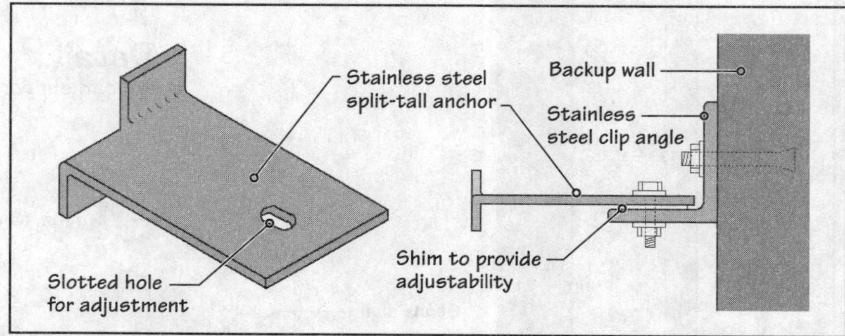

FIGURE 29.36 An adjustable split-tail tieback secured to the backup wall through a clip angle.

The kerfs in stone slabs must be filled completely with fast-curing sealant before inserting the anchors, Figure 29.37. An incomplete seal may lead to intrusion of rainwater in them, causing freeze-thaw damage to stone slabs.

The number of tieback anchors is determined by the load resistance of each anchor and the magnitude of lateral loads. However, a minimum of four anchors are required for a slab area of up to 12 ft^2, with additional anchors for a larger area, as needed.

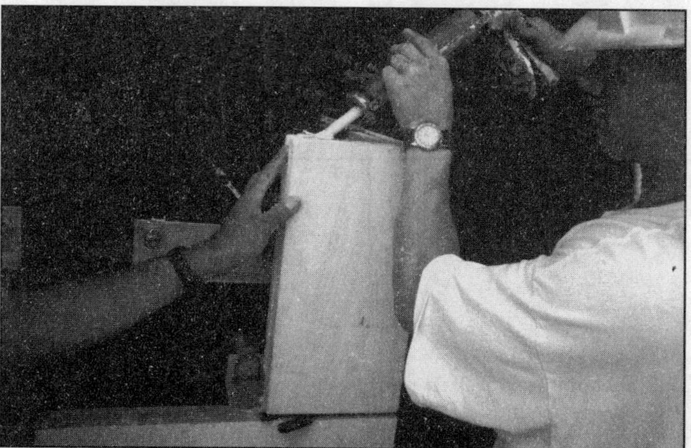

FIGURE 29.37 Applying sealant in kerfs of a stone slab before inserting tiebacks.

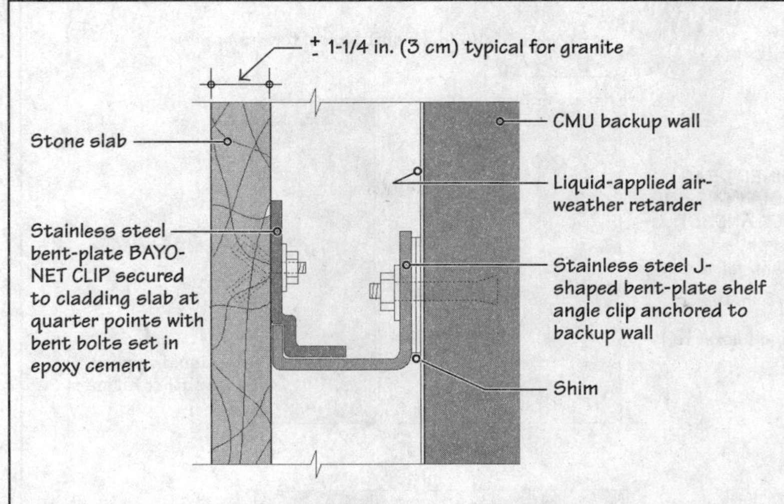

FIGURE 29.38 Bayonet anchor clips as an alternative to liner blocks for dead-load supports.

Bent-Plate Clips as an Alternative to Stone Liner Blocks

A commonly used substitute for a stone liner block is a stainless steel bent-plate clip, referred to as a *bayonet clip,* Figure 29.38. Slabs with bayonet clips have less stone and, therefore, a smaller dead load than slabs with liner blocks. Bayonet clips (two per slab) are bolted and epoxied to the cladding slab in the same way as liner blocks.

Dead-Load Support and Tieback Anchor in One Clip Anchor

The dead-load supports and tieback anchors for stone slabs can be combined in one clip anchor, which is fabricated from two stainless steel members—a bent plate and a flat plate, Figure 29.39. The use of one anchor substantially facilitates the installation of cladding. At locations where flashing is required, dead-load supports and tieback clip anchors need to be separated, as shown in Detail Q of Figure 29.40.

With rigid foam insulation provided between the cladding and the CMU backup wall, the insulation must be cut around dead-load supports and tieback anchors, Figure 29.41. In such a case, the anchorage clips may be required to be heavier or braced.

The details shown previously can be modified if the backup wall consists of a cold-formed steel-stud wall instead of a CMU backup wall, Figure 29.42.

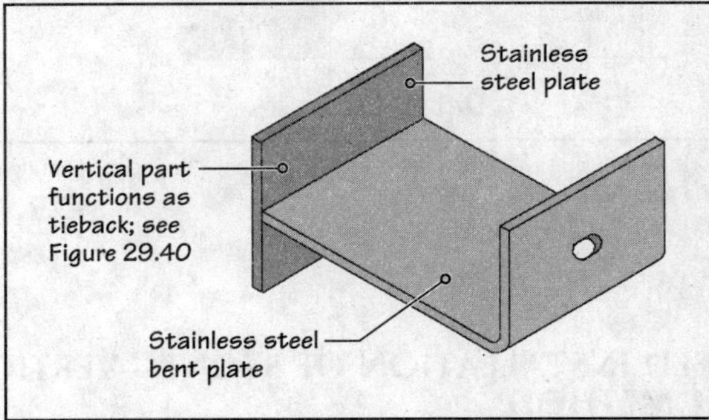

FIGURE 29.39 A combined dead-load support and tieback anchor formed by welding a stainless steel (flat) plate to a stainless steel bent plate.

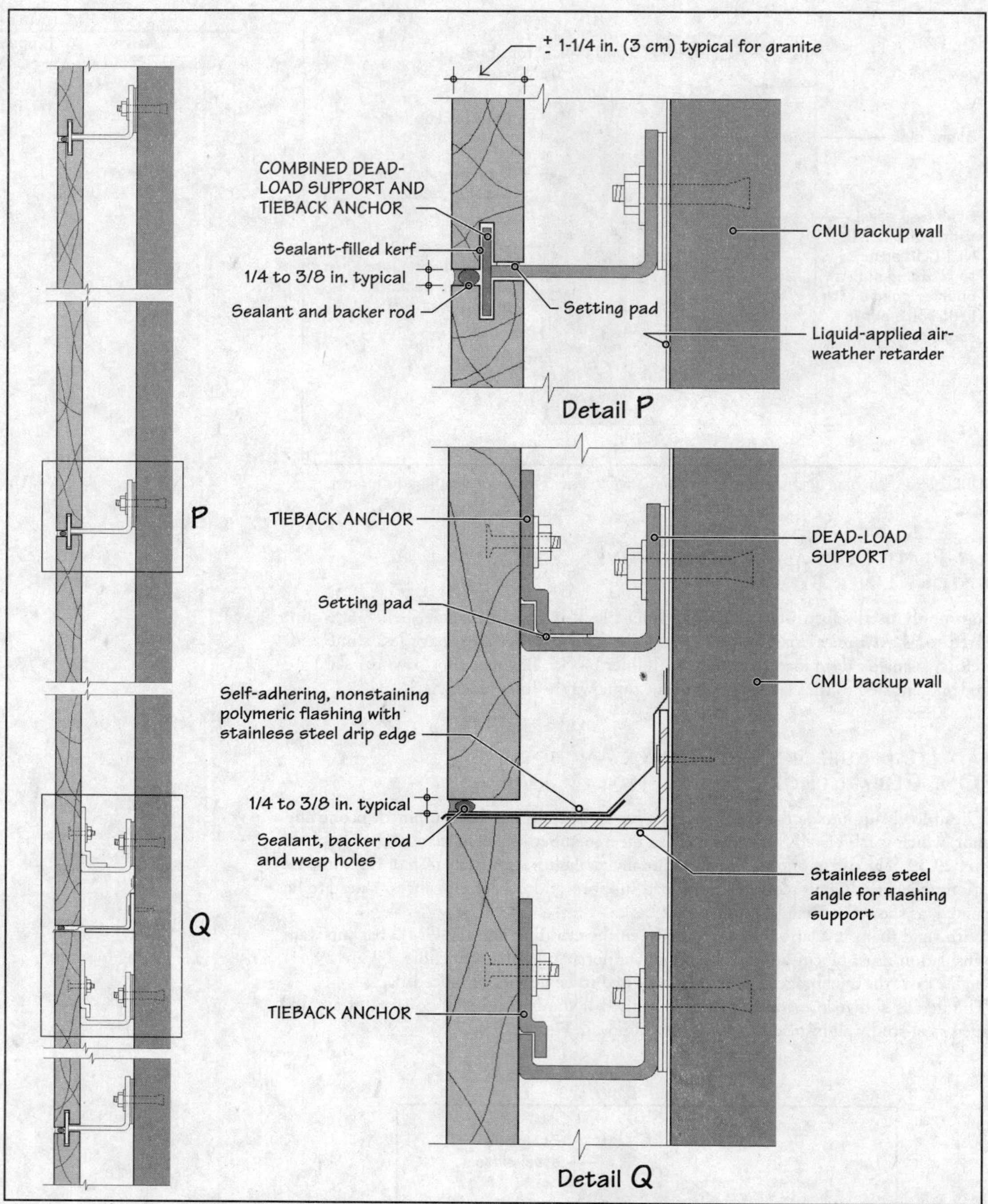

FIGURE 29.40 A typical section through a stone-clad exterior wall using combined dead-load and tieback anchors. Separate dead-load supports and tiebacks have been used at locations of flashing and weep holes.

29.11 FIELD INSTALLATION OF STONE—VERTICAL CHANNEL METHOD

The anchorage of stone slabs is substantially simplified by using continuous vertical support channels, Figure 29.43. The manufacturers of channels provide various accessories to anchor the channels to the backup wall.

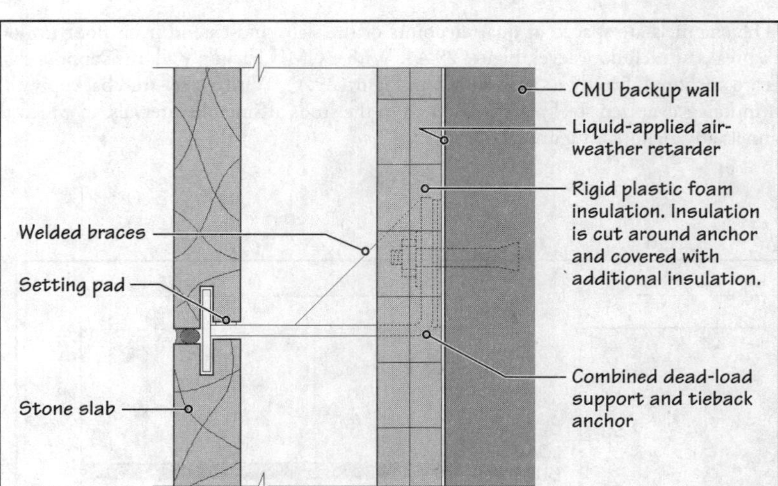

FIGURE 29.41 An alternative to Detail P in Figure 29.40.

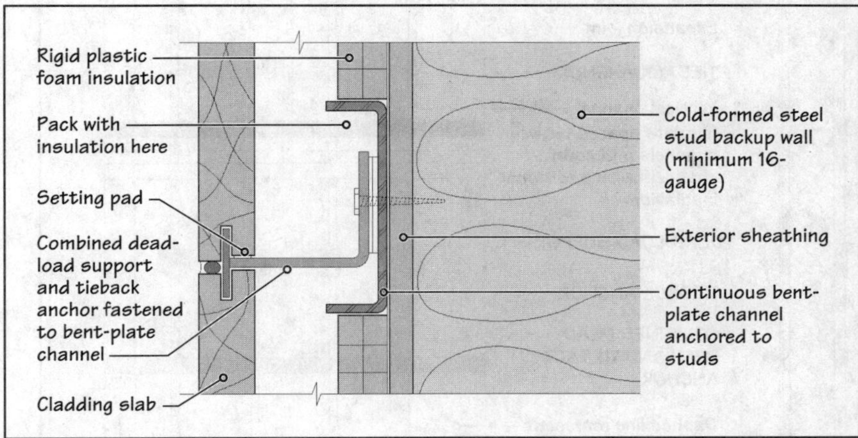

FIGURE 29.42 A typical stone cladding detail with a steel-stud backup wall.

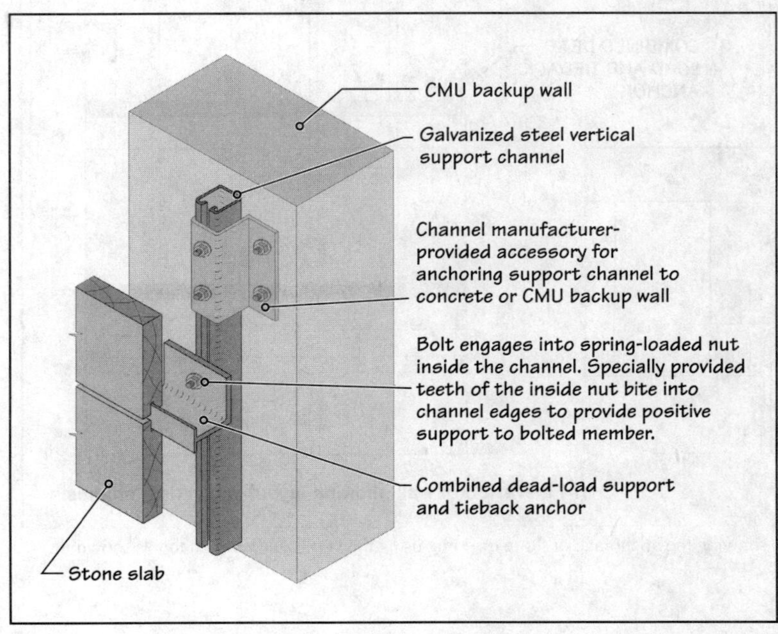

FIGURE 29.43 Adjustable anchorage of a vertical channel to a concrete or CMU backup wall; see also Figure 29.44.

The channels are spaced at quarter points of the slabs and extend from floor to floor, with breaks at each floor level, Figure 29.44. With a CMU backup wall, the support channels are anchored directly to it, as shown in Figure 29.43. With a steel-stud backup wall, a continuous galvanized steel plate is fastened to the studs at suitable intervals, to which the channels are anchored, Figure 29.45.

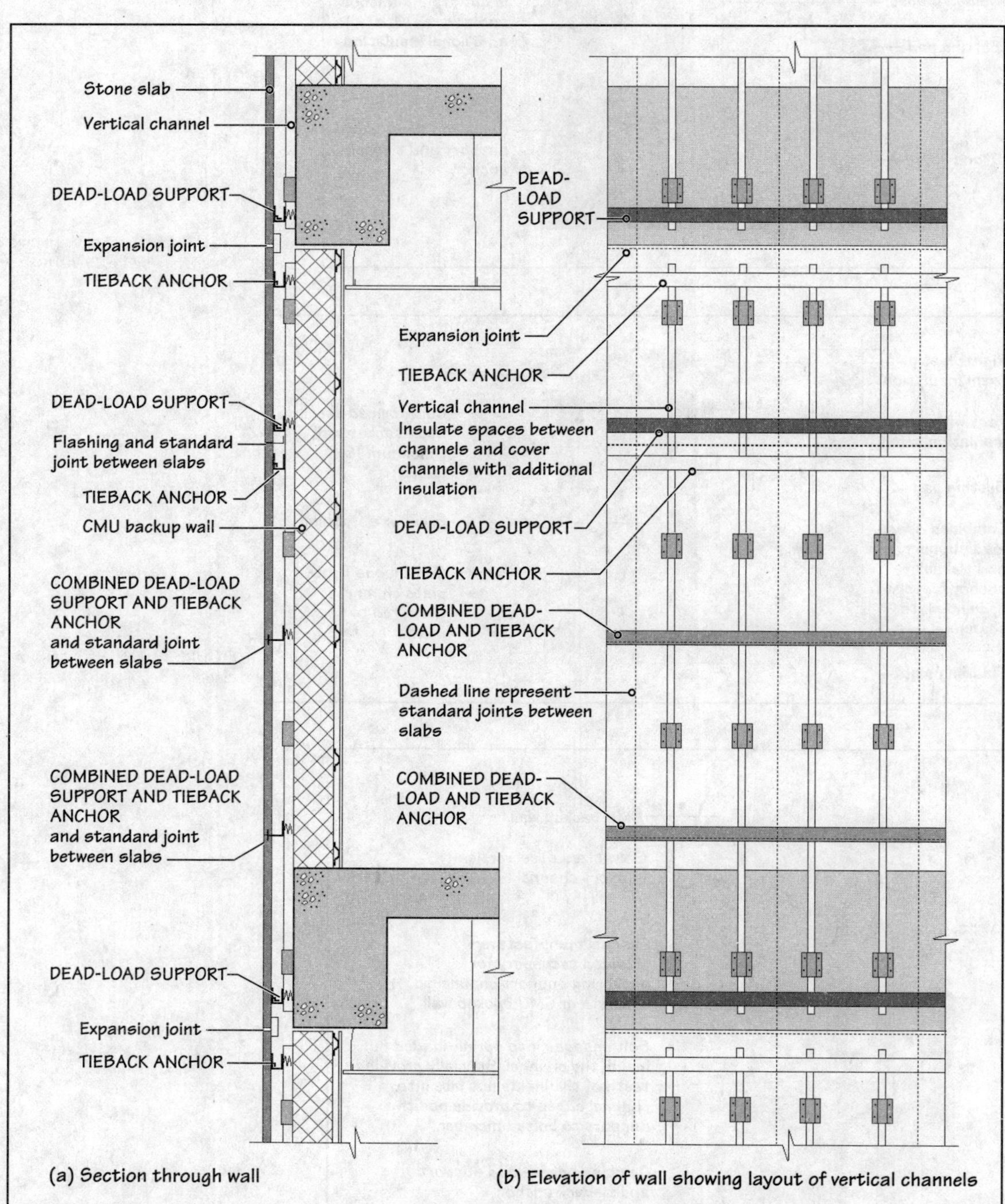

FIGURE 29.44 A typical section and elevation showing the anchorage of stone cladding using the vertical support channels shown in Figure 29.43.

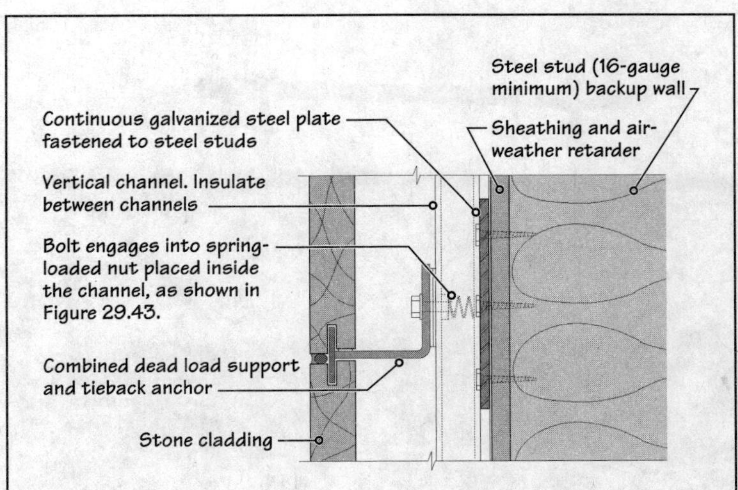

FIGURE 29.45 Anchorage of a vertical channel to a steel-stud backup wall.

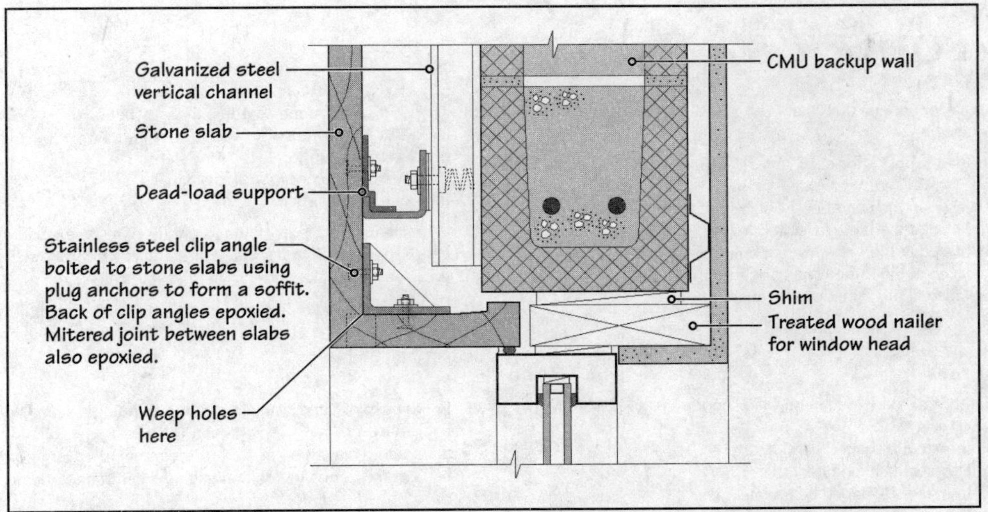

FIGURE 29.46 One of the several ways of detailing a stone-clad wall to create a return (soffit) at a window head.

Where a short length of stone return (e.g., at a soffit) is required, it is obtained through the use of stainless steel clip angles bolted to slabs, Figure 29.46.

29.12 PREFABRICATED STONE CURTAIN WALLS

Instead of installing stones slab by slab to a backup wall, they can be anchored to a steel truss frame, generally at the construction site. The stone-frame assembly forms a panel, which is lifted into position by a crane and hung from the building's structure like any other curtain wall panel (precast concrete or GFRC panel).

Generally, each panel extends from column to column and is supported on them. The panelized system is preferred for use in locations where labor costs are high, unfavorable weather conditions exist, or the site is unsuitable for the construction of scaffolding. Figure 29.47 shows a stone cladding panel.

FIGURE 29.47 A prefabricated stone curtain wall panel. (Photo courtesy of Dee Brown, Inc.)

PRACTICE QUIZ

Each question has only one correct answer. Select the choice that best answers the question.

25. The dead-load supports for each stone slab (in a stone-clad wall) are generally
    a. two, placed in the center of the slab, one above the other.
    b. two, placed at the same level at one-third points.
    c. three, placed at the same level at one-third points.
    d. two, placed at the same level at quarter points.
    e. four, placed at the same level at quarter points.

26. A bayonet clip in the stone cladding industry refers to
    a. dead-load support.
    b. lateral-load support.
    c. flashing retainer.
    d. none of the above.

27. A combined dead-load and lateral-load support anchor for stone cladding consists of
    a. a stainless steel L-shaped bent plate.
    b. a stainless steel J-shaped bent plate.
    c. a stainless steel H-shaped bent plate.
    d. a stainless steel L-shaped bent plate welded to a stainless steel flat plate.
    e. none of the above.

28. The term *kerf* in stone cladding industry refers to
    a. a slot in a stone slab.
    b. dead-load support for a stone slab.
    c. a tieback anchor for a stone slab.
    d. none of the above.

29. When vertical support channels are used in stone cladding, dead-load supports and tiebacks are not required for stone slabs.
    a. True
    b. False

30. The thickness of granite used in wall cladding is generally
    a. $1\frac{1}{4}$ in.
    b. $1\frac{1}{2}$ in.
    c. 2 in.
    d. $2\frac{1}{4}$ in.
    e. none of the above.

31. In the prefabricated, panelized stone cladding system, stone slabs are backed by
    a. a frame consisting of cold-formed steel members.
    b. a frame consisting of laminated veneer lumber members.
    c. a ribbed steel deck.
    d. a truss consisting of structural steel members.
    e. any one of the above, depending on the building.

29.13 THIN STONE CLADDING

Another form of panelized stone cladding uses an extremely thin (nearly $\frac{1}{4}$ in. thick) stone veneer bonded to an aluminum honeycomb backing. The panels are manufactured by epoxy-cementing aluminum honeycomb on both sides of an approximately $\frac{3}{4}$-in.-thick stone slab (generally granite or marble). The honeycomb-stone-honeycomb combination is sawn through the middle, producing two identical panels, Figure 29.48.

After sawing, the stone facing on each panel is finished as needed (e.g., polished, honed, abrasive-blasted, bush-hammered, etc.). The honeycomb backing is $\frac{3}{4}$ in. thick, giving an overall finished panel thickness of approximately 1 in. (For a panel intended for interior use, e.g., for lining the walls of elevator lobbies, foyers, and ceilings, the honeycomb backing is only $\frac{3}{8}$ in. thick.)

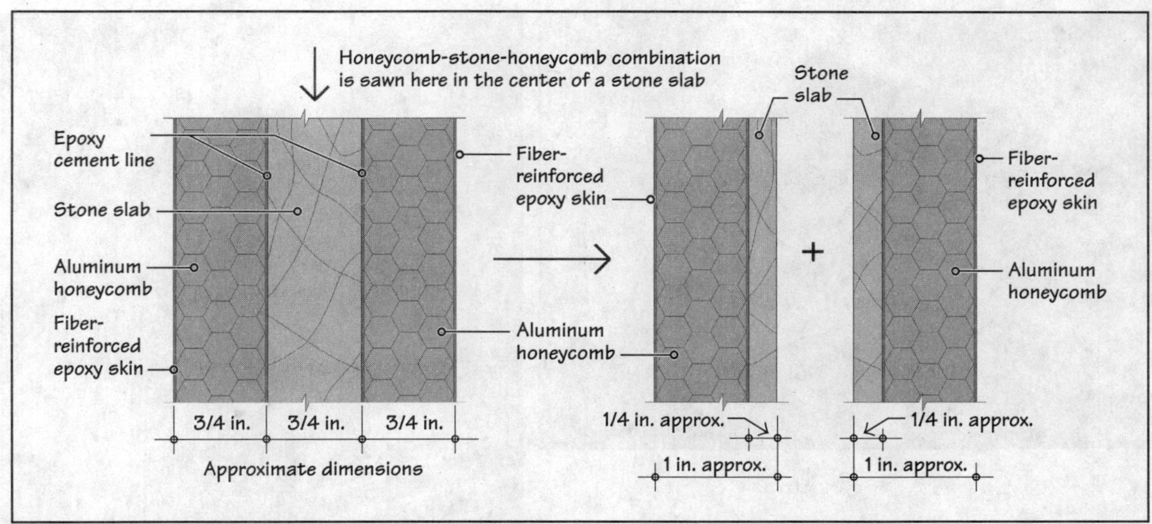

FIGURE 29.48 Making of a thin stone panel.

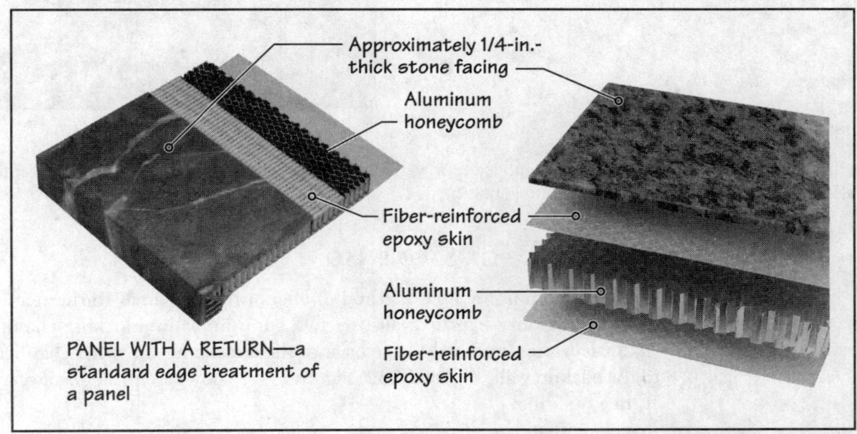

FIGURE 29.49 Anatomy of a stone-honeycomb panel showing the standard treatment of its exposed edge. (Photos courtesy of Stone Panels Inc.)

The standard treatment on an exposed edge of the panel is a small return, as shown in Figure 29.49. Where a larger return is required (e.g., to obtain a deeper soffit at a window head), it is obtained by cementing a continuous aluminum angle to the honeycomb, Figure 29.50.

The standard size of the panels is 4 ft × 8 ft, but other sizes are available, with a maximum of 5 ft × 10 ft. The light weight of the panels makes their installation convenient, particularly in high-labor areas. A 1-in.-thick stone-honeycomb panel weighs only 3.3 psf, which is approximately the weight of $\frac{1}{4}$-in.-thick glass.

The bending strength of a stone-honeycomb panel is fairly high because of the honeycomb backing and the fiber-reinforced epoxy skin bonded to it. The composition also gives a great deal of ductility to the panel so that it is able to flex under lateral loads.

The panels' light weight, high ductility, and bending strength make it ideal for use in seismic areas, where the aesthetics of natural stone cladding without its heavy weight are required. These are some of the reasons cited for their use in the courthouse in Anchorage, Alaska, and the International Business Center, Moscow, Russia, Figure 29.51.

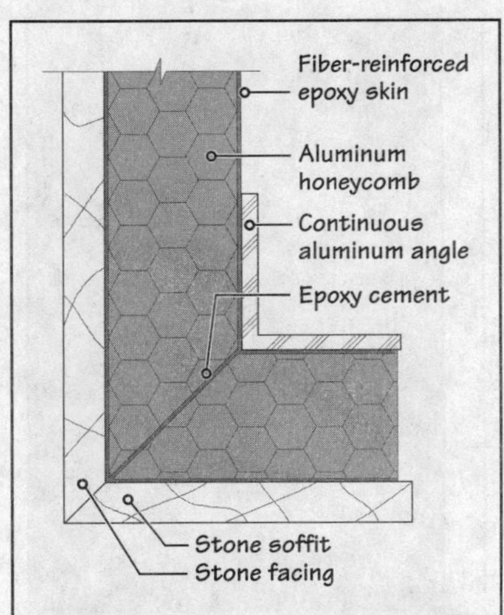

FIGURE 29.50 Anatomy of a stone-honeycomb panel with a deep soffit.

(a)

(b)

FIGURE 29.51 Examples of buildings with stone-honeycomb exterior wall cladding. **(a)** Courthouse, Anchorage, Alaska. **(b)** International Business Center, Moscow, Russia. (Photos courtesy of Stone Panels Inc.)

Anchorage of Stone-Honeycomb Panels

The most commonly used method of anchoring the panels to the steel stud, concrete, or concrete masonry backup wall uses two continuous interlocking channels. One of these channels is shop installed to the back of the panel, and the other channel is field anchored to the backup wall, Figure 29.52. Figure 29.53 shows a typical detail of the use of the panels in a building.

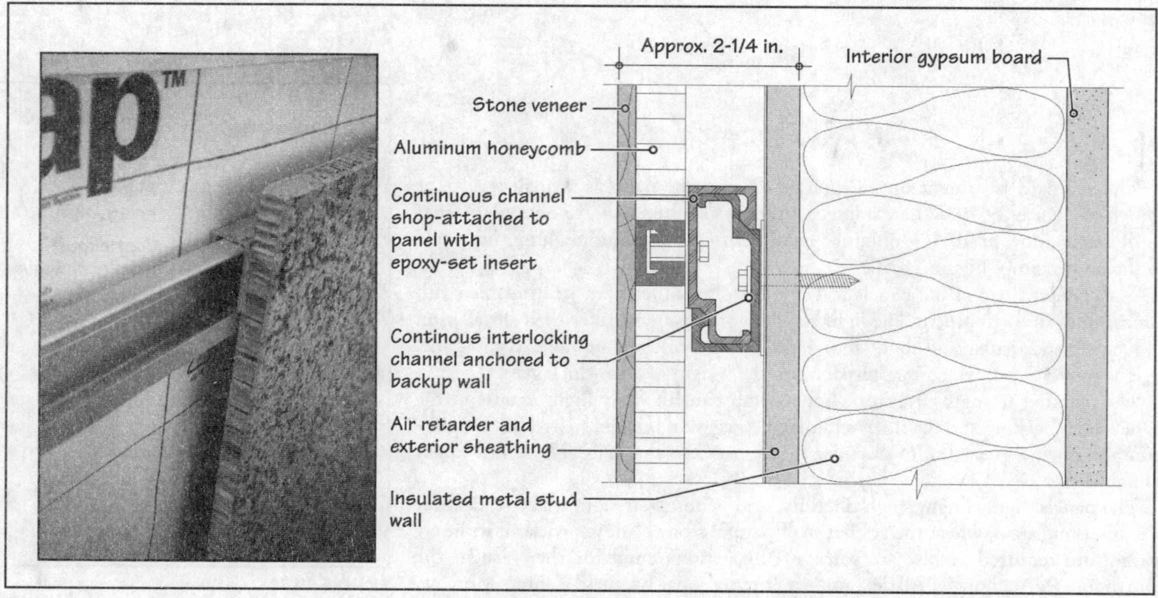

FIGURE 29.52 A commonly used method of anchoring stone-honeycomb panels to a steel-stud backup wall.

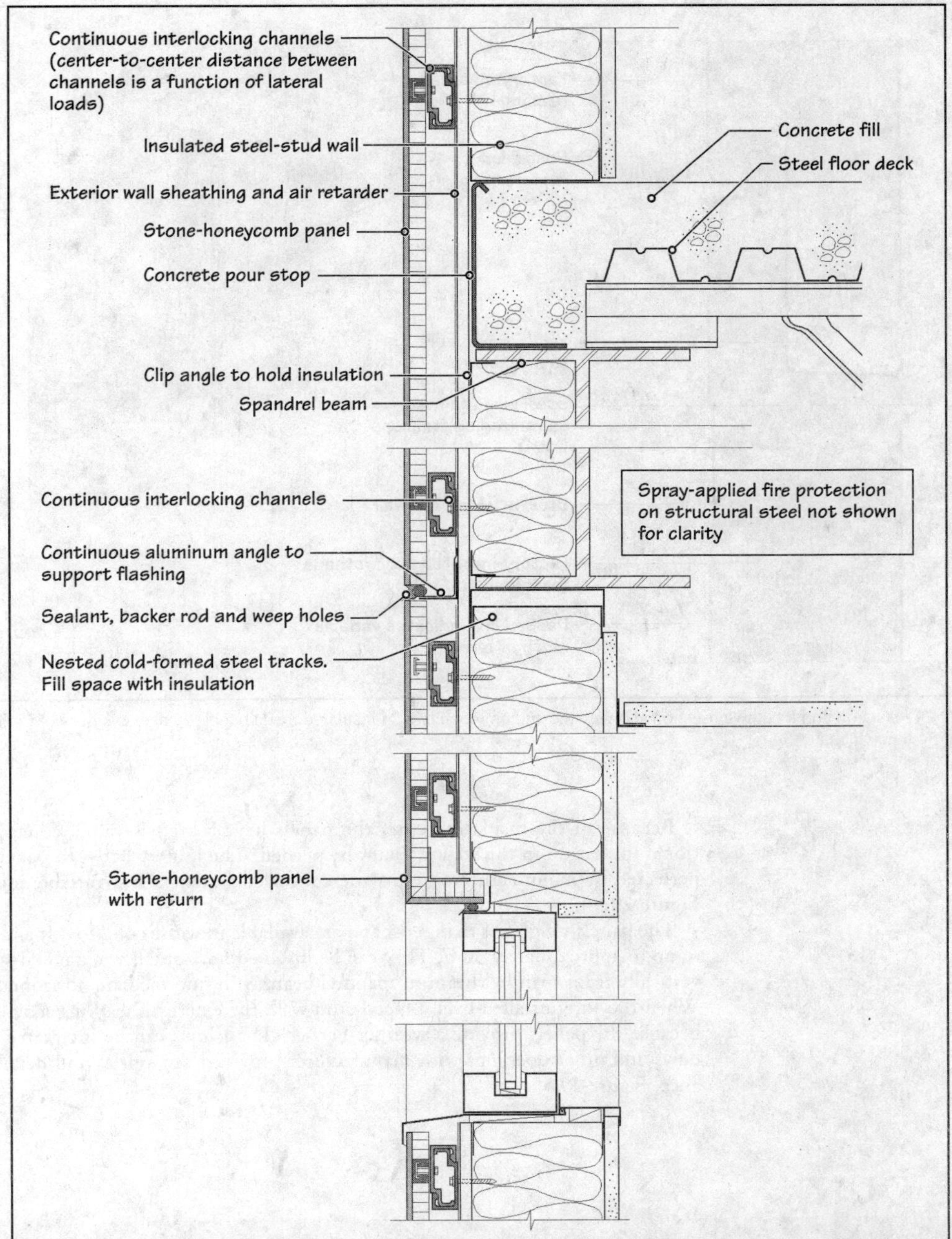

FIGURE 29.53 A typical wall section through a steel-frame building clad with stone-honeycomb panels.

Prefabricated Stone-Honeycomb Curtain Wall Panels

Stone-honeycomb panels can also be prefabricated into curtain wall panels, which generally extend from column to column and are hung from the building's structure like other curtain wall panels.

29.14 INSULATED METAL PANELS

Another lightweight exterior cladding system consists of metal (typically steel) panels with 2- to 3-in. factory-installed polyurethane foam insulation between metal skins. The panels are available factory painted in various colors, in galvanized steel and in stainless steel. Several surface finishes, such as smooth, embossed, and a precast concrete-like texture, are available.

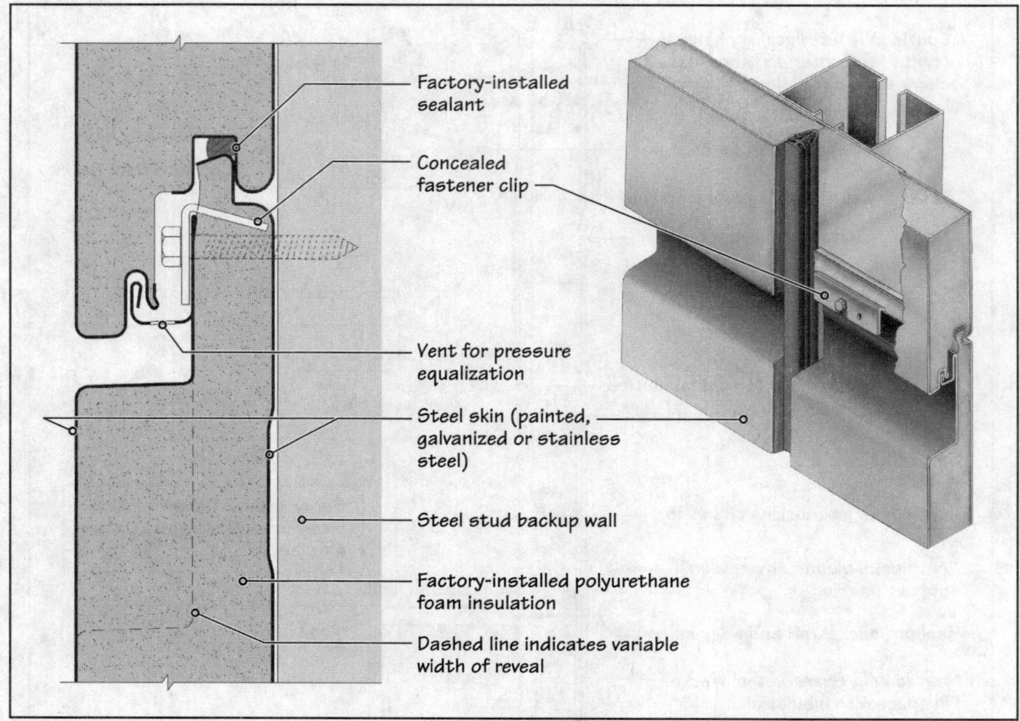

FIGURE 29.54 A section and a cutaway section showing the anatomy of a typical insulated metal panel. (Images courtesy of Centria Architectural Systems, Inc.)

Because of the insulating core, the panels have a high R-value. Therefore, additional insulation in the wall may not be needed. The joinery between panels has been perfected by manufacturers to provide concealed fasteners and variable joint widths, Figure 29.54.

The panels weigh less than 3 psf and are available in widths of 2 to 4 ft and in lengths of up to approximately 30 ft. They can be installed horizontally on a metal-stud wall or vertically from spandrel beam to spandrel beam with intermediate horizontal supports. When they are installed over a metal-stud wall, the exterior sheathing may be omitted because the panels provide a weather barrier. The panels can be integrated with windows and other openings with manufacturer-provided accessories and detailing assistance, Figure 29.55.

PRACTICE **QUIZ**

Each question has only one correct answer. Select the choice that best answers the question.

32. In a stone-honeycomb wall cladding panel, the thickness of the stone is approximately

- **a.** 1 in.
- **b.** $\frac{3}{4}$ in.
- **c.** $\frac{1}{2}$ in.
- **d.** $\frac{1}{4}$ in.
- **e.** $\frac{1}{8}$ in.

33. In a stone-honeycomb wall cladding panel, the honeycomb is generally made of

- **a.** stainless steel.
- **b.** aluminum.
- **c.** titanium.
- **d.** copper.
- **e.** none of the above.

34. A stone-honeycomb panel is anchored to the backup wall using

- **a.** epoxy cement.
- **b.** separate dead-load supports and tiebacks.
- **c.** two interlocking metal channels.
- **d.** stainless steel bolts.

35. In an insulated wall metal panel, the insulation is sandwiched between

- **a.** a metal sheet in the front and a fiberglass scrim at the back of the panel.
- **b.** a metal sheet on both sides.
- **c.** a metal sheet in the front and an acrylic sheet at the back of the panel.
- **d.** none of the above.

36. The weight of insulated metal panels and stone-honeycomb panels is approximately the same.

- **a.** True
- **b.** False

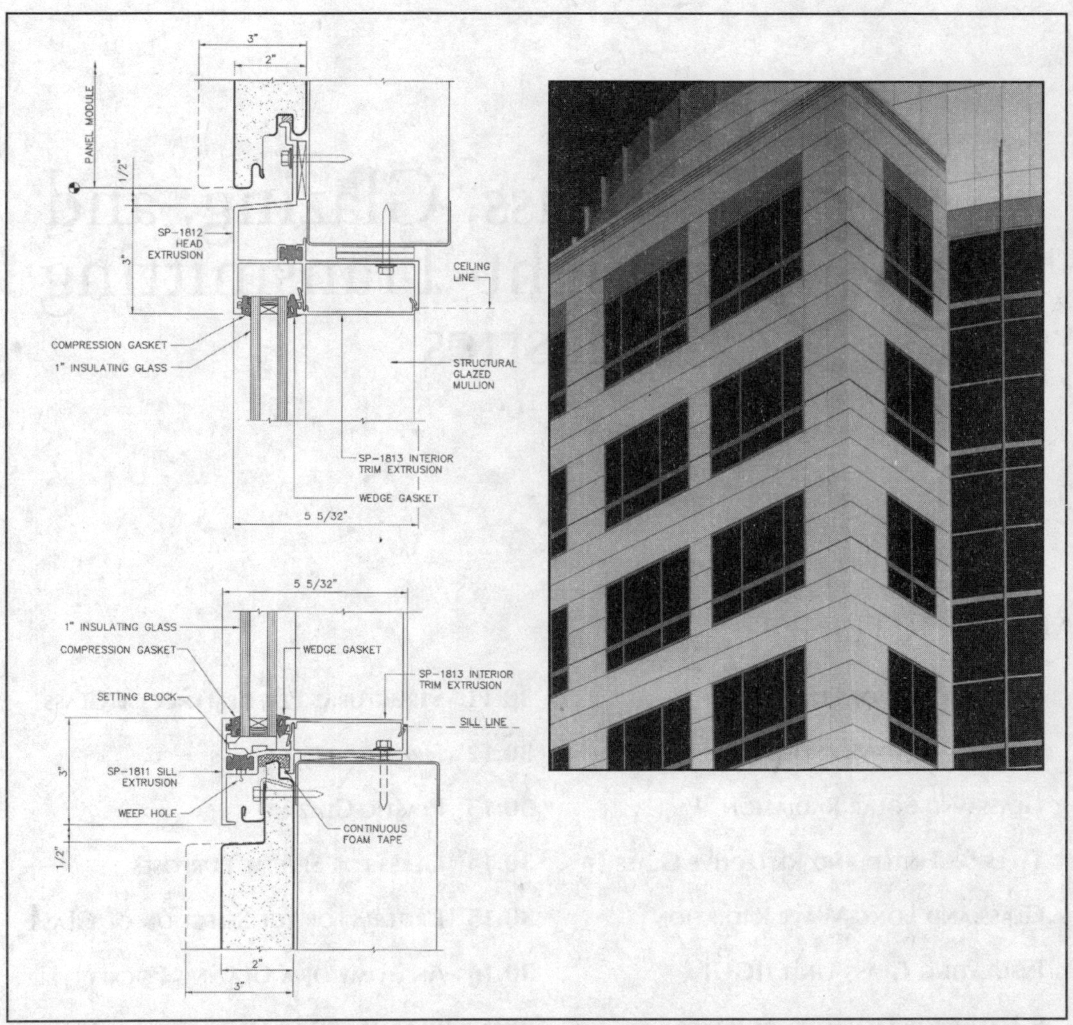

FIGURE 29.55 Insulated metal panels and their integration with windows and other openings as per manufacturer-provided details and accessories. (Photo and details courtesy of Centria Architectural Systems, Inc.)

REVIEW QUESTIONS

1. Using a sketch, explain the anatomy of stucco applied over a metal-stud wall assembly, showing all components.
2. What are the essential differences between stucco applied on a stud wall versus that applied on a CMU wall?
3. Discuss the difference between control joints and expansion joints in a stucco-clad wall.
4. Explain the difference between PB EIFS and PM EIFS. Which of the two methods is more commonly used?
5. With the help of a sketch, illustrate the anatomy of PB EIFS. Also, explain what back-wrapping is and where it is used in EIFS cladding.
6. Discuss the pros and cons of EIFS cladding compared with stucco cladding.
7. Sketch a split-tail anchor, showing the material it is made of and its function in a stone-clad wall.
8. Sketch a dead-load support of stone cladding with a bayonet clip.
8. Sketch a combined dead-load support and tieback anchor used in a stone-clad wall.
9. Explain the anatomy of a stone-honeycomb panel. With the help of a sketch, explain how the panels are anchored to a backup wall.

CHAPTER 30

Glass, Glazing, and Light-Transmitting Plastics

CHAPTER OUTLINE

The introduction of daylight into buildings has always been an important design requirement. In early buildings, this requirement was satisfied by providing voids in walls. Subsequently, in an effort to control daylight, ventilation, and security, windows with wood shutters were developed. The use of oiled-paper panels or muslin cloth in wooden frames, which provided some light without opening the windows, was the next step in providing transparency in buildings.

With the discovery of glass, the window's performance (as well as its appearance) changed dramatically. It was now possible to obtain daylight effectively without admitting other environmental elements, such as wind, rain, dust, and insects. The earliest use of glass in openings is usually traced to the Roman Empire, where it was employed in small sizes.

The first significant use of glass, however, occurred during the Middle Ages as stained glass windows gained popularity in churches. Staining (or painting) of the glass was used as decoration to hide defects such as air bubbles and colored impurities, which were integral to the glass made at that time. During the Middle Ages, church windows were particularly large. They not only provided daylight, but also used dramatic pictures made from brightly stained glass to explain biblical stories to people who could not read.

The transparency of glass improved greatly with the perfection of the *crown glass process* around the fourteenth century [30.1]. This method provided glass with fewer air bubbles and other defects. In this process, molten glass was first converted to a globe by blowing through a long pipe, then transferring the globe from the blow pipe to a rod, and subsequently cutting the globe open. The cut globe was finally flattened into a disc by repeated heating and vigorous spinning, Figure 30.1.

In a similar process, called the *cylindrical process,* molten glass was first blown into a globe and subsequently enlarged into a cylinder by swinging the pipe back and forth. The cylinder was later cut along its length, flattened by heating, and converted to sheets, Figure 30.2.

Clear glass, as we know it today, came into commercial use at the end of the seventeenth century with the invention of the glassmaking process by Bernard Perrot [30.2]. In Perrot's process, molten ingredients of glass were first cast in a mold and later spread into sheets by rollers. When the glass solidified fully, it was ground and polished on both sides with abrasives. The mechanization of the rolling and grinding operations brought down the price of glass sufficiently so that it could be used commercially in mirrors and storefronts. Although the process of manufacturing modern glass is substantially different from that of the cast or cylindrical glass of several centuries ago, the ingredients are virtually the same.

The transparency of glass, coupled with modern manufacturing techniques that provide large sheets of glass inexpensively, is responsible for the extensive use of glass in modern buildings. These developments have also led to the use of glass as the primary material for the cladding of many building types, Figures 30.3 and 30.4.

In this chapter, the modern process of manufacturing glass, as well as its various types and their properties, are described. Also discussed are light-transmitting plastics, which are sometimes used in lieu of glass.

FIGURE 30.1 Crown glassmaking process. (Courtesy of the Collection of the Rakow Research Library, The Corning Museum of Glass, Corning, NY.)

FIGURE 30.2 Cylindrical glassmaking process. (Courtesy of the Collection of the Rakow Research Library, The Corning Museum of Glass, Corning, NY.)

FIGURE 30.3 A large glass pyramid admits light into an underground addition to the Louvre museum, Paris, France. Architect: I. M. Pei and Partners. (Photo by Zoran Karapancev of Shutterstock)

FIGURE 30.4 A transparent cube of glass, supported by a light steel structural backup, encloses the Rose Center for Earth and Space, New York City (see also Chapter 32, Figure 32.17). Architects: Polshek Partnership.

EXPAND YOUR KNOWLEDGE

Why Glass Is Transparent

Although glass is a solid substance, its molecular structure resembles that of a liquid. A typical solid is composed of numerous tiny crystals. Each crystal has a definite geometric form at the atomic level, which reflects the arrangement of its constituent atoms. The crystals are packed together in a regular manner to form a repeating network or lattice.

Crystals are formed when the material changes from its molten state to a solid state. Crystallization occurs abruptly at a specific temperature for a given material, called the *freezing point* (or the *crystallization temperature*). In the molten (liquid) state, a material has an amorphous (noncrystalline) structure. An amorphous structure is one in which the constituent atoms are joined to one another but not in a regular three-dimensional pattern—rather in a random pattern.

Glass is a material that has never crystallized. It becomes hard while still retaining its liquid structure. Thus, glass is sometimes called an *amorphous solid* or a *supercooled liquid*. Because of its amorphous structure, glass does not have a definite melting point like a crystalline solid.

The amorphous structure of glass is responsible for its transparency. All solids except glass and clear plastics are opaque to light. The opaqueness of a crystalline solid is explained by the fact that light is reflected at each crystal boundary as it passes through the solid. At each reflection, some light is lost.

Because there are numerous such crystal boundaries, even within a small thickness of a solid material, light goes through a large number of reflections, losing some light at each reflection. This makes the material behave as an opaque material. Glass is virtually one large crystal containing no internal boundaries, and that is why it is transparent. Several crystalline impurities, however, can make glass translucent or opaque.

The amorphous structure of glass is not merely a property of the materials of which the glass is made because many other materials can also be made to produce an amorphous structure on solidification. The amorphous structure of a solid material is also a property of the rate at which molten ingredients are cooled. If cooled slowly, the atoms in the molten mass have sufficient time to organize into a regular pattern to become a crystalline solid.

Therefore, in the manufacturing of glass, the molten ingredients must be cooled rapidly to below the crystallization temperature to prevent crystallization.

30.1 MANUFACTURE OF FLAT GLASS

The primary raw material for making *flat glass* (i.e., window glass, to distinguish it from glass block and other glass products) is sand. Chemically, sand is silicon dioxide, also called *silica.* Sand used for glass manufacturing is obtained from sandstone deposits. (Seashore sand is unsuitable for glassmaking because it has too many impurities.)

Although silica is all that is needed to make glass, other ingredients are added to modify several of its properties. The two major ingredients added to silica for making flat glass are sodium oxide and calcium oxide (lime). That is why flat glass is sometimes referred to as *soda-lime glass.* Soda-lime glass consists of approximately 72% silica, 15% sodium oxide, 9% calcium oxide, and 4% other minor ingredients.

One of the minor impurities in the ingredients for making glass is iron oxide, which occurs naturally in sand. Iron oxide gives a clear glass sheet its bluish-green tinge when viewed from its edge.

Sodium oxide works as a *flux.* A flux is an additive that lowers the melting point of the main ingredient—in this case, silica. The melting point of pure silica is very high—nearly 3,100°F (1720°C). When sodium oxide is added to silica, the mixture melts at a much lower temperature, which reduces the cost of glassmaking.

Pure silica in its molten form is highly unworkable. It is so viscous and tacky that any bubbles of air or gas produced during melting do not readily escape from the molten mass. Sodium oxide also makes the molten mixture less viscous and, hence, more workable.

The mixture of silica and sodium oxide yields glass that is not too durable. It slowly dissolves in water and has low resistance to chemical attack. The addition of calcium oxide stabilizes the mix, resulting in a glass that is durable and more easily worked.

NOTE

Silica Glass

Glass made only from silica (silicon dioxide), with no other additives, is used in situations where its greater resistance to heat and higher transmissivity to radiation are required, such as in mercury vapor lamps and telescope mirrors. Glass made with pure silica is called *silica glass.*

THE FLOAT GLASS PROCESS

Although the raw materials for making flat glass have remained virtually unchanged, the manufacturing process has evolved considerably over the years. In the process that is commonly used today, the raw materials are granulated, mixed together, and loaded into a furnace, where they melt.

Broken pieces of glass from an earlier batch and scrap glass are also loaded into the furnace at this stage. From the furnace, the molten material goes to a molten glass tank and then to a bed of molten tin. Because the density of molten tin is higher than that of molten glass, the latter floats on the molten tin bed.

The process is controlled so that a predetermined thickness of molten glass travels continuously over the tank of molten tin, called the *float bath,* Figure 30.5. As the molten glass moves to the end of the float tank, it is cooled rapidly to prevent crystallization. There, it solidifies into a sheet and then travels over rollers to the annealing chamber, called the *annealing lehr.*

The glass that enters the annealing lehr has internal stresses locked within its body. This is due to the thermal gradient that is set up between the external surfaces and the interior of the glass due to rapid cooling in the float bath.

In the annealing lehr, the glass is first heated sufficiently to relieve any stresses created during the solidifying process. It is then cooled very slowly so that every glass particle cools at the same rate to ensure that no stresses are frozen in the glass. The annealing lehr is, therefore, several hundred feet long.

NOTE

Invention of the Float Glass Process

The float glass process was first used in 1959 by Pilkington Brothers Limited near Liverpool, England, and has since become a worldwide standard for the manufacturing of glass.

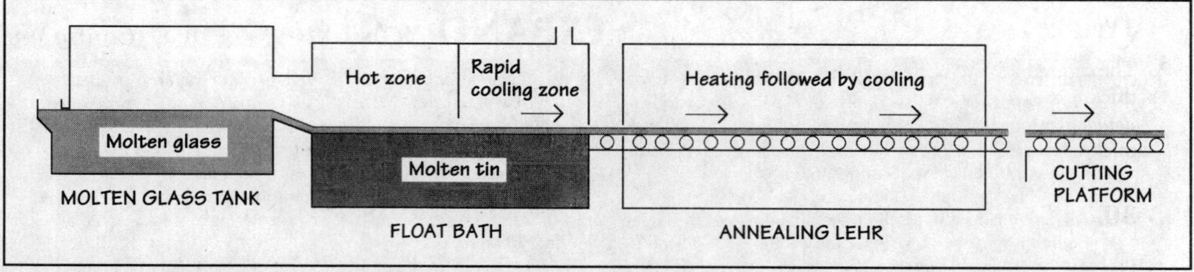

FIGURE 30.5 Outline of the float glass manufacturing process.

At the end of the annealing lehr, the glass sheet emerges as a continuous ribbon at room temperature, free of any internal stresses, Figure 30.6. It is then cut into desired lengths by automatic cutters, packed, and transported to its destination, Figure 30.7.

Because the top surface of a liquid must be uniformly horizontal due to gravity, the floating of molten glass over molten tin ensures that the top and bottom surfaces of glass are parallel. This provides a glass sheet of uniform thickness—an important requirement to ensure distortion-free vision through glass.

The glass obtained from this process is called *float glass* to distinguish it from sheet glass and plate glass, which were the two types of flat glass used before the discovery of the float glass manufacturing process. Virtually all flat glass manufactured today is produced by the float glass process. Sheet glass and plate glass are no longer commercially produced.

FIGURE 30.6 In the float glass manufacturing process, glass emerges as an endless ribbon at room temperature. (Photo courtesy of PPG Industries Inc.)

FIGURE 30.7 Glass is cut by automatic cutters to the required sizes before packing and shipping. (Photo courtesy of PPG Industries Inc.)

TYPICAL FLAT GLASS THICKNESS

The commonly used nominal thicknesses of flat glass range from $\frac{3}{32}$ in. to 1 in. Glass of $\frac{3}{32}$-in. thickness is referred to as *single-strength* (SS) *glass,* and that of $\frac{1}{8}$-in. thickness is called *double-strength* (DS) *glass.* Manufacturers must be consulted for the availability of certain thicknesses before specifying it for a project.

30.2 TYPES OF HEAT-MODIFIED GLASS

Flat glass obtained from the float glass process, without any further treatment, is the basic glass, referred to as *annealed glass.* Annealed glass, which may be clear or tinted, is the most commonly used glass type in buildings. However, where a stronger glass is required, annealed glass is heat treated before use. Heat treatment increases the bending strength and the temperature resistance of glass and makes it more suitable for applications where annealed glass is inadequate. There are three types of heat-modified glass:

- Tempered glass, also referred to as *fully tempered* (FT) *glass*
- Heat-soaked tempered glass
- Heat-strengthened (HS) glass

NOTE

Nominal versus Actual Thickness of Flat Glass

Glass thickness is given in terms of its nominal thickness. The actual glass thickness is its nominal thickness plus or minus a tolerance. For example, the actual thickness of a $\frac{1}{4}$-in. (6-mm)-thick glass sheet must lie between 0.219 in. (5.56 mm) and 0.244 in. (6.20 mm).

TEMPERED GLASS

Tempering is the exact opposite of annealing. Whereas annealing (the process of slow cooling after heating) reduces or eliminates locked-in stresses, tempering produces them. Tempering involves heating the glass below its softening point, to approximately 1,300°F (700°C), and suddenly cooling (quenching) it by blowing a jet of cold air on all surfaces of the glass simultaneously. This causes the the outer layers of the glass to harden quickly while the interior of the glass is still soft. As the interior begins to cool, it tends to shrink but is prevented from doing so by the already-hardened outer glass surfaces, Figure 30.8. Consequently, the exterior of the glass comes under a state of compression and the interior under a state of compensating tension.

Like all brittle materials, glass is weak in tension. Therefore, a glass sheet is weak in bending also, because bending creates tensile (and compressive) stresses in the sheet. The locked-in compressive stresses in the outer layers of tempered glass cancel (or reduce) the tensile stresses produced by bending. Consequently, tempered glass is approximately four times stronger than annealed glass in bending. It can also withstand greater deflection than annealed glass of the same thickness, Figure 30.9, and is far more resistant to impact and thermal stresses.

Tempering does not affect other properties of glass, such as solar heat gain, U-value, or the color of glass. However, because of the bowing and warping caused by the shrinkage of glass during heat treatment, tempered glass may contain a slightly noticeable distortion of view under some light conditions, which is generally not objectionable.

Tempering must be done after the glass has been cut to size. A tempered glass sheet must not be cut, drilled, or edged because these processes release the locked-in stresses, causing the glass to disintegrate abruptly. Abrasive blasting and etching may be done with some care. However, both abrasive blasting and etching reduce the thickness of the outer compressed layers, reducing the effectiveness of tempering.

NOTE

Annealed, Tempered, Heat-Soaked, and Heat-Strengthened Glasses

Annealed glass is the basic glass, obtained directly from the float glass process. *Tempered glass, heat-soaked glass,* and *heat-strengthened glass* are obtained by heat-treating annealed glass.

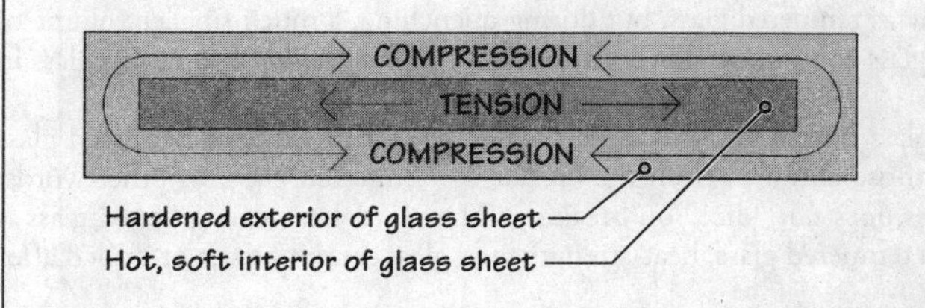

FIGURE 30.8 When the heated glass is quenched with cold air during the tempering process, the outside layers of glass become hard while the interior remains hot and soft. As the interior of the glass cools, it shrinks and tends to pull the hardened outside layers inward, producing compressive stresses in them. Because the outside layers have already hardened, they resist the inward pull. Consequently, the interior of the glass comes under tension. These stresses are permanently locked in a tempered glass sheet.

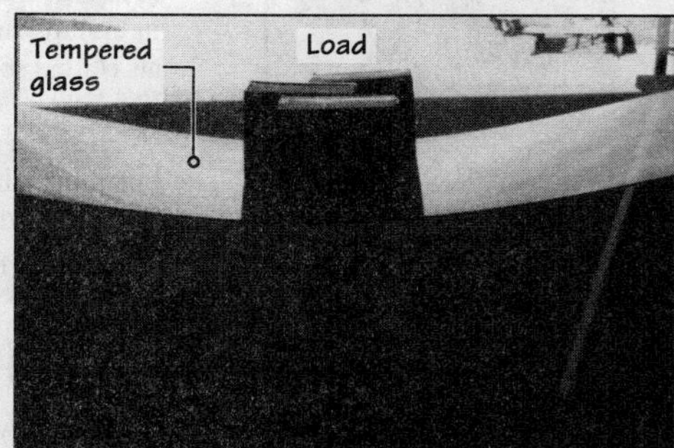

FIGURE 30.9 Deflection of a tempered glass sheet under load. Observe the ductility of tempered glass, that is, its ability to deflect substantially prior to failure.

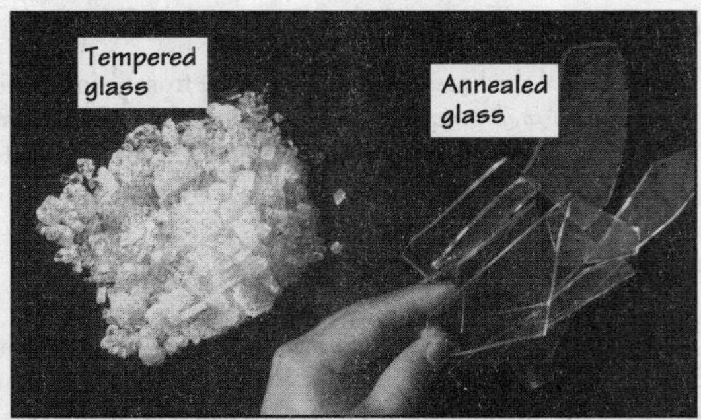

FIGURE 30.10 Breakage patterns of tempered glass and annealed glass. Tempered glass breaks into tiny square-edged granules. Annealed glass breaks into, large sharp-ended pieces.

USE OF TEMPERED GLASS

When tempered glass breaks, it breaks into tiny square-edged, cubicle-shaped granules—a breakage pattern usually referred to as *dicing,* Figure 30.10. Annealed glass, on the other hand, breaks into long, sharp-edged pieces. Tempered glass is, therefore, used in hazardous locations, provided that it meets the requirements of safety glazing, as described in Section 30.9.

SPONTANEOUS BREAKAGE OF TEMPERED GLASS

Nickel sulfide and certain other impurities present in glass ingredients do not fully melt during glass manufacturing. They are known to expand after days or even years of glass manufacturing when subjected to extreme thermal stresses or thermal shock (e.g., sudden cooling of sun-heated glass by rainwater). When this expansion occurs in tempered glass, it creates excessive tensile stress in the glass, resulting in its spontaneous breakage.

This breakage, although rare, occurs abruptly and can be a safety hazard. Building codes require that if there is a walking or other usable surface under a tempered glass opening, a protective screen must be provided below it to prevent injuries to humans.

HEAT-SOAKED TEMPERED GLASS

When it is necessary to reduce or eliminate spontaneous breakage of tempered glass, heat-soaked tempered glass is used. In the heat-soaking process, tempered glass panes are subjected to cyclical heating and cooling to simulate long-term field conditions. Glass that does not break in the process is safe against spontaneous breakage and is called *heat-soaked tempered glass.*

Because the testing is expensive and results in the destruction of a certain percentage of tempered glass panes, heat-soaked glass is expensive and is used only in projects where its higher cost can be justified.

HEAT-STRENGTHENED GLASS

Heat-strengthened glass falls in between annealed glass and tempered glass. It is heat treated in exactly the same way as tempered glass, but during quenching, a much smaller volume of air is used so that the glass cools more slowly. It is nearly twice as strong as annealed glass in bending.

Heat-strengthend glass breaks into pieces that are sharper than those of tempered glass but more blunt than those obtained from the breakage of annealed glass. In other words, heat-strengthened glass does not "dice" on breaking. Therefore, heat-strengthened glass is not a safety glass. Like tempered glass, heat-strengthened glass cannot be cut or drilled after heat treatment.

FIGURE 30.11 An example of a typical all-glass curtain wall building.

USE OF HEAT-STRENGTHENED GLASS

The primary use of heat-strengthened glass is in situations where the glass is subjected to large thermal stresses. One such situation is the spandrel area of an all-glass curtain wall, Figure 30.11. A spandrel area is an area of a wall that is between the head of a window on one floor and the sill of the window on the floor above. This is the part of the building

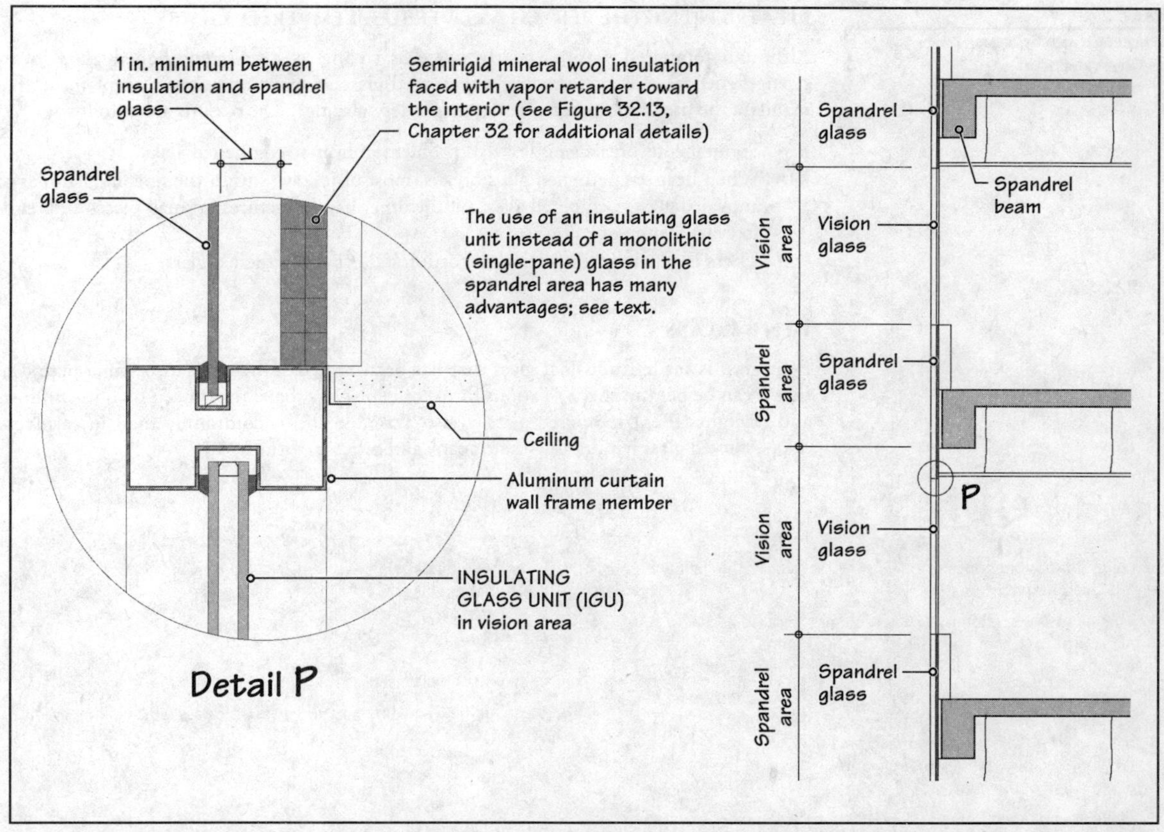

FIGURE 30.12 A section through an all-glass curtain wall showing spandrel glass and vision glass areas (see also Chapter 7, Figure 7.14). Although monolithic glass has been shown in the spandrel area (Detail P), the same insulating glass unit used in the vision area may be used in the spandrel area.

facade that includes edge beams and the floor slab. The purpose of spandrel glass panes is to hide the structural components behind them. Thus, an all-glass curtain wall consists of two distinct areas of glass: vision glass and spandrel glass, Figure 30.12.

In colder regions of North America, it is common practice to use monolithic (single) sheets of glass in spandrel areas and insulating glass units in vision areas. (Refer to Section 30.6 for the description of an insulating glass unit.) To prevent see-through, the spandrel glass is opacified either with a ceramic frit coating or a polyester film.

To improve the energy performance of the glass curtain wall, spandrel areas are generally insulated. At least a 1-in. space is generally recommended between the insulation and the glass to simulate the depth of view of vision glass. Using the same type of glass in both the vision and spandrel areas of the wall (particularly a tinted-reflective glass) adds to the blend between the vision and the spandrel panels.

Because of the presence of insulation, the spandrel glass in an all-glass curtain wall does not dissipate the sun's heat as readily as the vision glass. Consequently, the spandrel glass is subjected to much greater thermal stress than the vision glass. Because of its greater strength compared to annealed glass, heat-strengthened glass is usually specified for the spandrel panels.

The space between the insulation and spandrel glass is particularly susceptible to condensation, which may lead to wetting of insulation and corrosion of metals. Therefore, the entry of water vapor into this space should be prevented through the use of a good vapor retarder that is fully sealed toward the interior (see Chapter 32 for additional details). Condensation effects can be reduced by the use of insulating glass units in spandrel areas.

The use of insulating glass units in spandrel areas in place of monolithic glass is fairly common in milder climates of North America, which gives a better blend between vision and spandrel areas and may preclude the use of spandrel insulation. The inward glass pane of such an insulating glass unit in the spandrel is opacified and heat strengthened. In situations where a glass pane is subjected to uneven solar shading, the vision areas may also require the use of a heat-strengthened glass.

NOTE

Ceramic Frit on Glass

A ceramic frit coating is deposited on glass in individual dots or lines and then fused (baked) into the glass under high temperature. The coating can be of different colors and can be opaque or translucent. Fritted glass can use any pattern to produce images or graphic designs on glass.

HEAT-STRENGTHENED GLASS VERSUS TEMPERED GLASS

Although tempered glass is nearly twice as strong as heat-strengthened glass, heat-strengthened glass is preferred over tempered glass in most situations, except those that mandate the use of tempered glass (e.g., as safety glazing). The reasons are as follows:

- Spontaneous breakage is less of a problem in heat-strengthened glass.
- When heat-strengthened glass breaks, most of it stays within the opening, similar to annealed glass. Tempered glass, on the other hand, fractures in small pieces and tends to evacuate the opening.
- There is generally less optical distortion in heat-strengthened glass.

BENT GLASS

Bent glass is made from float glass that has been heated sufficiently to become plastic so that it can be bent to shape. Various forms of bent-glass shapes have been used in windows and skylights. Bent tempered glass is also available and is commonly used in revolving doors, curved glass handrails, elevator cars, and office partitions.

PRACTICE QUIZ

Each question has only one correct answer. Select the choice that best answers the question.

1. The microscopic (atomic) structure of glass resembles that of
   a. solids.
   b. liquids.
   c. gases.
   d. none of the above.
2. To make glass transparent, the molten glass ingredients must be
   a. cooled very slowly during manufacture.
   b. cooled slowly during manufacture.
   c. cooled rapidly during manufacture.
   d. any one of the above.
3. The glass obtained from the float glass process without any further treatment is called
   a. sheet glass.
   b. annealed glass.
   c. tempered glass.
   d. heat-strengthened glass.
   e. plate glass.
4. If glass is $\frac{1}{8}$ in. (3 mm) thick, it is also called
   a. single-layer glass.
   b. single-strength glass.
   c. double-strength glass.
   d. basic glass.
   e. monolithic glass.
5. Which of the following types of glass is generally used as safety glass?
   a. Sheet glass
   b. Annealed glass
   c. Tempered glass
   d. Heat-strengthened glass
   e. Plate glass
6. Which glass type is commonly specified for the spandrel areas of an all-glass curtain wall?
   a. Sheet glass
   b. Annealed glass
   c. Tempered glass
   d. Heat-strengthened glass
   e. Plate glass
7. Which of the following glass types has locked-in stresses?
   a. Tempered glass
   b. Heat-soaked tempered glass
   c. Heat-strengthened glass
   d. All of the above
   e. None of the above
8. To obtain heat-soaked tempered glass, the glass must be tempered first and then
   a. annealed.
   b. etched.
   c. heat strengthened in a furnace.
   d. subjected to cyclic heating and cooling.
   e. all of the above.
9. Which of the following glass types, when broken, breaks into tiny square-edged granules?
   a. Tempered glass
   b. Heat-strengthened glass
   c. Annealed glass
   d. None of the above
10. Spontaneous breakage generally occurs in
    a. annealed glass.
    b. heat-strengthened glass.
    c. heat-soaked tempered glass.
    d. tempered glass.
    e. laminated glass.

30.3 GLASS AND SOLAR RADIATION

When solar radiation (i.e., direct solar beam) falls on a glass surface, part of it is transmitted through glass, part is reflected, and part is absorbed by the glass. A $\frac{1}{8}$-in.-thick clear glass sheet transmits approximately 85%, reflects 10%, and absorbs nearly 5% of solar radiation. In other words, the transmissivity, reflectivity, and absorptivity of clear glass for solar radiation are 0.85, 0.10, and 0.05, respectively. Thus, if 100 units of solar energy fall on a clear glass sheet, 85 units are transmitted to the interior, 10 units are reflected to the exterior, and 5 units are absorbed by the sheet, Figure 30.13.

The 5 units of solar radiation absorbed by a clear glass sheet increase its temperature, and the glass becomes a low-temperature (long-wave) radiator. It releases (by radiation and convection) R_i units of absorbed heat to the inside and R_o units to the outside, so that $R_i + R_o = 5$ units.

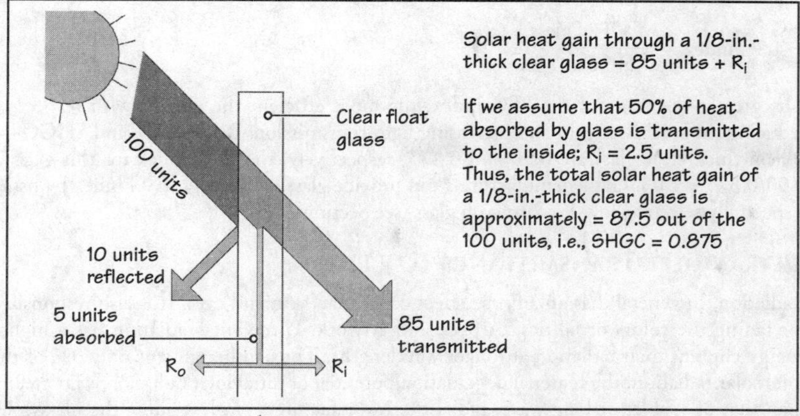

FIGURE 30.13 Properties of a $\frac{1}{8}$-in.-thick clear glass with respect to solar radiation.

The relative values of R_i and R_o depend on the internal and external air temperatures. A greater amount of heat is released toward the cooler side. Thus, if the inside air is cooler than the outside air, $R_i > R_o$ (e.g., R_i may equal 3.0 units, so $R_o = 2.0$ units). Conversely, if the outside air is cooler than the inside air, $R_o > R_i$.

When inside and outside air temperatures are the same, $R_i = R_o = 2.5$ units. In such a case, the total amount of solar radiation penetrating the glass will equal $85 + 2.5 = 87.5$ units. In other words, 87.5% of solar radiation incident on the glass penetrates the glass.

SOLAR HEAT-GAIN COEFFICIENT (SHGC)

A commonly used measure of how well a glass sheet performs with respect to direct solar radiation is the *solar heat-gain coefficient* (SHGC). It is defined as the amount of solar radiation that penetrates a given glass (including the absorbed radiation that is subsequently released to the interior) divided by the solar radiation incident on the glass:

$$\text{SHGC} = \frac{\text{solar energy gain through a glass}}{\text{solar energy incident on glass}}$$

Thus, as shown in Figure 30.13, SHGC = 0.875 for a $\frac{1}{8}$-in.-thick clear glass. The smaller the value of SHGC, the smaller the heat gain through the building. For warm (southern) regions of the United States, SHGC should be as low as possible. For cold (northern) regions of the United States, a higher SHGC value is desirable because it may provide daytime heating of the building through glazed openings.

SHADING COEFFICIENT (SC)

Another measure closely related to SHGC is the *shading coefficient* (SC). SC is defined as the solar heat gain through a given glass divided by the solar heat gain through an unshaded, clear $\frac{1}{8}$-in.-thick glass under the same internal and external conditions. In other words, the SC of glass is a ratio of the SHGC of glass and the SHGC of clear $\frac{1}{8}$-in.-thick clear glass:

$$\text{SC} = \frac{\text{SHGC of glass in question}}{\text{SHGC of } \frac{1}{8}\text{-in.-thick clear glass}}$$

From the preceding definition, the SC of a $\frac{1}{8}$-in.-thick clear glass is 1.0. This is practically the maximum possible value of the shading coefficient. The shading coefficient of a clear $\frac{3}{32}$-in.-thick glass is nearly 1.01.

VISIBLE TRANSMITTANCE (VT) AND THE LIGHT-TO-SOLAR GAIN (LSG) INDEX

Because transparency is the primary reason for using glass in buildings, another important property of glass is its ability to transmit light (the visible part of solar radiation; see the first Principles in Practice section at the end of this chapter). *Visible transmittance* (VT) is defined as the percentage of the visible part of solar radiation transmitted through glass. For $\frac{1}{8}$-in.-thick clear glass, VT is approximately 0.90.

NOTE

SHGC Value and the Energy Code

The International Energy Conservation Code (IECC) places restrictions on the SHGC values of glass used in building fenestration. For example, in a residential building, the maximum SHGC value of glass used in windows and skylights is limited to 0.30 for climate zones 1 to 3 (warmer climates). The value of 0.3 (in the 2009 version of the IECC) represents a reduction of 25% from 0.4 (in the 2006 version of the IECC), highlighting the importance of SHGC in affecting energy use (and, hence, sustainability) in warm climates.

There is no limit placed on the SHGC value for cooler climates (climate zones 4 to 8) because in these climates, limiting the maximum value of SHGC may increase energy consumption during the winters. In extremely cold climates (climate zones 7 and 8), a high SHGC value is obviously desirable.

Note that this information applies to compliance with the prescriptive provisions of the IECC. If the simulated performance approach is used, the designer has the freedom to trade off between various factors that affect energy use in the building.

The LSG (light-to-solar gain) index of a glass is the ratio of VT and SHGC:

$$\text{LSG index} = \frac{\text{VT of glass}}{\text{SHGC of glass}}$$

The greater the value of the LSG index, the more efficient the glass is with respect to reducing the solar heat gain and increasing light transmission. Because VT and SHGC for a $\frac{1}{8}$-in.-thick clear glass are 0.90 and 0.875, respectively, the LSG index for this glass is 0.90/0.875 = 1.03. Glass manufacturers can provide glass with a high LSG index by using a spectrally selective, low-E coating on glass (see Section 30.5).

ULTRAVIOLET TRANSMITTANCE (UVT)

Radiation, in general, has an adverse effect on human skin and eyes. It is also responsible for fading the colors of fabrics, carpets, and artwork. Ultraviolet radiation has a higher energy content than radiation at longer wavelengths. Thus, although it is only 3% of the total solar radiation, the material degradation potential of ultraviolet radiation is far greater than that of visible, or long-wave, radiation. Manufacturers of glass quote the ultraviolet transmittance (UVT) of their products. A glass with a lower UVT is generally preferable.

30.4 TYPES OF TINTED AND REFLECTIVE GLASS

Two commonly used glass products that are specially produced to lower solar heat gain are tinted glass and reflective glass. A glass may be either tinted, reflective, or tinted and reflective. Tinted and reflective glass is commonly used in the curtain walls of an all-glass building to reduce unwanted heat gain through the building.

TINTED (OR HEAT-ABSORBING) GLASS

Glass can be tinted to a desired color by adding metallic pigments to the molten constituents of glass during its manufacture. Tinted glass is also called *heat-absorbing glass* because it absorbs more solar radiation than clear glass under identical conditions.

A thicker glass, made with exactly the same concentration of pigment and the same batch of molten constituents as a thinner glass, appears to be deeper in color because a greater amount of light is absorbed by a thicker glass. For the same reason, a thicker tinted glass absorbs more heat than a thinner tinted glass.

As a result of greater heat absorption, the temperature of tinted glass is higher than that of clear glass when direct solar radiation is incident on the glass. A thick tinted glass sheet exposed to intense solar heat should be heat strengthened to withstand greater thermal stresses.

The most commonly used colors in tinted glass are bronze, gray, and blue-green. If everything else is the same, a green-tinted glass has a higher LSG index (i.e., is more efficient in reducing the solar heat gain and increasing light transmission) than other tinted glass.

REFLECTIVE GLASS

Reflective glass works as a mirror from the outside during the day, hiding interior activity. At night, however, the interior activity is visible from the outside when the interior space is lit.

Reflective glass is made by bonding metal or metal oxide coatings to one surface of a clear or tinted glass. Some of the metals used are chrome, stainless steel, titanium, and copper oxide. The coating is extremely thin so that sufficient light can pass through the glass and is deposited on the glass by one of two methods:

- Magnetic sputtering
- Pyrolytic deposition

Magnetic sputtering requires a large vacuum chamber in which atoms of metal or metal oxide are dislodged from the original material by an induced electrical charge. These atoms impinge on the surface of the glass placed in the chamber, creating a thin layer of coating. If the glass is required to be tempered or heat strengthened, it is heat treated prior to the application of the sputtered coating.

In the pyrolytic deposition process, a metal or metal oxide coating is applied to hot glass. This is accomplished either in the heat-strengthening furnace (for a heat-treated glass), at the hot end of the annealing lehr, or at the cool end of the float bath for annealed glass. Thus, the reflective coating is virtually fused into the glass sheet because the glass constituents are in a semisolid state when the coating is applied. Consequently, a pyrolytically deposited coating is more durable than a magnetically sputtered coating.

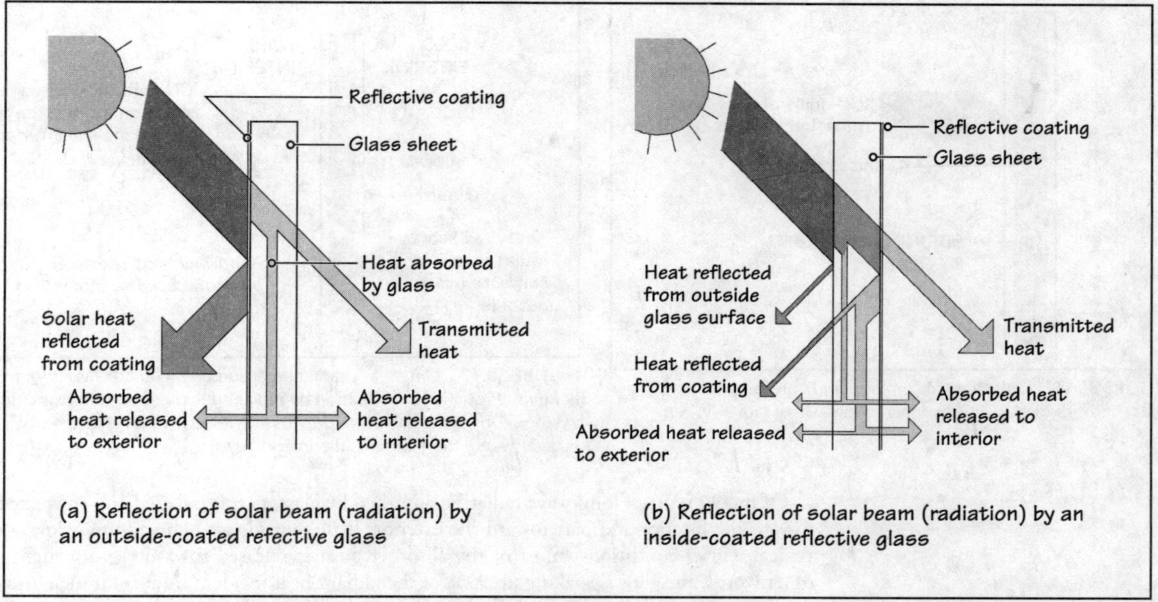

FIGURE 30.14 An outside-coated reflective glass reflects a greater amount of solar radiation than an inside-coated reflective glass but is less durable.

Reflective Glass and Solar Radiation

A reflective coating on the outside surface of a glass is more effective in reflecting solar radiation compared with the coating on the inside surface, Figure 30.14. This is due to the fact that in an inside-coated glass, the solar radiation passes through the glass twice, and hence it goes through two absorption cycles instead of one (in an outside-coated glass). Thus, an inside-coated glass will become warmer (and be subjected to greater thermal stress) than an outside-coated glass under the same conditions and will generally require heat strengthening. This is particularly true if the glass is also tinted.

The durability of the coating in an outside-coated glass must be examined. A sputtered coating is too soft to withstand the abrasive effects of wind and rain. Therefore, a sputtered coating should be exposed only to the interior. A more protected location, such as surface 2 or surface 3 of an insulating glass unit, is preferred for sputtered coating (see Section 30.6 for the description of an insulating glass unit). A pyrolytic coating may be exposed to the exterior, but the cleaning of such a glass must be done per the manufacturer's recommendations.

Reflective glass reduces solar heat gain but also reduces visible transmittance (VT). Because lower VT is generally undesirable, the use of reflective glass has decreased since the introduction of new spectrally selective, low-E coatings for glass, which provide low solar heat gain without greatly reducing VT (see Section 30.5).

Another disadvantage of reflective glass is that it reflects the solar radiation toward the street and the surrounding buildings. The glare created by such reflection is known to have temporarily blinded passing motorists. Several high-rise, all-glass buildings located on the opposite sides of a narrow street with reflective glass facades may make a street (particularly a narrow street) unduly warm in the summer due to multiple reflections of solar radiation, Figure 30.15.

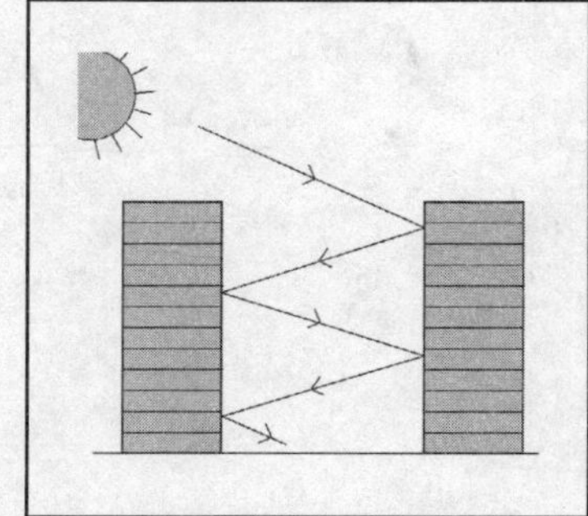

FIGURE 30.15 The use of reflective glass on the opposite facades of all-glass high-rise buildings can make a narrow street become unduly warm, particularly during summertime.

30.5 GLASS AND LONG-WAVE RADIATION

The properties of glass (SHGC, VT, and UVT) discussed previously are relevant only with respect to solar radiation (see the first Principles in Practice section at the end of the chapter). The properties of glass with respect to long-wave radiation are shown diagrammatically in Figure 30.16. Of the 100 units of long-wave radiation incident on glass, 90 units are absorbed, 10 units are reflected, and practically nothing is transmitted through glass. In other words, the long-wave absorptivity of glass is 0.90, reflectivity is 0.10, and transmissivity is zero. Because absorptivity is equal to emissivity (see Section 5.5), the emissivity of glass is also 0.90.

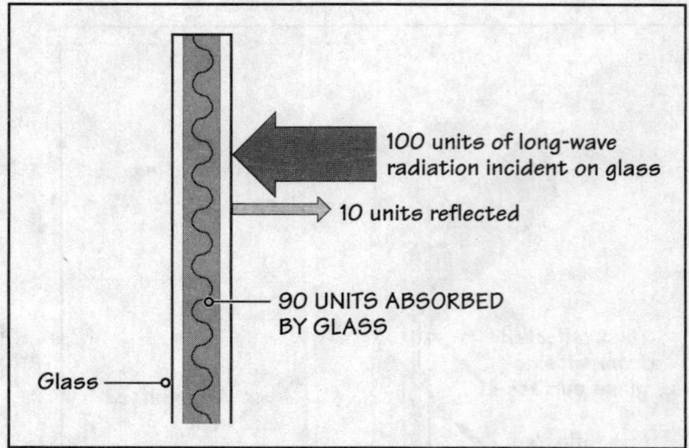

FIGURE 30.16 Reflection, absorption, and transmission characteristics of glass with respect to long-wave radiation.

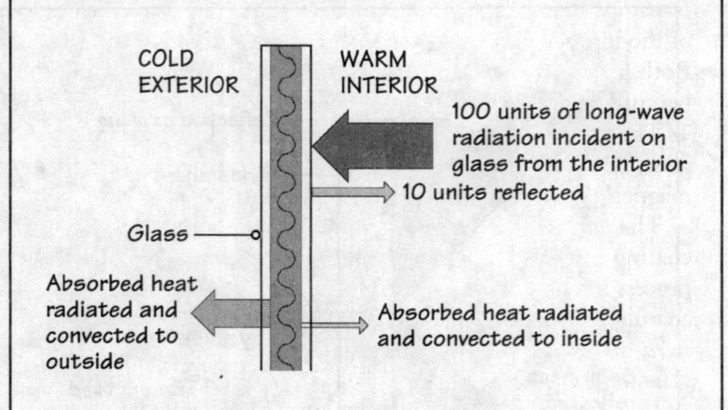

FIGURE 30.17 Long-wave radiation absorbed by (float) glass from a warm interior and its dissipation by radiation and convection to a cold exterior. Observe that the majority of interior heat is dissipated to the outside; very little returns to the interior.

Of the 90 units of long-wave radiation absorbed by glass, part is radiated and convected toward the interior and part toward the exterior. If the outside air temperature is low—a typical winter condition—most of the absorbed heat is released toward the outside. In other words, the glass is soaking up 90% of the radiant heat incident on it from the enclosure, most of which is being dissipated to the outside, Figure 30.17.

At this stage, it is important to appreciate that this mode of heat transfer is not the only mode by which heat from the inside is being lost to the outside. The other important mode of heat transfer is due to the temperature difference between the inside and outside air.

Low-Emissivity (Low-E) Glass

A simple way to reduce the loss of interior heat absorbed by glass to the exterior is to coat the surface of the glass with a low-emissivity (low-E) film. Because the emissivity of a surface is equal to its absorptivity, lowering the emissivity implies lowering its (long-wave) absorptivity and, hence, increasing the surface's (long-wave) reflectivity.

Consider once again 100 units of interior heat incident on a glass. If the interior surface of this glass is coated with a low-E coating with an emissivity of 0.1, 10% of the heat from the room interior will be absorbed and 90% will be reflected back to the interior, Figure 30.18. A glass with a low-emissivity coating is called a *low-E glass.*

Thus, low-E glass considerably reduces radiant heat loss through the glass and is extremely useful in cold climates. In other words, the behavior of low-E glass to long-wave radiation is similar to the behavior of reflective glass to solar radiation. Just as reflective glass reflects solar radiation back to the exterior, low-E glass reflects long-wave interior heat back to the interior. Low-E glass is also effective in warm climates, although less so (see the box entitled "Location of a Low-E Coating in an IGU" in Section 30.7).

A low-E coating is almost invisible to the eye because it transmits most of the visible solar radiation. A *spectrally selective low-E coating* is similar to a low-E coating in that it

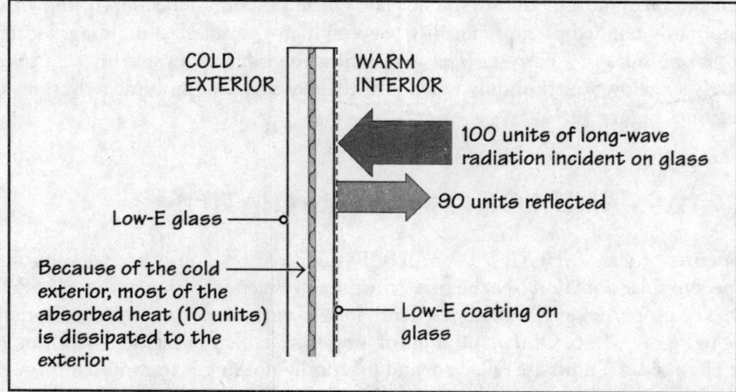

FIGURE 30.18 Long-wave radiation absorbed by low-E glass from a warm interior and its dissipation. Observe that the majority of interior heat returns to the interior.

transmits most of the visible radiation, but it is more efficient in blocking the transmission of the infrared and ultraviolet parts of solar radiation by absorbing and/or reflecting them. Both a reflective coating and tinting of the glass, on the other hand, are clearly visible because they reflect or absorb the visible solar radiation, respectively.

The process of producing low-E glass is exactly the same as that of producing a reflective glass, that is, by depositing a metal or a metal oxide coating on one of its surfaces, either by magnetic sputtering or pyrolytic deposition.

The sputtered coating has lower abrasion resistance than the pyrolytically deposited coating but gives a lower E value—nearly 0.10. Low-E glass produced by the pyrolytic process has an E value of nearly 0.3. Because of its lower abrasion resistance, the sputtered coating is referred to as a *soft-coat low-E* coating and the pyrolytic coating is referred to as a *hard-coat low-E* coating. Sputtered-coated low-E glass can be used only in an insulating glass unit where the coating is applied to either surface 2 or surface 3 of the unit.

30.6 INSULATING GLASS UNIT (IGU)

In Section 5.4, it was noted that the R-value of a glass sheet is approximately 1.0, which is provided almost entirely by the inside and outside surface resistances. This is an extremely low value compared to the value of the opaque portions of a wall. That is why most heat loss or gain in a building occurs through glazed areas.

Because air is a good insulator, the R-value of glass can be increased by providing an (air or gas-filled) space between two sheets of glass. To avoid problems with cleaning the surfaces of glass facing into this space, it is necessary that the intervening space be hermetically sealed. An unsealed space traps dust and moisture, both of which fog the view through the glass.

A sealed assembly consisting of two glass sheets with an intervening space is called an *insulating glass unit* (IGU). The most commonly used IGU in commercial construction is a 1-in. unit, in which the total width of the unit (the thickness of the two glass sheets plus the width of the space) is 1 in. In this unit, the width of the intervening space varies, depending on the thickness of the glass. If the two sheets of glass are each $\frac{1}{4}$ in. thick, the width of the intervening space is $\frac{1}{2}$ in. If the thickness of each glass sheet is $\frac{3}{16}$ in., the width of the intervening space is $\frac{5}{8}$ in. In residential windows, a $\frac{5}{8}$-in.-thick IGU (instead of a 1-in.-thick IGU) is generally used.

The construction of an IGU is shown in Figure 30.19. It consists of two sheets of glass with a spacer, which is sealed to the glass at the sides—the primary seal. The entire assembly is further sealed around all four edges—the secondary seal. The intervening space is filled with dry air or a gas.

FIGURE 30.19 Anatomy of an insulating glass unit (IGU). Although an aluminum spacer has been shown here, less conducting spacers are available that increase the R-value of the IGU.

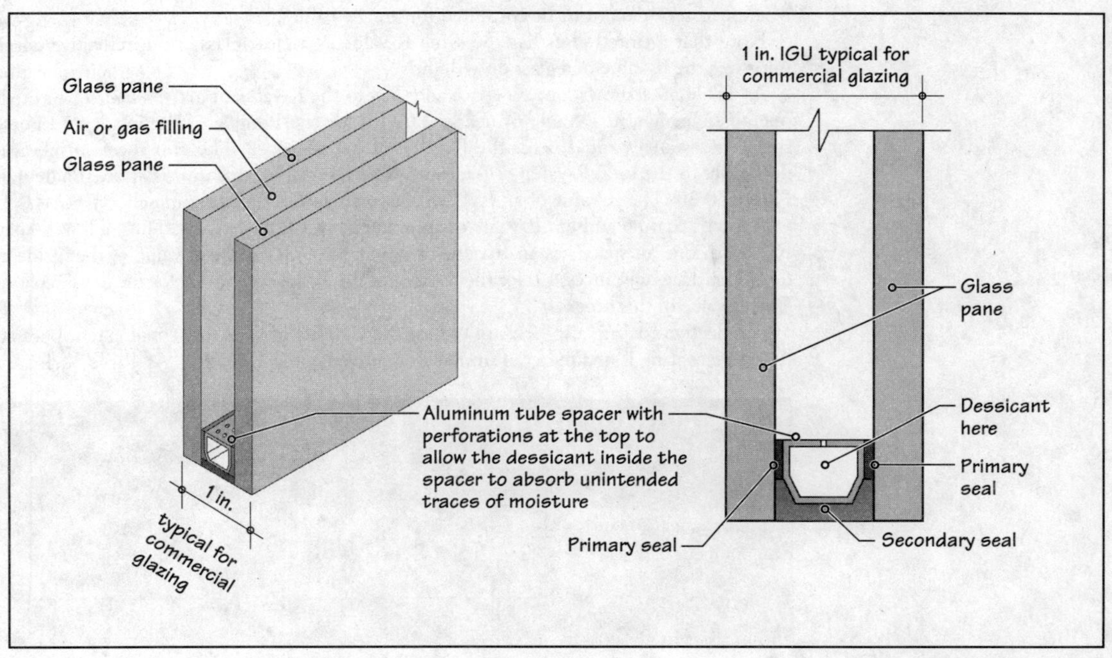

EXPAND YOUR KNOWLEDGE

Standard Surface Designation of Glass

Because contemporary window and curtain wall glass is generally not a single sheet of glass but a fabricated product with coatings, laminations, and gas or air filling, it is necessary to have a standard numbering system for various surfaces. The glass industry has adopted a numbering system for the surfaces of various glass products, as shown in the following images. Note that the number sequence starts from the exterior surface of glass. Only the glass surfaces carry a number. The surfaces of laminations or coatings are not numbered.

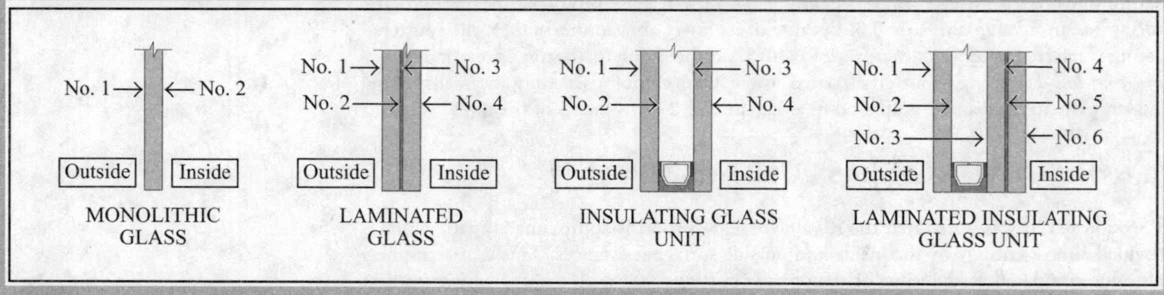

However, as an extra precaution, the metal spacer is filled with a desiccant to absorb any incidental moisture that may not have fully evacuated from the intervening space. The spacer may be of a clear silvery finish or any other required color.

Because argon has a higher insulating value than air (see Table 5.3), many manufacturers provide argon-filled units. Krypton is another gas that is used as a filling in IGUs. Krypton has better insulating properties than argon but is more expensive. Some IGU manufacturers use a krypton-argon mixture.

Air, argon, or krypton is filled in the intervening space with the pressure of 1 atm to prevent any loading of glass due to pressure differences. Note that if a vacuum were to replace the air or gas filling, the glass would be subjected to the atmospheric pressure (nearly 2,100 psf), which is several times greater than the wind load expected on a window glass, even in the worst possible hurricane.

30.7 R-VALUE (OR U-VALUE) OF GLASS

Knowing that the R-value of a single sheet of glass is approximately 1.0, the approximate R-values (or U-values) of other glass products can be obtained with reference to Table 30.1. Manufacturers' data must be consulted for precise values.

Note that a tinted glass has the same R-value as a clear glass and a reflective coating improves the R-value of a glass only slightly.

As Table 30.1 shows, one air space adds 1.0 to the R-value of glass. Thus, an IGU with one air space has an R-value of nearly 2.0 (1.0 for the R-value of the inside and outside surface resistances and 1.0 for the R-value of the air space. The two sheets of glass add negligibly to the total R-value of the unit). The U-value of this unit is approximately 0.5, Figure 30.20. The R-value of an IGU with two air spaces is approximately 3.0 ($U = 0.33$).

A low-E coating adds an R-value of approximately 1.0 to the unit. Thus, a low-E-coated IGU with one air space has an R-value of nearly 3.0 (1.0 for the R-value of the inside and outside surface resistances, 1.0 for the R-value of the air space, and 1.0 for the low-E coating). The U-value of this unit is 0.33.

A reflective coating also helps to reduce the U-value of glass to a small extent because a reflective coating is of a material similar to that of the low-E coating. As Table 30.1 shows,

TABLE 30.1 ADDITION TO THE R-VALUE OF GLASS BY VARIOUS ITEMS

Item	Contribution to R-value of glass
Air-filled space	1.0
Argon-filled space	1.6
Low-E coating	1.0
Reflective coating	0.2
Glass tint	0.0

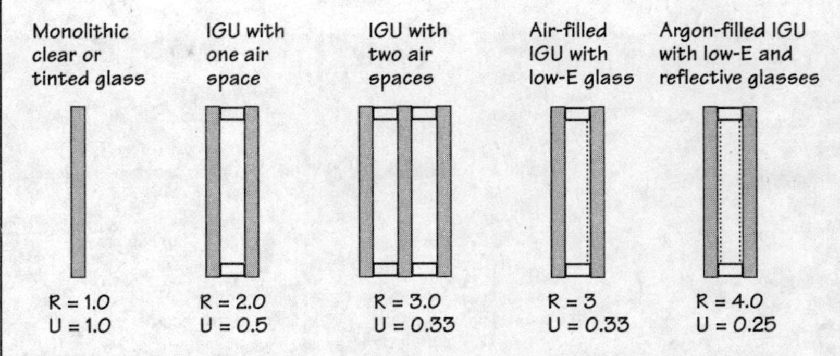

FIGURE 30.20 Approximate R-values (and U-values) of various types of IGUs. Note that the U-value is the inverse of the R-value.

the contribution of a reflective coating to the R-value is approximately 0.2. Thus, a reflective-coated monolithic glass will have an R-value of nearly 1.2 (U-value = 0.83). Note that the tinting of glass does not change its U-value.

Argon-filled units have lower U-values compared with units that contain air. Argon filling adds an R-value of approximately 1.6 to the unit. Thus, an argon-filled, low-E-coated reflective IGU has a total R-value of nearly 3.8 (1.0 for the internal and external surface resistances, 1.6 for the argon-filled space, 1.0 for the low-E coating, and 0.2 for the reflective coating).

An increase in the R-value of a glass product beyond 4.0 can be achieved only by adding a second intervening space in the IGU, but if this is accomplished by adding a third sheet of glass, it makes the unit heavier and costlier. One manufacturer makes a low-E IGU by using a patented system in which a clear plastic film is suspended halfway into the intervening space, Figure 30.21.

The film divides one space into two spaces, decreasing the U-value of the unit without making it heavier. The low-E coating is placed on the plastic film. Being transparent, the film is not visible, so that the unit is virtually indistinguishable from the standard one-intervening-space unit. The R-value of this unit with an argon filling is reported to be approximately 5.0, and it is 6.0 with a krypton filling.

Because of the plastic in the intervening space, this unit has a limited ability to withstand high temperatures. IGUs with two low-E-coated plastic films and krypton filling (giving three spaces) are available with an R-value of approximately 8.0.

U-VALUES FOR SUMMER AND WINTER CONDITIONS

In the previous discussion, we have assumed that the U-value of glass is constant for a given type of glass or an IGU. In fact, the U-value of glass depends, to some extent, on the external and internal environmental conditions. Thus, glass manufacturers quote separate U-values for summer and winter conditions. The difference between the two values is significant only if a detailed analysis of the energy performance of a building is needed.

NOTE

Location of a Low-E Coating in an IGU

A low-E coating is generally placed on the interior glass pane of an IGU (on surface 3). This location works well for cold climates because the low-E coating reflects the interior heat back.

For warm climates, a low-E coating is generally placed on the exterior glass pane of an IGU (on surface 2), which reduces the inward emission of solar heat absorbed by the exterior pane. However, if the IGU must also have a reflective coating, it is placed on surface 2 and the low-E coating is placed on surface 3. In this location, the low-E coating reflects the heat from the intervening space back to the exterior.

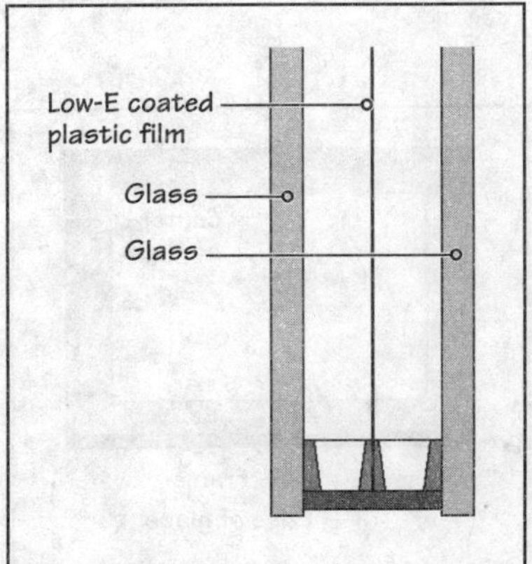

FIGURE 30.21 An IGU with a low-E-coated plastic film in air-filled or argon-filled space.

PRACTICE QUIZ

Each question has only one correct answer. Select the choice that best answers the question.

11. Tinted glass absorbs more solar radiation than clear glass.
a. True **b.** False

12. If the reflective coating on tinted glass is on the interior surface of the glass, the solar radiation absorbed by the glass is greater than if the same coating is on the outside surface.
a. True **b.** False

13. When solar radiation falls on a glass, a certain amount of solar energy enters the enclosure through the glass. A commonly used measure of solar heat gain through the glass is called
a. VT. **b.** SHGC.
c. UVT. **d.** CRF.
e. none of the above.

14. Before a reflective coating is applied on a glass, the glass must first be cut to size.
a. True **b.** False

15. Which of the following indices measures the effectiveness of the transparency of glass in relation to the solar heat gain through glass?
a. UVT **b.** SC
c. SHGC **d.** LSG
e. None of the above

16. A reflective coating increases the R-value of glass only marginally.
a. True **b.** False

17. An IGU is
a. an assembly of two glass sheets.
b. an assembly of two glass sheets with an intervening space.
c. a sealed assembly of two glass sheets with an intervening space.
d. a sealed assembly of two glass sheets with an intervening space and a spacer.
e. none of the above.

18. Which type of gas is used in an IGU?
a. Air **b.** Argon
c. Krypton **d.** All of the above
e. None of the above

19. To obtain an IGU with a high R-value, the space between the glass sheets should be
a. air filled. **b.** argon filled.
c. hydrogen filled. **d.** none of the above.

20. Surface 3 in an IGU is
a. the surface of the exterior glass sheet facing the exterior.
b. the surface of the interior glass sheet facing the interior.
c. the surface of the exterior glass sheet facing the intervening space.
d. the surface of the interior glass sheet facing the intervening space.

21. The approximate R-value of a clear glass sheet is
a. 1.0. **b.** 2.0.
c. 3.0. **d.** 4.0.
e. 5.0.

22. The approximate R-value of a tinted glass sheet is
a. 1.0. **b.** 2.0.
c. 3.0. **d.** 4.0.
e. 5.0.

23. The approximate R-value of an air-filled IGU is
a. 1.0. **b.** 2.0.
c. 3.0. **d.** 4.0.
e. none of the above.

24. The approximate R-value of an air-filled IGU with a low-E-coated glass is
a. 1.0. **b.** 2.0.
c. 3.0. **d.** 4.0.
e. none of the above.

30.8 GLASS AND GLAZING

The discussion of R-values given in the previous sections refers to glass only. In practice, however, we are interested in the R-value of glazing, which includes the glass and the frame in which the glass is held. The R-value of framing material (aluminum, vinyl, or wood) is not the same as that of glass. Therefore, the R-value of a glazing (a window or a curtain wall) is different from that of glass alone.

The overall R-value of glazing is obtained by averaging the R-values of its three parts: the center of the glass, the edge of the glass, and the frame. The center of the glass constitutes most of the glazing, Figure 30.22. The R-value of the center of the glass is the R-value of the glass itself. The approximate R-values given in Figure 30.20 refer to the R-values of the center of the glass.

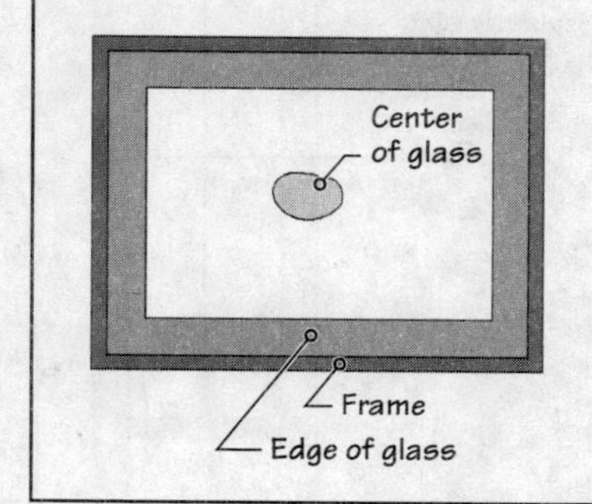

FIGURE 30.22 Three parts of a typical glazing (center of the glass, edge of the glass, and frame) that contribute differently to heat flow.

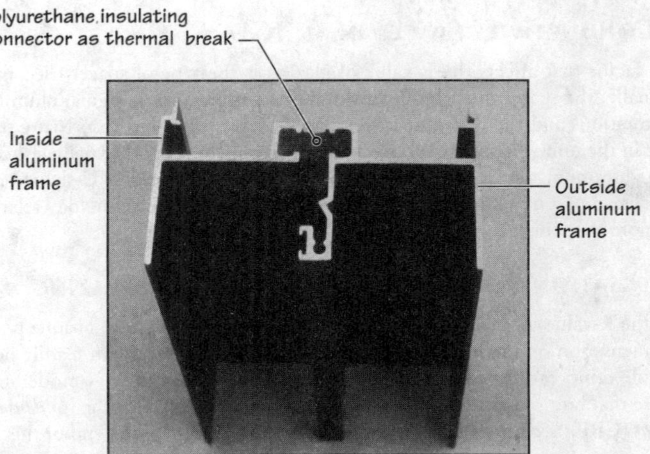

FIGURE 30.23 Polyurethane insulating connector (as a thermal break) between the inside and outside parts of an aluminum window frame.

The edge of the glass is an area surrounding the frame, which has a width of approximately 2.5 in. A greater amount of heat is conducted through this area than through the center of the glass because of its proximity to the frame (and the presence of the aluminum spacer in an IGU).

Therefore, the size of the glazing affects its R-value. A small aluminum window generally has a lower R-value than a large window, although the materials used in both windows may be the same. This is due to the relatively larger area of the frame and edges in a small window. In commercial curtain walls, where the framing members are spaced farther apart, the difference between the R-values of the glass and the glazing is smaller than that of a window.

Thermal Break in Aluminum Frames

The R-value of the frame depends on the dimensions of the frame, its material, and its configuration (i.e., whether or not there is a thermal break in the aluminum frame). A thermal break consists of an insulating connector (usually polyurethane) that connects the two parts of the aluminum frame, Figure 30.23. Polyurethane has excellent rigidity and strength to function as framing material.

The frames of most aluminum windows used in contemporary commercial construction are thermally broken. The use of a thermal break in aluminum window and glass curtain wall frames has improved the R-values of glazing in addition to reducing the condensation potential, Figure 30.24. Thermally broken window (or glass curtain wall) frames are more important for cold northern U.S. climates than for warm southern climates.

FIGURE 30.24 Illustration showing the effectiveness of a thermal break in an aluminum frame. (Photo courtesy of Vistawall Architectural Products)

LOW-CONDUCTIVITY SPACER IN AN IGU

Another factor that affects the R-value of glazing is the type of spacer used in an IGU. Traditionally, the spacer in an IGU consisted of a perforated rectangular aluminum tube. The perforations and the dessicant material within the tube help absorb any unintended moisture in the unit (Figure 30.19). Recently, spacers, which are thermally less conducting than the aluminum tube, have been introduced. These consist of a U-shaped aluminum spacer (instead of a tube), plastic, or a plastic foam spacer. The technology related to producing more efficient spacers is called *warm edge technology* (WET).

CONDENSATION RESISTANCE FACTOR (CRF) OF GLAZING

Because the R-value of glazing is fairly low compared with that of the opaque portions of a wall, condensation on the interior surface of the glass or its frame can readily occur when the outside temperature is low and the interior relative humidity is moderate to high. A measure that rates a glazing for its condensation potential is called the *condensation resistance factor* (CRF); see the second Principles in Practice section at the end of this chapter.

30.9 SAFETY GLASS

Because of the transparency of glass, a large glazed opening without an intermediate horizontal framing member may be mistaken for a clear (unglazed) opening. In this situation, there is a likelihood that a person may unwittingly attempt to walk through the glass, thinking that it is a clear opening. In this and several other situations, the use of annealed glass is hazardous because of the sharp pieces it produces on breaking.

Building codes require the use of safety glass in locations where accidental human impact is expected. Such locations are called *hazardous locations*. Some examples of hazardous locations are shown in the adjacent note. Two types of glass are considered to be safety glass:

- Fully tempered glass
- Laminated glass

They provide safety in different ways. On impact, tempered glass breaks into small, blunt, relatively harmless pieces. Laminated glass, on the other hand, is safe because it stays in place after breaking due to the plastic interlayer.

Not all tempered or laminated glass is safety glass. To be recognized as such, the glass must comply with the appropriate test standards. In this test, the glass specimen is mounted in a test frame, Figure 30.25. The test frame has an impactor suspended from the top. The impactor, which weighs 100 lb, consists of a leather punching bag filled with lead shots. The impact on the test specimen is produced by moving the impactor away from its rest position and releasing it.

The test simulates the impact of a person running into a glazing. Although 100 lb is less than the average weight of a human being, it is considered adequate because, in an impact with a glazing, the entire weight of the person is not likely to be involved. The glass is

NOTE

Hazardous Locations

Some examples of hazardous locations in buildings requiring safety glass are given here. For a comprehensive list and description, refer to the building code.

- Glass in a swinging door
- Glass in panels of a sliding door and a bifold closet door
- Glass in a door and enclosure for a hot tub, whirlpool, sauna, steam room, bathtub, and shower
- Glass in some fixed conditions adjacent to a glass door
- Glass in guardrails

(a) Specimen glass in the test frame, which is about to receive the impact

(b) Glass after the test. The size of broken glass pieces (which must be small enough and within the limits specified in the test standard) indicate that the glass has passed the test.

FIGURE 30.25 Testing a tempered glass specimen to verify that it conforms to the requirements for a safety glass.

assumed to have passed the test if the specimen satisfies the breakage criteria outlined in the test standard. Building codes require that every pane of tempered glass used in hazardous locations bear a label identifying its compliance with the test.

30.10 LAMINATED GLASS

In its simplest form, laminated glass is made from two layers of glass laminated under heat and pressure to a plastic interlayer so that all three layers are fused together, Figure 30.26. In the event of an impact, the glass layer in laminated glass may break, but it will not fall out of the frame. The glass tends to remain bonded to the interlayer, minimizing the hazard of shattered glass.

Laminated glass has been used for automobile windshields for a long time. In architectural applications, it is the product of choice for skylights, sloped and overhead glazing, zoos, and aquariums. Because the glass remains in the opening even after breakage, laminated glass is also recommended for windows in earthquake-prone areas to reduce injury from abrupt shattering of glass.

The plastic interlayer, generally polyvinyl butyryl (PVB), may be clear, patterned, or tinted. It comes in three standard thicknesses: 15 mil, 30 mil, and 60 mil. Where greater safety or security is required, a thicker interlayer (90 mil or thicker) or multiple layers of thinner sheets may be used.

A major advantage of the plastic interlayer is that it also blocks out most of the ultraviolet radiation. The plastic interlayer also improves the sound-transmission properties of glass. Any type of glass—annealed, heat-strengthened, or tempered—can be used in making laminated glass. Laminated glass can also be used in IGUs.

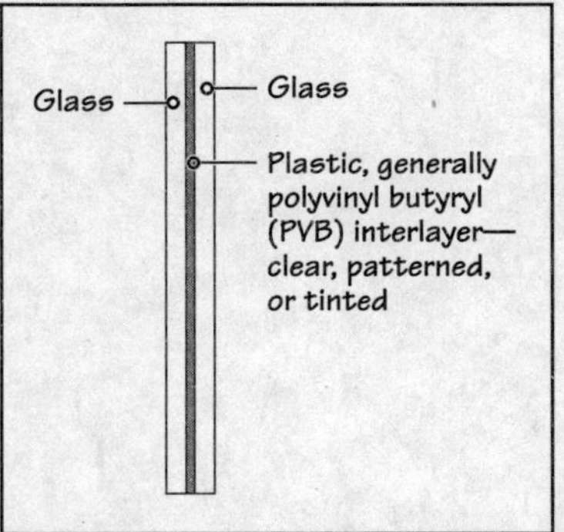

FIGURE 30.26 Section through a laminated glass sheet.

Laminated Glass as Safety Glass in Hazardous Locations

If used in hazardous locations, laminated glass must pass the same impact-resistance test as tempered glass. Additionally, it must pass the boil test. The boil test determines the ability of glass to withstand resistance to high temperatures. In this test, 12-in. by 12-in. specimens of laminated glass are immersed in boiling water for 2 h. The specimens are deemed to have passed the test if no bubbles or other defects more than $\frac{1}{2}$ in. from the outer edges of the glass develop.

Hurricane-Resistant and Blast-Resistant Glass

With a suitable thickness of PVB lamination, laminated glass can be made to resist the impact of wind-borne missiles and is mandated for openings in hurricane-prone coastal regions. Blast (explosion)-resistant laminates are also available, and blast-resistant glass is commonly specified to meet higher security requirements. Tempered glass can be laminated, and laminated-tempered glass can be used in IGUs.

30.11 STRUCTURAL PERFORMANCE OF GLASS

Glass is a brittle material; that is, a glass pane will fail at a relatively low deflection. As previously stated, glass is weak in tension and bending but relatively strong in compression.

Although glass is sometimes used in structural applications, such as small floor slabs and stair treads (see Chapter 35), its primary use is in nonstructural applications, such as windows, skylights, and guardrails. Therefore, glass is not considered a structural material like steel, wood, concrete, and masonry.

Glass in windows is required to resist wind, earthquake, and thermal loads. Glass in skylights must be able to withstand snow loads, loads due to repair personnel, and wind and earthquake loads. Glass in the rear walls of squash and racquetball courts is subjected to impact loads, and the glass used in aquariums must be able to withstand hydrostatic pressure. All of these loads cause bending in glass. Thus, the most important structural property of glass that is of interest to us is its bending strength.

Unlike other materials, the strength of glass is statistically complex because every glass sheet has numerous randomly distributed microscopic internal cracks and other flaws. Furthermore, the edge conditions of glass also affect its strength. Clean-cut edges provide the maximum strength. If the edges are damaged, the ability of the glass to resist loads is reduced. Thus, the actual breakage of glass results from a complex interaction between the size, orientation, distribution, and severity of cracks and edge defects.

Consequently, the strength of an individual glass pane cannot be predicted with reasonable accuracy. In other words, if several identical panes of glass are tested, there will be an unusually large variation in their breaking loads. Therefore, the strength of a glass pane is expressed in terms of the *probability of breakage* under a given load, that is, the number of panes that are likely to break per 1,000 panes. Thus, a probability of breakage of $\frac{3}{1,000}$ (three per thousand) means that 3 out of a total of 1,000 panes will probably break when a specified load is placed on the panes for the first time.

The maximum probability of breakage allowed by building codes for determining the glass thickness in windows and curtain walls to withstand wind loads is $\frac{8}{1,000}$. A lower probability of breakage may be required for glass in high-rise and other critical buildings.

ASTM Standard E 1300, "Standard Practice for Determining Load Resistance of Glass in Buildings," provides an authoritative procedure for determining the glass thickness to resist a given wind load.

HIGHER STRENGTHS OF HEAT-STRENGTHENED AND TEMPERED GLASS

As noted previously, tempered glass is nearly four times stronger than annealed glass. This implies that a tempered glass pane can withstand four times as much load as an annealed glass pane of the same thickness and size. However, the modulus of elasticity of tempered glass (and heat-strengthened glass) is the same as that of annealed glass. This implies that the deflection of tempered glass and annealed glass panes of the same size and thickness will be equal under a given load. Thus, if tempered glass is made to withstand a greater wind load because of its higher strength, greater deflection will be produced in the panes. This will place greater pressure on gaskets and seals, leading to their premature failure. It may also expel the glass out of the opening.

Therefore, in practice, tempered glass and heat-strengthened glass are designed to withstand the same wind load as annealed glass. In other words, the higher strengths of tempered glass and heat-strengthened glass are usually not exploited. Tempered glass and heat-strengthened glass are used primarily for their greater impact resistance, greater thermal stress resistance, and safer breakage pattern.

30.12 FIRE-RESISTANT GLASS

Ordinary glass (whether annealed, tempered, or laminated) is not resistant to fire. When subjected to a typical building fire, it shatters in 2 or 3 min. The types of glass commonly used as a fire-resistant glass are

- Wired glass
- Intumescent multilaminate glass
- Intumescent gel-filled glass units

WIRED GLASS

Wired glass is made by the rolling process, not by the float glass process. During the rolling process, welded wire mesh (which looks like chicken wire mesh) is embedded in the middle of the glass thickness so that the resulting product is a steel wire–reinforced glass, Figure 30.27. The wire diameter is nearly 0.020 in.

FIGURE 30.27 Wired glass in a fire-rated window between a parking garage and an office room.

The minimum thickness of glass recognized as fire-rated glass is $\frac{1}{4}$ in. When subjected to fire, wired glass also breaks in 2 to 3 min, but unlike annealed glass, which falls out of the frame, wired glass is held within the opening because of the wire mesh, and hence contains the spread of smoke and other fire gases.

Because of the embedded wires, wired glass is often confused with safety glass, which it is not. Unlike tempered glass, wired glass is not impact resistant. Additionally, its bending strength is only half that of annealed glass. On impact, wired glass breaks into sharp pieces, and the broken wires will generally project out of the glass, causing injury to a person. The wires can even act as a spider web on human impact and catch a victim rather than permit the person to pass through safely. Thus, although wired glass is fire resistant, it is not a safety glass.

INTUMESCENT MULTILAMINATE GLASS

Intumescent multilaminate glass (IMG) consists of multiple sheets of annealed glass alternated with intumescent interlayers. The number of glass and intumescent layers determine the fire rating. A fire rating of up to 2 h is attainable. When fire occurs, the intumescent layers become opaque and expand to provide the fire resistance.

INTUMESCENT GEL-FILLED GLASS UNITS

Intumescent gel-filled glass is an IGU in which the intervening space is filled with a clear gel. In the case of fire, the gel absorbs the heat, transforming the glass into a heat-insulating opaque crust.

A unit consisting of two sheets of $\frac{1}{4}$-in.-thick tempered glass with $\frac{3}{8}$-in. gel provides a $\frac{1}{2}$-h fire rating. The same construction with $1\frac{1}{8}$-in. gel provides a 1-h rating. A 2-h rating is obtained by a unit with two gel spaces of $1\frac{1}{4}$ in. and three sheets of glass. Under normal conditions (when there is no fire), this product looks like any other IGU with clear glass panes, Figure 30.28. These units can also be obtained to serve as safety glazing.

30.13 PLASTIC GLAZING

Clear plastic sheets (referred to as *light-transmitting plastics*) have gained popularity in some applications as an alternative glazing material. Three important advantages of plastic over glass are as follows:

- Plastic can be bent to curves far more easily than glass. Plastic is the material of choice in curved skylights, particularly doubly curved shapes such as domes.
- Plastic is several times stronger than glass of the same thickness and is also more impact resistant. It does not shatter or crack like glass. It is, therefore, specified in fenestration where glass breakage due to vandalism is a concern or where a high degree of security is required.
- Plastic glazing is lighter than glass.

NOTE

Fire-Rating Limitations of Wired Glass

Building codes place severe limitations on the use of wired glass as a fire-rated glass: $\frac{1}{4}$-in.-thick wired glass can be used in an opening whose fire rating is between 60 and 90 min, provided that the area of wired glass is less than 100 in.^{2} If the fire rating of the opening is 45 min, the maximum permissible area of wired glass is 9 ft^2 (1,296 in.^{2}). If the fire rating of a window is 20 min or less, the area of wired glass is not limited by the building codes.

For a fire-rated opening, both the wired glass and the opening must be so labeled by the manufacturers. Not every $\frac{1}{4}$-in.-thick wired glass is a fire-rated glass.

NOTE

Building Codes and Plastic Glazing

Building codes place stringent restrictions on the use of light-transmitting plastics in wall and roof openings. Less stringent restrictions apply to buildings of Type V(B) construction (see Section 7.5 for an explanation of the types of construction).

FIGURE 30.28 Lobby at the Red Lion Inn, Costa Mesa, California. Glazed walls provide a $1\frac{1}{2}$-h fire rating. (Photo courtesy of SAFTI FIRST, a division, O'Keeffe's Inc.)

However, the disadvantages of plastic glazing far outweigh its advantages. That is why its use is limited to applications where glass cannot be used. Some of the disadvantages of plastic glazing are as follows:

- Plastic has a much greater coefficient of thermal expansion than glass. If the movement of plastic glazing is restricted, it will visibly bow out in the direction of higher temperature. Framing details must, therefore, allow linear movement as well as the rotation of glazing. Consequently, framing details are more complex with plastic glazing, which further adds to its cost.
- Humidity changes also affect the dimensional stability of plastics. Plastic expands with increasing humidity, creating the same problems as thermal expansion.
- Plastic is not abrasion resistant, although abrasion-resistant coatings on plastic glazing considerably improve the abrasion resistance of plastics.
- Plastic is vapor permeable, so that sealed insulating units are not possible.
- Plastic yellows with age (ultraviolet degradation), reducing its clarity and light transmission.
- The most serious disadvantage of plastic is its combustibility. It will contribute fuel to a building fire. Therefore, building codes place severe limitations on the area of plastic glazing and the height of the building above which it cannot be used.

NOTE

Acrylic and Polycarbonate

Acrylic—more weather and chemical resistant; commonly specified for skylights and greenhouses

Polycarbonate—more vandal and heat resistant; commonly specified in bus shelters and glazing subjected to vandalism

ACRYLIC VERSUS POLYCARBONATE GLAZING

At the present time, two types of plastic are used for glazing: acrylic and polycarbonate. Acrylic has better resistance to ultraviolet radiation than polycarbonate. It yellows more slowly and has greater resistance to abrasion and scratching than polycarbonate. It also has greater resistance to chemicals such as household and glass cleaners.

Polycarbonate, on the other hand, has greater impact resistance than acrylic. It can withstand 30 times as great an impact as acrylic and 250 times as great an impact as glass, making it the most vandal-resistant glazing. Polycarbonate is, therefore, commonly specified for bus shelters and school windows where vandalism is a problem.

Polycarbonate can also withstand higher temperatures than acrylic. Therefore, polycarbonate is commonly used in signs that are exposed to high temperatures from light sources. Acrylic is commonly used in curved skylights and greenhouses, where greater durability and weatherability are required.

30.14 GLASS FOR SPECIAL PURPOSES

There have been major advances in glass and glazing technology in recent years. A few of the glass types used for special purposes are described next.

SECURITY GLAZING

Security glass refers to a glass that is resistant to burglary, forced entry, and bullet penetration. Because the safety against these threats is a function of both the glass and the frame in which the glass is held, the term *security glazing* is commonly used rather than *security glass.* Typically, security glazing is composed of single or multiple panes of glass fused with single or multiple sheets of plastic, usually held in a steel frame.

Security glazing can be designed to withstand different levels of threat, including repeat assaults. For example if the threat level is burglary, the design must consider repeated impacts from hand-held weapons such as hammers, crowbars, or bricks.

Ballistic threats of various levels can be accommodated, depending on the power of the firearms to be resisted. Ballistic gazing is designed to be shatterproof on the victim's side, because the shattering of glass on the victim's side can cause serious injuries. Ballistic glazing is generally specified for drive-in bank windows, armored vehicles, and any other application where a high degree of security is required.

RADIATION-SHIELDING GLASS AND PLASTIC

The ability of a building material to shield against radiation is generally a function of its density. The greater the density and the thicker the material, the greater its radiation-shielding capability. Metals, concrete, and masonry materials (provided that there are no cracks in concrete and masonry) provide excellent radiation protection.

However, lead—because of its highest density among building materials and its flexibility—is the most commonly used radiation-shielding material. Where transparency is not required, lead curtains are commonly used, and lead aprons are worn for additional protection.

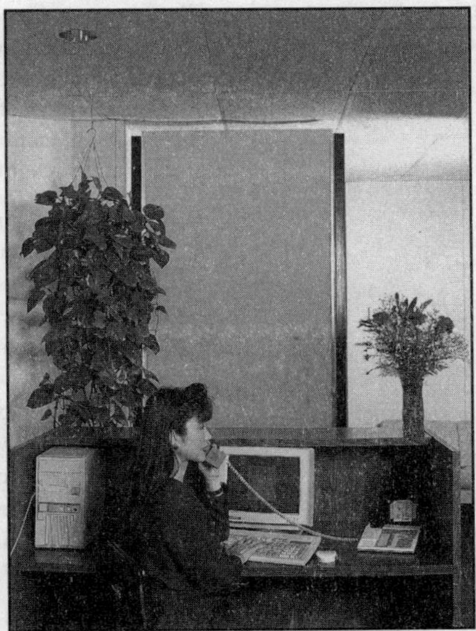

FIGURE 30.29 An example of the use of dimmable glass showing a glass pane that is transparent in the left-hand image and opaque in the right-hand image. (Photo courtesy of Polytronix, Inc.)

Radiation-shielding glass is made by adding lead (usually in the form of lead oxide) to the same ingredients as those of soda-lime glass. Lead acrylic is made from lead-acrylic polymer. Lead glass and lead acrylic have much greater radiation-shielding potential than ordinary soda-lime glass.

Dimmable Glass

Dimmable glass refers to glass in which the light-transmission characteristics change from an opaque to a transparent glass with the flick of a switch. It consists of two layers of glass with a plastic interlayer that encapsulates polymer-dispersed liquid crystals (PDLC). When an electric field is applied to PDLC, the crystals orient themselves to let the light through. In its natural state (i.e., without the electric current), the glass is opaque, Figure 30.29.

Also referred to as *electrochromic glass,* dimmable glass is used in situations where privacy requirements are changeable, such as conference rooms or executive offices. Although far more expensive than normal glass, it may be used in residential and office glazing for privacy and energy-conservation reasons. The switch mechanism can also be made automatic by a photovoltaic mechanism that responds to the ambient light environment. If the ambient light is excessive (e.g., due to direct solar radiation on the glass), the glass turns opaque. When the ambient light is low, the glass turns transparent.

Self-Cleaning Glass

A special oxide coating applied to one surface of the glass can make it self-cleaning. The coating breaks down organic dirt (due to trees, leaves, bird droppings, etc.) on the glass so that the glass is cleaned when rainwater falls on it. The coating does not affect inorganic dirt (e.g., plaster and paint droppings), which must be removed by conventional means. (The coating becomes active due to ultraviolet rays and may take nearly 1 week of ultraviolet exposure for a complete breakdown of organic dirt.)

30.15 CRITERIA FOR THE SELECTION OF GLASS

With so many factors to consider, the selection of glass, particularly for windows and curtain walls, is not an easy task. The selected glass must obviously be structurally adequate on its own. Additionally, the frame in which the glass is held must be adequate so that the entire glazing system is structurally adequate.

The safety of glass against human impact, the resistance of glass against wind-borne missiles, and the security of glass against blasts or bullets are other important considerations. These considerations apply to the glass as well as to the frame.

The energy performance of glass also plays an important role in its selection. As with structural and safety considerations, the energy performance of the entire glazing must be evaluated. To help designers, manufacturers of windows and curtain walls provide energy performance ratings of their products (see Chapters 31 and 32).

A checklist of factors that must be considered in the selection of glass for windows, curtain walls, and skylights is as follows:

- Structural adequacy—wind-load resistance, earthquake resistance (where needed), snow-load resistance (for skylights)
- Energy performance (U-value, SHGC, VT, LSG index, air infiltration, and CRF, as they apply to the entire glazing, are some of the important factors)
- Resistance to temperature-related stress
- Fire resistance, where needed
- Safety glass, where needed
- Blast resistance, where needed
- Missile-impact resistance and cyclic wind-load resistance in hurricane-prone areas
- Impact resistance in sports facilities (e.g., racquetball and squash courts)
- Radiation shielding, where needed

30.16 ANATOMY OF A GLAZING POCKET

In most glazing systems, the glass is captured by the framing members within a U-shaped pocket that is continuous on all four edges of the glass, referred to as the *glazing pocket,* Figure 30.30. Glazing pockets provide bearing and lateral load supports to glass. Additionally, they collect any incidental rainwater that might go past the glazing seals and weep it out.

Because the coefficient of thermal expansion of glass is different from that of the framing members, sufficient space must be left between the edges of the glass and the edges of the glazing pocket to allow the glass to move. Additionally, as the glass moves within the opening, it must not directly touch the frame but must be able to "float" within the opening. Important components of a glazing pocket are

- Setting blocks
- Edge blocks and (in seismic zones) corner blocks
- Glazing seals
- Weep holes

SETTING BLOCKS

Each pane of glass is supported on two identical *setting blocks,* generally at quarter points, Figure 30.31. Setting blocks serve two important functions: (a) they provide a hard, but resilient, bearing support to the glass, and (b) they lift glass up from the bottom of the glazing pocket to allow water to drain out from underneath and to prevent the glass from coming in direct contact with water. (The seals in an IGU and the PVB layer in a laminated glass are susceptible to damage by sustained contact with water.)

Neoprene, EPDM, or silicone are the materials commonly used for setting blocks. Generally, the setting blocks are rectangular in shape, but other profiles may be used as long as they allow sufficient clearance around the blocks for the passage of water.

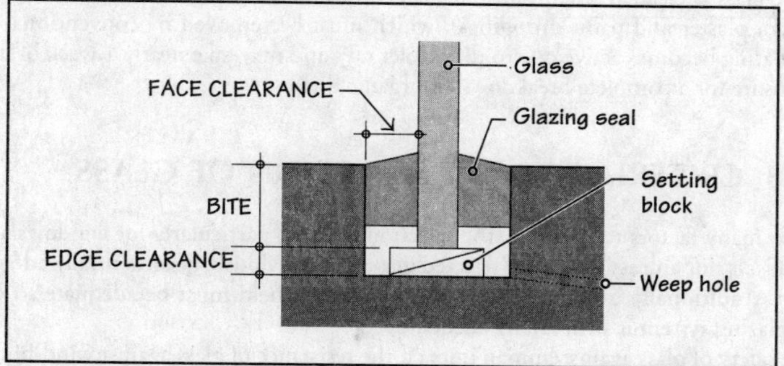

FIGURE 30.30 Glazing pocket and its components. Note that the dimensions of face clearance, edge clearance, and bite are controlled by industry standards.

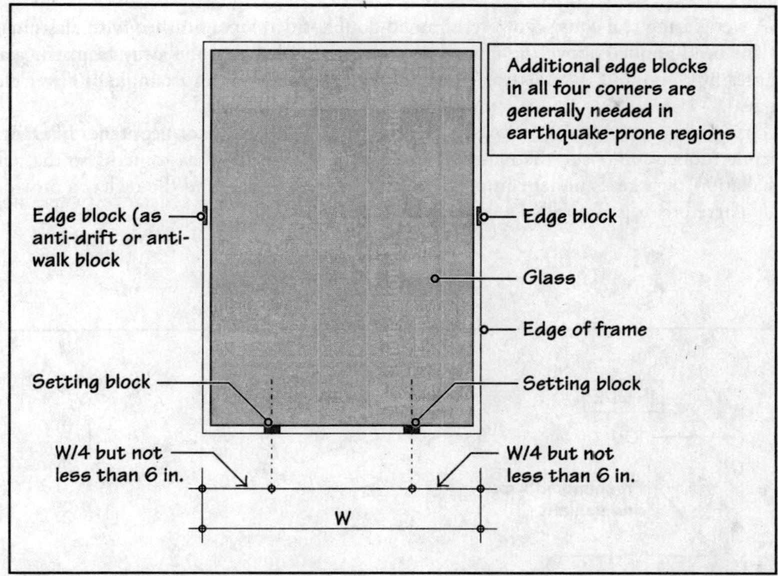

FIGURE 30.31 Setting blocks and edge blocks in a glazing. Both blocks are made of the same material.

EDGE BLOCKS

Edge blocks are used to separate the vertical edges of glass from the vertical frame members. Also referred to as *antiwalk blocks,* they cushion the glass from the frame and allow the glass to move within the opening without touching the frame. The movement of glass may be due to temperature changes or the effect of wind or earthquake forces. Corner blocks are required in addition to edge blocks in seismic zones.

GLAZING SEALS—WET-GLAZED AND DRY-GLAZED SEALS

Glazing seals are used within the glazing pockets to cushion the glass and help seal the glazing against air and water infiltration. One of the earliest glazing seals was *glazing putty,* also called *glazing compound.* Glazing putty is made from linseed oil and powdered limestone. It was generally used in a rabbeted frame to function both as a seal as well as a stop, Figure 30.32.

Glazing putty has largely been replaced by better glazing seals. Based on the glazing seal used, contemporary glazing is referred to as one of the following systems:

- Wet-glazed system
- Dry-glazed system
- Wet/dry-glazed system

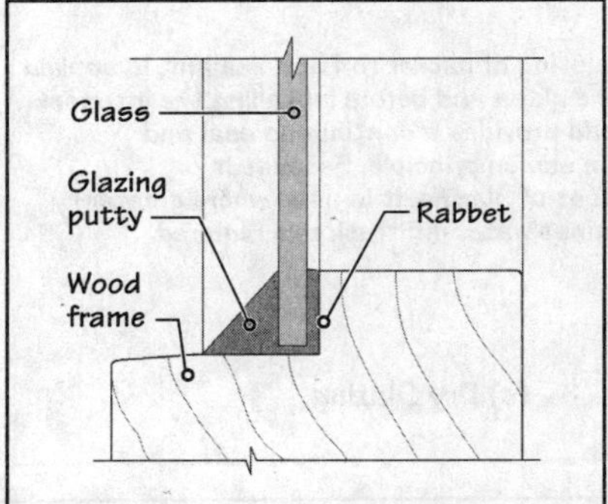

FIGURE 30.32 In earlier wood and steel windows, glazing putty served both as a stop and also as a seal.

The term *stop* in glazing literature is used for a member that holds the glass against the rabbet in the frame, providing lateral support to the glass. Thus, in this illustration, glazing putty serves as the stop.

The recess in the frame against which the glass is held is called a *rabbet.* The space between the stop and the rabbet constitutes the glazing pocket.

A wet glazing seal consists of preshimmed double-sided tape, finished with elastomeric sealant bead applied above the tape. The sealant is tooled to slope away from the glass, Figures 30.33(a) and (b). Glazing putty (of earlier times) was an example of a wet glazing seal.

Dry glazing seals consist of *gaskets.* A gasket is generally made of neoprene, EPDM, or silicone rubber composite. Its material, size, and profile make it elastomeric, so that after installation, a gasket is under compression, clinging to the glass and the reglet to provide a seal. Therefore, it is also called a *compression gasket.*

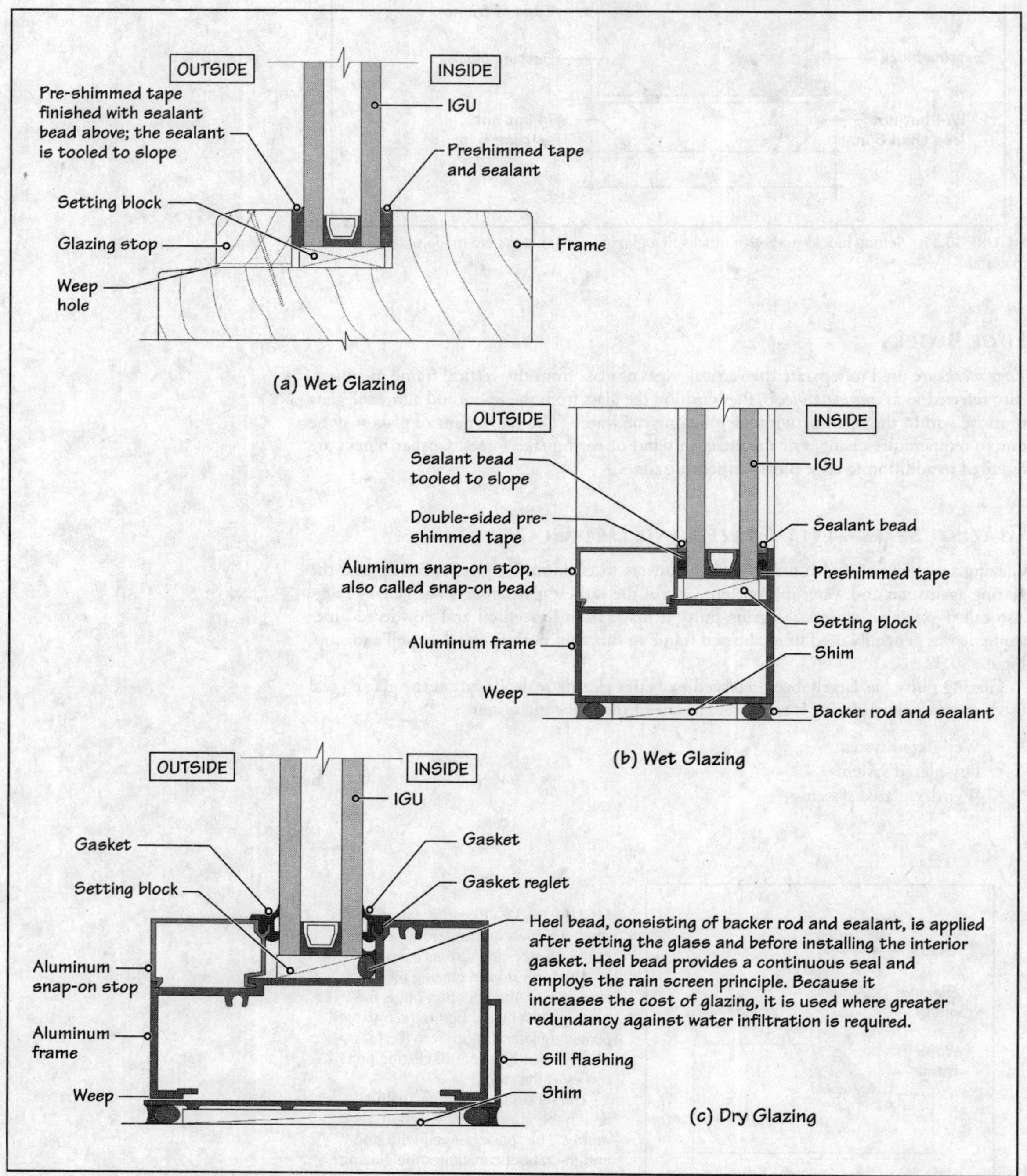

FIGURE 30.33 Examples of wet-glazed and dry-glazed seals.

TABLE 30.2 RELATIVE ADVANTAGES AND DISADVANTAGES OF DRY AND WET GLAZING SEALS

Type of seal	Advantages	Disadvantages
Wet glazing seal	Better resistance to water infiltration	Requires exterior and interior installation Performance dependent on workmanship Installation weather dependent More expensive
Dry glazing seal	Can be installed from the interior Less weather dependent Less dependent on workmanship Less expensive	Less watertight Gasket can shrink and create openings

Gaskets need some form of engagement into the frame. Therefore, gaskets are generally used with aluminum frames because, aluminum being a soft metal, can be easily profiled into a complex section to capture the gaskets. The part of the aluminum section that captures the gasket is referred to as a *gasket reglet,* Figure 30.33(c).

Dry glazing is more commonly used in contemporary glazing than wet glazing because the performance of dry glazing is less affected by workmanship, weather, and compatibility issues, Table 30.2. A wet/dry-glazed system is a combination of the two systems.

RAIN-SCREEN PRINCIPLE APPLIED TO GLAZING POCKET

Regardless of whether dry glazing or wet glazing is used, some water may enter the glazing pocket due to poor workmanship or the degradation of the seals. Some window and curtain wall manufacturers recommend the use of the rain-screen principle to glazing pockets.

Because an air seal in the backup is an important component of a rain screen, the use of an additional continuous sealant bead and a backer rod in the glazing pocket, referred to as a *heel bead,* is recommended, Figure 30.35(c). A heel bead is applied in the glazing pocket after setting the glass and before applying the interior gasket.

PRACTICE QUIZ

Each question has only one correct answer. Select the choice that best answers the question.

25. The R-value of an aluminum-framed glazing is generally
- **a.** higher than that of glass.
- **b.** lower than that of glass.
- **c.** equal to that of glass.

26. Laminated glass is
- **a.** an assembly of two glass sheets welded (fused) together into one sheet.
- **b.** an assembly of two glass sheets with an intervening space.
- **c.** a sealed assembly of two glass sheets with an intervening space.
- **d.** an assembly of two glass sheets fused together with a plastic interlayer.
- **e.** none of the above.

27. Which of the following statements is true about plastic glazing?
- **a.** It is stronger than glass.
- **b.** It is more vapor permeable than glass.
- **c.** It is less scratch resistant than glass.
- **d.** All of the above.
- **e.** None of the above.

28. The primary reason for the limited use of plastic glazing in buildings is that it is more expensive than glass.
- **a.** True
- **b.** False

29. Which type of glass is commonly used in drive-in bank windows?
- **a.** Annealed glass
- **b.** Tempered glass
- **c.** Laminated glass
- **d.** Heat-strengthened glass

30. Glass can be tempered as well as laminated; that is, laminated, tempered glass is available.
- **a.** True
- **b.** False

31. A glazing stop is used
- **a.** as a bearing support for glass.
- **b.** to seal the glass against water and air infiltration.
- **c.** to provide lateral support to glass.
- **d.** to provide a cushion between the glass and the glazing pocket.
- **e.** none of the above.

32. If the glass is sealed with preshimmed tape and sealant, the glazing is referred to as
- **a.** dry glazed.
- **b.** wet glazed.
- **c.** dry/wet glazed.
- **d.** wet/wet glazed.
- **e.** dry/dry glazed.

33. A heel bead is used in a glazing pocket.
- **a.** True
- **b.** False

34. A heel bead is used to
- **a.** increase the R-value of glazing.
- **b.** decrease the solar heat gain through glass.
- **c.** decrease water infiltration through glazing.
- **d.** increase the wind-load resistance of glazing.
- **e.** none of the above.

PRINCIPLES IN PRACTICE

Important Facts About Radiation

Radiation is emitted by all objects. Consequently, all objects receive radiation from other objects in their surroundings. In other words, a constant radiation exchange occurs between objects that have a direct line of sight between them (i,e., those that can see each other).

As the temperature of an object increases, the wavelength of radiation emitted by it becomes smaller. An object at a high temperature emits most of its radiation in short wavelengths. Conversely, an object at a low temperature radiates in long wavelengths.

A type of high-temperature radiation of interest is solar radiation. Although the actual temperature of the interior of the sun is several million degrees Fahrenheit, solar radiation received on the earth's surface has the same characteristics as the radiation from an object whose temperature is approximately 11,000°F (6,000°C).

Low-temperature radiation is present everywhere because it is emitted by objects on the surface of the earth, such as the ground, sky, landscape elements, building surfaces, and furniture. The temperature of these objects seldom exceeds 150°F (60°C).

Because solar radiation is high-temperature radiation, it is referred to as *shortwave radiation.* The radiation from earthly objects, on the other hand, is *long-wave radiation.* This difference is shown in the adjacent illustration, which gives the wavelength composition of solar radiation and radiation from an object at 100°F (38°C).

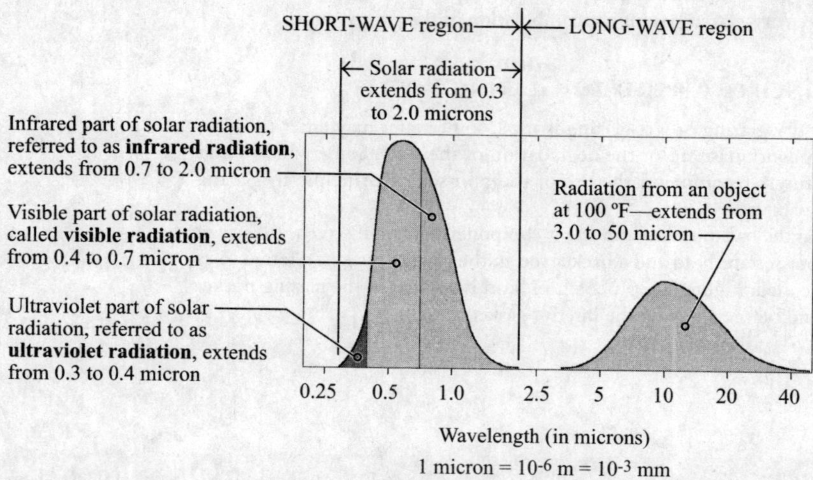

COMPONENTS OF SOLAR RADIATION (ULTRAVIOLET, VISIBLE, AND INFRARED RADIATION)

Observe that solar radiation extends from a wavelength of 0.3 μm to 2.0 μm. Approximately 3% of solar radiation lies in the ultraviolet region (wavelength less than 0.4 μm), 50% lies in the visible region (0.4 to 0.7 μm), and the remaining 47% lies in the infrared region. Radiation from an object at approximately 100°F (38°C) extends from a wavelength of nearly 3 to 50 μm.

The wavelength region up to 2.0 μm is generally referred to as the *short-wave region,* and beyond 2.0 μm, it is called the *long-wave region.* Building surfaces receive short-wave radiation only from the sun. However, they receive long-wave radiation from all surrounding objects, including the sky, because they are at low temperature.

The radiation from the sky is low-temperature radiation because it is due to the heat contained in the atmosphere (air, dust, and water vapor particles). It is also referred to as *diffuse solar radiation.* The term *solar radiation* refers to the radiation contained in direct solar beams.

SURFACE EMISSIVITY, SURFACE COLOR, AND LONG-WAVE RADIATION

Building surfaces and objects behave differently to short-wave and long-wave radiation. As shown in Figure 30.13, a clear glass sheet transmits virtually all (85%) short-wave solar radiation. Having penetrated the glass, solar radiation heats room surfaces and the contents of the room. These objects, being at low temperature, emit long-wave radiation. Glass is virtually

PRINCIPLES IN PRACTICE

Important Facts About Radiation (*Continued*)

opaque to long-wave radiation because it absorbs 90% of it (Figure 30.16). Having absorbed the interior heat, glass dissipates part of it to the interior and part to the exterior by convection and radiation. In other words, glass does not allow the interior heat to escape as readily as it admitted the solar heat. This is the well-known *greenhouse effect.*

From Section 5.5, we know that the emissivity of a surface—a concept that applies only to long-wave radiation—is equal to its absorptivity. Thus, the reason glass absorbs nearly 90% of long-wave radiation is that its emissivity is 0.9.

A low-E surface absorbs little and reflects most long-wave radiation. A low-E coating on glass changes the behavior of glass in response to long-wave radiation. A reflective coating on glass, on the other hand, changes the behavior of glass with respect to solar radiation. It does not greatly affect the behavior of glass with regard to long-wave radiation.

Similarly, long-wave radiation does not discriminate between the colors of surfaces. Solar radiation does. A white or light-colored surface is more reflective of solar radiation than a dark-colored surface. Long-wave radiation does not differentiate between the various colors.

PRINCIPLES IN PRACTICE

Condensation-Resistance Factor

The condensation-resistance factor (CRF) has been defined such that the higher its value, the better the performance of glazing with respect to condensation. The CRF of a glazing is obviously related to its R-value. The higher the R-value of glazing, the higher its CRF. Thus, a window with an IGU has a higher CRF than one with a single sheet of glass.

The CRF of a window or curtain wall is obtained by measuring it in a test chamber. It is given in whole numbers. Manufacturers provide CRF values for their products. In choosing a glazing based on the CRF value, Table 1, which gives the minimum recommended CRF value of glazing based on the interior humidity and outside winter air temperature, is generally invoked. The inside temperature is assumed to be 68°F (20°C).

For example, according to this table, if the minimum winter air temperature at a location is 20°F (−11°C) and if the interior relative humidity of a building is 40% (typical of an office building), a minimum CRF value of 48 should be specified for the glazing in that building. In obtaining CRF values from Table 1, the outside air temperature is the design winter temperature of the location, defined as the temperature that is exceeded more than 97.5% of the time during December, January, and February.

Specifying a CRF value based on Table 1 does not ensure that no condensation will occur on the interior surface of glazing. It simply means that condensation will be minimal and within acceptable limits. CRF values below 35 are generally not specified.

TABLE 1 RECOMMENDED MINIMUM CRF VALUES OF GLAZING

This table is based on American Architectural Manufacturers Association (AAMA) Standard 1503-98 (revised 2004), "Voluntary Test Method for Thermal Transmittance and Condensation Resistance Factor of Windows, Doors, and Glazed Wall Sections."

The entry *X* in the table indicates that no test data are available in this standard. Underlined numbers indicate extrapolated values from the test data.

The table is based on the interior temperature of 68°F (20°C) and outside wind speed of 15 mph.

	Outside air temperature	Inside relative humidity 15%	20%	25%	30%	35%	40%
		Minimum CRF values					
Minneapolis →	−20°F	45	52	57	62	X	X
Chicago →	−10°F	40	47	53	58	65	X
Boston →	0°F	32	40	47	52	58	63
New York →	10°F	17	30	37	45	52	57
Atlanta →	20°F	X	X	23	33	40	48

PRACTICE QUIZ

Each question has only one correct answer. Select the choice that best answers the question.

35. Building surfaces emit

a. short-wave radiation.
b. ultraviolet radiation.
c. long-wave radiation.
d. all of the above.
e. none of the above.

36. Glass is virtually opaque to

a. short-wave radiation.
b. ultraviolet radiation.
c. long-wave radiation.
d. all of the above.
e. none of the above.

37. Solar radiation is

a. short-wave radiation.
b. long-wave radiation.

38. A low-E glass is a poor reflector of long-wave radiation.

a. True
b. False

39. The condensation resistance factor (CRF) of glazing is directly related to

a. the material used in framing members of glazing.
b. whether or not reflective glass has been used in the glazing.
c. whether or not tempered glass has been used in the glazing.
d. the R-value of the glazing.
e. all of the above.

40. The minimum value of the CRF generally specified is

a. 55.
b. 45.
c. 35.
d. 25.
e. 15.

REVIEW QUESTIONS

1. Using sketches and notes, explain the process of manufacturing flat glass.
2. Using a sketch, explain why a tempered glass sheet is stronger than an annealed glass sheet.
3. What is heat-soaked glass and where is it used?
4. Using a sketch, explain the anatomy of an IGU.
5. Using a sketch, explain the standard surface designation system for (a) monolithic glass, (b) laminated glass, and (c) an IGU.
6. What is a low-E glass, and what are its benefits?
7. If low-E glass is used in an IGU, on which surface is the low-E coating placed, and why? Explain.
8. Explain how a low-E coating increases the R-value of glass.
9. Which type of glass would you recommend for the following locations?
   a. Sliding-glass entrance door
   b. Spandrel area of an all-glass curtain wall
   c. Teller window of a drive-in bank

CHAPTER 31 Windows and Doors

CHAPTER OUTLINE

Not long ago, windows and doors could be made only from wood. Ornate carpentry was often used, and architects spent an inordinate amount of time and effort detailing the cross-sectional profiles of window and door frames, window sashes, door panels, and the related trims.

Because of the lack of automation and manufacturing technology, the carpentry and joinery related to window and door fabrication were carried out at the construction site. The performance requirements of windows and doors were modest at best and were directly related to the skill of the craftsperson building the window or door; good workmanship yielded higher-performing windows or doors. For windows, letting the daylight in and keeping the rainwater out were the only requirements. Because large sheets of glass were not available at that time, window glazing consisted of small panes of glass held by narrow horizontal and vertical wood members. For doors, the primary performance requirement was security.

Like most other building components, windows and doors have evolved over the years. Windows are now made not only from wood but also from aluminum, steel, vinyl, and fiberglass. Additionally, two materials are sometimes combined in the same window to take advantage of their different properties.

Complex cross-sectional profiles for frames and sashes, sophisticated glazing, gaskets, weatherstripping, and hardware are required to satisfy more-demanding performance requirements for windows. Consequently, contemporary windows are almost always shop fabricated, and, in some cases, they are shipped preglazed to the construction site. This means that the only operation required at the site is to secure them in the opening and seal the joints between the window and the wall opening.

Due to the sophisticated fabrication process for windows and the rigorous performance testing to which they are subjected, the architect's design of contemporary windows is generally limited to establishing their size, shape, and relationship with the building facade and selecting the manufacturers and window material that will satisfy the performance and aesthetic requirements. The detailing of windows is limited to designing their surrounds, that is, detailing at the jambs, head, and sill.

This chapter begins with an introduction to the commonly used window styles, materials used in window and sash frames, performance ratings of windows, and window installation.

A similar discussion of door types follows. Because of their immense variety, this chapter focuses on the more commonly used door types (interior wood and steel doors). The fire rating of windows and doors is discussed next, and the chapter concludes with a presentation of frequently used window and door terminology. The reader may choose to browse through the terminology before reading the main text.

31.1 WINDOW STYLES

Figure 31.1 shows various commonly used window styles that are classified into two broad categories:

- Fixed windows
- Operable windows

Fixed windows are the simplest of all windows. Because they do not have (operable) sashes, they are more economical and less likely to leak air or water. They are commonly used in nonresidential buildings with highly controlled heating, cooling, and ventilating systems. A fixed window is also referred to as a *direct-glaze window.* The term *picture window* is used for a fixed window whose width is larger than its height to provide a panoramic view.

Operable windows are commonly used in residential occupancies (homes, apartments, and hotels), sometimes in patient rooms of a hospital, and so on. In many situations, fixed and operable units are combined in the same window.

NOTE

Window Terminology

The reader may wish to brush up on the window terminology given at the end of the chapter before proceeding with this section.

31.2 WINDOW MATERIALS

As stated previously, five materials are used in contemporary windows: wood, aluminum, steel, vinyl, and fiberglass. The strengths and weaknesses of these materials are discussed in this section.

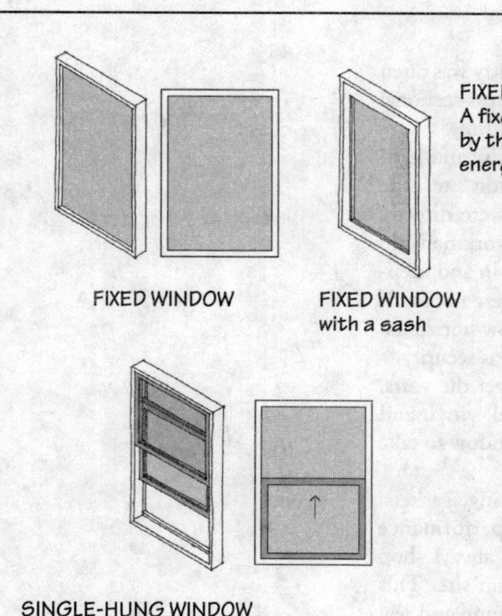

FIXED WINDOW
A fixed window generally does not have a sash because the glass is held directly by the window frame. A fixed window gives 0% ventilation, and is generally more energy efficient because of its lower air leakage rate.
FIXED WINDOW
FIXED WINDOW with a sash
SINGLE-HUNG WINDOW
A single-hung window consists of one sash that can slide vertically over a fixed lite of glass. It can provide up to 50% ventilation. An insect screen unit can be easily included on the outside.

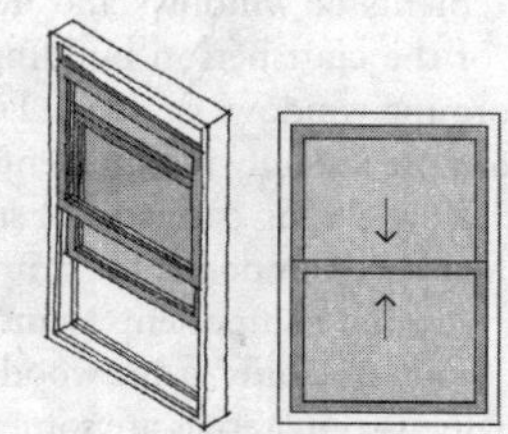

DOUBLE-HUNG WINDOW
A double-hung window consists of two sashes, both of which can slide up and down. The maximum amount of ventilation provided by a double-hung window is 50%. An insect screen can be included on the outside.
Some contemporary double-hung windows include a "tilt-and-turn" mechanism that allows the cleaning of the outside of the window from the inside.
TRIPLE-HUNG WINDOW
A triple-hung window consists of three sliding sashes, all of which can slide up and down. Like a double-hung window, a triple-hung window provides ventilation from both the top and bottom of the window.

FIGURE 31.1 Commonly used window styles.

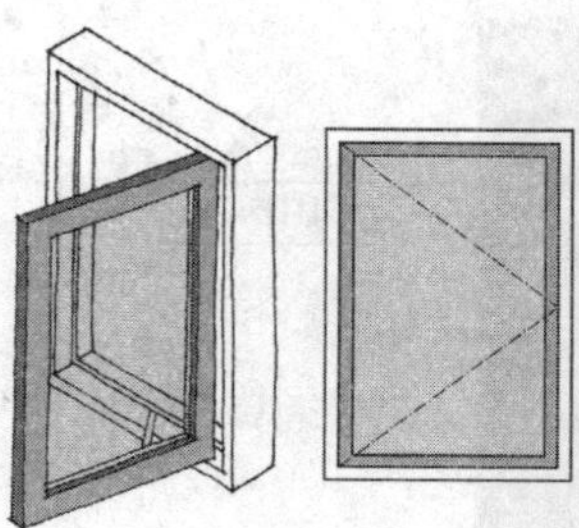

CASEMENT WINDOW
A casement window (also called a side-hung window) may consist of one operable sash, as shown here, or two sashes that close on each other with or without a center mullion. Because the sash closes on the frame with pressure, providing a compression seal, a casement window is generally less prone to air leakage and, hence, can be more energy efficient than single-, double-hung or sliding windows.

A casement window can provide up to 100% ventilation. Screen units with a casement window are generally provided on the inside with sashes opening out.

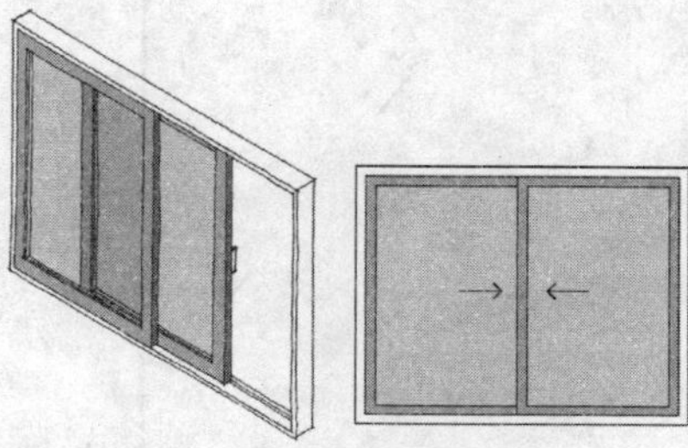

HORIZONTAL SLIDING WINDOW
A sliding window consists of one sash that slides horizontally over a fixed sash. Alternatively, both sashes may slide with respect to each other, as shown here.

A sliding window can provide up to 50% ventilation. A sliding window is more prone to air leakage than other window styles. An insect screen unit can be easily included toward the exterior.

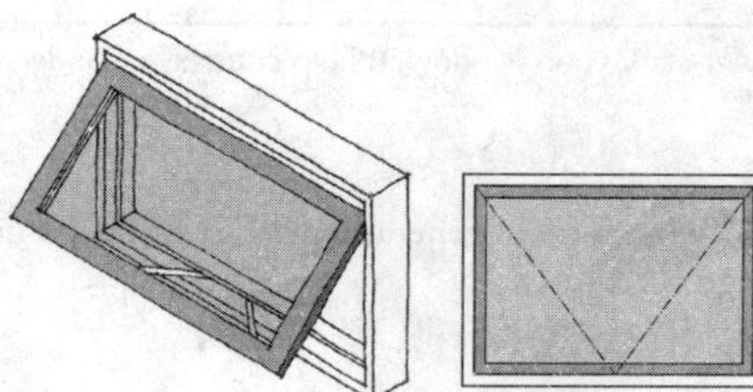

AWNING WINDOW
An awning window (also called a top-hung window) is similar to a casement window but provides a degree of rain protection when the window is partially open. It can provide up to 100% ventilation. An insect screen unit can only be provided toward the interior of an awning window.

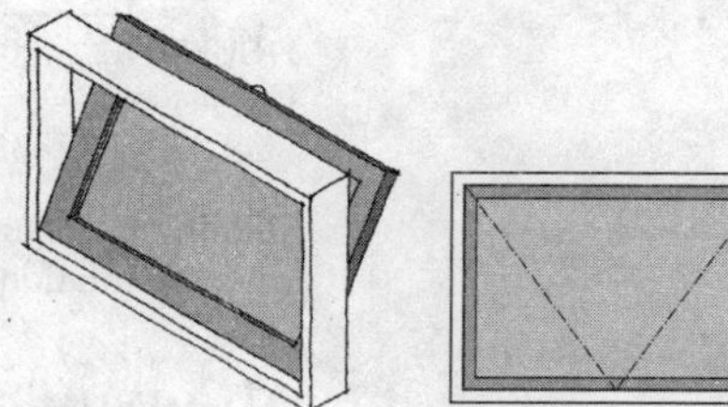

HOPPER WINDOW
A hopper window (also called a bottom-hung window) is similar to an awning window but opens inward at the top. Like an awning window, it can provide up to 100% ventilation. An insect screen unit can be provided toward the outside.

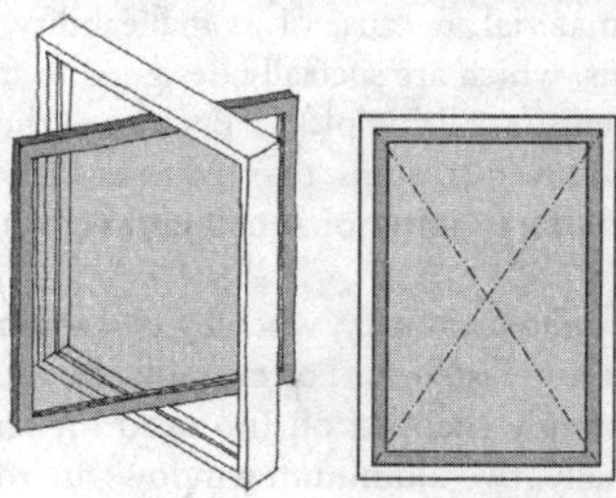

PIVOTING WINDOW
A pivoting window may be pivoted at the center or off center. It allows easy cleaning of the window from the inside and can provide up to 100% ventilation. It has the ability to direct the flow of ventilation. A screen unit cannot be provided with a pivoting window.

FIGURE 31.1 ***(Continued)*** Commonly used window styles.

Wood

Wood is the oldest material used for window frames and sashes. However, because of its swelling and shrinkage problems, fungal decay, and termite vulnerability, as well as the need for periodic staining or painting, all-wood windows are not often used today. The advantage of wood is its high R-value. Additionally, its warmth and beauty are unmatched by any other window material. Therefore, wood windows are generally used in high-end homes and offices requiring cozy and homey interiors.

To improve their durability and eliminate the need for periodic painting, manufacturers provide wood windows whose frames and sashes are clad on the exterior with vinyl or

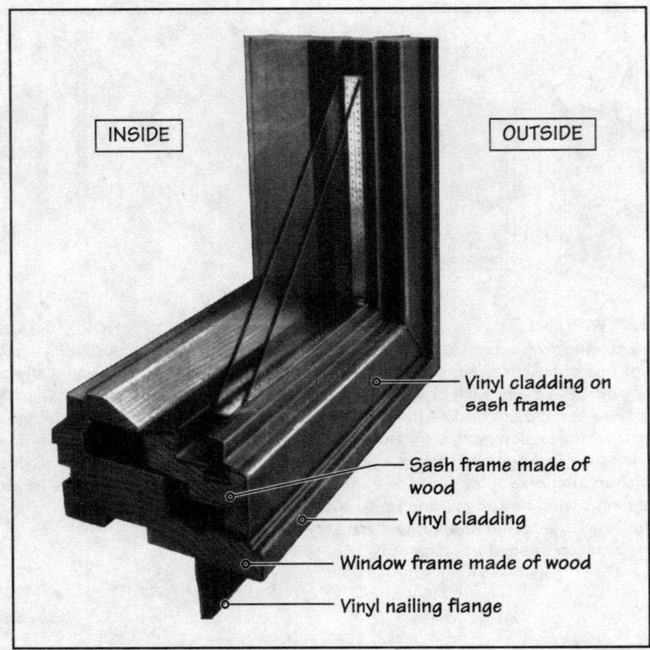

FIGURE 31.2 Cross section through a vinyl-clad wood window. (Photo courtesy of Andersen Corporation)

aluminum, Figure 31.2. Unclad wood windows are generally used in historic buildings where restoration of their original design is required.

ALUMINUM

Aluminum is by far the most commonly used window material today. It is the only window material that is also used in glass curtain walls and storefronts. It is not subject to moisture shrinkage or swelling. Because the anodized or painted finish on aluminum is virtually permanent, aluminum windows require almost no maintenance.

The malleability and flexibility of aluminum are perhaps the most important reasons for its extensive use as a window and curtain wall material. Because of its malleability, aluminum can be extruded into complex cross sections, which are specially designed to facilitate the joinery of window and curtain wall components (see Principles in Practice at the end of this chapter). For example, continuous round, hollow intrusions, referred to as *screw splines,* are commonly provided in aluminum cross sections to join horizontal and vertical frame members, Figure 31.3.

The flexibility of aluminum allows the use of *snap-on* joinery, whereby two separate aluminum sections can be joined together without any fasteners. For example, in aluminum windows and curtain walls, glazing stops are simply snapped on (no need for screws or other fastening material; Figure 31.3). Additionally, two aluminum windows are routinely snap-connected by using mating mullion profiles to provide a larger window with a common mullion—without the use of fasteners, Figure 31.4.

The disadvantage of aluminum is its low R-value. However, with the introduction of *thermally improved*—also called *thermally broken* (see Figure 30.23)—aluminum frames, the situation has substantially improved.

STEEL

Before the introduction of aluminum for windows, steel was commonly used as an alternative to wood. Steel corrosion and the need for frequent painting have substantially reduced the use of steel windows. However, because steel is a strong material, the frame members can be narrower than those of other materials, giving a sleeker window appearance. Because of steel's strength, steel windows are often specified for prisons and other locations where greater durability and greater resistance to lateral loads (e.g., blast-resistant glazing) are required. Steel windows are also used in the renovation work of historic buildings.

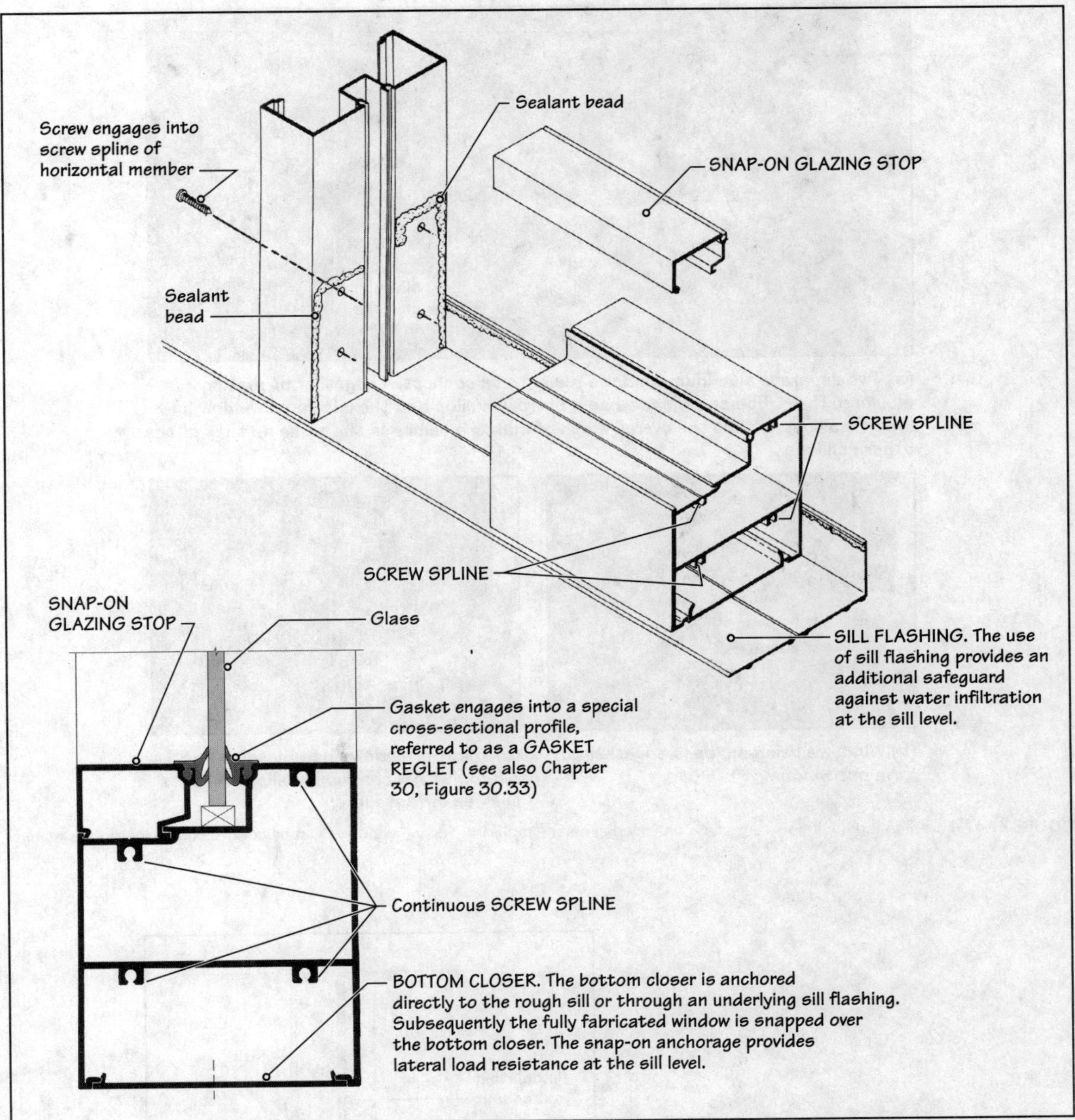

FIGURE 31.3 The cross-sectional profile of a typical aluminum window or curtain wall frame member looks complex, but each intrusion or protrusion has a special purpose. Observe the use of screw splines to connect horizontal and vertical frame members and snap-on cross-sectional profiles that allow members to be connected together without any fasteners.

Vinyl and Fiberglass

Vinyl (an abbreviation for polyvinyl chloride, PVC) frames and sashes are almost maintenance free because vinyl does not corrode or degrade due to fungal or termite attack. Vinyl windows require no painting and are available in a variety of nonfading colors. Vinyl frame and sash cross sections are made by the extrusion process and are generally hollow, Figure 31.5. They provide an excellent R-value.

A major disadvantage of vinyl, and of plastics in general, is their high coefficient of thermal expansion. The introduction of fiberglass in windows has substantially reduced this problem. The coefficient of thermal expansion of fiberglass is compatible with that of glass. Thus, fiberglass frames and sashes are dimensionally stable, strong, and available in different colors.

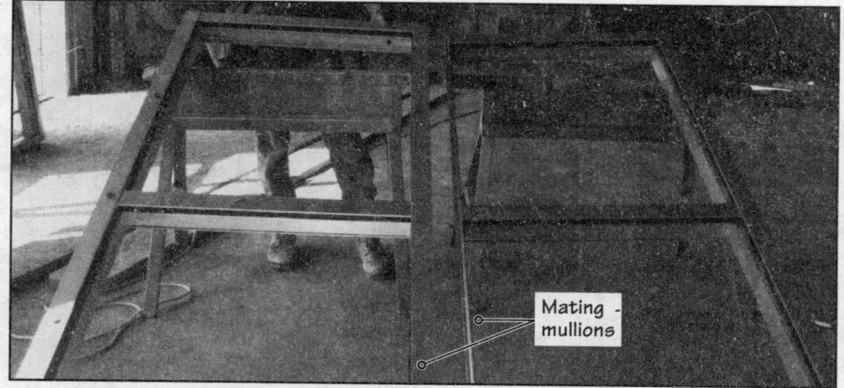

(a) Two separate aluminum windows ready to be connected together at mating mullions. The right-hand window has a narrow mullion and the left-hand window has a broad mullion, so that the overall width of mated mullions is the same as that of the other mullions.

(b) Windows being snapped together to form one window.

(c) The two windows have become one window with a common mullion without the use of fasteners.

FIGURE 31.4 Two aluminum windows are snap-connected at mating mullions. Several windows can be connected together in this manner.

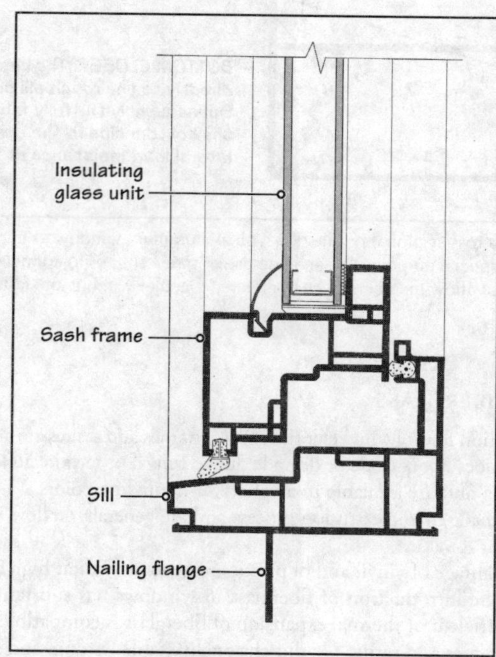

FIGURE 31.5 Typical cross-sectional profile of a vinyl window.

PRACTICE QUIZ

Each question has only one correct answer. Select the choice that best answers the question.

1. The term *direct-glaze window* implies
   a. an operable window.
   b. a fixed window.
   c. a fixed window whose length is greater than its height.
   d. a fixed window whose height is greater than its length.
   e. none of the above.

2. The term *picture window* implies
   a. an operable window.
   b. a fixed window.
   c. a fixed window whose length is greater than its height.
   d. a fixed window whose height is greater than its length.
   e. none of the above.

3. Which of the following window types can provide up to 100% ventilation when fully open?
   a. Single-hung window
   b. Double-hung window
   c. Sliding window
   d. Casement window
   e. All of the above

4. Which of the following windows is top-hung?
   a. Awning window
   b. Hopper window
   c. Casement window
   d. Pivoting window
   e. None of the above

5. In vinyl- (or aluminum)-clad windows, the cladding is used on exterior
   a. window frame members.
   b. sash frame members.
   c. both window frame and sash frame members.

6. Screw splines are provided in
   a. window and curtain wall frames made from wood.
   b. window and curtain wall frames made from steel.
   c. window and curtain wall frames made from vinyl.
   d. window and curtain wall frames made from aluminum.
   e. all of the above.

7. Screw splines are round hollow intrusions in frame members that facilitate the connection of horizontal and vertical members of a(n)
   a. wood window.
   b. aluminum window.
   c. steel window.
   d. vinyl window.
   e. all of the above.

8. Snap-on joinery is generally used in
   a. wood windows.
   b. aluminum windows.
   c. steel windows.
   d. all of the above.
   e. none of the above.

31.3 PERFORMANCE RATINGS OF WINDOWS

Apart from aesthetic considerations such as shape, style, frame color, and finish, several performance characteristics must be considered in selecting a window for a project. Two rating systems, one relating to a window's energy performance and the other relating to a window's structural and water-leakage performance, help architects in this task. Both systems are material neutral—that is, they apply uniformly to windows made of different materials. Additionally, they apply to the entire window assembly, consisting of the frame, sash, glass, gaskets, and so on.

ENERGY-PERFORMANCE RATING OF A WINDOW

The energy-performance rating of a window assembly is provided by the National Fenestration Rating Council (NFRC), which is backed by the U.S. government's Energy Star initiative. NFRC-rated windows are labeled for their U-values and solar heat-gain coefficient (SHGC), Figure 31.6. Labeling for other energy-related criteria, such as visible transmittance (VT), air leakage, and the condensation resistance factor (CRF), is at the discretion of the manufacturer. The current energy-conservation code requires conformance only with the U-value, air leakage, and SHGC of windows.

STRUCTURAL, WATER-LEAKAGE, AND OTHER PERFORMANCE RATINGS OF A WINDOW

In addition to being energy efficient, a window assembly must be strong enough to resist wind loads, it should not leak water or air, and the window's locking mechanism should prevent forced entry. The window manufacturing industry (comprising the manufacturers of wood, aluminum, vinyl, and fiberglass windows) has jointly established standards by which their windows are tested. All windows must be tested for

- Resistance to wind loads
- Resistance to water leakage
- Resistance to air leakage
- Resistance to forced entry

CM NFRC National Fenestration Rating Council CERTIFIED	World's Best Window Co. Millennium 2000+ Vinyl-Clad Wood Frame Double Glazing • Argon Fill • Low E Product Type: **Vertical Slider** (per NFRC 100-97)
ENERGY PERFORMANCE RATINGS	
U-Factor (U.S./I-P) **0.35**	Solar Heat Gain Coefficient **0.32**
ADDITIONAL PERFORMANCE RATINGS	
Visible Transmittance **0.51**	Air Leakage (U.S./I-P) **0.2**
Manufacturer stipulates that these ratings conform to applicable NFRC procedures for determining whole product performance. NFRC ratings are determined for a fixed set of environmental conditions and a specific product size. Consult manufacturer's literature for other product performance information. www.nfrc.org	

FIGURE 31.6 A typical label on an NFRC-rated window; see also Figure 31.9(a). (Illustration courtesy of National Fenestration Rating Council)

TABLE 31.1 ABBREVIATIONS FOR COMMONLY USED WINDOW STYLES IN THE WINDOW-RATING DESIGNATION

Window style	Code
Fixed	F
Single/double/triple hung	H
Casement	C
Horizontal sliding	HS
Awning/hopper	AP
Vertically pivoted	VP
Horizontally pivoted	HP

Based on the test results, a window is assigned a rating designation that includes the following information:

- *Window style:* whether the window is a fixed, casement, sliding, or awning window. The acronyms for the commonly used window styles are given in Table 31.1.
- *Window performance class:* the five performance classes, coded as R, LC, C, HC, or AW, Table 31.2.
- *Window performance grade:* the allowable (or design) wind load on the window in psf (pounds per square foot). A factor of safety of 1.5 is used so that if the window assembly resists a wind load of 45 psf, its design wind load is 30 psf (i.e., the window's performance grade is 30).

  The window is also tested for water leakage under a static water pressure given in Table 31.2. Under this pressure, the assembly should not leak any water. Thus, water leakage is a pass/fail test. The window's air leakage should not exceed 0.3 cfm/ft^2 (cubic feet per minute per square foot)—another pass/fail test. The window should also pass the forced-entry resistance test as specified in the standard, once again a pass/fail test.
- *Window size:* an important part of the designation system. It is stated as "window width × window height" in inches. Generally, if everything else is the same, a smaller window generally performs better than a larger window under the tests. Therefore, the standard specifies a certain minimum window size for each performance class and performance grade. The manufacturer can exceed the specified size requirement, and if that is the case, it can be so included in the designation, as explained in Figure 31.7.

TABLE 31.2 PERFORMANCE-CLASS AND PERFORMANCE-GRADE REQUIREMENTS OF WINDOWS

Performance class and abbreviation	Performance grade (design wind load) (psf)	Minimum wind load resistance (psf)*	Water-leakage test pressure (psf)†	Air-leakage test pressure (psf)	Air-leakage rate ft^3/(min. ft^2)
Residential (R)	15	22.5	2.86	1.57	0.3
Light commercial (LC)	25	37.5	3.75	1.57	0.3
Commercial (C)	30	45.0	4.50	1.57	0.3
Heavy commercial (HC)	40	60.0	6.00	1.57	0.3
Architectural (AW)	40	60.0	8.00	6.24	0.3

*The design wind load equals two-thirds of the load that a window resists under the test. For example, if the window resists 60 psf under the test, the design wind load for the window is 40 psf. In other words, the window may be subjected to a maximum wind load of 40 psf in service. Hence, the window's performance grade is 40.

†The water leakage of a window is determined by subjecting it to a static pressure that is 15% of its design wind load. Thus, if the design wind load for a window is 40 psf, it must pass a water-leakage test under a pressure of 6.0 psf. AW windows are tested under a pressure that is 20% of the design wind load (in place of 15%), and R windows are tested under 19% design wind-load pressure.

Observe that HC and AW windows have the same performance grade of 40, but AW windows must resist water leakage and air leakage under higher pressures.

HS-LC25 120 x 59

Window style
(see Table 29.1)

HS = horizontal sliding window

Window performance class and grade:
(see Table 29.2)

LC25 indicates that

(a) the window is classified as light commercial

(b) the design wind load on window should not exceed 25 psf

(c) it has passed the water leakage test under its grade

(d) its air infiltration rate is ≤ 0.3 cfm

(e) it has passed the forced entry test

Manufacturer's maximum window size

This dimension indicates that the manufacturer has tested the maximum window size for this class and grade at 120 in. × 59. As stated previously, each performance class and grade must be tested to a minimum window size. For example, HS-LC25 must be tested to a minimum window size of 69 in. × 54 in.

The manufacturer can exceed the size requirement of the standard, as indicated by this label, since 120 in. × 59 in. is greater than the required minimum size of 69 in. × 54 in.

As per the standard, the designation system recognizes a manufacturer exceeding the minimum requirements for a performance class and grade. For example, for a horizontal sliding window to receive a grade of HS C30 must be at least 71 in. × 59 in. and must resist a design wind load of at least 30 psf. If a man ufacturer's window of this size tests to a design wind load of 35 psf, it can be designated as HS C35. This higher test wind load of 35 × 1.5 = 52.5 psf also requir es the window to pass under a correspondingly higher water leakage test pressure.

FIGURE 31.7 Explanation of a typical label on a window tested under AAMA/WDMA/CSA101/I.S.2/A440.

Other Performance Characteristics of Windows

The four performance characteristics just described are the minimum requirements for windows. Generally, a manufacturer will have its windows tested for several other characteristics, such as the operating force test (force required to open and close windows), life-cycle test (which determines the damage over time to hardware parts used in the window), and the vertical deflection test (which determines the deflection of a sash in a casement window). Acoustical windows are tested for their sound insulation property.

Windows and Building Codes

Windows in habitable rooms of dwelling units must satisfy the building code's requirement for a certain minimum amount of natural light and ventilation. They are also required to provide a means of emergency egress from the dwelling in the event of fire. For example, building codes generally require the clear area of at least one window in each sleeping room of a dwelling, when open, to be at least 5.7 sq ft.

NOTE

Building Fenestration and the Energy Code

The International Energy Conservation Code (IECC) places restrictions on various energy-related properties of building fenestration (glazed and nonglazed doors, windows, skylights, etc.). For example, the maximum U-value of fenestration in climate zones 4 to 8 for residential buildings is limited to 0.35 (which approximates to a minimum R-value of 3); the maximum U-value for climate zone 3 is 0.5; for climate zone 2, the corresponding value is 0.65; and for climate zone 1, it is 1.20.

The fenestration leakage rate is limited to a maximum of 0.3 cfm/ft^2 for all climate zones. The maximum value of SHGC is also regulated and is a function of the climate zone (see Chapter 30).

Note that the preceding information applies to compliance with the prescriptive provisions of the IECC. IF the simulated performance approach is used, the designer has the freedom to trade off between various factors that affect energy use in the building.

31.4 WINDOW INSTALLATION AND SURROUNDING DETAILS

FIGURE 31.8 Aluminum windows were installed in this high-rise condominium building before the exterior (EIFS) wall finish was applied. The windows were, therefore, protected from damage, as shown here.

Contemporary windows are generally brought to the site fully assembled and glazed, which simplifies the installation process. On-site glazing is generally limited to fixed windows with large glass panes.

A window may be installed before or after the exterior wall cladding is in place. Most commercial window manufacturers recommend the installation of their windows after the exterior wall cladding is in place to prevent physical damage to windows, such as breakage of glass and scratching of coatings. Chemical damage is also a concern. For example, fresh mortar can adversely react with aluminum. If window installation must precede the exterior wall finish, adequate precautions must be taken to protect the windows from damage, Figure 31.8.

INSTALLATION OF WINDOWS WITH NAILING FLANGES

Windows with nailing flanges are generally used in buildings with exterior wood siding, stucco, and EIFS cladding. They can also be used in buildings with masonry veneer cladding, but the detailing is more complicated. The installation of a window with nailing flanges is relatively simple because the window is anchored to the wall through the flanges.

Because the flanges project over the entire window perimeter, they seal the rough opening, providing a built-in flashing. However, additional waterproofing of the rough opening is generally recommended to provide a second line of defense against water and air leakage, Figure 31.9.

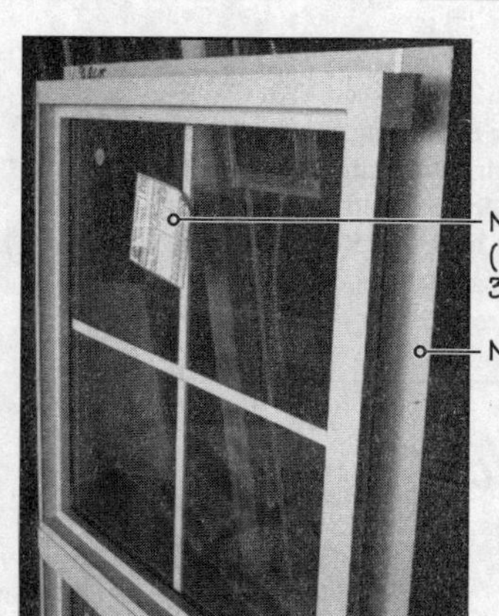

(a) A factory-glazed aluminum window with nailing flanges and simulated divided lites (muntins) within insulating glass unit

(b) Window being brought into the opening. Observe the self-adhering waterproof tape around the rough opening.

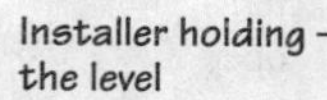

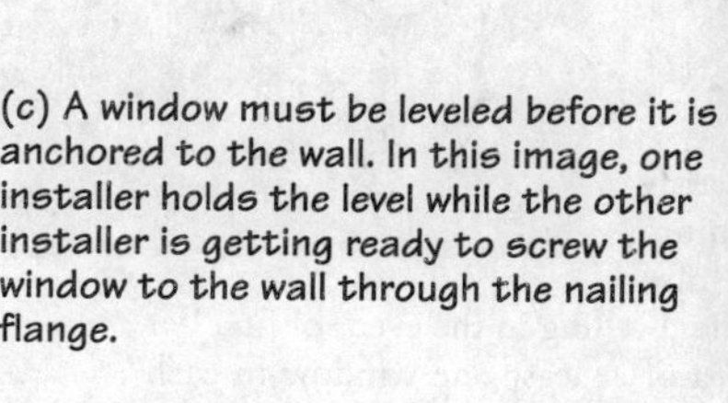
(c) A window must be leveled before it is anchored to the wall. In this image, one installer holds the level while the other installer is getting ready to screw the window to the wall through the nailing flange.

FIGURE 31.9 A few steps in the anchorage of a window with nailing flanges into a stud wall.

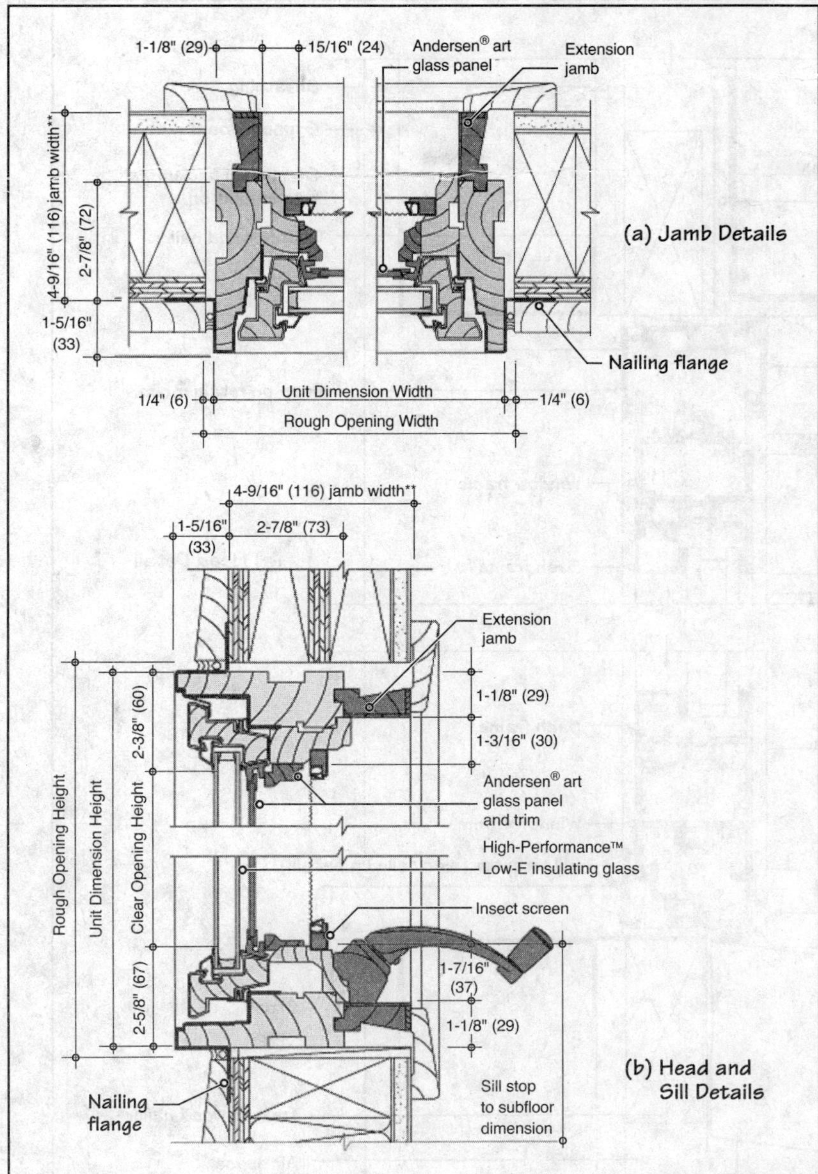

FIGURE 31.10 Suggested jamb, sill, and head details of a vinyl-clad wood window by Andersen Corporation. Observe that the window is anchored to the wall through vinyl nailing flanges. (Illustration courtesy of Andersen Corporation)

Manufacturers typically provide recommendations for anchorage and installation of their windows and detailing at the jambs, head, and sill, which an architect can use as a guide, Figure 31.10.

Installation of Windows without Nailing Flanges

Windows without nailing flanges can be installed in several ways. Two commonly used methods for aluminum windows are shown in Figures 31.11 and 31.12. Figure 31.11 shows the use of a perimeter receptor frame comprising a sill receptor (also called a *subsill*), two jamb receptors, and a head receptor. The receptors are first anchored to the rough opening with anchorage points fully sealed. Subsequently, the window (complete with frame, sashes, glass, hardware, etc.) is placed within the receptor frame.

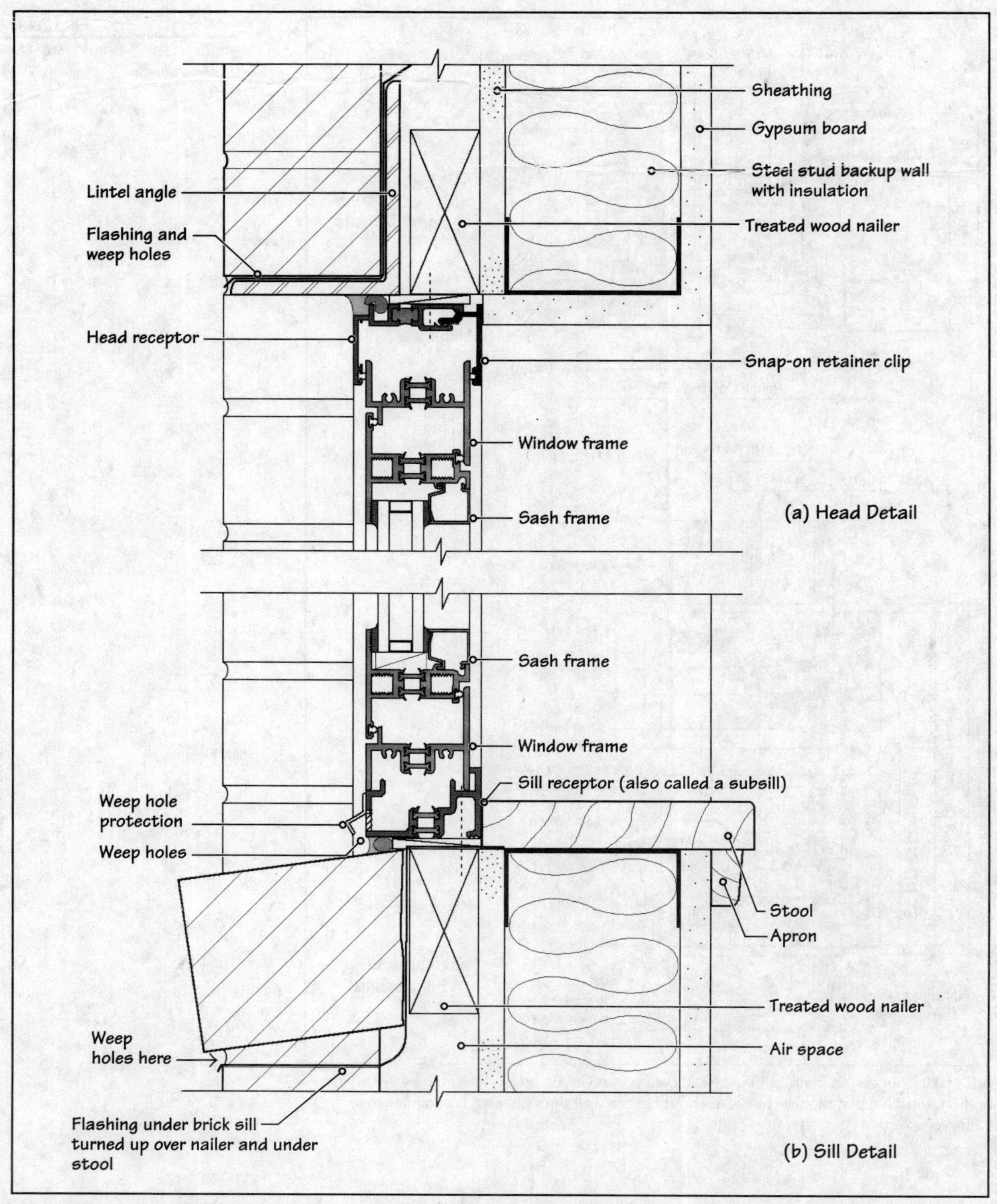

FIGURE 31.11 A preglazed aluminum window anchored to a steel stud–backed brick veneer wall using receptors and retainer clips.

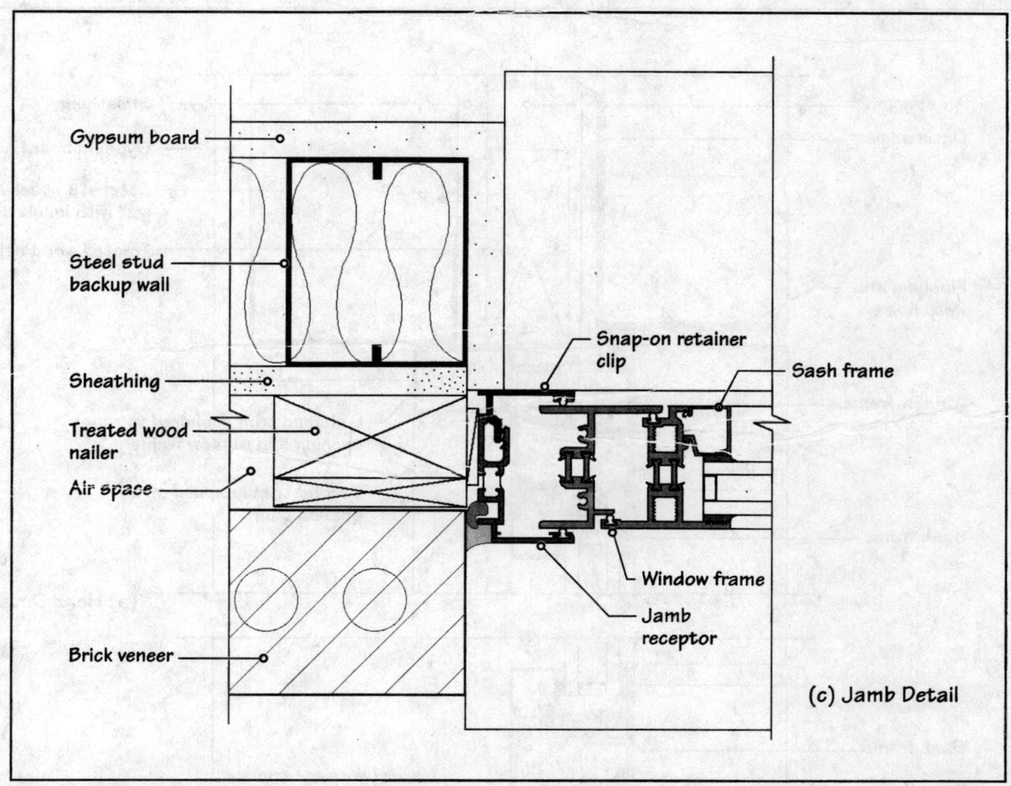

FIGURE 31.11 *(Continued)* A preglazed aluminum window anchored to a steel stud–backed brick veneer wall using receptors and retainer clips.

Lateral load resistance to the window is provided by aluminum retainer clips that are snapped onto the receptors from the inside. If their thickness, cross-sectional profile, and locking striations are changed, retainer clips can provide the required lateral load resistance. Interior finishes around the jambs, sill, and head are now installed. In this method of installation, the window is not directly anchored to the rough opening.

Any incidental water that leaks into the window is collected in the subsill. Therefore, the subsill is provided with end dams to prevent its runoff from the ends and also with weep holes to let the water drain out.

Figure 31.12 shows an alternative anchorage method, which utilizes two-piece interior perimeter trims. In this method, the window is first placed in the opening, squared, and leveled. The first (L-shaped) trim piece is then anchored to the rough opening on all four sides and also to the window. Subsequently, the second trim piece is snapped on the first trim piece. Interior finishes can then be applied to the opening.

The details shown in Figures 31.11 and 31.12 can be modified in several ways, depending on the wall construction and the exterior finish. Manufacturers have several standard components for use with different interior and exterior requirements. They can also provide custom-made trims and accessories at little or no additional cost.

31.5 CLASSIFICATION OF DOORS

Because of the large variety of doors used in contemporary buildings, they can be classified in several ways. The simplest classification is to group them as *interior* and *exterior doors.* Exterior doors include entry doors and garage and overhead doors. Entry doors in residential buildings are generally insulated and rated for energy performance.

Classification Based on Door Material

Doors can also be classified based on the materials of which they are made. Wood and metal (steel, stainless steel, and aluminum) are the commonly used door materials. Fiberglass-reinforced plastic (FRP) is a more recently introduced material.

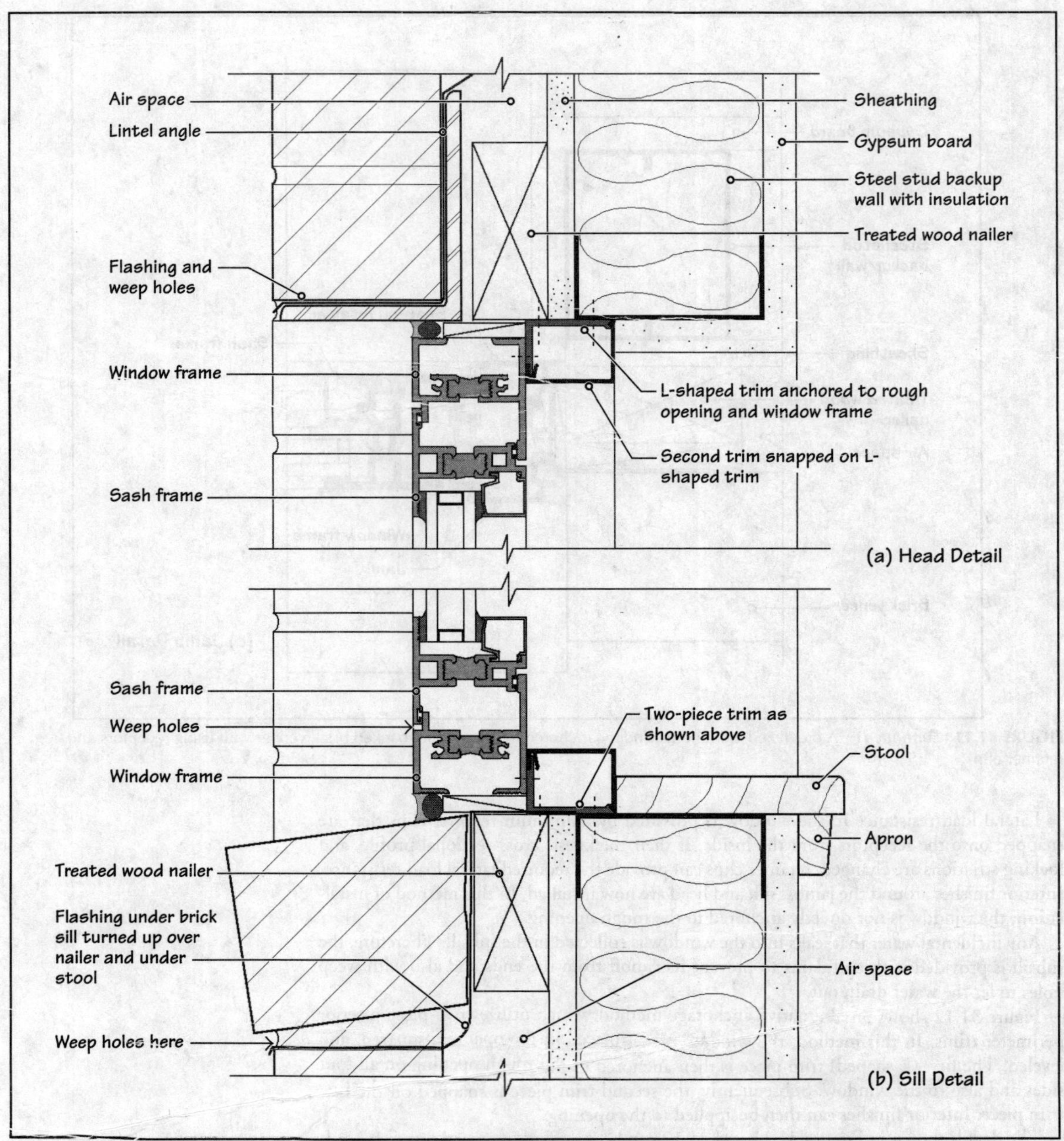

FIGURE 31.12 A preglazed aluminum window anchored using two-piece trims.

Wood doors are the most popular interior doors. Among the metal doors, steel doors are used where greater security, fire resistance, rot resistance, blast resistance, and wind-load resistance are required. Stainless steel doors are preferred for food-processing plants, freezer rooms, commercial kitchens, and so on. (Note: steel doors are commonly made from sheet steel and are generally referred to as *hollow metal doors.*)

Aluminum doors are generally glazed and are commonly used in public buildings. Aluminum flush doors and panel doors are specified where corrosion resistance is important, such as in water-treatment plants, swimming pools, and pumping stations.

Classifications Based on Door Operation and Door Styles

A helpful classification of doors is based on their mode of operation, as shown in Figure 31.13. Of these, the single-leaf hinged door is most commonly used. Another classification is based on door style (i.e., door elevation), Figure 31.14.

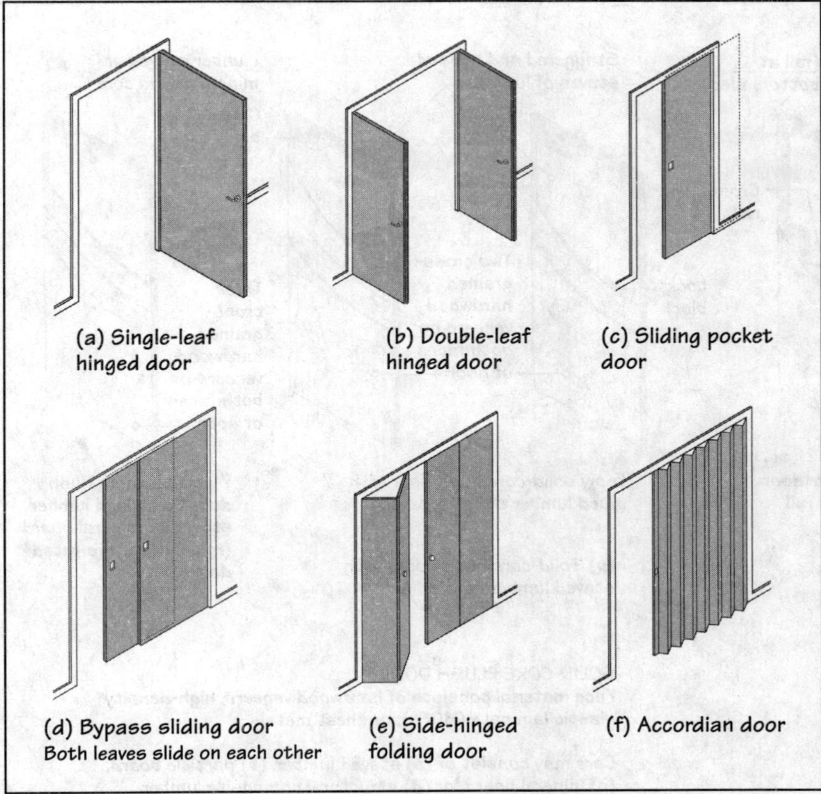

FIGURE 31.13 Classification of doors based on their mode of operation.

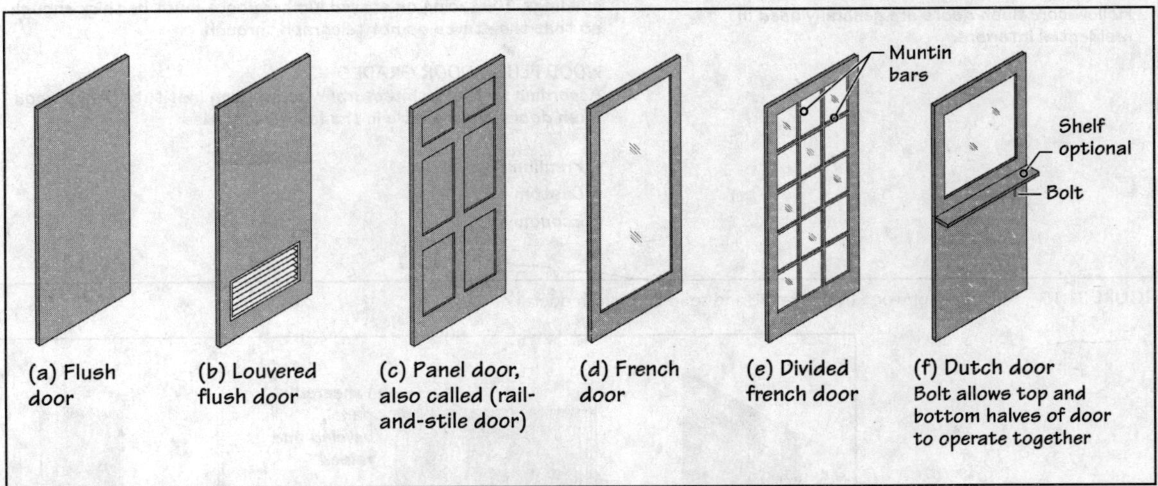

FIGURE 31.14 Common wood door styles.

Of the various styles shown in Figure 31.14, flush doors and panel doors are most popular. Construction details of wood flush doors and panel doors are illustrated in Figures 31.15 and 31.16, respectively. Figure 31.17 gives standard door sizes for wood doors, their hardware, and their hand conventions.

Metal (steel) doors are generally hollow-core doors and are available in both flush and panel door styles. The types and construction of metal hollow-core flush doors are illustrated in Figures 31.18 and 31.19, respectively.

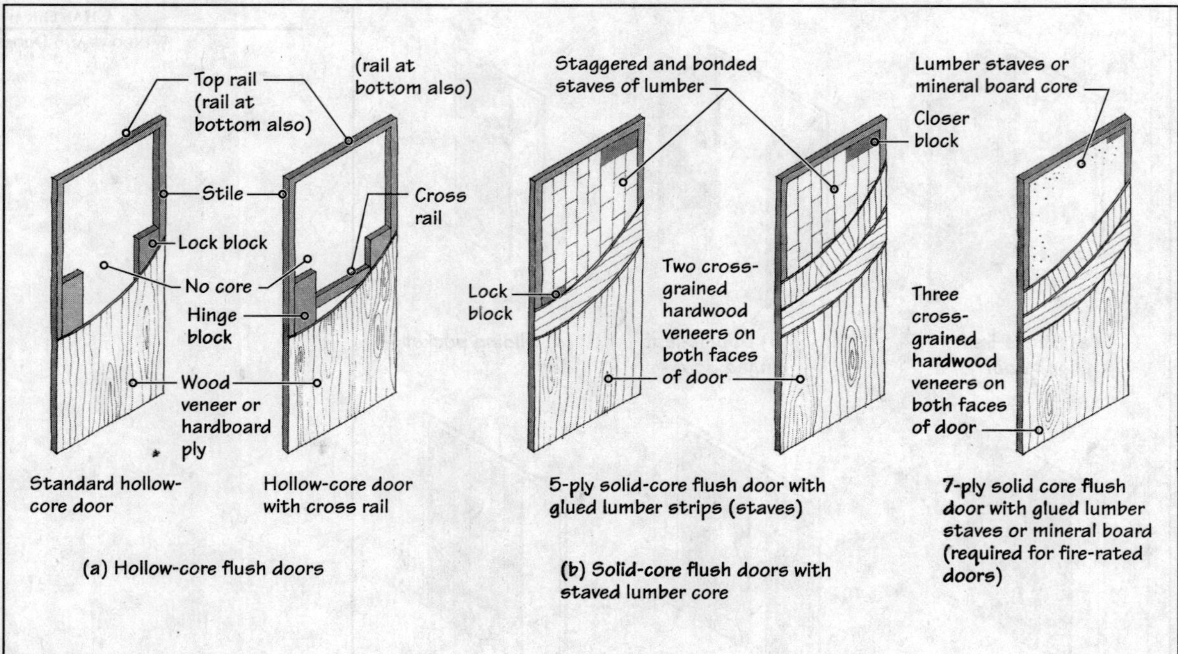

WOOD FLUSH DOOR ANATOMY

Flush doors consist of
(a) Face material
(b) Core

HOLLOW-CORE FLUSH DOORS
Face materials generally wood veneer or hardboard.
Core is usually hollow but can also be of honeycomb or cellular fiberboard.
Hollow-core flush doors are generally used in residential interiors.

SOLID-CORE FLUSH DOORS
Face material consists of hardwood veneers, high-density plastic laminate (HDPL), or sheet metal.

Core may consist of (a) staved lumber, (b) particle board, (c) mineral board, or (d) structural composite lumber.

Solid-core flush doors are more durable, secure, and provide greater sound insulation and fire resistance than hollow-core flush doors. They are commonly used as interior doors in public buildings. The facing on staved lumber doors must be thick enough so that the staves do not telegraph through.

WOOD FLUSH DOOR GRADES
According to the Architectural Woodworking Institute (AWI), wood flush doors are available in the following grades:

- Premium
- Custom
- Economy

FIGURE 31.15 Anatomies of wood (hollow-core and solid-core) flush doors.

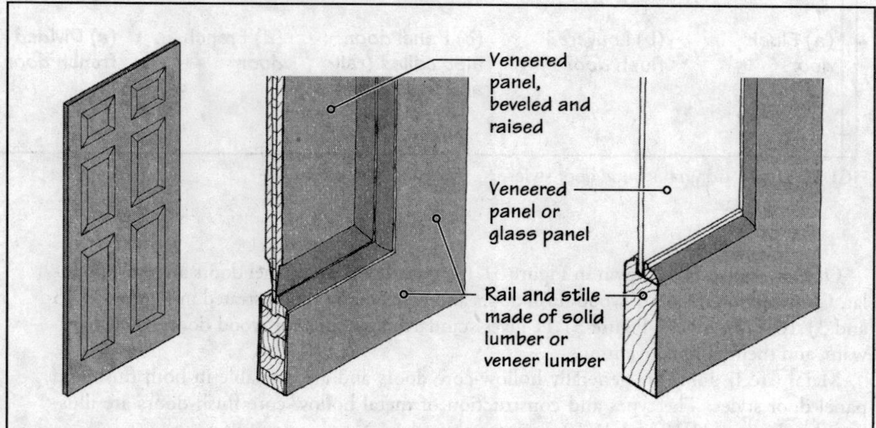

FIGURE 31.16 Details of a typical wood panel door. Panel doors are also imitated using other materials.

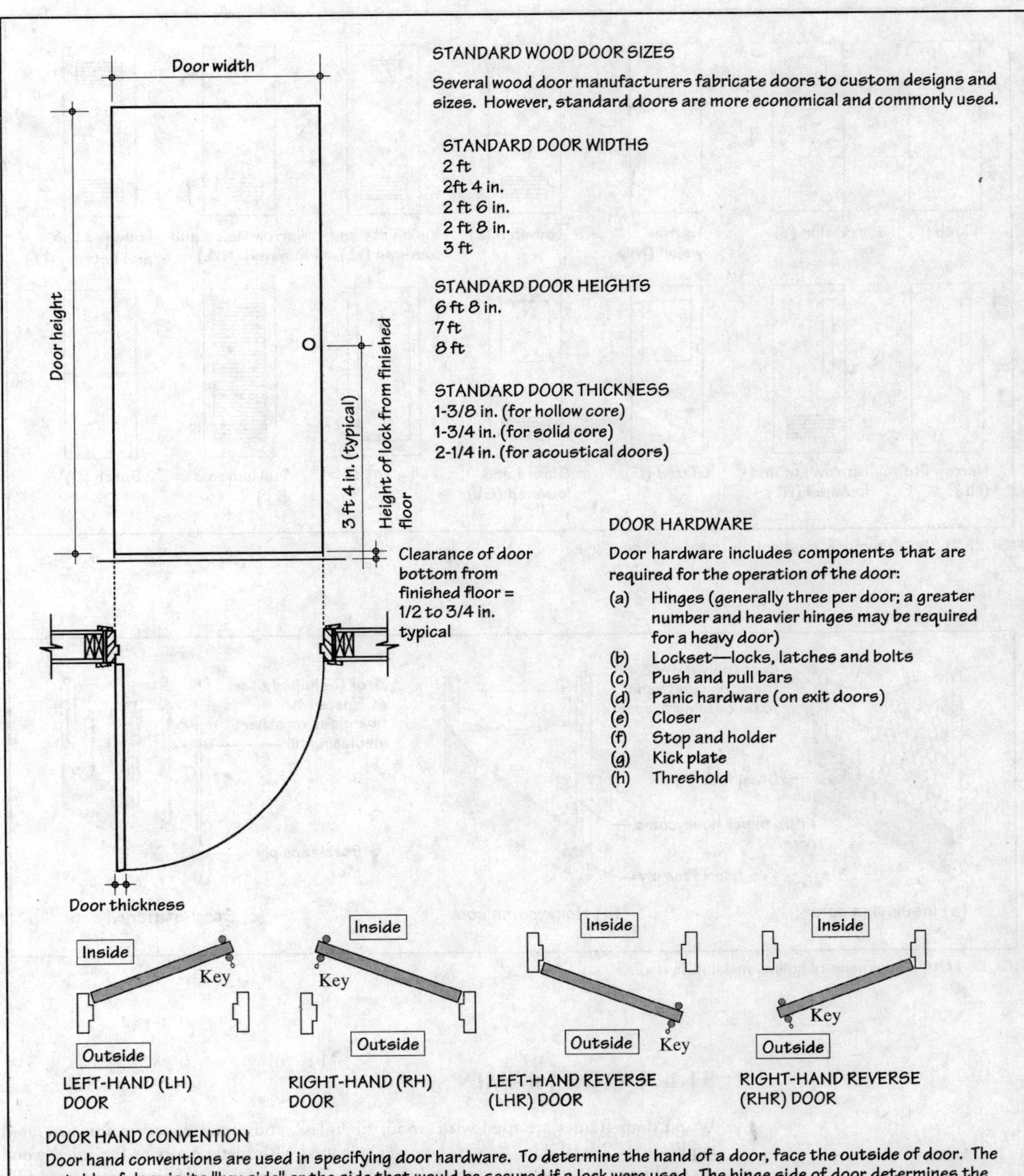

Door width
Door height
3 ft 4 in. (typical)
Height of lock from finished floor
Clearance of door bottom from finished floor = 1/2 to 3/4 in. typical
Door thickness
STANDARD WOOD DOOR SIZES
Several wood door manufacturers fabricate doors to custom designs and sizes. However, standard doors are more economical and commonly used.
STANDARD DOOR WIDTHS
2 ft
2ft 4 in.
2 ft 6 in.
2 ft 8 in.
3 ft
STANDARD DOOR HEIGHTS
6 ft 8 in.
7 ft
8 ft
STANDARD DOOR THICKNESS
1-3/8 in. (for hollow core)
1-3/4 in. (for solid core)
2-1/4 in. (for acoustical doors)
DOOR HARDWARE
Door hardware includes components that are required for the operation of the door:
(a) Hinges (generally three per door; a greater number and heavier hinges may be required for a heavy door)
(b) Lockset—locks, latches and bolts
(c) Push and pull bars
(d) Panic hardware (on exit doors)
(e) Closer
(f) Stop and holder
(g) Kick plate
(h) Threshold
Inside
Key
Outside
LEFT-HAND (LH) DOOR
Inside
Key
Outside
RIGHT-HAND (RH) DOOR
Inside
Outside
Key
LEFT-HAND REVERSE (LHR) DOOR
Inside
Key
Outside
RIGHT-HAND REVERSE (RHR) DOOR
DOOR HAND CONVENTION
Door hand conventions are used in specifying door hardware. To determine the hand of a door, face the outside of door. The outside of door is its "key side" or the side that would be secured if a lock were used. The hinge side of door determines the door's hand. The hand is referred to as"reverse", if the door swings toward the key side, i.e., toward the outside.

FIGURE 31.17 Standard wood door sizes, hardware, and door hand conventions.

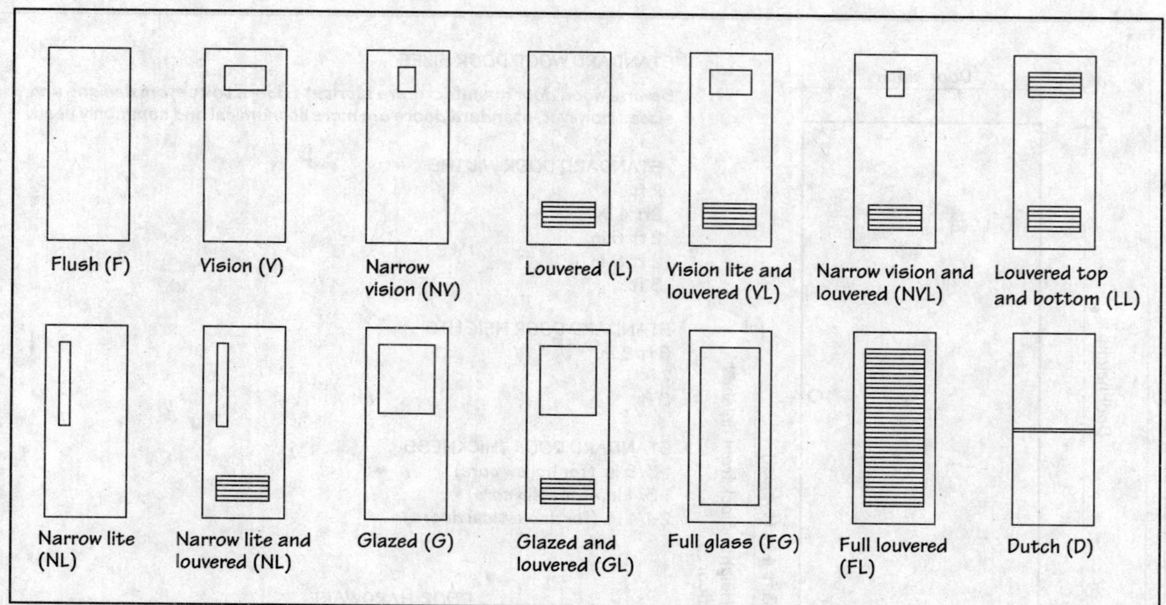

FIGURE 31.18 Commonly used hollow metal flush door styles.

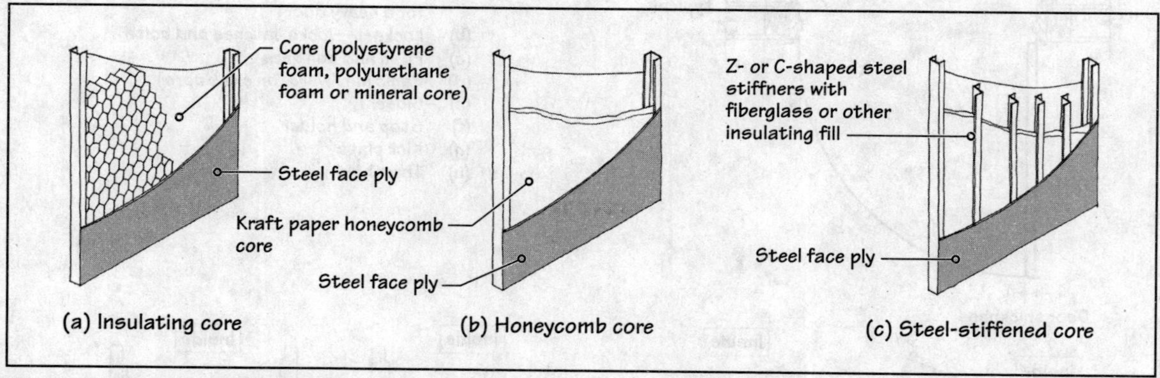

FIGURE 31.19 Anatomies of hollow metal flush doors.

31.6 DOOR FRAMES

Wood door frames are used with wood, fiberglass, and metal doors in non-fire-rated applications. Metal door frames, however, are used with metal doors or fire-rated wood doors. Figure 31.20 shows commonly used wood and metal door frame shapes, and Figures 31.21 and 31.22 show typical construction details of wood and metal door frames, respectively.

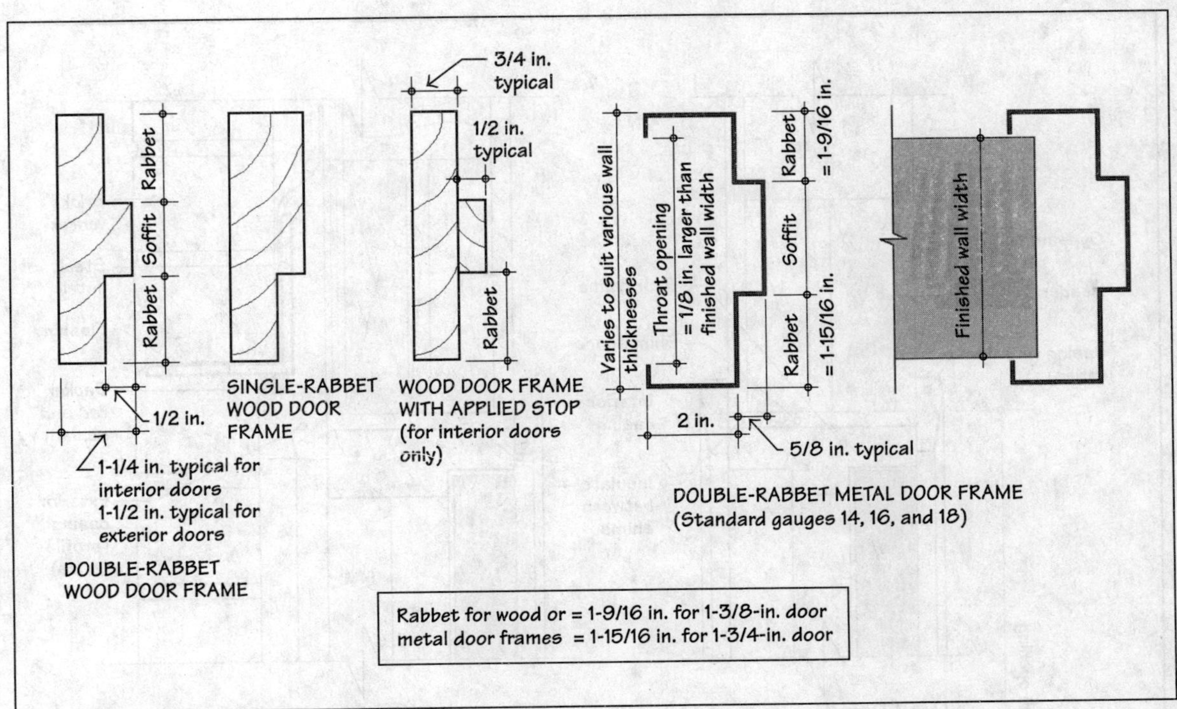

FIGURE 31.20 Typical wood and metal door frame shapes.

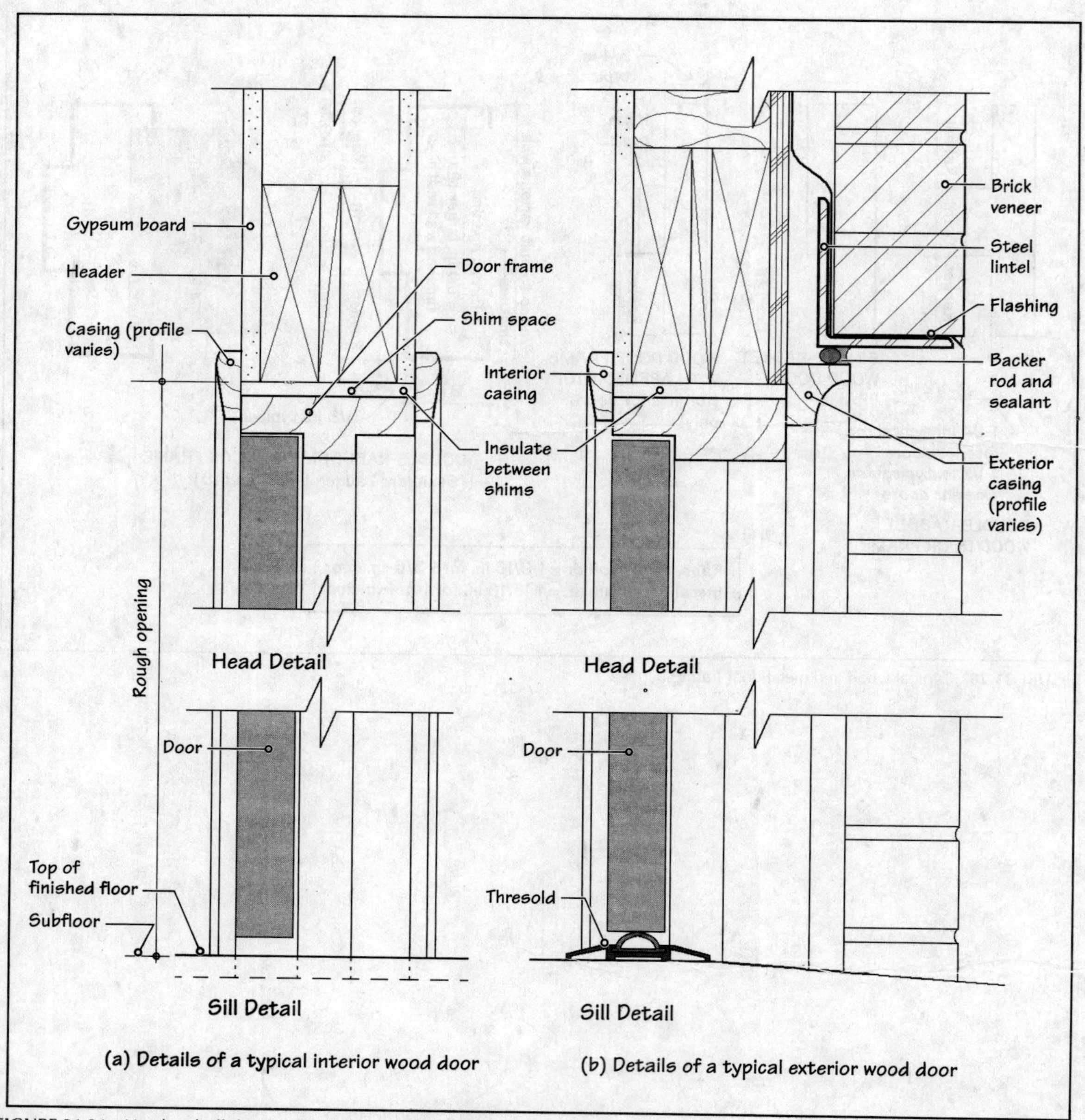

FIGURE 31.21 Head and sill details of typical interior and exterior wood doors.

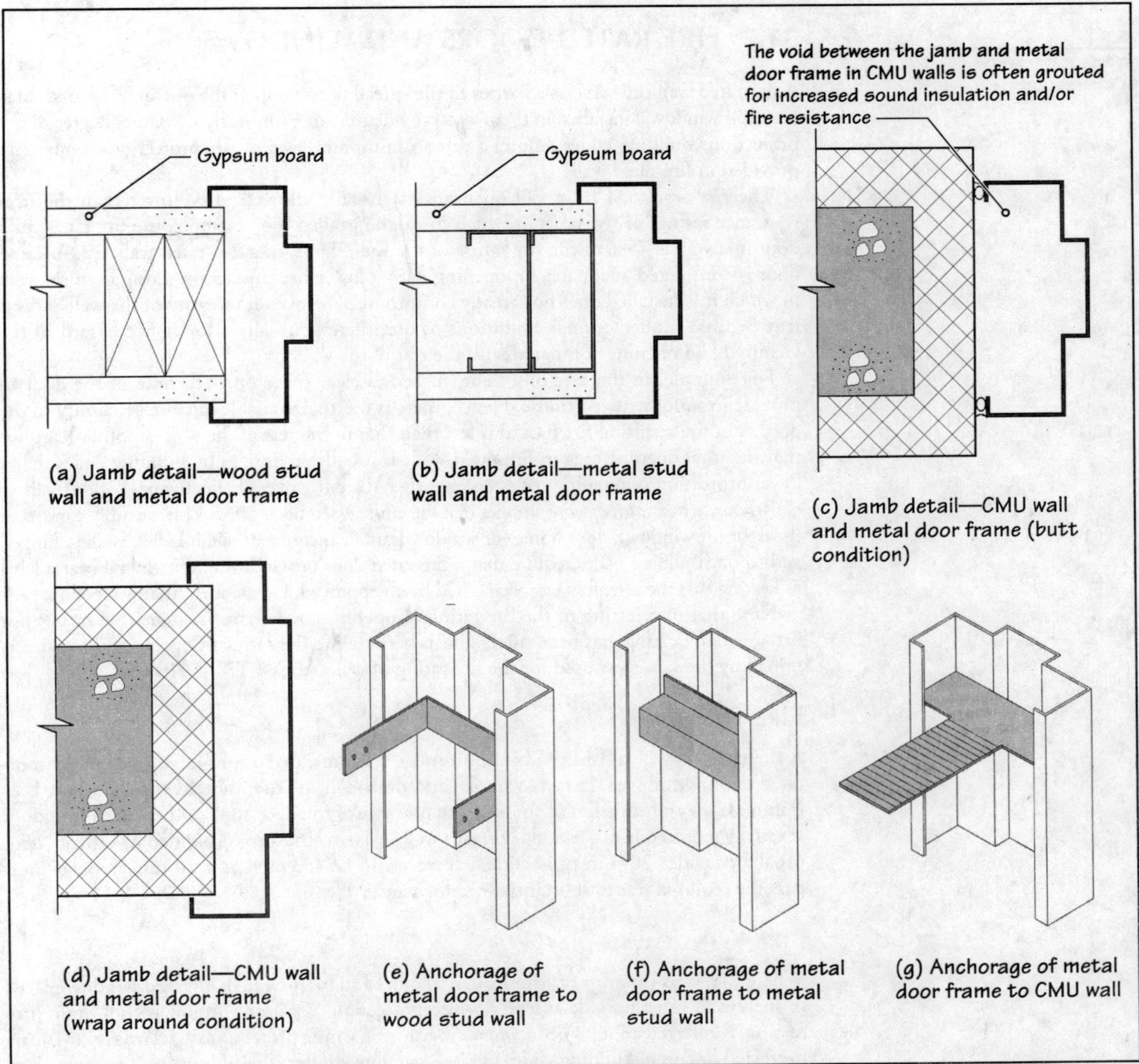

FIGURE 31.22 Details of the anchorage of a metal door frame to various types of walls.

PRACTICE QUIZ

Each question has only one correct answer. Select the choice that best answers the question.

9. The label on an NFRC-rated window must include the
- **a.** U-value and shading coefficient.
- **b.** U-value and air infiltration.
- **c.** U-value and solar heat gain coefficient.
- **d.** U-value and visible transmittance.
- **e.** Solar heat–gain coefficient and air infiltration.

10. In a window that carries the performance label "HS-C30 71 × 59", the number 30 indicates that the window
- **a.** will leak a maximum of 30% of the room's air per hour.
- **b.** conforms to the applicable standard up to a size of 30 in. × 30 in.
- **c.** should not be subjected to a maximum wind load of 30 psf.
- **d.** should not be subjected to a maximum combined wind load and vapor pressure of 30 psf.
- **e.** none of the above.

11. In a window that carries the performance label "HS-C30 71 × 59", the letter C indicates that the window
- **a.** belongs to a performance class called *commercial*.
- **b.** is a casement window.
- **c.** is a C-grade window.
- **d.** none of the above.

12. Nailing flanges are only provided in steel windows.
- **a.** True
- **b.** False

13. The use of nailing flanges in windows also requires the use of
- **a.** retainer clips.
- **b.** two L-shaped trims.
- **c.** none of the above.
- **d.** both (a) and (b).
- **e.** either (a) or (b).

14. Most windows are shop-glazed so that they arrive at the construction site in a fully assembled condition.
- **a.** True
- **b.** False

15. A hollow metal door frame is obtained from
- **a.** tubular steel.
- **b.** sheet steel that is pressed into shape.
- **c.** tubular steel and sheet steel welded together.
- **d.** none of the above.

16. A hollow metal door frame is installed so that it
- **a.** butts against the jambs of a CMU wall.
- **b.** wraps around the jambs of a CMU wall.
- **c.** (a) or (b).
- **d.** (a) and (b).

31.7 FIRE-RATED DOORS AND WINDOWS

A fire-rated wall can serve as a barrier to the spread of fire only if the openings (doors, shutters, and windows) installed in them also provide the same (or nearly the same) degree of fire protection. Building codes require a certain minimum level of fire protection of openings provided in fire-rated walls.

The *fire-protection ratings* of openings are usually allowed to be lower than the fire-resistance ratings of the walls in which they are installed. For example, the fire rating of a door in a wall is 75% of the fire rating of the wall. Thus, in a 2-h-rated wall, a $1\frac{1}{2}$-h-rated door is considered adequate. An opening with a lower fire-protection rating than the wall in which it is installed does not greatly compromise the overall integrity of the wall. This is true because, under normal conditions of use, there is usually a lower fire hazard in the vicinity of an opening compared with the rest of the wall.

For example, in the case of a door, there is a clear space on both sides of the door to provide for unobstructed traffic. Hence, there is less fuel available in close proximity to the door, and the resulting fire hazard is less than that in the rest of the wall. Another reason is that the total opening area in fire-rated walls is usually so limited by building codes that a lower protection of openings does not greatly reduce the overall effectiveness of the wall.

Reference to a *rated opening* implies that the entire assembly is rated. The assembly consists of the door (or window), door frame (or window frame), anchorages, sill, handles, latches, hinges, and so on. Building codes require that a fire-rated door or window and its frame bear a label indicating that the assembly has been tested by an approved, independent testing laboratory.

Note that in referring to the fire rating of openings, the term *fire-protection rating*, not fire-resistance rating, has been used. This is because the fire rating of openings is not determined by the same test used for the fire rating of walls (the ASTM E 119 test).

FIRE DOORS

A fire door may be a 3-h-, 90-min-, 60-min-, 45-min-, or 20-min-rated door. A 20-min-rated door assembly is, in fact, a smoke-and-draft-control assembly. Its main purpose is to minimize the transmission of smoke from one side of the door to the other. A 20-min door is usually required in a 1-h-rated corridor wall. A $1\frac{3}{4}$-in.-thick solid-bonded wood-core door usually provides 20-min protection. However, to be accepted as a 20-min door, it must pass the required test to substantiate performance.

FIRE WINDOWS

Where the walls (interior or exterior) are required to be fire rated, a glazed window placed in such walls must provide at least 45-min protection. A glazed window with 45-min protection is referred to as a *fire window*. Because 45-min protection is relatively low for a protected opening, building codes place restrictions on the allowable areas of fire windows.

EXPAND YOUR KNOWLEDGE

Window Terminology

Two important components of a window are (a) window frame and (b) window sash. The terminology related to these and other components of a window is given next.

Frame (the Fixed Part of the Window Assembly)

- *Head*—the top horizontal member of the frame
- *Jamb*—a side vertical member of the frame
- *Sill*—the bottom horizontal member of the frame
- *Mullion*—an intermediate vertical member of the frame
- *Rail*—an intermediate horizontal member of the frame
- *Nailing flange*—Some manufacturers provide windows with a nailing flange. A nailing flange runs continuously around the outside of a window frame and functions as a flashing. It also provides a means of anchoring the window to the wall opening.

Sash (the Operable Part of the Window Assembly)

- *Stile*—a vertical sash member
- *Rail*—a horizontal sash member
- *Muntins*—thin horizontal and vertical dividers, commonly used in early windows when large sheets of good-quality glass were not available. (In some contemporary windows, large sheets of glass are used, but to simulate a divided window, grilles are used within the cavity of an insulating glass unit. Alternatively, a grill may be placed over the glass pane, either on the inside face, the outside face, or both faces. The grilles are removable for easy cleaning of glass. A window with imitation muntins is referred to as comprising *simulated divided lites*. A window with true dividers is referred to as a window with *authentic, or true, divided lites*.)
- *Weatherstripping*—a strip of resilient material that provides a seal between the sash and the frame to reduce air and water leakage
- *Glazing stop*—a feature that holds the glass against the rabbet in the sash or frame
- *Rabbet*—a step in a sash or frame cross section against which the glass is held

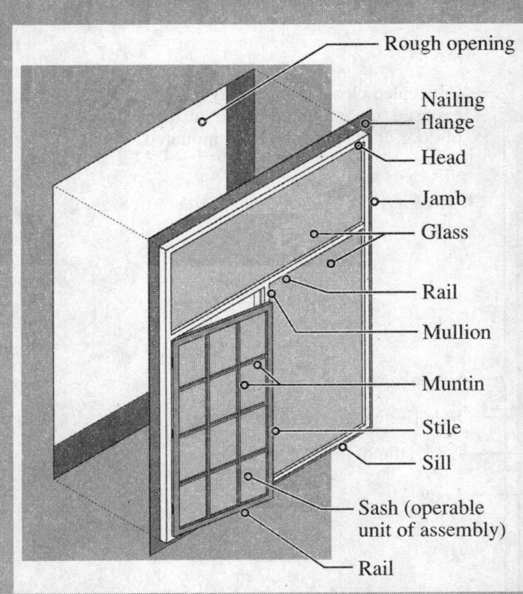

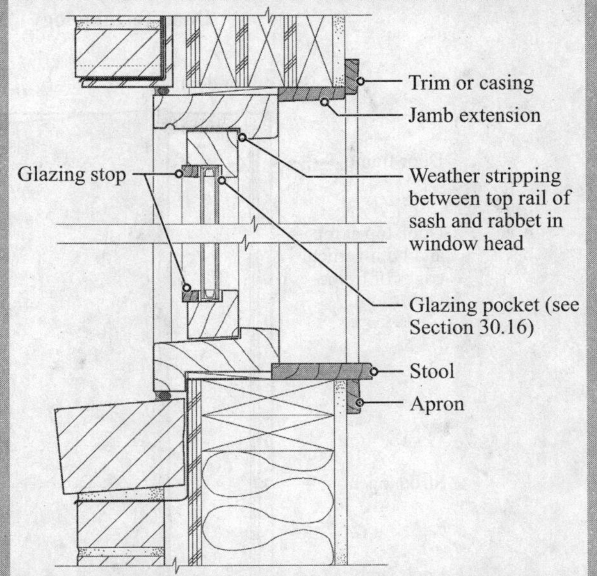

- *Gasket*—a strip of resilient material between the glass and the glazing pocket (Double-sided tape is used by some window manufacturers instead of a gasket.)
- *Daylight opening (DLO)*—visible glass area in a window (It is the total area of glass minus the invisible area, i.e., the area of glass that is buried within glazing pockets.)
- *Hardware*—hinges, latches, locks, levers, and so on

Window Surrounds (Interior)

- *Stool*—a horizontal trim member that abuts against the window sill and covers the rough sill
- *Jamb extension*—a horizontal or vertical trim that covers the rough head and jamb and extends the depth of the window frame (Note that the term *jamb extension* also refers to the extension of the head because the head is sometimes referred to as the *head jamb*.)
- *Apron*—the trim installed on the wall under the stool
- *Casing* (or trim)—decorative members that cover the joints between adjacent materials

Window-Opening Configurations

- *Rough opening*—the opening within which the window is placed (Rough opening dimensions are at least $\frac{1}{2}$ in. greater than the actual window unit dimensions to allow an easy fit and alignment of the window in the opening. The terms *rough sill*, *rough head*, and *rough jamb* refer to the respective areas of a rough opening.)
- *Punched window* and *punched glazing*—a window with an opaque wall around it (The term *punched glazing* is generally used for large, fixed glazing in which the glass is site installed. In terms of appearance, both punched windows and punched glazing are the same.)
- *Strip windows* and *glazing*—an array of windows placed side by side to form a horizontal strip (or ribbon) window system (The term *strip glazing* is used with large, fixed glazing units in which the glass is site installed.)
- *Window wall* and *glass curtain wall*—a wall in which the windows extend from floor to roof, or from floor to floor, called a *window wall* (A window wall must be distinguished from a glass curtain wall, which hangs as a curtain from the building's structure and bypasses the intermediate floors or roof; see Chapters 27 and 32.)
- *Projected window*—includes windows whose sashes project out of the window plane when open, such as casement, awning, hopper, and pivoted windows (Projected windows are generally more airtight in the closed position than hung or sliding windows because of the pressure seal between the frame and the sash.)

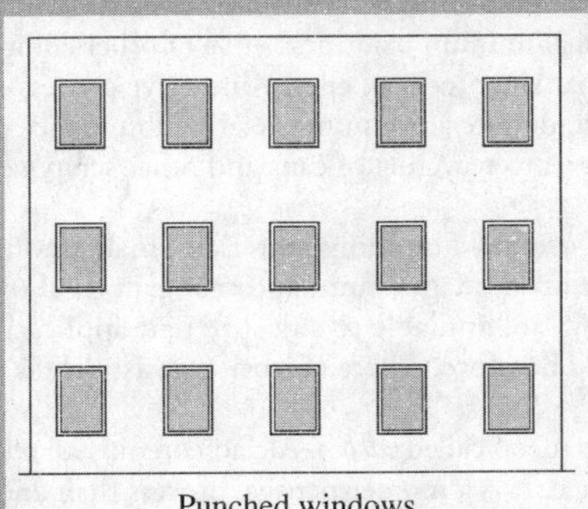
Punched windows

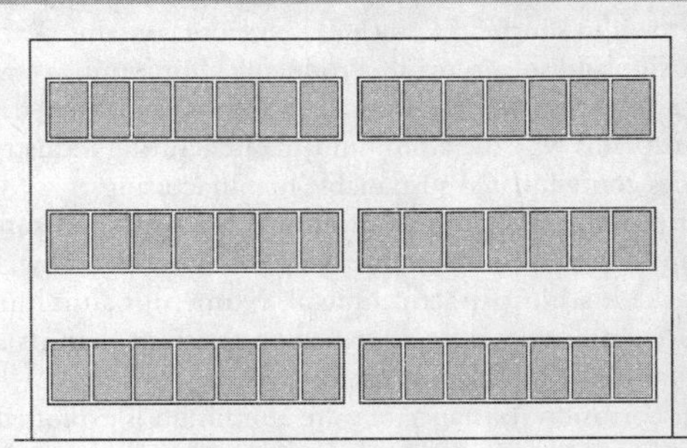
Strip (or ribbon) windows

EXPAND YOUR KNOWLEDGE

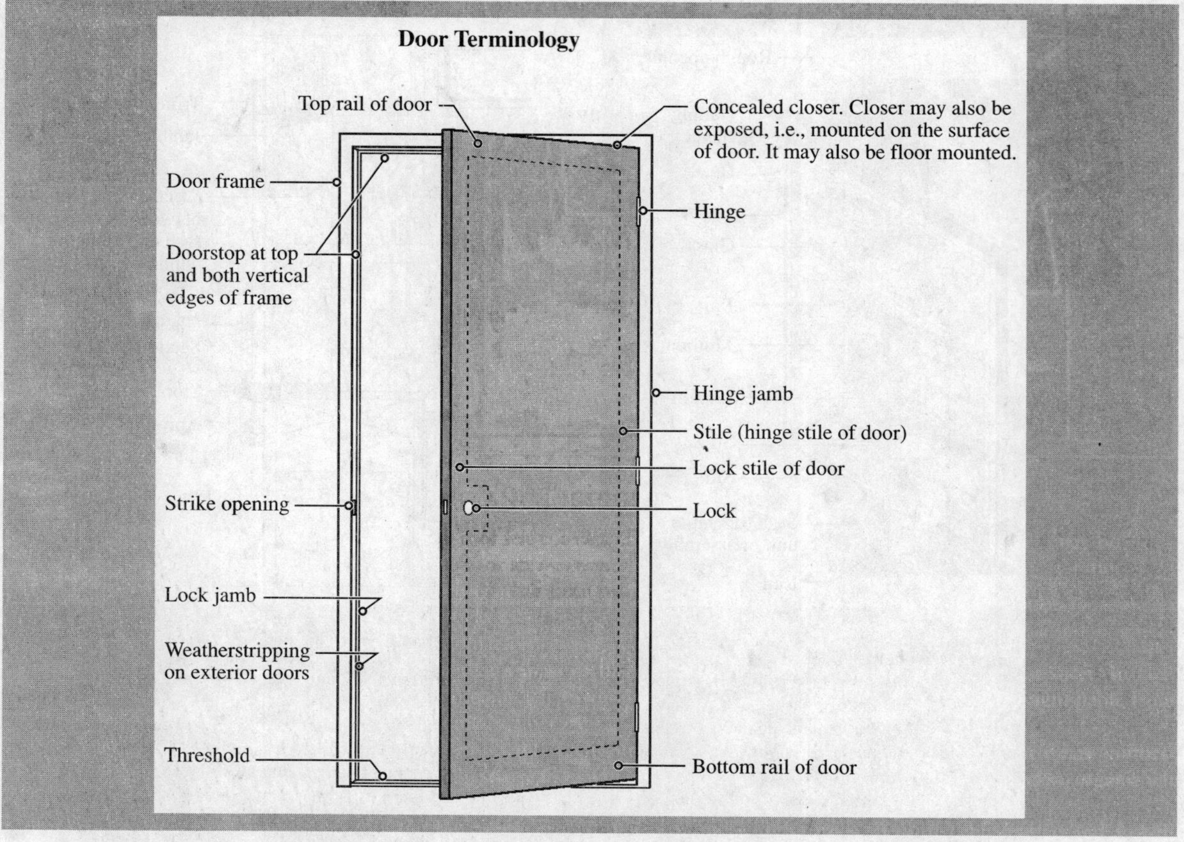

PRINCIPLES IN PRACTICE

A Note on Aluminum

Aluminum is a metal whose appearance resembles and whose corrosion resistance matches that of stainless steel. It is, however, much lighter, softer, and more flexible than steel or stainless steel. Its softness allows it to be easily extruded into complex cross-sectional profiles. Its flexibility allows aluminum components to be joined together by the snap-on technique, requiring no fasteners. Its corrosion resistance implies that it can be left exposed to the atmosphere without any protective coating.

Corrosion resistance, flexibility, and softness are the reasons aluminum is the preferred metal for glass curtain walls, storefronts, and windows. Nearly one-third of the entire production of aluminum is used in these components [31.1]. Aircrafts and automobiles also consume a large portion of the aluminum produced.

Aluminum, in the form of aluminum oxide, is far more abundant on the earth's crust than any other metal oxide, and most rocks and soils contain some aluminum oxide. However, the energy required to extract aluminum from common rocks and soils is very high. The only economical method of producing aluminum is from the ore called *bauxite,* named after Baux, a town in southern France. Bauxite ore contains nearly 60% aluminum oxide and 40% of other compounds, such as iron oxide and silicon oxide. Producing aluminum, even from bauxite ore, is an energy-intensive process. This explains why, weight for weight, aluminum is more expensive than steel, despite aluminum's relative abundance on the earth. It also explains why the aluminum-manufacturing industry resorted to recycling of cans and other scrap well before other industries contemplated sustainable manufacturing.

Like other pure metals, pure aluminum is too weak and soft to be used as a building material. Small amounts (altogether up to 5%) of other elements, such as copper, zinc, silicon, iron, magnesium, and manganese, alloyed with aluminum make it a more practical metal. Numerous aluminum alloys are available to suit different applications. However, pure aluminum is more corrosion resistant than its alloys. Therefore, where corrosion resistance is critical, anodized or painted aluminum is used.

The higher corrosion resistance of pure aluminum is exploited in the metal called *alclad.* Alclad contains an aluminum alloy core that is metallurgically bonded to pure aluminum on the surface. It is a more expensive process than anodizing and is commonly used in aircraft.

PRINCIPLES IN PRACTICE

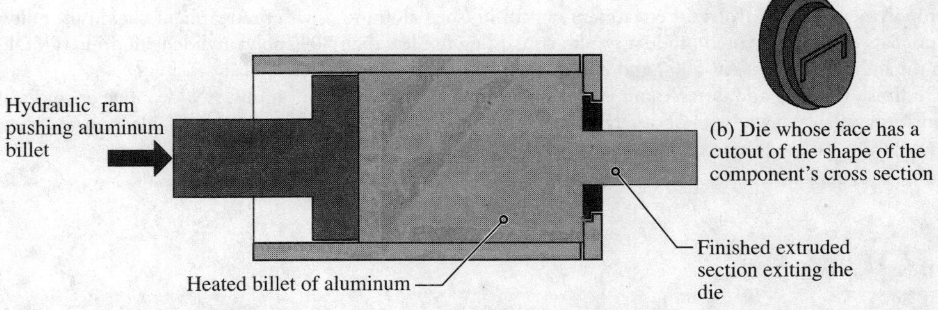

FIGURE 1 Aluminum window and curtain wall sections are made by pushing a heated aluminum billet through a die, a process called *extrusion*. Extrusion is similar to pushing toothpaste through the opening of a toothpaste tube.

ALUMINUM EXTRUSIONS

Manufacturing aluminum from bauxite ore has several similarities to manufacturing iron from its ore. The end product is cast pig aluminum, which is converted to aluminum billets, bloom, or slabs. Aluminum sections can be made by hot rolling billets in the same way as steel sections.

Sheets obtained from aluminum slabs can be cold-formed into simpler profiles, similar to light-gauge steel members. However, extrusion is a cheaper and more versatile process for aluminum than hot rolling or cold forming. Additionally, fairly complex cross-sectional shapes that cannot be obtained from hot rolling can be produced by extruding. Therefore, except for sheets, most architectural aluminum components are produced from the extrusion process, Figure 1.

The melting point of most aluminum alloys is nearly 1,100°F (600°C). Therefore, extrusion requires aluminum billets to be heated to a fairly low temperature—nearly 800°F (450°C). In this process, a billet of heated aluminum is pressed through a die, whose face has a cutout that matches the cross-sectional profile of the desired component. A different cross-sectional profile needs only a different die, which is not too expensive to fabricate.

ANODIZED ALUMINUM

Aluminum obtains its corrosion resistance because of the aluminum oxide film it develops on its surface on exposure to the atmosphere. This occurs fairly quickly. In fact, the extrusion emerging out of the die mouth is already covered with an oxide film. Like the chromium oxide film on stainless steel, aluminum oxide film is also self-repairing.

The aluminum surface obtained from the die is referred to as *mill finish.* Mill-finished aluminum has a bright, silvery (the familiar aluminum foil) finish, which turns dull gray after a few years. The number of years depends on the amount of pollution and the amount of rainfall. If water from rain or condensation is allowed to lodge on the surface, black streaks may develop there.

Most architectural aluminum is, therefore, either anodized or painted. Anodization essentially preoxidizes the aluminum. The only difference between natural mill-finished aluminum and anodized aluminum is in the thickness of the oxide film. The natural oxide film is nearly one-millionth of an inch (0.001 mil) thick. The anodic coating is several thousand times thicker—nearly 0.2 to 0.7 mil thick. A thicker coating is selected for exterior exposures. Coatings greater than 3 mil fracture and chip easily.

The anodic coating may either be clear or colored. A clear coating leaves aluminum looking its natural silvery color. It is also the more common and economical finish and has been used in several outstanding building facades.

The colors available in anodized aluminum are fairly limited. Generally, they are restricted to black, dark bronze, medium bronze, light bronze, and white, and there can be slight variations between pieces.

PAINTED ALUMINUM

Two types of paint finishes are commonly used on aluminum:

- Baked-enamel coating
- Fluoropolymer coating

A baked-enamel coating requires pretreatment of mill-finished aluminum with required chemical(s), followed by spray application of a thermosetting, modified acrylic enamel coating, followed by oven baking. A baked-enamel finish is generally recommended for interior applications. The colors obtained are uniform and opaque.

(Continued)

PRINCIPLES IN PRACTICE

A Note on Aluminum (*Continued*)

A fluoropolymer coating involves pretreatment of mill-finished aluminum with pretreatment chemicals, followed by one to four coats of a thermocured product system containing not less than 70% polyvinylidene fluoride (PVDF) resin, known in the industry as KYNAR 500® and HYLAR 5000.

A PVDF finish is strong and durable and is primarily suited for exterior applications. A wide range of colors, including exotic colors, is available. Two types of durability ratings with respect to PVDF finish are available; one gives 5 years of performance at the test requirements and another gives 10 years.

PRACTICE **QUIZ**

Each question has only one correct answer. Select the choice that best answers the question.

17. Which of the following terms is used with fire-rated windows or doors?
- **a.** *Fire-resistance rating*
- **b.** *Fire-protection rating*

18. Fire-rated doors are rated as
- **a.** 3 h, 90 min, 60 min, 45 min, or 20 min.
- **b.** 3 h, 120 min, 60 min, or 30 min.
- **c.** 4 h, 3 h, 2 h, 1 h, or 30 min.
- **d.** 4 h, 3 h, 2 h, or 1 h.
- **e.** none of the above.

19. Which of the following terms describes an intermediate vertical member of a window frame?
- **a.** *Mullion*
- **b.** *Stile*
- **c.** *Rail*
- **d.** *Jamb*
- **e.** *Muntin*

20. Which of the following terms describes an intermediate horizontal member of a window frame?
- **a.** *Mullion*
- **b.** *Stile*
- **c.** *Rail*
- **d.** *Jamb*
- **e.** *Muntin*

21. Which of the following terms refers to thin and narrow horizontal and vertical members of a window sash?
- **a.** *Mullion*
- **b.** *Stile*
- **c.** *Rail*
- **d.** *Jamb*
- **e.** *Muntin*

22. A stool is used in a window in the building's interior.
- **a.** True
- **b.** False

REVIEW QUESTIONS

1. Sketch the following window styles: (a) casement window, (b) double-hung window, (c) single-hung window, (d) awning window, and (e) hopper window.
2. A window has been rated as HS-C35 71 × 59. Explain each term of the rating.
3. Which materials are commonly used for cladding wood windows? What is the purpose of cladding, and which parts of a window are clad? Explain how the cladding helps to secure a window to the opening.
4. What does the acronym *NFRC* stand for? List the information provided in an NFRC label.
5. Explain the purpose and anatomy of a nailing flange. Which types of windows are typically provided with nailing flanges?
6. Sketch in plan various types of wood door frames, and give the terminology used to describe various parts of a door frame.
7. Sketch in plan a hollow metal door frame and its junction with (a) a metal stud wall and (b) a CMU wall.
8. Using sketches, explain the following terms: (a) *punched windows*, (b) *strip windows*, (c) *window wall*, and (d) *curtain wall*.

CHAPTER 32 Exterior Wall Cladding–IV (Wall Systems in Glass)

CHAPTER OUTLINE

Transparency, luminosity, and elegance are the reasons for the popularity of glass walls in modern architecture. Most glass walls are constructed with aluminum sections to support the glass. In other words, the glass panes (also called *lites*) are held within vertical and horizontal aluminum framing members. Therefore, they share some of the characteristics of their smaller counterparts—the aluminum windows, discussed in Chapter 31. However, there are many differences between the two: scale, aesthetic character, performance properties, design, detailing, and installation.

Three commonly used glass-aluminum wall system systems are

- Glass-aluminum curtain walls
- Punched and strip glazing systems
- Storefront systems

The vast majority of contemporary buildings include one or more of these systems in the same building. The reasons include the unparalleled opportunity provided by them to obtain the maximum amount of daylight and view, the cost savings compared with other exterior wall cladding systems, and the recent technological advances in the thermal and structural performance of glass wall systems.

Of the three systems listed above, the most frequently used and the most complex is the glass-aluminum curtain wall system, which is presented here in detail. The other two systems (strip system and storefront system) are discussed to the extent that they differ from the curtain wall system. Finally, the chapter deals with nontraditional glass wall systems—systems that do not include aluminum sections to support the glass.

32.1 GLASS-ALUMINUM CURTAIN WALLS

Because of their common use, glass-aluminum curtain walls (or simply *glass curtain walls*) are constantly evolving in their design and performance. Therefore, a succinct classification that includes all contemporary glass curtain walls is impossible. The American Architectural Manufacturers Association (AAMA), an association of the manufacturers of windows and curtain walls, however, classifies glass curtain wall systems into five types based on their anatomy:

- Stick-built (or, simply, stick) systems
- Unitized systems
- Unit and mullion systems
- Panel systems
- Column cover and spandrel systems

These systems are illustrated in Figure 32.1. The stick system is the oldest and the most widely used system. The remaining four systems are different from the stick system because they consist of prefabricated wall units similar to the (opaque) curtain wall panels.

STANDARD AND CUSTOM CURTAIN WALL SYSTEMS

Most major glass curtain wall manufacturers have their own facility for extruding the aluminum sections. Walls constructed from a manufacturer's commonly used and pretested aluminum sections are referred to as *standard walls*.

Custom curtain walls utilize cross-sectional shapes extruded specifically for a project in response to an architect's design. Because the cost of dies and other equipment required to extrude custom cross sections can be recovered from just one fair-size project, custom curtain walls are fairly common. Custom walls should, however, be tested for performance before they are used in a project. Performance data for standard walls are available from the manufacturers.

Walls made from standard components are obviously more economical. However, this does not imply that the standard components yield only one type of wall design. In fact, the components are generally quite adaptable, and manufacturers can provide a few custom components for a standard system, so that the facade expressions obtained from the use of standard components can be numerous. If the number of custom components in a wall becomes excessive, the cost of a standard wall may approach that of a custom wall.

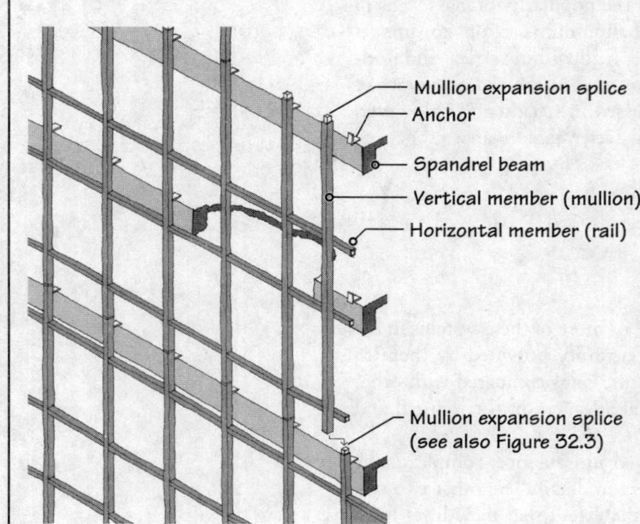

(a) STICK SYSTEM

In the stick system, the curtain wall is installed piece by piece at the site. Generally, the mullions are installed first, followed by the rails. Subsequently the glass panes are installed within the mullion-rail framework.

The anchorage of the wall to the structural frame is through the mullions. The mullions may span from floor to floor or over two floors. Thermal expansion and contraction of mullions are accommodated by expansion joints in mullions.

The system components are shop-fabricated and shipped to the construction site in a knocked-down (KD) version. Therefore, the system has relatively low shipping costs and also permits a greater degree of on-site adjusment as compared to the other systems.

Its disadvantages include longer on-site assembly time and more on-site labor than the other systems.

FIGURE 32.1 Types of glass curtain walls—the stick system. (Illustration adapted from AAMA, *Curtain Wall Design Guide*, 1996, with permission)

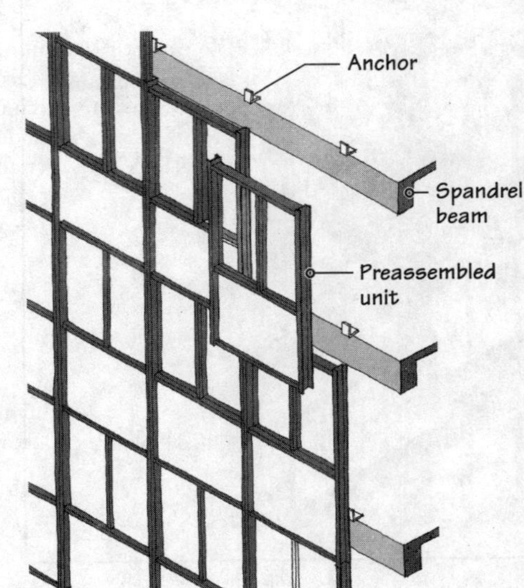

(b) UNITIZED SYSTEM
A unitized system consists of framed wall units that are shop-fabricated, preassembled, and generally preglazed. The units are designed so that the vertical and horizontal members in adjacent units interlock to form common mullions and rails. The units may be one or two stories high. They are anchored to the building's structural frame in essentially the same way as the mullions in the stick system.

The advantage of this system is its greater degree of quality control resulting from shop fabrication. Its disadvantages are the greater shipping cost because of the added bulk from assembled units, the need of a greater degree of protection of units during transporation, and a lower degree of field adjustment.

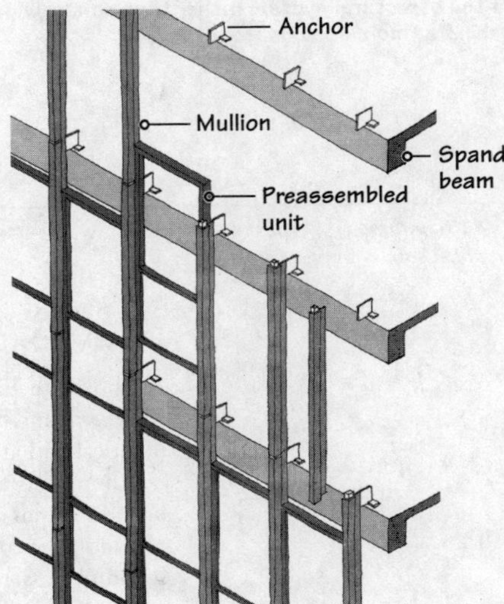

(c) UNIT AND MULLION SYSTEM
The unit and mullion system combines the advantages of both the stick system as well as the unitized system. It is constructed by first installing the mullions; subsequently, factory-assembled units are placed between the mullions.

Because the system is a compromise between the stick and unitized systems, it has the advantages and disadvantages of both, i.e., its transportation cost is lower than that of the unitized system but greater than that of the stick system. A greater degree of site adjustability is available in the unit and mullion system, but it is less that that of the stick system.

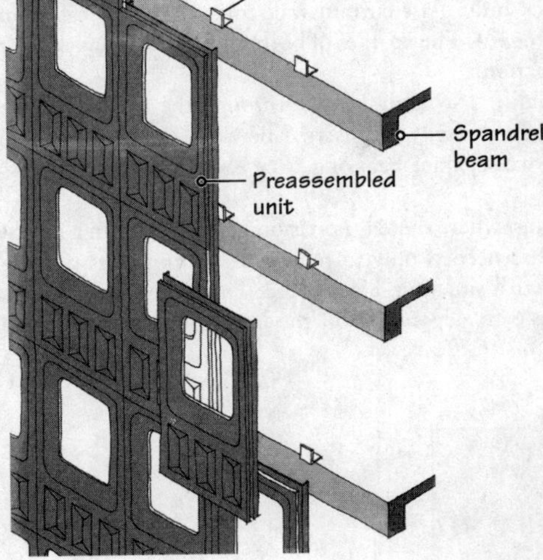

(d) PANEL SYSTEM
The panel system consists of preassembled (and sometimes preglazed) homogeneous sheet metal panels with glass infills that generally span from floor to floor. The curtain wall's appearance is more integrated and comprehensive rather that a grid pattern of horizontal and vertical elements.

The panels can be formed by stamping or casting. The casting system is economical only where a large number of identical panels are needed.

FIGURE 32.1 *(continued)* Types of glass curtain walls—unitized system, unit and mullion system, and panel system. (Illustrations adapted from AAMA, *Curtain Wall Design Guide*, 1996, with permission)

32.2 ANCHORAGE OF A STICK-BUILT GLASS CURTAIN WALL TO A STRUCTURE

Like other curtain walls, a glass curtain wall must be spaced away from the building's structural frame to account for the small dimensional variations (within the allowed tolerances) in the structural frame. A 2-in. space is generally the minimum requirement. A wider space may be required for tall buildings.

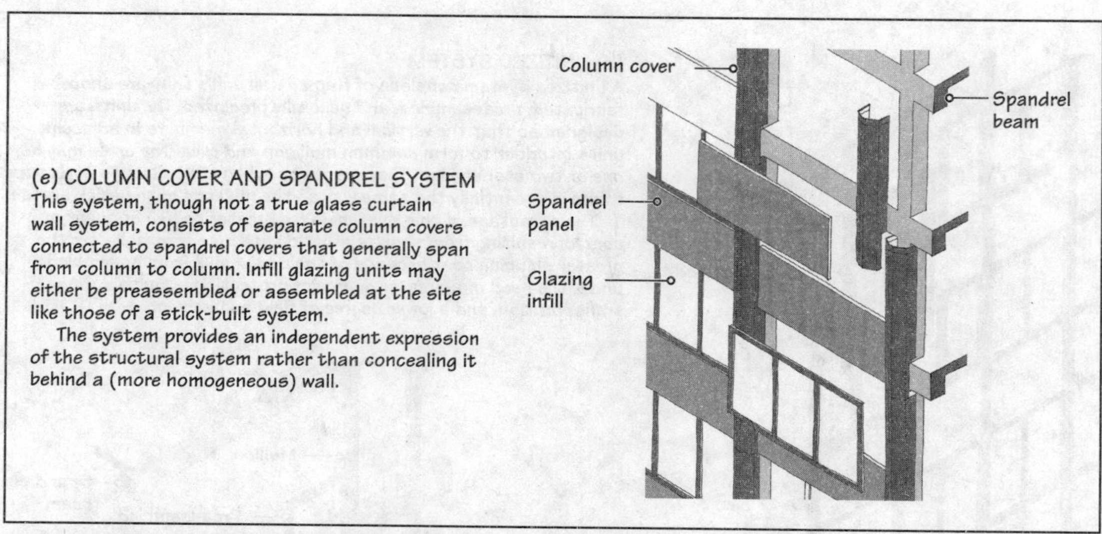

The system provides an independent expression of the structural system rather than concealing it behind a (more homogeneous) wall.

Glazing infill

FIGURE 32.1 (*continued*) Types of glass curtain walls—column cover and spandrel system. (Illustration adapted from AAMA, *Curtain Wall Design Guide*, 1996, with permission)

Dead-Load Anchors and Expansion Anchors

As shown in Figure 32.1(a), a stick-built glass curtain wall consists of vertical members (*mullions*) and horizontal members (*rails*). The profiles of both mullions and rails are almost identical and are tubular in cross section.

The wall is anchored to the building's structural frame through the mullions. All mullions in a wall are installed first; then the rails are inserted between them. Three rails are commonly used per floor to create two separate areas of glass at each floor—vision glass and spandrel glass.

In a building (or part of a building) where there is no vision glass, such as in a multistory parking garage, intermediate rails are needed only to reduce the size of glass panes. Two rails per floor are commonly used in that situation, Figure 32.2. The center-to-center spacing between mullions is generally 4 to 6 ft, depending on the lateral load intensity and the desired appearance of the facade.

10
Office floors
Parking floors
After the framing of the curtain wall (mullions and rails) for the parking floors is complete, its glazing has started, while the framing for the office floors is yet to begin. Because there are no vision areas on the parking floors, the glazing on each floor has been divided into two parts by one intermediate rail. The purpose of the rail is merely to reduce the size of the glass.

FIGURE 32.2 (a) The progress in the installation of a stick-built glass curtain wall on an office building in which the lower floors are parking floors and the upper floors are office floors. See also Figure 32.2(b).

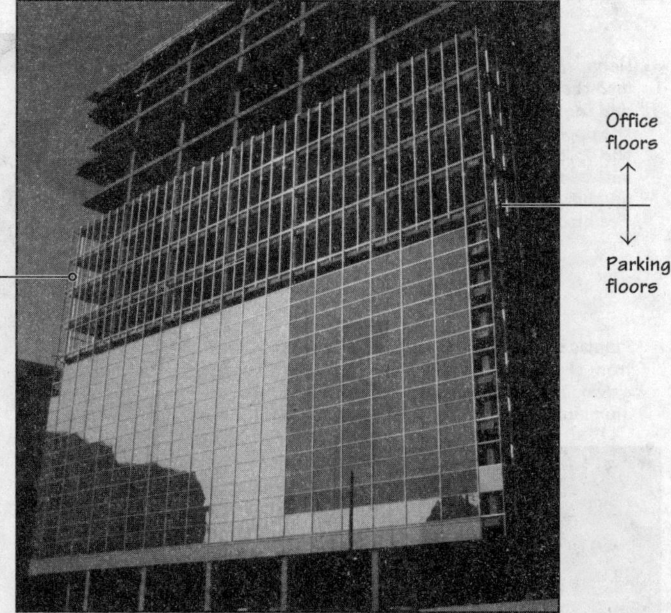
In this photograph, the glazing of the curtain wall on the parking floors of the building shown in Figure 32.2(a) is almost complete. Now the framing for the curtain wall on the office floors has begun. As shown here, the mullions are installed first, followed by the rails.
This (part of the) photograph shows the progress in the installation of rails on the office floors. Because an office floor has separate spandrel and vision glass areas, there will be three rails per floor.
Office floors
Parking floors

FIGURE 32.2 (b) The progress in the installation of a stick-built glass curtain wall on an office building in which the lower floors are parking floors and the upper floors are office floors; see also Figure 32.2(a).

To allow for the expansion and contraction of mullions caused by temperature changes, each mullion must be provided with expansion joints. Thus, the mullions consist of short lengths (one or two floors tall) that terminate in expansion joints at both ends, Figure 32.3.

An expansion joint also absorbs the creep in concrete columns and the live-load deflection of the spandrel beam to which the mullions are anchored. Therefore, the expansion joint width must be determined on a project-by-project basis. Note that an expansion joint allows movement in the vertical direction only.

Because all loads on a wall are transferred to the structural frame through the mullions, each mullion is provided with a dead-load support anchor (or, simply, a *DL anchor*) designed to carry the weight of the respective portion of the curtain wall.

A DL anchor fully restrains the movement of a mullion; that is, the mullion is immobile in all three principal directions at a dead-load support. Therefore, a DL anchor transfers both the dead loads and the lateral loads on a mullion to the building's structural frame.

Two types of mullion spans are generally used in a stick-built glass curtain wall, Figure 32.4:

- Single-span mullion systems
- Twin-span mullion systems

In a single-span mullion system, each mullion extends only over one floor. DL anchors are, therefore, required at every floor, except at the ground floor, where the building's foundation provides dead-load support to the first mullion length, Figure 32.4(a).

In a twin-span mullion system, the mullions extend over two floors. Because a mullion can have only one dead-load support, DL anchors are provided at alternate floors, Figure 32.4(b). Another difference between a single-span and a twin-span system is that in a twin-span system, expansion anchors (or, simply, *EX anchors*) are also required at alternate floors. In a single-span system, an EX anchor is required only at the second floor of the building.

DL anchors and EX anchors are steel (or aluminum) members to which the mullions are bolted. As shown in Figure 32.4(b), they are almost identical. The only difference between them is that in a DL anchor, the upper pair of holes is round, and in an EX anchor, the upper pair of holes has vertically slotted holes that allow vertical movement.

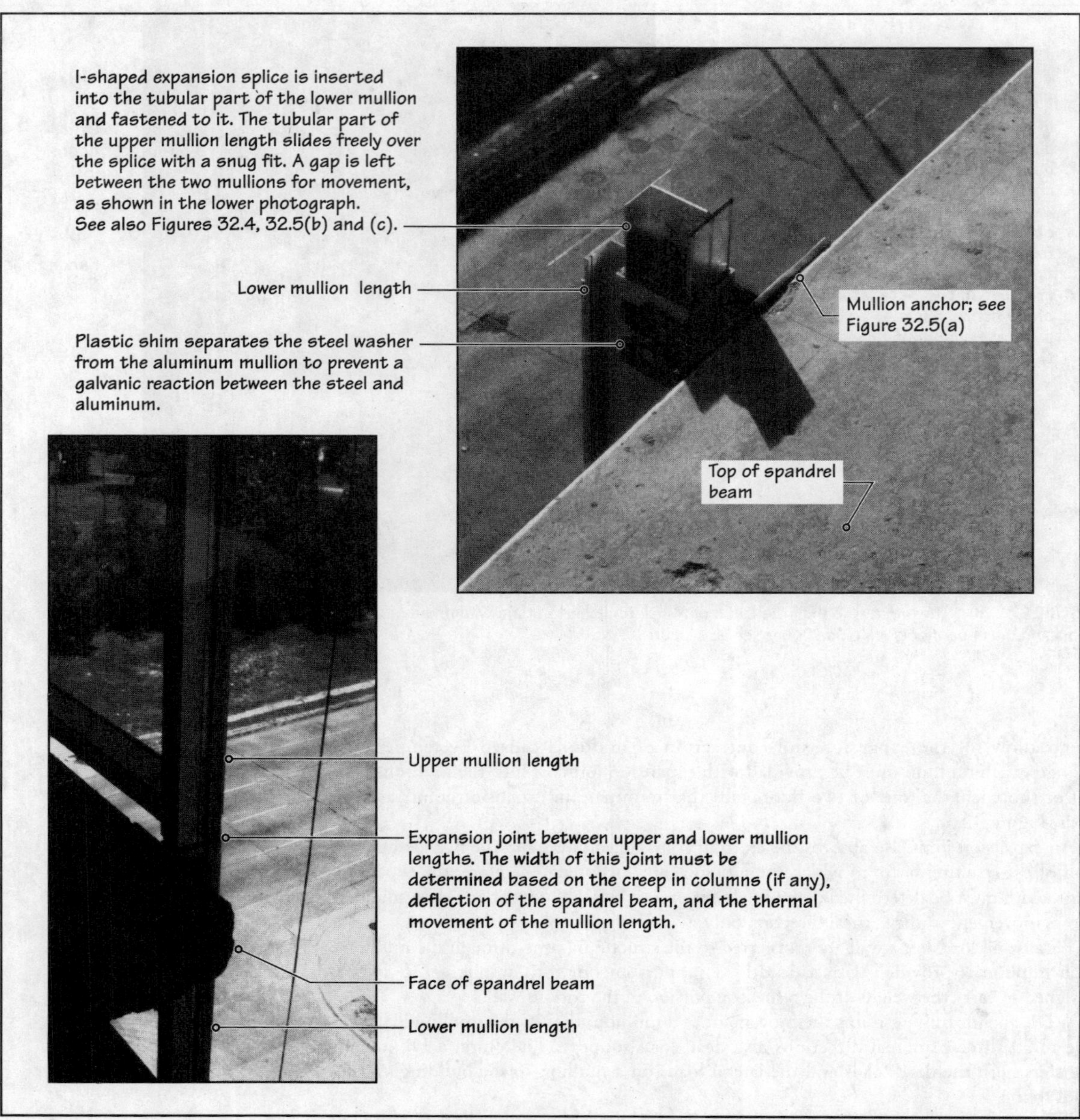
I-shaped expansion splice is inserted into the tubular part of the lower mullion and fastened to it. The tubular part of the upper mullion length slides freely over the splice with a snug fit. A gap is left between the two mullions for movement, as shown in the lower photograph. See also Figures 32.4, 32.5(b) and (c).
Lower mullion length
Plastic shim separates the steel washer from the aluminum mullion to prevent a galvanic reaction between the steel and aluminum.
Mullion anchor; see Figure 32.5(a)
Top of spandrel beam
Upper mullion length
Expansion joint between upper and lower mullion lengths. The width of this joint must be determined based on the creep in columns (if any), deflection of the spandrel beam, and the thermal movement of the mullion length.
Face of spandrel beam
Lower mullion length

FIGURE 32.3 A typical expansion joint between two mullion lengths.

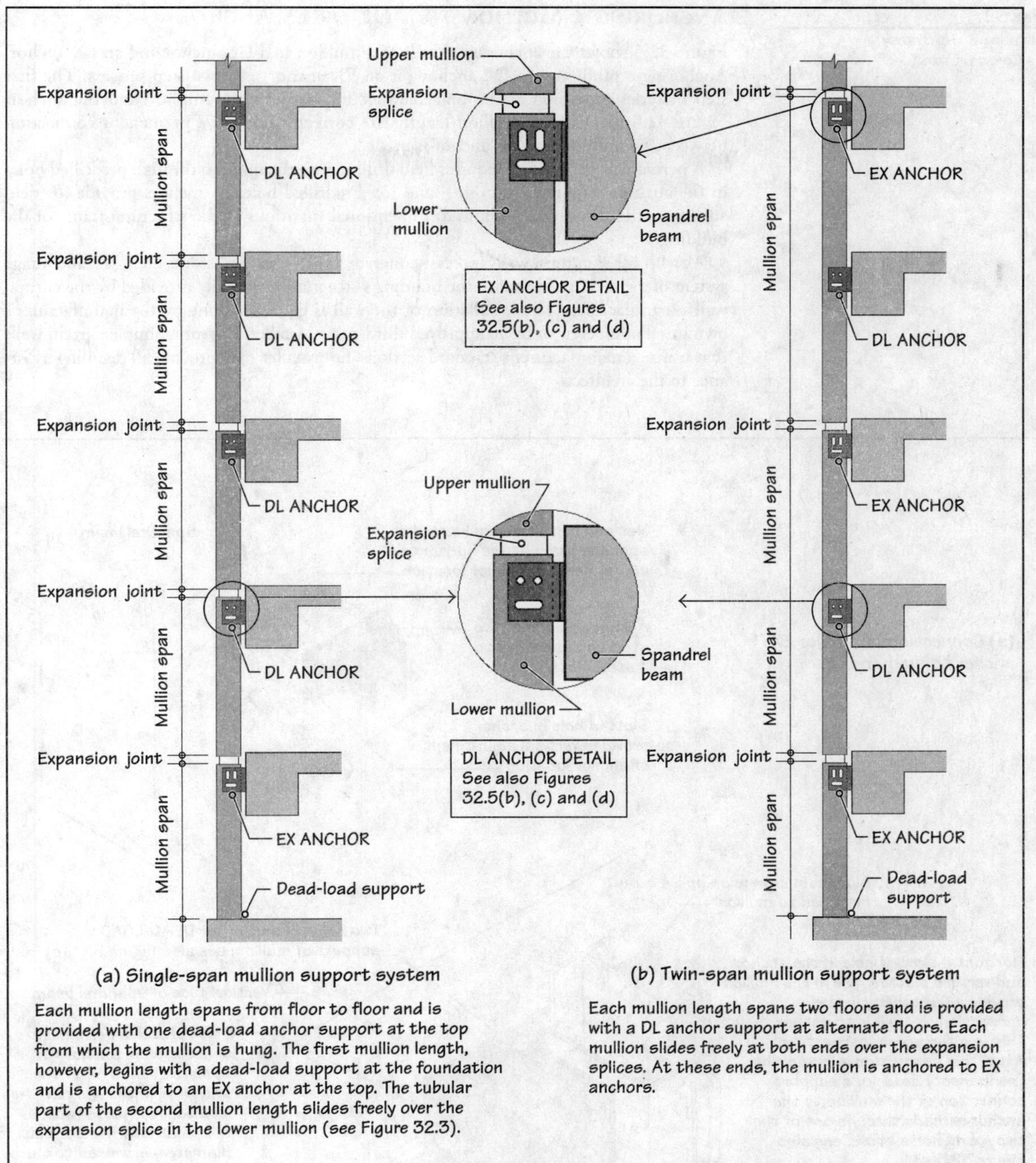

Upper mullion
Expansion splice
Lower mullion
Spandrel beam
EX ANCHOR DETAIL
See also Figures 32.5(b), (c) and (d)
Upper mullion
Expansion splice
Spandrel beam
Lower mullion
DL ANCHOR DETAIL
See also Figures 32.5(b), (c) and (d)
Expansion joint
DL ANCHOR
Mullion span
EX ANCHOR
Dead-load support
(a) Single-span mullion support system
Each mullion length spans from floor to floor and is provided with one dead-load anchor support at the top from which the mullion is hung. The first mullion length, however, begins with a dead-load support at the foundation and is anchored to an EX anchor at the top. The tubular part of the second mullion length slides freely over the expansion splice in the lower mullion (see Figure 32.3).
(b) Twin-span mullion support system
Each mullion length spans two floors and is provided with a DL anchor support at alternate floors. Each mullion slides freely at both ends over the expansion splices. At these ends, the mullion is anchored to EX anchors.

FIGURE 32.4 Support systems for single-span and twin-span curtain wall mullions. Observe that each mullion has only one dead-load support.

Anchoring a Mullion to a DL or EX Anchor

Figure 32.5 shows the anchorage details of a mullion to a DL anchor and an EX anchor. Anchoring a mullion to a DL anchor (or an EX anchor) is a two-step process. The first step includes providing a temporary connection between the mullion and the anchor, Figure 32.5(b). After all mullion lengths are correctly aligned, a permanent connection between the mullion and the anchor is made.

A permanent connection requires field drilling into the mullion through predrilled holes in the anchors, Figure 32.5(c), (d), and (e). Predrilled holes in anchors provide for field adjustment to cater to the (allowed) dimensional variations in the structural frame of the building.

As with other curtain walls (precast concrete, GFRC, natural stone, etc.), the anchorage system of a glass curtain wall to the building's structure is typically provided by the curtain wall's manufacturer. The installation of the wall is generally done by the manufacturer's own installation crew or by an approved third-party installer. For some simple curtain walls that utilize a manufacturer's standard sections, an installer may provide all detailing assistance to the architect.

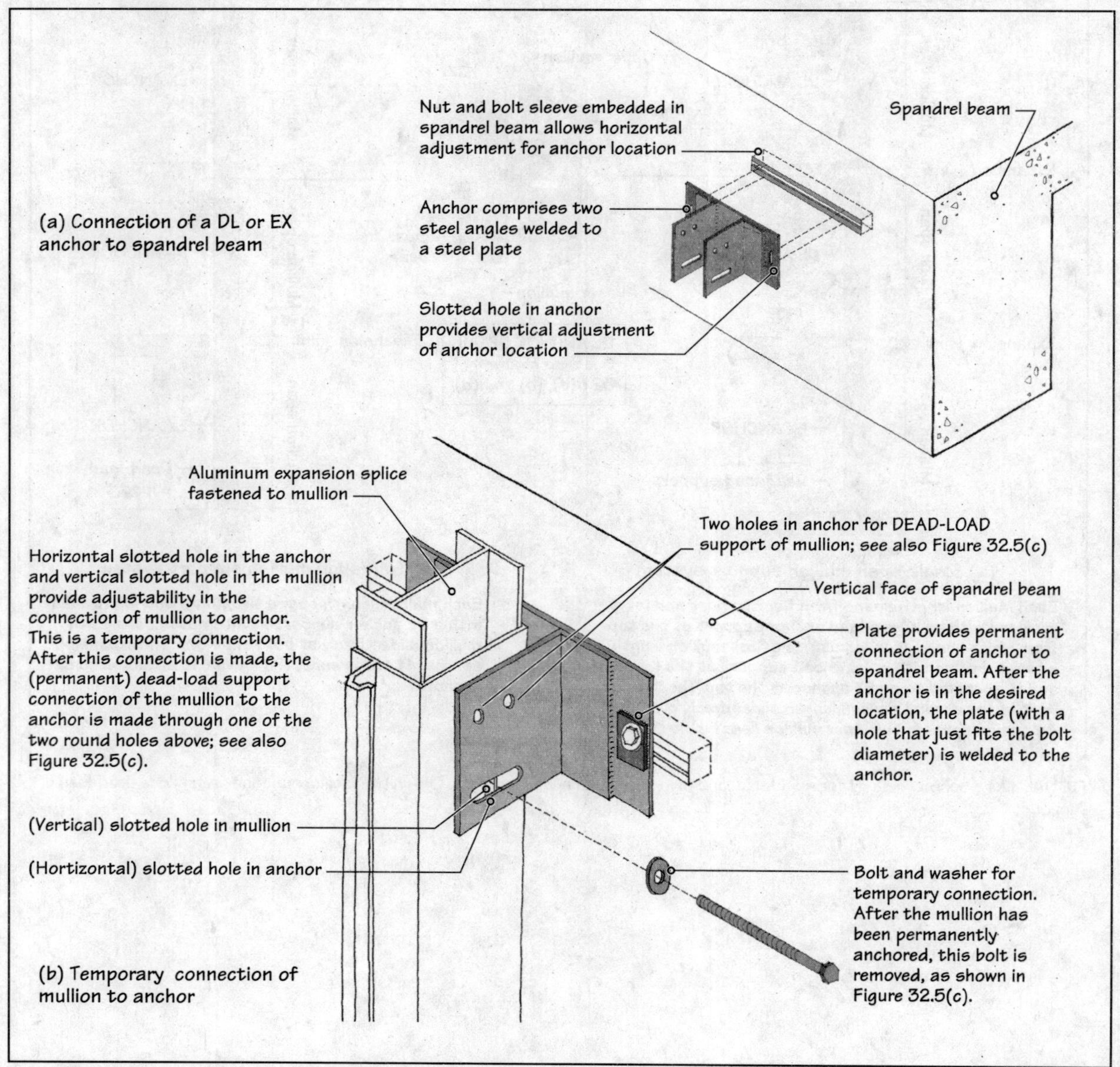
Nut and bolt sleeve embedded in spandrel beam allows horizontal adjustment for anchor location
Spandrel beam
Anchor comprises two steel angles welded to a steel plate
(a) Connection of a DL or EX anchor to spandrel beam
Slotted hole in anchor provides vertical adjustment of anchor location
Aluminum expansion splice fastened to mullion
Two holes in anchor for DEAD-LOAD support of mullion; see also Figure 32.5(c)
Horizontal slotted hole in the anchor and vertical slotted hole in the mullion provide adjustability in the connection of mullion to anchor. This is a temporary connection. After this connection is made, the (permanent) dead-load support connection of the mullion to the anchor is made through one of the two round holes above; see also Figure 32.5(c).
Vertical face of spandrel beam
Plate provides permanent connection of anchor to spandrel beam. After the anchor is in the desired location, the plate (with a hole that just fits the bolt diameter) is welded to the anchor.
(Vertical) slotted hole in mullion
(Hortizontal) slotted hole in anchor
Bolt and washer for temporary connection. After the mullion has been permanently anchored, this bolt is removed, as shown in Figure 32.5(c).
(b) Temporary connection of mullion to anchor

FIGURE 32.5 Typical anchorage details of a mullion to a spandrel beam.

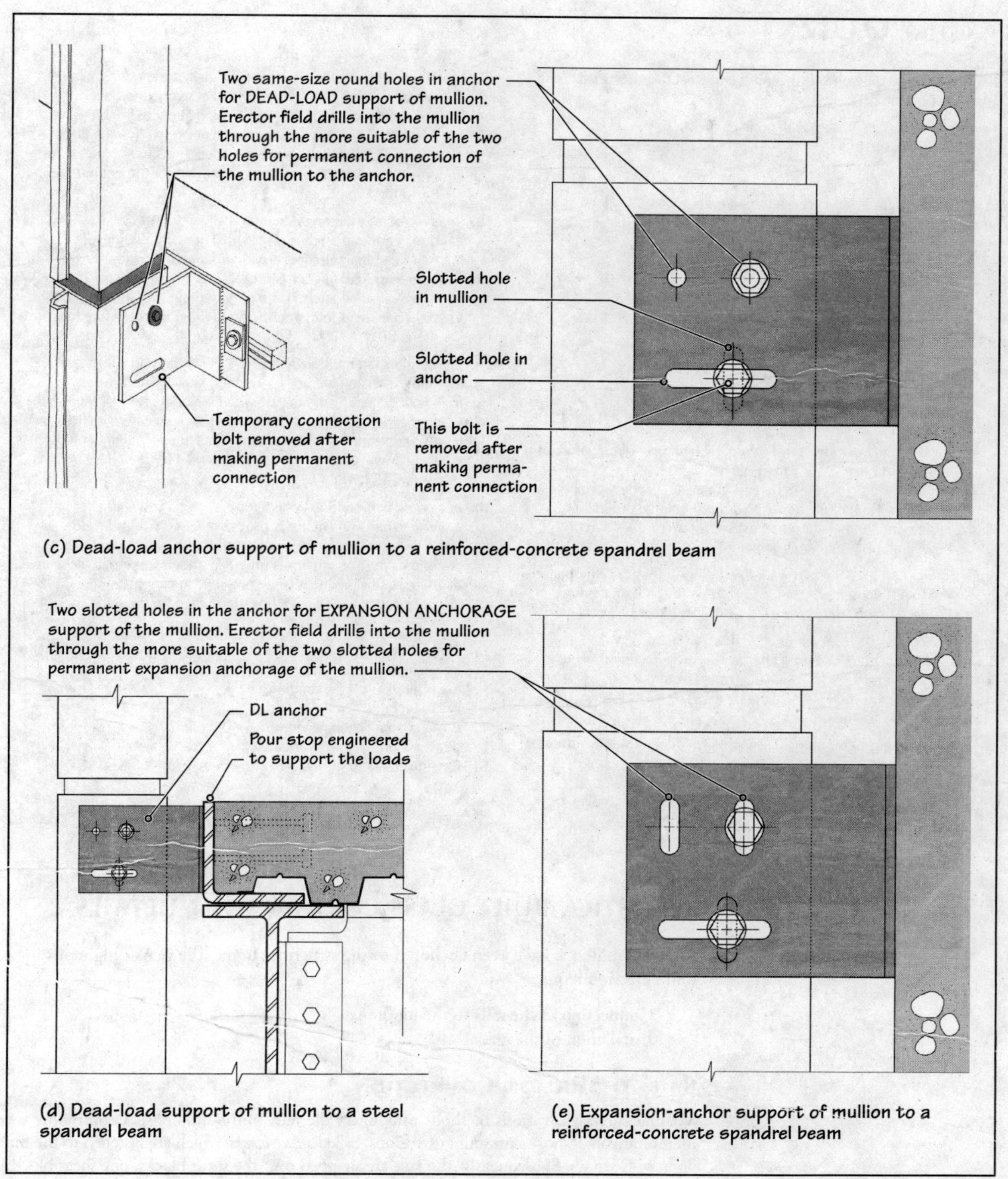

Two same-size round holes in anchor for DEAD-LOAD support of mullion. Erector field drills into the mullion through the more suitable of the two holes for permanent connection of the mullion to the anchor.
Slotted hole in mullion
Slotted hole in anchor
Temporary connection bolt removed after making permanent connection
This bolt is removed after making perma-nent connection
(c) Dead-load anchor support of mullion to a reinforced-concrete spandrel beam
Two slotted holes in the anchor for EXPANSION ANCHORAGE support of the mullion. Erector field drills into the mullion through the more suitable of the two slotted holes for permanent expansion anchorage of the mullion.
DL anchor
Pour stop engineered to support the loads
(d) Dead-load support of mullion to a steel spandrel beam
(e) Expansion-anchor support of mullion to a reinforced-concrete spandrel beam

FIGURE 32.5 ***(continued)*** Typical anchorage details of a mullion to a spandrel beam.

PRACTICE QUIZ

Each question has only one correct answer. Select the choice that best answers the question.

1. A stick-built glass curtain wall consists of
   a. mullions and glass.
   b. rails and glass.
   c. mullions, rails, and glass.
   d. mullions, rails, and preassembled units.
   e. preassembled units and glass.

2. A unitized glass curtain wall consists of
   a. mullions and glass.
   b. rails and glass.
   c. mullions, rails, and glass.
   d. mullions, rails, and preassembled units.
   e. preassembled units and glass.

3. Glass curtain walls in which aluminum framing sections are specially profiled for a particular project are
   a. rare because of the prohibitive cost of manufacturing custom profiles.
   b. uncommon because of the extremely high cost of manufacturing custom profiles.
   c. not uncommon because the cost of custom profiles can be recovered from a few repeat mid-sized to large projects.
   d. fairly common because the cost of custom profiles can be recovered from one large project.

4. In a stick-built glass curtain wall, the mullions are typically spaced at
   a. 2 ft to 4 ft on center.
   b. 4 ft to 6 ft on center.
   c. 6 ft to 10 ft on center.
   d. 10 ft to 15 ft on center.
   e. as needed for the project.

5. In a stick-built glass curtain wall, the rails are typically spaced at
   a. 2 ft to 4 ft o.c.
   b. 4 ft to 6 ft o.c.
   c. 6 ft to 10 ft o.c.
   d. 10 ft to 15 ft o.c.
   e. as needed for the project.

6. A stick-built glass curtain wall is anchored to the building's structure through
   a. mullions.
   b. rails.
   c. both mullions and rails.
   d. none of the above.

7. In a single-span mullion system for a glass curtain wall,
   a. one dead-load anchor is provided at every floor level.
   b. two dead-load anchors are provided at every floor level.
   c. one dead-load anchor is provided at every alternate floor level.
   d. two dead-load anchors are provided at every alternate floor level.
   e. only one dead-load anchor is provided for the entire height of the wall.

8. In a twin-span mullion system for a glass curtain wall,
   a. one dead-load anchor is provided at every floor level.
   b. two dead-load anchors are provided at every floor level.
   c. one dead-load anchor is provided at every alternate floor level.
   d. two dead-load anchors are provided at every alternate floor level.
   e. only one dead-load anchor is provided for the entire height of the wall.

9. In a single-span mullion system for a glass curtain wall,
   a. one expansion anchor is provided at every floor level.
   b. two expansion anchors are provided at every floor level.
   c. one expansion anchor is provided at every alternate floor level.
   d. two expansion anchors are provided at every alternate floor level.
   e. only one expansion anchor is provided for the entire height of the wall.

10. In a twin-span mullion system for a glass curtain wall,
    a. one expansion anchor is provided at every floor level.
    b. two expansion anchors are provided at every floor level.
    c. one expansion anchor is provided at every alternate floor level.
    d. two expansion anchors are provided at every alternate floor level.
    e. only one expansion anchor is provided for the entire height of the wall.

11. The width of an expansion joint between adjacent mullion lengths in a typical stick-built glass curtain wall
    a. is generally 1 in. standard.
    b. is generally $\frac{1}{2}$ in. standard.
    c. is generally $\frac{1}{4}$ in. standard.
    d. is generally $\frac{1}{16}$ in. standard.
    e. must be determined on a project-by-project basis.

32.3 STICK-BUILT GLASS CURTAIN WALL DETAILS

After the mullions have been anchored to the structural frame, the remaining items in the wall's erection require

- Connection of the rails to the mullions
- Installation of the glass.

RAIL-TO-MULLION CONNECTION

Manufacturers use various methods to connect the rails to the mullions. A commonly used method involves short aluminum extrusions, called *shear blocks,* which are fastened to the mullions with screws. Subsequently, the rails are snapped over the shear blocks, one shear block at each end of a rail, Figure 32.6. Thus, no fasteners are used between the rail and the shear blocks.

Because the length of each rail is small (4 ft to 6 ft is typical), a fairly small space is required for the expansion or contraction of a rail. In general, the length of a rail is $\frac{1}{16}$ in. less than the clear distance between mullions.

OUTSIDE-GLAZED AND INSIDE-GLAZED CURTAIN WALLS

One of the factors that determines the cross-sectional shapes of mullions and rails is whether the glass in the wall is to be installed from the outside or the inside of the building, referred to, respectively, as

- Outside-glazed curtain walls
- Inside-glazed curtain walls

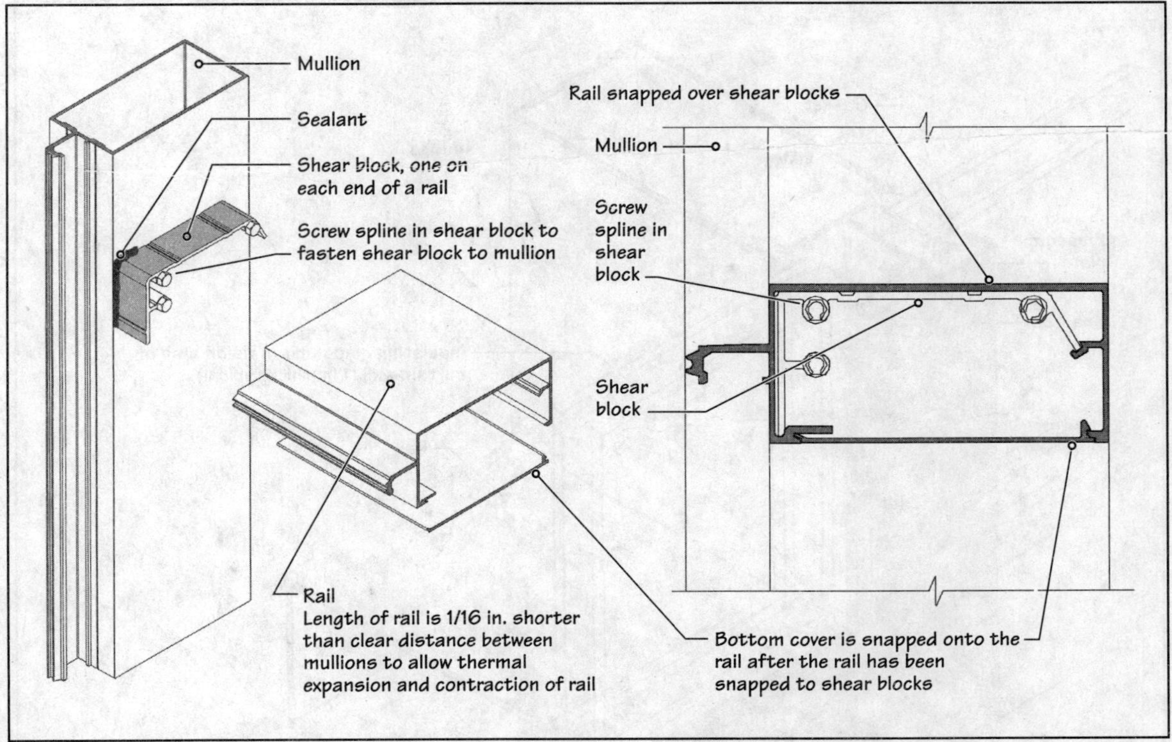

FIGURE 32.6 Typical connection between rails and mullions of a stick-built glass curtain wall.

In an *outside-glazed* wall, the glass panes are installed from the outside of the building by workers standing on a scaffold or staging. This method of installing glass is less efficient and more expensive due to the cost of scaffolding or staging. It is generally used for low- to mid-rise buildings. The glass in an outside-glazed wall can be secured in two ways:

- Pressure plate–captured glass (Figures 32.7 to 32.9)
- Structural silicone sealant–adhered glass (Figure 32.10)

In an *inside-glazed* wall, the glass is installed by workers standing on the appropriate floor of the building. The system is more efficient because it does not require scaffolding or staging. It is the system of choice for high-rise buildings. However, the cross-sectional shapes of mullions and rails for the inside-glazed system are more complex than the corresponding shapes for the outside-glazed system.

Outside-Glazed Walls (Pressure Plate–Captured Glass)

In an outside-glazed curtain wall, the glass is held by horizontal and vertical pressure plates, which are fastened to the mullions and rails with screws. A plastic insert is used between the pressure plate and the mullion (or the rail), which functions as a thermal separator. The pressure plates are finally covered with snap-on covers, Figure 32.7.

Because the covers are the only externally visible part of the curtain wall frame, they have a major influence on the curtain wall's appearance. The covers can be profiled into various shapes, Figure 32.8.

The exterior and interior gaskets should prevent water from leaking through the wall. However, a curtain wall system typically includes accommodations for the drainage of water, should it penetrate beyond the gaskets. This is accomplished through drainage weep holes in the pressure plates and the covers. Thus, in a typical curtain wall, each glass-pane frame is drained independently. Figure 32.9 shows typical sections through a pressure plate–captured outside-glazed curtain wall.

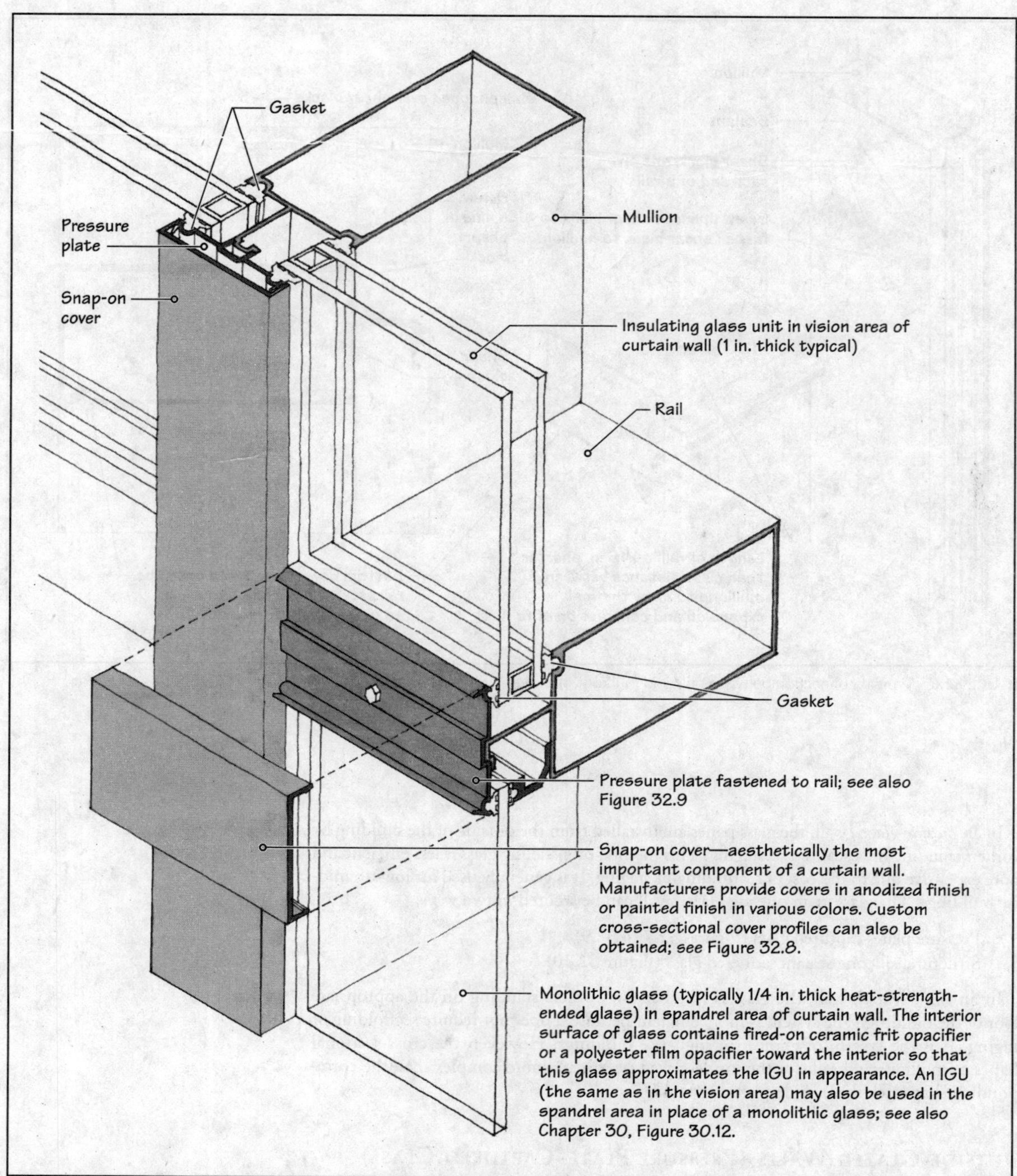

Gasket
Pressure plate
Snap-on cover
Mullion
Insulating glass unit in vision area of curtain wall (1 in. thick typical)
Rail
Gasket
Pressure plate fastened to rail; see also Figure 32.9
Snap-on cover—aesthetically the most important component of a curtain wall. Manufacturers provide covers in anodized finish or painted finish in various colors. Custom cross-sectional cover profiles can also be obtained; see Figure 32.8.
Monolithic glass (typically 1/4 in. thick heat-strengthended glass) in spandrel area of curtain wall. The interior surface of glass contains fired-on ceramic frit opacifier or a polyester film opacifier toward the interior so that this glass approximates the IGU in appearance. An IGU (the same as in the vision area) may also be used in the spandrel area in place of a monolithic glass; see also Chapter 30, Figure 30.12.

FIGURE 32.7 (a) Anatomy of an outside-glazed glass curtain wall (pressure plate–captured glass).

FIGURE 32.7 (b) In fastening the pressure plates to curtain wall framing members, the glass is temporarily held by small pressure plate members. After several glass panes are in position, the temporary pressure plates are removed and replaced by full-length pressure plates.

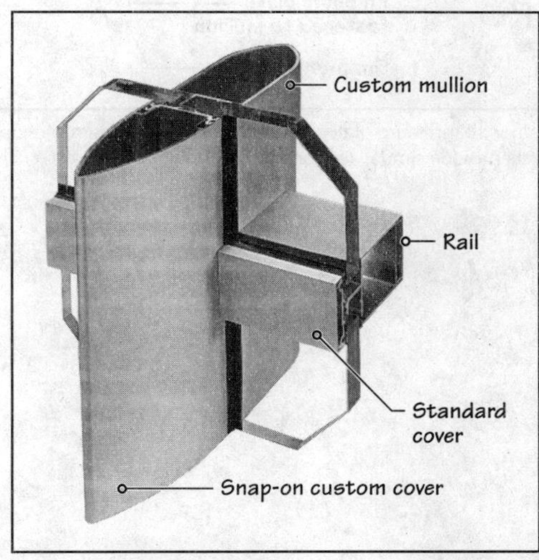

FIGURE 32.8 A custom cover and custom mullion for an outside-glazed glass curtain wall. (Photo courtesy of Vistawall Architectural Products)

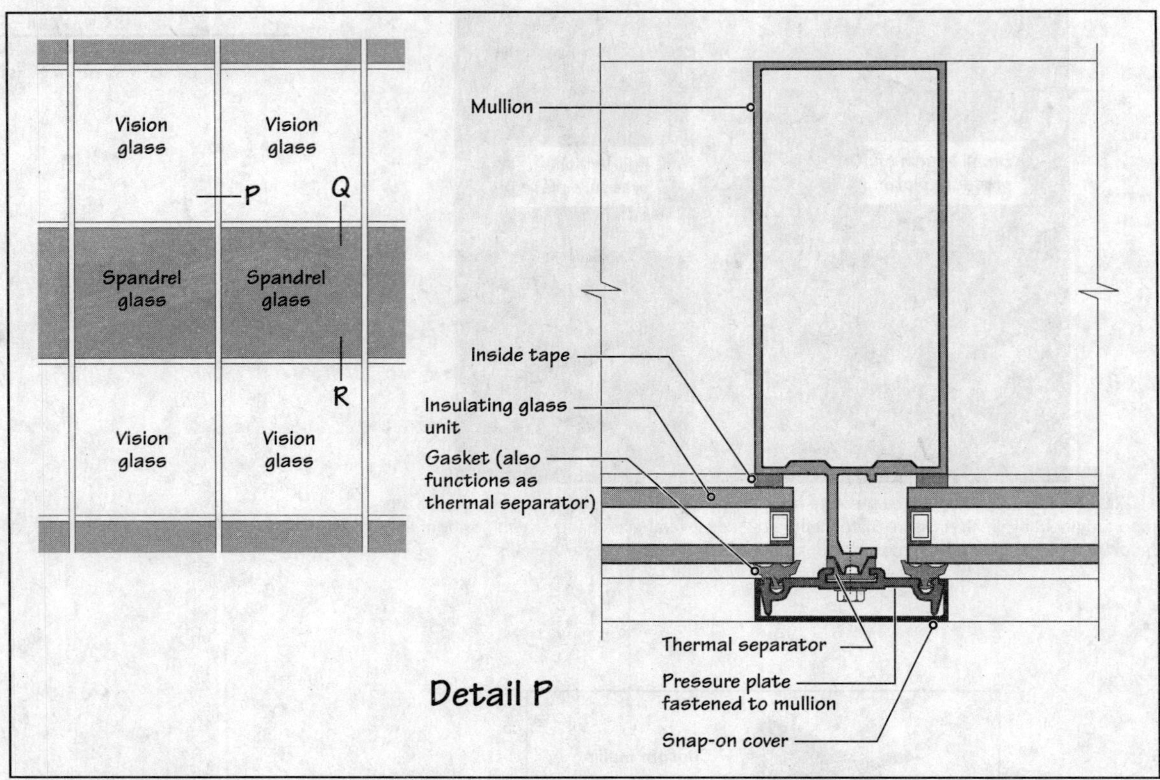

FIGURE 32.9 (a) Typical details of an outside-glazed glass curtain wall (pressure plate–captured glass). Aluminum sections used in the details are by Vistawall Architectural Products. Other manufacturers provide similar sections.

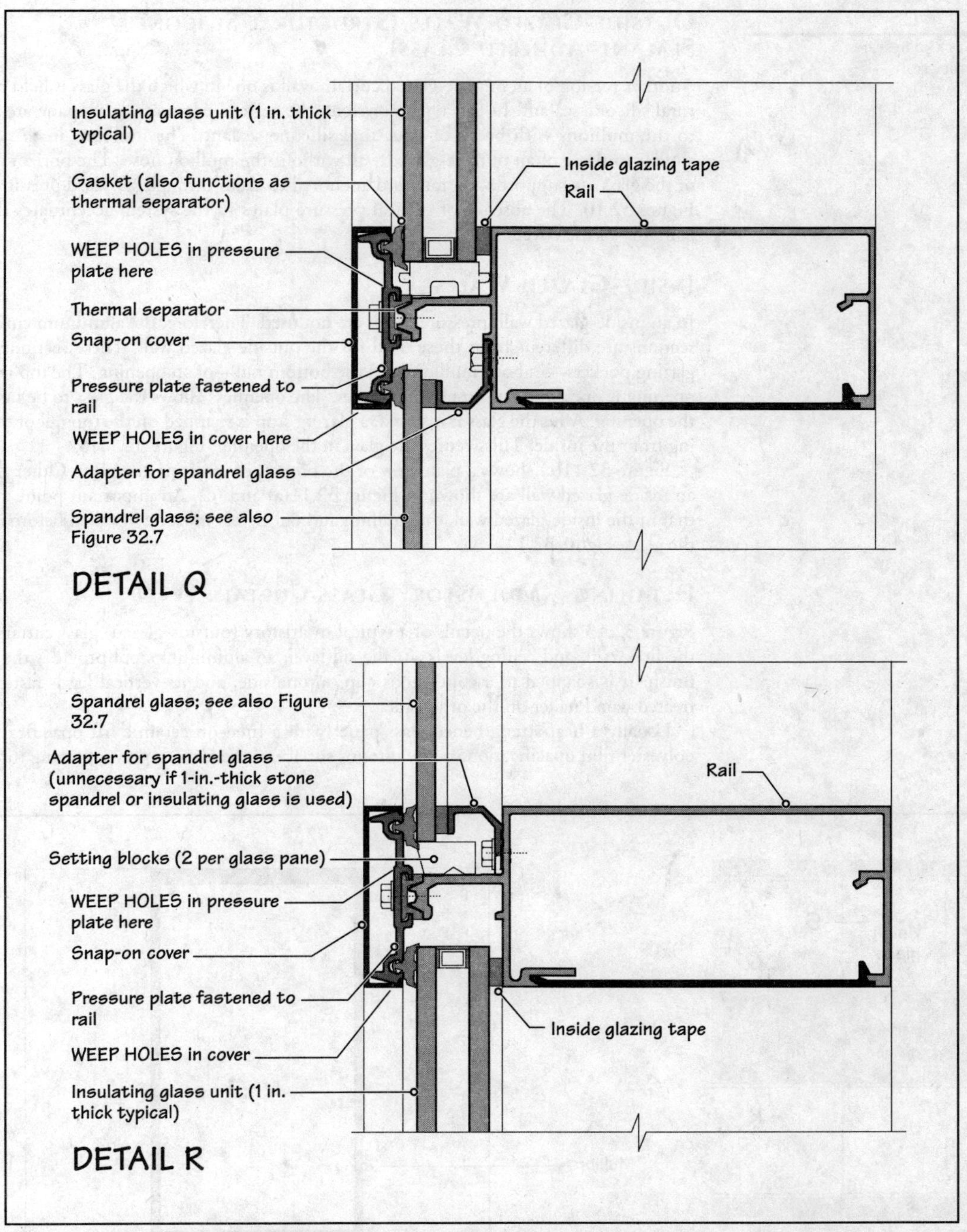

FIGURE 32.9 (b) Typical details of an outside-glazed glass curtain wall (pressure plate–captured glass). Aluminum sections used in the details are by Vistawall Architectural Products. Other manufacturers provide similar sections.

Outside-Glazed Walls (Structural Silicone Sealant–Adhered Glass)

Another version of an outside-glazed curtain wall is one in which the glass is held by structural silicone sealant. In this type of system, the vertical edges of a glass pane are adhered to the mullions with beads of structural silicone sealant. The mullions in this wall are similar to those of an outside-glazed wall without the mullion nose. The horizontal edges of the glass are supported on rails and anchored to them through standard pressure plates, Figure 32.10. The absence of vertical pressure plates in the system accentuates the horizontality of the covers.

Inside-Glazed Walls

In an inside-glazed wall, pressure plates are not used. Therefore, the aluminum curtain wall sections are different from those used for the outside-glazed wall. These sections include glazing pockets—in both mullions and the bottom rail—of an opening. The top rail of the opening is open and has no glazing pocket. The openness allows the glass to be inserted in the opening. After the glass is inserted, a glazing stop is snapped on the top rail of the opening from the inside. This secures the glass in the opening, Figure 32.11(b).

Figure 32.11(c) shows a plan view of the process of inserting the glass. Other details of an inside-glazed wall are shown in Figure 32.11(a) and (d). An important point to note is that in the inside-glazed wall, the mullion and rail covers must be installed before inserting the glass, Figure 32.12.

Detailing a Multistory Glass Curtain Wall

Figure 32.13 shows the details of a typical multistory (outside-glazed) glass curtain wall at the floor, sill, and ceiling levels. At the sill level, an aluminum stool provides the interior finish. It is snapped to a continuous clip on one side, and its vertical leg is fastened to a treated wood nailer on the other side.

Usually a heat-strengthened glass pane (with a fired-on ceramic frit opacification or a polyester film opacification on the interior surface of the glass to prevent seeing through it)

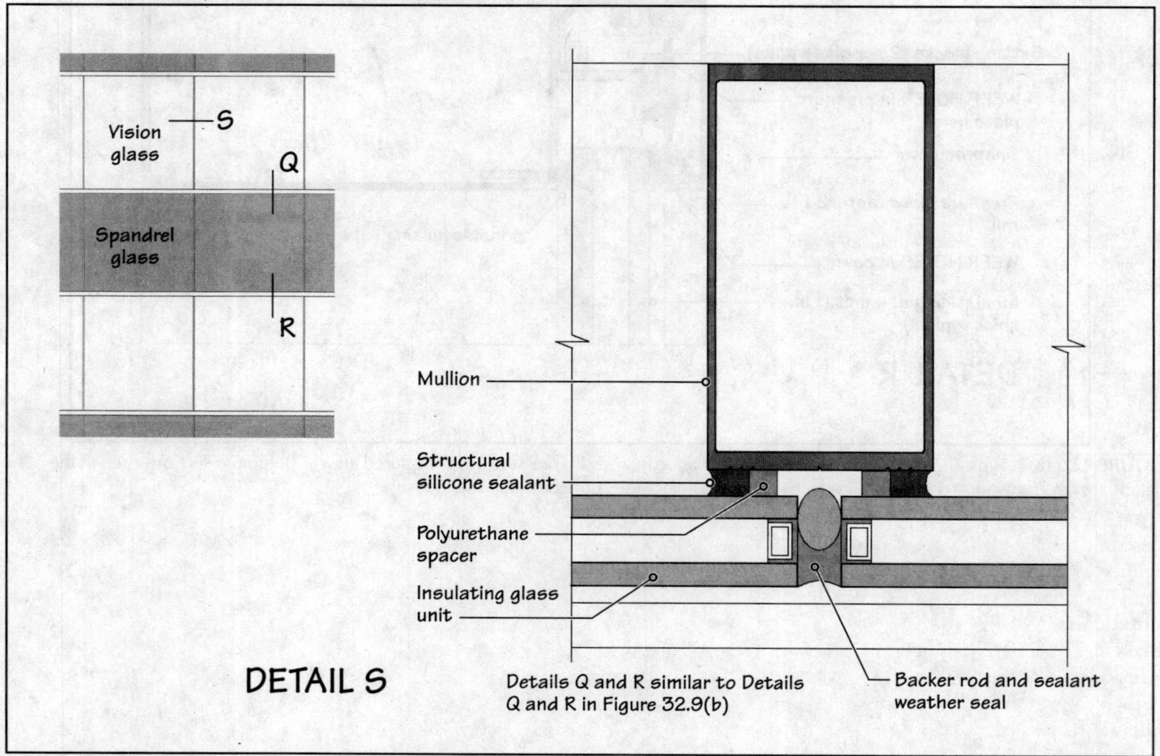

FIGURE 32.10 Typical details of an outside-glazed curtain wall (structural silicone sealant–adhered glass). Aluminum section used in the detail is by Vistawall Architectural Products. Other manufacturers provide a similar section.

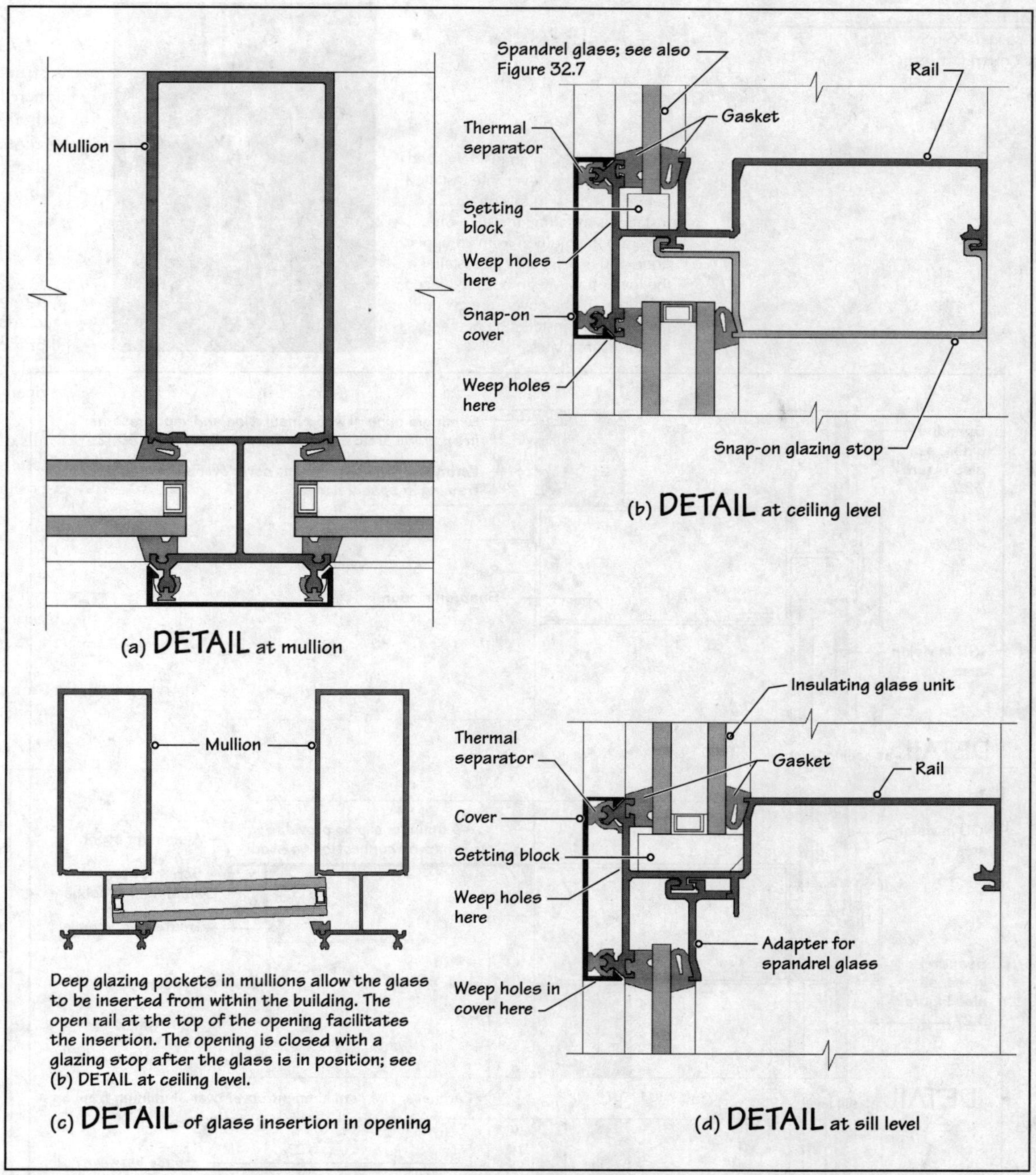

FIGURE 32.11 Typical details of an inside-glazed curtain wall. Aluminum sections used in the details are by Vistawall Architectural Products. Other manufacturers provide similar sections.

is used in the spandrel area (see Figure 32.7(a)). A fire-containment assembly is generally required to prevent the passage of fire and smoke between the adjacent floors of the building. This assembly consists of semirigid mineral wool insulation pressure-fitted (and supported on metal clips) in the space between the curtain wall and the spandrel beam. To obtain a good seal, the insulation is topped with a liquid-applied, fire-resistive sealant.

In addition to the fire-containment assembly, the entire spandrel area of the curtain wall is provided with mineral wool insulation placed behind the spandrel glass. To prevent condensation between the glass and the insulation, a vapor retarder is provided in this area. This generally consists of an aluminum foil lamination on the insulation's interior face. Where a high degree of condensation potential exists, a metal panel (referred to as a *metal back pan*) should be specified in place of an aluminum foil lamination.

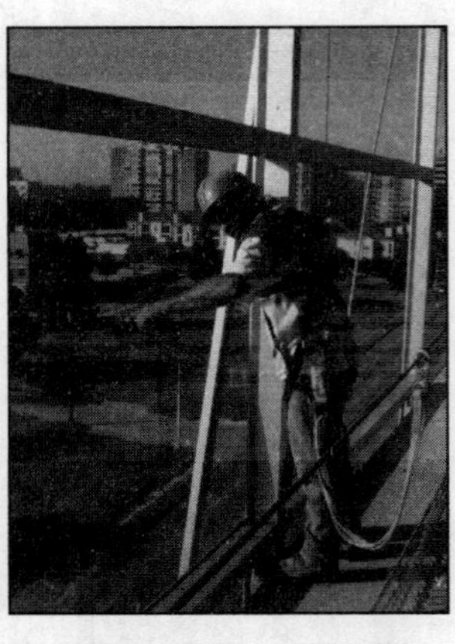

FIGURE 32.12 In an inside-glazed curtain wall, the covers are installed before the glass. In this photo, an installer is installing the snap-on cover on the mullion from an upper floor, and is helped by an installer at the lower floor. In an outside-glazed curtain wall, the covers are generally installed after installation of the glass and pressure plates.

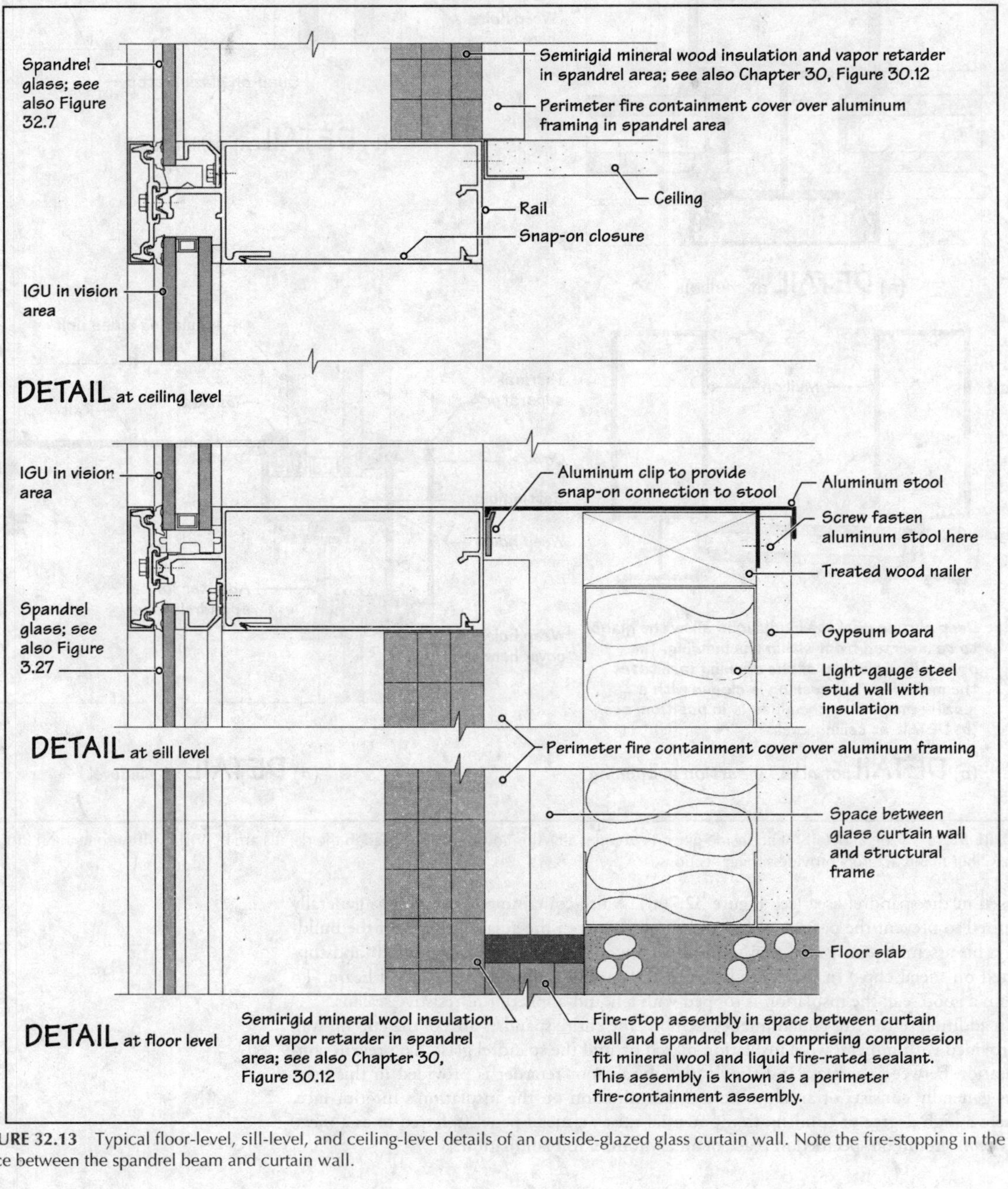

FIGURE 32.13 Typical floor-level, sill-level, and ceiling-level details of an outside-glazed glass curtain wall. Note the fire-stopping in the space between the spandrel beam and curtain wall.

PRACTICE QUIZ

Each question has only one correct answer. Select the choice that best answers the question.

12. The width of an expansion joint in the rails of a typical stick-built glass curtain wall
    a. is 1 in. standard.
    b. is $\frac{1}{2}$ in. standard.
    c. is $\frac{1}{4}$ in. standard.
    d. is $\frac{1}{32}$ in. standard.
    e. must be determined on project-by-project basis.

13. In a stick-built glass curtain wall, the mullions are erected first and the rails are inserted between them.
    a. True
    b. False

14. A shear block is used
    a. at a mullion expansion joint to allow the mullion to move.
    b. at the dead-load anchor of a mullion.
    c. at the expansion anchor of a mullion.
    d. to connect a rail to adjacent mullions.
    e. none of the above.

15. A glass curtain wall in which the glass is pressure plate captured is glazed from the
    a. building's interior.
    b. building's exterior.
    c. either (a) or (b).

16. As shown in this text, a glass curtain wall in which the glass is structural silicone adhered is glazed from the
    a. building's interior.
    b. building's exterior.
    c. (a) or (b).

17. In a glass curtain wall with pressure plate–captured glass, there are
    a. no covers.
    b. covers in both horizontal and vertical directions.
    c. covers only in the horizontal direction.
    d. covers only in the vertical direction.

18. As shown in the text, a glass curtain wall with structural silicone–adhered glass
    a. has no exterior covers.
    b. has exterior covers in both horizontal and vertical directions.
    c. has exterior covers only in the horizontal direction.
    d. has exterior covers only in the vertical direction.

19. A typical glass curtain wall is provided with weep holes to drain infiltrating water even though it is sealed from both the inside and the outside with gaskets or tapes.
    a. True
    b. False

32.4 UNITIZED GLASS CURTAIN WALL

As shown in Figure 32.1(b), in a unitized glass curtain wall, the wall units are preassembled (and generally preglazed) in a fabrication shop and brought to the site for installation, so that the wall is assembled at the site, unit by unit, instead of assembling sticks of mullions and rails. Figure 32.14(a) to (d) show the important steps in the installation of a typical unitized wall to a building structure.

The units are designed to mate with the adjacent units at the mullions, and at the top and bottom rails. The bottom rail of the upper unit connects to the top rail of the lower unit. As shown in Figure 32.14(b), the splices projecting from the top rail of the lower unit fit snugly into the void in the bottom rail of the upper unit. This detail provides the lateral-load resistance and is similar to the detail used in a stick-built wall (see Figure 32.3). The dead-load resistance of the unit is provided through its anchorage to the floor, Figure 32.14(d). Two adjacent units generally share the same dead-load anchorage.

32.5 STRUCTURAL PERFORMANCE OF A GLASS-ALUMINUM WALL

The most important structural requirement of a glass-aluminum wall is its ability to resist lateral loads (particularly wind loads), including missile-impact resistance in hurricane-prone regions. Just as the design of a glass-aluminum wall's anchorage to the structure is accomplished by the wall manufacturer (or installer, see Section 32.2), its lateral-load-resistance design is also provided by the manufacturer (or installer) based on the lateral-load intensities provided by the project architect or structural engineer.

Manufacturers generally have several standard sections designed to suit various lateral-load intensities. For high lateral-load intensities, a strategy often used is to enclose structural steel (or aluminum) sections within the mullions, Figure 32.15. The enclosed steel sections and the mullions are fastened together to produce a composite action between them. Structural C- or I-sections are commonly used as enclosed sections. Channels provide the advantage of nesting, so that two or three channels may be used within the same mullion. The enclosed steel sections are suitably coated to prevent galvanic action between the aluminum and steel.

An alternative to enclosed steel sections is to anchor the mullions to an independent steel structural frame, Figure 32.16. This strategy is generally used in a tall glass-aluminum wall where the mullions do not have intermediate supports to reduce their span, such as those provided by the floor structure in a multistory curtain wall.

(a) Using a crane, unitized curtain wall elements are carried up from the delivery truck to the building facade

(b) The new unit is installed by placing both of its mullions over the (projecting) splices in the lower unit. The splices function as expansion splices (see Figure 32.3) and also provide lateral-load support to the unit. The lower unit is anchored to the floor structure through a dead-load support to the unit.

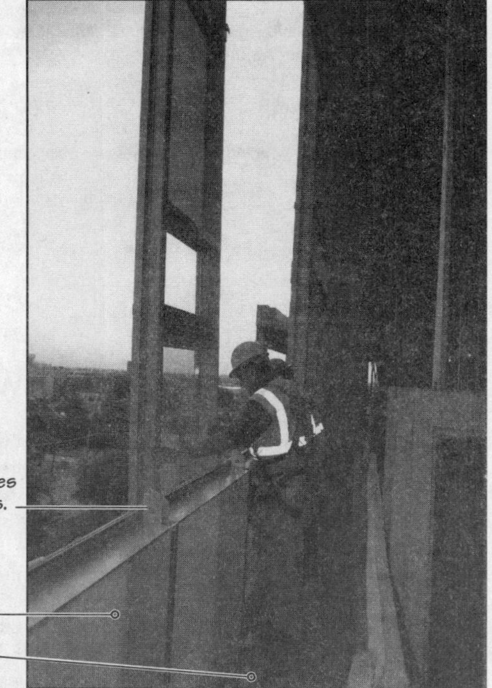

Projecting splices (one on each side of the lower unit) function as expansion splices and also as lifting elements. Observe holes in splices to which lifting cables are attached.
Unit already installed
Dead-load anchor for the unit here; see (d) below

(c) The new unit being forced over the (projecting) splice

(d) Anchorage of unit to the floor structure, providing dead-load support

FIGURE 32.14 Installation of a unitized curtain wall element.

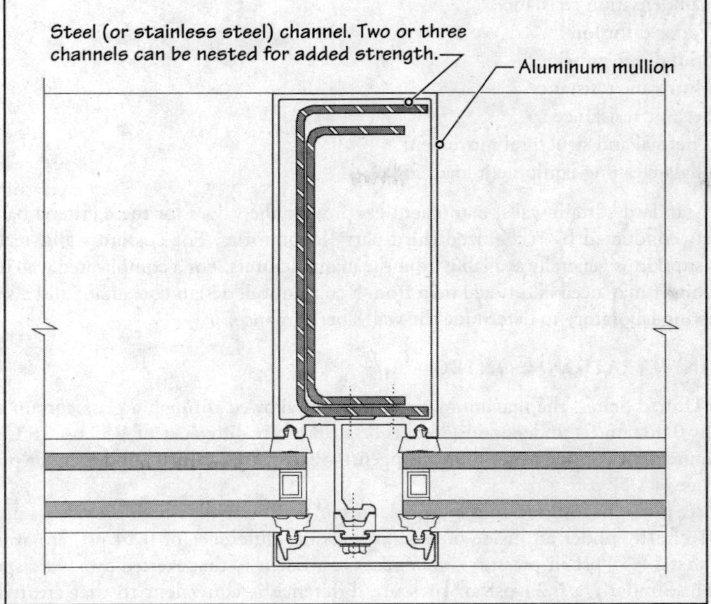

Steel (or stainless steel) channel. Two or three channels can be nested for added strength.
Aluminum mullion

FIGURE 32.15 One of the ways to increase the lateral-load resistance of aluminum mullions is to enclose structural steel sections within them.

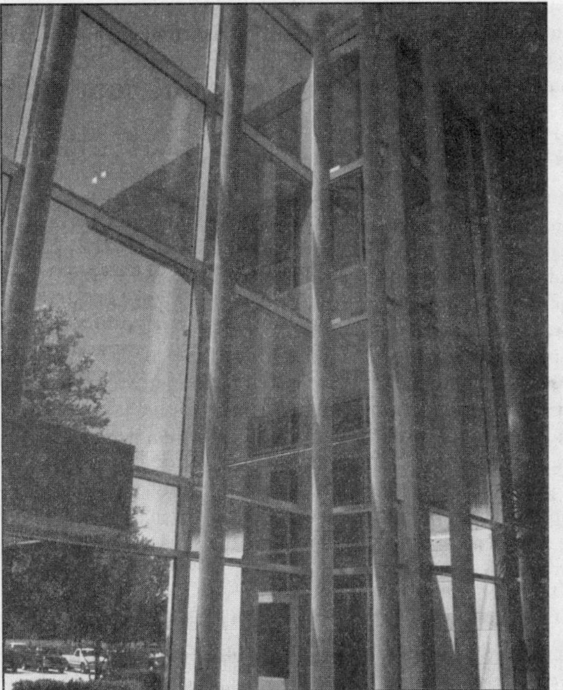

In this glass-aluminum wall, the supporting structure for aluminum curtain wall sections consists of steel pipe verticals. In a very tall curtain wall, vertical steel trusses and horizontal steel members or a steel space frame may be used.

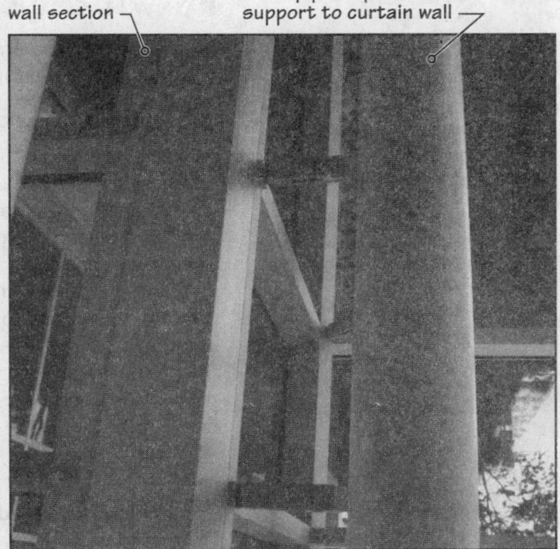

FIGURE 32.16 A tall glass wall with standard curtain wall sections anchored to an interior structural steel vertical member to provide lateral-load support to aluminum mullions. In this building, a steel pipe support is used. In taller walls, vertical steel trusses are common.

32.6 ENVIRONMENTAL PERFORMANCE CRITERIA FOR A GLASS CURTAIN WALL

The nonstructural performance of a glass curtain wall is just as important as its structural performance. Among the important nonstructural design criteria for a glass curtain wall are

- Air-infiltration control
- Rainwater- and meltwater-penetration control
- U-value
- Solar heat gain

- Condensation resistance
- Vapor diffusion
- Sound transmission
- Hurricane resistance
- Seismic resistance
- Thermal and structural movement
- Glass-cleaning-equipment load

For standard curtain walls, manufacturers provide the values for these criteria based on the tests conducted by recognized third-party laboratories. For custom walls, technical design support is generally available from the manufacturers. For a complicated wall design, the architect may need additional help from a curtain wall design consultant and a specialized testing laboratory to determine the wall's performance.

AIR-INFILTRATION CONTROL

In the United States, the maximum air infiltration allowed through a glass curtain wall is typically 0.06 cfm/ft^2 under an inside-outside air pressure difference of 1.57 psf. In Canada, the requirement is three times more stringent—that is, 0.02 cfm/ft^2 under an air pressure difference of 1.57 psf.

Where lower air infiltration is required, curtain wall systems, which provide a rate of up to 0.01 cfm/ft^2 under an inside-outside air pressure difference of 6.24 psf, are available. (Note that a 1.57-psf air pressure difference is equivalent to that exerted by a wind speed of 25 mph. Similarly a 6.24-psf air pressure difference is equivalent to that created by a 50-mph wind speed; see Chapter 3.

Air-infiltration control not only conserves energy but also reduces ice buildup on the exterior of curtain wall components. Ice buildup is caused by the condensation of water vapor that escapes from the building's interior along with air. When the ice melts, the meltwater may leak into the building's interior. Therefore, a more stringent air-infiltration-control criterion is generally needed in colder climates.

NOTE

cfm is an acronym for cubic feet per minute.

RAINWATER- AND MELTWATER-PENETRATION CONTROL

Water-penetration control (of both rainwater and meltwater) is perhaps the most important nonstructural performance requirement of a glass curtain wall. Glass curtain wall systems are generally designed to ensure no water penetration when tested under a static air pressure difference (between the inside and the outside) that is at least 20% of the inward structural design wind load on the curtain wall. Thus, if the inward structural design wind load on the wall is 50 psf, the system is tested for water penetration under a static air pressure difference of at least 10 psf. A more stringent water-penetration criterion is required for buildings located in areas subjected to frequent and intense wind-driven rain. In addition to conforming to the static pressure criterion, a glass curtain wall is required to conform to dynamic pressure test criterion.

Water-penetration control is accomplished in different ways by system manufacturers; it typically includes adequate drainage in the aluminum joinery and glazing pockets. For example, in the stick-built glass curtain wall described earlier, weep holes are provided in the pressure plates and snap-on covers that drain the water to the outside. The architect's details must also ensure the management of water entering the curtain wall from a nonglass facade that is above the curtain wall, where such a facade exists.

NOTE

AAMA and Water Penetration

The American Architectural Manufacturers Association (AAMA) defines water penetration as the appearance of uncontrolled water other than condensation on the interior face of any part of the curtain wall.

U-VALUE, SOLAR HEAT GAIN, AND CONDENSATION RESISTANCE

These three interrelated criteria are a function of the type of glass, the type of aluminum framing (thermally improved or not), and the center-to-center spacing of framing members, as explained in Chapter 30. The architect must specify their values in consultation with the HVAC consultant (who also needs these values to design the building's HVAC system and to meet the energy code requirements).

VAPOR DIFFUSION ANALYSIS

As previously stated, interior water vapor may result in ice buildup on curtain wall framing and condensation of water vapor in the building's interior. These issues are particularly critical in cold climates, where a vapor analysis of the curtain wall system is generally required.

SOUND TRANSMISSION

Glass curtain walls in buildings located in areas where high levels of exterior noise are present (e.g., near airports or busy highways) may need a higher sound-transmission-loss specification than other areas.

HURRICANE IMPACT RESISTANCE AND SEISMIC RESISTANCE

Because of Florida's experience with extensive wind damage to glass curtain walls from hurricanes, several coastal cities in the United States are requiring that glass curtain walls be missile-impact resistant, particularly in the lower floors of the building. Similarly, because of California's experience with earthquake damage to glass curtain walls, resistance to shaking, glass drifting, and horizontal movement of components is required for buildings in seismically active areas.

THERMAL AND STRUCTURAL MOVEMENT

Aluminum-framing members and glass expand and contract due to temperature changes and the sudden cooling effects of precipitation. Glass curtain walls require sufficient expansion and contraction control built into them to allow thermal movement and the movement of spandrel beams due to live-load deflection. The expansion splice required in a stick-built curtain wall (Figure 32.3) accounts for this requirement.

GLASS-CLEANING-EQUIPMENT LOAD

High-rise curtain walls include provisions for periodic cleaning. This means that they must include anchorage points for the staging of cleaning equipment that is lowered from the roof to the front of the curtain wall. These anchors add point loads on the wall, and this information needs to be communicated by the architect to the system manufacturer.

32.7 OTHER GLASS-ALUMINUM WALL SYSTEMS

In addition to curtain walls, two additional glass-aluminum wall systems commonly used are

- Punched and strip glazing
- Storefront system

PUNCHED AND STRIP GLAZING

Punched glazing is similar to a punched window (see the window terminology in Chapter 31), except that the glass in punched glazing is generally fixed and site installed due to its large size. By contrast, a punched window is generally shop glazed and may contain operable sashes. The frame for punched glazing may either be shop assembled or stick-built on site. The frame is anchored to the opening (jambs, head, and sill) instead of being anchored to the building's structural frame (as in a curtain wall).

Strip (also called *ribbon*) glazing is similar to punched glazing, with several glazing units placed in a linear alignment. Strip glazing is also anchored to the head and sill of an opening.

STOREFRONT SYSTEM

A storefront is a large glass-aluminum wall that is generally one story tall and extends from the ground to the second floor of the building. Three major differences distinguish a curtain wall from a storefront. They are:

(a) The storefront wall lies under the second-floor structure of the building and (unlike a curtain wall), it is not spaced away from the structural frame of the building. Often, a storefront is protected by an overhang to control water leakage through the system.
(b) The glazing system's performance for structural and nonstructural criteria (air infiltration, water penetration, CRF, etc.) is lower than that of a curtain wall.
(c) Rainwater that enters a curtain wall is drained by the (horizontal) rail support of each individual lite. In a storefront system, the water entering through a lite must travel vertically down the mullion and be drained at the weep holes at the ground level.

32.8 NONTRADITIONAL GLASS WALLS

The glass-aluminum wall types discussed so far are traditional types that have evolved over a long period. Three recently introduced systems that do not rely on aluminum sections for support are

- Glass wall supported by cable trusses
- Glass wall supported by a pretensioned cable net
- Double-skin glass walls

GLASS WALL SUPPORTED BY CABLE TRUSSES

A sophisticated and unique version of a glass wall that is relatively uncommon (due to its cost) is the mullionless glass wall developed by the Pilkington Company. In this system, each glass pane is suspended at four corners by stainless steel *spider-shaped connectors.* The connectors are held by a horizontal truss consisting of stainless steel tension cables and compression struts. Several such trusses are anchored to vertical steel frames, Figure 32.17.

The glass used in the system is generally a high-R-value, insulating glass unit with laminated and heat-soaked tempered glass (see Section 30.2). Only elastomeric, silicone sealant separates a glass pane from the adjacent panes.

GLASS WALL SUPPORTED BY A PRETENSIONED CABLE NET

Another recently introduced glass wall system is one that uses a grid of pretensioned (stretched) cables to support the glass. Conceptually, the support system resembles a two-directional grid of strings in a tennis racket. Because the strings in a tennis racket are highly stretched, they create a stiff plane. When held vertically, the stringed plane of the racket resembles a stiff wall that can resist lateral loads and can be used to suspend planar elements from it.

Similar to the stringed plane of a tennis racket, a cable-net-supported glass wall consists of pretensioned (approximately 1-in.-diameter) stainless steel cables arrayed in both the horizontal and vertical directions. The vertical cables are stretched between the top and bottom of an opening—generally between the spandrel beams at the upper and lower floors. The horizontal cables are stretched between the sides of the opening—generally

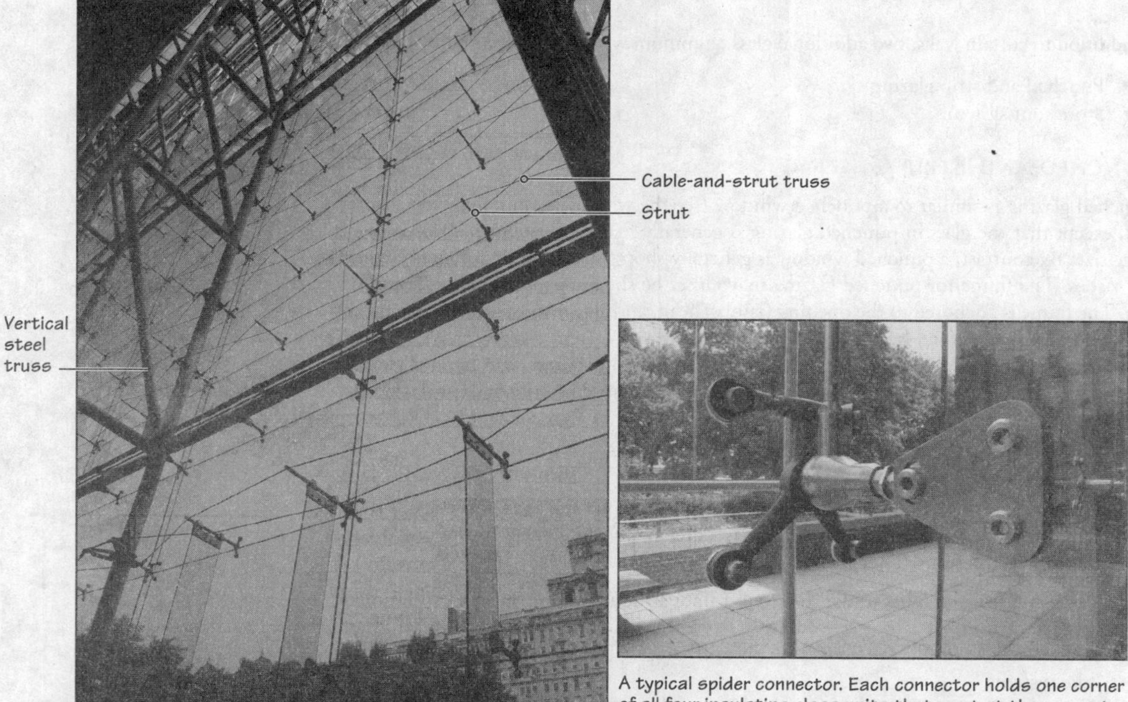

This glass wall is supported by a two-directional (consisting of horizontal and vertical elements) backup structure. The horizontal elements of the backup structure consist of trusses with tension cables and compression struts. The vertical elements consist of trusses made of steel pipes. The glass is held by spider connectors, which are connected to the compression struts. At a level close to the floor where tension cables and compression struts cannot be used, glass fins have been used to provide lateral support to the glass.

A typical spider connector. Each connector holds one corner of all four insulating glass units that meet at the connector. Elastomeric silicone sealant seals the edges between the glass units.

FIGURE 32.17 Pilkington's Planar Glass Wall System used in the Rose Center for Earth and Space, New York City (see also Chapter 30, Figure 30.4). Architects: Polshek Partnership.

between the columns or walls supporting the building. Two examples of cable-net-supported glass walls are shown in Figure 32.18(a) and (b).

The horizontal and vertical cables are held together at intersections through special stainless steel connectors, which also serve as points for securing the glass to the grid. A typical connector is shown in Figure 32.18(c). The glass panels generally consist of insulating glass units with laminated and heat-soaked tempered glass.

Because the cables are highly stressed, they impose a large load on the boundary elements of the opening—the two spandrel beams and the columns (or walls)—which must be designed to resist this additional load.

(a) Cable-net glass wall in the entrance lobby of the One North Wacker Building, Chicago, Architects: Goettsch Partners.

(b) Cable-net glass in the atrium of the Time Warner Center, New York City. Architects: Skidmore, Owing and Merril.

(c) Typical connector at a node—the point of intersection between cables. The connector connects the cables together and also provides suppor to the glass. Four glass corners meet at each connector.

FIGURE 32.18 Two examples of cable-net-supported glass walls. Note that all photographs have been taken from inside the buildings.

DOUBLE-SKIN GLASS WALLS

Another innovative glass wall system is the double-skin wall system, also referred to as a *bioclimatic glass wall.* In this system, two glass walls, separated by 1 ft to 5 ft of air space, are used. The air space serves as a buffer between the two skins, tempering the outdoor air, and also serves as a plenum for the building's HVAC system. The outer skin may include computer-controlled operable glazing and solar shading devices.

Although the primary benefit of a double-skin system is energy conservation, it also provides effective water-penetration control, better air-infiltration control, higher sound insulation, and so on. The proponents of the system, which has been used extensively in Europe, claim it to be more sustainable under the present energy prices in Europe, which are considerably higher than those in the United States.

As smart glazing systems with environment-adaptive technologies (e.g., electrochromic glass) and energy-generating capabilities (e.g., photovoltaic glass) evolve further, double-skin glass wall systems may become more popular. In that scenario, the outer glass skin may not only help conserve energy but also generate some energy to power the building.

PRACTICE QUIZ

Each question has only one correct answer. Select the choice that best answers the question.

20. The air infiltration rate through a glass curtain wall is specified in terms of
- **a.** pounds per square feet.
- **b.** cubic feet per minute.
- **c.** cubic feet under a given inside-outside air pressure difference.
- **d.** cubic feet per minute under a given inside-outside air pressure difference.
- **e.** none of the above.

21. In a glass wall supported by cable trusses, the glass
- **a.** bears on (horizontal) rails.
- **b.** is connected to (vertical) mullions.
- **c.** both (a) and (b) above.
- **d.** none of the above.

22. In a cable-net-supported glass wall, the glass
- **a.** bears on (horizontal) rails.
- **b.** is connected to (vertical) mullions.
- **c.** both (a) and (b) above.
- **d.** is supported by a spider-shaped connector.
- **e.** none of the above.

23. The environmental performance requirements for a storefront are generally much higher than those for a glass curtain wall.
- **a.** True
- **b.** False

REVIEW QUESTIONS

1. Using sketches and notes, explain the differences between a single-span mullion support system and a twin-span mullion support system for a glass curtain wall.
2. Use three-dimensional sketches to illustrate each of the following:
   a. A typical dead-load anchor used in a glass curtain wall
   b. A typical expansion anchor used in a glass curtain wall
3. Sketch in three dimensions a typical spider-shaped connector.
4. List the major differences between a glass curtain wall and a storefront.
5. Explain the purpose of the following items and state where they are used: (a) shear block, (b) pressure plate, and (c) adapter for spandrel glass.
6. Using sketches and notes, explain how we increase the lateral load-bearing capacity of a standard glass-aluminum wall.

CHAPTER 33 Roofing–I (Low-Slope Roofs)

CHAPTER OUTLINE

The roof is one of the most critical components of a building. It is one of the the primary sources of construction litigation and building owners' complaints. One estimate indicates that nearly 65% of all lawsuits brought against architects originate in roofing problems [33.1]. This statistic becomes significant considering that the roof covering of a typical building is replaced every few years—12 to 14 years on the average. Would we accept replacement of other components of a building envelope—doors, windows, curtain walls, and so on—so frequently? A major reason for this situation is the severity of physical and chemical degradation, by solar radiation, rainwater, snow, hail, and windstorm, to which a roof is subjected. Another reason is the ever-increasing complexity of contemporary buildings and, more specifically, of contemporary roofs.

Until the mid-twentieth century, roofing alternatives were few and roof design was relatively simple. For a flat roof, the roofing membrane consisted of either a coal tar or an asphalt built-up roof on a concrete or wood deck. For a steep roof, clay tiles or wood shingles were the major alternatives. This is not true today. The relatively sudden spurt in the gross national products of Western economies after World War II introduced large industrial buildings with relatively light steel roof decks. A roof that performed well on a stiff concrete or wood deck failed prematurely on a light, flexible steel deck.

The advent of air conditioning and more efficient heating systems led to urban development in extremely warm and cold climates. Roof insulation that was seldom used earlier became the norm. The greater degradation of roofs in extremely hot and cold climates, and the effect of roof insulation, introduced factors whose impact was not realized until several years later.

The oil embargo of 1973 led to significant increases in the cost of asphalt-based built-up roofs, which made single-ply roofs economically competitive for the first time. This, combined with several other market forces, resulted in a significant increase in the market share of single-ply roofs. Some of the early single-ply roofs had a fairly brief history of use, which further increased roofing problems. Although roofing products are generally subjected to several laboratory tests before being introduced in the market, none of these tests can fully simulate the actual field conditions.

This chapter is the first of two chapters on roofing. It begins with a general introduction to roofing, including the basis for distinguishing between low-slope and steep roofs. Subsequently, various aspects of low-slope roofs are covered. Steep roofs are discussed in the following chapter.

33.1 LOW-SLOPE AND STEEP ROOFS DISTINGUISHED

A discussion of roofing must begin with its classification into two types:

- Low-slope roof
- Steep roof

A low-slope roof is a *water-resisting* roof and consists of a continuous roof membrane over a relatively flat roof surface. Although water ponding is generally to be avoided, a low-slope roof has to be designed for a certain depth of rainwater accumulation that might occur in the event of blockage of the roof's drainage system. The depth of rainwater accumulation is predetermined based on the roof's geometry and the design of its drainage system. The roof membrane in a low-slope roof must, therefore, act as a waterproofing membrane and be able to resist the rainwater load and water leakage until the drainage system is able to function again. This period may range from a few hours to a few days.

NOTE

Low-Slope Roof and Steep Roof

Low-slope roof	Slope < 3:12
Steep roof	Slope ≥ 3:12

A steep roof, also called a *water-shedding* roof, typically consists of small individual roofing units (shingles) that overlap each other. The roof surface must be sloped so that the water is shed off the roof by gravity. The slope must generate adequate gravitational force to overcome the forces produced by wind, head pressure, and capillary action that might push the water up the slope between adjacent shingles and cause the roof to leak. Double coverage in shingles and adequate overlap are necessary to prevent roof leakage.

The rate of discharge of water from the roof is not only critical in keeping the roof and the building's interior dry, it is also one of the most important factors in establishing the durability of the roof. There are sufficient long-term data to confirm that—everything else being the same—the durability of a roof increases directly with the slope of the roof. Steep roofs, in general, outsurvive low-slope roofs.

A low-slope roof is defined as a roof with a slope of less than 3:12 (25% slope) and greater than $\frac{1}{4}$:12 (2% slope). A dead-flat roof is not permitted by the building codes. A steep roof is defined as a roof with a slope greater than or equal to 3:12.

Despite the greater durability of a steep roof, the market share of low-slope roofs is nearly twice as large as that of steep roofs, Figure 33.1 [33.2]. Low-slope roofs are generally used for commercial and industrial buildings and are mostly constructed to a minimum mandated

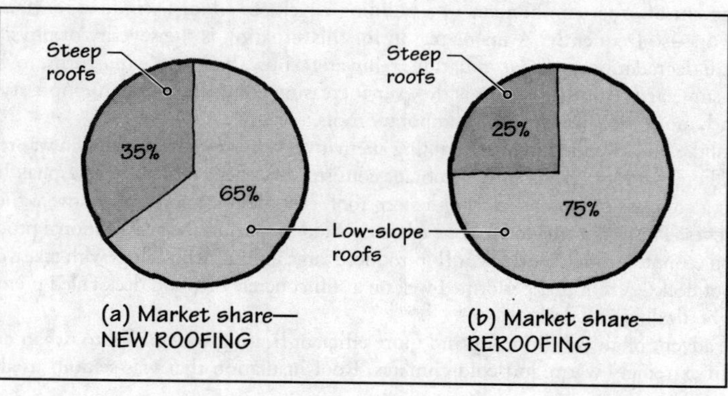

FIGURE 33.1 Approximate market shares of low-slope and steep roofs.

slope of $\frac{1}{4}$:12. This slope, which yields an almost flat roof, makes the structural framing of the roof and the supporting structure most economical because an increase in roof slope increases the roof area, the dimensions of framing members, and the related structure. Additionally, a flat roof provides an easily accessible space for installing HVAC and other equipment. Buildings with flat roofs generally have their HVAC equipment mounted on the roofs. Steep roofs, which are generally used for low-rise residential buildings (single-family homes, apartment buildings, motels, etc.) have their HVAC equipment placed on the ground.

A low-slope roof also makes maintenance and reroofing, particularly on high-rise buildings, relatively easy. Imagine reroofing or repairing a steep roof a few hundred feet above the ground!

33.2 LOW-SLOPE ROOF FUNDAMENTALS

A low-slope roof has the following components, Figure 33.2:

- Roof membrane, including a protective cover or coating above the membrane, where needed
- Insulation
- Roof deck
- Flashings

In some roofs, particularly in cold regions or over high-humidity enclosures, a vapor retarder may also be needed.

ROOF MEMBRANES

The roof membrane is the most important component of a roof because it is the waterproofing layer. Being the top layer, it is constantly subjected to the weathering effects of sun, rain, snow, hail, and wind. Roof membranes are divided into three general categories:

- Built-up roof membrane
- Modified bitumen roof membrane
- Single-ply roof membrane

A *built-up roof membrane* consists of three to five plies (layers) of felt with intervening moppings of bitumen (asphalt or coal tar). Because both asphalt and coal tar are adversely affected by ultraviolet radiation, a built-up roof membrane is generally protected by an aggregate cover. The aggregate cover also increases the roof's fire resistance. Of the two, asphalt built-up membranes are far more common than coal tar built-up membranes.

A *modified bitumen roof membrane* is similar to a built-up roof membrane and comprises two to three plies of modified bitumen sheets with intervening moppings of bitumen. A protective aggregate cover or a mineral granule–surfaced cap sheet is required.

A *single-ply roof membrane* consists of only one sheet of a synthetic polymer (plastic) and does not require any protective cover. Although slightly variable, the relative market shares of built-up, modified bitumen, and single-ply roof membranes are approximately the same—nearly one-third each.

ROOF INSULATION

In a typical low-slope roof, the insulation is sandwiched between the membrane and the deck. Apart from reducing energy costs, insulation provides a suitable substrate for the roof membrane. Over a steel roof deck, insulation also provides the necessary surface flatness.

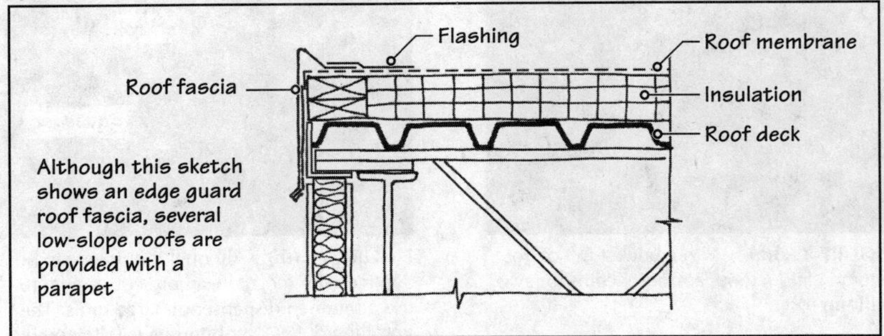
Flashing
Roof membrane
Roof fascia
Insulation
Roof deck
Although this sketch shows an edge guard roof fascia, several low-slope roofs are provided with a parapet

FIGURE 33.2 Components of a low-slope roof.

NOTE

Metal Panel Roofs and Protected Membrane Roofs

In a typical membrane-covered low-slope roof assembly, the membrane deteriorates more rapidly due to its exposure to weathering elements. In response to this weakness, two other low-slope roofs have gradually increased their market shares:

- Metal panel roofs
- Protected membrane roofs

A metal panel roof consists of thin sheets of metal bent into panels that are laid over roof insulation and anchored to the deck. They may be used in low-slope or steep roofs, although they function better on steep slopes (slope greater than 3:12). The initial cost of metal panel roofs (a nonmembrane roof assembly) is higher than that of membrane roofs. Metal panel roofs are discussed in Chapter 34.

In a protected membrane roof, the insulation is loose laid over the deck to which the membrane is attached directly. The protected membrane roof is discussed later in this chapter.

Fluid-Applied Membrane Roofs

Fluid-applied membrane roofs consist of polyurethane foam insulation sprayed over a low-slope roof deck. The insulation is covered with a (fluid) coating that cures into a membrane. It is commonly used in the southern and southwestern United States and has a relatively small overall market share.

ROOF DECK

A low-slope roof deck may be of steel, wood, cast-in-place concrete, precast concrete, gypsum concrete, or wood fiber–cement deck. Steel and concrete roof decks are more commonly used in low-slope roofs.

ROOF FLASHINGS

Flashings are an integral part of a low-slope roof. They are required at roof terminations, such as the free edges of a roof, at parapets, and at roof expansion joints. Flashings are also required at penetrations, such as around interior roof drains, and around pipe or tubular supports for rooftop equipment.

LOW-SLOPE ROOF—A SYSTEM OF COMPATIBLE COMPONENTS

Each roof component must obviously serve its own function, but it must also be compatible with other roof components. For instance, the insulation must not only give the required R-value, it must also provide a rigid substrate against hail and foot traffic. It should also have long-term chemical compatibility with the deck and the membrane.

Similarly, the roof deck must not only provide the required structural support for the entire assembly and other loads, it must also be dimensionally stable to prevent the overstressing of the insulation and the membrane. It must provide a suitable base for the anchorage of insulation and the membrane and have adequate fire resistance.

With several factors that affect its performance, and with several components and subcomponents that constitute a low-slope roof, roof design must be considered a system design. It should consider the interaction between roof components and the effects of external factors such as the external climate, solar radiation, wind, rain, hailstorms, fire, and rooftop traffic (necessitated by the equipment's and roof's maintenance operations). The selection of all components of the system, including fasteners and adhesives, from the same manufacturer is, therefore, important in the case of a low-slope roof. Where a manufacturer does not make all components of the system, components approved by the primary manufacturer must be used to ensure their compatibility.

33.3 BUILT-UP ROOF MEMBRANE

A typical built-up roof membrane consists of a felt laid over a mopping of bitumen, followed by a second mopping of bitumen, then by the second felt, and so on. In other words, a number of felt layers (called plies), separated by interply moppings of bitumen, constitute a built-up roof membrane.

The greater the number of plies, the thicker and, hence, stronger and more durable the membrane. Three to five plies is generally the norm. The last ply is typically covered with a surfacing material to protect the entire membrane from the effects of weather and external fire. The most common surfacing material is stone aggregate laid over a flood coat of bitumen. Because the quantity of bitumen for the flood coat is much greater than that required for interply moppings, it is poured over the roof, whereas the interply bitumen is simply mopped on. Figures 33.3(a) to (f) illustrate a few stages in the laying of a typical built-up roof membrane.

FIGURE 33.3(a) With a bucket full of hot bitumen and a mop, a roofer begins to lay a built-up roof.

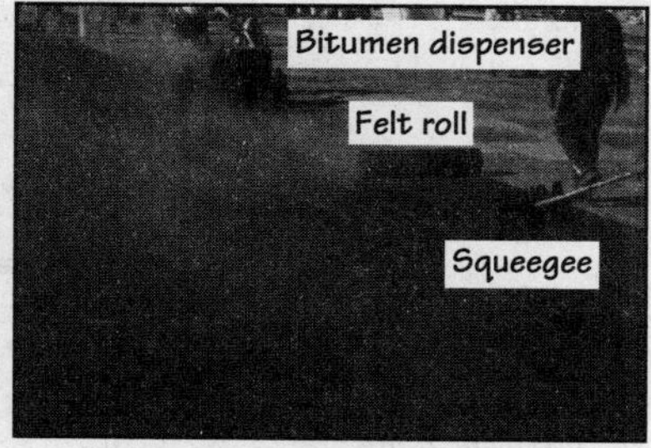

FIGURE 33.3(b) Although bitumen can be applied with mops, it is more convenient to use a bitumen dispenser on large roofs. The unrolling of felt over bitumen is followed by pressing the felt down with a squeegee.

FIGURE 33.3(c) This image shows the roof after all built-up felt plies have been applied. The roof is now ready to receive the flood coat of bitumen and the stone aggregate surfacing, as shown in the following two images.

FIGURE 33.3(d) The bitumen under aggregate surfacing is poured over the deck, not mopped. That is why it is called a flood coat of bitumen.

FIGURE 33.3(e) The bitumen flood coat is covered with stone aggregate. For a small roof, shovels are used to spread the aggregate.

FIGURE 33.3(f) Although the bitumen for a flood coat can be poured with buckets, as shown previously, it is usually poured on large roofs using a bitumen dispenser followed by an aggregate spreader.

In the alternate layers of felt and bitumen in a built-up roof, the bitumen is the waterproofing material. However, bitumen alone cannot be used, because it is a thermoplastic material. It becomes soft at high temperatures and begins to flow. At low temperatures, it becomes hard and brittle and cracks. Thus, bitumen does not have the requisite tensile strength to withstand the stresses imposed by the changes in temperature, deck movement, foot traffic, hailstorms, and so on.

The felts work as reinforcing material, giving the required tensile strength to the membrane, Figure 33.4. The felts also stabilize the bitumen against flow, because the interwoven felt fibers form mini-receptacles within which the bitumen is held, preventing its flow, Figure 33.5. A heavy mopping of bitumen without felt simply cracks in cold weather, due to the lack of tensile strength, and flows like a thick paste during hot weather, due to the lack of containment. Thus, the felts allow for a more significant buildup of bitumen, which increases the weatherproofing and waterproofing of the membrane.

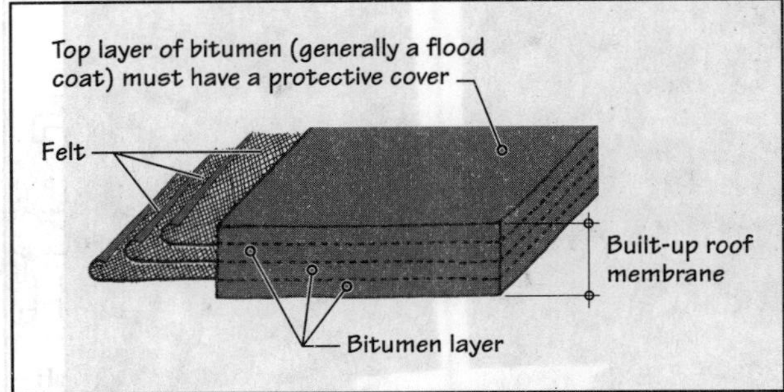

FIGURE 33.4 A built-up roof membrane consists of alternate layers of bitumen moppings and felts. The felts provide tensile strength to the membrane, in which the bitumen is the waterproofing agent. The greater the number of felts, the stronger the roof membrane.

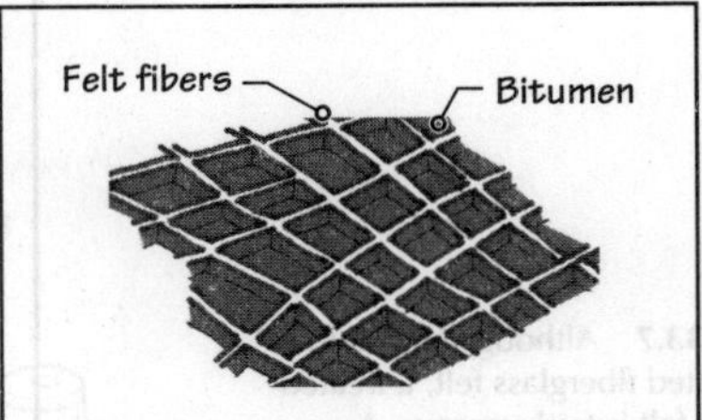

FIGURE 33.5 Interwoven felt fibers form miniature containers to hold the bitumen, preventing its flow.

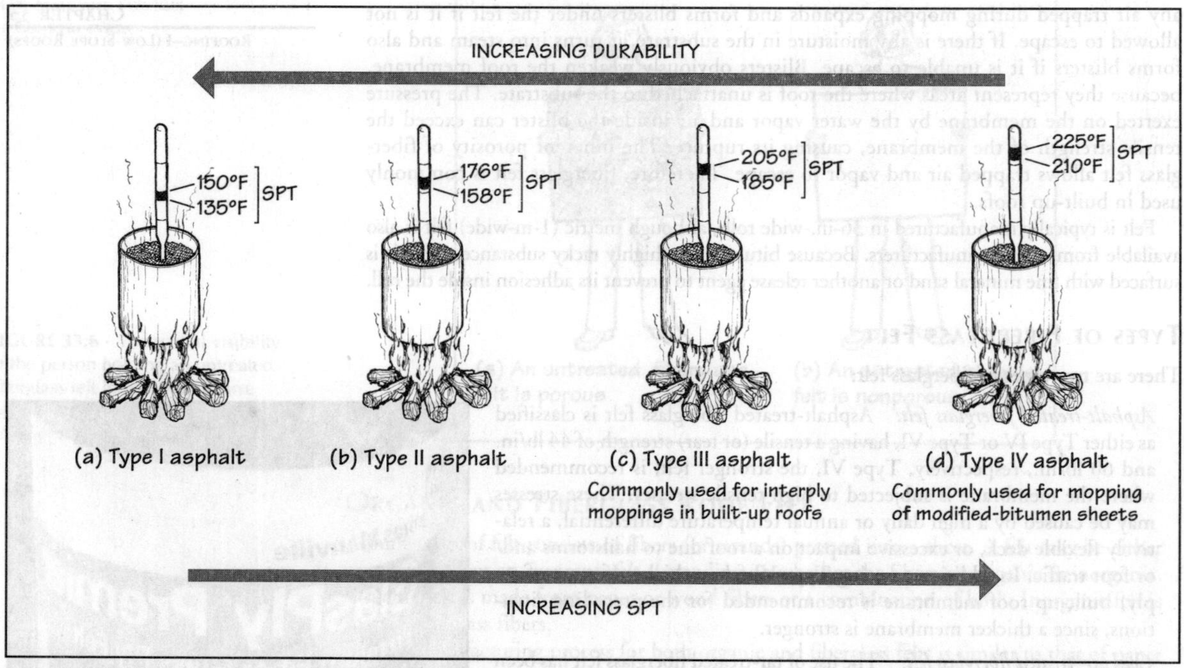

FIGURE 33.8 Types of roofing asphalt and their softening point temperature (SPT) and relative durability. Type III asphalt is most commonly used in built-up roofs and Type IV in mop-applied modified bitumen roofs.

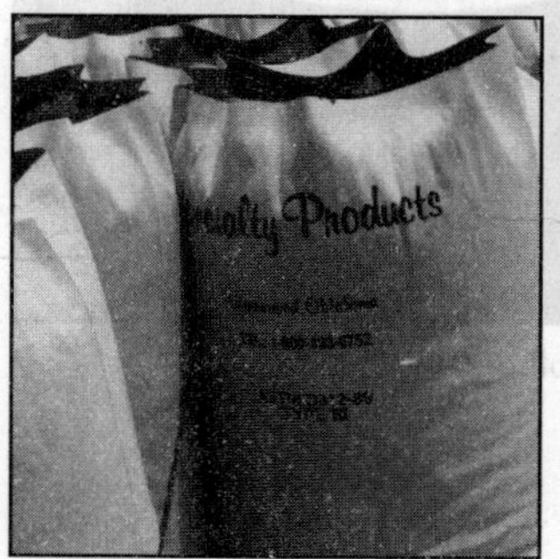

FIGURE 33.9 Types III and IV asphalt are available in paper-wrapped kegs, as shown.

Types of Roofing Asphalt

ASTM specifications divide roofing asphalts into four types: Types I, II, III, and IV. The most important property that distinguishes one type from the other is the *softening-point temperature* (SPT) of asphalt. SPT is the temperature at which asphalt begins to flow. It is directly related to the weathering characteristics of asphalt. The lower the SPT, the more durable the asphalt and the more easily the asphalt will heal any cracks caused by expansion or contraction in roof membrane. Figure 33.8 gives the SPT ranges of the four asphalt types.

The most commonly used asphalt is Type III. It is used as an interply adhesive and as an adhesive to bond insulation to the deck or to bond two layers of insulation together. Because of its lower viscosity, Type III (and also Type IV) can be used to adhere bituminous flashings on vertical surfaces.

Type IV asphalt performs better than Type III, but its lower durability is a deterrent. Type III asphalt is also specified for the flood coat for roofs in warm climates. However, in colder climates (and roof slope permitting), Type I or Type II asphalt should be considered for the flood coat due to its better weathering characteristics. Types III and IV asphalts are available in paper-wrapped kegs, Figure 33.9. Types I and II are available in metal containers because of their lower SPT.

NOTE

Roof Square

Built-up roof felt rolls are 36 in. wide and 36 ft long. Thus, the area of a roll is 108 ft^2. With a 2-in. head lap and a 6-in. lap at the end of a roll, one roll covers a roof area of 34 in. × 35 ft 6 in., which is approximately equal to 100 ft^2.

In the roofing industry, 108 ft^2 is referred to as a *factory square* and 100 ft^2 is called a *roof square*.

Amount of Asphalt Required

Approximately 25 lb of asphalt per roof square is recommended for interply moppings, and 60 lb per roof square is suggested for the flood coat. The flood coat is poured on the roof, not mopped on, to provide the large quantity of asphalt required.

Built-Up Roof Surfacing

Because both coal tar and asphalt degrade over time due to their exposure to ultraviolet radiation, the topmost layer in a built-up roof must be protected. Stone aggregate is by far the most commonly used protective covering. Apart from protecting the bitumen from ultraviolet radiation, an aggregate surface increases the fire resistance of the roof, protects the membrane from hail impact, and adds weight over the membrane, increasing its resistance to wind uplift.

The aggregate must be a graded aggregate so that most of the particles sink into the bitumen. The average recommended aggregate size is $\frac{3}{8}$ in., with a maximum aggregate size of $\frac{3}{4}$ in. A minimum aggregate weight of 400 lb per roof square is generally recommended.

Water-worn river gravel is ideal surfacing material because it does not damage the membrane under occasional foot traffic, but crushed stone may be used if the traffic on the roof is minimal. Blast-furnace slag (see Section 18.1) also makes a good surfacing material. The use of lightweight aggregate, which may be dislodged under high wind or rainstorm conditions, is generally discouraged. It may clog the drains and disintegrate under freeze-thaw action.

NOTE

Weight of Aggregate Surfacing

The weight of aggregate surfacing on a built-up roof is generally 400 lb per roof square, that is, 4 lb/ft^2.

PRACTICE **QUIZ**

Each question has only one correct answer. Select the choice that best answers the question.

1. A low-slope roof is defined as a roof whose slope is less than
   a. 1:12.
   b. 1.5:12.
   c. 2:12.
   d. 2.5:12.
   e. 3:12.
2. Low-slope roofs are more commonly used than steep roofs.
   a. True
   b. False
3. Because of the general trend toward *sustainable* and *organic* construction, organic felts are increasingly replacing fiberglass felts in built-up roofs.
   a. True
   b. False
4. Bitumen-treated fiberglass felt
   a. is less porous than bitumen-treated organic felt.
   b. is more porous than bitumen-treated organic felt.
   c. has the same porosity as bitumen-treated organic felt.
   d. none of the above.
5. Which of the following represents an asphalt-treated fiberglass felt?
   a. Type II
   b. Type III
   c. Type IV
   d. No. 15
   e. No. 30
6. A built-up roof membrane is generally finished with
   a. a single-ply roof membrane.
   b. aggregate surfacing.
   c. a flood coat of bitumen.
   d. all of the above.
   e. (b) and (c).
7. A roof square implies
   a. 1 ft^2 of roof area.
   b. 10 ft^2 of roof area.
   c. 100 ft^2 of roof area.
   d. 1 m^2 of roof area.
   e. 10 m^2 of roof area.
8. Asphalt is a waste product obtained from
   a. the manufacturing of coal tar.
   b. the manufacturing of rubber.
   c. petroleum refining.
   d. the manufacturing of steel.
   e. none of the above.
9. The weight of aggregate surfacing on a built-up roof is generally
   a. 5 lb per roof square.
   b. 15 lb per roof square.
   c. 25 lb per roof square.
   d. 40 lb per roof square.
   e. 400 lb per roof square.
10. Which of the following built-up roof membranes is more commonly used?
    a. Asphalt built-up roof membrane
    b. Tar built-up roof membrane
11. Which of the following roofing asphalts is commonly used for interply mopping?
    a. Type I
    b. Type II
    c. Type III
    d. Type IV
    e. None of the above

33.4 MODIFIED BITUMEN ROOF MEMBRANE

A modified bitumen sheet is similar to built-up roof felt because the waterproofing agent is the bitumen, to which a polymer has been added to modify the bitumen's properties. The polymer's addition to the bitumen improves its characteristics so that the modified bitumen is more pliable and elastomeric compared to the (unmodified) bitumen. The polymer also increases the ultraviolet radiation resistance of the bitumen. Modified bitumen is, therefore, more resistant to cold temperatures and to ultraviolet radiation than the bitumen used in a built-up roof membrane.*

Of the two modified bitumens—modified asphalt and modified coal tar—modified asphalt is far more commonly used. The two most common asphalt modifiers are

- Styrene butadiene styrene (SBS)
- Attactic polypropylene (APP)

NOTE

SBS- and APP-Modified Bitumen Sheets

SBS-Modified Bitumen Sheet

- Greater flexibility
- Greater resistance to a cold environment

APP-Modified Bitumen Sheet

- Greater resistance to ultraviolet radiation

*Regular (unmodified) ashphalt and coal tar become relatively brittle in a cold environment. However, this fact does not greatly influence the performance of a built-up roof in cold climates, as is sometimes believed. For instance, the long-term performance of built-up roofs in Canada and the northern United States compares favorably with that of modified bitumen or single-ply roofs.

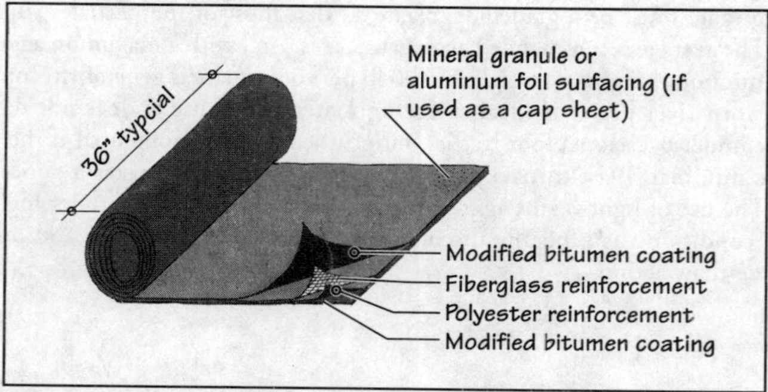

FIGURE 33.10 Anatomy of a modified bitumen membrane with fiberglass and polyester scrims.

The SBS modifier is a synthetic rubber. Therefore, an SBS-modified sheet is more flexible and more resistant to cold temperatures than an APP-modified sheet. APP-modified asphalt has greater resistance to ultraviolet radiation, which gives an APP sheet greater weatherability than an SBS sheet.

A modified bitumen sheet consists of a reinforcing mat, called a *carrier,* which is impregnated and coated with modified bitumen on both sides. The carriers commonly used are polyester or fiberglass scrims or both. A sheet that has both fiberglass and polyester reinforcements is referred to as a *dual-carrier* sheet, Figure 33.10. Most manufacturers make modified bitumen sheets in 36-in. widths, like the built-up roof felt, although the use of 1-m-wide felt is gradually increasing.

NOTE

$1 \text{ mil} = \frac{1}{1{,}000} \text{ in.}$

Polyester, which is itself a polymer, has high elongation and thus gives pliability to the sheet. It also adds puncture and tear resistance to the sheet, which is important in resisting rooftop traffic and damage caused by roofers' and repairpersons' tools. Fiberglass has very little elongation, but it provides tensile strength and increases the sheet's fire resistance.

The thickness of a modified bitumen sheet varies from about 100 to 200 mil. Therefore, a modified bitumen sheet is much thicker than an asphalt felt, which is only about 30 mil thick. It is also nonporous, unlike the asphalt-treated felt. A thicker modified bitumen sheet is selected if greater strength is required to withstand greater impact and foot traffic. Because of its greater thickness, a modified bitumen roof needs to consist of only two to three sheets to give almost the same performance as four to five plies of a built-up roof.

A modified bitumen sheet may either be smooth or surfaced with mineral granules or metal foil. A smooth-surfaced sheet is generally used as an intermediate ply or as a base sheet in a two- or three-ply modified bitumen membrane. A mineral granule or foil-faced sheet is used as a cap sheet. The commonly used metal foils are aluminum, copper, and stainless steel.

A cap sheet performs the same function as the flood coat and surfacing in a built-up roof. Mineral granules (typically $\frac{1}{10}$-in. stone chips) and metal foil function as protective covers in a cap sheet. If a smooth-surfaced modified bitumen sheet is used as a cap sheet, it must be protected by a bitumen flood coat and aggregate, like a typical built-up roof.

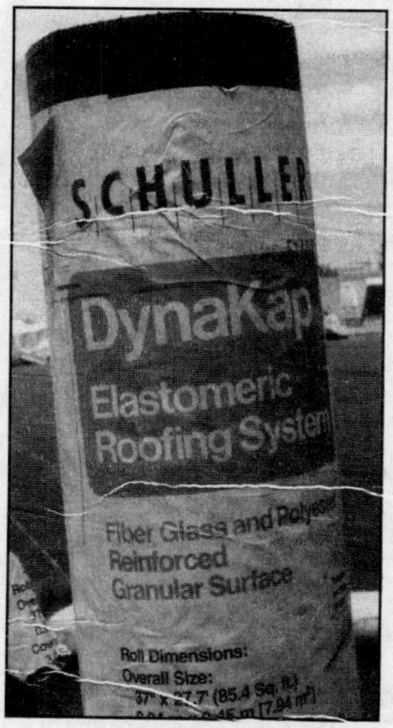

A typical SBS-modified granule-surfaced cap sheet roll.

SBS-MODIFIED BITUMEN MEMBRANE

An SBS sheet is typically mop applied with hot asphalt. SBS sheets requiring torch application and, more recently, self-adhering membranes are also available. The low-temperature flexibility (elastomeric nature) of an SBS sheet provides some advantage in colder climates over a conventional built-up roof or an APP-modified bitumen roof.

With hot-asphalt-mopped SBS sheets, Type IV roofing asphalt is commonly used. Type IV asphalt has an application temperature of nearly 500°F—a temperature high enough to soften the SBS-modified asphalt in SBS sheets. (SBS-modified asphalt softens at approximately 350°F.) The amount of asphalt required is the same as for interply moppings in a built-up roof, that is, nearly 25 lb per roof square.

A two- or three-ply SBS membrane gives good performance under most conditions. A two-ply membrane may consist of a smooth-surfaced SBS base sheet covered by a granule-surfaced

The Terms: Felt, Ply, Membrane, and Sheet

As mentioned in Section 33.3, the individual sheets between bitumen moppings in a built-up roof are referred to as *felts*. A felt and the corresponding layer of bitumen (adhesive) is referred to as a *ply*. An assembly of two or more plies is referred to as a *membrane*. When modified bitumen sheets were first introduced (in the mid-1960s), it was thought that only one modified bitumen sheet would perform as well as an assembly of felts and bitumen moppings in a built-up roof—the built-up roof membrane. Therefore, modified bitumen sheets were referred to as *membranes* and were even included in the single-ply membrane family.

It was subsequently realized that a single sheet of modified bitumen is inadequate and that two or more sheets (plies) are generally needed to provide a reasonable roof. However, each such sheet is referred to as a *membrane* in some roofing literature, which is an incorrect use of the term *roof membrane*. A membrane is the entire waterproofing layer on the roof. It may consist of one ply, as in a single-ply roof membrane; an assembly of four to five plies, as in a built-up roof; or two to three plies, as in a modified bitumen roof.

Similarly, the term *felt*, in roofing vernacular, is reserved for either an organic or fiberglass sheet, typically used in a built-up roof, which together with a bitumen mopping constitutes one built-up roof ply. Other roofing sheets are not referred to as felts. For instance, in a commonly used modified bitumen membrane assembly, consisting of two plies, the first ply is called a *base sheet* and the second ply is called a *cap sheet*. The base sheet is usually a smooth-surfaced modified bitumen sheet, and the cap sheet is a granule-surfaced modified bitumen sheet.

SBS cap sheet. A foil-faced SBS cap sheet may be substituted for a granule-surfaced cap sheet. A three-ply SBS membrane consists of two smooth-surfaced sheets, followed by a mineral or foil-faced SBS cap sheet, Figure 33.11.

Built-Up and SBS Hybrid Membrane

An SBS-modified bitumen sheet can also be combined with a built-up roof membrane to give an SBS and built-up hybrid. Such a membrane may consist of a two- to four-ply conventional built-up roof followed by a granule or foil-faced SBS cap sheet. In place of an SBS cap sheet, a smooth-surfaced SBS sheet with an asphalt flood coat and gravel surfacing may be used, as in a built-up roof, Figure 33.12.

The asphalt for the flood coat is generally the regular (unmodified) roofing asphalt, although SBS-modified asphalt can also be used. SBS-modified asphalt is compatible with (regular) roofing asphalt and can be used with built-up roof felts or with SBS sheets.

APP-Modified Bitumen Membrane

Just as the addition of an SBS modifier raises the softening-point temperature of asphalt, so does the addition of an APP modifier. In fact, the addition of an APP modifier raises the softening-point temperature of the APP-modified asphalt so much that an APP-modified membrane cannot be hot mopped like an SBS sheet or a built-up roof felt. The application temperature of even Type IV asphalt (nearly 500°F) is not high enough to melt the asphalt in an APP sheet for good adhesion.

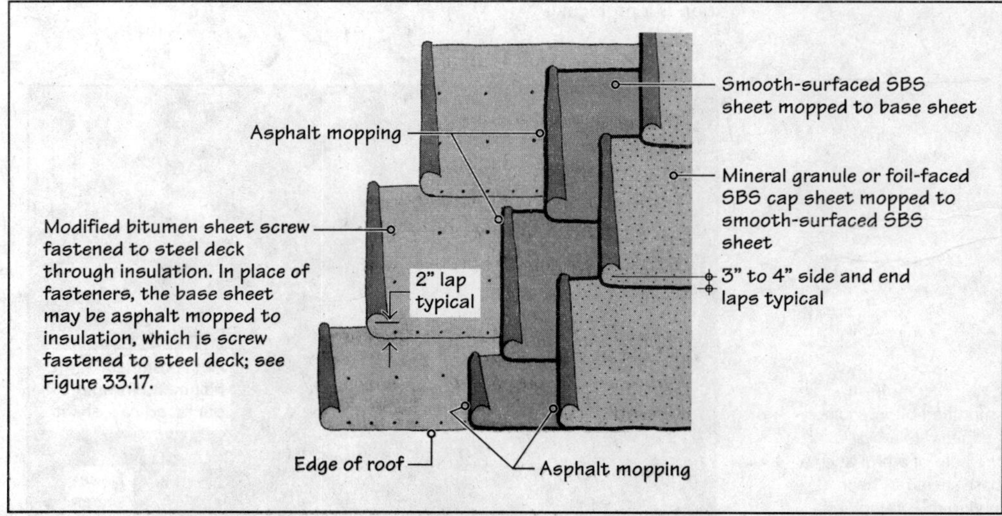

FIGURE 33.11 A three-ply SBS-modified roof membrane consisting of a granule- or foil-faced, SBS-modified bitumen cap sheet and two smooth-surfaced SBS sheets.

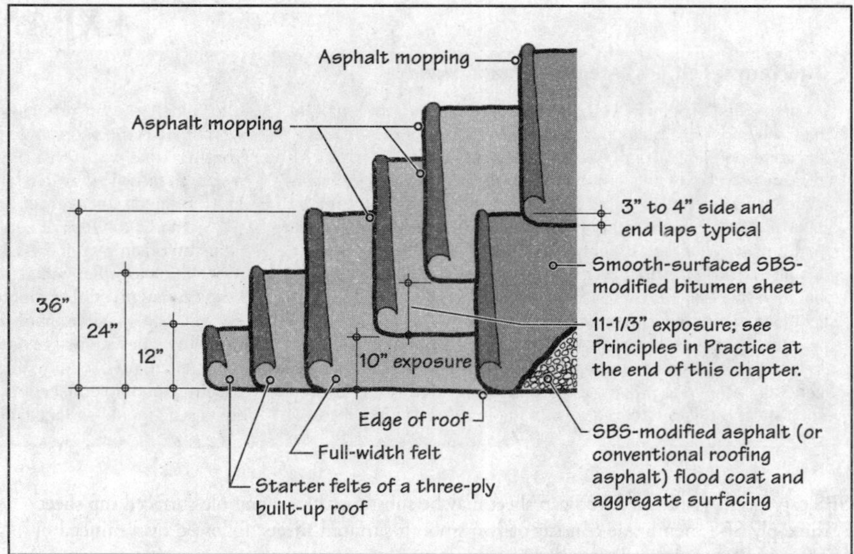

FIGURE 33.12 A built-up and SBS hybrid consisting of a smooth-surfaced SBS-modified bitumen sheet and an aggregate-covered flood coat over a three-ply built-up roof. Note that built-up roof felts are mopped in the shingled format (see Principles in Practice at the end of this chapter).

The most common method of installing an APP sheet is through the open flame of a handheld. propane torch, Figure 33.13, which has a temperature of more than 1,000°F. In this application, the flame is applied to the sheet roll from one end to the other. As the flame melts the modified asphalt on the underside of the sheet, the roll is unwound and pressed down. In this way, the sheet is fully adhered to the substrate. The sheets are usually factory laminated with a thin polyethylene release sheet to prevent adhesion in the roll. The torch burns the polyethylene sheet. Some manufacturers use sand dusting in place of a polyethylene sheet.

An alternative to the hand-held torch is a *dragon wagon,* which consists of a series of propane torch heads mounted on a wheelable cart that also carries the APP sheet roll, Figure 33.14. A dragon wagon speeds up the installation of an APP sheet but should be used only on a large open roof with few penetrations.

Torched Membrane's Suitability

A torched membrane is suitable in situations where the hot bitumen used in a conventional built-up roof (or an SBS roof) cannot be pumped from the ground to the roof, such as on the roof of a high-rise building. The torched system also has an advantage in cold climates, where keeping the asphalt hot enough between the kettle and the point of application is a problem.

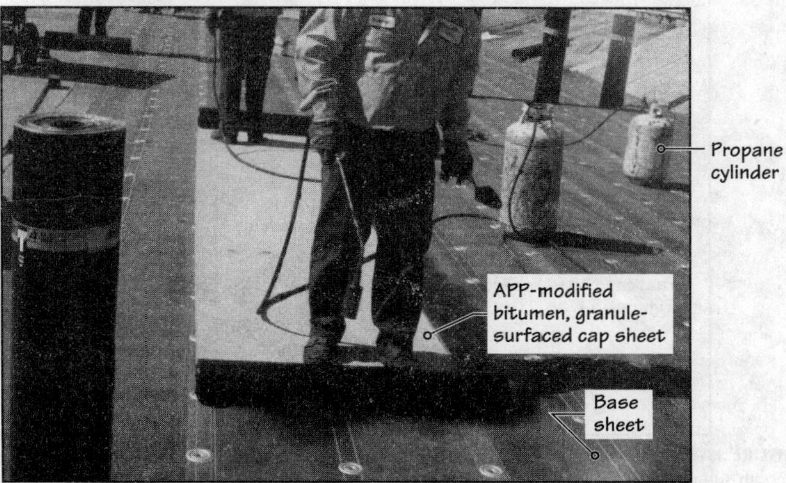

FIGURE 33.13 A granule-surfaced, APP-modified bitumen cap sheet being adhered to a base sheet with the help of a propane torch. The base sheet has been screw fastened to the underlying steel deck through rigid board foam insulation.

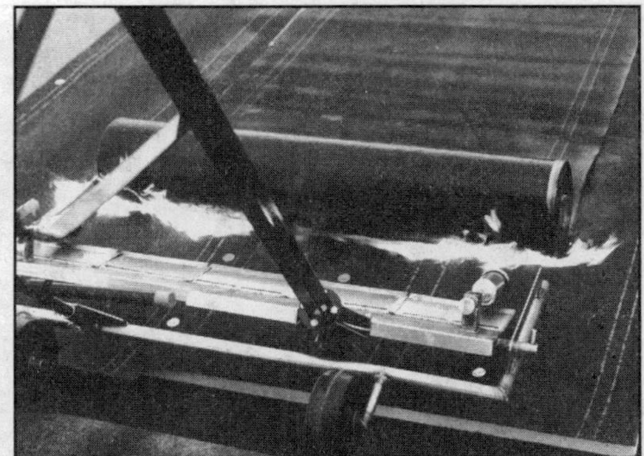

FIGURE 33.14 A dragon wagon with its multiple torches. (Photo courtesy of Siplast Inc.)

In fact, the torch-adhered, SBS-modified bitumen roof may be preferred for cold regions for the preceding reasons and also because of the greater flexibility of an SBS roof compared with an APP or a built-up roof. However, the open torch used in laying a torchable membrane presents a fire hazard, which must be considered before specifying a torched membrane. Fire extinguishers must be available on the roof for any emergency, and personnel need to be left on the roof for a "fire watch" for a few hours after the day's work.

Although an APP sheet has a degree of ultraviolet resistance, a surfacing over it gives additional protection. Mineral granule or foil surfacing on an APP cap sheet adds to its durability. Where practical, an APP roof should be flood coated and covered with gravel for greater durability and fire resistance.

APP AND BUILT-UP ROOF HYBRID

A two-layer APP membrane—the first layer, a smooth-surfaced APP sheet, and the top layer, a mineral-surfaced APP cap sheet—can provide an adequate membrane in most situations. Alternatively, an APP cap sheet may be used over two to three plies of built-up roof felt, giving an APP and built-up roof hybrid membrane.

33.5 SINGLE-PLY ROOF MEMBRANE

Although a built-up roof membrane (and, to a smaller extent, a modified bitumen membrane) is constructed on the roof felt by felt, a single-ply roof has only a one-ply membrane and does not require the cumbersome use of hot bitumen. It is easier to lay because it has only one ply.

A single-ply membrane is made of a polymeric material. Because a polymer is a synthetic material, it is capable of a great deal of chemical manipulation by minor changes in one or more of its constituents. Therefore, a large number of polymers are used as roof membranes. However, polymers (plastics), in general, can be divided into two broad categories:

- *Thermosetting polymer (unweldable plastic):* a polymer that does not soften on heating once it is cured into hardness, like a boiled egg. Therefore, the seams in a thermosetting membrane cannot be heat welded but must be adhered together by an adhesive or a double-sided seam tape. The most commonly used thermosetting membrane is EPDM, an acronym that stands for ethylene propylene diene monomer.
- *Thermoplastic polymer (heat-weldable plastic):* a polymer that softens on heating and hardens on cooling over and over again, like butter. Asphalt and coal tar behave like a thermoplastic. A thermoplastic material is heat weldable, which makes the seaming operation relatively easier. However, thermoplastic polymer is not as stretchable as thermosetting polymer. Currently, the most commonly used thermoplastic roof membranes are PVC (polyvinyl chloride) and TPO (thermoplastic polyolefin); TPO is more commonly used than PVC.

ADVANTAGES AND DISADVANTAGES OF A SINGLE-PLY MEMBRANE

By virtue of the material of which it is made, a single-ply membrane is highly flexible and can withstand large elongation before failure. For example, a typical EPDM membrane will

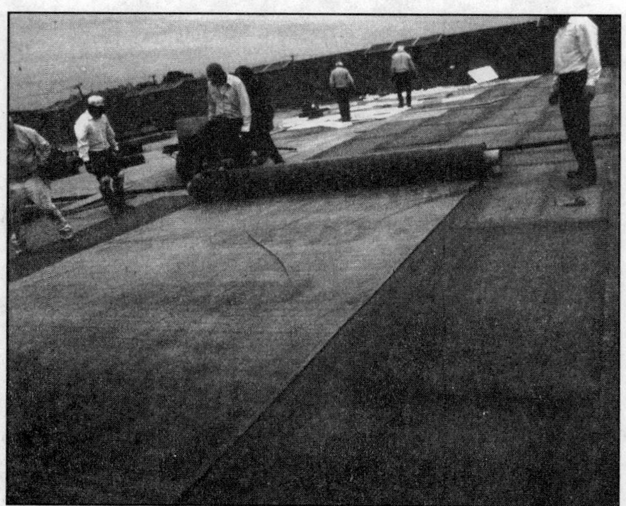

FIGURE 33.15 Although this is a 30-ft-wide EPDM membrane (with a center fold so that the roll is only 15 ft wide), single-ply membrane rolls are available in widths larger than 30 ft, shown here.

stretch 300% to 500% of its original length before tearing apart, compared to 50% for an SBS-modified bitumen sheet and 4% for a fiberglass-reinforced built-up roof felt.

For a variety of reasons, a typical single-ply membrane comes in much wider and longer rolls than a built-up roof felt or a modified bitumen sheet, Figure 33.15. Although advantageous (because it reduces the number of lap splices, which reduces lap-failure possibilities), a large roll size also means that roof penetrations are more difficult to work around. Therefore, a single-ply membrane has an advantage on a large roof with few penetrations.

The relative flexibility and the elongation characteristic of a single-ply membrane are useful in climates with a large annual or daily temperature variation. A single-ply membrane (particularly EPDM) retains its elastomeric character even in subzero (Fahrenheit) temperatures. Another advantage of a single-ply membrane is that it does not require aggregate surfacing.

A major disadvantage of a single-ply roof membrane is the absence of second, third, or additional plies. This implies a lack of redundancy in a single-ply roof—a redundancy that is helpful in any waterproofing application. A puncture or split results in a roof leak more readily in a single-ply roof than in a built-up roof because of the greater thickness, multiple plies, and self-healing property of bitumen in a built-up roof.

As stated previously, the overall market share of single-ply membranes is approximately the same as that of built-up or modified bitumen membranes—that is, roughly one-third each. Because EPDM, PVC, and TPO are the more commonly used single-ply membranes, the following discussion relates to them only, although it is general enough to apply to other single-ply membranes as well.

EPDM MEMBRANE

Because EPDM is a synthetic rubber, it is essentially similar to the material used in automobile tires. Firestone is, therefore, one of the major EPDM membrane manufacturers. The typical thickness of an EPDM membrane is 45 to 60 mil, although some manufacturers make a thicker membrane for use where higher puncture and tear resistance are needed.

EPDM polymer lacks inherent fire resistance, but fire-resistive EPDM membranes are available. However, with a loose-laid, ballasted EPDM roof, a non-fire-resistive membrane may be specified because the ballast will provide the fire resistance. Because EPDM is a synthetic rubber (with its consequent flexibility), its use is more common in colder climates of North America. In warmer climates and in response to sustainability considerations, several architects and roof consultants specify white membranes, which are produced more economically in PVC or TPO.

PVC MEMBRANE

PVC, the most versatile plastic, is available in two types. Rigid PVC is used for pipes, siding, and window sashes. Soft, pliable PVC is used in shower curtains, raincoats, electrical

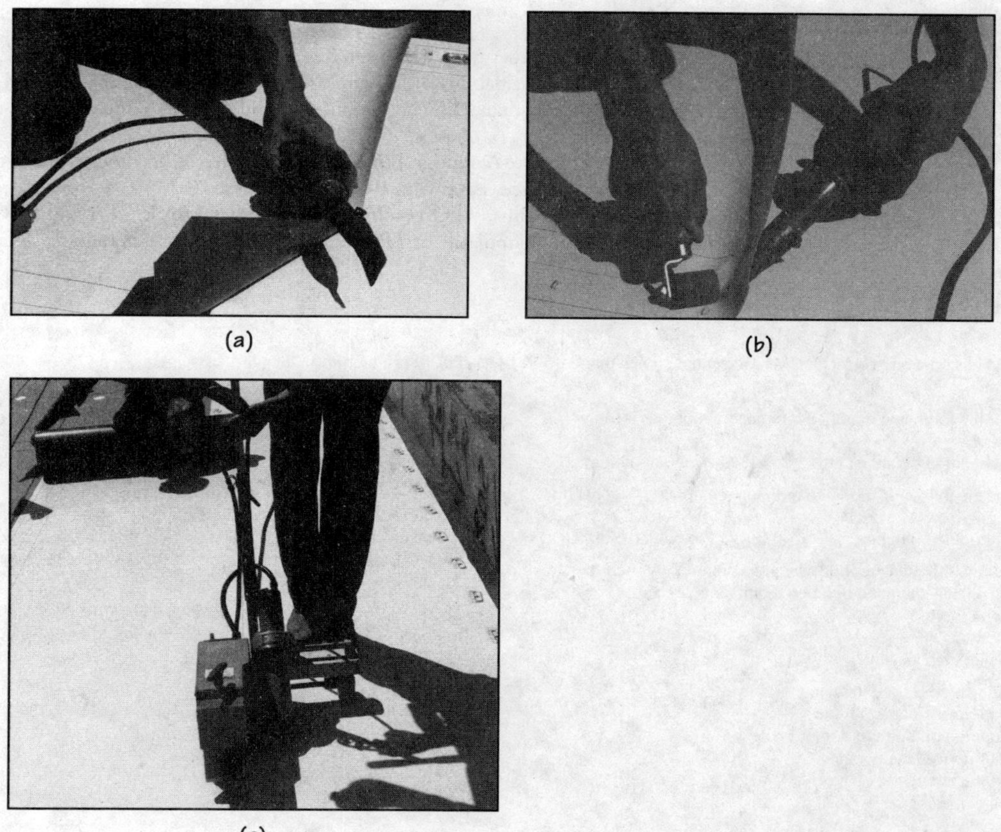

FIGURE 33.16 **(a)** A hand-held welding tool that supplies hot compressed air through a flat nozzle. **(b)** A roofer welds the seams of a white PVC membrane and presses down the fused seams with a roller. **(c)** A self-propelled hot-air welding machine that welds and presses the seams—an alternative to the hand-held welding tool. The machine speeds the welding process.

wire insulation, and roof membranes. Softness and flexibility—properties that are necessary for a roofing membrane—are obtained through the addition of a plasticizer. However, despite the addition of a plasticizer, a PVC membrane, though flexible, is far less stretchable than an EPDM membrane.

The primary advantage of a PVC membrane is that the seams in the membrane can be heat fused. A simple tool, such as that shown in Figure 33.16, which supplies hot compressed air, is all that is needed to fuse the seams. This makes roof installation much easier and more economical compared to EPDM roofs, in which the seams must be joined together with a double-sided tape.

Another advantage of PVC is its inherent fire resistance, which is provided by its chlorine atom. PVC membranes are generally used in 45- to 80-mil thickness. They are available in several colors, although white and gray are most common.

EXPAND YOUR KNOWLEDGE

EPDM, PVC, and TPO Membranes Compared

Criterion	EPDM	PVC	TPO
Lap-seam attachment	Two-sided tape or contact adhesive	Seams heat welded	Seams heat welded
Fire resistance	Fire-resistive EPDM available	Inherently good	Fire-resistive TPO available
White-sheet availability	Available	More easily made	More easily made
Resistance to ultraviolet radiation	Good for black EPDM	Good	Good
Compatibility with insulation	Compatible with most insulations	Incompatible with polystyrene foam due to plasticizers	Compatible with most insulation
Compatibility with asphalt	Incompatible	Incompatible	Incompatible

TPO MEMBRANE

Like PVC, TPO is also thermoplastic. Therefore, it has the same advantage over EPDM as PVC, that is, the heat weldability of its seams. Although less flexible than EPDM, TPO is inherently more flexible than PVC, not requiring the addition of plasticizers in its chemical formulation.

This makes TPO more economical than PVC; also, there is no concern about the gradual evaporation of plasticizers (as with PVC), which causes the PVC membrane to lose its flexibility over time. Thus, TPO combines the advantages of EPDM and PVC, which explains the increasing popularity of TPO among the single-ply membranes.

PRACTICE QUIZ

Each question has only one correct answer. Select the choice that best answers the question.

12. Which of the following modified bitumen membranes is more elastomeric?
a. SBS-modified bitumen
b. APP-modified bitumen

13. Which of the following modified bitumen membranes is generally torch applied?
a. SBS-modified bitumen
b. APP-modified bitumen

14. Which of the following roofing asphalts is commonly used for mopping an SBS-modified bitumen membrane?
a. Type I
b. Type II
c. Type III
d. Type IV
e. SBS-modified asphalt

15. A dragon wagon is used in laying
a. a single-ply roof membrane.
b. a modified bitumen roof membrane.
c. a built-up roof membrane.
d. all of the above.
e. none of the above.

16. EPDM is a thermosetting plastic.
a. True
b. False

17. The lap seams in an EPDM roof membrane are typically sealed by
a. heat welding.
b. liquid adhesives.
c. single-sided tape.
d. double-sided tape.
e. any one of the above.

18. The lap seams in a PVC roof membrane are typically sealed by
a. heat welding.
b. liquid adhesives.
c. single-sided tape.
d. double-sided tape.
e. any one of the above.

19. The lap seams in a TPO roof membrane are typically sealed by
a. heat welding.
b. liquid adhesives.
c. single-sided tape.
d. double-sided tape.
e. any one of the above.

20. Of the two single-ply roof membranes, EPDM and PVC, PVC is generally
a. more flexible and more fire resistant.
b. less flexible but more fire resistant.
c. more flexible but less fire resistant.
d. less flexible and less fire resistant.

33.6 RIGID-BOARD INSULATION AND MEMBRANE ATTACHMENT

A type of insulation commonly used in a low-slope roof consists of semirigid boards. Batt and blanket insulations are not suitable for low-slope roofs because of their low compressive strength. Two commonly used semirigid board insulations (both are plastic foams; see Principles in Practice, Chapter 5) are

- Polyisocyanurate boards (also called *iso boards*)
- Extruded polystyrene boards (also called *XPS boards*)

Iso boards have much wider applicability than XPS boards. An XPS board is chemically incompatible with some membranes, particularly a PVC membrane.

Between the roof membrane and the insulation, an intervening medium (called a *cover board*) is commonly used to provide compatibility between the membrane and the insulation. With hot-applied (hot asphalt or torched) membranes, such as built-up and modified bitumen membranes, the most commonly used cover board is perlite board (typically 1 in. thick). Perlite board also adds to the R-value of the system, although it is not as effective an insulator as an iso board or an XPS board.

With single-ply membranes, fiberglass-reinforced gypsum board (typically $\frac{1}{4}$-in.-thick) is commonly used. Apart from providing compatibility, gypsum board adds to the rigidity of the roof system, providing a better bearing surface for foot traffic and greater resistance to hail damage.

NOTE

Cover Board

With hot-applied membranes, the cover board is perlite board. If the asphalt is applied directly over iso board or XPS board, both of which are sensitive to the high temperature of asphalt, the membrane tends to blister. Perlite board, which has good high-temperature resistance, has been found to be a good cover board with hot-applied membranes.

With single-ply membranes, the use of a cover board is not required but it is preferred. The commonly used cover board with single-ply membranes is fiberglass-reinforced gypsum board. It is generally not recommended under a hot-applied membrane because of its chemically combined water (see Chapter 16). Under the high temperature of asphalt (approximately 400°F), the water in gypsum converts to steam. This causes blisters to form under the membrane. Additionally, the high temperature of asphalt may result in the disintegration of the gypsum.

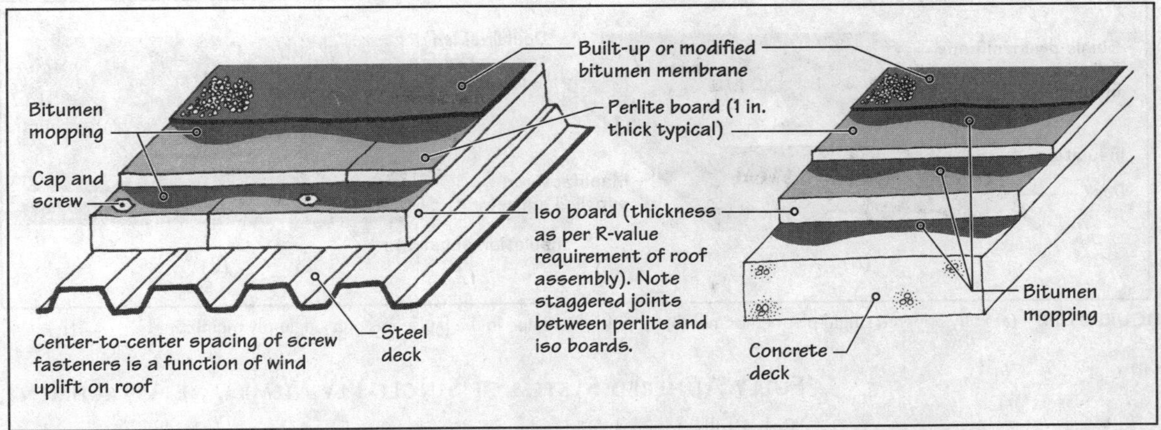

FIGURE 33.17 Attachment of insulation to steel and (dry) concrete decks—built-up and modified bitumen membranes.

Insulation Attachment—Built-Up and Modified Bitumen Membranes

The insulation used with built-up and modified bitumen membranes is iso board covered with perlite board. The insulation and cover-board combination also helps prevent thermal bridging through insulation joints. Therefore, the joints between the two layers are generally staggered.

The typical thickness of perlite board is 1 in. The thickness of iso board varies, depending on the required R-value of the roof assembly. Figure 33.17 illustrates the arrangements for the attachment of insulation to steel and concrete decks and that of membrane to the insulation.

As Figure 33.17 shows, the insulation is attached to a steel deck with screws. For a concrete deck, the insulation may be adhered with bitumen mopping, provided that the deck is dry. If the deck is not fully dry, the moisture in the deck will blister through bitumen as steam, reducing attachment strength. Nonbituminous adhesives are available that do not require a fully dry concrete deck. Spot or strip mopping is generally used with such adhesives, Figure 33.18.

Insulation Attachment—Single-Ply Membranes

With a single-ply membrane, the combination of iso board and cover board in staggered layers is generally used. Whereas built-up and modified bitumen membranes are attached through bitumen mopping, a single-ply membrane (EPDM, PVC, and TPO) is attached using one of the following three systems:

- Fully adhered
- Mechanically fastened
- Ballasted

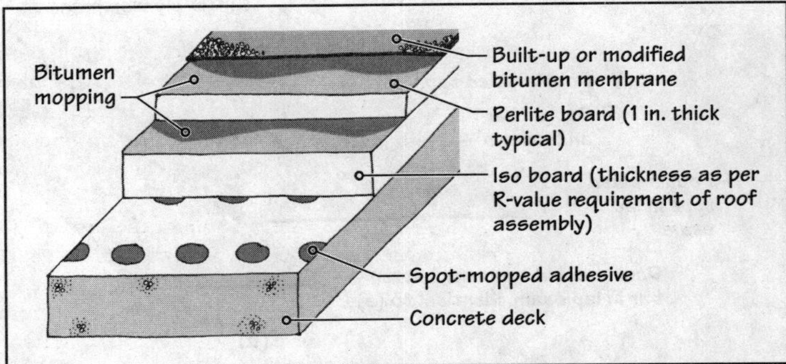

FIGURE 33.18 Attachment of insulation to a concrete deck with nonbituminous spot- or strip-mopped adhesive, generally used with a concrete deck that may not be fully dry at the time of insulation attachment.

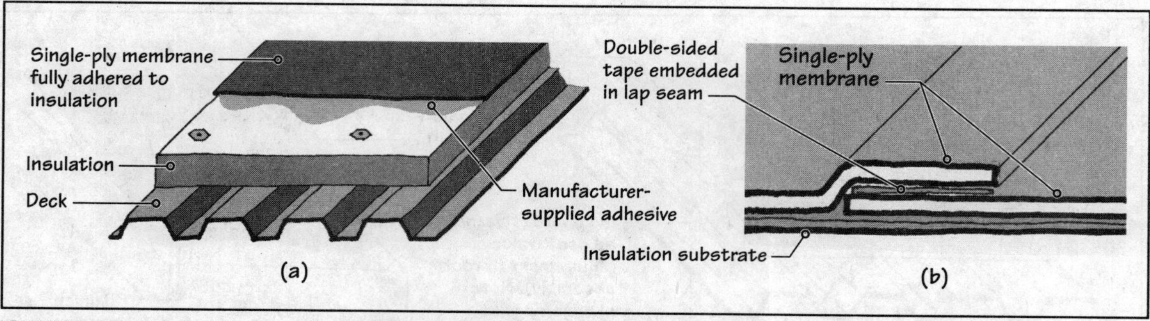

FIGURE 33.19 **(a)** Fully adhered single-ply membrane. **(b)** Double-sided tape in the lap seams of a single-ply membrane.

Fully Adhered System of Single-Ply Membrane Attachment

With a fully adhered single-ply membrane, manufacturer-provided adhesive is applied to the underside of the entire membrane, Figure 33.19(a). After the membrane is adhered, the laps between adjacent membranes are joined using a double-sided tape embedded in lap seams, Figure 33.19(b).

A fully adhered single-ply membrane system is labor intensive compared to a mechanically fastened or ballasted single-ply system. However, the system is particularly well suited to high-wind regions.

Mechanically Fastened Attachment System for a Single-Ply Membrane

In a mechanically fastened system, the membrane is laid over the insulation, which has already been fastened to the deck. (The fastening of insulation to the deck is nominal—to hold the insulation in place until the membrane has been anchored.) The spacing of fasteners for the anchorage of the membrane to the deck is a function of the wind uplift on the roof.

The membrane-fastening system consists of screws and a continuous bar (referred to as a *batten bar* in the industry) through which the screws are applied, Figure 33.20(a). The batten bar, typically made of metal or plastic, is covered over by double-sided tape in a lap seam in a thermosetting membrane, such as EPDM. (In a thermoplastic membrane such as PVC and TPO, the lap seams are heat welded.)

Intermediate fasteners and the batten bar, which are not within the seams, are covered with self-adhesive splice tape, Figure 33.20(b). Some manufacturers provide individual plates instead of batten bars for their systems.

Under the effect of wind, a mechanically fastened system is subjected to concentrated loads at the fasteners. This is unlike a fully adhered system, in which the loads are distributed over the entire membrane. Wind may make the mechanically fastened membrane flutter, causing its premature failure at the fasteners. Therefore, care should be taken in specifying a mechanically fastened system in a high-wind location.

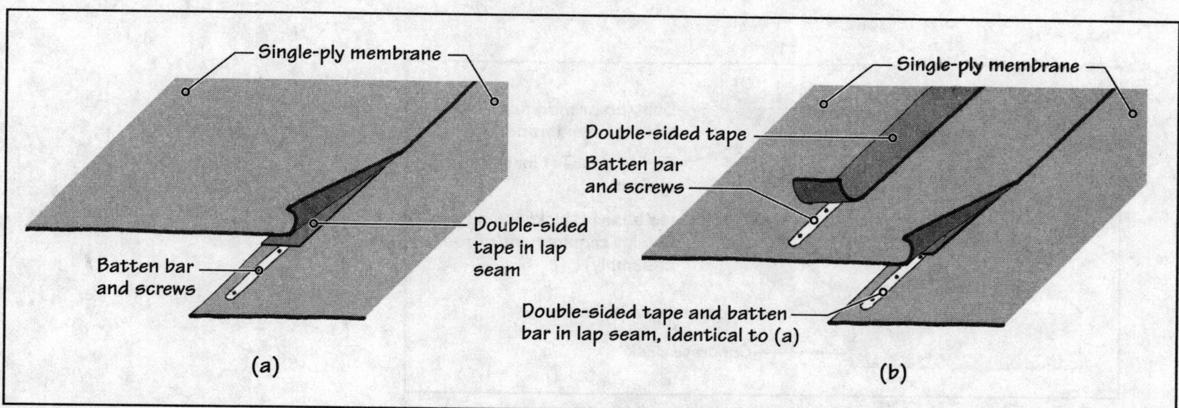

FIGURE 33.20 **(a)** The anchorage of a mechanically fastened single-ply membrane utilizes a batten bar and screws in the lap seams of membranes. **(b)** Where the membrane requires fastening in areas other than the seams, a batten bar, screws, and double-sided tape are used.

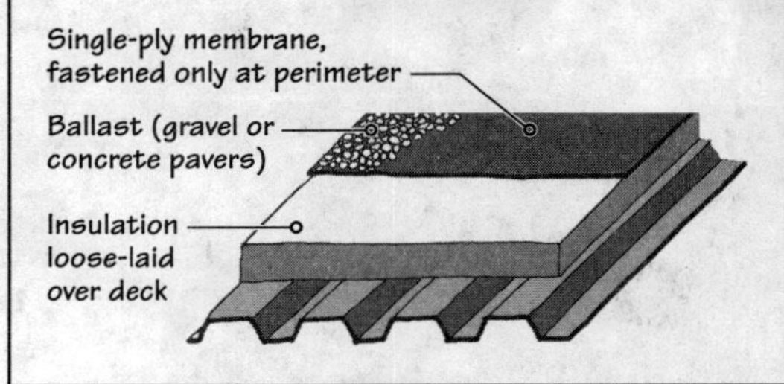

FIGURE 33.21 Loose-laid, ballasted single-ply membrane system. Note that the insulation is not fastened to the roof deck.

FIGURE 33.22 In a loose-laid, ballasted single-ply membrane system, the ballast is generally brought onto the roof using a crane and buckets. Ballast spreaders are then used to spread the ballast on the roof.

LOOSE-LAID BALLASTED SYSTEM OF SINGLE-PLY MEMBRANE ATTACHMENT

In a loose-laid ballasted system, the entire membrane is laid loose over the insulation, followed by a loose-laid single-ply membrane, Figure 33.21. The lap seams are adhered together using a double-sided tape (thermosetting membrane) or a heat-welded thermoplastic membrane. The membrane is anchored to the deck only at the roof perimeter, at curbs and penetrations. The entire membrane is then covered with ballast, Figure 33.22.

Ballast generally consists of river-worn gravel ($1\frac{1}{2}$-in. to 2-in. particle size). The use of crushed stone is discouraged because it can puncture the membrane under foot traffic. The weight of the gravel required varies with the wind uplift on the membrane. Concrete pavers may also be used instead of gravel.

The ballasted system is the most economical single-ply system. The ballast adds to fire resistance and weatherability. The system is recommended only for roofs of low-rise buildings in low-wind regions. Because the dead load of ballast is high, the deck must be structurally adequate to withstand the ballast load.

33.7 INSULATING CONCRETE AND MEMBRANE ATTACHMENT

A low-slope roof insulation (other than the rigid boards) is lightweight insulating concrete or foamed concrete. As described in in Chapter 5 (Principles in Practice), insulating concrete or foamed concrete is poured over a roof deck in wet format. When the concrete hardens, it becomes monolithic with the deck.

INSULATING CONCRETE

Where insulating concrete is used over a steel deck, the use of a slotted (i.e., perforated) deck is recommended. The perforations in the deck allow the moisture in the concrete to gradually vent out from the underside of the deck by the building's heating and cooling system. Therefore, the roof membrane can be installed as soon as the concrete has become sufficiently hard to hold the fasteners, even though it may not have fully dried.

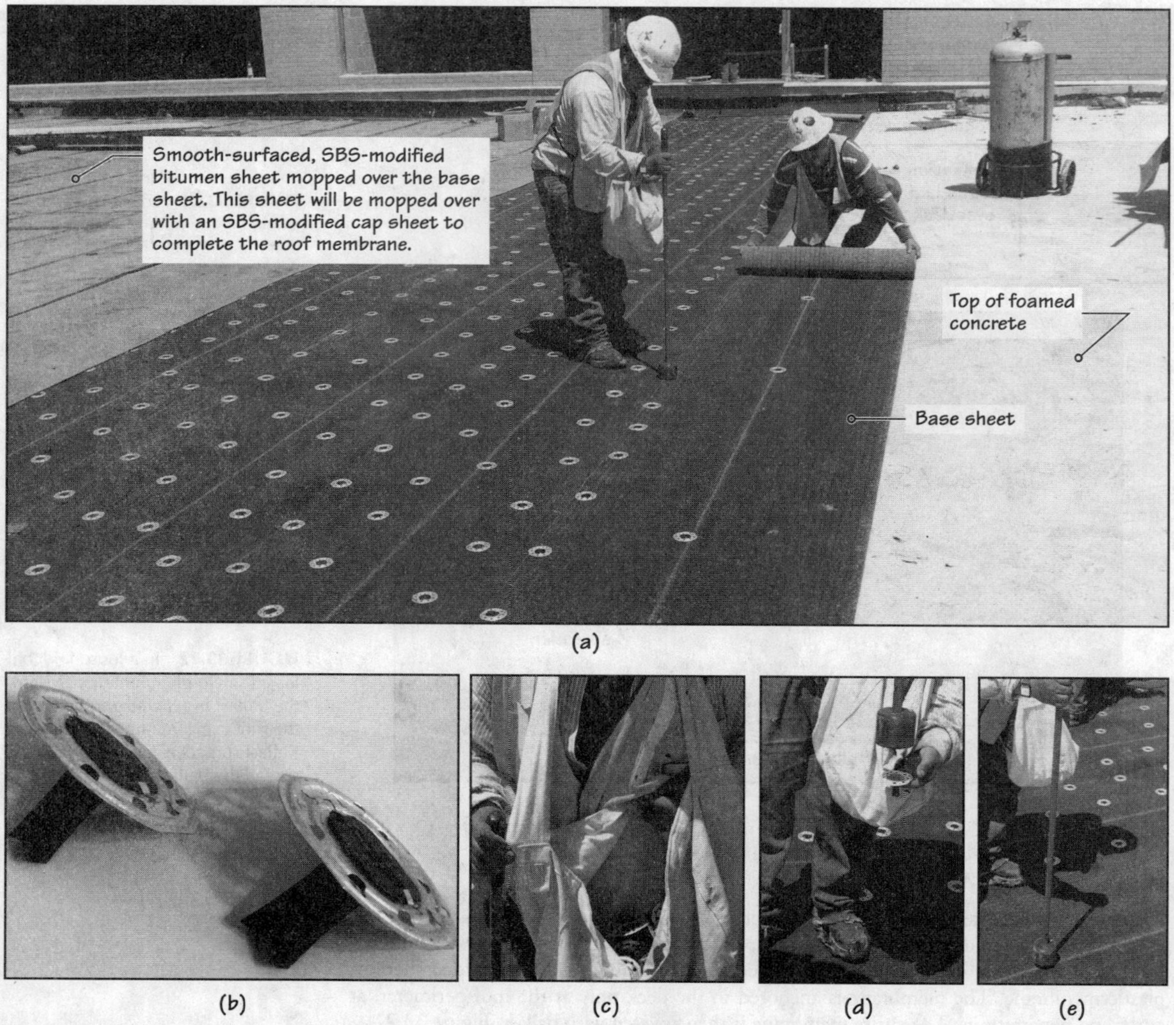

FIGURE 33.23 **(a)** One roofer unrolls the base sheet, aligning it with the adjacent, already laid base sheet, and another roofer drives the fasteners into the foamed concrete. **(b)** Close-up of a fastener. **(c)** A roofer with a pouch full of fasteners. **(d)** The roofer attaches the fastener to the head of the magnetic driver, then orients and holds the fastener perpendicular to the roofing surface. **(e)** A vertical tamping motion sets the fastener securely into the concrete.

After the insulating concrete has hardened (typically after 3 to 5 days), an asphalt-treated base sheet (preferably a mineral-granule-faced base sheet) is fastened to the concrete as a substrate for a built-up or modified bitumen roof. Special fasteners have been developed by the industry to fasten the base sheet, which can simply be driven into the insulating concrete, Figure 33.23. The fasteners flare as they penetrate into the concrete, Figure 33.24.

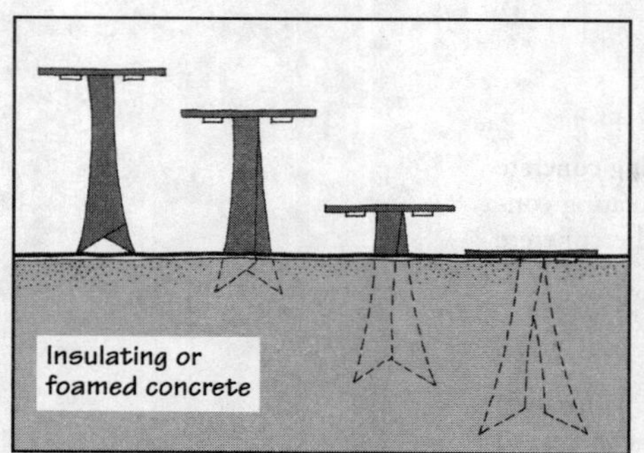

FIGURE 33.24 The fastener flares as it is driven into the concrete. (Sketch adapted from that of ES Products, Inc.)

After the base sheet has been fastened, the built-up roof or modified bitumen membrane is mopped to the base sheet. A single-ply membrane may also be installed over insulating concrete, but the manufacturers of insulating concrete recommend consulting with the manufacturers of single-ply membranes for compatibility, and so on.

Because below-the-deck venting cannot be provided in a reinforced-concrete or precast concrete deck, the use of insulating concrete on such decks requires that the insulating concrete be so dry that it does not create water vapor problems in the roofing membrane.

A major advantage of insulating concrete is that roof slope can easily be provided in the insulation, even if the roof deck is dead level. This is done by using expanded polystyrene (EPS) boards of different thicknesses to create a stairstepped pattern, Figure 33.25. Providing slope over a dead-level deck with

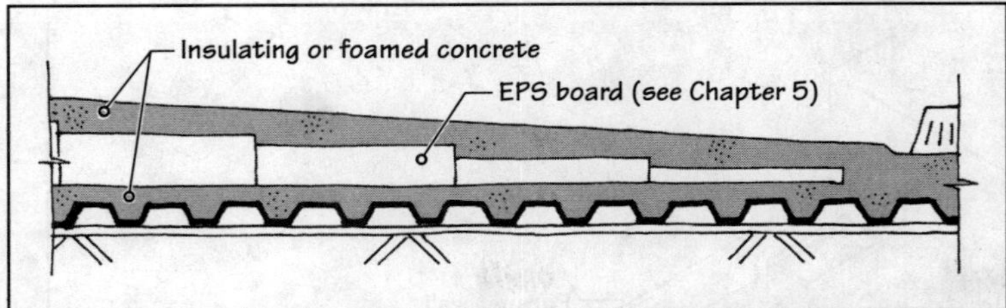

FIGURE 33.25 Section through a roof showing a stairstepped EPS board pattern to achieve a slope on a dead-level roof deck. See also Figures 9 to 11 of Principles in Practice, Chapter 5.

tapered rigid board insulation is expensive and complicated. Insulating concrete is also well suited for decks that have a complex slope geometry.

FOAMED CONCRETE

As explained in Chapter 5 (Principles in Practice), foamed concrete, also called *cellular concrete,* is a nonaggregate (insulating) concrete that consists of portland cement, water, and a liquid foaming concentrate. Foamed concrete requires only one-third the amount of water used in insulating concrete. Therefore, foamed concrete does not need a slotted metal deck, although its use is preferred. The other details, such as fastening a base sheet, stairstepping EPS boards for roof slope, and so on, are similar to those used with insulating concrete.

PRACTICE QUIZ

Each question has only one correct answer. Select the choice that best answers the question.

21. Perlite board (as cover board), sandwiched between the insulation and the roof membrane, is generally used with
a. an asphalt-mopped membrane.
b. a fully adhered single-ply membrane.
c. a mechanically fastened single-ply membrane.
d. a loose-laid, ballasted single-ply membrane.

22. The cover board generally used between a single-ply roof membrane and rigid board insulation is
a. iso board.
b. specially treated perlite board.
c. glass-fiber-reinforced gypsum board.
d. extruded polystyrene board.
e. none of the above.

23. A mechanically fastened single-ply roof membrane is generally used with
a. steel roof decks.
b. reinforced-concrete decks.
c. precast-concrete decks.
d. none of the above.

24. A loose-laid ballasted roof is generally used with
a. a built-up roof membrane.
b. a modified bitumen membrane.
c. a single-ply membrane.
d. all of the above.

25. For a building located in a high-wind region, which of the following roof systems is most suitable?
a. Loose-laid ballasted single-ply system
b. Fully adhered single-ply system
c. Mechanically fastened single-ply system
d. Gravel-covered built-up roof
e. Protected membrane system

26. With a steel deck topped with lightweight insulating concrete, the commonly used board insulation is
a. iso board.
b. polyurethane board.
c. expanded polystyrene (EPS) board.
d. not used with insulating concrete.

27. A slotted steel roof deck is mandated for use with
a. all rigid-board insulations.
b. lightweight insulating concrete.
c. foamed concrete.
d. none of the above.

33.8 LOW-SLOPE ROOF FLASHINGS

The termination of a roof, which generally occurs over a wall, requires careful detailing. The wall either stops below the roof, in which case the roof terminates in a free edge (requiring a guard edge), or the wall projects above the roof as a parapet wall. In either case, the differential movement between the roof and the wall, and the higher wind loads at roof terminations, increase the stresses in a roof membrane at a wall-roof junction. Additional strength is, therefore, required at a roof-wall junction, which is provided by a special sheet material called *flashing.* Flashing also helps to integrate the roof membrane with the wall and further waterproofs the wall-roof junction, which is more vulnerable to leakage than the field-of-roof, which is the area of the roof that is not a roof perimeter, corner, or termination.

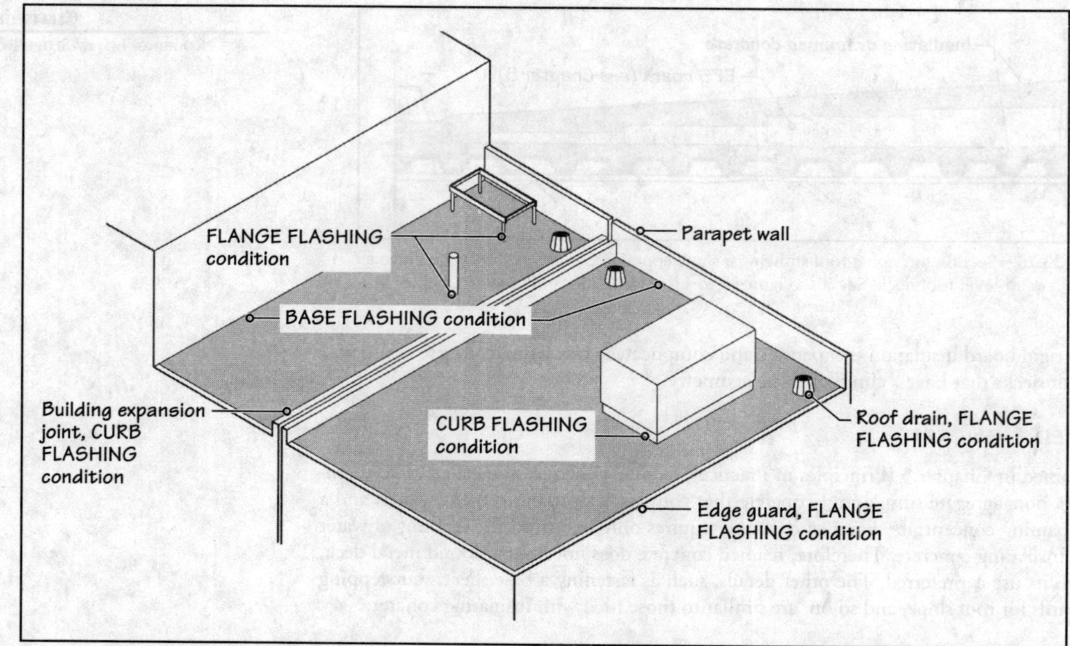

FIGURE 33.26 Three types of roof flashing: base flashing, curb flashing, and flange flashing.

Flashing is also required around penetrations in the roof, such as roof drains, gutters, vent and plumbing stacks, curbs below skylights and rooftop equipment, expansion joints, and area dividers.

Flashings may be divided into three basic types, Figure 33.26:

- Base flashings
- Curb flashings
- Flange flashings

33.9 BASE FLASHING DETAILS

Base flashing occurs at the junction of a roof and a wall that rises above the roof, such as a parapet wall. Five important components of base flashing details are

- Membrane base flashing
- Perimeter wood nailers
- Cants (required only with built-up and modified bitumen membranes)
- Counterflashing
- Parapet coping

Figure 33.27 illustrates the principles of detailing a base flashing against a low-height parapet (up to about 2 ft high) where the roof deck is supported by a bearing wall.

Figure 33.28 illustrates the same principles, but the roof deck is supported on a structural frame (and not by a bearing wall). In this case, the detail must allow differential movement between the exterior wall and the roof, which is provided for by separating the roof from the wall at the base flashing level. Base flashing details for a high parapet are essentially similar but are a little more complex [33.1].

Membrane Base Flashing

Membrane base flashing provides additional reinforcement to the roof membrane and extends the waterproofing ability of the roof to the highest anticipated water level (8 in. minimum is recommended). Membrane flashing also provides additional waterproofing at the roof-wall junction, which, being under greater stress, is more prone to leakage than the field-of-roof.

Membrane base flashing must have adequate strength and pliability to withstand minor movement between the wall and the roof. Therefore, the most commonly used membrane base flashing material with a built-up roof is an SBS-modified bitumen sheet. An SBS

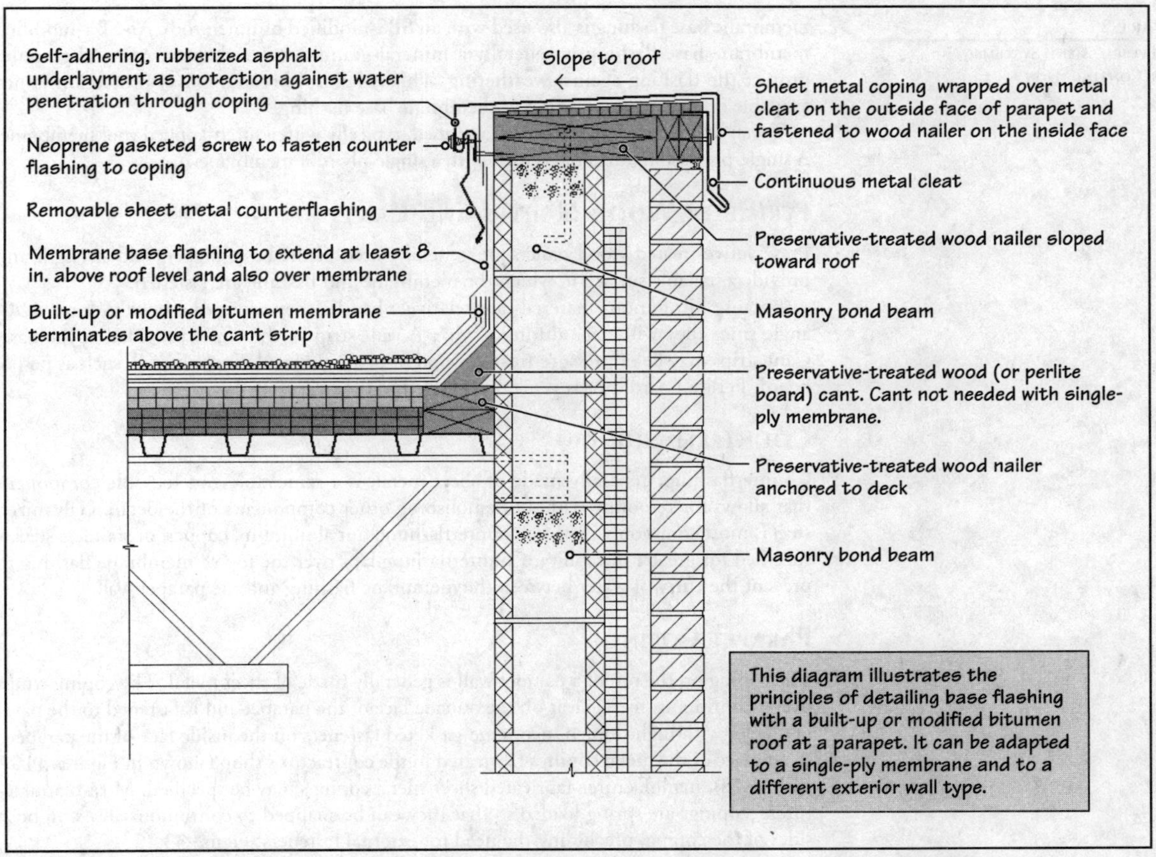

Self-adhering, rubberized asphalt underlayment as protection against water penetration through coping
Slope to roof
Sheet metal coping wrapped over metal cleat on the outside face of parapet and fastened to wood nailer on the inside face
Neoprene gasketed screw to fasten counter flashing to coping
Continuous metal cleat
Removable sheet metal counterflashing
Membrane base flashing to extend at least 8 in. above roof level and also over membrane
Preservative-treated wood nailer sloped toward roof
Built-up or modified bitumen membrane terminates above the cant strip
Masonry bond beam
Preservative-treated wood (or perlite board) cant. Cant not needed with single-ply membrane.
Preservative-treated wood nailer anchored to deck
Masonry bond beam
This diagram illustrates the principles of detailing base flashing with a built-up or modified bitumen roof at a parapet. It can be adapted to a single-ply membrane and to a different exterior wall type.

FIGURE 33.27 Base flashing and parapet coping details where the roof deck is supported by an exterior bearing wall.

Flexible insulation held within vapor retarder
Sheet metal coping wrapped over metal cleat on the outside face of parapet and fastened to wood nailer on the inside face
Slope to roof
Self-adhering, rubberized asphalt underlayment as protection against water penetration through coping
Neoprene gasketed screw to fasten reglet to coping
Continous metal cleat
Sheet metal reglet
Removable sheet metal counterflashing
Treated wood curb
Membrane base flashing
Multiply built-up roof membrane
Treated wood cant and nailer below
This diagram illustrates the principles of detailing base flashing with a built-up or modified bitumen roof at a parapet. It can be adapted to a single-ply membrane and to a different exterior wall type.

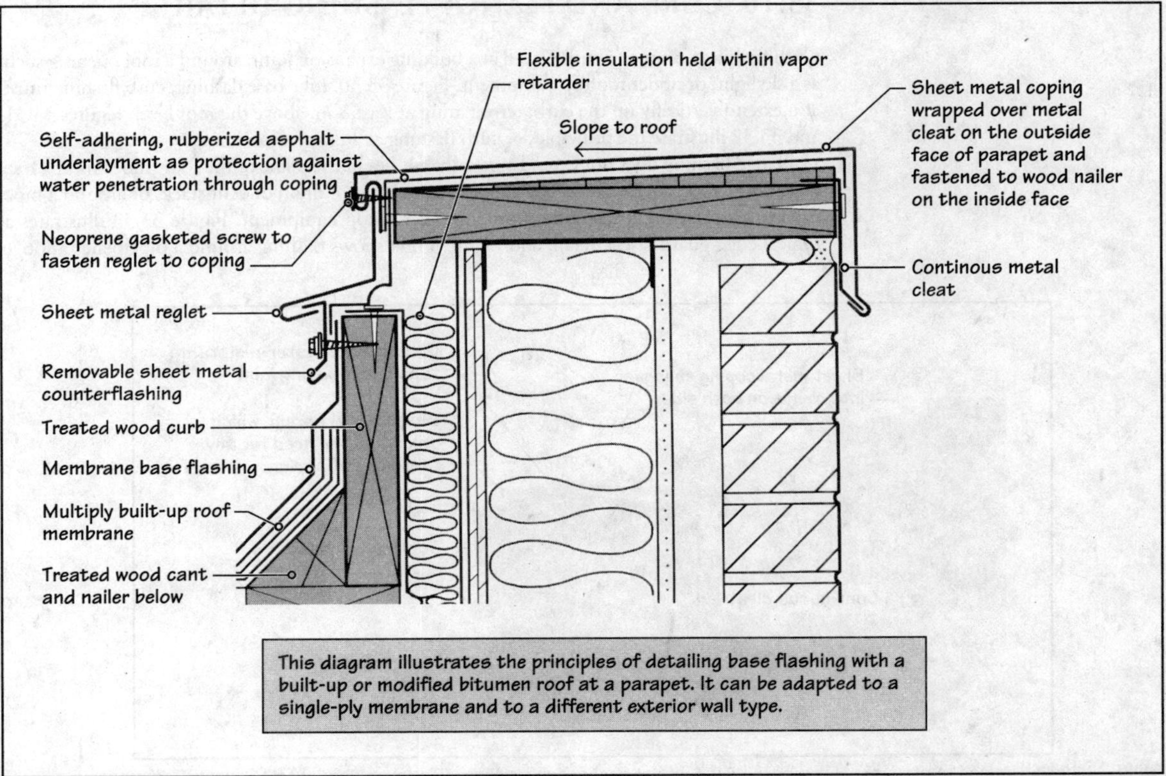

FIGURE 33.28 Base flashing and parapet coping details where the roof deck is supported by the structural frame and the exterior wall is nonbearing.

membrane base flashing is also used with an SBS-modified bitumen roof. An SBS-modified membrane base flashing is generally a mineral-granule-surfaced sheet. Mineral granules protect the flashing against weathering. This is helpful because gravel protection is not available on the vertical portion of membrane base flashing.

Torch-applied flashing sheets are also used, typically with a torch-applied roof membrane. A single-ply flashing material is used with a single-ply roof membrane.

PERIMETER WOOD NAILERS AND CANTS

Preservative-treated wood nailers are generally used at the base. Nailers add strength and provide a nailable surface to which the membrane and flashing are fastened.

Because bituminous materials are relatively brittle, they cannot be bent to form a 90° angle since they will crack during service. A cant strip is, therefore, provided at the base. Cant strips may be of pressure-treated wood or of a rigid insulating material, such as perlite board. Perlite-board cant is generally adhered to the nailer.

COUNTERFLASHING

Counterflashing, generally made of sheet metal, is a removable and reusable component that allows for reroofing without demolishing other components of the detail. Galvanized steel is most commonly used for counterflashing, but aluminum, copper, or stainless steel is specified for greater durability. Counterflashing laps over the top of membrane flashing to prevent the entry of water between the membrane flashing and the parapet wall.

PARAPET COPING

The coping on the top of a parapet wall is generally made of sheet metal. The coping wraps over a continuous metal cleat on the outside face of the parapet and is fastened to the parapet nailer with high-domed, neoprene-gasketed fasteners on the inside face of the parapet.

In place of the metal copings fabricated in the contractor's shop (shown in Figures 33.27 and 33.28), manufacturer-fabricated sheet-metal copings may be specified. Most manufacturers' copings are spring loaded so that they can be snapped to continuous cleats on both sides of the coping, precluding the need for external fasteners, Figure 33.29.

33.10 CURB AND FLANGE FLASHING DETAILS

Curb flashing occurs at a curb placed at a building expansion joint, around a roof opening such as a skylight, or under rooftop equipment, Figure 33.30. Like base flashing, curb flashing must also extend vertically up the curb, terminating at least 8 in. above the roof level. Figures 33.31 and 33.32 illustrate the principles of curb flashing at an expansion joint.

Flange flashing does not extend vertically but lies in the plane of the roof. It occurs at a free edge of the roof, such as at an edge guard, around a roof drain or vent stack, or around a pipe used for supporting a structural frame under rooftop equipment. Figure 33.33 illustrates a typical edge-guard fascia detail, and Figure 33.34 shows flashing around a pipe penetration.

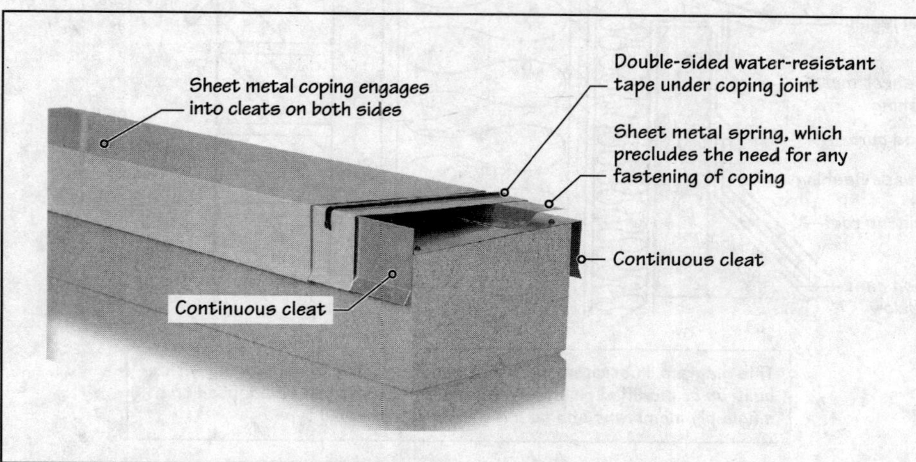

FIGURE 33.29 Snappped-in-place metal coping, which does not require any fasteners. (Photo courtesy of MM Systems Corporation.)

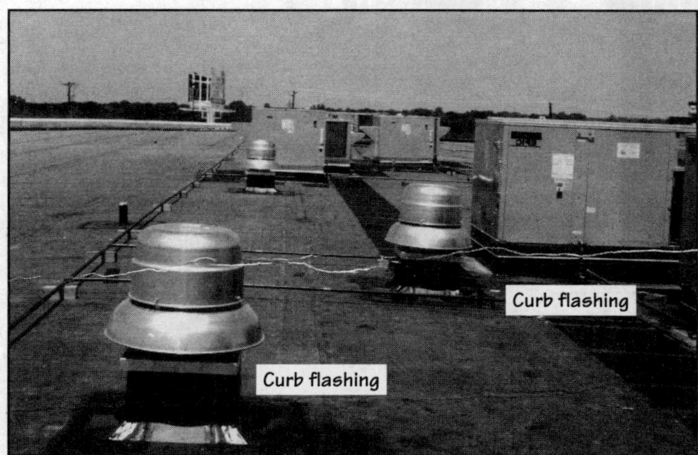

FIGURE 33.30 Examples of curb flashing.

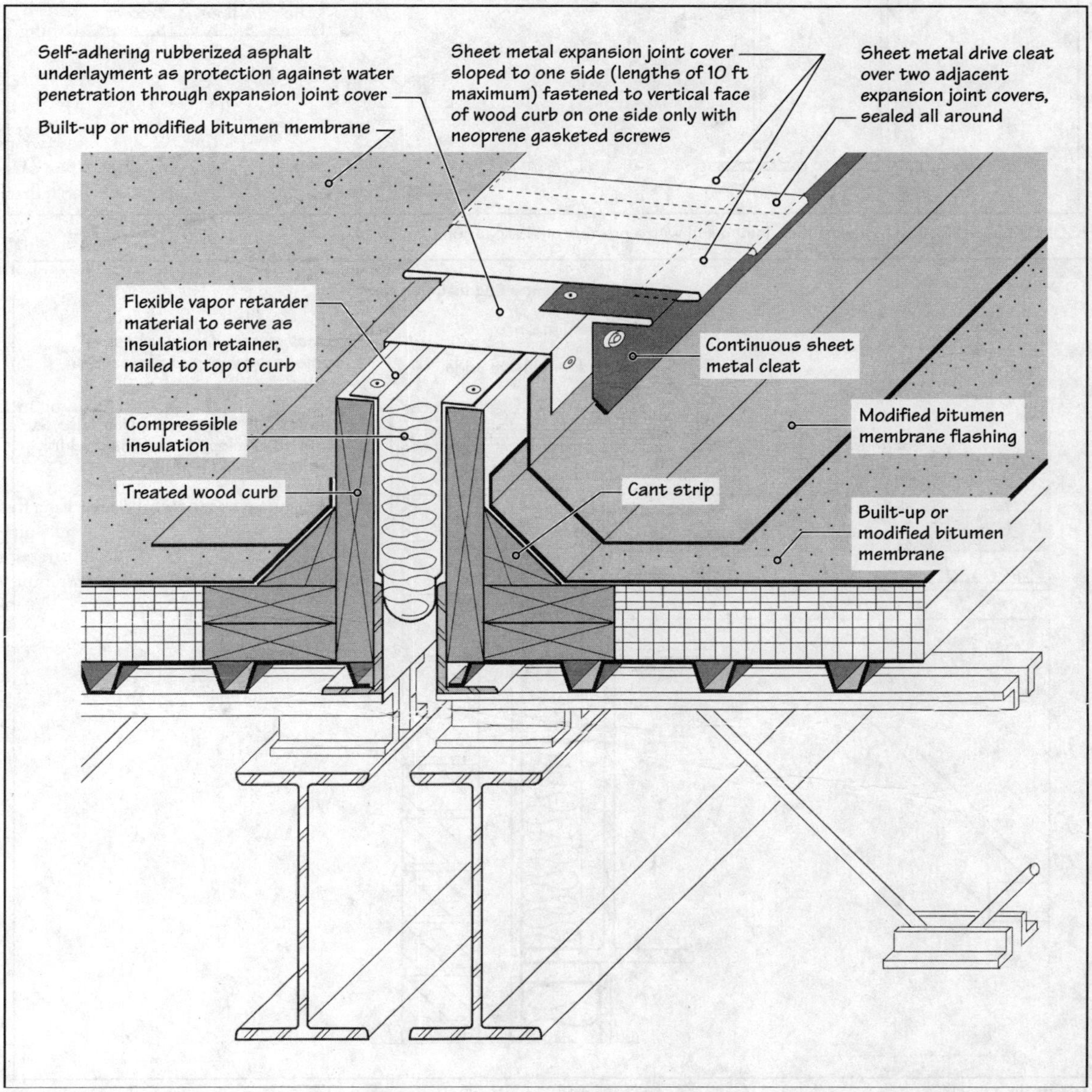

FIGURE 33.31 A typical roof expansion joint detail with a site-fabricated sheet metal expansion joint cover.

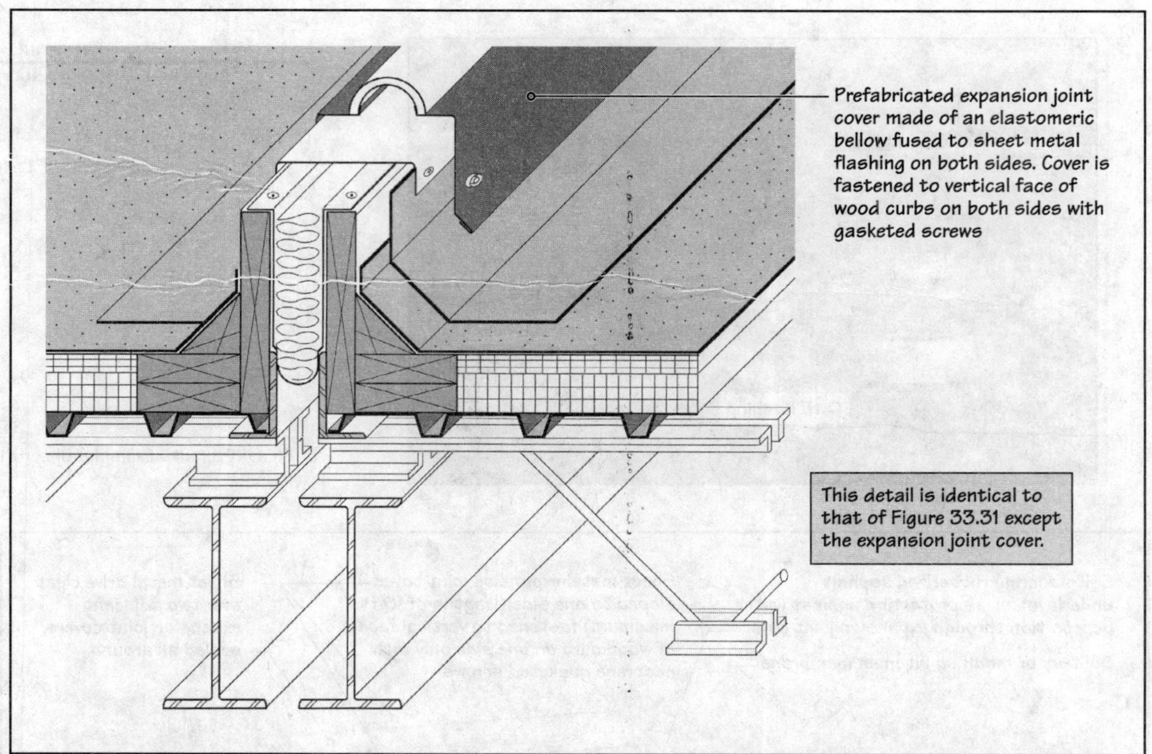

Prefabricated expansion joint cover made of an elastomeric bellow fused to sheet metal flashing on both sides. Cover is fastened to vertical face of wood curbs on both sides with gasketed screws
This detail is identical to that of Figure 33.31 except the expansion joint cover.

FIGURE 33.32 A typical roof expansion joint detail with a prefabricated expansion joint cover.

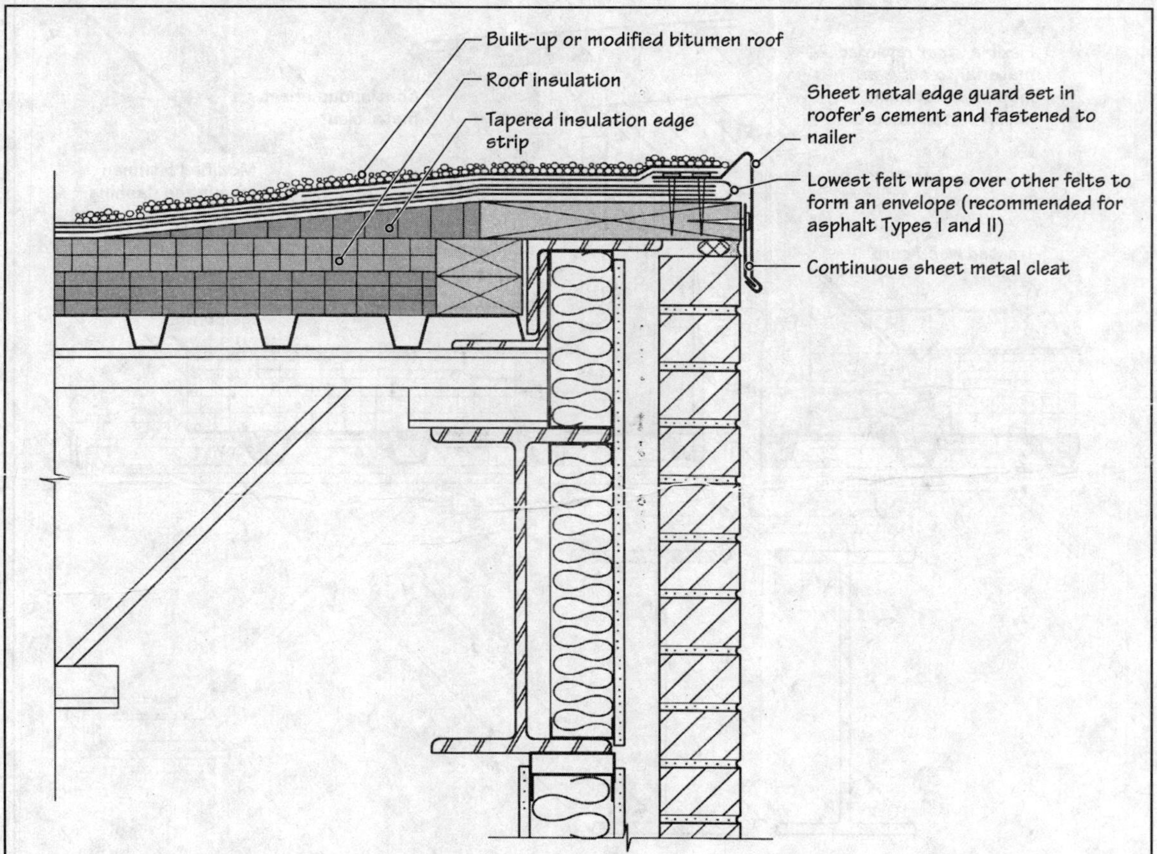

Built-up or modified bitumen roof
Roof insulation
Tapered insulation edge strip
Sheet metal edge guard set in roofer's cement and fastened to nailer
Lowest felt wraps over other felts to form an envelope (recommended for asphalt Types I and II)
Continuous sheet metal cleat

FIGURE 33.33 Flange flashing at an edge guard. This detail shows the use of a built-up roof membrane. Detail using a single-ply membrane is similar.

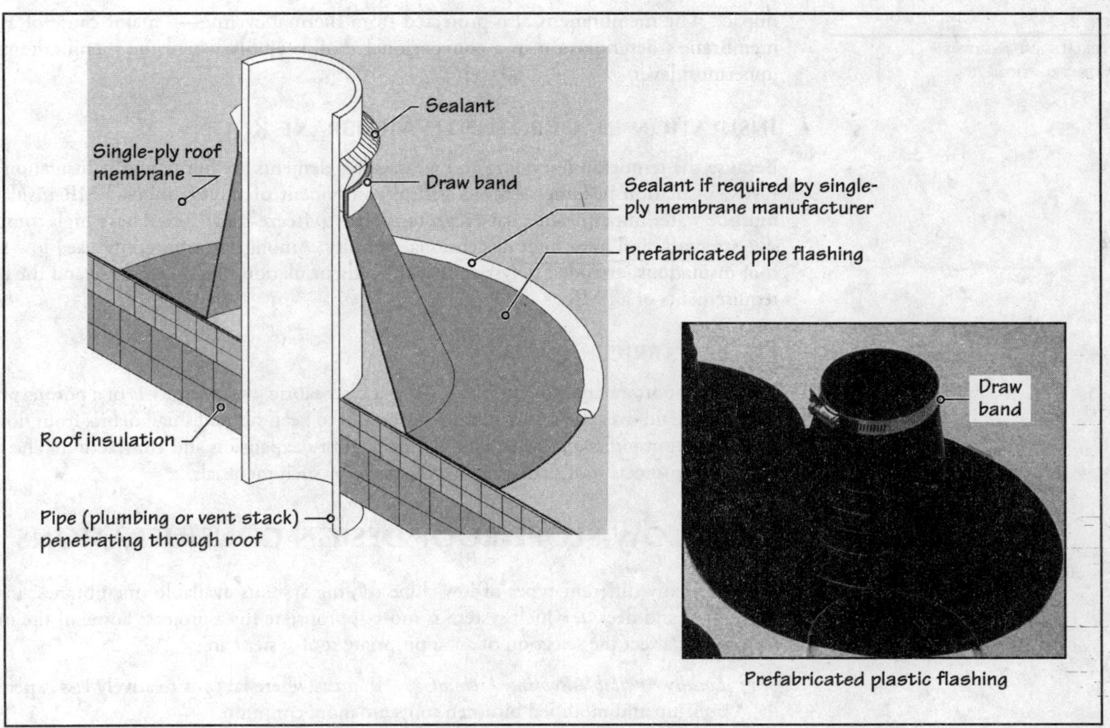

FIGURE 33.34 Flashing around a pipe penetration with a single-ply roof membrane.

33.11 PROTECTED MEMBRANE ROOF

In a protected membrane roof (PMR), the insulation is laid loose over the roof membrane, which is generally installed straight on the deck, Figure 33.35. In the case of a steel deck, a leveling board (typically a low-insulation board) is used under the membrane to provide a flat, rigid surface. The type of membrane and its installation are similar to those in a conventional roof assembly. Any one of the various membranes (built-up roof, modified bitumen, or single-ply membrane) can be used in a protected membrane roof.

After the installation of the membrane and the associated flashings at the roof perimeter and penetrations, the membrane is covered with insulation, which is then covered with ballast. The ballast may consist of river gravel or precast-concrete pavers. Pavers make the entire roof surface walkable, in addition to providing an easily maintainable roof. Pavers also make the roof look cleaner compared to gravel or crushed stone. Because the order of insulation and roof membrane is reversed in a PMR, it is also referred to as an *inverted roof.*

With the protection provided to it by the insulation and the ballast cover, the roof membrane is shielded from weathering elements, such as ultraviolet radiation and hail and traffic

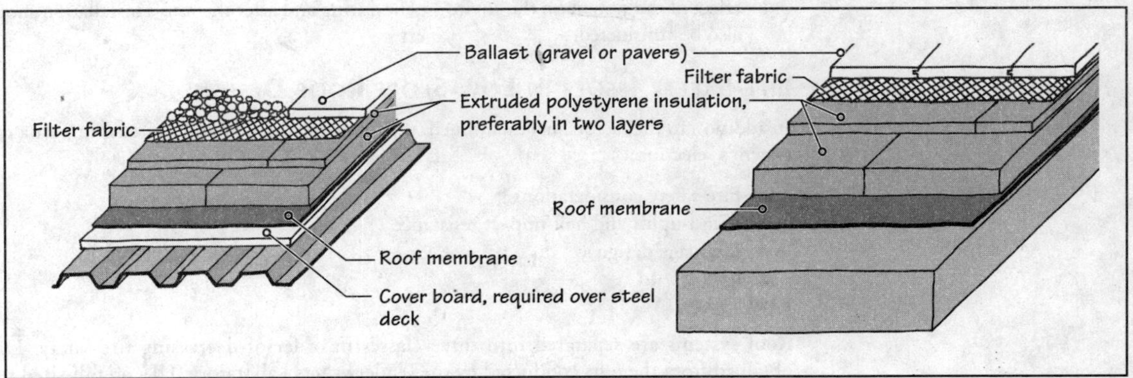

FIGURE 33.35 Components of a protected roof membrane.

impact. The membrane is also protected from thermal cycling—a major cause of a roof membrane's deterioration in a conventional roof assembly where the membrane is the uppermost layer.

Insulation in a Protected Membrane Roof

Because the temperature cycling and weathering elements are borne by the insulation in a PMR, insulation becomes a more critical component of the assembly. PMR insulation must be water impermeable, have high resistance to freeze-thaw cycles, have high compressive strength, and have high dimensional stability. Among the commonly used low-slope roof insulations, extruded polystyrene is the only insulation that meets these and the other requirements of a PMR.

Filter Fabric

Another important component in a PMR is a filter fabric, which consists of a porous plastic membrane laid over the insulation. Its purpose is to keep silt and small debris from flowing into insulation joints and restricting the insulation's expansion and contraction. The filter fabric also protects roof drains against blockage by such materials.

33.12 LOW-SLOPE-ROOF DESIGN CONSIDERATIONS

With so many different types of low-slope roofing systems available (membranes, insulation types, and decks), which system is most appropriate for a project? Some of the major factors that affect the selection of an appropriate roof system are:

- *Locally Available Roofing Technology* In areas where labor is relatively less expensive, built-up and modified bitumen roofs are more common.
- *System's Suitability to the Local Climate* In extremely cold climates, a mechanically fastened or ballasted single-ply system may be more appropriate than a fully adhered single-ply or hot-asphalt-applied membrane system because both asphalt and single-ply membrane adhesives require additional precautions when applied in a cold environment.

  In a high-wind region, fully adhered systems are more appropriate because the uplift stresses are distributed over the entire roof membrane instead of being concentrated at the fasteners, as in mechanically fastened systems. A ballasted system should be avoided, and a mechanically fastened system should be detailed with care.
- *Curved Roofs and Roofs with Large Slopes* Such roofs do not lend themselves to the application of hot bitumen. A fully adhered single-ply membrane is generally more appropriate for such projects.
- *Chemical Environment on the Roof* Roof membranes on industrial facilities and commercial kitchens must be examined to ensure the membrane's compatibility with the environment on the roof. Roofing-system manufacturers generally provide the required information.
- *Roof Warranty* Everything else being the same, the manufacturer's warranty may impact the system's selection.
- *Long-Term Cost and Sustainability Aspects of the System* Although the initial cost of the system is important, its long-term life-cycle cost should not be overlooked. The effect of the system on the environment during and after the roof's installation should also be considered.

Life-Safety Issues in Low-Slope Roof Design

In addition to the factors just mentioned, the following three life-safety issues impact a roof system's selection:

- Fire-safety considerations
- Wind-uplift and hail-impact resistance
- Drainage design

Fire Safety

Roof systems are separated into three classes in order of decreasing fire safety. Data obtained from the tests conducted by the Underwriters Laboratory (UL) on full-sized roof assembly specimens are considered to be most authoritative. UL classification is based on the exposure of a roof to an exterior fire. (The fire-resistance rating of construction assemblies [see Chapter 7] are based on fire assumed to originate from the interior of the building.) The three UL roof classes are:

- Class A roof
- Class B roof
- Class C roof

Class A is the most fire-resistive and Class C is the least fire-resistive roof. Generally, the fire rating of a given roof system decreases with increasing slope because the exterior fire spreads more easily on a roof with a larger slope. Additionally, gravel-covered, ballast-covered, or foil-covered roofs are generally more fire resistive than those without such covers.

WIND-UPLIFT AND HAIL-IMPACT RESISTANCE

A low-slope roof system must be designed to withstand twice the maximum expected design wind uplift on a roof. In other words, if the maximum expected wind uplift on a roof is 75 psf, the roof system must be able to withstand at least 150 psf, which implies a minimum safety factor of 2.0.

The Factory Mutual Research Corporation (FMRC) is the most widely recognized establishment for rating roof assemblies for wind-uplift and hail-impact resistance. FMRC data, obtained from tests conducted on full-sized roof-assembly specimens, are provided in the following format:

Class 1-60 and Class 1-SH

In this classification, 1 implies that the assembly has passed the calorimeter test—a pass-fail test that measures the amount of combustible content in the assembly. An assembly that exceeds the combustible content limit fails the test and is not tested for any other performance (wind or hail). An assembly that passes the calorimeter test is called an *approved assembly*.

Thus, Class 1-60 implies an approved assembly with wind-uplift resistance of 60 psf. Such an assembly can be used on a roof designed to withstand a maximum wind uplift of 30 psf. Similarly, 1-240 means an assembly can be used where the design wind uplift on roof is 120 psf.

Class 1-SH implies an approved assembly that can withstand severe hail impact. In terms of hail resistance, a roof assembly is rated either as 1-SH or 1-MH. A 1-MH rating implies an approved assembly that has medium hail-impact resistance.

Manufacturers of roof systems provide UL and FMRC ratings of their assemblies—important considerations in roof system's selection.

DRAINAGE DESIGN

After the appropriate roof system has been selected, the roof's drainage design is completed. Like fire and wind, drainage design is an important life-safety issue because a poor drainage design can lead to the collapse of a low-slope roof, which is generally an abrupt collapse.

FOCUS ON SUSTAINABILITY

Sustainability Considerations in Low-Slope Roofs

In response to sustainability concerns, manufacturers of roofing systems have introduced a number of ecofriendly initiatives.

Recycling of Roofing Materials and Reuse of Industrial Wastes

Recycling has been initiated throughout the roofing industry. Built-up roofing and modified bitumen roofing products are being recycled for use in asphalt pavements for roads and highways. Prior to this initiative, nearly 9 million tons of asphalt roofing ended up in landfills in the United States.

New roofing products are being developed from waste obtained from old single-ply roofs, automobile tires, shoes, and several other products. The industry is conscious of the need to make recycling and reuse economically viable.

Roof Surface Reflectivity and Global Warming—The Cool Roof Approach

Another effective means of increasing a roof's sustainability is to improve its energy efficiency by increasing the reflectivity of its surface, thus reducing its solar heat absorption. The commonly used black roofs absorb approximately 80% of the solar energy incident on the roof. If the roof surface is white, its solar energy absorption is reduced to about 20%, which can reduce the roof surface temperature by as much as 15°F to 20°F, reducing the peak cooling loads substantially during the summers.

(Continued)

FOCUS ON SUSTAINABILITY (Continued)

As one might expect, the use of a reflective roof does not significantly increase the heating load in winters for the following reasons:

- On an annual basis, the heat loss through a roof is typically much smaller than the heat gain. For example, when it is 30°F outside and 70°F inside, the temperature differential between the inside and the outside is 40°F. (It may even be less on a sunny day because the roof's surface temperature will be greater than 30°F.) In a warm climate, a dark roof's surface temperature may be 140°F or more, resulting in a temperature differential of 70°F.
- In the wintertime, the sun is much lower in the sky and solar radiation is less intense. Winter days are also shorter. Therefore, the sun's heating potential is lower in winter than in summer. (When passive solar heating is used in buildings in cold climates, we generally make use of sunshine through the windows.) Additionally, the winter days are shorter.
- There are more cloudy days and, in some regions, the roofs may be covered with snow for extended periods during the winter.

The most important advantage of reflective roofs is their contribution to moderating the urban heat island effect and, hence, global warming. LEED 2009 offers one credit when roof membranes with a solar reflectance index (SRI) of 0.78 or greater are used in low-slope roofs. (According to the Cool Roof Rating Council, the SRI of a "standard black" surface is zero and that of a "standard white" surface is 1.0)

A reflective roof's lower surface temperature also decreases the membrane's degradation, increasing the service life of the roof. Roofing membrane manufacturers are developing and promoting white roofs through white single-ply membranes or membranes coated with reflective surfaces. More and more cities now require the use of white roofs to obtain LEED certification and lower utility bills. White aggregate- or ballast-covered roofs generally do not qualify as white roofs because they lose their reflectivity fairly rapidly.

Landscaped (Green) Roof Approach

Landscaped roofs, also called *green roofs,* containing vegetation planted on a lightweight soil or another growth medium, have received considerable attention. Like cool roofs, landscaped roofs reduce energy consumption in the building and the urban heat island effect. Unlike cool roofs, however, they also absorb air pollution, control stormwater runoff by temporarily storing rainwater on the roof, and provide wildlife habitat. Their higher initial cost (because of the greater roof load, enhanced waterproofing requirements, and larger number of roof system components) and higher maintenance cost are deterrents against their widespread use.

PRINCIPLES IN PRACTICE

Shingling of Built-Up Roof Felts

The recommended application procedure of a built-up roof membrane involves the shingling of felts, Figure 1. Layer-by-layer application (i.e., phased application) of felts is discouraged.

In a shingled application, the upper felt overlaps the lower felt by a constant dimension. In a three-ply roof membrane, there are at least three felts under any roof point. Similarly, a four-ply roof must have four felts at every point.

FIGURE 1 Successive built-up roof felts are generally laid in shingle fashion, with the upper felt overlapping the lower felt by a constant dimension. The alternative, applying the felts layer by layer (i.e., in a phased manner), has several disadvantages, and is not used.

In a shingled built-up roof (BUR), the lap at the exposed end of a felt with respect to the lowest felt under that end, referred to as the *head lap,* is 2 in. as the industry standard. Under the head lap, there is one additional felt. Thus, there are four felts under a head lap in a three-ply membrane, Figure 2. The dimension EXP is called the *exposure* of the felt. EXP is related to the number of plies by the following relationship:

$$\text{EXP} = \frac{\text{felt width} - \text{head lap}}{\text{number of BUR plies}}$$

PRINCIPLES IN PRACTICE

Shingling of Built-Up Roof Felts (*Continued*)

Thus, with a felt width of 36 in., EXP (in a two-ply membrane) = 17 in. In a three-ply membrane, EXP = $11\frac{1}{3}$ in., and so on. Built-up roof felts are manufactured with two-, three-, and four-ply lines marked on them, as shown in Figure 3, to help roofers lay the felts correctly.

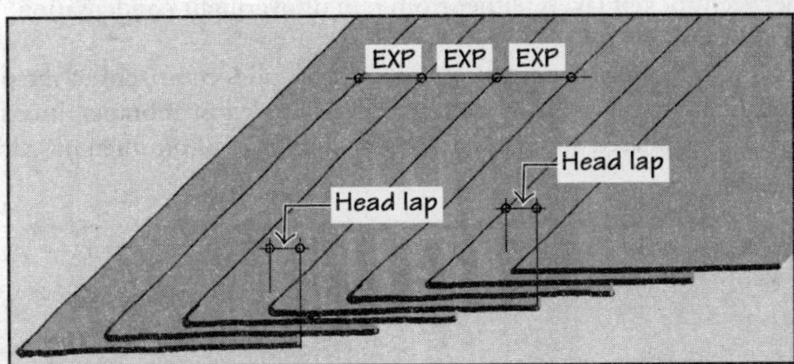

FIGURE 2 Shingling of felts in a built-up roof, with terms exposure (EXP) and head lap indicated. The industry standard for the head lap is 2 in. This sketch shows a three-ply built-up roof membrane.

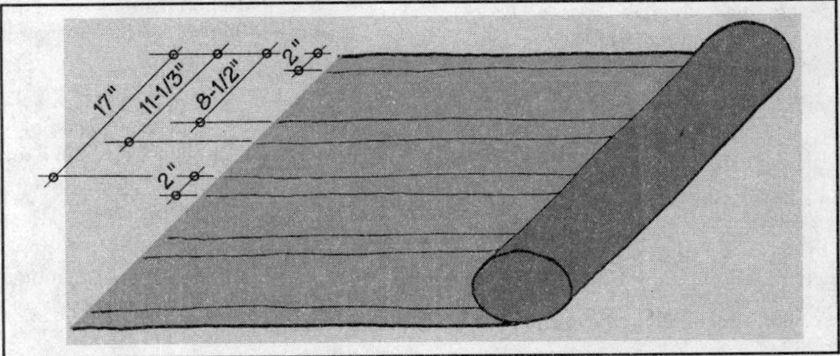

FIGURE 3 Factory-applied lines on a built-up roof felt. The spacings between lines on the left side of the felt are the same as on the right side.

To provide the necessary number of felts at the starting edge of the roof (usually the roof's lowest edge), the first few felts are not the full width of 36 in. The width of the starter felts is obtained simply by dividing the full width by the number of built-up roof plies.

Thus, in a four-ply membrane, there are three partial-width felts at the starting edge—9 in., 18 in., and 27 in. wide, Figure 4. In a three-ply membrane, there are two partial-width felts—12 in. and 24 in. wide. In a two-ply built-up roof membrane (not usually specified), there is only one partial width felt—18 in. wide.

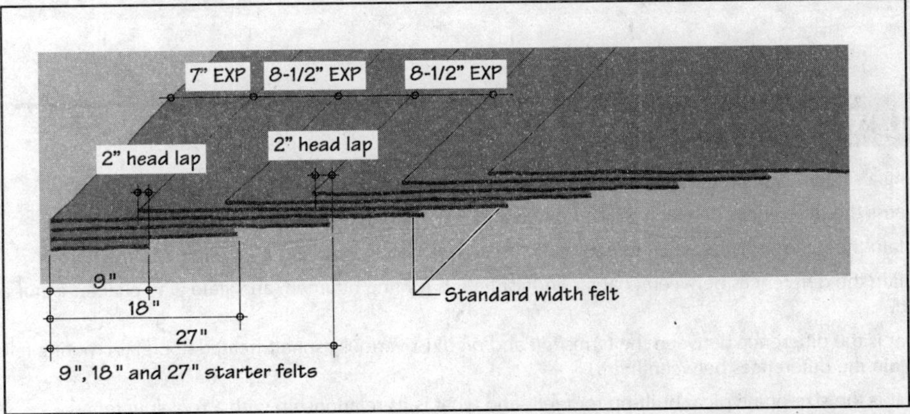

FIGURE 4 The use of starter felts at the starting edge of a built-up roof membrane. This illustration shows a four-ply membrane.

(Continued)

PRINCIPLES IN PRACTICE

Shingling of Built-Up Roof Felts (*Continued*)

ADVANTAGES OF SHINGLING

Shingling of felts results in completed sections of the roof at the end of each day. Thus, the membrane remains protected from interply moisture penetration during interruptions in roofing operations. If the felts were laid in a phased manner, moisture might be trapped between the felt layers, either from rain or overnight condensation. The entrapment of debris in a phased application is also of concern.

Another advantage of shingling is that it provides better adhesion of plies compared to the phased application. This is due to the slower cooling of bitumen in a shingled membrane. In a shingled membrane, three to four layers of bitumen are applied within a short time span. In a phased application, the previous layer of bitumen has virtually cooled off when the subsequent layer is applied.

PRACTICE QUIZ

Each question has only one correct answer. Select the choice that best answers the question.

28. The flashing around a pipe penetrating through the roof is
a. base flashing. **b.** curb flashing.
c. flange flashing. **d.** none of the above.

29. The flashing around a roof drain is
a. base flashing. **b.** curb flashing.
c. flange flashing. **d.** none of the above.

30. A cant strip is not required in a base flashing detail with a single-ply roof membrane.
a. True **b.** False

31. Counterflashing is generally
a. made of PVC and is permanently anchored to the deck or parapet wall.
b. made of PVC and is a removable item.
c. made of sheet metal and is permanently anchored to the deck or parapet wall.
d. made of sheet metal and is a removable item.
e. none of the above.

32. In a protected membrane roof, the insulation is
a. below the deck.
b. between the deck and the membrane.
c. above the membrane and fully adhered to it.
d. above the membrane but loose laid over it.
e. not used.

33. FM 1-120 represents
a. the fire rating of a roof assembly.
b. the wind-uplift resistance of a roof assembly.
c. the hail-impact resistance of a roof assembly.
e. none of the above.

34. UL Class B represents:
a. the fire rating of a roof assembly.
b. the wind-uplift resistance of a roof assembly.
c. the hail-impact resistance of a roof assembly.
e. none of the above.

35. FM MH represents
a. the fire rating of a roof assembly.
b. the wind-uplift resistance of roof assembly.
c. the hail-impact resistance of roof assembly.
d. none of the above.

36. The exposure (EX) of felts in a four-ply built-up roof is
a. 2 in. **b.** 4 in.
c. 6 in. **d.** 8 in.
e. none of the above.

37. The head lap in a four-ply built-up roof is
a. 2 in. **b.** 4 in.
c. 5 in. **d.** 6 in.
e. none of the above.

REVIEW QUESTIONS

1. Using a sketch, explain the essential components of a low-slope roof.
2. Discuss the differences between SBS-modified and APP-modified bitumen roof membranes.
3. Explain the differences between an inorganic (fiberglass) roofing felt and an organic roofing felt.
4. Explain the differences between coal tar and asphalt as roofing bitumen, and state why coal tar is not commonly used.
5. What is the difference between the terms *felt* and *ply*? List various asphalt-treated fiberglass roofing felts and briefly explain the differences between them.
6. What is the size of a typical built-up roof felt, and what is its relationship with a roof square?
7. Explain the difference between thermosetting and thermoplastic roof membranes, and give at least one example of each. How does the difference between the two types influence the sealing of lap seams between adjacent membranes?

8. Using sketches, explain the anatomies of the following roofs. Show how the insulation and roof membrane are attached.
   a. Steel deck, rigid-board insulation, and built-up roof membrane
   b. Reinforced-concrete deck, rigid-board insulation, and SBS-modified bitumen membrane
   c. Steel deck, foamed-concrete insulation, and built-up roof membrane
   d. Concrete deck, rigid-board insulation, and loose-laid, ballasted single-ply membrane
9. Explain what a cool roof is, and state its advantages and limitations.
10. List and briefly explain the life-safety issues in roofing design.
11. With the help of a sketch, describe a protected membrane roof.

CHAPTER 34

Roofing–II (Steep Roofs)

CHAPTER OUTLINE

As a continuation of the previous chapter, in which low-slope roofs were discussed, this chapter deals with steep roofs—roofs that have a slope of greater than or equal to 3:12. It begins with an introduction to the components of a steep roof and then discusses the types of steep roofs commonly used. Of the various types of steep roofs, asphalt shingle roofs are most commonly used. Therefore, asphalt shingle roofs are covered in greater detail than the other roof types.

As mentioned in Chapter 33, a roof type that is used for both low-slope and steep roofs is the architectural sheet-metal roof. Architectural sheet-metal roofs are becoming increasingly popular because of their greater durability and the availability of a vast array of nonfading colors. Architectural sheet-metal roofs are covered at the end of the chapter.

34.1 STEEP-ROOF FUNDAMENTALS

The basic components of a steep roof are illustrated in Figure 34.1:

- Roof shingles
- Roof underlayment
- Ice dam protection membrane, where needed
- Roof deck
- Flashings

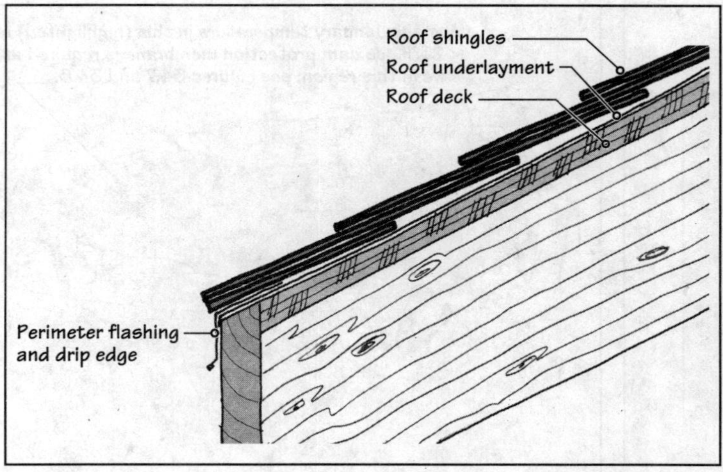

FIGURE 34.1 Components of a steep roof. (Insulation in the attic under a steep roof and the attic's ventilation may also be considered as components of a steep roof; see Chapter 5 and Section 6.8.)

ROOF SHINGLES

Roof shingles can be of various types. The more commonly used shingle varieties are

- Asphalt shingles
- Slate shingles
- Concrete and clay tiles
- Wood shingles
- Metal shingles

As stated earlier, asphalt shingles (also called *composition shingles*) have the largest steep-roof market share because of their low cost and acceptable life span. They are available in different colors and textures and have aesthetic appeal. With shadow lines added to their surface, some varieties can mimic a wood shingle roof. Another advantage of asphalt shingles is their light weight, which reduces labor and transportation costs and also the cost of the structural components of the roof.

Slate shingles have one of the longest recorded histories of use. They have been used in many buildings of historic importance in Europe and the United States. Because slate is a hard, dense natural stone, slate roofs last a long time. They perform quite well when subjected to repetitive freeze-thaw cycles. Well constructed and detailed, slate roofs can last for 75 years or more. Obviously, they are expensive. Imitation slates are also available, but their performance records must be checked before specifying them.

Clay tiles create a handsome roof due to the availability of a large variety of tile shapes. Like slate roofs, clay tile roofs are quite durable. However, it is important to understand that there can be significant differences in the quality of tiles, which can affect the durability of a clay tile roof.

Concrete tile roofs, though fairly durable, are generally less so than clay tile roofs. There can also be problems with long-term fading of colors of concrete tiles.

Wood shingles and shakes, once extensively used in the United States, have a decreasing market share today because of their fire hazard. Although fire-retardant-treated wood shingles are available, their long-term performance history is uncertain. A large variety of metal shingles are used in contemporary steep roofs. Like wood shingles, their market share is also small.

ROOF UNDERLAYMENT

In a properly designed and constructed steep roof, rainwater should normally not get under the shingles. However, a water-resistant layer, called a *roof underlayment,* is required under the shingles throughout the roof as a second line of protection against water leakage. The requirements for the type of roof underlayment varies with the type of shingles and the roof slope. (Wood shingles do not require underlayment in some situations.)

A valley requires an additional roofing layer, referred to as a *valley underlayment.* A valley is more vulnerable to water leakage than the field of the roof because its slope is smaller than that of the roof itself. For example, the slope of a valley created by two intersecting planes, each with a slope of 4:12, is only 2.8:12—a 30% reduction in slope, Figure 34.2.

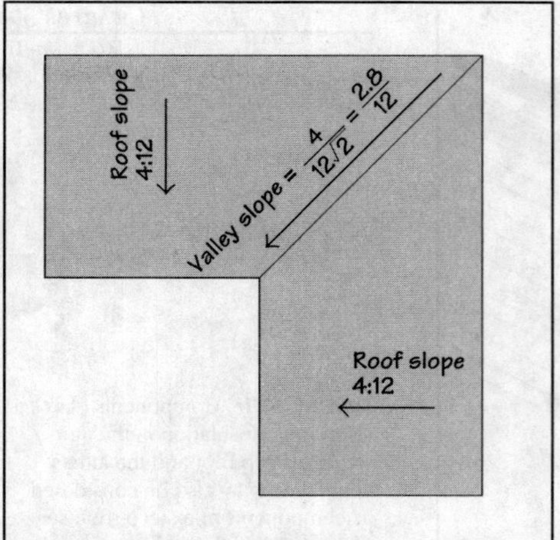

FIGURE 34.2 The roof slope in a valley is much lower than that in the rest of the roof. Therefore, a valley requires additional (valley) underlayment; see Figure 34.14.

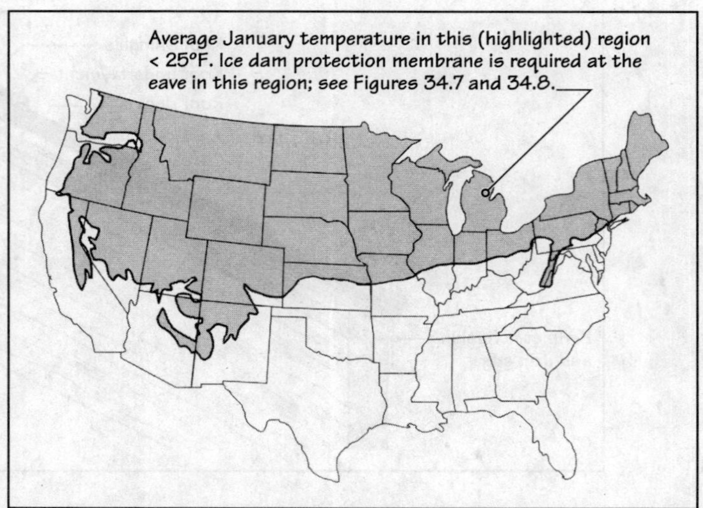

FIGURE 34.3 Approximate areas of the United States where the use of an ice dam protection membrane at the eave is recommended (see [33.1] in Chapter 33). The local building code should be consulted for precise information.

A significant head of water can be present in a long, flat valley. In addition to a smaller slope, the water running through the valley experiences more turbulence because it arrives there from two opposite directions.

Ice Dam Protection Membrane

Another special type of underlayment is called an *ice dam protection membrane*. It is required for use at the eaves in climates where there is a potential for the formation of ice at an overhanging eave.

The formation of an ice dam can make water back up under the shingles (see Figure 6.11). Ice dam protection is recommended in regions where the average daily temperature in January is less than 25°F, Figure 34.3. However, many roofers and roofing consultants recommend the provision of an ice dam protection membrane at the eaves in all locations.

Roof Deck

The deck, being a structural component of the roof, is its most important element. In a contemporary steep roof, the deck generally consists of plywood or oriented strandboard panels. A spaced deck, consisting of wood lath (typically 1 × 4 or 1 × 6 wood members spaced according to the length of the shingles), may be used with some steep-roof types, but its use is limited today.

Flashings

As in low-slope roofs, flashings are additional waterproofing components used to create a watertight seal at terminations and transitions in the roof. Critical areas for flashings are valleys, eaves, rakes, chimneys, and other penetrations.

Flashings around roof penetrations, such as a chimney or a pipe, are particularly critical. The water is not only interrupted by a penetration but is redirected and compressed on a smaller area of the roof as it flows around a penetration.

34.2 ASPHALT SHINGLES AND ROOF UNDERLAYMENT

Asphalt shingles are manufactured in strips measuring approximately 36 in. × 12 in., with up to five cutouts. The cutouts separate a shingle strip into tabs, making the strip appear to be composed of several smaller shingles, which improves the shingle's appearance.

Asphalt shingles vary in weight from about 200 lb to nearly 400 lb per roof square. The heavier varieties, which consist of two or more layers of shingles laminated together, are generally more durable and aesthetically more pleasing.

FIGURE 34.4 Asphalt shingles with color variations in their granule surface that give shadow lines to mimic wood or slate roofing.

Each shingle strip is made from a fiberglass mat, which is saturated and coated with asphalt and surfaced with mineral granules. The granules give the desired surface color to a shingle, which may vary from white, gray, and black to various shades of red, brown, and green, Figure 34.4.

Three of the most commonly used asphalt shingle profiles are shown in Figure 34.5. Each shingle is coated with a line of self-sealing adhesive, which helps to adhere that shingle with the overlying shingle. The adhesion is activated by the sun's heat after the shingles have been installed on the roof. The adhesive line increases the wind resistance of shingles. Of the three self-sealing shingle varieties shown in Figure 34.5, the three-tab shingle is the most commonly specified.

ROOF UNDERLAYMENT AND ICE DAM PROTECTION FOR A STANDARD-SLOPE ASPHALT SHINGLE ROOF

The recommended roof slope for asphalt shingles is between 4:12 and 21:12. This is referred to as the *standard slope* for asphalt shingles. A slope lower than the standard slope may also be used, but it requires special treatment, Figure 34.6.

The recommended roof underlayment for a standard-slope roof is No. 15 or heavier asphalt-treated organic felt. The underlayment is applied with 2-in. laps at the edges of the roll and 6-in. laps at the ends of the roll, Figure 34.7. It is nominally nailed or stapled to

NOTE

Asphalt-Treated Organic Felts as Underlayment

As stated in Chapter 33, asphalt-treated organic felts are used as steep-roof underlayment. They are available in two types:

- No. 15 felt
- No. 30 felt

When first developed, No. 15 felt weighed 15 lb per roof square and No. 30 felt weighed 30 lb per roof square. Currently, as per the ASTM D226 Standard, No. 15 felt and No. 30 felt weigh 11.5 lb and 26 lb per roof square, respectively.

Because of its greater thickness and weight, a No. 30 felt is more durable and performs better than a No. 15 felt.

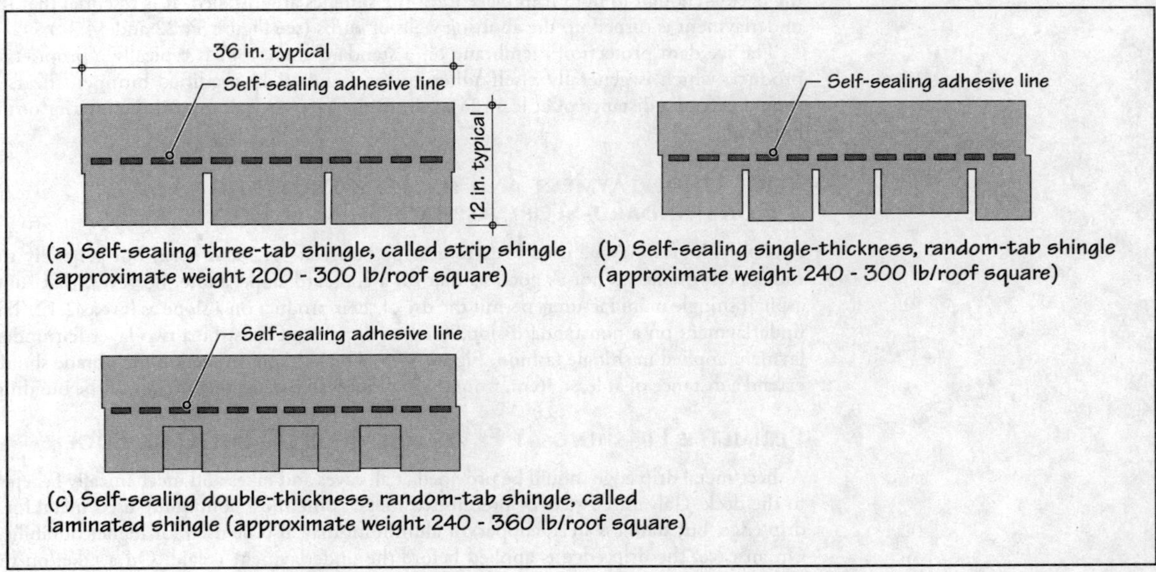

FIGURE 34.5 Commonly used asphalt shingle profiles.

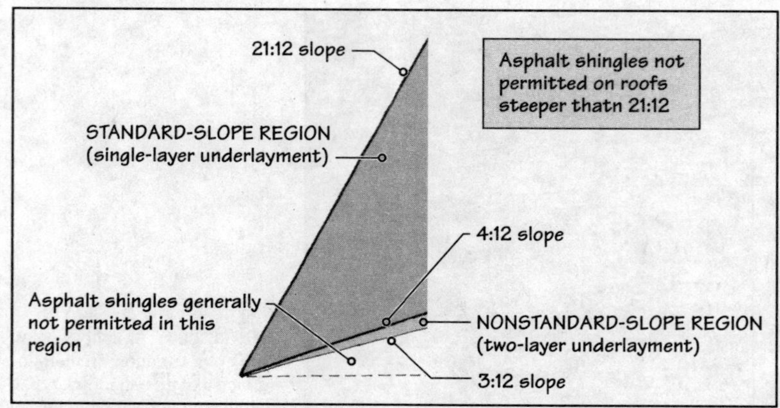

FIGURE 34.6 Slope and underlayment requirements for asphalt shingle roofs.

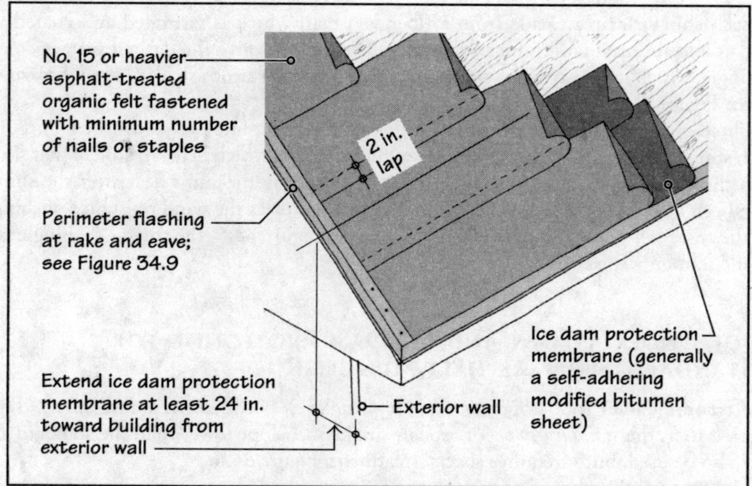

FIGURE 34.7 Roof underlayment and ice dam protection membrane on a standard-slope (≥ 4:12) asphalt shingle roof.

the deck—enough to hold it in place until the shingles are installed. It is essential that the underlayment is turned up the abutting walls or curbs (see Figure 34.22 and 34.24).

The ice dam protection membrane on a standard-slope roof is typically a proprietary product, which is generally a self-adhering (peel-and-stick) modified bitumen sheet. It should extend a distance of at least 24 in. from the exterior wall toward the interior of the building.

Roof Underlayment and Ice Dam Protection for a Nonstandard-Slope Asphalt Shingle Roof

Asphalt shingles may be used on a roof slope between 3:12 and 4:12, but generally the roof's performance is not as good as that on a standard slope. Note, however, that some asphalt shingle manufacturers permit the use of their product on a slope as low as 2:12. The underlayment on a nonstandard-slope (or low-slope) roof must be a two-layer felt underlayment applied in shingle fashion, Figure 34.8. The ice dam protection membrane should extend a distance of at least 36 in. from the exterior wall toward the interior of the building.

Perimeter Flashing at Eaves and Rakes—The Drip Edge

A sheet-metal drip edge should be provided at all eaves and rakes and mechanically fastened to the deck. Galvanized steel or prefinished metal is the most commonly used metal for a drip edge, but stainless steel, copper, or aluminum may also be used for higher durability. On an eave, the drip edge is applied before the underlayment is laid. On a rake, on the other hand, it is applied over the underlayment, Figure 34.9.

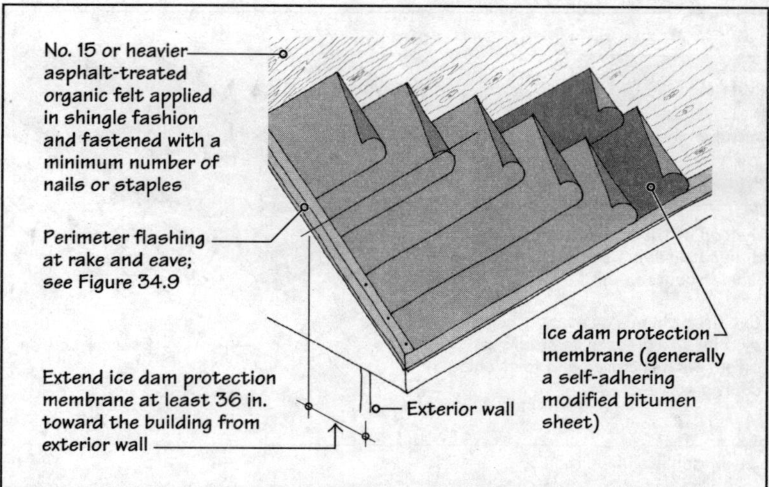

FIGURE 34.8 Roof underlayment and ice dam protection membrane on a nonstandard-slope asphalt shingle roof (slope between 3:12 and 4:12).

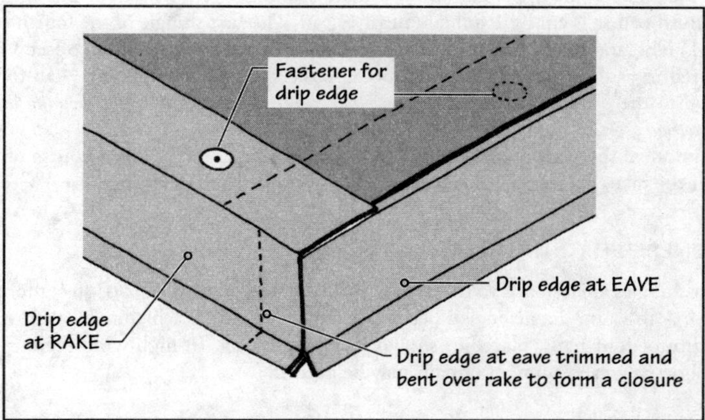

FIGURE 34.9 Drip edge at the eave and rake of an asphalt shingle roof.

34.3 INSTALLATION OF ASPHALT SHINGLES

After the roof underlayment and ice dam protection membrane have been installed, the roof is ready to receive the shingles. The procedures used to install various types of asphalt shingles (see Figure 34.5) is almost identical. Because three-tab shingles are most commonly used, the installation procedure described here refers to their installation.

The installation of three-tab shingles begins with the starter course at the eave, which is applied directly over the underlayment. A starter course consists of the standard three-tab shingles, whose tabs have been cut off, Figure 34.10(a). The remaining part of each shingle, which is nearly 7 in. wide, is nailed to the deck, with the self-sealing adhesive line placed along the outside edge of the eave, Figure 34.10(b).

The self-sealing adhesive line of the starter course bonds with the overlying shingles (of the first course), improving the wind-resistance performance of the roof. The starter course is laid from one end of the roof to the other. The first shingle in the starter course is cut by 3 in. along the shingle's length, so that the joints in the starter course are offset from the joints in the first course of shingles.

Next, the first course of shingles is laid. The first course of shingles fully covers the starter course, and each shingle is a full-length shingle. The second course of shingles is laid so that only the tabs of the first course are exposed, and so on. In other words, asphalt shingles have a double coverage—a two-shingle thickness under every point, except under a headlap, where there is a three-shingle thickness.

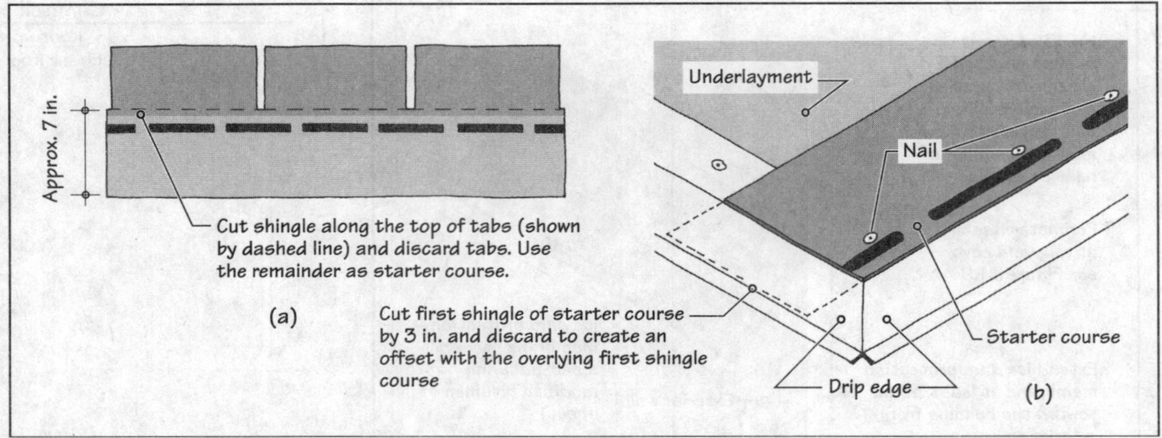

FIGURE 34.10 Installation of a starter course in an asphalt shingle roof.

The first shingle of the second course is cut in length by $\frac{1}{2}$ tab (nearly 6 in.). This cut breaks the joints in the upper and lower courses of shingles, Figure 34.11. The first shingle of the third course is cut by 1 full tab, nearly 12 in. The first shingle of the fourth course is cut by $1\frac{1}{2}$ tabs, and so on. The pattern of shingles generated by this half-tab offset procedure is referred to as the *6-in. offset method,* because the shingles are offset by $\frac{1}{2}$ tab (6 in.). In addition to the 6-in. offset method, the *3-in. offset method, 4-in. offset method,* and *5-in. offset method* are also used.

To improve the performance of the roof, it is desirable to include a course of bleeder strips at the rake, Figure 34.12. A bleeder strip is similar to the starter course.

Fasteners for Shingles

Asphalt shingles are fastened to the deck with 11- or 12-gauge galvanized steel roofing nails with $\frac{3}{8}$-in.- to $\frac{7}{16}$-in.-diameter nail heads (see Figure 14.34). Each shingle strip requires a minimum of four nails, placed as shown in Figure 34.13. In high-wind regions and on steep slopes, a greater number of nails may be needed.

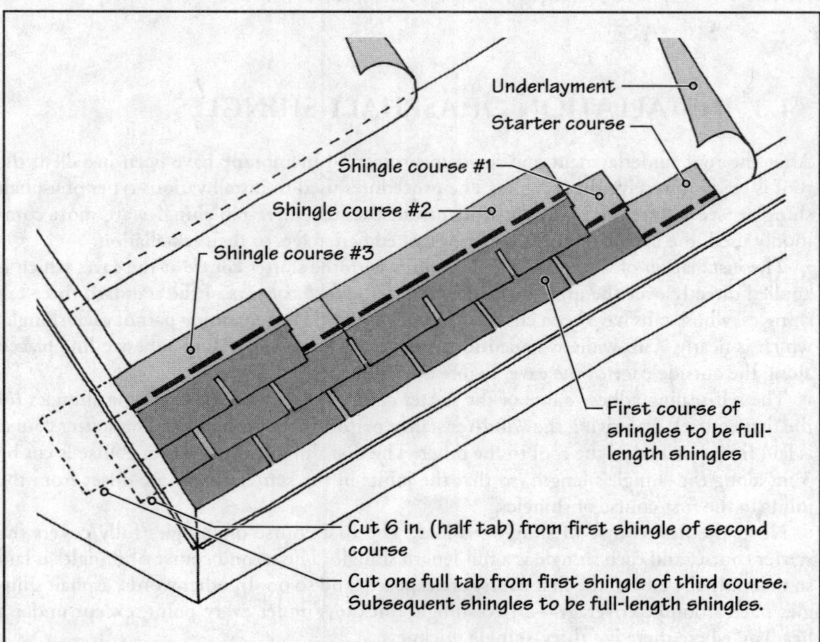

FIGURE 34.11 Layout of the starter course and the first few shingle courses. Note that asphalt shingles are laid with double coverage—two-shingle thickness under every point except under a headlap, where there is three-shingle thickness.

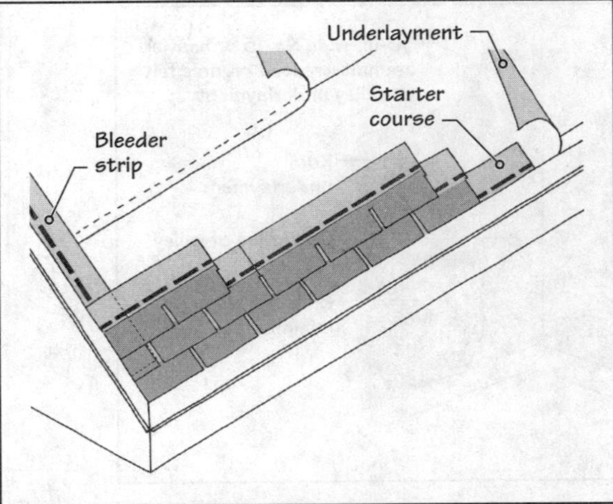

FIGURE 34.12 The use of a bleeder strip at the rake. A bleeder strip shingle is identical to the starter strip shingle at the eave.

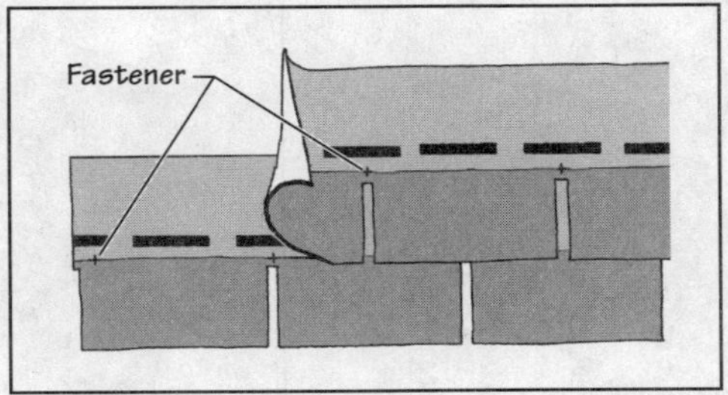

FIGURE 34.13 The standard fastening pattern requires four nails per shingle—one fastener at each end (nearly 1 in. from the end) and one nail between each tab. In high-wind regions, a larger number of nails are needed.

PRACTICE QUIZ

Each question has only one correct answer. Select the choice that best answers the question.

1. Which of the following steep-roof materials has the longest history of use?
   - a. Asphalt shingle
   - b. Clay tile
   - c. Concrete tile
   - d. Slate roofs
   - e. Wood shingle
2. Which of the following steep-roof materials is most commonly used in contemporary buildings?
   - a. Asphalt shingle
   - b. Clay tile
   - c. Concrete tile
   - d. Slate
   - e. Wood shingle
3. In a steep roof, the ice dam protection membrane is generally provided at the
   - a. ridge.
   - b. eave.
   - c. rake.
   - d. valley.
   - e. hip.
4. A double-layer roof underlayment is required in an asphalt shingle roof if the roof slope is
   - a. between 3:12 and 4:12.
   - b. between 3:12 and 5:12.
   - c. between 3:12 and 6:12.
   - d. between 3:12 and 9:12.
   - e. none of the above.
5. An asphalt shingle roof is generally not permitted if the roof slope exceeds
   - a. 12:12.
   - b. 15:12.
   - c. 18:12.
   - d. 21:12.
   - e. 24:12.
6. The roof underlayment commonly used in steep roofs is
   - a. No. 1 asphalt-treated fiberglass felt.
   - b. No. 15 asphalt-treated organic felt.
   - c. No. 30 asphalt-treated organic felt.
   - d. (a) and (b).
   - e. (b) and (c).
7. At the eave of an asphalt shingle roof,
   - a. a drip edge is installed before installing roof underlayment.
   - b. a drip edge is installed after installing roof underlayment.
   - c. a drip edge is not required.
8. On the rake of an asphalt shingle roof,
   - a. a drip edge is installed before installing roof underlayment.
   - b. a drip edge is installed after installing roof underlayment.
   - c. a drip edge is not required.
9. Asphalt shingles are generally laid in
   - a. single coverage with a 5-in. overlap between the top and bottom shingles.
   - b. double coverage with a 2-in. overlap between the top and bottom shingles.
   - c. double coverage with a 2-in. head lap between the top and bottom shingles.
   - d. triple coverage with no headlap.
   - e. none of the above.

34.4 VALLEY TREATMENT IN AN ASPHALT SHINGLE ROOF

A valley occurs where two pitched roofs meet and water converges. In asphalt shingle roofs, any one of the following three types of valley treatments may be used:

- Open valley
- Woven valley (also called *closed valley*)
- Closed-cut valley

Regardless of the type of valley, each valley requires

- Valley underlayment
- Valley flashing

Although the flashing material varies with the type of valley, the underlayment requirement for each valley is the same. A valley underlayment consists of a full-width (36-in.-wide) asphalt-treated organic felt (No. 15 or heavier) laid with the center of the roll laid on the

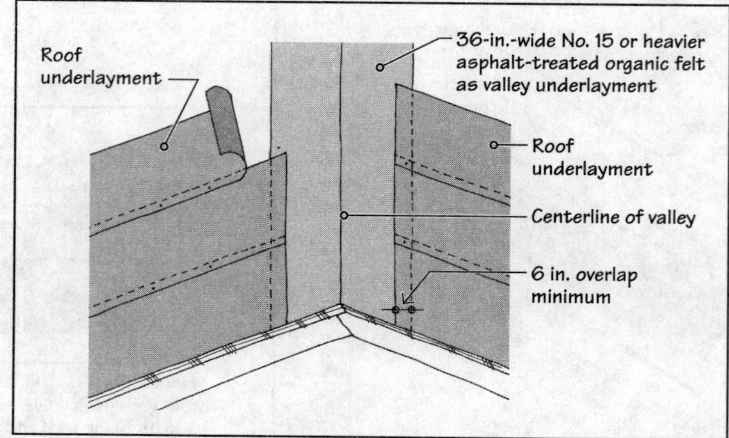

FIGURE 34.14 Valley underlayment and roof underlayment.

centerline of the valley, Figure 34.14. A self-adhering ice dam protection membrane below the valley underlayment is helpful because valleys are common sources of roof leaks.

Like the roof underlayment, the valley underlayment is also secured with a minimum number of nails. Roof underlayment from the field of the roof is cut so that it overlaps the valley underlayment by a minimum of 6 in. Some roofers prefer to use a self-adhering ice dam protection membrane as the valley underlayment.

OPEN VALLEY—ROLL ROOFING MATERIAL AS VALLEY FLASHING

In an open valley, the shingles are held away from the centerline of the valley so that the valley flashing is exposed to view. An open valley provides a relatively smooth and more rapid discharge of water.

The flashing in an open valley may consist of either roll roofing material or sheet metal. The construction of an open valley using roll roofing material as valley flashing is shown in Figure 34.15. Roll roofing material is an asphalt-treated organic felt with one smooth surface and the other surface covered with mineral granules.

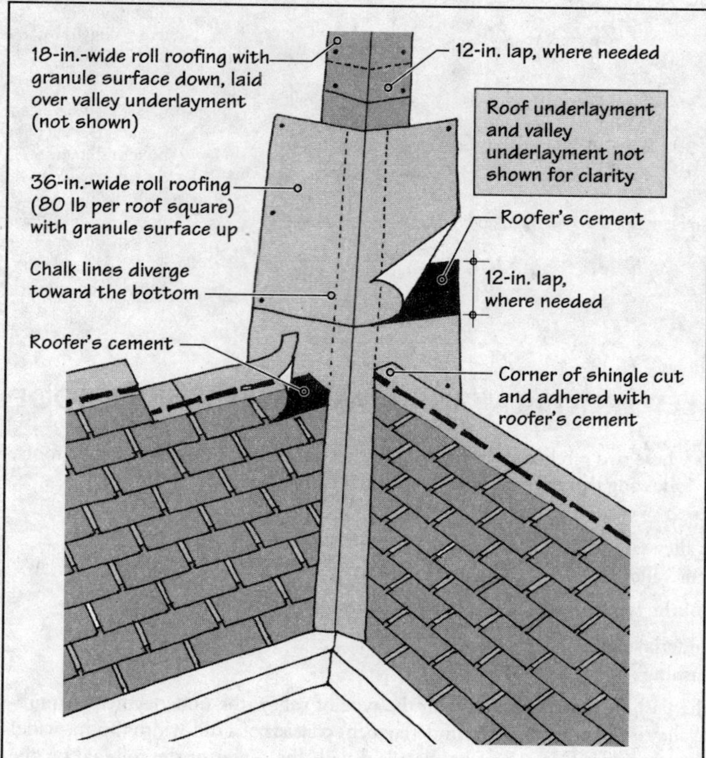

FIGURE 34.15 Open-valley flashing with roll roofing material used as valley flashing.

Roll roofing valley flashing consists of two layers of felt. The first layer, which is laid immediately over the valley underlayment, is an 18-in.-wide roll roofing felt with the mineral granule surface facing down. The second layer consists of the same material, but it is 36 in. wide. It is laid over the first layer with the mineral granule surface facing up. Both layers of roll roofing are secured with a minimum number of nails or staples.

The width of the exposed part of the valley flashing should increase at the rate of $\frac{1}{8}$ in. per ft toward the bottom of the valley. In practice, this is achieved by placing chalk lines over the valley flashing that converge upward. The minimum width of the exposed part of the valley flashing at the top is 6 in. Thus, if the valley is 24 ft long, the minimum width of the exposed part of the valley flashing at the bottom is 9 in.

The shingles are trimmed along the chalk lines. The end and corner of each shingle adjacent to the valley are cut and adhered with roofer's cement to prevent water from entering under the shingles.

OPEN VALLEY—SHEET-METAL VALLEY FLASHING

The construction of an open valley using sheet-metal flashing is similar to that using roll roofing flashing. The sheet metal generally consists of a minimum of 24-gauge galvanized steel and must be at least 24 in. wide. It is placed over the valley underlayment, Figure 34.16(a). It is typically not nailed, but secured with special metal clips that allow the sheet metal to expand and contract, Figure 34.16(b). Sheet-metal flashing should be used in lengths of no more than 12 ft to reduce their expansion and contraction.

The metal clips, which are located at 8 in. to 24 in. on center, grip the sheet-metal flashing at one end. The other end of the clip is nailed to the deck and then folded over to cover the nails. This protects the shingles against damage by any back-out of the nails.

Although the fastening of metal flashing with clips is generally recommended, the clips can telegraph through a lightweight shingle roof, adversely affecting the roof's appearance and performance. In such a case, sheet-metal flashing may be nailed at the roof's outer edges. The outer edges may then be stripped in with a self-adhering ice dam protection membrane.

However, nailing sheet-metal flashing can be problematic in regions with large daily or annual temperature variations. That is why an open-valley sheet-metal flashing is generally not recommended in such areas.

Sheet-metal flashing is generally profiled to form a W-shaped rib in the center. This reduces the crossover of water from one side of the roof to the other.

WOVEN VALLEY

In a woven valley, shingles from the two adjacent roofs weave into each other, covering the valley flashing, Figure 34.17. Because of the weaving of shingles, the thickness of roofing material in a woven valley is greater than in an open valley. Therefore, a woven valley is more durable. It can better withstand the traffic of roofers—who tend to walk in the valley because the valley's slope is lower than that of the adjoining areas of the roof.

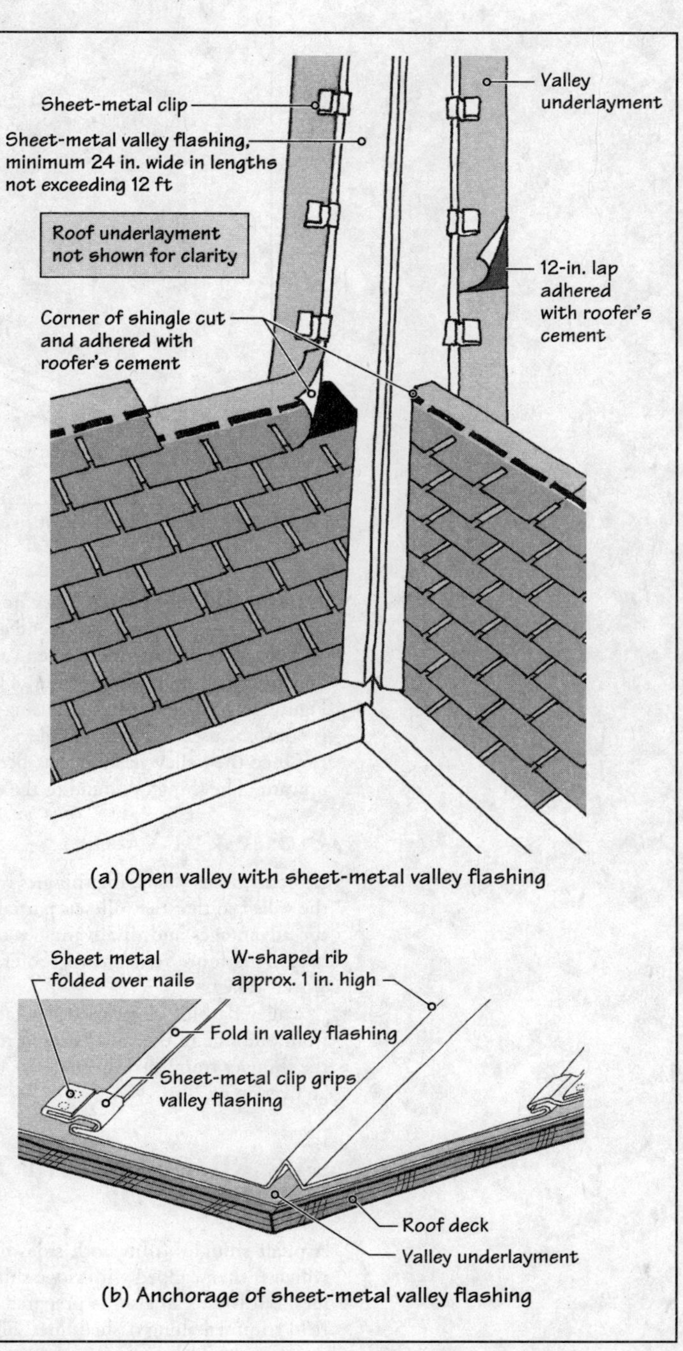

FIGURE 34.16 Open valley in an asphalt shingle roof with sheet-metal valley flashing.

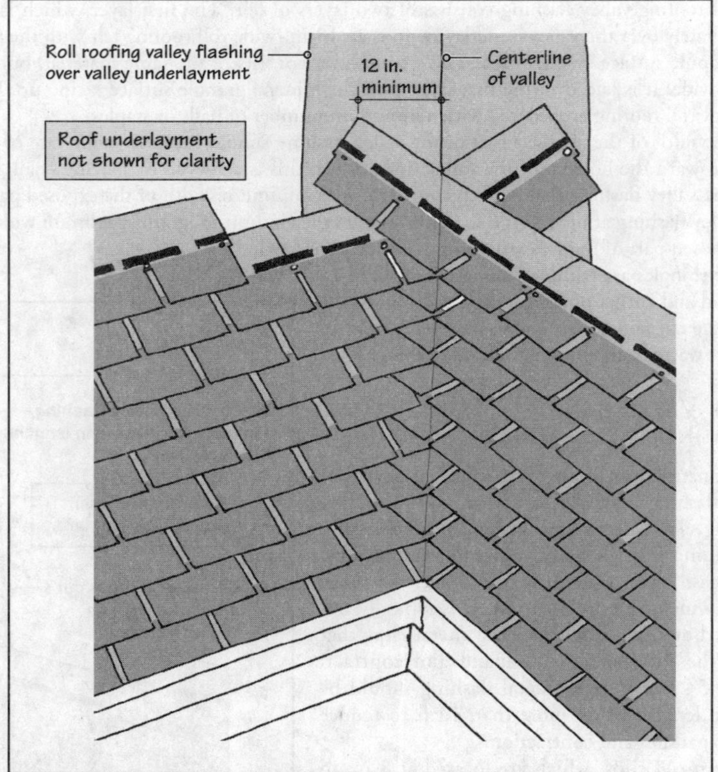

FIGURE 34.17 A woven valley in an asphalt shingle roof.

A woven valley, however, does not work well with heavier-weight laminated shingles because these shingles do not bend as much.

The valley flashing in a woven valley consists of roll roofing material and is identical to that provided in an open valley. However, the 18-in.-wide roll roofing sheet shown in Figure 34.15 may be omitted, using only the 36-in.-wide sheet. Sheet-metal valley flashing is not to be used in a woven valley.

Once the valley flashing has been installed, shingles are laid, starting from the eave upward. The shingles extend to the opposite side of the valley by at least 12 in.

Closed-Cut Valley

In a closed-cut valley, the shingles from one side of the valley are cut parallel to the line of the valley so that the valley is partially closed, Figure 34.18. A closed-cut valley combines the advantages and disadvantages of the other two valley types and is more commonly used. It is more resistant to roofers' traffic than an open valley but less than that of a woven valley.

Valley flashing in a closed-cut valley is identical to that in a woven valley. The shingles from one side of the valley extend to the opposite side by a minimum of 12 in. However, the shingles from the other side are trimmed parallel to the centerline of the valley, leaving an overlap of 2 in.

34.5 RIDGE AND HIP TREATMENT IN AN ASPHALT SHINGLE ROOF

Asphalt shingles from both sides of a ridge are butted against each other at the ridge, which is then capped with ridge shingles. Some manufacturers provide special ridge shingles, but these can also be prepared on site from standard shingles. To do so, a shingle is field trimmed along dashed lines, Figure 34.19. The dashed lines taper slightly away from the cutouts. The trimming converts a three-tab shingle into three nearly 12-in. × 12-in. shingles.

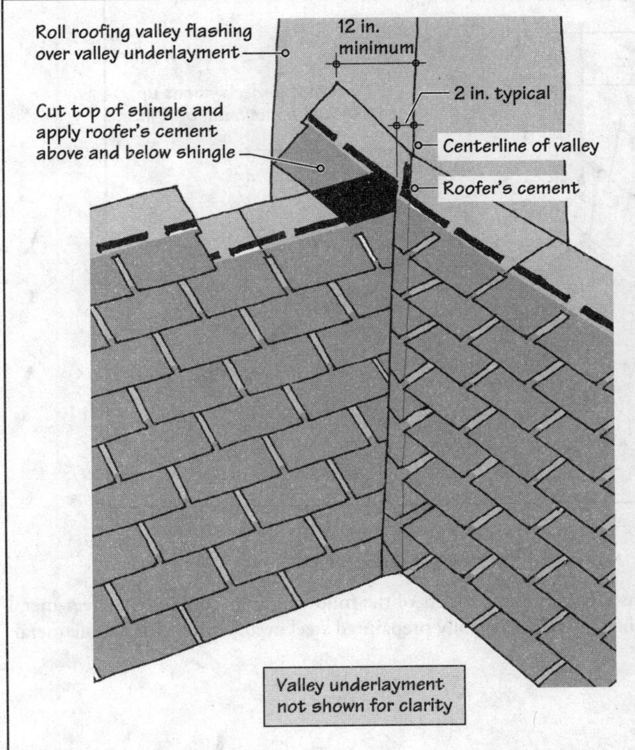

FIGURE 34.18 A closed-cut valley.

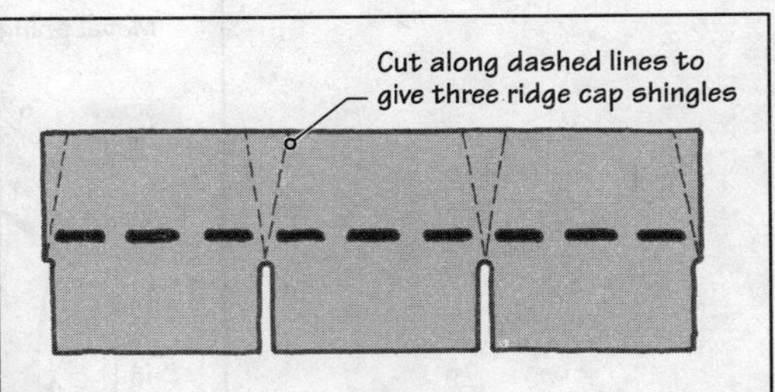

FIGURE 34.19 Creating ridge (or hip) caps from a standard three-tab shingle.

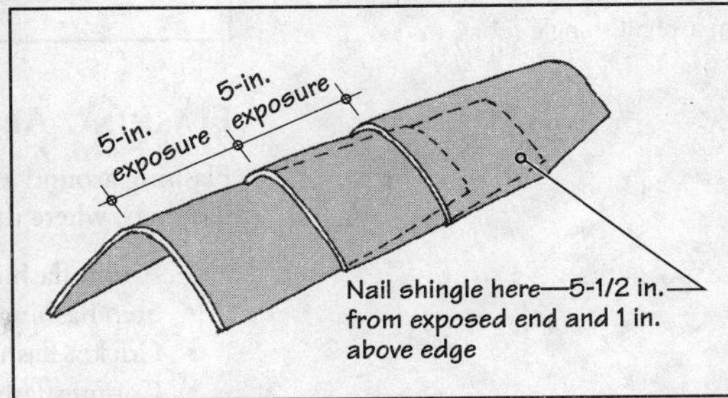

FIGURE 34.20 Layout and nailing of ridge cap shingles.

These smaller shingles are now bent and nailed over the ridge as caps, using a 5-in. exposure, Figure 34.20. In cold weather, the shingles may be warmed somewhat to increase their pliability, because fiberglass shingles are generally brittle.

The treatment of a hip is similar to that of a ridge. The hip is capped with the same shingles as a ridge, and the shingles are installed from the bottom of the hip to the top, Figure 34.21.

FIGURE 34.21 An asphalt shingle roof showing hip cap shingles and ridge cap shingles. Note that the two types of shingles are identical.

34.6 FLASHINGS IN AN ASPHALT SHINGLE ROOF

A sloping roof plane that abuts a vertical side wall is flashed with metal shingles placed over the end of each asphalt shingle course—a flashing system referred to as *step flashing.* Metal shingle flashing is made of galvanized or prepainted sheet steel.

The profile of a metal shingle flashing is an L shape, each leg at least 5 in., so that the shingle flashing extends 5 in. in both horizontal and vertical directions. The length of shingle flashing is 2 in. greater than the exposure of roof shingles. Thus, with the most commonly used three-tab roof shingles that have a 5-in. exposure, 7-in.-long metal shingle flashing is commonly used. With its 7-in. length and 5-in. exposure, each metal shingle flashing overlaps the underlying shingle flashing by 2 in., Figure 34.22.

Each shingle flashing piece is placed slightly upslope of the exposed edge of the roofing shingle that overlaps it. In other words, the horizontal surface of metal shingles is not visible, and each metal shingle flashing is completely covered by the overlying roof shingle. Each shingle flashing piece is secured to the deck by two nails placed at its top edge; the vertical leg of a shingle flashing is not fastened. Metal shingle flashing is counterflashed by siding that terminates 2 in. above the roof surface.

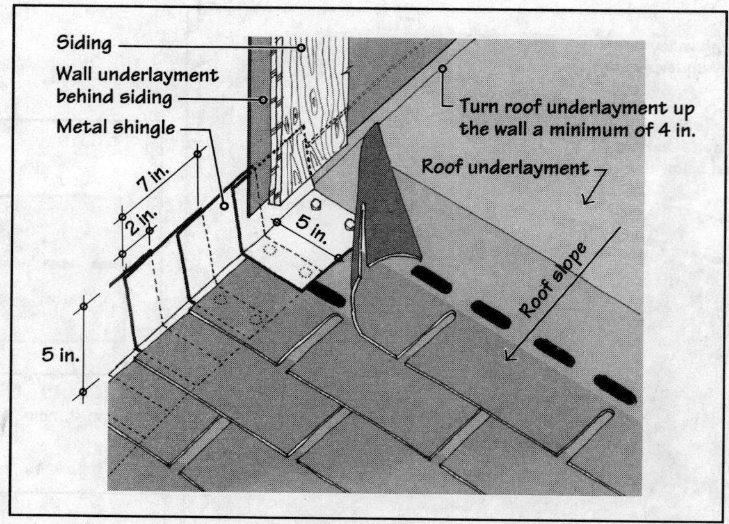

FIGURE 34.22 Step flashing against a wall that abuts the sloping side of an asphalt shingle roof.

Flashing Around a Masonry Chimney

Flashing around a masonry chimney consists of the following four-component sheet-metal flashings, where the sheet metal is generally prepainted steel or any other compatible metal:

- Apron flashing
- Step flashing
- Cricket flashing
- Counterflashing

Sheet-metal apron flashing is placed over the downslope side of chimney and tightly wrapped around it, Figure 34.23(a). This requires prefabricating the flashing to a profile, shown in Figure 34.23(b), and adhering it to the chimney with roofer's cement.

After the apron flashing has been applied, step flashing is installed on the side walls of the chimney in the same way shown in Figure 34.22. Note that the first step-flashing shingle wraps over the apron flashing at the corner by at least 2 in., Figure 34.24. Step flashing is also set with roofer's cement applied to chimney walls.

The upslope side (backside) of a chimney requires a cricket. Thus, before installing any metal flashing, roof shingles, or a roof underlayment around a chimney, a cricket is formed with plywood or oriented strandboard, Figure 34.25. Roof underlayment is subsequently installed on the deck and over the ice dam protection membrane. (Because a chimney is one of the primary sources of leaks on an asphalt shingle roof, a self-adhering ice dam protection membrane should preferably be installed all around the base of the chimney, with its edges turned up the chimney walls.)

The flashing over the cricket consists of a prefabricated sheet-metal flashing, Figure 34.26. The joints in cricket flashing are thoroughly soldered. The chimney is then ready to receive sheet-metal counterflashing, Figure 34.27.

FIGURE 34.23 Apron flashing on the front face of a chimney.

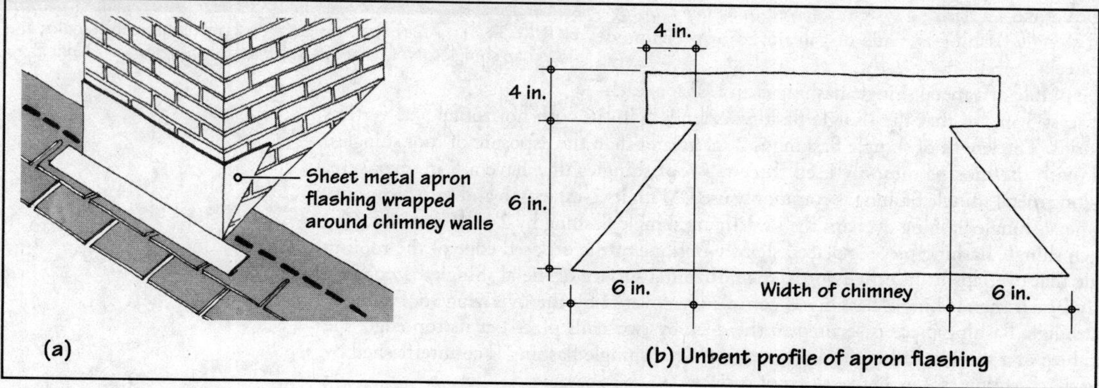

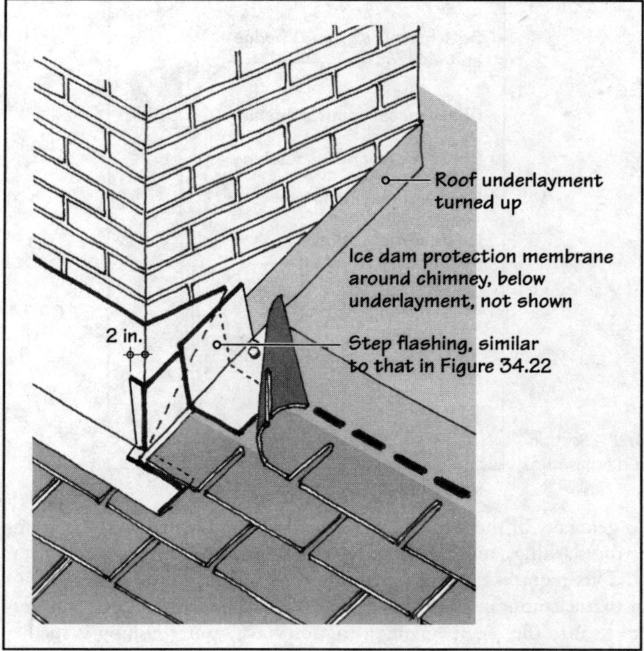

FIGURE 34.24 Step flashing on the side walls of a masonry chimney.

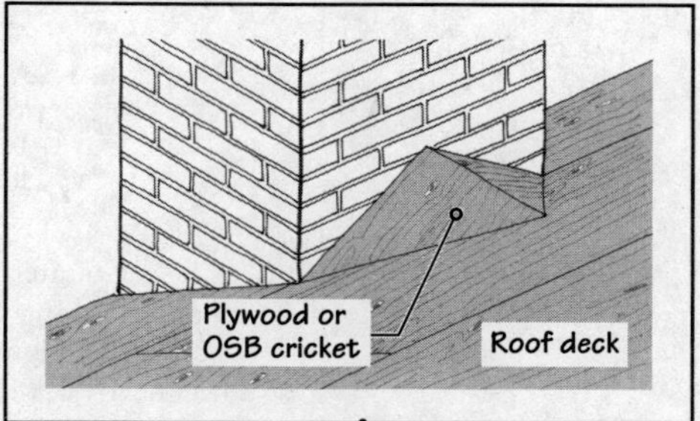

FIGURE 34.25 Formation of a plywood or OSB cricket on the upslope side (back) of a chimney wall.

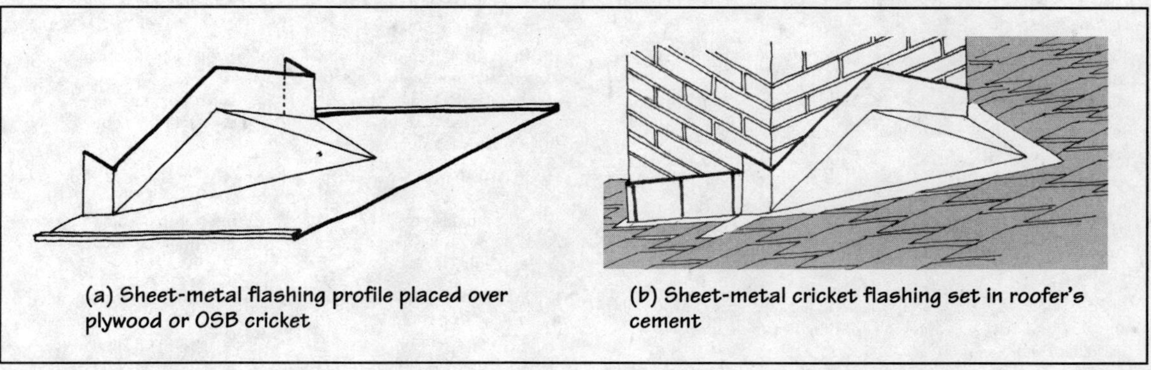

(a) Sheet-metal flashing profile placed over
plywood or OSB cricket
(b) Sheet-metal cricket flashing set in roofer's
cement

FIGURE 34.26 Sheet-metal cricket flashing.

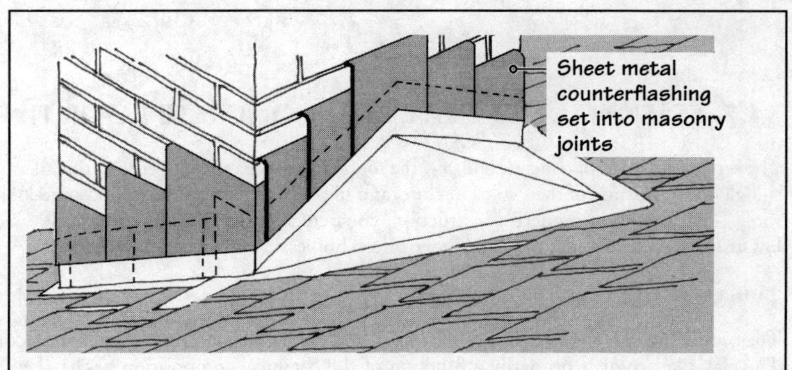

FIGURE 34.27 Counterflashing over a chimney wall, set into masonry joints as shown in Figure 34.28.

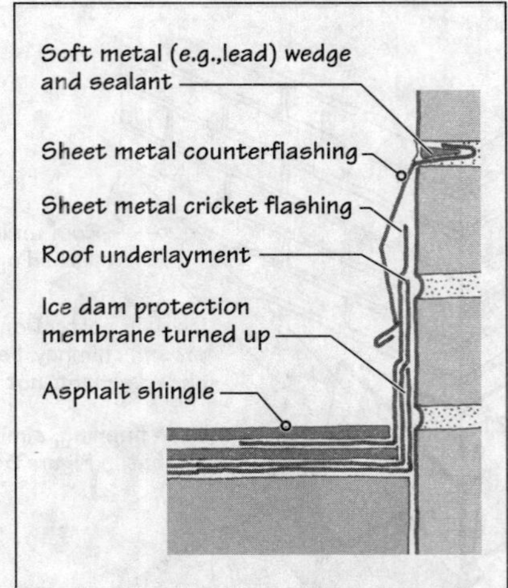

FIGURE 34.28 Section through counterflashing.

Counterflashing is generally of the same metal as the flashing. One edge of the counterflashing laps over the flashing, and the other edge is inset into the masonry mortar joints, Figure 34.28. This requires cutting the appropriate mortar joints to a depth of nearly 1.5 in.—a cut that is commonly referred to as a *raggle*. The inset edge of counterflashing is profiled to fit into the raggle, giving a friction fit. Counterflashing is further secured by driving a soft metal (e.g., lead) wedge into the raggle and finishing the joint with an elastomeric sealant.

PRACTICE **QUIZ**

Each question has only one correct answer. Select the choice that best answers the question.

10. In an asphalt shingle roof, roof underlayment is installed first, followed by valley underlayment.
a. True **b.** False

11. In an asphalt shingle roof, either valley underlayment or valley flashing is used.
a. True **b.** False

12. In an asphalt shingle roof, sheet-metal valley flashing is used in an
a. open valley. **b.** closed-cut valley.
c. woven valley. **d.** all of the above.
e. (a) and (b) only.

13. In an asphalt shingle roof, the ridge and hip are capped with shingles. The ridge cap and hip cap shingles can be used interchangeably.
a. True **b.** False

14. The junctions between the walls of a chimney and a steep roof need to be flashed. Which of the following flashings is required as chimney flashing?
a. Step flashing **b.** Apron flashing
c. Cricket flashing **d.** All of the above
e. Only (a) and (c)

15. In an asphalt shingle roof, which of the following flashings is interwoven with shingles?
a. Step flashing **b.** Apron flashing
c. Cricket flashing **d.** All of the above
e. Only (a) and (c)

16. In a steep roof, which of the following flashings is made from sheet metal?
a. Step flashing **b.** Apron flashing
c. Cricket flashing **d.** All of the above
e. Only (a) and (c)

17. Which of the following flashings in a steep roof has the shape of a small, steep roof?
a. Step flashing **b.** Apron flashing
c. Cricket flashing **d.** All of the above
e. Only (a) and (c)

34.7 ESSENTIALS OF CLAY AND CONCRETE ROOF TILES

The service life spans of concrete and clay tile roofs are significantly longer than that of asphalt shingle roofs. Because of their color, texture, and tile profile, they produce handsome-looking roofs. That is why in modern construction, concrete and clay tiles are extensively used in low-rise commercial structures and higher-priced homes and apartment buildings.

Types of Clay and Concrete Roof Tiles

Tiles come in a large variety of profiles, sizes, and colors, depending on the manufacturer. The clay tile's color is primarily a function of the chemical composition of the clay from which it is made. If any additional color-modifying agent is used, it is added to the clay before firing. Therefore, the color becomes integral to the tile. With its integral color, clay tile weathers only slightly with time, caused mainly by the pollutants in the air. Glaze-coated clay tiles are also available. The glaze is applied to the already-fired tiles, which are then refired to obtain a durable glaze.

The color of concrete tiles is a function of the color of the portland cement, aggregates, and any pigments used. The gradual erosion of the surface of the tile due to running water on the roof exposes the aggregates in a concrete tile, which can create a noticeable change in the tile's color over time. Initial and periodic sealing of the tiles' surface helps retain the color longer.

Concrete tiles can also be factory painted on their exposed surface. Several manufacturers make concrete tiles with the characteristic (brown) terra-cotta color painted on them to mimic clay tiles. Being an applied finish, the paint tends to fade over time.

CLAY AND CONCRETE ROOF TILE PROFILES

Because of the enormous variety in the profiles of concrete and clay tiles, it is impossible to discuss all available profiles. Although some profiles are used only with clay tiles and others only with concrete tiles, most tile profiles are essentially similar. Except for flat tile, all tile profiles provide for water channels that direct water down the roof. That is why roof tiles are generally laid in single coverage with a simple overlap of 3 in. between succeeding courses. This is unlike asphalt shingle roofs, which are laid in double coverage with a 2-in. headlap.

One of the oldest profiles, and a commonly used one in clay tiles, is the *mission tile,* which consists of an almost semicircular barrel profile. The tiles are laid alternately, one tile in a water-holding setting (pan) and the other tile in a water-shedding setting (cover), Figure 34.29. Both pan and cover tiles are provided with one hole to receive a fastener at the head of each tile.

Another commonly used clay tile profile is a derivative of the mission tile. Called an *S-shape tile,* a *one-piece barrel,* or an *S-shape mission tile,* it integrates the pan and cover of the mission tile into one unit, Figure 34.30. The S-shape tile generally comes with its two opposite ends chamfered to allow for easier drainage of water. The chamfer is generally of the same size as the overlap needed between the adjacent tiles to give a better, more exact fit between them. Each tile is provided with two fastener holes at its head.

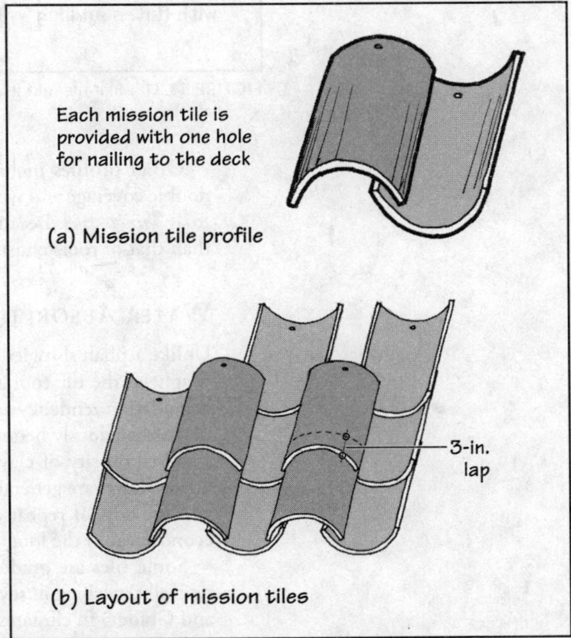

FIGURE 34.29 Mission tile profile and layout.

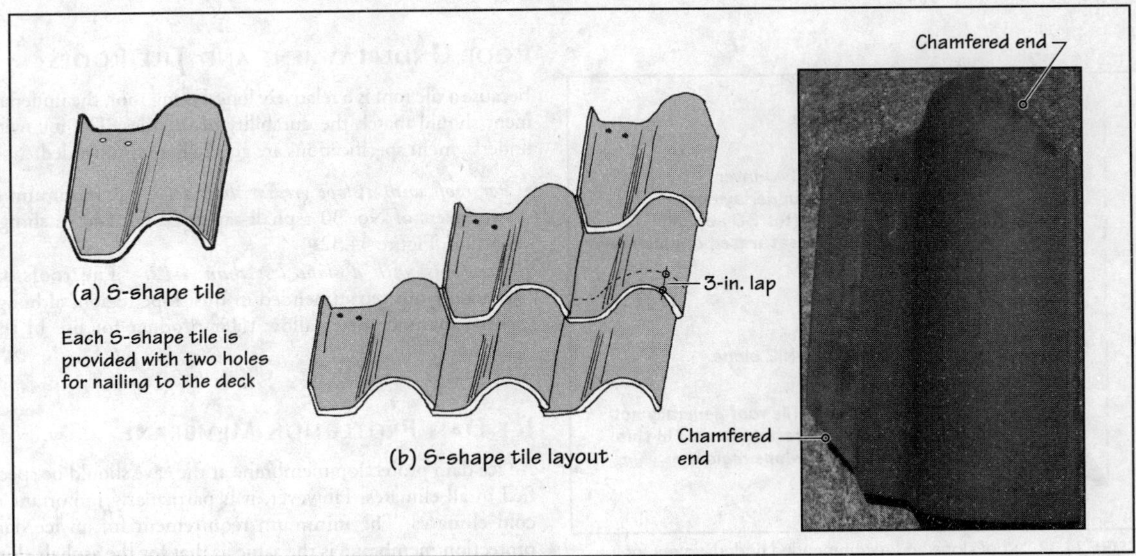

FIGURE 34.30 S-shape tile and its layout.

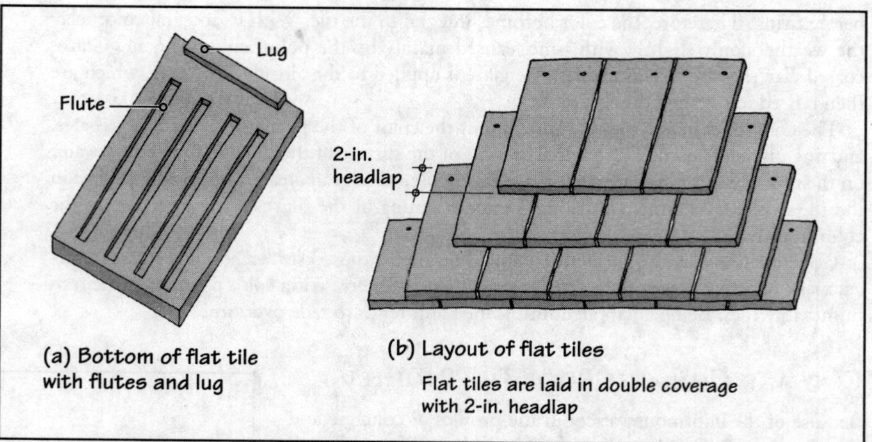

FIGURE 34.31 Flat tile and its layout.

Other profiles include a flat tile. Flat tiles are laid like asphalt shingles or slates—in double coverage and with a headlap of 2 in., Figure 34.31. That is why they are also referred to as *shingle tiles*. Because of its double coverage, the weight of a shingle tile roof is higher than that of roofs constructed of curved roof tiles.

WATER ABSORPTION AND STRENGTH OF TILES

Unlike asphalt shingles, concrete and clay tiles absorb water. Water absorption increases the weight of the tile roof and leads to freeze-thaw damage of tiles. It also increases the color-degradation tendency as the water-soluble pollutants enter the body of the tile. Less porous tiles are obviously better but are more expensive.

The porosity of clay tiles varies from 2% to 10%; that of concrete tiles, from 3% to 20%. Sealers are generally used to reduce the effect of porosity. The service life of the sealers and the cost of repeat applications must be considered in tile selection. It may be more economical in the long run to specify a denser, albeit more expensive, tile.

Some tiles are graded (Grades 1, 2, or 3) for their freeze-thaw resistance. Grade 1 is typically specified in severe freeze-thaw climates, Grade 2 in moderate freeze-thaw climates, and Grade 3 in climates with negligible freeze-thaw activity.

Porosity is not merely an index of the tile's water absorptivity; it also indicates the tile's strength. A more porous tile is weaker. Strength is an important selection criterion for concrete and clay tiles, because a stronger tile is generally more durable.

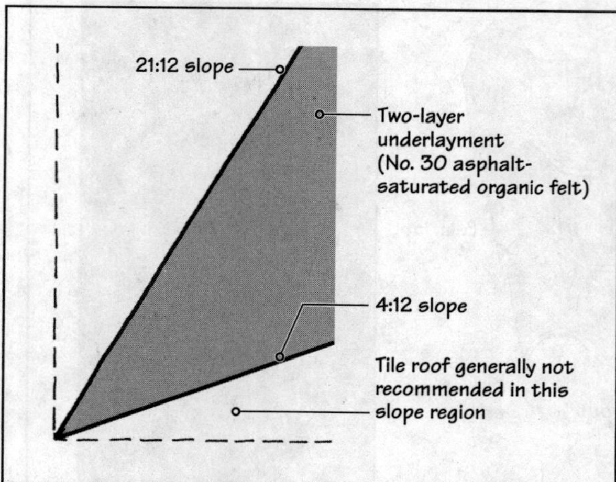

FIGURE 34.32 Roof slope and recommended underlayment for tile roofs.

ROOF UNDERLAYMENT AND TILE ROOFS

Because a tile roof is a relatively long-lasting roof, the underlayment should match the durability of the tiles. The following underlayment specifications are generally recommended:

- *For roofs with a slope greater than 4:12:* A minimum of two layers of No. 30 asphalt-saturated felt laid in shingle fashion, Figure 34.32.
- *For roofs with a slope less than 4:12:* Tile roofs are generally not recommended in this slope region, although some manufacturers allow their product for use in this region.

ICE DAM PROTECTION MEMBRANE

An ice dam protection membrane at the eave should be specified in all climates. However, it is particularly important in cold climates. The minimum requirement for an ice dam protection membrane is the same as that for the asphalt shingle roofs.

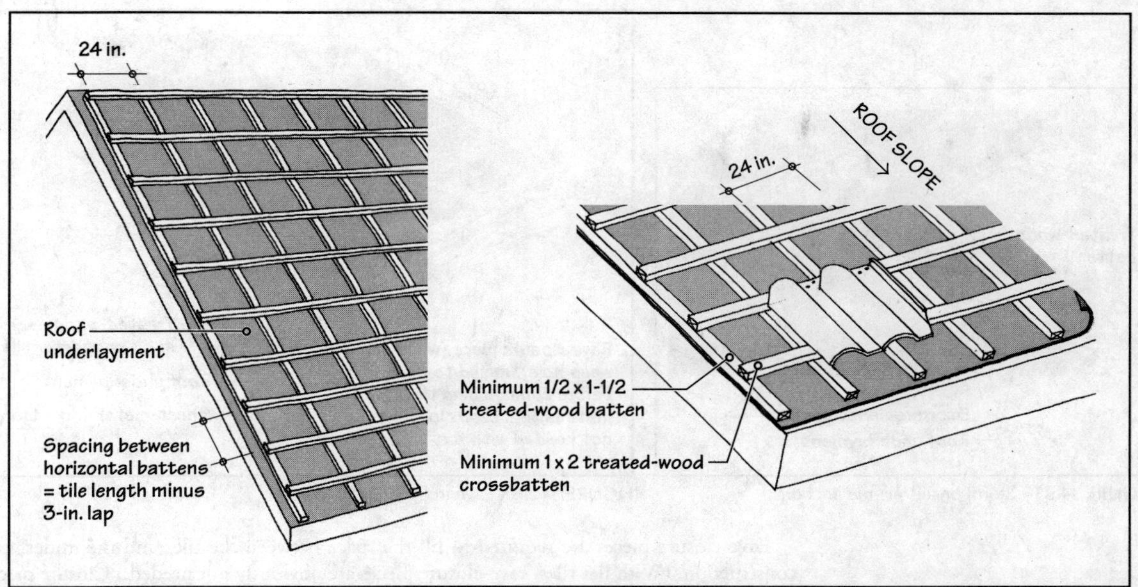

FIGURE 34.33 Installation of tiles on a network of preservative-treated wood battens and crossbattens.

34.8 CLAY AND CONCRETE TILE ROOF DETAILS

The easiest and most commonly used method of fastening the tiles is to nail them to the substrate. The nails should be of stainless steel or copper to match the long life of a tile roof. Except for mission tiles, other tiles can be nailed straight to the deck. However, the best installation of tiles is achieved if a network of battens and crossbattens of treated wood is nailed to the deck over the underlayment, Figure 34.33. The tiles are then nailed to the (top) battens, which run horizontally. The horizontal battens should be a minimum of 1 in. × 2 in. (nominal), and the vertical battens should be a minimum of $\frac{1}{2}$ in. × $1\frac{1}{2}$ in. nominal.

The vertical battens are generally laid 24 in. on center. The spacing of the horizontal battens is a function of the length of the tile. Thus, with a 3-in. lap between successive courses, the center-to-center spacing of horizontal battens equals the tile length minus 3 in.

The batten network creates a space under the tiles, which allows for clear drainage of water should the water seep under the tiles. Because the tiles are generally laid in single coverage with only a 3-in. lap, seepage of water under the tiles in severe weather is not uncommon.

The space under the tiles also permits more rapid drying of the underlayment. Battens also help to hang the tiles that consist of lugs. A major advantage of a batten and crossbatten network is that such a system can be used with slopes less than 4:12, provided that the underlayment is treated like a built-up roof. In that case, the tiles serve simply as a decorative roof covering.

In place of the batten and crossbatten network, a system of one-way horizontal battens may be used, Figure 34.34. The battens are generally laid in 4-ft lengths, separated by a minimum of a $\frac{1}{2}$-in. space. The $\frac{1}{2}$-in. separation allows the drainage of any water trapped by the battens. However, this system does not perform as well as the crossbatten system because the dust and debris block the drainage, causing leaks.

In addition to nailing each tile, additional anchorage may be required in a high-wind or high-seismic region, Figure 34.35.

EAVE TREATMENT

Unlike an asphalt shingle roof, the drip edge is not necessary with a tile roof because the eave tiles can be made to project sufficiently beyond the fascia. However, if a drip edge is used, it should be of a durable metal (e.g., 24-oz copper) to match the long life of a tile roof. A treated-wood edge strip is also required at the eave to give the correct alignment to the first tile, Figure 34.36. Tile manufacturers also make starter tiles that preclude the need of an edge strip.

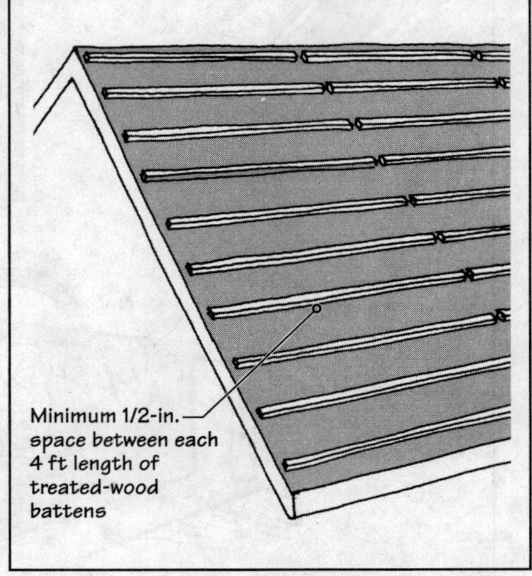

FIGURE 34.34 In place of the batten and crossbatten arrangement of Figure 34.33, one-way battens may be used.

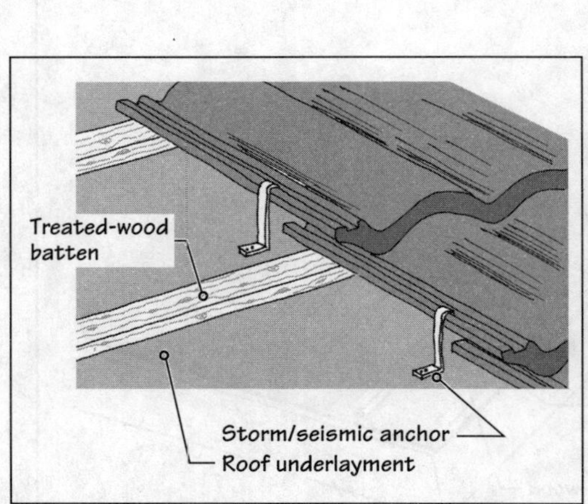

FIGURE 34.35 Storm or seismic tile anchors.

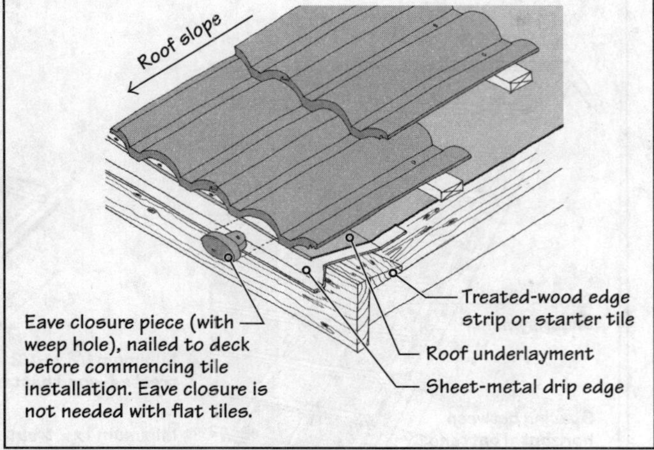

FIGURE 34.36 Eave treatment of a tile roof.

Eave-closure pieces are required to fill the space between the tiles and the underlying construction. (With flat tiles, eave-closure pieces are obviously not needed.) Closure pieces can be of clay (with clay tiles), concrete, or plastic. They are provided with weep holes to allow the escape of water that may seep under the tiles. The closure pieces are nailed to the deck before the tiles are installed. Closing the space under the tiles prevents the nesting of birds and insects. Portland cement mortar may also be used to fill the gap between the eave tiles and the underlying part of the roof if weep holes are provided in each mortar filling.

Rake, Ridge, and Hip Treatments

Specially prepared (curved) rake tiles are generally used to weatherproof the rake, Figure 34.37(a). The construction of a ridge and a hip is similar. With curved tiles, a treated-wood

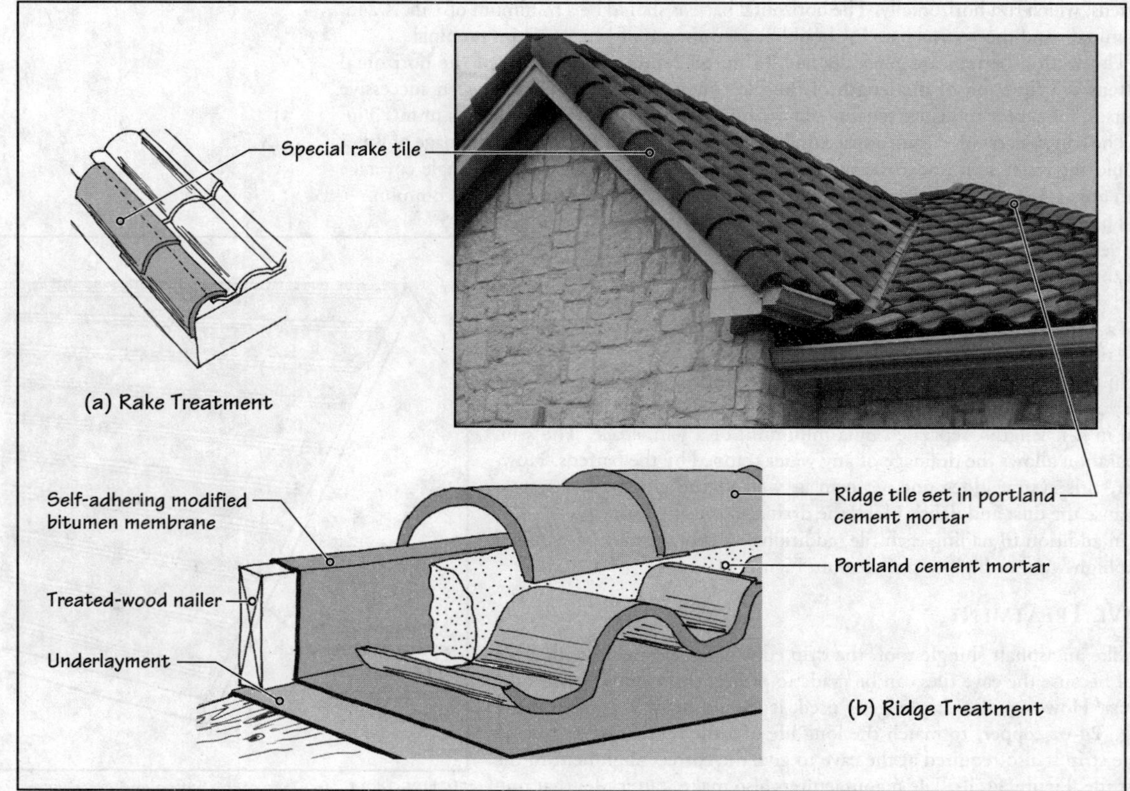

FIGURE 34.37 Rake and hip treatment of a tile roof. Ridge treatment is similar.

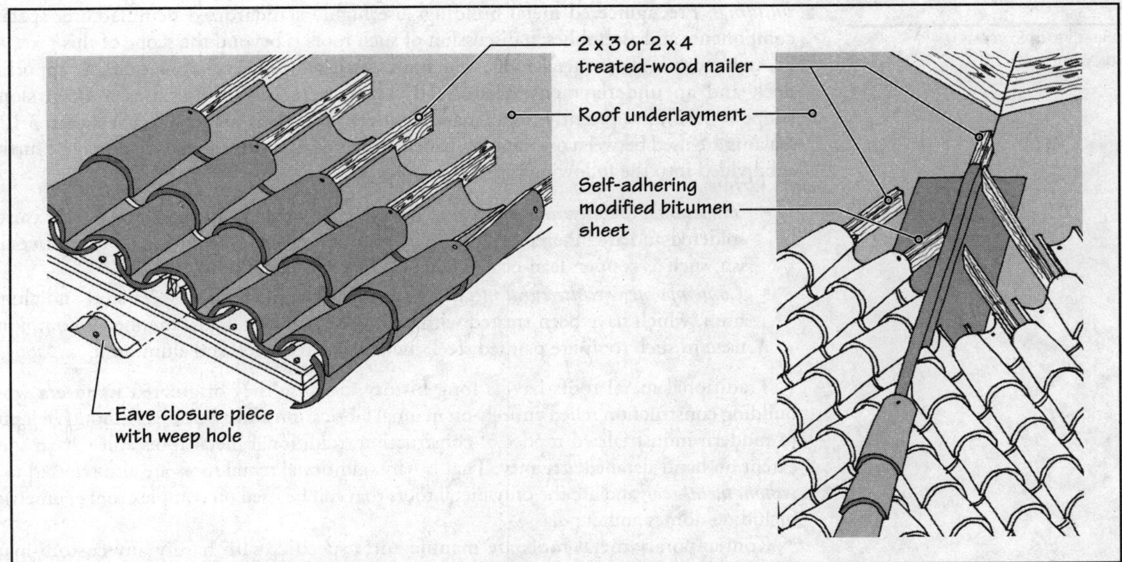

FIGURE 34.38 Installation details of mission tiles.

nailer is installed at the ridge (or the hip), and a self-adhering modified bitumen sheet is wrapped over the nailer. Field tiles are brought close to the nailer, and the nailer is covered with special ridge (or hip) tiles set in portland cement mortar. The mortar holds the ridge (or hip) tiles and also closes the gap between the ridge tiles (or hip tiles) and the field tiles, Figure 34.37(b). Each ridge (or hip) tile is nailed to the nailer. Although these nails are protected by the overlapping ridge (or hip) tile, it is necessary to seal the nail with roofer's cement or sealant.

Installation of Mission Tiles

Mission tiles require an entirely different installation system. A treated-wood nailer, aligned along the slope of the roof, is placed under every cover piece, and the cover piece is nailed to the nailer, Figure 34.38. The nailer is usually either 2 × 3 or 2 × 4 (nominal size), depending on the height of the cover piece. The nailer not only helps to secure the tile, it also reduces the possibility of tile breakage when roofers walk on the tiles. The tiles are laid with 3-in. standard overlap, as with other tiles.

Eave-closure pieces are required, and the rake is covered with the field tiles and rake cover tiles. Additional nailers are needed at the rake. The construction of a ridge or a hip is similar to that of other tile roofs.

34.9 TYPES OF SHEET-METAL ROOFS

Sheet-metal roofs are increasingly becoming part of the urban landscape. The development of durable coatings on sheet steel, coupled with a vast array of nonfading coating colors, has increased the popularity of metal roofs. Metal roofs are spared the rapid atmospheric and solar degradation to which other commonly used low-slope roofs (e.g., built-up, modified bitumen, or single-ply roofs) are subjected.

Some owners of commercial buildings, unhappy with the frequent replacement of their existing low-slope roofs, are switching to metal roofs when reroofing is needed. This is despite the fact that a change from a low-slope roof to a metal roof on an existing building requires many of additional under-roof components. Sheet-metal roofs can be divided into two types:

- Structural sheet-metal roofs
- Architectural sheet-metal roofs

Structural sheet-metal roofs double as a roof covering and also as a structural deck. A separate deck is not provided with such roofs, and the roof panels are supported directly on purlins, which are typically spaced 4 to 5 ft on center. Structural metal roofs are commonly used in low-cost industrial buildings in what are generally referred to as *preengineered metal buildings.* Preengineered metal buildings use highly standardized manufacturer-specific components and assemblies; a discussion of such roofs is beyond the scope of this text.

Architectural sheet-metal roofs, which are discussed further here, require a supporting deck and an underlayment. Historically, they were limited for use in steep slopes (slope ≥ 3:12), but recently, some manufacturers allow their use in slopes less than 3:12 if sealants are used between overlapping panels. Architectural sheet-metal roofs may be further subdivided into the following types:

- *Traditional architectural metal roofs* are typically made from sheet metals that can be soldered and are inherently durable in their natural state, requiring no protective finish, such as copper, lead-coated copper, or terne-coated stainless steel.
- *Contemporary architectural metal roofs* are typically made from sheet steel and aluminum, which have been treated with protective finishes. Thus, the metals commonly used in such roofs are painted steel, metallic-coated steel, and aluminum.

Traditional metal roofs have a long history, because they originated in an era when building construction relied entirely on manual fabrication and assembly. Although adapted to modern industrialized modes of construction, traditional metal roofs still rely to some extent on hand-detailed artisanry. That is why traditional metal roofs are also referred to as *custom metal roofs* and are the only metal roofs that can be used on complex roof geometries, including domes and cupolas.

Contemporary metal roofs are manufacturer specific, with hardly any custom-made components. The introduction of industrialized methods of metal forming and the discovery of durable paints and metallic coatings have made contemporary metal roofs an economical alternative to the traditional copper or lead-coated copper roofs. In fact, the durability and colorfastness of paints and coatings are responsible for the acceptance of sheet steel and aluminum as roofing material.

The discussion provided here is brief and applies to contemporary architectural metal roofs. For additional details, the reader should refer to a text that covers metal roofs in greater detail.

34.10 CONTEMPORARY ARCHITECTURAL METAL ROOFS

As stated earlier, architectural metal roofs require a deck and an underlayment. The deck can be any one of the deck types used in low-slope roof applications, such as concrete, steel, plywood, or oriented strandboard. In commercial buildings, in which steel or concrete decks are common, a typical metal roof assembly consists of the structural deck, one or two layers of insulation, a layer of plywood (or OSB), an underlayment, and the metal roof, Figure 34.39.

In this assembly, plywood (or OSB) works as a nail base to which the metal panels are fastened. The underlayment provides the secondary waterproofing and is generally one layer of No. 30 asphalt-saturated organic felt. In some applications, however, the nail base may be omitted, and the metal panels may be fastened to the deck through the insulation.

An ice dam protection membrane is also needed, as in other steep roofs. The recommendation for an ice dam protection membrane is the same as for other steep roofs. In areas with high snow loads, an ice dam protection membrane should be installed over the entire roof. Additionally, continuous ventilation under the plywood nail base should be provided, and the roof should be designed as a cold roof.

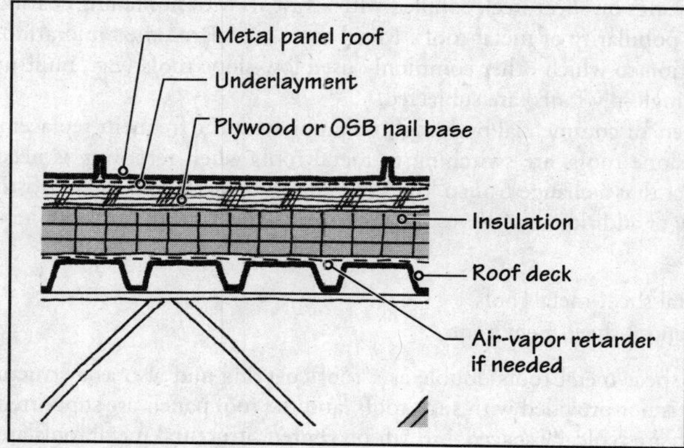

FIGURE 34.39 Typical anatomy of a metal panel roof.

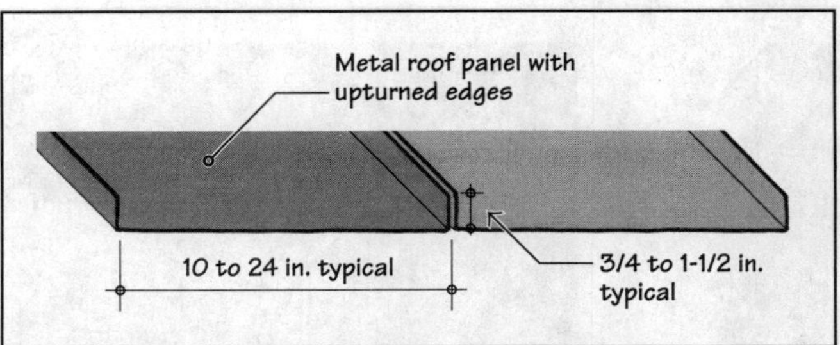

FIGURE 34.40 Typical metal panel roof profile.

FABRICATION OF CONTEMPORARY ARCHITECTURAL METAL ROOF PANELS

One of the several contemporary architectural metal roofs is the standing seam type, in which each panel has two upturned legs, one at each longitudinal edge of the panel. The upturn is generally $\frac{3}{4}$ in. to $1\frac{1}{2}$ in. high. A larger upturn is needed where rainfall intensity is high. The width of the panel varies from 10 in. to 24 in., Figure 34.40. The upturned edges of adjacent panels meet to form a seam, which can be waterproofed in several ways.

In addition to being shop fabricated, panels can be fabricated to the required profile at the construction site from a coil of (prefinished) sheet metal, Figure 34.41. This means that panels of almost any length can be used. The longer the panel, the smaller the number of transverse seams.

In many cases, only one length of panel on each side of a sloping roof is needed, completely precluding the use of transverse seams. In such a case, there is only one joint between the two panels from the two opposite slopes of the roof—at the ridge. The length of a panel is limited by its handleability. Panels 100 ft in length are not uncommon.

Site fabrication of panels reduces the cost of transporting long panels from the factory to the site. In tall buildings, the panel-fabricating equipment is generally placed on the flat portion of the roof, which avoids the need to lift panels from the ground to the roof.

ANCHORING THE PANELS

The panels are fastened to the substrate by metal cleats, whose design varies from manufacturer to manufacturer. One such fastening method is shown in Figure 34.42. The center-to-center distance between cleats is a function of the wind uplift on the roof and may vary from 12 in. (or less) to 5 ft. The cleat shown is of a *fixed type* that does not allow any movement in panels. Generally, if panels are longer than 40 ft, *expansion cleats* are required. Expansion cleats allow the panels to move—an important requirement for long panels.

After the panels have been fastened, each seam is covered with a snap-on cover piece, Figure 34.43. The snap-on cover piece has a factory-applied sealant throughout the length of the cover piece, which squeezes over and around the panel seams, providing additional waterproofing, Figure 34.44.

FIGURE 34.41 For some projects (particularly those requiring long panels), architectural metal roof panels are fabricated at the construction site from a sheet-metal roll using a simple panel-forming machine. Panel length is restricted only by the handling capability. Large lengths are particularly difficult to handle on a windy day.

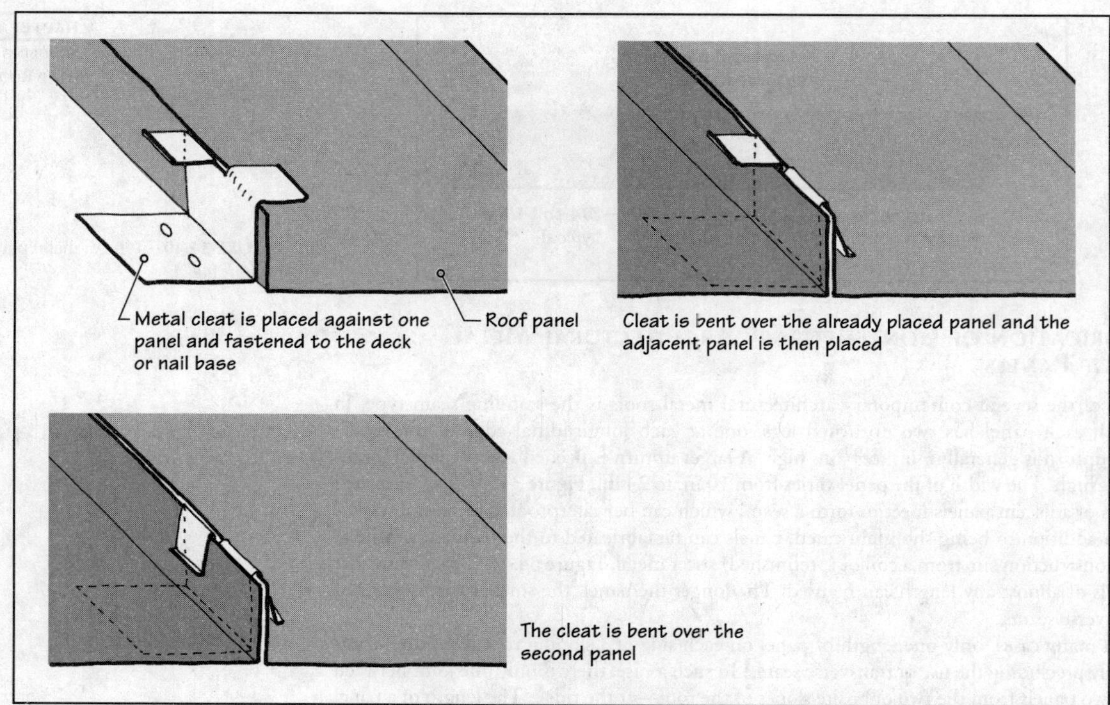

Cleat is bent over the already placed panel and the adjacent panel is then placed
The cleat is bent over the second panel

FIGURE 34.42 Details of the anchorage of metal panels to a roof deck (or nail base) using sheet metal cleats.

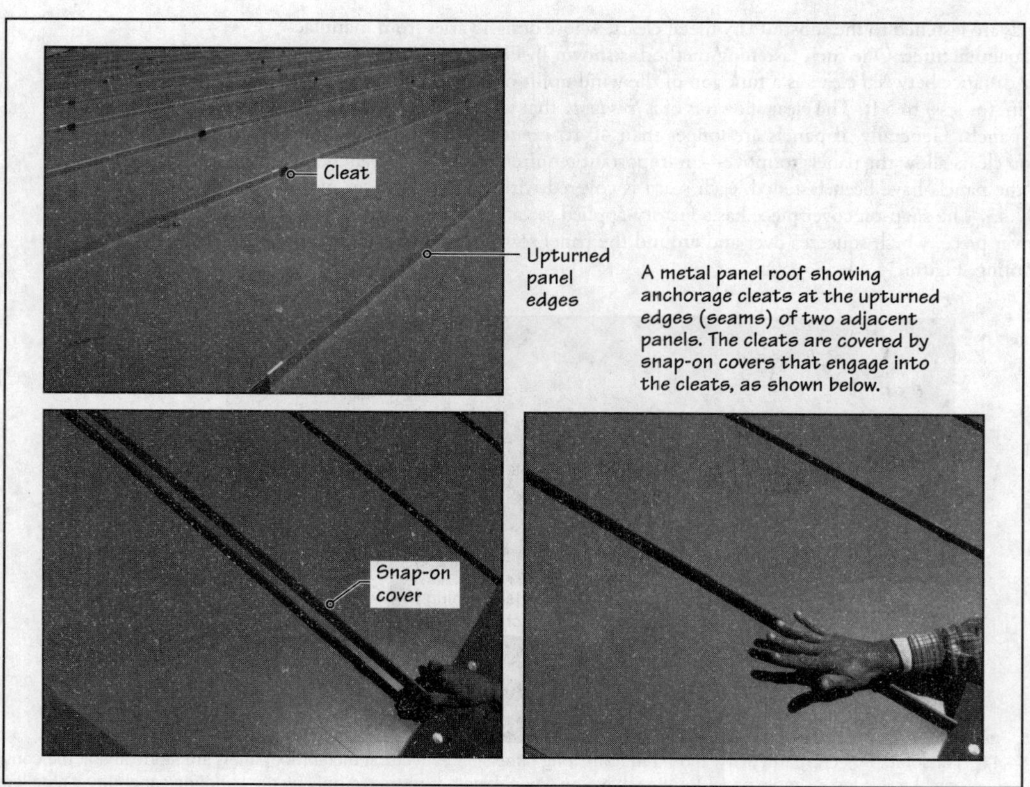

Cleat
Upturned panel edges
A metal panel roof showing anchorage cleats at the upturned edges (seams) of two adjacent panels. The cleats are covered by snap-on covers that engage into the cleats, as shown below.
Snap-on cover

FIGURE 34.43 Installation of snap-on cover pieces over panel seams.

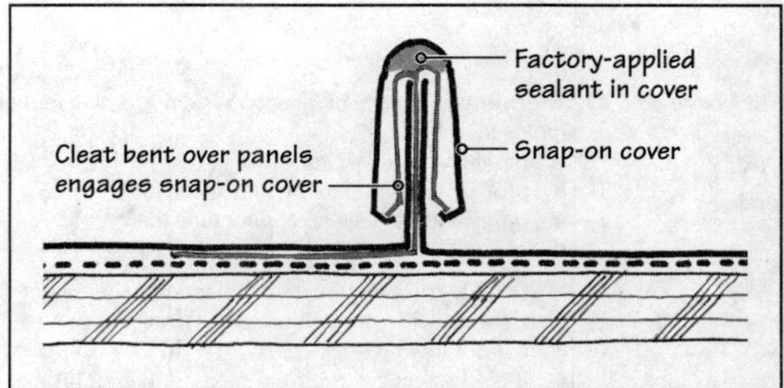

FIGURE 34.44 Section through a seam showing the cleat and the snap-on cover.

RIDGE AND EAVE DETAILS

A typical ridge detail, which consists of a continuous Z-shaped sheet-metal closure set in sealant and soldered to the underlying metal panels, is shown in Figure 34.45. A preformed ridge cap engages into the closure. An ice dam protection membrane is adhered over the ridge directly below the roof panels. Figure 34.46 shows a typical eave detail. Note that the flat portion of the roof panels is turned over the eave flashing.

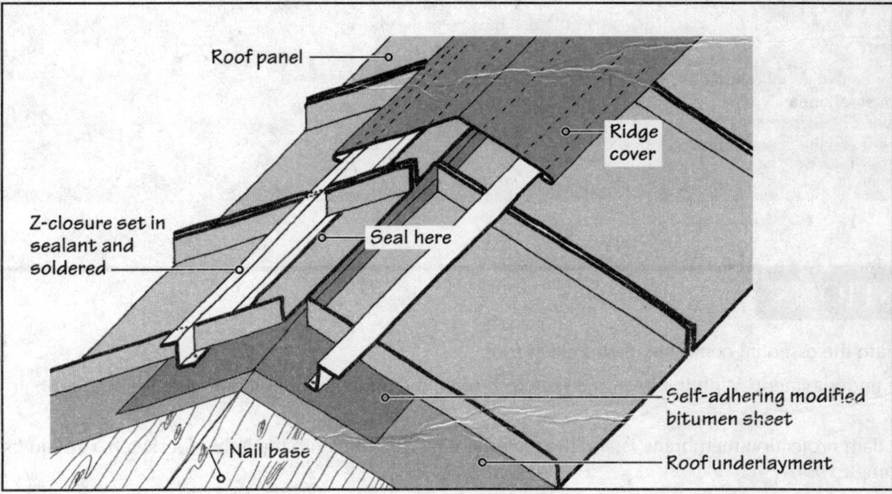

FIGURE 34.45 A typical ridge detail.

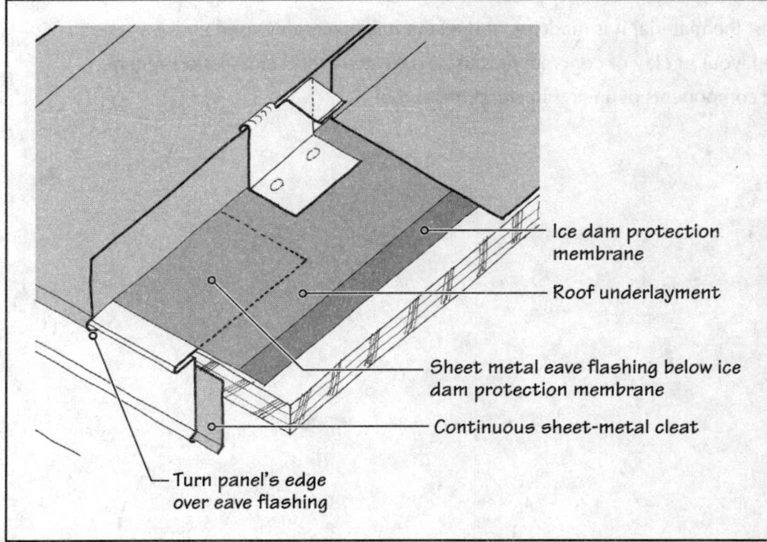

FIGURE 34.46 A typical eave detail.

PRACTICE QUIZ

Each question has only one correct answer. Select the choice that best answers the question.

18. Clay and concrete tiles are generally laid in
- **a.** single coverage with a 5-in overlap between the top and bottom shingles.
- **b.** double coverage with a 2-in. overlap between the top and bottom shingles.
- **c.** double coverage with a 2-in. headlap between the top and bottom shingles.
- **d.** triple coverage with no headlap.
- **e.** none of the above.

19. A roof tile profile that resembles a barrel is generally referred to as a
- **a.** *barrel tile.*
- **b.** *mission tile.*
- **c.** *semicircular tile.*
- **d.** all of the above.

20. Because a clay or concrete tile roof is more durable, the requirement for roof underlayment in a concrete or clay tile roof is less stringent than that in an asphalt shingle roof.
- **a.** True
- **b.** False

21. Everything else being the same, which of the following clay or concrete tile roofs gives the best performance?
- **a.** Tiles laid over a two-way batten system
- **b.** Tiles laid over a one-way batten system
- **c.** Tiles without any batten system

22. The batten system in a clay or concrete tile roof is installed
- **a.** between the roof deck and the roof underlayment.
- **b.** between the deck and the clay (or concrete) tiles.
- **c.** between the roof underlayment and the clay (or concrete) tiles.
- **d.** none of the above.

23. The primary purpose of a batten system in a clay (or concrete) tile roof is to
- **a.** provide a nail base for tiles.
- **b.** provide a cushion to the tiles.
- **c.** provide a drainage layer under the tiles.
- **d.** none of the above.

24. Mission tile roofs require a treated-wood nailer under each row of tiles. The nailer is generally
- **a.** 2 × 2 lumber.
- **b.** 2 × 3 lumber.
- **c.** 2 × 4 lumber.
- **d.** (a) or (b).
- **e.** (b) or (c).

25. An architectural metal roof serves both as roof deck and roof cover.
- **a.** True
- **b.** False

26. An architectural metal roof can be used as a low-slope or steep roof.
- **a.** True
- **b.** False

27. In a standing-seam metal roof, the roof panels are
- **a.** shop formed and site finished.
- **b.** shop formed and shop finished.
- **c.** site formed from a prefinished sheet metal roll.
- **d.** any one of the above.

REVIEW QUESTIONS

1. Using a sketch, explain the essential components of a steep roof.
2. What is a steep roof underlayment? With the help of a sketch, explain the underlayment requirements for an asphalt shingle roof.
3. Explain what an ice dam protection membrane is and the material it is made of. With the help of a sketch, explain its use on an asphalt shingle roof.
4. Sketch a typical three-tab asphalt shingle, showing the adhesive lines and their purpose.
5. Explain the pros and cons of using an open valley, a closed-cut valley, and a woven valley in an asphalt shingle roof.
6. Explain what step flashing is, the material it is made of, and where it is commonly used.
7. Using a sketch, explain the layout of clay or concrete tiles using a treated-wood crossbatten system.
8. Using a sketch, explain the components of an architectural metal roof.

CHAPTER 35

Stairs

CHAPTER OUTLINE

A stair, defined as a series of ascending (or descending) steps, is an important element that allows occupants to move vertically in a building. Architectural historians claim that the stair remained a purely functional element (without artistic overtones) until the end of the fifteenth century. The beginning of the sixteenth century, inspired by Leonardo da Vinci's sketches, however, "signaled a new era of expression for the staircase" [35.1]. From then on, the staircase played an increasingly important visual role, often becoming a sculptural feature in a space, an imperial entrance to a public building or a significant facade element.

The birth of the elevator—and, subsequently, the escalator—reduced the importance of the stair. More recently, the requirement to make buildings accessible to persons with disabilities further eroded its significance.

Because a stair cannot be used by people in a wheelchair, it is no longer a mandatory feature of an entrance lobby. (Increasingly, entrance lobbies in contemporary public buildings are designed without a stair.) Consequently, stairs are reverting to their purely functional role—fulfilling the requirement as exit stair or standby vertical circulation in the event of electrical outage or mechanical interruption.

However, despite the stair's decreasing significance, in many contemporary buildings the rhythm and repetitive features of a stair have been transformed into an important aesthetic component of the interior space, as shown by the images in Figure 35.1. This chapter begins with a general introduction to stairs, followed by the details of construction of simple wood, steel, and concrete stairs.

(a)

(b)

(c)

FIGURE 35.1 Images showing the aesthetic potential of stairs; see also Figure 35.20. **(a)** An entrance lobby stair with structural steel beams (stringers) and concrete-filled sheet-steel tread pans (see Figure 35.13). **(b)** A highly transparent glass wall showcasing the stair in the Madison Museum of Contemporary Art, Madison, Wisconsin. **(c)** Stair in the entrance lobby of the Madison Museum of Contemporary Art, Madison, Wisconsin with structural steel beams (stringers) and glass treads. Architect: Cesar Pelli and Associates.

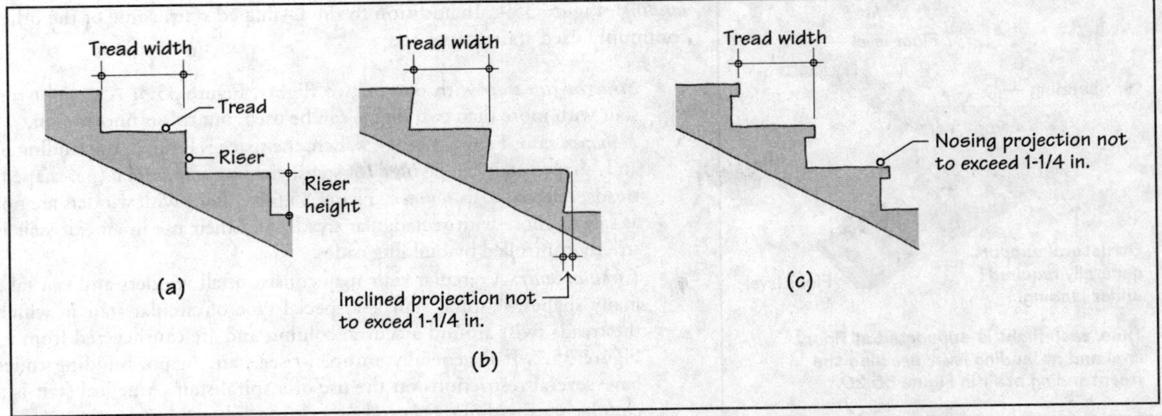

FIGURE 35.2 Tread, riser, and nosing configurations in a stair.

35.1 STAIR FUNDAMENTALS

Because a stair provides vertical transportation, it is part of the means-of-egress (exit) system of a building. It is also a relatively hazardous element because injuries due to falls from stairs are not uncommon. For this reason, stair design is stringently controlled by building codes.

TREAD, RISER, AND NOSING

There are two main components of a stair: *treads* and *risers.* A tread is the horizontal surface on which one walks. The riser is the vertical component that separates one tread from another. Generally, a stair has several treads and risers. For the sake of safety, the dimensions of treads and risers must be uniform in a stair. Building codes allow a small dimensional variation because perfect uniformity is unachievable.

In walking on a horizontal or an inclined surface, an average person can comfortably traverse a distance of 24 to 25 in. in one step. Therefore, a rule of thumb generally used in proportioning the treads and risers of a stair is

2(riser height) + tread width = 24 to 25 in.

Thus, if the risers in a stair are each 5 in. high, the tread width should lie between 14 and 15 in. The most commonly used dimensions for an interior public stair are 12- to 13-in. treads and 6-in. risers. Outdoor stairs generally have a smaller riser and hence a wider tread. Building codes generally require a riser height between 4 in. and 7 in. and a minimum tread width of 11 in.

In most stairs, the tread is a simple flat surface, and the riser is a solid vertical surface, Figure 35.2(a). Where space is limited, the effective tread width can be increased somewhat by inclining the risers, Figure 35.2(b), or by projecting the front edge of the tread beyond the riser, Figure 35.2(c). The front edge of a tread is referred to as the *nosing.*

When an inclined riser or a projected nosing is used, the code-required minimum width of a tread does not change. In other words, the width of a tread is considered the horizontal distance between the vertical planes of the foremost projections of adjacent treads, as shown in Figure 35.2(b) and (c).

The nosing of a tread is subjected to the maximum abrasion. In public stairs with heavy traffic, the treads should consist of a strong, dense material such as granite, high-strength concrete, or steel. Alternatively, a separate nosing (approximately $2\frac{1}{2}$ in. wide) consisting of an abrasion-resistant and skid-resistant material is epoxied or embedded into the tread.

Stairs can also be constructed without risers, referred to as *open-riser stairs,* Figure 35.3. Because of safety concerns, open-riser stairs are subject to more stringent code restrictions than stairs with solid risers. For example, open-riser stairs are generally not allowed as exit stairs. Additionally, the clear vertical distance between the treads of open-riser stairs cannot exceed 4 in.

STAIR SHAPES

The most commonly used stair shape is a U-shaped stair (in plan). It consists of two flights of stairs between floors with a midfloor landing (or simply a *midlanding* or

NOTE

Treads and Risers of a Stair

The treads and risers of a stair must meet the following dimensional requirements:

Minimum tread width = 11 in.

Riser height = 4 to 7 in.

Residential Stair

Building codes are less restrictive for a stair within a dwelling unit:

Minimum tread width = 10 in.

Riser height = 4 to $7\frac{3}{4}$ in.

A 10-in. tread is allowed in a dwelling unit stair, provided that a nosing projection of $\frac{3}{4}$ to $1\frac{1}{4}$ in. is used. If a nosing projection is not used, an 11-in. minimum tread width is required.

Some local building codes may allow a riser height of greater than $7\frac{3}{4}$ in. for a residential basement stair.

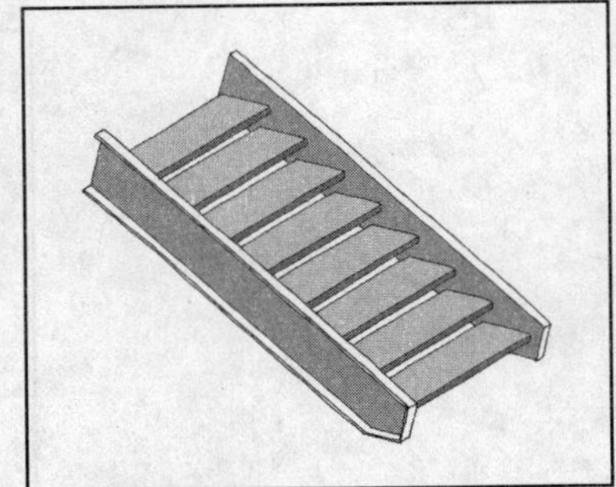

FIGURE 35.3 A stair with open risers.

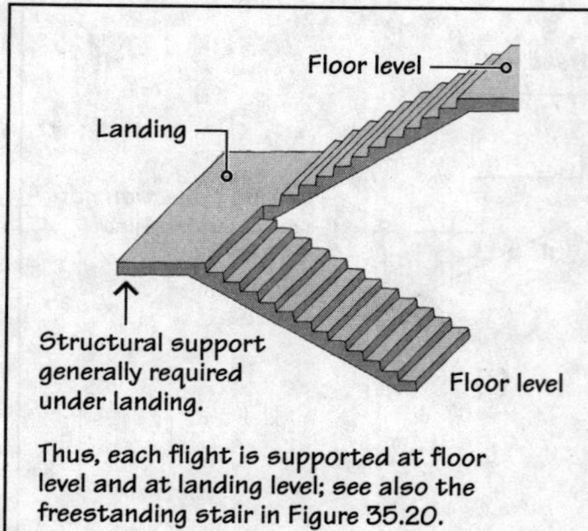
Floor level
Landing
Structural support
generally required
under landing.
Floor level
Thus, each flight is supported at floor
level and at landing level; see also the
freestanding stair in Figure 35.20.

FIGURE 35.4 A U-shaped stair.

landing), Figure 35.4. In addition to the U-shaped stair, some of the other commonly used stair shapes are

- *Straight-run stair* with one or two flights, Figure 35.5: A straight-run stair with more than two flights can be used, but this is uncommon.
- *L-shaped stair,* Figure 35.6(a): Where the space is limited, the landing of an L-shaped stair can be used for steps, yielding trapezoidal (pie-shaped) treads, referred to as *winders,* Figure 35.6(b). Stairs with winders are not as safe as those with rectangular treads, and their use in an exit stair is strictly controlled by building codes.
- *Circular stair:* A circular stair may consist of all winders and can take many shapes. A spiral stair is a special type of circular stair in which the treads twist around a central column and are cantilevered from it, Figure 35.7. It is generally an open-riser stair. Again, building codes have several restrictions on the use of a spiral stair. A helical stair is a circular stair without a central supporting column (see Section 35.5).

Open and Closed Stairs

Stairs are also described as either open or closed. An open stair is exposed to the area below on one or more sides, whereas a closed stair is fully enclosed with a stair enclosure (stair shaft) and is usually accessed through a doorway.

Width of Stair

The minimum width of a stair is determined by its purpose. When it is used as an exit stair, its width depends on the number of occupants it serves (occupant load) but is not less than 44 in. clear (between handrail and handrail) for an open exit stair or 48 in. for an enclosed exit stair. An exit stair for an occupant load of less than 50, or a stair within a dwelling unit, has a minimum width of 36 in.

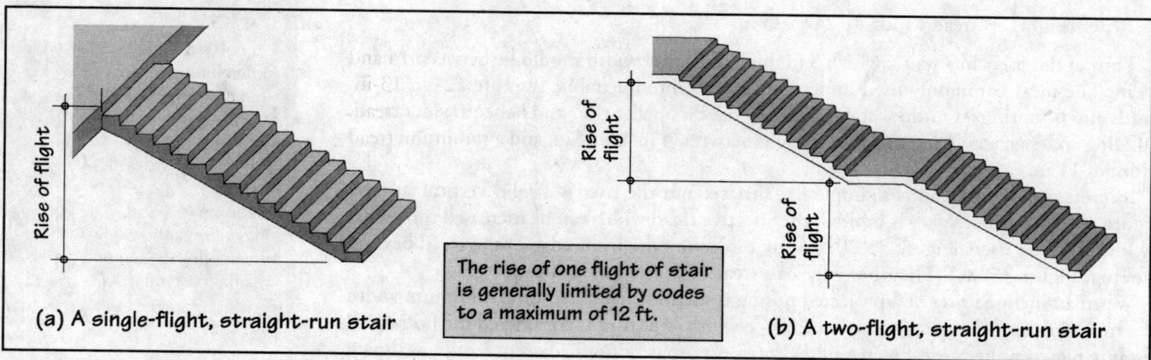
Rise of flight
(a) A single-flight, straight-run stair
The rise of one flight of stair
is generally limited by codes
to a maximum of 12 ft.
Rise of
flight
Rise of
flight
(b) A two-flight, straight-run stair

FIGURE 35.5 Straight-run stairs; see Fig. 35.1(a).

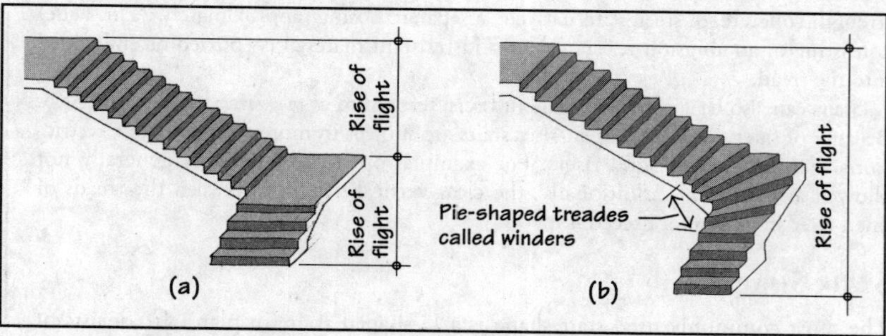
Rise of
flight
Rise of
flight
(a)
Pie-shaped treades
called winders
Rise of flight
(b)

FIGURE 35.6 L-shaped stairs **(a)** without winders and **(b)** with winders.

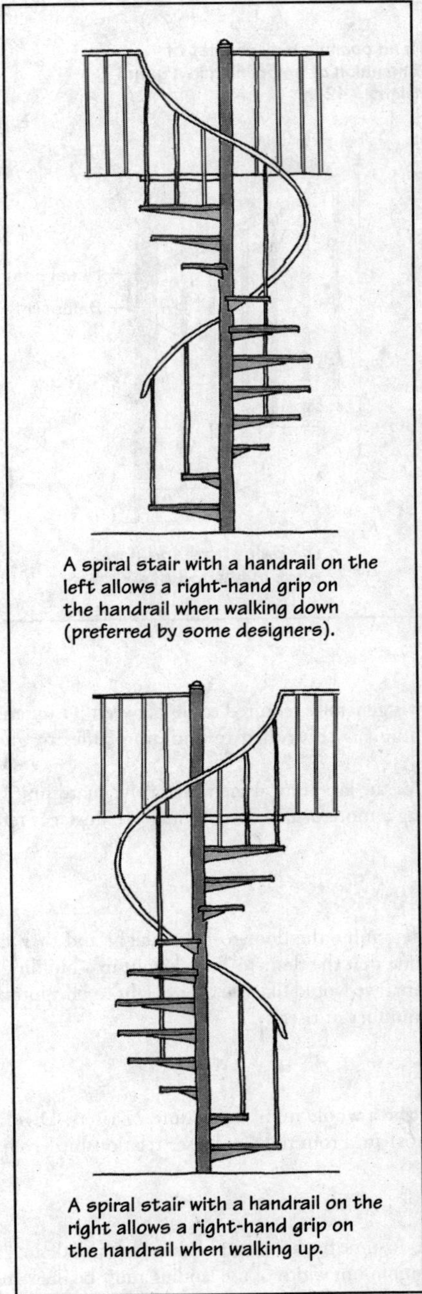
A spiral stair with a handrail on the left allows a right-hand grip on the handrail when walking down (preferred by some designers).
A spiral stair with a handrail on the right allows a right-hand grip on the handrail when walking up.

FIGURE 35.7 A spiral stair.

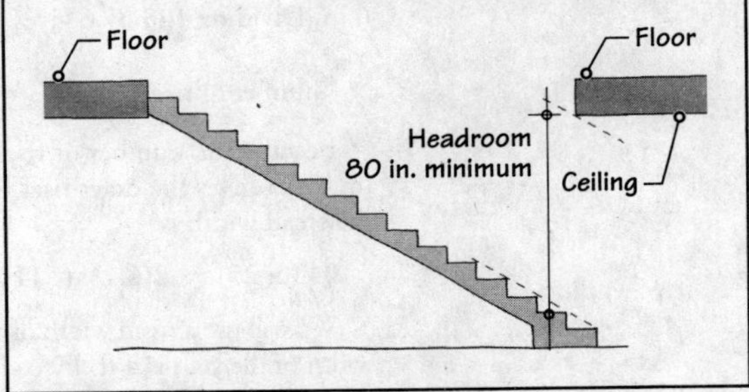
Floor
Floor
Headroom
80 in. minimum
Ceiling

FIGURE 35.8 Headroom in a stair.

HEADROOM

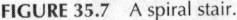

The headroom in a stair is the minimum clearance between a tread and a projection above, Figure 35.8. Building codes generally require the headroom to be a minimum of 80 in. at any point on the stair.

GUARD UNIT, HANDRAIL, BALUSTERS, AND NEWEL POST

The edge of a stair exposed to a change in height (i.e., not protected by the wall of the enclosure) must have a *guard unit* to protect against falling. The minimum height of a guard unit is 42 in., Figure 35.9(a). The clear distance of openings in a guard unit must not exceed 4 in.

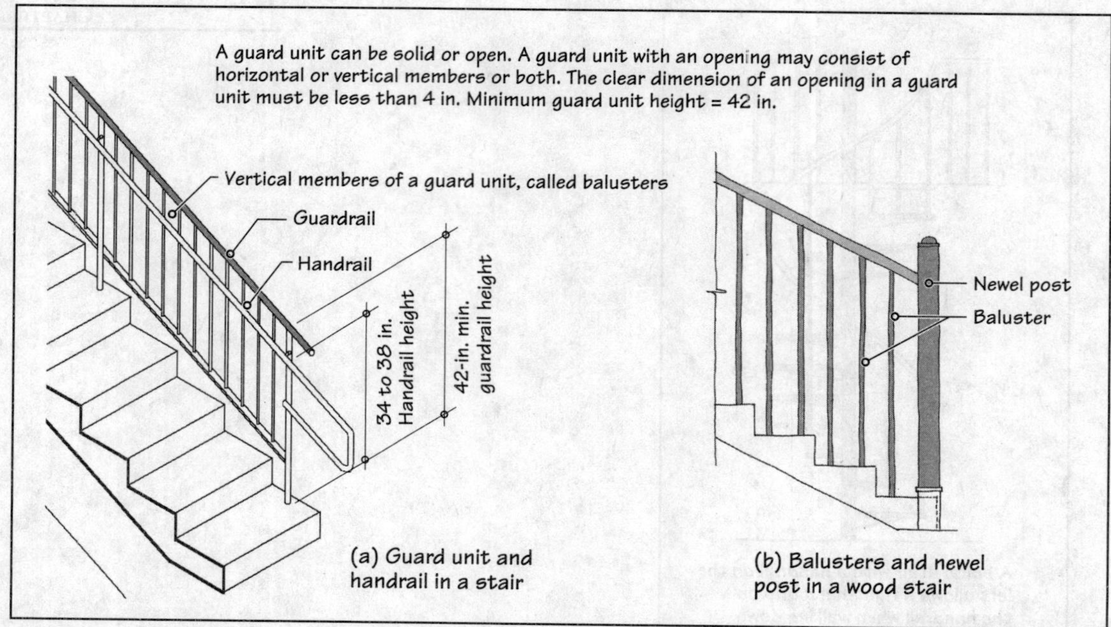
A guard unit can be solid or open. A guard unit with an opening may consist of horizontal or vertical members or both. The clear dimension of an opening in a guard unit must be less than 4 in. Minimum guard unit height = 42 in.
Vertical members of a guard unit, called balusters
Guardrail
Handrail
34 to 38 in.
Handrail height
42-in. min.
guardrail height
(a) Guard unit and handrail in a stair
Newel post
Baluster
(b) Balusters and newel post in a wood stair

FIGURE 35.9 Guard unit, handrail, balusters and newel post.

The height of a handrail in a stair is generally required to lie between 34 in. and 38 in. The cross-sectional profile of a handrail is controlled by building codes to give it the required graspability.

In some wood stairs, the first and/or the last vertical member of the guard unit (referred to as a *baluster*) is highlighted by using a more ornate design. Such a baluster is referred to as a *newel post,* Figure 35.9(b).

Stair Layout and Stair Plan

In preparing a stair layout, we first determine the floor-to-floor height and then calculate the number of risers and treads. Assume that the floor-to-floor height in a building is 10 ft 8 in., that is, 128 in. Assume further that we would like the riser height to be approximately 6 in. Dividing 128 by 6 gives us the number of risers:

$$\text{Number of risers} = \frac{128}{6} = 21.3$$

Because the number of risers must be a whole number, assume 21 risers. Dividing 128 in. by 21 gives the exact riser height, 6.1 in. From the tread-riser relationship given earlier, the tread width is

$$24 \text{ (or 25)} - 2(6.1) = 11.8 \text{ to } 12.8 \text{ in.}$$

We will use a tread width of 12.0 in. Assume further that a U-shaped stair is desired and the width of the stair is 4 ft. By code, the minimum width of the landing must be the same as the

EXPAND YOUR KNOWLEDGE

Summary of Stair-Design Criteria

1. Tread depth and riser height must be dimensionally uniform throughout.
2. Minimum tread width = 11 in. Riser height = 4 to 7 in. (See Section 35.1 for exceptions for residential stairs.)
3. Nosing projection $\leq 1\frac{1}{4}$ in.
4. Stair width = function of the occupant load, but not less than 48 in. for an enclosed exit stair, 44 in. for an open exit stair, or 36 in. for a stair serving an occupant load of less than 50 or a residential stair.
5. Width of landing $\geq$ stair width.
6. Rise of one flight $\leq$ 12 ft.
7. Headroom $\geq$ 80 in.
8. Use of winders is restricted.
9. Height of guardrail = 42 in. minimum. Height of handrail = 34 to 38 in.
10. Handgrip portion of the handrail must have a circular cross section between $1\frac{1}{4}$ in. and $2\frac{5}{8}$ in. Noncircular profiles must provide equivalent graspability.

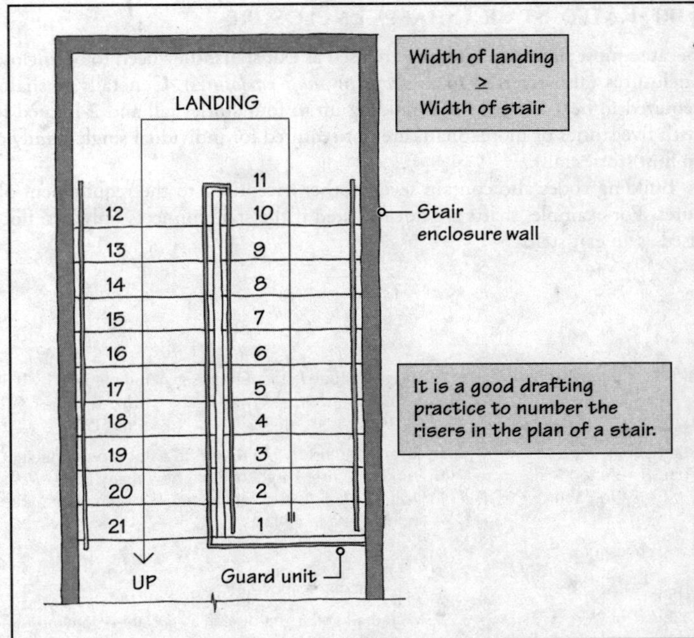

FIGURE 35.10 Plan of a U-shaped stair (at the second floor level) that extends only from the first floor to the second floor.

width of the stair, that is, 4 ft. With these data, a plan of the stair can be drawn, as shown in Figure 35.10.

STAIR-DRAWING CONVENTIONS

There are standard conventions for how stairs are shown in building plans. Figure 35.10 is the plan at the second-floor level of a U-shaped stair that extends from the first to the second floor only. The plan of the same stair at the first-floor level is shown in Figure 35.11(a). If the same stair were to extend over several floors, then the plan of the stair at a typical floor would generally be drawn as shown in Figure 35.11(b).

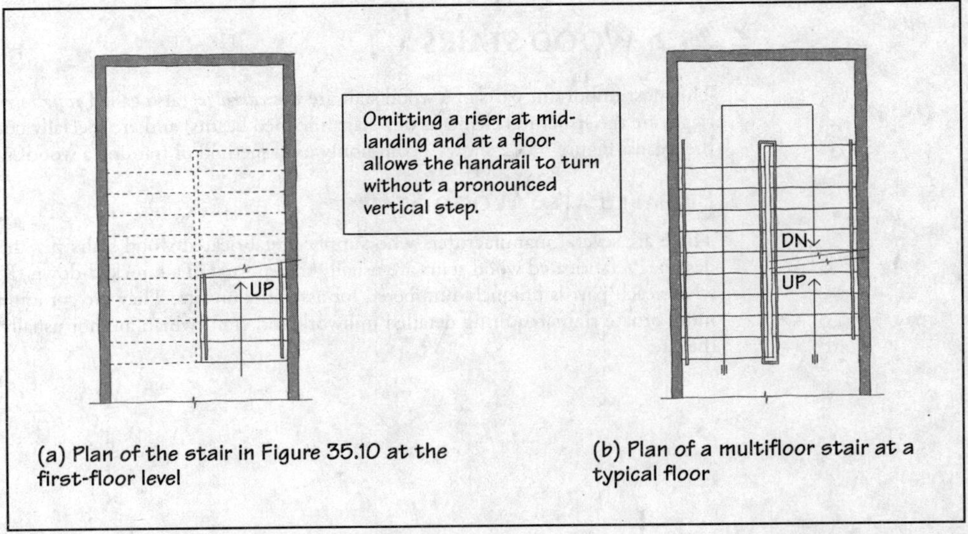

FIGURE 35.11 Stair plans at different levels.

FIRE-RATED STAIR (SHAFT) ENCLOSURE

Because most stairs in a building are used as exit stairs, they need to be enclosed by vertical enclosures (also referred to as *shafts* or *shaft enclosures*). Generally, a shaft enclosure is required to be 1-h rated for a building up to four stories tall and 2-h rated for a building with five stories or more. Shafts are not required for individual single-family dwellings (up to four stories tall).

Building codes also contain several other exceptions to the requirement of shaft enclosures. For example, shafts are not required if the stair connects only two floors and is not used as an exit stair.

PRACTICE QUIZ

Each question has only one correct answer. Select the choice that best answers the question.

1. An approximate formula generally used in determining the tread dimension (T) and riser dimension (R) in a stair is
- **a.** $2T + R = 24$ to 25 in.
- **b.** $2T + 2R = 24$ to 25 in.
- **c.** $2R + T = 24$ to 25 in.
- **d.** $R + T = 24$ to 25 in.
- **e.** none of the above.

2. The minimum tread width required by building codes for a nonresidential stair is
- **a.** 11.0 in.
- **b.** 11.5 in.
- **c.** 12.0 in.
- **d.** 12.5 in.
- **e.** 13.0 in.

3. The minimum riser height required by building codes for a nonresidential stair is
- **a.** 6.0 in.
- **b.** 5.5 in.
- **c.** 5.0 in.
- **d.** 4.5 in.
- **e.** 4.0 in.

4. The maximum nosing projection allowed for a stair is
- **a.** 3.0 in.
- **b.** $2\frac{1}{2}$ in.
- **c.** 2 in.
- **d.** $1\frac{1}{2}$ in.
- **e.** $1\frac{1}{4}$ in.

5. A riser must be vertical. It cannot be inclined.
- **a.** True
- **b.** False

6. A U-shaped stair has been provided between the first floor and the second floor of a building with a midlanding. This stair has
- **a.** one flight.
- **b.** two flights.
- **c.** three flights.
- **d.** four flights.

7. The rise of one flight of stair is generally limited by building codes to
- **a.** 7 ft.
- **b.** 8 ft.
- **c.** 10 ft.
- **d.** 12 ft.

8. Given a multistory building with a floor-to-floor height of 10 ft and an optimal riser height of 7 in., how many treads would you use for a U-shaped stair with a midlanding between floors? (The landing is not counted as a tread.)
- **a.** 15
- **b.** 16
- **c.** 17
- **d.** 18
- **e.** None of the above

9. A handrail and a guardrail in a stair are synonymous.
- **a.** True
- **b.** False

10. The minimum height of a guardrail in a stair is
- **a.** 34 in.
- **b.** 36 in.
- **c.** 38 in.
- **d.** 40 in.
- **e.** none of the above.

11. A stair constructed without risers is generally called a
- **a.** no-riser stair.
- **b.** closed-riser stair.
- **c.** open-riser stair.
- **d.** hollow stair.

12. A stair with treads cantilevered from a central column is a
- **a.** circular stair.
- **b.** U-shaped stair.
- **c.** L-shaped stair.
- **d.** spiral stair.
- **e.** helical stair.

13. The minimum width of a stair in a dwelling unit is
- **a.** 2 ft 6 in.
- **b.** 3 ft.
- **c.** 3 ft 6 in.
- **d.** 4 ft.
- **e.** none of the above.

35.2 WOOD STAIRS

The most important parts of a wood stair are the *carriages* (also called *rough stringers*). Carriages are the structural elements of a stair (inclined beams) and are specially cut to support the treads. Figure 35.12 shows a commonly used method of framing a wood stair.

PREFABRICATED WOOD STAIRS

There are several manufacturers who supply prefabricated wood stairs per the architect's design. Prefabricated wood stairs are usually transported in a knocked-down (KD) version, where each part is uniquely numbered for assembly on site. They are commonly used for more ornate stairs requiring detailed millwork and craft, which are not usually possible at the site.

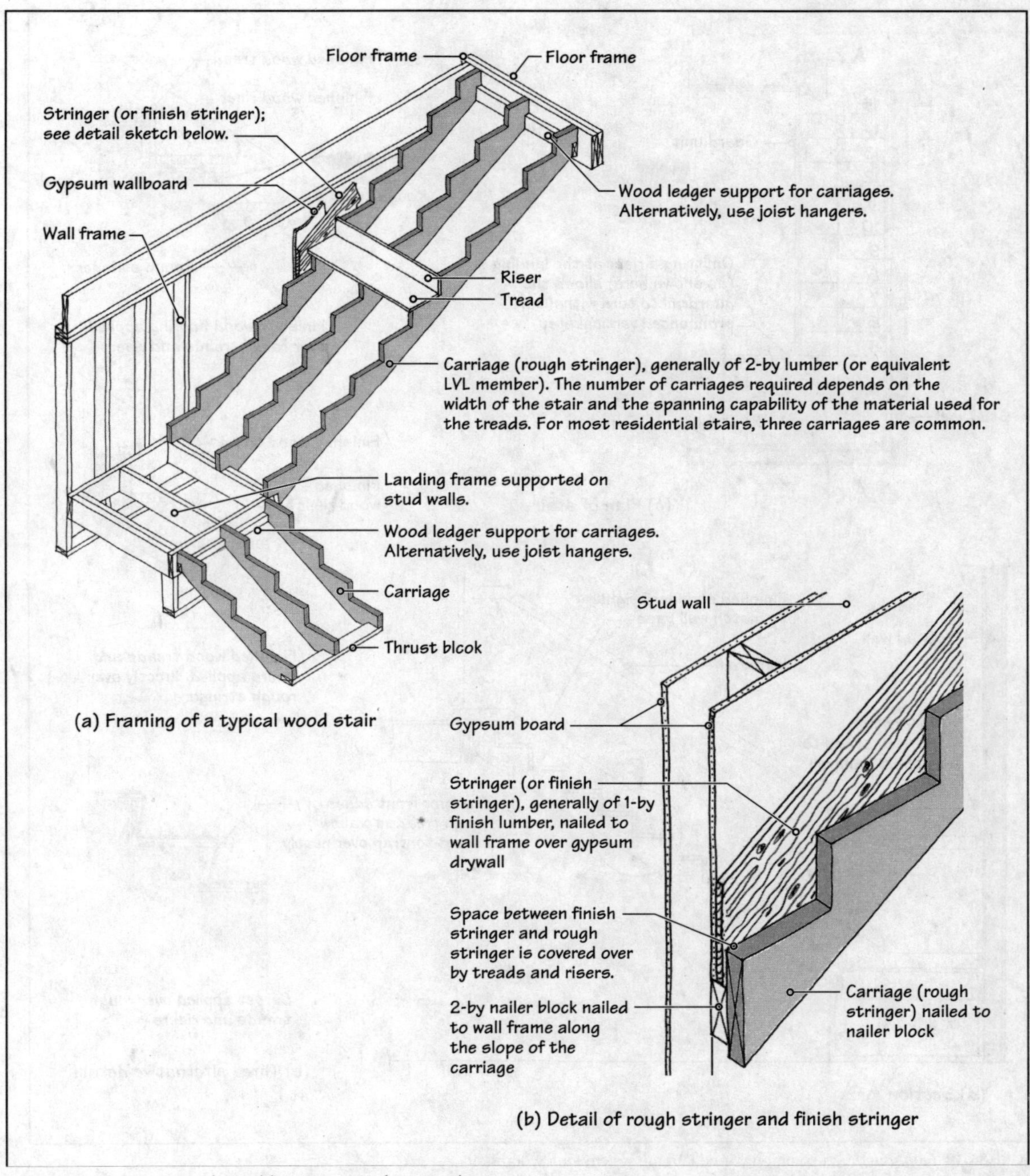
Floor frame
Floor frame
Stringer (or finish stringer); see detail sketch below.
Gypsum wallboard
Wall frame
Wood ledger support for carriages. Alternatively, use joist hangers.
Riser
Tread
Carriage (rough stringer), generally of 2-by lumber (or equivalent LVL member). The number of carriages required depends on the width of the stair and the spanning capability of the material used for the treads. For most residential stairs, three carriages are common.
Landing frame supported on stud walls.
Wood ledger support for carriages. Alternatively, use joist hangers.
Carriage
Thrust blcok
(a) Framing of a typical wood stair
Stud wall
Gypsum board
Stringer (or finish stringer), generally of 1-by finish lumber, nailed to wall frame over gypsum drywall
Space between finish stringer and rough stringer is covered over by treads and risers.
2-by nailer block nailed to wall frame along the slope of the carriage
Carriage (rough stringer) nailed to nailer block
(b) Detail of rough stringer and finish stringer

FIGURE 35.12 A commonly used framing system for a wood stair.

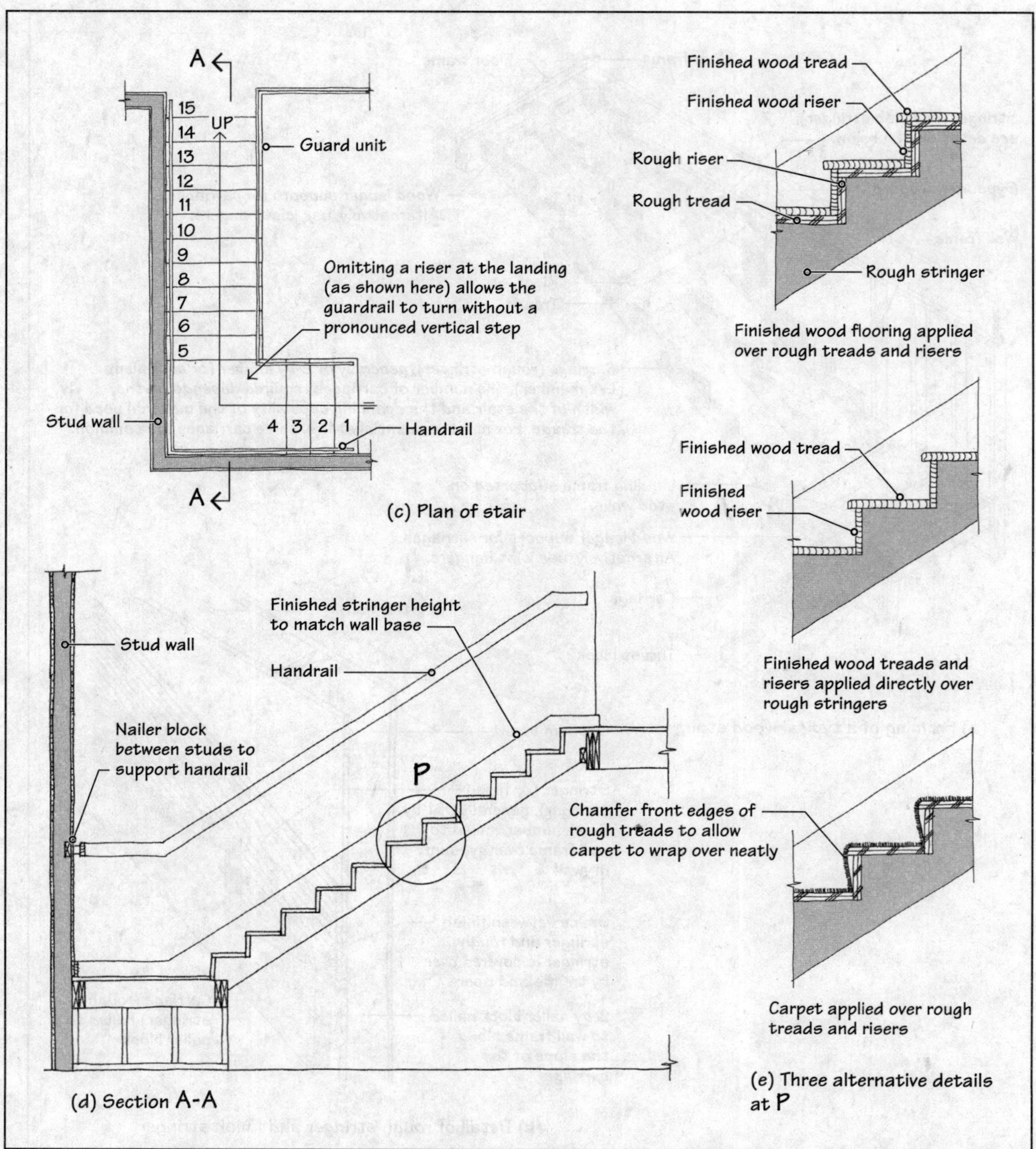

FIGURE 35.12 ***(continued)*** A commonly used framing system for a wood stair.

35.3 STEEL STAIRS

Stairs in public buildings are generally constructed of steel or concrete. Because steel stairs can be shop fabricated and brought to the site ready for installation, they are far more commonly used than concrete stairs. Another reason for the lack of use of concrete stairs is that their formwork is complicated and expensive.

Prefabricated steel stairs are used in all types of public buildings, that is, steel- and concrete-frame buildings and load-bearing masonry buildings. They are particularly popular for exit stairs.

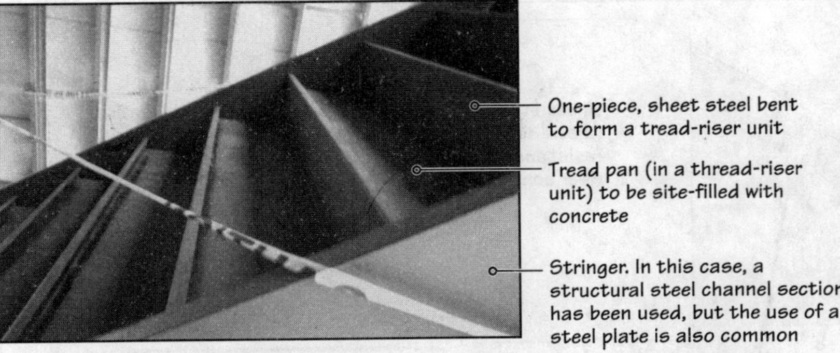

FIGURE 35.13 A typical prefabricated steel stair consists of two stringer beams (stringers) to which tread-riser units made from sheet steel are welded; see Figure 35.14 for details. (In this stair, the guard unit and handrail have not yet been installed.)

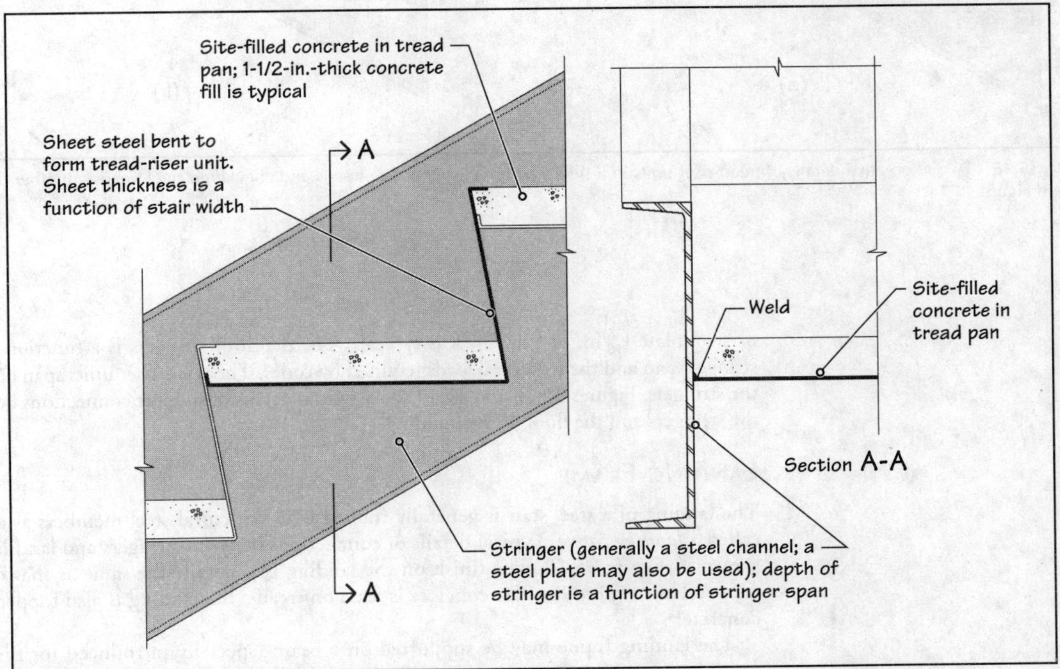

FIGURE 35.14 Typical detail of tread-riser units welded to stringers. In this detail, tread pans are site-filled with concrete; see Figure 35.15 for alternatives.

A typical prefabricated flight of a steel stair consists of two stringer beams (stringers) to which tread-riser units made of sheet steel are welded, Figure 35.13.

The tread pan is generally site filled with concrete, Figure 35.14. For good wear resistance, a concrete strength of 5,000 psi is generally specified. Other tread finishes include a precast-concrete *drop-in tread* with a slip-resistant broom finish, Figure 35.15(a), and sheet steel with a raised, diamond-shaped checkered pattern, Figure 35.15(b). Factory-installed epoxy-aggregate fill or wear- and slip-resistant coatings can also be used.

Stringers

Stringers in a steel stair function as inclined beams, spanning from the floor to the landing and from the landing to the next floor. They generally consist of a structural-steel channel

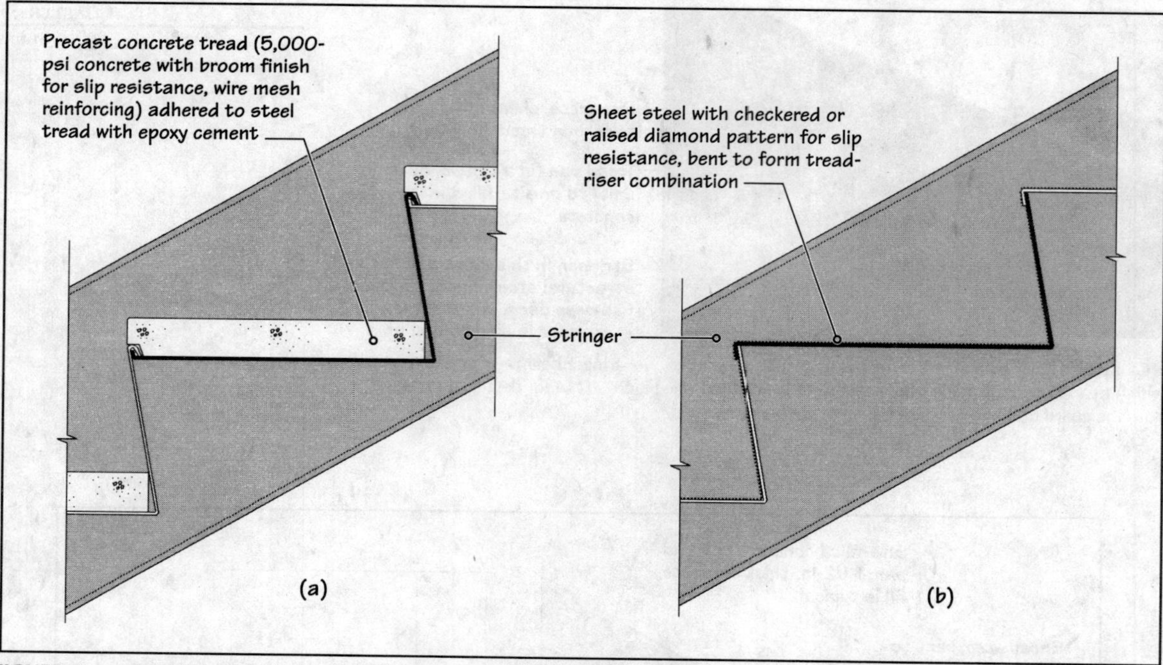

FIGURE 35.15 Two alternative details of tread-riser units in a steel stair (precast concrete treads and checkered steel treads); another detail is shown in Figure 35.14.

or steel plate ($\frac{3}{16}$ in. or $\frac{1}{4}$ in. thick is typical). The depth of stringers is a function of the stringer span and the structural loads required by codes. The tread-riser units span between the stringers. Figures 35.16 and 35.17 show typical details of support connections between the stringers and the floor of the building.

Landing Frame

The landing of a steel stair is generally framed with structural steel members as a unit, called a *landing frame.* Typical details of connections between stringers and landings are shown in Figure 35.18. The finish on the landing is generally the same as that on the treads. Thus, where site-cast concrete is used on treads, the landing is also topped with concrete.

The landing frame may be supported on a beam (specially introduced for the purpose) between the upper and lower floors of the building, on (masonry or concrete) stair-enclosure walls, or on columns independent of the structural frame of the building. In most buildings, however, the landing frame for a prefabricated steel stair is supported by suspending it from the upper-level floor beams with steel hanger bars, Figure 35.19.

A major advantage of a suspended landing is that it allows adjustment of the height of the landing with a few turns of the nuts. Additionally, the entire stair can be erected before constructing the walls of the stair enclosure.

35.4 CONCRETE STAIRS

Although concrete stairs can be precast and prefabricated, their use is limited because they are heavy, which increases the cost of transportation and installation. Most concrete stairs are site cast. As previously stated, the formwork for concrete stairs is intricate, which increases the cost and causes construction delays. Their use is, therefore, infrequent, even in buildings with a reinforced-concrete structural frame.

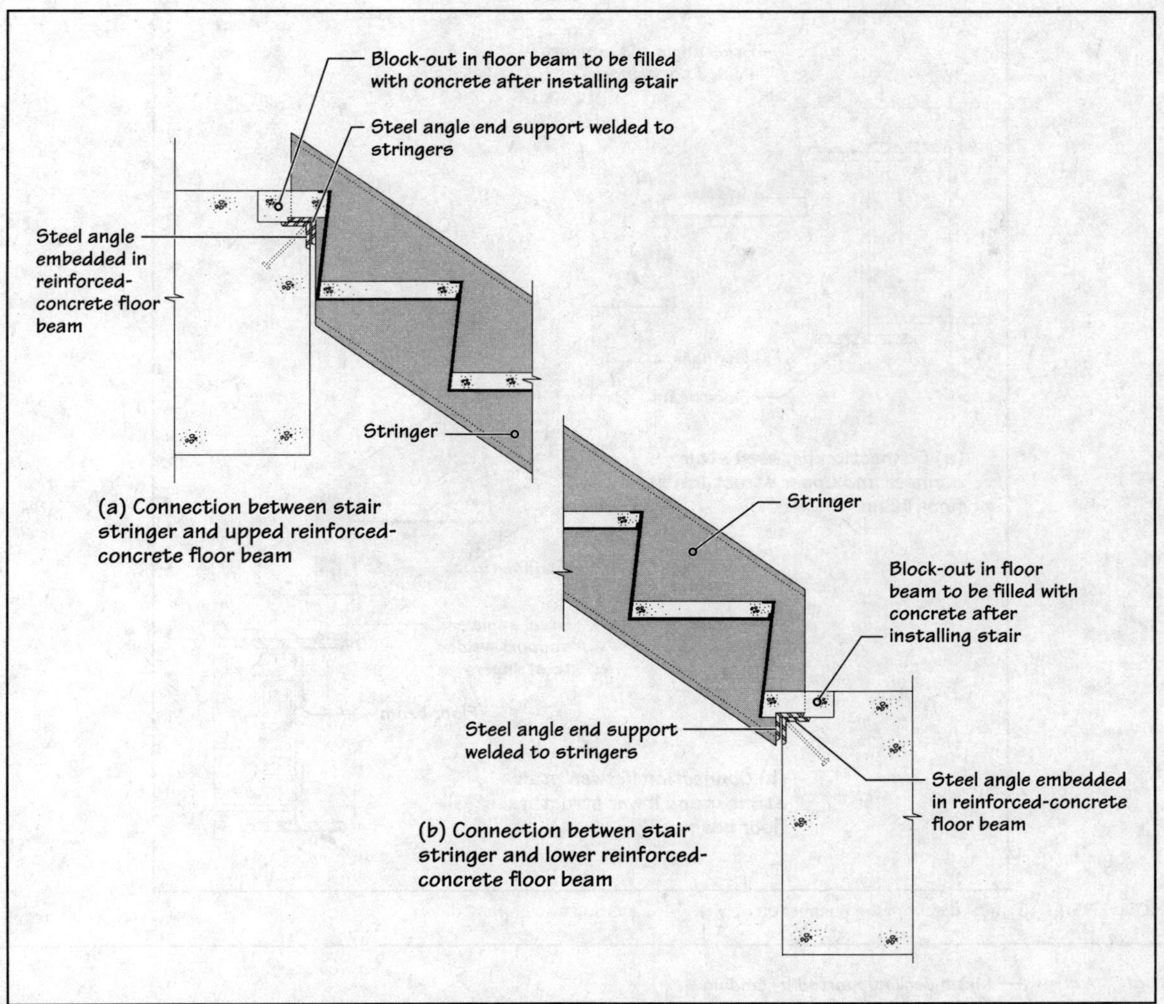

FIGURE 35.16 Typical details of the connection between stringers and a reinforced-concrete floor.

35.5 FREESTANDING CANTILEVERED STAIRS

In the various stair types discussed so far, each flight is supported at the floor and landing levels. The use of steel and reinforced concrete, however, allows the stairs to be constructed without any supports at the landings (designed as cantilevers and supported only at the floors). Cantilevered stairs, also referred to as *self-supporting stairs,* can either be U-shaped or circular in plan.

A cantilevered, self-supporting, U-shaped reinforced stair is shown in Figure 35.20(a). Figure 35.20(b) shows a cantilevered steel stair. In this stair, the stringers (of structural-steel channels) function as continuous spatially bent beams that are rigidly connected to the floor beams at both floors. Tread-riser units that span between stringer beams are made of structural-steel plate.

Self-supporting circular steel or concrete stairs can be constructed with or without landings. Called *helical* (or *helicoidal*) stairs, they are fairly common in steel, concrete, and wood. A helical stair is similar to a spiral stair but has no central column support.

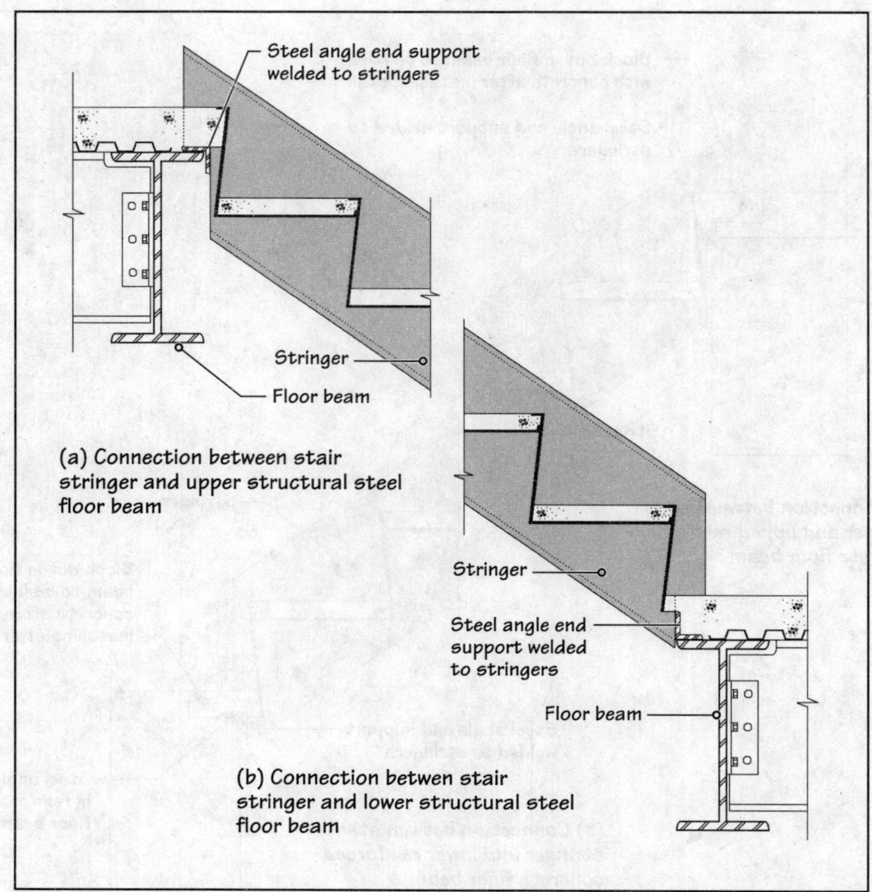

FIGURE 35.17 Typical details of the connection between stringers and a steel-framed floor.

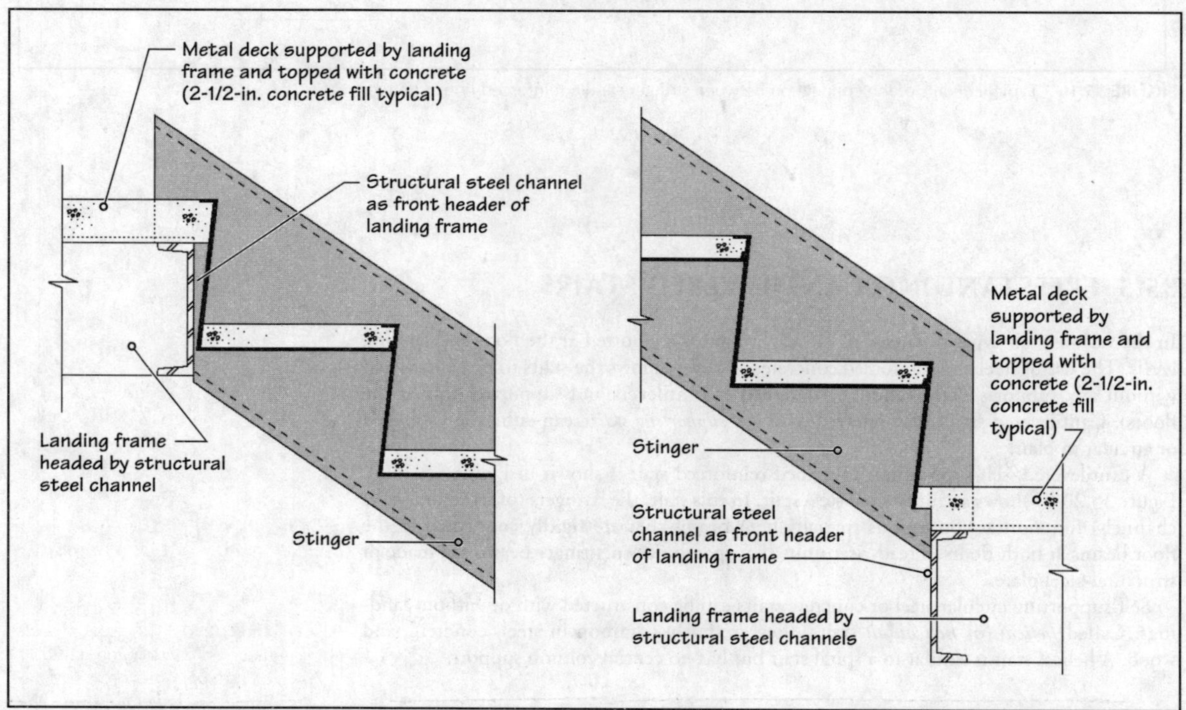

FIGURE 35.18 Typical details of the connection between stringers and the landing frame.

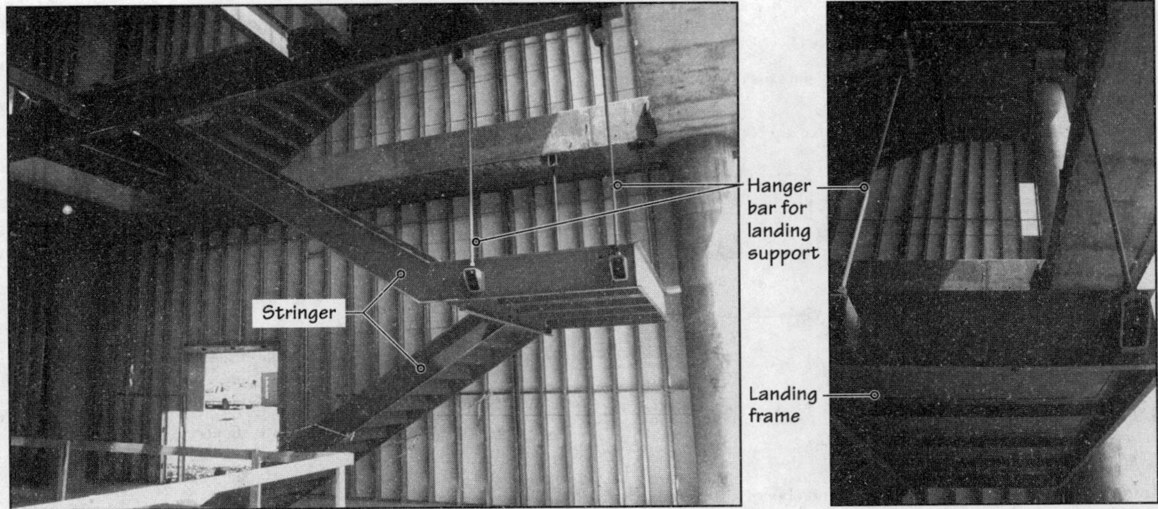

FIGURE 35.19 A typical steel stair with a suspended landing —a commonly used landing support system for exit stairs in concrete and steel-frame buildings. (Guard units and handrail have not yet been installed.)

(a)

(b)

FIGURE 35.20 Two examples of cantilevered freestanding stairs, which are supported on the upper and lower floor structures, with no supports provided at mid-landing levels.

PRACTICE **QUIZ**

Each question has only one correct answer. Select the choice that best answers the question.

14. In a typical wood stair, inclined beams that are cut to allow for the support of treads are called
a. *rough stringers.*
b. *finish stringers.*
c. *balusters.*
d. all of the above.
e. none of the above.

15. In a typical wood stair, the number of stringers required is determined by the
a. width of the stair.
b. spanning capacity of the carriage material.
c. floor-to-floor height.
d. thickness of the treads.
e. all of the above.

16. In a prefabricated steel stair, treads and risers are generally two separate components.
a. True
b. False

17. In a typical prefabricated steel stair, the number of stringers required is determined by the
a. width of the stair.
b. spanning capacity of the stringers.
c. floor-to-floor height.
d. spanning capacity of the tread-riser units.
e. all of the above.

18. In a typical prefabricated steel stair, the stringers are cut to accommodate treads and risers.
a. True
b. False

19. The stringers in a typical prefabricated steel stair are generally made of
a. wide-flange sections.
b. channel sections.
c. plates.
d. (a) and (b).
e. (b) and (c).

20. The landing frame in a typical prefabricated steel stair is generally hung from the building's structural frame.
a. True
b. False

21. A self-supporting cantilevered stair is supported on intermediate landings only.
a. True
b. False

22. A self-supporting cantilevered stair can be made only of reinforced concrete.
a. True
b. False

REVIEW QUESTIONS

1. Provide the approximate expression used in proportioning the dimensions of the treads and risers of a stair. What is the basis for this expression? List building codes' restrictions on the dimensions of treads and risers.
2. With the help of at least two sketches, explain what a flight of stairs implies. What is the code-mandated maximum height of a flight of stairs?
3. Using a sketch, explain the difference between a handrail and guardrail of a stair, and give their code-mandated heights.
4. List the factors that determine the width of a stair.
5. Using sketches, explain how a steel stair's landing frame can be supported. Which one of these support methods is most commonly used?
6. Explain why a prefabricated steel stair is most commonly used even in buildings that are built with a reinforced-concrete frame structure.
7. Using a sketch, describe a cantilevered freestanding stair.

CHAPTER 36 Floor Coverings

CHAPTER OUTLINE

Floors have had a broad range of functions and finishes throughout the history of architecture, ranging from basic and utilitarian to complex and ornate. They have been ignored, and they have been treated as the canvas for some of the most enduring artistic expressions. It should be noted that "looking down" after entering a building is an important consideration that should not be ignored.

The commitment to floors has always been based on the project budget. If funds are not available, floor finishes are left basic, and if funds are available, finishes tend to be of a higher quality. Because there are so many material, color, pattern, and covering options available, floors are still valued as surfaces worthy of financial investment and creative expression.

36.1 SUBFLOORS

The top surface of the structural floor is called the *subfloor*. Subfloors are constructed early in the building process. Therefore, it should be expected that the subfloor surface will become damaged due to the construction that takes place on or above it, necessitating some subfloor surface repairs before the floor covering.

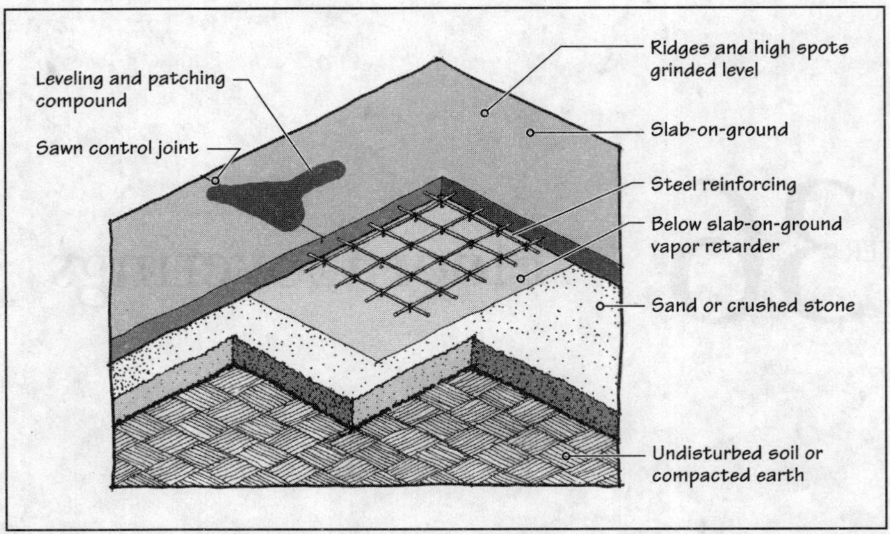

FIGURE 36.1 Typical concrete slab-on-ground components, including the leveling and patching compound used before laying the floor covering.

Types of Floor Structures

There are generally two types of subfloor structures over which floor coverings can be installed:

- *Concrete Subfloors:* Elevated concrete floors (cast-in-place concrete and concrete on metal deck) and concrete slabs-on-ground, Figure 36.1.
- *Wood Subfloors:* Wood panels supported by wood light-frame or cold-formed steel-frame members, Figure 36.2.

Subfloor Characteristics

The long-term durability of floor coverings depends on the quality of the subfloor conditions. A good subfloor will have some of the following characteristics:

- *Sound:* Concrete should be finished smooth, without ridges, ripples, and imperfections, and wood structures should be properly attached, solid and tight, with nails, screws, or adhesive.

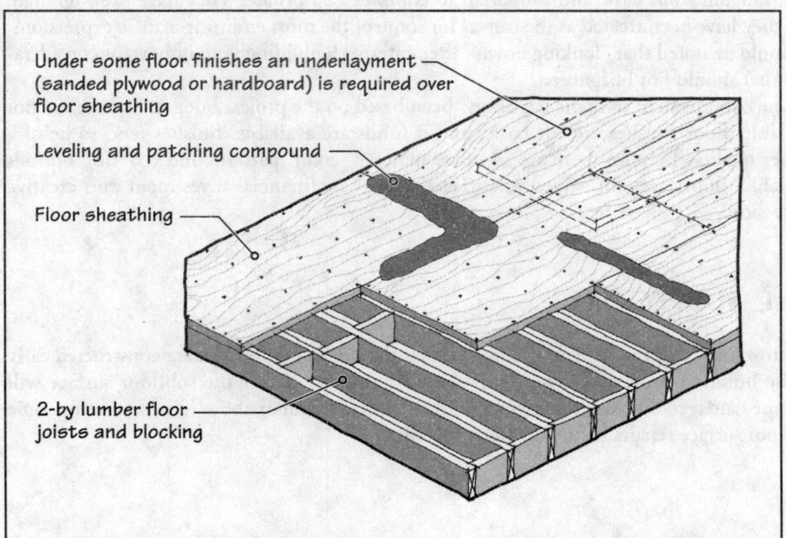

FIGURE 36.2 Typical wood-framed floor components, including the leveling and patching compound used before laying the floor covering.

- *Dry:* Subfloors must be sufficiently dry because adhesives will not adhere properly to a wet or partially wet surface. Additionally, excessive moisture in a concrete subfloor will damage floor coverings as the moisture leaves the subfloor.
- *Flat, Smooth, and Level:* Subfloors should be flat, smooth, level, and free from irregularities such as rough spots, cracks, depressions, ridges, and holes. Many floor coverings are thin and flexible, and imperfections in the subfloor will telegraph through the product and become permanent imperfections in the floor covering. Generally, subfloors should be level, with no more than $\frac{1}{8}$ to $\frac{3}{16}$ in. of vertical change in any 10-ft horizontal dimension. Deflection also contributes to surface irregularities.
- *Clean and Free of Foreign Materials:* There should be no foreign materials, including dust, solvents, paint, wax, oil, and so on, on the surface of the subfloor.

NOTE

The moisture content of a concrete slab-on-ground is particularly important if the floor covering is polymer-based (e.g., vinyl tiles, linoleum, thin-set terrazzo, and resinous floorings). Because these floor coverings are not vapor permeable, they do not allow water vapor to pass through them. Therefore, as the water in the slab converts to water vapor, it delaminates the floor covering from the slab. (Remember from Chapter 6 that water vapor pressure is generally very high.) The use of a vapor retarder under a concrete slab-on-gound is important for the same reason.

The dryness of a concrete slab is not of much concern if the floor covering and adhesive are portland cement–based, such as portland cement terrazzo.

36.2 SELECTION CRITERIA FOR FLOOR COVERINGS

An almost limitless number of floor covering products and materials is available today. The design and articulation of floor surfaces are limited only by the designer's imagination. In fact, it is not unusual for the floor coverings to be the dominant element in building interiors. The selection of materials must consider the design intent as well as the manner in which people will interact with the floors.

PERFORMANCE FACTORS

There are many performance factors that should be considered when selecting floor coverings, Some of these are related to the type of use that is anticipated for a given room or area within a building. Most materials will be tested according to ASTM standards and rated for their ability to perform under a variety of conditions.

Some of the following factors are critically important and necessary to consider, whereas others do not apply to every situation.

- *Slip Resistance:* Many injuries result annually from falls on slippery surfaces, particularly when wet. In areas where this is a hazard, surface conditions must provide adequate resistance to slipping. This is measured by laboratory testing that establishes a numerical value known as the *coefficient of friction.*
- *Durability and Longevity:* Floor coverings may be subjected to many types of use and exposure to environmental impacts. These include wear and damage due to the abrasive impact of pedestrian and other types of traffic; staining and damage from liquids, chemicals, and reagents; degradation from exposure to ultraviolet radiation from skylights or large windows coverings; and issues related to water and moisture.
- *Flammability:* Building codes require resistance to fire propagation (flame spread and smoke development) for some applications.
- *Sound:* Floor coverings impact the transmission of sound in several ways. Some types can absorb sound, lowering the airborne noise level within rooms. They can also be selected for their ability to provide sound isolation based on the impact insulation class (IIC); see Chapter 8.
- *Hygienic Qualities:* In some applications, such as a hospital or commercial kitchen, a floor covering should not give sanctuary to dirt and bacteria; resistance to mold and mildew growth is especially important.
- *Walking Comfort:* When people will be constantly standing on, or walking over, a floor covering, underfoot comfort is very important. Hard, unyielding surfaces can cause walker fatigue and, in some cases, lead to foot, leg, or back stress problems.
- *Sustainability:* Maintaining good indoor air quality is one of the criteria that contribute to a sustainable environment. Floor coverings and adhesives that do not outgas volatile organic compounds are important parts of the sustainability criteria included in the LEED rating system. Other considerations include the content and source of the material.
- *Maintenance Requirements:* Maintenance is an ongoing expense for every building. For every floor covering, there is a prescribed maintenance procedure to ensure the longest possible performance time. Some floor coverings are more time-consuming and expensive to maintain than others, a factor that must be considered in terms of long-term maintenance goals and budgets for the building.

36.3 CERAMIC AND STONE TILE FLOORING

The composition and installation of ceramic tile have changed little over the decades. Stone can be fabricated into either tiles, considered in this section, or panels, considered in Section 36.4. The tiles are adhered with mortar, and the voids between tile edges are then filled with grout.

CERAMIC TILE CHARACTERISTICS

Ceramic tile is made from natural clay, porcelain, or other ceramic materials. The exposed face is either glazed or left unglazed and then fired to a temperature sufficient to produce the necessary physical properties.

Glazed tiles have glassy, or glossy, exposed surfaces. The glaze protects the tile body against water absorption and provides for a wide range of colors. Glazes can be opaque, transparent clear, or colored clear, and sheens may be mat (low or no gloss), semimat (moderate gloss), or bright (high gloss). One of the weaknesses of glaze is that it can be scratched, which changes its appearance and performance characteristics.

STONE TILE CHARACTERISTICS

Stone tile, also known as *cut stone* tile, has many of the same characteristics as stone panel flooring, which is discussed in Section 36.4. In contrast to panels, the backs of tiles are gauged, or cut flat, to establish uniform thicknesses.

PHYSICAL PROPERTIES OF CERAMIC AND STONE TILE

Ceramic and stone tile have a variety of physical properties, which influence the selection of the appropriate tile for a particular use.

- *Quality and Uniformity:* Consistent tile size, color, pattern, and texture are important for the finished construction. Because ceramic tiles are fired products, facial and structural defects may develop that disqualify them for installation. Crazing, the development of tiny cracks in the surface of a tile or in the glazing process, also causes premature deterioration.
- *Shapes and Dimensions:* Dimensional consistency is important for proper installation. The width of the grout joint is determined by the amount of variation that is allowed in the width, length, and thickness of a group of tiles. Tight control over dimensional variation allows smaller grout joints, whereas looser control permits wider joints.
- *Warpage:* If tiles are not uniformly flat, they will not lay flat on the subfloor and will break under heavier loads because portions of the tile will not be adequately supported by the underlying mortar. In addition, after installation, the edges of warped tiles that are slightly higher than those of an adjoining tile, called *lippage,* could cause people to trip and fall.
- *Water Absorption:* Tiles do not absorb atmospheric moisture, as wood does; however, they absorb water when in direct contact with it. The tile industry classifies tiles according to permeability, Table 36.1.
- *Breaking Strength:* Tiles should maintain their integrity when subjected to loads or when objects are dropped on the floor. The breaking strength of a tile is a good indication of how well it will perform.
- *Abrasion Hardness:* Tiles should be hard enough to resist the abrasion that will occur over their service life.

TABLE 36.1 PERMEABILITY AND APPLICATION OF VARIOUS TILES

Tile type	Permeability	Application
Impervious	0.5% or less	Frequent or regular water contact
Vitreous	0.5 to 3.0%	Occasional water contact and/or
Semivitreous	3.0 to 7.0%	when water is quickly removed
Nonvitreous	Above 7.0%	No water contact except maintenance

TYPES OF CERAMIC TILES AND TRIMS

Ceramic tiles come in a variety of shapes with corresponding trims (transition pieces between the floor and the vertical surfaces, such as walls and columns). Trim shapes are shown in Figure 36.3. The Tile Council of North America (TCNA) publishes the TCA *Handbook for Ceramic Tile Installation,* which contains installation methods for every type of mortar and the various applications in which each can be used.

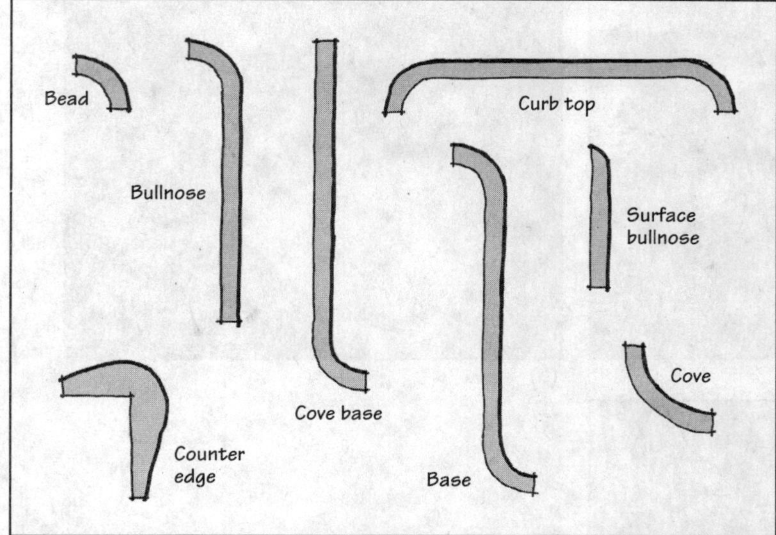

FIGURE 36.3 Ceramic trim shapes. The glazed side is indicated by the heavier line.

SETTING METHODS

As stated earlier, tiles are set using mortar. First, a mortar bed is set. If waterproofing is required, the mortar bed is placed in two applications with a waterproof membrane in the middle. If the floor will be subjected to a considerable load or if the floor structure deflection will be more than allowed by the tile installation method, the mortar bed should be reinforced with wire mesh or metal lath. Additionally, a cleavage (bond-breaking) membrane is sometimes necessary to prevent bonding with the subfloor. Reinforcing and cleavage membranes are not needed if the mortar bed is to be bonded with the subfloor.

Second, tile is set on the mortar bed while the mortar is still green, that is, not cured. If the tile is set after the bed has cured, a portland cement bond coat is required between the tile and the mortar bed.

The primary differences between the three primary tile-setting methods (described below) are the thickness of the mortar bed and the type of mortar used.

- *Thick-Set (or Thick-Bed):* Usually requiring a 2- to 3-in.-thick mortar bed, this method is necessary where (a) the floor tiles are large, generally more than 12 in. × 12 in., (b) the floor slopes to floor drains, (c) there is excessive variation in the thickness of tiles, as is generally the case with natural stone panels, or (d) the subfloor has surface irregularities.
- *Medium-Set (or Medium-Bed):* Although not officially recognized by the several industry standards, the medium-set method actually involves thin-set mortars that can be applied thicker than traditional thin-set mortars. Allowing a bed of $\frac{1}{4}$ to $\frac{3}{4}$ in. in thickness, this method provides extra setting space when the subfloor is not properly prepared or when large-format tiles (18 to 24 in. in one dimension) are used.
- *Thin-Set (or Thin-Bed):* More popular than the other methods (requiring less material and labor), this method is used where (a) the tiles are small, generally less than 12 in. × 12 in., (b) no slope to floor drains is required, (c) the tile thickness is relatively uniform, or (d) the subfloor does not have excessive surface irregularities. The mortar bed is generally $\frac{1}{8}$ in. thick and consists of polymer-based adhesives, Figures 36.4, 36.5, and 36.6.

SETTING MATERIALS—MORTARS, ADHESIVES, AND EPOXIES

A wide variety of setting materials are available to adhere ceramic and stone tile to the subfloor using any of the three setting methods.

- *Organic adhesives are usually* ready-to-use liquid or powdered water-emulsion latex products that cure by evaporation They are typically for light-duty, interior-use-only installations and are not suitable for high temperatures.

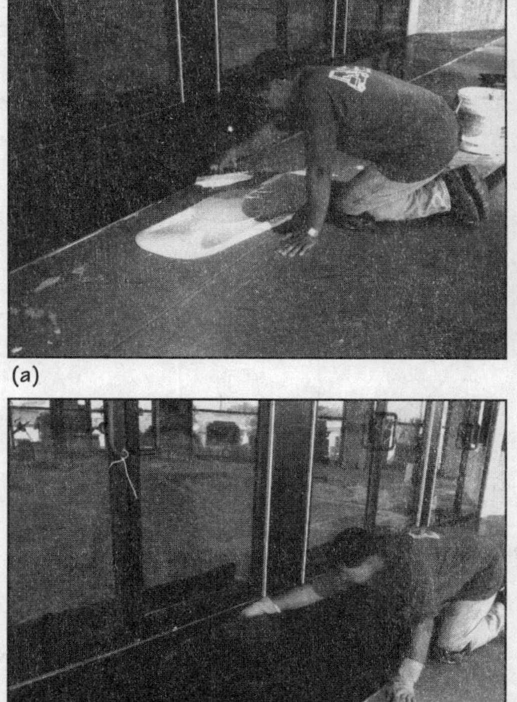

(a)

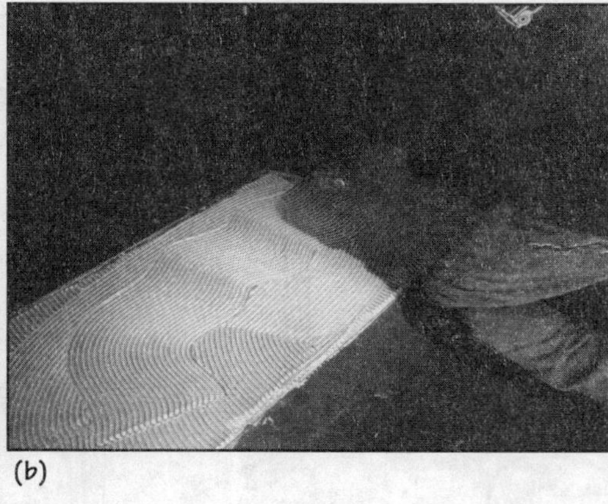

(b)

(c)

FIGURE 36.4 Installation of tile using the thin-set method at an exterior building entrance. **(a)** Latex-modified portland cement mortar being spread on a concrete slab using a notched trowel for a thin-set application. **(b)** A tile being set into the mortar bed. The purpose of the notched trowel is to ensure that the proper amount of mortar will be applied; when the tile is pressed down, the mortar becomes evenly distributed. **(c)** Grout being applied to the tile joints, usually 24 h after tile has been set.

FIGURE 36.5 A closer view of mortar being spread using a notched trowel (in this case, an epoxy mortar). (Photo courtesy of Mapei Corp.)

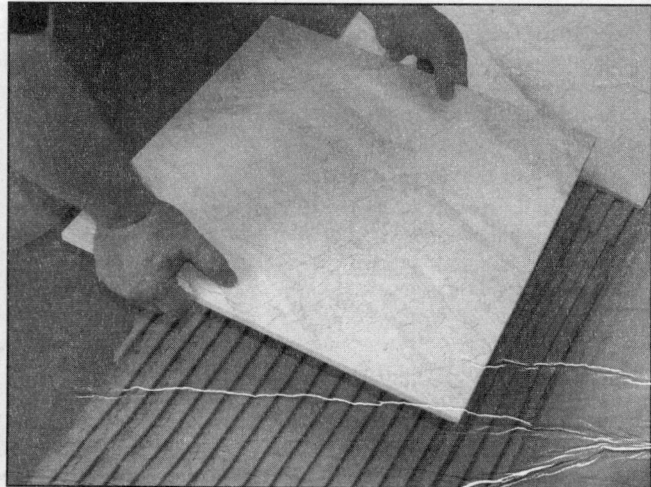

FIGURE 36.6 A closer view of setting large-format tile in mortar. The purpose of the notched trowel is that when the tile is set and pushed into place, the mortar will redistribute itself into a uniform bed. (Photo courtesy of Mapei Corp.)

- *Cement mortars* consist of mixtures of portland cement, sand, water, and water-retentive additives and are for general-duty installations.
- *Water-cleanable epoxies,* composed of an epoxy resin and a hardener, are suitable for heavy-duty installations, high-temperature conditions, and specific functions.
- *Furan resin mortars* consist of furan resin, powder containing carbon or silica fillers, and an acid catalyst and are formulated for resistance to chemicals.

TABLE 36.2 TILE TYPES, GENERAL DESCRIPTIONS, AND DIMENSIONAL CHARACTERISTICS OF THE VARIOUS CERAMIC TILE CLASSIFICATIONS

Tile type	General description of tile	Impervious tile	Vitreous tile	Semivitreous tile	Nonvitreous tile
UNGLAZED CERAMIC MOSAIC TILE	• Formed by the dust-pressed or plastic method • Either porcelain or natural clay composition • Plain or abrasive mixture throughout • Homogeneous color throughout the body	X			
QUARRY TILE	• Glazed or unglazed • Formed by the extrusion method • Either natural clay or shale composition	X	X	X	
PAVER TILE	• Glazed or unglazed • Formed by the dust-pressed method • Either porcelain or natural clay composition	X	X	X	
PORCELAIN TILE	• A ceramic mosaic or paver tile • Formed by the dust-pressed method • Composition resulting in a dense, smooth, fine-grained tile with a face that is sharply formed	X			
GLAZED WALL TILE	• Not expected to withstand excessive impact • Not subject to freezing and thawing conditions • Not suitable for floors				X
GLASS MOSAIC TILE		X			

Setting Materials—Grouts

After the tiles have been set and the mortar has partially cured, grout is used to fill the joints between the tile edges. As with mortars, there are several types of grout, each with specific application methods, described in the TCA *Handbook for Ceramic Tile Installation.*

- *Sand–portland cement grouts* are used for joints greater than $\frac{1}{8}$ in. wide, whereas *unsanded cement grouts* contain water-retentive additives and are for joints up to $\frac{1}{8}$ in. wide.
- *Polymer-modified cement grouts* tend to perform better than portland cement grouts. These grouts possess increased color stability, good flexural and bond strengths, stain resistance, and lower moisture absorption, so they resist frost damage.
- *Water-cleanable epoxy and furan resin grouts* are essentially the same as the mortars and are used with the appropriate mortar.

Installation

The following industry standards establish the appropriate installation method for an application:

- TCA *Handbook for Ceramic Tile Installation*
- ANSI A108/A118/A136 Series, *American National Standard Specification for the Installation of Ceramic Tile*

These standards provide details about installing tile on various subfloors, such as concrete, and wood panel floors. Other installation methods include those for ceilings, swimming pools, countertops, stairs, steam rooms, shower stalls, bathtub walls, refrigerated rooms, new tile over existing tile, window stools, and thresholds. The standards also contain guidelines for installing tile over sound-rated floors.

One of the most commonly overlooked aspects of successful tile installation for large floor areas is the need for movement joints. Movement joints are typically filled with a pedestrian traffic grade urethane or silicone sealant over either a foam backer rod or bond-breaking tape. The following general guidelines are helpful for locating movement joints in a tile installation:

- For interior installations, 20 to 25 ft in each direction when not exposed to direct sunlight and 8 to 12 ft in each direction when exposed. For exterior installations, 8 to 12 ft is required in each direction.

- When tile adjoins other restraining structural members, such as columns, curbs, walls, and ceilings; also at changes in floor elevation.
- At subfloor construction, contraction, and expansion joints.
- The widths of joints should never be less than the joint in the subfloor below.

36.4 STONE PANEL FLOORING

Stone panels for flooring, also known as *dimension stone*, are natural stones that have been selected and fabricated (cut and trimmed) to specific shapes and sizes, with or without mechanical dressing of one or more surfaces.

Many of the physical properties of natural stone are discussed in Chapter 25. It is important to note that the properties of stone that are critical for selecting exterior cladding may not be important (or as important) for the stone used in floors. For example, the weatherability and flexural strength of stone (important for exterior cladding) are less important for flooring, where stone's compressive strength and abrasion resistance are more important.

Natural stones used for panel flooring include granite, marble, limestone, slate, and other quartz-based stones such as sandstone, bluestone, and quartzite.

Stone panels are uniformly dressed on five faces, with the backs being left ungauged (cut in a way that does not allow for uniform thicknesses). The panels are actually slabs of stone and usually have large dimensions in one or both directions. Because the panels are not uniformly thick, they must be installed over a thick-set mortar bed.

Physical Properties of Stone Panels

Most stones are more than adequate for use as a floor covering because they have the ability to resist abrasion, wear, and absorption. When they are used in exterior applications, damage is possible due to water permeability, inelasticity, or low compressive strength. Some fabrication options, such as microfractures caused by thermal finishing, may render stone unsuitable for exterior applications.

Patterns and Finishes for Stone Panel Flooring

Virtually any stone panel floor pattern can be used. The choice is limited only by the shape and size of the individual stone panels that can be fabricated.

Although there are more suitable finishes for some stones, such as polished for granite and honed for marble, any of the following mechanical finishes can be used with most stones:

- *Polished:* Finished to a reflective sheen and resistant to wear (granites more so than marbles), a polished finish can be scratched and dulled by abrasive materials, such as sand on the shoes of people that walk over it.
- *Honed:* Finished to a uniformly matte sheen, a honed finish can be used to mask wear.
- *Thermal:* Exposure to an open flame essentially burns off the immediate surface, leaving a slightly roughened surface that has improved slip resistance.

Installation

The usual installation method for stone panel flooring is the thick-set (thick-bed) method, Figure 36.7. Because the backs of stone panels are usually ungauged, the panels are larger than tiles, and there is slight variation in thickness among the panels; therefore, the thin-set installation method cannot be used.

An important consideration is whether the flooring and its setting bed should bond to the subfloor or not. Because stone panel flooring is a rigid assembly, it is likely to develop cracks if the subfloor deflects excessively. If the stone panel flooring is installed on a concrete slab-on-ground that is not subject to deflection, a cleavage membrane is not needed. On the other hand, if the stone panel flooring is installed on a subfloor subjected to excessive deflection, an unbonded application using a cleavage membrane is required so that the setting bed does not bond to the subfloor.

A setting bed composed of a damp mix of portland cement and sand that is reinforced for unbonded assemblies or unreinforced for bonded assemblies is placed and then leveled. Prior to setting of the stone panels, the setting bed is sprinkled with water to start the cement-setting process.

(a) (b) (c) (d)

FIGURE 36.7 **(a)** Natural stone panel flooring being set by the thick-set method. Once the exact panel locations are determined, several rows are set as the guide for the remainder of the installation. Note the water-sprinkling can for sprinkling the setting bed to start the setting process. Also note that this is a bonded installation without reinforcing or a cleavage membrane. **(b)** The thick-set mortar is being readied for the installation of the next stone panel. The rubber mallet is used to adjust the panels. **(c)** Plastic shims are used to keep the joints between the stone panels the same width. The round stickers on the panel edges are numbers that are coordinated with the shop drawings. **(d)** Finished stone flooring after grouting and polishing.

While the setting bed is still plastic, the backs of the stone panels are fully buttered with a bond coat of cementitious materials and then set in place. Care must be taken to maintain uniform joint widths and consistent elevations of the various panels.

After the mortar has cured sufficiently, the joints are grouted. It should be noted that because of the uniformity of panel sizes, the joints of stone panel flooring may be as small as $\frac{1}{8}$ in. in width.

PRACTICE QUIZ

Each question has only one correct answer. Select the choice that best answers the question.

1. The architect should expect the subfloor to sustain some damage during the construction process.
   a. True **b.** False
2. The two most common materials used as a subfloor are
   a. steel decking and wood plank.
   b. concrete and steel.
   c. concrete and wood panel.
   d. wood light-frame and plywood panels.
   e. carpet and vinyl flooring.
3. Subfloors should be free of moisture because
   a. moisture can cause wood to swell, damaging floor coverings.
   b. moisture in the subfloor causes water-based adhesives to debond.
   c. moisture can make floor surfaces slippery.
   d. all of the above.
   e. (a) and (b).
4. Slip resistance is measured through laboratory tests that determine
   a. the coefficient of resistance. **b.** slip resistance.
   c. emulsification. **d.** the coefficient of friction.
   e. the impact frequency.
5. In applications where hygiene is the most important consideration, a floor covering that is resistant to mold and mildew growth is required. The floor covering should also
   a. provide sanctuary to dust and dander.
   b. resist penetration by dust and bacteria.
   c. have low impact resistance.
   d. be in a color that camouflages any dirt that may accumulate.
   e. be inexpensive.

PRACTICE QUIZ *(Continued)*

6. *Stone panel* is another name for *stone tile*.
   a. True
   b. False
7. The thick-set method of setting tile is used when
   a. a slope is required to drain water.
   b. thin ceramic tile is used.
   c. irregular tile is used.
   d. (a) and (b).
   e. (a) and (c).
8. Epoxy mortar and epoxy grout are similar in chemical composition.
   a. True
   b. False
9. Stone panel flooring is laid using the
   a. deep-set method.
   b. medium-set method.
   c. thick-set method.
   d. grout mix.
   e. thin-set method.
10. The stone panels used for flooring are left ungauged
    a. on the sides.
    b. at the bottom face.
    c. at the top face.
    d. on all sides and faces.

36.5 TERRAZZO FLOORING

Terrazzo flooring has much in common with concrete. Like concrete, a terrazzo binding matrix is mixed with several aggregates and placed, wet and plastic, in its final location. However, unlike concrete, the exposed surface is ground and polished after it cures to expose the binding matrix and aggregates, thus revealing a smooth and colorful finish. No longer limited to utilitarian purposes, as in the past, the use of terrazzo has become much more common; it can be placed in an explosive variety of creative designs, Figure 36.8.

CHARACTERISTICS OF TERRAZZO FLOORING

A binding matrix is the material that holds the aggregate chips in position. Until about the end of the twentieth century, the most common binding matrix was cementitious, combining portland cement (usually white), aggregate chips, and pigments if needed (either alkali-resistant or synthetic powdered, inorganic substances) to form a uniform matrix color. This binding matrix is now being replaced with a resinous epoxy, most commonly polyester, polyacrylate-modified cement, or polyurethane.

The creative design of terrazzo flooring is the primary criterion for selecting aggregate chips. The defining quality for these chips is their ability to be ground and polished. Marble chips are the most common aggregate; however, granite, onyx, travertine, or glass chips are also used, and mother-of-pearl is especially favored.

Metal divider and control strips are used to control cracking and to create designs. Cementitious terrazzo requires closely spaced strips to control cracking. Resinous terrazzo does not crack, so the strips are used to create decorative designs. Usually made of white zinc alloy, brass, or plastic, strips are available in standard, K, and L shapes.

TYPES AND INSTALLATION OF TERRAZZO FLOORING

Industry standards for the various types of terrazzo are defined in the National Terrazzo and Mosaic Association's (NTMA) *Terrazzo Ideas and Design Guide*. Terrazzo types are distinguished by the manner in which they are installed:

FIGURE 36.8 The creative use of divider strips and colored terrazzo at the entry of a hospital. (Photo courtesy of HKS Inc.)

- *Cementitious terrazzo* requires a recessed subfloor up to 3-in. deep, Figure 36.9, and has several variations.
- *Sand-cushion cementitous terrazzo* consists of a $\frac{1}{2}$-in.-thick standard terrazzo topping over a $2\frac{1}{2}$-in.-thick reinforced cementitious underbed that is separated from—and not allowed to bond with—the subfloor by a $\frac{1}{4}$-in-thick sand bed. The total thickness ranges from $2\frac{3}{4}$ to 3 in. and shrinkage is relieved by the divider strips, which are required at closely spaced intervals. In this system, substrate cracks do not telegraph through to the topping.
- *Bonded cementitious terrazzo* is not as deep, and consists of a $\frac{1}{2}$-in.-thick standard terrazzo topping over an unreinforced cementitious underbed that is bonded to the subfloor, resulting in a total thickness that ranges from $1\frac{3}{4}$ to $2\frac{1}{4}$ in. Crack control is very important, and the installation should include divider strips at closely spaced intervals.

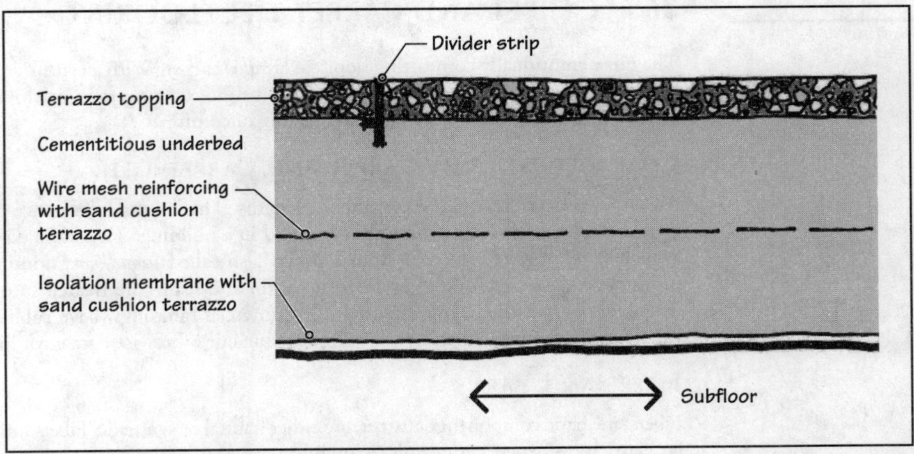

FIGURE 36.9 A section through sand-cushion cementitious terrazzo.

- *Structural terrazzo* combines a standard topping with a concrete slab-on-ground structural slab and is used when single responsibility for construction is desired. The total thickness ranges from $4\frac{1}{2}$ to 6 in.
- *Rustic terrazzo* is used when grinded surfaces are not required to be smooth, such as at exterior locations. Rather than being ground and polished, rustic terrazzo is washed with water or otherwise treated to expose the aggregate.
- *Epoxy terrazzo* combines the advantages of cementitious systems with the dramatic improvements made in epoxy resins. It is lighter and more flexible than cementitious terrazzo, Figure 36.10, it can develop higher bond strengths, and it is resistant to mild acids, impact indentations, concentrated load impressions, and staining. Epoxy terrazzo requires a thin veneer ($\frac{1}{4}$ to $\frac{3}{8}$ in.), eliminating the need for floor recesses. Unless required for a design, divider strips may be located at larger intervals than those used with cementitious systems. The time from placement to finish grinding is shorter than with cementitious systems, and when combined with crack isolation underlayment, epoxy terrazzo can have the highest resistance to crack telegraphing of any system.
- *Other thin-set terrazzo systems* may be used when specific performance characteristics are required. These include conductive, polyester-resin, and polyacrylate-modified cement types.
- *Precast terrazzo* shapes are made at the manufacturing plant and include floor tiles, sloped shower-stall floors, stair treads and risers, and wall base strips.

Installation

When the subfloor or cementitious underbed has been prepared (depending on the system used), including filling cracks, divider strips are attached and then the terrazzo topping is placed in a wet and plastic condition. It is leveled, rolled, compacted, and troweled to a dense, uniform, and flat surface that reveals the divider strips. The topping is allowed to cure until it develops sufficient strength to prevent chips from being lifted or pulled out by the grinding machine. It is then ground with a series of various grit stones, progressing from rough to fine; imperfections are grouted; and, finally, it is polished to the desired sheen.

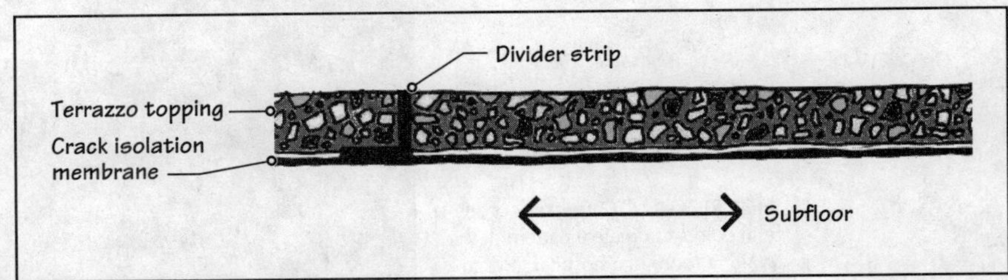

FIGURE 36.10 A section through epoxy terrazzo.

36.6 CARPET AND CARPET TILE FLOORING

The most emotionally comforting floor covering is carpet. With a centuries'-long history of exquisite rugs and carpets, carpet continues to be the covering of choice for many due to its softness, contribution to quietness, and feeling of comfort.

CHARACTERISTICS OF CARPET AND CARPET TILE

All carpet is manufactured in continuous lengths. The essential difference between carpets is in their construction and how they are used in a building. Two basic types of carpet are available: *rolled goods* and *tiles*. Rolled goods, also called *broadloom*, dominates the carpet industry (rolls are typically 6 or 12 ft in width, and seams are treated so that they appear invisible). Carpet tiles, typically used in commercial buildings, have rubber backings and can be easily changed when a tile becomes stained or otherwise damaged, Figure 36.11.

FIBERS AND YARNS

Fibers, the basic component of carpet, are either natural or synthetic. Fibers must be of particular elasticity, fineness, uniformity, durability, weight (called *denier*, which is a measure of weight to unit length), luster (brightness or reflectivity), cleanability, and soil hiding (ability to hide the presence of soil). Three fibers are used to make the yarn that is used to make carpet:

- *Nylon* has proven to be among the most durable fibers developed and is the most popular fiber commercially. Nylon can be dyed to achieve bright colors and is resistant to crushing, has good fade resistance, and can be easily cleaned. One of its weaknesses is that it is prone to static electricity buildup; however, that problem can be alleviated with antistatic treatments.
- *Polypropylene*, a by-product of the gasoline-refining process, has superior soil-resistant properties and low water-absorption rates. Weaknesses are that colors tend to be dull, because the fibers do not accept water-based dyes or stains, and that polypropylene has poor resiliency, a lower melting point than nylon, and poor texture retention.
- *Wool* is the original carpet fiber and is generally the most expensive. It has an outstanding ability to be dyed in deep, rich colors. It has exceptional memory and will return to its original state, without rupturing or tearing, after being stretched; it can also shed water. Although not commonly used in commercial buildings, wool is especially suited for hospitality and entertainment areas and is common in residential applications.

Fibers are grouped together in various ways to make yarn, which is then used to construct the exposed face of carpet, Table 36.3. There are several ways to achieve colors and patterns in carpet and carpet tile. In some cases, the fiber is dyed before being made into yarn (synthetic fibers are solution-dyed as the filament is extruded), in other cases the yarn is stock-dyed before spinning or yarn-dyed after spinning, and in still other cases the carpet is dyed after it is constructed (similar to printing).

CONSTRUCTION AND PHYSICAL PROPERTIES OF CARPET AND CARPET TILES

There are different ways of constructing carpet, each with particular advantages and disadvantages. As the yarns are constructed into a carpet, there are several face constructions that are available, Table 36.4.

FIGURE 36.11 Carpet tiles of several colors to create a pattern in the waiting room of a hospital. (Photo courtesy of HKS Inc.)

TABLE 36.3 CARPET CONSTRUCTION TYPES

Type		Description
TUFTED	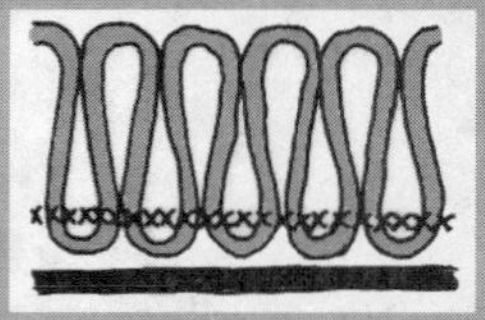	Accounting for the vast majority of broadloom carpets, this inexpensive and fast construction method involves hundreds of needles on tufting machines, which essentially sew, or punch, multiple rows of yarns into and through a primary backing. The back is coated with latex to stabilize the tufts, and is then backed with a secondary backing to give the completed construction dimensional stability.
FUSION BONDED	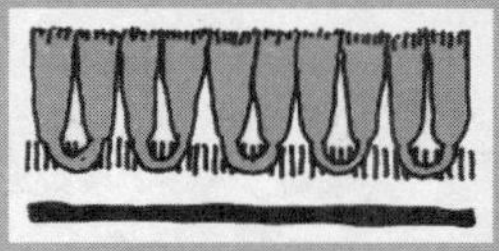	Dominating the carpet tile industry, fusion bonding consists of a spun yarn bundle that is heat-fused into a liquid vinyl compound in a sandwich-like configuration. A knife then splits the sandwich into two carpets; thus, cut pile is the only face construction that can be produced.
NEEDLE-PUNCHED	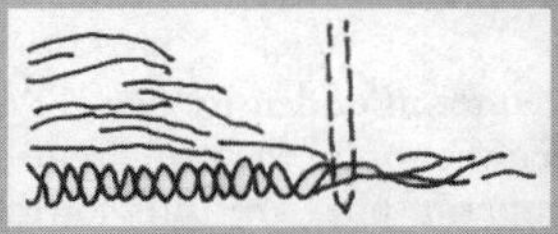	Hundreds of barbed needles punch individual loose synthetic fibers through a woven blanket of synthetic fibers that forms a dense, homogeneous sheet without pile, similar to a heavy felt.
KNITTED		Similar to textile knitting, various yarns are knitted into connected loops in differing directions to produce a sheet that does not require backings.
WOVEN	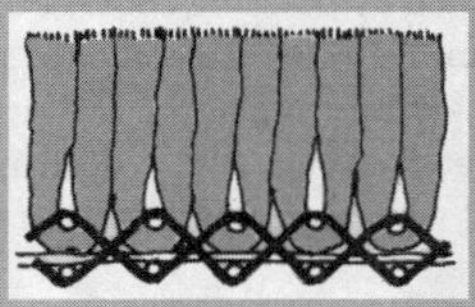	Historically, carpets and rugs were woven by hand on looms. Today, machines weave pile yarns and backing yarns simultaneously. Although requiring more time than the tufting process, woven carpet is distinguished by its tailored textures and intricate patterns. Two of the most important assets of woven carpets are that they are more dimensionally stable and more resistant to wear than other carpet constructions.

TABLE 36.4 CARPET FACE CONSTRUCTIONS

	Description
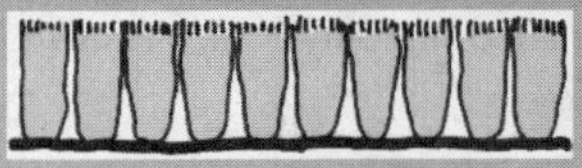	CUT PILE Tops of the yarn loops are cut to reveal the ends of fibers, creating a plush appearance.
	TEXTURED SAXONY Tops of the yarns are evenly cut across the top to reveal more tip definition.
	FRIEZE Yarn is tightly twisted to produce a curled, or kinked, effect.
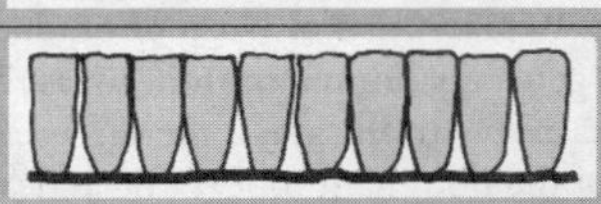	TIP-SHEARED Tops of tufted loops are shaved to create a cut or uncut texture or pattern.
	RANDOM-SHEARED Similar to the multilevel loop, but the higher loops are sheared off to reveal a combination cut and uncut textures.
	MULTILEVEL LOOP Yarns are looped at varying heights to create patterns or effects.
	LEVEL LOOP Tops of the yarn loops are level with each other.

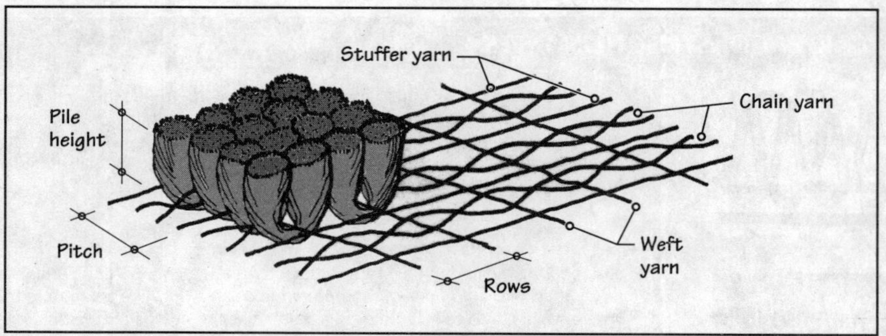

FIGURE 36.12 General characteristics of knitted and woven carpets.

NOTE

Critical radiant flux (CRF) refers to the rate of flow of radiant energy through a material. In the case of resilient flooring, it is measured in terms of watts per square centimeter (W/cm^2) because it is a direct reflection of resistance to the spread of fire in this type of material. Class I materials (the most fire resistant) have a 0.45 W/cm^2 CRF, and Class II materials are 0.22 W/cm^2.

Carpet construction is detailed, Figure 36.12, in terms of density (weight of the pile in a unit volume), pile height (dimension between the top of the primary backing and the finished top of the yarn), rows or wires (number of pile yarn tufts for each 1 in.), stitches (number of yarn tufts for each 1 in. of length), gauge (number of yarn ends counted across a specific dimension of the width), pitch (number of surface yarn ends in 27 in. of width), total weight, face weight, tuft density (total tufts for each 1 in.), and yarn count. Each of these properties affects the look, feel, and performance of a given carpet.

Carpets are also evaluated for flammability, as regulated by building codes. Flame spread is tested by the flooring radiant panel test, which determines a carpet's ability to spread fire; this is identified as the critical radiant flux (CRF). Another measure of the ability of a carpet to promulgate a flame from a small source, such as a match or cigarette, is determined from the pass-fail methenamine pill test, in which an ignited tablet is dropped on the carpet and flammability is then measured.

In addition, the performance of carpets can be enhanced by treating them to resist staining, static electricity, and bacteria growth.

Carpet Cushion

Some, but certainly not all, broadloom or rolled goods carpets require a cushion between the subfloor and the carpet to achieve their desired performance. Cushions tend to be used in hospitality areas, hotel rooms, and most residential buildings. Carpet cushions are not required for carpet tiles. Several types of carpet cushions are commonly available, including fiber, sponge rubber, and polyurethane foam.

Installation

There are various methods for installing carpet and carpet cushion; the choice depends on the carpet type, the end use of the areas, and the type of subfloor. The most commonly used methods are as follows:

- The *stretch-in* method is used for broadloom carpets. The carpet is placed over an independent cushion, and is stretched (placed in tension) and then securely hooked to tackless nail strips at the perimeter of the room. Replacement of worn or outdated carpet can be easily accomplished. This method is generally limited to those areas that experience low traffic. If it is used in heavy or extra-heavy traffic areas, the carpet and the cushion will tend to "travel" in the direction of the dominant foot traffic and will become wrinkled or detached from the nail strips, and the cushion may bunch up, or buckle, underneath the carpet.
- *Direct glue-down* is used for broadloom carpet and carpet tile. In this method, a compatible adhesive is used to attach the carpet to the subfloor, and it will prevent buckling under traffic load. It will also resist carpet movement due to rolling loads as well as changes due to temperature and humidity.
- *Double glue-down* is essentially the same as the direct glue-down method; however, the carpet cushion is also glued to the subfloor.

While not frequently used, several other methods are available, including the hook-and-loop method and a preapplied-adhesive system.

36.7 WOOD FLOORING

Wood and manufactured wood products are discussed in Chapters 13 and 14, and much of the same information also applies to wood flooring. Wood flooring is valued in residential applications for the warmth and beauty it brings to a room. It is also used extensively for athletic floors, such as basketball, volleyball, and racquetball courts, as well as for theater and dance floors, where durability, hardness, and resiliency are highly desired. Although most wood species can be fabricated for wood flooring, all of them are not equal in performance.

CHARACTERISTICS OF WOOD FLOORING

Virtually any hardwood or softwood species that can be domestically harvested or imported and provided to a fabricator can be used for wood flooring. However, some species may not be easily available, and some cannot be suitably finished for wood flooring applications. The characteristics of most importance for evaluating wood flooring are species, appearance (including grain, texture, and color), cut, durability, dimensional stability, and ability to accept finishes.

Oak and maple are the most commonly used species in the United States; however, currently, as many as 50 domestic and exotic species are available, including the following:

White ash
American and red beech
Yellow birch
African, black, and Brazilian cherry
Hickory
Black and hard maple
Northern red, red Southern, and white oak
Pecan
Eastern white, Southern yellow, and heart pine
African, black, Brazilian, and Peruvian walnut

Generally, wood floorings are classified as either solid wood flooring or engineered wood flooring. As the name implies, in *solid wood flooring* the same wood species is used throughout the entire piece. It is susceptible to changes in size due to moisture content. It can be flat sawn, quarter sawn, or rift sawn and is available in different appearance grades. Like plywood, *engineered wood flooring* is a combination of a surface veneer, usually a hardwood, which is laminated to one or more plies of a wood veneer from a less expensive wood species that provides dimensional stability and added strength.

Either material can be fabricated into a variety of shapes and sizes. They are categorized into three groups: strips (usually $1\frac{1}{2}$ to $2\frac{1}{4}$ in. in width and random in length), planks (usually 3 to 8 in. in width and random in length), and parquet (a patterned floor, which can be strips, planks, blocks, or square tiles).

The typical profile for wood flooring is known as a *tongue-and-groove profile*, Figure 36.13. This allows for accurate alignment during installation and—because of its uniformity—tight,

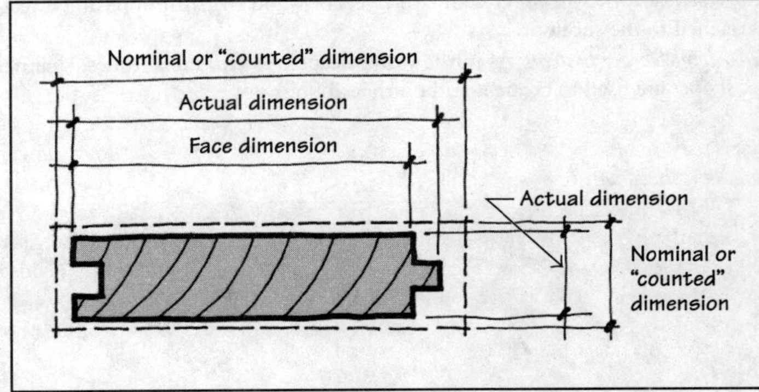

FIGURE 36.13 Solid wood flooring cross-sectional dimension terminology.

level joints. The thickness of solid wood flooring varies according to usage: $\frac{5}{16}$, $\frac{3}{8}$, $\frac{1}{2}$, and $\frac{5}{8}$ in. thick for light use; $\frac{3}{4}$ and $\frac{25}{32}$ in. thick for normal use; and $\frac{33}{32}$, $1\frac{1}{4}$ and $1\frac{1}{2}$ in. thick for heavy use.

Wood floorings are finished to protect the wood from dirt, abrasion, wear, oxidation, and moisture (to some extent). They can be finished by the manufacturer, resulting in higher-performing finishes than can be achieved when finished in place. When finished in place, the floor is sanded, stained, sealed, and waxed. Wood flooring can be factory impregnated with an acrylic, which increases its density, hardness, and resistance to wear. Because the acrylic seals the wood from moisture, the wood's dimensional stability is greatly improved.

PHYSICAL PROPERTIES OF WOOD FLOORING

The two most important physical properties to consider for wood flooring are density and hardness. A wood species with a high density is required for wood flooring. Hardness is necessary to resist indentation and marring. The side hardness of the wood species listed above range from 380 ft-lb for eastern white pine (softest) to 3,680 ft-lb for Brazilian walnut (hardest).

The species of the wood used and its grade and cut are important considerations that affect the dimensional stability, durability, and appearance of wood flooring.

Wood flooring used for athletic purposes should possess the physical properties of absorption of impact shock, ball rebound, deflection resistance, and control of surface friction. Hard maple is favored because it meets these requirements well.

INSTALLATION

Unlike the other floor coverings discussed in this chapter, controlling the moisture content of wood flooring during fabrication and installation is imperative. Wood is a hygroscopic material (each wood species and cut absorbs and releases moisture at unique rates) and, therefore, changes dimensionally by swelling and shrinking. Controlling moisture content is more important for solid wood flooring than for engineered wood flooring. To control moisture, wood is kiln dried to establish a baseline of moisture content; then, during fabrication and installation, it should be protected from changes in moisture. It is common to place wood flooring in the place where it will be installed for some time before installation to allow it to acclimate to the temperature and relative humidity.

Another unique factor concerning wood flooring is that an expansion space is necessary around the perimeter to accommodate the small amount of movement the flooring may encounter due to changes in temperature and relative humidity during the life of the installation.

There are several methods of installation for those wood floorings not installed for athletic purposes. Historically, strips and planks were installed over wood sleepers. Today, strips, planks, and parquet can be installed over a plywood subbase or directly to the subfloor. Nails, staples, and water-based adhesives are used to attach to wood subfloors, whereas adhesives are required for concrete.

A more sophisticated method of installation, in which an air space is established between the wood flooring and the substrate, is used for athletic wood flooring. Usually a vapor retarder is placed over the concrete substrate; then one of a variety of support systems may be used:

- *Floating Systems:* Resilient materials such as neoprene, rubber, or polyvinyl chloride sheets may be adhered to the substrate or a foam underlayment may be used to isolate the wood flooring from the subfloor.
- *Fixed-Wood-Sleeper Systems:* One or two layers of wood strips, at opposing directions, are attached to the subfloor.
- *Fixed-Metal Sleeper Systems:* As in the wood sleeper system, metal tubes, channels, or other shapes are used to create a grid on the subfloor.

PRACTICE **QUIZ**

Each question has only one correct answer. Select the choice that best answers the question.

11. Terrazzo flooring is like concrete in that it

a. is composed of a binding agent and aggregate.
b. is liquid when placed and dries hard.
c. is saw-cut to control cracking.
d. all of the above.
e. (a) and (b).

12. Cementitious terrazzo floor coverings are approximately $\frac{1}{4}$ in. thick.

a. True **b.** False

13. Carpet is a relatively new flooring material.
 a. True b. False

14. The major physical characteristic that differentiates carpets is
 a. color. b. fabric.
 c. construction. d. size.
 e. cost.

15. The desired carpet color is achieved by
 a. dying the fiber.
 b. dying the yarn.
 c. dying the carpet after it is constructed.
 d. all of the above.
 e. none of the above.

16. In the face construction of cut pile carpet,
 a. the yarn is twisted, curled, and kinked.
 b. a multilevel effect is created by shearing the yarn.
 c. the yarn is looped at varying heights.
 d. the tops of yarn loops are cut to reveal the tips of the fibers.

17. The majority of broadloom carpets are
 a. fusion bonded. b. woven.
 c. tufted. d. knitted.
 e. needle punched.

18. All carpet installations include a cushion between the subfloor and the carpet.
 a. True b. False

19. The standard size for wood floor pieces is
 a. strips $1\frac{1}{2}$ to $2\frac{1}{4}$ in. wide.
 b. planks 3 to 8 in. wide.
 c. parquet blocks or tiles.
 d. There is no standard size for wood flooring.

20. The highest-performing floor finishes are
 a. waxed in place.
 b. acrylic impregnated.
 c. finished in place with water-based sealant.
 d. applied by the manufacturer prior to placing the wood.

21. It is essential to control the moisture content of wood flooring, especially during installation and fabrication.
 a. True b. False

22. Wood flooring is installed using
 a. adhesives. b. nails.
 c. screws. d. all of the above.
 e. none of the above.

36.8 RESILIENT FLOORING

One of the two most commonly used floor coverings is resilient flooring (the other being carpet). Resilient flooring comprises those products that spring back into shape after being compressed as a result of impacts due to walking or other activities normally found in buildings. It can be easily cleaned without special knowledge or specialized products, and it can withstand general use without permanent deformations or damage. Resilient flooring is typically used for utilitarian purposes in rooms that require regular wet cleaning and where there is a probability that water will be on the floor.

Forms of Resilient Flooring

There are two forms of resilient flooring: tiles and sheets. Resilient tiles are uniformly sized after manufacturing, most commonly 12 in. square; however, other sizes and shapes are available. Tiles are usually more economical and are used when joints are not objectionable or when there is a desire to create a particular pattern with tiles of different colors. Resilient sheets are available in rolls of continuous length, which are typically 6 or 12 ft in width. Sheets are used when there is a desire either to cover as much of the floor as possible without joints or to create a larger pattern that cannot be accomplished by tiles.

Physical Properties of Resilient Flooring

As with the floor covering materials discussed previously, there are several physical properties that must be considered. Among all of the building materials and products available, few exceed the vast variety of colors, patterns, textures, styles, and designs that are commonly available in resilient flooring. Common appearances include the traditional speckled pattern, solid colors, and imitations of natural stone and transparent finished woods.

Many resilient flooring products are chemical resistant; however, solvents such as ketones, esters, and chlorinated and aromatic hydrocarbons may cause permanent damage, Resilient floorings also have varying degrees of burn resistance. (See the note on CRF in Section 36.6.)

Another consideration is that all resilient flooring is not equally resilient. Permanent indention of the surface can result from furniture and equipment loading if the appropriate floor covering is not selected. This is described in terms of *static and dynamic load resistance.*

Some resilient floorings are also measured for their hardness or softness. Soft materials may be required in some uses to resist cuts, punctures, tears, or heavy traffic. In other instances, the materials may need to be hard to endure their use.

Exterior application of resilient flooring is usually not recommended by manufacturers because ultraviolet light can cause fading, shrinking, and blistering. When it is used in rooms flooded with exterior light, the degradation potential should be evaluated. In some applications *electrostatic discharge resistance* is required, and product types include static-dissipative, static generation–resistant, and static decay–resistant resilient floorings.

NOTE

For many years, resilient flooring and adhesives contained a strong, fire-resistant mineral fiber called *asbestos*. This material, also previously use for ceiling tiles, insulation, roofing shingles, and other products, which is known to cause cancer, has been banned from all construction materials, including resilient flooring.

NOTE

In fire-rated rooms and corridors, the use of resilient floor coverings may be restricted based on the CRF (class).

NOTE

Resin is a naturally occurring sticky, organic substance that is discharged from pine, fir, or other plants. Rosin is a hard, amber-colored by-product of the distillation of turpentine or naptha extract from pine trees. The "resin" in resinous flooring is a liquid synthetic polymer or epoxy that forms into a sheet as it cures.

SOLID VINYL TILES

Solid vinyl tiles are composed of three primary ingredients—pigments, fillers, a vinyl chloride polymer or copolymer binder—and other modifying resins, plasticizers, and stabilizers. The thickness is approximately $\frac{3}{32}$ in. and sizes range from 12 to 36 in. square. Solid vinyl is also available in planks that replicate the appearance of transparent finished wood, which are 36 in. in length and 3 to 6 in. in width. Tile classifications include monolithic, surface-decorated, and printed film (all smooth or embossed surface).

VINYL COMPOSITION TILES

Widely popular and similar to solid vinyl tiles, vinyl composition tiles are less expensive, Figure 36.14. The thickness is usually $\frac{1}{8}$ in. However, $\frac{3}{32}$ in. is also available, and the size is usually 12 in. square. The classifications include solid-color, through-pattern, and surface-pattern tiles.

RUBBER TILES

Rubber tiles are manufactured from a vulcanized compound of natural or synthetic rubber (no minimum content required) along with pigments, fillers, and plasticizers. Thicknesses range from $\frac{3}{16}$ to $\frac{5}{16}$ in., and sizes range from 12 to 24 in. square. Rubber tiles are commonly used for stair treads and riser covers. Tile classifications include homogeneous (solid-colored and mottled) and laminated (solid-colored and mottled wear-layer). Recycled-rubber material content is very popular and is formed into mats with thicknesses ranging from 1 to $2\frac{1}{2}$ in.

SHEET VINYL

Similar to solid vinyl and vinyl composition tiles, sheet vinyl is ideally suited for situations where the floors are regularly or frequently cleaned with water and cleaners. It provides minimal joints, and the sheets can be bent up the wall to form an integral base, known as a *flash cove base.* Additionally, the joints may be sealed using a vinyl rod of the same color and pattern as the sheet vinyl. The joint is either chemically bonded or welded with hot air, making the joint watertight and giving a seamless appearance. The wear layer can be clear, translucent, or opaque, and it can have a background that is either printed or otherwise prepared. The surface can be smooth, embossed, or textured. The thickness is approximately $\frac{3}{32}$ in., and the widths are 6 ft 0 in., 6 ft 7 in., and 12 ft 0 in. Sheet classifications include backed (nonasbestos fibrous formulations, nonfoamed plastic, and foamed plastic, each graded according to application) and unbacked.

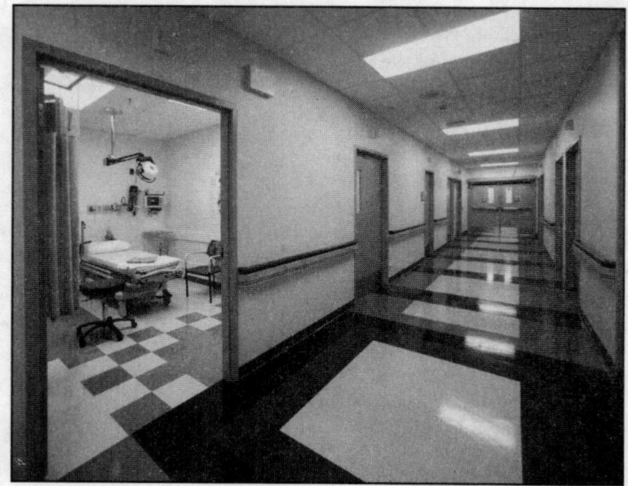

FIGURE 36.14 A multicolored pattern of vinyl composition tiles installed in a hospital. (Photo courtesy of HKS Inc.)

LINOLEUM TILES AND SHEETS

Linoleum is the original resilient sheet on which all other resilient floorings have been patterned. Invented in England in 1860, linoleum consists of a binding agent (a mixture of linseed oil combined with either fossil, pine, or other resins or rosins or an equivalent oxidized oleoresinous material) and a filler (ground cork, wood flour, mineral fillers, and pigments). This mixture is then solidified, bonded, and keyed to a fibrous backing such as burlap (jute). The joints of sheets and tiles can be made watertight by sealing them with heat-welded rods of linoleum. Thicknesses range from $\frac{3}{32}$ to $\frac{1}{8}$ in., and sheets are as wide as 6 ft 7 in.

CORK TILES

The bark of cork trees and recycled cork are ground into small granules to form a synthetic resin matrix, which, when formed into a tile, becomes a flexible product that is homogeneous and uniform in composition. Product options include solid and homogeneous tiles and engineered tiles. Engineered cork tile may be composed of a patterned cork veneer laminated to a cork base and combined with rubber (30% by volume) to form cork rubber, or it may be laminated around a medium-density fiberboard core to form planking with tongue-and-groove edges (that can be constructed into a floating floor). The surface can be sanded, left unsanded, waxed, stained, coated with polyurethane or acrylic, or finished in the factory or in the building with other finishes. Thicknesses range from $\frac{3}{16}$ to $\frac{5}{16}$ in., and sizes range from 12 to 24 in. square.

STATIC-CONTROL RESILIENT FLOORING

There are situations in which floor coverings are required to control electrostatic discharge, such as in sensitive manufacturing environments, computer and electronics rooms, and explosive environments. A static-control floor covering system is vital in resisting the electrostatic charges generated by people, furniture casters, and equipment movements. Essentially, static-control resilient flooring directs an electrical charge to a reliable grounding source. These systems are composed of floor-covering products (like those described earlier) with conductivity elements within the body of the material, and require the use of a static-control adhesive and grounding strips.

INSTALLATION

Preparation of the subfloor is critically important for a successful installation of resilient flooring, and smoothness is the most essential factor. Ridges, high spots, and other imperfections extending above the surrounding surface should be removed. Depressions, scratches, gouges, and other imperfections extending below the surrounding surface should be filled. Latex-modified portland cement–based products can be applied with a trowel to level and patch imperfections. In some instances, a self-leveling product may be necessary for an extensive area.

Once the imperfections are corrected and the subfloor is free of dust and debris, the resilient flooring is applied by a full spreading of adhesive using a notched trowel.

The layout of the floor covering should be considered in order to achieve the desired design expression and joint pattern and to minimize waste. Consideration should be given to alignment with the principal wall while discounting minor offsets. Floor coverings should be scribed and cut to fit around, and butt tightly to, vertical surfaces and permanent fixtures, furniture, and cabinetry. Cuts should be neat and straight, and joints should be tight.

36.9 RESINOUS FLOORING

In recent years, the dramatic advance in chemical technology has led to improvements in liquid-applied resinous flooring systems. Resinous flooring, also called *polymer flooring* or *epoxy flooring*, is applied in its liquid form; when cured, it provides a flexible, seamless, and uniform surface (applied as a liquid that cures into a sheet). The flooring is thin and has excellent bonding capability, mechanical strength, and abrasion and impact resistance. It may be used in decorative applications of different performance qualities. These combine decorative aggregates (ceramic-coated silica, known as *colored quartz*, marble, granite, dyed stone, or vinyl flakes) in an epoxy resin matrix. (Resinous [epoxy] terrazzo is actually resinous flooring.)

High-performance or *special-use* resinous systems may be formulated for specific environmental exposures or for specific purposes such as resistance to severe and corrosive chemicals, acids, and solvents; resistance to extreme cyclic changes in temperature; or for static-dissipative or conductive purposes. Epoxy, epoxy-novolac, urethane, and vinyl-ester systems are the primary systems available.

36.10 OTHER FLOOR-COVERING MATERIALS

In addition to the floor-covering materials discussed previously, there are other materials that are commonly available and can be used, including brick and concrete.

FIGURE 36.15 A concrete floor that has been highly polished with an abrasive medium. Notice the mirror-like reflection and the clarity of the reflected light fixtures. (Used with permission of the Concrete Polishing Association of America.)

Full-size brick or split-face bricks can be used for interior and exterior floor coverings. They can be thin-set or thick-set mortared for interior floors, with or without grouted joints, generally following the same methods used for ceramic and stone tile flooring or stone panel flooring.

Concrete is also becoming more popular as a finish material. It may be dyed or stained so that color penetrates through the surface and reduces the visibility of wear. The finish has a mottled appearance and can either be left as is or covered with a clear sealing product. Concrete can also be polished to various sheen levels. The surface is ground, honed, and polished using abrasive media, with finishes that range from matt to reflective sheens, Figure 36.15. It is also possible to expose the aggregate in the concrete matrix. A polished or high-gloss finish can also be created using clear or tinted sealers, Figure 36.16.

There are many floor-covering products that can be used for special purposes. These include rubber mats around swimming pools and in athletic locker rooms, and elevated access flooring used in computer equipment and television broadcast rooms where there is a considerable amount of wiring on the floor. Also increasingly popular in large office buildings is elevated access flooring combined with the heating-cooling systems to distribute air below the flooring instead of through overhead ducts.

FIGURE 36.16 A clear coating applied to exposed concrete flooring. (Used with permission of RetroPlate System, PSI Permanent Surfaces; photo by Rhonda Clinton.)

36.11 UNDERLAYMENTS

An underlayment is a thin material that is adhered or applied to the subfloor prior to installing the final floor covering. It either provides protection or prepares the subfloor to receive the flooring. There are three types of underlayments: membrane, fill, and solid.

Membrane underlayments can perform three different functions, and some products may simultaneously perform two or all three of them:

- *Waterproofing:* These products are single or multiple components (some contain internal reinforcing fabrics) that are dimensionally stable, load bearing, and resistant to mold growth and possess strength in breaking (at seams and in shear). Product types include liquid-applied components (which cure into a continuous membrane), plastic sheets (polyvinyl chloride [PVC], chlorinated polyethylene, and polyethylene), Figure 36.17, and self-adhering bituminous sheets.
- *Crack Isolation* (also known as *Crack Suppression*)*:* Because subfloor structures expand, contract, and deflect very slightly over the course of the life of a building, it is sometimes necessary to provide crack-isolation membranes. They are designed to either eliminate the transfer of cracks or greatly reduce the likelihood of cracks. They are fungus and microorganism resistant and have good shear strength, good point-load resistance, and crack resistance. It should be noted that crack-isolation membranes are not a substitute for properly located and designed expansion and control joints in the building structure.
- *Sound Reduction:* These membranes mitigate, or reduce, the noise between floors, such as between two stacked residential units. It is common for building codes to require certain performance levels.

Fill underlayments can be troweled over irregular surfaces, providing a smooth surface. They are also used for leveling or resurfacing a subfloor. If the surface of the subfloor has to be raised, the self-leveling versions can be used. Fill underlayments are not intended as wear surfaces. They are typically hydraulic cement based (with high compressive strength and used for commercial applications) and gypsum based (with low compressive strength and used for residential applications) materials.

When the elevation of the subfloor has to be raised slightly or the subfloor is significantly damaged, *solid underlayments* may be used. Common materials used include medium-density fiberboard, cement fiberboards, and cement backer units. These are commonly attached using an adhesive and/or mechanical fasteners.

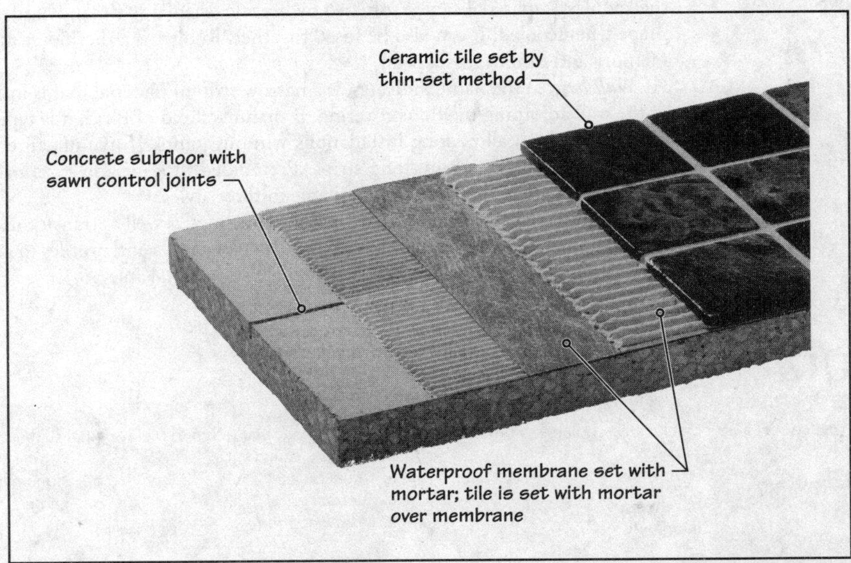

FIGURE 36.17 A waterproofing underlayment in a thin-set ceramic tile assembly. (Used with permission of Noble Company.)

TABLE 36.5 TYPICAL RESILIENT WALL BASE AND MOLDING PROFILES

	WALL BASE Left: Toeless, or straight, base Middle: Cove base Right: Trimmed base (where floor covering extends up to but not under the base) Typical heights are $2\frac{1}{2}$, 4, $4\frac{1}{2}$, and 6 in.
	STAIR TREADS AND RISER COVERS Left: Riser cover Middle: Tread cover Right: Alternate tread nosing Options include an abrasive strip just behind the nosing (shown); surface patterns on treads (smooth, dots, grooved diamond, longitudinal grooves) and glow-in-the-dark colored nosings
	STAIR NOSINGS Options include surface patterns on the tread surface abrasive strip, longitudinal grooves; glow-in-the-dark colors
	CARPET-TO-CARPET MOLDING Resilient exposed trim that snaps into an aluminum retainer; options include transitions into other floor-covering materials
	TRANSITION MOLDINGS

36.12 RESILIENT ACCESSORIES—WALL BASE AND MOLDINGS

Just as woodwork is trimmed with narrow strips of wood to close gaps between surfaces, narrow strips of resilient materials are used to trim around flooring installations and to cover small gaps, Table 36.5. Resilient accessories are typically manufactured of PVC. This thermoplastic material can be softened by heating, shaped to fit, and hardened. It retains its shape when cooled. It can also be fused together. Rubber is a thermoset material that has a memory and cannot be shaped.

Wall base, as the name indicates, is a narrow strip of material that is installed at the base of the wall adjoining the floor covering. If manufactured of PVC, it is typically available in long rolls, which allow long installations without joints. If manufactured of rubber, it is typically available in 48-in.-long strips. Premolded interior and exterior corners are also available. A variety of heights are available in three styles.

Moldings cover small gaps between floor coverings as well as transitions from one covering to another. They are available in a multitude of shapes and profiles to accommodate the many situations that occur, a few of which are shown in Table 36.5.

PRACTICE QUIZ

Each question has only one correct answer. Select the choice that best answers the question.

23. Resilient flooring is available in
- **a.** blankets.
- **b.** sheets.
- **c.** tiles.
- **d.** panels.
- **e.** (b) and (c).

24. Floor coverings that return to their original shape after normal impacts are called
- **a.** *ceramic tile.*
- **b.** *stone panels.*
- **c.** *terrazzo.*
- **d.** *carpet.*
- **e.** *resilient flooring.*

25. All vinyl tile is available only in 12-in. × 12-in. squares.
a. True **b.** False

26. The binding agent in linoleum is
a. portland cement. **b.** PVC (polyvinyl chloride).
c. linseed oil and organic resins. **d.** epoxy.

27. Electrostatic discharges generated by people, furniture casters, and the movement of equipment can be hazardous in
a. rooms or buildings that house computers.
b. some manufacturing facilities.
c. storage rooms for explosives.
d. all of the above.
e. none of the above.

28. Resinous flooring is a site-applied, synthetic seamless sheet material.
a. True **b.** False

29. The purpose of a membrane underlayment is to provide
a. resistance to cracking.
b. mitigation of the transmission of sound.
c. waterproofing.
d. any of the above.
e. none of the above.

30. A fill underlayment is a solid, hard material that is nailed directly to the subfloor.
a. True **b.** False

31. Most floor coverings require moldings or trim where the flooring meets the wall or where one type of flooring meets a different type of flooring.
a. True **b.** False

REVIEW QUESTIONS

1. Use a sketch and notes to illustrate the difference between the thin-set and thick-set methods of laying tiles. Describe when each of these systems should be used.
2. Describe the differences and similarities between terrazzo and resinous flooring.
3. Draw plans and cross sections of a ceramic tile, stone tile, and stone panel that illustrate the differences between the materials.
4. Use a sketch and notes to describe the difference between a conventional wood plank floor and a floating wood plank floor assembly.
5. Using the information provided in this chapter, select an appropriate type of flooring and state why it should be used for the following applications: busy corridor in a school, hospital kitchen, residential living room, and office with a raised-floor system.

CHAPTER 37

Ceilings

CHAPTER OUTLINE

Ceilings can be brought low and articulated to create intimate spaces, or they can be pushed high to create awe-inspiring and dramatic spaces. They can be subtle in design or they can be highly expressive. They can impede sound or amplify it, and they can direct light by reflection or they can black it out. Whatever the requirements, ceilings can be used to accomplish varied design intents.

Historically, ceilings were the underside of the floor or roof structure above; the ceilings were either finished or the structure was left exposed. The several architects of St. Peter's Cathedral in Rome, Figure 37.1, placed an exquisitely detailed dome at the center of the

FIGURE 37.1 The smooth suspended gypsum ceiling (curved in plan, narrows in width from one end to the other, and barrel vault in profile) in a contemporary foundation building on a college campus stands in contrast to, and is reminiscent of, St. Peter's Cathedral in Rome. (Photo of foundation building courtesy of HKS Inc.)

building and consequently created one of the most awe-inspiring buildings in the history of architecture. The ceiling of another awe-inspiring ancient building, the Pantheon in Rome, is coffered concrete that was originally covered in stucco (see Figure 21.5).

Buildings today are far more complex and require space for their utilities (ductwork for the distribution of heat and cooling, electrical power, lighting, communications wiring, fire protection piping, etc), as can be seen in Figure 37.2. In most buildings, the space overhead is the preferred location for many of these components; for that reason, ceiling systems were developed to conceal these from view. Today a vast number of products and materials are available for this purpose.

FIGURE 37.2 In most buildings, there are usually many utilities located overhead, as shown in this image. Because many of these utilities require periodic access, an acoustical suspended ceiling is frequently selected. (Photo courtesy of HKS Inc.)

In this chapter, we discuss the three primary strategies for constructing ceilings. First, and the simplest to accomplish, is the one in which the underside of the structure above forms the ceiling and any overhead mechanical and electrical components are left exposed. Second, the ceiling finish materials are attached directly to the overhead building structure to form a cover. Third, and most popular, ceiling finish materials are suspended from the overhead building structure. In the second and third strategies, the space between the ceiling and the floor or roof above is known as a *plenum.* When the ceiling is on the exterior of a building, it is called a *soffit.*

37.1 SELECTION CRITERIA FOR CEILING FINISH MATERIALS

Unless the ceiling is part of a fire-rated assembly (which is quite often the case), the reasons for the ceiling are discretionary. In residential buildings, the ceiling is attached directly to the structure, and in most commercial buildings the suspended ceiling strategy is used. When ceilings are not included, the reason may be to reduce costs or to express the building utilities as part of the design. At other times, suspended ceilings can be used creatively to enhance the building's design.

There are several factors that must be considered when determining the appropriate ceiling type and finishes, and many of the principles discussed in Part I of the text apply. Factors unique to ceilings include the following:

- *Aesthetic Expectations:* Ceiling finish materials, heights, and profiles can create inviting environments and can influence the way light interacts with the building's interior. Unlike floors, ceilings are not limited to a single horizontal plane; they can be articulated at several elevations or can include vertical or sloped surfaces. Ceilings are used to introduce a sense of scale and proportion to an interior space. For example, if the area of the room is large, a high ceiling can create a grand space; however, if the area is small and the ceiling is high, the room may simply feel cavernous.
- *Concealing the Building's Utilities Overhead (mechanical and electrical equipment and components):* As previously discussed, the most common reason for a ceiling is to conceal the building's structure and overhead utilities, Figure 37.2.
- *Wind Loading:* Wind can impose an uplift load on the exterior soffit surface; therefore, wind-uplift resistance becomes an important consideration. In circumstances when large areas of the building enclosure are regularly opened to the exterior, wind can enter the building and exert a force on the ceilings.
- *Volume of Occupied Space:* Ceiling height is related directly to the volume of space required to be heated and cooled.
- *Humidity:* Ceilings adjacent to openings in exterior walls, especially major openings such as at a bank of frequently used doors, may be subjected to higher levels of humidity than the ceilings in the remainder of the building. Because many materials, when oriented horizontally, are vulnerable to sagging due to moisture absorption, the choice of materials must be carefully considered.

FIGURE 37.3 The building structure, utilities, and lapidaries of a professional basketball and hockey stadium are exposed as part of the architectural design. (Photo courtesy of HKS Inc.)

- *Flammability:* Building codes sometimes require ceilings to be resistant to fire propagation and spread.
- *Seismic Activity:* In seismically active areas, building codes require products, materials, and equipment installed overhead to resist the movements caused by an earthquake. During seismic activity, support structures and ceiling finish materials must not fall down, which would cause injuries to people trying to vacate the building.
- *Sound Absorption:* Ceiling finish materials and their method of application can help absorb sound generated within a room or space.
- *Sound Isolation:* Building codes sometimes require a floor-ceiling assembly to help isolate sound from above. See the discussion of the impact isolation class (IIC) in Chapter 8.
- *Sustainability:* Many ceiling products contain recycled content and can make contributions to sustainability goals.
- *Antimicrobial Resistance:* In addition to normal maintenance, some applications may require ceilings to be resistant to bacterial growth and mold and mildew development.
- *Light Reflectance:* Ceilings can be used to reflect and/or diffuse light from other sources in order to distribute it uniformly throughout a space. Sometimes certain ceiling finish materials are selected specifically to give the space a brighter appearance because indirect lighting is the preferred lighting source. This is also an important consideration when daylighting of an interior space is part of the design.
- *Maintenance:* Some occupancies, such as healthcare facilities, require ceiling finishes that can be regularly cleaned and scrubbed to remove possible contaminants. Other occupancies may require ceilings to resist soiling, scratching, and impact.

37.2 NO CEILING FINISH—EXPOSED TO THE AREA ABOVE

One method of treating an interior space is not to have a ceiling, but rather to expose the building structure and utilities. Although this may seem simple, it requires careful design, detailing, and coordination of all exposed elements to produce a satisfactory result, Figure 37.3. This approach can make it easier to access certain elements for routine maintenance.

37.3 CEILINGS ATTACHED TO THE BUILDING STRUCTURE

In most residential buildings and some light-commercial buildings, lightweight ceiling finish materials are attached directly to the building structure. This is the most economical approach in residential applications, especially when there are few mechanical, plumbing, or electrical systems to be accommodated within the plenum. The most important consideration for this type of ceiling is that the underside of the building structure be specifically designed and located to provide surfaces for attaching the finished ceiling. Ceilings attached directly to the building structure do not usually require another structure, as is needed for suspended ceilings.

FIGURE 37.4 A typical gypsum board ceiling attached to the wood framing in a residential application.

The most common material attached to the building structure is gypsum board. It is either (a) nailed to the underside of the wood framing, Figure 37.4, or (b) screwed to light-gauge steel framing, as at the underside of the framing in Figure 20.24. The gypsum board is subsequently taped, bedded, textured, and painted. In addition, acoustical tiles can be attached with adhesive to the gypsum board finish.

PRACTICE QUIZ

Each question has only one correct answer. Select the choice that best answers the question.

1. All ceilings are formed by suspending a lightweight material from the underside of the structure to provide the interior finish.
   a. True **b.** False
2. The most common reason for a ceiling is to
   a. control heat and humidity.
   b. provide a sculpted or multilevel surface overhead.
   c. reflect light.
   d. conceal the building structure and overhead utilities.
   e. isolate sound.
3. Antimicrobial ceiling materials are most likely to be used in a
   a. university classroom.
   b. hospital or medical laboratory.
   c. department store.
   d. pharmacy.
   e. warehouse.
4. In a building where no ceiling is provided and the structure and utilities are exposed,
   a. mechanical and electrical subcontractors are allowed easy access and freedom of choice in locating their work.
   b. all systems must be carefully designed.
   c. on-site coordination of construction is critical.
   d. all of the above.
   e. (b) and (c).
5. The ceiling provides an excellent opportunity to modulate natural and artificial light.
   a. True **b.** False

37.4 CEILINGS SUSPENDED FROM THE BUILDING STRUCTURE

By far the most common ceiling in commercial buildings is one that is suspended from the building structure overhead.

PRINCIPLES OF CEILING SUSPENSION SYSTEMS

A suspended ceiling system is a relatively simple concept. A lightweight metal grid structure is suspended from the building's overhead structure to provide support for the ceiling finish products as well as to establish the ceiling height. This grid structure consists of main runners, suspended from above by steel wire, and interconnecting cross runners spaced uniformly to support the ceiling finish products.

There are three types of metal suspension systems, Figure 37.5. The type of suspension system used for a particular application depends on the structural requirements to support the ceiling finish products and other components that will be mounted on or attached to the grid structure, Table 37.1.

SUSPENDED ACOUSTICAL CEILINGS

The most common commercial ceiling consists of modular acoustical panels laid in a direct-hung suspension system known as an *inverted-tee grid.* The acoustical panels are typically 24 in. square, or 24 by 48 in., and are laid on the horizontal legs of the inverted tees, lower image in Figure 37.6.

This grid structure is composed of corrosion-resistant steel or extruded aluminum inverted tees (upside-down T-shape). The main runners and cross runners are in the same plane and are assembled by interlocking the pieces to form a unified, modular grid structure, usually on some increment of 12 in. The main runners have slots and holes that have been prepunched in the web at regular spacings to receive the end clips that are on each end of the cross tees.

TABLE 37.1 METAL SUSPENSION SYSTEMS—DUTY CLASSIFICATIONS AND STRUCTURAL CAPACITIES

		Minimum load-carrying capacity (pounds per square foot)		
Duty	Carrying capacity	Direct Hung	Indirect Hung	Furring Bar
Light	Supports only the finish material itself	5.0	2.0	4.5
Intermediate	For ordinary ceilings, supports finish material and lightweight components such as light fixtures and ceiling diffusers/grilles	12.0	3.5	6.5
Heavy	Greatest capacity to support finish material and other ceiling-mounted components	16.0	8.0	N/A

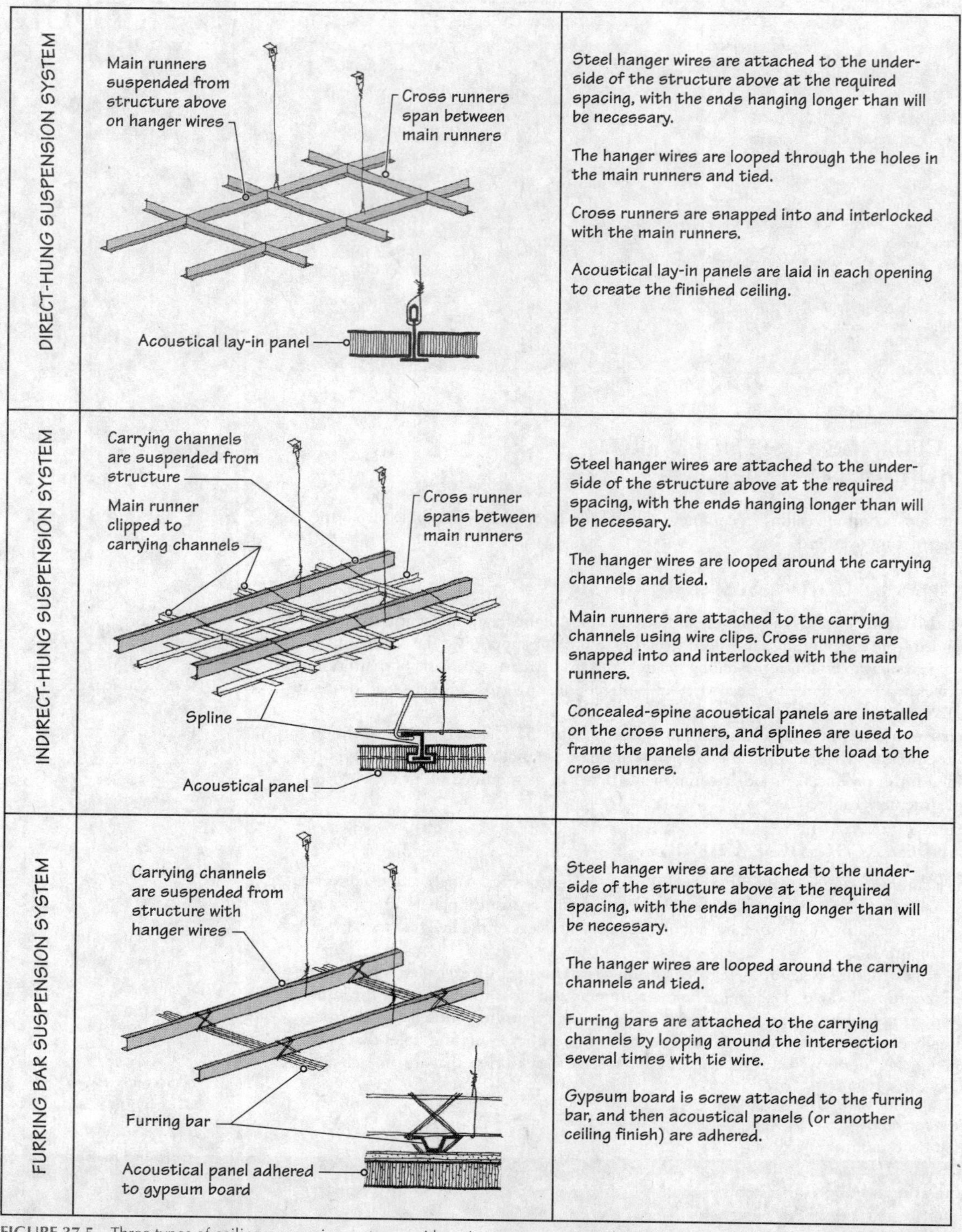

DIRECT-HUNG SUSPENSION SYSTEM
Main runners suspended from structure above on hanger wires
Cross runners span between main runners
Acoustical lay-in panel
Steel hanger wires are attached to the underside of the structure above at the required spacing, with the ends hanging longer than will be necessary.
The hanger wires are looped through the holes in the main runners and tied.
Cross runners are snapped into and interlocked with the main runners.
Acoustical lay-in panels are laid in each opening to create the finished ceiling.
INDIRECT-HUNG SUSPENSION SYSTEM
Carrying channels are suspended from structure
Main runner clipped to carrying channels
Cross runner spans between main runners
Spline
Acoustical panel
Steel hanger wires are attached to the underside of the structure above at the required spacing, with the ends hanging longer than will be necessary.
The hanger wires are looped around the carrying channels and tied.
Main runners are attached to the carrying channels using wire clips. Cross runners are snapped into and interlocked with the main runners.
Concealed-spine acoustical panels are installed on the cross runners, and splines are used to frame the panels and distribute the load to the cross runners.
FURRING BAR SUSPENSION SYSTEM
Carrying channels are suspended from structure with hanger wires
Furring bar
Acoustical panel adhered to gypsum board
Steel hanger wires are attached to the underside of the structure above at the required spacing, with the ends hanging longer than will be necessary.
The hanger wires are looped around the carrying channels and tied.
Furring bars are attached to the carrying channels by looping around the intersection several times with tie wire.
Gypsum board is screw attached to the furring bar, and then acoustical panels (or another ceiling finish) are adhered.

FIGURE 37.5 Three types of ceiling suspension systems, with various components indicated.

Acoustical panels and tiles are composed of a variety of materials, including mineral wool, mineral fibers, fiberglass, and/or perlite, which are combined with fillers, binders, and water and then formed into sheets (and cut into panels or tiles) or molded into pans. There are several surface finishes available, including paint, membrane overlay, fabric, and thin metal sheets. Surface textures and patterns include perforations, fissures, embossings, printings,

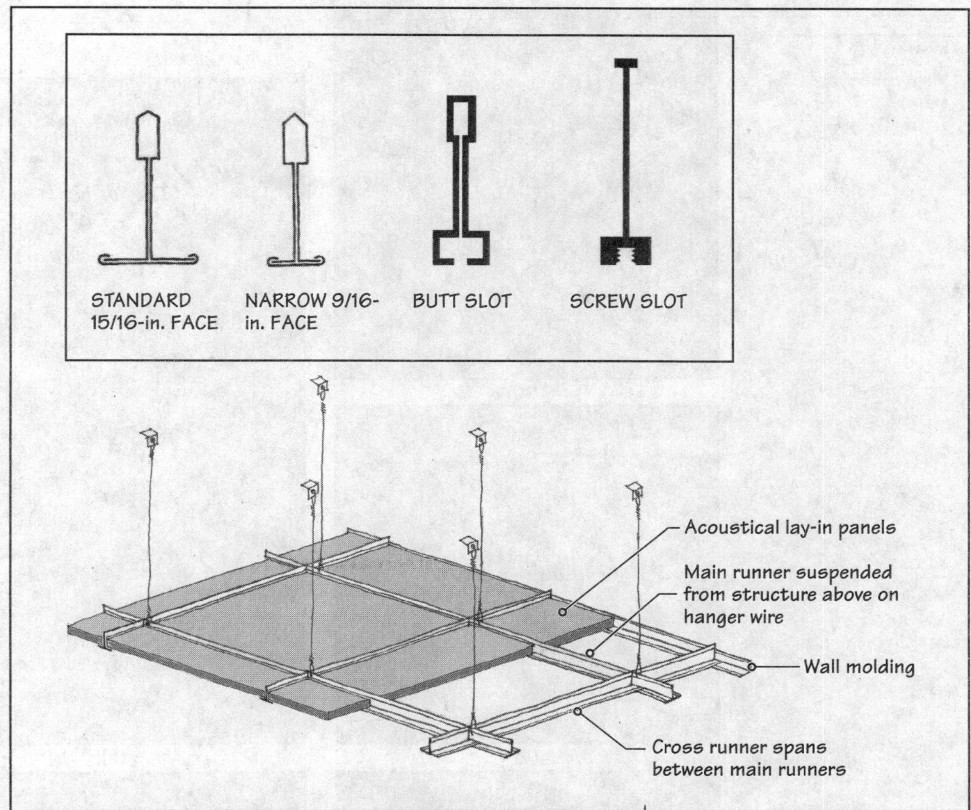

FIGURE 37.6 Suspended acoustical ceiling components and typical profiles for inverted tees.

and scorings. Finally, panels can have one of several edge treatments, Figure 37.7. Once the grid structure is assembled, Figure 37.8, and the panels and tiles are installed, no other finishing work is necessary.

When an acoustical ceiling without an exposed grid is desired, a variation of the inverted-tee grid structure is used. In addition to a suspended grid, metal splines are inserted in the edges of acoustical tiles, usually 12 in. square, which are, in turn, supported on a suspended grid.

Suspended Gypsum Board Ceilings

The second most common commercial suspended ceiling consists of gypsum board sheets screw-attached to an inverted-tee grid structure. The gypsum board is then finished like other gypsum board surfaces. Suspended gypsum board ceilings are versatile because they can be integrated with stud framing hung from the building structure above to form ceilings that have various profiles vertically and horizontally.

The suspended inverted-tee grid structure for gypsum board ceilings is very similar to the grid structure used for suspended acoustical ceilings, except that the bottom surface of the horizontal tee flange is embossed to improve the grip on the gypsum board, Figures 37.9 and 37.10.

Suspended Gypsum Plaster Ceilings and Portland Cement Plaster Soffits

Portland cement and gypsum plasters are heavier than acoustical products and materials and thus require the use of a channel grid structure. Wire lath is attached horizontally to the main and cross runners using steel wire, Figure 37.11; then the plaster is applied.

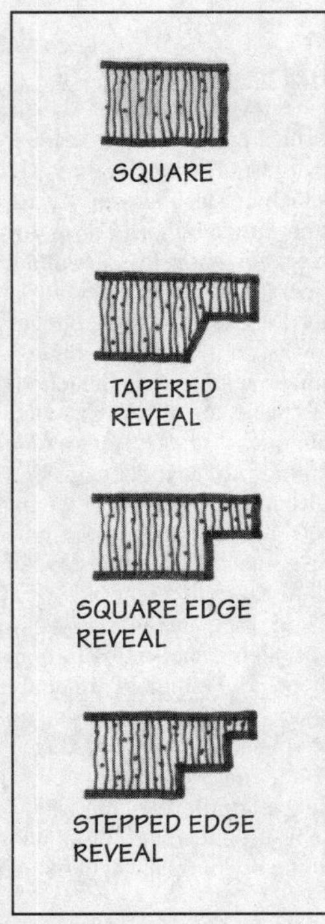

FIGURE 37.7 Various acoustical lay-in panel edge types.

FIGURE 37.8 Installation of a suspended grid. (Photos of Vaden's Acoustical & Drywall, Inc. by Gary Yancy of USG Building Systems). **(a)** Attaching steel hanger wire to the steel structural joist above. In this case, the hanger wire is tied to the steel joist. When a concrete slab or metal deck is present, a hanger wire, with an integral clip angle, is usually powder-fastener attached. Hanger wires are always attached to the building structure, not to building utilities. **(b)** Layout begins by establishing the location of the main runners. For most ceilings, installers will use stilts to assist in installation or will use a rolling scaffold. **(c)** Using string and a laser level (see (e)), installers establish main runner locations. **(d)** Until permanent attachments can be made, installers clamp intersecting tees together. **(e)** Continuing with installation; notice the laser level at the wall. **(f)** Final suspended grid installation, including openings for light fixtures.

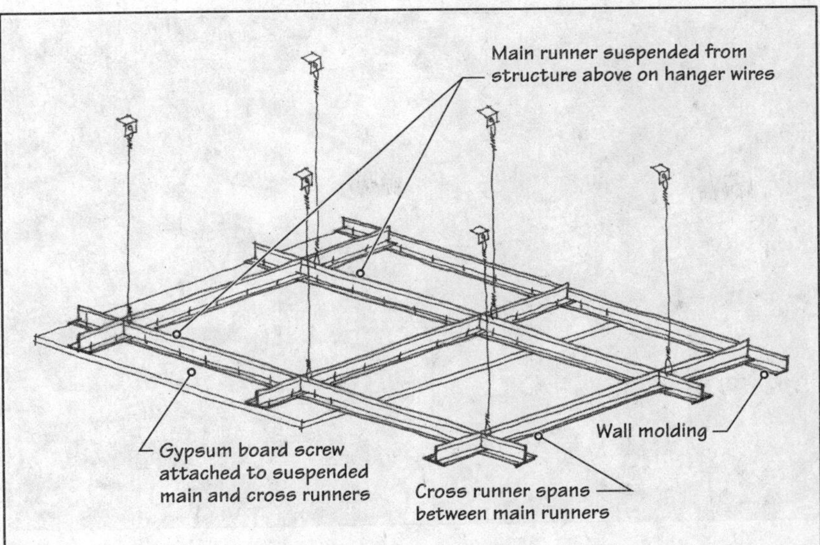

FIGURE 37.9 Suspended gypsum board ceiling components.

FIGURE 37.10 A sculpted suspended gypsum board ceiling in the dining room of an office building. (Photo courtesy of HKS Inc.)

FIGURE 37.11 Suspended portland cement plaster ceiling/soffit components. Gypsum plaster components would be similar; however, gypsum board lath would be used in lieu of wire mesh lath.

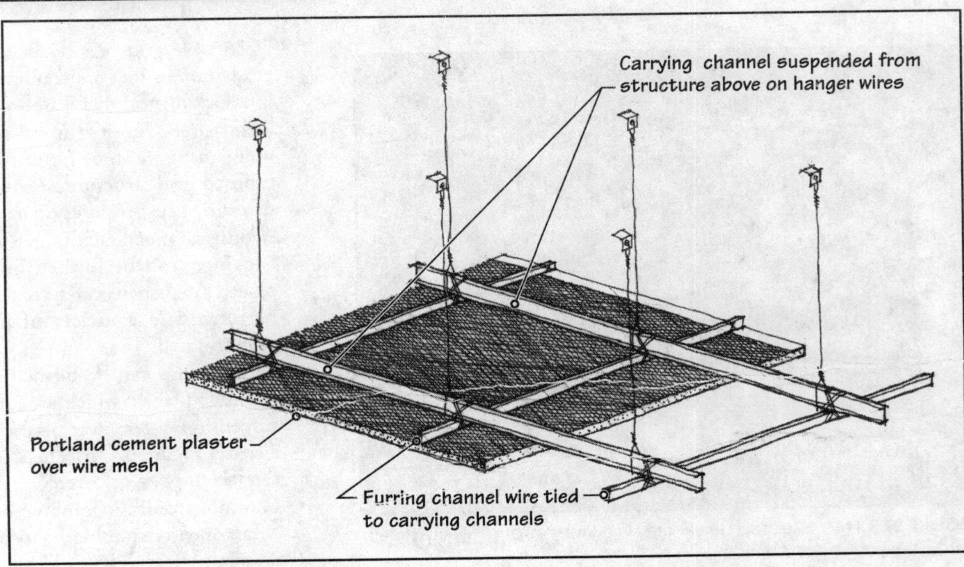

FIGURE 37.12 Suspended wood panel ceiling. (Photo courtesy of Armstrong Ceiling Systems.)

The channel grid structure is used for ceilings that are heavier than acoustical or gypsum board materials. Using galvanized-steel channel shapes, main runners and cross runners are overlapped, and each intersection is tied with steel wire. Unlike the inverted-tee grid, the spacing of the channels can be adjusted to fit the application conditions, the applied finish, and the component loading.

SUSPENDED SPECIALTY CEILINGS AND SOFFITS

As with so many other building products, a wide variety of specialty ceiling and soffit finishes and materials is available, limited only by the designer's imagination. These systems can be flat planes, squared with corners, curved vertically or horizontally, or formed into the appearance of waves. They can be continuous or broken into sections. Lighting fixtures can be above, within, or below the systems. The following discussion describes just a few suspended specialty ceilings and soffits.

- Canopy ceilings appear as floating ceilings because they emphasize the negative space above and at the sides of suspended ceiling shapes. The materials involved are usually metals, and an almost infinite number of shapes and combinations can be achieved.
- Decorative grids use rectangular main runners and cross runners that are combined and formed into an interlocking grid that is uniform in appearance.
- Transparent-finished wood panels (very much like wall paneling) can be formed into panels laid into an exposed grid structure or into planks attached to a specialized suspension system, Figure 37.12.
- Acoustical metal facings, with or without acoustical backings, can be formed into panels or tiles and placed in suspension systems. The surface can be perforated in a variety of patterns, or it can be unperforated.
- Sheet metals can be formed into linear strips that resemble planks, which are attached to a specialized suspension system to achieve a directional look within a space, Figure 37.13. The spaces between the strips can be left open or covered.
- Luminous ceilings combine overhead lighting with either open or translucent (opaque) ceilings to create dramatic effects, Figure 37.14.

FIGURE 37.13 A suspended linear metal ceiling in an area where streets and drives pass under a building.

FIGURE 37.14 A luminous ceiling, imitating a basketball, in a professional basketball team's locker room. (Photo courtesy of HKS Inc.)

PRACTICE QUIZ

Each question has only one correct answer. Select the choice that best answers the question.

6. In a suspended ceiling system, steel hanger wires are supplied by the manufacturer already cut to length.
  a. True **b.** False

7. The strongest metal suspension system is always a properly installed direct-hung system.
  a. True **b.** False

8. In suspended acoustical ceilings, the grid structure is made of
  a. a hardwood such as oak or maple.
  b. extruded aluminum tees.
  c. corrosion-resistant steel tees.
  d. all of the above.
  e. (b) and (c).

9. Acoustical panels and tiles are fabricated using
  a. mineral wood, mineral fibers, fiberglass, and/or perlite.
  b. fillers, binders, and water.
  c. (a) and (b).
  d. none of the above.

10. The edges of acoustical panels
  a. are designed to rest on the metal suspension system.
  b. are available in three different profiles.
  c. are snapped into extruded plastic shapes.
  d. do not affect the overall pattern of the ceiling.
  e. all of the above.

11. Gypsum board is a commonly used finish material. It is typically
  a. attached directly to a structural metal or wood light-frame.
  b. attached directly to a suspended metal frame.
  c. designed to rest on the metal suspension system.
  d. all of the above.
  e. (a) and (b).

12. The design of a ceiling in a large office building calls for a multilevel series of ceiling planes finished with gypsum board. The most likely method for attaching the ceiling to the structure is to
  a. nail gypsum board directly to the wood light-frame structure.
  b. screw the gypsum board directly to a suspended metal framework.
  c. use lay-in acoustical tile.

13. It is not possible to use lath and plaster with a suspended ceiling structure.
  a. True **b.** False

14. Ceiling materials are essentially limited to drywall, lath and plaster, and acoustical tiles.
  a. True **b.** False

REVIEW QUESTIONS

1. Diagram and describe the difference between a ceiling that is attached to the building structure and one that is suspended from the building structure.
2. Examine the ceiling in one of your classrooms. What type of ceiling is it? Do a section sketch illustrating the parts of the system.
3. Use a section sketch and notes to illustrate a direct-hung suspended ceiling.
4. Use a sketch and notes to illustrate two methods of attachment when gypsum board is used as a finish ceiling material.

APPENDIX A SI System and U.S. System of Units

APPENDIX OUTLINE

- Rules of Grammar in the SI System
- Length, Thickness, Area, and Volume
- Fluid Capacity
- Mass, Force, and Weight
- Pressure and Stress
- Unit Weight of Materials
- Temperature and Energy
- Conversion from the U.S. System to the SI System

The system of measurement (or units) commonly used at the present time for the design and construction of buildings in the United States is the foot-pound-second (FPS) system. In this system, length is measured using the foot or its multiples—the yard and the mile—or its submultiple—the inch. Weight is measured in pounds, kilopounds, or ounces. Time is measured in hours, minutes, or seconds. Although the United States was the first country to establish the decimal currency in 1785, it is one of the two (or three) countries where (the nondecimal) FPS system is still used.* Therefore, the FPS system, earlier known as the *Imperial System of Units*, is now commonly known as the *U.S. System of Units* or *U.S. Customary Units*.

The system of units used by the rest of the world is the meter-kilogram-second (MKS) system. A rationalized and more commonly used version of the MKS system is called the *SI system*, popularly known as the *International System of Units*.

The advantage of the SI system (or the MKS system) lies primarily in the fact that the multiples and submultiples of each unit (the secondary units) have a decimal relationship with each other, which makes computations easier and less susceptible to errors. For instance, the secondary units of length, the centimeter and the kilometer, are related to the base unit, the meter, by 10^{-2} and 10^{3}, respectively. By contrast, the length unit in the U.S. system, the foot, does not bear a decimal relationship with its secondary units, the inch, the yard, and the mile.

Twelve prefixes have been standardized in the SI system for use with the base unit to make the secondary units. These prefixes, along with their symbols, are given in Table 1. Note that the prefixes are uppercase for magnitudes 10^{6} and greater and lowercase for magnitudes 10^{3} and lower.

NOTE

The SI system, although commonly referred to as the *International System*, is, in fact, an acronym for *Le Système International d'Unités*, a name given by the thirty-six nations meeting at the eleventh General Conference on Weights and Measures (CGPM) held in Paris in 1960. (*CGPM* is an acronym for *Conférence Générale des Poids et Mesures*.)

*Although not yet officially adopted, the SI system is being increasingly used (together with the U.S. System—in a dual-unit format) in several important U.S. publications that regulate building design and construction.

TABLE 1 PREFIXES IN THE SI SYSTEM

Factor	Prefix	Symbol
10^{12}	tera	T
10^{9}	giga	G
10^{6}	mega	M
10^{3}	kilo	k
10^{2}	hecto	h
10	deka	da
10^{-1}	deci	d
10^{-2}	centi	c
10^{-3}	milli	m
10^{-6}	micro	μ
10^{-9}	nano	n
10^{-12}	pico	p

The SI system uses seven base quantities—length, mass, time, temperature, electric current, luminous intensity, and amount of substance. The units corresponding to the base quantities and their symbols are listed in Table 2. Of these, only the first four quantities are of interest to design and construction professionals—length, mass, time, and temperature.

Quantities other than the base quantities, such as force, stress, pressure, velocity, acceleration, and energy, and their units are derived from a combination of two or more base quantities. For example, acceleration is the rate of change of velocity, which, in turn, is the rate of change of distance. Consequently, the unit for acceleration is meters per second squared (m/s^2).

Another advantage of the SI system is that there is one and only one unit for each quantity—meter for length, kilogram for mass, second for time, and so on. This is not so in the U.S. system, where multiple units are often used. For instance, power is measured in Btu per hour and also in horsepower.

RULES OF GRAMMAR IN THE SI SYSTEM

The symbols for units in the SI system are always lowercase unless the unit is named after a person, such as Newton, Pascal, Hertz, or Kelvin. In this case, the first letter of the symbol is uppercase and the second letter is lowercase, as in Pa. An exception is, however, made in the case of the liter, which is given the symbol L.

Multiples of base units are given with a single prefix. Double prefixes are not allowed. Thus, megakilometer (Mkm) is incorrect; instead, we use gigameter (Gm). No space is left between a prefix and a symbol. Thus, we use km, not k m.

The product of two or more units in symbolic form is given by using a multiplication dot between individual symbols (e.g., N · m). Mixing symbols and names of units is incorrect (e.g., N · meter).

A space must be left between the numerical value and the unit symbol. Thus, we use 300 m, not 300m. No space, however, is left between the degree symbol and C (Celsius). Plurals are not used in symbols. For instance, we use 1 m and 50 m. Periods are not used after symbols except at the end of a sentence.

In architectural and engineering drawings, dimensions are generally given in millimeters, and when that is done, the use of mm is avoided. For instance, the measurements of a floor tile are given as 300 × 300, not as 300 mm × 300 mm. Larger dimensions may be given in

TABLE 2 BASE QUANTITIES AND THEIR SYMBOLS IN THE SI SYSTEM

Quantity	Unit name	Symbol
Length	meter	m
Mass	kilogram	kg
Time	second	s
Temperature	kelvin	K
Electric current	ampere	A
Luminous intensity	candela	cd
Amount of substance	mole	mol

meters, if necessary. For instance, a building may be dimensioned as 30.800 m $\times$ 50.600 m in plan (elevation or section), but it is preferable to dimension it as 30,800 $\times$ 50,600.

LENGTH, THICKNESS, AREA, AND VOLUME

In the U.S. system, the standard unit of length is the foot: 1 ft = 12 in. Long distances are measured in miles. In the SI system, the standard unit of length is the meter. Most dimensions in the SI system are given in millimeters (mm), where 1 mm = 10^{-3} m. Long distances are given in kilometers (km).

1 ft = 0.3048 m; 1 m = 3.281 ft = 39.37 in.; 1 in. = 25.4 mm

1 mi = 1.609 km

FLUID CAPACITY

Fluid capacity is usually given in gallons in the U.S. system and in liters (L) in the SI system. The imperial gallon is slightly larger than the U.S. gallon, but the imperial gallon is no longer used as a unit.

1 gal (U.S.) = 4 qt = 3.7854 L

1 L = 0.001 m^3 (by definition) = 0.264 gal

MASS, FORCE, AND WEIGHT

In the U.S. system, the unit of weight (and mass) is the pound (lb); the corresponding unit of mass in the SI system is the kilogram (kg).

1 lb = 16 oz = 7,000 gr (grains) = 0.4536 kg

1 kg = 2.205 lb

Force is defined as mass times acceleration. Because the unit of acceleration in the U.S. system is feet/second^2 (ft/s^2), the unit of force is (lb-ft/s^2). This unit is called *pound-force,* usually referred to as the *pound.* In the SI system, the unit of force is $(kg \cdot m)/s^2$. This complex unit is called the *newton* (N), after the famous physicist Isaac Newton (1642–1727).

Because the weight of an object is the force exerted on it by the gravitational pull of the earth, the weight is equal to mass $\times$ acceleration due to gravity. The acceleration due to gravity on the earth's surface is 9.8 m/s^2. Therefore, the weight of an object whose mass is 1 kg is 9.8 $(kg \cdot m)/s^2$, or 9.8 N.

We see that in the SI system, there is a clear distinction between the units of mass and the weight of an object. The distinction between the mass and the weight in the U.S. system is rather obscure because the pound is used as the unit for both the mass and the weight of an object.

1 kilogram-force = 9.8 N; 1 lb = 4.448 N; 1 N = 0.2248 lb

1 kilopound (kip) = 1,000 lb = 4.448 kN; 1 kN = 0.2248 kip

NOTE

Some countries that have not fully adopted the SI system use kilogram as the unit for mass as well as force (in the same way as pound is used as a unit for mass and weight). In these countries, the unit for stress is kilogram per square centimeter (kg/cm^2) instead of Pa. Weight density is expressed as kilograms per cubic centimeter (kg/cm^3) instead of N/m^3.

PRESSURE AND STRESS

Because pressure, or stress, is defined as force per unit area, the unit of pressure, or stress, in the U.S. system is lb/ft^2 (psf). Other units commonly used are pounds per square inch (psi) and kilopounds per square inch (ksi). In the SI system, the unit of pressure, or stress, is N/m^2. This unit is called the *pascal* (Pa) after the physicist Blaise Pascal (1623–1662). Thus, 1 N/m^2 = 1 Pa.

1 psf = 47.880 Pa; 1 Pa = 0.208 85 psf

1 psi = 6.895 kPa; 1 kPa = 0.1450 psi

In weather-related topics, the unit of pressure is the atmosphere (atm); 1 atm is the standard atmospheric pressure at sea level.

1 atm = 760 mm of mercury (Hg) = 29.92 in. of Hg = 14.69 psi = 2,115.4 psf = 101.3 kPa. For all practical purposes, the atmospheric pressure may be taken as 2,100 psf or 101 kPa.

UNIT WEIGHT OF MATERIALS

Density is defined as the mass per unit volume. Its units are lb/ft^3 in the U.S. system and kg/m^3 in the SI system, respectively.

$$1 \text{ lb/ft}^3 = 16.018 \text{ kg/m}^3; \quad 1 \text{ kg/m}^3 = 0.064\ 243 \text{ lb/ft}^3$$

In building construction, however, we are more interested in the weight density of materials instead of the mass density. Weight density (or simply unit weight) is defined as the weight per unit volume, and its units are lb/ft^3 and N/m^3 in the U.S. system and SI system, respectively.

$$1 \text{ lb/ft}^3 = 157.1 \text{ N/m}^3$$

TEMPERATURE AND ENERGY

In the U.S. system, temperature is measured in degrees Fahrenheit (°F). This scale was introduced during the early eighteenth century. Zero on the Fahrenheit scale was established based on the lowest obtainable temperature at the time, and 100°F was established on the basis of human body temperature as it was then considered to be.

On the Celsius scale (°C), earlier known as the *centigrade scale,* 0°C refers to the freezing point of water and 100°C to the boiling point of water.

The unit of temperature in the SI system is the kelvin (K). The preference for the Kelvin scale over the Celsius scale is due to the fact that on the Kelvin scale, the temperature is always positive. This is not so on the Celsius or the Fahrenheit scale, on which the temperature may be positive or negative. In other words, 0 K (as we know it now) is the lowest obtainable temperature and is, therefore, called ***absolute zero.*** The relationship between Kelvin and Celsius temperatures is T°C = (T + 273.15) K. In other words, 20°C = 293.15 K. Other relationships are

$$T°F = [(1.8)T + 32]°C; \quad T°C = [(0.555\ldots)(T - 32)]°F$$

The intervals in both the Kelvin and Celsius scales are equal. Therefore, the Celsius and Kelvin scales start at different points, but their subdivisions are equal. Note that the word *degree* is not used on the Kelvin scale. (That is, we say "8 Kelvins", not "8 degrees Kelvin".) Although Kelvin is the appropriate scale to use in the SI system, degrees Celsius are also used because of the smaller numbers associated with the Celsius scale.

CONVERSION FROM THE U.S. SYSTEM TO THE SI SYSTEM

The conversion of a physical quantity such as length, weight, or stress from the U.S. system to the SI system simply involves using the appropriate conversion factor. A comprehensive list of conversion factors used to convert from the U.S. system to the SI system is given in Table 3. However, when it comes to converting building products' sizes from the U.S. system to the SI system, three types of conversion are possible:

- *Exact Conversion:* This conversion is made simply by multiplying a value given in the U.S. system by the appropriate conversion factor to obtain the corresponding value in the SI system. For example, 12 in. is exactly equal to 304.8 mm.
- *Soft Conversion:* In this conversion, a product's size is not converted, only its description. For example, a manufacturer may decide to continue making 12 in. × 12 in. floor tiles but market them as 305 mm × 305 mm tiles. During the initial period of changeover to the SI system in the building industry, most product sizes will be soft-converted.
- *Hard Conversion:* In hard conversion, the physical size of the product is changed to a new metric equivalent. For example, 12 in. × 12 in. floor tiles will probably be changed to 300 mm × 300 mm tiles. Hard conversion of a product requires a change in the manufacturing equipment and a great deal of coordination among various related products but has the advantage that the product sizes are rationalized.

TABLE 3 UNIT CONVERSION FACTORS

Quantity	To convert from	To	Multiply by
Length	mi	km	1.609 344*
	yd	m	0.9144*
	ft	m	0.304 8*
	ft	mm	304.8*
	in.	mm	25.4*
Area	mi^2	km^2	2.590 00*
	acre	m^2	4 046.87
	acre	ha**	0.404 687
	ft^2	m^2	0.092 903 04*
	in.^{2}	mm^2	645.16*
Volume	yd^3	m^3	0.764 555
	ft^3	m^3	0.028 3168
	100 board feet	m^3	0.235 974
	gal	L	3.785 41
	in.^{3}	cm^3	16.387 064
	in.^{3}	mm^3	16 387.064
Velocity	ft/s	m/s	0.3048
Rate of fluid flow, infiltration	ft^3/s	m^3/s	0.028 3168
	gal/h	mL/s	1.051 50
Acceleration	ft/s^2	m/s^2	0.3048
Mass	lb	kg	0.453 59
Mass per unit area	psf	kg/m^2	4.882 43
Mass density	pcf	kg/m^3	16.018 5
Force	lb	N	4.448 22
Force per unit length	plf	N/m	14.593 9
Pressure, stress	psf	Pa	47.880 26
	psi	kPa	6.894 76
	in. of mercury (in. Hg)	kPa	3.386 38
	in. of Hg (in. Hg)	psf	70.72
	atm***	kPa	101.325
Temperature	°F	°C	5/[9(°F − 32)]
	°F	K	(°F + 459.7)/1.8
Quantity of heat	Btu	J	1055.056
Power	ton (refrigeration)	kW	3.517
	Btu/h	W	0.293 07
	hp	W	745.7
	Btu/(h-ft^2)	W/m^2	3.154 59
Thermal conductivity	Btu-in/(ft^2-h-°F)	W/(m · °C)	0.144 2
Thermal conductance, or thermal transmittance, U	Btu/(ft^2-h-°F)	W/(m^2 · °C)	5.678 263
Thermal resistance	(ft^2-h-°F)/Btu	(m^2 · °C)/W	0.176 110
Thermal capacity	Btu/(ft^2-°F)	kJ/(m^2 · °C)	20.44
Specific heat	Btu/(lb-°F)	J/(kg · °C)	4.186 8
Vapor permeability	perm-in	ng/(Pa · m · s)	1.459 29
Vapor permeance	perm	ng/(Pa · m^2 · s)	57.213 5
Angle	degree	radian	0.017 453

*Denotes exact conversion.
**1 hectare (ha) = 100 m × 100 m.
***1 atmosphere (atm) = 29.92 in. of mercury.

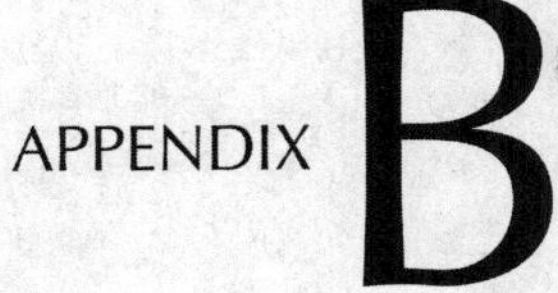

APPENDIX B Preliminary Sizing of Structural Members

APPENDIX OUTLINE

- **Conventional Wood Light Frame (WLF) Buildings**
- **Conventional Cold-Formed Steel Frame (CFSF) Buildings**
- **Structural Steel Frame Buildings**
- **Site-Cast Concrete Frame Buildings**
- **Precast, Prestressed Concrete Members**
- **Load-Bearing Masonry and Concrete Buildings**

The information provided in this appendix is valid only for approximate sizing of structural members during the sketch design (SD) or initial design development (DD) stages of conventional buildings. A structural engineer should be consulted for final member sizes.

CONVENTIONAL WOOD LIGHT FRAME (WLF) BUILDINGS

Wall Framing

Approximate Stud Size and Spacing

Number of stories	Stud size	Stud spacing
3 stories	2 × 6	16 in. o.c.
2 stories	2 × 6 2 × 4	16 in. o.c. 12 in. o.c.
1 story	2 × 4	24 in. o.c.

1. Stud size and spacing are for approximately 10-ft-high studs. For tall walls, such as those used in double-height spaces, doubled studs and (or) closer stud spacing may be required.
2. In cold climates, the stud size may be governed by insulation requirements. For example, 2 × 6 studs may be necessary where 2 × 4 studs are structurally adequate.
3. In high-wind regions, the exterior walls may require larger studs and (or) closer spacing or doubled studs.

FLOOR FRAMING

Lumber Joists—Approximate Span Capabilities

	Joist spacing		
Joist size	**12 in. o.c.**	**16 in. o.c**	**24 in. o.c.**
2 × 6	11 ft	10 ft	9 ft
2 × 8	15 ft	13 ft	11 ft
2 × 10	19 ft	17 ft	14 ft
2 × 12	23 ft	20 ft	16 ft

I-Joists—Approximate Span Capabilities

	I-joist spacing		
I-joist depth	**12 in. o.c.**	**16 in. o.c**	**24 in. o.c.**
$9\frac{1}{2}$ in.	18 ft	16 ft	14 ft
$11\frac{7}{8}$ in.	21 ft	19 ft	15 ft
14 in.	24 ft	20 ft	17 ft
16 in.	28 ft	24 ft	19 ft

Trussed Joists—Span Capabilities

Unlike lumber or I-joists, trussed joists are custom manufactured for a project and are not trimmable. Typical spacing of trussed joists = 24 in. o.c.

$$\textbf{Approx. joist depth} = \frac{\textbf{joist span}}{\textbf{18}}$$

Example

If joist span = 30 ft, approximate joist depth = (30 × 12)/18 = 20 in.
Because trussed joists are made of 2 × 4 lumber, the width of joists is $3\frac{1}{2}$ in.

ROOF FRAMING

Sawn Lumber Rafters—Approximate Span Capabilities

	Rafter spacing		
Rafter size	**12 in. o.c.**	**16 in. o.c.**	**24 in. o.c.**
2 × 6	14 ft	13 ft	12 ft
2 × 8	19 ft	18 ft	15 ft
2 × 10	25 ft	22 ft	18 ft

The above table applies to a roof live load (or snow load) ≤20 psf and a light roof cover (such as asphalt shingles). If the snow load exceeds 20 psf and (or) the roof cover is heavier (such as clay or concrete tiles), the span capability of a given rafter size will be smaller.

Sawn Lumber Ceiling Joists—Approximate Span Capabilities

	Joist spacing		
Joist size	**12 in. o.c.**	**16 in. o.c.**	**24 in. o.c.**
2×6	18 ft	16 ft	14 ft
2×8	24 ft	22 ft	18 ft
2×10	—	—	22 ft

The above table applies to an uninhabitable attic without storage.

A dash (—) indicates that the span capability exceeds 26 ft—the maximum sawn lumber length available.

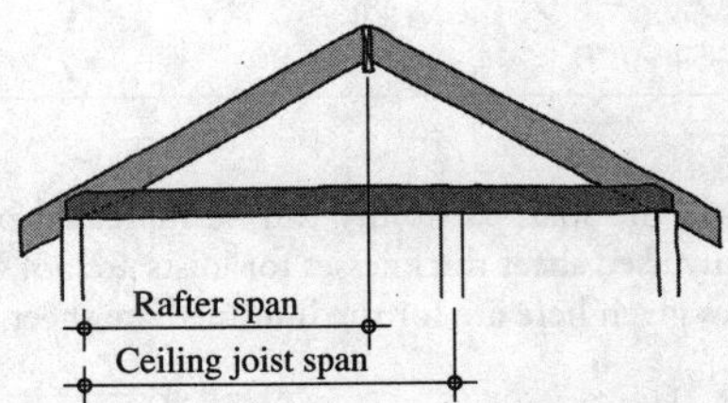

Because the span of a hip or valley rafter is larger than the common rafters, hip or valley rafters are generally one size larger than the common rafters. Thus, for 2 x 8 common rafters, a hip or valley rafter is generally 2 x 10.

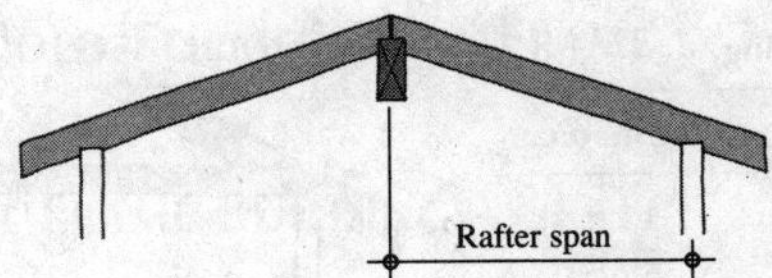

CONVENTIONAL COLD-FORMED STEEL FRAME (CFSF) BUILDINGS

WALL FRAMING

Approximate Stud Size and Spacing

Number of stories	**Stud size**	**Stud spacing**
2 stories	550S162–33	24 in. o.c.
	350S162–43	16 in. o.c.
1 story	350S162–33	24 in. o.c.

Stud size and spacing are based on sheet steel with a yield strength of 33 ksi. The other commonly used yield strength is 50 ksi (see Chapter 17).

550S162–33 implies a stud (joist or rafter) with a web depth of 5.5 in. and a flange width of 1.625 in., made of 33-mil (0.033-in.)-thick sheet steel (see Chapter 17).

1. Stud size and spacing are for approximately 10-ft-high studs. For tall walls, such as those used in double-height spaces, larger stud size and (or) thicker sheet steel or closer spacing of studs may be required.

2. In cold climates, the stud size may be governed by insulation requirements. For example, 550S162 studs may be necessary where 350S162 studs are structurally adequate.
3. In high-wind regions, the exterior walls may require larger studs, studs made of thicker sheets, or higher yield strength.

FLOOR FRAMING

Joists—Approximate Span Capabilities

Joist size	Joist spacing 16 in. o.c.	24 in. o.c.
550S162–54	11 ft	10 ft
800S162–54	15 ft	13 ft
1000S162–54	18 ft	17 ft
1200S162–54	21 ft	18 ft

1. For a given joist size, the span capability can be increased by increasing the sheet thickness. Commonly used sheet thicknesses for joists are 33, 43, 54, 68, and 97 mil.
2. The span capabilities given here are for the intermediate sheet thickness of 54 mil.

ROOF FRAMING

Rafters—Approximate Span Capabilities

Rafter size	Rafter spacing 16 in. o.c.	24 in. o.c.
550S162–54	18 ft	15 ft
800S162–54	24 ft	20 ft
1000S162–54	28 ft	23 ft
1200S162–54	30 ft	25 ft

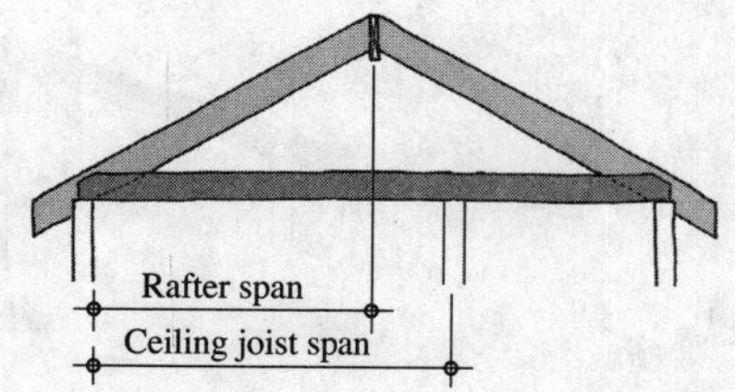

1. The above table applies to a roof live load (or snow load) ≤20 psf and a light roof cover (such as asphalt shingles). If the snow load exceeds 20 psf and (or) the roof cover is heavier (such as clay or concrete tiles), the span capability of a given rafter size will be smaller.
2. For a given rafter size, the span capability can be increased by increasing the sheet thickness. Commonly used sheet thicknesses for rafters are 33, 43, 54, 68, and 97 mil.
3. The span capabilities given here are for the intermediate sheet thickness of 54 mil.

Ceiling Joists—Approximate Span Capabilities

Joist size	Joist spacing 16 in. o.c.	24 in. o.c.
550S162–54	18 ft	15 ft
800S162–54	20 ft	18 ft
1000S162–54	21 ft	19 ft
1200S162–54	23 ft	20 ft

1. The above table applies to an uninhabitable attic without storage.
2. For a given joist size, the span capability can be increased by increasing the sheet thickness. Commonly used sheet thicknesses for rafters are 33, 43, 54, 68, and 97 mil.
3. The span capabilities given here are for the intermediate sheet thickness of 54 mil.

STRUCTURAL STEEL FRAME BUILDINGS

FLOORS AND ROOFS

W-section Girders and Beams

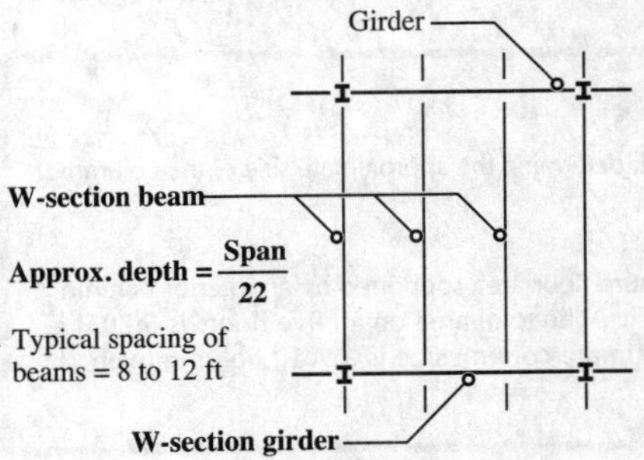

Approx. depth $= \frac{\textbf{Span}}{\textbf{22}}$

Typical spacing of beams = 8 to 12 ft

Approx. depth $= \frac{\textbf{Span}}{\textbf{16}}$

but not less than beam's depth

Nominal Depths of W-Sections in inches
44, 40, 36 33, 30, 27, 24, 21 18, 16, 14, 12, 10, 8, 6, 4

Girders and K-series joists

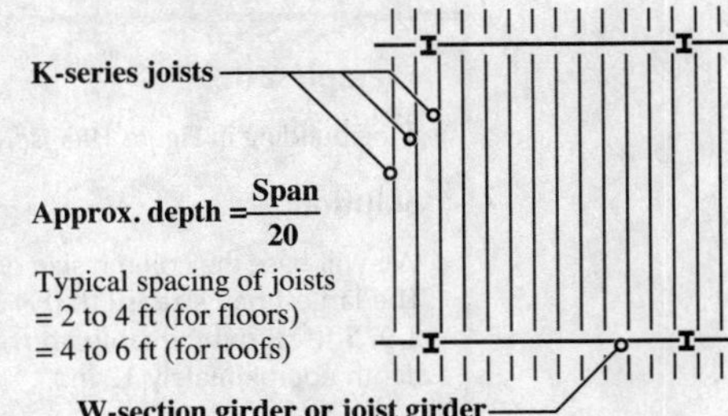

Approx. depth $= \frac{\textbf{Span}}{\textbf{20}}$

Typical spacing of joists
= 2 to 4 ft (for floors)
= 4 to 6 ft (for roofs)

Approx. depth $= \frac{\textbf{Span}}{\textbf{16}}$ (for W-section girder)

Approx. depth $= \frac{\textbf{Span}}{\textbf{14}}$ (for joist girder)

Standard Depths of K-Series Joists in inches
8, 10, 12, 14, 16, ...30

Example 1

Determine the approximate depths of the beams and girders for the floor of Figure 19.5.

Solution

The maximum span of the girder is 26.5 ft. Therefore, approximate girder depth = $(26.5 \times 12)/16 = 19.9$ in., or 20 in. The nearest W-section is W21. Hence, use a W21 girder.

The maximum beam span is 30 ft. Therefore, approximate beam depth = $(30 \times 12)/22 = 16.4$ in. Hence, use a W16 beam.

COLUMNS

Interior Columns

To approximate the size of an interior column, compute the total floor area supported by the column on all floors and then select the column size from the following table.

Column size	Maximum floor area on all floors
8 in. x 8 in.	2,000 sq. ft
10 in. x 10 in.	3,000 sq. ft
12 in. x 12 in.	4,500 sq. ft
14 in. x 14 in.	6,000 sq. ft

Composite Floor Decks

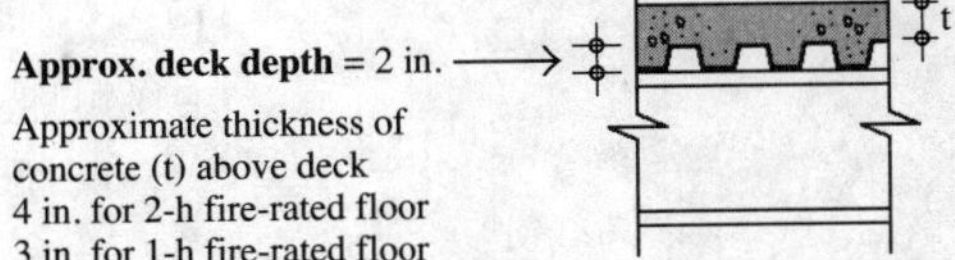

Approx. deck depth = 2 in.

Approximate thickness of concrete (t) above deck
4 in. for 2-h fire-rated floor
3 in. for 1-h fire-rated floor

Roof Decks

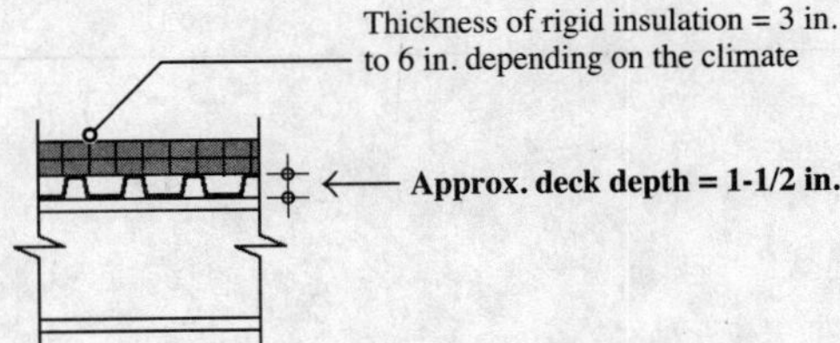

Thickness of rigid insulation = 3 in. to 6 in. depending on the climate

Approx. deck depth = 1-1/2 in.

Exterior Columns

Although the floor area supported by an exterior column is less than that of an interior column, exterior columns will generally support (exterior) walls, which generally are not supported by interior columns. Therefore, exterior columns may be assumed to be the same size as interior columns.

> **Example 2**
>
> If the building in Figure 19.5 is five stories tall, determine the approximate size of the columns.
>
> **Solution**
>
> We will base the column size on the maximum floor area supported by an interior column. The largest bay size is (30.0 × 26.5) = 795 ft^2. The total area on all five floors is 5(795) = 3,975 ft^2. From the table above, an approximate column size is a W12 column (web depth approximately 12 in.).

SITE-CAST CONCRETE FRAME BUILDINGS

Beam and Girder-Supported One-Way Solid Slab

Reinforced Concrete

Girder

Approx. depth $= \dfrac{\textbf{span}}{\textbf{12}}$

Width = 0.6 (depth)

Beam

Approx. depth $= \dfrac{\textbf{span}}{\textbf{15}}$

Width = 0.6 (depth)

Slab

Approx. thickness $= \dfrac{\textbf{span}}{\textbf{24}}$

Typical distance between beams = 8 ft to 15 ft

Posttensioned Concrete

Girder

Approx. depth $= \dfrac{\textbf{span}}{\textbf{18}}$

Width = 0.6 (depth)

Beam

Approx. depth $= \dfrac{\textbf{span}}{\textbf{20}}$

Width = 0.6 (depth)

Slab

Approx. thickness $= \dfrac{\textbf{span}}{\textbf{40}}$

Typical distance between beams = 15 ft to 25 ft

Span for girder is clear distance between columns, and span for beams is clear distance between girders.

Depth of girder or beam includes thickness of slab.

Span for slab is clear distance between beams.

As far as possible, girder and beam depths should be the same. Therefore, the girder should span along the shorter direction.

Round beam and girder widths and depths to whole inches. Round slab thickness to $\frac{1}{2}$ inch, not less than 4 in. Slab thickness may be governed by fire-resistance requirements.

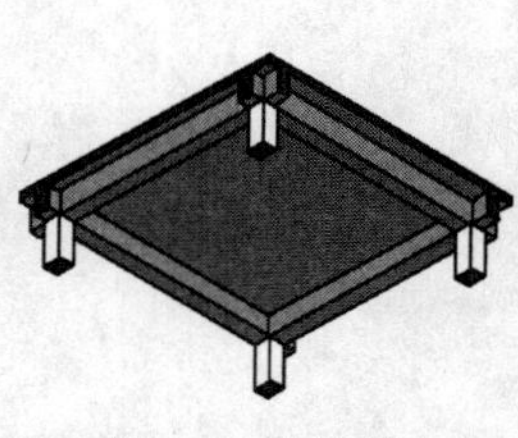

Two-way Solid Slab

Reinforced Concrete

Beam

Approx. depth $= \dfrac{\textbf{span}}{\textbf{15}}$

Width = 0.6 (depth)

Slab

Approx. thickness $= \dfrac{\textbf{span}}{\textbf{36}}$

Typical column spacing = 8 ft to 25 ft

Posttensioned Concrete

Beam

Approx. depth $= \dfrac{\textbf{span}}{\textbf{20}}$

Width = 0.6 (depth)

Slab

Approx. thickness $= \dfrac{\textbf{span}}{\textbf{48}}$

Typical column spacing = 25 ft to 30 ft

Span for beams is clear distance between columns.

Span for slab is the longer of the two clear distances between beams.

Round beam width and depth to whole inches. Round slab thickness to $\frac{1}{2}$ inch, not less than 4 in. Slab thickness may be governed by fire-resistance requirements.

As far as possible, distance between columns in both directions should be the same.

One-Way Band Beam Slab	*Reinforced Concrete* *Beam* **Approx. depth = 2.0 to 2.5 (slab thickness)** Width = 0.25 to 0.3 (center-to-center beam spacing) *Slab* **Approx. thickness** $= \frac{\textbf{span}}{\textbf{24}}$ Typical column spacing = 25 ft to 30 ft	*Posttensioned Concrete* *Beam* **Approx. depth = 2.0 to 2.5 (slab thickness)** Width = 0.25 to 0.3 (center-to-center beam spacing) *Slab* **Approx. thickness** $= \frac{\textbf{span}}{\textbf{40}}$ Typical column spacing = 30 ft to 40 ft	Span for slab is clear distance between beams. Round beam width and depth to whole inches. Round slab thickness to $\frac{1}{2}$ inch. Slab thickness should not be less than 4 in. Slab thickness may be governed by fire-resistance requirements.
One-Way Joist Slab	*Reinforced Concrete* *Beam* **Depth = same as joists** Width = 1.75 (joist depth) *Joist* **Approx. depth** $= \frac{\textbf{span}}{\textbf{18}}$ Width = 5 in. (standard-module pans), 7 in. for wide-module pans *Slab* **Thickness = 3 in.** (standard-module pans), **4 in.** (wide-module pans) Typical column spacing = 25 ft to 40 ft	*Post-tensioned Concrete* *Beam* **Depth same as joists =** Width = 1.25 (joist depth) *Joist (Reinforced Concrete)* **Approx. depth** $= \frac{\textbf{span}}{\textbf{18}}$ Width = 5 in. (standard-module pans), 7 in. for wide-module pans *Slab* **Thickness 3 in.** (standard-module pans), **4 in.** (wide-module pans) Typical column spacing = 35 ft to 50 ft	Span for joists is clear distance between beams. Depth of joists or beams includes slab thickness. Round joist depth to pan depth + slab thickness. Slab thickness may be governed by fire-resistance requirements.
Two-Way Joist (Waffle) Slab	*Reinforced Concrete* *Joist* **Approx. depth** $= \frac{\textbf{span}}{\textbf{22}}$ **Width = 5 in. to 6 in.,** depending on dome size *Slab* **Thickness = 3 in. to 4 in.** Typical column spacing = 30 ft to 45 ft	*Posttensioned Concrete* *Joist* **Approx. depth** $= \frac{\textbf{span}}{\textbf{30}}$ **Width = 5 in. to 6 in.,** depending on dome size *Slab* **Thickness = 3 in. to 4 in.** Typical column spacing = 40 ft to 60 ft	Depth of joists includes slab thickness. Round joist depth to dome depth + slab thickness. Slab thickness may be governed by fire-resistance requirements. The number of domes filled around columns is a function of column spacing, floor load, and dome size. Beams may be used in place of filled domes.
Flat Plate	*Reinforced Concrete* *Slab* **Approx. thickness** $= \frac{\textbf{span}}{\textbf{30}}$ Typical column spacing = 15 ft to 20 ft	*Posttensioned Concrete* *Slab* **Approx. thickness** $= \frac{\textbf{span}}{\textbf{40}}$ Typical column spacing = 20 ft to 25 ft	Slab span is (longer) clear distance between columns. As far as possible, distance between columns in both directions should be the same. Slab thickness may be governed by fire-resistance requirements.
Flat Slab	*Reinforced Concrete* *Slab* **Approx. thickness** $= \frac{\textbf{span}}{\textbf{35}}$ Typical column spacing = 20 ft to 30 ft	*Posttensioned Concrete* *Slab* **Approx. thickness** $= \frac{\textbf{span}}{\textbf{45}}$ Typical column spacing = 30 ft to 45 ft	Slab span is (longer) clear distance between columns. As far as possible, distance between columns in both directions should be the same. Slab thickness may be governed by fire-resistance requirements.

Columns

Column size depends on various factors, such as the total floor area supported by the column, concrete strength, amount of reinforcement, column height, and whether the column is part of the lateral load resistance system of the building. For conventional buildings in which the lateral load is resisted by shear walls, the following rule of thumb may be used for the approximate size of an interior column:

$$\text{Area of column} = \frac{\text{total floor area supported by column}}{10 \text{ to } 20(\text{depending on concrete strength})} \text{ not less than 10 in. in any direction}$$

As far as possible, column size should be the same for interior and exterior columns and from floor to floor. Note that the amount of reinforcement and the concrete strength in a column can be increased toward the lower floors.

Example: Determine the column size required for a square reinforced-concrete column at the ground floor of a building supporting 1,000 ft^2 at each floor. Number of floors = 4.

Solution:

$$\text{Area of column} = \frac{4{,}000}{10} = 400 \text{ in.}^2$$

Hence, approx. column size = 20 in. × 20 in.

PRECAST, PRESTRESSED CONCRETE MEMBERS

Double-T Units

Approx. depth, $h = \frac{\text{Span}}{28}$

Hollow-core slabs

Approx. depth, $h = \frac{\text{Span}}{40}$

LOAD-BEARING MASONRY AND CONCRETE BUILDINGS

Reinforced-Concrete Masonry Bearing Walls in Residential (e.g., Apartment and Hotel) Buildings

8-in.-thick, CMU walls for up to 8-floor-high buildings.

10-in.-thick CMU walls for 11- to 15-floor-high buildings.

12-in.-thick CMU walls for 16- to 20-floor-high buildings.

Upper floors in 10-in.- or 12-in.-thick wall structures may be constructed of 8-in.-thick walls.

Reinforced-Concrete Masonry Bearing Walls in Single-Story, Long-Span Structures (e.g., Gymnasiums)

10-in.- or 12-in.-thick CMU walls, depending on the span.

Site-Cast Reinforced-Concrete Bearing Walls in Residential Buildings

6-in.-thick site-cast reinforced-concrete walls for up to 20-floor-high buildings.

Precast-Concrete Tilt-Up Walls

The thickness of tilt-up walls in a single-story building can be approximated by dividing the wall height by 48 (not less than 6 in.). For two- or three-story buildings, slightly thicker walls may be needed.

GLOSSARY

A

Abrasive blasting A method of roughening a concrete or steel surface by an abrasive medium, such as sand, under high air pressure.
Acid etching A method used to expose the aggregate on a concrete surface through the use of a dilute acid wash.
Acoustical materials See *sound-absorbing materials.*
ACQ See *alkaline copper quat.*
ADA Americans with Disabilities Act.
Adhered veneer A thin finish material, such as thin brick or tile, that is adhered directly to a backup wall with mortar or adhesive.
Adobe Sun-dried clay masonry units, generally molded on site.
Aggregate Granular materials, including gravel, sand, crushed stone, expanded shale, and slag, that serve as filler in a concrete mix.
Air retarder A microperforated plastic sheet that prevents the passage of outside air through a wall or roof assembly but allows water vapor to pass through.
Air seasoning A method of drying lumber in which lumber members are stacked to allow free circulation of air around them.
Alkaline copper quat (ACQ) A chemical used for wood preservation.
Alloy A chemical combination of two or more metals.
Anchored veneer A cladding system (usually exterior) in which masonry units are mechanically anchored to the backup wall.
Anisotropic material A material that has different properties in different directions (e.g., wood). See also *isotropic material.*
Annealed glass Flat glass obtained by heating and then gradually cooling it to relieve internal stresses that develop during the early stage of its manufacturing process. Annealed glass is the basic form of flat glass.
Arch A curved structural member that spans over a wall opening and carries loads in compression.
Arching action A structural action in which a vertical load on a wall or beam translates into two inclined loads; generally occurs in a deep beam or wall.
Architectural concrete Concrete that is obtained from a specially designed formwork to give a smooth, patterned, textured, or otherwise detailed surface finish on concrete.
Architectural precast concrete Nonstructural precast concrete, as in a concrete curtain wall.
Architectural sheet-metal roof A roof covering made of sheet metal (e.g., standing seam roof, batten seam roof), applied over a structural roof deck and underlayment.
Ashlar Stone dressed with square-cut edges and laid in a coursed or random pattern in a wall.
Asphalt A tacky black liquid derivative of the petroleum-distillation process, used in roofing and waterproofing applications.
Asphalt-saturated felt An organic paper felt that is saturated with asphalt, typically used as an underlayment in steep roofs (sometimes in walls).
Asphalt shingles A roofing unit composed of heavy organic paper or fiberglass felt saturated (or treated and coated) with asphalt and faced with mineral granules.
Asphalt-treated felt A fiberglass felt that is treated with asphalt, typically used in built-up roofs.
ASTM American Society for Testing and Materials.
Attactic polypropylene (APP) A polymer used to obtain an APP modified-bitumen roof membrane.
Auger A tool for drilling holes in wood or soil. Large mechanized augers drill holes in the ground for piers, piles, or caissons.
Axial A direction that is parallel to the length of a structural member. For a column, the axial direction is vertical; for a horizontal beam, the axial direction is horizontal.

B

Backer rod A compressible spherical foam rod used to control the depth of an elastomeric sealant, which allows the tooling of the sealant.
Backfill Soil used to fill the space between the foundation and the boundaries of an excavation. Generally, the soil excavated from foundation trenches and pits is used as backfill unless it is unsuitable.
Backup wall Load-bearing or non-load-bearing wall to which exterior cladding is adhered or anchored.
Balance point temperature Exterior temperature at which heat loss through the building envelope equals heat gain from interior activities such as cooking, lighting, and human occupancy.
Ballast (in roofing) Aggregate or concrete pavers used over a loose-laid, single-ply roof membrane to resist wind uplift and protect the membrane from degradation by solar radiation.
Balloon frame One of the two wood light frame construction systems in which the studs extend from the foundation to the roof, bypassing the intermediate floor(s). See also *platform frame.*
Ballooning Interior pressurization of a building due to high wind. It typically occurs in buildings that have large openings on only one exterior face, such as an aircraft hangar.
Baluster Vertical support of a staircase handrail.
Barrier wall An exterior wall that resists water leakage by providing an impervious barrier or by serving as a water reservoir.
Basement Below-ground portion of a building.
Basic wind speed Peak 3-s gust wind speed with a 50-year recurrence interval, used to determine wind loads on a building.
Batt Precut units of fiberglass insulation sized to fit between wall studs, ceiling joists, or rafters.
Batten A long, narrow strip of wood.
Batten bar A long metal or plastic bar used in a mechanically fastened single-ply roof membrane.
Bay In a building structure, regular, repeated space defined by four adjacent columns and beams (and/or girders) that span between them.

Beamless floor A reinforced-concrete slab supported directly on columns (without supporting beams).
Bearing pad A block of high-strength plastic, metal, or rubber placed under a beam or curtain wall to distribute the load on a larger area of the supporting member.
Bearing plate A steel plate welded to the base of a steel column to distribute the column load on a larger area of the footing. Also a steel plate welded to the ends of a steel joist or beam.
Bearing wall A wall that carries gravity loads from the floor(s) and/or roof.
Bed joint Horizontal mortar joint in a masonry wall. See also *head joint.*
Bending Deformation of a structural member caused by a force acting perpendicular to the member axis.
Bent glass Flat glass bent to shape by heating.
Bentonite clay A highly expansive clay typically used as slurry to fill the excavation for a concrete wall.
Billet A large rectangular bar of cast steel used to roll finished shapes, such as smaller bars and rods.
Bioclimatic glass curtain wall See *double-skin glass curtain wall.*
Bitumen A waterproofing compound (asphalt or coal tar) used for roofing and waterproofing.
Blast furnace A furnace used to melt iron ore.
Blocking Short pieces of lumber that fit between adjacent wood joists, rafters, or studs to provide lateral stability and additional nailing members.
Bloom Steel cast in a large rectangular section, which is then used to make wide-flange sections, angles, and channels by the hot-rolling process.
Board foot The amount of lumber contained in a 1-in.-thick board that measures 1 ft by 1 ft.
Board lumber Lumber that is less than 2 in. (nominal) thick.
Bolt A threaded steel rod with a fixed head at one end and closed with a nut at the other end.
Bond beam A continuous reinforced beam embedded in a masonry wall (generally at the floor and roof levels) formed either by bond beam masonry units or site-cast concrete.
Bond breaker A material used to prevent the adhesion of an elastomeric sealant to a backup surface.
Bond pattern The pattern used to arrange masonry units in a wall.
Breathable material A material that allows water vapor to pass through but is resistant to the passage of (bulk) water. An air retarder (sheet or liquid-applied film) is required to be breathable.
Brick A small masonry unit, generally manufactured from clay. Concrete bricks are also available.
Brick ledge Depressed portion of a concrete foundation to support the first story of brick veneer.
Bridging Structural members, laid perpendicular to wood or steel framing members, to stabilize them against overturning and brace them against buckling.
Brittle material A material that deforms little before failure. It is generally stronger in compression than in tension.
Brown coat The second layer of base coat in portland cement plaster (stucco) applied on metal lath.
Buckling A type of failure that results in the sudden bending of a slender structural member subjected to excessive axial loading.
Building brick A type of clay brick recommended for interior applications, with less-stringent durability specifications than facing brick.
Building code A legal document that regulates the design and construction of buildings to ensure that the buildings meet minimum standards of health, safety, and welfare.
Building component A part of a building that performs a specific function, that is, a window, door, or wall assembly.
Building envelope All building components that separate the building's interior from the exterior environment.
Building expansion joint See *building separation joint.*
Building information modeling (BIM) A three-dimensional software tool for designing buildings in real time, which facilitates design and construction and also allows the extraction of conventional two-dimensional plans, elevations, and sections. The model contains information about the attributes of building materials and assemblies, and therefore can be used after construction for repairs, alterations, and maintenance over the building's entire life.
Building movement joint See *building separation joint.*
Building separation joint A continuous joint that extends through all floors and the roof of a building, dividing the building into smaller buildings that can move independently of each other, generally $1\frac{1}{2}$ to 2 in. wide.
Built-up beam A beam made by combining two or more standard structural members.
Built-up roof (BUR) membrane A multi-ply roof membrane composed of alternating layers of felt and moppings of bitumen with a top cover of aggregate.

C

Caisson A large-diameter, deep reinforced-concrete foundation element made by drilling a hole into the ground and filling it with concrete; an enclosure that permits excavation work to be carried out underwater.
Calcium silicate masonry unit Sand-lime bricks or blocks cured in an autoclave to enhance the pozzolanic reaction between sand and lime.
Camber An upward curvature introduced in a beam to ensure that it will be flat under dead loads.
Cant strip A triangular strip of perlite board or pressure-treated wood to provide a smooth transition between a horizontal and a vertical surface on a roof, required with a built-up or modified-bitumen roof membrane.
Capital Upper part of a column, generally of a larger cross section than the column.
Carriage An inclined structural member (beam) that supports a stair.
Cast-in-place (CIP) concrete See *site-cast concrete.*
Cast iron An iron-carbon alloy formed by casting molten iron in a sand mold and milling to the final shape—relatively brittle and nonmalleable.
Caulk Soft, pliable material (such as polyisobutyline and acrylic) used to fill narrow, nonmovement joints between building components.
Ceiling The visible overhead interior surface of a room.
Cells Voids in concrete masonry units; voids in foamed insulation; microscopic voids in wood.
Centering Temporary formwork for constructing an arch, vault, or dome.
Ceramic material A material produced by firing clay in a high-temperature kiln, such as brick, tile, or porcelain.
Certified wood Wood obtained from sustainable forestry practices.

Chair A small support to raise steel reinforcing bars above the surface of the formwork or ground.

Charcoal The heating of wood in the absence of oxygen drives water and other volatile compounds (a process called pyrolysis) leaving a dry, porous, soft black substance referred to as charcoal because it resembles coal.

Chimney A hollow vertical structure lined with an internal flue to carry smoke and other effluents from wood or charcoal fires used for heating a building.

Chord Member of a truss; can be the top or bottom chord of a truss.

CIP concrete See *cast-in-place concrete.*

Cladding Exterior weather-resistant layer of a wall assembly.

Clad window Framing members of a wood window clad in aluminum, polyvinyl chloride, or fiberglass on the outside to increase wood's durability.

Clapboard A type of horizontal lap siding

CMU See *concrete masonry unit.*

Coal tar A tacky black liquid derivative of the coal-distillation process, used in roofing and waterproofing applications.

Coefficient of friction A measure of the resistance to sliding between the contact surfaces of two components.

Cohesive soil A soil, such as clay, whose particles tend to adhere (cohere) to each other in the presence of water.

Cold-formed metal framing See *light-gauge steel framing.*

Cold joint See *construction joint.*

Column An upright vertical structural member that supports a slab, beam, or truss.

Column cover Preformed exterior cladding element that covers a column for aesthetic purposes.

Column splice A method of connecting the lower part of a tall steel column with the upper part. Column splices are typically provided every two stories.

Combustible A material that will ignite in the presence of an open flame or high temperature.

Commissioning A process of adjusting the performance of individual components of the mechanical and electrical systems of a building to achieve energy efficiency.

Compaction A decrease in the amount of void space in a soil mass brought about by mechanical action (e.g., tamping) or a natural process, resulting in a densified soil mass.

Composite deck Corrugated-steel floor deck, which acts as formwork and primary reinforcement for a concrete slab.

Compression Stress created in a structural member as a result of bending or axial compressive force.

Compressive strength Measure of the ability of a material to resist compressive force.

Concave joint A type of mortar joint profile that is most weather resistant.

Concrete A composite material consisting of portland cement, coarse aggregate (crushed stone), fine aggregate (sand), and water.

Concrete admixture Material added to a concrete mix to influence its performance.

Concrete block See *concrete masonry unit.*

Concrete masonry unit Standardized masonry unit made of concrete and consisting of face shells and webs surrounding two or three voids, called cells.

Condensation resistance factor A measure of the potential for condensation to occur on a glazed assembly (window or glass curtain wall).

Conduction A mode of heat transfer in a solid.

Conduit A hollow metal or plastic tube used as protective cover for electric wires.

Construction documents Documents used by a contractor to construct a building, including drawings and specifications.

Construction drawings Drawings used by a contractor to construct a building.

Construction joint A nonmovement joint resulting when fresh concrete is placed against previously placed concrete.

Construction manager A manager hired by the owner to coordinate and supervise the work of multiple contractors. A project with a construction manager does not generally have a general contractor.

Control joint A sawed or tooled joint on the top surface of a concrete slab-on-ground; a continuous vertical joint in a concrete masonry wall.

Convection A mode of heat transfer in liquids and gases.

Conveying equipment Mechanical equipment that is used to move people or products vertically and/or horizontally through a building.

Coping A protective cap or cover used on the exposed top of a wall, typically sloped to shed water.

Core holes Holes in an extruded brick to provide uniform drying and firing.

Corrosion An electrochemical process that causes deterioration of metal surfaces exposed to air, water, or excessive humidity.

Counterflashing A removable sheet-metal flashing that laps over membrane flashing to prevent water from penetrating between the membrane and the wall.

Course A single horizontal layer of units in a masonry wall.

Coursed rubble Irregularly shaped and sized stone laid with periodically aligned bed joints.

Crane A type of construction equipment used to hoist and place heavy building components or materials during construction.

Crawl space Air space between an elevated ground floor and the ground.

Creep Permanent deformation of a component under sustained loads, generally of concern in concrete and wood.

CRF See *condensation resistance factor.*

Cricket A pyramidal formation on a roof to divert water to drains or on an easily drained area of the roof.

CSI Construction Specification Institute.

CSMU See *calcium silicate masonry unit.*

Curtain wall Exterior wall cladding system suspended from or supported by the structural frame of the building.

D

Dampproofing A material applied to the exterior of a wall to resist the intrusion of capillary water.

Dead load Loads created by the weight of building components.

Decibel (dB) A measure of the loudness of sound.

Deep foundation A foundation element that extends deep below the ground to reach bedrock or higher-bearing-capacity soil.

Deformation Change in the size or shape of a structural member as a result of an applied load.

Dew point The temperature at which the relative humidity of an air mass reaches 100% and the water vapor in air converts to (liquid) water, that is, condenses.

Diagonal brace A linear, diagonal stiffening element against lateral loads.

Die A tool with an orifice (hole) through which a soft, heated metal or plastic or a column of wet clay is pushed to give a long, continuous element.

Dimension lumber Lumber that ranges from 2 in. to 4 in. (nominal dimensions) thick.

Dimension stone Stone that has been fabricated to specific dimensions, texture, and finish for use in buildings.

Dimmable glass A glass that changes from a transparent to a translucent or opaque condition when exposed to sun (photochromic glass) or electrical current (electrochromic glass).

Dome A curved roof structure over a round or rectangular space.

Double-skin glass curtain wall A glass curtain wall assembly composed of two layers of glazing separated by 1 to 5 ft of air space, also called a double-skin facade or bioclimatic glass curtain wall.

Double-strength (DS) glass Flat glass that is $\frac{1}{8}$ in. thick.

Double tee A precast concrete floor or roof element to span long distances.

Dowel A steel or wood pin used with an adhesive to connect adjacent pieces of wood; a steel reinforcing bar that projects from a foundation to form a splice with reinforcing bars in a concrete column or wall above the foundation.

Drainage wall A cavity wall with an air space between the exterior cladding and the inner backup wall, allowing water to drain out through weep holes.

Drilled pier A reinforced-concrete column constructed by drilling a round hole (of required diameter) in the soil, reinforcing it, and filling the hole with concrete. Drilled piers are used as foundation elements where near-surface soil is unsuitable for building foundations.

Drip edge Perimeter or edge flashing with an angled lip to keep water away from a vertical surface below.

Drop panels A widening of the underside of a concrete slab at a column location.

Dry glazing Use of preformed compression gaskets to seal the glass against the metal frame of a window or metal-glass curtain wall.

Drywall See *gypsum board.*

Ductile material A material that produces large deformations under load before failure.

Dynamic joint A joint that allows a predetermined amount of unrestrained movement between components.

E

Eave The low end of a sloped roof.

Edge block A resilient material used to prevent the vertical edge of glass from touching the frame of a window or curtain wall.

Efflorescence A white deposit of water-soluble salts on an exterior masonry or concrete wall caused by the wetting of the wall.

Egress window A window in a habitable room of a dwelling that opens to a minimum size for use as a fire escape.

EIFS See *exterior insulation and finish system.*

Elasticity The ability of a deformed material to return to its original shape and size after the removal of a load, such as rubber.

Elastomeric A material that exhibits elastic (rubberlike) properties.

Elastomeric sealant A synthetic semiliquid material used to seal the joints between adjacent building components, which becomes an elastic material after curing.

Elevated slab An above-ground floor or roof slab, supported on columns and/or beams and forming an integral part of a structural frame.

Embodied energy Total amount of energy required to extract raw materials and manufacture a building material.

Emissivity A property of the surface of a material that governs its potential to emit radiation; the value lies between 0 and 1.0.

End-bearing pile A deep foundation element within the ground that transfers load to bedrock, in contrast to a friction pile that transfers load to the surrounding soil. See also *friction pile.*

Engineered fill Soil and/or aggregate with specified proportions and properties compacted into place so that the resulting product has predictable physical characteristics.

Engineered wood products Manufactured wood products rated for structural applications, such as plywood, oriented strandboard, or laminated veneer lumber; see also *manufactured wood products* and *industrial wood products.*

Epoxy A polymer-based adhesive.

Ethylene propylene diene monomer (EPDM) A stretchable polymeric membrane used as a single-ply roof membrane.

Expanded polystyrene Open-cell foamed plastic insulating material formed into rigid boards.

Expansion anchor An anchor that allows expansion and contraction while connecting the vertical framing members of a glass curtain wall to the building structure.

Expansion joint A narrow space between two building components that allows their unrestrained expansion and contraction.

Exposed aggregate A type of concrete finish with coarse aggregate exposed at the surface.

Exposed ceiling A ceiling where all structural components and mechanical and electrical services are left exposed to the area below.

Exposure The exposed portion of lapped roofing felt or shingles.

Exterior insulation and finish system A stuccolike exterior finish that includes a layer of foam insulation, fiberglass reinforcing mesh, and one or two coats of a polymer-based finish; also called synthetic stucco.

Extruded brick Wet clay extruded through a die and cut to size before drying and firing. See also *molded brick.*

Extruded polystyrene Closed-cell foamed plastic insulating material formed as rigid boards by an extrusion process.

Extrusion See *die.*

F

Face shell Longitudinal walls of a concrete masonry unit.

Facing brick Brick used in an exposed exterior masonry wall, controlled for dimensional variations, warpage, and durability.

Factor of safety A safety margin added to structural calculations to account for our imprecise knowledge of material strength, loads, and theories pertaining to structural analysis and design, defined as the actual strength of a component divided by the strength required to carry the load.

Ferrous metal A metal that contains mainly iron.

Ferrule A threaded cylindrical sleeve that accepts a bolt, typically embedded in a precast-concrete member.

Fiberboard A product made from wood fibers (softwood or hardwood). In terms of increasing density, particle board, medium-density fiberboard (MDF) and high-density fiberboard (hardboard) are three types of fiberboard commonly used.

Fiberglass Mineral fiber spun from glass.

Fibrous insulation An insulating material that has a high R-value due to the air contained between the fibers.

Fill Soil or aggregate that fills the area under the foundations or between the exterior of a foundation and the boundaries of an excavation. See also *engineered fill and backfill.*

Filter fabric A porous filter fabric for filtering out silt, sand, or debris in a protected membrane roof, French drain, or foundation drain.
Finger joint A method of joining short lengths of lumber using interlocking "fingers" at the end of each piece and gluing the members together.
Finish coat The last layer of material that provides the final color and texture on a surface, such as the portland cement–based layer on stucco or the polymer-based layer on the exterior insulation and finish system.
Fire-rated glazing A glazing system that will resist a standard fire for the measured duration.
Fire rating The ability of a building assembly to endure fire, measured in hours or minutes of time and determined from standardized full-scale tests.
Fire-resistance rating See *fire rating.*
Fire-retardant lumber Wood that is pressure treated with chemicals to increase its fire resistance.
Fire-stopping A noncombustible material used to seal the space around a penetration or joint in a fire-rated wall, floor, or roof.
Fixed window A window unit where the glass is permanently fixed into the frame.
Flame finish A type of finish on a stone surface (generally granite) using a torch on wetted stone to break the surface, creating a rough, slip-resistant finish; also called thermal finish.
Flame spread index (FSI) A measure of the rate at which flames spread on the surface of an interior finish.
Flange Top and bottom components of an I-section or C-section steel beam.
Flashing A flexible metal or plastic used at roof terminations, edges, or penetrations to increase the strength and water resistance of the roof; a flexible metal or plastic sheet used as a drainage channel at the base of a cavity wall.
Flat glass Planar glass sheet (window glass), as opposed to glass block or other glass products.
Flat plate A reinforced-concrete slab of uniform thickness supported directly on columns.
Flat-sawn lumber See *plain-sawn lumber.*
Flat slab A reinforced-concrete slab, like flat plate, but the slab is thickened at the columns.
Flex anchor An L-shaped steel pin that connects to the GFRC skin with a bonding pad and is welded to the supporting light-gauge steel frame.
Float glass Flat glass manufactured using the float glass process.
Floating The process of smoothing a freshly placed concrete surface after it has been struck (leveled). See also *striking.*
Floating floor A floor assembly consisting of a resilient underlayment to isolate the finished floor from the subfloor; can also be a double floor with an intervening air space.
Fluid-applied roof membrane A combination of sprayed polyurethane foam insulation topped by a fluid that cures to a membrane.
Flush joint Troweled masonry mortar joint, where the mortar is finished flush with the masonry face.
Flux A material added to the primary material to lower the primary material's melting-point temperature, saving energy.
Fly ash A waste product obtained from the combustion of coal in a thermal power plant (a pozzolanic material), used as a concrete admixture.
Flying form A large, prefabricated formwork for concrete floors, which can be lifted (flown) into position by a crane and reused multiple times.
Foamed concrete See *insulating concrete.*
Foamed plastic Plastic insulating material that contains small volumes of air or another gas.
Footing The bottom portion of a foundation resting directly on the supporting soil with an area larger than that of the supported wall or column.
Form ties Specially shaped steel wires that are used to connect and separate opposing panels of formwork for a concrete wall.
Formwork Temporary support system for fresh concrete to give it the desired shape (form) until it cures to hardness.
Frame structure Building structure that is composed of columns and beams, called a skeleton frame in the case of a steel frame structure.
Framing plan Architectural plan that delineates the organization and dimensions of columns, beams, and load-bearing walls (if any).
Freeze-thaw A cycle of freezing and thawing of water within water-absorptive materials, such as soil, masonry walls, or concrete.
Friction pile A load-bearing pile that carries load by friction developed between the surface area of the pile and the soil. See also *end-bearing pile.*
Fritted glass Glass made semiopaque or opaque by the application of patterns of tiny dots or lines of ceramic material on one surface of flat glass.
Frog A small depression in a molded brick.
Frost line The depth of soil below which groundwater will not freeze.
Furring Narrow wood or formed sheet-metal sections applied to a wall or ceiling, used for supporting and attaching a finish material on the wall.

G

Gable Triangular portion of an exterior wall between the eave and the ridge in a gable roof.
Gable roof A roof with two sloping planes that meet at a point at the top, called a ridge.
Gable vent Passive exhaust ventilation unit located high on the gable end, used to exhaust attic air.
Galvanic corrosion Corrosion that occurs when two dissimilar metals are in direct contact with each other.
Galvanizing Method of coating steel with zinc to provide a corrosion-resistant (sacrificial) coating.
Gasket A shaped piece of resilient material that provides a weatherproof seal between the glass and the frame in a window or metal-glass curtain wall. See also *dry glazing.*
General contractor A contractor who has the final responsibility for the construction of the project including site safety, supervision, and coordination of the work of all subcontractors.
Geodesic dome A hemispheric dome formed by short, crisscross linear members that run along great circles of the sphere, creating triangles or polygons on the surface of the dome.
Geotechnical report A document prepared by a geotechnical (soils) engineer that identifies and describes the properties of soil and the underlying geology of the building site.
GFRC Glass fiber–reinforced concrete.
GFRP Glass fiber–reinforced plastic.
Girder A large beam (also called a primary beam) that carries the loads from secondary beams or joists to columns in a frame structure.
Glass block See *glass masonry unit.*
Glass fiber–reinforced concrete Glass fibers reinforce a mix of portland cement, sand, and water that is sprayed on formwork

to form a panel that is integrated with a light-gauge steel frame and used in curtain wall assemblies.

Glass fiber–reinforced plastic A plastic material reinforced with fiberglass, molded to the desired shape.

Glass masonry unit A hollow glass unit used in non-load-bearing wall applications to provide a translucent wall.

Glazing An assembly of flat glass and its supporting frame.

Glazing compound A semisolid compound mastic, which cures to hardness, used to bed small panes of glass in a window frame; not commonly used in contemporary buildings.

Glue laminated Lengths of dimension lumber, glued and laminated together to create a structural member of a large cross section.

Glulam See *glue laminated.*

GMU Glass masonry unit.

Gradation Distribution by size of soil particles within a soil mass; also applies to aggregates in a concrete mix.

Grade beam A reinforced-concrete beam constructed at ground level.

Graded lumber Lumber with a grade stamp.

Grade stamp Stamp affixed by a grading (inspection) agency on lumber or a wood panel that describes its in-service performance.

Grading Mechanically moving soil on a site to predetermined ground elevations.

Grain structure Pattern and density of fibers that compose wood.

Granite Strongest and densest of stones; it takes good polish and weathers more slowly than other stones; obtained from igneous rock.

Gravity load Building load caused by gravity and acting in the vertical direction; also called vertical load.

Green roof A landscaped roof.

Grille A grating or screen that protects an opening while allowing the passage of air.

Grout A high-slump (semiliquid) mortar that flows well enough to be placed or pumped between the voids in a masonry wall; a mortarlike, high-strength cementitious material for filling the space between the steel base plate of a steel column and the concrete footing; a cementitious material used to fill joints between ceramic or quarry floor tiles or panels after they have been laid.

Growth ring A nearly circular ring of wood fibers; a tree adds approximately one ring per year.

Guard rail A horizontal rail and associated supports to prevent people from falling from a stair or balcony.

Gunite See *shotcrete.*

Gusset plate A plate that connects two structural steel members, commonly used in a large steel truss.

Gypsum board A building panel faced on two sides by paper (or vinyl) with a gypsum core. Typically used on fire-rated interior walls. It is also called drywall, wallboard, or gypsum wallboard.

H

Handrail Diagonal rail at hand level adjacent to a stairway to provide support for people using the stairs, as distinguished from a guard rail.

Hardboard A product made from wood fibers and a resin binder under heat and pressure. It is denser than medium-density fiberboard (MDF). See also *fiberboard.*

Hardwood Wood obtained from trees that are deciduous and have broad leaves. See also *softwood.*

Head Top of a window or door.

Header The short face of a brick when laid in a horizontal position in a course of bricks, historically used to tie together two wythes of a thick brick wall; a beam (lintel) above a door or window opening in a wood frame or light-gauge steel construction.

Head joint Vertical mortar joint in a masonry wall. See also *bed joint.*

Headroom Minimum code-required clearance between a floor and an overhead projection or ceiling.

Heartwood The central portion of a tree trunk that no longer conducts nutrients, generally darker than sapwood. See also *sapwood.*

Heat-absorbing glass See *tinted glass.*

Heat-soaked glass A type of tempered glass obtained from a process that reduces (or eliminates) the possibility of spontaneous breakage of tempered glass during its service life. See also *tempered glass.*

Heat-strengthened glass A glass obtained by heating annealed (basic) glass to a high temperature and then suddenly cooling it; approximately twice as strong as annealed glass. See also *annealed glass* and *tempered glass.*

Heavy timber Sawn lumber with both cross-sectional dimensions greater than 5 in. (nominal).

Heavy-timber construction A construction type in which the floors, roofs, and structural frame are composed of heavy sections of wood without added fire protection.

Hip The edge of a hip roof where adjacent slopes meet; shown as a diagonal line in the plan of the roof.

Hip roof A roof formed by four sloping planes that intersect to form a pyramidal or elongated pyramidal shape.

Hollow-core door Wood or metal veneer placed on both sides of a frame to build a door that is partially or almost entirely hollow within.

Hollow structural section (HSS) Square or rectangular tubular steel section used as columns or beams in a steel-frame structure or as components of a steel truss.

Honed finish A type of smooth finish on stone or metal created by using abrasives.

Horizontal diaphragm A structural engineering term that refers to the floors and roof in a building when they function as part of the lateral load–resisting system.

Hydrated lime Calcium hydroxide, produced by combining quicklime and water to produce a relatively benign lime, used in powder form in masonry mortar. See also *slaked lime.*

Hydration Chemical reaction between portland cement and water that causes the mix to become hard.

Hydraulic cement A cement (glue) that, after setting, does not react chemically with or dissolve in water, such as portland cement.

I

IBC International Building Code.

Ice dam protection membrane A self-adhering waterproof membrane that prevents water from backing up under roof shingles when an ice dam forms at the eave.

IGU Insulating glass unit.

IMG Intumescent multilaminate glass

Impact insulation class (IIC) Measure of the ability of an assembly to resist the transmission of structure-borne sounds.

Impact sound Structure-borne sound produced by an impact on a building component.

Impact wrench A tool for tightening bolts and nuts through rapidly repeated torque impulses produced by electric or pneumatic power.

Industrial wood products Manufactured wood products used for nonstructural applications, including particle board, medium-density fiberboard (MDF), and high-density fiberboard (HDF). See also *manufactured wood products* and *engineered wood products.*

Infrared radiation Long-wave radiation emitted by objects at low temperature.

Infrastructure Basic organizational systems that provide services to a building, including roads, power, water, and electricity.

Initial rate of absorption (IRA) A measure of the ability of clay bricks to absorb water, used to determine if bricks need to be wetted before being laid.

Inside glazed Glazing system in which the glass is set into the frame from the inside of a building. See also *outside glazed.*

Insulated metal panels Metal panels consisting of polyurethane foam sandwiched between and bonded to two metal sheets, used in curtain wall applications.

Insulating concrete Lightweight concrete consisting of portland cement, water, and expanded aggregate (perlite or vermiculite), primarily used as low-slope roof insulation. A type of insulating concrete (called foamed concrete) does not contain aggregate but does contain air particles. A fire-resistant insulation material for roofs and floors.

Insulating glass unit A sealed assembly of two layers of glass separated by an air- or gas-filled chamber. The assembly has a higher R-value than a single sheet of glass.

Insulation See *thermal insulation* and *sound insulation.*

Integrated project delivery (IPD) A project delivery method in which all people connected with the design and construction of the building, including the owner and fabricators of assemblies, collaborate as a team in a no-blame contractual agreement as if they all belonged to the same organization. The profits are generally shared at the completion of the project by team members on a predetermined basis.

Interior finish class Rating of interior finish materials as Class A, B, or C based on the flame-spread index (FSI) and the smoke-developed index (SDI).

Intumescent paint A paint that swells when exposed to heat, creating an insulating layer that protects steel from short-term exposure to fire.

Inverted roof See *protected membrane roof.*

Inverted T-beam A precast concrete beam with flanges at the bottom, generally used to support double-tee concrete floors.

Isolation joint A joint in a concrete slab-on-ground that penetrates the entire thickness of the slab, used to separate the slab from the structure (walls and columns).

Isotropic material A material that has the same properties in all directions, such as steel. See also *anisotropic material.*

J

Jack rafter A shorter rafter (than a common rafter) that joins to a hip or valley rafter.

Jack stud A stud of a shorter length attached to a longer king stud to support a header in the opening of a wood light frame wall.

Jamb Side of a window, door opening, or frame.

Joint compound Soft material used together with tape to finish a joint or cover nail dents in a gypsum wallboard assembly.

Joint cover Cover for an expansion or seismic joint in buildings to provide a continuous protective surface.

Joint reinforcement Continuous horizontal steel wire reinforcement laid in mortar joints of concrete masonry to control shrinkage cracks.

Joint sealants See *elastomeric sealants.*

Joist floor An elevated reinforced concrete slab with integral narrow ribs (like wood floor joists), also called a one-way joist floor or ribbed floor.

Joists Slender, closely spaced, parallel beams in a wood light (or light-gauge steel) frame floor.

Junction box Rectangular or circular metal or plastic box used to protect connections between electrical wires.

K

Kerf Slit cut into a surface, used to form a capillary break (drip edge) at the bottom edge of an exterior surface; a slit cut into building stone (used as wall cladding) to serve as a receptacle for a mechanical attachment system.

Key See *shear key.*

Keyway A gap between two adjacent precast-concrete floor units, typically reinforced and filled with portland cement grout.

Kiln A furnace used to dry materials such as green lumber; a furnace for drying and firing bricks; a furnace for making quicklime.

Kilopascal (kPa) A measure of load (force) in the SI system of units.

Knot An approximately circular formation of wood fibers formed where a limb branches from a tree trunk. A knot generally lowers the strength of wood.

Kraft paper Asphalt-treated paper that serves as a vapor retarder, commonly used as lamination on fiberglass insulation.

L

Laminated glass Two pieces of glass laminated under heat and pressure to a plastic interlayer to form a fused unit.

Laminated veneer lumber (LVL) Dried wood veneers laminated in layers, all oriented in the same direction, to form a large structural member.

Landing A horizontal platform between two flights of stairs.

Landing frame A structural frame that forms a landing in a prefabricated steel stair.

Lap splice A connection between two steel or wood members, where the ends of each member are overlapped and connected so that the two members function as one continuous member.

Lateral load Load that acts predominantly in the horizontal direction, such as wind or earthquake load or soil pressure on a basement wall.

Lath Expanded sheet metal attached to a backup wall and used as a mechanical key to bond and reinforce portland cement plaster (stucco).

LEED Leadership in Energy and Environmental Design.

Light-gauge steel framing A framing system that mimics wood light frame construction, but the elements are made of cold-formed, galvanized sheet steel.

Light-gauge stud Generally a C-shaped member made of light-gauge steel.

Light-transmitting plastic Transparent or opaque polycarbonate or acrylic sheet used as glazing in some limited situations.

Lime See *hydrated lime* and *quicklime.*

Limestone A sedimentary rock composed primarily of calcium carbonate and used as exterior and interior wall cladding.

Linoleum A resilient sheet or tile flooring made from linseed oil, organic resins, and fillers bonded to a fibrous backing.

Lintel A beam that spans over a door or window opening.

Lite A pane or sheet of glass.

Live load A type of gravity load that changes in magnitude and placement over time due to factors such as human occupancy or storage of roof repair or maintenance materials, subdivided into floor live load and roof live load. See also *dead load.*

Load-bearing-wall A wall that supports superimposed gravity loads, such as from floors and the roof. See also *non-load-bearing wall.*

Longitudinal bars Steel reinforcing bars placed in the long direction of a reinforced-concrete beam or column.

Louvers Fixed or operable, closely spaced, angled slats attached to a frame to direct the passage of air or light.

Low-emissivity (low-E) coating A surface coating that reflects most of the long-wave radiation.

Low-emissivity glass A glass treated with a low-E coating, called low-E glass.

Low-emissivity (low-E) material A material that is an efficient reflector (poor absorber) of long-wave radiation (heat), generally obtained through a low-E lamination (such as aluminum foil) or a low-E coating.

Low-slope roof A roof with a slope less than 3:12.

Lumber Solid wood products derived directly from logs by sawing and planing only, with no additional machining.

LVL See *laminated veneer lumber.*

M

Machine grading A method of grading lumber using a machine, also called machine stress-rated lumber (MSR), in contrast with the more commonly used method of visual grading.

Machine stress-rated (MSR) lumber See *machine grading.*

Malleability A property of metals that indicates their ability to be formed to shape by hammering, bending, forging, pressing, rolling, and so on.

Manufactured wood products Any of a variety of wood products made by bonding smaller pieces of wood veneers, fibers, strands, wafers, or particles to produce a composite wood material. See also *engineered wood products* and *industrial wood products.*

Marble A metamorphic rock geologically formed from limestone under high pressure and heat.

Masonry Materials such as brick, stone, concrete masonry units (concrete blocks), and glass masonry units that are stacked and adhered using mortar.

Masonry cement A preblended mix of portland cement, lime, and/or pulverized limestone (typically manufacturer specific) used in masonry mortar in place of portland cement and lime.

Masonry grout A high-slump concrete used to fill the voids in a masonry wall.

Masonry unit Single brick, concrete (or glass) block, or stone used in masonry construction.

MasterFormat The standard organizational format for construction specifications (note the absence of space between *r* and *F* in "MasterFormat").

Mat foundation A type of concrete foundation where one large, combined footing is used for several columns and load-bearing walls, often for the entire building. A concrete slab-on-ground used as a foundation for light frame buildings is the simplest type of mat foundation.

MDF An acronym for medium-density fiberboard. It is made from wood fibers and resin binder under heat and pressure. It is denser than plywood and is commonly used for cabinets and other furniture. See also *fiberboard.*

Means of egress The route by which one exits a building in the case of a fire.

Metal A highly refined (generally solid) material that is typically hard, malleable, ductile, and capable of carrying an electrical current. Steel, copper, and aluminum are commonly used metals in building construction.

Metal panel roof A roof covering composed of thin sheets of metal, laid over rigid insulation and mechanically attached to the roof structure.

Mil A unit of thickness typically used to express the thickness of sheet plastic or sheet metal. 1 mil = 1/1,000 in.

Mission tile A clay or concrete tile with a half-barrel profile.

Model code Building code developed and updated regularly by an independent agency and adopted and adapted for use by federal, state, or local jurisdictions. The International Building Code, International Residential Code, and International Plumbing Code are examples of model codes.

Modified bitumen A type of polymer-modified bitumen (generally asphalt) used in making modified-bitumen roof membranes.

Modular brick The most commonly used size of extruded brick (4 in. × $2\frac{2}{3}$ in. × 8 in. nominal dimensions).

Modulus of elasticity A property of a material that measures its stiffness against deformation under loads.

Moisture movement Expansion or contraction of a material that occurs as a result of a changing moisture content.

Moisture retarder A membrane that retards (prevents) liquid water from passing through.

Molded brick A brick formed by pressing small clumps of wet clay into molds. See also *extruded brick.*

Moment connection A connection between a beam and its support (e.g., a wall or column) that transfers the bending of the beam to the support, or vice versa.

Mortar A mixture of portland cement, lime, sand, and water used to bond masonry units; cementitious mixture used to adhere tiles to a subfloor.

Mortar-capturing device A synthetic mesh located just above the flashing in the air space in a brick veneer wall used to prevent mortar from clogging the weep holes.

Mortar cement A preblended (typically manufacturer-specific) mix of portland cement, lime, and admixtures.

Mortar joint The layer of mortar placed between masonry units, generally about $\frac{3}{8}$ in. thick.

Movement joint Same as *dynamic joint.*

Mud sill See *sill plate.*

Mullion A vertical member in a metal-glass curtain wall; a vertical member between two adjacent windows or doors.

N

Nail Metal fastener made from a wire with a pointed tip (bottom) and a (generally flat) head.

Nailing flange Continuous flat metal or plastic extrusion from a window frame used to nail the unit to the exterior of a wall opening in a wood frame building.

Nail plate A metal plate with several sharp, flat, naillike projections.

Neoprene A synthetic rubber primarily used in glazing gaskets.

NFRC rating Window-performance rating system developed by the National Fenestration Rating Council (NFRC).

Noise-reduction coefficient (NRC) A measure of the ability of a material to absorb sound; the value lies between 0 and 1.0.

Noncombustible A material that will not ignite when subjected to an open flame or high temperature. See also *combustible.*

Non-load-bearing wall A wall that does not support floor or roof loads.

Nonmovement joint Same as *static joint.* See also *dynamic joint* or *movement joint.*

Nosing The leading edge of a tread of a stair.

Nylon A versatile polymer whose fibers are used in carpet and fabrics; also produced in sheets or molded into fittings used in mechanical devices.

O

Occupancy load Floor live load based on its occupancy.

One-way joist floor See *joist floor.*

One-way slab An elevated reinforced concrete slab where most of the load on the slab is carried to the supporting beams in one direction; a four-sided, supported rectangular slab whose length is greater than or equal to twice its width.

Open-web steel joist A standardized, prefabricated steel parallel chord truss used to span between beams, larger joists, or trusses.

Operable window A window unit that opens to allow the passage of air, sound, and so on. See also *fixed window.*

Organic soil A soil containing decayed plant material.

Oriented strandboard A wood-based panel made by gluing several layers of wood strands under heat and pressure so that the adjacent layers are oriented in opposite directions.

OSB See *oriented strandboard.*

Outside glazed Glazing system in which the glass is set into the frame from the outside of a building.

Oxidation Chemical weathering process by which the source material combines with atmospheric oxygen, generally leading to the material's degradation, such as oxidation of iron or steel, called corrosion, or rusting.

P

Pan Glass-reinforced plastic pans used as an economical formwork for one-way concrete joist floors.

Pane A sheet of glass.

Panel A sheet of plywood, oriented strandboard, particle board, and so on; a prefabricated building component whose surface dimensions are much greater than its thickness, such as a curtain wall panel.

Panel door A wood door constructed by inserting solid wood panels or glass panes in horizontal and vertical frame members (rails and stiles, respectively).

Panelized stone cladding Multiple stone slabs attached edge to edge to a prefabricated steel-frame work, which is then attached to the exterior of a steel- or concrete-frame building.

Panel points Points on a truss where members of a truss connect.

Parallel-strand lumber Manufactured wood product composed of narrow strands of veneered lumber glued together, all oriented in the same direction to form a member of large cross section.

Parapet Portion of an exterior wall that extends above the roof line.

Parging A thin, rough (unfinished) coat of portland cement plaster on a masonry wall. The purpose of parging is to increase the water resistance of the wall and to provide a smoother wall surface. Parging is not intended to be finished plaster.

Partition wall A non-load-bearing interior wall that separates spaces but does not carry floor or roof loads.

Patching compound A liquid or semisolid compound that can be troweled on to correct surface irregularities.

Paving bricks Bricks used for paving, graded for abrasion resistance and freeze-thaw damage.

Permanent formwork Material used as formwork (typically, metal deck) and retained as a permanent part of a reinforced-concrete floor slab.

Permeability Property of a material (e.g., soil) or assembly that allows water or water vapor to pass through.

Permeance A measure of the rate at which water vapor flows through a material or assembly.

Pier A relatively short column or a deep vertical foundation element (typically of cast-in-place concrete). A pier is not the same as a drilled pier (see *drilled pier*).

Pilaster A column formed by thickening an area of a masonry or concrete wall.

Pile Driven or drilled long, slender foundation element.

Pile cap A concrete cap that transfers foundation loads to multiple piles beneath.

Placing concrete The operation of pouring fresh concrete from a concrete mixer into the formwork.

Plain concrete Concrete without steel reinforcement.

Plain masonry A masonry wall without (vertical) steel reinforcement.

Plain-sawn lumber Lumber produced by sawing a log in one or two directions only. The grain pattern varies from nearly parallel to the wide face to perpendicular.

Planing Operation that uses high-speed knives to smooth the surface of rough-sawn lumber.

Plaster Pastelike cementitious material that, when applied to the surface of a wall or ceiling, cures to a hard surface.

Plasterboard Specific type of gypsum board used as a substrate for plaster.

Plastic A synthetic material composed of polymers that can be molded or shaped when soft; it cures into a rigid or semielastic form and does not return to its original shape after being deformed.

Plate girder A heavy steel beam fabricated from steel plates.

Platform frame One of the two types of wood light frame constructions, the other being a balloon frame. In a platform frame, the studs are one floor high and extend between adjacent floors or from the floor or roof. See also *balloon frame.*

Plywood A panel made from multiple thin layers of wood veneer adhered with glue under heat and pressure so that the grain direction of a veneer is perpendicular to that of the adjacent veneer.

Pointing Raking an existing (generally defective) masonry mortar joint to sufficient depth and then finishing it with new mortar.

Point-supported curtain wall Mullionless glass curtain wall with glass panes supported at its corners by a metal connector and glass panes sealed at vertical and horizontal joints with a sealant. See also *spider connector.*

Polished finish A type of stone finish, attained by buffing to give a reflective sheen.

Polyethylene A polymer, typically used in sheets as a vapor retarder.

Polyisocyanurate Closed-cell foamed plastic insulating material, sandwiched between two facing layers, available as rigid (flat or tapered) boards, typically used as roof insulation.

Polypropylene Synthetic fiber, commonly used in carpets and fabrics.

Polyvinyl chloride (PVC) A versatile thermoplastic polymer, used in single-ply roof membranes, rigid plastic pipes, and so on.

Porosity The property of a material that allows liquid or air to pass through.

Portland cement A noncombustible hydraulic cement produced by burning limestone and clay. See also *hydraulic cement.*

Portland cement plaster Multilayered application of a cementitious mix of portland cement, lime, sand, and water to a thickness ranging from $\frac{5}{8}$ in. to 1 in., commonly used on exterior walls.

Posttensioning Subjecting a concrete or masonry member to compressive stresses by tensioning high-strength steel strands (wires) after the concrete has developed sufficient strength.

Pour To cast or place concrete; an increment of concrete placement carried out without a long interruption.

Pour stop An L-shaped steel member welded to a steel spandrel or edge beam to stop fresh concrete from flowing beyond the edge of a floor or roof.

Pozzolanic reaction Reaction between lime and amorphous silica (such as in fly ash, silica fume, etc.) that converts the mixture into a hydraulic cement. See also *hydraulic cement.*

Precast concrete Concrete members typically cast in a precasting plant and transported to the construction site for erection. Precasting can also be done at the construction site, such as with concrete tilt-up walls.

Precast, prestressed concrete A precast-concrete member that has been subjected to compressive stresses by tensioning high-strength steel wires before casting concrete for the member.

Preformed tape An elastomeric material (available in rolls) that provides a seal between the glass and the supporting frame in a window or metal-glass curtain wall.

Pressure-equalized wall A drainage wall where the air space between the exterior cladding and the backup wall is at the same pressure as the outside air so that water penetration through the wall is minimized.

Pressure-treated wood Wood into which preservatives have been pressure injected to retard termite infestation and fungal decay. Using a different preservative, the pressure treatment can also be used for increasing the fire resistance of wood.

Prestressed concrete See *precast, prestressed concrete.*

Pretensioning Same as prestressing. See also *posttensioning.*

Primary reinforcement Steel reinforcement in a one-way concrete slab oriented in the direction that carries most of the loads; see also *one-way slab.*

Primer A liquid that improves the adhesion of a sealant or paint to the substrate.

Probability of breakage The probability that a pane of glass will break unpredictably, measured as the number of broken panes (out of 1,000) when the panes are subjected to a given load.

Protected membrane roof (PMR) A roof membrane directly attached to the roof deck and covered with rigid insulation held in place by ballast.

PSL See *parallel-strand lumber.*

Punched window A window surrounded by opaque portions of the wall.

Punch list A list of outstanding problems that must be corrected before the architect certifies that the building is complete.

Putty See *glazing compound.*

Pylon A heavy vertical support member, commonly used in suspension bridges.

Q

Quarry An excavation from which stone is obtained.

Quarry tile A clay floor tile, generally unglazed.

Quarter-sawn lumber Lumber produced by cutting the log radially into four quarters and then sawing it along radial lines.

Quicklime Lime obtained by heating limestone. Chemically, it is calcium oxide, which is a fairly caustic substance and is, therefore, not used in construction. When quicklime is mixed with water, it becomes calcium hydroxide, also called hydrated lime. Hydrated lime is used in building construction.

R

Racking Angular deformation of a structure under lateral loads, such as the deformation caused by a horizontal force on a book rack.

Radiant barrier Layer of material that slows the transmission of heat due to its low emissivity. Aluminum foil is a commonly used radiant barrier.

Radiation Emission of electromagnetic waves by an object. All objects emit radiation at all times.

Rain screen wall Same as *pressure-equalized wall.*

Rake Sloping edge of a gable roof.

Raked joint Tooled, inset masonry mortar joint that creates a strong shadow line.

Random rubble Irregularly shaped and sized stone laid in a random pattern in a wall.

Rebar An abbreviation for reinforcing bar (a deformed steel bar used as concrete reinforcement).

Reciprocating saw Fixed or handheld saw that uses a linear blade, which moves horizontally and/or vertically at the same time.

Reflective glass A glass that reflects incoming visible radiation due to a very thin metal oxide coating on one surface.

Reglet A slot in a concrete or masonry wall in which a flashing or roof membrane is inserted and held in place by inserting a compressible material in the slot.

Reinforced concrete Concrete with integral steel reinforcing bars.

Reinforced masonry A masonry wall with vertical reinforcing bars; see also *plain masonry.*

Reinforcement cage A preassembled unit of reinforcing bars used in concrete members.

Relative humidity Amount of water vapor present in the air, measured as a percentage of the maximum water vapor the air can hold at the same temperature.

Reshoring A method of supporting a concrete floor after the formwork has been stripped and before it has gained the strength required to support the superimposed loads.
Resilient channel Light-gauge steel channel used for attaching gypsum boards so that the gypsum boards are acoustically decoupled from the backup wall on which the channel is attached.
Resilient flooring Synthetic flooring material, such as linoleum, vinyl, or rubber tiles or sheets, that returns to its original shape after being compressed by impacts caused by walking, dropped objects, and so on.
Resinous flooring A liquid-applied flooring material that dries to form a thin, strong floor surface; also called epoxy flooring. Includes epoxy-based terrazzo.
Ribbed floor See *joist floor.*
Ribbed slab A concrete slab that is stiffened by the edge and interior ribs (beams) in both directions, used as an elevated or ground-supported slab. Where the ribs are closely spaced (in both directions), it is also called a waffle slab or two-way joist floor.
Ridge beam A structural beam that supports the top ends of roof rafters at the ridge line and forms a triangular shape without creating lateral thrust.
Ridge board A nonstructural board used to align and join rafters at the ridge line of a sloping roof.
Ridge line The line of intersection of two oppositely sloping roof planes.
Ridge vent Passive air-exhaust unit located along the ridge of a sloping roof to exhaust attic air.
Rigid connection See *moment connection.*
Rigid insulation Rigid boards of foamed, granular, or cellular materials.
Rim joist Outer joist that surrounds the wood frame floor structure.
Riser Vertical component of a stair.
Roof deck A load-bearing surface on which a roof membrane is attached.
Roofing felt Organic (typically paper-based) or inorganic (typically fiberglass) fibers pressed into a thin, flat sheet and saturated with bitumen (asphalt or coal tar).
Roof membrane The entire waterproofing layer on the roof.
Rubber flooring Rubber sheets or tiles used as floor finish.
Running bond Masonry bond pattern that staggers the units in each course, creating staggered head joints.
R-value See *thermal resistance.*

S

Safety glass Glass that can resist impact (per the safety glass test standard) and, upon breaking, falls into pieces that are small and blunt enough not to cause injury.
Safety margin See *factor of safety.*
Sapwood The outer portion of a tree trunk that conducts nutrients, typically lighter in color than heartwood. See also *heartwood.*
Sash The (operable) part of a window, which holds the glass.
SBS See *styrene butadiene styrene.*
Scab A small piece of lumber nailed to two butt-jointed lumber members to give them continuity.
Scaffold An elevated temporary platform (movable or unmovable) on which workers (such as bricklayers, plasterers, and painters) stand to perform their work. The term scaffolding is used for a system of scaffolds.
Scratch coat First of two base coats in stucco.
Seasoning Process of drying of lumber until the moisture content reaches the desired percentage.
Secondary reinforcement Reinforcement in a one-way concrete slab placed perpendicular to primary reinforcement. See *one-way slab.*
Security glazing Glazing system that can withstand various levels of assault from handheld weapons, ballistic weapons, and so on.
Seismic joint Continuous joint that divides large or complex buildings into smaller buildings that can move independently of each other to reduce damage from an earthquake; generally wider than a building separation joint. See also *building separation joint.*
Self-drilling fastener A fastener that drills its own hole.
Self-furring metal lath A metal lath with dimples that holds the lath away from the substrate to which the lath is applied, allowing the plaster to go through the lath and embed the lath in the plaster.
Setting block A strong, resilient block that supports the glass pane in its frame and cushions it against damage; located at the lower edge of a glazing pocket.
Setting material A material used to adhere ceramic and stone floor tiles to the subfloor.
Settlement Natural compaction of a soil mass, or shifts in a structure resulting from compaction or instability of soil underlying the foundation.
S4S A lumber member surfaced on all four sides.
Shading coefficient Solar heat gain through a glass divided by the solar heat gain through clear $\frac{1}{8}$-in.-thick glass.
Shaft A continuous vertical enclosure around stairs, elevators, ducts, pipes, and so on.
Shear Stress that acts tangential to the cross section of a member, causing various planes of the member to slide relative to each other.
Shear block A small metal block used to attach horizontal rails to mullions in a metal-glass curtain wall.
Shear key A continuous groove in the first concrete pour that is filled with concrete during the second pour to prevent differential movement between the two elements, such as between two adjacent concrete slabs or a foundation wall and its footing.
Shear stud A short, headed steel rod welded to the top of a steel beam (girder or joist) that is embedded in the concrete slab over the beam so that the beam and the slab act as one structural unit.
Shear wall See *vertical diaphragm.*
Sheathing A panelized material applied to the exterior surfaces of wood or light-gauge steel frame members to add rigidity to the frame and to serve as a base for (wall) cladding or roofing.
Shed roof A roof that slopes to one side.
Sheet bracing A panelized material applied to a structural frame to prevent the frame from racking. See *racking.*
Sheeting Sheet piles; a thin sheet material such as polyethylene sheet.
Sheet pile Piles made of interlocking sheet steel driven into the ground to support an excavation.
Shelf angle Steel angle attached to a spandrel beam or a load-bearing wall to support masonry veneer.
SHGC See *solar heat-gain coefficient.*
Shim A thin, flat piece of metal, plastic, or wood placed between two components to adjust their relative positions during construction.

Shingles Small roofing units of asphalt-treated felts, wood, metal, concrete, or clay tiles, slate, and so on, that lap over each other to waterproof steep roofs.
Shop drawings Detailed drawings of a building component developed by its fabricator based on construction drawings and specifications.
Shores Temporary vertical or inclined supports used in concrete formwork or excavation.
Shotcrete Concrete mix that is deposited through a nozzle at high pressure in combination with a stream of compressed air; also called gunite.
Siding Exterior wall-finish material applied to a wood light frame or light-gauge steel-frame building.
Sieve An open-top container with a wire mesh bottom used for screening particles. Sieves are used in laboratories for analyzing the particle size distribution of soils and concrete aggregates.
Sill Bottom horizontal portion of a window, the exterior part of which is typically sloped away from the window.
Sill plate The bottom, horizontal member in a wood light frame (or light-gauge steel) wall that is directly in contact with the foundation.
Simple (shear) connection A connection between a beam and its support (e.g., a wall or column) that allows the beam to bend without transferring its bending to the support.
Simply supported beam A single-span beam with simple connections at both supports.
Single-ply roof (SPR) membrane A single-layer synthetic roof membrane.
Single-strength (SS) glass A $\frac{3}{32}$-in.-thick flat glass.
SIPS See *structural insulated panels.*
Site-cast concrete Concrete cast in the position where it will remain in the final structure.
Slab-on-grade A concrete slab that is supported directly on the ground.
Slab-on-ground Same as *slab-on-grade.*
Slaked lime Same as *hydrated lime*
Slag A noncombustible waste product obtained from the manufacture of iron in a blast furnace, commonly used as aggregate on built-up roofs.
Slag wool Slag converted to fibrous insulation.
Slip connection A connection that allows movement between connected elements in at least one direction.
Slip-critical connection A connection between two structural steel members bolted together in which the load on one member is transferred to the other member mainly by friction.
Slump test A test that measures the workability of fresh concrete by filling a cone-shaped mold with concrete, removing the mold, and measuring the height to which the concrete settles below its original height.
Slurry A liquid mixture of an insoluble material and water, such as portland cement and water (called portland cement slurry) or bentonite clay and water (called bentonite slurry).
Slurry wall A method of constructing a reinforced-concrete basement wall by temporarily stabilizing the wall excavation with bentonite slurry and then pumping concrete into it. As the concrete is pumped, the slurry is displaced.
Smoke-developed index (SDI) A measure of the visibility through smoke in a building fire.
Smooth, off-the-form concrete The smooth finish on concrete revealed when formwork is removed, without any further surface treatment.
Snap tie A steel wire tie used in formwork for concrete walls.
Soffit The underside of a horizontal building element, such as that of a roof overhang or of a stair.
Soffit vent Air intake ventilation unit located in the soffit of a roof overhang.
Softwood Wood obtained from trees that have thin conical leaves and are typically evergreen. See *hardwood.*
Solar heat-gain coefficient Solar heat transmitted through a glass divided by the solar heat that falls on the glass.
Solid-core door Flush wood or metal veneers laminated to both sides of a solid interior, in contrast to a hollow-core door.
S1S2E A lumber member surfaced on one side and two edges.
Sound-absorbing material A material that absorbs more sound than most other materials.
Sound frequency The number of to-and-fro movements of air particles in 1 s in a sound wave.
Sound insulation A material that retards the transmission of sound.
Space truss A three-dimensional truss.
Spall Loss of the surface of a material, such as concrete or masonry, resulting from stresses caused by expansion or contraction of the material, such as during freeze-thaw cycles.
Spandrel area The area of the exterior facade of a building at the level of the spandrel beam.
Spandrel beam A beam that spans between columns on the exterior face of a frame structure.
Spandrel glass Glass (usually opaque) used in a metal-glass curtain wall in the spandrel area of the facade.
Spandrel panel A panel used in the spandrel area of a glass curtain wall.
Specific gravity Density of a material divided by the density of water.
Specifications Written technical description of materials and assemblies shown in construction drawings, organized in three parts: general, products, and execution.
Spider connector Four-pronged stainless steel connector used at the corner of four lites of glass in a mullionless glass curtain wall.
SPR See *single-ply roof (SPR) membrane.*
Spray-applied fire protection A mixture composed of portland cement, water, and mineral fibers (or lightweight aggregate) sprayed on a steel member to increase its fire resistance.
Stack bond Masonry bond pattern in which units are stacked vertically with continuous head joints.
Standing-seam metal roof Sheet metal roof panels joined by folding and/or interlocking seams.
Static joint Same as *nonmovement joint.*
Steel A strong, malleable metal (alloy of iron and carbon) used for structural and nonstructural shapes.
Steel decking A ribbed-sheet steel formed into panels, commonly used in floors and roofs.
Steep roof A roof with a slope greater than or equal to 3:12. See *low-slope roof.*
Stem wall A short foundation wall, generally used to form a crawl space below the ground floor of a light-frame building.
Step flashing Small lengths of metal sheets placed between roof shingles, used at locations where a sloping roof meets a wall.
Stick-built curtain wall A metal-glass curtain wall whose framing members are installed at the site member by member.
Stiffback A member attached to a frame to increase its stiffness.
Stirrup A loop or U-shaped steel reinforcement used to increase the shear resistance of a concrete beam.
Stone cladding Stone panels attached to a backup wall or curtain wall frame.

Stone honeycomb panel A lightweight panel made of thin stone veneer laminated to an aluminum honeycomb backing.
Stone masonry Stone laid with mortar unit by unit.
Storefront system Glazed facade, generally one or two stories high from the ground, with a framing system similar to that of a metal-glass curtain wall but with less stringent performance requirements.
Story drift Horizontal deflection of an upper-story floor with respect to a lower floor caused by lateral loads as a result of racking.
Strain Change in the dimension of a member divided by its original dimension.
Strands High-strength steel wires used in prestressing cables; thin wafers of wood.
Stress Internal resistance created by a member in response to an applied external force, expressed as force divided by the cross-sectional area of the member.
Stretcher Long face of a brick laid in its usual way in a wall.
Striking Leveling a freshly placed concrete surface, generally followed by floating. See *floating.*
Stringer See *carriage.*
Strip flooring Wood finish flooring made of long, narrow tongue-and-groove boards.
Stripping time Length of time between placing concrete and removing the formwork.
Strip windows An array of continuous window units placed side by side horizontally or vertically.
Structural composite lumber Manufactured wood products produced by gluing together longitudinally oriented wood veneers or stands.
Structural insulated panels Structural panels composed of rigid insulation glued on each side with oriented strandboard or plywood panels.
Stucco See *portland cement plaster.*
Studs Closely spaced vertical members that constitute a wood light frame or light-gauge steel frame wall.
Styrene butadiene styrene A polymer with rubberlike properties generally used as a modifier of asphalt, giving an SBS-modified roof membrane.
Subfloor Structural (rough) floor beneath a floor finish, such as carpet or floor tiles.
Substrate Underlying member on which other material is placed.
Substructure Below-ground portion of a structure, including the basement and foundation.
Sump A pit used to collect water, from which it is pumped out.
Superstructure Above-ground portion of a structure.
Surety bond A type of insurance required of a contractor by the owner to ensure that the obligations of the construction contract will be met.
Suspended ceiling A ceiling hung from the overlying floor or roof structure.
Sustainable Comprehensively addressing aspects of building design and construction in addition to health, safety, welfare, and aesthetic aspects that account for resource consciousness and stewardship of the environment and ensure that the needs of future generations are not compromised by today's generation.
Synthetic stucco Another name for exterior insulation and finish system (EIFS).

T

T&G See *tongue-and-groove.*
Tabular area Allowable area per floor as obtained from the building code table.
Tempered glass A glass obtained by heating annealed glass to a high temperature and then suddenly cooling it, which makes it four times stronger than annealed (basic) glass; used as safety glass because it breaks into pieces that are small and blunt enough not to cause injury.
Tendon The combination of high-strength steel strands, sleeves, and end anchorages used for posttensioning concrete.
Terrazzo A concrete-like material (binder, filler, and aggregate mix) that is poured over a subfloor in a thin layer, ground, and polished after it cures to form a floor finish.
Thermal break A section of a material with high thermal resistance epoxied between two parts of a material with low thermal resistance.
Thermal bridge A portion of a building envelope or a building assembly with a much lower thermal resistance than the rest.
Thermal capacity Ability of a building material or component to store heat.
Thermal conductor Material that conducts heat rapidly
Thermal finish See *flame finish.*
Thermal insulation A material or assembly that retards the flow of heat.
Thermal movement Changes in the dimensions of a material that occur as a result of a change in its temperature.
Thermal resistance (R-value) Measure of the ability of a building material or assembly to resist the flow of heat.
Thermal transmittance Inverse of thermal resistance.
Thermoplastic polymer A polymer that softens when heated and hardens when cooled.
Thermoplastic polyolefin A thermoplastic polymer used as a single-ply roof membrane.
Thermosetting polymer A polymer that will not soften when heated.
Thick-set mortar Thick bed of cementitious mortar on which floor tiles are set, used where the subfloor surface or the bottom of the floor tiles (or slabs) are irregular or too large for the use of thin-set mortar.
Thin-set mortar Method for attaching tile to the subfloor or underlayment using a thin layer of adhesive.
Threshold Floor surface immediately below an exterior door, typically sloped to the exterior.
Tieback A fastener that connects an exterior cladding to the supporting frame to resist lateral loads.
Ties Closed-loop reinforcement used in columns to prevent the buckling of longitudinal bars.
Tile Small, thin flooring or ceiling unit.
Tilt-up wall Reinforced-concrete wall, precast on the ground-floor slab of the building and lifted by a crane to its required position.
Timber See *heavy timber.*
Tinted glass A type of glass made by adding a metallic pigment during its manufacture.
Tongue-and-groove A joint between two members in which the edge of one member is milled with a groove and that of the other member is milled with a projecting tongue.
Tooled joint Joint in masonry mortar or sealant that is compressed by a specific tool.

Torched membrane SBS modified-bitumen roofing sheet adhered using an open-flame torch.
TPO See *thermoplastic polyolefin.*
Travertine A type of natural stone pitted with voids.
Tread Horizontal (walking) surface of a stair.
Tread pan A steel pan in a prefabricated steel stair that is later filled with concrete at the site to give a finished tread.
Tributary area Area of a building that contributes to the load on a building component.
Truss A structural member with triangulated, linear elements, typically used for large spans.
Trussed joist Parallel chord truss used as a floor joist.
Two-way joist floor See *ribbed slab* and *waffle slab.*
Two-way slab An elevated reinforced-concrete slab in which the load on the slab is carried to the supporting beams in both directions; a four-side-supported rectangular slab whose length is less than twice its width.

U

Ultraviolet radiation (UV) Short-wave solar radiation that can damage human skin and eyes as well as fade the colors of materials and art.
Ultraviolet transmittance A measure of the transmission of ultraviolet radiation through glass.
Underlayment (floor) A panel attached to the subfloor to create a smooth, rigid surface for the application of finish flooring.
Underlayment (roof) Water-resistant sheet applied as a second layer of defense under shingles in a steep roof.
Unit-and-mullion curtain wall Preassembled metal-glass curtain units attached to mullions that have already been attached to the building frame.
Unitized curtain wall Preassembled metal-glass curtain wall units attached to the building frame.
UVT See *ultraviolet transmittance.*

V

Valley A trough formed at the intersection of two adjacent sloping roofs.
Valley flashing Flashing material used in a steep roof valley to increase the water resistance of a valley trough.
Vapor drive analysis A numerical analysis that determines the amount of water vapor flow through the building envelope to ascertain the condensation potential.
Vapor pressure Pressure exerted by water vapor on the building envelope. Vapor pressure is independent of air pressure.
Vapor retarder A material that restricts the flow (transmittance) of water vapor.
Vault A continuous, curved roof member.
Vaulted ceiling High ceiling that follows the shape of the roof structure. If the gypsum board is attached directly to the rafters in a wood frame building (with no attic space), such a ceiling is called a vaulted ceiling.
Veneer A thin layer of material over a backup component. See also *adhered veneer* and *anchored veneer.*
Vertical diaphragm Walls that serve to transfer lateral loads in a building to the foundation by interacting with horizontal diaphragms (floor and roof structures).
Vinyl Abbreviation for polyvinyl chloride, used in window frames, plumbing pipes, and so on; sheet vinyl used in floor tiles, siding, trim, and so on.
Visible transmittance A measure of the transmission of visible light through glass.
Vision glass Transparent glass used between the spandrel areas of a curtain wall.
Visual grading A method of grading lumber based on visual inspection of each piece by trained (graders) inspectors.
Vitreous Glasslike substance with zero vapor permeability.
VT See *visible transmittance.*

W

Waffle slab Also called a two-way joist floor or ribbed slab that has closely spaced ribs (beams) in both directions. See also *ribbed slab.*
Waler Horizontal member used to stiffen and support concrete formwork.
Wall anchor A steel fastener that connects a veneer wall to the backup wall.
Wallboard See *gypsum board.*
Wall cavity Air space between the veneer and the backup wall.
Waterproofing Liquid or sheet material applied to the exterior of a basement wall or floor to resist the intrusion of water under hydrostatic pressure, as distinct from dampproofing.
Waterstop A rubber or polymeric strip of material, used to seal joints between two concrete pours.
Water table The level below ground where the water pressure due to the water in the soil equals the atmospheric pressure; the level below ground to which water will fill an excavation.
Wavelength Distance between adjacent air pressure maxima (compression peaks) or adjacent minima (rarefaction crests) at a given point in time caused by sound.
Weatherstripping Strip of material (wood, metal, foam, etc.) applied to an exterior door or window, which allows the unit to open and close but prevents the passage of air and rainwater when it is in the closed position.
Web Transverse element in a concrete masonry unit; the vertical part of a steel wide-flange section; members in a steel truss between the top and bottom chords.
Weep hole A small opening in a veneer wall directly above the flashing to drain water.
Weld Method of attaching two steel members by heating them until they fuse together.
Welded-wire reinforcement (WWR) A prefabricated rectangular grid of steel wires spot-welded together at intersections, used as reinforcement in concrete slabs.
Wet glazing A semiliquid elastomeric sealant used to set glass in a frame. See also *dry glazing.*
White cement White portland cement, commonly used in architectural concrete, stucco, and terrazzo flooring.
Wind load Difference between inside and outside air pressures on a building component caused by the wind.
Wind uplift Upward pressure caused by wind on a building component, generally on the roof.
Wired glass Rolled glass sheet with embedded thin steel wire mesh, used in fire-rated glazed openings.
Wood flooring Solid or manufactured wood available in strips, planks, and tilelike units and used as a floor finish.

Wood light frame A structural frame assembly composed primarily of dimension lumber studs, floor joists, and roof rafters and panels of wood-based sheathing materials.
Workability Ease with which fresh concrete can be placed and compacted, roughly measured by a slump test; ease with which fresh masonry mortar or portland cement plaster can be troweled on.
Worked lumber Lumber that has been machined beyond surfacing to obtain a specific profile or cross section.
Wrought iron Iron-carbon alloy that is hammered into shape.
Wythe Vertical section of masonry that is 1 unit thick.

Y

Yield strength Maximum stress in steel before it experiences excessive deformation.

Z

Zero-slump concrete A stiff concrete that will not settle at all in a slump test, generally used in making concrete masonry units.
Zoning ordinance A document that describes regulations for the use of land in a particular jurisdiction.

REFERENCES

2.1 Sanderson, R. "Code and Code Administration." Country Club Hills, IL: Building Officials and Code Administrators, 1969, p. 102.

3.1 Walker, B. *Earthquake.* Alexandria, VA: Time-Life Books, 1982, p. 7.

5.1 Baird, George, et al. *Energy Performance of Buildings.* Boca Raton, FL: CRC Press,1984, p. 2.

7.1 National Fire Protection Association, Fire Analysis and Research Division. "A Few Facts at the Household Level." Quincy, MA: July 2009. (http://www.facebook.com/topic.php?uid=205062782061&topic=12838)

10.1 World Commission on Environment and Development (WCED), Melbourne. *Our Common Future* (The Brundtland Report). Oxford: Oxford University Press, 1987.

10.2 Kats, Gregory, et al. "The Costs and Benefits of Green Buildings." A Report to California's Sustainable Building Task Force, October 2003.

10.3 McLaughlin, Lisa. "Technology, Health, Money, Fitness, Home Trends." *Time,* March 2005, p. 74.

10.4 Alternative Technology and Lifestyle Association (ATLA). "Embodied Energy of Materials." Wellington, New Zealand: Newsletter Issue 7, No. 4, 1998. (http://www.converge.org.nz/atla/new-11-98-p1.html)

12.1 National Association of Home Builders. *Revised Builder's Guide to Frost Protected Shallow Foundations.* Upper Marlboro, MD: 2004. (http://www.toolbase.org/Design-Construction-Guides/Foundations/Design-Guide-Frost-Protected-Shallow-Foundation)

12.2 International Code Council. *2009 International Residential Code.* Washington, DC: 2009.

13.1 Hoadley, Bruce. *Understanding Wood.* Newtown, CT: Taunton Press, 1980, p. 69.

14.1 Giedion, Sigfried. *Space, Time and Architecture.* Cambridge, MA: Harvard University Press, 1971, p. 350.

14.2 Bear, James. "Mr. Jefferson's Nails." *The Magazine of Albermarle County History,* 16, pp. 47–53.

15.1 Giedion, Sigfried. *Space, Time and Architecture.* Cambridge, MA: Harvard University Press, 1971, p. 350.

15.2 Williams, Henry, and Williams, Ottalie. *Old American Houses (1700–1850)—How to Restore, Remodel and Reproduce Them.* New York: Bonanza Books, 1946, p. 27.

15.3 Isham, Norman. *Early American Houses and a Glossary of Colonial Architectural Terms.* New York: Da Capo Press, 1967, p. 26.

18.1 Reid, Esmond. *Understanding Buildings: A Multidisciplinary Approach.* Cambridge, MA: MIT Press, 1984, p. 21.

18.2 American Institute of Steel Construction. *A Guide for Architects.* Chicago, IL: 1988, p. 85.

19.1 Zahner, L. W. *Architectural Metals.* New York: John Wiley, 1995, p. 30.

21.1 Kosmatka, S. H., and Panarese, W. C. *Design and Control of Concrete Mixtures,* 13th ed. Skokie, IL: Portland Cement Association, 1994, pp. 21, 68.

21.2 Harriman, M. S. "Structural Concrete: High Strength." *Architectural Record,* October 1990, pp. 85–92.

21.3 Godfrey, K. A. "Concrete Strength Record Jumps 36%." *Civil Engineering,* October 1987, pp. 82–97.

21.4 Kosmatka, S. H., and Panarese, W. C. *Design and Control of Concrete Mixtures,* 13th ed. Skokie, IL: Portland Cement Association, 1994, p. 69.

22.1 Building Research Advisory Board (BRAB), National Academy of Sciences/National Research Council. *Criteria for Selection and Design of Residential Slabs-on-Ground.* Washington, DC: 1968.

22.2 American Concrete Institute (ACI). *Design and Construction of Concrete Slabs on Grade.* Seminar Course Manual/SCM-11(86). Farmington Hills, MI: 1986.

23.1 Alsamsan, I., and Kamara, M. *Simplified Design of Reinforced Concrete Buildings of Moderate Size and Height.* Skokie, IL: Portland Cement Association, 2004.

26.1 International Masonry Institute (IMI). *The Planning and Design of Loadbearing Masonry Buildings.* Annapolis, MD.

28.1 American Concrete Institute. *Building Code Requirements and Specifications for Masonry Structures and Related Commentaries,* ACI530-530.1(08). Farmington Hills, MI: 2008.

28.2 Brick Industry Association (BIA). "Brick Veneer Steel Stud Walls." *Technical Notes on Brick Construction* (28B). Reston, VA: 2005.

30.1 Newman, Harold. *An Illustrated Dictionary of Glass.* London: Thames and Hudson, 1977, p. 82.

30.2 Peter, John. *Design with Glass,* New York: Reinhold, 1964, p. 9.

31.1 Zahner, L. W. *Architectural Metals.* New York: John Wiley, 1995, p. 30.

33.1 Patterson, S., and Mehta, M. *Roofing Design and Practice.* Upper Saddle River, NJ: Prentice Hall, 2001.

33.2 Kane, K: "Another Year for the Industry." *Professional Roofing,* April 1999, p. 14.

35.1 Spens, Michael. *Staircases.* London: Academy Editions, 1995, p. 7.

INDEX